Dictionary of Organic Compounds

FIFTH EDITION

SIXTH SUPPLEMENT

Dictionary
of
Organic
Compounds

FIFTH EDITION

SIXTH SUPPLEMENT

LONDON   NEW YORK
CHAPMAN AND HALL

The Fifth Edition of the Dictionary of Organic Compounds
in seven volumes published 1982
The First Supplement published 1983
The Second Supplement published 1984
The Third Supplement published 1985
The Fourth Supplement published 1986
The Fifth Supplement (in two volumes) published 1987
This Sixth Supplement published 1988
Chapman and Hall
11 New Fetter Lane, London EC4P 4EE
29 West 35th Street, New York, NY 10001

Printed in Great Britain at
the University Press, Cambridge

ISBN 0 412 17060 4
ISSN 0264-1100

© 1988 Chapman and Hall

British Library Cataloguing in Publication Data

Buckingham, J.
    Dictionary of organic compounds – 5th ed.
    Sixth supplement
    1. Chemistry, Organic – Dictionaries
    I. Title
    547′.003′21   QD246

    ISBN 0-412-17060-4

Library of Congress Cataloging in Publication Data

Main entry under title:

Dictionary of organic compounds.

    Bibliography: p. 000
    Includes index.
    1. Chemistry, Organic – Dictionaries.
    QD246.D5   1982   547′.003′21   82-2280
    ISSN 0264-1100
    ISBN 0-412-17060-4 (Sixth supplement)   AACR2

Sixth Supplement

Introduction

For detailed information about how to use DOC 5, see the Introduction in Volume 1 of the Main Work.

1. Using DOC 5 Supplements

As in the Main Work volumes, every Entry is numbered to assist ready location. The DOC Number consists of a letter of the alphabet followed by a five-digit number. In this sixth supplement the first digit is invariably 6. Cross-references within the text to Entries having numbers beginning with zero refer to Main Work Entries and with 1, 2, 3, 4 or 5 refer to the first five supplements.

Where a Supplement Entry contains additional or corrected information referring to an Entry in the Main Work or earlier supplements, the whole Entry is reprinted, with the accompanying statement "Updated Entry replacing . . .". In such cases, the new Entry contains all of the information which appeared in the former Entry, except for any which has been deliberately deleted. In such cases there is therefore no necessity for the user to consult the Main Work or previous supplements.

2. Literature Coverage

In compiling this Supplement the primary literature has been surveyed to mid-1987. A considerable number of compounds from the older literature have also been included for the first time.

3. Indexes

The indexes in the Supplement cover this Sixth Supplement only. A cumulative index volume to Supplements 1–5 inclusive was issued as part of the Fifth Supplement. In order to find a compound in DOC, look first in the Sixth Supplement, then in the Fifth Supplement cumulative indexes, then in the Main Work indexes.

Note to Readers

Always use the latest Supplement

Supplements are published in the middle of each year and contain new and updated Entries derived from the primary literature of the preceding year. Searching the entire Supplement series is facilitated by consulting first the indexes in the latest Supplement. The Supplement indexes are cumulative to facilitate the rapid location of data.

For full information on Supplements please write to:

The Marketing Manager, DOC 5   *or*   Chapman and Hall
Chapman and Hall   29 West 35th Street
11 New Fetter Lane   New York, NY 10001
London
EC4P 4EE

New Compounds for the DOC 5

The Editor is always pleased to receive comments on the selection policy of DOC 5, and in particular welcomes specific suggestions for compounds or groups of compounds to be considered for inclusion in the annual Supplements.

Write to:

The Editor, DOC 5
Chapman and Hall
11 New Fetter Lane
London
EC4P 4EE

Specialist Dictionaries

Four important specialist publications are now available which greatly extend the coverage of the DOC databank in key specialist areas. These are as follows:

Dictionary of Alkaloids, 1989, 2 volumes, ISBN 0412 24910 3

Dictionary of Antibiotics and Related Substances, 1987, ISBN 0412 25450 6

Dictionary of Organophosphorus Compounds, 1987, ISBN 0412 25790 4

Carbohydrates (a Chapman and Hall Chemistry Sourcebook), 1987, ISBN 0412 26960 0

Caution

Contents

A

Abbreviatin PB   A-60001

[84633-06-7]

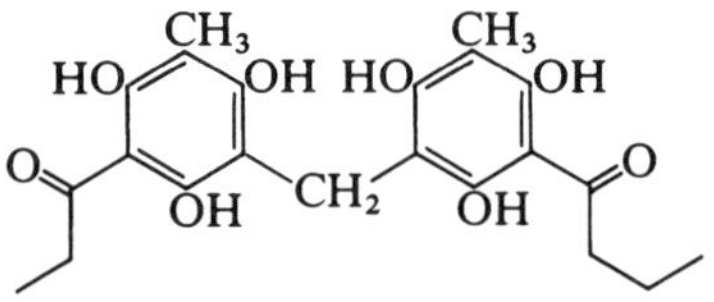

$C_{22}H_{26}O_8$ M 418.443

Isol. from *Dryopteris abbreviata*. Pale-yellow powder (CHCl₃). Mp 206-208°.

Coskun, M. *et al*, *Chem. Pharm. Bull.*, 1982, **30**, 4102 (*isol*)

Abeoanticopalic acid   A-60002

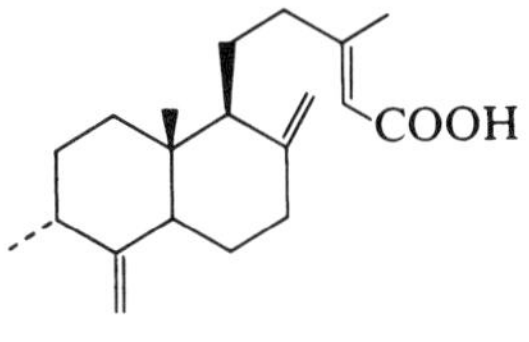

$C_{20}H_{30}O_2$ M 302.456

Constit. of *Pinus strobus*.

Me ester: Oil. $[\alpha]_D^{20}$ +41.5° (c, 1.3 in CHCl₃).

Zinkel, D.F. *et al*, *Phytochemistry*, 1987, **26**, 769.

19(4→3)-Abeo-11,12-dihydroxy-4(18),8,11,13-abietatetraen-7-one   A-60003
19(4→3)-Abeo-O-demethylcryptojaponol

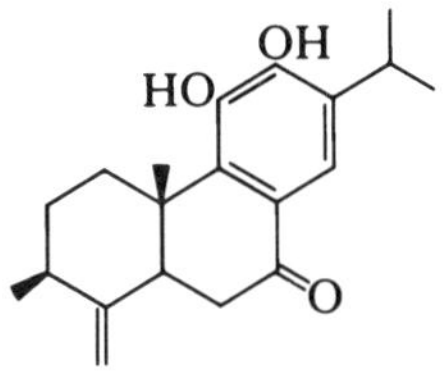

$C_{20}H_{26}O_3$ M 314.424

Constit. of *Salvia pubescens*. Oil.

Di-Ac: Cryst. (CH₂Cl₂/pet. ether). Mp 135-136°. $[\alpha]_D^{20}$ +126° (c, 0.44 in CHCl₃).

Galicia, M.A. *et al*, *Phytochemistry*, 1988, **27**, 217.

19(4β→3β)-Abeo-6,11-epoxy-6,12-dihydroxy-6,7-seco-4(18),8,11,13-abietatetraen-7-al   A-60004

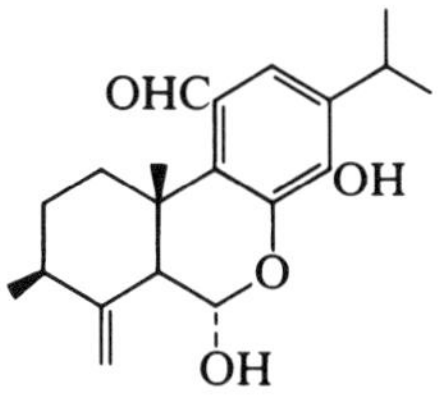

$C_{20}H_{26}O_4$ M 330.423

6α-form [111508-79-3]

Constit. of *Coleus barbatus*.

Di-Ac: [111508-81-7]. Cryst. Mp 250-252°. $[\alpha]_D^{25}$ +155.3° (c, 0.36 in CHCl₃).

Kelecom, A. *et al*, *Phytochemistry*, 1987, **26**, 2337.

Abiesonic acid   A-60005

[107195-86-8]

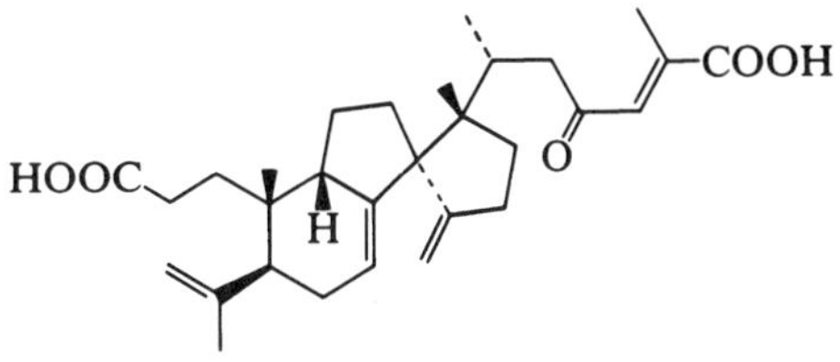

$C_{30}H_{42}O_5$ M 482.659

Constit. of the oleoresin of *Abies sibirica*.

Di-Me ester: [107195-81-3]. Cryst. (pentane). Mp 93-94°. $[\alpha]_D^{20}$ −37.3° (c, 4.84 in CHCl₃).

Raldugin, V.A. *et al*, *Khim. Prir. Soedin.*, 1986, **22**, 548 (*isol, cryst struct*)

6,8,11,13-Abietatetraene-11,12,14-triol   A-60006
Cryptanol

[110209-97-7]

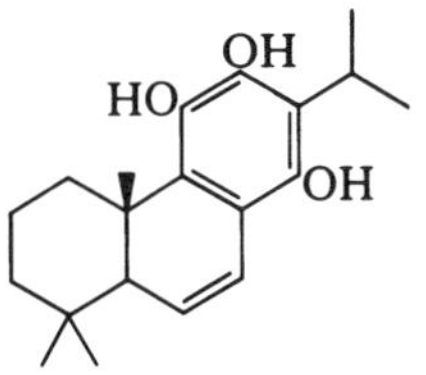

$C_{20}H_{28}O_3$ M 316.439

Constit. of *Salvia cryptantha*. Cryst. Mp 138-142°.

Ulubelen, A. *et al*, *Phytochemistry*, 1987, **26**, 1534.

8,11,13-Abietatrien-19-ol   A-60007
Dehydroabietinol

[24035-43-6]

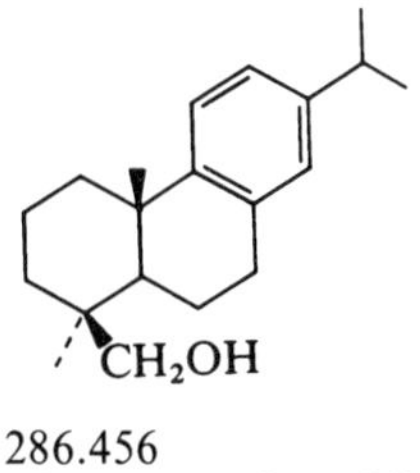

$C_{20}H_{30}O$ M 286.456

Constit. of *Calceolaria ascendens*. Oil. $[\alpha]_D^{25}$ +43.4° (c, 2 in CHCl₃).

Ac: [24462-15-5]. *Dehydroabietinol acetate.*
$C_{22}H_{32}O_2$ M 328.494
Constit. of *Pinus silvestris*. Mp 60-61°. $[\alpha]_D^{25}$ +54.3° (c, 1.1 in CHCl₃).

19-(Carboxyacetyl):
$C_{23}H_{32}O_4$ M 372.503
Constit. of *C. ascendens*.

19-(Carboxyacetyl), Me ester: Oil. $[\alpha]_D^{25}$ +32.5° (c, 0.6 in CHCl₃).

Carman, R.M. *et al*, *Aust. J. Chem.*, 1967, **20**, 2789 (*synth*)
Schmidt, E.N. *et al*, *Khim. Prir. Soedin.*, 1969, **5**, 187 (*deriv*)
Chamy, M.C. *et al*, *Phytochemistry*, 1987, **26**, 1763 (*isol*)

8,11,13-Abietatrien-12,16-oxide   A-60008

Updated Entry replacing A-10001
12,16-Epoxy-8,11,13-abietatriene
[37842-29-8]

Absolute
configuration

$C_{20}H_{28}O$ M 284.441

Constit. of *Thujopsis dolabrata*. Cryst. Mp 49-52°. $[\alpha]_D^{27}$
+18.9° (c, 1.06 in $CHCl_3$). Has been synthesised
apparently as a single stereoisomer of undetd. config.

Burnell, R.H. *et al*, *Can. J. Chem.*, 1978, **56**, 517 (*synth*)
Hasegawa, S. *et al*, *Phytochemistry*, 1982, **21**, 643 (*isol, struct*)
Matsumoto, T. *et al*, *Bull. Chem. Soc. Jpn.*, 1987, **60**, 2401
 (*synth, abs config*)

8(14)-Abieten-18-oic acid 9,13-endoperoxide   A-60009

(9α,3α)-*form*

$C_{20}H_{30}O_4$ M 334.455

(9α,13α)-*form*
 Constit. of *Elodea canadensis*.
(9β,13β)-*form*
 Constit. of *E. canadensis*.

 Monaco, P. *et al*, *Tetrahedron Lett.*, 1987, **28**, 4609.

Acalycixeniolide *A*   A-60010

$C_{19}H_{28}O_2$ M 288.429

Constit. of gorgonian *Acalycigorgia inermis*. Inhibits cell
 division in fertilised starfish eggs. Amorph. powder.
 $[\alpha]_D$ +143° (c, 0.31 in $CHCl_3$).

 Fusetani, N. *et al*, *Tetrahedron Lett.*, 1987, **28**, 5837.

Acalycixeniolide *B*   A-60011

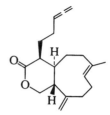

$C_{19}H_{26}O_2$ M 286.413

Constit. of gorgonian *Acalycigorgia inermis*. Inhibits cell
 division in fertilised starfish eggs.

 Fusetani, N. *et al*, *Tetrahedron Lett.*, 1987, **28**, 5837.

Acanthospermal *A*   A-60012

Updated Entry replacing A-00044
Acanthoglabrolide
[56689-33-9]

$R^1 = CH_3$, $R^2 = OC$

$C_{23}H_{30}O_8$ M 434.485

Constit. of *Acanthospermum australe*. Gum. $[\alpha]_{Hg}^{25}$−54°
 (c, 0.33 in $CHCl_3$).

De-Ac: **Desacetylacanthospermal A**.
 $C_{19}H_{24}O_6$ M 348.395
 Constit. of *Milleria quinqueflora*. Oil.

 Herz, W. *et al*, *J. Org. Chem.*, 1975, **40**, 3486 (*isol*)
 Jakupovic, J. *et al*, *Phytochemistry*, 1987, **26**, 2011 (*deriv*)

3,4-Acenaphthenedicarboxylic acid   A-60013

1,2-Dihydro-3,4-acenaphthylenedicarboxylic acid, 9CI
[13055-36-2]
$C_{14}H_{10}O_4$ M 242.231
Cryst. Mp 244-246°.

Di-Et ester: [108665-29-8].
 $C_{18}H_{14}O_4$ M 294.306
 Yellow cryst. (MeOH). Mp 94-95°.
Anhydride: [108665-25-4].
 $C_{14}H_8O_3$ M 224.215
 Cryst. (Ac_2O). Mp 256-257°.

 Sangaiah, R. *et al*, *J. Org. Chem.*, 1987, **52**, 3205 (*synth, deriv,*
  uv, ir, pmr)

Acenaphtho[5,4-*b*]furan, 9CI   A-60014

[19536-50-6]

$C_{14}H_8O$ M 192.217
Yellow cryst. Mp 46-48°.

 Lee-Ruff, E. *et al*, *J. Heterocycl. Chem.*, 1986, **23**, 1551 (*synth,*
  pmr, uv, ms)

Acenaphtho[5,4-*b*]thiophene, 9CI   A-60015
[58427-00-2]

$C_{14}H_8S$ M 208.277
Yellow-orange cryst. Mp 71-73°.

Lee-Ruff, E. *et al, J. Heterocycl. Chem.*, 1986, **73**, 1551 (*synth, pmr, uv, ms*)

Acetamidoxime, 8CI   A-60016
N-*Hydroxyethanimidamide*, 9CI
[22059-22-9]

$$H_3CC(NH_2)=NOH$$

$C_2H_6N_2O$ M 74.082
Cryst. Mp 140°. Hygroscopic.
▷AC6900000.
B,HCl: [5426-04-0]. Cryst. Mp 140-142°.
O-Formyl:
   $C_3H_6N_2O_2$ M 102.093
   Cryst. (hexane). Mp 29°.
O-Ac:
   $C_4H_8N_2O_2$ M 116.119
   Cryst. (C_6H_6). Mp 96°.
O-Benzyl:
   $C_9H_{12}N_2O$ M 164.207
   Pale-yellow oil. Bp 200°.
O-Benzyl; B,HCl: Cryst. Mp 163°.
N-Ph: [5661-30-3].
   $C_8H_{10}N_2O$ M 150.180
   Large brown-yellow plates. Mp 121°.

Nordmann, E., *Ber.*, 1884, **17**, 2746 (*synth, deriv*)
Lenaers, R. *et al, Helv. Chim. Acta*, 1962, **45**, 441 (*synth*)
Eloy, F. *et al, Bull. Soc. Chim. Belg.*, 1964, **73**, 518 (*deriv*)
Bedford, C.D. *et al, J. Med. Chem.*, 1986, **29**, 2174.

Acetophenone, 8CI   A-60017
Updated Entry replacing A-00121
1-Phenylethanone, 9CI. *Methyl phenyl ketone. Hypnone*
[98-86-2]

$$PhCOCH_3$$

C_8H_8O M 120.151
Has soporific props. Solvent, resin intermed.
   Photochemical sensitiser. Plates (freq. obt. as liq.). d_4^{20}
   1.029. Mp 20°. Bp 202°, Bp_5 67°. n_D^{20} 1.5342. For
   enol form see 1-Phenylethenol, P-60092 .
▷AM5250000.
(*E*)-*Oxime:* [10341-75-0]. syn-*Oxime.*
   C_8H_9NO M 135.165
   Mp 59.5-60.5°.
(*Z*)-*Oxime:* [50314-86-8]. anti-*Oxime.* Cryst.
   ($CHCl_3$/pet. ether). Mp 81-83°.
Hydrazone: [13466-30-3].
   $C_8H_{11}N_2$ M 135.188
   Cryst. (pet. ether). Mp 22°.
Phenylhydrazone: [583-11-9]. Needles (EtOH). Mp
   106°.
(*E*)-*2,4-Dinitrophenylhydrazone:* [23245-99-0]. syn-*2,4-*
   Dinitrophenylhydrazone. Yellow cryst. Mp 147-
   147.5°.

(*Z*)-*2,4-Dinitrophenylhydrazone:* anti-*2,4-*
   Dinitrophenylhydrazone. Red cryst. Mp 244-246°.

Org. Synth. Coll. Vol., **1**, 109 (*synth*)
Aldrich Library of IR Spectra, 2nd Ed., 746H (*ir*)
Aldrich Library of NMR Spectra, **6**, 6A (*pmr*)
Sadtler Standard C-13 NMR Spectra, 550 (*cmr*)
Sadtler Standard Ultraviolet Spectra, 953 (*uv*)
Registry of Mass Spectral Data, Wiley-Interscience, 180 (*ms*)
Fieser, M. *et al, Reagents for Organic Synthesis*, Wiley, 1967-
   84, **4**, 5.
Sax, N.I., *Dangerous Properties of Industrial Materials*, 5th
   Ed., Van Nostrand-Reinhold, 1979, 905.

4-Acetoxy-6-(4-hydroxy-2-cyclohexenyl)-   A-60018
2-(4-methyl-3-pentenyl)-2-heptenoic acid

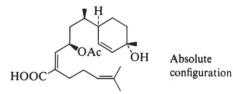

Absolute
configuration

$C_{22}H_{34}O_5$ M 378.508
Compd. not named in the reference. Constit. of *Eremo-*
   phila foliosissima.
Me ester: Unstable oil.

Forster, P.G. *et al, Tetrahedron*, 1987, **43**, 2999.

6-Acetyl-2-amino-1,7,8,9-tetrahydro-4*H*-   A-60019
pyrimido[4,5-*b*][1,4]diazepin-4-one, 9CI
2-Amino-6-acetyl-3,4,7,8-tetrahydro-4-oxo-3H,9H-
pyrimido[4,5-b][1,4]diazepine. Quench spot
[80003-63-0]

$C_9H_{11}N_5O_2$ M 221.218
Isol. from eye tissue of *Drosphila melanogaster.*
   Intermed. in biosynth. of Drosopterin and other eye
   pigments.

Jacobson, K.B. *et al, Biochemistry*, 1982, **21**, 5700 (*isol, uv,
   pmr, cmr, ms, struct*)

[3][14-Acetyl-14-azacyclohexacosan-   **A-60020**
one][25,26,53,54,55,56-hexaacetoxytri-
cyclo[49.3.1.1^{24,28}]hexapentaconta-
1(55),24,26,28(56),51,53-hexaene][14-
acetyl-14-azacyclohexacosanone]-
catenane

C$_{122}$H$_{210}$N$_2$O$_{16}$ M 1961.005

First [3]catenane. Directed synthesis produces 3 isomers, that in which the 2 smaller rings encircle different sides of the large ring (illus.), that in which they both encircle the bridge adjacent to the more highly substd. sides of the benzene rings, and that in which they both encircle the bridge adjacent to the less highly substd. sides. Corresponding hexa-Me-ether with OMe replacing OAc also prepd.

Rissler, K. *et al, Chem. Ber.*, 1986, **119**, 1374 (*synth, ir, cmr, pmr, ms*)

1-Acetyl-3-bromonaphthalene   **A-60021**
1-(3-Bromo-1-naphthalenyl)ethanone, 9CI. 3-Bromo-1-naphthyl methyl ketone, 8CI
[58149-65-8]

C$_{12}$H$_9$BrO   M 249.107
Liq. Bp$_1$ 137-138°.

Tsuno, Y. *et al, Bull. Chem. Soc. Jpn.*, 1975, **48**, 3347 (*synth*)

1-Acetyl-4-bromonaphthalene   **A-60022**
1-(4-Bromo-1-naphthalenyl)ethanone, 9CI. 4-Bromo-1-naphthyl methyl ketone, 8CI
[46258-62-2]
C$_{12}$H$_9$BrO   M 249.107
Cryst. (hexane). Mp 47-47.5°. Bp$_{0.7}$ 141-143°.
Oxime: Cryst. (EtOH aq.). Mp 143-144°.

Jacobs, T.L. *et al, J. Org. Chem.*, 1946, **11**, 27 (*synth*)
Tsuno, Y. *et al, Bull. Chem. Soc. Jpn.*, 1975, **48**, 3347 (*synth*)

1-Acetyl-5-bromonaphthalene   **A-60023**
1-(5-Bromo-1-naphthalenyl)ethanone, 9CI. 5-Bromo-1-naphthyl methyl ketone, 8CI
[58149-87-4]

C$_{12}$H$_9$BrO   M 249.107
Pale-yellow cryst. (pet. ether). Mp 55.5-56.5°. Bp$_7$ 179-182°.

Tsuno, Y. *et al, Bull. Chem. Soc. Jpn.*, 1975, **48**, 3356 (*synth*)

1-Acetyl-7-bromonaphthalene   **A-60024**
1-(7-Bromo-1-naphthalenyl)ethanone, 9CI. 7-Bromo-1-naphthyl methyl ketone, 8CI
[1590-24-5]
C$_{12}$H$_9$BrO   M 249.107
Cryst. (pet. ether). Mp 63-64°.

Girdler, R.B. *et al, J. Chem. Soc. (C)*, 1966, 518 (*synth, ir, pmr*)
Tsuno, Y. *et al, Bull. Chem. Soc. Jpn.*, 1975, **48**, 3347 (*synth*)

2-Acetyl-6-bromonaphthalene   **A-60025**
1-(6-Bromo-2-naphthalenyl)ethanone, 9CI. 6-Bromo-2-naphthyl methyl ketone, 8CI
[1590-25-6]
C$_{12}$H$_9$BrO   M 249.107
Cryst. (pet. ether). Mp 102°.
2,4-Dinitrophenylhydrazone: [5020-69-9]. Red cryst. (xylene). Mp 274°.

Girdler, R.B. *et al, J. Chem. Soc. (C)*, 1966, 518 (*synth, ir, pmr*)

1-Acetylcycloheptene   **A-60026**
1-(1-Cyclohepten-1-yl)ethanone, 9CI
[14377-11-8]

C$_9$H$_{14}$O   M 138.209
Liq. Bp$_{14}$ 110° (Bp$_{16}$ 97°).
Semicarbazone: Needles (EtOH). Mp 197°.
2,4-Dinitrophenylhydrazone: Crimson plates (EtOH). Mp 178°.

Heilbron, I. *et al, J. Chem. Soc.*, 1949, 1827 (*synth*)
Tamura, R. *et al, J. Org. Chem.*, 1986, **51**, 4368 (*synth, ir, pmr*)

1-Acetylcyclooctene, 9CI   **A-60027**
1-(1-Cyclooctenyl)ethanone, 9CI. (1-Cyclooctan-1-yl) methyl ketone
[17339-74-1]

C$_{10}$H$_{16}$O   M 152.236
(*E*)-*form* [60727-70-0]
Liq. Bp$_{0.15}$ 60°.

Wroble, R.R. *et al, J. Org. Chem.*, 1976, **41**, 2939 (*synth*)
Tamura, R. *et al, J. Org. Chem.*, 1986, **51**, 4368 (*synth, ir, pmr*)

1-Acetylcyclopentene   A-60028

1-(1-Cyclopenten-1-yl)ethanone, 9CI. 1-Cyclopenten-1-ylmethyl ketone, 8CI

[16112-10-0]

$C_7H_{10}O$ M 110.155

Oil. Bp_{22} 75-78°, Bp_{16} 67°.

Semicarbazone: Needles (EtOH). Mp 211-212°.
2,4-Dinitrophenylhydrazone: Bright-red needles. Mp 203°.

Rapson, W.G. *et al, J. Chem. Soc.,* 1935, 1285 (*synth*)
Heilbron, I. *et al, J. Chem. Soc.,* 1949, 1827 (*synth*)

5-Acetyl-2,2-dimethyl-1,3-dioxane-4,6-dione, 9CI   A-60029

Acetylmeldrum's acid

[72324-39-1]

$C_8H_{10}O_5$ M 186.164

Intermed. for heterocyclic compds. Cryst. (Et_2O). Mp 83-84°. Other acylmeldrum's acids also prepd.

Yamamoto, Y. *et al, Chem. Pharm. Bull.,* 1987, **35**, 1860, 1871 (*synth, pmr, use*)

3-Acetyl-4-hydroxy-2(5H)-furanone, 9CI   A-60030

3-Acetyltetronic acid

[22621-26-7]

$C_6H_6O_4$ M 142.111

Mp 77-79°. General method for 3-acyltetronic acids given by Nomura *et al.*

Tanaka, K. *et al, Chem. Pharm. Bull.,* 1979, **27**, 1901 (*synth*)
Nomura, K. *et al, Chem. Pharm. Bull.,* 1986, **34**, 5188 (*synth*)

6-Acetyl-5-hydroxy-2-hydroxymethyl-2-methylchromene   A-60031

1-[5-Hydroxy-2-(hydroxymethyl)-2-methyl-2H-1-benzopyran-6-yl]ethanone, 9CI

$C_{13}H_{14}O_4$ M 234.251

Constit. of *Blepharispermum subsessile.* Cryst. Mp 108°. $[\alpha]_D$ +5.7° (c, 0.35 in $CHCl_3$).

Kulkarni, M.M. *et al, Phytochemistry,* 1987, **26**, 2971.

2-Acetylindole   A-60032

1-(1H-Indol-2-yl)ethanone, 9CI. 2-Indolyl methyl ketone

[4264-35-1]

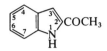

$C_{10}H_9NO$ M 159.187

Needles (EtOAc). Mp 154-155°.

Wong, L.C.H. *et al, J. Chem. Soc., Chem. Commun.,* 1980, 200 (*synth*)
Beugelmans, R. *et al, Tetrahedron, Suppl. 9,* 1981, 393 (*synth*)
Murakawa, Y. *et al, Synthesis,* 1984, 738 (*synth*)
Akimoto, H. *et al, Bull. Chem. Soc. Jpn.,* 1985, **58**, 123 (*synth*)

4-Acetylindole   A-60033

Updated Entry replacing A-40020
1-(1H-Indol-4-yl)ethanone, 9CI. 4-Indolyl methyl ketone

[50614-86-3]

$C_{10}H_9NO$ M 159.187

Cryst. (EtOAc). Mp 163-164° (159-160°).

1-Ac: [83188-13-0]. *1,4-Diacetylindole.*
    $C_{12}H_{11}NO_2$ M 201.224
    Cryst. (cyclohexane). Mp 110-113°.

Baron, M. *et al, Bull. Soc. Chim. Fr.,* 1982, 249 (*synth*)
Clark, R.D., *J. Heterocycl. Chem.,* 1983, **20**, 1393 (*synth*)
Barrett, A.G.M. *et al, J. Org. Chem.,* 1984, **49**, 4409 (*synth, ir, pmr, ms*)

5-Acetylindole   A-60034

1-(1H-Indol-5-yl)ethanone, 9CI. 5-Indolyl methyl ketone

[53330-94-2]

$C_{10}H_9NO$ M 159.187

Mp 94-95°.

▷OB4525000.

Oxime:
    $C_{10}H_{10}N_2O$ M 174.202
    Mp 163.5-164.5°.
Semicarbazone: Mp 200.5-201°.

Terent'ev, A.P. *et al, Zh. Obshch. Khim.,* 1959, **29**, 2875; *CA,* **54**, 12098d (*synth*)

6-Acetylindole   A-60035

1-(1H-Indol-6-yl)ethanone, 9CI. 6-Indolyl methyl ketone

[81223-73-6]

$C_{10}H_9NO$ M 159.187

Cryst. (EtOAc). Mp 122-123°.

Kim, P.T. *et al, J. Heterocycl. Chem.,* 1981, **18**, 1365 (*synth*)
Baron, M. *et al, Bull. Soc. Chim. Fr.,* 1982, 249 (*synth*)

7-Acetylindole   A-60036

1-(1H-Indol-7-yl)ethanone, 9CI. 7-Indolyl methyl ketone

[104019-20-7]

$C_{10}H_9NO$ M 159.187

Eur. Pat., 181 136 (*1986*); *CA,* **105**, 152922 (*synth*)

1-Acetyl-7-iodonaphthalene   A-60037

1-(7-Iodo-1-naphthalenyl)ethanone, 9CI. 7-Iodo-1-naphthyl methyl ketone, 8CI

[1590-23-4]

$C_{12}H_9IO$ M 296.107
Cryst. (pet. ether). Mp 65-66°.

Harnik, M. *et al, Israel J. Chem.*, 1965, **3**, 13 (*synth*)
Endo, K. *et al, Bull. Chem. Soc. Jpn.*, 1971, **44**, 2465.

1-Acetyl-8-iodonaphthalene   A-60038

1-(8-Iodo-1-naphthalenyl)ethanone, 9CI. 8-Iodo-1-naphthyl methyl ketone, 8CI

[32141-12-1]
$C_{12}H_9IO$ M 296.107
Cryst. (EtOH). Mp 55-57°.

Hellwinkel, D. *et al, Chem. Ber.*, 1987, **120**, 1151 (*synth, ir, pmr*)

1-Acetylisoquinoline   A-60039

1-(1-Isoquinolinyl)ethanone, 9CI. 1-Isoquinolinyl methyl ketone, 8CI

[58022-21-2]

$C_{11}H_9NO$ M 171.198
Pale-yellow cryst. (pet. ether). Mp 14-15°. Bp₁ 115-118°.
Oxime:
   $C_{11}H_{10}N_2O$ M 186.213
   Rhomboids (EtOH aq.). Mp 211.5-213.5°.
Semicarbazone: Fine needles (EtOH aq.). Mp 208°.
Phenylhydrazone: Orange needles (EtOH). Mp 165-167° dec.

Sakamoto, T. *et al, Synthesis*, 1984, 245 (*synth*)

3-Acetylisoquinoline   A-60040

1-(3-Isoquinolinyl)ethanone, 9CI. 3-Isoquinolinyl methyl ketone

[91544-03-5]
$C_{11}H_9NO$ M 171.198
Plates (pet. ether). Mp 88°.
Semicarbazone: Rods (EtOH aq.). Mp 216°.
Phenylhydrazone, picrate: Orange-red needles (EtOH). Mp 209°.

Sakamoto, T. *et al, Synthesis*, 1984, 245 (*synth*)

4-Acetylisoquinoline   A-60041

1-(4-Isoquinolinyl)ethanone, 9CI. 4-Isoquinolinyl methyl ketone

[40570-74-9]
$C_{11}H_9NO$ M 171.198
Needles (pet. ether). Mp 71-72°. Bp₈ 160-164°.
Oxime:
   $C_{11}H_{10}N_2O$ M 186.213
   Needles (EtOH aq.). Mp 194.5-195.5°.
Semicarbazone: Fine cryst. (EtOH aq.). Mp 211.5-212°.

Phenylhydrazone: Orange plates (EtOH). Mp 173.5-176° dec.

Abramovitch, R.A. *et al, Can. J. Chem.*, 1963, **41**, 2265 (*synth*)
Sakamoto, T. *et al, Synthesis*, 1984, 245 (*synth*)
Baradarani, M.M. *et al, J. Chem. Soc., Perkin Trans. 1*, 1985, 1503 (*synth*)

5-Acetylisoquinoline   A-60042

1-(5-Isoquinolinyl)ethanone, 9CI. 5-Isoquinolinyl methyl ketone

[54415-44-0]
$C_{11}H_9NO$ M 171.198
No phys. props. reported.

Cohylakis, D. *et al, J. Chem. Soc., Perkin Trans. 1*, 1974, 1518 (*synth*)

1-Acetyl-2-methylcyclopentene   A-60043

1-(2-Methyl-1-cyclopenten-1-yl)ethanone, 9CI

[3168-90-9]

$C_8H_{12}O$ M 124.182
Bp₄ 58-60°, Bp₀.₈ 44°.

Hudlicky, T. *et al, Tetrahedron Lett.*, 1981, **22**, 3351 (*synth, ir, pmr, cmr, ms*)
Hudlicky, T. *et al, Synthesis*, 1986, 716 (*synth, ir, pmr, cmr, ms*)

2-Acetylpyrimidine   A-60044

1-(2-Pyrimidinyl)ethanone, 9CI. Methyl 2-pyrimidinyl ketone

[53342-27-1]

$C_6H_6N_2O$ M 122.126
Needles. Mp 52°.

Alexander, E.C. *et al, J. Org. Chem.*, 1975, **40**, 1500 (*ms*)
Naumenko, I.I. *et al, Khim. Geterotsikl. Soedin.*, 1981, 958; *Chem. Heterocycl. Compd.*, 710 (*synth, uv, pmr*)

4-Acetylpyrimidine   A-60045

1-(4-Pyrimidinyl)ethanone, 9CI. Methyl 4-pyrimidinyl ketone

[39870-05-8]
$C_6H_6N_2O$ M 122.126
Light-yellow cryst. Mp 67°.

Robba, M., *Ann. Chim. (Paris)*, 1960, **5**, 380 (*synth*)
Alexander, E.C. *et al, J. Am. Chem. Soc.*, 1974, **96**, 5663 (*synth*)
Alexander, E.C. *et al, J. Org. Chem.*, 1975, **40**, 1500 (*ms*)
Ger. Pat., 2 800 443, (*1978*); *CA*, **89**, 163595 (*synth*)
Naumenko, I.I. *et al, Khim. Geterotsikl. Soedin.*, 1981, 958; *Chem. Heterocycl. Compd.*, 710 (*synth, pmr*)

5-Acetylpyrimidine   A-60046

1-(5-Pyrimidinyl)ethanone, 9CI. Methyl 5-pyrimidinyl ketone

[10325-70-9]
$C_6H_6N_2O$ M 122.126
Mp 88-89°.

Zymalkowski, F. *et al, Arch. Pharm. (Weinheim, Ger.)*, 1966, **299**, 362.

Alexander, E.C. *et al*, *J. Am. Chem. Soc.*, 1974, **96**, 5663
   (*synth*)
Alexander, E.C. *et al*, *J. Org. Chem.*, 1975, **40**, 1500 (*ms*)
Naumenko, I.I. *et al*, *Khim. Geterotsikl. Soedin.*, 1981, 958;
   Chem. Heterocycl. Compd., 710 (*synth, uv, pmr*)

5-Acetyl-2,4(1*H*,3*H*)-pyrimidinedione, 9CI   A-60047

5-Acetyluracil

[6214-65-9]

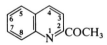

$C_6H_6N_2O_3$ M 154.125
Prisms (EtOH). Mp 283-285° (294°) dec.

2,4-Dinitrophenylhydrazone: Orange-red needles
   (AcOH). Mp 276° dec.
1-Ph: [36980-84-4].
   $C_{12}H_{10}N_2O_3$ M 230.223
   Needles (EtOH). Mp 269° dec.

Bergmann, W. *et al*, *Ber.*, 1933, **66**, 1492 (*synth*)
Dewar, J.H. *et al*, *J. Chem. Soc.*, 1961, 3254 (*synth*)

5-Acetyl-2(1*H*)-pyrimidinone, 9CI   A-60048

[87573-88-4]

$C_6H_6N_2O_2$ M 138.126
Mp 170°. $Bp_{0.03}$ 140-150° subl.

Undheim, K. *et al*, *Acta Chem. Scand.*, *Ser. B*, 1983, **37**, 235;
   1986, **40**, 764 (*synth, pmr, ms*)

2-Acetylquinoline   A-60049

1-(2-Quinolinyl)ethanone, 9CI. *Methyl 2-quinolinyl ketone*, 8CI

[1011-47-8]

$C_{11}H_9NO$ M 171.198
Cryst. (EtOH aq.). Mp 50-52° (80-81°). Bp_1 117-119°,
   $Bp_{0.45}$ 93-95°.

Oxime:
   $C_{11}H_{10}N_2O$ M 186.213
   Mp 143-145°.
Phenylhydrazone: [72404-94-5]. Cryst. (MeOH). Mp
   154°.

Campbell, K.N. *et al*, *J. Am. Chem. Soc.*, 1946, **68**, 1840
   (*synth*)
Capuano, L., *Chem. Ber.*, 1959, **92**, 2670 (*synth*)
Yamazaki, T. *et al*, *Heterocycles*, 1976, **4**, 713 (*synth*)
Case, F.H. *et al*, *J. Heterocycl. Chem.*, 1979, **16**, 1135 (*synth*)
Citterio, A. *et al*, *J. Chem. Res. (M)*, 1982, 2801 (*synth*)
Sakamoto, T. *et al*, *Synthesis*, 1984, 245 (*synth*)

3-Acetylquinoline   A-60050

1-(3-Quinolinyl)ethanone, 9CI. *Methyl 3-quinolinyl ketone*

[33021-53-3]

$C_{11}H_9NO$ M 171.198
Cryst. (EtOH). Mp 101-102°.

Semicarbazone: Mp 236-236.5°.
2,4-Dinitrophenylhydrazone: Red cryst. Mp 236-236.5°.
Picrate: Mp 215-215.5°.

Haug, U. *et al*, *Chem. Ber.*, 1960, **93**, 593 (*synth*)
Biellmann, J.F. *et al*, *Tetrahedron*, 1971, **27**, 1789 (*synth*)
Sakamoto, T. *et al*, *Synthesis*, 1984, 245 (*synth*)
Miura, M. *et al*, *J. Chem. Soc., Chem. Commun.*, 1986, 241
   (*synth*)

4-Acetylquinoline   A-60051

1-(4-Quinolinyl)ethanone, 9CI. *Methyl 4-quinolinyl
ketone*

[60814-30-4]

$C_{11}H_9NO$ M 171.198
Mp 33-34°. Bp_1 118-121°.

1-Oxide:
   $C_{11}H_9NO_2$ M 187.198
   Needles (C_6H_6). Mp 103°.
Picrate: Yellow needles (EtOH). Mp 165-170° dec.

Elliott, I.W., *J. Org. Chem.*, 1960, **25**, 1256 (*synth*)
van Tamelen, E.E. *et al*, *Tetrahedron Lett.*, 1961, 390 (*synth*)
Ochiai, E. *et al*, *Chem. Pharm. Bull.*, 1963, **11**, 137 (*synth*)
Citterio, A. *et al*, *J. Chem. Res. (M)*, 1982, 2081 (*synth*)
Sakamoto, T. *et al*, *Synthesis*, 1984, 245 (*synth*)
Ihara, M. *et al*, *Tetrahedron*, 1985, **41**, 2109 (*synth*)

5-Acetylquinoline   A-60052

1-(5-Quinolinyl)ethanone, 9CI. *Methyl 5-quinolinyl
ketone*

$C_{11}H_9NO$ M 171.198
Yellow oil. $Bp_{0.4}$ 124-127°.

Oxime: [30074-90-9].
   $C_{11}H_{10}N_2O$ M 186.213
   Mp 153-154°.
Phenylhydrazone: Mp 168-173°.

Haug, U. *et al*, *Chem. Ber.*, 1960, **93**, 593 (*synth*)
Gregory, B.J. *et al*, *J. Chem. Soc. (B)*, 1970, 1687 (*synth*)

6-Acetylquinoline   A-60053

1-(6-Quinolinyl)ethanone, 9CI. *Methyl 6-quinolinyl
ketone*

[73013-68-0]

$C_{11}H_9NO$ M 171.198
Cryst. (EtOH). Mp 75-76°. Bp_2 145°.

Oxime: [30074-91-0].
   $C_{11}H_{10}N_2O$ M 186.213
   Mp 169-170°.
Phenylhydrazone: Mp 157-159°.
Picrate: Mp 242°.

Waley, S.G., *J. Chem. Soc.*, 1948, 2008 (*synth*)
Pugin, A. *et al*, *Helv. Chim. Acta*, 1952, **35**, 2322 (*synth*)
Haug, U. *et al*, *Chem. Ber.*, 1960, **93**, 593 (*synth*)
Ferles, M. *et al*, *Collect. Czech. Chem. Commun.*, 1979, **44**,
   2672 (*synth*)

7-Acetylquinoline   A-60054

1-(7-Quinolinyl)ethanone, 9CI. *Methyl 7-quinolinyl
ketone*

$C_{11}H_9NO$ M 171.198
Cryst. (EtOH). Mp 76.5-77.5°.

Oxime: [31189-39-6].
   $C_{11}H_{10}N_2O$ M 186.213

Mp 183-184°.
Phenylhydrazone: Mp 198-201°.

Haug, U. *et al, Chem. Ber.,* 1960, **93,** 593 (*synth*)

8-Acetylquinoline   A-60055
1-(8-Quinolinyl)ethanone, 9CI. Methyl 8-quinolinyl ketone, 8CI

[56234-20-9]

$C_{11}H_9NO$ M 171.198
Mp 42.5-44°. Bp_{12} 176-180°.
Oxime: [30197-64-9].
   $C_{11}H_{10}N_2O$ M 186.213
   Mp 144-145°.
2,4-Dinitrophenylhydrazone: Mp 253°.

Campbell, K.N. *et al, J. Am. Chem. Soc.,* 1946, **68,** 1844 (*synth*)
Haug, U. *et al, Chem. Ber.,* 1960, **93,** 593 (*synth*)

2-Acetylquinoxaline   A-60056
Updated Entry replacing A-50037
1-(2-Quinoxalinyl)ethanone, 9CI. Methyl 2-quinoxalinyl ketone, 8CI

[25594-62-1]

$C_{10}H_8N_2O$ M 172.186
Cryst. (EtOH aq.). Mp 76.5-77.5°.
Oxime: [78583-89-8].
   $C_{10}H_9N_3O$ M 187.201
   Cryst. (Py or DMF). Mp 217-218°.
Phenylhydrazone: Small yellow prisms. Mp 222° dec.
1,4-Dioxide: [79441-12-6].
   $C_{10}H_8N_2O_3$ M 204.185
   Mp 185-188° dec.

Henseke, G. *et al, Chem. Ber.,* 1958, **91,** 1605 (*synth*)
Gardini, G.P. *et al, J. Chem. Soc.* (*C*), 1970, 929 (*synth*)
Usta, J.A. *et al, J. Heterocycl. Chem.,* 1981, **18,** 655 (*deriv*)
Sarodnick, G. *et al, Z. Chem.,* 1982, **22,** 300 (*synth*)

5-Acetylquinoxaline   A-60057
1-(5-Quinoxalinyl)ethanone, 9CI. Methyl 5-quinoxalinyl ketone

[89334-34-9]

$C_{10}H_8N_2O$ M 172.186
Pale-yellow plates. Mp 75-76°.

Sudoh, Y. *et al, Bull. Chem. Soc. Jpn.,* 1983, **56,** 3358 (*synth*)

6-Acetylquinoxaline   A-60058
1-(6-Quinoxalinyl)ethanone, 9CI. Methyl 6-quinoxalinyl ketone

[83570-42-7]

$C_{10}H_8N_2O$ M 172.186
Cryst. (cyclohexane). Mp 106-108°.

Silk, J.A., *J. Chem. Soc.,* 1956, 2058 (*synth*)
McNab, H., *J. Chem. Soc., Perkin Trans. 1,* 1982, 1941 (*synth*)

1-Acetyltriphenylene   A-60059
1-(1-Triphenylenyl)ethanone, 9CI. Methyl 1-triphenylenyl ketone, 8CI

[74732-99-3]

$C_{20}H_{14}O$ M 270.330
Cryst. Mp 122°.

Gore, P.H. *et al, J. Chem. Res.* (*S*), 1980, 40 (*synth, ir, pmr*)

2-Acetyltriphenylene   A-60060
1-(2-Triphenylenyl)ethanone, 9CI. Methyl 2-triphenylenyl ketone

[74733-00-9]

$C_{20}H_{14}O$ M 270.330
Cryst. (EtOH). Mp 155-156°.
Oxime:
   $C_{20}H_{15}NO$ M 285.345
   Needles (EtOH). Mp 202°.

Barker, C.C. *et al, J. Chem. Soc.,* 1955, 4482 (*synth*)
Gore, P.H. *et al, J. Chem. Res.* (*S*), 1980, 40 (*synth, ir, pmr*)
Tanga, M.J. *et al, J. Heterocycl. Chem.,* 1987, **24,** 39 (*synth, uv, ir, pmr, cmr*)

Achillolide *B*   A-60061

$C_{19}H_{24}O_7$ M 364.394
Constit. of *Achillea fragrantissima.* Noncryst. $[\alpha]_D^{22}$ −1.4° (c, 0.5 in EtOH).

1-Deacetoxy, 1-oxo, 6-epimer: **Achillolide A.**
   $C_{17}H_{20}O_6$ M 320.341
   Constit. of *A. fragrantissima.* Cryst. (Me_2CO/pet. ether). Mp 144-145°. $[\alpha]_D^{22}$ −37.4° (c, 0.5 in EtOH).

Segal, R. *et al, Tetrahedron,* 1987, **43,** 4125.

Acuminatin   A-60062

$C_{20}H_{22}O_6$ M 358.390
See also under Licarin *A,* L-20040 . Constit. of *Helichrysum acuminatum.* Oil.

Jakupovic, J. *et al, Phytochemistry,* 1987, **26,** 803.

Acuminatolide   A-60063
[108645-28-9]

$C_{13}H_{12}O_5$ M 248.235

Constit. of *Helichrysum acuminatum*. Cryst. Mp 118°.

Jakupovic, J. *et al, Phytochemistry*, 1987, **26**, 803.

Agasyllin   A-60064
[23402-19-9]

$C_{19}H_{20}O_5$ M 328.364

Isol. from *Agasyllis latifolia, Eryngium campestre, Seseli tortuosum, Xanthogalum purpurascens* and *Zosima korovinii*. Mp 78-80°. $[\alpha]_D^{24}$ −44° (CHCl$_3$).

Nikonov, G.K. *et al, Khim. Prir. Soedin.*, 1969, **119**, 317 (*isol, struct*)
Medvedev, F.A. *et al, Khim. Prir. Soedin.*, 1977, 508 (*ms*)
Gonzalez, A.G. *et al, An. Quim., Ser. C*, 1982, **78**, 184, 407 (*isol*)
Sklyar, Y.E. *et al, Khim. Prir. Soedin.*, 1982, 779 (*isol*)
Erdelmeier, C.A.J. *et al, Planta Med.*, 1985, 407 (*isol*)

Ageratone   A-60065
Updated Entry replacing A-00640
1-[2-[1-[(Acetyloxy)methyl]ethenyl]-5-hydroxy-6-benzofuranyl]ethanone, 9CI. 2-Acetoxyisopropenyl-6-acetyl-5-hydroxybenzofuran
[28915-02-8]

$C_{15}H_{14}O_5$ M 274.273

Constit. of roots of *Ageratum houstonianum*. Mp 122-124°.

De-O-Ac: 6-Acetyl-5-hydroxy-2-(1-hydroxymethylvinyl)benzo[b]furan.
$C_{13}H_{12}O_4$ M 232.235
Constit. of *A. houstonianum*.

Deacetoxy: 6-Acetyl-5-hydroxy-2-isopropenylbenzo[b]-furan.
$C_{13}H_{12}O_3$ M 216.236
Constit. of *A. houstonianum*.

Anthonsen, T. *et al, Acta Chem. Scand.*, 1970, **24**, 721 (*isol, uv, ir, pmr, ms*)
Breuer, M. *et al, Phytochemistry*, 1987, **26**, 3055 (*isol, struct*)

Agrimophol   A-60066
[65792-05-4]

$C_{26}H_{34}O_8$ M 474.550

Isol. from *Agrimonia pilosa*. Possess anthelmintic props. Bp 138.5-139°.

Anon, *CA*, 1978, **88**, 104842; **89**, 20275 (*isol, synth*)
Tang, Y. *et al, CA*, 1982, **97**, 22919 (*cryst struct*)

Agrostistachin   A-60067

$C_{20}H_{30}O_3$ M 318.455

Constit. of *Agrostistachys hookeri*. Cytotoxic. Cryst. (EtOAc). Mp 155-157°. $[\alpha]_D$ +248° (c, 0.18 in CHCl$_3$).

Choi, Y.-H. *et al, Tetrahedron Lett.*, 1986, **27**, 5795 (*cryst struct*)

Ajafinin   A-60068
Updated Entry replacing A-40039

$C_{15}H_{20}O_6$ M 296.319

Constit. of *Ajania fastigiata*. Cryst. (C$_6$H$_6$/EtOAc). Mp 162-164°.

Yusupov, M.I. *et al, Khim. Prir. Soedin.*, 1983, **19**, 650.
Tashkhodzhaev, B. *et al, Khim. Prir. Soedin.*, 1986, **22**, 270 (*cryst struct*)

N-Alanylcysteine, 9CI   A-60069

$H_3CCH(NH_2)CONHCH(COOH)CH_2SH$

$C_6H_{12}N_2O_3S$ M 192.232

D-L-form [33512-01-5]
Isol. from *Lactobacillus plantarum*.

L-L-form [2490-72-4]
Cryst. (MeOH).

Boc-Ala-Cys-OMe: [86810-04-0]. Mp 110-112°. $[\alpha]_D^{26}$ −35.0° (c, 1 in EtOH).

Z-Ala-Cys-OMe: [34804-98-3]. Cryst. (EtOAc). Mp 116.5-118°. $[\alpha]_D^{26}$ −26.5° (c, 1.27 in MeOH).

Wieland, T. *et al, Justus Liebigs Ann. Chem.*, 1954, **588**, 15 (*synth*)

Leahy, J. *et al*, *Arch. Biochem., Biophys.*, 1965, **109**, 449 (*isol*)
Neuenschwander, M. *et al*, *Helv. Chim. Acta*, 1978, **61**, 2437.
Ueki, M. *et al*, *Bull. Chem. Soc. Jpn.*, 1983, **56**, 1187 (*synth*)

Albidin   A-60070

6-Methoxy-3-methyl-4,5-isobenzofurandione
[1402-69-3]

$C_{10}H_8O_4$ M 192.171
Red pigment from *Penicillium albidum*. Red needles
(EtOH). Mp >380°.

Curtis, P.J. *et al*, *Nature (London)*, 1947, **160**, 574 (*isol*)
Grove, J.F. *et al*, *J. Chem. Soc., Perkin Trans. 1*, 1986, 1145
(*isol, struct*)

6-Aldehydoisoophiopogone *A*   A-60071

RR′ = —OCH$_2$O—

$C_{19}H_{14}O_7$ M 354.315
Constit. of *Ophiopogon japonicus*. Orange needles
(MeOH/CHCl$_3$). Mp 170-172°.

Zhu, Y. *et al*, *Phytochemistry*, 1987, **26**, 2873.

6-Aldehydoisoophiopogone *B*   A-60072

As 6-Aldehydoisoophiopogone *A*, A-60071 with

R = H, R′ = OMe

$C_{19}H_{16}O_6$ M 340.332
Constit. of *Ophiopogon japonicus*. Cryst.
(MeOH/CHCl$_3$). Mp 144-145°.

Zhu, Y. *et al*, *Phytochemistry*, 1987, **26**, 2873.

Alethine   A-60073

N,N′-(*Dithiodi-2,1-ethanediyl*)*bis*[*3-aminopropana-
mide*], *9CI*. N,N′-(*Dithiodiethylene*)*bis*[*3-aminopropion-
amide*], *8CI*. [N-(β-Alanyl)-2-aminoethyl] disulfide. N-
(β-Alanyl)cystamine
[646-08-2]

[H$_2$NCH$_2$CH$_2$CONHCH$_2$CH$_2$S]$_2$

$C_{10}H_{22}N_4O_2S_2$ M 294.429
B,2HCl: [14307-88-1]. Needles (EtOH aq.). Mp 221-
222°.
B,(*COOH*)$_2$: Cryst. (EtOH aq.). Mp 186°.

Wittle, E.L. *et al*, *J. Am. Chem. Soc.*, 1953, **75**, 1694 (*synth*)
Bowman, R.E. *et al*, *J. Chem. Soc.*, 1954, 1171 (*synth*)
Viscontini, M. *et al*, *Helv. Chim. Acta*, 1954, **37**, 375 (*synth*)
Kopelevich, V.M. *et al*, *Bioorg. Khim.*, 1979, **5**, 254 (*synth*)

Alloaromadendrane-4,10-diol   A-60074

$C_{15}H_{26}O_2$ M 238.369
(*4α,10β*)-*form* [109360-94-3]
Constit. of *Ambrosia peruviana*. Cryst. Mp 112-113°.
$[\alpha]_D^{25}$ +7° (c, 0.45 in CHCl$_3$).

Goldsby, G. *et al*, *Phytochemistry*, 1987, **26**, 1059.

Allopteroxylin   A-60075

Updated Entry replacing A-20069
*5-Hydroxy-2,8,8-trimethyl-4*H,8H-*benzo[1,2-b:3,4-b′]-
dipyran-4-one*, *9CI*
[4670-29-5]

$C_{15}H_{14}O_4$ M 258.273
Found in roots of *Spathelia sorbifolia*. Yellow cryst. Mp
159-160°.
Ac: [27305-38-0]. Cryst. (EtOAc/pet. ether). Mp 145-
146°.
Me ether: [35930-31-5]. **Perforatin A**. O-
Methylalloptaeroxylin.
  $C_{16}H_{16}O_4$ M 272.300
  Constit. of *Harrisonia abyssinica*, *H. perforata* and
  Ptaeroxylon obliquum. Cryst. (EtOAc/pet. ether). Mp
  152.5-154°.
6-(3-Methyl-2-butenyl): **6-(3-Methyl-2-butenyl)-
allopteroxylin**.
  $C_{20}H_{22}O_4$ M 326.391
  Constit. of *Dictyoloma incanescens*. Yellow cryst.
  (MeOH aq.). Mp 101-103°.
6-(3-Methyl-2-butenyl), Me ether: 6-(3-Methyl-2-
butenyl)allopteroxylin methyl ether.
  $C_{21}H_{24}O_4$ M 340.418
  Constit. of *D. incanescens*. Cryst. (Et$_2$O). Mp 102-
103°.

Dean, F.M. *et al*, *J. Chem. Soc. (C)*, 1969, 114 (*isol, struct*)
Bajwa, B.S. *et al*, *Indian J. Chem.*, 1971, **9**, 17 (*synth, uv, pmr*)
Taylor, D.R. *et al*, *J. Chem. Soc., Perkin Trans. 1*, 1977, 397
(*isol, synth, ir, uv, pmr, ms*)
Ahluwalia, V.K. *et al*, *Bull. Chem. Soc. Jpn.*, 1982, **55**, 2649
(*synth*)
Prasad, K.J.R. *et al*, *Indian J. Chem., Sect. B*, 1982, **21**, 570
(*synth*)
Campos, A.M. *et al*, *Phytochemistry*, 1987, **26**, 2819 (*deriv*)

Allosamidin   A-60076
[103782-08-7]

$C_{25}H_{42}N_4O_{14}$ M 622.625
Aminoglycoside antibiotic. Prod. by *Streptomyces* sp. In-
secticide, by chitinase inhibition. Powder. $[\alpha]_D$ −24.8°
(c, 0.5 in 0.1M AcOH).

6″-Me ether: [107395-29-9]. **Methylallosamidin**.
$C_{26}H_{44}N_4O_{14}$ M 636.652
From *S.* sp. Insect chitinase inhibitor.

Sakuda, S. *et al, Tetrahedron Lett.,* 1986, **27**, 2475 (*isol, struct*)
Koga, D. *et al, Agric. Biol. Chem.,* 1987, **51**, 471 (*props*)
Sakuda, S. *et al, CA,* 1987, **106**, 134812 (*isol, struct, props*)
Sakuda, S. *et al, J. Antibiot.,* 1987, **40**, 296 (*props*)

Aloenin B   A-60077
[106533-41-9]

$C_{34}H_{38}O_{17}$ M 718.664
Constit. of Kenyan aloe. Cryst. Mp 186-188°. $[\alpha]_D^{30}$
−24.9° (c, 0.22 in MeOH).

Speranz, G. *et al, J. Nat. Prod.,* 1986, **49**, 800.

Alterperylenol   A-60078
Updated Entry replacing A-30073
*1,2,12a,12b-Tetrahydro-1,4,9,12a-tetrahydroxy-3,10-
perylenedione, 9CI. Alteichin*
[88899-62-1]

$C_{20}H_{14}O_6$ M 350.327
Same struct. detd. crystallographically for Alterperylenol
and Alteichin but widely differing phys. consts.
reported. Pigment from *Alternaria* sp. Antifungal and
phytotoxic substance. Cryst. (MeOH/CHCl₃). Mp
182-185°, Mp >350°. $[\alpha]_D^{26}$ +699° (c, 0.26 in CHCl₃),
$[\alpha]_D$ +90°.

5,6-Dihydro: [88899-63-2]. **Dihydroalterperylenol**. *Altertoxin I.*
$C_{20}H_{16}O_6$ M 352.343

Constit. of *A.* sp. Antifungal and cytotoxic. Cryst.
(MeOH/CHCl₃). Mp 147-150°. $[\alpha]_D^{26}$ +380° (c, 0.20
in Me₂CO).

Okuno, T. *et al, Tetrahedron Lett.,* 1983, **24**, 5653 (*cryst struct*)
Robeson, D. *et al, Experientia,* 1984, **40**, 1248 (*isol, cryst
struct*)
Stack, M.E. *et al, J. Nat. Prod.,* 1986, **49**, 866 (*isol, bibl*)

Alterporriol B   A-60079
Updated Entry replacing A-40045
*5,6,7,8-Tetrahydro-4,4′,5,6,6′,7,8-heptahydroxy-2,2′-di-
methoxy-7,7′-dimethyl-[1,1′-bianthracene]-9,9′,10,10′-
tetrone, 9CI*
[88901-69-3]

$C_{32}H_{26}O_{13}$ M 618.550
Metab. of *Alternaria porri*. Dark-red cryst. Mp >300°.
$[\alpha]_D^{25}$ −310° (c, 0.05 in EtOH).

Atropisomer: **Alterporriol A**.
$C_{32}H_{26}O_{13}$ M 618.550
Metab. of *A. porri*. Dark-red amorph. solid. Mp 300°
dec. $[\alpha]_D^{25}$ −235° (c, 0.05 in EtOH).

Suemitsu, R. *et al, Agric. Biol. Chem.,* 1984, **48**, 2611 (*isol*)
Suemitsu, R. *et al, Phytochemistry,* 1987, **26**, 3221 (*isol*)

Altersolanol A   A-60080
Updated Entry replacing A-00870
Stemphylin
[22268-16-2]

R¹ = R² = OH

$C_{16}H_{16}O_8$ M 336.298
Pigment produced by *Alternaria solani, A. porri,
Dactylaria lutea, Phomopsis juniperivora* and
Stemphylium botryosum. Active against gram-positive
bacteria. Phytotoxic. Red cryst. Mp 218° dec. $[\alpha]_D^{27}$
−290° (c, 0.25 in Py).

Tetra-Ac: Cryst. (Et₂O/pet. ether). Mp 188-192° with
prior charring.
Penta-Ac: Cryst. (EtOH). Mp 169-173°.
4-Deoxy: [67022-41-7]. **Altersolanol C**. *Dactylariol.*
$C_{16}H_{16}O_7$ M 320.298
From *A. porri* and *D. lutea*. Phytotoxic. Red prisms
(EtOH). Mp 219-223° and 304-306° (double Mp).
$[\alpha]_D$ −33° (c, 0.032 in EtOH).
1,4-Dideoxy: [22350-90-9]. **Altersolanol B**. *Dactylarin.*
$C_{16}H_{16}O_6$ M 304.299

Pigment from *A. solani*, *A. porri*, *D. lutea* and *P. juniperivora*. Active against gram-positive bacteria and protozoa. Phytotoxic. Red-brown plates (EtOH). Mp 228-230° (201-205° dec.).

1,4-Dideoxy, *tri-Ac*: Mp 182-185°. $[\alpha]_D^{23}$ −44° (c, 0.25 in CHCl$_3$).

Stoessl, A., *Can. J. Chem.*, 1969, **47**, 767, 777 (*isol, struct*)
Gordon, M. *et al*, *Can. J. Chem.*, 1972, **50**, 122 (*stereochem*)
Barash, I. *et al*, *Plant Physiol.*, 1975, **55**, 646 (*isol*)
Wheeler, M.M. *et al*, *Phytochemistry*, 1975, **14**, 288 (*isol*)
Bugle, R.C., *Diss. Abstr. Int. B*, 1977, **38**, 192 (*synth*)
Becher, A.M. *et al*, *J. Antibiot.*, 1978, **31**, 324 (*isol*)
Kelly, T.R. *et al*, *Tetrahedron Lett.*, 1978, 4309 (*synth*)
Suemitsu, R. *et al*, *Agric. Biol. Chem.*, 1978, **42**, 1801; 1981, **45**, 2363; 1982, **46**, 1693; 1984, **48**, 2383 (*isol, uv, ir, pmr, props*)
Stoessl, A. *et al*, *Tetrahedron Lett.*, 1979, 2481; *Can. J. Chem.*, 1983, **61**, 372 (*biosynth*)
Arnone, A. *et al*, *J. Chem. Soc.*, *Perkin Trans. 1*, 1986, 525 (*struct*)
Assante, G. *et al*, *Phytochemistry*, 1987, **26**, 703 (*isol, struct*)

Altertoxin II   A-60081

[56257-59-1]

$C_{20}H_{14}O_6$ M 350.327

Constit. of *Alternaria alternata*. Cytotoxic agent. Cryst. (CHCl$_3$/hexane). Mp 245-250°. $[\alpha]_D$ +636° (c, 0.001 in CHCl$_3$).

▷BC9625300.

Stack, M.E. *et al*, *J. Nat. Prod.*, 1986, **49**, 866.

Altertoxin III   A-60082

[105579-74-6]

$C_{20}H_{12}O_6$ M 348.311

Constit. of *Alternaria alternata*. Cryst. Mp 175-230°. $[\alpha]_D$ +845° (c, 0.0004 in CHCl$_3$).

Stack, M.E. *et al*, *J. Nat. Prod.*, 1986, **49**, 866.

Altholactone   A-60083

Updated Entry replacing A-00876
2,3,3a,7a-Tetrahydroxy-3-hydroxy-2-phenyl-5H-furo[3,2-b]pyran-5-one, *9CI*. Goniothalenol
[65408-91-5]

$C_{13}H_{12}O_4$ M 232.235

Constit. of *Polyalthia* spp. and *Goniothalamus giganteus*. Cytotoxic agent. Cryst. (H$_2$O or C$_6$H$_6$). Mp 75°. $[\alpha]_D^{20}$ +188° (c, 0.5 in EtOH).

Ac:
$C_{15}H_{14}O_5$ M 274.273
Mp 142°.

Loder, J.W. *et al*, *Heterocycles*, 1977, **7**, 113 (*isol, struct*)
El-Zayat, A. *et al*, *Tetrahedron Lett.*, 1985, **26**, 955 (*isol, cryst struct*)
Gesson, J.-P. *et al*, *Tetrahedron Lett.*, 1987, **28**, 3945, 3949 (*synth*)

Amanitins   A-60084

Updated Entry replacing A-00888
Amantins

α-Amanitin, R = NH$_2$, R′ = R″ = OH
β-Amanitin, R = R′ = R″ = OH
γ-Amanitin, R = NH$_2$; R′ = OH, R″ = H
ε-Amanitin, R = R′ = OH, R″ = H
Amanullin, R = NH$_2$, R′ = R″ = H
Amanullinic acid, R = R″ = OH, R′ = H

$C_{39}H_{54}N_{10}O_{14}S$ M 918.974

A family of highly toxic constits. of the Green Deathcap Toadstool, *Amanita phalloides*. Highly toxic to humans, *via* cytotoxicity to hepatocytes and kidney secretory cells.

α-**Amanitin** [23109-05-9]
Cyclic (L-asparaginyl-4-hydroxy-L-prolyl-(R)-4,5-dihydroxy-L-isoleucyl-6-hydroxy-2-mercapto-L-tryptophylglycyl-L-isoleucylglycyl-L-cysteinyl) cyclic (4→8)-sulfide (R)-S-oxide, *9CI*
$C_{39}H_{54}N_{10}O_{14}S$ M 918.974
Toxic constit. of *Amanita phalloides*. Needles. Mp 254-255° dec. $[\alpha]_D^{20}$ +191° (H$_2$O).

▷V. poisonous (LD$_{50}$ ca. 3µg/kg). BD6195000.

β-**Amanitin** [21150-22-1]
1-L-Aspartic acid-α-amanitin, *9CI*
$C_{39}H_{53}N_9O_{15}S$ M 919.959
Toxic constit. of several spp. of *Amanita*, notably *A. phalloides* and also found in *Galerina autumnalis*. Cryst. (EtOH). Sol. H$_2$O, EtOH. $C_{39}H_{53}N_9O_{15}S$, M 919.

▷V. poisonous (LD$_{50}$ ca. 10µg/Kg). NJ8324000.

Deoxy: [21150-21-0]. **Amanine**. *1-L- Aspartic acid-4-(2-mercapto-L-trytophan)-α-amanitin*, *9CI*.
$C_{39}H_{53}N_9O_{14}S$ M 903.960
Constit. of *A. phalloides* toxin. Lacks the OH-group on the tryptophan ring.

▷Highly toxic. NJ8326000.

γ-**Amanitin** [21150-23-2]
3-(4-Hydroxy-L-isoleucine)-α-amanitin, *9CI*
$C_{39}H_{54}N_{10}O_{13}S$ M 902.975
Constit. of *A. phalloides*.

▷NJ8323000.

ε-Amanitin [21705-02-2]
1-L-Aspartic acid-3-(S)-4-hydroxy-L-isoleucine-α-
amanitin, 9CI
$C_{39}H_{53}N_9O_{14}S$ M 903.960
▷NJ8325000.
Amanullinic acid [54532-45-5]
1-L-Aspartic acid-3-isoleucine-α-amanitin, 9CI
$C_{39}H_{53}N_9O_{13}S$ M 887.960
Constit. of *A. phalloides.*
Amanullin
$C_{39}H_{54}N_{10}O_{12}S$ M 886.975

Constit. of *A. phalloides.* Nontoxic.

Wieland, Th., *Fortschr. Chem. Org. Naturst.*, 1967, **25**, 214
(*rev*)
Wieland, T., *Justus Liebigs Ann. Chem.*, 1974, 1561, 1570,
1580, 1587 (*bibl, abs config*)
Yocum, R.R. *et al*, *Biochemistry*, 1978, **17**, 3786, 3790 (*isol,
purifn, cryst struct*)

Amarolide   A-60085
Updated Entry replacing A-00894
[29913-86-8]

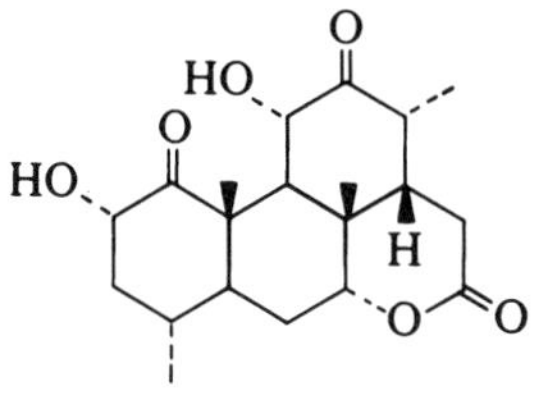

$C_{20}H_{28}O_6$ M 364.438
Constit. of *Ailanthus glandulosa.* Cryst. Mp 253-255°.
Stöcklin, W. *et al*, *Tetrahedron Lett.*, 1970, 2399.
Hirota, H. *et al*, *Tetrahedron Lett.*, 1987, **28**, 435 (*synth*)

Amentadione   A-60086

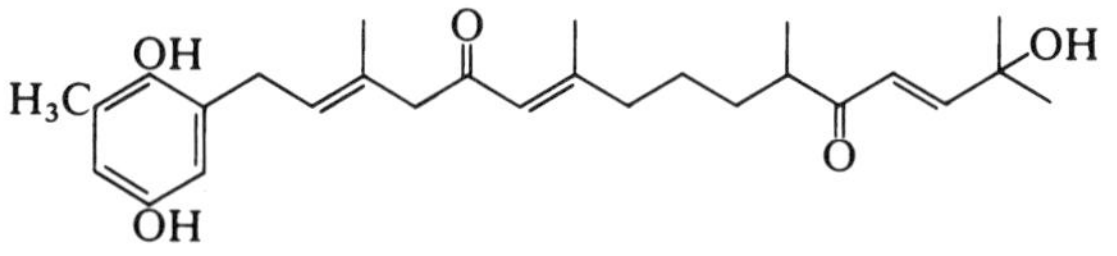

$C_{27}H_{38}O_5$ M 442.594
Constit. of *Cystoseira stricta.* Oil. $[\alpha]_D^{20}$ −0.44° (c, 1.6 in
EtOH).
Amico, V. *et al*, *Phytochemistry*, 1987, **26**, 1715.

Amentaepoxide   A-60087

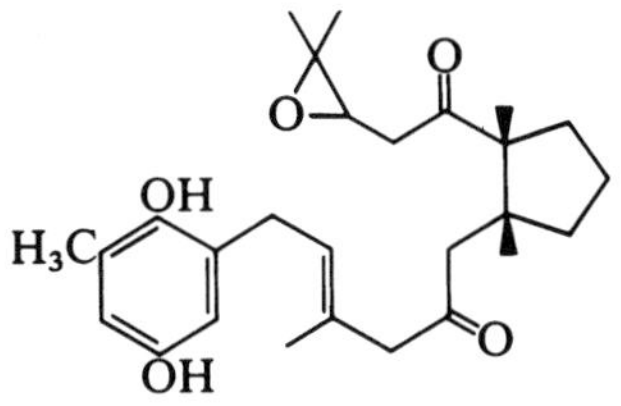

$C_{27}H_{38}O_5$ M 442.594
Constit. of *Cystoseira stricta.* Oil. $[\alpha]_D^{20}$ +14.6° (c, 1.25
in EtOH).
Amico, V. *et al*, *Phytochemistry*, 1987, **26**, 1715.

Amentol   A-60088

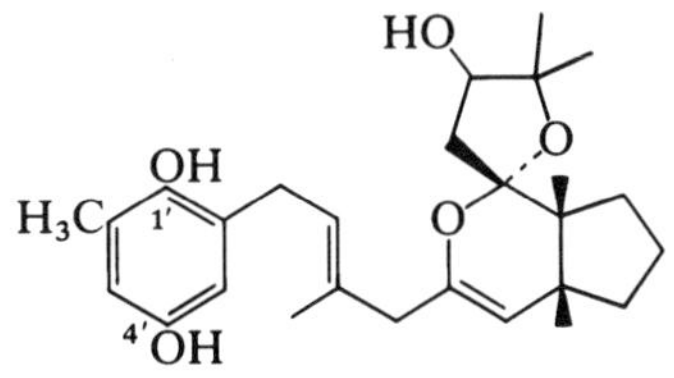

$C_{27}H_{38}O_5$ M 442.594
Constit. of brown alga *Cystoseira stricta.* Oil. $[\alpha]_D$ +7.3°
(c, 1.6 in EtOH).
1′-Me ether: Amentol 1′-methyl ether.
$C_{28}H_{40}O_5$ M 456.621
Constit. of *C. stricta.* Oil. $[\alpha]_D^{20}$ +2.4° (c, 5.7 in
EtOH).
Amico, V. *et al*, *Tetrahedron*, 1986, **42**, 6015.

Amino-1,4-benzoquinone   A-60089
Updated Entry replacing A-01035
2-Amino-2,5-cyclohexadiene-1,4-dione, 9CI.
Aminoquinone
[2783-57-5]

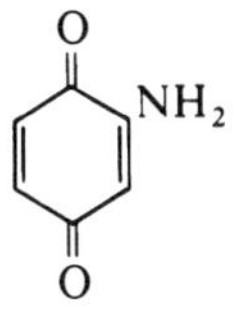

$C_6H_5NO_2$ M 123.111
N-Ac:
$C_8H_7NO_3$ M 165.148
Orange cryst. (C_6H_6). Mp 142°.
Dianil:
$C_{18}H_{15}N_3$ M 273.337
Red cryst. Mp 167°.
Orlov, E.I., *CA*, 1929, **23**, 1125.
Buttrus, N.H. *et al*, *J. Chem. Soc., Perkin Trans. 1*, 1987, 851
(*synth*)

α-Aminobenzo[*b*]thiophene-3-acetic acid,   A-60090
9CI
3-Benzothienylglycine
[95834-94-9]

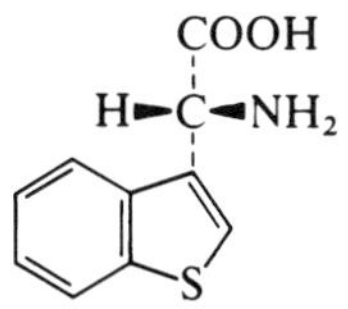

(R)-form

$C_{10}H_9NO_2S$ M 207.247
(R)-form [95909-98-1]
Mp 203-207°. $[\alpha]_D$ −182.6° (c, 1 in 1M HCl).
Me ester: [95909-97-0]. $[\alpha]_D$ −173.8° (c, 1 in EtOH).
Isopropyl ester: [98819-36-4]. Gum. $[\alpha]_D$ −35.6° (c, 1 in
EtOH).
(S)-form
Me ester: [98819-33-1]. $[\alpha]_D$ +165.4° (c, 1 in EtOH).
Isopropyl ester: [98819-35-3]. $[\alpha]_D$ +48.6° (c, 1 in
EtOH).
(±)-form [95834-55-2]
Mp 195-198°.
Me ester: [95835-03-3].
$C_{11}H_{11}NO_2S$ M 221.273

Yellow oil.
Me ester; B,HCl: [98760-30-6]. Mp 180-182°.

Kukolja, S. *et al, J. Med. Chem.*, 1985, **28**, 1903 (*synth*)

3-Amino-2-benzo[*b*]thiophenecarboxylic acid, 9CI   A-60091

[40142-71-0]

C₉H₇NO₂S   M 193.220
Cryst. (EtOH or EtOAc). Mp 140-146° dec.
▷Blue-violet fluor. in soln.
K salt: [40139-58-0]. Mp >300°.
Me ester: [35212-85-2].
 C₁₀H₉NO₂S   M 207.247
 Cryst. (EtOH aq.). Mp 110-111°.
Et ester:
 C₁₁H₁₁NO₂S   M 221.273
 Needles (EtOH aq.). Mp 86.5-87°.
Amide: [37839-59-1].
 C₉H₈N₂OS   M 192.235
 Cryst. (EtOH). Mp 218-220°.
Nitrile: [34761-14-3]. *3-Amino-2-cyanobenzo*[b]-*thiophene.*
 C₉H₆N₂S   M 174.220
 Cryst. (EtOH). Mp 155-156°.

Friedlaender, P. *et al, Justus Liebigs Ann. Chem.*, 1907, **351**, 416 (*synth*)
Carrington, D.E. *et al, J. Chem. Soc.* (*C*), 1971, 3903 (*synth*)
Beck, J.R. *et al, J. Org. Chem.*, 1972, **37**, 3224; 1973, **38**, 2450; 1974, **39**, 3440; 1976, **41**, 1733 (*derivs*)

4-Amino-1,2,3-benzotriazine   A-60092

Updated Entry replacing A-20089
1,2,3-Benzotriazin-4-amine, 9CI

Amine-*form*   2*H*-*form*   3*H*-*form*

C₇H₆N₄   M 146.151
Amine-form
 Mp 284-285° (266° dec.). Major tautomer.
B,HCl: Mp 160-163° dec.
N(4)-Benzyl: [26944-65-0].
 C₁₄H₁₂N₄   M 236.276
 Mp 204-205°.
N(4)-Ph: [888-35-7]. *4-Anilino-1,2,3-benzotriazine.*
 C₁₃H₁₀N₄   M 222.249
 Mp 201° dec.
N(4)-OH: 4-Hydroxylamino-1,2,3-benzotriazine.
 C₇H₆N₄O   M 162.151
 Cream needles (MeOH aq. or MeNO₂). Mp 175° dec.
2-Oxide:
 C₇H₆N₄O   M 162.151
 Cryst. (MeOH). Mp 314-319° dec.
3-Oxide: [52745-08-1].
 C₇H₆N₄O   M 162.151
 Yellow solid, pale-yellow needles + 1EtOH (EtOH). Mp 181°.

N(4)-Butyl: [25465-37-6]. *4-(Butylimino)-3,4-dihydro-1,2,3-benzotriazine (incorr.).*
 C₁₁H₁₄N₄   M 202.258
 Cryst. (EtOH). Mp 153-155°.
N,N(4)-Di-Me: 4-Dimethylamino-1,2,3-benzotriazine.
 C₉H₁₀N₄   M 174.205
 Plates (CH₂Cl₂/pet. ether). Mp 113-114°.
N,N(4)-Di-Me, 2-Oxide:
 C₉H₁₀N₄O   M 190.204
 Cryst. (toluene). Mp 182.5-184°.
2H-form
 Minor tautomer.
2-Me, N(4)-Ph: [29984-72-3].
 C₁₄H₁₂N₄   M 236.276
 Red needles (C₆H₆/pet. ether). Mp 130-131°.
2-Et, N(4)-Ph: [29984-74-5].
 C₁₅H₁₄N₄   M 250.302
 Red prisms (pet. ether). Mp 64-65°.
2-Propyl, N(4)-butyl; B,HI: [90522-71-7]. Ochre solid. Mp 162-163°.
3H-form
 Minor tautomer.
3-Ph: [954-63-2].
 C₁₃H₁₀N₄   M 222.249
 Yellow flakes (2-methoxyethanol). Mp 112-114°.
3-Me, N(4)-Ph: [29372-44-9].
 C₁₄H₁₂N₄   M 236.276
 Yellow needles (EtOH). Mp 131°.
3-Et, N(4)-Ph: [29372-45-0].
 C₁₅H₁₄N₄   M 250.302
 Yellow cryst. Mp 83-84°.
3,N(4)-Di-Ph: [29980-84-5].
 C₁₉H₁₄N₄   M 298.346
 Mp 139-140°.
3-Benzyl: [958-17-8].
 C₁₄H₁₂N₄   M 236.276
 Cryst. (C₆H₆/pet. ether). Mp 119-120°.

Grundmann, C. *et al, J. Org. Chem.*, 1959, **24**, 272 (*synth*)
Parnell, E.W., *J. Chem. Soc.*, 1961, 4930 (*synth*)
Partridge, M.W. *et al, J. Chem. Soc.*, 1964, 3663 (*derivs*)
Gilbert, E.E. *et al, J. Heterocycl. Chem.*, 1969, **6**, 779 (*derivs*)
Stevens, H.N.E. *et al, J. Chem. Soc.* (*C*), 1970, 765, 2289 (*derivs*)
Clark, B.J. *et al, J. Chem. Res.* (*S*), 1984, 62, 64 (*N mmr, derivs, tautom*)

2-Aminobenzoxazole, 8CI   A-60093

2-Benzoxazolamine, 9CI
[4570-41-6]

C₇H₆N₂O   M 134.137
Leaflets (C₆H₆). Mp 129-130°.
▷DM4500000.

Skraup, S., *Justus Liebigs Ann. Chem.*, 1919, **419**, 1 (*synth*)
Reid, W. *et al, Justus Liebigs Ann. Chem.*, 1964, **676**, 114 (*synth*)
Weidinger, H. *et al, Chem. Ber.*, 1964, **97**, 1599 (*synth*)

Handle all chemicals with care

2-Amino-2-benzylbutanedioic acid   A-60094
2-(Phenylmethyl)aspartic acid, 9CI. α-Benzylaspartic acid

$$H_2N-\overset{\underset{|}{CH_2COOH}}{\overset{|}{\underset{}{C}}}-CH_2Ph \qquad (S)\text{-}form$$

$C_{11}H_{13}NO_4$ M 223.228

(S)-form
Mp 235° dec. $[\alpha]_D^{20}$ +50.88° (c, 0.8 in H_2O).
(±)-form [70398-11-7]
Powder. Mp 275° dec.

Kuebel, B. *et al, Chem. Ber.*, 1979, **112**, 128 (*synth, ir, pmr*)
Fadel, A. *et al, Tetrahedron Lett.*, 1987, **28**, 2243 (*synth, ir, pmr*)

7-Aminobicyclo[4.1.0]heptane-7-carboxylic acid   A-60095

$C_8H_{13}NO_2$ M 155.196

trans-form [103348-91-0]
exo-*form*
Mp 185° dec.

Schöllkopf, U. *et al, Angew. Chem., Int. Ed. Engl.*, 1986, **25**, 754 (*synth, pmr*)

7-Amino-3,10-bisaboladiene   A-60096
7-Amino-7,8-dihydro-α-bisabolene

$C_{15}H_{27}N$ M 221.385

(6R,7R)-form [105281-34-3]
Constit. of *Ciocalypta* sp. Oil. $[\alpha]_D$ −15° (c, 0.4 in MeOH).
B,HCl: [105281-33-2]. Yellow oil. $[\alpha]_D^{23}$ −8.3° (c, 0.5 in MeOH).
(6R,7S)-form [105281-43-4]
Constit. of a *Halichondria* sp. Oil. $[\alpha]_D^{20}$ +59.9° (c, 3 in $CHCl_3$).

Sullivan, B.W. *et al, J. Org. Chem.*, 1986, **51**, 5134 (*isol*)
Gulavita, N.K. *et al, J. Org. Chem.*, 1986, **51**, 5136 (*isol*)

2-Amino-3-bromobutanoic acid, 9CI   A-60097
3-Bromobutyrine
[68942-37-0]

$C_4H_8BrNO_2$ M 182.017

(2R,3R)-form
(−)-threo-*form*
B,HCl: Cryst. (MeOH/Et_2O). Mp 185° dec. $[\alpha]_D^{20}$ −26.3° (c, 0.5 in H_2O).

(2R,3S)-form [64187-50-4]
(+)-erythro-*form*
Cryst. (Me_2CO aq.). Mp 198°. $[\alpha]_D^{21}$ +15° (c, 1 in H_2O).
B,HCl: Cryst. (MeOH/Et_2O). Mp 185° dec. $[\alpha]_D^{20}$ +12.7° (c, 0.44 in H_2O).
Me ester, N-benzyloxycarbonyl: Cryst. (EtOAc/pet. ether). Mp 73-74° (53-54°). $[\alpha]_D^{20}$ +31.5° (c, 0.47 in $CHCl_3$).

Wieland, T. *et al, Justus Liebigs Ann. Chem.*, 1977, 806 (*synth*)
Akhtar, M. *et al, Tetrahedron*, 1987, **43**, 5341 (*synth*)

2-Amino-4-bromobutanoic acid, 9CI   A-60098
4-Bromobutyrine

$$H_2N-\overset{\underset{|}{CH_2CH_2Br}}{\overset{|}{\underset{}{C}}}-H \qquad (S)\text{-}form$$

$C_4H_8BrNO_2$ M 182.017

(S)-form [92136-58-8]
B,HBr: [15159-65-6]. Cryst. (Et_2O/MeOH). Mp 188-190°. $[\alpha]_D$ +5.3° (c, 0.4 in DMF).
N-tert-*Butyloxycarbonyl, Me ester:* [76969-87-4]. Cryst. (Et_2O/pet. ether). Mp 62-64°. $[\alpha]_D$ −46° (c, 3 in MeOH).
(±)-form [63038-23-3]
Mp 214-216°.
B,HBr: Cryst. (AcOH). Mp 172°.
tert-*Butyl ester:*
$C_8H_{16}BrNO_2$ M 238.124
Mp 118-121° (as hydrochloride).

Ermolaev, K.M. *et al, CA*, 1966, **65**, 9013 (*synth*)
Nollet, A.J.H. *et al, Tetrahedron*, 1969, **25**, 5971 (*synth*)
DuBois, G.E. *et al, J. Agric. Food Chem.*, 1982, **30**, 676 (*synth*)
Prochazka, Z. *et al, Collect. Czech. Chem. Commun.*, 1982, **47**, 2291 (*pmr, cmr*)
Bajgrowicz, J.A. *et al, Tetrahedron*, 1985, **41**, 1833 (*synth*)

4-Amino-2-bromobutanoic acid, 9CI   A-60099

$$H_2NCH_2CH_2CHBrCOOH$$

$C_4H_8BrNO_2$ M 182.017

B.P., 728 970, (*1955*); *CA*, **50**, 7848 (*synth*)
Fowden, L., *Biochem. J.*, 1956, **64**, 323 (*synth*)

6-Amino-2-bromohexanoic acid, 9CI   A-60100

$$H_2NCH_2(CH_2)_3CHBrCOOH$$

$C_6H_{12}BrNO_2$ M 210.070

(±)-form
B,HBr: [77300-37-9]. Cryst. (EtOH/EtOAc). Mp 118-119.5°.
N-*Ac:*
$C_8H_{14}BrNO_3$ M 252.108
Cryst. (EtOAc). Mp 92-93°.

Wineman, R.J. *et al, J. Am. Chem. Soc.*, 1958, **80**, 6233 (*synth*)
Fukumoto, T. *et al, CA*, 1964, **60**, 653 (*synth*)

2-Amino-5-bromo-3-hydroxypyridine   **A-60101**
2-Amino-5-bromo-3-pyridinol, 9CI
[39903-01-0]

$C_5H_5BrN_2O$ M 189.011
Mp 204-207°.

Mattern, G., *Helv. Chim. Acta*, 1977, **60**, 2062 (*synth*)

2-Amino-3-butylbutanedioic acid   **A-60102**
3-Butylaspartic acid, 9CI. 2-Amino-3-butylsuccinic acid

(2S,3S)-form

$C_8H_{15}NO_4$ M 189.211
(±)-*form* [15383-91-2]
Mp 230-233°.

Winkler, M.F. *et al*, *Biochim. Biophys. Acta*, 1967, **146**, 287 (*synth*)
Akhtar, M. *et al*, *Tetrahedron*, 1987, **43**, 5899.

2-Amino-3-chlorobutanoic acid, 9CI   **A-60103**
3-Chlorobutyrine
[14561-56-9]

(2R,3S)-form

$C_4H_8ClNO_2$ M 137.566
(**2R,3S**)-*form* [59980-98-2]
D-threo-*form*
Me ester:
    $C_5H_{10}ClNO_2$ M 151.593
    Cryst. (MeOH/EtOAc) (as hydrochloride). Mp 119-120° (hydrochloride). $[\alpha]_D$ +22° (c, 0.9 in H_2O).
(**2S,3S**)-*form*
L-erythro-*form*
Plates (H_2O). Mp 176°.
B,HCl: Needles (MeOH aq.). Mp 220° dec.
Me ester: Needles (as hydrochloride). Mp 187° dec. (hydrochloride).
(**2RS,3RS**)-*form* [14561-37-6]
    (±)-threo-*form*
Me ester: [62076-81-7]. Cryst. (MeOH/EtOAc) (as hydrochloride). Mp 159-161° (hydrochloride).
(**2RS,3SR**)-*form* [5856-47-3]
    (±)-erythro-*form*
B,HCl: Mp 196-197°.
Me ester: [62076-67-9]. Cryst. (MeOH/EtOAc) (as hydrochloride). Mp 169-172° (hydrochloride).
N-*Benzoyl:* [62076-72-6]. Cryst. (EtOAc/pet. ether). Mp 77-78°.

Plattner, P.A. *et al*, *Helv. Chim. Acta*, 1957, **40**, 1531 (*synth*)
Tanaka, T. *et al*, *CA*, 1961, **55**, 25962 (*synth*)
Walsh, C.T. *et al*, *J. Biol. Chem*, 1973, **248**, 1946 (*synth*)
Srinivasan, A. *et al*, *J. Org. Chem.*, 1977, **42**, 2256 (*synth*)

2-Amino-4-chlorobutanoic acid, 9CI   **A-60104**
4-Chlorobutyrine
[3157-50-4]

$$ClCH_2CH_2CH(NH_2)COOH$$

$C_4H_8ClNO_2$ M 137.566
(±)-*form*
Mp 192-193°.
B,HCl: Mp 152-155°.
Me ester:
    $C_5H_{10}ClNO_2$ M 151.593
    Mp 115-120° (as hydrochloride).

Frankel, M. *et al*, *J. Am. Chem. Soc.*, 1958, **80**, 3147 (*synth*)
Borcsok, E. *et al*, *Arch. Biochem. Biophys.*, 1982, **213**, 695.

4-Amino-2-chlorobutanoic acid, 9CI   **A-60105**
[4219-23-2]

$$H_2NCH_2CH_2CHClCOOH$$

$C_4H_8ClNO_2$ M 137.566
(±)-*form* [39919-02-3]
Mp 172-174°.

Fowden, L., *Biochem. J.*, 1956, **64**, 323 (*synth*)
Japan. Pats., 71 03 767, 03 964, (*1971*); *CA*, **74**, 124854, 140928 (*synth*)

4-Amino-3-chlorobutanoic acid, 9CI   **A-60106**
[32954-39-5]

$$H_2NCH_2CHClCH_2COOH$$

$C_4H_8ClNO_2$ M 137.566
(±)-*form* [71388-73-3]
    B,HCl: [79958-32-0]. Cryst. (AcOH/EtOAc). Mp 158-159°.

Japan. Pat., 71 08 682, (*1971*); *CA*, **75**, 36680 (*synth*)
Buu, N.T. *et al*, *Br. J. Pharmacol.*, 1974, **52**, 401 (*synth, props*)
Silverman, R.B. *et al*, *J. Biol. Chem*, 1981, **256**, 11565 (*synth, props*)

2-Amino-4-chloro-4,4-difluorobutanoic   **A-60107**
acid, 9CI

$$ClF_2CCH_2CH(NH_2)COOH$$

$C_4H_6ClF_2NO_2$ M 173.547
(±)-*form* [80262-07-3]
    Cryst. Mp 217-218° dec.
    B,HCl: [80386-32-9]. Cryst. Mp 172-176°.
    Et ester: [80262-04-0].
        $C_6H_{10}ClF_2NO_2$ M 201.600
        Cryst. (as hydrochloride). Mp 144-147° (hydrochloride).

Heinzer, F. *et al*, *Helv. Chim. Acta*, 1981, **64**, 1379 (*synth*)

2-Amino-4-chloro-4-fluorobutanoic acid,   **A-60108**
9CI

$$ClFCHCH_2CH(NH_2)COOH$$

$C_4H_7ClFNO_2$ M 155.556
1:1 mixt. of diastereoisomers.
B,HCl: Mp 165-166° dec.
Et ester:
    $C_6H_{11}ClFNO_2$ M 183.610

Mp 148-150° (as hydrochloride).

Heinzer, F. *et al, Helv. Chim. Acta*, 1981, **64**, 1379 (*synth*)

6-Amino-2-chlorohexanoic acid, 9CI   A-60109

$$H_2NCH_2(CH_2)_3CHClCOOH$$

$C_6H_{12}ClNO_2$ M 165.619

(±)-form

B,HCl: [43084-29-3]. Cryst. (EtOH/EtOAc). Mp 83.5-85°.

N-*Ac:*
$C_8H_{14}ClNO_3$ M 207.657
Cryst. (C_6H_6/pet. ether). Mp 76-78°.

Wineman, R.J. *et al, J. Am. Chem. Soc.*, 1958, **80**, 6233 (*synth*)
Effenberger, F. *et al, Chem. Ber.*, 1981, **114**, 173 (*synth*)

2-Amino-5-chloro-6-hydroxy-4-hexenoic acid   A-60110

COOH
H₂N—C—H
CH₂   Cl
C=C
H   CH₂OH

$C_6H_{10}ClNO_3$ M 179.603

(S,Z)-form [108101-63-9]

Isol. from the mushroom *Amanita abrupta*. Mp 200-212° dec. $[\alpha]_D^{20}$ −17.9° (c, 0.19 in H_2O).

Ohta, T. *et al, Phytochemistry*, 1987, **26**, 565 (*isol*)

2-Amino-5-chloro-3-hydroxypyridine   A-60111

2-Amino-5-chloro-3-pyridinol, 9CI

[40966-87-8]

Cl   OH
      NH₂

$C_5H_5ClN_2O$ M 144.560

Mp 198-201°.

Mattern, G., *Helv. Chim. Acta*, 1977, **60**, 2062 (*synth*)

2'-Amino-2-chloro-3'-methylacetophenone, 8CI   A-60112

1-(2-Amino-3-methylphenyl)-2-chloroethanone, 9CI. 2-Chloroacetyl-6-methylaniline

[109532-22-1]

COCH₂Cl
       NH₂
       CH₃

$C_9H_{10}ClNO$ M 183.637

Intermediate for indole synthesis. Yellow cryst. ($CHCl_3$/hexane). Mp 57-58°.

Nimtz, M. *et al, Justus Liebigs Ann. Chem.*, 1987, 765 (*synth, pmr, cmr, ms*)

2'-Amino-2-chloro-4'-methylacetophenone, 8CI   A-60113

1-(2-Amino-4-methylphenyl)-2-chloroethanone, 9CI. 2-Chloroacetyl-5-methylaniline

[109532-23-2]

$C_9H_{10}ClNO$ M 183.637

Yellow cryst. ($CHCl_3$/hexane). Mp 99°.

Nimtz, M. *et al, Justus Liebigs Ann. Chem.*, 1987, 765 (*synth, pmr, cmr, ms*)

2'-Amino-2-chloro-5'-methylacetophenone, 8CI   A-60114

1-(2-Amino-5-methylphenyl)-2-chloroethanone, 9CI. 2-Chloroacetyl-4-methylaniline

[61871-80-5]

$C_9H_{10}ClNO$ M 183.637

Yellow cryst. ($CHCl_3$/hexane). Mp 127°.

Nimtz, M. *et al, Justus Liebigs Ann. Chem.*, 1987, 765 (*synth, pmr, cmr, ms*)

2'-Amino-2-chloro-6'-methylacetophenone, 8CI   A-60115

1-(2-Amino-6-methylphenyl)-2-chloroethanone, 9CI. 2-Chloroacetyl-3-methylaniline

[109532-24-3]

$C_9H_{10}ClNO$ M 183.637

Yellow needles (CCl_4/hexane). Mp 48°.

Nimtz, M. *et al, Justus Liebigs Ann. Chem.*, 1987, 765 (*synth, pmr, cmr, ms*)

2-Amino-6-chloro-1H-purine   A-60116

6-Chloro-1H-purin-2-amine, 9CI

[10310-21-1]

$C_5H_4ClN_5$ M 169.573

Cryst. (H_2O). Mp >275° dec.

▷UO7502000.

Daves, G.D. *et al, J. Am. Chem. Soc.*, 1960, **82**, 2633 (*synth*)

1-Amino-1,4-cyclohexanedicarboxylic acid   A-60117

H₂N COOH

COOH

$C_8H_{13}NO_4$ M 187.195

(1RS,4SR)-form [21146-28-1]

trans-*form*

Mp 247° dec.

B,HCl: [21146-26-9]. Mp 244-248° dec.

Di-Me ester; B,HCl: [21146-34-9]. Mp 173-175.5°.

B.P., 1 115 817, (1968); CA, **70**, 28834 (*synth*)
Ledford, N.D. *et al, Org. Prep. Proced. Int.*, 1987, **19**, 209 (*synth, cmr*)

2-Amino-1,4-cyclohexanedicarboxylic acid   A-60118

(1RS,2RS,4RS)-form

C$_8$H$_{13}$NO$_4$ M 187.195

(1RS,2RS,4RS)-form

(±)(1α,2β,4β)-form

Di-Et ester: [53978-11-3].
   C$_{12}$H$_{21}$NO$_4$ M 243.302
   Mp 43.5-45°. Bp$_{0.05}$ 70-80°.

Di-Et ester; B,HCl: [53978-31-7]. Cryst. (CHCl$_3$/Et$_2$O).
   Mp 188-189°.

Di-Et ester, N-benzoyl: Cryst. (Me$_2$CO). Mp 161-162°.

(1RS,2SR,4RS)-form

(±)(1α,2α,4β)-form

Di-Et ester, N-benzoyl: [53978-17-9]. Cryst.
   (CH$_2$Cl$_2$/hexane). Mp 122-123°.

(1RS,2RS,4SR)-form

(±)(1α,2β,4α)-form

Di-Et ester: [53978-12-4]. Bp$_{0.05}$ 70-90°.

Di-Et ester; B,HCl: [53978-32-8]. Cryst.
   (CHCl$_3$/Et$_2$O/hexane). Mp 132-133°.

Di-Et ester, N-benzoyl: Cryst. (CH$_2$Cl$_2$/Et$_2$O/hexane).
   Mp 150-151°.

(1RS,2SR,4SR)-form

(±)(1α,2α,4α)-form

Di-Et ester: [53978-13-5]. Bp$_{0.0005}$ 75-83°.

Di-Et ester; B,HCl: [53978-33-9]. Cryst.
   (CHCl$_3$/Et$_2$O/hexane). Mp 161-163°.

Di-Et ester, N-benzoyl: Cryst. (Me$_2$CO). Mp 124-125°.

Skaric, V. *et al, J. Chem. Soc., Perkin Trans. 1*, 1974, 1406.

3-Amino-1,2-cyclohexanedicarboxylic acid   A-60119

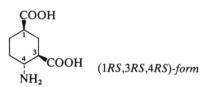

(1RS,2RS,3RS)-form

C$_8$H$_{13}$NO$_4$ M 187.195

(1RS,2RS,3RS)-form

(±)(1α,2α,3β)-form

Di-Et ester: [61883-14-5].
   C$_{12}$H$_{21}$NO$_4$ M 243.302
   Bp$_{0.05}$ 75-77°.

Di-Et ester, N-benzoyl: Cryst. (Et$_2$O/hexane). Mp 130-
   135°.

(1RS,2SR,3SR)-form

(±)(1α,2β,3α)-form

Di-Et ester: [61915-22-8]. Bp$_{0.05}$ 75-80°.

Di-Et ester, N-benzoyl: Cryst. (Et$_2$O/hexane). Mp 165-
   167°.

(1RS,2SR,3RS)-form

(±)(1α,2β,3β)-form

Di-Et ester: [61915-20-6]. Bp$_{0.05}$ 70-75°.

Di-Et ester, N-benzoyl: Cryst. (Et$_2$O/hexane). Mp 119-
   121°.

(1RS,2RS,3SR)-form

(±)(1α,2α,3α)-form

Di-Et ester, N-benzoyl: Cryst. (Et$_2$O/hexane). Mp 96-
   98°.

Skaric, V. *et al, Croat. Chim. Acta*, 1976, **48**, 341.

4-Amino-1,1-cyclohexanedicarboxylic acid,   A-60120
9CI

[57899-74-8]

C$_8$H$_{13}$NO$_4$ M 187.195
Cryst. (Et$_2$O/EtOH aq.). Mp 230-235°.

Di-Et ester: [57899-68-0].
   C$_{12}$H$_{21}$NO$_4$ M 243.302
   Bp$_{0.01}$ 80°.

Di-Et ester; B,HCl: [21896-87-7]. Cryst. (CHCl$_3$/Et$_2$O-
   /hexane). Mp 190-194°.

Fetizon, M. *et al, Bull. Soc. Chim. Fr.*, 1969, 194 (*ester*)
Skaric, V. *et al, Croat. Chim. Acta*, 1975, **47**, 145 (*synth*)

4-Amino-1,3-cyclohexanedicarboxylic acid,   A-60121
9CI

(1RS,3RS,4RS)-form

C$_8$H$_{13}$NO$_4$ M 187.195

(1RS,3RS,4RS)-form

(±)(1α,3α,4β)-form

Di-Et ester: [58660-75-6].
   C$_{12}$H$_{21}$NO$_4$ M 243.302
   Bp$_{0.05}$ 75-80°.

Di-Et ester, N-benzoyl: Cryst. (Et$_2$O/hexane). Mp 140-
   143°.

(1RS,3RS,4SR)-form

(±)(1α,3α,4α)-form

Di-Et ester: [58660-74-5]. Bp$_{0.05}$ 70-75°.

Di-Et ester, N-benzoyl: Cryst. (Et$_2$O/hexane). Mp 115-
   117°.

Skaric, V. *et al, J. Chem. Soc., Perkin Trans. 1*, 1975, 1959.

1-Amino-2-cyclopropene-1-carboxylic acid,   A-60122
9CI

[110374-54-4]

C$_4$H$_5$NO$_2$ M 99.089
Inhibitor of ethylene biosynth. in plants.

B,HCl: [110374-53-3]. Cryst. Mp 164° dec.

N-tert-*Butyloxycarbonyl:* [110374-52-2]. Cryst. (EtOA-
   c/hexane). Mp 139° dec.

N-tert-*Butyloxycarbonyl, Me ester:* [110374-51-1].
   Light-yellow solid. Mp 83-84.5° dec.

Wheeler, T.N. *et al, J. Org. Chem.*, 1987, **52**, 4875 (*synth*)

3-(1-Aminocyclopropyl)-2-propenoic acid   A-60123

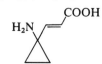

$C_6H_9NO_2$ M 127.143

(E)-form [98212-93-2]

Cryst. Mp 168-171°. Could not be recryst.

B,HCl: [103500-27-2]. Cryst. (EtOH/EtOAc). Mp 195-200° dec.

N-Benzyloxycarbonyl: [103500-26-1]. Fine needles (EtOAc/hexane). Mp 188-189°.

N-Benzyloxycarbonyl, Et ester: Cryst. (Et₂O). Mp 103-104°.

Silverman, R.B. *et al, J. Med. Chem.,* 1986, **29**, 1840 (*synth, ir, pmr*)

6-Amino-2,2-dibromohexanoic acid, 9CI   A-60124

$$H_2NCH_2(CH_2)_3CBr_2COOH$$

$C_6H_{11}Br_2NO_2$ M 288.966

B,HCl: Mp 187-188° dec.

N-Ac:
  $C_8H_{13}Br_2NO_3$ M 331.004
  Mp 105-110°.

Wineman, R.J. *et al, J. Am. Chem. Soc.,* 1958, **80**, 6233 (*synth*)

2-Amino-4,4-dichloro-3-butenoic acid, 9CI   A-60125

[80745-01-3]

$$Cl_2C=CHCH(NH_2)COOH$$

$C_4H_5Cl_2NO_2$ M 169.995

(±)-form [80300-23-8]

Mp 164-165° dec.

Heinzer, F. *et al, Helv. Chim. Acta,* 1981, **64**, 1379, 2279 (*synth*)

6-Amino-2,2-dichlorohexanoic acid, 9CI   A-60126

[22993-77-7]

$$H_2NCH_2(CH_2)_3CCl_2COOH$$

$C_6H_{11}Cl_2NO_2$ M 200.064

Cryst. (H₂O). Mp 181.5-182.5°.

B,HCl: [3080-67-9]. Cryst. Mp 189.5-190.5° dec.

N-Ac:
  $C_8H_{13}Cl_2NO_3$ M 242.102
  Cryst. Mp 86-91°.

N-Benzoyl: [24769-97-9].
  $C_{13}H_{15}Cl_2NO_3$ M 304.172
  Cryst. (pet. ether). Mp 124.5-125.5°.

Wineman, R.J. *et al, J. Am. Chem. Soc.,* 1958, **80**, 6233 (*synth*)
Reinisch, G. *et al, J. Prakt. Chem.,* 1969, **311**, 455 (*synth*)

2-Amino-3-(2,3-dichlorophenyl)propanoic acid   A-60127

2,3-Dichlorophenylalanine, 9CI

$C_9H_9Cl_2NO_2$ M 234.082

(±)-form [110300-04-4]

Mp 238-240° dec.

Taylor, D.C. *et al, Bioorg. Chem.,* 1987, **15**, 335 (*synth*)

2-Amino-3-(2,4-dichlorophenyl)propanoic acid   A-60128

2,4-Dichlorophenylalanine, 9CI

$C_9H_9Cl_2NO_2$ M 234.082

(±)-form [5472-68-4]

Mp 239-241° dec.

Burckhalter, J.H. *et al, J. Am. Chem. Soc.,* 1951, **73**, 56 (*synth*)
Yurkevich, A.M. *et al, Zh. Obshch. Khim.,* 1958, **28**, 227; *CA*, **52**, 12797g (*synth*)
Taylor, D.C. *et al, Bioorg. Chem.,* 1987, **15**, 335 (*synth*)

2-Amino-3-(2,5-dichlorophenyl)propanoic acid   A-60129

2,5-Dichlorophenylalanine, 9CI

$C_9H_9Cl_2NO_2$ M 234.082

(±)-form [110351-36-5]

Mp 237-238° dec.

U.S.P., 3 005 021, (*1961*); *CA*, **56**, 12801g (*synth*)
Taylor, D.C. *et al, Bioorg. Chem.,* 1987, **15**, 335 (*synth*)

2-Amino-3-(2,6-dichlorophenyl)propanoic acid   A-60130

2,6-Dichlorophenylalanine, 9CI

$C_9H_9Cl_2NO_2$ M 234.082

(±)-form [110300-03-3]

Mp 257-259° dec.

Taylor, D.C. *et al, Bioorg. Chem.,* 1987, **15**, 335 (*synth*)

2-Amino-3-(3,4-dichlorophenyl)propanoic acid   A-60131

3,4-Dichlorophenylalanine, 9CI

[59331-62-3]

$C_9H_9Cl_2NO_2$ M 234.082

(±)-form [5472-67-3]

Mp 229-234° dec. (213-216°).

Burckhalter, J.H. *et al, J. Am. Chem. Soc.,* 1951, **73**, 56 (*synth*)
U.S.P., 3 790 679, (*1974*); *CA*, **81**, 126797 (*synth*)
Taylor, D.C. *et al, Bioorg. Chem.,* 1987, **15**, 335 (*synth*)

2-Amino-3-(3,5-dichlorophenyl)propanoic acid   A-60132

3,5-Dichlorophenylalanine, 9CI

$C_9H_9Cl_2NO_2$ M 234.082

(±)-form [93930-25-7]

Mp 229-230° dec.

Taylor, D.C. *et al, Bioorg. Chem.*, 1987, **15**, 335 (*synth*)

2-Amino-3,3-difluorobutanoic acid, 9CI   A-60133

$$H_3CCF_2CH(NH_2)COOH$$

$C_4H_7F_2NO_2$ M 139.102

(±)-*form*
Et ester: [73757-45-6].
  $C_6H_{11}F_2NO_2$ M 167.155
  Mp 156° (as hydrochloride).
Wade, T.M. *et al, J. Org. Chem.*, 1980, **45**, 5333.

3-Amino-4,4-difluorobutanoic acid, 9CI   A-60134
[77162-46-0]

$$F_2CHCH(NH_2)CH_2COOH$$

$C_4H_7F_2NO_2$ M 139.102
Inhibitor of GABA transaminase.
Eur. Pat., 24 965, (*1981*); *CA*, **95**, 62710 (*synth, props*)
U.S.P., 4 435 425, (*1984*); *CA*, **100**, 209153 (*synth, props*)

2-Amino-3,3-difluoropropanoic acid   A-60135
3,3-Difluoroalanine, 9CI
[97995-82-9]

$$H_2N\!\!-\!\!\overset{\displaystyle COOH}{\underset{\displaystyle CHF_2}{C}}\!\!\!-\!\!H \qquad (S)\text{-}form$$

$C_3H_5F_2NO_2$ M 125.075
Kollonitsch, J. *et al, J. Org. Chem.*, 1976, **41**, 3107 (*synth*)
Wang, E.A. *et al, J. Biol. Chem*, 1981, **256**, 6917.
Tsushima, T. *et al, Tetrahedron Lett.*, 1985, **26**, 2445 (*synth*)

9-Amino-9,10-dihydroanthracene   A-60136
9,10-Dihydro-9-anthracenamine, 9CI. 9,10-Dihydro-9-anthrylamine
[97825-91-7]

$C_{14}H_{13}N$ M 195.263
N-*Ac:*
  $C_{16}H_{15}NO$ M 237.301
  Cryst. (C_6H_6). Mp 206.5-207°.
Yasuda, M. *et al, J. Org. Chem.*, 1985, **50**, 3667; 1987, **52**, 753
  (*synth, deriv, ir, pmr, ms*)

4-Amino-5,6-dihydro-4H-cyclopenta[b]-thiophene   A-60137
5,6-Dihydro-4H-cyclopentathiophen-4-amine, 9CI
[108046-24-8]

C_7H_9NS M 139.215

(±)-*form*
Bp$_{0.5}$ 70°.
B,HCl: [108046-27-1]. Mp 172°.
Dallemagne, P. *et al, Tetrahedron Lett.*, 1986, **27**, 2607 (*synth, ir, pmr*)

3-Aminodihydro-2(3H)-furanone, 9CI, 8CI   A-60138
Homoserine lactone. 2-Amino-4-butanolide. 2-Amino-γ-butyrolactone. α-Aminobutyrolactone. 2-Amino-4-hydroxybutanoic acid lactone
[1192-20-7]

(*S*)-*form*

$C_4H_7NO_2$ M 101.105
(*S*)-*form* [2185-02-6]
L-form
B,HBr: [6305-38-0]. Mp 242-244° (198-201°). [α]$_D$
  −21.0° (c, 0.38 in H_2O).
N-*Hexadecanoyl:* [87206-01-7]. N-(*Tetrahydro-2-oxo-3-furanyl)hexadecanamide, 9CI. N-Hexadecanoylhomoserine lactone.*
  $C_{20}H_{37}NO_3$ M 339.517
  Isol. from methanol-utilising bacteria. Shows herbicidal props. Needles. Mp 137-138°.
N-*Benzoyl:* [87219-30-5].
  $C_{11}H_{11}NO_3$ M 205.213
  Mp 123.5-126°. [α]$_D^{20}$ −97.0° (c, 1.4 in $CHCl_3$).
(±)-*form*
B,HCl: [42417-39-0]. Mp 198°.
B,HBr: Mp 224-246°.
N-*Ac:*
  $C_6H_9NO_3$ M 143.142
  Bp$_1$ 160-180°.
Weiss, S. *et al, J. Am. Chem. Soc.*, 1951, **73**, 2497.
Jošt, K. *et al, Collect. Czech. Chem. Commun.*, 1967, **32**, 2365
  (*synth*)
Japan. Pat., 83 096 079, (*1983*); *CA*, **99**, 174212 (*isol, uv, ir, pmr, cmr, deriv*)
McGarvey, G.J. *et al, J. Am. Chem. Soc.*, 1986, **108**, 4943
  (*synth, pmr, ir*)

1-Amino-1,4-dihydronaphthalene   A-60139
1,4-Dihydro-1-naphthalenamine, 9CI. 1,4-Dihydro-1-naphthylamine

$C_{10}H_{11}N$ M 145.204
Liq.
(±)-*form* [97825-92-8]
N-*Ac:*
  $C_{12}H_{13}NO$ M 187.241
  Cryst. (C_6H_6/hexane). Mp 156-157°.
Yasuda, M. *et al, J. Org. Chem.*, 1985, **50**, 3667; 1987, **52**, 753
  (*synth, deriv, ir, pmr, ms*)

1-Amino-5,8-dihydronaphthalene   A-60140
5,8-Dihydro-1-naphthalenamine, 9CI. 5,8-Dihydro-1-naphthylamine
[32666-56-1]

$C_{10}H_{11}N$ M 145.204
Cryst. (pet. ether). Mp 37-38°. Bp$_7$ 123-129°.

N-Ac:
$C_{12}H_{13}NO$ M 187.241
Cryst. (EtOH). Mp 164°.

Watt, G.W. et al, J. Am. Chem. Soc., 1947, **69**, 1657 (synth)
Plieninger, H. et al, Chem. Ber., 1956, **89**, 270 (synth)

2-Amino-9,10-dihydrophenanthrene   A-60141

9,10-Dihydro-2-phenanthrenamine, 9CI. 9,10-Dihydro-2-phenanthrylamine, 8CI

[76302-58-4]

$C_{14}H_{13}N$ M 195.263
Cryst. (Et$_2$O). Mp 52°. Bp$_{0.005}$ 204-210°.

N-Ac:
$C_{16}H_{15}NO$ M 237.301
Cryst. (MeOH). Mp 173-174°.

Kreuger, J.W. et al, J. Org. Chem., 1939, **3**, 340 (synth)
Wirth, H.O. et al, Makromol. Chem., 1963, **63**, 53 (synth)
Byron, D.J. et al, J. Chem. Soc., Perkin Trans. 2, 1983, 197 (synth, deriv)
Sugiura, M. et al, Nippon Kagaku Kaishi, 1986, 197; CA, **105**, 236275f (synth)

4-Amino-9,10-dihydrophenanthrene   A-60142

9,10-Dihydro-4-phenanthrenamine. 9,10-Dihydro-4-phenanthrylamine

[83527-88-2]

$C_{14}H_{13}N$ M 195.263
Prisms (EtOAc). Mp 53-54°.

Krueger, J.W. et al, J. Org. Chem., 1939, **3**, 340 (synth)
Abramovitch, R.A. et al, J. Org. Chem., 1982, **47**, 4818.

9-Amino-9,10-dihydrophenanthrene   A-60143

9,10-Dihydro-9-phenanthrenamine, 9CI. 9,10-Dihydro-9-phenanthrylamine, 8CI

$C_{14}H_{13}N$ M 195.263

(±)-*form* [97825-83-7]

N-Ac:
$C_{16}H_{15}NO$ M 237.301
Cryst. (MeOH). Mp 155.5-156.5°.

Yasuda, M. et al, J. Org. Chem., 1985, **50**, 3667; 1987, **52**, 753 (synth, ir, pmr, ms)

6-Amino-1,3-dihydro-2H-purin-2-one, 9CI   A-60144

Updated Entry replacing A-01495
Isoguanine
[3373-53-3]

$C_5H_5N_5O$ M 151.127
Several tautomers possible. Aglycone from *Croton tiglium*. Amorph. powder. Mp >360°.

1-Me: [73691-67-5]. *6-Amino-1,3-dihydro-1-methyl-2H-purine-2-one, 9CI.*
$C_6H_7N_5O$ M 165.154

Powder. Mp >300°.
9-(β-D-Arabinofuranosyl): [38819-11-3].
$C_{10}H_{13}N_5O_5$ M 283.243
Light-yellow powder. Mp 269-272° dec. [α]$_D^{24}$ +29.5° (c, 0.5 in H$_2$O).

6-Ac, 1-Me: [98933-71-2]. N-(2,9-Dihydro-1-methyl-2-oxo-1H-purin-6-yl)acetamide, 9CI.
$C_8H_9N_5O_2$ M 207.191
Cryst. (MeOH). Mp >274° dec.

6,9-Di-Ac, 1-Me: [98933-69-8].
$C_{10}H_{11}N_5O_3$ M 249.229
Cryst. Mp 222-223° dec. Exists as 6-imino tautomer.

9-β-D-Arabinofuranosyl, 1-Me: [77856-33-8]. *Aradoridosine.*
$C_{11}H_{15}N_5O_5$ M 297.270
Shows antiviral props. Plates (H$_2$O). Mp 223-225° dec. [α]$_D^{24}$ +25.5° (c, 0.5 in H$_2$O).

Spies, J.R., J. Am. Chem. Soc., 1939, **61**, 350 (isol)
Taylor, E.C. et al, J. Am. Chem. Soc., 1959, **81**, 2442 (synth, uv)
Yamazaki, A. et al, Nucleic Acids Res., 1976, **3**, 251 (synth)
Nachman, R.J. et al, J. Chem. Soc., Perkin Trans. 1, 1985, 1315 (deriv, synth, cmr, pmr, ms, uv, ir, cryst struct)

1-Amino-4,5-dihydroxy-7-methoxy-2-methylanthraquinone   A-60145

4-Aminophyscion

$C_{16}H_{13}NO_5$ M 299.282
Metab. of *Dermocybe canaria*. Red needles (EtOAc). Mp 216-217°.

Keller, G. et al, Phytochemistry, 1987, **26**, 2119.

2-Amino-6-(1,2-dihydroxypropyl)-3-methylpterin-4-one   A-60146

Absolute configuration

$C_{10}H_{13}N_5O_3$ M 251.244
Metab. from the marine anthozoan *Astroides calycularis*. Appears to possess cell-growth inhibiting activity. Pale-yellow cryst. powder (MeOH). Mp 229-231°. [α]$_D$ −60° (c, 0.03 in 0.1M HCl).

N,O,O-*Tri-Ac:* [α]$_D$ −67° (c, 0.04 in CHCl$_3$).

Aiello, A. et al, Experientia, 1987, **43**, 950 (isol, uv, pmr, cmr, ms, struct, synth)

2-Amino-3,3-dimethylbutanedioic acid   A-60147

3,3-Dimethylaspartic acid, 9CI

HOOCC(CH$_3$)$_2$CH(NH$_2$)COOH

$C_6H_{11}NO_4$ M 161.157

(±)-*form* [61521-39-9]
Cryst. (H$_2$O). Mp 267-272° dec.

Jönsson, A. et al, CA, 1956, **50**, 10709b (synth)
Bochenska, M. et al, Pol. J. Chem., 1978, **52**, 2355 (synth)
Wanner, M.J. et al, Recl. Trav. Chim. Pays-Bas, 1980, **99**, 20 (synth)

2-Amino-2,3-dimethylbutanoic acid   A-60148

2-Methylvaline, 9CI

[4378-19-2]

$$H_3C - \underset{\underset{CH(CH_3)_2}{|}}{\overset{\overset{COOH}{|}}{C}} \blacktriangleright NH_2 \qquad (R)\text{-}form$$

$C_6H_{13}NO_2$ M 131.174

(*R*)-form
D-form
Me ester: [71785-68-7].
   $C_7H_{15}NO_2$ M 145.201
   Bp$_{4-6}$ 45-55°.

(*S*)-form
L-form
B,HCl: [73473-42-4]. $[\alpha]_D$ +4.6° (c, 2 in H_2O).

Bey, P. *et al*, *Tetrahedron Lett.*, 1977, 1455 (*synth*)
Kolb, M. *et al*, *Tetrahedron Lett.*, 1979, 2999 (*synth*)
Schoellkopf, U. *et al*, *Justus Liebigs Ann. Chem.*, 1981, 696
   (*synth*)

2-Amino-1,3-diphenyl-1-propanone   A-60149

$$PhCOCH(NH_2)CH_2Ph$$

$C_{15}H_{15}NO$ M 225.290

(±)-form
B,HBr: [102831-11-8]. Cryst. (MeOH/Et$_2$O). Mp 249-
250°.
N-Formyl: [102831-18-5].
   $C_{16}H_{15}NO_2$ M 253.300
   Oil.

Muchowski, J.M. *et al*, *J. Org. Chem.*, 1986, **51**, 3374 (*synth*,
   deriv)

2-Amino-2-ethyl-3-butenoic acid, 9CI   A-60150

2-Ethyl-2-vinylglycine

$$H_3CCH_2 - \underset{\underset{CH=CH_2}{|}}{\overset{\overset{COOH}{|}}{C}} \blacktriangleright NH_2$$

$C_6H_{11}NO_2$ M 129.158

(*R*)-form [109918-71-0]
Cryst. + 1H$_2$O (EtOH aq.). Mp 246-247°. $[\alpha]_D^{20}$ +27.7°
(c, 0.365 in H_2O).

Weber, T. *et al*, *Helv. Chim. Acta*, 1986, **69**, 1365 (*synth, pmr,
ms, ir*)

1-Amino-2-ethylcyclopropanecarboxylic acid   A-60151

[87480-58-8]

$C_6H_{11}NO_2$ M 129.158

(*1R,2R*)-form [63393-57-7]
(−)-*Coronamic acid*
$[\alpha]_D^{25}$ −14.2° (c, 1.67 in H_2O).
(*1S,2S*)-form [63393-56-6]
(+)-*Coronamic acid*
$[\alpha]_D^{20}$ +14.7° (c, 1.67 in H_2O).

(*1RS,2RS*)-form [63364-56-7]
(±)-*Coronamic acid*
Trifluoroacetate salt: Mp 168-169°.
(*1S,2R*)-form [65878-52-6]
(+)-*Allocoronamic acid*
$[\alpha]_D^{23}$ +65.0° (c, 1.83 in H_2O).
(*1R,2S*)-form [65878-53-7]
(−)-*Allocoronamic acid*
$[\alpha]_D^{21}$ −68.4° (c, 1.15 in H_2O).
(*1RS,2SR*)-form [65878-54-8]
(±)-*Allocoronamic acid*
Glassy solid. Mp 184-185° subl.

Shiraishi, K. *et al*, *Agric. Biol. Chem.*, 1977, **41**, 2497 (*synth,
   resoln, abs config*)
Ichihara, A. *et al*, *Tetrahedron Lett.*, 1979, 365 (*abs config*)
Suzuki, M. *et al*, *Tetrahedron Lett.*, 1983, **24**, 3839 (*synth*)
Baldwin, J.E. *et al*, *Tetrahedron Lett.*, 1985, **26**, 481, 485
   (*synth, pmr, resoln*)

S-(2-Aminoethyl)cysteine, 9CI   A-60152

3-[(2-Aminoethyl)thio]alanine, 8CI. *Thialysine*.
Thiosine
[617-71-0]

$$H_2N - \underset{\underset{CH_2SCH_2CH_2NH_2}{|}}{\overset{\overset{COOH}{|}}{C}} \blacktriangleright H$$

$C_5H_{12}N_2O_2S$ M 164.222

(*R*)-form [2936-69-8]
L-form
Isol. from the mushroom *Rozites caperta*. Possesses
antibacterial props. Antimetabolite of lysine.
B,HCl: [4099-35-8]. Mp 204-205° dec. (194-195°). $[\alpha]_D^{20}$
−9.0° (c, 1 in 1M HCl).
▷HA1690000.
N$^\alpha$-Ac:
   $C_7H_{14}N_2O_3S$ M 206.259
   Cryst. (Me$_2$CO aq.). Mp 198-200°. $[\alpha]_D^{26}$ −3.5° (c, 4.9
   in H$_2$O).
N$^\epsilon$-Ac:
   $C_7H_{14}N_2O_3S$ M 206.259
   Cryst. (Me$_2$CO aq.). Mp 208-210° dec. $[\alpha]_D^{24}$ −18.5°
   (c, 3 in H$_2$O).
N$^\alpha$,N$^\epsilon$-Di-Ac:
   $C_9H_{16}N_2O_4S$ M 248.296
   Cryst. (MeOH/Et$_2$O). $[\alpha]_D^{21}$ −1.5° (c, 4 in AcOH).
S-Oxide:
   $C_5H_{12}N_2O_3S$ M 180.221
   Mp 183-184° dec. (as hydrochloride). $[\alpha]_D^{25}$ +3.75° (c,
   3.74 in H$_2$O).
S,S-Dioxide:
   $C_5H_{12}N_2O_4S$ M 196.221
   Cryst. (EtOH) (as hydrochloride). Mp 178-180° dec.
   (hydrochloride). $[\alpha]_D^{22}$ −3.7° (c, 4 in H$_2$O).

Hope, D.B. *et al*, *J. Chem. Soc. (C)*, 1966, 1098 (*synth, bibl*)
Hermann, P. *et al*, *J. Prakt. Chem.*, 1969, **311**, 1018 (*synth*)
Matsumoto, N., *CA*, 1985, **103**, 3724 (*isol*)

4-Amino-5-ethynyl-2(1*H*)-pyrimidinone, A-60153
9CI

5-Ethynylcytosine

[65223-79-2]

$C_6H_5N_3O$ M 135.125

Solid. No Mp given.

1-β-D-Ribofuranosyl: [65223-78-1]. *5-Ethynylcytidine,*
9CI.
$C_{11}H_{13}N_3O_5$ M 267.241
Cryst. (MeOH). No Mp given.
1-β-D-(2′-Deoxyribofuranosyl: [69075-47-4]. *2′-Deoxy-*
5-ethynylcytidine, 9CI.
$C_{11}H_{13}N_3O_4$ M 251.241
Mp 155°.

Barr, P.J. *et al, J. Chem. Soc., Perkin Trans. 1*, 1978, 1263
(*synth, deriv, uv, ir, pmr*)

2-Amino-3-fluorobutanoic acid, 9CI A-60154

3-Fluorobutyrine

[50885-01-3]

              (2R,3R)-form

$C_4H_8FNO_2$ M 121.111

(2R,3R)-form [58960-35-3]
Mp 193-194° dec. $[\alpha]_D$ −28.4° (c, 1 in H_2O).
(2R,3S)-form [68781-14-6]
Mp 197-198° dec. $[\alpha]_D$ +12.5° (c, 1 in H_2O).
(2S,3R)-form [68781-15-7]
Mp 197-198° dec. $[\alpha]_D$ −26° (c, 1 in 1M HCl).
(2S,3S)-form [68781-16-8]
Mp 194-195° dec. $[\alpha]_D$ +18° (c, 1 in 1M HCl).

Gershon, H. *et al, J. Med. Chem.*, 1973, **16**, 1407 (*synth*)
Loy, R.S. *et al, J. Fluorine Chem.*, 1976, **7**, 421 (*synth*)
Kollonitsch, J. *et al, J. Org. Chem.*, 1979, **44**, 771 (*synth*)
Pansare, S.V. *et al, J. Org. Chem.*, 1987, **52**, 4804 (*synth*)

2-Amino-4-fluorobutanoic acid, 9CI A-60155

4-Fluorobutyrine

[401-53-6]

$FCH_2CH_2CH(NH_2)COOH$

$C_4H_8FNO_2$ M 121.111

(±)-form [16652-35-0]
B,HCl: [16699-76-6]. Cryst. (EtOH/Et_2O). Mp 110-
112°.

Bergmann, E.D. *et al, Isr. J. Chem.*, 1967, **5**, 15 (*synth*)
Lettre, H. *et al, Justus Liebigs Ann. Chem.*, 1967, **708**, 75
(*synth*)

3-Amino-4-fluorobutanoic acid, 9CI A-60156

[77162-47-1]

$FCH_2CH(NH_2)CH_2COOH$

$C_4H_8FNO_2$ M 121.111
Inactivator of GABA transaminase.

(±)-form
B,HCl: [78347-59-8]. Flaky cryst. (AcOH/EtOAc). Mp
151-152°.

Mathew, J. *et al, Synth. Commun.*, 1985, **15**, 377 (*synth, props*)

4-Amino-2-fluorobutanoic acid, 9CI A-60157

[5130-17-6]

$H_2NCH_2CH_2CHFCOOH$

$C_4H_8FNO_2$ M 121.111

(±)-form
Cryst. + $1H_2O$ (EtOH aq.). Mp 201-203°.

Buchanan, R.L. *et al, Can. J. Chem.*, 1965, **43**, 3466 (*synth*)
Japan. Pat., 73 52 721, (*1973*); *CA*, **80**, 47465 (*synth*)
Borthwick, P.W. *et al, J. Mol. Struct.*, 1977, **41**, 253 (*conformn*)

4-Amino-3-fluorobutanoic acid, 9CI A-60158

$H_2NCH_2CHFCH_2COOH$

$C_4H_8FNO_2$ M 121.111

(±)-form
B,HCl: [68781-12-4]. Cryst. (EtOH). Mp 159.5-160.5°.

Kollonitsch, J. *et al, J. Org. Chem.*, 1979, **44**, 771 (*synth*)

2-Amino-3-fluoro-3-methylbutanoic acid A-60159

3-Fluorovaline, 9CI

[43163-94-6]

            (R)-form

$C_5H_{10}FNO_2$ M 135.138

(R)-form [59752-74-8]
D-form
Plates (2-Propanol aq.).
(±)-form
Cryst. (EtOH aq.). Mp 204° (193-194°).
Nitrile: [79205-57-5].
$C_5H_9FN_2$ M 116.138
No phys. props. reported.

Gershon, H. *et al, J. Med. Chem.*, 1973, **16**, 1407 (*synth*)
Kollonitsch, J. *et al, J. Org. Chem.*, 1976, **41**, 3107 (*synth*)
Ayi, A.I. *et al, J. Fluorine Chem.*, 1984, **24**, 137 (*synth*)

2-Amino-3-fluoropentanoic acid A-60160

3-Fluoronorvaline, 9CI

[43163-95-7]

$H_3CCH_2CHFCH(NH_2)COOH$

$C_5H_{10}FNO_2$ M 135.138
Mp 188-190° dec. Data given is for a 1:1 mix of
diastereoisomers.

Gershon, H. *et al, J. Med. Chem.*, 1973, **16**, 1407 (*synth*)
Pansare, S.V. *et al, J. Org. Chem.*, 1987, **52**, 4804 (*synth*)

4-Amino-5-fluoro-2-pentenoic acid A-60161

$FCH_2CH(NH_2)CH=CHCOOH$

$C_5H_8FNO_2$ M 133.122

(±)-(E)-form [102491-83-8]

Irreversible inhibitor of 4-aminobutyrate-2-oxoglutarate aminotransferase (GABA-T). Cryst. (EtOH aq.). Mp 168°.

Bey, P. *et al, J. Org. Chem.*, 1986, **51**, 2835 (*synth, pmr, use*)

3-Amino-2-fluoropropanoic acid   A-60162

Updated Entry replacing A-01687

2-Fluoro-β-alanine

[3821-81-6]

$C_3H_6FNO_2$ M 107.084

(R)-form [88099-66-5]

Needles (Me$_2$CO aq.). Mp 250-253° dec. $[\alpha]_D^{20}$ +28.4° (c, 0.51 in H$_2$O).

B,HCl: Mp 190-195° dec.

(±)-form

B,HCl: Mp 190-195° dec.

Bergmann, E.D. *et al, J. Chem. Soc.*, 1961, 4669 (*synth*)
Tolman, V. *et al, Collect. Czech. Chem. Commun.*, 1964, **29**, 234 (*synth*)
Abraham, J. *et al, Tetrahedron*, 1977, **33**, 1227 (*nmr*)
Somekh, L. *et al, J. Am. Chem. Soc.*, 1982, **104**, 5836 (*synth*)
Gani, D. *et al, J. Chem. Soc., Perkin Trans. 1*, 1985, 1363 (*synth, pmr*)

8-Amino-6-fluoro-9H-purine   A-60163

$C_5H_4FN_5$ M 153.119

9-β-D-Ribofuranosyl:
$C_{10}H_{12}FN_5O_4$ M 285.234
Cytotoxic agent. Several forms include solid and needles (MeOH). Usually slow dec. >200°; one sample had Mp 231-233° dec.

Secrist, J.A. *et al, J. Med. Chem.*, 1986, **29**, 2069 (*synth, pmr, cmr, F nmr, ms, uv, biochem*)

2-Amino-3-heptenedioic acid, 9CI   A-60164

2-Amino-3,4-dehydropimelic acid

$C_7H_{11}NO_4$ M 173.168

(R,E)-form [89648-25-9]

Isol. from *Asplenum unilalerale* and *A. wilfordii*. Cryst. Mp 172-177°. $[\alpha]_D^{20}$ −70° (c, 0.3 in H$_2$O).

Murakami, N. *et al, Phytochemistry*, 1983, **22**, 2735; 1985, **24**, 2291.

2-Amino-5-heptenedioic acid, 9CI   A-60165

2-Amino-4,5-dehydropimelic acid

$C_7H_{11}NO_4$ M 173.168

(2S,5E)-form

L-trans-form
Mp 208-209°. $[\alpha]_D$ +39.6° (c, 1 in 5M HCl).

Barton, D.H.R. *et al, Tetrahedron*, 1987, **43**, 4297 (*synth*)

4-Amino-2,5-hexadienoic acid   A-60166

$$H_2C{=}CHCH(NH_2)CH{=}CHCOOH$$

$C_6H_9NO_2$ M 127.143

(±)-(E)-form [102420-40-6]

Irreversible inhibitor of 4-aminobutyrate-2-oxoglutarate aminotransferase (GABA-T). Cryst. (EtOH aq.). Mp 134°.

Bey, P. *et al, J. Org. Chem.*, 1986, **51**, 2835 (*synth, pmr, use*)

2-Amino-2-hexenoic acid, 9CI   A-60167

2,3-Didehydronorleucine

$C_6H_{11}NO_2$ M 129.158
Free acid unknown.

(E)-form

N-*Ac, Me ester:* [79357-45-2].
$C_9H_{15}NO_3$ M 185.222
Oil. Bp$_{0.5}$ 85°.

(Z)-form

Me ester: [95824-80-9].
$C_7H_{13}NO_2$ M 143.185
Oil. Bp$_{0.5}$ 56-59°.
N-*Ac, Me ester:* [79357-44-1]. Cryst. (Et$_2$O/hexane). Mp 52-53°.

Scott, J.W. *et al, J. Org. Chem.*, 1981, **46**, 5086.
Effenberger, F. *et al, Chem. Ber.*, 1984, **117**, 1497.
Shin, C.G. *et al, Chem. Pharm. Bull.*, 1984, **32**, 3934.

2-Amino-3-hexenoic acid, 9CI   A-60168

2-Aminohydrosorbic acid

$$H_3CCH_2CH{=}CHCH(NH_2)COOH$$

$C_6H_{11}NO_2$ M 129.158

(±)-(E)-form [80745-00-2]

Cryst. Mp 210-211° dec.

B,HCl: [80744-88-3]. Cryst. Mp 163-166° dec.

Heinzer, F. *et al, Helv. Chim. Acta*, 1981, **64**, 2279 (*synth*)

2-Amino-5-hexenoic acid, 9CI

A-60169

[16258-05-2]

$$\begin{array}{c} \text{COOH} \\ | \\ \text{H} \blacktriangleright \text{C} \blacktriangleleft \text{NH}_2 \\ | \\ \text{CH}_2\text{CH}_2\text{CH}=\text{CH}_2 \end{array} \quad (R)\text{-}form$$

$C_6H_{11}NO_2$ M 129.158
Cryst. (H_2O). Mp 252-255°.
Nitrile: [84673-59-6].
   $C_6H_{10}N_2$ M 110.158
   No phys. props. reported.

Smith, P.W.G. *et al, J. Chem. Soc. (C)*, 1971, 1305 (*synth*)
Kennewell, P.D. *et al, J. Chem. Soc., Perkin Trans. 1*, 1982, 2553 (*synth*)
Bajgrowicz, J.A. *et al, Tetrahedron*, 1985, **41**, 1833 (*nitrile*)
Baldwin, J.E. *et al, J. Chem. Soc., Chem. Commun.*, 1986, 273 (*synth*)

3-Amino-2-hexenoic acid, 9CI

A-60170

$$\text{H}_3\text{CCH}_2\text{CH}_2\text{C(NH}_2)=\text{CHCOOH}$$

$C_6H_{11}NO_2$ M 129.158
Free acid unknown.
Et ester: [58096-02-9].
   $C_8H_{15}NO_2$ M 157.212
   Bp_5 115°, Bp_2 88°.
tert-*Butyl ester:* [87512-32-1].
   $C_{10}H_{19}NO_2$ M 185.266
   $Bp_{0.3}$ 78°.
Amide: [95395-77-0].
   $C_6H_{12}N_2O$ M 128.174
   No phys. props. reported.
Nitrile: [95882-39-6].
   $C_6H_{10}N_2$ M 110.158
   No phys. props. reported.

Celerier, J.P. *et al, Synthesis*, 1981, 130 (*ester*)
Slopianka, M. *et al, Justus Liebigs Ann. Chem.*, 1981, 2258 (*ester*)
Hiyama, T. *et al, Tetrahedron Lett.*, 1982, **23**, 1597 (*ester*)
Sato, M. *et al, Chem. Pharm. Bull.*, 1984, **32**, 3848 (*amide*)
B.P., 2 141 712, (*1985*); *CA*, **103**, 5909 (*nitrile*)

3-Amino-4-hexenoic acid, 9CI

A-60171

$$\begin{array}{c} \text{CH}_2\text{COOH} \\ | \\ \text{H} \blacktriangleright \text{C} \blacktriangleleft \text{NH}_2 \\ | \\ \text{H} \diagdown \text{C} \diagup \text{H} \\ \| \\ \text{C} \\ | \\ \text{CH}_3 \end{array} \quad (S,E)\text{-}form$$

$C_6H_{11}NO_2$ M 129.158
Free acid unknown.
(3S,4E)-form
N-*Benzoyl, Me ester:* [75812-86-1].
   $C_{14}H_{17}NO_3$ M 247.293
   Cryst. (Me_2CO/hexane). Mp 89°. $[\alpha]_D^{23}$ +5.4° (c, 10 in MeOH).
(±)-(E)-form
Me ester: [89889-13-4].
   $C_7H_{13}NO_2$ M 143.185
   Needles (Me_2CO/hexane) (as hydrochloride). Mp 117-120° (hydrochloride).
N-*Benzoyl, Me ester:* [89921-31-3]. Needles. Mp 74-76°.

Hauser, F.M. *et al, J. Org. Chem.*, 1984, **49**, 2236.

3-Amino-5-hexenoic acid, 9CI

A-60172

[87255-31-0]

$$\begin{array}{c} \text{CH}_2\text{COOH} \\ | \\ \text{H} \blacktriangleright \text{C} \blacktriangleleft \text{NH}_2 \\ | \\ \text{CH}_2\text{CH}=\text{CH}_2 \end{array} \quad (R)\text{-}form$$

$C_6H_{11}NO_2$ M 129.158
Intermediate in carbapenem synth.
(R)-form [82448-92-8]
   β-*Lactam:* [82468-77-7]. 4-(2-*Propenyl*)-2-*azetidinone*, 9CI. 4-*Allyl-2-azetidinone*.
   C_6H_9NO M 111.143
   $[\alpha]_D$ −3.41° (c, 8.41 in $CHCl_3$).

Takano, S. *et al, Chem. Lett.*, 1982, 631 (*synth*)
Hua, D.H. *et al, Tetrahedron Lett.*, 1985, **26**, 547 (*lactam*)

6-Amino-2-hexenoic acid, 9CI

A-60173

$$\text{H}_2\text{NCH}_2\text{CH}_2\text{CH}_2\text{CH}=\text{CHCOOH}$$

$C_6H_{11}NO_2$ M 129.158
(E)-form
   B,HCl: [19991-88-9]. Cryst. (EtOH). Mp 156-159°.
   N-*Phthalimido:* [19991-86-7].
     $C_{14}H_{13}NO_4$ M 259.261
     Cryst. (EtOH). Mp 156-158°.
   N-*Phthalimido, Me ester:* [19991-87-8].
     $C_{15}H_{15}NO_4$ M 273.288
     Cryst. Mp 90-92°.

Stammer, C.H. *et al, J. Org. Chem.*, 1969, **34**, 2306 (*synth*)

1-Amino-2-hydroxyanthraquinone

A-60174

1-Amino-2-hydroxy-9,10-anthracenedione, 9CI
[568-98-9]

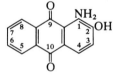

$C_{14}H_9NO_3$ M 239.230
Cryst. (EtOH). Mp 264-266°.
N-*Ac:*
   $C_{16}H_{11}NO_4$ M 281.267
   Golden needles (EtOH). Mp 170°.
Me ether: [10165-33-0]. *1-Amino-2-methoxyanthraquinone*.
   $C_{15}H_{11}NO_3$ M 253.257
   Red cryst. Mp 221-222°.
▷CB5735000.

Liebermann, C. *et al, Justus Liebigs Ann. Chem.*, 1876, **183**, 206 (*synth*)
Lagodzinski, K., *Justus Liebigs Ann. Chem.*, 1905, **342**, 85 (*synth*)
Dokunikhin, N.S. *et al, CA*, 1950, **44**, 1948 (*synth*)
Hida, M. *et al, CA*, 1962, **57**, 16514e (*synth*)
Ger. Pat., 2 531 259, (*1976*); *CA*, **84**, 166267 (*synth*)
Allen, N.S. *et al, Chem. Ind. (London)*, 1979, 214 (*props*)

1-Amino-4-hydroxyanthraquinone

A-60175

1-Amino-4-hydroxy-9,10-anthracenedione, 9CI. *C.I. Disperse Red 15. C.I. Solvent Red 53. Celliton Fast Pink B*
[116-85-8]
$C_{14}H_9NO_3$ M 239.230

Dyestuff. Pink plates (C_6H_6), violet needles (C_6H_6/pet. ether). Mp 215° (207-209°).
▷CB5600000.
Me ether: [116-83-6]. *1-Amino-4-methoxyanthraquinone.*
$C_{15}H_{11}NO_3$ M 253.257
Needles (toluene). Mp 186°.
▷CB5735100.
N-*Me:* [6373-16-6]. *1-Hydroxy-4-(methylamino)-anthraquinone. C.I. Disperse Blue 22.*
$C_{15}H_{11}NO_3$ M 253.257
Dyestuff. Violet needles (butanol). Mp 171.5-172°.

Garg, S.P. *et al, J. Org. Chem.,* 1973, **38**, 1247 (*synth, bibl*)
Morley, J.O. *et al, J. Chem. Technol. Biotechnol.,* 1980, **30**, 409 (*synth*)
Flemming, C.A. *et al, Can. J. Chem.,* 1982, **60**, 624 (*synth*)

1-Amino-5-hydroxyanthraquinone   A-60176
1-Amino-5-hydroxy-9,10-anthracenedione, 9CI
[71502-46-0]
$C_{14}H_9NO_3$ M 239.230
Red needles (EtOH aq.). Mp 216° (208-209°).
N-*Ac:*
$C_{16}H_{11}NO_4$ M 281.267
Mp 228.2-228.6°.
Me ether: [17734-85-9]. *1-Amino-5-methoxyanthraquinone.*
$C_{15}H_{11}NO_3$ M 253.257
Mp 227°.

Ger. Pat., 2 013 790, (*1971*); *CA,* **76**, 101212 (*deriv*)
Ruediger, E.H. *et al, J. Org. Chem.,* 1980, **45**, 1974 (*synth*)

1-Amino-8-hydroxyanthraquinone   A-60177
1-Amino-8-hydroxy-9,10-anthracenedione, 9CI
[63572-76-9]
$C_{14}H_9NO_3$ M 239.230
Violet cryst. (EtOH). Mp 230-231.4°.
N-*Ac:*
$C_{16}H_{11}NO_3$ M 265.268
Mp 216.8-218°.

Schrobsdorff, H., *Ber.,* 1903, **36**, 2936 (*synth*)
Egorova, L.M. *et al, CA,* 1958, **52**, 1613 (*synth*)

2-Amino-1-hydroxyanthraquinone   A-60178
2-Amino-1-hydroxy-9,10-anthracenedione, 9CI
[568-99-0]
$C_{14}H_9NO_3$ M 239.230
Red-brown cryst. (AcOH). Mp 257-258°.

Liebermann, C. *et al, Justus Liebigs Ann. Chem.,* 1876, **183**, 210 (*synth*)
Tanaka, M. *et al, J. Chem. Soc. Jpn.,* 1935, **56**, 192 (*synth*)
Dokunikhin, N.S. *et al, CA,* 1954, **48**, 13669i (*synth*)

2-Amino-3-hydroxyanthraquinone   A-60179
2-Amino-3-hydroxy-9,10-anthracenedione, 9CI
[117-77-1]
$C_{14}H_9NO_3$ M 239.230
Red-brown cryst. Mp >300°.

Hardacre, R.W. *et al, J. Chem. Soc.,* 1929, 180.
U.S.P., 1 922 480, (*1933*); *CA,* **27**, 5083 (*synth*)
Ger. Pat., 605 125, (*1934*); *CA,* **29**, 1834 (*synth*)
Hida, M. *et al, CA,* 1962, **56**, 5901i (*synth*)

3-Amino-1-hydroxyanthraquinone   A-60180
3-Amino-1-hydroxy-9,10-anthracenedione, 9CI
[27165-32-8]
$C_{14}H_9NO_3$ M 239.230
Red needles. Mp 310°.

Scholl, R. *et al, Ber.,* 1904, **37**, 4436 (*synth*)
U.S.S.R. Pat., 364 593, (*1972*); *CA,* **78**, 147685 (*synth*)

9-Amino-10-hydroxy-1,4-anthraquinone   A-60181
9-Amino-10-hydroxy-1,4-anthracenedione, 9CI. 1,4-Dihydroxy-10-imino-9(10H)-anthracenone, 9CI
[96423-72-2]

$C_{14}H_9NO_3$ M 239.230
Tautomeric. Violet cryst. Mp 278° subl.
10-Me ether: [96423-70-0]. *9-Amino-10-methoxy-1,4-anthraquinone, 9CI.*
$C_{15}H_{11}NO_3$ M 253.257
Mp 175-178°.

Farina, F. *et al, Tetrahedron Lett.,* 1985, **26**, 111 (*synth*)

3-Amino-5-hydroxybenzyl alcohol   A-60182
3-Amino-5-hydroxybenzeneethanol, 9CI
$C_7H_9NO_2$ M 139.154
B,HCl: Pale-cream solid. Mp 146-149°.

Rickards, R.W. *et al, Aust. J. Chem.,* 1987, **40**, 1011 (*synth, pmr*)

3-Amino-2-hydroxybutanoic acid, 9CI   A-60183
Updated Entry replacing A-01762
3-Methylisoserine
[565-81-1]

$$\begin{array}{c} COOH \\ H\!\!\blacktriangleright\!\!\underset{|}{C}\!\!\blacktriangleleft\!\!OH \\ H_2N\!\!\blacktriangleright\!\!\underset{|}{C}\!\!\blacktriangleleft\!\!H \\ CH_3 \end{array}$$

$C_4H_9NO_3$ M 119.120
(2R,3S)-form is illus.
(2R,3S)-form
D-*Isothreonine*
Fine needles (EtOH aq.). Mp 215-217°. $[\alpha]_D^{20}$ +22.2° (c, 0.46 in H_2O).
(2S,3S)-form
L-*Alloisothreonine*
Fine needles. Mp 242-243°. $[\alpha]_D^{20}$ −26.15° (c, 1.1 in H_2O).
(2RS,3SR)-form
(±)-*Isothreonine*
Plates (EtOH aq.). Mp 250° dec.
N-*Benzoyl:*
$C_{11}H_{14}NO_3$ M 208.237
Needles (EtOH aq.). Mp 235-236°.

Balenović, K. *et al, Croat. Chim. Acta,* 1956, **28**, 279; *CA,* **51**, 12000e (*synth*)
Liwschitz, Y. *et al, Isr. J. Chem.,* 1963, **1**, 441 (*synth*)
Ger. Pat., 2 632 396, (*1977*); *CA,* **78**, 160113c
Wolf, J.-P. *et al, Helv. Chim. Acta,* 1987, **70**, 116 (*synth, pmr, ms*)

2-Amino-1-hydroxy-1-cyclobutaneacetic acid, 9CI   A-60184

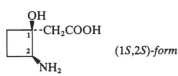

(1S,2S)-form

$C_6H_{11}NO_3$ M 145.158

(**1S,2S**)-*form* [108428-42-8]
Shows antibacterial activity. $[\alpha]_D^{20}$ +8.4° (c, 0.13 in H_2O).

(**1RS,2RS**)-*form* [108507-92-2]
Mp 173-181°.
tert-*Butyl ester:*
$C_{10}H_{19}NO_3$ M 201.265
Solid (EtOAc). Mp 68-69°.

(**1RS,2SR**)-*form* [108507-93-3]
Mp 195-204°.
tert-*Butyl ester:* Solid (EtOAc). Mp 85-86°.

Baldwin, J.E. *et al, Tetrahedron,* 1986, **42**, 2575 (*synth, ir, pmr, cmr, ms*)

2-Amino-3-hydroxy-4-(4-hydroxyphenyl)-butanoic acid   A-60185

α-*Amino-β,4-dihydroxybenzenebutanoic acid, 9CI. 3-Hydroxyhomotyrosine*

$C_{10}H_{13}NO_4$ M 211.217

(**2S,3R**)-*form* [104241-29-4]
Constituent amino acid of the Echinocandins. Cryst. Mp 200-203° dec. $[\alpha]_D^{25}$ +54.1° (c, 1.22 in 1M HCl).

Kurokawa, N. *et al, J. Am. Chem. Soc.,* 1986, **108**, 6041 (*synth, pmr*)

2-Amino-4-hydroxy-2-methylbutanoic acid   A-60186

4-Hydroxyisovaline, 9CI. Methylhomoserine

$C_5H_{11}NO_3$ M 133.147

(**R**)-*form* [109918-65-2]
Cryst. Mp 199-200°. $[\alpha]_D^{25}$ +21.8° (c, 1 in H_2O).

Weber, T. *et al, Helv. Chim. Acta,* 1986, **69**, 1365 (*synth, ir, pmr, ms*)

2-Amino-2-(hydroxymethyl)-3-butenoic acid   A-60187

α-*Vinylserine*

$C_5H_9NO_3$ M 131.131

(±)-*form*
Weak inhibitor of hydroxymethyltransferase. Needles (EtOH aq.). Mp 173-175° dec.

Tendler, S.J.B. *et al, J. Chem. Soc., Perkin Trans. 1,* 1987, 2617 (*synth*)

3-Amino-2-hydroxy-5-methylhexanoic acid, 9CI   A-60188

[62023-30-7]

$C_7H_{15}NO_3$ M 161.200

(**2R,3R**)-*form* [70853-17-7]
(+)-erythro-*form*
Needles. Mp 195-197°. $[\alpha]_D^{24}$ +34.0° (c, 0.5 in AcOH).
N-*Benzyloxycarbonyl:* Mp 116-118°. $[\alpha]_D$ +24.0° (c, 1 in MeOH).

(**2R,3S**)-*form* [73397-21-4]
(+)-threo-*form*
Cryst. (1-Propanol aq.). Mp 228-229°. $[\alpha]_D^{25}$ +26.9° (c, 0.32 in AcOH).
N-*Benzyloxycarbonyl:* Mp 54-56°. $[\alpha]_D$ −17.0° (c, 1 in MeOH).

(**2S,3R**)-*form* [70853-11-1]
(−)-threo-*form*
Component of Amastatin. Needles (2-Propanol aq.). Mp 188-189°. $[\alpha]_D^{22}$ −28.0° (c, 0.5 in AcOH).
N-*Benzyloxycarbonyl:* Mp 78-79°. $[\alpha]_D$ +42.0° (c, 1 in MeOH).

(**2S,3S**)-*form* [73397-20-3]
(−)-erythro-*form*
Mp 280-282° (237-239.5° dec.). $[\alpha]_D^{25}$ −30.6° (c, 0.53 in AcOH).
N-*Benzyloxycarbonyl:* Mp 94-96°. $[\alpha]_D$ −7.0° (c, 1 in MeOH).

(**2RS,3SR**)-*form* [75556-30-8]
(±)-threo-*form*
Mp 243-245° dec.

Johnson, R.L., *J. Med. Chem.,* 1982, **25**, 605 (*synth*)
Tobe, H. *et al, Agric. Biol. Chem.,* 1982, **46**, 1865 (*synth*)

2-Amino-3-hydroxy-4-methyl-6-octenoic acid, 9CI   A-60189

4-(2-Butenyl)-4-methylthreonine

$C_9H_{17}NO_3$ M 187.238

(**2S,3R,4R,6E**)-*form*
N-*Me:* [59865-23-5].
$C_{10}H_{19}NO_3$ M 201.265
Component of Cyclosporine. Cryst. (EtOH aq.). Mp 242-243°. $[\alpha]_D^{20}$ +13.5° (c, 0.5 in H_2O at pH7, phosphate buffer).

Wenger, R., *Helv. Chim. Acta,* 1983, **66**, 2308 (*synth*)
Evans, D.A. *et al, J. Am. Chem. Soc.,* 1986, **108**, 6757 (*synth*)
Tung, R.D. *et al, Tetrahedron Lett.,* 1987, **28**, 1139 (*synth*)

2-Amino-2-(hydroxymethyl)-4-pentenoic acid　　A-60190

α-Allylserine

$$H_2C=CHCH_2 \quad \overset{\displaystyle H_2N \quad COOH}{\underset{\displaystyle CH_2OH}{C}}$$

$C_6H_{11}NO_3$　　M 145.158

(±)-*form*

Cryst. (EtOH aq.). Mp 215-220° dec.

Tendler, S.J.B. *et al, J. Chem. Soc., Perkin Trans. 1*, 1987, 2617 (*synth*)

2-Amino-2-(hydroxymethyl)-4-pentynoic acid　　A-60191

α-Propargylserine

$$HC{\equiv}CCH_2 \quad \overset{\displaystyle COOH}{\underset{\displaystyle NH_2}{C}} \quad HOH_2C$$

$C_6H_9NO_3$　　M 143.142

(±)-*form*

Cryst. (EtOH aq.). Mp 157-160°.

Tendler, S.J.B. *et al, J. Chem. Soc., Perkin Trans. 1*, 1987, 2617 (*synth*)

2-Amino-3-hydroxy-6-methylpyridine　　A-60192

2-Amino-6-methyl-3-pyridinol, 9CI

[20348-16-7]

$C_6H_8N_2O$　　M 124.142

Mp 150-153°. Darkens in air.

Greuter, H. *et al, J. Heterocycl. Chem.*, 1977, **14**, 203 (*synth*)

5-Amino-3-hydroxy-2-methylpyridine　　A-60193

5-Amino-2-methyl-3-pyridinol, 9CI

[57183-27-4]

$C_6H_8N_2O$　　M 124.142

Cryst. (EtOAc/hexane). Mp 189-191°.

Morisawa, Y. *et al, Agric. Biol. Chem.*, 1975, **39**, 1275 (*synth*)

3-Amino-5-hydroxy-7-oxabicyclo[4.1.0]-hept-3-en-2-one, 9CI　　A-60194

2-Amino-5,6-epoxy-3-hydroxy-2-cyclohexen-1-one

[89020-30-4]

(1S,5R,6S)-form

$C_6H_7NO_3$　　M 141.126

(1S,5R,6S)-*form*

N-*Ac*: [93752-54-6]. N-(*5-Hydroxy-2-oxo-7-oxobicyclo[4.1.0]hept-3-en-3-yl)acetamide*. **Antibiotic LL C10037α**. *LL C10037α*.
$C_8H_9NO_4$　　M 183.163
Isol. from *Streptomyces* spp. Shows antitumour props. Weakly active against bacteria. Fluffy powder. $[\alpha]_D^{26}$ −155° (c, 0.1 in H_2O).

N-*Ac, 5-Ketone*: [89020-31-5]. *Antibiotic 36531*.

(1S,5S,6S)-*form* [89020-30-4]
Antibiotic MM 14201. *MM 14201*
Isol. from *S.* spp.

N-*Ac*: Antibiotic MT 35214. *MT 35214*.
$C_8H_9NO_4$　　M 183.163
Semisynthetic. Weakly active against gram-positive and -negative bacteria and *Candida*. Prisms (Me_2CO/Et_2O). Mp 149-151° dec. $[\alpha]_D^{20}$ +104° (c, 1 in MeOH).

5-*Ketone*: *3-Amino-7-oxabicyclo[4.1.0]hept-3-ene-2,5-dione. 2-Amino-5,6-epoxy-2-cyclohexene-1,4-dione. Antibiotic MT 36531. MT 36531.*
$C_8H_7NO_4$　　M 181.148
Semisynthetic. Shows weak antibacterial activity. Needles (EtOAc/Et_2O). Mp 144-145°. $[\alpha]_D^{20}$ −99° (c, 0.5 in MeOH).

Box, S.J. *et al, J. Antibiot.*, 1983, **36**, 1631 (*isol, props*)
Lee, M.D. *et al, J. Antibiot.*, 1984, **37**, 1149 (*isol, struct*)
Whittle, Y.G. *et al, J. Am. Chem. Soc.*, 1987, **109**, 5043 (*biosynth, struct*)

2-Amino-4-hydroxypentanedioic acid　　A-60195

Updated Entry replacing A-20117
4-Hydroxyglutamic acid, 9CI

$$\begin{array}{c} COOH \\ H{\blacktriangleright}\overset{}{C}{\blacktriangleleft}NH_2 \\ CH_2 \\ HO{\blacktriangleright}\overset{}{C}{\blacktriangleleft}H \\ COOH \end{array} \quad \begin{array}{l} (2R,4R)\text{-}form \\ \text{Absolute} \\ \text{configuration} \end{array}$$

$C_5H_9NO_5$　　M 163.130

(2R,4R)-*form*
D-threo-*form*
$[\alpha]_D$ +1.32° (c, 1 in H_2O), $[\alpha]_D$ −0.306° (c, 1 in 5M HCl).

(2S,4S)-*form* [3913-68-6]
L-allo-*form*. L-threo-*form*
Present in the green parts of *Phlox decassata, Linaria vulgaris* and *Hemerocallis* spp. Mp 183-185°. $[\alpha]_D^{26}$ −1.38° (c, 1 in H_2O), $[\alpha]_D$ +0.306° (c, 1 in 5M HCl).

(2R,4S)-*form*
D-erythro-*form*
$[\alpha]_D^{26}$ +1.91° (c, 1 in H_2O), $[\alpha]_D$ −3.8° (c, 1 in 5M HCl).

(2S,4R)-*form* [2485-33-8]
L-erythro-*form*
Intermediate in hydroxyproline metab. in mammals.
$[\alpha]_D^{26}$ −1.95° (c, 1 in H_2O), $[\alpha]_D$ +3.78° (c, 1 in 5M HCl).

(2RS,4RS)-*form* [17093-75-3]
(±)-threo-*form*
Lactone:
$C_5H_7NO_4$　　M 145.115
Mp 228-230° dec. (as hydrochloride).
Lactone, N-chloroacetyl: Cryst. (Me_2CO/Et_2O) (as hydrochloride). Mp 183-185° (hydrochloride).

(2RS,4SR)-*form* [38523-30-7]
(±)-erythro-*form*
Cryst. (EtOH aq.). Mp 165-166° dec.
Lactone, N-chloroacetyl: Mp 172-173°.

Biochem. Prep., 1962, **9**, 69, 74 (*isol*)
Kusami, T. *et al, Bull. Chem. Soc. Jpn.*, 1978, **51**, 1261 (*synth*)
Bjerg, B. *et al, Acta Chem. Scand., Ser. B*, 1983, **37**, 321 (*abs config*)

Passerat, N. *et al*, *Tetrahedron Lett.*, 1987, **28**, 1277 (*synth*)

5-Amino-4-hydroxypentanoic acid   A-60196
[25635-44-3]

$$CH_2CH_2COOH$$
$$H\text{—}C\text{—}OH$$
$$CH_2NH_2$$

$C_5H_{11}NO_3$ M 133.147
(*S*)-*form* [102774-90-3]
Powder (MeOH). Mp 129°. $[\alpha]_D^{20}$ +12.4° (c, 0.4 in H_2O).
Lactone: [102774-96-9]. *5-(Aminomethyl)dihydro-2(3H)-furanone, 9CI.*
$C_5H_9NO_2$ M 115.132
Needles (MeOH) (as hydrochloride). Mp 169-170° (hydrochloride). $[\alpha]_D^{20}$ +81° (c, 1 in MeOH).

Hergeis, C., *Synthesis*, 1986, 232 (*synth*)

1-Amino-2-(4-hydroxyphenyl)-   A-60197
cyclopropanecarboxylic acid, 9CI
Cyclopropyltyrosine

$C_{10}H_{11}NO_3$ M 193.202
(*1RS,2SR*)-*form* [74214-39-4]
(±)-*cis-form*
Inhibitor of dopa decarboxylase.
B,HCl: [111314-74-0]. Prisms + ½H_2O. Mp 208-209° dec.
Me ester: [111314-77-3].
$C_{11}H_{13}NO_3$ M 207.229
Yellow cryst. (Et_2O/2-Propanol) (as hydrochloride). Mp 191-196° (hydrochloride).
Me ether: [95474-41-2].
$C_{11}H_{13}NO_3$ M 207.229
Inhibits dopa decarboxylase, cryst. (EtOH/Et_2O) (as hydrochloride). Mp 198-199° dec.

Bernabe, M. *et al*, *Eur. J. Med. Chem.-Chim. Ther.*, 1979, **14**, 33 (*synth, props*)
Arenal, I. *et al*, *Tetrahedron*, 1985, **41**, 215 (*synth*)
Suzuki, M. *et al*, *Bioorg. Chem.*, 1987, **15**, 43 (*synth*)

2-Amino-3-(4-hydroxyphenyl)-3-methylbu-   A-60198
tanoic acid
β,β-Dimethyltyrosine, 9CI, 8CI. α-Amino-β-(p-hydroxyphenyl)isovaleric acid. Neotyrosine
[565-12-8]

$C_{11}H_{15}NO_3$ M 209.244
(±)-*form*
Cryst. (H_2O). Dec. at ca. 245°.
N-Ac:
$C_{13}H_{17}NO_4$ M 251.282
Prisms (H_2O). Mp 181-181.5°.
Me ether: 2-Amino-3-(4-methoxyphenyl)-3-methylbutanoic acid.
$C_{12}H_{17}NO_3$ M 223.271

Cryst. (H_2O). Dec. at ca. 220°.

Jönsson, A. *et al*, *Acta Chem. Scand.*, 1954, **8**, 1203, 1492 (*synth*)
Jönsson, A. *et al*, *Acta Pharm. Suec.*, 1976, **13**, 75 (*synth*)

4-Amino-3-hydroxy-5-phenylpentanoic acid   A-60199
γ-Amino-β-hydroxybenzenepentanoic acid, 9CI
[105114-21-4]

$$CH_2COOH$$
$$HO\text{—}\overset{3}{C}\text{—}H$$
$$H_2N\text{—}\overset{4}{C}\text{—}H$$
$$CH_2Ph$$ (*3R,4S*)-*form*

$C_{11}H_{15}NO_3$ M 209.244
Component of Ahpatinins.
(*3R,4S*)-*form* [72155-51-2]
Mp 170°. $[\alpha]_D^{24}$ −42.7° (c, 0.12 in H_2O).
N-tert-Butyloxycarbonyl: Mp 187.5°. $[\alpha]_D^{24}$ −16.1° (c, 1.1 in MeOH).
N-tert-Butyloxycarbonyl, Et ester: Mp 140-140.5°. $[\alpha]_D^{24}$ −14.2° (c, 1 in MeOH).
(*3S,4S*)-*form* [72155-50-1]
Component of Pepstatin. Mp 193°. $[\alpha]_D^{24}$ −24.8° (c, 0.44 in H_2O).
N-tert-Butyloxycarbonyl: Mp 148-148.5°. $[\alpha]_D^{24}$ −37.0° (c, 1.1 in MeOH).
N-tert-Butyloxycarbonyl, Et ester: Mp 88-89°. $[\alpha]_D^{24}$ −35.9° (c, 1 in MeOH).

Rich, D.H. *et al*, *J. Med. Chem.*, 1980, **23**, 27 (*synth*)
Omura, S. *et al*, *J. Antibiot.*, 1986, **39**, 1079 (*isol*)

2-Amino-3-hydroxypyridine   A-60200
2-Amino-3-pyridinol, 9CI, 8CI. 3-Hydroxy-2-pyridinamine
[16867-03-1]

$C_5H_6N_2O$ M 110.115
Cryst. (MeOH). Mp 172°.
Picrate: Cryst. (EtOH aq.). Mp 256° dec.

Frazer, J. *et al*, *J. Chem. Soc.*, 1957, 4625 (*synth*)
Boyland, E. *et al*, *J. Chem. Soc.*, 1958, 4198 (*synth*)
Lewicka, K. *et al*, *Recl. Trav. Chim. Pays-Bas*, 1959, **78**, 644 (*synth*)
Greuter, H. *et al*, *J. Heterocycl. Chem.*, 1977, **14**, 203 (*synth*)

2-Amino-5-hydroxypyridine   A-60201
6-Amino-3-pyridinol, 9CI. 5-Hydroxy-2-pyridinamine
[55717-46-9]
$C_5H_6N_2O$ M 110.115
B,HCl: Pale-yellow cryst. (EtOH/Et_2O). Mp 124-126°.

Scudi, J.V. *et al*, *J. Biol. Chem*, 1956, **218**, 587 (*synth*)

4-Amino-3-hydroxypyridine   A-60202
4-Amino-3-pyridinol, 9CI. 3-Hydroxy-4-pyridinamine
[52334-53-9]
$C_5H_6N_2O$ M 110.115

Cryst. (EtOH/pet. ether). Mp 240-242°.
Boyland, E. *et al, J. Chem. Soc.*, 1958, 4198.

5-Amino-3-hydroxypyridine   A-60203
5-Amino-3-pyridinol, 9CI. 5-Hydroxy-3-pyridinamine
[3543-01-9]
$C_5H_6N_2O$ M 110.115
Needles (MeOH/C_6H_6). Mp 116-117°.
B,HBr: Mp 120-125°.
Picrate: Yellow needles. Mp 225-227° dec.

Moore, J.A. *et al, J. Am. Chem. Soc.*, 1959, **81**, 6049 (*synth*)
Roelfsema, W.A., *CA*, 1977, **77**, 19492 (*synth*)
Tamura, Y. *et al, Heterocycles*, 1981, **15**, 871 (*synth*)

α-Amino-1*H*-imidazole-1-propanoic acid,   A-60204
9CI
β-(Imidazol-1-yl)-α-alanine. 1-Imidazolealanine. 1-Isohistidine
[501-32-6]

$C_6H_9N_3O_2$ M 155.156
(±)-*form* [68068-70-2]
Cryst. (EtOH aq.). Mp 231° (212°).
Picrolonate: Cryst. + 2H_2O. Dec. at 250°.

Wieland, T. *et al, Chem. Ber.*, 1957, **90**, 194 (*synth*)
Trout, G.E., *J. Med. Chem.*, 1972, **15**, 1259 (*synth*)
Draminski, M. *et al, Makromol. Chem.*, 1978, **179**, 2195 (*synth*)

α-Amino-1*H*-imidazole-2-propanoic acid,   A-60205
9CI
β-(Imidazol-2-yl)-α-alanine. 2-Imidazolealanine. 2-Isohistidine
[34175-33-2]
$C_6H_9N_3O_2$ M 155.156
(±)-*form* [95462-97-8]
Mp 254-255° dec. Enantiomers, together with many derivs., are descr. in the patent lit.
N^α-*Ac:* [71239-87-7].
  $C_8H_{11}N_3O_3$ M 197.193
  Cryst. (Me_2CO/MeOH/$CHCl_3$). Mp 234.5° dec.
N^α-*Ac*, N^1-*benzyl:* [71239-86-6]. Cryst.
  (Me_2CO/MeOH/$CHCl_3$/Et_2O). Mp 136.5-138°.
N^α-tert-*Butoxycarbonyl*, N^1-*benzyl:* [71239-88-8].
  Cryst. (MeOH/$CHCl_3$/Et_2O/heptane). Mp 198-199°.

Trout, G.E., *J. Med. Chem.*, 1972, **15**, 1259 (*synth*)
Hsieh, K.-H. *et al, J. Med. Chem.*, 1979, **22**, 1199 (*derivs*)
Eur. Pat., 118 787, (*1984*); *CA*, **102**, 149782 (*synth*)
Lloyd, M.J.B. *et al, J. Chromatogr.*, 1986, **351**, 219 (*resoln*)

9-Aminoimidazo[4,5-*f*]quinazoline   A-60206
1H-Imidazo[4,5-f]quinazolin-9-amine. prox-*Benzoadenine*
[53449-43-7]

$C_9H_7N_5$ M 185.188

Cryst. (EtOH). Mp >320°.

Morrice, A.G. *et al, J. Org. Chem.*, 1975, **40**, 363 (*synth, pmr, uv*)

8-Aminoimidazo[4,5-*g*]quinazoline   A-60207
1H-Imidazo[4,5-g]quinazolin-8-amine, 9CI. lin-*Benzoadenine*
[53449-12-0]

$C_9H_7N_5$ M 185.188
Beige cryst.
B,HCl: [53449-13-1]. Cryst. (H_2O/EtOH/Et_2O). Mp >320°.
3-(β-D-Ribofuranosyl): [60189-62-0]. lin-*Benzoadenosine.*
  $C_{14}H_{15}N_5O_4$ M 317.304
  Needles (H_2O). Mp 294-296° dec.
1-(β-D-Ribofuranosyl): [60189-88-0].
  $C_{14}H_{15}N_5O_4$ M 317.304
  Needles (H_2O). Mp 277-280° dec.

Leonard, N.J. *et al, J. Org. Chem.*, 1975, **40**, 356; *J. Am. Chem. Soc.*, 1976, **98**, 3987 (*synth, pmr, uv, deriv*)

7-Aminoimidazo[4,5-*f*]quinazolin-9(8*H*)-   A-60208
one
prox-*Benzoguanine*
[103884-21-5]

$C_9H_7N_5O$ M 201.187
Solid.
B,2HCl: [103884-30-6]. Cryst. Mp >390° dec.

Schneller, S.W. *et al, J. Org. Chem.*, 1986, **51**, 4067 (*synth, ir, pmr*)

6-Aminoimidazo[4,5-*g*]quinolin-8(7*H*)-one   A-60209
lin-*Benzoguanine*
[60064-29-1]

$C_9H_7N_5O$ M 201.187
Solid. Mp >300°.

Keyser, G.E. *et al, J. Org. Chem.*, 1976, **41**, 3529 (*synth, pmr, uv, ir*)

Aminoiminomethanesulfonic acid, 9CI   A-60210
Thiourea S-trioxide. Guanylsulfonic acid
[1184-90-3]

$$HN{=}C(NH_2)SO_3H$$

$CH_4N_2O_3S$ M 124.114

Zwitterionic. Cryst. (AcOH). Mp 131-131.5° (112-115°). Stable at 0°.

N-*Me:* [25343-47-9].
$C_2H_6N_2O_3S$　　M 138.141
Cryst. Mp 154-156° dec.

N,N-*Di-Me:* [25343-55-9].
$C_3H_8N_2O_3S$　　M 152.168
Rhombic cryst. Mp 169-173° dec.

N-tert-*Butyl:* [25343-56-0].
$C_5H_{12}N_2O_3S$　　M 180.221
Prisms. Mp 199-202°.

N-*Benzyl:* [101030-88-0].
$C_8H_{10}N_2O_3S$　　M 214.239
Cryst. Mp 168° dec.

N-*Ph:* [25343-52-6].
$C_7H_8N_2O_3S$　　M 200.212
Solid. Mp 161-161.5° (171-172°).

N,N-*Diisopropyl:* [25348-84-9].
$C_7H_{16}N_2O_3S$　　M 208.275
Prisms. Mp 210.5-212° dec.

N,N-*Di-Ph:* [107678-85-3].
$C_{13}H_{12}N_2O_3S$　　M 276.309
Cryst. (MeOH/Et$_2$O). Mp 209-209.5°.

Debowski, Z., *Przem. Chem.*, 1965, **44**, 82; *CA*, **63**, 479 (*synth*)
Walter, W. *et al, Justus Liebigs Ann. Chem.*, 1969, **722**, 98 (*deriv, synth*)
Miller, A.E. *et al, Synthesis*, 1986, 777 (*synth, ir*)

3-Amino-1-indanone, 8CI　　　　　A-60211
3-Amino-2,3-dihydro-1H-inden-1-one, 9CI

C_9H_9NO　　M 147.176

(±)-*form*
Unstable oil.
B,HCl: Cryst. (2-propanol). Mp >260°.

Rault, S. *et al, Bull. Soc. Chim. Fr.*, 1987, 1079 (*synth, ir, pmr*)

2-Amino-2-isopropylbutanedioic acid　　　A-60212
2-Amino-2-(1-methylethyl)butanedioic acid, 9CI. α-Iso-propylaspartic acid

$C_7H_{13}NO_4$　　M 175.184

(*R*)-*form*
Mp 190°. [α]$_D^{20}$ +55.38° (c, 0.86 in H$_2$O). Incorr. descr. as the (*S*)-form.

Fadel, A. *et al, Tetrahedron Lett.*, 1987, **28**, 2243 (*synth, ir, pmr*)

2-Amino-3-mercapto-3-phenylpropanoic　　A-60213
acid

Updated Entry replacing A-01925
β-Mercaptophenylalanine, 9CI. β-Phenylcysteine
[4371-55-5]

$$PhCH(SH)CH(NH_2)COOH$$

$C_9H_{11}NO_2S$　　M 197.251

Leucine aminopeptidase inhibitor.
B,HCl: [59779-79-2]. Cryst. Mp 222-223° (202-203°) dec.

Heilbron, I. *et al, J. Chem. Soc.*, 1948, 1060 (*synth*)
Kollonitsch, J. *et al, J. Org. Chem.*, 1976, **41**, 3107 (*synth*)
Kraus, J.L., *Pharmacol. Res. Commun.*, 1984, **16**, 533 (*props*)

2-Amino-3-(4-mercaptophenyl)propanoic　　A-60214
acid
4-Mercaptophenylalanine, 9CI

$C_9H_{11}NO_2S$　　M 197.251

(*S*)-*form* [84053-10-1]
No phys. props. reported. V. labile against air oxidation. Forms a symmetrical disulfide.

N,S-*Bis*(tert-*butyloxycarbonyl*): Mp 112.5-114°. [α]$_D$ +8.95° (EtOH).

N-tert-*Butyloxycarbonyl*, S-*Me:* Mp 119.5-121°. [α]$_D$ +6.73° (EtOH).

Escher, E. *et al, Helv. Chim. Acta*, 1983, **66**, 1355 (*synth*)

Aminomethanesulfonic acid, 9CI　　　A-60215
[13881-91-9]

$$H_2NCH_2SO_3H$$

CH_5NO_3S　　M 111.115
Cryst. (H$_2$O). Mp 184-185° dec.

N-*Me:* [23592-45-2]. (*Methylamino*)*methanesulfonic acid, 9CI.*
$C_2H_7NO_3S$　　M 125.142
Cryst. (H$_2$O). Mp 165.5-167.5° dec.

N,N-*Di-Me:* [68507-34-6]. (*Dimethylamino*)-*methanesulfonic acid, 9CI.*
$C_3H_9NO_3S$　　M 139.169
Cryst. Mp 155°.

Backer, H.J. *et al, Recl. Trav. Chim. Pays-Bas*, 1933, **52**, 454 (*derivs*)
Hartough, H.D. *et al, J. Am. Chem. Soc.*, 1950, **72**, 1572 (*synth*)
Lacoste, R.G. *et al, J. Am. Chem. Soc.*, 1955, **77**, 5512 (*synth*)
Fujii, A. *et al, J. Med. Chem.*, 1975, **18**, 502 (*synth, props*)
Boehme, H. *et al, Chem. Ber.*, 1978, **111**, 3294 (*deriv*)
Cassidei, L. *et al, J. Magn. Reson.*, 1985, **62**, 529 (^{33}S nmr)
King, J.F. *et al, Phosphorus Sulfur*, 1985, **25**, 11 (*deriv*)

1-Amino-3-methyl-2,3-butanediol　　　A-60216
Updated Entry replacing A-20123

$C_5H_{13}NO_2$　　M 119.163

(*S*)-*form*
N-*Ac:* [81892-89-9]. N-(*2,3-Dihydroxy-3-methylbutyl*)acetamide, 9CI.
$C_7H_{15}NO_3$　　M 161.200
Isol. from *Verbesina enceloides.* [α]$_D^{25}$ −20.7° (c, 0.15 in CHCl$_3$).

Eichholzer, J.V. *et al, Phytochemistry*, 1982, **21**, 97 (*isol, struct, synth*)

Eichholzer, J.V. *et al*, *Aust. J. Chem.*, 1986, **39**, 1907 (*synth*, *abs config, pmr, cmr*)

3-Amino-3-methyl-2-butanol   A-60217

[13325-14-9]

$$(H_3C)_2C(NH_2)CH(OH)CH_3$$

$C_5H_{13}NO$ M 103.164

(±)-*form*

Oil. d_{25}^{25} 0.92. Bp_{15} 63-65°.

B,HCl: Cryst. Mp 172-174°.

Picrate: [101210-95-1]. Yellow needles (EtOH). Mp 164-166°.

Jones, G.D., *J. Org. Chem.*, 1944, **9**, 484 (*synth*)
Al-Hassan, S.S. *et al*, *J. Chem. Soc., Perkin Trans. 1*, 1985, 1645 (*synth, pmr*)

2-Amino-3-methyl-3-butenoic acid, 9CI   A-60218

Isodehydrovaline

[23311-86-6]

(R)-*form*

$C_5H_9NO_2$ M 115.132

(R)-*form* [60103-01-7]

D-form

Microprisms (EtOH aq.). Mp 210-213° dec. $[\alpha]_D^{27}$ −165° (c, 3.32 in H_2O).

B,HCl: [61348-76-3]. Mp 202-205° dec. $[\alpha]_D^{27}$ −112° (c, 3.44 in H_2O).

Me ester: [61348-78-5].

$C_6H_{11}NO_2$ M 129.158

Cryst. (MeOH/Et_2O) (as hydrochloride). Mp 157-159.5° (hydrochloride).

Benzyl ester: [75520-53-5].

$C_{12}H_{15}NO_2$ M 205.256

Cryst. (MeOH/CH_2Cl_2) (as hydrochloride). Mp 166-168° dec. (hydrochloride). $[\alpha]_D^{25}$ −44.0° (c, 1.052 in MeOH).

(S)-*form* [61376-23-6]

L-form

Microprisms. Mp 212-215° dec. $[\alpha]_D^{27}$ +165° (c, 4.6 in H_2O).

B,HCl: [61376-24-7]. Mp 201-204° dec. $[\alpha]_D^{27}$ +113° (c, 3.64 in H_2O).

(±)-*form* [60049-36-7]

Cryst. (EtOH aq.). Mp 212-215° (217-218° dec.).

B,HCl: [61348-75-2]. Mp 206-208° dec.

Baldwin, J.E. *et al*, *J. Org. Chem.*, 1977, **42**, 1239 (*synth*, *resoln*)
Nunami, K. *et al*, *J. Chem. Soc., Perkin Trans. 1*, 1979, 2224 (*synth*)
Uyeo, S. *et al*, *Chem. Pharm. Bull.*, 1980, **28**, 1563 (*deriv*)
Heinzer, F. *et al*, *Helv. Chim. Acta*, 1981, **64**, 2279 (*synth*)

2-Amino-3-methylenepentanedioic acid   A-60219

3-Methyleneglutamic acid, 9CI

$$H_2C{=}C(CH_2COOH)CH(NH_2)COOH$$

$C_6H_9NO_4$ M 159.141

(±)-*form* [97402-98-7]

Cryst. Mp 187-193°.

B,HCl: [102831-40-3]. Mp 125° dec.

Amide: [97402-99-8].

$C_6H_{10}N_2O_3$ M 158.157

Solid. Mp 130-144° dec.

Amide; B,HCl: [102831-41-4]. Solid. Mp 123-131° dec.

Dowd, P. *et al*, *J. Org. Chem.*, 1986, **51**, 2910 (*synth, deriv, ir, pmr, cmr*)

2-Amino-2-methyl-3-(4-hydroxyphenyl)-   A-60220
propanoic acid

α-*Methyltyrosine*, 9CI. 4-Hydroxy-α-methylphenylalanine

[658-48-0]

(S)-*form*

$C_{10}H_{13}NO_3$ M 195.218

Tyrosine hydroxylase inhibitor.

(S)-*form* [672-87-7]

L-form. Metyrosine. Demser. MK 781

Antihypertensive agent. Cryst. (H_2O). Mp 310-315°. $[\alpha]_D^{20}$ −4.4° (c, 1 in 1M HCl).

(±)-*form* [620-30-4]

Cryst. (H_2O). Mp 318-320° dec.

Potts, K.T., *J. Chem. Soc.*, 1955, 1632 (*synth*)
Stein, G.A. *et al*, *J. Am. Chem. Soc.*, 1955, **77**, 700 (*synth*)
Netherlands Pat., 6 607 757, (1966); *CA*, **67**, 91108 (*synth*)
Saari, W.S., *J. Org. Chem.*, 1967, **32**, 4074 (*synth*)
Weinges, K. *et al*, *Chem. Ber.*, 1971, **104**, 3594.
Gaudestad, O. *et al*, *Acta Chem. Scand., Ser. B*, 1976, **30**, 501 (*cryst struct*)
Robinson, C., *Med. Actual.*, 1980, **16**, 343 (*rev*)
O'Donnell, M.J. *et al*, *Tetrahedron Lett.*, 1982, **23**, 4259 (*synth*)
Bjorkling, F. *et al*, *Tetrahedron Lett.*, 1985, **26**, 4957 (*synth*)

4-Amino-5-methyl-3-isoxazolidinone, 9CI,   A-60221
8CI

Cyclothreonine

[433-69-2]

(4R,5R)-*form*

$C_4H_8N_2O_2$ M 116.119

Possesses antibacterial props.

(4R,5R)-*form*

D-trans-form

$[\alpha]_D$ +159° (c, 1 in 1M NaOH). Browns >205°, does not melt.

(4S,5S)-*form*

L-trans-form

$[\alpha]_D$ −157° (c, 1 in 1M NaOH). Browns >205°, does not melt.

(4RS,5RS)-*form*

(±)-trans-*form*

Mp ca. 160° dec.

(4RS,5SR): (±)-cis-*form*. Mp ca. 160° dec.

Plattner, P.A. *et al*, *Helv. Chim. Acta*, 1957, **40**, 1531 (*synth*)
Stammer, C.H. *et al*, *J. Am. Chem. Soc.*, 1957, **79**, 3236 (*synth*)

3-Amino-4-methyl-5-nitrophenol A-60222

$C_7H_8N_2O_3$ M 168.152

Me ether: [16024-30-9]. *5-Methoxy-2-methyl-3-ni-troaniline. 2-Amino-4-methoxy-6-nitrotoluene.*
$C_8H_{10}N_2O_3$ M 182.179
Mp 87-88°.

Büchi, G. *et al, J. Am. Chem. Soc.,* 1986, **108**, 4115 *(deriv, synth, uv, pmr, ms)*

2-Amino-4-methylpentanedioic acid A-60223

Updated Entry replacing A-02067
4-Methylglutamic acid, 9CI. γ-Methylglutamic acid
[2596-04-5]

$$H_2N\overset{COOH}{\underset{CH_2}{\underset{}{\overset{2}{C}}}}\!\!\blacktriangleleft H$$

H₃C►C◄H
COOH   *(2S,4R)-form*

$C_6H_{11}NO_4$ M 161.157
Cryst. (H_2O). Mp 155-158°.

(2S,4R)-form [33511-70-5]
Found in the ferm *Phyllitis scolopendrium* and in tulips.
$[\alpha]_D^{25}$ +21.4° (c, 2.62 in 6*M* DCl).

(2S,4S)-form [33511-69-2]
$[\alpha]_D^{25}$ +36.3° (c, 2.58 in 6*M*HCl).

Smrt, J. *et al, Collect. Czech. Chem. Commun.,* 1953, **18**, 131 *(synth)*
Blake, J. *et al, Biochem. J.,* 1964, **92**, 136 *(isol, abs config)*
Bory, S. *et al, J. Chem. Soc., Perkin Trans. 1,* 1984, 475 *(resoln)*
Belokon, Y.N. *et al, J. Chem. Soc., Perkin Trans. 1,* 1986, 1865 *(synth, pmr)*

2-Aminomethyl-1,10-phenanthroline A-60224

1,10-Phenanthroline-2-methanamine, 9CI
[111622-25-4]

$C_{13}H_{11}N_3$ M 209.250
Chelating agent.
B,AcOH: Mp 170-172°.

Engberson, J.F.J. *et al, J. Heterocycl. Chem.,* 1986, **23**, 989 *(synth, pmr)*

2-Amino-2-methyl-3-phenylpropanoic acid A-60225

α-Methylphenylalanine, 9CI. 2-Methyl-3-phenylalanine, 8CI

$$H_2N\blacktriangleright\overset{COOH}{\underset{CH_2Ph}{C}}\!\!\blacktriangleleft CH_3 \quad \textit{(S)-form}$$

$C_{10}H_{13}NO_2$ M 179.218
(S)-form [23239-35-2]
L-form
Cryst. (MeOH). Mp 308-309° dec. $[\alpha]_D^{22}$ −22° (c, 1 in H_2O).
B,HCl: [30990-90-0]. Mp 249° dec. $[\alpha]_D^{20}$ −8.6° (c, 1 in H_2O).

Me ester: [28385-45-7].
$C_{11}H_{15}NO_2$ M 193.245
Bp₀.₀₁ 90°. $[\alpha]_D^{20}$ +2.4° (c, 0.8 in EtOH).

(±)-form [1132-26-9]
Mp 294-295° dec.
B,HCl: Dec. at 236°.
Et ester:
$C_{12}H_{17}NO_2$ M 207.272
Bp₉ 152-154°.
N-Ac:
$C_{12}H_{15}NO_3$ M 221.255
Cryst. (EtOH). Mp 203-204°.

Terashima, S. *et al, Chem. Pharm. Bull.,* 1966, **14**, 1138 *(abs config)*
Torchiana, M.L. *et al, Biochem. Pharmacol.,* 1970, **19**, 1601 *(props)*
Hoppe, I. *et al, Synthesis,* 1983, 789 *(synth)*
Kolb, M. *et al, Justus Liebigs Ann. Chem.,* 1983, 1668 *(synth, bibl)*
Karady, S. *et al, Tetrahedron Lett.,* 1984, **25**, 4337 *(synth)*
Belokon, Y.N. *et al, J. Chem. Soc., Chem. Commun.,* 1985, 171 *(synth)*
Bjorkling, F. *et al, Tetrahedron,* 1985, **41**, 1347 *(synth)*
Bjorkling, F. *et al, Tetrahedron Lett.,* 1985, **26**, 4957 *(synth)*

6-Amino-7-methylpurine A-60226

Updated Entry replacing A-02108
7-Methyl-7H-purin-6-amine, 9CI. 7-Methyladenine
[935-69-3]

$C_6H_7N_5$ M 149.155
Powder (H_2O), minute cryst. Mp 349-350° dec.

Fischer, E., *Ber.,* 1898, **31**, 111 *(synth)*
Chenon, M.T. *et al, J. Am. Chem. Soc.,* 1975, **97**, 4627 *(cmr)*
Kistenmacher, T.J. *et al, Acta Crystallogr., Sect. B,* 1975, **31**, 211 *(cryst struct)*
Yamauchi, K. *et al, J. Org. Chem.,* 1976, **41**, 3691 *(synth)*
Leonard, N.J. *et al, Chem. Pharm. Bull.,* 1986, **34**, 2037 *(synth, uv, bibl)*

2-(Aminomethyl)pyridine A-60227

Updated Entry replacing A-02112
2-Pyridinemethanamine, 9CI. 2-Pyridylmethylamine. α-Picolylamine. Monopicolylamine
[3731-51-9]

$C_6H_8N_2$ M 108.143
Sol. H_2O. Bp₁₅ 91°, Bp₁₂ 82°.
▷US1840000.

B,2HBr: Mp 234°.
B,AcOH: Mp 127°.
Picrate: Needles (EtOH). Mp 162° dec.
N-Benzoyl: [35854-47-8].
$C_{13}H_{12}N_2O$ M 212.251
Rosettes (pet. ether). Mp 53°. Bp₁₅ 235°.

Kolloff, H.G. *et al, J. Am. Chem. Soc.,* 1941, **63**, 490.
Biniecki, S. *et al, CA,* 1970, **72**, 66760.
Volkova, L.D., *CA,* 1974, **80**, 120705 *(synth)*
Anderegg, G. *et al, Helv. Chim. Acta,* 1986, **69**, 329 *(nmr)*
Engbersen, J.F.J. *et al, J. Heterocycl. Chem.,* 1986, **23**, 989 *(synth)*

3-(Aminomethyl)pyridine   A-60228

Updated Entry replacing A-02113
3-Pyridinemethanamine, 9CI. *3-Pyridylmethylamine. β-Picolylamine*
[3731-52-0]
$C_6H_8N_2$ M 108.143
Oil. Misc. H_2O, EtOH, Et_2O. Bp_{18} 112°.
B,AcOH: Mp 113°.
Picrate: Cryst. (EtOH). Mp 211°.

Adkins, H. *et al, J. Am. Chem. Soc.*, 1944, **66**, 1293.
Fischer, E. *et al, Chimia*, 1969, **23**, 155.
Engbersen, J.F.J. *et al, J. Heterocycl. Chem.*, 1986, **23**, 989 (*synth*)

4-(Aminomethyl)pyridine   A-60229

Updated Entry replacing A-02114
4-Pyridinemethanamine, 9CI. *4-Pyridylmethylamine. γ-Picolylamine*
[3731-53-1]
$C_6H_8N_2$ M 108.143
Oil. Bp_{11} 103°.
B,AcOH: Mp 95°.
N-*Benzoyl:* [3820-26-6].
    $C_{13}H_{12}N_2O$ M 212.251
    Mp 108°. Bp_{12} 240°.

Graf, R. *et al, J. Prakt. Chem.*, 1936, **146**, 88 (*synth*)
Biniecki, S. *et al, Acta Pol. Pharm.*, 1969, **26**, 277; *CA*, **72**, 66760 (*synth*)
Engbersen, J.F.J. *et al, J. Heterocycl. Chem.*, 1986, **23**, 989 (*synth*)

2-Amino-6-methyl-4(1*H*)-pyrimidinone, 9CI   A-60230

Updated Entry replacing A-02136
6-Methylisocytosine
[3977-29-5]
$C_5H_7N_3O$ M 125.130
1*H*-form is present in cryst. state; tautom. to 3*H*-form possible in soln. Needles (H_2O). Spar. sol. H_2O, org. solvs., sol. acids, alkalis. Mp 297-299° dec. Aq. soln. reacts alkaline.
B,HCl: Mp 295°.
B,H_2SO_4: Mp 180°.
2-N-Me:
    $C_6H_9N_3O$ M 139.157
    Cryst. (H_2O). Mp 202°.

Gabriel, S. *et al, Ber.*, 1899, **32**, 2924.
Iwakwa, Y. *et al, J. Chem. Soc. Jpn.*, 1954, **57**, 947 (*synth*)
Lowe, P.R. *et al, Acta Crystallogr., Sect. C*, 1987, **43**, 330 (*cryst struct*)

2-Amino-5-(methylthio)pentanoic acid   A-60231

Updated Entry replacing A-30133
5-Methylthionorvaline, 9CI, 8CI. *Homomethionine*
[5632-95-1]

COOH
|
H_2N►C◄H   (*S*)-*form*
|
$CH_2CH_2CH_2SMe$

$C_6H_{13}NO_2S$ M 163.234
(*S*)-*form* [25148-30-5]
    L-*form*
    Isol. from cabbage and horseradish. Plates (EtOH aq.).
    Mp 235°, 247-248° dec. $[\alpha]_D^{25.5}$ +21° (c, 0.3 in 6M HCl).

(±)-*form* [6094-76-4]
    Hexagonal plates (EtOH). Mp 247-248°.

Kjaer, A. *et al, Acta Chem. Scand.*, 1955, **9**, 721 (*synth*)
Sugii, M. *et al, Chem. Pharm. Bull.*, 1964, **12**, 1115 (*isol*)
Suketa, Y. *et al, Chem. Pharm. Bull.*, 1970, **18**, 249 (*isol, ord, biosynth*)
Vriesema, B.K. *et al, Tetrahedron Lett.*, 1986, **27**, 2045 (*resoln, cd, abs config*)

3-Amino-5-nitro-2,1-benzisothiazole   A-60232

5-Nitro-2,1-benzisothiazol-3-amine, 9CI
[14346-19-1]

$C_7H_5N_3O_2S$ M 195.195
Dye intermed. Red needles (AcOH or MeOH). Mp >300°.

Belg. Pat., 670 652, (*1966*); *CA*, **66**, 65467

3-Amino-5-nitrobenzyl alcohol   A-60233

3-Amino-5-nitrobenzenemethanol, 9CI
[90390-46-8]
$C_7H_8N_2O_3$ M 168.152
Mp 91.5°.

Meindl, W.R. *et al, J. Med. Chem.*, 1984, **27**, 1111 (*synth, pmr*)

6-Amino-8-nitroquinoline   A-60234

8-Nitro-6-quinolinamine, 9CI
[107678-82-0]
$C_9H_7N_3O_2$ M 189.173
Red needles (toluene). Mp 162°.

Schofield, J. *et al, Chem. Ind. (London)*, 1986, 587 (*synth, pmr*)

8-Amino-6-nitroquinoline   A-60235

Updated Entry replacing A-02425
6-Nitro-8-quinolinamine, 9CI
[88609-21-6]
$C_9H_7N_3O_2$ M 189.173
Red cryst. (EtOH). Mp 194°.
B,MeI: Red cryst. (EtOH). Mp 176°.
8-N-Ac: [88609-21-6].
    $C_{11}H_9N_3O_3$ M 231.210
    Light-yellow cryst. Mp 224°. Sublimes.

Schofield, J. *et al, Chem. Ind. (London)*, 1986, 587 (*synth*)

3-Amino-4-oxo-4-phenylbutanoic acid   A-60236

3-Amino-3-benzoylpropanoic acid

$PhCOCH(NH_2)CH_2COOH$

$C_{10}H_{11}NO_3$ M 193.202
(±)-*form*
    B,HBr: Cryst. (Me_2CO/Et_2O). Mp 177-180°.

Seki, M. *et al, Chem. Pharm. Bull.*, 1986, **34**, 4516 (*synth, pmr*)

2-Amino-3*H*-phenoxazin-3-one, 9CI, 8CI   A-60237

Updated Entry replacing A-02477
Questiomycin A. *AV toxin* C
[1916-59-2]

$C_{12}H_8N_2O_2$ M 212.207

Isol. from *Acrospermum viticola, Brevibacterium iodinum, Calocybe gambosa, Microbispora aerata, Streptomyces thioluteus, Pycnoporus* sp. and *Waksmania* sp. Active against gram-positive bacteria, mycobacteria, *Candida albicans* and shows antitumour activity. Phytotoxin. Dark-brown or red cryst. (EtOH). Mp 255-257° subl. Sometimes occurs in amorph. form with Mp 296-297°.

▷SP7695000.

N-*Ac:* [1916-55-8]. N-*(3-Oxo-3*H*-phenoxazin-2-yl)-acetamide, 9CI. 2-Acetamido-3*H*-phenoxazin-3-one.
$C_{14}H_{10}N_2O_3$ M 254.245
From *B. iodinum, M. aerata, S. thioluteus* and *W.* sp. Active against *Sarcina lutea* and *Trichophyton* sp. Orange cryst. Mp 165° subl.

Gerber, N.N. *et al, Biochemistry*, 1964, **3**, 598; 1966, **5**, 3824; *J. Org. Chem.*, 1967, **32**, 4055 (*isol, uv, ir, bibl*)
Ikekawa, T. *et al, Chem. Pharm. Bull.*, 1968, **16**, 1705 (*synth, ir*)
Baer, H. *et al, Pharmazie*, 1971, **26**, 108, 314 (*isol*)
Sullivan, G. *et al, J. Pharm. Sci.*, 1971, **60**, 1097 (*isol*)
Hishida, T. *et al, Chem. Lett.*, 1974, 293 (*synth*)
Schlunegger, U.V. *et al, Helv. Chim. Acta*, 1976, **59**, 1383 (*isol*)
Motohashi, N., *Yakugaku Zasshi*, 1983, **103**, 364 (*synth, props*)
Bolognese, A. *et al, J. Heterocycl. Chem.*, 1986, **23**, 1003 (*synth, pmr, uv, deriv*)
Kinjo, J. *et al, Tetrahedron Lett.*, 1987, **28**, 3697 (*isol*)

2-Amino-3-phenyl-3-butenoic acid   A-60238

β-Methylenephenylalanine, 9CI
[71028-59-6]

(R)-form

$C_{10}H_{11}NO_2$ M 177.202

(R)-form

Me ester: [79435-73-7].
$C_{11}H_{13}NO_2$ M 191.229
Bp$_{0.1}$ 100-110°. $[\alpha]_D^{20}$ −62.1° (c, 0.6 in EtOH).

(±)-form [80875-71-4]
Cryst. (MeOH). Mp 173-174° dec. (165-166°).
B,HCl: [80875-72-5]. Cryst. Mp 170-172° dec.

Chari, R.V.J. *et al, Tetrahedron Lett.*, 1979, 111 (*synth*)
Heinzer, F. *et al, Helv. Chim. Acta*, 1981, **64**, 2279 (*synth*)
Schoellkopf, U. *et al, Angew. Chem., Int. Ed. Engl.*, 1981, **20**, 977 (*ester*)

2-Amino-4-phenyl-3-butenoic acid, 9CI   A-60239

α-Amino-β-benzalpropionic acid
[72764-82-0]

$PhCH=CHCH(NH_2)COOH$

$C_{10}H_{11}NO_2$ M 177.202

(−)-(*E*)-form [27038-07-9]
Cryst. (MeOH aq.). $[\alpha]_D^{20}$ −20° (c, 1 in 0.1*M* HCl).
(±)-(*E*)-form [90528-92-0]
Mp 198-200° dec.

Sakota, N. *et al, Bull. Chem. Soc. Jpn.*, 1970, **43**, 1138 (*synth*)
Hines, J.W. *et al, J. Org. Chem.*, 1976, **41**, 1466 (*synth*)
Greenlee, W.J. *et al, J. Org. Chem.*, 1984, **49**, 2632 (*synth*)

5-Amino-3-phenylisoxazole   A-60240

Updated Entry replacing A-02514
3-Phenyl-5-isoxazolamine
[4369-55-5]
$C_9H_8N_2O$ M 160.175
Cryst. Mp 110-112°.

N-*Formyl:* [86685-13-4].
$C_{10}H_8N_2O_2$ M 188.185
Cryst. + ½H$_2$O (H$_2$O). Mp 115-117° (112°).
N-*Ac:* [31301-40-3].
$C_{11}H_{10}N_2O_2$ M 202.212
Cryst. Mp 163-164°.
N-*Me:* [86685-14-5].
$C_{10}H_{10}N_2O$ M 174.202
Cryst. (C$_6$H$_6$). Mp 110-112°.
N-*Et:* [86685-93-0].
$C_{11}H_{12}N_2O$ M 188.229
Cryst. (hexane/toluene). Mp 115-116°.
N-*Ph:* [15055-49-9].
$C_{15}H_{12}N_2O$ M 236.273
Cryst. (MeOH). Mp 143-144° (131-133°).

Boulton, A.J. *et al, Tetrahedron*, 1961, **17**, 51 (*ir, uv, struct*)
Iwai, I. *et al, Chem. Pharm. Bull.*, 1966, **14**, 1277 (*synth, pmr*)
Griss, G. *et al, Justus Liebigs Ann. Chem.*, 1970, **738**, 60 (*synth*)
Tatee, T. *et al, Chem. Pharm. Bull.*, 1986, **34**, 1643 (*derivs*)

2-Amino-3-phenylpentanedioic acid   A-60241

Updated Entry replacing P-01195
3-Phenylglutamic acid, 9CI

2S,3S-form

$C_{11}H_{13}NO_4$ M 223.228

(2S,3S)-form [109905-96-6]
Mp 278-280°. $[\alpha]_D^{25}$ +11.1° (c, 0.81 in 6*M*DCl).
N-*Benzoyl:*
$C_{18}H_{17}NO_5$ M 327.336
Needles (H$_2$O). Mp 171-172°.

(2S,3R)-form
Mp 214-216°. $[\alpha]_D^{25}$ +19.15° (c, 0.87 in 6*M* DCl).

Harrington, C.R., *J. Biol. Chem.*, 1925, **64**, 29 (*synth*)
Pachaly, P., *Arch. Pharm. (Weinheim, Ger.)*, 1972, **305**, 176 (*synth*)
Belokon, Y.N. *et al, J. Chem. Soc., Perkin Trans. 1*, 1986, 1865 (*synth, ir, pmr*)

The symbol ▷ *in Entries highlights hazard or toxicity information*

2-Amino-4(1*H*)-pteridinone, 9CI   A-60242
2-Aminopteridinol. 2-Amino-4-hydroxypteridine. Pterin
[2236-60-4]

$C_6H_5N_5O$ M 163.138
Form shown predominates. Isol. from numerous natural sources.
N^2-*Ac*:
  $C_8H_7N_5O_2$ M 205.176
  Cryst. (propanol). Mp >350°. Browns at 270°.
1-Me: [13005-86-2]. *2-Amino-1-methyl-4(1H)-pteridinone, 9CI*.
  $C_7H_7N_5O$ M 177.165
  Mp 335-337° dec.
N^2-*Me*: [13005-84-0]. *2-(Methylamino)-4(1H)-pteridinone, 9CI*.
  $C_7H_7N_5O$ M 177.165
  Pale-yellow cryst. Mp >350°.
8-Oxide: [42346-89-4].
  $C_6H_5N_5O_2$ M 179.138
  Bright-yellow cryst. Mp >360°.

Dick, G.P.G. *et al, J. Chem. Soc.*, 1955, 1379 (*synth*)
Pfleiderer, W. *et al, Chem. Ber.*, 1960, **93**, 2015 (*synth, struct*)
Brown, D.J. *et al, J. Chem. Soc.*, 1961, 4413 (*tautom*)
Williams, V.P. *et al, J. Heterocycl. Chem.*, 1973, **10**, 827 (*ms*)
Taylor, E.C. *et al, J. Org. Chem.*, 1975, **40**, 2341 (*oxide*)
Tobias, S. *et al, Chem. Ber.*, 1985, **118**, 354 (*cmr*)

2-Aminopurine, 8CI   A-60243
1H-Purin-2-amine, 9CI. Isoadenine
[452-06-2]

$C_5H_5N_5$ M 135.128
Cryst. (H_2O). Mp 277-278°.
▷UO7475000.
N^2-*Me*: [1931-01-7]. *2-(Methylamino)purine, 8CI*.
  $C_6H_7N_5$ M 149.155
  Cryst. (H_2O). Mp 278-280°.
N^2,N^2-*Di-Me*: [23658-61-9]. *2-(Dimethylamino)purine, 8CI*.
  $C_7H_9N_5$ M 163.182
  Cryst. (H_2O). Mp 222-223°.
N^2,N^2-*Di-Et*: [5167-17-9]. *2-(Diethylamino)purine*.
  $C_9H_{13}N_5$ M 191.235
  Cryst. (EtOH aq.). Mp 228-230°.

Robins, R.K. *et al, J. Am. Chem. Soc.*, 1953, **75**, 263 (*synth*)
Albert, A. *et al, J. Chem. Soc.*, 1954, 2060 (*synth*)
Brown, D.J. *et al, J. Chem. Soc.*, 1965, 3770 (*deriv*)
Coburn, W.C. *et al, J. Org. Chem.*, 1965, **30**, 1114 (*pmr*)
Fr. Pat., 1 415 224, (*1965*); *CA*, **64**, 5116 (*synth*)
Cerami, A., *Diss. Abstr. Int. B*, 1968, **29**, 39 (*biochem*)
Fr. Pat., 1 514 638, (*1968*); *CA*, **70**, 95464 (*manuf*)
Thorpe, M.C. *et al, J. Magn. Reson.*, 1974, **15**, 98 (*cmr*)

8-Aminopurine, 8CI   A-60244
9H-Purin-8-amine, 9CI
[20296-09-7]

$C_5H_5N_5$ M 135.128
Small needles (H_2O). Mp >360°.
N^8-*Me*: [23658-67-5]. *8-(Methylamino)purine, 8CI*.
  $C_6H_7N_5$ M 149.155
  Cryst. (H_2O). Mp 332-334° dec.
N^8,N^8-*Di-Me*: [23687-23-2]. *8-(Dimethylamino)purine, 8CI*.
  $C_7H_9N_5$ M 163.182
  Needles (H_2O). Mp 292° dec.

Albert, A. *et al, J. Chem. Soc.*, 1954, 2060 (*synth*)
Lewis, A.F. *et al, Can. J. Chem.*, 1963, **41**, 1807 (*synth*)
Lonnberg, H. *et al, Acta Chem. Scand., Ser. A*, 1985, **39**, 171 (*deriv*)

9-Aminopurine   A-60245
9H-Purin-9-amine, 9CI
[6313-13-9]
$C_5H_5N_5$ M 135.128
Prisms (MeOH). Mp 222-223°.
N^9-*Benzyloxycarboxyl*: Needles (Et_2O). Mp 142.5-143.5°.

Somei, M. *et al, Chem. Pharm. Bull.*, 1978, **26**, 2522 (*synth*)

5-Aminopyrazolo[4,3-*d*]pyrimidin-7(1*H*,6*H*)-one   A-60246
5-Amino-1,4-dihydro-7H-pyrazolo[4,3-d]pyrimidin-7-one, 9CI
[41535-76-6]

$C_5H_5N_5O$ M 151.127
Cryst. Mp >300° dec. Slowly hydrates in air.

Lewis, A.F. *et al, J. Am. Chem. Soc.*, 1982, **104**, 1073 (*synth, deriv, pmr, uv*)

3-Amino-2(1*H*)-pyridinethione   A-60247
3-Amino-2-pyridinethiol. 3-Amino-2-mercaptopyridine
[38240-21-0]

$C_5H_6N_2S$ M 126.176
Greenish-yellow needles (MeOH or C_6H_6). Mp 133-134°.
▷Prep. descr. in earlier ref. can be explosive
N^3-*Ac*:
  $C_7H_8N_2O_5$ M 200.151
  Mp 196-197°.
B,HCl: Yellow needles (EtOH/Et_2O). Mp 227-228° dec.

Rodig, O.R. *et al, J. Org. Chem.*, 1964, **29**, 2652 (*synth*)
Hagen, S. *et al, Acta Chem. Scand., Ser. B*, 1974, **28**, 523 (*synth*)
Krowicki, K., *Pol. J. Chem.*, 1978, **52**, 2039 (*synth*)
Okafor, C.O., *J. Org. Chem.*, 1982, **47**, 592 (*synth*)

1-Amino-2(1H)-pyridinone, 9CI　　　　A-60248
1-Amino-2(1H)-pyridone
[54931-11-2]

$C_5H_6N_2O$　　M 110.115
Rods (C_6H_6/pet. ether). Mp 64-66°.
B,HCl: [62438-05-5]. Needles (2-propanol). Mp 175-177°.
N-Ac: [66193-78-0].
　$C_7H_8N_2O_2$　　M 152.152
　Cryst. (C_6H_6). Mp 159-161°. Softens at 126-130°.

Hoegerle, K., *Helv. Chim. Acta*, 1956, **39**, 1203; 1958, **41**, 539 (*synth*)
Katritzky, A.R. *et al*, *An. Quim.*, 1974, **70**, 994 (*synth*)
Boyers, J.T. *et al*, *J. Chem. Soc., Perkin Trans. 1*, 1977, 1960 (*synth*)

3-Amino-2(1H)-pyridinone, 9CI　　　　A-60249
3-Amino-2-pyridinol. 3-Amino-2-hydroxypyridine
$C_5H_6N_2O$　　M 110.115
NH-form predominates.
NH-form [33630-99-8]
Cryst. Mp 134-137° (132-133°).
B,HCl: [33631-21-9]. Cryst. Mp 240-242°.
Picrate: [33631-22-0]. Cryst. (H_2O). Mp 217-218° dec.
1-Me: [33631-01-5]. *3-Amino-1-methyl-2(1H)-pyridinone.*
　$C_6H_8N_2O$　　M 124.142
　Oil.
1-Me, picrate: Cryst. (EtOH). Mp 204° (199-200° dec.).
OH-form [59315-44-5]
Me ether: [20265-38-7]. *3-Amino-2-methoxypyridine, 8CI.*
　$C_6H_8N_2O$　　M 124.142
　Cryst. (pet. ether). Mp 67-69.5°.

Barlin, G.B. *et al*, *J. Chem. Soc.* (*B*), 1971, 1425 (*synth, uv*)
Takeuchi, I. *et al*, *Chem. Pharm. Bull.*, 1976, **24**, 1813 (*synth*)
Hwang, D.R. *et al*, *J. Pharm. Sci.*, 1980, **69**, 1074 (*synth, props*)
Oklobdzija, M. *et al*, *J. Heterocycl. Chem.*, 1983, **20**, 1329 (*synth*)

4-Amino-2(1H)-pyridinone, 9CI　　　　A-60250
4-Amino-2-pyridinol. 4-Amino-2-hydroxypyridine
$C_5H_6N_2O$　　M 110.115
NH-form predominates.
NH-form [38767-72-5]
Cryst. (Me_2CO). Mp 214°.
N-Oxide: [13602-69-2].
　$C_5H_6N_2O_2$　　M 126.115
　Cryst. (EtOH). Mp 230-233° dec.
N-Oxide; B,HCl: [13602-68-1]. Cryst. (EtOH/Et_2O).
　Mp 191-193° dec.
OH-form [33630-97-6]
Me ether: [20265-39-8]. *4-Amino-2-methoxypyridine, 8CI.*
　$C_6H_8N_2O$　　M 124.142
　Needles (pet. ether). Mp 92°.

Barlin, G.B. *et al*, *J. Chem. Soc.* (*B*), 1971, 1425 (*synth, uv*)

Hung, N.C. *et al*, *Synthesis*, 1984, 765 (*synth*)

5-Amino-2(1H)-pyridinone, 9CI　　　　A-60251
5-Amino-2-pyridinol. 5-Amino-2-hydroxypyridine
$C_5H_6N_2O$　　M 110.115
NH-form predominates.
NH-form [33630-94-3]
B,HCl: Cryst. (EtOH). Mp 236-237°.
1-Me: [33630-96-5]. *5-Amino-1-methyl-2(1H)-pyridinone.*
　$C_6H_8N_2O$　　M 124.142
　Cryst. (H_2O) (as picrate). Mp 216-217° dec. (204°) (picrate).
OH-form
Me ether: [6628-77-9]. *5-Amino-2-methoxypyridine, 8CI.*
　$C_6H_8N_2O$　　M 124.142
　$Bp_{1.5}$ 79-80°.

Barlin, G.B. *et al*, *J. Chem. Soc.* (*B*), 1971, 1425 (*synth, uv*)

6-Amino-2(1H)-pyridinone, 9CI　　　　A-60252
6-Amino-2(1H)-pyridone, 8CI
$C_5H_6N_2O$　　M 110.115
Cryst. (H_2O). Mp 214°.
NH-form [5154-00-7]
Cryst. (H_2O). Mp 214°. Major tautomer.
1-Me: [17920-37-5]. *6-Amino-1-methyl-2(1H)-pyridinone, 9CI.*
　$C_6H_8N_2O$　　M 124.142
　Cryst. (H_2O). Mp 166-169°.
OH-form
6-Amino-2-pyridinol. 2-Amino-6-hydroxypyridine
Me ether: [17920-35-3]. *2-Amino-6-methoxypyridine, 8CI.*
　$C_6H_8N_2O$　　M 124.142
　$Bp_{12.5}$ 114-115°.
Me ether, picrate: Cryst. (EtOH). Mp 213-214°.

Sharma, B.D. *et al*, *Acta Crystallogr.*, 1966, **20**, 921 (*cryst struct*)
Barlin, G.B. *et al*, *J. Chem. Soc.* (*B*), 1971, 1425 (*synth, uv*)
Dorie, J. *et al*, *Org. Magn. Reson.*, 1979, **12**, 229 (*N nmr*)

2-Amino-4(1H)-pyridinone　　　　A-60253
[33623-18-6]
$C_5H_6N_2O$　　M 110.115
Cryst. Mp 181.5-182.5°.
OH-form [33631-05-9]
2-Amino-4-pyridinol. 2-Amino-4-hydroxypyridine
Cryst. Mp 181.5-182.5°.
Picrate: Cryst. (H_2O). Mp 224-227°.
Me ether: [10201-73-7]. *2-Amino-4-methoxypyridine.*
　$C_6H_8N_2O$　　M 124.142
　Cryst. (pet. ether). Mp 116-117°.

Barlin, G.B. *et al*, *J. Chem. Soc.* (*B*), 1971, 1425 (*synth, uv*)
Lombardino, J.G., *J. Med. Chem.*, 1981, **24**, 39 (*synth*)

3-Amino-4(1H)-pyridinone　　　　A-60254
3-Amino-4-pyridinol. 3-Amino-4-hydroxypyridine
$C_5H_6N_2O$　　M 110.115

NH-form predominates.

NH-form [15590-89-3]

Prisms. Oxidises in air.

B,2HCl: Cream cryst. Mp 228-230° dec.

1-Me: [33631-10-6]. *3-Amino-1-methyl-4(1*H)-*pyridinone.*
$C_6H_8N_2O$ M 124.142
Cryst. (EtOAc). Mp 147-148°.

1-Me, N^3-Ac:
$C_8H_{10}N_2O_2$ M 166.179
Long needles (C_6H_6). Mp 185-186°.

1-N-Me, N^3,N^3-di-Ac:
$C_{10}H_{12}N_2O_3$ M 208.216
Needles (C_6H_6). Mp 150-152°.

1-Me, picrate: Yellow needles (MeOH). Mp 209-210° dec.

OH-form [6320-39-4]

Me ether: [33631-09-3]. *3-Amino-4-methoxypyridine.*
$C_6H_8N_2O$ M 124.142
Mp 83.5-85°. Bp$_{1.25}$ 104°.

Jones, R.A. *et al, J. Chem. Soc.* (*B*), 1967, 84 (*tautom, bibl*)
Barlin, G.B. *et al, J. Chem. Soc.* (*B*), 1971, 1425 (*synth, bibl*)

2-Aminopyrrolo[2,3-*d*]pyrimidin-4-one   A-60255

*2-Amino-1,7-dihydro-4*H-*pyrrolo*[2,3-d]*pyrimidin-4-one, 9CI. 2-Amino-4-hydroxypyrrolo*[2,3-d]*pyrimidine. 7-Deazaguanine*

[7355-55-7]

$C_6H_6N_4O$ M 150.140
Needles (H_2O). Mp 323-324° dec.

Davoll, J., *J. Chem. Soc.*, 1960, 131 (*synth*)

3-Amino-2(1*H*)-quinolinone   A-60256

Updated Entry replacing A-02670
3-Aminocarbostyril, 8CI. 3-Amino-2-hydroxyquinoline. 3-Amino-2-quinolinol

[5873-00-7]

$C_9H_8N_2O$ M 160.175
Tautomeric. Cryst. (EtOH). Mp 285° (211-213°).

3-N-Ac:
$C_{11}H_{10}N_2O_2$ M 202.212
Cryst. (EtOH). Mp 262-264°.

Hashimoto, T. *et al, J. Pharm. Soc. Jpn.*, 1960, **80**, 1806.
Bowman, R.G. *et al, J. Chem. Soc.*, 1965, 1080.
Leclerc, G. *et al, J. Med. Chem.*, 1986, **29**, 2427 (*synth*)

4-Amino-2,3,5,6-tetrafluoropyridine, 8CI   A-60257

2,3,5,6-Tetrafluoro-4-pyridinamine, 9CI

[1682-20-8]

$C_5H_2F_4N_2$ M 166.078
Cryst. (pet. ether). Mp 85-86°.

Chambers, R.D., *J. Chem. Soc.*, 1964, 3736 (*synth*)
Banks, R.E. *et al, J. Chem. Soc.*, 1965, 575 (*synth*)
Lee, J. *et al, J. Chem. Soc.*, 1965, 582 (*F nmr*)

2-Amino-4,5,6,7-tetrahydrobenzo[*b*]-thiophene-3-carboxylic acid, 9CI   A-60258

2-Amino-4,5-tetramethylene-3-thiophenecarboxylic acid

[5936-58-3]

$C_9H_{11}NO_2S$ M 197.251
Cryst. (propanol). Mp 158-160°.

Et ester: [4506-71-2].
$C_{11}H_{15}NO_2S$ M 225.305
Mp 115°.

Et ester; B,HCl: [76488-06-7]. Mp 137-138°.

Et ester, N-Ac: [5919-29-9].
$C_{13}H_{17}NO_3S$ M 267.342
Mp 111-113°.

Et ester, N-benzoyl: [52535-73-6].
$C_{18}H_{19}NO_3S$ M 329.413
Mp 166-168°.

Amide: [4815-28-5].
$C_9H_{12}N_2OS$ M 196.267
Possesses analgesic props. Cryst. (EtOH). Mp 189-190°.

Nitrile: [4651-91-6].
$C_9H_{10}N_2S$ M 178.251
Mp 147-148°.

Hydrazide: [22721-28-4]. Mp 165-167°.

Gewald, K. *et al, Chem. Ber.*, 1966, **99**, 94.
Gewald, K., *Z. Chem.*, 1967, **7**, 186.
Perrissin, M. *et al, Eur. J. Med. Chem.-Chim. Ther.*, 1980, **15**, 413 (*synth, props*)
Achakzi, D. *et al, Chem. Ber.*, 1981, **114**, 3188 (*deriv*)

3-Amino-1,2,3,4-tetrahydrocarbazole   A-60259

*1,2,3,4-Tetrahydro-9*H-*carbazol-3-amine*

$C_{12}H_{14}N_2$ M 186.256
Rigid tryptophan analogue.

(±)-form

Mp 176-177° (170-172°). Bp$_{0.1}$ 180° subl.

N^3-*Benzoyl:*
$C_{20}H_{18}N_2O$ M 302.375
Mp 138-140°.

Bird, C.W. *et al, J. Heterocycl. Chem.*, 1985, **22**, 191 (*synth, ir, pmr, bibl*)

5-Amino-2-thiazolecarbothioamide, 9CI   A-60260

5-Aminothiazole-2-thiocarboxamide. 5-Aminothio-2-thiazolecarboxamide. 5-Amino-2-thiocarbamoylthiazole. Chrysean

[535-67-1]

$C_4H_5N_3S_2$ M 159.224
Cryst. (AcOH aq.). Mp 215-216°.
N^5-*Ac:*
   $C_6H_7N_3OS_2$ M 201.261
   Yellow needles (AcOH aq.). Mp 250° dec.

Adams, A. *et al, J. Chem. Soc.*, 1956, 1870 (*synth, bibl*)
Runge, F. *et al, J. Prakt. Chem.*, 1962, **16**, 297 (*synth*)
Arias, P. *et al, CA*, 1974, **80**, 82775 (*synth*)

5-Amino-4-thiazolecarbothioamide   A-60261

5-Aminothiazole-4-thiocarboxamide. 5-Aminothio-4-thiazolecarboxamide. 5-Amino-4-thiocarbamoylthiazole. Isochrysean

$C_4H_5N_3S_2$ M 159.224
Needles (EtOH). Mp 155°.
N^5-*Ac:*
   $C_6H_7N_3OS_2$ M 201.261
   Pale-yellow needles (AcOH aq.). Mp 208-210°.

Adams, A. *et al, J. Chem. Soc.*, 1956, 1870 (*synth*)

2-Amino-3-(2,3,4-trichlorophenyl)-propanoic acid   A-60262

2,3,4-Trichlorophenylalanine, 9CI

$C_9H_8Cl_3NO_2$ M 268.527
(±)-*form* [110300-02-2]
Mp 254-257° dec.

Taylor, D.C. *et al, Bioorg. Chem.*, 1987, **15**, 335 (*synth*)

2-Amino-3-(2,3,6-trichlorophenyl)-propanoic acid   A-60263

2,3,6-Trichlorophenylalanine, 9CI
$C_9H_8Cl_3NO_2$ M 268.527
(±)-*form* [110300-01-1]
Mp 261-263° dec.

Taylor, D.C. *et al, Bioorg. Chem.*, 1987, **15**, 335 (*synth*)

2-Amino-3-(2,4,5-trichlorophenyl)-propanoic acid   A-60264

2,4,5-Trichlorophenylalanine, 9CI
$C_9H_8Cl_3NO_2$ M 268.527
(±)-*form* [110300-00-0]
Mp 254-255° dec.

Taylor, D.C. *et al, Bioorg. Chem.*, 1987, **15**, 335 (*synth*)

7-Amino-4-(trifluoromethyl)-2H-1-benzo-pyran-2-one, 9CI   A-60265

7-Amino-4-(trifluoromethyl)coumarin

[53518-15-3]

$C_{10}H_6F_3NO_2$ M 229.158
Laser dye. Fluorescent marker in proteinase detection.
Mp 222°.
N-*Ac:* [78277-38-0].
   $C_{12}H_8F_3NO_3$ M 271.195
   Solid (MeOH/Et$_2$O). Mp 184°.
N-*Benzoyl:* [78277-39-1].
   $C_{17}H_{10}F_3NO_3$ M 333.266
   Solid (EtOAc). Mp 229.9°.
N-*Et:* [52840-38-7].
   $C_{12}H_{10}F_3NO_2$ M 257.212
   Solid (EtOH aq.). Mp 161.7°.
N-*Ph:* [78277-34-6].
   $C_{16}H_{10}F_3NO_2$ M 305.256
   Solid (EtOH aq.). Mp 161°.

Bissel, E.R. *et al, J. Org. Chem.*, 1980, **45**, 2283 (*synth, pmr, ir*)
Smith, R.E. *et al, Thromb. Res.*, 1980, **17**, 393; *CA*, **92**, 159413 (*derivs*)
Bissel, E.R. *et al, J. Chem. Eng. Data*, 1981, **26**, 348 (*derivs*)

3-Amino-2,4,6-trimethylbenzoic acid   A-60266

3-Aminomesitoic acid

[106567-40-2]

$C_{10}H_{13}NO_2$ M 179.218
Cryst. (EtOH aq.). Mp 209-210°. pK_a 5.39 (55% EtOH aq., 22°).

Beringer, F.M. *et al, J. Am. Chem. Soc.*, 1955, **75**, 3319 (*synth, props*)
Cuyegkeng, M.A. *et al, Chem. Ber.*, 1987, **120**, 803 (*synth, pmr*)

5-Amino-2,3,4-trimethylbenzoic acid   A-60267

[20804-96-0]
$C_{10}H_{13}NO_2$ M 179.218
Cryst. (H$_2$O). Mp 252-254°.

Okukado, N. *et al, CA*, 1969, **70**, 3406t (*synth*)

3-Amino-2-ureidopropanoic acid   A-60268

2-[(Aminocarbonyl)amino]-β-alanine. Isoalbizziine

$C_4H_9N_3O_3$ M 147.133
(R)-form
D-form
Cryst. (EtOH). Mp 204°. $[\alpha]_D^{22}$ +44.6° (c, 1 in 0.1M HCl).
(±)-form
Cryst. (H$_2$O). Mp 201°.

Kjaer, A. *et al, Acta Chem. Scand.*, 1959, **13**, 1565; 1960, **14**, 961 (*synth*)

2-Amino-5-vinylpyrimidine
A-60269

5-Ethenyl-2-pyrimidinamine, 9CI

[108444-56-0]

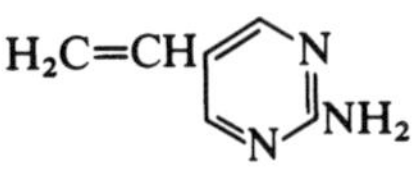

$C_6H_7N_3$ M 121.141
Cryst. by subl. Mp 136°.
N^2,N^2-*Di-Me:* [108461-92-3].
  $C_8H_{11}N_3$ M 149.195
  Cryst. by subl. Mp 67°.

Kvita, V. *et al, Synthesis*, 1986, 786 (*synth, pmr*)

Anamarine
A-60270

[73413-69-1]

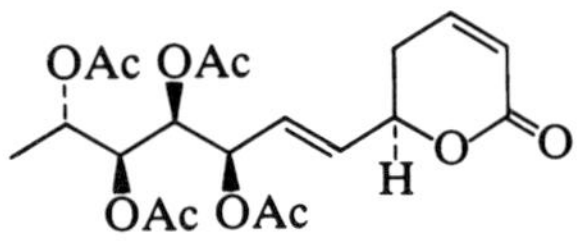

Absolute configuration

$C_{20}H_{26}O_{10}$ M 426.419
Constit. of flowers and leaves of *Hyptis* sp. Cryst. Mp
110-112°. $[\alpha]_D$ +18.8° (c, 0.75 in $CHCl_3$).

Alemany, A. *et al, Tetrahedron Lett.*, 1979, 3579, 3583 (*isol, cryst struct*)
Lichtenthaler, F.W. *et al, Tetrahedron Lett.*, 1987, **28**, 47
  (*synth, abs config*)
Lorenz, K. *et al, Tetrahedron Lett.*, 1987, **28**, 6437 (*synth*)

Andirolactone
A-60271

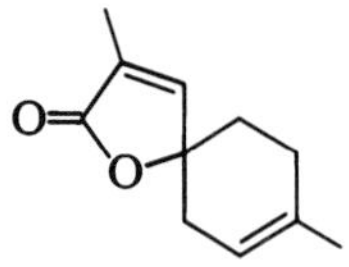

$C_{11}H_{14}O_2$ M 178.230
Constit. of *Cedrus libanotica*. Oil. $[\alpha]_D^{25}$ +3.2° (c, 2.1 in
  $CHCl_3$).

Avcibasi, H. *et al, Phytochemistry*, 1987, **26**, 2852.

Andrimide
A-60272

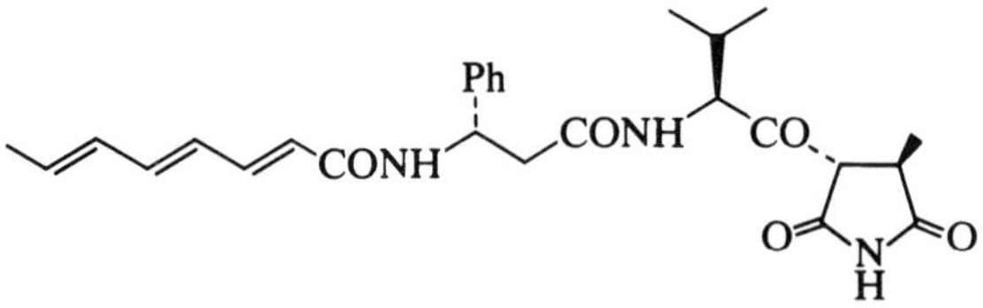

$C_{27}H_{33}N_3O_5$ M 479.575
Peptide antibiotic. Prod. by a bacterial symbiont from
  Nilaparvata lugens. Active against *Xanthomonas* sp.
  Mp 172-173.5°.

Fredenhagen, A. *et al, J. Am. Chem. Soc.*, 1987, **109**, 4409.

Anhydrohapaloxindole
A-60273

*8-Chloro-9-ethenyl-6,6a,7,8,9,10-hexahydro-10-iso-
cyano-6,6,9-trimethylnaphth[1,2,3-cd]indol-1(2H)-one,
9CI*

[109217-16-5]

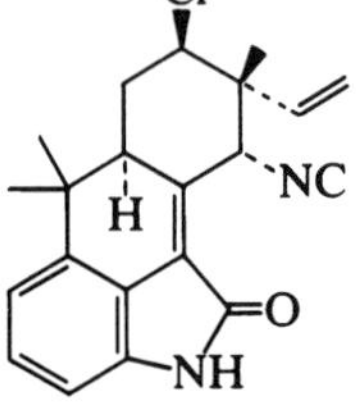

$C_{21}H_{21}ClN_2O$ M 352.863
Isol. from blue-green alga *Hapalosiphon fontinalis*. Mp
123° dec. $[\alpha]_D$ +150° (c, 0.4 in EtOH). Related to
Hapalindoles.

Moore, R.E. *et al, J. Org. Chem.*, 1987, **52**, 3773 (*isol, struct*)

15-Anhydrothyrsiferol
A-60274

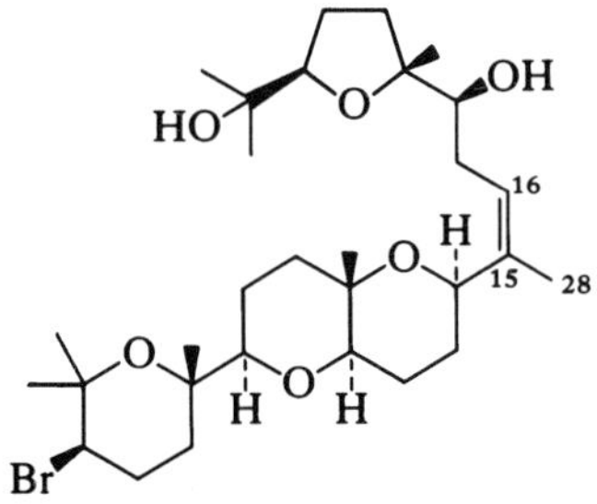

$C_{30}H_{51}BrO_6$ M 587.633
Di-Ac: [107040-99-3]. *15-Anhydrothyrsiferyl diacetate.*
  $C_{34}H_{55}BrO_8$ M 671.708
  Constit. of red alga *Laurencia obtusa*. Cytotoxic.
  Cryst. Mp 163-164°. $[\alpha]_D$ +12.9° (c, 1.00 in $CHCl_3$).
$\Delta^{15(28)}$-*Isomer, di-Ac:* [107065-86-1]. *15(28)-Anhy-
  drothyrsiferyl diacetate.*
  $C_{34}H_{55}BrO_8$ M 671.708
  Constit. of *L. obtusa*. Cytotoxic. Cryst. Mp 94-95°.
  $[\alpha]_D$ +8.8° (c, 1.00 in $CHCl_3$).

Suzuki, T. *et al, Chem. Lett.*, 1987, 361.

Anisomelic acid
A-60275

Updated Entry replacing A-50281
*2,3,3a,4,5,8,9,12,13,15a-Decahydro-6,14-dimethyl-3-
methylene-2-oxocyclotetradeca[b]furan-10-carboxylic
acid, 9CI. Anisomelolide*

[59632-76-7]

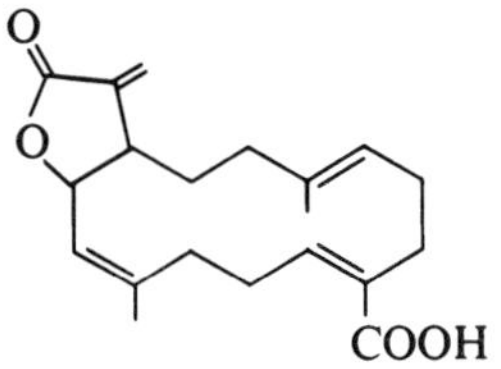

$C_{20}H_{26}O_4$ M 330.423
Constit. of *Anisomeles malabarica*. Cryst. Mp 155°
  (142-143°). $[\alpha]_D^{25}$ −35° ($CHCl_3$).

Purushothaman, K.K. *et al, Indian J. Chem.*, 1975, **13**, 1357
  (*isol*)
Devi, G. *et al, Indian J. Chem., Sect. B*, 1978, **16**, 441 (*isol,
  struct*)

Marshall, J.A. *et al*, *Tetrahedron*, 1987, **43**, 4849 (*synth*)

Annulin *A*   A-60276

[105335-73-7]

$C_{19}H_{20}O_7$ M 360.363

Constit. of *Garveia annulata*. Shows antibacterial activity. Orange cryst. (EtOH). Mp 174-176°.

Fahy, E. *et al*, *J. Org. Chem.*, 1986, **51**, 5145 (*isol, cryst struct*)

Annulin *B*   A-60277

[105335-74-8]

$C_{21}H_{22}O_7$ M 386.401

Constit. of *Garveia annulata*. Orange oil. $[\alpha]_D$ +8.0° (c, 0.2 in CHCl₃).

Fahy, E. *et al*, *J. Org. Chem.*, 1986, **51**, 5145.

Antheridic acid   A-60278

Antheridiogen A_n

$C_{19}H_{22}O_6$ M 346.379

Constit. of fern *Anemia phyllitidis*. Antheridium-inducing factor.

Corey, E.J. *et al*, *Tetrahedron Lett.*, 1986, **27**, 5083 (*struct*)
Corey, E.J. *et al*, *J. Am. Chem. Soc.*, 1985, **107**, 5574 (*synth*)

Anthra[9,1,2-*cde*]benzo[*rst*]pentaphene, 9CI   A-60279

*Dinaphtho[1,2,3-*cd:3',2',1'*-lm]perylene, 8CI. Violanthrene* A

[188-87-4]

$C_{34}H_{18}$ M 426.516
Red cryst. Mp 506.5°.

5,10-Dihydro: [81-31-2]. *5,10-Dihydroanthra[9,1,2-*cde]benzo[rst]pentaphene. Violanthrene.
$C_{34}H_{20}$ M 428.532

Red platelets.

Clar, E., *Ber.*, 1943, **76**, 458 (*synth*)
Aoki, J., *Bull. Chem. Soc. Jpn.*, 1961, **34**, 1817; 1964, **37**, 1356 (*synth, deriv*)
Ueda, T. *et al*, *Org. Mass Spectrom.*, 1983, **18**, 105 (*ms*)

9-Anthraceneacetaldehyde   A-60280

9-Anthrylacetaldehyde

[84332-58-1]

$C_{16}H_{12}O$ M 220.270
Light-green cryst. (CH₂Cl₂/hexane). Mp 144-145°.

Becker, H.-D. *et al*, *J. Org. Chem.*, 1986, **51**, 2956 (*synth, pmr, ir*)

1,8-Anthracenedicarboxylic acid, 9CI   A-60281

Updated Entry replacing A-50283

[38378-77-7]
$C_{16}H_{10}O_4$ M 266.253
Gold needles (EtOH or dioxan aq.). Mp 353° (345°).

Di-Me ester: [93655-34-6].
    $C_{18}H_{14}O_4$ M 294.306
    Yellow needles (MeOH). Mp 104-105°.
Dichloride: [90885-92-0].
    $C_{16}H_8Cl_2O_2$ M 303.144
    Orange needles (toluene). Mp 220°, 255-257°.
Dinitrile: [92967-66-3]. *1,8-Dicyanoanthracene*.
    $C_{16}H_8N_2$ M 228.253
    Yellow needles (AcOH). Mp 304-306°.

Waldmann, H., *Chem. Ber.*, 1950, **83**, 167 (*synth*)
Akiyama, S. *et al*, *Bull. Chem. Soc. Jpn.*, 1962, **35**, 1829 (*synth, derivs*)
Golden, R., *J. Am. Chem. Soc.*, 1972, **94**, 3080 (*synth*)
Rogers, M.E. *et al*, *J. Org. Chem.*, 1986, **51**, 3308 (*nitrile, synth, pmr, ir, ms*)

2-(9-Anthracenyl)ethanol   A-60282

9-Anthraceneethanol, 9CI. 2-(9-Anthryl)ethanol

[54060-73-0]

$C_{16}H_{14}O$ M 222.286
Pale-yellow cryst. Mp 106-108°.

Becker, H.-D. *et al*, *J. Org. Chem.*, 1986, **51**, 2956 (*synth, pmr*)

Antiarrhythmic peptide (ox atrium), 9CI   A-60283

N-[N-[N-[1-(1-Glycylproline)-4-hydroxyprolyl]-glycyl]alanyl]glycine. AAP

H-Gly-Pro-*trans*-Hyp-Gly-Ala-Gly-OH

$C_{19}H_{30}N_6O_8$ M 470.481

L-L-L-form [81771-37-1]

Posseses antiarrhythmic and antithrombotic activities.
$[\alpha]_D^{20}$ −133° (c, 1 in 0.5M HCl).

Boc-Gly-Pro-Hyp-Gly-Ala-Gly-OBzl: Cryst. (Et₂O).
Mp 164-165°. $[\alpha]_D^{25}$ −91.2° (c, 0.25 in MeOH).

Aonuma, S. *et al, Chem. Pharm. Bull.,* 1980, **28**, 3332; 1983, **31**, 612; 1984, **32**, 219 (*isol, synth, props*)

Kundu, B. *et al, Indian J. Chem., Sect. B,* 1986, **25**, 930 (*synth*)

Antibiotic FR 900452   A-60284

FR 900452

[101706-33-6]

$C_{22}H_{25}N_3O_3S$ M 411.518

Prod. by *Streptomyces phaeofaciens.* Inhibits PAF-induced platelet aggregation. Pale-yellow powder. Mp 112-120° dec. $[\alpha]_D^{23}$ +97.0° (c, 0.5 in CHCl₃).

Okamoto, M. *et al, J. Antibiot.,* 1986, **39**, 198 (*isol*)

Takase, S. *et al, J. Org. Chem.,* 1987, **52**, 3485 (*cryst struct*)

Antibiotic FR 900482   A-60285

FR 900482

[102363-08-6]

$C_{14}H_{15}N_3O_6$ M 321.289

Exhibits epimerisation at anomeric centre. Prod. by *Streptomyces sandaensis.* Active against gram-positive and -negative bacteria and tumours. Powder. Mp 175° dec. $[\alpha]_D^{23}$ +8° (c, 1 in H₂O). Synthetic derivs. showed antileukaemic activity.

Kiyoto, S. *et al, J. Antibiot.,* 1987, **40**, 589, 594, 600, 607 (*isol, struct, props*)

Uchida, I. *et al, J. Am. Chem. Soc.,* 1987, **109**, 4108 (*struct*)

Antibiotic PDE I   A-60286

PDE I

[62497-62-5]

R = −CONH₂

$C_{13}H_{13}N_3O_5$ M 291.263

Isol. from culture closely related to *Streptomyces griseoflavus.* Shows antitumour activity. Inhibitor of cyclic adenosine-3′,5′-monophosphate phosphodiesterase. Cryst. (MeOH). Mp 235° dec.

Enomoto, Y. *et al, Agric. Biol. Chem.,* 1978, **42**, 1331, 1337; 1979, **43**, 559 (*isol, uv, ir, pmr, cmr, cryst struct, synth*)

Bolton, R.E. *et al, J. Chem. Soc., Perkin Trans. 1,* 1987, 931 (*synth*)

Carter, P. *et al, J. Am. Chem. Soc.,* 1987, **109**, 2711 (*synth, bibl*)

Boger, D.L. *et al, J. Am. Chem. Soc.,* 1987, **109**, 2717 (*synth*)

Antibiotic Sch 38519   A-60287

Sch 38519

$C_{24}H_{25}NO_8$ M 455.463

Isol. from a *Thermomonospora* sp. Active against gram-positive and -negative bacteria. Related to Medermycin.

B,HCl: Red needles. Mp 215-220° dec. $[\alpha]_D^{22}$ +74.5° (c, 0.5 in MeOH).

Hegde, V.R. *et al, Tetrahedron Lett.,* 1987, **28**, 4485 (*pmr, cmr, abs config*)

Antibiotic SF 2312   A-60288

(1,5-Dihydroxy-2-oxo-3-pyrrolidinyl)phosphonic acid, 9CI. SF 2312

$C_4H_8NO_6P$ M 197.084

Prod. by *Micromonospora* sp. Active against *Pseudomonas* and *Proteus* spp. Isol. as Na salt.

Mono-Na salt: Cryst. + ½H₂O. Mp 160-164° dec.

Japan. Pat., 85 224 493, (*1985*); *CA,* **104**, 107918

Watabe, H. *et al, CA,* 1987, **106**, 134903, 152686 (*isol, struct, props*)

Antibiotic SF 2339   A-60289

Valyl-N-(1,2-dicarboxy-2-hydroxyethoxy)valinamide, 9CI. SF 2339

[103528-06-9]

$C_{14}H_{25}N_3O_8$ M 363.367

Oligopeptide antibiotic. Prod. by *Dactylosporangium* sp. Active against gram-positive and -negative bacteria in synthetic medium. Powder.

Japan. Pat., 85 258 200, (*1985*); *CA,* **105**, 59450 (*isol, struct*)

Yashida, S. *et al, CA,* 1987, **106**, 135131.

Polyangium Antibiotics
A-60290

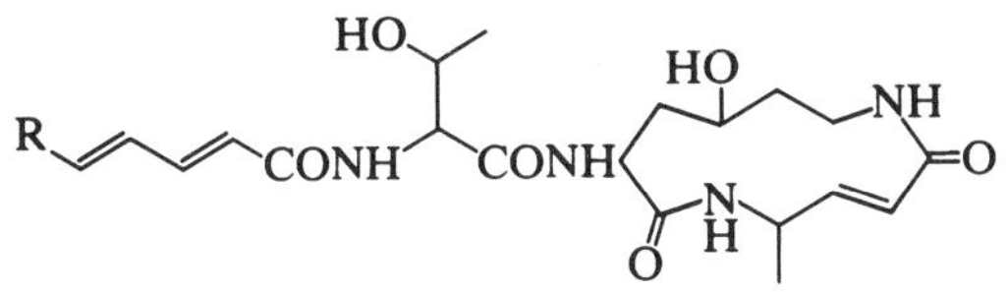

R = $-(CH_2)_6CH_3$, $-CH_2CH_2CH=CH(CH_2)_{14}CH_3$

or $-(CH_2)_8CH_3$

Lipopeptide antibiotics. Prod. by myxobacterium *Polyangium branchysporum*. MF's: $C_{27}H_{44}N_4O_6$, $C_{29}H_{46}N_4O_6$, $C_{29}H_{48}N_4O_6$.

Ger. Pat., 3 629 465, (*1987*); *CA*, **106**, 212569 (*isol, props*)

Aparjitin
A-60291

6-Eicosyl-4-methyl-2H-pyran-2-one. 6-Icosyl-4-methyl-2H-pyran-2-one

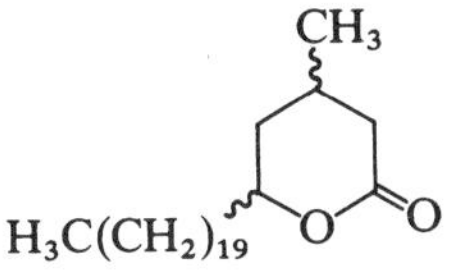

$C_{26}H_{50}O_2$ M 394.680
Isol. from leaves of *Clitoria ternatea*. Mp 92-93°.

Tiwari, R.D. *et al, J. Indian Chem. Soc.*, 1959, **36**, 243 (*isol*)

Araneophthalide
A-60292

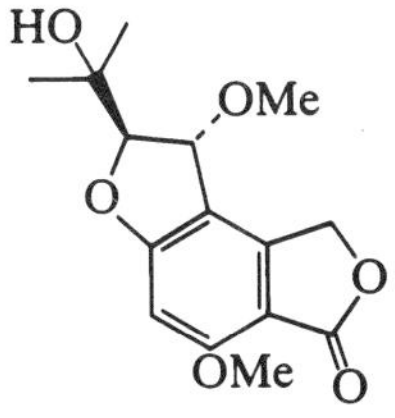

$C_{15}H_{18}O_6$ M 294.304
Constit. of *Anaphalis araneosa*. Gum.

Jakupovic, J. *et al, Phytochemistry*, 1987, **26**, 580.

Aranochromanophthalide
A-60293

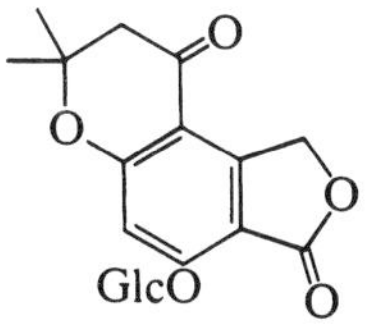

$C_{19}H_{22}O_{10}$ M 410.377
Constit. of *Anaphalis araneosa*. Amorph.

Jakupovic, J. *et al, Phytochemistry*, 1987, **26**, 580.

Arborone
A-60294

[108069-03-0]

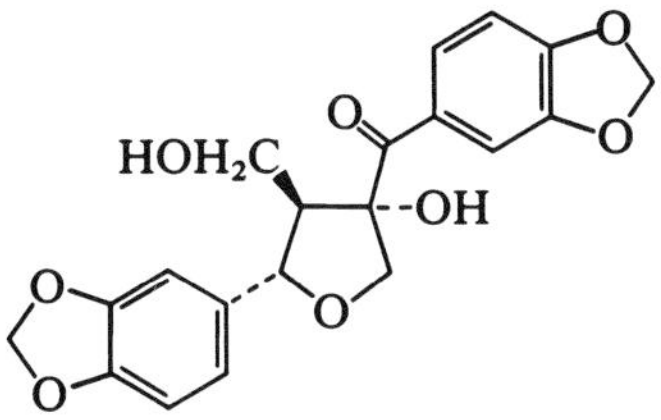

$C_{20}H_{18}O_8$ M 386.357

Lignan from *Gmelina arborea*. Cryst. (EtOAc). Mp 216-217°. $[\alpha]_D^{28}$ +81.48° (c, 0.054 in $CHCl_3$).

Satyanarayana, P. *et al, J. Nat. Prod.*, 1986, **49**, 1061.

Argiopine
A-60295

Argiotoxin 636

[105029-41-2]

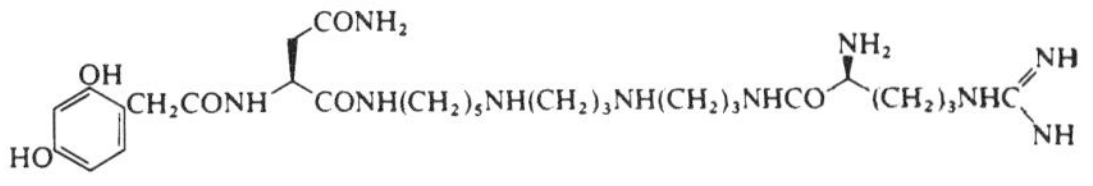

$C_{29}H_{52}N_{10}O_6$ M 636.793
Isol. from venom of the spider *Argiope lobata*. Glutamate receptor channel activator.

Shih, T.L. *et al, Tetrahedron Lett.*, 1987, **28**, 6015 (*struct, bibl*)

Aristolactone
A-60296

Updated Entry replacing A-30198
5-Methyl-8-(1-methylethenyl)-10-oxabicyclo[7.2.1]-dodeca-1(12),4-dien-11-one, 9CI

[6790-85-8]

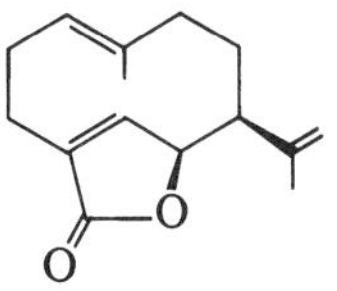

$C_{15}H_{20}O_2$ M 232.322
Constit. of *Aristolochia reticulata* and *A. serpentaria*. Cryst. (Me_2CO aq.). Mp 110-111°. $[\alpha]_D^{14}$ +156.4° (c, 1 in EtOH).

Martin-Smith, M. *et al, Tetrahedron Lett.*, 1964, 2391 (*isol*)
Ferguson, G. *et al, J. Chem. Res. (S)*, 1982, 304 (*cryst struct*)
Marshall, J.A. *et al, J. Org. Chem.*, 1987, **52**, 3883 (*synth*)
Marshall, J.A. *et al, Tetrahedron Lett.*, 1987, **28**, 723, 3323 (*synth, abs config*)

Aristolignin
A-60297

[110268-34-3]

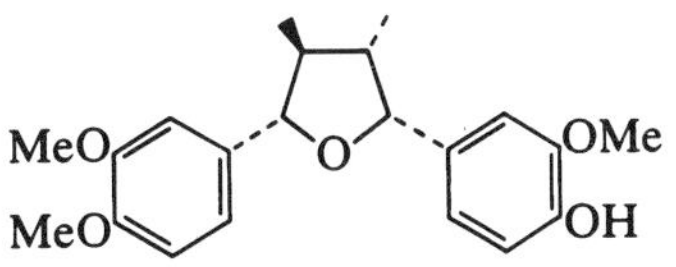

$C_{21}H_{26}O_5$ M 358.433
Constit. of *Aritolochia chilensis*. Oil. $[\alpha]_D$ +24.3° (c, 1.35 in $CHCl_3$).

Urzúa, A. *et al, Phytochemistry*, 1987, **26**, 1509.

Aristotetralone
A-60298

[111188-75-1]

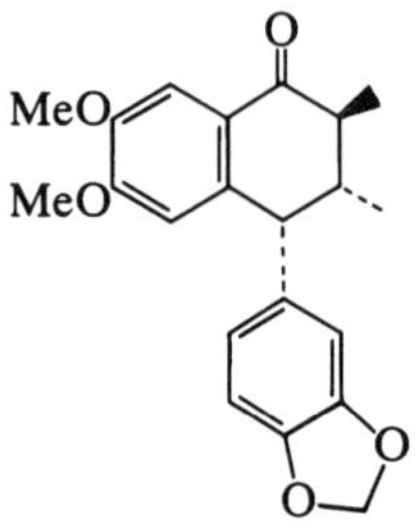

C$_{21}$H$_{22}$O$_5$ M 354.402

Constit. of *Aristolochia chilensis*. Amorph. solid. [α]$_D$ −164.3° (c, 1.54 in CHCl$_3$).

Urzua, A. *et al*, *Phytochemistry*, 1987, **26**, 2414.

10(14)-Aromadendrene-4,8-diol
A-60299

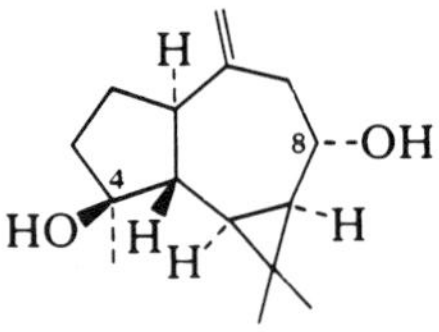

C$_{15}$H$_{24}$O$_2$ M 236.353

(4β,8α)-form [85526-81-4]

Constit. of *Parthenium argentatum*. Cryst. Mp 156-157.5°. [α]$_D^{26}$ +7.1° (c, 1.71 in CHCl$_3$).

8-Cinnamoyl: [107812-57-7]. **Guayulin C.**
C$_{24}$H$_{30}$O$_3$ M 366.499
Constit. of *P. argentatum*. Cryst. Mp 111-112°. [α]$_D^{26}$ −7.8° (c, 2.95 in CHCl$_3$).

8-(4-Methoxybenzoyl): [107812-58-8]. **Guayulin D.**
C$_{23}$H$_{30}$O$_4$ M 370.488
From *P. argentatum*. Oil.

Crevoisier, M. *et al*, *Acta Crystallogr.*, *Sect. C*, 1984, **40**, 979 (*cryst struct*)
Martinez, M. *et al*, *J. Nat. Prod.*, 1986, **49**, 1102 (*isol, deriv*)

Artanomaloide
A-60300

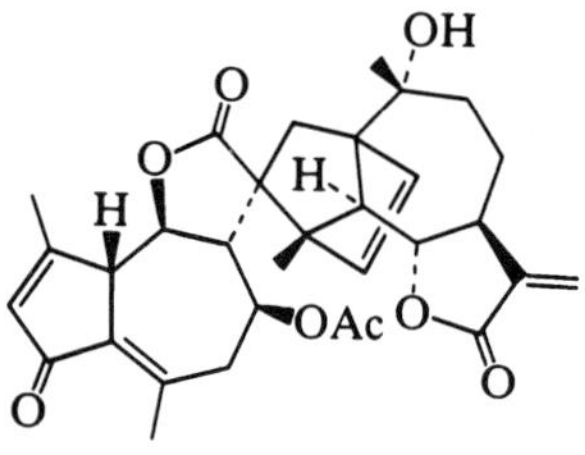

C$_{32}$H$_{36}$O$_8$ M 548.632

Constit. of *Artemisia anomala*. Gum. [α]$_D^{24}$ −17° (c, 0.2 in CHCl$_3$).

Jakupovic, J. *et al*, *Phytochemistry*, 1987, **26**, 2777.

Consult the Dictionary of Antibiotics and Related Substances for a fuller treatment of antibiotics and related compounds.

Artapshin
A-60301

Updated Entry replacing A-40173

Decahydro-4,6-dihydroxy-3,5a-dimethyl-9-methylenenaphtho[1,2-b]furan-2(3H)-one, 9CI

[80377-69-1]

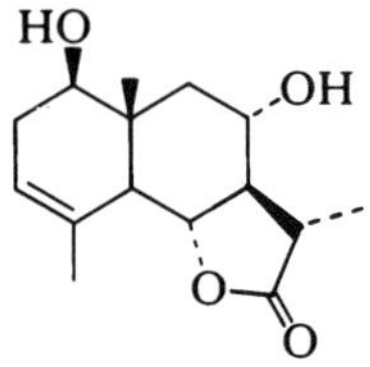

C$_{15}$H$_{22}$O$_4$ M 266.336

Constit. of *Artemisia fragrans*. Oil.

Serkerov, S.V. *et al*, *Khim. Prir. Soedin.*, 1983, **19**, 578 (*isol*)
Fernandez, J. *et al*, *Tetrahedron*, 1987, **43**, 805 (*synth*)

Artecalin
A-60302

Dehydroarsanin

[24778-20-9]

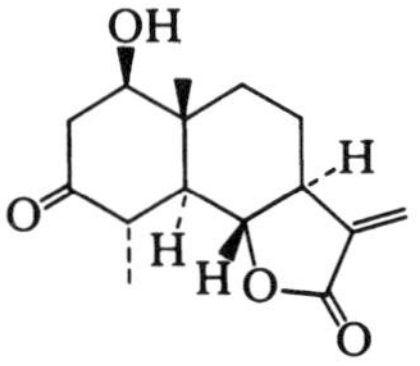

C$_{15}$H$_{20}$O$_4$ M 264.321

Isol. from *Artemisia californica*, *A. tripartita* and many other plants. Cryst. (EtOAc). Mp 225-227°. [α]$_D^{23}$ +45° (c, 0.01 in CHCl$_3$).

Geissman, T.A. *et al*, *Phytochemistry*, 1969, **8**, 1297 (*isol*)
Yamakawa, K. *et al*, *Tetrahedron Lett.*, 1975, 2829 (*synth*)
Abduazimov, B.K. *et al*, *Khim. Prir. Soedin.*, 1980, 633 (*isol*)
Yunusov, A.I. *et al*, *Khim. Prir. Soedin.*, 1983, 532 (*isol*)

Artelein
A-60303

[103654-30-4]

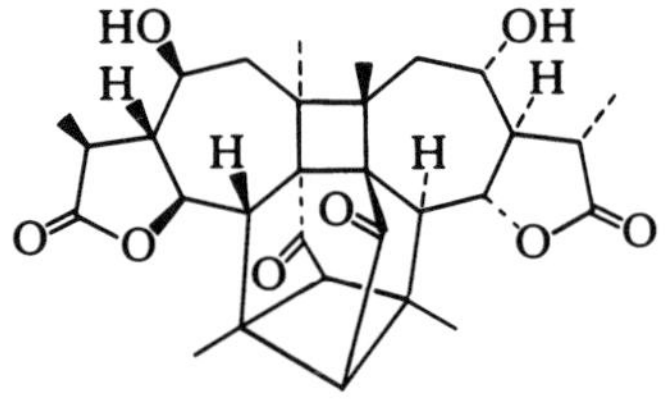

C$_{30}$H$_{36}$O$_8$ M 524.610

Constit. of *Artemisia leucodes*. Cryst. Mp 465-466°.

Mallabaev, A. *et al*, *Khim. Prir. Soedin.*, 1986, **22**, 42 (*isol, cryst struct*)

Artelin
A-60304

[66173-52-2]

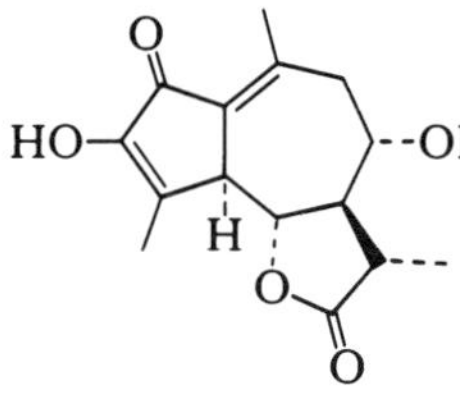

C$_{15}$H$_{18}$O$_5$ M 278.304

See also Artelin under 5,6,7,8-Tetrahydroxy-2*H*-1-benzo-
pyran-2-one, T-20082 . Constit. of *Artemisia leucodes*.
Cryst. (EtOH). Mp 228-229°.

Saitbaeva, I.M. *et al*, *Khim. Prir. Soedin.*, 1986, **22**, 112.

Artemone   A-60305

Updated Entry replacing A-03496

2-(5-Ethenyltetrahydro-5-methyl-2-furanyl)-4,4-di-
methyl-5-hexen-3-one, 9CI

[30925-48-5]

$C_{15}H_{24}O_2$ M 236.353

Constit. of *Artemisia pallens*. Liq. $[\alpha]_D$ +41.4° (c, 1.9 in
CHCl$_3$).

Naegeli, P. *et al*, *Tetrahedron Lett.*, 1970, 5021 (*isol, struct*)
Akhila, A. *et al*, *Tetrahedron Lett.*, 1986, **27**, 5885 (*biosynth*)

Artesovin   A-60306

$C_{15}H_{20}O_3$ M 248.321

Constit. of *Artemisia szowitsiana*. Cryst. (EtOH aq.).
Mp 142-144°.

Serkerov, S.V. *et al*, *Khim. Prir. Soedin.*, 1986, **22**, 609.

Articulin   A-60307

[75478-98-7]

$C_{20}H_{26}O_5$ M 346.422
Constit. of *Baccharis articulata*.

Ac: Articulin acetate.
  $C_{22}H_{28}O_6$ M 388.460
  Constit. of *B. articulata*.
2-Deoxy: Desoxyarticulin.
  $C_{20}H_{26}O_4$ M 330.423
  Constit. of *B. pedicellata* and *B. marginalis*. Cryst.
  (Et$_2$O). Mp 130-132°. $[\alpha]_D^{25}$ −146.4° (c, 1.1 in
  CHCl$_3$).

Stapel, G. *et al*, *Planta Med.*, 1980, **39**, 366 (*isol, struct*)
Faini, F. *et al*, *Phytochemistry*, 1987, **26**, 3281 (*deriv*)

Asadanin   A-60308

3,8,9,12,17-Pentahydroxytricyclo[12.3.1.1^{2,6}]nonadeca-
1(18),2,4,6(19),14,16-hexaen-10-one, 9CI

[22756-44-1]

$C_{19}H_{20}O_6$ M 344.363
Isol. from *Ostrya japonica*. Mp 236-239°. $[\alpha]_D^{17}$ +84°.
Penta-Ac: Mp 133-135°.
10-Alcohol: [32479-45-1]. *Asadanol*.
  $C_{19}H_{22}O_6$ M 346.379
  From *O. japonica*.

Yasue, M. *et al*, *CA*, 1965, **63**, 13553g; 1966, **64**, 6591dg (*isol,
struct, derivs*)

Ascorbigen   A-60309

[8075-98-7]

$C_{15}H_{15}NO_6$ M 305.287
Present in plants, esp. cabbage and other crucifers. Bound
form of ascorbic acid.

(*2R*)-*form* [26548-49-2]
  Ascorbigen B. *2-C-(1H-Indol-3-ylmethyl)-β-L-lyxo-3-*
  hexulofuranosonic acid γ-lactone, 9CI
  Light-yellow amorph. powder. Mp ca. 70° (sinters). $[\alpha]_D^{25}$
  +12.5° (c, 1.0 in MeOH).
(*2S*)-*form* [26676-89-1]
  Ascorbigen A. *2-C-(1H-Indol-3-ylmethyl)-β-L-xylo-3-*
  hexulofuranosonic acid γ-lactone, 9CI
  Amorph. powder. Mp ca. 65° (sinters). $[\alpha]_D^{25}$ +11.0° (c,
  2.0 in EtOH).

Piironen, E. *et al*, *Acta Chem. Scand.*, 1962, **16**, 1286 (*synth,
bibl*)
Kiss, G. *et al*, *Helv. Chim. Acta*, 1966, **49**, 989 (*synth, ir, uv,
pmr, struct*)

N-Asparaginylalanine, 9CI   A-60310

$$H_2NCOCH_2CH(NH_2)CONHCH(COOH)CH_3$$

$C_7H_{13}N_3O_4$ M 203.197
L-L-form [20917-57-1]
  N^2-*Benzyloxycarbonyl:* [74216-21-0]. *Z-Asn-Ala-OH*.
    Mp 217-219° dec. $[\alpha]_D^{21}$ +° 1.1 (c, 1 in DMF).
  N^2-*(p-Methoxybenzyloxycarbonyl), Me ester:* [38428-
    17-0]. Mp 172-174°. $[\alpha]_D^{26}$ −8.5° (MeOH).

Kiso, Y. *et al*, *Chem. Pharm. Bull.*, 1973, **21**, 2507 (*deriv*)
Van Nispen, J.W. *et al*, *Recl. Trav. Chim. Pays-Bas*, 1980, **99**,
57 (*synth*)
Kiyama, S. *et al*, *Int. J. Pept. Protein Res.*, 1984, **23**, 174 (*ester*)

Aspartame
A-60311

Updated Entry replacing M-00852

N-α-*Aspartylphenylalanine 1-methyl ester, 9CI.* N-α-*Aspartyl-3-phenylalanine methyl ester. Methyl aspartylphenylalanine.* α-*APM*

$$HOOCCH_2CH(NH_2)CONHCH(CH_2Ph)COOMe$$

$C_{14}H_{18}N_2O_5$ M 294.307

D-D-form

Mp 190° and 244-245° (double Mp). $[\alpha]_D \pm 0°$ (H_2O).

D-L-form

Cryst. (H_2O). Mp 159-160° and 212-213° (double Mp). $[\alpha]_D -18°$ (c, 1 in H_2O).

L-D-form [22839-65-2]

Cryst. (H_2O). Mp 157-159° and 212-213° (double Mp). $[\alpha]_D +19°$ (c, 1 in H_2O).

L-L-form [22839-47-0]

Canderel

Compd. with 100 times the sweetness of sucrose. Artifical sweetener. Cryst. (EtOH aq. or H_2O). Mp 190° and 245-247° (double Mp) (235-236° dec.). $[\alpha]_D \pm 0°$ (H_2O), $[\alpha]_D^{22} +32.0°$ (c, 1 in AcOH).

▷WM3407000.

B,HCl: [5910-52-1]. Mp 127-128° dec. (partly melts at 103°). $[\alpha]_D^{25} +1.3°$ (c, 2 in H_2O).

B,HBr: [36771-92-3]. Mp 155° dec. $[\alpha]_D^{25} +1.0°$ (c, 2 in H_2O). Mp 180-181°.

Neuman, H., *Med. Actual.*, 1980, **16**, 63 (*rev*)
Shvachkin, Y.P. *et al, Zh. Obshch. Khim.*, 1982, **52**, 2791 (*synth*)
Stegink, L.D. *et al, Food Science and Technol. Vol. 12*, Dekker, N.Y., 1984 (*book*)
Hatada, M. *et al, J. Am. Chem. Soc.*, 1985, **107**, 4279 (*cryst struct*)
Renwick, A.G., *Food Chem.*, 1985, **16**, 281 (*metab*)
Tou, J.S. *et al, J. Org. Chem.*, 1985, **50**, 4982 (*synth*)
Fuganti, C. *et al, J. Org. Chem.*, 1986, **51**, 1126 (*synth*)
Görbitz, C.H., *Acta Crystallogr., Sect. C*, 1987, **41**, 87 (*cryst struct*)

Aspicillin
A-60312

Updated Entry replacing A-03571

5,6,7-Trihydroxy-18-methyloxacyclooctadec-3-en-2-one, 9CI. 4,5,6-Trihydroxyoctadec-2-en-1,17-olide. Aspicilin

[52461-05-9]

(−)-*form*

$C_{18}H_{32}O_5$ M 328.448

Obt. from the lichens *Aspicilia calcarea* and *A. gibbosa*. Platelets (MeOH). Mp 153-154°. $[\alpha]_D^{20} +32°$ (c, 2.31 in $CHCl_3$).

Tri-Ac: [52461-07-1]. Needles (MeOH aq.). Mp 118-119°.

Dihydro: [52461-08-2].
$C_{18}H_{34}O_5$ M 330.464
Needles. Mp 168°.

Huneck, S. *et al, Tetrahedron*, 1973, **29**, 3687 (*ir, ms, pmr, cd*)
Waanders, P.P. *et al, Tetrahedron Lett.*, 1987, **28**, 2409 (*synth, struct*)

Aurachin *A*
A-60313

α-*(4,8-Dimethyl-3,7-nonadienyl)-1,2-dihydro-α,4-dimethylfuro[2,3-c]quinoline-2-methanol 5-oxide, 9CI*

[108354-15-0]

$C_{25}H_{33}NO_3$ M 395.541

Prod. by *Stigmatella aurantiaca*. Active against gram-positive bacteria and a few yeasts and molds. Cryst. (Et_2O). Mp 111-112°. $[\alpha]_D -49.2°$ (c, 0.4 in MeOH).

Kunze, B. *et al, J. Antibiot.*, 1987, **40**, 258 (*isol, props*)

Aurachin *B*
A-60314

[108354-12-7]

$C_{25}H_{33}NO_2$ M 379.541

Prod. by *Stigmatella aurantiaca*. Active against gram-positive bacteria and a few yeasts and molds. Cryst. (Et_2O). Mp 93-94°.

Kunze, B. *et al, J. Antibiot.*, 1987, **40**, 258 (*isol, props*)

Aurachin *D*
A-60315

[108354-13-8]

$C_{25}H_{33}NO$ M 363.542

Prod. by *Stigmatella aurantiaca*. Active against gram-positive bacteria and a few yeasts and molds. Cryst. (Et_2O). Mp 165-168°.

N-*Hydroxy:* [108354-14-9]. **Aurachin C**.
$C_{25}H_{33}NO_2$ M 379.541
From *S. aurantiaca*. Active against gram-positive bacteria and a few yeasts and molds. Cryst. (Et_2O). Mp 124-125°.

Kunze, B. *et al, J. Antibiot.*, 1987, **40**, 258 (*isol, props*)

Aurantioclavine
A-60316

*3,4,5,6-Tetrahydro-6-(2-methyl-1-propenyl)-1*H-*aze-pino[5,4,3-*cd*]indole, 9CI*

[80152-02-9]

$C_{15}H_{18}N_2$ M 226.321

(−)-*form*

Isol. from *Penicillium aurantio-virens*. No opt. rotn. reported.

(±)-*form* [99211-67-3]

Off-white powder. Mp 194-195°.

Hegedus, L.S. *et al*, *J. Org. Chem.*, 1987, **52**, 3319 (*synth*, *bibl*)

Auricularic acid
A-60317

13(15),16-Cleistanthadien-18-oic acid

$C_{20}H_{30}O_2$ M 302.456

Constit. of *Pogostemon auricularis*.

Prakash, O. *et al*, *Tetrahedron Lett.*, 1987, **28**, 685.

Aurintricarboxylic acid
A-60318

5-[(3-Carboxy-4-hydroxyphenyl)(3-carboxy-4-oxo-2,5-cyclohexadien-1-ylidene)methyl]-2-hydroxybenzoic acid, 9CI

[4431-00-9]

$C_{22}H_{14}O_9$ M 422.347

Effective inhibitor of complement activity in pharmaceuticals.

Tri-NH$_4$ salt: [569-58-4]. *Aluminon.*

$C_{22}H_{25}N_3O_{10}$ M 491.454

Reagent for Al. Yellow-brown powder.

▷GU4800000.

Tri-Na salt: [13186-45-3]. *C.I. Mordant Violet 39.* Dyestuff.

Org. Synth., Coll. Vol., **1**, 54.
Smith, W.H. *et al*, *Anal. Chem.*, 1949, **21**, 1334 (*synth*, *use*)
U.S.P., 4 007 270, (*1977*); *CA*, **86**, 177318 (*use*)
Janowski, A. *et al*, *Pol. J. Chem.*, 1982, **56**, 451 (*ir*)
Pannell, L.K. *et al*, *Anal. Chem.*, 1985, **57**, 1060 (*ms*)
Sax, N.I., *Dangerous Properties of Industrial Materials*, 6th Ed., Van Nostrand-Reinhold, 1984, 174.

Auropolin
A-60319

Updated Entry replacing A-03653

[77984-53-3]

$C_{24}H_{30}O_9$ M 462.496

Constit. of *Teucrium polium*. Cryst. (EtOAc/EtOH). Mp 170-172°. $[\alpha]_D^{20}$ +26.0° (c, 0.33 in CHCl$_3$).

Ac: Cryst. (EtOAc/hexane). Mp 186°. $[\alpha]_D^{19}$ +63.6° (c, 0.23 in CHCl$_3$).

Eguren, L. *et al*, *J. Org. Chem.*, 1981, **46**, 3364 (*isol*)
Camps, F. *et al*, *Phytochemistry*, 1987, **26**, 1475 (*cryst struct*)

Austrobailignan-6
A-60320

Updated Entry replacing A-03670

As Austrobailignan-5, A-03669 with

$$R = OH, R' = Me$$

$C_{20}H_{24}O_4$ M 328.407

(**2R,3R**)-*form* [55890-24-9]

Lignan from *Austrobaileya scandens*. Oil. Bp$_{0.02}$ 100-120°. $[\alpha]_D^{25}$ −32° (c, 1.3 in CHCl$_3$).

(**2R,3S**)-*form*

Macelignan

Constit. of *Myristica fragrans*. Cryst. (hexane). Mp 70-72°. $[\alpha]_D^{20}$ +5.28° (c, 1.8 in CHCl$_3$).

(**2RS,3RS**)-*form* [68964-82-9]

Oil.

Murphy, S.T. *et al*, *Aust. J. Chem.*, 1975, **28**, 81 (*isol*, *struct*)
Biftu, T. *et al*, *J. Chem. Soc., Perkin Trans. 1*, 1978, 1147 (*synth*)
Woo, W.S. *et al*, *Phytochemistry*, 1987, **26**, 1542 (*isol*, *cryst struct*)

Austrocortilutein
A-60321

Updated Entry replacing A-40187

1,2,3,4-Tetrahydro-1,3,8-trihydroxy-6-methoxy-3-methyl-9,10-anthracenedione, 9CI

[97400-70-9]

$C_{16}H_{16}O_6$ M 304.299

Constit. of fruit bodies of *Cortinarius* sp. Cryst. Mp 183-185°. $[\alpha]_D^{20}$ +62° (c, 0.546 in EtOH).

1-Deoxy: **Deoxyaustrocortilutein**.

$C_{16}H_{16}O_5$ M 288.299

Constit. of *C.* sp. Orange plates (CHCl$_3$/pet. ether). Mp 206-212°. $[\alpha]_D^{20}$ −78° (c, 0.05 in CHCl$_3$).

Gill, M. *et al*, *Tetrahedron Lett.*, 1985, **26**, 2593 (*isol*)
Gill, M. *et al*, *Phytochemistry*, 1987, **26**, 2999 (*deriv*)

Austrocortirubin   A-60322

Updated Entry replacing A-40188

5,6,7,8-Tetrahydro-5,7,9,10-tetrahydroxy-2-methoxy-7-methyl-1,4-anthracenedione, 9CI

[97400-69-6]

$C_{16}H_{16}O_7$ M 320.298

Constit. of fruit bodies of *Cortinarius* sp. Cryst. Mp 193-195°. $[\alpha]_D^{20}$ +109° (c, 0.824 in EtOH).

1-Deoxy: **Deoxyaustrocortirubin**.

$C_{16}H_{16}O_6$ M 304.299

Constit. of *C.* sp. Red needles (C_6H_6/pet. ether). Mp 211-216°. $[\alpha]_D^{20}$ −59° (c, 0.05 in $CHCl_3$).

Gill, M. *et al*, *Tetrahedron Lett.*, 1985, **26**, 2593 (*isol, struct*)
Gill, M. *et al*, *Phytochemistry*, 1987, **26**, 2999 (*deriv*)

Avellanin *A*   A-60323

[110297-47-7]

R = −CH₂CH₃

$C_{31}H_{39}N_5O_5$ M 561.680

Isol. from *Hamigera avellanea*. Possesses pressor props. Plates (Me₂CO/hexane). Mp 202-204°. $[\alpha]_D^{24}$ +161.5° (c, 0.5 in $CHCl_3$). Similar to Cycloaspeptides.

Yamazaki, M. *et al*, *Chem. Pharm. Bull.*, 1987, **35**, 2122 (*isol, struct*)

Avellanin *B*   A-60324

[110297-46-6]

As Avellanin *A*, A-60323 with

R = CH₃

$C_{30}H_{37}N_5O_5$ M 547.653

Stereochem. not verified. Isol. from *Hamigera avellanea*. Possesses pressor props. Amorph. $[\alpha]_D^{24}$ +281.9° (c, 0.17 in $CHCl_3$).

Yamazaki, M. *et al*, *Chem. Pharm. Bull.*, 1987, **35**, 2122 (*isol, struct*)

Avenaciolide   A-60325

Updated Entry replacing A-30220

Dihydro-3-methylene-4-octylfuro[3,4-b]furan-2,6(3H,4H)-dione, 9CI, 8CI

[16993-42-3]

(−)-*form*
Absolute
configuration

$C_{15}H_{22}O_4$ M 266.336

Bislactone antibiotic.

(+)-form

Mp 54-56°. $[\alpha]_D^{25}$ +41.3° (c, 1 in EtOH).

(−)-form [20223-76-1]

Metab. of *Aspergillus avenaceus* and *A. fischeri* var. *glaber*. Antifungal agent. Cryst. (Et₂O/pet. ether). Mp 49-50° and 54-56° (double Mp). $[\alpha]_D^{25}$ −41.6° (c, 1.0 in EtOH).

(±)-form [26057-70-5]

Cryst. Mp 54-57°.

Brookes, D. *et al*, *J. Chem. Soc.*, 1963, 5385 (*isol, uv, ir, nmr, cryst struct*)
Brookes, D. *et al*, *Aust. J. Chem.*, 1965, **18**, 373 (*pmr*)
Herrmann, J.L. *et al*, *J. Am. Chem. Soc.*, 1973, **95**, 7923 (*synth*)
Tanabe, M. *et al*, *J. Chem. Soc., Chem. Commun.*, 1973, 212 (*cmr, biosynth*)
Ohrui, H. *et al*, *Tetrahedron Lett.*, 1975, 3657 (*synth, pmr, ir*)
Hughes, D.L., *Acta Crystallogr., Sect. B*, 1978, **34**, 3674 (*cryst struct*)
Kido, F. *et al*, *Chem. Lett.*, 1983, 881 (*synth*)
Schreiber, S.L. *et al*, *J. Am. Chem. Soc.*, 1984, **106**, 7200 (*synth*)
Kallmerten, J. *et al*, *J. Org. Chem.*, 1985, **50**, 1128 (*synth*)
Anderson, R.C. *et al*, *J. Org. Chem.*, 1985, **50**, 4781 (*synth, pmr*)
Burke, S.D. *et al*, *Tetrahedron Lett.*, 1986, **27**, 3345 (*synth, bibl*)
Kotsuki, H. *et al*, *Bull. Chem. Soc. Jpn.*, 1986, **59**, 3881 (*synth*)
Suzuki, K. *et al*, *Tetrahedron Lett.*, 1986, **27**, 6237 (*synth*)

Avicennioside   A-60326

$C_{15}H_{24}O_{11}$ M 380.348

Constit. of *Avicennia officinalis*. Oil.

König, G. *et al*, *Phytochemistry*, 1987, **26**, 423.

Axamide-2   A-60327

Updated Entry replacing A-03702

[56012-89-6]

$C_{16}H_{27}NO$ M 249.395

Constit. of *Axinella cannabina*. Oil. $[\alpha]_D$ +37.5° (c, 0.9 in $CHCl_3$).

1-Epimer: [108739-41-9]. ***10α-Formamidoalloaromadendrane***.
$C_{16}H_{27}NO$ M 249.395
Constit. of *A. cannabina*. Oil.

Fattorusso, E. *et al, Tetrahedron*, 1975, **31**, 269 (*isol*)
Ciminiello, P. *et al, Can. J. Chem.*, 1987, **65**, 518 (*isol, struct*)

Axisonitriles   A-60328

Updated Entry replacing A-03711

Sesquiterpenes having structures identical with Axamides 1-4 in which the isocyano group —N≡C replaces —NHCHO

Axisonitrile-1 [53822-96-1]
$C_{16}H_{25}N$ M 231.380

Constit. of *Axinella cannabina*. Cryst. Mp 43-45°. $[\alpha]_D$ +22.6° (c, 1 in $CHCl_3$).

Axisonitrile-2 [55907-33-0]
$C_{16}H_{25}N$ M 231.380

Constit. of *A. cannabina*. Oil. $[\alpha]_D$ +29° (c, 0.5 in $CHCl_3$).

1-Epimer: [108739-39-5]. ***10α-Isocyanoalloaromadendrane***.
$C_{16}H_{25}N$ M 231.380
Constit. of *A. cannabina*. Oil. $[\alpha]_D$ −17.21° (c, 0.7 in $CHCl_3$).

Axisonitrile-3 [59633-83-9]
$C_{16}H_{25}N$ M 231.380

Constit. of *A. cannabina*. Cryst. (pet. ether). Mp 101-103°. $[\alpha]_D$ +68.44° (c, 1 in $CHCl_3$).

Axisonitrile-4 [62078-10-8]
$C_{16}H_{23}N$ M 229.364

Constit. of *A. cannabina*. Cryst. Mp 56-58°. $[\alpha]_D$ +51.4° (c, 1 in $CHCl_3$).

Cafieri, F. *et al, Tetrahedron*, 1973, **29**, 4259 (*isol*)
Fattorusso, E. *et al, Tetrahedron*, 1974, **30**, 3911 (*isol*)
Blasio, B.Di *et al, Tetrahedron*, 1976, **32**, 473 (*isol*)
Adinolfi, N. *et al, Tetrahedron Lett.*, 1977, 2815 (*struct*)
Iengo, A. *et al, Experientia*, 1977, **33**, 11 (*isol*)
Piers, E. *et al, Can. J. Chem.*, 1986, **64**, 2475 (*synth*)
Ciminiello, P. *et al, Can. J. Chem.*, 1987, **65**, 518 (*isol, struct*)

Axisothiocyanates   A-60329

Updated Entry replacing A-03712
Sesquiterpenes having structures identical with Axamides 1-4 in which the thiocyanato group —NCS replaces —NHCHO. See Axamide-1, A-03701 and following entries.

Axisothiocyanate-1 [53822-97-2]
$C_{16}H_{25}NS$ M 263.440

Constit. of *Axinella cannabina*. Oil. $[\alpha]_D$ +5.9° (c, 2.5 in $CHCl_3$).

Axisothiocyanate-2 [56012-90-9]
$C_{16}H_{25}NS$ M 263.440

Constit. of *A. cannabina*. Oil. $[\alpha]_D$ +12.8° (c, 1.5 in $CHCl_3$).

1-Epimer: [108739-40-8]. ***10α-Isothiocyanatoalloaromadendrane***.
$C_{16}H_{25}NS$ M 263.440

Constit. of *A. cannabina*. Oil. $[\alpha]_D$ −6.95° (c, 0.9 in $CHCl_3$).

Axisothiocyanate-3 [59633-81-7]
$C_{16}H_{25}NS$ M 263.440

Constit. of *A. cannabina*. Oil. $[\alpha]_D$ +165.2° (c, 1 in $CHCl_3$).

Axisothiocyanate-4 [62078-11-9]
$C_{16}H_{23}NS$ M 261.424

Constit. of *A. cannabina*. Oil. $[\alpha]_D$ −35.9° (c, 1.2 in $CHCl_3$).

Cafieri, F. *et al, Tetrahedron*, 1973, **29**, 4259 (*isol*)
Fattorusso, E. *et al, Tetrahedron*, 1975, **31**, 269.
Blasio, B.Di *et al, Tetrahedron*, 1976, **32**, 473.
Adinolfi, M. *et al, Tetrahedron Lett.*, 1977, 2815 (*struct*)
Iengo, A. *et al, Experientia*, 1977, **33**, 11 (*isol*)
Ciminiello, P. *et al, Can. J. Chem.*, 1987, **65**, 518 (*isol, struct*)

1-Azabicyclo[3.2.0]heptane-2,7-dione   A-60330

$C_6H_7NO_2$ M 125.127

(±)-*form*
Possesses cognition enhancing props. Mp 61-63°.

Drummond, J.T. *et al, Tetrahedron Lett.*, 1987, **28**, 5245 (*synth, props*)

7-Azabicyclo[4.2.0]oct-3-en-8-one, 9CI   A-60331

C_7H_9NO M 123.154

(1S,6R)-*form* [98856-65-6]
Intermed. for carbapenems. Prisms. Mp 163-164°. $[\alpha]_D^{25}$ −28.6° (c, 0.585 in $CHCl_3$).

Tamura, N. *et al, Chem. Pharm. Bull.*, 1987, **35**, 996 (*synth, pmr*)

14-Azaprostanoic acid   A-60332

$C_{19}H_{37}NO_2$ M 311.507
Cryst. ($Et_2O/CHCl_3$). Mp 86-88°.

Cho, B.P. *et al, J. Org. Chem.*, 1986, **51**, 4279 (*synth, pmr, ms*)

4-Azatricyclo[4.3.1.1³·⁸]undecane, 9CI   A-60333
Updated Entry replacing A-03827
4-Azahomoadamantane
[22776-74-5]

$C_{10}H_{17}N$ M 151.251
Derivs. show antiviral, antiarryhthmic, antiinflammatory or cardiovascular activities. Solid (hexane). Mp 300° (290°).

B,HCl: Mp 374-378° dec.
N-(p-*Nitrobenzoyl*): Cryst. Mp 150°.

Decherches, E. *et al, Synthesis*, 1974, 812 (*synth*)
Quast, H. *et al, Justus Liebigs Ann. Chem.*, 1974, 1727 (*synth*)
St. Georgiev, V. *et al, J. Heterocycl. Chem.*, 1986, **23**, 1023 (*synth, bibl*)

2'-Azidoacetophenone, 8CI   A-60334
1-(2-Azidophenyl)ethanone, 9CI. 1-Acetyl-2-azidobenzene. 2-Azidophenyl methyl ketone
[16714-26-4]

$C_8H_7N_3O$ M 161.163
Cryst. (pet. ether). Mp 23°.
2,4-Dinitrophenylhydrazone: Red-brown platelets. Mp 182-183° dec.

Meisenheimer, J. *et al, Ber.*, 1927, **60**, 1736 (*synth*)
Adger, B.M. *et al, J. Chem. Soc., Perkin Trans. 1*, 1975, 31 (*synth*)

3'-Azidoacetophenone, 8CI   A-60335
1-(3-Azidophenyl)ethanone, 9CI. 1-Acetyl-3-azidobenzene. 3-Azidophenyl methyl ketone
[70334-60-0]
$C_8H_7N_3O$ M 161.163
Yellow oil.

Ohba, Y. *et al, Bull. Chem. Soc. Jpn.*, 1986, **59**, 2317 (*synth, ir, pmr*)

4'-Azidoacetophenone, 8CI   A-60336
1-(4-Azidophenyl)ethanone, 9CI. 1-Acetyl-4-azidobenzene. 4-Azidophenyl methyl ketone
[20062-24-2]
$C_8H_7N_3O$ M 161.163
Yellow needles. Mp 44°.

Ohba, Y. *et al, Bull. Chem. Soc. Jpn.*, 1986, **59**, 2317 (*synth, ir, pmr*)

2-Azidoadamantane   A-60337
2-Azidotricyclo[3.3.1.1³·⁷]decane, 9CI
[34197-88-1]
$C_{10}H_{15}N_3$ M 177.249
Mp 40-41°.

Sasaki, T. *et al, Heterocycles*, 1977, **7**, 315 (*synth*)
Prakash, G.K.S. *et al, J. Org. Chem.*, 1986, **51**, 3215 (*synth, pmr*)

2-Azidobenzophenone   A-60338
(2-Azidophenyl)phenylmethanone, 9CI. 2-Azidophenyl phenyl ketone
[16714-27-5]

$C_{13}H_9N_3O$ M 223.234
Pale-yellow needles (pet. ether). Mp 36-37°.

Smith, P.A.S. *et al, J. Am. Chem. Soc.*, 1953, **75**, 6335 (*synth*)
Adger, B.M. *et al, J. Chem. Soc., Perkin Trans. 1*, 1975, 31 (*synth*)

3-Azidobenzophenone   A-60339
(3-Azidophenyl)phenylmethanone, 9CI. 3-Azidophenyl phenyl ketone
[107128-58-5]
$C_{13}H_9N_3O$ M 223.234
Yellow oil.

Ohba, Y. *et al, Bull. Chem. Soc. Jpn.*, 1986, **59**, 2317 (*synth, ir, pmr*)

4-Azidobenzophenone   A-60340
(4-Azidophenyl)phenylmethanone, 9CI. 4-Azidophenyl phenyl ketone
[36210-71-6]
$C_{13}H_9N_3O$ M 223.234
Yellow needles. Mp 74°.

Dimroth, O. *et al, Ber.*, 1910, **43**, 2757.
Ohba, Y. *et al, Bull. Chem. Soc. Jpn.*, 1986, **59**, 2317 (*synth, ir, pmr*)

1-Azidodiamantane   A-60341
1-Azidodecahydro-3,5,1,7-[1,2,3,4]-butanetetraylnaphthalene, 9CI
[87999-44-8]

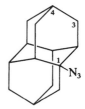

$C_{14}H_{19}N_3$ M 229.324
Mp 114°.

Sasaki, T. *et al, J. Org. Chem.*, 1984, **49**, 444 (*synth, ir, pmr, ms*)
Prakash, G.K.S. *et al, J. Org. Chem.*, 1986, **51**, 3215 (*synth, cmr*)

3-Azidodiamantane   A-60342
2-Azidodecahydro-3,5,1,7-[1,2,3,4]-butanetetraylnaphthalene, 9CI. 3-Azidopentacyclo[7.3.1.1⁴·¹².0²·⁷.0⁶·¹¹]tetradecane
[102920-05-8]
$C_{14}H_{19}N_3$ M 229.324
Cryst. Mp 46-47°.

Prakash, G.K.S. *et al, J. Org. Chem.*, 1986, **51**, 3215 (*synth, cmr*)

4-Azidodiamantane A-60343

3-Azidodecahydro-3,5,1,7-[1,2,3,4]-butanetetraylnaphthalene, 9CI

[87999-45-9]

$C_{14}H_{19}N_3$ M 229.324
Cryst. Mp 85°.

Sasaki, T. *et al*, *J. Org. Chem.*, 1984, **49**, 444 (*synth, ir, pmr, ms*)
Prakash, G.K.S. *et al*, *J. Org. Chem.*, 1986, **51**, 3215 (*synth, cmr*)

1-Azido-2,2-dimethylpropane A-60344

[102711-08-0]

$$(H_3C)_3CCH_2N_3$$

$C_5H_{11}N_3$ M 113.162
Bp_{55} 37-39°.

Quast, H. *et al*, *Justus Liebigs Ann. Chem.*, 1986, 1891 (*synth, pmr*)

4-Azido-1*H*-imidazo[4,5-*c*]pyridine A-60345

1H-Imidazo[4,5-c]tetrazolo[1,5-a]pyridine, 9CI

[98858-06-1]

$C_6H_4N_6$ M 160.138
Tautomeric system. Yellow solid. Mp >250°. Tetrazolo tautomer dominant in solid.

Krenitsky, T.A. *et al*, *J. Med. Chem.*, 1986, **29**, 138 (*synth, uv, pmr, ir*)

1-Azido-2-iodoethane, 9CI A-60346

2-Iodoethyl azide

[42059-30-3]

$$ICH_2CH_2N_3$$

$C_2H_4IN_3$ M 196.978
Reagent for electrophilic aminoethylation. Yellowish oil. Bp_{20} 70-75°. Stable at 0° in dark.

Khoukhi, M. *et al*, *Tetrahedron Lett.*, 1986, **27**, 1031 (*synth, use*)

1-Azido-3-iodopropane, 9CI A-60347

3-Iodopropyl azide

[58503-62-1]

$$ICH_2CH_2CH_2N_3$$

$C_3H_6IN_3$ M 211.005
Reagent for electrophilic aminopropylation. Yellowish oil. $Bp_{0.01}$ 38-40°. Stable at 0° in dark.

Khoukhi, M. *et al*, *Tetrahedron Lett.*, 1986, **27**, 1031 (*synth, use*)

3-Azido-3-methylbutanoic acid A-60348

[105090-72-0]

$$(H_3C)_2C(N_3)CH_2COOH$$

$C_5H_9N_3O_2$ M 143.145

Oil. $Bp_{0.017}$ 92-95°.

Nagarajan, S. *et al*, *J. Org. Chem.*, 1986, **51**, 4856 (*synth, pmr, ir, ms*)

2-Azido-2-nitropropane, 9CI A-60349

[85620-94-6]

$$(H_3C)_2C(N_3)(NO_2)$$

$C_3H_6N_4O_2$ M 130.106
$Bp_{0.1}$ 35-38°.

Al-Khalil, S.I. *et al*, *J. Chem. Soc., Perkin Trans. 1*, 1986, 555 (*synth, pmr, ir*)

Azidopentafluorobenzene, 9CI, 8CI A-60350

[1423-15-0]

$$(C_6F_5)N_3$$

$C_6F_5N_3$ M 209.078
Yellow oil. Bp_5 35°. Dec. smoothly at 80-120°.

Birchall, J.M. *et al*, *J. Chem. Soc.*, 1962, 4966 (*synth*)
Banks, R.E. *et al*, *Tetrahedron Lett.*, 1973, 99 (*use*)
Kanjia, D.M. *et al*, *J. Chem. Soc., Perkin Trans. 2*, 1981, 975 (*N nmr*)

3-Azido-1,2,4-thiadiazole A-60351

[105376-48-5]

C_2HN_5S M 127.123

Butler, R.N. *et al*, *J. Chem. Soc., Chem. Commun.*, 1986, 800 (*synth*)

Azidotriphenylmethane, 8CI A-60352

1,1′,1″-(Azidomethylidyne)trisbenzene, 9CI. *Triphenyl-methyl azide*

[14309-25-2]

$$Ph_3CN_3$$

$C_{19}H_{15}N_3$ M 285.348
Cryst. (hexane). Mp 64-65°.

Wieland, H., *Ber.*, 1909, **42**, 3027 (*synth*)
Senior, J.K., *J. Am. Chem. Soc.*, 1916, **38**, 2718 (*synth, props*)
Saunders, W.H. *et al*, *J. Am. Chem. Soc.*, 1958, **80**, 3328; 1964, **86**, 861 (*synth, props*)
Preston, P.N. *et al*, *Spectrochim. Acta, Part A*, 1972, **28**, 197 (*pmr*)

2,2′-Azodiquinoxaline A-60353

Updated Entry replacing T-50234

[93764-49-9]

$C_{16}H_{10}N_6$ M 286.295
Originally incorrectly descr. as 1,2,4,5-Tetrazino[1,6-a:4,3-a′]diquinoxaline.

(*E*)-*form* [97274-08-3]
Deep-red monoclinic needles (EtOH). Mp 250-252°.

Koçak, A. *et al*, *Helv. Chim. Acta*, 1984, **67**, 1503 (*synth, uv, ir, pmr*)

Krieger, C. *et al*, *Helv. Chim. Acta*, 1985, **68**, 581 (*cryst struct*)

4-Azoniaspiro[3.3]heptane-2,6-diol

A-60354

Charamin

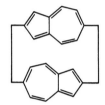

C$_6$H$_{12}$NO$_2$$^{\oplus}$ M 130.166 (ion)

Isol. from the green alga *Chara globularis*. Possesses antibacterial props. No phys. props. reported.

Anthoni, U. *et al*, *J. Org. Chem.*, 1987, **52**, 694 (*isol, synth*)

[2.2](2,6)Azulenophane

A-60355

Updated Entry replacing A-03950

Pentacyclo[10.4.4.4^{4,9}.0^{6,23}.0^{14,19}]tetracosa-4,6,8,12,14,16,17,19,21,23-decaene, 9CI

C$_{24}$H$_{20}$ M 308.422

anti-form [60549-48-6]

Blue-green plates (CHCl$_3$). Spar. sol. most solvs. Mp >340°. λ_{max} 605 (ϵ 418), 277 nm (69 700) (CHCl$_3$).

Kato, N. *et al*, *Tetrahedron Lett.*, 1976, 2045 (*synth, struct*)

Luhowy, R. *et al*, *J. Am. Chem. Soc.*, 1977, **99**, 3797 (*synth, struct, spectra*)

Koenig, T. *et al*, *J. Org. Chem.*, 1987, **52**, 641 (*synth, props*)

Azuleno[1,2-*d*]thiazole

A-60356

C$_{11}$H$_7$NS   M 185.243

Parent compd. currently unknown. Substd. derivs. known.

Fujimori, K. *et al*, *Bull. Chem. Soc. Jpn.*, 1986, **59**, 3320.

Azuleno[2,1-*d*]thiazole

A-60357

C$_{11}$H$_7$NS   M 185.243

Blue plates. Mp 113-114°.

Yamane, K. *et al*, *Chem. Lett.*, 1982, 707 (*synth, ir, pmr, cmr*)

B

Baccharinoid B9 B-60001

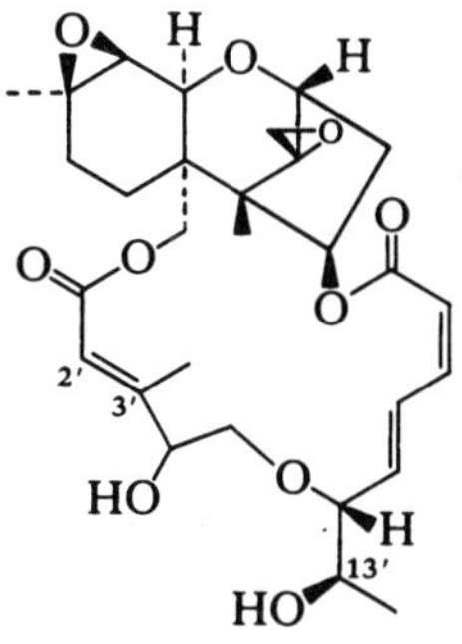

C$_{29}$H$_{38}$O$_{10}$ M 546.613
Constit. of *Baccharis megapotamica*. Cryst.
(CH$_2$Cl$_2$/Et$_2$O). Mp 216-218°. [α]$_D^{25}$ +2.4° (c, 0.68 in CH$_2$Cl$_2$).

13′-Epimer: **Baccharinoid B10**.
C$_{29}$H$_{38}$O$_{10}$ M 546.613
From *B. megapotamica*. Cryst. (CH$_2$Cl$_2$/Et$_2$O). Mp 157-158°. [α]$_D^{25}$ +8.1° (c, 0.62 in CHCl$_3$).

13′-Epimer, 2′,3′-Dihydro: **Baccharinoid B20**.
C$_{29}$H$_{40}$O$_{10}$ M 548.629
From *B. megapotamica*. Cryst. (Me$_2$CO/hexane). Mp 170-172°. [α]$_D^{25}$ +39.9° (c, 1.44 in CHCl$_3$).

Jarvis, B.B. *et al*, *J. Org. Chem.*, 1987, **52**, 45.

Baccharinoid B13 B-60002

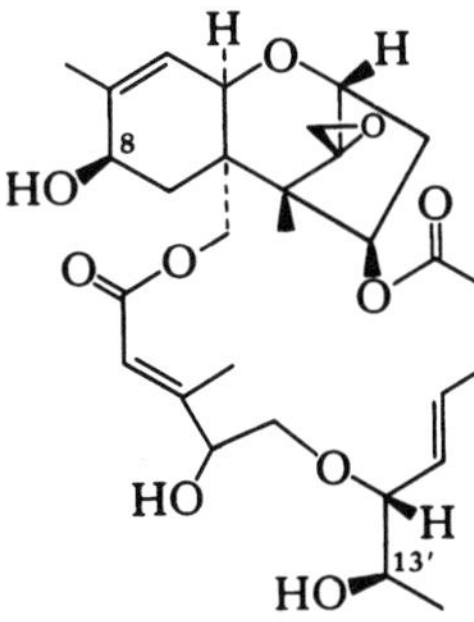

C$_{29}$H$_{38}$O$_{10}$ M 546.613
Constit. of *Baccharis megapotamica*. Cryst. (EtOAc).
Mp 218-219°. [α]$_D^{25}$ +130° (c, 0.74 in MeOH).

13′-Epimer: **Baccharinoid B14**.
C$_{29}$H$_{38}$O$_{10}$ M 546.613
Constit. of *B. megapotamica*. Cryst. (EtOAc/hexane). Mp 149-151°. [α]$_D^{25}$ +72.4° (c, 0.76 in MeOH).

13′-Epimer, 8-Ketone: **Baccharinoid B27**.
C$_{29}$H$_{36}$O$_{10}$ M 544.597
Constit. of *B. megapotamica*. Cryst. (CH$_2$Cl$_2$/hexane). Mp 165°. [α]$_D^{25}$ +5.4° (c, 0.4 in MeOH).

Jarvis, B.B. *et al*, *J. Org. Chem.*, 1987, **52**, 45.

Baccharinoid B16 B-60003

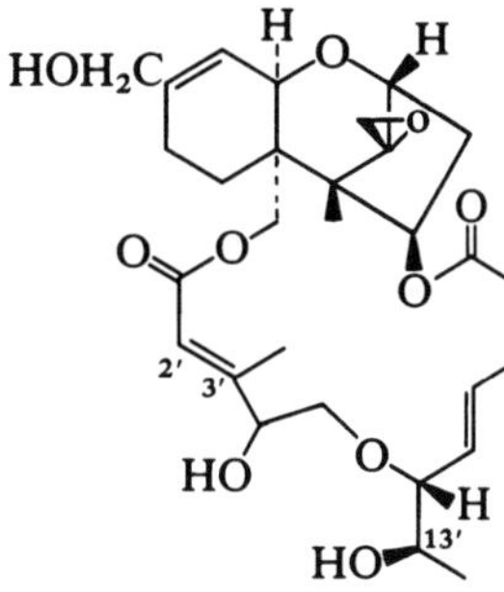

C$_{29}$H$_{38}$O$_{10}$ M 546.613
Constit. of *Baccharis megapotamica*. Cryst. (EtOH/hexane). Mp 160-161°. [α]$_D^{25}$ +63.5° (c, 0.47 in MeOH).

2′,3′-Dihydro: **Baccharinoid B23**.
C$_{29}$H$_{40}$O$_{10}$ M 548.629
From *B. megapotamica*. Glass. [α]$_D^{25}$ +113.5° (c, 0.9 in MeOH).

2′,3′-Dihydro, 13′-epimer: **Baccharinoid B24**.
C$_{29}$H$_{40}$O$_{10}$ M 548.629
From *B. megapotamica*. Glass. [α]$_D^{25}$ +90.8° (c, 1.36 in MeOH).

Jarvis, B.B. *et al*, *J. Org. Chem.*, 1987, **52**, 45.

Baccharinoid B25 B-60004

C$_{27}$H$_{32}$O$_{10}$ M 516.544
Constit. of *Baccharis megapotamica*. Cryst. (CH$_2$Cl$_2$/hexane). Mp 205°. [α]$_D^{25}$ +214° (c, 0.7 in MeOH).

Jarvis, B.B. *et al*, *J. Org. Chem.*, 1987, **52**, 45.

Balearone B-60005

Updated Entry replacing B-50009
[92675-09-7]

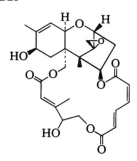

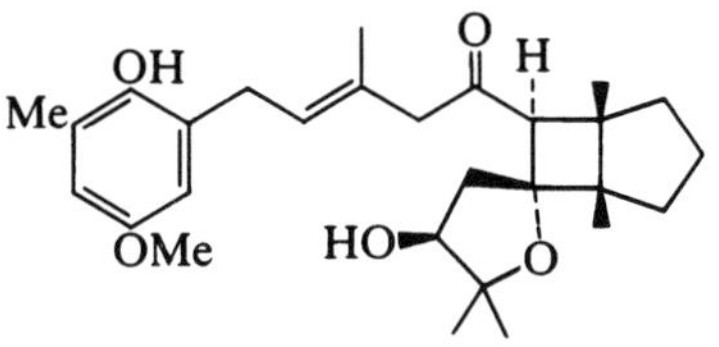

Relative configuration

C$_{28}$H$_{40}$O$_5$ M 456.621
Metabolite of brown alga *Cystoseira balearica*. Cryst.
(hexane). Mp 94-96°. [α]$_D^{20}$ +52°.

Z-Isomer: [110351-75-2]. **Isobalearone**.
C$_{28}$H$_{40}$O$_5$ M 456.621

Constit. of *Cystoseira stricta*. Oil. $[\alpha]_D^{20}$ +38.3° (c, 1.2 in EtOH).

Amico, V. *et al*, *Tetrahedron*, 1984, **40**, 1721 (*isol, struct*)
Amico, V. *et al*, *Phytochemistry*, 1987, **26**, 1719 (*isol*)

ψ-Baptigenin   B-60006

Updated Entry replacing B-00072
3-(1,3-Benzodioxol-5-yl)-7-hydroxy-4H-1-benzopyran-4-one, 9CI. 7-Hydroxy-3-(3,4-methylenedioxyphenyl)-chromone. Pseudobaptigenin
[90-29-9]

$C_{16}H_{10}O_5$ M 282.252
Constit. of *Baptisia* spp. and *Pterocarpus erinaceous*. Mp 296-298°. λ_{max} 226 (log ϵ 4.4), 248 sh and 294 nm (4.4) (MeOH).

Ac: Mp 173°.
Me ether: Mp 180-182°.
7-O-D-Glucosylrhamnoside: ψ-**Baptisin**. *Pseudobaptisin*.
   $C_{22}H_{20}O_{10}$ M 444.394
   Cryst. + $3H_2O$. Mp 148-150° (resolidifies at 180-210° and remelts at 249-251°). $[\alpha]_D$ −98.1°.
O^7-*(3-Methyl-2-butenyl):* **Maximaisoflavone** B.
   $C_{21}H_{18}O$ M 286.373
   Constit. of *Tephrosia maxima*. Cryst. (C_6H_6/pet. ether). Mp 132-133° (126-128°).

Schmidt, O. *et al*, *Monatsh. Chem.*, 1929, **53**, 454 (*isol, struct*)
Baker, W. *et al*, *J. Chem. Soc.*, 1953, 1582 (*synth*)
Kukla, A.S. *et al*, *Tetrahedron*, 1962, **18**, 1443 (*deriv*)
Bevan, C.W.L. *et al*, *J. Chem. Soc. (C)*, 1966, 509 (*pmr*)
Dhoubhadel, S.P. *et al*, *J. Indian Chem. Soc., Sect. B*, 1975, **52**, 440 (*synth*)
Schuda, P.F. *et al*, *J. Org. Chem.*, 1987, **52**, 1972 (*synth*)

Barbatusol   B-60007

(5S)-9(10→20)-Abeo-1(10),8,11,13-abietatetraene-11,12-diol

Absolute configuration

$C_{20}H_{28}O_2$ M 300.440
Constit. of *Coleus barbatus*. Amorph. $[\alpha]_D$ −102.5° (c, 1.88 in CCl_4).

Kelecom, A., *Tetrahedron*, 1983, **39**, 3603 (*struct*)
Koft, E.R. *et al*, *Tetrahedron*, 1987, **43**, 5775 (*synth*)

Bedfordiolide   B-60008

[108885-55-8]

$C_{15}H_{22}O_3$ M 250.337
Constit. of *Bedfordia arborescens*. Oil.

Zdero, C. *et al*, *Phytochemistry*, 1987, **26**, 1207.

Benz[*d*]aceanthrylene   B-60009

[19770-52-6]

$C_{20}H_{12}$ M 252.315
Violet cryst. (C_6H_6/hexane). Mp >300°.

Sangaiah, R. *et al*, *J. Org. Chem.*, 1987, **52**, 3205 (*synth, uv, pmr*)

Benz[*k*]aceanthrylene   B-60010

[16683-64-0]

$C_{20}H_{12}$ M 252.315
Violet needles (MeOH). Mp >300°.

Sangaiah, R. *et al*, *J. Org. Chem.*, 1987, **52**, 3205 (*synth, uv, pmr*)

Benz[*a*]azulene-1,4-dione, 9CI   B-60011

[87121-83-3]

$C_{14}H_8O_2$ M 208.216
Black-green cryst. ($CHCl_3$/pet. ether). Mp 197-198°.

Bindl, J. *et al*, *Chem. Ber.*, 1983, **116**, 2408 (*synth, ms, ir, uv, pmr*)

Benzenehexamine, 9CI   B-60012

Updated Entry replacing B-40012
Hexaaminobenzene
[4444-26-2]

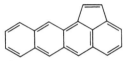

$C_6H_{12}N_6$ M 168.201
Needles. Mp ca. 255°. Formerly descr. as a dark-brown cryst. powder. Prepn. under Ar is necessary to obt. a colourless prod. V. air- and light-sensitive.
▷Potentially hazardous synth.

Kohne, B. *et al*, *Justus Liebigs Ann. Chem.*, 1987, 265 (*synth, ir, pmr, cmr, ms*)

Benzenesulfenyl thiocyanate   B-60013

Cyano phenyl disulfide, 9CI
[3153-52-4]

PhSSCN

$C_7H_5NS_2$ M 167.243
Yellow oil, cryst. (Me$_2$CO/pet. ether at low temp.). Dec. on dist.

Lecher, H. *et al*, *Ber.*, 1922, **55**, 1474 (*synth*)

1,2,3,4-Benzenetetrol, 9CI   B-60014

Updated Entry replacing B-00217
1,2,3,4-Tetrahydroxybenzene. Apionol. Phenetrol
[642-96-6]

$C_6H_6O_4$ M 142.111
Needles (EtOAc). Mp 160°.
Tetra-Ac:
  $C_{14}H_{14}O_8$ M 310.260
  Mp 139-141°.
1,4-Di-Me ether: 3,6-Dimethoxy-1,2-benzenediol.
  $C_8H_{10}O_4$ M 170.165
  Prisms (CH$_2$Cl$_2$/pet. ether). Mp 105-106°.
Tetra-Me ether: 1,2,3,4-Tetramethoxybenzene.
  $C_{10}H_{14}O_4$ M 198.218
  Mp 89°.

Bogert, M.T. *et al*, *J. Am. Chem. Soc.*, 1915, **37**, 2723.
Mayer, W. *et al*, *Chem. Ber.*, 1956, **89**, 511.
Rizzacasa, M.A. *et al*, *J. Chem. Soc., Perkin Trans. 1*, 1987, 2017 (*synth, deriv, pmr*)

Benz[f]indane   B-60015

2,3-Dihydro-1H-benz[f]indene, 9CI. 2,3-Cyclopentenon-aphthalene. 5,6-Benzohydrindene. 5,6-Benzindane
[1624-26-6]

$C_{13}H_{12}$ M 168.238
Flakes (EtOH). Mp 94° (84-85°).
Picrate: Golden-yellow needles. Mp 120-121°.

Sen Gupta, S.C., *J. Indian Chem. Soc.*, 1939, **16**, 89 (*synth*)
El-Abbady, A.M. *et al*, *J. Am. Chem. Soc.*, 1957, **79**, 1757 (*synth*)
Christol, H. *et al*, *Bull. Soc. Chim. Fr.*, 1960, 1576 (*synth, uv*)
Tius, M.A. *et al*, *Tetrahedron Lett.*, 1986, **27**, 2571 (*synth, ir, pmr, cmr, ms*)

Benzo[a]biphenylene, 9CI   B-60016

1,2-Benzobiphenylene
[252-47-1]

$C_{16}H_{10}$ M 202.255
Bright-yellow needles (MeOH). Mp 72-73°.
2,4,7-Trinitrofluorenone complex: [15300-62-6]. Black needles (C$_6$H$_6$/MeOH). Mp 201.5-202.5°.

Cava, M.P. *et al*, *J. Am. Chem. Soc.*, 1955, **77**, 6022; 1957, **79**, 1701 (*synth, uv, ir*)
Barton, J.W. *et al*, *J. Chem. Soc. (C)*, 1967, 1276 (*synth, pmr*)

Barton, J.W. *et al*, *Tetrahedron*, 1985, **41**, 1323 (*pmr*)

Benzo[b]biphenylene, 9CI   B-60017

2,3-Benzobiphenylene
[259-56-3]

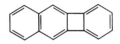

$C_{16}H_{10}$ M 202.255
Pale-yellow cryst. (C$_6$H$_6$/cyclohexane). Mp 242-243°.
2,4,7-Trinitrofluorenone complex: Red cryst. (C$_6$H$_6$/MeOH). Mp 214-216°.

Jensen, F.R. *et al*, *Tetrahedron Lett.*, 1959, 7 (*synth*)
Baker, W. *et al*, *J. Chem. Soc.*, 1962, 2633 (*synth, uv, ir*)
Figeys, H.P. *et al*, *Tetrahedron*, 1976, **32**, 2571 (*pmr*)
Barton, J.W. *et al*, *Tetrahedron*, 1985, **41**, 1323 (*pmr*)
Barton, J.W. *et al*, *J. Chem. Soc., Perkin Trans. 1*, 1986, 967 (*synth*)

Benzo[1,2-b:4,5-b']bis[1]benzothiophene, 9CI   B-60018

[241-34-9]

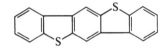

$C_{18}H_{10}S_2$ M 290.397
Cryst. by subl. Mp 313-315°.

Ahmed, M. *et al*, *J. Chem. Soc., Perkin Trans. 1*, 1973, 1099 (*synth, pmr*)

Benzo[1,2-b:5,4-b']bis[1]benzothiophene, 9CI   B-60019

[241-37-2]

$C_{18}H_{10}S_2$ M 290.397
Cryst. (C$_6$H$_6$). Mp 216-218°.

Ahmed, M. *et al*, *J. Chem. Soc., Perkin Trans. 1*, 1973, 1099 (*synth, pmr*)

Benzo[c]cinnoline, 9CI   B-60020

Updated Entry replacing B-00279
Phenazone. 9,10-Diazaphenanthrene
[230-17-1]

$C_{12}H_8N_2$ M 180.209
Yellow needles (EtOH aq.). V. sol. AcOH, CHCl$_3$, sol. Et$_2$O, EtOH, C$_6$H$_6$, spar. sol. H$_2$O, pet. ether. Mp 156°. Bp >360°.
N-Oxide: [6141-98-6].
  $C_{12}H_8N_2O$ M 196.208
  Cryst. (EtOH aq.). Mp 140-142° (139-140°).
5,6-Di-N-oxide:
  $C_{12}H_8N_2O_2$ M 212.207
  Cryst. (MeOH aq. or Me$_2$CO aq.). Mp 246-248° (240°). A higher-melting modification Mp 262-264° was obt. on one occasion.

Badger, G.M. *et al, J. Chem. Soc.*, 1951, 3199, 3207.
Corbett, J.F. *et al, J. Chem. Soc.*, 1961, 3695.
Étienne, A. *et al, Bull. Soc. Chim. Fr.*, 1962, 292.
Dewar, M.J.S. *et al, J. Chem. Soc.*, 1963, 2201.
Lewis, G.E. *et al, Aust. J. Chem.*, 1963, **16**, 1042; 1966, **19**, 1445 (*synth*)
Budzikiewicz, H. *et al, Z. Naturforsch., B*, 1970, **25**, 178 (*ms*)
van der Meer, H., *Acta Crystallogr., Sect. B*, 1972, **28**, 367 (*cryst struct*)
Barton, J.W., *Adv. Heterocycl. Chem.*, 1978, **24**, 151 (*rev*)
Smith, W.B., *J. Heterocycl. Chem.*, 1987, **24**, 745 (*synth, pmr, cmr, oxides*)

Benzo[*a*]coronene, 9CI   B-60021

1:2-Benzocoronene

[190-70-5]

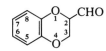

C$_{28}$H$_{14}$ M 350.419

Pale-yellow needles (xylene). Mp 292-294° (290°).

Clar, E. *et al, Tetrahedron*, 1959, **6**, 358 (*synth, uv*)
Bunte, R. *et al, Chem. Ber.*, 1986, **119**, 3521 (*synth, pmr*)

1,4-Benzodioxan-2-carboxaldehyde, 8CI   B-60022

2,3-Dihydro-1,4-benzodioxin-2-carboxaldehyde, 9CI. 2-Formyl-1,4-benzodioxan

[64179-67-5]

C$_9$H$_8$O$_3$ M 164.160

(±)-*form*

Oil. Bp$_{0.09}$ 63°.

2,4-Dinitrophenylhydrazone: Cryst. (MeOH). Mp 154-156° dec.

Semicarbazone: Cryst. (EtOH). Mp 183-185° dec.

Rosnati, V. *et al, Tetrahedron*, 1962, **18**, 289 (*synth, props*)
Petragnani, N. *et al, Farmaco, Ed. Sci.*, 1977, **32**, 512 (*synth, props*)

1,4-Benzodioxan-6-carboxaldehyde, 8CI   B-60023

2,3-Dihydro-1,4-benzodioxin-6-carboxaldehyde, 9CI. 6-Formyl-1,4-benzodioxan. 3,4-Ethylenedioxybenzaldehyde

[29668-44-8]

C$_9$H$_8$O$_3$ M 164.160

Cryst. (hexane). Mp 52.5°.

Oxime: [31127-39-6].
　C$_9$H$_9$NO$_3$ M 179.175
　Mp 54-59°. Bp$_{0.16}$ 130°.

Sugasawa, S. *et al, Chem. Pharm. Bull.*, 1956, **4**, 406 (*synth*)
Sasamoto, M. *et al, Chem. Pharm. Bull.*, 1960, **8**, 324 (*synth*)
Corsano, S. *et al, Farmaco, Ed. Sci.*, 1983, **38**, 265 (*synth*)

1,4-Benzodioxan-2-carboxylic acid, 8CI   B-60024

2,3-Dihydro-1,4-benzodioxin-2-carboxylic acid, 9CI

[3663-80-7]

(*R*)-*form*

C$_9$H$_8$O$_4$ M 180.160

(*R*)-*form* [70918-53-5]

Cryst. (toluene). Mp 98-99°. [α]$_D^{20}$ +62.1° (c, 1 in CHCl$_3$).

(±)-*form* [34385-93-8]

Cryst. (C$_6$H$_6$). Mp 125-126°.

K salt: Cryst. (EtOH). Mp 198-202° dec.

Me ester: [3663-79-4].
　C$_{10}$H$_{10}$O$_4$ M 194.187
　Mp 51-53°. Bp$_3$ 136-137°.

Chloride: [3663-81-8].
　C$_9$H$_7$ClO$_3$ M 198.606
　Needles. Mp 56-57°. Bp$_2$ 122-123°.

Amide: [33070-04-1].
　C$_9$H$_9$NO$_3$ M 179.175
　Cryst. (EtOH aq.). Mp 142-144°.

Nitrile: [1008-92-0]. *1,4-Benzodioxan-2-carbonitrile. 2-Cyano-1,4-benzodioxan.*
　C$_9$H$_7$NO$_2$ M 161.160
　Cryst. (toluene/pet. ether). Mp 57°. Bp$_{0.2}$ 93-94°.

Koo, J. *et al, J. Am. Chem. Soc.*, 1955, **77**, 5373 (*synth*)
Mndzhoyen, A.L. *et al, Izv. Akad. Nauk SSSR, Ser. Khim.*, 1965, **18**, 297 (*synth*)
Cook, M.J. *et al, J. Chem. Soc. (B)*, 1970, 1207 (*conformn*)
Martin, A.R. *et al, J. Org. Chem.*, 1974, **39**, 1808 (*synth*)
Campbell, S.F. *et al, J. Med. Chem.*, 1987, **30**, 49 (*resoln*)

1,4-Benzodioxan-6-carboxylic acid   B-60025

2,3-Dihydro-1,4-benzodioxin-6-carboxylic acid. 3,4-Ethylenedioxybenzoic acid

[4442-54-0]

C$_9$H$_8$O$_4$ M 180.160

Needles (H$_2$O). Mp 138-139°.

Me ester: [20197-75-5].
　C$_{10}$H$_{10}$O$_4$ M 194.187
　Bp$_{1-2}$ 135-140°.

Chloride: [6761-70-2].
　C$_9$H$_7$ClO$_3$ M 198.606
　Mp 102°.

Hydrazide:
　C$_9$H$_{10}$N$_2$O$_3$ M 194.190
　Needles (H$_2$O). Mp 141°.

Heertjes, P.M. *et al, J. Chem. Soc.*, 1957, 3445 (*synth*)
Lipp, M. *et al, Chem. Ber.*, 1958, **91**, 2247 (*deriv*)
Byrne, M.M. *et al, J. Chem. Soc. (B)*, 1968, 809 (*synth*)
Coudert, G. *et al, Tetrahedron Lett.*, 1978, 1059 (*synth*)

4*H*-1,3-Benzodioxin-6-carboxaldehyde, 9CI   B-60026

6-Formyl-4H-1,3-benzodioxin

[92607-80-2]

C$_9$H$_8$O$_3$ M 164.160
Mp 58°.

Denis, A. *et al*, *J. Heterocycl. Chem.*, 1984, **21**, 517 (*synth*)

4H-1,3-Benzodioxin-6-carboxylic acid, 9CI   B-60027

[33835-87-9]

$C_9H_8O_4$ M 180.160
Cryst. (MeOH aq.). Mp 197° (193°).

Me ester: [33835-89-1].
  $C_{10}H_{10}O_4$ M 194.187
  Cryst. (MeOH aq.). Mp 65.5°.

Kaemmerer, H. *et al*, *Monatsh. Chem.*, 1971, **102**, 946 (*synth*)
Denis, A. *et al*, *J. Heterocycl. Chem.*, 1984, **21**, 517 (*synth*)

4H-1,3-Benzodioxin-2-one   B-60028

$C_8H_6O_3$ M 150.134
Mp 72-73°.

Podraza, K.F., *J. Heterocycl. Chem.*, 1987, **24**, 801 (*synth, pmr, cmr*)

1,3-Benzodioxole-4-acetic acid   B-60029

2,3-Methylenedioxyphenylacetic acid
[100077-49-4]

$C_9H_8O_4$ M 180.160
Needles. Mp 103-104°.

Nitrile: [84434-78-6]. *1,3-Benzodioxole-4-acetonitrile, 9CI. 2,3-Methylenedioxyphenylacetonitrile.*
  $C_9H_7NO_2$ M 161.160
  Needles (pet. ether). Mp 75-76°. Bp_1 115°.

Govindachari, T.R. *et al*, *Chem. Ber.*, 1958, **91**, 36 (*deriv*)
Hanaoka, M. *et al*, *Chem. Pharm. Bull.*, 1985, **33**, 2273 (*synth*)

Benzo[1,2-b:4,5-b']dithiophene, 9CI   B-60030

Updated Entry replacing B-00343
[267-65-2]

$C_{10}H_6S_2$ M 190.277
Flakes (C_6H_6 or pet. ether). Mp 198°.

▷Earlier reported syntheses are said to be irreproducible or difficult

Nair, P.J. *et al*, *Indian J. Chem.*, 1974, **12**, 589 (*pmr*)
Beimling, P. *et al*, *Chem. Ber.*, 1986, **119**, 3198 (*synth, ir, ms*)

Benzo[1,2-b:4,5-b']dithiophene-4,8-dione, 9CI   B-60031

[32281-36-0]

$C_{10}H_4O_2S_2$ M 220.260

Olive-green microcryst. Mp 260-262° (258-260°).

Beimling, P. *et al*, *Chem. Ber.*, 1986, **119**, 3198 (*synth, pmr, bibl*)

3-Benzofurancarboxylic acid, 9CI   B-60032

Updated Entry replacing B-00360
[26537-68-8]

$C_9H_6O_3$ M 162.145
Cryst. (C_6H_6/pet. ether). Mp 159-160°.

Me ester: [4687-24-5].
  $C_{10}H_8O_3$ M 176.171
  Oil. $Bp_{0.2}$ 142-143°.

Hallmann, G. *et al*, *Ann. Chim.* (*Paris*), 1963, **662**, 147
Capuano, L., *Chem. Ber.*, 1965, **98**, 3659.
Robba, M. *et al*, *Bull. Soc. Chim. Fr.*, 1977, 142.
Chou, C.-H. *et al*, *J. Org. Chem.*, 1986, **51**, 4208 (*synth, pmr, cmr*)

Benzo[b]naphtho[2,1-d]furan-5,6-dione   B-60033

α-Brazanquinone

$C_{16}H_8O_3$ M 248.237

Quinoxaline deriv.: Benzo[a]benzofuro[2,3-c]phenazine.
  $C_{22}H_{12}N_2O$ M 320.350
  Mp 250-253°.

Chatterjea, J.N., *Experientia*, 1956, **12**, 18 (*synth*)

Benzo[b]naphtho[2,3-d]furan-6,11-dione, 9CI   B-60034

β-Brazanquinone
[479-11-8]

$C_{16}H_8O_3$ M 248.237
Pale-yellow needles (EtOH). Mp 243°.

Chatterjee, J.N. *et al*, *J. Indian Chem. Soc.*, 1957, **34**, 155, 347 (*synth*)
Shand, A.J. *et al*, *Tetrahedron*, 1963, **19**, 1919 (*synth*)
Stadlbauer, W. *et al*, *Z. Naturforsch., B*, 1975, **30**, 139 (*synth*)

Benzonitrile N-oxide, 9CI   B-60035

[873-67-6]

$$PhC \equiv \overset{\oplus}{N} - O^{\ominus}$$

C_7H_5NO M 119.123
Undergoes many cycloaddn. reacns. Needles or oil with characteristic odour. Mp 15°. Readily dimerises and isomerises.

Wieland, H., *Ber.*, 1907, **40**, 1667 (*synth*)
Huisgen, R. *et al*, *Chem. Ber.*, 1972, **105**, 2805.

Benzo[rst]phenaleno[1,2,3-de]pentaphene,   **B-60036**
9CI
Violanthrene C
[51958-76-0]

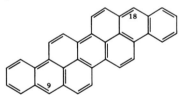

C₃₄H₁₈   M 426.516
Li, J.H. *et al, CA*, 1975, **83**, 58510 (*synth*)

Benzo[rst]phenanthro[1,10,9-cde]-   **B-60037**
pentaphene, 9CI
Dibenzo[a,cd]*naphtho*[1,2,3-lm]*perylene. Isoviolanth-rene* B
[190-93-2]

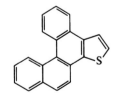

C₃₄H₁₈   M 426.516
Brown-yellow needles. Mp 318.3°.

Aoki, J. *et al, Bull. Chem. Soc. Jpn.*, 1977, **50**, 1017 (*synth*)
Ueda, T. *et al, Org. Mass. Spectrom.*, 1983, **18**, 105 (*ms*)

Benzo[rst]phenanthro[10,1,2-cde]-   **B-60038**
pentaphene, 9CI
Dinaphtho[1,2,3-cd:1',2',3'-lm]*perylene. Isoviolanthrene* A
[188-84-1]

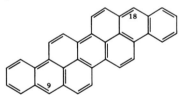

C₃₄H₁₈   M 426.516
Deep-red cryst. Mp 526.5°.

9,18-Dihydro: [4430-29-9]. *9,18-Dihydrobenzo*[rst]-*phenanthro*[10,1,2-cde]*pentaphene,* 9CI.
Isoviolanthrene.
C₃₄H₂₀   M 428.532

Clar, E., *Ber.*, 1939, **72**, 1645 (*synth*)
Parkyns, N. *et al, J. Chem. Soc.*, 1960, 4188 (*synth*)
Ueda, T. *et al, Org. Mass Spectrom.*, 1983, **18**, 105 (*ms*)

Benzo[3,4]phenanthro[1,2-b]thiophene, 9CI   **B-60039**
[107971-15-3]

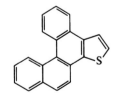

C₂₀H₁₂S   M 284.375

Mp 101-102°.

Stuart, J.G. *et al, J. Heterocycl. Chem.*, 1986, **23**, 1215 (*synth, pmr, cmr, ms, cryst struct*)

Benzo[3,4]phenanthro[2,1-b]thiophene, 9CI   **B-60040**
[107971-19-7]

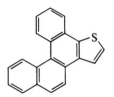

C₂₀H₁₂S   M 284.375
Mp 101.5°.

Stuart, J.G. *et al, J. Heterocycl. Chem.*, 1986, **23**, 1215 (*synth, pmr, cmr, ms*)

1H-2-Benzopyran-1-one, 9CI   **B-60041**
Updated Entry replacing B-00474
Isocoumarin
[491-31-6]

C₉H₆O₂   M 146.145
Cryst. (C₆H₆). Mp 47°. Bp₇₁₉ 285-286°. Steam-volatile.

Narasimhan, N.S. *et al, Synthesis*, 1975, 797 (*synth*)
Korte, D.E. *et al, J. Org. Chem.*, 1977, **42**, 1329 (*synth*)
Sakamoto, T. *et al, Chem. Pharm. Bull.*, 1986, **34**, 2754 (*synth*)

10H-[1]Benzopyrano[3,2-c]pyridin-10-one,   **B-60042**
9CI
Updated Entry replacing B-00480
2-Azaxanthone
[54629-30-0]

C₁₂H₇NO₂   M 197.193
V. pale-greenish needles (EtOH). Mp 183-185°.

Villani, F.J. *et al, J. Org. Chem.*, 1975, **40**, 1734.
Sliwa, H. *et al, J. Heterocycl. Chem.*, 1977, **14**, 169.
Cordonnier, G. *et al, J. Heterocycl. Chem.*, 1987, **24**, 111 (*synth, ir, pmr, ms*)

[1]-Benzopyrano[2,3-d]-1,2,3-triazol-   **B-60043**
9(1H)-one, 9CI
9-Oxo-1H,9H-benzopyrano[2,3-d]-v-*triazole*
[75020-20-1]

C₉H₅N₃O₂   M 187.157
Cryst. (EtOH). Mp 250-251° dec.

Buckle, D.R. *et al, J. Heterocycl. Chem.*, 1981, **18**, 1117 (*synth, ir, pmr, uv*)

Buckle, D.R. *et al*, *J. Med. Chem.*, 1983, **26**, 251 (*biochem*)

Benzo[g]quinazoline-6,9-dione, 9CI   B-60044

[107427-67-8]

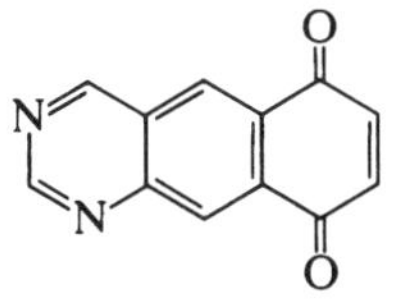

$C_{12}H_6N_2O_2$ M 210.192
Yellow cryst. (CHCl$_3$/EtOH). Mp >160° dec.

Shepherd, M.K., *J. Chem. Soc.*, *Perkin Trans. 1*, 1986, 1495 (*synth, ir, ms, pmr*)

Benzo[g]quinoline-6,9-dione, 9CI   B-60045

[107427-65-6]

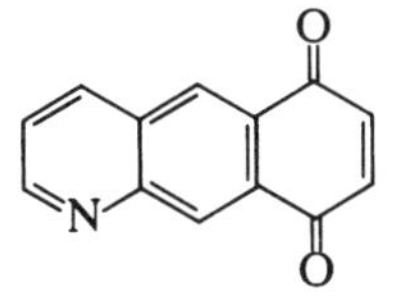

$C_{13}H_7NO_2$ M 209.204
Yellow cryst. (CHCl$_3$). Mp >160° dec.

Shepherd, M.K., *J. Chem. Soc.*, *Perkin Trans. 1*, 1986, 1495 (*synth, ir, pmr, ms*)

Benzo[g]quinoxaline-6,9-dione, 9CI   B-60046

[107427-66-7]

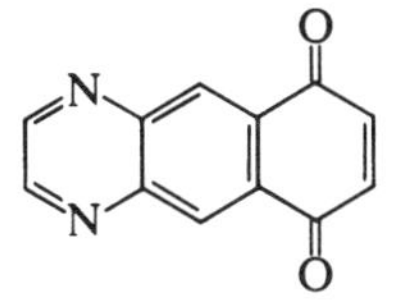

$C_{12}H_6N_2O_2$ M 210.192
Pale-yellow cryst. (CHCl$_3$). Mp >200° dec.

Shepherd, M.K., *J. Chem. Soc.*, *Perkin Trans. 1*, 1986, 1495 (*synth, pmr, ir, ms*)

1-Benzothiepin-5(4H)-one, 9CI   B-60047

[107798-86-7]

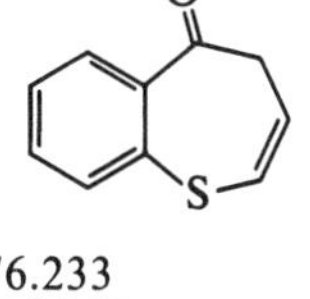

$C_{10}H_8OS$ M 176.233
Cryst. (Et$_2$O). Mp 35-36°.

Hofmann, H. *et al*, *Justus Liebigs Ann. Chem.*, 1987, 505 (*synth, pmr*)

Benzo[b]thiophen-3(2H)-one   B-60048

Thioindoxyl. 1-Thianaphthen-3-one. 3(2H)-Thian-aphthenone. 3-Keto-2,3-dihydrothianaphthene
[130-03-0]

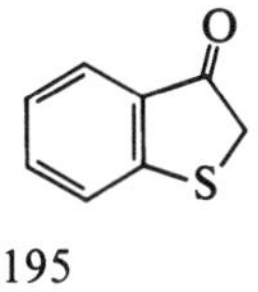

C_8H_6OS M 150.195

Major tautomer of 3-Hydroxybenzo[b]thiophene, H-01342 . Needles (H$_2$O). Mp 71°. Readily oxidised to thioindigo.
1,1-Dioxide: [1127-35-1].
  $C_8H_6O_3S$ M 182.194
  Cryst. (EtOH). Mp 136-137°.
Hydrazone, 1,1-dioxide:
  $C_8H_8N_2O_2S$ M 196.223
  Needles (EtOH). Mp 188-189°.

Friedländer, P. *et al*, *Justus Liebigs Ann. Chem.*, 1907, **351**, 390 (*synth*)
Weston, A.W. *et al*, *J. Am. Chem. Soc.*, 1939, **61**, 389 (*deriv, synth*)
Kloosterziel, H. *et al*, *Recl. Trav. Chim. Pays-Bas*, 1952, **71**, 368 (*deriv, synth*)
Bolssens, J. *et al*, *Recl. Trav. Chim. Pays-Bas*, 1954, **73**, 819 (*deriv, synth*)

[1]Benzothiopyrano[6,5,4-def][1]-benzothiopyran, 9CI   B-60049

1,6-Dithiapyrene. Naphtho[1,8-bc:5,4-b',c']bisthiopyran
[194-07-0]

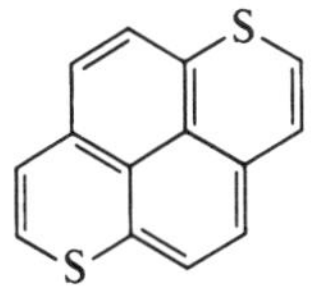

$C_{14}H_8S_2$ M 240.337
Orange needles (C$_6$H$_6$). Mp 228-229° dec.

Nakasuji, K. *et al*, *J. Am. Chem. Soc.*, 1986, **108**, 3460 (*synth, uv, pmr*)

1,2,3-Benzotriazine-4(3H)-thione, 9CI   B-60050

4-Mercapto-1,2,3-benzotriazine. 1,2,3-Benzotriazine-4-thiol
[25465-34-3]

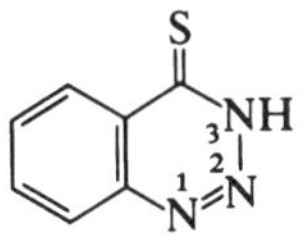

$C_7H_5N_3S$ M 163.197
3H-thione-form
  Yellow needles (EtOH). Mp 187.5° dec. Major tautomer.
  N-*Ac:*
    $C_9H_7N_3OS$ M 205.234
    Golden-yellow needles (CCl$_4$). Mp 144° dec.
  N-*Benzoyl:*
    $C_{14}H_9N_3OS$ M 267.305
    Orange prisms (CCl$_4$). Mp 163° dec.
  3-N-*Me:* [22305-48-2].
    $C_8H_7N_3S$ M 177.223
    Yellow needles (MeOH aq.). Mp 107-108°.
Thiol-form [2536-88-1]
  Minor tautomer.
  S-*Me:* [22305-56-2].
    $C_8H_7N_3S$ M 177.223
    Prisms (EtOH or pet. ether). Mp 101-103°.
(2H)-form
  Minor tautomer.
  2-N-*Me:* [22305-49-3]. *4-Mercapto-2-methyl-1,2,3-ben-zotriazinium hydroxide inner salt, 9CI.*
    $C_8H_7N_3S$ M 177.223

Red prisms (propanol). Mp 197-199°.
2-N-Et: [22298-37-9].
  $C_9H_9N_3S$ M 191.250
  Red needles (EtOH). Mp 190-192°.

Reissert, A. *et al, Ber.,* 1909, **42**, 3710 (*synth, derivs*)
Wagner, G. *et al, Arch. Pharm. (Weinheim, Ger.),* 1968, **301**, 923 (*derivs, uv*)
Wagner, G. *et al, Pharmazie,* 1968, **23**, 629 (*synth, derivs, ms*)
Gilbert, E.E. *et al, J. Heterocycl. Chem.,* 1969, **6**, 779 (*synth, tautom*)

Benzo[1,2-*b*:3,4-*b*′:5,6-*b*″]tripyrazine-2,3,6,7,10,11-tetracarboxylic acid   B-60051

1,4,5,8,9,12-Hexaazatriphenylenehexacarboxylic acid
[105598-29-6]

$C_{18}H_6N_6O_{12}$ M 498.278
Cryst. + $1\frac{1}{2}H_2O$. Mp >350°.
Hexa-Me ester:
  $C_{24}H_{18}N_6O_{12}$ M 582.439
  Cryst. (MeCN). Mp >350°.
Hexaamide:
  $C_{18}H_{12}N_{12}O_6$ M 492.370
  Grey-black solid + $1H_2O$. Mp >350°.
Hexanitrile: [105598-27-4].
  $C_{18}N_{12}$ M 384.278
  Brown-black insol. solid. Mp >350°.
Trianhydride: [105618-29-9].
  $C_{18}N_6O_9$ M 444.233
  Needles (MeCN/C_6H_6). Mp >350°. Moisture sensitive.

Kanakarajan, K. *et al, J. Org. Chem.,* 1986, **51**, 5241 (*synth, pmr, ir, uv*)

Benzo[1,2-*b*:3,4-*b*′:5,6-*b*″]tris[1]-benzothiophene, 9CI   B-60052

s-*Tris-2,3-thiocoumaronobenzene* (*obsol.*)
[199-14-4]

$C_{24}H_{12}S_3$ M 396.539
Cryst. (xylene). Mp 422° (>360°).

Dalgliesh, C.E. *et al, J. Chem. Soc.,* 1945, 910 (*synth*)
Proetzsch, R. *et al, Z. Naturforsch., B,* 1976, **31**, 529 (*synth*)
Bergman, J. *et al, Tetrahedron,* 1986, **42**, 763 (*synth, ir, ms*)

Benzo[1,2-*b*:3,4-*b*′:6,5-*b*″]tris[1]-benzothiophene, 9CI   B-60053

[106020-39-7]

$C_{24}H_{12}S_3$ M 396.539
Cryst. (MeCN). Mp 253-255°.

Bergman, J. *et al, Tetrahedron,* 1986, **42**, 763 (*synth, ir, ms*)

[1,4]Benzoxaselenino[3,2-*b*]pyridine, 9CI   B-60054

10-Oxa-9-selena-1-azaanthracene. 1-Azaphenoxaselenine
[109681-73-4]

$C_{11}H_7NOSe$ M 248.142
Cryst. (Et$_2$O/hexane). Mp 67.5-68.5°.

Smith, K.E. *et al, J. Chem. Soc., Perkin Trans. 1,* 1986, 2075 (*synth, ir, uv, pmr, ms*)

[1,4]Benzoxazino[3,2-*b*][1,4]benzoxazine, 9CI   B-60055

[258-19-5]

$C_{14}H_8N_2O_2$ M 236.229
Light-yellow needles (toluene). Mp 290-291°. Obt. in only 2% yield. An earlier reported synth. (1925) was erroneous.

Tetrahydro: see 5a,6,11a,12-Tetrahydro[1,4]-benzoxazino[3,2-b] [1,4]benzoxazine, T-60059

Tauer, E. *et al, Chem. Ber.,* 1986, **119**, 3316 (*synth, uv, pmr*)

1-[*N*-[3-(Benzoylamino)-2-hydroxy-4-phenylbutyl]alanyl]proline   B-60056

[95549-39-6]

$C_{25}H_{31}N_3O_5$ M 453.537
Potent inhibitor of angiotensin converting enzyme.
(2S,8S,11R,12S)-form [95589-34-7]
  B,HCl: Mp 140-165°. $[\alpha]_D^{25}$ −84.2° (c, 1.13 in MeOH).
(2S,8S,11S,12S)-form [95589-35-8]
  B,HCl: Mp 234-237°. $[\alpha]_D^{25}$ −110.2° (c, 1.12 in MeOH).

Gordon, E.M. *et al, Biochem. Biophys. Res. Commun.*, 1985, **126**, 419.
Godfrey, J.D. *et al, J. Org. Chem.*, 1986, **51**, 3073; *Tetrahedron Lett.*, 1987, **40**, 1603 (*synth*)

4(5)-Benzoylimidazole   B-60057

1H-Imidazol-4-ylphenylmethanone, 9CI
[61985-32-8]

$C_{10}H_8N_2O$ M 172.186
Needles (Me$_2$CO). Mp 142°.

Iwasaki, S., *Helv. Chim. Acta*, 1976, **59**, 2738 (*synth, uv, ir, pmr, cmr*)

6-Benzoylindole   B-60058

1H-Indol-6-ylphenylmethanone. 6-Indolyl phenyl ketone
[105205-50-3]
$C_{15}H_{11}NO$ M 221.258
Moyer, M.P. *et al, J. Org. Chem.*, 1986, **51**, 5106 (*synth, pmr*)

1-Benzoylisoquinoline   B-60059

1-Isoquinolylphenylmethanone, 9CI. 1-Isoquinolyl phenyl ketone
[16576-23-1]

$C_{16}H_{11}NO$ M 233.269
Needles. Mp 76-77°.
Oxime:
   $C_{16}H_{12}N_2O$ M 248.284
   Mp 214-216°.
Hydrazone: [58022-24-5]. Needles (cyclohexane). Mp 108°.

Boekelheide, V. *et al, J. Am. Chem. Soc.*, 1952, **74**, 660 (*synth*)
Chazerain, J., *Ann. Chim. (Paris)*, 1963, **8**, 255 (*synth*)

3-Benzoylisoquinoline   B-60060

3-Isoquinolylphenylmethanone, 9CI
[83629-95-2]
$C_{16}H_{11}NO$ M 233.269
Mp 79-81°.

Yamamoto, Y. *et al, Chem. Pharm. Bull.*, 1982, **30**, 2003 (*synth, pmr*)

4-Benzoylisoquinoline   B-60061

4-Isoquinolylphenylmethanone
[20335-71-1]
$C_{16}H_{11}NO$ M 233.269
Needles (hexane). Mp 76-78°. Bp$_{0.3}$ 157-160°.

Tachibana, S. *et al, Chem. Pharm. Bull.*, 1968, **16**, 414 (*synth, uv*)

4-Benzoylisoxazole   B-60062

4-Isoxazolylphenylmethanone, 9CI

$C_{10}H_7NO_2$ M 173.171
Could not be cryst.
Oxime:
   $C_{10}H_8N_2O_2$ M 188.185
   Mp 133-135°.
p-*Nitrophenylhydrazone:* Brown-orange cryst. Mp 204°.

Panizzi, L., *Gazz. Chim. Ital.*, 1947, **77**, 283 (*synth*)

5-Benzoylisoxazole   B-60063

5-Isoxazolylphenylmethanone, 9CI
$C_{10}H_7NO_2$ M 173.171
Plates (pet. ether). Mp 51°.
p-*Nitrophenylhydrazone:* Yellow prisms (EtOAc). Mp 198°.
Semicarbazone: Needles (EtOAc). Mp 190°.

Mina, G.A. *et al, J. Chem. Soc.*, 1962, 4234 (*synth*)

3-Benzoyl-2-methylpropanoic acid   B-60064

Updated Entry replacing B-00670
α-Methyl-γ-oxobenzenebutanoic acid, 9CI. 3-Benzoylisobutyric acid. 2-Phenacylpropionic acid
[1771-65-9]

$$PhCOCH_2CH(CH_3)COOH$$

$C_{11}H_{12}O_3$ M 192.214
(±)-***form***
   Cryst. (C$_6$H$_6$). Mp 139-140.5°.
   Et ester: [6938-44-9].
      $C_{13}H_{16}NO_3$ M 234.274
      Bp$_{0.2}$ 98-100°.

Alexander, E.R. *et al, J. Am. Chem. Soc.*, 1950, **72**, 3194 (*synth*)
Scarpati, R. *et al, Gazz. Chim. Ital.*, 1967, **97**, 654 (*synth*)
Kurihara, T. *et al, Chem. Pharm. Bull.*, 1986, **34**, 4620 (*synth, pmr*)

4-Benzoylpyrazole   B-60065

Phenyl-1H-pyrazol-4-ylmethanone, 9CI
[37687-16-4]

$C_{10}H_8N_2O$ M 172.186
Cryst. (EtOAc/CHCl$_3$). Mp 156°.

Zbiral, E. *et al, Tetrahedron*, 1972, **28**, 4189 (*synth*)

5-Benzoyl-2(1H)-pyrimidinone   B-60066

$C_{11}H_8N_2O_2$ M 200.196
Cryst. (MeOH). Mp 244°.

Undheim, K. *et al, Acta Chem. Scand., Ser. B*, 1986, **40**, 588, 764 (*synth, pmr, ms*)

1-Benzyl-1,4-dihydronicotinamide   B-60067

1,4-Dihydro-1-(phenylmethyl)-3-pyridinecarboxamide,
9CI

[952-92-1]

$C_{13}H_{14}N_2O$ M 214.266

Reducing agent for many substrates; a model for the reducing co-enzyme NADH. Cryst. (EtOH). Mp 105-122°, Mp 120-122° dec.

Karrer, P. *et al, Helv. Chim. Acta*, 1937, **20**, 418 (*synth*)
Mauzerall, D. *et al, J. Am. Chem. Soc.*, 1955, **77**, 2261 (*synth, use*)
Brown, A. *et al, J. Am. Chem. Soc.*, 1976, **98**, 5682 (*synth, pmr, ir*)
Stout, D.M. *et al, Chem. Rev.*, 1982, **82**, 223.

Benzylidenecycloheptane   B-60068

(Phenylmethylene)cycloheptane, 9CI

[15537-53-8]

$C_{14}H_{18}$ M 186.296
Bp$_2$ 100-110°.

Ohta, S. *et al, Chem. Pharm. Bull.*, 1986, **34**, 5144 (*synth, ir, pmr*)

9-Benzylidene-1,3,5,7-cyclononatetraene   B-60069

9-(Phenylmethylene)-1,3,5,7-cyclononatetraene, 9CI. 10-*Phenylnonafulvene*

[104170-63-0]

$C_{16}H_{14}$ M 206.287
Light-yellow cryst. Handled below −20°.

Sabbioni, G. *et al, Helv. Chim. Acta*, 1985, **68**, 1543 (*synth, uv, pmr, cmr*)
Furrer, J. *et al, Helv. Chim. Acta*, 1987, **70**, 862 (*pmr*)

N-Benzylidenemethylamine   B-60070

N-*(Phenylmethylene)methanamine, 9CI*

[622-29-7]

PhCH=NMe

C_8H_9N M 119.166.
Oil. Bp$_{34}$ 92-93°, Bp$_{20}$ 78-83°. n_D^{20} 1.5528.

Org. Synth., Coll. Vol., **4**, 605 (*synth*)
Anderson, W.K. *et al, J. Med. Chem.*, 1986, **29**, 2241 (*synth, ir, pmr*)

Benzyloxycarbonyl chloride   B-60071

Phenylmethyl carbonochloridate. Carbobenzoxy chloride. Benzyl chloroformate

[501-53-1]

PhCH$_2$OCOCl

$C_8H_7ClO_2$ M 170.595
Widely used means of *N*-protection in peptide synth. Oil with acrid odour. Mp 0°. Bp$_{20}$ 103°, Bp$_7$ 85-87°.
▷Lachrymator. LQ5860000.

Org. Synth., Coll. Vol., **3**, 167 (*synth*)
Bergmann, M. *et al, Ber.*, 1932, **65**, 1192 (*synth, use*)
Farthing, A.C., *J. Chem. Soc.*, 1950, 3213 (*synth*)
Fieser, M. *et al, Reagents for Organic Synthesis*, Wiley, 1967-84, **1**, 109, 856 (*synth, use*)

2-(Benzyloxy)phenol   B-60072

2-(Phenylmethoxy)phenol, 9CI. Benzyl o-hydroxyphenyl ether

[6272-38-4]

$C_{13}H_{12}O_2$ M 200.237
Bp$_{13}$ 173-174°, Bp$_{0.1}$ 133-135°.
Me ether: [835-79-0]. *1-Methoxy-2-(phenylmethoxy)-benzene, 9CI.* o-*Benzyloxyanisole.*
$C_{14}H_{14}O_2$ M 214.263
Mp 54-55°.

Jones, J.H. *et al, J. Chem. Soc. (C)*, 1968, 436 (*synth*)
Colquhoun, H.M. *et al, J. Chem. Soc., Perkin Trans. 2*, 1985, 607 (*synth*)

3-(Benzyloxy)phenol   B-60073

3-(Phenylmethoxy)phenol, 9CI. Resorcinol benzyl ether. Benzyl m-hydroxyphenyl ether

[3769-41-3]
$C_{13}H_{12}O_2$ M 200.237
Needles (pet. ether). Mp 52°. Bp$_{11}$ 202-210°.
Benzoyl:
$C_{20}H_{16}O_3$ M 304.345
Mp 81-82°.
Me ether: [21144-16-1]. *1-Methoxy-3-(phenylmethoxy)benzene.* m-*Benzyloxyanisole.*
$C_{14}H_{14}O_2$ M 214.263
Mp 32°. Bp$_4$ 166-168°, Bp$_{0.2}$ 120-125°.

Freundenberg, K. *et al, Justus Liebigs Ann. Chem.*, 1958, **612**, 78 (*synth*)
Fitton, A.O. *et al, J. Chem. Soc.*, 1962, 4870 (*synth*)
Kametani, T. *et al, J. Chem. Soc. (C)*, 1969, 4 (*deriv*)
Ward, R.S. *et al, J. Am. Chem. Soc.*, 1969, **91**, 2727 (*ms*)

4-(Benzyloxy)phenol, 8CI   B-60074

4-(Phenylmethoxy)phenol, 9CI. Benzyl p-hydroxyphenyl ether. Monobenzone. Agerite. Benoquin. Benzoquin. Depigman. Pigmex

[103-16-2]
$C_{13}H_{12}O_2$ M 200.237
Depigmenting agent. Leaflets (H$_2$O). Mp 122°.
▷SJ7700000.
Ac:
$C_{15}H_{14}O_3$ M 242.274
Cryst. (CHCl$_3$/pet. ether). Mp 104-107°.
Me ether: [6630-18-8]. *1-Methoxy-4-(phenylmethoxy)-benzene, 9CI.* p-*Benzyloxyanisole.*
$C_{14}H_{14}O_2$ M 214.263

Cryst. (MeOH). Mp 74°.

Rowe, E.J. et al, J. Org. Chem., 1958, **23**, 1622 (synth)
Zaslow, B. et al, Mol. Cryst., 1967, **3**, 297 (cryst struct)
Ward, R.S. et al, J. Am. Chem. Soc., 1969, **91**, 2727 (ms)
Leznoff, C.C. et al, Can. J. Chem., 1977, **55**, 3351 (synth)
Al-Hamdany, R. et al, J. Chem. Soc., Perkin Trans. 2, 1985, 1395 (synth)

Biadamantylideneethane   B-60075
2,2'-(1,2-Ethanediylidene)bistricyclo[3.3.1.1³,⁷]decane, 9CI

[94052-77-4]

$C_{22}H_{30}$ M 294.479
Mp 310° dec. subl. Bp$_{0.1}$ 140° subl.

Nelsen, S.F. et al, J. Am. Chem. Soc., 1986, **108**, 5503 (synth, uv, pmr, cmr)

Bianthrone A2a   B-60076

R =

$C_{45}H_{46}O_8$ M 714.854
Constit. of *Psorospermum tenuifolium*. Cryst. (Et$_2$O-/hexane). Mp 81-83°.

Delle Monache, G. et al, Phytochemistry, 1987, **26**, 2611.

Bianthrone A2b   B-60077
As Bianthrone A2a, B-60076 with

R = H

$C_{40}H_{38}O_8$ M 646.735
Constit. of *Psorospermum tenuifolium*. Cryst. (Et$_2$O-/hexane). Mp 178-179°.

Delle Monache, G. et al, Phytochemistry, 1987, **26**, 2611.

7,7'-Bi(bicyclo[2.2.1]heptylidene)   B-60078
7-Bicyclo[2.2.1]hept-7-ylidenebicyclo[2.2.1]heptane, 9CI. 7,7'-Binorbornylidene

[51689-29-3]

$C_{14}H_{20}$ M 188.312
Cryst. (MeOH/pet. ether). Mp 140.5-141°.

Bartlett, P.D. et al, J. Am. Chem. Soc., 1974, **96**, 627 (synth, ir, pmr, ms)

Kabe, Y. et al, J. Am. Chem. Soc., 1984, **106**, 8174 (synth, pmr, cmr)

9,9'-Bi(bicyclo[3.3.1]nonylidene)   B-60079
[55993-21-0]

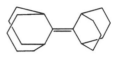

$C_{18}H_{28}$ M 244.419
Mp 144-146°. Bp$_{0.05}$ 100° subl.

Keul, H., Chem. Ber., 1975, **108**, 1207 (synth, ir, pmr)
Gerson, F. et al, J. Am. Chem. Soc., 1981, **103**, 6716 (synth, esr)
Kabe, Y. et al, J. Am. Chem. Soc., 1984, **106**, 8174 (synth, pmr, cmr)

1,1'-Bibiphenylene, 9CI   B-60080
1,1'-Bisbiphenylenyl

[96694-90-5]

$C_{24}H_{14}$ M 302.375
Pale-yellow needles (hexane). Mp 162-163°.

Cracknell, M.E. et al, J. Chem. Soc., Perkin Trans. 1, 1985, 115 (synth, uv, pmr)

2,2'-Bibiphenylene, 9CI   B-60081
2,2'-Bisbiphenylenyl

[96694-92-7]
$C_{24}H_{14}$ M 302.375
Yellow needles (EtOH). Mp 246-248°.

Baker, W. et al, J. Chem. Soc., 1958, 2666 (synth)
Cracknell, M.E. et al, J. Chem. Soc., Perkin Trans. 1, 1985, 115 (synth)

Bicoumol   B-60082
7,7'-Dihydroxy-[6,6'-bi-2H-benzopyran]-2,2'-dione, 9CI. 7,7'-Dihydroxy-6,6-bicoumarin, 8CI

[15575-52-7]

$C_{18}H_{10}O_6$ M 322.273
Isol. from ladino clover and *Ruta* sp. Cryst. (EtOH/Me$_2$CO). Mp 293-294°.

Di-Ac: [18304-03-5]. Cryst. (Me$_2$CO aq.). Mp 228.5-229.5°.

Di-Me ether: [3153-73-9]. *7,7'-Dimethoxy-[6,6'-bi-2H-benzopyran]-2,2'-dione, 9CI. 7',7'-Dimethoxy-6,6-bicoumarin, 8CI.*
$C_{20}H_{14}O_6$ M 350.327
Platelets (MeOH). Mp 266.5-267.5°.

Spencer, R.R. et al, J. Agric. Food Chem., 1967, **15**, 536 (isol, uv, ir, pmr)
Gonzalez, A.G. et al, An. Quim., 1977, **73**, 1015 (isol)

Bicyclo[2.2.1]hepta-2,5-diene-2-carboxaldehyde   B-60083

2-Formylnorbornadiene

[5212-50-0]

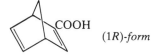

C_8H_8O M 120.151

(±)-**form**

Bp$_{15}$ 70°.

Verkruijsse, H.D. *et al, Recl. Trav. Chim. Pays-Bas*, 1986, **105**, 66.

Bicyclo[2.2.1]hepta-2,5-diene-2-carboxylic acid, 9CI   B-60084

2,5-Norbornadiene-2-carboxylic acid, 8CI. 2,5-Endomethylenebenzoic acid

[698-40-8]

(1R)-*form*

$C_8H_8O_2$ M 136.150

(1R)-**form**

Me ester: [3604-36-2].
  $C_9H_{10}O_2$ M 150.177
  $[\alpha]_D^{22}$ −39.22° (c, 0.2 in CHCl$_3$).

(±)-**form**

Cryst. Mp 93-94°.

Me ester: [99946-16-4]. Bp$_{0.1}$ 28°.

Alder, K. *et al, Justus Liebigs Ann. Chem.*, 1936, **525**, 183 (*synth*)
Fienemann, H. *et al, J. Org. Chem.*, 1979, **44**, 2802 (*synth*)
Baumgaertel, O. *et al, Chem. Ber.*, 1983, **116**, 2180 (*synth*)
De Lucchi, O. *et al, J. Org. Chem.*, 1986, **51**, 1457 (*synth, pmr*)

Bicyclo[2.2.1]heptane-1-carboxylic acid, 9CI   B-60085

Updated Entry replacing B-01036
Norbornane-1-carboxylic acid, 8CI

[18720-30-4]

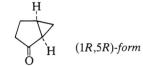

$C_8H_{12}O_2$ M 140.182

Cryst. (pentane) with odour similar to that of butyric acid. Mp 111-112° (109.5-110.5°). Sol. Et$_2$O. Mp 108-110° (111-112°). Bp 113.8-115.5°, Bp$_{10}$ 80° (subl.).

Me ester: [2287-57-2].
  $C_9H_{14}O_2$ M 154.208
  Liq. Sol. Et$_2$O. Bp$_{1.8}$ 52-53°.

Amide: [69095-03-0].
  $C_8H_{13}NO$ M 139.197
  Cryst. (H$_2$O). Mp 234-236°.

Bixler, P.L. *et al, J. Org. Chem.*, 1958, **23**, 742 (*synth*)
Kwart, H. *et al, J. Am. Chem. Soc.*, 1958, **80**, 248; 1959, **81**, 2765 (*synth, ester*)
Herwig, K. *et al, Chem. Ber.*, 1972, **105**, 363 (*synth*)

Riecke, R.D. *et al, Chem. Comm.*, 1973, 879 (*synth*)
Poindexter, G.S. *et al, J. Org. Chem.*, 1976, **41**, 1215 (*cmr*)
Della, G.W. *et al, Aust. J. Chem.*, 1986, **39**, 2061 (*synth, pmr*)

Bicyclo[3.1.1]heptane-1-carboxylic acid   B-60086

[91239-72-4]

$C_8H_{12}O_2$ M 140.182
Cryst. (pentane). Mp 61-62°.

Della, E.W. *et al, Aust. J. Chem.*, 1986, **39**, 2061 (*synth, pmr, cmr*)

Bicyclo[3.1.1]hept-2-ene, 9CI   B-60087

Norpinene

[7095-82-1]

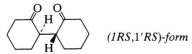

C_7H_{10} M 94.156
Liq. Bp 102°. n_D^{20} 1.4713.

Grychtol, K. *et al, Chem. Ber.*, 1972, **105**, 1798 (*synth, pmr*)
Herzog, C. *et al, Chem. Ber.*, 1986, **119**, 3027 (*synth, pmr, cmr*)

Bicyclo[3.1.0]hexan-2-one   B-60088

[4160-49-0]

(1R,5R)-*form*

C_6H_8O M 96.129

(1R,5R)-**form** [58001-78-8]
No phys. props. given.

(1RS,5RS)-**form**
Bp$_{20}$ 69°. n_D^{25} 1.4747.

2,4-Dinitrophenylhydrazone: Cryst. (EtOH/C$_6$H$_6$). Mp 170.6-171.9°.

Nelson, W.A. *et al, J. Org. Chem.*, 1957, **22**, 1146 (*synth, uv*)
Lightner, D.A. *et al, Tetrahedron Lett.*, 1975, 3051 (*synth, abs config*)

[1,1'-Bicyclohexyl]-2,2'-dione   B-60089

[32673-76-0]

(1RS,1'RS)-*form*

$C_{12}H_{18}O_2$ M 194.273

(1RS,1'RS)-**form**
C$_2$-form
Mp 70-71°.

(1RS,1'SR)-**form**
C$_S$-form
Oil. Stereochemical assignments reversed in 1986.

Denmark, S.E. *et al, Tetrahedron Lett.*, 1986, **27**, 3693 (*synth, pmr, cmr, ir, stereochem, bibl*)

Bicyclo[4.1.1]octa-2,4-diene, 9CI   B-60090

Updated Entry replacing B-01102

[61885-53-8]

C_8H_{10} M 106.167

Oil. λ_{max} 266 (ϵ 2 800), 277 (3 800), 288 (3 750), 301 nm (2 000) (hexane).

Gleiter, R. *et al, J. Am. Chem. Soc.*, 1977, **99**, 8 (*synth*)
Yin, T.-K. *et al, J. Org. Chem.*, 1985, **50**, 531 (*synth*)
Christl, M. *et al, Chem. Ber.*, 1986, **119**, 3059, 3067 (*synth, pmr, cmr*)

Bicyclo[4.1.1]octane   B-60091

Updated Entry replacing B-01118

[7078-34-4]

C_8H_{14} M 110.199

Low-melting solid. Mp ca. 20°.

Gleiter, R. *et al, J. Am. Chem. Soc.*, 1977, **99**, 8 (*synth, pmr*)
Christl, M. *et al, Chem. Ber.*, 1986, **119**, 3059, 3067 (*synth, cmr*)

Bicyclo[2.2.2]octane-1-carboxylic acid   B-60092

Updated Entry replacing B-01120

[699-55-8]

$C_9H_{14}O_2$ M 154.208

Platelets (H_2O). Sol. EtOH, C_6H_6, Et_2O, hexane. Mp 140.8-141.3°.

Et ester: [31818-12-9].
    $C_{11}H_{18}O_2$ M 182.262
    Bp_3 75-76°.

Roberts, J.D. *et al, J. Am. Chem. Soc.*, 1953, **75**, 637 (*synth*)
Cross, A.D. *et al, CA*, 1972, **76**, 140343h (*use*)
Iwao, T. *et al, Bull. Chem. Soc. Jpn.*, 1974, **47**, 3079 (*synth*)
Mornz, D.G. *et al, J. Chem. Soc., Perkin Trans. 2*, 1975, 734 (*cmr*)
Della, E.W. *et al, Aust. J. Chem.*, 1986, **39**, 2061 (*synth, pmr, cmr*)

Bicyclo[4.2.0]octane-2,5-dione, 9CI   B-60093

[68882-69-9]

$C_8H_{10}O_2$ M 138.166

(*1RS,2SR*)-*form* [54338-82-8]
    cis-*form*
    Oil. $Bp_{0.15}$ 68-70°.

Iyoda, M. *et al, Synthesis*, 1986, 322 (*synth, pmr*)

Bicyclo[3.2.1]octan-8-one   B-60094

[55679-31-7]

$C_8H_{12}O$ M 124.182

Mp 140-141°.

Semicarbazone: Cryst. Mp 189.8-190.5°.
2,4-Dinitrophenylhydrazone: Flat needles. Mp 175.4-176.2°.
Ethylene ketal: [337-65-5]. *Spiro[bicyclo[3.2.1]octane-8,2'-[1,3]dioxolane], 9CI.*
    $C_{10}H_{16}O_2$ M 168.235
    Oil. $Bp_{0.1}$ 60°.

Mayer, R. *et al, Chem. Ber.*, 1958, **91**, 1616 (*synth*)
Cope, A.C. *et al, J. Am. Chem. Soc.*, 1960, **82**, 4299 (*synth, ir*)
Foote, C.S. *et al, Tetrahedron*, 1964, **20**, 687 (*synth*)
Nelsen, S.F. *et al, J. Am. Chem. Soc.*, 1986, **108**, 1265 (*synth, pmr*)

Bicyclo[4.1.1]oct-2-ene, 9CI   B-60095

[16544-26-6]

C_8H_{12} M 108.183

Liq.

Christl, M. *et al, Chem. Ber.*, 1986, **119**, 3059, 3067 (*synth, ir, pmr, cmr*)

Bicyclo[4.1.1]oct-3-ene   B-60096

Updated Entry replacing B-01138

[61885-54-9]
    C_8H_{12} M 108.183
    Liq.

Gleiter, R. *et al, J. Am. Chem. Soc.*, 1977, **99**, 8 (*synth, pmr*)
Christl, M. *et al, Chem. Ber.*, 1986, **119**, 3059, 3067 (*synth, cmr*)

Bicyclo[4.2.0]oct-7-ene-2,5-dione, 9CI   B-60097

$C_8H_8O_2$ M 136.150

(*1RS,5SR*)-*form* [54338-83-9]
    cis-*form*
    Oil. $Bp_{0.15}$ 61-64°.

Iyoda, M. *et al, Synthesis*, 1986, 322 (*synth, pmr*)

Bicyclo[3.3.0]oct-1(2)-en-3-one   B-60098

Updated Entry replacing B-10154

$C_8H_{10}O$ M 122.166
Liq.
Dimer:
   C_8H_8O M 120.151
   Cryst. (hexane). Mp 88-89°.

Begley, M.G. *et al*, *Tetrahedron Lett.*, 1981, 257 (*synth, spectra*)
Klipa, D.K. *et al*, *J. Org. Chem.*, 1981, **46**, 2815 (*synth*)
Davidsen, S.K. *et al*, *Synthesis*, 1986, 842 (*synth, pmr, cmr, ir*)

Bicyclo[3.3.0]oct-6-en-2-one   B-60099

$C_8H_{10}O$ M 122.166
(*RS,SR*)-*form* [32405-38-2]
cis-*form*
Intermed. for polycyclopentenoid nat. prods. No phys. props. given.

Hudlicky, T. *et al*, *J. Org. Chem.*, 1980, **45**, 5020 (*synth, ir, pmr, cmr, ms*)
Hashimoto, S. *et al*, *Tetrahedron Lett.*, 1986, **27**, 2885 (*synth*)

Bi-2,4-cyclopentadien-1-yl, 9CI   B-60100

9,10-Dihydropentafulvalene
[21423-86-9]

$C_{10}H_{10}$ M 130.189
Almost colourless oil. Tautomerises slowly at 0°, rapidly at r.t.

Escher, A. *et al*, *Helv. Chim. Acta*, 1986, **69**, 1644 (*synth, uv, pmr*)

Bicyclo[1.1.1]pentane-1-carboxylic acid, 9CI, 8CI   B-60101

Updated Entry replacing B-01149
[22287-28-1]

$C_6H_8O_2$ M 112.128
Mp 59-59.7°.

Wiberg, K.B. *et al*, *J. Org. Chem.*, 1970, **35**, 369 (*synth*)
Della, E.W. *et al*, *Aust. J. Chem.*, 1986, **39**, 2061 (*synth, pmr*)

Bicyclo[5.3.1]undeca-1,3,5,7,9-pentaene, 9CI   B-60102

Benzo[c]*-1,7-methano*[*12*]*annulene. Benz*[b]*-homoheptalene*
[65754-71-4]

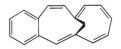

$C_{17}H_{14}$ M 218.298
Stable orange cryst. Mp 53-54°.

Scott, L.T. *et al*, *J. Am. Chem. Soc.*, 1983, **105**, 1372 (*synth, pmr, cmr, ir, uv*)
Günther, M.-G. *et al*, *Chem. Ber.*, 1986, **119**, 2942 (*pmr*)

2,2'-Bi-1,3-dithiolane   B-60103

[6784-47-0]

$C_6H_{10}S_4$ M 210.385
Plates (EtOH or by subl.). Mp 135-136°.

Arbuzov, B.A. *et al*, *Izv. Akad. Nauk SSSR, Ser. Khim.*, 1973, 2422 (*Engl. transl.* 2368) (*synth, conformn*)
Chastrette, F. *et al*, *Bull. Soc. Chim. Fr.*, 1976, 601 (*synth*)
Pericas, M.A. *et al*, *Tetrahedron*, 1986, **42**, 2717 (*conformn*)

9,9'-Bifluorenylidene   B-60104

Updated Entry replacing B-01189
9-(9H-Fluoren-9-ylidene)-9H-fluorene, 9CI. $\Delta^{9,9'}$-Bi-9H-fluorene. $\Delta^{9,9'}$-Bi-9-fluorenyl
[746-47-4]

$C_{26}H_{16}$ M 328.412
Red cryst. (EtOH). Mp 182-183°. Forms a dianion stable in soln. at −30°.

Fukunaga, K., *Synthesis*, 1975, 442.
Bailey, N.A. *et al*, *Acta Crystallogr., Sect. B*, 1978, **34**, 3287.
Newkome, G.R. *et al*, *J. Org. Chem.*, 1979, **44**, 5025.
Lemmen, P. *et al*, *Chem. Ber.*, 1984, **117**, 2300 (*synth, ir, pmr, ms*)
Cohen, Y. *et al*, *J. Chem. Soc., Chem. Commun.*, 1986, 1071 (*dianion*)
Kikuchi, O. *et al*, *Chem. Pharm. Bull.*, 1986, **59**, 3043 (*conformn, struct*)

Bifurcarenone   B-60105

Updated Entry replacing B-01196

(*2E*)-*form*

$C_{27}H_{38}O_5$ M 442.594
(*2E*)-*form*
   E-*Bifurcarenone*
   Constit. of *Cystoseira stricta*. Oil. $[\alpha]_D^{20}$ +5.5° (c, 8 in EtOH).

2'-Me ether:
  $C_{28}H_{40}O_5$ M 456.621
  Constit. of *C. stricta*. Oil. $[\alpha]_D^{20}$ +5.3° (c, 1.15 in EtOH).

(2Z)-form [75872-68-3]
  Z-Bifurcarenone
  Constit. of *Bifurcaria galapagensis*. Antibacterial. Oil. $[\alpha]_D$ −5.7° (c 0.34 in CHCl₃).

  Sun, H.H. *et al*, *Tetrahedron Lett.*, 1980, 3123 (*isol*, *struct*, *spectra*)
  Amico, V. *et al*, *Phytochemistry*, 1987, **26**, 1715 (*isol*)

3,3'-Bi[5-hydroxy-2-methyl-1,4-naphtho-quinone]   B-60106

Updated Entry replacing B-01202
  8,8'-Dihydroxy-3,3'-dimethyl[2,2'-binaphthalene]-1,1',4,4'-tetrone, 9CI. **3,3'-Biplumbagin**
  [34341-27-0]

$C_{22}H_{14}O_6$ M 374.349
Constit. of roots of *Plumbago zeylanica* and leaves and rhizomes of *Aristea eclclonii*. Orange cryst. Mp 214-216°.

Sankaram, A.V.B. *et al*, *Tetrahedron Lett.*, 1971, 2385 (*struct*)
Sankaram, A.V.B. *et al*, *Indian J. Chem., Sect. B*, 1974, **12**, 519 (*synth*)
Laatsch, H., *Justus Liebigs Ann. Chem.*, 1983, 299 (*synth*)
Sankaram, A.V.B. *et al*, *Phytochemistry*, 1986, **25**, 2867 (*cmr*)

2,2'-Bi-1H-imidazole, 9CI   B-60107

Updated Entry replacing B-01203
  Glycosine. 2,2'-Diiminazole
  [492-98-8]

$C_6H_6N_4$ M 134.140
Prisms (AcOH or H₂O). V. spar. sol. org. solvs., insol. cold H₂O. Mp >350°.

B,2HCl: Needles (1M HCl/EtOH). Mp 307°.

Duranti, E. *et al*, *Synthesis*, 1974, 815.
Matthews, D.P. *et al*, *Synthesis*, 1986, 336 (*synth*)
Matthews, D.P. *et al*, *J. Heterocycl. Chem.*, 1987, **24**, 689 (*bibl*)

1,1'-Bi-1H-indene, 9CI   B-60108

1,1'-Biindenyl
[2177-49-3]

(1RS,1'RS)-form

$C_{18}H_{14}$ M 230.309

(1RS,1'RS)-form [81523-14-0]
  (±)-form
  Needles (EtOH). Mp 99-100°.
(1RS,1'SR)-form [74339-76-7]
  meso-*form*
  Platelets (EtOH). Mp 77.5°.

  Maréchal, E. *et al*, *Bull. Soc. Chim. Fr.*, 1964, 1740 (*synth*)
  Escher, A. *et al*, *Helv. Chim. Acta*, 1986, **69**, 1644 (*synth, uv, pmr, ms*)

Δ^{1,1'}-Biindene, 9CI   B-60109

Updated Entry replacing B-01207
  sym-*Dibenzfulvalene. 1,2;5,6-Dibenzopentafulvalene. 1,1'-Biindenylidene*
  [27949-41-3]

$C_{18}H_{12}$ M 228.293

(E)-(?)-form
  Red platelets (EtOH). Mp >300°.
(Z)-(?)-form
  Orange leaflets. Mp >300° (darkens at 220°). Stable.

  Anastassiou, A.G. *et al*, *J. Org. Chem.*, 1966, **31**, 2705 (*synth, ir, pmr, uv*)
  Lacy, P.H. *et al*, *J. Chem. Soc. (C)*, 1971, **41**, (*synth, ms*)
  Escher, A. *et al*, *Helv. Chim. Acta*, 1986, **69**, 1644 (*synth, uv, pmr, cmr, ms*)

Bindschedler's green   B-60110

N-[4-[[4-(*Dimethylamino*)*phenyl*]*imino*]-2,5-cyclohex-adien-1-ylidene]-N-*methylmethanaminium, 9CI*

$C_{16}H_{20}N_3^{\oplus}$ M 254.354 (ion)
Chloride: [4486-05-9].
  $C_{16}H_{20}ClN_3$ M 289.807
  Used in anal. of hydrazo compds. Dyestuff. Green solid. V. sol. H₂O.

Bindschedler, R., *Ber.*, 1883, **16**, 864 (*synth*)
Wieland, H., *Ber.*, 1915, **48**, 1078 (*synth*)
Shine, H.J. *et al*, *Anal. Chem.*, 1958, **30**, 383 (*synth, use*)

Biopterin   B-60111

Updated Entry replacing B-01229
  2-Amino-6-(1,2-dihydroxypropyl)-4(1H)-pteridinone, 9CI. 2-Amino-4-hydroxy-6-(1',2'-dihydroxypropyl)-pterin

(1'S, 2'S)-form
Absolute configuration

$C_9H_{11}N_5O_3$ M 237.218
(1'R,2'R)-form [13019-52-8]
  D-threo-*form*
  Mp >300°. pK_{a1} 2.20, pK_{a2} 7.92.
(1'S,2'R)-form [22150-76-1]
  L-erythro-*form*
  Widely distributed in microorganisms, insects, algae, amphibia and mammals. Found in urine. Growth factor. Pale-yellow cryst. (AcOH aq.). Mp 250-280° dec. $[\alpha]_D^{20}$ −66° (c, 0.2 in 0.1M HCl). pK_{a1} 2.23, pK_{a2} 7.89.

(*1′S,2′S*)-*form* [13039-82-2]
L-threo-*form*. Ciliapterin
Mp >300°. pK_{a1} 2.24, pK_{a2} 7.87.
(*1′R,2′S*)-*form* [13039-62-8]
D-erythro-*form*
Mp >300°. pK_{a1} 2.23, pK_{a2} 7.90.

Rembold, H. *et al*, *Angew. Chem., Int. Ed. Engl.*, 1972, **11**, 1061 (*biochem*)
Sugimoto, T. *et al*, *Bull. Chem. Soc. Jpn.*, 1975, **48**, 3767 (*synth*)
Taylor, E.C. *et al*, *J. Am. Chem. Soc.*, 1976, **98**, 2301 (*synth*)
Schircks, B. *et al*, *Helv. Chim. Acta*, 1977, **60**, 211; 1985, **68**, 1639 (*synth, pmr, cmr*)
Armarego, W.L.F. *et al*, *Aust. J. Chem.*, 1982, **35**, 785 (*synth*)
Kappel, M. *et al*, *Justus Liebigs Ann. Chem.*, 1984, 1815 (*synth*)

α-Biotol   B-60112

Updated Entry replacing B-01231
2,3,4,7,8,8a-Hexahydro-3,6,8,8-tetramethyl-1H-3a,7-methanoazulen-1-ol, 9CI
[19902-30-8]

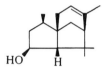

$C_{15}H_{24}O$ M 220.354
Constit. of the essential oil of the wood of *Biota orientalis*. Cryst. Mp 78°. $[\alpha]_D^{28}$ −27.3° (CCl₄).

Tomita, B. *et al*, *Tetrahedron Lett.*, 1968, 843 (*isol, struct*)
Trifilieff, E. *et al*, *Tetrahedron Lett.*, 1975, 4307 (*synth*)
Brun, P., *Tetrahedron Lett.*, 1977, 2269 (*synth*)
Grewal, R.S. *et al*, *J. Chem. Soc., Chem. Commun.*, 1987, 1290 (*synth*)

β-Biotol   B-60113

Updated Entry replacing B-01232
Octahydro-3,8,8-trimethyl-6-methylene-1H-3a,7-methanoazulen-1-ol, 9CI
[19902-26-2]

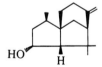

$C_{15}H_{24}O$ M 220.354
Constit. of the essential oil of *Biota orientalis*. Cryst. Mp 84°. $[\alpha]_D^{24}$ +43.8° (CCl₄).

Tomita, B. *et al*, *Tetrahedron Lett.*, 1968, 843 (*isol, struct*)
Grewal, R.S. *et al*, *J. Chem. Soc., Chem. Commun.*, 1987, 1290 (*synth*)

Biphenomycin *A*   B-60114
WS 43708A. Antibiotic WS 43708A
[95485-50-0]

$C_{23}H_{28}N_4O_8$ M 488.496
Cyclic peptide antibiotic. Isol. from *Streptomyces griseorubiginosus*. Active against bacteria; esp. potent against gram-positive bacteria.
B,2HCl: Needles + 1H₂O. Mp 205-209° dec. $[\alpha]_D^{20}$ −22.5° (c, 0.1 in 1M HCl).
7-Deoxy: [100217-74-1]. **Biphenomycin B**. *WS 43708B. Antibiotic WS 43708B.*
$C_{23}H_{28}N_4O_7$ M 472.497
From *S. griseorubiginosus*. Shows sl. weaker activity than Biphenomycin *A*.
7-Deoxy; B,2HCl: Needles. Mp 206-209° dec. $[\alpha]_D^{20}$ −10.6° (c, 0.1 in 1M HCl).

Ezaki, M. *et al*, *J. Antibiot.*, 1985, **38**, 1453, 1462 (*isol, uv, ir, pmr, cmr, struct*)
Uchida, I. *et al*, *J. Org. Chem.*, 1985, **50**, 1341 (*struct*)
Kannon, R. *et al*, *J. Org. Chem.*, 1987, **52**, 5435 (*stereochem*)

[1,1′-Biphenyl]-2,2′-diacetic acid, 9CI   B-60115
2,2′-Bis(carboxymethyl)biphenyl

HOOCCH₂CH₂COOH

$C_{16}H_{14}O_4$ M 270.284
Prisms (H₂O). Mp 152°.
Dinitrile: [3526-27-0]. *2,2′-Bis(cyanomethyl)biphenyl.*
$C_{16}H_{12}N_2$ M 232.284
Cryst. (EtOH). Mp 76-78°.

Kenner, J. *et al*, *J. Chem. Soc.*, 1911, 2101 (*synth, deriv*)
Cope, A.C. *et al*, *J. Am. Chem. Soc.*, 1956, **78**, 1012 (*synth, deriv*)

[1,1′-Biphenyl]-4,4′-diacetic acid, 9CI   B-60116
4,4′-Bis(carboxymethyl)biphenyl
[19806-14-5]
$C_{16}H_{14}O_4$ M 270.284
Cryst. (AcOH). Mp 282-284°.

Dinitrile: [7255-83-6]. *4,4′-Bis(cyanomethyl)biphenyl.*
$C_{16}H_{12}N_2$ M 232.284
Cryst. Mp 185-187°.

Schwenk, E. *et al*, *J. Org. Chem.*, 1946, **11**, 798 (*synth*)
Addison, A.W. *et al*, *Can. J. Chem.*, 1977, **55**, 4191 (*dinitrile, pmr*)
Ishibashi, H. *et al*, *Chem. Pharm. Bull.*, 1985, **33**, 5310 (*synth*)
Gotthardt, H. *et al*, *Chem. Ber.*, 1987, **120**, 411 (*synth*)

[1,1'-Biphenyl]-4,4'-diylbis[diphenylmethyl], B-60117
9CI

α,α,α',α'-*Tetraphenyldi*-p-*xylylene. Chichibabin's hydrocarbon*

[6418-52-6]

$C_{38}H_{28}$ M 484.639

Shiny metallic-green cryst. Very oxygen- and moisture-sensitive. Gives blue-violet solns.

Chichibabin, A.E., *Ber.*, 1907, **40**, 1810 (*synth*)
Montgomery, L.K. *et al*, *J. Am. Chem. Soc.*, 1986, **108**, 6004 (*synth, uv, cryst struct, bibl*)

1,8-Biphenylenedicarboxaldehyde B-60118
[58746-94-4]

$C_{14}H_8O_2$ M 208.216

Bright-yellow needles (cyclohexane). Mp 151-152°.

Wilcox, C.F. *et al*, *Tetrahedron*, 1975, **31**, 2889 (*synth, pmr, ir, ms*)
Wilcox, C.F. *et al*, *J. Am. Chem. Soc.*, 1986, **108**, 7693 (*synth*)

1,3(5)-Bi-1H-pyrazole B-60119
[71426-39-6]

$C_6H_6N_4$ M 134.140

Cryst. (diisopropyl ether or by subl.). Mp 91-92°.

2-Me: [104384-54-5]. *1'-Methyl-1,5'-bi-1H-pyrazole,* 9CI.

$C_7H_8N_4$ M 148.167
Bp_2 85°.

Cohen-Fernandes, P. *et al*, *J. Org. Chem.*, 1979, **44**, 4156 (*synth, cmr, pmr, ir*)
Bruix, M. *et al*, *Tetrahedron Lett.*, 1985, **26**, 5485 (*synth*)

2,3'-Bipyridine, 9CI B-60120

Updated Entry replacing B-01354

2,3'-Dipyridyl. **Isonicoteine**

[581-50-0]

$C_{10}H_8N_2$ M 156.187

Occurs in tobacco (*Nicotiana tabacum*) (Solanaceae) and the hoplonemertine *Amphiporus angulatus*. Liq. Insol. H_2O. Bp 298°.

▷Neurotoxin

B,2HClO₄: Mp 215-216°.
Picrate: Mp 153-154°.
Dipicrate: Mp 165-168°.
1'-Oxide:
    $C_{10}H_8N_2O$ M 172.186
    Cryst. + $1H_2O$ (Me_2CO/hexane). Mp 78°.
1,1'-Dioxide:
    $C_{10}H_8N_2O_2$ M 188.185
    Cryst. (MeOH). Mp 240-243°.

Smith, J.R., *J. Am. Chem. Soc.*, 1930, **52**, 397 (*synth*)

Morgan, G.T. *et al*, *J. Chem. Soc.*, 1932, 20 (*synth*)
Späth, E. *et al*, *Ber.*, 1936, **69**, 2448 (*isol*)
Leete, E., *J. Am. Chem. Soc.*, 1969, **91**, 1697 (*synth*)
Kem, W.R. *et al*, *Experientia*, 1976, **32**, 684 (*isol, ms*)
Leete, E. *et al*, *J. Am. Chem. Soc.*, 1976, **98**, 6326 (*biosynth*)
Moran, D.B. *et al*, *J. Heterocycl. Chem.*, 1986, **23**, 1071 (*oxides*)

2,4'-Bipyridine, 9CI B-60121

Updated Entry replacing B-01355

2,4'-Bipyridyl

[581-47-5]

$C_{10}H_8N_2$ M 156.187
Mp 61.5°. Bp_{11} 148-150°.

B,MeI: Mp 188-190°.
B,2MeI: Mp 225°.
1-Oxide:
    $C_{10}H_8N_2O$ M 172.186
    Cryst. (Me_2CO/hexane). Mp 119-121°.
1,1'-Dioxide:
    $C_{10}H_8N_2O_2$ M 188.185
    Mp 240-242°.

Homer, R.F., *J. Chem. Soc.*, 1958, 1574.
Moran, D.B. *et al*, *J. Heterocycl. Chem.*, 1986, **23**, 1071 (*oxides*)

3,3'-Bipyridine, 9CI B-60122

Updated Entry replacing B-01356

3,3'-Bipyridyl

[581-46-4]

$C_{10}H_8N_2$ M 156.187
Misc. H_2O, EtOH. d^{20} 1.1635. Mp 68°. Bp 300-301°, Bp_{25} 190-192°.

Mono-N-oxide:
    $C_{10}H_8N_2O$ M 172.186
    Off-white cryst. Mp 151-153°.

Morgan, G.T. *et al*, *J. Chem. Soc.*, 1932, 20.
Moran, D.B. *et al*, *J. Heterocycl. Chem.*, 1986, **23**, 1071 (*oxide*)

3,4'-Bipyridine, 9CI B-60123

3,4'-Bipyridyl

[4394-11-0]

$C_{10}H_8N_2$ M 156.187
Bp_{15} 144-146°, Bp_3 115°.

Picrate: Mp 204-205°.

Otroshchenko, O.S. *et al*, *Chem. Heterocycl. Compd. Engl. Trans.*, 1969, **5**, 277 (*synth, bibl*)
Yamamoto, Y. *et al*, *Synthesis*, 1986, 564 (*synth*)

[2,2'-Bipyridine]-3,3'-dicarboxylic acid, 9CI B-60124

2,2'-Bipyridyl-3,3'-dicarboxylic acid. 2,2'-Binicotinic acid

[4433-01-6]

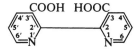

$C_{12}H_8N_2O_4$ M 244.206
Needles. Mp 262°.

Mono-Me ester: [99165-99-8].
    $C_{13}H_{10}N_2O_4$ M 258.233
    Cryst. (CH_2Cl_2). Mp 182-185°.

Di-Me ester: [39775-31-0].
$C_{14}H_{12}N_2O_4$ M 272.260
Needles. Mp 152°.

Wimmer, F. *et al*, *Org. Prep. Proced. Int.*, 1983, **15**, 368 (*synth*)
Dholakia, S. *et al*, *Polyhedron*, 1985, **4**, 791 (*synth*)
Rebek, J. *et al*, *J. Am. Chem. Soc.*, 1985, **107**, 7487 (*ester*)

[2,2′-Bipyridine]-3,5-dicarboxylic acid, 9CI   **B-60125**
2,2′-Bipyridyl-3,5′-dicarboxylic acid. 2,6′-Binicotinic acid
[74357-37-2]
$C_{12}H_8N_2O_4$ M 244.206
Di-Me ester:
  $C_{14}H_{12}N_2O_4$ M 272.260
  Cryst. Mp 115-116°.

Sugimori, A. *et al*, *Bull. Chem. Soc. Jpn.*, 1981, **54**, 2068 (*synth*)

[2,2′-Bipyridine]-4,4′-dicarboxylic acid, 9CI   **B-60126**
2,2′-Bipyridyl-4,4′-dicarboxylic acid. 2,2′-Biisonicotinic acid
[6813-38-3]
$C_{12}H_8N_2O_4$ M 244.206
Mp 384-388°.
Di-Me ester: [71071-46-0].
  $C_{14}H_{12}N_2O_4$ M 272.260
  Cryst. (Me₂CO/CHCl₃). Mp 208-210°.
Diamide:
  $C_{12}H_{10}N_4O_2$ M 242.237
  Cryst. (1,2-ethanediol). Mp 251° dec.
Dichloride: [72460-28-7].
  $C_{12}H_6Cl_2N_2O_2$ M 281.098
  Cryst. (heptane). Mp 141-143°.

Maerker, G. *et al*, *J. Am. Chem. Soc.*, 1958, **80**, 2745 (*synth, amide*)
Bos, K.D. *et al*, *Synth. Commun.*, 1979, **9**, 497 (*synth*)
Elliott, C.M. *et al*, *J. Am. Chem. Soc.*, 1982, **104**, 7519 (*synth*)
Evers, R.C. *et al*, *J. Polym. Sci. Polym. Chem.*, 1986, **24**, 1863 (*synth*)

[2,2′-Bipyridine]-5,5′-dicarboxylic acid, 9CI   **B-60127**
2,2′-Bipyridyl-5,5′-dicarboxylic acid. 6,6′-Binicotinic acid
[1802-30-8]
$C_{12}H_8N_2O_4$ M 244.206
Mp >360°.
Di-Me ester: [1762-45-4].
  $C_{14}H_{12}N_2O_4$ M 272.260
  Cryst. (dioxane). Mp 261-262°.
Diamide: [4444-36-4].
  $C_{12}H_{10}N_4O_2$ M 242.237
  Cryst. Mp >310°.
Dinitrile: [1802-29-5]. *5,5′-Dicyano-2,2′-bipyridine.*
  $C_{12}H_6N_4$ M 206.206
  Cryst. (DMF). Mp 272-273°.
Dihydrazide: [63361-67-1].
  $C_{12}H_{12}N_6O_2$ M 272.266
  Mp >360°.

Badger, G.M. *et al*, *J. Chem. Soc.*, 1956, 616 (*synth*)
Whittle, C.P. *et al*, *J. Heterocycl. Chem.*, 1977, **14**, 191 (*synth, derivs*)

[2,2′-Bipyridine]-6,6′-dicarboxylic acid, 9CI   **B-60128**
2,2′-Bipyridyl-6,6′-dicarboxylic acid. 6,6′-Bipicolinic acid
[4479-74-7]
$C_{12}H_8N_2O_4$ M 244.206
Needles (H₂O). Mp 288° dec.
Di-Et ester: [65739-40-4].
  $C_{16}H_{16}N_2O_4$ M 300.313
  Cryst. (MeOH). Mp 140-141°.
Dichloride: [65739-39-1].
  $C_{12}H_6Cl_2N_2O_2$ M 281.098
  Mp 177-178°.

Burstall, F.H., *J. Chem. Soc.*, 1938, 1662 (*synth*)
Buhleier, E. *et al*, *Chem. Ber.*, 1978, **111**, 200 (*synth, pmr*)

[2,3′-Bipyridine]-2,3′-dicarboxylic acid, 9CI   **B-60129**
2,3′-Bipyridyl-2′,3-dicarboxylic acid
$C_{12}H_8N_2O_4$ M 244.206
Mp 214-215° dec.

Skraup, S. *et al*, *Monatsh. Chem.*, 1882, **3**, 588, 591 (*synth*)

[2,4′-Bipyridine]-2′,6′-dicarboxylic acid, **B-60130**
9CI
2,4′-Bipyridyl-2′,6′-dicarboxylic acid
$C_{12}H_8N_2O_4$ M 244.206
Mp 240° dec.

Kuffner, F. *et al*, *Monatsh. Chem.*, 1957, **88**, 793 (*synth*)

[2,4′-Bipyridine]-3,3′-dicarboxylic acid, 9CI   **B-60131**
2,4′-Bipyridyl-3,3′-dicarboxylic acid. 2,4′-Binicotinic acid
$C_{12}H_8N_2O_4$ M 244.206
Cryst. (H₂O). Mp 275°.

Homer, R.F., *J. Chem. Soc.*, 1958, 1574 (*synth*)

[2,4′-Bipyridine]-3′,5-dicarboxylic acid, 9CI   **B-60132**
2,4′-Bipyridine-3,5′-dicarboxylic acid
[50722-63-9]
$C_{12}H_8N_2O_4$ M 244.206

Ziyaev, A.A. *et al*, *Khim. Geterotsikl. Soedin.*, 1973, 1076.

[3,3′-Bipyridine]-2,2′-dicarboxylic acid, 9CI   **B-60133**
3,3′-Bipyridyl-2,2′-dicarboxylic acid. 3,3′-Bipicolinic acid
[3723-32-8]
$C_{12}H_8N_2O_4$ M 244.206
Fine cryst. (DMSO aq.). Mp 220°.
Di-Me ester: [39770-63-3].
  $C_{14}H_{12}N_2O_4$ M 272.260
  Cryst. (MeOH). Mp 156-157°.
Dihydrazide: [39770-64-4].
  $C_{12}H_{12}N_6O_2$ M 272.266
  Cryst. (EtOH). Mp 217°.

Kuhn, R. *et al*, *J. Prakt. Chem.*, 1965, **98**, 2139 (*synth*)
Williams, R.L. *et al*, *J. Heterocycl. Chem.*, 1972, **9**, 1021 (*synth*)

[3,3'-Bipyridine]-4,4'-dicarboxylic acid, 9CI   B-60134

3,3'-Bipyridyl-4,4'-dicarboxylic acid. 3,3'-Biisonicotinic acid

$C_{12}H_8N_2O_4$ M 244.206

Di-Me ester: [82828-86-2].
  $C_{14}H_{12}N_2O_4$ M 272.260
  Mp 184-187° subl.

Sugiyama, T. *et al, Bull. Chem. Soc. Jpn.*, 1984, **57**, 1882.

[3,4'-Bipyridine]-2',6'-dicarboxylic acid,   B-60135
9CI

3,4'-Bipyridyl-2',6'-dicarboxylic acid

$C_{12}H_8N_2O_4$ M 244.206
Mp 250° dec.

Kuffner, F. *et al, Monatsh. Chem.*, 1957, **88**, 793 (*synth*)

[4,4'-Bipyridine]-2,2'-dicarboxylic acid, 9CI   B-60136

4,4'-Bipyridyl-2,2'-dicarboxylic acid. 4,4'-Bipicolinic acid

[85531-49-3]
$C_{12}H_8N_2O_4$ M 244.206
Needles. Mp 247.5°.

Heuser, A. *et al, J. Prakt. Chem.*, 1886, **44**, 405 (*synth*)

[4,4'-Bipyridine]-3,3'-dicarboxylic acid, 9CI   B-60137

4,4'-Bipyridyl-3,3'-dicarboxylic acid. 4,4'-Binicotinic acid

[23245-77-4]
$C_{12}H_8N_2O_4$ M 244.206
Mp >250° dec.

Ziyaev, A.A. *et al, CA*, 1976, **84**, 30823 (*synth*)
Rebek, J. *et al, J. Am. Chem. Soc.*, 1985, **107**, 7487 (*synth*)

Bisabolangelone   B-60138

Updated Entry replacing B-01386
3,3a,7,7a-Tetrahydro-3-hydroxy-3,6-dimethyl-2-(3-methyl-2-butenylidene)-4(2H)-benzofuranone, 9CI.
Angelikoreanol

[30557-81-4]

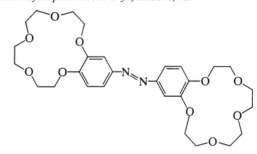

$C_{15}H_{20}O_3$ M 248.321
Constit. of *Angelica silvestris* seeds. Cryst. (C_6H_6). Mp 148-149°. $[\alpha]_D$ +49.3° (c, 0.073 in dioxan).

Novotný, L. *et al, Tetrahedron Lett.*, 1966, 3541 (*isol, struct*)
Hata, K. *et al, Chem. Pharm. Bull.*, 1971, **19**, 1963 (*isol, struct*)
Riss, B.P. *et al, Tetrahedron Lett.*, 1986, **27**, 4979 (*synth*)

1,2-Bis(2-aminoethoxy)ethane   B-60139

2,2'-[1,2-Ethanediylbis(oxy)]bisethanamine, 9CI. 1,8-Diamino-3,6-dioxaoctane

[929-59-9]

$$H_2NCH_2CH_2OCH_2CH_2OCH_2CH_2NH_2$$

$C_6H_{16}N_2O_2$ M 148.205
Reagent for prepn. of diazo-crown ethers. Liq. Bp_1 95°, $Bp_{0.2}$ 77-78°. Hygroscopic.

Dwyer, F.P. *et al, J. Am. Chem. Soc.*, 1953, **75**, 1526 (*synth*)
Dietrich, B. *et al, Tetrahedron*, 1973, **29**, 1629 (*synth, pmr*)
Gatto, V.J. *et al, J. Org. Chem.*, 1986, **51**, 5373 (*synth*)
Kulstad, S. *et al, Acta Chem. Scand., Ser. B*, 1979, **33**, 469 (*synth*)

3,3-Bis(aminomethyl)oxetane   B-60140

3,3-Oxetanedimethanamine, 9CI. 3,3-Oxetanebis(methylamine), 8CI

[23500-57-4]

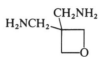

$C_5H_{12}N_2O$ M 116.163
Monohydrate. Bp_{15} 121°.

N,N'-*Di-Me, N,N'-dinitro:* 3,3-Bis(methylnitraminomethyl)oxetane.
$C_7H_{14}N_4O_5$ M 234.211
Monomer for explosive propellant polymers.

Salvador, R.L. *et al, Can. J. Chem.*, 1968, **46**, 751 (*synth*)
George, C. *et al, Acta Crystallogr., Sect. C*, 1986, **42**, 1161 (*cryst struct, deriv*)

1,2-Bis(4'-benzo-15-crown-5)diazene   B-60141

Bis(2,3,5,6,8,9,11,12-octahydro-1,4,7,10,13-benzopentaoxacyclopentadecin-5-yl)diazene, 9CI

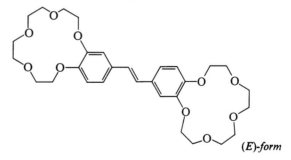

$C_{28}H_{38}N_2O_{10}$ M 562.616
Photoresponsive crown ether.

(*E*)-*form* [73491-37-9]
Yellow needles. Mp 187-188°.

Shinkai, S. *et al, J. Am. Chem. Soc.*, 1981, **103**, 111.

1,2-Bis(4'-benzo-15-crown-5)ethene   B-60142

15,15'-(1,2-Ethenediyl)bis[2,3,5,6,8,9,11,12-octahydro-1,4,7,10,13-benzopentaoxacyclopentadecin], 9CI. 1',2'-Bis(2,5,8,11,14-pentaoxabicyclo[13.4.0]nonadeca-1(15),16,18-trien-17-yl)ethene. Stilbenobis-15-crown-5

(E)-form

$C_{30}H_{40}O_{10}$ M 560.640

(*E*)-*form* [107825-31-0]
Cryst. (CH$_2$Cl$_2$/MeOH). Mp 190-192°.
(*Z*)-*form* [107825-33-2]
Oil. The higher homologue 1,2-Bis(4'-benzo-18-crown-6) was also prepd.

Lindsten, G. *et al*, *Acta Chem. Scand., Ser. B*, 1986, **40**, 545 (*synth, pmr, ms*)

Bisbenzo[3,4]cyclobuta[1,2-c;1',2'-g]-phenanthrene   B-60143
[96695-04-4]

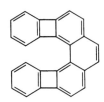

C$_{26}$H$_{14}$ M 326.397
Bright-yellow cryst. (hexane). Mp 176-179°:

Cracknell, M.E. *et al*, *J. Chem. Soc., Perkin Trans. 1*, 1985, 115 (*synth, ir, uv, pmr*)

2,3-Bis(bromomethyl)-1,4-dibromobutane, 8CI   B-60144
1,1,2,2-Tetrakis(bromomethyl)ethane
[19760-94-2]

$$(BrH_2C)_2CHCH(CH_2Br)_2$$

C$_6$H$_{10}$Br$_4$ M 401.761
Cryst. (pet. ether). Mp 88-89°.

Weinges, K., *Chem. Ber.*, 1968, **101**, 3010 (*synth*)

1,3-Bis(bromomethyl)-5-nitrobenzene, 9CI   B-60145
*ω,ω'-Dibromo-5-nitro-*m*-xylene*
[51760-20-4]

C$_8$H$_7$Br$_2$NO$_2$ M 308.957
Cryst. (pet. ether). Mp 105-106.5°. Lachrymator and skin irritant.

Reich, W.S. *et al*, *J. Chem. Soc.*, 1947, 1234 (*synth*)
Sherrod, S.A. *et al*, *J. Am. Chem. Soc.*, 1974, **96**, 1565 (*synth, pmr*)
Vögtle, F. *et al*, *Chem. Ber.*, 1979, **112**, 1400 (*synth*)

2,6-Bis(bromomethyl)pyridine   B-60146
[7703-74-4]

BrH$_2$C〔pyridine〕CH$_2$Br

C$_7$H$_7$Br$_2$N   M 264.947
Cryst. (pet. ether). Mp 84-87°.

▷Lachrymator

Baker, W. *et al*, *J. Chem. Soc.*, 1958, 3594 (*synth*)
Offerman, W. *et al*, *Synthesis*, 1977, 272 (*synth, pmr*)
Šket, B. *et al*, *J. Org. Chem.*, 1986, **51**, 929 (*synth*)

3,5-Bis(bromomethyl)pyridine   B-60147
[35991-75-4]

C$_7$H$_7$Br$_2$N   M 264.947
Cryst. Mp 97-98° dec.

Beeby, P.J. *et al*, *J. Am. Chem. Soc.*, 1972, **94**, 2128 (*synth*)

2,3-Bis(chloromethyl)-1,4-dichlorobutane, 8CI   B-60148
1,1,2,2-Tetrakis(chloromethyl)ethane
[19869-19-3]

$$(ClH_2C)_2CHCH(CH_2Cl)_2$$

C$_6$H$_{10}$Cl$_4$ M 223.957
Cryst. (MeOH). Mp 52-53°.

Weinges, K., *Chem. Ber.*, 1968, **101**, 3010 (*synth*)

3,3-Bis(chloromethyl)oxetane   B-60149
BCMO
[78-71-7]

C$_5$H$_8$Cl$_2$O   M 155.024
Polymer intermed. Liq. d$_{25}^{25}$ 1.295. Mp 18.7°. Bp$_{27}$ 101°. n$_D^{20}$ 1.4858.

▷RQ6826000.

Farthing, A.C., *J. Chem. Soc.*, 1955, 3648 (*synth*)
Campbell, T.W., *J. Org. Chem.*, 1957, **22**, 1029 (*synth, bibl*)
Ratz, L. *et al*, *Z. Naturforsch., A*, 1968, **23**, 2100 (*ir*)
Kirk-Othmer Encycl. Chem. Technol., 3rd Ed., 1982, **18**, 646 (*use*)

2,3-Bis(dibromomethyl)pyrazine   B-60150
[107427-52-1]

C$_6$H$_4$Br$_4$N$_2$ M 423.727
Cryst. (CHCl$_3$/EtOH). Mp 168-170°.

Shepherd, M.K., *J. Chem. Soc., Perkin Trans. 1*, 1986, 1495 (*synth, pmr, ms*)

2,3-Bis(dibromomethyl)pyridine, 9CI   B-60151
[107427-51-0]

C$_7$H$_5$Br$_4$N   M 422.739
Cryst. (EtOH). Mp 158-159°.

Shepherd, M.K., *J. Chem. Soc., Perkin Trans. 1*, 1986, 1495 (*synth, pmr, ms*)

3,4-Bis(dibromomethyl)pyridine   B-60152
[107445-08-9]
C$_7$H$_5$Br$_4$N   M 422.739
Cryst. solid. Dec. within hrs. at 0°, rapidly >30°.

Shepherd, M.K., *J. Chem. Soc., Perkin Trans. 1*, 1986, 1495 (*synth, pmr, ms*)

4,5-Bis(dibromomethyl)pyrimidine   B-60153
[107427-53-2]

$C_6H_4Br_4N_2$ M 423.727
Mp ca. 160°. Dec. > ca. 140°.

Shepherd, M.K., *J. Chem. Soc., Perkin Trans. 1*, 1986, 1495 (*synth, pmr, ms*)

Bis(1,1-dicyclopropylmethyl)diazene   B-60154

$C_{14}H_{22}N_2$ M 218.341
(**E**)-**form** [100515-66-0]
   Needles (pentane). Mp 39-40°.
(**Z**)-**form** [100515-65-9]
   Yellowish cryst. (pet. ether). Mp 116-117°.

Bruch, M. *et al, J. Org. Chem.*, 1986, **51**, 2969 (*synth, pmr, ir, uv*)

1,5-Bis(3,4-dihydroxyphenyl)-4-pentyne-1,2-diol   B-60155
1,5-Bis(3,4-dihydroxyphenyl)-1,2-dihydroxy-4-pentyne

$C_{17}H_{16}O_6$ M 316.310
2-O-β-D-Glusopyranoside: **Nyasicaside**.
   $C_{23}H_{26}O_{11}$ M 478.452
   Constit. of *Hypoxis nyasica*. Cryst. (EtOH/EtOAc).
   Mp 120-122°. $[\alpha]_D^{20}$ +14.7° (c, 0.9 in MeOH).
Galeffi, C. *et al, Tetrahedron*, 1987, **43**, 3519.

3,6-Bis(3,4-dimethoxyphenyl)tetrahydro-1*H*,3*H*-furo[3,4-*c*]furan-1,4-diol, 9CI   B-60156
[111614-78-9]

$C_{22}H_{26}O_8$ M 418.443
Constit. of *Eremophila dalyana*. Cryst. (EtOH-/Me₂CO). Mp 207-209°. $[\alpha]_D$ −16.9° (c, 0.2 in MeOH).

Ghisalberti, E.L. *et al, Aust. J. Chem.*, 1987, **40**, 405 (*cryst struct*)

3,6-Bis(diphenylmethylene)-1,4-cyclohexadiene   B-60157
*1,1′,1″,1‴-(2,5-Cyclohexadiene-1,4-diylidenedimethanetetrayl)tetrakisbenzene, 9CI.
α,α,α′,α′-Tetraphenyl-p-xylylene. Thiele's hydrocarbon*
[26392-12-1]

$C_{32}H_{24}$ M 408.542
Orange-yellow cryst. (hexane). Mp 234-241°.

Thiele, J. *et al, Ber.*, 1904, **37**, 1463 (*synth*)
Montgomery, L.K. *et al, J. Am. Chem. Soc.*, 1986, **108**, 6004 (*synth, pmr, cmr, cryst struct, bibl*)

1,4-Bis(ethylthio)-3,6-diphenylpyrrolo[3,4-c]pyrrole, 9CI   B-60158
1,4-Bis(ethylthio)-3,6-diphenyl-2,5-diazapentalene
[108472-40-8]

$C_{22}H_{20}N_2S_2$ M 376.533
Simplest 2,5-diazapentalene (pyrrolo[3,4-*c*]pyrrole) so far obt. Violet cryst. Mp 223°. Dec. by acids or on htg. in polar solvs. The *p,p′*-dimethoxy analogue is stable to protonation.

Closs, F. *et al, Angew. Chem., Int. Ed. Engl.*, 1987, **26**, 552 (*synth, uv*)

2,3-Bis(hydroxymethyl)-1,4-butanediol, 8CI   B-60159
1,1,2,2-Tetrakis(hydroxymethyl)ethane
[5373-21-7]

$$(HOH_2C)_2CHCH(CH_2OH)_2$$

$C_6H_{14}O_4$ M 150.174
Cryst. (MeOH). Mp 110°.

Weinges, K. *et al, Chem. Ber.*, 1968, **101**, 3010 (*synth*)

2,5-Bis(hydroxymethyl)furan   B-60160
2,5-Furandimethanol, 9CI, 8CI. 5-(Hydroxymethyl)-furfuryl alcohol
[1883-75-6]

$C_6H_8O_3$ M 128.127
Plates (EtOAc). Mp 76°.
Di-Ac: [5076-10-8].
   $C_{10}H_{12}O_5$ M 212.202
   Needles (pet. ether). Mp 64°.
Dibenzoyl: [94465-43-7].
   $C_{20}H_{16}O_5$ M 336.343
   Prisms (MeOH). Mp 76-77°.

Finan, P.A., *J. Chem. Soc.*, 1963, 3917 (*synth*)
Gagnaire, D. *et al, Bull. Soc. Chim. Fr.*, 1965, 474 (*synth*)

Timko, J.M. *et al*, *J. Am. Chem. Soc.*, 1977, **99**, 4207 (*synth*)

3,4-Bis(hydroxymethyl)furan   B-60161

3,4-Furandimethanol, 9CI

[14496-24-3]

$C_6H_8O_3$ M 128.127

Oil. Bp_2 129-130°, $Bp_{0.1}$ 110°.

Di-Ac: [30614-73-4].

$C_{10}H_{12}O_5$ M 212.202

Liq. $Bp_{0.1}$ 88-90°.

Cook, M.J. *et al*, *Tetrahedron*, 1968, **24**, 4501 (*synth*)
Koenig, H. *et al*, *Justus Liebigs Ann. Chem.*, 1981, 668 (*synth*)
Jenneskens, L.W. *et al*, *J. Chem. Soc., Perkin Trans. 1*, 1985, 2119 (*synth, pmr, ms*)

1,5-Bis(4-hydroxyphenyl)-1,4-pentadiene   B-60162

Di-(p-hydroxystyryl)methane

$C_{17}H_{16}O_2$ M 252.312

(1Z,4Z)-form

Constit. of *Alpinia galanga* rhizomes.

Di-Ac: Cryst. (C_6H_6/pet. ether). Mp 140-142°.

Barik, B.R. *et al*, *Phytochemistry*, 1987, **26**, 2126.

1,2-Bis(2-iodoethoxy)ethane, 9CI   B-60163

[36839-55-1]

$$ICH_2CH_2OCH_2CH_2OCH_2CH_2I$$

$C_6H_{12}I_2O_2$ M 369.969

Intermed. for crown ether synth. Liq. $Bp_{0.5}$ 126-129°.

Kulstad, S. *et al*, *Acta Chem. Scand., Ser. B*, 1979, **33**, 469 (*synth, pmr, ms, use*)

2,2-Bis(mercaptomethyl)-1,3-propanedithiol, 9CI   B-60164

Tetrakis(mercaptomethyl)methane. Pentaerythritol tetramercaptan

[4720-60-9]

$$C(CH_2SH)_4$$

$C_5H_{12}S_4$ M 200.390

Na deriv.: Tetragonal plates (EtOH). Mp 73-73.5°. Dec. by acids.

Tetra-S-Me: [65654-51-5].

Tetrakis(methylthiomethyl)methane.

$C_9H_{20}S_4$ M 256.497

Liq.

Tetra-S-Me, tetrakis-S-oxide:

Tetrakis(methylsulfinylmethyl)methane.

$C_9H_{20}O_4S_4$ M 320.495

Phase-transfer catalyst. Mp 205-215° (205°).

Backer, H.J. *et al*, *Recl. Trav. Chim. Pays-Bas*, 1933, **52**, 701; 1937, **56**, 681 (*synth*)

Fujihara, H. *et al*, *Heterocycles*, 1981, **16**, 1701 (*synth*)
Fujihara, H. *et al*, *J. Chem. Soc., Perkin Trans. 1*, 1986, 333 (*deriv, ir, pmr, use*)

1,6:7,12-Bismethano[14]annulene   B-60165

*Tricyclo[8.4.1.1^{2,7}]hexadeca-2,4,6,8,10,12,14-heptaene,
9CI. 1,6:7,12-Dimethano[14]annulene*

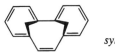

syn-form

$C_{16}H_{14}$ M 206.287

syn-form [85385-68-8]

Air-stable orange rhombs (pentane). Mp 93-94°.

anti-form [85440-62-6]

Pale-yellow liq.

Vogel, E. *et al*, *Angew. Chem., Int. Ed. Engl.*, 1986, **25**, 720, 723 (*synth, uv, pmr, cmr, cryst struct*)

1,2-Bis(methylene)cyclohexane, 9CI   B-60166

1,2-Dimethylenecyclohexane. 1,2-Dimethylidenecyclohexane

[2819-48-9]

C_8H_{12} M 108.183

Bp_{740} 124°, Bp_{90} 60-61°. n_D^{25} 1.4718.

Bailey, W.J. *et al*, *J. Am. Chem. Soc.*, 1953, **75**, 4780 (*synth, uv*)
Blomquist, A.T. *et al*, *J. Am. Chem. Soc.*, 1957, **79**, 3916 (*synth*)
Pfeffer, H.U. *et al*, *Org. Magn. Reson.*, 1977, **9**, 121 (*cmr*)
v. Straten, J.W. *et al*, *Recl. Trav. Chim. Pays-Bas*, 1978, **92**, 105 (*synth, uv, pmr*)
Bates, R.B. *et al*, *J. Org. Chem.*, 1984, **49**, 2981 (*synth, pmr, uv*)
Müller, P. *et al*, *Helv. Chim. Acta*, 1986, **69**, 1546.

1,3-Bis(methylene)cyclohexane, 9CI   B-60167

1,3-Dimethylenecyclohexane. 1,3-Dimethylidenecyclohexane

[52086-82-5]

C_8H_{12} M 108.183

Bp 122°. n_D^{25} 1.4697.

Bailey, W.J. *et al*, *J. Org. Chem.*, 1958, **23**, 1002 (*synth*)
Block, E. *et al*, *J. Am. Chem. Soc.*, 1983, **105**, 6165 (*synth*)

1,4-Bis(methylene)cyclohexane, 9CI   B-60168

1,4-Dimethylenecyclohexane, 8CI. 1,4-Dimethylidenecyclohexane

[5664-20-0]

C_8H_{12} M 108.183

Has been used in polymerisations. Bp_{98} 60-65°. $n_D^{19.5}$ 1.4734.

Butler, G.B. *et al*, *J. Macromol. Sci. Chem.*, 1974, **A8**, 1139 (*synth, uv*)

1,2-Bis(methylene)cyclopentane

B-60169

1,2-Dimethylenecyclopentane, 8CI. 1,2-
Dimethylidenecyclopentane
[20968-70-1]

C_7H_{10} M 94.156
d_4^{25} 0.821. Bp_{100} 45°. n_D^{25} 1.4750.

Blomquist, A.T. *et al, J. Am. Chem. Soc.*, 1956, **78**, 6057
   (*synth*)
Bartlett, P.D. *et al, J. Am. Chem. Soc.*, 1968, **90**, 6067 (*synth*)
v. Straten, J.W. *et al, Recl. Trav. Chim. Pays-Bas*, 1978, **97**,
   105 (*synth, uv, pmr*)
Müller, P. *et al, Helv. Chim. Acta*, 1986, **69**, 1546 (*synth*)

1,3-Bis(methylene)cyclopentane

B-60170

1,2-Dimethylenecyclopentane. 1,2-
Dimethylidenecyclopentane

[59219-48-6]
C_7H_{10} M 94.156
Bp 102°.

Wiberg, K.B. *et al, J. Org. Chem.*, 1976, **41**, 2711 (*synth, ir,*
   pmr)

3,6-Bis(2-methylpropyl)-2,5-piperazine-dione

B-60171

3,6-Diisobutylpiperazinedione. Cyclo(leucylleucyl)
[1436-27-7]

(3R,6R)-form

$C_{12}H_{22}N_2O_2$ M 226.318
(3R,6R)-form
   D-D-form
   $[\alpha]_D^{20}$ +46.02°.
(3S,6S)-form [952-45-4]
   L-L-form
   Cryst. (MeOH aq.). $[\alpha]_D^{20}$ −47.1° (c, 0.7 in AcOH).
(3RS,6SR)-form [16679-66-6]
   meso-*form*
   Mp 287-289°.

Fischer, E. *et al, Justus Liebigs Ann. Chem.*, 1907, **354**, 39
   (*synth*)
Exner, M.M. *et al, Biopolymers*, 1977, **16**, 1387 (*synth, cmr*)
Kanmera, T. *et al, Int. J. Pept. Protein Res.*, 1980, **16**, 280
   (*synth*)
Kricheldorf, H., *Org. Magn. Reson.*, 1980, **13**, 52 (*cmr, N nmr*)
Ueda, T. *et al, Bull. Chem. Soc. Jpn.*, 1983, **56**, 568 (*hplc*)
Tanihara, M. *et al, Bull. Chem. Soc. Jpn.*, 1983, **56**, 1155.

1,4-Bis(2-pyridylamino)phthalazine

B-60172

Updated Entry replacing B-20132
N,N′-*Di-2-pyridinyl-1,4-phthalazinediamine, 9CI. 1,4-*
Di(2-pyridylamino)phthalazine. pap
[25535-53-9]

$C_{18}H_{14}N_6$ M 314.349
Complexing agent forming binuclear complexes with
transition-metal salts. Cryst. (EtOH or by subl.). Mp
210°. Several analogues substd. in the pyridine rings
e.g. papf, paps, papfs are also employed.

B,2HNO₃: Mp 187° dec.

Thompson, L.K. *et al, Can. J. Chem.*, 1969, **47**, 4141 (*synth*)
Dewan, J.C. *et al, Can. J. Chem.*, 1982, **60**, 121 (*cryst struct*)
Bullock, G. *et al, Can. J. Chem.*, 1983, **61**, 57.
Mandal, S.K. *et al, Aust. J. Chem.*, 1986, **39**, 1007 (*complexes*)

Bis(quinuclidine)bromine(1+)

B-60173

Bis(1-azabicyclo[2.2.2]octane)bromine(1+), 9CI

$C_{14}H_{26}BrN_2^{\oplus}$ M 302.277 (ion)
Tetrafluoroborate: [85282-86-6].
   $C_{14}H_{26}BBrF_4N_2$ M 389.080
   Source of $Br^{\oplus}$, oxidising agent. Cryst. (MeCN). Mp
   >130° dec.
Bromide: [85282-84-4].
   $C_{14}H_{26}Br_2N_2$ M 382.181
   Solid. Mp >137° dec.

Blair, L.K. *et al, J. Am. Chem. Soc.*, 1983, **105**, 3649 (*synth,*
   cryst struct)
Blair, L.K. *et al, J. Org. Chem.*, 1986, **51**, 5454 (*pmr, use*)

Bissetone

B-60174

$C_9H_{14}O_5$ M 202.207
Constit. of gorgonian *Briareum polyanthes*. Antimicrobial agent. Oil. $[\alpha]_D$ −43.6° (c, 4.25 in EtOH).

Cardellina, J.H. *et al, Tetrahedron Lett.*, 1987, **28**, 727 (*cryst*
   struct)

1,2-Bis(trifluoromethyl)benzene, 9CI   B-60175

α,α,α,α′,α′,α′-Hexafluoro-o-xylene
[433-95-4]

$C_8H_4F_6$ M 214.110
Liq. Bp 143°.

Hasek, W.R. et al, J. Am. Chem. Soc., 1960, **82**, 543 (synth)
Kobayashi, Y. et al, Chem. Pharm. Bull., 1970, **18**, 2334 (synth)
Takahashi, K. et al, Bull. Chem. Soc. Jpn., 1985, **58**, 755 (pmr, cmr, F nmr)

1,3-Bis(trifluoromethyl)benzene, 9CI   B-60176

α,α,α,α′,α′,α′-Hexafluoro-m-xylene
[402-31-3]

$C_8H_4F_6$ M 214.110
Liq. d 1.378. Bp 116-116.3°.

Kobayashi, Y. et al, Chem. Pharm. Bull., 1970, **18**, 2334 (synth)
Takahashi, K. et al, Bull. Chem. Soc. Jpn., 1985, **58**, 755 (pmr, cmr, F nmr)

1,4-Bis(trifluoromethyl)benzene, 9CI   B-60177

α,α,α,α′,α′,α′-Hexafluoro-p-xylene
[433-19-2]

$C_8H_4F_6$ M 214.110
Liq. d 1.381. Bp 116° (113-115°).

Hasek, W.R. et al, J. Am. Chem. Soc., 1960, **82**, 543 (synth)
Takahashi, K. et al, Bull. Chem. Soc. Jpn., 1985, **58**, 755 (pmr, cmr, F nmr)

3,3-Bis(trifluoromethyl)-3H-diazirine   B-60178

[3024-50-8]

$C_3F_6N_2$ M 178.037
Source of $(CF_3)_2\ddot{C}$ carbene. Gas. Bp −12°.

Gale, D.M. et al, J. Am. Chem. Soc., 1966, **88**, 3617 (synth, F nmr)

3,4-Bis(trifluoromethyl)-1,2-dithiete, 9CI   B-60179

[360-91-8]

$C_4F_6S_2$ M 226.154
Malodorous yellow liq. Mp ca. −79°. Bp 99° (93-94°). Dimerises over several days.

Krespan, C.G. et al, J. Am. Chem. Soc., 1961, **83**, 3434, 3438 (synth, ir, props)

N,N-Bis(trifluoromethyl)hydroxylamine, 8CI   B-60180

1,1,1-Trifluoro-N-hydroxy-N-(trifluoromethyl)-methanamine, 9CI

[359-63-7]

$(F_3C)_2NOH$

C_2HF_6NO M 169.026
Precursor to the nitroxide. Liq. d_4^4 1.606. Bp 36° (32.5°). n_D^4 1.2600.

Me ether: [22264-96-6]. 1,1,1,1′,1′,1′-Hexafluoro-N-methoxydimethylamine, 8CI.
$C_3H_3F_6NO$ M 183.053
Liq. Bp 19°.

Et ether: [32872-29-0]. N-Ethoxy-1,1,1,1′,1′,1′-hexa-fluorodimethylamine, 8CI.
$C_4H_5F_6NO$ M 197.080
Liq. Bp not recorded.

O-Nitroso: [359-75-1]. 1,1,1-Trifluoro-N-(nitrosooxy)-N-(trifluoromethyl)methanamine, 9CI. O-Nitroso-N,N-bis(trifluoromethyl)hydroxylamine, 8CI.
$C_2F_6N_2O_2$ M 198.025
Red-brown liq. Bp 10°.

Blackley, W.D. et al, J. Am. Chem. Soc., 1965, **87**, 802 (synth, F nmr)
Makarov, S.P. et al, Dokl. Akad. Nauk SSSR, Ser. Sci. Khim., 1965, **160**, 1319 (synth)
Banks, R.E. et al, J. Chem. Soc., Chem. Commun., 1967, 413 (synth, use)
Banks, R.E. et al, J. Fluorine Chem., 1978, **12**, 27 (synth)
Compton, D.A. et al, J. Phys. Chem., 1981, **85**, 3093 (ir, raman, conformn)

2,3-Bis(trifluoromethyl)pyridine, 8CI   B-60181

[1644-68-4]

$C_7H_3F_6N$ M 215.098
Mp −2°. Bp 158°. V. low basicity. Does not form a hydrochloride.

Soon, N.G. et al, J. Chem. Phys., 1964, **40**, 2090 (F nmr)
Kobayashi, Y. et al, Chem. Pharm. Bull., 1967, **15**, 1896, 1901 (synth, ir, ms)

2,4-Bis(trifluoromethyl)pyridine, 8CI   B-60182

[454-99-9]
$C_7H_3F_6N$ M 215.098
Bp 121-122°. V. weak base, does not form a hydrochloride.

1-Oxide: [22253-61-8].
$C_7H_3F_6NO$ M 231.097
Oil. Bp_{20} 140°.

Kobayashi, Y. et al, Chem. Pharm. Bull., 1967, **15**, 1896, 1901; 1969, **15**, 510 (synth, ir, ms, uv)

2,5-Bis(trifluoromethyl)pyridine, 8CI   B-60183

[20857-44-7]
$C_7H_3F_6N$ M 215.098
Mp 27-28°. Bp 130-131°.

1-Oxide: [22253-60-7].
$C_7H_3F_6NO$ M 231.097
Prisms (C_6H_6/hexane). Mp 90-91°.

Kobayashi, Y. et al, Chem. Pharm. Bull., 1967, **15**, 1896, 1901; 1969, **17**, 510 (synth, uv, pmr, ir, ms)

2,6-Bis(trifluoromethyl)pyridine, 9CI   B-60184

[455-00-5]

$C_7H_3F_6N$ M 215.098

Solid (EtOH aq.). Mp 55-56°.

1-Oxide: [22245-81-4].

$C_7H_3F_6NO$ M 231.097

Solid (C_6H_6/pet. ether). Mp 100°.

Kobayashi, Y. *et al, Chem. Pharm. Bull.,* 1967, **15**, 1896, 1901 (*synth, ir, ms*)
Kobayashi, Y. *et al, Chem. Pharm. Bull.,* 1969, **17**, 510 (*oxide*)
Shustov, L.D. *et al, J. Gen. Chem. USSR (Engl. Transl.),* 1983, **53**, 85 (*synth*)

3,4-Bis(trifluoromethyl)pyridine, 8CI   B-60185

[20857-46-9]

$C_7H_3F_6N$ M 215.098

Bp 132-133°. V. weak base, does not form a hydrochloride.

Kobayashi, Y. *et al, Chem. Pharm. Bull.,* 1967, **15**, 1896, 1901 (*synth, ir, ms*)

3,5-Bis(trifluoromethyl)pyridine, 9CI   B-60186

[20857-47-0]

$C_7H_3F_6N$ M 215.098

Mp 35-36°. Bp 117-118°. V. weak base, does not form hydrochloride.

Kobayashi, Y. *et al, Chem. Pharm. Bull.,* 1967, **15**, 1896, 1901 (*synth, ir, ms*)

Bis(trifluoromethyl)thioketene   B-60187

3,3,3-Trifluoro-2-(trifluoromethyl)-1-propene-1-thione, 9CI. Thiobis(trifluoromethyl)ketene, 8CI

[7445-60-5]

$$(F_3C)_2C{=}C{=}S$$

C_4F_6S M 194.094

Stable, reactive intermediate. Red-orange liq., freezes to a yellow solid. d_4^{25} 1.462. Mp −55°. Bp 52°. n_D^{25} 1.3495. Dimerises with trace of Lewis base. Polymerises at < −20°.

Raasch, M.S., *J. Org. Chem.,* 1970, 3470 (*synth, use, ir, F nmr*)
Gleiter, R. *et al, Chem. Ber.,* 1983, **116**, 2888 (*pe*)

2,4-Bis[2,2,2-Trifluoro-1-(trifluoromethyl)-ethylidene]-1,3-dithietane, 9CI   B-60188

[7445-61-6]

$C_8F_{12}S_2$ M 388.189

Precursor to, and dimer of, Bis(trifluoromethyl)-thioketene, B-60187 . Mp 85°. Bp 170°.

Raasch, M.S., *J. Chem. Soc., Chem. Commun.,* 1966, 577 (*synth, F nmr, ir, props*)

Birum, G.H. *et al, J. Org. Chem.,* 1967, **32**, 3554 (*synth*)
Krespan, C.G. *et al, J. Org. Chem.,* 1968, **33**, 1850 (*synth*)
Raasch, M.S., *J. Org. Chem.,* 1970, **35**, 3470 (*synth, F nmr*)
Sterling, S.R. *et al, Izv. Akad. Nauk SSSR, Ser. Khim.,* 1971, 2517 (*Engl. transl.* p. 2368) (*synth*)

4,4′-Bi-4*H*-1,2,4-triazole, 8CI   B-60189

[16227-15-9]

$C_4H_4N_6$ M 136.116

Cryst. (EtOH). Mp 268° dec.

Bartlett, R.K. *et al, J. Chem. Soc. (C),* 1967, 1664 (*synth*)

Biuret   B-60190

Updated Entry replacing B-01765

Imidodicarbonic diamide, 9CI. Ureidoformamide. Allophanamide

[108-19-0]

$$H_2NCONHCONH_2$$

$C_2H_5N_3O_2$ M 103.080

Cryst. from water as $5C_2H_5N_3O_2.4H_2O$ which dehydrates at 110°. Foaming agent for plastics, fireproofing for synthetic fibres. Cryst. (H_2O, EtOH). Mp 193° dec. Gives violet biuret reacn.

N-*Ac:*

$C_4H_7N_3O_2$ M 129.118

Cryst. (EtOH). Mp 193-194°.

N-*Benzoyl:*

$C_9H_9N_3O_3$ M 207.188

Cryst. (H_2O). Mp 223-224°.

N-*Me:* [6937-91-3]. *Allophanic methylamide.*

$C_3H_7N_3O_2$ M 117.107

Mp 167-168°.

1,3,5-Tri-Me: [816-00-2].

$C_5H_{11}N_3O_2$ M 145.161

Solid. Mp 125-126°.

1,1,3,5-Tetra-Me: [54070-65-4].

$C_6H_{13}N_3O_2$ M 159.188

Yellow oil.

Thiele, J. *et al, Justus Liebigs Ann. Chem.,* 1898, **303**, 93 (*synth*)
Biltz, H. *et al, Ber.,* 1923, **56**, 1914.
Kurzer, F., *Chem. Rev.,* 1956, **56**, 95 (*rev*)
Piskala, A., *Tetrahedron Lett.,* 1964, 2587 (*derivs*)
D'Silva, T.D.J. *et al, J. Org. Chem.,* 1986, **51**, 3781 (*derivs, ir, pmr, cmr*)

Blastmycetin *D*   B-60191

[110064-64-7]

$C_{29}H_{45}N_3O_3$ M 483.693

Isol. from *Streptoverticillium blastmyceticum*. Amorph. $[\alpha]_D^{23}$ −51° (c, 0.27 in MeOH).

Irie, K. *et al*, *Agric. Biol. Chem.*, 1987, **51**, 1733 (*isol, struct*)

Bonducellin
B-60192

Updated Entry replacing B-20137

2,3-Dihydro-7-hydroxy-3-[4-methoxyphenyl]-methylene]-4H-1-benzopyran-4-one, 9CI

[83162-84-9]

$C_{17}H_{14}O_4$ M 282.295

Homoisoflavone from *Caesalpinia bonducella* and *C. pulcherrima*. Yellow needles (CHCl$_3$/MeOH). Mp 208° (205°).

8-Methoxy: 8-Methoxybonducellin.
$C_{18}H_{16}O_5$ M 312.321
Isol. from *C. pulcherrima*. Yellow needles (CHCl$_3$/pet. ether). Mp 124-125°.

Purushothaman, K.K. *et al*, *Indian J. Chem., Sect. B*, 1982, **21**, 383 (*isol*)
McPherson, D.D. *et al*, *Phytochemistry*, 1983, **22**, 2835 (*isol*)
Parmar, V.S. *et al*, *Acta Chem. Scand., Ser. B*, 1987, **41**, 267 (*synth*)

Boronolide
B-60193

Updated Entry replacing B-50299

5,6-Dihydro-6-[1,2,3-tris(acetyloxy)heptyl]-2H-pyran-2-one, 9CI

[33903-83-2]

$C_{18}H_{26}O_8$ M 370.399

Constit. of *Tetradenia fruticosa*. Mp 90°. $[\alpha]_D^{26}$ +28° (c, 0.08 in EtOH).

Franca, N.C. *et al*, *C. R. Hebd. Seances Acad. Sci.*, 1971, **273**, 439 (*isol, struct*)
Kjaer, A. *et al*, *Acta Chem. Scand., Ser. B*, 1985, **39**, 745 (*cryst struct*)
Davies-Coleman, M.T. *et al*, *Phytochemistry*, 1987, **26**, 3047 (*pmr, cmr, struct*)

Botryodiplodin
B-60194

Updated Entry replacing B-50300

1-(Tetrahydro-5-hydroxy-4-methyl-3-furanyl)-ethanone, 9CI. 4-Acetyl-2-hydroxy-3-methyltetrahydrofuran

[27098-03-9]

Absolute configuration

$C_7H_{12}O_3$ M 144.170

Antibiotic metab. of *Botryodiplodia theobromae* and of *Penicillium carneoluteus* and *P. roquefortii*. Shows antifungal and antileukaemic activity. Mycotoxin. Cryst. (Et$_2$O). Mp 50-52°. Anomerises in soln. to a mixt. of 2-epimers.

▷OB6005000.

Ac: Cryst. (Et$_2$O). Mp 45-47°. $[\alpha]_D$ −104° (c, 0.09 in CHCl$_3$).

McCurry, P.M. *et al*, *J. Am. Chem. Soc.*, 1973, **95**, 5824; *Tetrahedron Lett.*, 1973, 4103 (*synth, stereochem*)
Wilson, S.R. *et al*, *J. Org. Chem.*, 1975, **40**, 3309 (*synth*)
Sakai, K. *et al*, *CA*, 1979, **90**, 186692p (*synth, abs config*)
Moreau, S. *et al*, *J. Org. Chem.*, 1982, **47**, 2358 (*isol, cryst struct*)
Renauld, F. *et al*, *Tetrahedron*, 1984, **40**, 1823; 1985, **41**, 955 (*biosynth, props*)
Kurth, M.J. *et al*, *J. Org. Chem.*, 1985, **50**, 1840 (*synth*)
Rehnberg, N. *et al*, *Tetrahedron Lett.*, 1987, **28**, 3589 (*synth*)

Brachycarpone
B-60195

Updated Entry replacing B-50303

[103223-12-7]

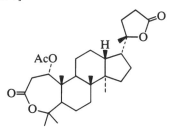

$C_{29}H_{44}O_6$ M 488.663

Constit. of *Cleome brachycarpa*. Cryst. (Et$_2$O). Mp 228°. $[\alpha]_D$ +25.6° (c, 1 in CHCl$_3$).

Deacetoxy: [106709-91-5]. **Deacetoxybrachycarpone.**
$C_{27}H_{42}O_4$ M 430.626
Constit. of *C. brachycarpa*. Cryst. Mp 185-186°. $[\alpha]_D$ +47° (CHCl$_3$).

Ahmad, V.U. *et al*, *J. Nat. Prod.*, 1986, **49**, 249 (*isol*)
Ahmad, V.U. *et al*, *Phytochemistry*, 1987, **26**, 315 (*deriv*)

Brachynereolide
B-60196

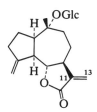

$C_{21}H_{30}O_8$ M 410.463

Constit. of *Brachylaena nereifolia*. Gum.

Tetra-Ac: Cryst. Mp 83°.
*11β,13-Dihydro: **11β,13-Dihydrobrachynereolide.***
$C_{21}H_{32}O_8$ M 412.479
Constit. of *B. nereifolia*. Gum.
11β,13-Dihydro, Tetra-Ac: Cryst. Mp 185-186°.

Zdero, C. *et al*, *Phytochemistry*, 1987, **26**, 2597.

Brasudol
B-60197

5-Bromodecahydro-α,α,4a-trimethyl-8-methylene-2-naphthalenemethanol, 9CI. 1β-Bromo-4(15)-eudesmen-11-ol

[72154-35-9]

$C_{15}H_{25}BrO$ M 301.266

Isol. from *Aplysia brasiliana*. Exhibits potent fish anti-feedant activity. Mp 105-106°. $[\alpha]_D^{26}$ +16.5°.

2-Epimer: [72154-34-8]. **Isobrasudol**.
$C_{15}H_{25}BrO$ M 301.266
From *A. brasiliana*. Mp 105-107°. $[\alpha]_D^{26}$ +10.3°.

Dieter, R.K. *et al*, *Tetrahedron Lett.*, 1979, 1645 (*isol, struct*)

Brevifloralactone   B-60198

[110668-26-3]

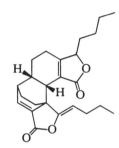

$C_{20}H_{28}O_4$ M 332.439
Constit. of *Salvia breviflora*. Cryst. Mp 126°. $[\alpha]_D^{20}$ +0.232° (c, 2.3 in MeOH).

Ac: Brevifloralactone acetate.
$C_{22}H_{30}O_5$ M 374.476
Constit. of *S. breviflora*. Cryst. Mp 112°.

Cuevas, G. *et al*, *Phytochemistry*, 1987, **26**, 2019.

2-Bromo-1,3,5-benzenetriol, 9CI   B-60199

2-Bromophloroglucinol
[84743-77-1]

$C_6H_5BrO_3$ M 205.008
Isol. from *Rhabdonia verticillata*.

Tri-Ac: [96820-10-9].
$C_{12}H_{11}BrO_6$ M 331.119
Isol. from *Eisenia arborea*. No phys. props. reported.
Tri-Me ether: [1131-40-4]. *2-Bromo-1,3,5-trimethoxy-benzene*, 9CI.
$C_9H_{11}BrO_3$ M 247.088
Cryst. (Et$_2$O). Mp 99°.

Blackman, A.J. *et al*, *Phytochemistry*, 1982, **21**, 2141 (*isol*)
Fischer, A. *et al*, *Can. J. Chem.*, 1983, **61**, 1045 (*deriv*)
Glombitza, K.W. *et al*, *Phytochemistry*, 1985, **24**, 543 (*deriv*)

4-Bromo-1,2,3-benzenetriol, 9CI   B-60200

4-Bromopyrogallol, 8CI
[17345-72-1]
$C_6H_5BrO_3$ M 205.008
Mp 119-120°.

1,3-Di-Me ether: [18111-34-7]. *3-Bromo-2,6-dimethox-yphenol, 8CI.*
$C_8H_9BrO_3$ M 233.061
Bp$_{0.4}$ 88°.
Tri-Me ether: [10385-36-1]. *1-Bromo-2,3,4-trimethoxy-benzene*, 9CI.
$C_9H_{11}BrO_3$ M 247.088
Bp$_{13}$ 148°, Bp$_{0.5}$ 97°.

Friedman, D. *et al*, *J. Org. Chem.*, 1958, **23**, 16 (*synth*)
Goldman, J. *et al*, *Tetrahedron*, 1973, **29**, 3833 (*deriv*)

5-Bromo-1,2,3-benzenetriol, 9CI   B-60201

5-Bromopyrogallol, 8CI
[16492-75-4]
$C_6H_5BrO_3$ M 205.008
Off-white needles (CHCl$_3$). Mp 148°.

1,3-Di-Me ether: [70654-71-6]. *4-Bromo-2,6-dimethox-yphenol, 9CI.*
$C_7H_7BrO_3$ M 219.035
Mp 99.5-100°.
Tri-Me ether: [2675-79-8]. *5-Bromo-1,2,3-trimethoxy-benzene, 9CI.*
$C_9H_{11}BrO_3$ M 247.088
Cryst. (hexane). Mp 78-81°.

Critchlow, A. *et al*, *Tetrahedron*, 1967, **23**, 2829 (*synth*)
Jung, M.E. *et al*, *J. Org. Chem.*, 1985, **50**, 1087 (*derivs*)

2-Bromobenzocyclopropene   B-60202

2-Bromobicyclo[4.1.0]hepta-1,3,5-triene

C_7H_5Br M 169.021
Oil.

Halton, B. *et al*, *Aust. J. Chem.*, 1987, **40**, 475 (*synth, ir, pmr, cmr*)

3-Bromobenzocyclopropene   B-60203

Updated Entry replacing B-01938
3-Bromobicyclo[4.1.0]hepta-1,3,5-triene, 9CI
[63370-07-0]
C_7H_5Br M 169.021
λ_{max} 273, 278, 285 nm (EtOH).

Billups, W.E. *et al*, *Tetrahedron Lett.*, 1977, 571 (*synth, pmr, ir, uv*)
Halton, B. *et al*, *Aust. J. Chem.*, 1987, **40**, 475 (*synth, ir, pmr, cmr*)

2-Bromobenzo[*b*]thiophene, 9CI   B-60204

[5394-13-8]

C_8H_5BrS M 213.092
Pale buff plates (EtOH aq.). Mp 41-42°. Bp$_{20}$ 135-138°.

1-Oxide: [57147-27-0].
C_8H_5BrOS M 229.091
Mp 79°.
1,1-Dioxide: [5350-05-0].
$C_8H_5BrO_2S$ M 245.090
Needles (EtOH). Mp 150-151°.

Faller, P., *Bull. Soc. Chim. Fr.*, 1966, 3618 (*synth*)
Acheson, R.M. *et al*, *J. Chem. Soc. (C)*, 1970, 1764 (*synth*)
Geneste, P. *et al*, *Bull. Soc. Chim. Fr.*, 1977, 271 (*oxides*)
Geneste, P. *et al*, *J. Org. Chem.*, 1979, **44**, 2887 (*cmr*)

3-Bromobenzo[*b*]thiophene, 9CI   B-60205

[7342-82-7]
C_8H_5BrS M 213.092
Pink liq. Bp$_{13}$ 137-138°.

Picrate: Long yellow needles (EtOH). Mp 115-116°.

1-Oxide: [57147-26-9].
  C_8H_5BrOS M 229.091
  Mp 141°.
1,1-Dioxide: [16957-97-4].
  $C_8H_5BrO_2S$ M 245.090
  Needles (EtOH). Mp 183.5-184°.

Faller, P., *Bull. Soc. Chim. Fr.*, 1966, 3618 (*synth*)
Acheson, R.M. *et al*, *J. Chem. Soc. (C)*, 1970, 1764 (*synth, bibl*)
Geneste, P. *et al*, *Bull. Soc. Chim. Fr.*, 1977, 271 (*derivs*)
Geneste, P. *et al*, *J. Org. Chem.*, 1979, **44**, 2887 (*cmr*)

4-Bromobenzo[*b*]thiophene, 9CI   B-60206

[5118-13-8]
C_8H_5BrS M 213.092
Mp 31-32°. Bp_{20} 144°, Bp_5 115.5-117°.
Picrate: Mp 135-136°.
1,1-Dioxide:
  $C_8H_5BrO_2S$ M 245.090
  Cryst. (EtOH). Mp 147°.

Faller, P., *Bull. Soc. Chim. Fr.*, 1966, 3667 (*synth*)
Titus, R.L. *et al*, *J. Heterocycl. Chem.*, 1967, **4**, 651 (*synth*)
Matsuki, Y. *et al*, *CA*, 1968, **69**, 59018 (*synth*)

5-Bromobenzo[*b*]thiophene, 9CI   B-60207

[4923-87-9]
C_8H_5BrS M 213.092
Plates. Mp 47-48°. Bp_{12} 142-143°, $Bp_{0.5}$ 50-52°.
Picrate: Yellow needles (EtOH). Mp 83.5°.
1,1-Dioxide:
  $C_8H_5BrO_2S$ M 245.090
  Needles (EtOH aq.). Mp 144-144.5°.

Banfield, J.E. *et al*, *J. Chem. Soc.*, 1956, 2603 (*synth*)
Badger, G.M. *et al*, *J. Chem. Soc.*, 1957, 2624 (*synth*)
Faller, P., *Bull. Soc. Chim. Fr.*, 1966, 3667 (*synth*)
Amin, H.B. *et al*, *J. Chem. Soc., Perkin Trans. 2*, 1982, 1489 (*synth*)

6-Bromobenzo[*b*]thiophene, 9CI   B-60208

[17347-32-9]
C_8H_5BrS M 213.092
Mp 57°.
Picrate: Mp 69-72°.
1,1-Dioxide:
  $C_8H_5BrO_2S$ M 245.090
  Cryst. (EtOH). Mp 139°.

Titus, R.L. *et al*, *J. Heterocycl. Chem.*, 1967, **4**, 651 (*synth*)
Amin, H.B. *et al*, *J. Chem. Soc., Perkin Trans. 2*, 1982, 1489 (*synth*)

7-Bromobenzo[*b*]thiophene, 9CI   B-60209

[1423-61-6]
C_8H_5BrS M 213.092
Bp_{10} 108-109°, $Bp_{0.4}$ 70-72°.
Picrate: Mp 144-145°.

Faller, P., *Bull. Soc. Chim. Fr.*, 1966, 3667 (*synth*)
Amin, H.B. *et al*, *J. Chem. Soc., Perkin Trans. 2*, 1982, 1489 (*synth*)

2-Bromobicyclo[2.2.1]heptane-1-carboxylic acid   B-60210

2-Bromo-1-norbornanecarboxylic acid

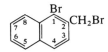

$(1R,2R)$-*form*

$C_8H_{11}BrO_2$ M 219.078
(**1R,2R**)-*form* [104113-37-3]
  (−)-exo-*form*
Mp 114°. $[\alpha]_D^{20}$ −72.5°.
(**1S,2S**)-*form* [104113-38-4]
  (+)-exo-*form*
Mp 112-114°. $[\alpha]_D^{20}$ +74.9°.

Müller, S. *et al*, *J. Chem. Soc., Chem. Commun.*, 1986, 297 (*synth, abs config*)

3-Bromo[1,1′-biphenyl]-4-carboxylic acid, 9CI   B-60211

Updated Entry replacing B-01975
2-Bromo-4-phenylbenzoic acid
$C_{13}H_9BrO_2$ M 277.117
Cryst. (MeOH/pet. ether). Mp 179-180°.
Nitrile: 3-Bromo-4-cyanobiphenyl.
  $C_{13}H_8BrN$ M 258.117
  Yellow needles (EtOH). Bp_3 180-188°.

Case, F.H., *J. Am. Chem. Soc.*, 1936, **58**, 1249 (*synth*)
Hori, M. *et al*, *J. Chem. Soc., Perkin Trans. 1*, 1987, 187 (*synth, deriv, ir*)

1-Bromo-2-(bromomethyl)naphthalene   B-60212

[37763-43-2]

$C_{11}H_8Br_2$ M 299.992
Cubes (hexane). Mp 107-109°.
▷QJ1550000.

Weber, E. *et al*, *J. Am. Chem. Soc.*, 1984, **106**, 3297 (*synth, ir, pmr*)
Smith, J.G. *et al*, *J. Org. Chem.*, 1986, **51**, 3762 (*synth, ir, pmr*)
Abeywickrema, A.N. *et al*, *J. Org. Chem.*, 1987, **52**, 4072 (*synth, pmr*)

1-Bromo-4-(bromomethyl)naphthalene   B-60213

[79996-99-9]
$C_{11}H_8Br_2$ M 299.992
Cryst. Mp 103-104°.

Boekelheide, V. *et al*, *J. Am. Chem. Soc.*, 1954, **76**, 604 (*synth*)
Mamalis, P. *et al*, *J. Chem. Soc.*, 1962, 3915 (*synth*)
Dixon, E.A. *et al*, *Can. J. Chem.*, 1981, **59**, 2629 (*synth*)

1-Bromo-5-(bromomethyl)naphthalene   B-60214

$C_{11}H_8Br_2$ M 299.992
Needles (pet. ether). Mp 101°.

Shoesmith, J.B. *et al*, *J. Chem. Soc.*, 1928, 3098 (*synth*)

1-Bromo-7-(bromomethyl)naphthalene   B-60215

[98331-27-2]

$C_{11}H_8Br_2$ M 299.992

Duchêne, K.H. *et al*, *Angew. Chem., Int. Ed. Engl.*, 1985, **24**, 885 (*synth*)

1-Bromo-8-(bromomethyl)naphthalene   B-60216

[72758-17-9]

$C_{11}H_8Br_2$ M 299.992
Cryst. (Et$_2$O/pentane). Mp 76-77°.

Karabatsos, G.J. *et al*, *Tetrahedron Lett.*, 1964, 2113.
Kiely, J.S. *et al*, *J. Organomet. Chem.*, 1979, **182**, 173 (*synth, ir, pmr, ms*)

2-Bromo-3-(bromomethyl)naphthalene   B-60217

[38399-20-1]

$C_{11}H_8Br_2$ M 299.992
Leaflets (EtOH). Mp 111-112°.

Staab, H.A. *et al*, *Chem. Ber.*, 1972, **105**, 2290 (*synth*)
Smith, J.G. *et al*, *J. Org. Chem.*, 1986, **51**, 3762 (*synth*)

2-Bromo-6-(bromomethyl)naphthalene   B-60218

$C_{11}H_8Br_2$ M 299.992
Cryst. Mp 124-125°.

Jones, R.G. *et al*, *J. Am. Chem. Soc.*, 1948, **70**, 2843 (*synth*)

3-Bromo-1-(bromomethyl)naphthalene   B-60219

[80063-50-9]

$C_{11}H_8Br_2$ M 299.992
Cryst. Mp 94-96°.

Dixon, E.A. *et al*, *Can. J. Chem.*, 1981, **59**, 2629 (*synth*)

6-Bromo-1-(bromomethyl)naphthalene   B-60220

[86456-69-1]

$C_{11}H_8Br_2$ M 299.992
Cryst. (C$_6$H$_6$/pet. ether). Mp 109-110.5°.

Newman, M.S. *et al*, *J. Org. Chem.*, 1983, **48**, 2926 (*synth*)

7-Bromo-1-(bromomethyl)naphthalene   B-60221

[81830-68-4]

$C_{11}H_8Br_2$ M 299.992

Newman, M.S. *et al*, *J. Org. Chem.*, 1962, **27**, 76; 1982, **47**, 2837 (*synth*)

5-(3-Bromo-4-chloro-4-methylcyclohexyl)-   B-60222
5-methyl-2(5H)-furanone

Absolute configuration

$C_{12}H_{16}BrClO_2$ M 307.614
Metabolite of red alga *Laurencia caespitosa*. Mp 122-123°. [α]$_D$ +16.0° (c, 0.23 in CHCl$_3$).

Estrada, D.M. *et al*, *Tetrahedron Lett.*, 1987, **28**, 687 (*cryst struct*)

3-Bromo-5-chloropyridine   B-60223

[73583-39-8]

C_5H_3BrClN M 192.442

Mallet, M. *et al*, *Tetrahedron*, 1979, **35**, 1625 (*synth*)

4-Bromo-3-chloropyridine   B-60224

[73583-41-2]

C_5H_3BrClN M 192.442
Cryst. Mp 71-72°.
Picrate: Cryst. (EtOH). Mp 126°.

Talik, T. *et al*, *Rocz. Chem.*, 1962, **36**, 417, 1049 (*synth*)

5-Bromo-5-decene, 9CI   B-60225

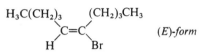

(*E*)-*form*

$C_{10}H_{19}Br$ M 219.164
(*E*)-*form* [72612-74-9]
Bp$_{0.9}$ 64-66°. n_D^{20} 1.4693.
(*Z*)-*form* [72612-75-0]
Bp$_{19}$ 88°, Bp$_{0.9}$ 62-64°. n_D^{22} 1.4645.

Brown, H.C. *et al*, *Synthesis*, 1986, 480 (*synth, pmr*)
Tamao, K. *et al*, *J. Org. Chem.*, 1987, **52**, 1100 (*synth, pmr, ms*)

2-Bromo-3,5-dichlorophenol   B-60226

[13659-22-8]

$C_6H_3BrCl_2O$ M 241.899
Prisms. Mp 80°.

Soma, T. *et al*, *CA*, 1967, **66**, 65223 (*synth*)
Tuzun, C. *et al*, *CA*, 1980, **93**, 7761c (*deriv*)

2-Bromo-4,5-dichlorophenol   B-60227

[2316-56-5]

$C_6H_3BrCl_2O$ M 241.899
Mp 70-71°.

Gump, W.S. *et al*, *CA*, 1965, **62**, 14546d (*synth*)

3-Bromo-2,4-dichlorophenol   B-60228

[13659-21-7]

$C_6H_3BrCl_2O$ M 241.899
Needles. Mp 85° (78°).

Soma, T. *et al*, *CA*, 1967, **66**, 65223g (*synth*)
Tashiro, M. *et al*, *J. Org. Chem.*, 1980, **43**, 196 (*synth*)

4-Bromo-2,3-dichlorophenol   B-60229

[1940-44-9]

$C_6H_3BrCl_2O$ M 241.899
Me ether: [109803-52-3]. *1-Bromo-2,3-dichloro-4-methoxybenzene. 4-Bromo-2,3-dichloroanisole.*
$C_7H_5BrCl_2O$ M 255.926
Mp 74-76°.

Soma, T. *et al*, *CA*, 1967, **66**, 65223g (*synth*)
Wyrick, S.D. *et al*, *J. Med. Chem.*, 1987, **30**, 1798 (*synth, deriv, pmr*)

4-Bromo-3,5-dichlorophenol   B-60230

[1940-28-9]

$C_6H_3BrCl_2O$ M 241.899

Mp 121-123°.

Gump, W.S. *et al, CA*, 1965, **62**, 14546d (*synth*)
Soma, T. *et al, CA*, 1967, **66**, 65223g (*synth*)
Eur. Pat., 8 061, (*1980*); *CA*, **93**, 114125v

1-Bromo-1,4-dihydro-1,4-methanonaphtha-   B-60231
lene, 9CI

1-Bromobenzonorbornadiene. 1-Bromo-2,3-benzobicyclo[2.2.1]hepta-2,5-diene

[23537-80-6]

$C_{11}H_9Br$ M 221.096

(±)-*form*

Bp$_{1.5}$ 90-93°.

Wilt, J.W. *et al, J. Org. Chem.*, 1970, **35**, 1562 (*synth, ir, pmr*)
Chenier, P.J. *et al, J. Org. Chem.*, 1973, **38**, 4350 (*synth*)

2-Bromo-1,4-dihydro-1,4-methanonaphtha-   B-60232
lene, 9CI

2-Bromobenzonorbornadiene

[23537-79-3]

$C_{11}H_9Br$ M 221.096

(±)-*form*

Yellow oil. Bp$_{0.9}$ 80.5-81°.

Wilt, J.W. *et al, J. Org. Chem.*, 1970, **35**, 1562 (*synth, ir, pmr*)

9-Bromo-1,4-dihydro-1,4-methanonaphtha-   B-60233
lene, 9CI

9-Bromobenzonorbornadiene. 7-Bromobenzonorbornadiene

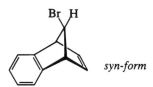

syn-form

$C_{11}H_9Br$ M 221.096

syn-form [22436-26-6]
Cryst. (Et$_2$O/hexane). Mp 133.5-135°.
anti-form [7605-10-9]
Cryst. (EtOH). Mp 58-59°. Bp$_1$ 94-96°.

Wilt, J.W. *et al, J. Org. Chem.*, 1967, **32**, 893; 1970, **35**, 1562, 1571 (*synth, uv, ir, pmr*)
Cristol, S.J. *et al, J. Org. Chem.*, 1967, **32**, 3727 (*synth*)

7-Bromo-3,4-dihydro-1(2H)-naphthalen-   B-60234
one, 9CI

Updated Entry replacing B-40172
7-Bromo-1-tetralone

[32281-97-3]

$C_{10}H_9BrO$ M 225.085

Cryst. (pet. ether). Mp 77-78.5°.

Newman, M.S. *et al, J. Org. Chem.*, 1962, **27**, 76 (*synth*)
Griffin, R.W. *et al, J. Org. Chem.*, 1964, **29**, 2109 (*synth*)

Adachi, K. *et al, CA*, 1972, **76**, 141055c (*synth*)
Buckle, D.R. *et al, J. Med. Chem.*, 1977, **20**, 1059 (*synth*)

1-Bromo-2,3-dimethylcyclopropene   B-60235

[87619-34-9]

C_5H_7Br M 147.014

(±)-*form*

Liq. Bp$_{14}$ 19°.

Baird, M.S. *et al, J. Chem. Soc., Perkin Trans. 1*, 1986, 1845 (*synth, pmr, ir*)

4-Bromodiphenylmethanol   B-60236

4-Bromo-α-phenylbenzenemethanol, 9CI. 4-Bromophenylphenylmethanol. p-Bromobenzhydrol

[29334-16-5]

(S)-form

$C_{13}H_{11}BrO$ M 263.133

(*S*)-*form* [73773-07-6]
Needles (hexane). $[\alpha]_D^{20}$ +21.0° (c, 0.8 in C$_6$H$_6$).
(±)-*form*
Cryst. (pet. ether). Mp 63-65°.

Bachmann, W.E. *et al, J. Org. Chem.*, 1948, **13**, 916 (*synth*)
Capillon, J. *et al, Tetrahedron*, 1979, **35**, 1801 (*synth, abs config*)
Hügel, H.M. *et al, Aust. J. Chem.*, 1979, **32**, 1511 (*cmr*)
de Vries, J.G. *et al, J. Org. Chem.*, 1980, **45**, 4126 (*synth*)
Wu, B. *et al, J. Org. Chem.*, 1986, **51**, 1904 (*synth, abs config*)

1-Bromo-1,2-diphenylpropene   B-60237

1,1'-(1-Bromo-2-methyl-1,2-ethenediyl)bisbenzene, 9CI

(E)-form

$C_{15}H_{13}Br$ M 273.172

(*E*)-*form* [63904-72-3]
Cryst. (EtOH). Isom. to *Z-form* at ca. 140°.
(*Z*)-*form* [63904-73-4]
Cryst. (EtOH). Mp 158-159°.

Koelsch, C.F. *et al, J. Org. Chem.*, 1941, **6**, 602 (*synth*)
Friedrich, L.E. *et al, J. Org. Chem.*, 1978, **43**, 34 (*synth, uv, ir, pmr*)
Kitamura, T. *et al, J. Am. Chem. Soc.*, 1986, **108**, 2641 (*synth, pmr*)

6-Bromo-6-dodecene, 9CI   B-60238

(E)-form

$C_{12}H_{23}Br$ M 247.218

(E)-form [106924-75-8]
Bp$_{0.01}$ 70-72°. n$_D^{20}$ 1.4667.
(Z)-form [106924-82-7]
Bp$_{0.01}$ 68-70°. n$_D^{22}$ 1.4664.
Brown, H.C. *et al*, *Synthesis*, 1986, 480 (*synth, pmr*)

2-Bromo-5,6-epoxy-4-hydroxy-2-cyclo-hexen-1-one   B-60239
3-Bromo-5-hydroxy-7-oxabicyclo[4.1.0]hept-3-en-2-one, 9CI

C$_6$H$_5$BrO$_3$ M 205.008
(4S,5R,6R)-form [110786-63-5]
Metabolite of sea acorn *Ptychodera* sp. Cryst.
(CHCl$_3$/hexane). Mp 123-127°. [α]$_D^{22}$ +220° (c, 0.09 in CHCl$_2$).
Ac: [110786-62-4]. *4-Acetoxy-2-bromo-5,6-epoxy-2-cyclohexen-1-one.*
C$_8$H$_7$BrO$_4$ M 247.045
Metabolite of *P.* sp. Needles (EtOAc/hexane). Mp 93-94°. [α]$_D^{19}$ +265° (c, 0.12 in CHCl$_3$).
Higa, T. *et al*, *Tetrahedron*, 1987, **43**, 1063 (*cryst struct*)

2-Bromo-4-fluorobenzoic acid   B-60240
[1006-41-3]

C$_7$H$_4$BrFO$_2$ M 219.010
Cryst. (EtOH aq.). Mp 171-172.5°.
Me ester: [653-92-9].
C$_8$H$_6$BrFO$_2$ M 233.037
Liq. Bp$_{10}$ 122°.

Dewar, M.J.S. *et al*, *J. Org. Chem.*, 1963, **28**, 1759 (*synth*)
Quang, N.N. *et al*, *Recl. Trav. Chim. Pays-Bas*, 1964, **83**, 1142 (*synth*)
Rajšner, M. *et al*, *Collect. Czech. Chem. Commun.*, 1975, **40**, 719 (*synth*)

2-Bromo-5-fluorobenzoic acid   B-60241
[394-28-5]
C$_7$H$_4$BrFO$_2$ M 219.010
Microcryst. (C$_6$H$_6$). Mp 151°.

Varma, P.S. *et al*, *J. Indian Chem. Soc.*, 1944, **21**, 112.
Quang, N.N. *et al*, *Recl. Trav. Chim. Pays-Bas*, 1964, **83**, 1142 (*synth*)

2-Bromo-6-fluorobenzoic acid   B-60242
[2252-37-1]
C$_7$H$_4$BrFO$_2$ M 219.010
Needles (EtOH aq.). Mp 155°.
Dewar, M.J.S. *et al*, *J. Org. Chem.*, 1963, **28**, 1759 (*synth*)

3-Bromo-4-fluorobenzoic acid   B-60243
[1007-16-5]

C$_7$H$_4$BrFO$_2$ M 219.010
Needles (C$_6$H$_6$). Mp 156° (143°).

Varma, P.S. *et al*, *J. Indian Chem. Soc.*, 1944, **21**, 112 (*synth*)
Quang, N.N. *et al*, *Recl. Trav. Chim. Pays-Bas*, 1964, **83**, 1142 (*synth*)

3-Bromo-4-fluorobenzyl alcohol   B-60244
3-Bromo-4-fluorobenzenemethanol, 9CI
[77771-03-0]

C$_7$H$_6$BrFO   M 205.026
Eur. Pat., 24 612, (*1981*); *CA*, **95**, 42620 (*synth*)

5-Bromo-2-fluorobenzyl alcohol   B-60245
5-Bromo-2-fluorobenzenemethanol, 9CI
[99725-13-0]
C$_7$H$_6$BrFO   M 205.026
Oil. Bp$_{0.4}$ 82-87°.
Stokker, G.E. *et al*, *J. Med. Chem.*, 1986, **29**, 170 (*synth, pmr*)

1-Bromo-2-fluoro-3,5-dinitrobenzene, 9CI   B-60246
[2367-75-1]

C$_6$H$_2$BrFN$_2$O$_4$ M 264.995
Pale-yellow needles (EtOH aq.). Mp 69°.
Channing, D.M. *et al*, *J. Chem. Soc.*, 1953, 2481 (*synth*)

2-Bromo-3-fluorophenol, 9CI, 8CI   B-60247

C$_6$H$_4$BrFO   M 190.999
Liq. Mp <−20°. Bp$_{80}$ 123°.
Me ether: 2-Bromo-1-fluoro-3-methoxybenzene. *2-Bromo-3-fluoroanisole, 8CI.*
C$_7$H$_6$BrFO   M 205.026
Bp$_{755}$ 220°.
Hodgson, H.H. *et al*, *J. Chem. Soc.*, 1929, 1632; 1931, 981.

2-Bromo-4-fluorophenol, 9CI, 8CI   B-60248
[496-69-5]
C$_6$H$_4$BrFO   M 190.999
Mp 42-43°. Bp$_1$ 89°.
Me ether: [452-08-4]. *2-Bromo-4-fluoro-1-methoxybenzene, 9CI. 2-Bromo-4-fluoroanisole, 8CI.*
C$_7$H$_6$BrFO   M 205.026
Liq. Bp$_5$ 79°. n$_D^{20}$ 1.5447.
Finger, G.C. *et al*, *J. Am. Chem. Soc.*, 1959, **81**, 94 (*synth, deriv*)
Nesmeyanov, A.N. *et al*, *Dokl. Chem.*, 1968, **183**, 1098 (*F nmr*)

2-Bromo-6-fluorophenol, 9CI　　　　　　**B-60249**
[2040-89-3]
C_6H_4BrFO　　M 190.999
Mp 47-48°.

Baker, A.W. *et al, Can. J. Chem.*, 1965, **43**, 650 (*synth*)

3-Bromo-4-fluorophenol, 9CI, 8CI　　　　**B-60250**
[27407-11-0]
C_6H_4BrFO　　M 190.999
Liq. Bp$_{0.5}$ 78°.

B.P., 1 265 212, (*1972*); *CA*, **76**, 140531 (*synth*)

4-Bromo-2-fluorophenol, 9CI　　　　　　**B-60251**
[2105-94-4]
C_6H_4BrFO　　M 190.999
Herbicide precursor. Oil. Bp$_{15}$ 90°, Bp$_7$ 79°. n_D^{24} 1.5645.
Methanesulfonate: [15149-00-5]. Bp$_{0.3}$ 115-120°.
Me ether: [2357-52-0]. *4-Bromo-2-fluoro-1-methoxy-*
benzene, 9CI. *4-Bromo-2-fluoroanisole,* 8CI.
C_7H_6BrFO　　M 205.026
Liq. Mp 16°. Bp 207-208°, Bp$_{14}$ 90°. n_D^{20} 1.5448.

Finger, G.C. *et al, J. Am. Chem. Soc.*, 1959, **81**, 94 (*deriv*)
Aymes, D.J. *et al, Bull. Soc. Chim. Fr.*, 1980, 175 (*synth*)
Eur. Pat., 50 219, (*1982*); *CA*, **97**, 127815 (*synth*)
Eur. Pat., 138 359, (*1985*); *CA*, **103**, 195896 (*synth, use*)

2-Bromo-5-fluoropyridine　　　　　　　**B-60252**
[41404-58-4]

C_5H_3BrFN　　M 175.988
Mp 27.5-29°. Bp$_{44}$ 80-83°.

Abramovitch, R.A. *et al, J. Org. Chem.*, 1974, **39**, 1802 (*synth*)

3-Bromo-2-fluoropyridine　　　　　　　**B-60253**
[36178-05-9]
C_5H_3BrFN　　M 175.988
Liq. Bp$_{20}$ 76°.

Mallet, M. *et al, C.R. Hebd. Seances Acad. Sci., Ser. C*, 1972,
　274, 719 (*synth*)

3-Bromo-5-fluoropyridine　　　　　　　**B-60254**
C_5H_3BrFN　　M 175.988
Mp 24.5-25°.

Wibaut, J.P. *et al, Recl. Trav. Chim. Pays-Bas*, 1955, **74**, 1062
　(*synth*)

4-Bromo-3-fluoropyridine　　　　　　　**B-60255**
[2546-52-3]
C_5H_3BrFN　　M 175.988
Bp 163°.
Picrate: [1799-37-7]. Mp 115°.

Talik, T. *et al, Rocz. Chem.*, 1964, **38**, 777; 1968, **42**, 1861
　(*synth*)
Marsaio, F. *et al, Tetrahedron*, 1983, **39**, 2009 (*synth*)

5-Bromo-2-fluoropyridine　　　　　　　**B-60256**
[766-11-0]

C_5H_3BrFN　　M 175.988
Bp$_{15}$ 63°.

Finger, G.C. *et al, J. Org. Chem.*, 1963, **28**, 1666 (*synth*)
Giam, C.-S. *et al, J. Chem. Soc. (B)*, 1970, 1516 (*F nmr*)
Wielgat, J., *Rocz. Chem.*, 1971, **45**, 931 (*synth*)

3-Bromo-2-furancarboxaldehyde　　　　**B-60257**
3-Bromo-2-formylfuran. 3-Bromofurfural
[14757-78-9]

$C_5H_3BrO_2$　　M 174.982
Bp$_{13}$ 97-98°, Bp$_{0.7}$ 60°.

Robba, M. *et al, C.R. Hebd. Seances Acad. Sci., Ser. C*, 1967,
　264, 413 (*synth, ir, pmr*)
Gronowitz, S. *et al, Ark. Kemi*, 1970, **32**, 283 (*synth, pmr*)

4-Bromo-2-furancarboxaldehyde　　　　**B-60258**
4-Bromo-2-formylfuran. 4-Bromofurfural
[21921-76-6]
$C_5H_3BrO_2$　　M 174.982
Cryst. (pentane). Mp 52°. Bp$_4$ 80-83°.

Sornay, R. *et al, Bull. Soc. Chim. Fr.*, 1971, 990 (*synth*)
Antonioletti, R. *et al, J. Chem. Soc., Perkin Trans. 1*, 1985,
　1285 (*synth*)

5-Bromo-2-furancarboxaldehyde　　　　**B-60259**
2-Bromo-5-formylfuran. 5-Bromofurfural
[1899-24-7]
$C_5H_3BrO_2$　　M 174.982
Needles (EtOH aq.). Mp 82°.

Nazarova, Z.N., *Zh. Obshch. Khim.*, 1954, **24**, 575 (*Engl.
　transl. p. 589*) (*synth*)

7-Bromo-1-heptene　　　　　　　　　　**B-60260**
[4117-09-3]

$$BrCH_2(CH_2)_4CH{=}CH_2$$

$C_7H_{13}Br$　　M 177.084
Bp$_{20}$ 77-81°, Bp$_{19}$ 73-74°.

Gaubert, P. *et al, J. Chem. Soc.*, 1937, 1971 (*synth*)
Hideg, K. *et al, J. Chem. Soc., Perkin Trans. 1*, 1986, 1431
　(*synth*)

7-Bromo-5-heptynoic acid　　　　　　　**B-60261**
[41300-60-1]

$$BrCH_2C{\equiv}C(CH_2)_3COOH$$

$C_7H_9BrO_2$　　M 205.051
Intermed. in prostagladin synth.
Me ester: [41349-38-6].
　$C_8H_{11}BrO_2$　　M 219.078
　Oil. Bp$_{1.0}$ 90-94°.

Corey, E.J. *et al, J. Am. Chem. Soc.*, 1973, **95**, 8483 (*synth*)
Theil, F. *et al, J. Prakt. Chem.*, 1985, **327**, 917 (*synth*)
Casy, G. *et al, Tetrahedron*, 1986, **42**, 5849 (*synth, ir, pmr*)

5-Bromo-2-hexanane　　　　　　　　　　**B-60262**
[52355-85-8]

$$H_3CCHBrCH_2CH_2COCH_3$$

$C_6H_{11}BrO$ M 179.056

(±)-*form*

Liq. Bp$_{14}$ 102-104°, Bp$_1$ 56°.

Ethylene acetal:
  $C_8H_{15}BrO_2$ M 223.109
  Yellow-orange liq. Bp$_{11}$ 126-128°.

Mihailovic, M.L. *et al, Tetrahedron,* 1973, **29**, 3675 (*synth, pmr, ir, bibl*)
Cornish, C.A. *et al, J. Chem. Soc., Perkin Trans. 1,* 1985, 2585 (*synth, ir, pmr, ms*)

5-Bromo-1-hexene   B-60263

[4558-27-4]

$$H_2C{=}CHCH_2CH_2CHBrCH_3$$

$C_6H_{11}Br$ M 163.057

(±)-*form*

Liq. Bp 141-143°.

Wood, H.B. *et al, J. Am. Chem. Soc.,* 1953, **75**, 5511 (*synth*)

6-Bromo-1-hexene   B-60264

[2695-47-8]

$$BrCH_2(CH_2)_3CH{=}CH_2$$

$C_6H_{11}Br$ M 163.057
Bp 149-150°, Bp$_{20}$ 47-48°.

Kraus, G.A. *et al, Synthesis,* 1984, 885 (*synth*)
Hideg, K. *et al, J. Chem. Soc., Perkin Trans. 1,* 1986, 1431 (*synth*)

5′-Bromo-2′-hydroxyacetophenone, 8CI   B-60265

Updated Entry replacing B-02490
1-(5-Bromo-2-hydroxyphenyl)ethanone, 9CI
$C_8H_7BrO_2$ M 215.046
Cryst. (EtOH aq.). Mp 62°. Bp$_{16}$ 135-143°.

Me ether: [16740-73-1]. *1-(5-Bromo-2-methoxyphenyl)-
  ethanone, 9CI. 5′-Bromo-2′-methoxyacetophenone.*
  $C_9H_9BrO_2$ M 229.073
  Mp 39°. Bp$_{24}$ 173°.
Me ether, semicarbazone: Mp 205°.

Bégué, J. *et al, Bull. Soc. Chim. Fr.,* 1969, 781.
Srebuik, M. *et al, J. Chem. Soc., Perkin Trans. 1,* 1987, 1423 (*synth, deriv, pmr*)

3-Bromo-5-hydroxy-2-methyl-1,4-naphtho-   B-60266
quinone

*3-Bromo-5-hydroxy-2-methyl-1,4-naphthalenedione,
9CI. 3-Bromoplumbagin*

$C_{11}H_7BrO_3$ M 267.079
Isol. from *Diospyros maritima.* Orange needles (hexane).
Mp 121-122°.

Higa, M. *et al, Chem. Pharm. Bull.,* 1987, **35**, 4366 (*isol, synth,
  uv, ir, pmr, ms*)

2-Bromo-3-hydroxy-6-methylpyridine   B-60267

*2-Bromo-6-methyl-3-pyridinol, 9CI, 8CI. 6-Bromo-5-
hydroxy-2-picoline*

[23003-35-2]

C_6H_6BrNO M 188.024
Cryst. (MeOH aq.). Mp 191-192°.

N-Oxide: [24207-04-3].
  $C_6H_6BrNO_2$ M 204.023
  Cryst. (AcOH/EtOH). Mp 194-196°.
Me ether: [24207-22-5]. *2-Bromo-3-methoxy-6-methyl-
  pyridine. 6-Bromo-5-methoxy-2-picoline, 8CI.*
  C_7H_8BrNO M 202.050
  Cryst. (H$_2$O). Mp 54°.

Undheim, K. *et al, Acta Chem. Scand., Ser. B,* 1969, **23**, 1704,
  2065, 2075 (*synth*)
Reistad, K.J. *et al, Acta Chem. Scand., Ser. B,* 1974, **28**, 667
  (*synth*)
Weis, C.D., *J. Heterocycl. Chem.,* 1976, **13**, 145 (*synth*)

2-Bromo-3-hydroxy-1,4-naphthoquinone,   B-60268
8CI

Updated Entry replacing B-02542
2-Bromo-3-hydroxy-1,4-naphthalenedione, 9CI
[1203-39-0]

$C_{10}H_5BrO_3$ M 253.052
Yellow prisms (EtOH). Spar. sol. EtOH. Mp 202°.
  Sublimes (Mp 196° after subl.).

Me ether: [26037-61-6]. *2-Bromo-3-methoxy-1,4-
  naphthoquinone.*
  $C_{11}H_7BrO_3$ M 267.079
  Needles (MeOH). Mp 163-164°.
Et ether: [62452-74-8]. *2-Bromo-3-ethoxy-1,4-
  naphthoquinone.*
  $C_{12}H_9BrO_3$ M 281.105
  Yellow needles (EtOH). Mp 118°.

Dimroth, O. *et al, Justus Liebigs Ann. Chem.,* 1916, **411**, 345.
Read, G. *et al, J. Chem. Soc., Perkin Trans. 1,* 1973, 368
  (*spectra*)
Otsuki, T. *et al, Bull. Chem. Soc. Jpn.,* 1976, **49**, 3713; *J.
  Heterocycl. Chem.,* 1980, **17**, 695 (*synth, deriv*)

2-Bromo-5-hydroxy-1,4-naphthoquinone,   B-60269
8CI

Updated Entry replacing B-02543
*2-Bromo-5-hydroxy-1,4-naphthalenedione, 9CI. 2-
Bromojuglone*
[69008-03-3]
$C_{10}H_5BrO_3$ M 253.052
Orange needles (pet. ether). Mp 135-136°.

Ac: [77189-69-6].
  $C_{12}H_7BrO_4$ M 295.089
  Yellow needles (EtOH). Mp 154.5-156°.
Me ether: [69833-09-6]. *2-Bromo-5-methoxy-1,4-
  naphthoquinone.*
  $C_{11}H_7BrO_3$ M 267.079
  Yellow needles (EtOH). Mp 133-134°.

Hannan, R.L. *et al, J. Org. Chem.,* 1979, **44**, 2153 (*synth, deriv,
  chromatog*)

Heinzman, S.W. *et al, Tetrahedron Lett.*, 1980, **21**, 4305 (*synth*)
Jung, M.E. *et al, J. Org. Chem.*, 1987, **52**, 1889 (*synth, deriv, ir, pmr, ms*)

2-Bromo-6-hydroxy-1,4-naphthoquinone   B-60270
2-Bromo-6-hydroxy-1,4-naphthalenedione
$C_{10}H_5BrO_3$ M 253.052
Yellow needles (AcOH). Mp 193-194° (170° dec.).
Ac:
$C_{12}H_7BrO_4$ M 295.089
Pale-yellow needles (MeOH aq.). Mp 102°.
Me ether: [69008-15-7]. *2-Bromo-6-methoxy-1,4-naphthaquinone.*
$C_{11}H_7BrO_3$ M 267.079
Yellow needles (EtOAc/pet. ether). Mp 155-156.5°.

Lyons, J.M. *et al, J. Chem. Soc.*, 1953, 2910 (*synth, deriv*)
Cameron, D.W. *et al, Aust. J. Chem.*, 1978, **31**, 1335 (*synth, deriv, uv, ir, pmr*)

2-Bromo-7-hydroxy-1,4-naphthoquinone   B-60271
$C_{10}H_5BrO_3$ M 253.052
Me ether: [69008-16-8]. *2-Bromo-7-methoxy-1,4-naphthoquinone.*
$C_{11}H_7BrO_3$ M 267.079
Orange-yellow needles (EtOAc/pet. ether). Mp 133-133.5°.

Cameron, D.W. *et al, Aust. J. Chem.*, 1978, **31**, 1335 (*synth, deriv, uv, ir, pmr*)

2-Bromo-8-hydroxy-1,4-naphthoquinone   B-60272
2-Bromo-8-hydroxy-1,4-naphthalenedione. 3-Bromojuglone
[52431-65-9]
$C_{10}H_5BrO_3$ M 253.052
Orange-brown needles (AcOH). Mp 172°.
Ac: [77197-58-1].
$C_{12}H_7BrO_4$ M 295.089
Yellow plates (EtOH). Mp 151.5°.
Me ether: [69833-10-9]. *2-Bromo-8-methoxy-1,4-naphthoquinone.*
$C_{11}H_7BrO_3$ M 267.079
Yellow needles (EtOAc/pet. ether). Mp 154-155°.

Bowden, B.F. *et al, Aust. J. Chem.*, 1979, **32**, 769 (*cmr*)
Hannan, R.L. *et al, J. Org. Chem.*, 1979, **44**, 2153 (*synth, deriv, chromatog*)
Cameron, D.W. *et al, Aust. J. Chem.*, 1981, **34**, 1513 (*synth, deriv, uv, ir*)
Parker, K.A. *et al, J. Org. Chem.*, 1981, **46**, 3218 (*deriv, pmr*)

7-Bromo-2-hydroxy-1,4-naphthoquinone   B-60273
7-Bromo-2-hydroxy-1,4-naphthalenedione
[58472-34-7]
$C_{10}H_5BrO_3$ M 253.052
Cryst. (EtOH). Mp 216°.

Buckle, D.R. *et al, J. Med. Chem.*, 1977, **20**, 1059 (*synth*)

8-Bromo-2-hydroxy-1,4-naphthoquinone   B-60274
8-Bromo-2-hydroxy-1,4-naphthalenedione
[54808-16-1]
$C_{10}H_5BrO_3$ M 253.052
Yellow cryst. (C_6H_6). Mp 185-187°.

Hewgill, F.R. *et al, Aust. J. Chem.*, 1975, **28**, 355 (*synth, ir*)

2-Bromo-3-hydroxypyridine   B-60275
2-Bromo-3-pyridinol, 9CI
[6602-32-0]

C_5H_4BrNO M 173.997
Mp 184-185°.
▷UU7705000.
B,HBr: Mp >360° dec.
N-Oxide: [6602-29-5].
$C_5H_4BrNO_2$ M 189.996
Cryst. (AcOH aq.). Mp 188-193° (180-185°).
Me ether: [24100-18-3]. *2-Bromo-3-methoxypyridine.*
C_6H_6BrNO M 188.024
Cryst. (C_6H_6). Mp 43-45°. Bp$_9$ 123°.
Me ether, N-Oxide:
$C_6H_6BrNO_2$ M 204.023
Mp 159-161°.
Et ether: 2-Bromo-3-ethoxypyridine.
C_7H_8BrNO M 202.050
Cryst. (EtOH/Et$_2$O). Mp 187°.
Et ether; B,HBr: Mp 183°.

Den Hertog, H.J. *et al, Recl. Trav. Chim. Pays-Bas*, 1949, **68**, 275 (*synth*)
Lewicka, K. *et al, Rocz. Chem.*, 1966, **40**, 405 (*synth*)
Nedenskov, P. *et al, Acta Chem. Scand.*, 1969, **23**, 1971 (*deriv, synth*)
Undheim, K. *et al, Acta Chem. Scand.*, 1969, **23**, 2075 (*deriv*)
Katritzky, A.R. *et al, J. Chem. Soc., Perkin Trans. 1*, 1979, 2528 (*synth*)
Clark, G.J. *et al, Aust. J. Chem.*, 1981, **34**, 927 (*synth*)
Tiecco, M. *et al, Tetrahedron*, 1986, **42**, 1475 (*synth, pmr*)

2-Bromo-5-hydroxypyridine   B-60276
6-Bromo-3-pyridinol
[55717-45-8]
C_5H_4BrNO M 173.997
Cryst. (C_6H_6). Mp 135.5-136.5°.

Den Hertog, H.J. *et al, Recl. Trav. Chim. Pays-Bas*, 1950, **69**, 1281 (*synth*)

3-Bromo-5-hydroxypyridine   B-60277
5-Bromo-3-pyridinol, 9CI
[74115-13-2]
C_5H_4BrNO M 173.997
Needles (H$_2$O). Mp 166.5-167.5°.
Me ether: [50720-12-2]. *3-Bromo-5-methoxypyridine, 9CI.*
C_6H_6BrNO M 188.024
Mp 33.5-34°. Bp$_2$ 80-82°.

Den Hertog, H.J. *et al, Recl. Trav. Chim. Pays-Bas*, 1950, **69**, 1281 (*synth*)
Czuba, W., *Rocz. Chem.*, 1960, **34**, 1639 (*synth*)
Ziegler, F.E. *et al, J. Am. Chem. Soc.*, 1973, **95**, 7458 (*synth*)
B.P., 2 025 953, (*1980*); *CA*, **93**, 114330 (*synth*)

4-Bromo-3-hydroxypyridine   B-60278
4-Bromo-3-pyridinol
C_5H_4BrNO M 173.997
Cryst. (C_6H_6/pet. ether). Mp 123.5-124°.

Den Hertog, H.J. *et al*, *Recl. Trav. Chim. Pays-Bas*, 1950, **69**, 1281 (*synth*)

5-Bromo-2-iodoaniline   B-60279
5-Bromo-2-iodobenzenamine, 9CI
C_6H_5BrIN M 297.921
Needles (hexane/Et_2O). Mp 55°.

Sakamoto, T. *et al*, *Chem. Pharm. Bull.*, 1987, **35**, 1823 (*synth*, *pmr*)

2-Bromo-4-iodopyridine   B-60280

C_5H_3BrIN M 283.894
Mp 61°.

Talik, T., *Rocz. Chem.*, 1957, **31**, 569 (*synth*)

2-Bromo-5-iodopyridine   B-60281
[73290-22-9]
C_5H_3BrIN M 283.894
Mp 122-125°.

Wibaut, J.P. *et al*, *CA*, 1927, **21**, 3619 (*synth*)
Spiers, C.W.F. *et al*, *Recl. Trav. Chim. Pays-Bas*, 1937, **56**, 573.

3-Bromo-4-iodopyridine   B-60282
C_5H_3BrIN M 283.894
Mp 112°.

Talik, T., *Rocz. Chem.*, 1962, **36**, 1049 (*synth*)

3-Bromo-5-iodopyridine   B-60283
C_5H_3BrIN M 283.894
Mp 132-134°.

Wibaut, J.P. *et al*, *Recl. Trav. Chim. Pays-Bas*, 1955, **74**, 1062 (*synth*)

4-Bromo-2-iodopyridine   B-60284
C_5H_3BrIN M 283.894
Mp 46°.

Talik, T., *Rocz. Chem.*, 1957, **31**, 569 (*synth*)

4-Bromo-3-iodopyridine   B-60285
C_5H_3BrIN M 283.894
Mp 77-78°.

Talik, T., *Rocz. Chem.*, 1962, **36**, 1049 (*synth*)

5-Bromo-2-iodopyridine   B-60286
C_5H_3BrIN M 283.894
Mp 117°.

Wibaut, J.P. *et al*, *CA*, 1927, **21**, 3619 (*synth*)
Spiers, C.W.F. *et al*, *Recl. Trav. Chim. Pays-Bas*, 1937, **56**, 573.

2-Bromo-4-iodothiazole   B-60287
[41731-34-4]

$C_3HBrINS$ M 289.916
Mp 93-95°.

Dondoni, A. *et al*, *Synthesis*, 1986, 757 (*synth*, *ir*, *pmr*)

2-Bromo-5-iodothiazole   B-60288
[108306-63-4]
$C_3HBrINS$ M 289.916
Mp 110-112°.

Dondoni, A. *et al*, *Synthesis*, 1986, 757 (*synth*, *ir*, *pmr*)

4-Bromo-2-iodothiazole   B-60289
[108306-56-5]
$C_3HBrINS$ M 289.916
Mp 80-82°.

Dondoni, A. *et al*, *Synthesis*, 1986, 757 (*synth*, *ir*, *pmr*)

5-Bromo-2-iodothiazole   B-60290
[108306-64-5]
$C_3HBrINS$ M 289.916
Oil.

Dondoni, A. *et al*, *Synthesis*, 1986, 757 (*synth*, *ir*, *pmr*)

Bromomethanesulfonic acid   B-60291

$BrCH_2SO_3H$

CH_3BrO_3S M 174.997
Na salt: [34239-78-6]. Cryst. (H_2O). Mp 277-281°.
Chloride: [10099-08-8].
     CH_2BrClO_2S M 193.443
     Bp$_{15}$ 87-89°.
Bromide: [54730-18-6].
     $CH_2Br_2O_2S$ M 237.894
     Light-yellow oil. Bp 60-68°.

Truce, W.E. *et al*, *J. Org. Chem.*, 1967, **32**, 990 (*synth*, *ir*, *pmr*)
Petigara, R.B. *et al*, *J. Heterocycl. Chem.*, 1974, **11**, 331 (*synth*)
Block, E. *et al*, *J. Am. Chem. Soc.*, 1986, **108**, 4568 (*deriv*, *synth*, *pmr*, *ir*)

2-Bromo-3-methylbenzaldehyde, 9CI   B-60292
[109179-31-9]

C_8H_7BrO M 199.047
Needles. Mp 53-54°. Purity ~95%.

Miyano, S. *et al*, *Bull. Chem. Soc. Jpn.*, 1986, **59**, 3285 (*synth*, *pmr*)

2-Bromo-3-methylbenzoic acid, 9CI   B-60293

Updated Entry replacing B-02681
2-Bromo-m-toluic acid, 8CI

$C_8H_7BrO_2$ M 215.046
Needles. Mp 135-137°. pK_a 2.9.

Org. Synth., Coll. Vol., **4**, 114 (*synth*)
Miyano, S. *et al*, *Bull. Chem. Soc. Jpn.*, 1986, **59**, 3285 (*synth*)

3-Bromo-4-methylbenzoic acid, 9CI   B-60294

Updated Entry replacing B-02686
3-Bromo-p-toluic acid, 8CI
[7697-26-9]
$C_8H_7BrO_2$ M 215.046
Mp 204°. pK_a 3.96. Sublimes, steam-volatile.

Pearson, D.E. *et al*, *J. Org. Chem.*, 1958, **23**, 1412 (*synth*)
Fleifel, A.M., *J. Org. Chem.*, 1960, **25**, 1024 (*synth*)
Hands, D. *et al*, *J. Heterocycl. Chem.*, 1986, **23**, 1333 (*synth*)

3-(Bromomethyl)-2,3-dihydrobenzofuran   B-60295

[78739-85-2]

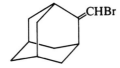

C_9H_9BrO M 213.074
(±)-*form*
Oil. Bp$_{0.7}$ 85°.

Meijs, G.F. *et al*, *J. Am. Chem. Soc.*, 1986, **108**, 5890 (*synth, ms, pmr, ir*)

(Bromomethylidene)adamantane   B-60296

(*Bromomethylene*)*tricyclo[3.3.1.1³,⁷]decane*, 9CI
[68251-94-5]

$C_{11}H_{15}Br$ M 227.144
Oil.

Nelsen, S.F. *et al*, *J. Am. Chem. Soc.*, 1986, **108**, 5503 (*synth, pmr, cmr*)

4-(Bromomethyl)-2-nitrobenzoic acid, 9CI   B-60297

4-Carboxy-3-nitrobenzyl bromide
[100466-27-1]

$C_8H_6BrNO_4$ M 260.044
Cryst. (CHCl₃). Mp 134-136°.
tert-Butyl ester:
     $C_{12}H_{14}BrNO_4$ M 316.151

Cryst. (Et₂O/hexane). Mp 64°.

Gelb, M.H. *et al*, *J. Med. Chem.*, 1986, **29**, 585 (*synth, pmr*)

1-Bromo-4-methyl-2-pentanone   B-60298

Bromomethyl isobutyl ketone
[29585-02-2]

$$(H_3C)_2CHCH_2COCH_2Br$$

$C_6H_{11}BrO$ M 179.056
Bp₄₀ 89-91°.

Elderfield, R.C. *et al*, *J. Am. Chem. Soc.*, 1950, **72**, 4059 (*synth*)
Ogawa, K. *et al*, *Chem. Pharm. Bull.*, 1986, **34**, 3252 (*synth, pmr*)

5-Bromo-3-methyl-1-pentene, 9CI   B-60299

$$H_2C{=}CHCH(CH_3)CH_2CH_2Br$$

$C_6H_{11}Br$ M 163.057
(±)-*form* [59822-10-5]
Liq. Bp 130-140°.

Bromidge, S.M. *et al*, *J. Chem. Soc., Perkin Trans. 1*, 1985, 1725 (*synth, ir, pmr*)

Bromomethyl phenyl selenide   B-60300

[(*Bromomethyl*)*seleno*]*benzene*, 9CI
[60466-50-4]

$$PhSeCH_2Br$$

C_7H_7BrSe M 249.996
Bp₀.₂₅ 82-86°.

Petragnani, N. *et al*, *J. Organomet. Chem.*, 1976, **114**, 281 (*synth*)
Reich, H.J. *et al*, *J. Org. Chem.*, 1986, **51**, 2985 (*synth, ir, pmr*)

(3-Bromomethyl)-2,4,10-trioxatricyclo[3.3.1.1³,⁷]decane   B-60301

3-(Bromomethyl)-2,4,10-trioxadamantane

$C_8H_{11}BrO_3$ M 235.077
Masked deriv. of bromoacetic acid of use in synth. Cryst. (ligroin). Mp 95°.

Stetter, W. *et al*, *Chem. Ber.*, 1953, **86**, 790 (*synth*)
Voss, G. *et al*, *Helv. Chim. Acta*, 1983, **66**, 2294 (*use*)

1-Bromo-2-naphthalenecarboxaldehyde, 9CI   B-60302

Updated Entry replacing B-30297
1-Bromo-2-naphthaldehyde, 8CI
[3378-82-3]

$C_{11}H_7BrO$ M 235.080
Needles (AcOH). Mp 118°.

Di-Me acetal: [103668-59-3]. *1-Bromo-(2-dimethoxymethyl)naphthalene.*
$C_{13}H_{13}BrO_2$ M 281.149
Cryst. (hexane). Mp 55-56°.

Mayer, F. *et al*, *Chem. Ber., B*, 1922, **55**, 1835 (*synth*)
Trost, B.M. *et al*, *J. Org. Chem.*, 1981, **46**, 4617 (*synth, ir, pmr*)
Koppenhoefer, B. *et al*, *Acta Crystallogr., Sect. C*, 1986, **42**, 1612 (*cryst struct*)
Smith, J.G. *et al*, *J. Org. Chem.*, 1986, **51**, 3762 (*synth, deriv, ir, pmr*)

1-Bromo-2-nitroethane   **B-60303**
[10524-56-8]

$$BrCH_2CH_2NO_2$$

$C_2H_4BrNO_2$ M 153.963
Liq. d_4^{20} 1.81. Bp_2 46-47°.

Shul'man, M.L. *et al*, *J. Org. Chem. USSR (Engl. Transl.)*, 1967, **3**, 840 (*synth*)

4-Bromo-3(5)-nitro-1*H*-pyrazole   **B-60304**
[89717-64-6]

$C_3H_2BrN_3O_2$ M 191.972
Cryst. (EtOH aq.). Mp 197-198°.
1-Me: [89607-12-5].
    $C_4H_4BrN_3O_2$ M 205.999
    Cryst. (EtOH aq.). Mp 158-160°.

Newton, C.G. *et al*, *J. Chem. Soc., Perkin Trans. 1*, 1984, 63 (*synth, ir, pmr*)

3(5)-Bromo-5(3)-nitro-1*H*-pyrazole   **B-60305**
[104599-38-4]
$C_3H_2BrN_3O_2$ M 191.972
Cryst. (EtOH aq.). Mp 150-151°.

Juffermans, J.P.H. *et al*, *J. Org. Chem.*, 1986, **51**, 4656 (*synth, pmr*)

1-Bromo-2-nitro-3-(trifluoromethyl)-   **B-60306**
benzene, 9CI
3-Bromo-2-nitrobenzotrifluoride. 3-Bromo-α,α,α-trifluoro-2-nitrotoluene

$C_7H_3BrF_3NO_2$ M 270.005
Cryst. Mp 29-30°. Bp_{14} 138-140°.

Forbes, E.J. *et al*, *Tetrahedron*, 1960, **8**, 73 (*synth*)

1-Bromo-4-nitro-2-(trifluoromethyl)-   **B-60307**
benzene, 9CI
2-Bromo-5-nitrobenzotrifluoride. 2-Bromo-5-nitro-α,α,α-trifluorotoluene
[367-67-9]
$C_7H_3BrF_3NO_2$ M 270.005
Cryst. (EtOH). Mp 46-48°. $Bp_{2.5-3}$ 87-88°.

McBee, E.T. *et al*, *J. Am. Chem. Soc.*, 1951, **73**, 3932 (*synth*)

Filler, R. *et al*, *J. Org. Chem.*, 1961, **26**, 2707 (*synth*)

2-Bromo-1-nitro-3-(trifluoromethyl)-   **B-60308**
benzene, 9CI
2-Bromo-α,α,α-trifluoro-3-nitrotoluene, 8CI. 2-Bromo-3-nitrobenzotrifluoride
[24034-22-8]
$C_7H_3BrF_3NO_2$ M 270.005
Cryst. (MeOH). Mp 61.5-62°.

Saggiomo, A.J. *et al*, *J. Heterocycl. Chem.*, 1969, **6**, 631 (*synth*)

4-Bromo-1-nitro-2-(trifluoromethyl)-   **B-60309**
benzene
5-Bromo-α,α,α-trifluoro-2-nitrotoluene. 5-Bromo-2-nitrobenzotrifluoride
[344-38-7]
$C_7H_8BrF_3NO_2$ M 275.045
Cryst. (EtOH). Mp 40-44°. Bp_5 99-100°.

McBee, E.T. *et al*, *J. Am. Chem. Soc.*, 1951, **73**, 3932 (*synth*)

1-Bromo-3-nonene, 9CI   **B-60310**

$$H_3C(CH_2)_4CH=CHCH_2CH_2Br$$

$C_9H_{17}Br$ M 205.137
(*Z*)-form [60705-54-6]
    $Bp_{6.5}$ 90°.

Rosini, G. *et al*, *Synthesis*, 1986, 46 (*synth, pmr*)

1-Bromo-1,3-pentadiene   **B-60311**

$$BrCH=CHCH=CHCH_3$$

C_5H_7Br M 147.014
(*E,E*)-form [103980-84-3]
    Bp_{45} 49-51°.

Hayashi, T. *et al*, *J. Org. Chem.*, 1986, **51**, 3772 (*synth, pmr*)

Bromopentamethylbenzene   **B-60312**
[5153-40-2]

$C_{11}H_{15}Br$ M 227.144
Cryst. (EtOH). Mp 162-163°.

Baciocchi, E. *et al*, *J. Am. Chem. Soc.*, 1965, **87**, 3953 (*synth*)
Charbonneau, G. *et al*, *J. Chim. Phys.*, 1967, **64**, 273 (*cryst struct*)
Kohl, F.X. *et al*, *Chem. Ber.*, 1987, **120**, 1539 (*pmr, cmr*)

5-Bromo-2-pentanol   **B-60313**
[62957-46-4]

$$H_3CCH(OH)CH_2CH_2CH_2Br$$

$C_5H_{11}BrO$ M 167.045
(±)-form
    $Bp_{0.05}$ 49-50°.

Audin, P. *et al*, *Bull. Soc. Chim. Fr.*, 1984, 297 (*synth, ir*)
Yadav, V.K. *et al*, *J. Org. Chem.*, 1986, **51**, 3372 (*synth, ir, pmr*)

3-Bromo-1-phenyl-1-butyne   B-60314
3-Bromo-1-butynylbenzene, 9CI
[27975-80-0]

$$PhC{\equiv}CCHBrCH_3$$

$C_{10}H_9Br$ M 209.085
(±)-*form* [104519-25-7]

Mannschreck, A. *et al*, *Tetrahedron*, 1986, **42**, 399 (*synth*, *ir*, *pmr*, *ms*)

3-Bromo-2(1*H*)-pyridinethione, 9CI   B-60315

C_5H_4BrNS M 190.057
NH-form is said to predominate in soln., *SH*-form in solid state, but the evidence for this is weak.
NH-form [65938-86-5]
SH-form
　3-Bromo-2-pyridinethiol. 3-Bromo-2-mercaptopy•idine
　Major tautomer in solid state. Yellow solid (EtOH). Subl. without melting.
　Me thioether: [51933-77-8]. *3-Bromo-2-(methylthio)-pyridine*, 9CI.
　C_6H_6BrNS M 204.084
　Liq. Bp_{22} 131-132°.

Trecourt, F. *et al*, *J. Chem. Res.* (*M*), 1979, 536 (*synth*, *tautom*)

3-Bromo-4(1*H*)-pyridinethione   B-60316
3-Bromo-4-pyridinethiol. 3-Bromo-4-mercaptopyridine
[82257-10-1]
C_5H_4BrNS M 190.057
NH-form predominates.

Trécourt, F. *et al*, *J. Chem. Res.* (*S*), 1982, 76 (*synth*)

5-Bromo-2(1*H*)-pyridinethione, 9CI   B-60317
5-Bromo-2-pyridinethiol. 5-Bromo-2-mercaptopyridine
[56673-34-8]
C_5H_4BrNS M 190.057
B.P., 791 190, (*1958*); *CA*, **52**, 15597 (*synth*)
Davies, J.S. *et al*, *Tetrahedron Lett.*, 1979, 5035 (*synth*)

5-Bromo-2,4-(1*H*,3*H*)-pyrimidinedione   B-60318
5-Bromouracil, 8CI
[51-20-7]

$C_4H_3BrN_2O_2$ M 190.984
Prisms (H_2O). Mp 293°.
▷Mutagen. YQ9060000.

Wheeler, H.L. *et al*, *Am. Chem. J.*, 1903, **29**, 478 (*synth*)

Levene, P.A. *et al*, *Ber.*, 1912, **45**, 608 (*synth*)
Merck Index, 10th Ed., no. 1414.

6-Bromo-2,4-(1*H*,3*H*)-pyrimidinedione, 9CI   B-60319
6-Bromouracil, 8CI
[4269-93-6]
$C_4H_3BrN_2O_2$ M 190.984
Cryst. (MeOH). Mp >340° (darkens at 270°).

Horwitz, J.P. *et al*, *J. Org. Chem.*, 1961, **26**, 3392 (*synth*, *uv*)

1-Bromo-2,3,4,5-tetrafluorobenzene, 9CI, 8CI   B-60320
[1074-91-5]

C_6HBrF_4 M 228.972
Liq. Bp 140-142° (136°), Bp_{25} 48-55°. η_D^{22} 1.4650.

Burdon, J. *et al*, *Tetrahedron*, 1966, **22**, 2541 (*synth*)
Netherlands Pat., 6 577 222, (*1966*); *CA*, **65**, 18525 (*synth*)
Belf, L.J. *et al*, *Tetrahedron*, 1967, **23**, 4719 (*synth*)

2-Bromo-1,3,4,5-tetrafluorobenzene, 9CI, 8CI   B-60321
1-Bromo-2,3,4,6-tetrafluorobenzene
[1559-86-0]
C_6HBrF_4 M 228.972
Liq. Bp 141-142°. n_D^{19} 1.4682.

Tilney-Bassett, J.F., *Chem. Ind.* (*London*), 1965, 693 (*synth*)
Burdon, J. *et al*, *Tetrahedron*, 1966, **22**, 2541 (*synth*)
Netherlands Pat., 6 577 222, (*1966*); *CA*, **65**, 18525 (*synth*)

3-Bromo-1,2,4,5-tetrafluorobenzene, 9CI, 8CI   B-60322
1-Bromo-2,3,5,6-tetrafluorobenzene
[1559-88-2]
C_6HBrF_4 M 228.972
Liq. d_4^{20} 1.883. Bp 145-147°. η_D^{20} 1.4688.

Netherlands Pat., 6 517 222, (*1966*); *CA*, **65**, 18525 (*synth*)
Aldrich Library of FT-IR Spectra, 1st Ed., **1**, 1013B (*ir*)
Aldrich Library of NMR Spectra, 2nd Ed., **1**, 812B (*pmr*)
Sigma-Aldrich Library of Chemical Safety Data, 1st Ed., 288D (*haz*)

1-Bromo-2,3,4,5-tetrafluoro-6-nitrobenzene, 9CI, 8CI   B-60323
[5580-83-6]

$C_6BrF_4NO_2$ M 273.969
Liq. Bp 200°, Bp_{20} 94°. n_D^{24} 1.4970.

Coe, P.L. *et al*, *J. Chem. Soc.* (*C*), 1966, 2323 (*synth*)
Belf, L.J. *et al*, *Tetrahedron*, 1967, **23**, 4719 (*synth*)
Yakobson, G.G. *et al*, *Zh. Obshch. Khim.*, 1967, **37**, 1289.

1-Bromo-2,3,4,6-tetrafluoro-5-nitrobenzene, 9CI   B-60324

[17826-68-5]

$C_6BrF_4NO_2$ M 273.969
Bp 213-214° dec., Bp_{10} 90-91°.

Netherlands Pat., 6 613 001, (*1967*); *CA*, **68**, 68762 (*synth*)
Burdon, J. *et al*, *J. Fluorine Chem.*, 1981, **18**, 507 (*synth, F nmr*)

1-Bromo-2,3,5,6-tetrafluoro-4-nitrobenzene, 9CI   B-60325

[17823-37-9]

$C_6BrF_4NO_2$ M 273.969
Solid by subl. Mp 50-51°. Bp_{20} 90-100°.

Yakobson, G.G. *et al*, *Zh. Obshch. Khim.*, 1967, **37**, 1289 (*synth*)
Coe, P.L. *et al*, *Tetrahedron*, 1968, **24**, 5913 (*synth*)

2-Bromo-3,4,5,6-tetrafluoropyridine, 9CI   B-60326

[20973-46-0]

C_5BrF_5N M 248.958
Liq. Bp_{769} 140-142°.

Anderson, L.P. *et al*, *J. Chem. Soc.* (*C*), 1969, 2559 (*synth, F nmr*)

3-Bromo-2,4,5,6-tetrafluoropyridine, 9CI   B-60327

[40392-86-7]

C_5BrF_5N M 248.958
Liq.

Furin, G.G. *et al*, *Izv. Sib. Otd. Akad. Nauk SSSR, Ser. Khim. Nauk*, 1972, 128 (*synth*)
Ger. Pat., 2 241 562, (*1973*); *CA*, **78**, 159438f (*synth*)

4-Bromo-2,3,5,6-tetrafluoropyridine, 9CI   B-60328

[3511-90-8]

C_5BrF_4N M 229.959
Liq. Bp 134-135°.

Chambers, R.D. *et al*, *J. Chem. Soc.*, 1964, 5040; *J. Chem. Soc.* (*C*), 1969, 1700 (*synth*)

3-Bromo[1,2,3]triazolo[1,5-a]pyridine, 9CI   B-60329

[106911-04-0]

$C_6H_4BrN_3$ M 198.022
Cryst. (pet. ether). Mp 91°.

Jones, G. *et al*, *J. Chem. Soc., Perkin Trans. 1*, 1985, 2719 (*synth, uv, ms, pmr*)

7-Bromo[1,2,3]triazolo[1,5-a]pyridine, 9CI   B-60330

[107465-26-9]

$C_6H_4BrN_3$ M 198.022
Reagent for prepn. of 2,6-disubst. pyridines. Mp 95-95.5°.

Abarca, B. *et al*, *Tetrahedron Lett.*, 1986, **27**, 3543 (*synth, use*)

3-Bromo-2,4,6-trimethylbenzoic acid   B-60331

3-Bromomesitoic acid

[5333-13-1]

$C_{10}H_{11}BrO_2$ M 243.100
Needles (EtOH aq.). Mp 168° (162-162.5°). pK_a 4.84 (55% EtOH aq., 22°).
Me ester: [26584-20-3].
    $C_{11}H_{13}BrO_2$ M 257.127
    Cryst. (pet. ether). Mp 42.5-43°.

Shildneck, P.R. *et al*, *J. Am. Chem. Soc.*, 1931, **53**, 343 (*synth*)
Beringer, F.M. *et al*, *J. Am. Chem. Soc.*, 1953, **75**, 3319 (*synth*)
Hart, H. *et al*, *J. Org. Chem.*, 1970, **35**, 3637 (*synth, deriv, ir, pmr*)
Cuyegkeng, M.A. *et al*, *Chem. Ber.*, 1987, **120**, 803 (*synth, pmr*)

2-Bromotryptophan   B-60332

Updated Entry replacing B-40220
2-Amino-3-(2-bromo-1H-indol-3-yl)propanoic acid

$C_{11}H_{11}BrN_2O_2$ M 283.124
(**S**)-*form* [89311-52-4]
    L-form
    Plates (EtOH). $[\alpha]_D^{20}$ +1.2° (c, 0.52 in H_2O). Darkens at 170°, dec. at 195-200°.

Phillips, R.S. *et al*, *J. Am. Chem. Soc.*, 1986, **108**, 2023 (*synth, uv, pmr, ms*)

5-Bromotryptophan, 9CI, 8CI   B-60333

$C_{11}H_{11}BrN_2O_2$ M 283.124
(**R**)-*form* [93299-40-2]
    D-form
    Cryst. (MeOH aq.). Mp 289-290°. $[\alpha]_D$ +27.1° (c, 1.1 in AcOH).
    N^α-*Ac:* [75816-16-9].
        $C_{13}H_{13}BrN_2O_3$ M 325.161
        Cryst. + $\frac{1}{2}H_2O$ (MeOH aq.). Mp 120-122° dec. $[\alpha]_D^{26}$ −15.1° (c, 2 in MeOH).
    Me ester: [93299-39-9].
        $C_{12}H_{13}BrN_2O_2$ M 297.151
        Cryst. (2-propanol/diisopropyl ether). Mp 98-100°. $[\alpha]_D^{20}$ −49.2° (c, 1 in MeOH).
(**S**)-*form* [25197-99-3]
    L-form
    Mp 288-291° dec. $[\alpha]_D^{20}$ −28.5° (c, 1 in AcOH), $[\alpha]_D^{20}$ +31.8° (c, 1 in 1M HCl).

Me ester: [93299-38-8]. Cryst. (2-propanol/diisopropyl ether). Mp 100-102°. $[\alpha]_D^{20}$ +48.6° (c, 1 in MeOH).

(±)-form [6548-09-0]
Mp 273-276° dec. Sinters at 258°.
N^α-*Ac:* [75816-15-8]. Cryst. (MeOH aq.). Mp 212-214°.
Raverty, W.D. *et al, J. Chem. Soc., Perkin Trans. 1*, 1977, 1204 (*synth*)
Allen, M.C. *et al, J. Chem. Soc., Perkin Trans. 1*, 1980, 1928 (*synth*)
Irie, K. *et al, Chem. Pharm. Bull.*, 1984, **32**, 2126 (*synth, derivs*)
Allen, M.C. *et al, J. Chem. Soc., Perkin Trans. 1*, 1986, 989 (*synth*)

6-Bromotryptophan, 9CI　　　　B-60334
$C_{11}H_{11}BrN_2O_2$　　M 283.124
(S)-form [52448-17-6]
$[\alpha]_D^{20}$ −12.1° (c, 0.38 in AcOH).
N^α-tert-*Butyloxycarbonyl:* [97444-12-7]. $[\alpha]_D^{20}$ +27.8° (c, 0.84 in CHCl$_3$).
(±)-form [33599-61-0]
Mp 275-280° dec.
N^α-*Ac:* [85515-06-6].
　$C_{13}H_{13}BrN_2O_3$　　M 325.161
　Cryst. (EtOAc/pet. ether). Mp 140-142°.
N^α-*Ac, Et ester:* [97444-11-6].
　$C_{15}H_{17}BrN_2O_3$　　M 353.215
　Cryst. (Et$_2$O). Mp 138-140°.
Schreier, E., *Helv. Chim. Acta*, 1976, **59**, 585 (*synth*)
Schmidt, U. *et al, Justus Liebigs Ann. Chem.*, 1985, 785 (*synth*)

7-Bromotryptophan, 9CI　　　　B-60335
$C_{11}H_{11}BrN_2O_2$　　M 283.124
(S)-form [75816-19-2]
L-form
Mp 276-280° dec. $[\alpha]_D^{20}$ −7.6° (c, 1 in AcOH).
(±)-form
N^α-*Ac:* [75816-18-1].
　$C_{13}H_{13}BrN_2O_3$　　M 325.161
　Cryst. (MeOH aq.). Mp 216-218°.
Allen, M.C. *et al, J. Chem. Soc., Perkin Trans. 1*, 1980, 1928 (*synth*)

11-Bromo-1-undecene　　　　B-60336
[7766-50-9]

$$BrCH_2(CH_2)_8CH=CH_2$$

$C_{11}H_{21}Br$　　M 233.191
Liq. d_{20}^{20} 1.03. Bp$_4$ 103°, Bp$_1$ 83-84°.
Tomecko, C.G. *et al, J. Am. Chem. Soc.*, 1927, **49**, 522 (*synth*)
Hideg, K. *et al, J. Chem. Soc., Perkin Trans. 1*, 1986, 1431 (*synth*)

Bullerone　　　　B-60337

$C_{15}H_{18}O_3$　　M 246.305

Constit. of *Cyathus bulleri*. Yellow oil.
Ayer, W.A. *et al, Can. J. Chem.*, 1987, **65**, 15.

Bursatellin　　　　B-60338
[75347-13-6]

$C_{13}H_{16}N_2O_4$　　M 264.280
Constit. of *Bursatella leachii pleii, B. leachii leachii* and *B. leachii savignyana.* Oil. $[\alpha]_D$ −8.8° (c, 2.4 in MeOH).
Gopichand, Y. *et al, J. Org. Chem.*, 1980, **45**, 5383 (*isol*)
Cimino, G. *et al, J. Org. Chem.*, 1987, **52**, 2301 (*struct*)

1,3-Butanedithiol　　　　B-60339
1,3-Dimercaptobutane
[24330-52-7]

$$H_3CCH(SH)CH_2CH_2SH$$

$C_4H_{10}S_2$　　M 122.243
(±)-form
Liq. Bp$_{10}$ 64-67°. n_D^{20} 1.5208.
Eliel, E.L. *et al, J. Am. Chem. Soc.*, 1969, **91**, 2703 (*synth*)

2,2-Butanedithiol　　　　B-60340
2,2-Dimercaptobutane
[15089-43-7]

$$H_3CC(SH)_2CH_2CH_3$$

$C_4H_{10}S_2$　　M 122.243
Liq. Bp$_{19}$ 46°, Bp$_{13}$ 37-38°.
Magnusson, B., *Acta Chem. Scand.*, 1963, **17**, 273 (*synth*)
Demuynck, M. *et al, Bull. Soc. Chim. Fr.*, 1967, 1213 (*synth, pmr*)

1-*tert*-Butylcyclohexene, 8CI　　　　B-60341
Updated Entry replacing B-03631
1-(1,1-Dimethylethyl)cyclohexene, 9CI
[3419-66-7]

$C_{10}H_{18}$　　M 138.252
Liq. Bp$_{14}$ 62-65°.
Goering, H.L. *et al, J. Am. Chem. Soc.*, 1956, **78**, 4926 (*synth*)
Filler, R. *et al, J. Am. Chem. Soc.*, 1959, **81**, 658 (*synth*)
Benkeser, R.A., *Synthesis*, 1971, 147 (*synth*)
Corona, T. *et al, J. Chem. Soc., Perkin Trans. 1*, 1985, 1607 (*synth*)

3-*tert*-Butylcyclohexene, 8CI　　　　B-60342
3-(1,1-Dimethylethyl)cyclohexene, 9CI
[14072-87-8]

(S)-form

$C_{10}H_{18}$　　M 138.252

(**S**)-*form* [77242-41-2]
  $[\alpha]_D^{20}$ +93.2° (c, 0.144 in CHCl$_3$).
(±)-*form*
  Bp 170°, Bp$_{17}$ 64°.

Benkeser, R.A. *et al*, *J. Org. Chem.*, 1964, **29**, 1313.
Richer, J.-C. *et al*, *Can. J. Chem.*, 1968, **46**, 3709.
Bellucci, G. *et al*, *J. Org. Chem.*, 1977, **42**, 1079 (*synth, abs config*)
Sadozai, S.K. *et al*, *Bull. Soc. Chim. Belg.*, 1980, **89**, 637 (*synth, ord, cmr*)
Grundy, S.L. *et al*, *J. Organomet. Chem.*, 1984, **272**, 265 (*synth*)
Goering, H.L. *et al*, *J. Org. Chem.*, 1986, **51**, 2884 (*synth*)
Tseng, C.C. *et al*, *J. Org. Chem.*, 1986, **51**, 2884 (*synth, pmr, cmr*)

4-*tert*-Butylcyclohexene, 8CI   B-60343

4-(1,1-Dimethylethyl)cyclohexene, 9CI
[2228-98-0]

(**R**)-*form*

C$_{10}$H$_{18}$ M 138.252
(**R**)-*form* [61062-50-8]
  $[\alpha]_D^{25}$ +82.8° (CHCl$_3$).
(**S**)-*form* [77242-40-1]
  $[\alpha]_D^{20}$ −75.9° (c, 1.14 in CHCl$_3$).
(±)-*form* [92619-42-6]
  Oil. Bp 172°. n_D^{20} 1.4583.

Winstein, S. *et al*, *J. Am. Chem. Soc.*, 1955, **77**, 5562 (*synth*)
Benkeser, R.A. *et al*, *J. Org. Chem.*, 1964, **29**, 1313 (*synth*)
Bellucci, G. *et al*, *J. Org. Chem.*, 1977, **42**, 1079 (*synth, abs config*)
Sadozai, S.K. *et al*, *Bull. Soc. Chim. Belg.*, 1980, **89**, 637 (*synth, ord, cmr*)
Caputo, R. *et al*, *Tetrahedron Lett.*, 1981, **22**, 3551 (*synth*)
Scott, W.J. *et al*, *J. Am. Chem. Soc.*, 1986, **108**, 3033 (*synth*)

1-*tert*-Butyl-1,2-cyclooctadiene   B-60344

1-(1,1-Dimethylethyl)-1,2-cyclooctadiene, 9CI
[108186-13-6]

C$_{12}$H$_{20}$ M 164.290
First stable 1,2-cyclooctadiene. Distillable liq.

Price, J.D. *et al*, *Tetrahedron Lett.*, 1986, **27**, 4679 (*synth, cmr, pmr, ir*)

2-*tert*-Butyl-2,4-dihydro-6-methyl-1,3-dioxol-4-one   B-60345

2-tert-Butyl-6-methyl-1,3-dioxan-4-one

C$_9$H$_{14}$O$_3$ M 170.208
(**R**)-*form*
  Chiral acetoacetic acid enol acetal with synthetic uses. Enolate stable at low temps. and can be alkylated stereospecifically. Mp 48.5°. $[\alpha]_D$ −217.7° (c, 1.2 in CHCl$_3$).

Seebach, D. *et al*, *Helv. Chim. Acta*, 1986, **69**, 1147 (*synth, pmr, use*)

1-*tert*-Butyl-2-iodobenzene   B-60346

1-(1,1-Dimethylethyl)-2-iodobenzene, 9CI
[62171-59-9]

C$_{10}$H$_{13}$I   M 260.117
Liq. Bp$_{10}$ 116-118°, Bp$_3$ 94-96°.

Lesslie, M.S. *et al*, *J. Chem. Soc.*, 1961, 611 (*synth*)
Berger, S., *Tetrahedron*, 1976, **32**, 2451 (*cmr*)
Akermark, B. *et al*, *J. Organomet. Chem.*, 1978, **149**, 97 (*synth*)

1-*tert*-Butyl-3-iodobenzene   B-60347

1-(1,1-Dimethylethyl)-3-iodobenzene, 9CI
[58164-02-6]
C$_{10}$H$_{13}$I   M 260.117
Liq. Bp$_9$ 106-108°.

Shoesmith, J.B. *et al*, *J. Chem. Soc.*, 1928, 2334 (*synth*)
Berger, S., *Tetrahedron*, 1976, **32**, 2451 (*cmr*)

1-*tert*-Butyl-4-iodobenzene   B-60348

1-(1,1-Dimethylethyl)-4-iodobenzene, 9CI
[35779-04-5]
C$_{10}$H$_{13}$I   M 260.117
Liq. Bp$_{0.05}$ 114°.

Shoesmith, J.B. *et al*, *J. Chem. Soc.*, 1928, 2334 (*synth*)
Berger, S., *Tetrahedron*, 1976, **32**, 2451 (*cmr*)
Fields, E.K. *et al*, *J. Org. Chem.*, 1978, **43**, 4705 (*synth*)
Barluenga, J. *et al*, *J. Chem. Soc., Perkin Trans. 1*, 1984, 2623 (*synth*)
Ranganathan, S. *et al*, *Tetrahedron Lett.*, 1985, **26**, 4955.

2-*tert*-Butyl-4(3*H*)-pyrimidinone   B-60349

C$_8$H$_{12}$N$_2$O   M 152.196
Off-white solid. Mp 150-151° (147-148°).

Tielemans, M. *et al*, *J. Heterocycl. Chem.*, 1987, **24**, 705 (*synth, pmr*)

3-*tert*-Butyl-2,2,4,5,5-tetramethyl-3-hexene   B-60350

3-(1,1-Dimethylethyl)-2,2,4,5,5-pentamethyl-3-hexene, 9CI. Tri-tert-butylmethylethylene
[89046-74-2]

C$_{15}$H$_{30}$ M 210.402
Oil.

Krebs, A. *et al*, *Tetrahedron*, 1986, **42**, 1693 (*synth, ir, pmr, cmr*)

2-Butynedial, 9CI   B-60351

Acetylenedicarboxaldehyde

[21251-20-7]

OHCC≡CCHO

$C_4H_2O_2$ M 82.059

Forms Diels-Alder adducts. Obt. in soln. only, readily polymerises.

Mono(diethyl acetal): [74149-25-0]. *4,4-Diethoxy-2-butynal.*
$C_8H_{12}O_3$ M 156.181
Pale-yellow liq. Bp_4 73-74°. Stable stored at −20°.

Bis(diethyl acetal): [3975-08-4]. *1,1,4,4-Tetraethoxy-2-butyne.*
$C_{12}H_{22}O_4$ M 230.303
Cryst. at low temp. Mp 18-19°. $Bp_{0.6}$ 91-3°.

Henkel, C. *et al*, *Ber.*, 1943, **76**, 812 (*deriv*)
Gorgues, A. *et al*, *J. Chem. Soc., Chem. Commun.*, 1979, 765 (*synth, ir, pmr, use*)
Gorgues, A. *et al*, *Tetrahedron*, 1986, **42**, 351 (*synth, ir, pmr, props*)

C

ε-Cadinene
C-60001

Updated Entry replacing C-50010
Decahydro-1,6-bis(methylene)-4-(1-methylethyl)-naphthalene, 9CI. Decahydro-4-isopropyl-1,6-dimethy-lenenaphthalene. 4(14),10(15)-Cadinadiene
[25548-04-3]

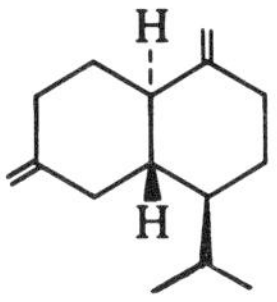

Absolute configuration

$C_{15}H_{24}$ M 204.355
Constit. of *Mentha arvensis* and *Juniperus communis.*
Oil. Bp_2 99-100°. $[\alpha]_D^{22}$ +50.7° (c, 1.5 in $CHCl_3$).

Burk, L.A. *et al, Tetrahedron,* 1976, **32**, 2083 (*bibl*)
Koster, F.-H. *et al, Justus Liebigs Ann. Chem.,* 1986, 78 (*synth, cmr*)
Hagiwara, H. *et al, J. Chem. Soc., Chem. Commun.,* 1987, 1333 (*synth*)

Caesalpin *F*
C-60002

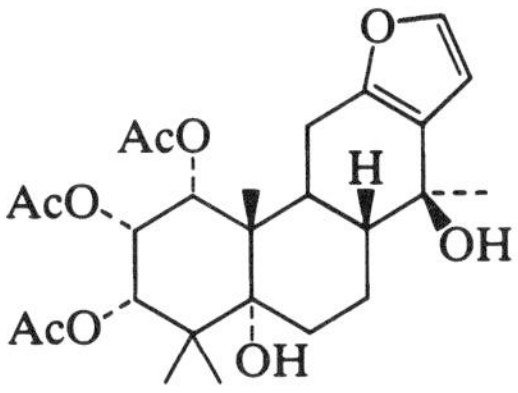

$C_{26}H_{36}O_9$ M 492.565
Constit. of *Caesalpinia bonducella.* Cryst. (EtOAc). Mp 236.5-237.5°.

Pascoe, K.O. *et al, J. Nat. Prod.,* 1986, **49**, 913.

Caffeidine
C-60003

N,1-Dimethyl-4-(methylamino)-1H-imidazole-5-carboxamide, 9CI. 1-Methyl-4-methylamino-5-(methylaminocarbonyl)imidazole
[20041-90-1]

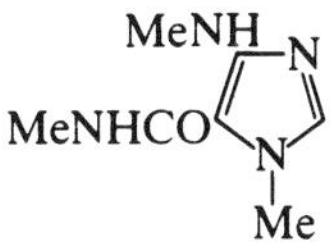

$C_7H_{12}N_4O$ M 168.198
Hydrol. prod. of Caffeine, C-00038 . Cryst. (Et_2O). Mp 93° (83-85°).

Ac:
   $C_9H_{14}N_4O_3$ M 226.235
   Cryst. (C_6H_6). Mp 143-145°.

Biltz, H. *et al, Ber.,* 1928, **61**, 1409.
Hoskinson, R.M., *Aust. J. Chem.,* 1968, **21**, 1913 (*synth, pmr, uv*)
Ohsaki, T. *et al, Chem. Pharm. Bull.,* 1986, **34**, 36.

Cajaflavanone
C-60004

Updated Entry replacing C-00047
Erythrisenegalone
[68236-12-4]

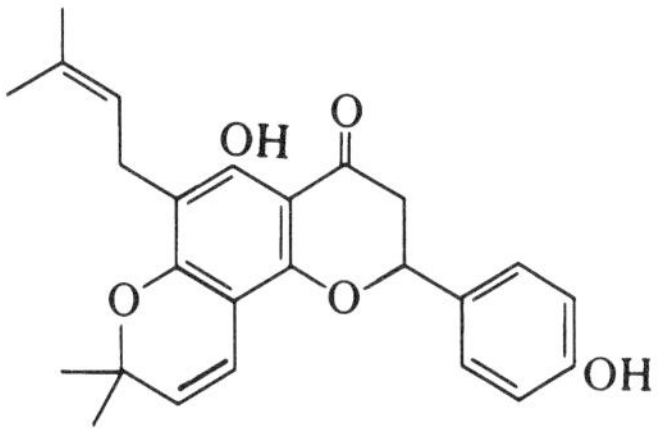

$C_{25}H_{26}O_5$ M 406.477
Constit. of *Cajanus cajan* and *Erythrisia senegalensis.*
Straw-coloured cryst. Mp 129-130° (122-124°). $[\alpha]_D$ −66.6° (c, 1 in $CHCl_3$), $[\alpha]_D$ −5° (c, 1 in $CHCl_3$).

(±)-*form* [68682-03-1]
Light-yellow cryst. (C_6H_6/pet. ether). Mp 138-140°.

Bhanumati, S. *et al, Phytochemistry,* 1978, **17**, 2045 (*isol*)
Jain, A.C. *et al, Tetrahedron,* 1978, **34**, 2607 (*synth*)
Nagar, A. *et al, Tetrahedron Lett.,* 1978, 2031 (*synth*)
Fomum, Z.T. *et al, Phytochemistry,* 1985, **24**, 3075 (*isol*)

Caleine *E*
C-60005

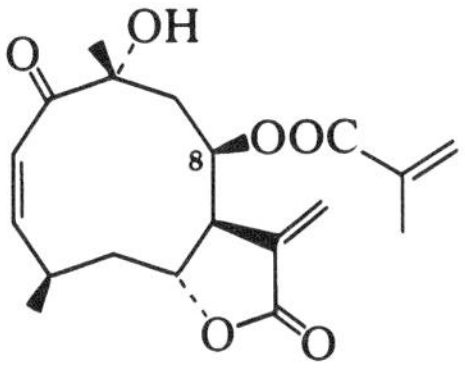

$C_{19}H_{24}O_6$ M 348.395
Constit. of *Calea zacatechichi.* Cryst. (Me_2CO/diisopropyl ether). Mp 150-151°. $[\alpha]_D$ +47.4° ($CHCl_3$).

8-Deacyl, 8-tigloyl: Caleine F.
   $C_{20}H_{26}O_6$ M 362.422
   From *C. zacatechichi.* Cryst. Mp 141-143°.

Martinez, M. *et al, Phytochemistry,* 1987, **26**, 2104.

Calopogonium isoflavone *B*
C-60006

3-(3,4-Methylenedioxyphenyl)-8,8-dimethyl-4H,8H-benzo[1,2-b:3,4-b']dipyran-4-one
[62502-14-1]

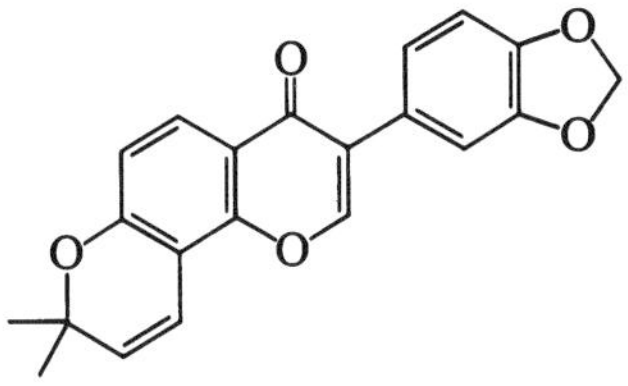

$C_{21}H_{16}O_5$ M 348.354
Constit. of *Calopogonium mucunoides.* Pale-yellow cryst. (Et_2O). Mp 169-171°.

Vilain, C. *et al, Bull. Soc. R. Sci. Liege,* 1976, **45**, 468 (*isol, struct*)

Vilain, C. *et al*, *Bull. Soc. Chim. Belg.*, 1977, **86**, 237 (*synth*)
Schuda, P.F. *et al*, *J. Org. Chem.*, 1987, **52**, 1972 (*synth*)

Caloverticillic acid *A*   C-60007

$C_{34}H_{50}O_6$ M 554.765
Constit. of *Calophyllum verticillatum*.
Me ether, Me ester: Amorph. $[\alpha]_D$ −78.5° (c, 0.5 in CHCl$_3$).
Diastereoisomer: **Caloverticillic acid B**.
   $C_{34}H_{50}O_6$ M 554.765
   Constit. of *C. verticillatum*.
Diastereoisomer, Me ether, Me ester: Amorph. $[\alpha]_D$ 0° (c, 0.5 in CHCl$_3$).

Ravelonjato, B. *et al*, *Phytochemistry*, 1987, **26**, 2973.

Caloverticillic acid *C*   C-60008

$C_{34}H_{50}O_6$ M 554.765
Constit. of *Calophyllum verticillatum*.
Me ether, Me ester: Amorph. $[\alpha]_D$ +6° (c, 1 in CHCl$_3$).
Ravelonjato, B. *et al*, *Phytochemistry*, 1987, **26**, 2973.

Cantabradienic acid   C-60009
[106941-41-7]

$C_{15}H_{20}O_2$ M 232.322
Constit. of *Artemisia cantabrica*.
Me ester: [106941-33-7]. Pale-yellow oil. $[\alpha]_D^{24}$ −22.3° (c, 0.5 in CHCl$_3$).
San Feliciano, A. *et al*, *J. Nat. Prod.*, 1986, **49**, 845.

Cantabrenolic acid   C-60010
[106941-43-9]

$C_{15}H_{22}O_3$ M 250.337
Constit. of *Artemisia cantabrica*.
Me ester: [106941-36-0]. Viscous oil. $[\alpha]_D^{24}$ −47.9° (c, 1.2 in CHCl$_3$).
5-Ketone: [106941-40-6]. **Cantabrenonic acid**.
   $C_{15}H_{20}O_3$ M 248.321
   From *A. cantabrica*.
5-Ketone, Me ester: [106941-35-9]. Cryst. (hexane). Mp 70°. $[\alpha]_D^{24}$ −36.9° (c, 1 in CHCl$_3$).
5-Ketone, 6α,7α-epoxide: [106941-42-8]. **Epoxycantabronic acid**.
   $C_{15}H_{20}O_4$ M 264.321
   From *A. cantabrica*.
5-Ketone, 6α,7α-epoxide, Me ester: [106941-34-8]. Cryst. (hexane). Mp 63°. $[\alpha]_D^{24}$ −5.2° (c, 0.8 in CHCl$_3$).

San Feliciano, A. *et al*, *J. Nat. Prod.*, 1986, **49**, 845.

Capsenone   C-60011
[41720-93-8]

$C_{15}H_{22}O_2$ M 234.338
Detoxification product produced from capsidiol by *Botrytis cinerea* and *Fusarium oxysporum*.
9,10β-Dihydro: [109986-01-8]. cis-*9,10-Dihydrocapsenone*.
   $C_{15}H_{24}O_2$ M 236.353
   Constit. of *Capsicum annuum*.

Ward, E.W.B. *et al*, *Phytopathology*, 1972, **62**, 1186 (*isol*)
Whitehead, I.M. *et al*, *Phytochemistry*, 1987, **26**, 1367 (*isol, deriv, pmr*)

1*H*-Carbazole-1,4(9*H*)-dione, 9CI   C-60012
1,4-Dihydro-1,4-carbazoledione
[70377-05-8]

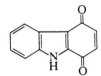

$C_{12}H_7NO_2$ M 197.193
Mp 136°.

Benzies, D.W.M. *et al*, *J. Chem. Soc., Chem. Commun.*, 1986, 1019 (*synth, uv, ir, pmr, cmr*)

Carbonochloridothioic acid, 9CI   C-60013
Chlorothioformic acid, 8CI. Thiochloroformic acid
[16890-85-0]

$$ClC(S)OH \rightleftharpoons ClC(O)SH$$

CHClOS   M 96.531
Derivs. only known. Monochloride of Thiocarbonic acid, T-02027 . O-Derivs. are alkoxythiocarbonylating agents.

O-*Me:* [2812-72-8].
C_2H_3ClOS M 110.558
Bp 105-107°.
O-*Et:* [2812-73-9].
C_3H_5ClOS M 124.585
Bp_{40} 52-55°, Bp_{33} 46°.
▷LQ6950000.
O-*Propyl:* [2812-74-0].
C_4H_7ClOS M 138.612
Bp_{12} 42°.
O-*Isopropyl:* [38363-18-7].
C_4H_7ClOS M 138.612
Bp_{10} 34° (Bp_4 35°).
O-*Butyl:* [2812-75-1].
C_5H_9ClOS M 152.639
Bp_{12} 62°.
O-tert-*Butyl:*
C_5H_9ClOS M 152.639
Bp_{10} 54°.
S-*Et:* [2941-64-2].
C_3H_5ClOS M 124.585
Bp 136°.
S-*Propyl:* [13889-92-4].
C_4H_7ClOS M 138.612
Bp_{26} 59-60°.
▷LQ7005000.
S-*Butyl:* [13889-94-6].
C_5H_9ClOS M 152.639
Bp 179°.
S-*Ph:* [13464-19-2].
C_7H_5ClOS M 172.629
Bp_{13} 104°.

Reimschneider, R. *et al, Monatsh. Chem.,* 1953, **84**, 518 (*synth*)
Tilles, H., *J. Am. Chem. Soc.,* 1959, **81**, 714 (*synth*)
Martin, D. *et al, Chem. Ber.,* 1965, **98**, 2059 (*synth, use*)
McKinnon, D.M. *et al, Can. J. Chem.,* 1972, **50**, 1401 (*props*)
Barany, G. *et al, J. Org. Chem.,* 1983, **48**, 4750 (*synth, props, bibl*)
Martinez, M. *et al, Synthesis,* 1986, 760 (*synth, ir, pmr, use*)

1,1'-Carbonothioylbis-2(1H)pyridinone, 9CI   C-60014
1,1'-Thiocarbonyl-2,2'-pyridone
[102368-13-8]

$C_{11}H_8N_2O_2S$ M 232.256
Reagent for prepn. of nitriles, carbodiimides, cyclic thionocarbonates and isothiocyanates and deoxygenation of alcohols under neutral conditions. Dark-orange cryst. solid. Mp 162-164°.
Kim, S. *et al, J. Org. Chem.,* 1986, **51**, 2613 (*synth, ir, pmr, use*)

O,O'-Carbonylbis(hydroxylamine)   C-60015
$CO(ONH_2)_2$

$CH_4N_2O_3$ M 92.054
N,N'-*Di-tert*-butyl: [103258-83-9]. N,N'-
[*Carbonylbis(oxy)*]*bis[2-methyl-2-propanamine], 9CI.*
$C_9H_{20}N_2O_3$ M 204.269

Cryst. (hexane). Mp 82.5-84.0°.
Stowell, J.C. *et al, J. Org. Chem.,* 1986, **51**, 3355 (*synth, ir, pmr*)

3-(Carboxymethylamino)propanoic acid   C-60016
N-(*Carboxymethyl*)-β-*alanine, 9CI.* N-(*Carboxyethyl*)-*glycine. Iminopropionicacetic acid*
[505-72-6]

$$HOO^1CCH_2CH_2NHCH_2COOH$$

$C_5H_9NO_4$ M 147.130
Isol. from *Phaseolus radiatus* var. *typicus.* Cryst. (H_2O or EtOH aq.). Mp 195-198° (191-192°) dec.
B,HCl: Mp 120-122°.
Di-Et ester: [3783-61-7].
$C_9H_{17}NO_4$ M 203.238
Oil. Bp 253° sl. dec., Bp_1 98-99°.
Di-Et ester; B,HCl: Mp 85-87°.
1-Nitrile: [3088-42-4]. N-(*2-Cyanoethyl*)*glycine, 9CI. 3-*
(*Carboxymethylamino*)*propionitrile.*
$C_5H_8N_2O_2$ M 128.130
Cryst. Mp 193-195°.
▷MB9275000.

Cocker, W. *et al, J. Chem. Soc.,* 1952, 1182 (*deriv*)
McKinney, L.L. *et al, J. Am. Chem. Soc.,* 1952, **74**, 1942, 5183 (*synth*)
Coburn, M.D. *et al, J. Heterocycl. Chem.,* 1965, **2**, 308 (*deriv*)
Kasai, T. *et al, Agric. Biol. Chem.,* 1971, **35**, 1603 (*isol, synth*)
Kawashiro, K. *et al, Bull. Chem. Soc. Jpn.,* 1984, **57**, 1097, 2871 (*synth, ms*)

2-Carboxy-3-thiopheneacetic acid   C-60017
[57279-42-2]

$C_7H_6O_4S$ M 186.182
Mp 189-191°.
Anhydride: [104292-92-4]. *5H-Thieno[2,3-c]pyran-5,7(4H)dione, 9CI.*
$C_7H_4O_3S$ M 168.167
Cryst. (C_6H_6). Mp 149-153°.
Ames, D.E. *et al, J. Chem. Soc., Perkin Trans. 1,* 1975, 1390 (*synth*)
Kita, Y. *et al, J. Org. Chem.,* 1986, **51**, 4150 (*deriv, synth, ir, pmr*)

4-Carboxy-3-thiopheneacetic acid   C-60018
[57279-54-6]

$C_7H_6O_4S$ M 186.182
Cryst. (H_2O). Mp 210-213°.
Anhydride: [57279-38-6]. *4H-Thieno[3,4-c]pyran-4,6(7H)-dione, 9CI.*
$C_7H_4O_3S$ M 168.167
Cryst. (C_6H_6). Mp 157-158°.
Acetamide:
$C_7H_7NO_3S$ M 185.197
Cryst. (EtOH). Mp 193-194°.
N-*Methylacetamide:*
$C_8H_9NO_3S$ M 199.224

Cryst. (H$_2$O). Mp 162-164°.

Ames, D.L. *et al*, *J. Chem. Soc., Perkin Trans. 1*, 1975, 1390 (*synth, pmr*)

3-Caren-5-one   C-60019
4,7,7-Trimethylbicyclo[4.1.0]heptan-2-one

C$_{10}$H$_{14}$O   M 150.220
Constit. of *Kaempferia galanga*. Oil.

Kiuchi, F. *et al*, *Phytochemistry*, 1987, **26**, 3350.

Carnosine   C-60020
Updated Entry replacing C-00360
N-*β-Alanylhistidine*, *9CI*

H$_2$NCH$_2$CH$_2$CONH−C−H

(S)-form
Absolute
configuration

C$_9$H$_{14}$N$_4$O$_3$ M 226.235
(R)-form [5853-00-9]
D-form
No depressor activity on blood pressure. Mp 260°. $[\alpha]_D^{28}$ −20.4° (c, 1.5 in H$_2$O).
B,HCl: Mp 245°.
Benzyloxycarbonyl: Mp 161°. $[\alpha]_D^{24}$ −11° (H$_2$O).
(S)-form [305-84-0]
L-form
Occurs in the skeletal muscles of many animals and man. No biochemical role has yet been assigned although it exerts a depressor effect on blood pressure. Needles. Mp 246-250° dec. $[\alpha]_D^{20}$ +24.1° (c, 1.5 in H$_2$O).
▷MS3080000.
B,HCl: [5852-99-3]. Mp 245°.
B,HNO$_3$: [5852-98-2]. Needles. Mp 219° (222° dec.).
B$_2$,H$_2$SO$_4$: Dec. at 238-40°.

Rinderknecht, H. *et al, J. Org. Chem.*, 1964, **29**, 1968 (*synth*)
Pietta, P.G. *et al, Ann. Chim.* (*Rome*), 1968, **58**, 1431 (*synth*)
Barrons, Y. *et al, J. Mol. Struct.*, 1976, **30**, 225 (*cryst struct*)
Itoh, H. *et al, Acta. Crystallogr., Sect. B*, 1977, **33**, 2959 (*cryst struct*)
Friedrich, J.O. *et al, Can. J. Chem.*, 1986, **64**, 2132 (*cmr, pmr*)

8-Carotene-4,6,10-triol   C-60021
8-Daucene-4,6,10-triol
[103654-31-5]

C$_{15}$H$_{26}$O$_3$ M 254.369

(4β,6α,10α)-form
Pallinol
Oil. $[\alpha]_D^{20}$ +37.3° (c, 0.1 in CHCl$_3$).
O^6,O^{10}-*Diangeloyl:* [90695-04-8]. **Pallinin**.
C$_{25}$H$_{38}$O$_5$ M 418.572
Constit. of *Ferula pallida*. Cryst. (EtOAc/pet. ether). Mp 79-80°. $[\alpha]_D^{20}$ −148.5° (c, 0.1 in CHCl$_3$).

Kushmuradov, A.Yu. *et al, Khim. Prir. Soedin.*, 1986, **22**, 48.

Carriomycin   C-60022
Updated Entry replacing C-00389
6,27-Didemethoxyantibiotic A 204A. T 42082. Antibiotic T 42082
[65978-43-0]

C$_{47}$H$_{80}$O$_{15}$ M 885.140
Polyether antibiotic. Isol. from *Streptomyces hygroscopicus*. Ionophore. Needles. Mp 120-122°. Structurally related to Septamycin, S-00300 .
▷CB9376436.
Na salt: Mp 180-182°. $[\alpha]_D^{25}$ −4.5° (c, 1 in CHCl$_3$).

Imada, A. *et al, J. Antibiot.*, 1978, **31**, 7.
Radios, N.A. *et al, Bull. Soc. Chim. Belg.*, 1978, **87**, 437 (*conformn*)
Nakayama, H. *et al, J. Chem. Soc., Perkin Trans. 2*, 1979, 293 (*cryst struct*)
Can. Pat., 1 091 175, (*1980*); *CA*, **94**, 154981 (*isol*)

[2',2']-Catechin-taxifolin   C-60023

C$_{30}$H$_{24}$O$_{13}$ M 592.512
Constit. of willow bark (*Salix* spp.).

Kolodziej, H. *et al, J. Chem. Soc., Chem. Commun.*, 1987, 205.

Cavoxinine   C-60024
3-Hydroxy-5-methoxy-4-(1-oxo-2,4-octadienyl)-benzeneacetic acid, 9CI
[109517-71-7]

C$_{17}$H$_{20}$O$_5$ M 304.342
Metab. of *Phoma cava*. Pale-yellow oil.

Evidente, A., *J. Nat. Prod.*, 1987, **50**, 173.

Cavoxinone   C-60025

3,4-Dihydro-5-methoxy-4-oxo-2-(1-pentenyl)-2H-1-benzopyran-7-acetic acid, 9CI

[109517-70-6]

$C_{17}H_{20}O_5$ M 304.342

Metab. of *Phoma cava*. Oil.

Evidente, A., *J. Nat. Prod.*, 1987, **50**, 173.

Cedrelopsin   C-60026

7-Hydroxy-6-methoxy-8-(3-methyl-2-butenyl)-2H-1-benzopyran-1-one, 9CI. 8-Dimethylally-7-hydroxy-6-methylcoumarin

[19397-28-5]

$C_{15}H_{16}O_4$ M 260.289

Constit. of *Cedrelopsis grevei* timber. Pale-yellow needles (MeOH). Mp 170-174°.

Me ether: [73815-13-1]. *O-Methylcedrelopsin*.
$C_{16}H_{18}O_4$ M 274.316
Constit. of *Zanthoxylum usambarense*. Cryst. Mp 66-68°.

O-(3-Methyl-2-butenyl): **Brayleanin**.
$C_{20}H_{24}O_4$ M 328.407
Isol. from *Flindersia brayleana*. Cubes. Mp 95°.

Anet, F.A. *et al*, *Aust. J. Sci. Res., Sect. A*, 1949, **2**, 608 (*Brayleanin*)
Eshiett, I.T. *et al*, *J. Chem. Soc. (C)*, 1968, 481 (*isol, struct*)
Kokworo, J.O. *et al*, *Planta Med.*, 1983, **47**, 251 (*deriv*)

Cedronellone   C-60027

$C_{20}H_{28}O_2$ M 300.440

Constit. of *Cedronella canariensis*. Cryst. (MeOH). Mp 106-107°. $[\alpha]_D^{20}$ +79.3° (c, 0.916 in $CHCl_3$).

Carreiras, M.C. *et al*, *Phytochemistry*, 1987, **26**, 3351.

3,7,11,15(17)-Cembratetraen-16,2-olide   C-60028

Updated Entry replacing C-10052

$C_{20}H_{28}O_2$ M 300.440

(1S,2S,3E,7E,11E)-form

Constit. of *Sinularia mayi* and *Lobophytum michaelae*. Cryst. Mp 98-99°. $[\alpha]_D$ +89.9° (c, 1.9 in $CHCl_3$).

(1S,2R,3E,7E,11E)-form

Constit. of *S. mayi*. Oil. $[\alpha]_D$ −29.0° (c, 3.40 in $CHCl_3$).

Kodama, M. *et al*, *Tetrahedron Lett.*, 1982, **23**, 5175 (*synth*)
Uchio, Y. *et al*, *Chem. Lett.*, 1982, 277 (*isol*)
Aoki, M. *et al*, *Chem. Lett.*, 1984, 695 (*synth*)
Marshall, J.A. *et al*, *Tetrahedron Lett.*, 1987, **28**, 5081 (*synth*)

Cephalochromin   C-60029

Updated Entry replacing C-00507

2,2',3,3'-Tetrahydro-5,5',6,6',8,8'-hexahydroxy-2,2'-dimethyl[9,9'-bi-4H-naphtho[2,3-b]pyran]-4,4'-dione, 9CI

[25908-26-3]

$C_{28}H_{22}O_{10}$ M 518.476

Constit. of *Cephalosporium* spp. and *Verticillium* spp. Also isol. from *Nectria* spp. Active against grampositive bacteria. Orange cryst. (hexane). Mp >300°. $[\alpha]_D$ +510° ($CHCl_3$).

3-Me: **Chaetochromin C**.
$C_{29}H_{24}O_{10}$ M 532.503
From *C. gracile*. Yellow cryst. ($CHCl_3$/hexane). Mp 214-217°. $[\alpha]_D^{20}$ +454° (c, 0.08 in dioxan).

3,3'-Di-Me: [75514-37-3]. **Chaetochromin**. *Chaetochromin A*.
$C_{30}H_{26}O_{10}$ M 546.529
Prod. by *Chaetomium thielavioideum*. Mycotoxin; cytoxic and mutagenic. Yellow needles. Mp 222-224°. $[\alpha]_D$ +634° (c, 1 in $CHCl_3$).

3,3'-Di-Me, 2,3-didehydro: **Chaetochromin D**.
$C_{30}H_{24}O_{10}$ M 544.514
From *C. gracile*. Orange cryst. (EtOAc/hexane). Mp >300°. $[\alpha]_D^{20}$ +411° (c, 0.06 in dioxan).

3,3'-Di-Me, stereoisomer: **Chaetochromin B**.
$C_{30}H_{26}O_{10}$ M 546.529
From *C. gracile*. Yellow cryst. ($CHCl_3$/hexane). Mp 204-206°. $[\alpha]_D^{20}$ +524° (c, 0.05 in dioxan).

Tertzakian, G. *et al*, *Proc. Chem. Soc., London*, 1964, 195 (*isol*)
Matsumoto, M. *et al*, *J. Antibiot.*, 1975, **28**, 602 (*isol*)
Nair, M.S.R. *et al*, *Lloydia*, 1975, **38**, 448 (*isol*)
Sekita, K. *et al*, *Chem. Pharm. Bull.*, 1980, **28**, 2428; *Can. J. Microbiol.*, 1981, **27**, 766 (*isol, struct, deriv*)
Ito, Y. *et al*, *CA*, 1983, **98**, 138826 (*derivs*)
Koyama, K. *et al*, *Chem. Pharm. Bull.*, 1987, **35**, 578 (*deriv, bibl*)

> *Consult the Dictionary of Alkaloids for a comprehensive treatment of alkaloid chemistry.*

Ceratenolone

C-60030

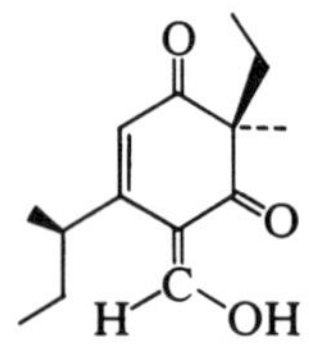

$C_{14}H_{20}O_3$ M 236.310

Metab. of *Ceratocystis minor*. Blue oil. $[\alpha]_D$ 0°, $[\alpha]_{360}$+210° (c, 0.33 in MeOH).

Ayer, W.A. *et al*, *Can. J. Chem.*, 1987, **65**, 765.

Chalepin

C-60031

Updated Entry replacing C-00562

6-(1,1-Dimethyl-2-propenyl)-2,3-dihydro-2-(1-hydroxy-1-methylethyl)-7H-furo[3,2-g] [1]benzopyran-7-one, 9CI. Heliettin

[13164-04-0]

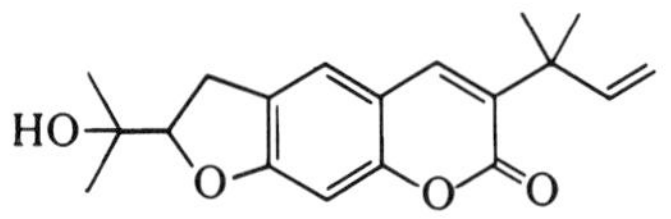

$C_{19}H_{22}O_4$ M 314.380

Constit. of *Ruta chalepensis*. Cryst. Mp 118°, 165-166°. $[\alpha]_D^{24}$ +31° (c, 1.4 in CHCl$_3$) (±0°).

Ac: [14882-94-1]. **Rutamarin**.
$C_{21}H_{24}O_5$ M 356.418
Constit. of *R. chalepensis* and *R. graveolens*. Cryst. Mp 107-108°. $[\alpha]_D^{22}$ +14° (c, 2.3 in CHCl$_3$).

Brooker, R.M. *et al*, *Lloydia*, 1967, **30**, 73 (*isol*)
Pozzi, H. *et al*, *Tetrahedron*, 1967, **23**, 1129 (*isol*)
Reisch, J. *et al*, *Acta Pharm. Suec.*, 1967, **4**, 179 (*isol*)
Sharma, R.B. *et al*, *Indian J. Chem., Sect. B*, 1983, **22**, 538 (*synth*)
Massonet, G.M. *et al*, *Heterocycles*, 1987, **26**, 1541 (*synth*)

Chamaedroxide

C-60032

Updated Entry replacing C-20084

4β,6β;15,16-Diepoxy-2β-hydroxy-13(16),14-clerodadiene-18,19;20,12S-diolide

[82679-43-4]

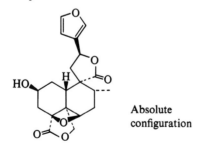

Absolute configuration

$C_{20}H_{22}O_7$ M 374.390

Constit. of *Teucrium chamaedrys*. Cryst. (Me$_2$CO/Et$_2$O). Mp 255-257°. $[\alpha]_D^{20}$ +37.1° (c, 0.42 in Py).

2-Deoxy: **2-Deoxychamaedroxide**.
$C_{20}H_{22}O_6$ M 358.390
Constit. of *T. divaricatum*. Amorph. powder. Mp 85-89°.

Eguren, L. *et al*, *J. Org. Chem.*, 1982, **47**, 4157 (*isol, struct*)
Bruno, M. *et al*, *Phytochemistry*, 1987, **26**, 2859 (*deriv*)

Chamaepitin

C-60033

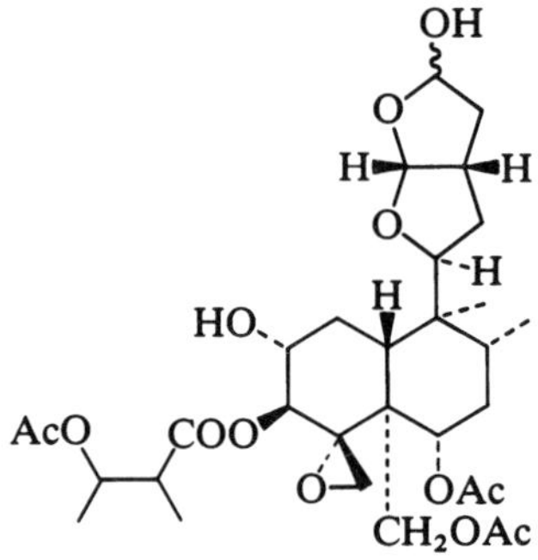

$C_{31}H_{46}O_{13}$ M 626.697

Constit. of *Ajuga chamaepitys*. Amorph.

Camps, F. *et al*, *Phytochemistry*, 1987, **26**, 1475.

Chilenone *B*

C-60034

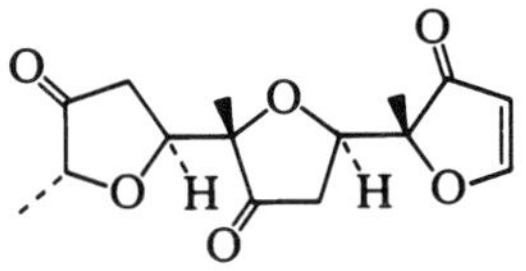

$C_{15}H_{18}O_6$ M 294.304

Constit. of red alga *Laurencia chilensis*. Cryst. (Et$_2$O-/hexane). Mp 111°. Racemate.

San-Martin, A. *et al*, *Tetrahedron Lett.*, 1987, **28**, 6013 (*cryst struct*)

Chinensin I

C-60035

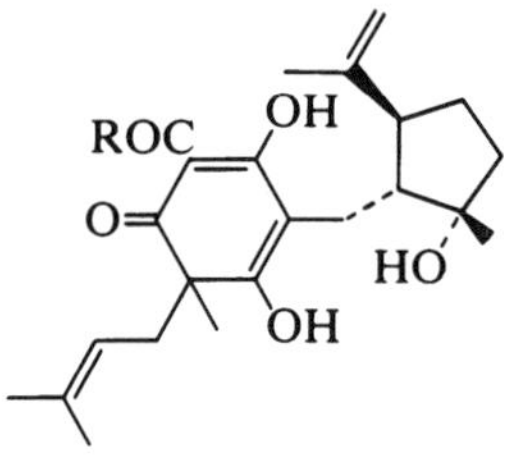

R = −CH(CH$_3$)CH$_2$CH$_3$

$C_{27}H_{40}O_5$ M 444.610

Constit. of flowers of *Hypericum chinense*. Antimicrobial agent. Oil. $[\alpha]_D$ +69° (c, 0.12 in MeOH).

Nagai, M. *et al*, *Chem. Lett.*, 1987, 1337.

Chinensin II

C-60036

As Chinensin I, C-60035 with

R = −CH(CH$_3$)$_2$

$C_{26}H_{38}O_5$ M 430.583

Constit. of flowers of *Hypericum chinense*. Antimicrobial agent. Oil.

Nagai, M. *et al*, *Chem. Lett.*, 1987, 1337.

Chiratanin   C-60037

[109237-38-9]

$C_{30}H_{22}O_{13}$ M 590.496
Constit. of *Swertia chirata*. Yellow needles. Mp 152°.

Mandal, S. *et al, Tetrahedron Lett.*, 1987, **28**, 1309.

Chlamydocin   C-60038

Updated Entry replacing C-30064
SL 3440. Antibiotic SL 3440
[53342-16-8]

$C_{28}H_{38}N_4O_6$ M 526.631
Cyclic peptide antibiotic. Metab. of *Diheterospora chlamydosporia*. Shows antifungal and cytostatic activity. Foam. $[\alpha]_D^{20}$ −147.5° (c, 0.3 in C_6H_6).

Zeller, E.A., *Adv. Enzymol.*, 1948, **8**, 459 (*isol*)
Ger. Pat., 2 011 982, (*1970*); *CA*, **73**, 129583 (*isol*)
Closse, A. *et al, Helv. Chim. Acta*, 1974, **57**, 533 (*struct*)
Kawai, M. *et al, J. Am. Chem. Soc.*, 1983, **105**, 4456 (*conformn*)
Rich, D.H. *et al, Tetrahedron Lett.*, 1983, **24**, 5305 (*synth*)
Schmidt, U. *et al, Tetrahedron Lett.*, 1983, **24**, 3573 (*synth*)
Schmidt, U. *et al, Angew. Chem., Int. Ed. Engl.*, 1984, **23**, 318 (*synth*)
Schmidt, U. *et al, Synthesis*, 1986, 361 (*synth*)

1-Chloroadamantane   C-60039

1-Chlorotricyclo[3.3.1.1³,⁷]decane, 9CI
[935-56-8]

$C_{10}H_{15}Cl$ M 170.682
Cryst. (MeOH aq. or by subl.). Mp 165°.

Stetter, H. *et al, Chem. Ber.*, 1959, **92**, 1629 (*synth*)

2-Chloroadamantane   C-60040

2-Chlorotricyclo[3.3.1.1³,⁷]decane, 9CI
[7346-41-0]
$C_{10}H_{15}Cl$ M 170.682
Waxy cryst. Mp 190-191.8° (186-188°).

Hoek, W. *et al, Recl. Trav. Chim. Pays-Bas*, 1966, **85**, 1045 (*synth*)

Kovacic, P. *et al, J. Org. Chem.*, 1971, **36**, 3138 (*synth*)

2-Chloro-1,4,9,10-anthracenetetrone, 9CI   C-60041

[109141-97-1]

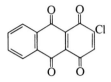

$C_{14}H_5ClO_4$ M 272.644
Cryst. (nitrobenzene/CS_2). Mp 206-208°.

Cano, P. *et al, J. Chem. Soc., Perkin Trans. 1*, 1986, 1923 (*synth, pmr*)

2-Chlorobenzo[*b*]thiophene, 9CI   C-60042

[7342-85-0]

C_8H_5ClS M 168.641
Mp 34°. Bp$_{0.1}$ 60-70°.
1-Oxide: [57147-28-1].
     C_8H_5ClOS M 184.640
     Mp 67-68°.
1,1-Dioxide: [10133-41-2].
     $C_8H_5ClO_2S$ M 200.639
     Mp 142-143°.

Van Zyl, G. *et al, Can. J. Chem.*, 1966, **44**, 2283 (*synth*)
Geneste, P. *et al, Bull. Soc. Chim. Fr.*, 1977, 271 (*synth*)
Geneste, P. *et al, J. Org. Chem.*, 1979, **44**, 2887 (*cmr*)

3-Chlorobenzo[*b*]thiophene, 9CI   C-60043

[7342-86-1]
C_8H_5ClS M 168.641
Bp$_5$ 96-101°.
1-Oxide: [63724-95-8].
     C_8H_5ClOS M 184.640
     Mp 113-114°.
1,1-Dioxide: [21211-29-0].
     $C_8H_5ClO_2S$ M 200.639
     Mp 170°.

Schlesinger, A.H. *et al, J. Am. Chem. Soc.*, 1951, **73**, 2614 (*synth*)
Schmitt, J. *et al, Bull. Soc. Chim. Fr.*, 1968, 4575 (*deriv*)
Geneste, P. *et al, Bull. Soc. Chim. Fr.*, 1977, 271 (*synth*)
Geneste, P. *et al, J. Org. Chem.*, 1979, **44**, 2887 (*cmr*)

4-Chlorobenzo[*b*]thiophene, 9CI   C-60044

[66490-33-3]
C_8H_5ClS M 168.641
Pale-yellow oil. Bp$_{15}$ 123-125°.
Picrate: Cryst. (EtOH). Mp 135-136°.

Hansch, C. *et al, J. Org. Chem.*, 1955, **20**, 1056 (*synth*)
Clark, P.D. *et al, J. Chem. Res. (S)*, 1978, 10 (*synth*)

5-Chlorobenzo[*b*]thiophene, 9CI   C-60045

[20532-33-6]
C_8H_5ClS M 168.641
Mp 34-36°. Bp$_4$ 84° (bath).

Picrate: Cryst. (EtOH). Mp 71-72°.

El Shanta, M.S. *et al, J. Chem. Soc. (C),* 1967, 2084 (*synth*)
Caddy, B. *et al, Aust. J. Chem.,* 1968, **21**, 1853 (*pmr*)

6-Chlorobenzo[*b*]thiophene, 9CI   C-60046

[66490-20-8]

C_8H_5ClS M 168.641
Cryst. (EtOH). Mp 42-43°.
Picrate: Cryst. (EtOH). Mp 74-75°.

Hansch, C. *et al, J. Org. Chem.,* 1955, **20**, 1056 (*synth*)

7-Chlorobenzo[*b*]thiophene, 9CI   C-60047

[90407-14-0]

C_8H_5ClS M 168.641
Oil. Bp_{10} 115°, $Bp_{0.7}$ 90-94°.

1,1-Dioxide:
    $C_8H_5ClO_2S$ M 200.639
    Cryst. (AcOH aq.). Mp 79-80°.

Mustafa, A. *et al, J. Am. Chem. Soc.,* 1956, **78**, 6174 (*synth*)
Rahman, L.K.A. *et al, J. Chem. Soc., Perkin Trans. 1,* 1984, 385 (*synth, pmr*)

2-Chlorobenzoxazole   C-60048

[615-18-9]

C_7H_4ClNO M 153.568
Bp 200°, Bp_{10} 90-96°.
▷DM4680000.

B,HCl: Cryst. Mp 57-58° dec. Unstable.
B,HBr: Cryst. Mp 155° dec.

McCoy, H.N., *Am. Chem. J.,* 1899, **21**, 122 (*synth*)
Profft, E. *et al, Z. Chem.,* 1965, **5**, 178 (*synth*)

2-Chloro-2-cyclohepten-1-one, 9CI   C-60049

[67382-69-8]

C_7H_9ClO M 144.601
Liq. $Bp_{0.05}$ 49-53°. Rapid dec. at r.t.

2,4-Dinitrophenylhydrazone: Red-purple cryst. Mp 198-199°.

Ohno, M., *Tetrahedron Lett.,* 1963, 1753 (*synth, ir, uv, pmr*)
Buckley, D.J. *et al, J. Chem. Soc., Perkin Trans. 1,* 1985, 2193 (*synth, ir, pmr, cmr*)

2-Chloro-2-cyclohexen-1-one, 9CI   C-60050

Updated Entry replacing C-10094

[3400-88-2]

C_6H_7ClO M 130.574

Mp 72°.

Laryutina, E.A. *et al, CA,* 1972, **76**, 59035n (*synth*)
Grenier-Loustalot, M.F. *et al, Synthesis,* 1976, 33 (*synth*)
Ley, S.V. *et al, Tetrahedron Lett.,* 1981, **22**, 3301 (*synth*)
Buckley, D.J. *et al, J. Chem. Soc., Perkin Trans. 1,* 1985, 2193 (*synth, ir, pmr, cmr*)

2-Chloro-2-cyclopenten-1-one, 9CI   C-60051

Updated Entry replacing C-10096

[3400-89-3]

C_5H_5ClO M 116.547
Oil.

Ley, S.V. *et al, Tetrahedron Lett.,* 1981, **22**, 3301 (*synth*)
Buckley, D.J. *et al, J. Chem. Soc., Perkin Trans. 1,* 1985, 2193 (*synth, ir, pmr, cmr*)

2-Chloro-4,6-difluoro-1,3,5-triazine, 9CI   C-60052

[696-85-5]

$C_3ClF_2N_3$ M 151.503
d_4^{20} 1.632. Mp 23.5°. Bp 113°. V. sensitive to hydrolysis.

Maxwell, A.F. *et al, J. Am. Chem. Soc.,* 1958, **80**, 548 (*synth*)
Sawodny, W. *et al, Spectrochim. Acta, Part A,* 1967, **23**, 1327 (*ir, raman*)
Young, J.A. *et al, J. Org. Chem.,* 1967, **32**, 2237 (*F nmr*)

6-Chloro-3,5-difluoro-1,2,4-triazine   C-60053

[82736-94-5]
$C_3ClF_2N_3$ M 151.503
Solid. Mp 28-30°.

Barlow, M.G. *et al, J. Chem. Soc., Perkin Trans. 1,* 1982, 1251 (*synth, F nmr*)

2-Chloro-2,3-dihydro-1*H*-imidazole-4,5-dione   C-60054

2-Chloro-4,5-imidazolidinedione

$C_3H_3ClN_2O_2$ M 134.522

(±)-*form*

1,3-Di-Ph: [104716-67-8].
    $C_{15}H_{11}ClN_2O_2$ M 286.717
    Solid which partially dec. on recryst. Mp 152-160° dec.
    Readily hydrolysed.

Barsa, E.A. *et al, J. Org. Chem.,* 1986, **51**, 4483 (*deriv, synth, ir, pmr, cmr*)

4-Chloro-2,3-dihydro-1*H*-indole, 9CI   C-60055
4-Chloroindoline
[41910-64-9]

C_8H_8ClN M 153.611
Liq. Bp_{35} 160-162°, Bp_{10} 135°.
Picrate: Yellow needles (EtOH). Mp 182-183°.
1-Me:
   $C_9H_{10}ClN$ M 167.638
   Needles (cyclohexane). Mp 90-91°.
1-Benzyl: [102493-70-9].
   $C_{15}H_{14}ClN$ M 243.735
   Mp 49-51°. Bp_1 155-160°.

Huisgen, R. *et al, Chem. Ber.*, 1960, **93**, 1496 (*synth*)
Florvall, L. *et al, J. Med. Chem.*, 1986, **29**, 1406 (*synth*)

2-Chloro-3,6-dihydroxybenzaldehyde, 9CI   C-60056
6-Chlorogentisaldehyde, 8CI
[32744-84-6]
$C_7H_5ClO_3$ M 172.568
Orange-yellow needles (CH_2Cl_2). Mp 144-147°.

Séquin-Frey, M. *et al, Helv. Chim. Acta*, 1971, **54**, 851 (*synth*)

3-Chloro-2,5-dihydroxybenzaldehyde   C-60057
3-Chlorogentisaldehyde, 8CI
[32744-83-5]
$C_7H_5ClO_3$ M 172.568
Light-yellow cryst. (CH_2Cl_2). Mp 120-129°.

Kurdukar, R. *et al, CA*, 1964, **60**, 11972 (*deriv*)
Séquin-Frey, M. *et al, Helv. Chim. Acta*, 1971, **54**, 851 (*synth*)

3-Chloro-2,6-dihydroxybenzaldehyde   C-60058
3-Chloro-γ-resorcylaldehyde
$C_7H_5ClO_3$ M 172.568
2-Me ether: [84290-29-9]. *3-Chloro-6-hydroxy-2-
   methoxybenzaldehyde.*
$C_8H_7ClO_3$ M 186.595
No phys. props. in abstract.
Eur. Pat., 54 924, (*1982*); *CA*, **98**, 53404 (*deriv*)

3-Chloro-2,6-dihydroxy-4-methylbenzalde-   C-60059
hyde, 9CI
Chloratranol. Chloroatranol
[57074-21-2]

$C_8H_7ClO_3$ M 186.595
Constit. of oakmoss *Evernia prunastri* and the lichen
   Platismatia glauca. Cryst. (Me_2CO). Mp 139-140.5°.

Ter Heide, R. *et al, J. Agric. Food Chem.*, 1975, **23**, 950 (*isol*)
Hveding-Bergseth, N. *et al, Phytochemistry*, 1983, **22**, 1826
   (*isol*)

3-Chloro-4,6-dihydroxy-2-methylbenzalde-   C-60060
hyde, 9CI
5-Chloroorsellinaldehyde. 4-Chloro-6-formylorcinol
[83324-59-8]
$C_8H_7ClO_3$ M 186.595
Powder. Mp 168-170°.

Chen, K.M. *et al, J. Org. Chem.*, 1985, **50**, 3997 (*synth*)

2-Chloro-4,6-dimethylbenzaldehyde   C-60061
[88174-22-5]

C_9H_9ClO M 168.623
Liq., cryst. on chilling.
Oxime:
   $C_9H_{10}ClNO$ M 183.637
   Cryst.
Semicarbazone: Mp 219-220°.

v. Auwers, K., *Ber.*, 1911, **44**, 793, 808.

4-Chloro-2,6-dimethylbenzaldehyde, 9CI   C-60062
C_9H_9ClO M 168.623
Cryst. Mp 55.5-56°.
2,4-Dinitrophenylhydrazone: Red cryst. Mp 269°.

Hjeds, H. *et al, Acta Chem. Scand.*, 1965, **19**, 2966 (*synth*)
Jerslev, B. *et al, Acta Crystallogr., Sect. C*, 1983, **39**, 1134
   (*cryst struct*)

4-Chloro-3,5-dimethylbenzaldehyde, 9CI   C-60063
[51719-64-3]
C_9H_9ClO M 168.623
Ger. Pat., 2 336 407, (*1974*); *CA*, **80**, 108534 (*synth*)

5-Chloro-2,4-dimethylbenzaldehyde   C-60064
C_9H_9ClO M 168.623
Mp 53°.
Semicarbazone: Mp 254°.

v. Auwers, K., *Ber.*, 1911, **44**, 793.

4-Chlorodiphenylmethanol   C-60065
*4-Chloro-α-phenylbenzenemethanol, 9CI. 4-Chlorophen-
ylphenylmethanol.* p-*Chlorobenzhydrol*
[119-56-2]

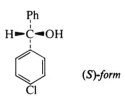

(S)-form

$C_{13}H_{11}ClO$ M 218.682
(S)-form [101402-04-4]
Needles (hexane). Mp 60-61°. $[\alpha]_D^{20}$ +22.0° (c, 0.9 in
   $CHCl_3$).

(±)-form

Cryst. Mp 59-61°.

Bachmann, W.E. *et al, J. Org. Chem.*, 1948, **13**, 916 (*synth*)
Green, G.H. *et al, J. Chem. Soc.*, 1950, 751 (*synth, resoln*)
Hügel, H.M. *et al, Aust. J. Chem.*, 1979, **32**, 1511 (*cmr*)
Wu, B. *et al, J. Org. Chem.*, 1986, **51**, 1904 (*synth, abs config*)

2-Chloro-3,7-epoxy-9-chamigranone   C-60066

[108906-79-2]

C_{15}H_{23}ClO_{2}   M 270.799
Constit. of *Laurencia obtusa*. Oil.

Brennan, M.R. *et al, Phytochemistry*, 1987, **26**, 1053.

(1-Chloroethyl) ethyl carbonate   C-60067

Carbonic acid 1-chloroethyl ethyl ester, 9CI

[50893-36-2]

EtOCOOCHClCH₃

C_{5}H_{9}ClO_{3}   M 152.577

(±)-form

Reagent for protection of primary and secondary amines.
Liq. Bp₂₂ 67°.

Barcelo, G. *et al, Synthesis*, 1986, 627 (*synth, pmr, use*)

1-Chloro-1-ethynylcyclopropane, 9CI   C-60068

[38387-33-6]

Cl C≡CH

C_{5}H_{5}Cl   M 100.548
Bp 89°.

Liese, T. *et al, Chem. Ber.*, 1986, **119**, 2995 (*synth, ir, pmr*)

3-Chloro-4-fluorobenzenethiol, 9CI   C-60069

3-Chloro-4-fluorothiophenol. 2-Chloro-1-fluoro-4-mercaptobenzene

[60811-23-6]

C_{6}H_{4}ClFS   M 162.609
Bp₂₀ 91-92°.

Cerevena, I. *et al, Collect. Czech. Chem. Commun.*, 1976, **41**, 881 (*synth*)

4-Chloro-2-fluorobenzenethiol, 9CI   C-60070

4-Chloro-2-fluorothiophenol. 4-Chloro-2-fluoro-1-mercaptobenzene

[73129-12-1]

C_{6}H_{4}ClFS   M 162.609
Liq. Bp₅ 104-106°.

Cerevena, I. *et al, Collect. Czech. Chem. Commun.*, 1979, **44**, 2139 (*synth*)

4-Chloro-3-fluorobenzenethiol, 9CI   C-60071

4-Chloro-3-fluorothiophenol. 1-Chloro-2-fluoro-4-mercaptobenzene

[60811-22-5]

C_{6}H_{4}ClFS   M 162.609
Liq. Bp₁₄ 88°.

Cerevena, I. *et al, Collect. Czech. Chem. Commun.*, 1976, **41**, 881 (*synth*)

2-Chloro-4-fluoro-5-nitrobenzoic acid   C-60072

C_{7}H_{3}ClFNO_{4}   M 219.556
Pale-yellow solid. Mp 183-185°.

Wheeler, T.N. *et al, Synthesis*, 1987, 883 (*synth, pmr*)

2-Chloro-4-fluorophenol, 9CI, 8CI   C-60073

[1996-41-4]

C_{6}H_{4}ClFO   M 146.548
Liq. Fp 20°. Bp₇ 64°. n_D^{20} 1.5173.
Me ether: [2267-25-6]. *2-Chloro-4-fluoro-1-methoxybenzene, 9CI. 2-Chloro-4-fluoroanisole.*
C_{7}H_{6}ClFO   M 160.575
Mp −17°. Bp₁₅ 67°. n_D^{20} 1.5173.

Finger, G.C. *et al, J. Am. Chem. Soc.*, 1959, **81**, 94.

2-Chloro-5-fluorophenol, 9CI   C-60074

C_{6}H_{4}ClFO   M 146.548
Liq. Bp 185°.
Me ether: [450-89-5]. *1-Chloro-4-fluoro-2-methoxybenzene, 9CI. 2-Chloro-5-fluoroanisole.*
C_{7}H_{6}ClFO   M 160.575
Liq. Bp₇₅₇ 195°.

Hodgson, H.H. *et al, J. Chem. Soc.*, 1931, 981 (*deriv*)
Finger, G.C. *et al, J. Am. Chem. Soc.*, 1959, **81**, 94 (*synth*)

2-Chloro-6-fluorophenol, 9CI   C-60075

[2040-90-6]
C_{6}H_{4}ClFO   M 146.548
Plates. Mp 63-64°.

Finger, G.C. *et al, J. Am. Chem. Soc.*, 1959, **81**, 5904 (*synth*)
Baker, A.W. *et al, Can. J. Chem.*, 1965, **43**, 650 (*synth*)
Watson, W.D. *et al, J. Org. Chem.*, 1985, **50**, 2145 (*synth*)

3-Chloro-2-fluorophenol, 9CI   C-60076

C_{6}H_{4}ClFO   M 146.548
Mp 38°. Bp₃₁ 96°.

Finger, G.C. *et al, J. Am. Chem. Soc.*, 1959, **81**, 94 (*synth*)

3-Chloro-4-fluorophenol, 9CI   C-60077

[2613-23-2]

C_6H_4ClFO M 146.548
Mp 42-44°. Bp_{11} 104°.
▷Irritant
Finger, G.C. *et al, J. Am. Chem. Soc.*, 1959, **81**, 94 (*synth*)

3-Chloro-5-fluorophenol, 9CI   C-60078
C_6H_4ClFO M 146.548
Bp_{11} 104°.
Finger, G.C. *et al, J. Am. Chem. Soc.*, 1959, **81**, 94 (*synth*)

4-Chloro-2-fluorophenol, 9CI   C-60079
[348-62-9]
C_6H_4ClFO M 146.548
Liq. Mp 23°. Bp_{40} 88°, Bp_7 64°.
Me ether: [452-09-5]. *4-Chloro-2-fluoro-1-methoxyben-*
zene, 9CI. 4-Chloro-2-fluoroanisole.
C_7H_6ClFO M 160.575
Mp −6°. Bp_{48} 102-103°. n_D^{25} 1.5163.
U.S.P., 2 606 183, (*1952*); *CA*, **47**, 3875 (*synth, deriv*)
Finger, G.C. *et al, J. Am. Chem. Soc.*, 1959, **81**, 94 (*synth*)
Watson, W.D. *et al, J. Org. Chem.*, 1985, **50**, 2145 (*synth*)

4-Chloro-3-fluorophenol, 9CI   C-60080
C_6H_4ClFO M 146.548
$Bp_{4.4}$ 84°.
Me ether: 1-Chloro-2-fluoro-4-methoxybenzene. 4-
Chloro-3-fluoroanisole.
C_7H_6ClFO M 160.575
Liq. Bp_{757} 196°.
Hodgson, H.H. *et al, J. Chem. Soc.*, 1931, 981 (*deriv*)
Finger, G.C. *et al, J. Am. Chem. Soc.*, 1959, **81**, 94 (*synth*)

2-Chloro-3-fluoropyridine   C-60081
[17282-04-1]

C_5H_3ClFN M 131.537
Liq. Bp_{55} 84-85°, Bp_{28} 62-64.5°.
N-Oxide: [85386-94-3].
   C_5H_3ClFNO M 147.536
   Mp 145-147°.
Link, W.J. *et al, J. Heterocycl. Chem.*, 1967, **4**, 641 (*synth*)
Thomas, W.A. *et al, Org. Magn. Reson.*, 1970, **2**, 503 (*synth, F
   nmr*)
Boudakian, M.M., *J. Fluorine Chem.*, 1981, **18**, 497 (*synth*)
Dehmlow, E.V. *et al, Tetrahedron Lett.*, 1985, **26**, 4903 (*synth,
   pmr*)

2-Chloro-4-fluoropyridine   C-60082
[34941-91-8]
C_5H_3ClFN M 131.537
Ger. Pat., 2 128 540, (*1971*); *CA*, **76**, 59469 (*synth*)

2-Chloro-5-fluoropyridine   C-60083
[31301-51-6]
C_5H_3ClFN M 131.537
No phys. props. reported.
Thomas, W.A. *et al, Org. Magn. Reson.*, 1970, **2**, 503 (*synth*)

2-Chloro-6-fluoropyridine   C-60084
[20885-12-5]
C_5H_3ClFN M 131.537
Needles. Mp 34-36° (24-25°). Bp_{51} 81-82°. Subl. at
   64°/0.5mm.
Boudakian, M.M., *J. Heterocycl. Chem.*, 1968, **5**, 683 (*synth*)
Thomas, W.A. *et al, Org. Magn. Reson.*, 1970, **2**, 503 (*synth, F
   nmr*)

3-Chloro-2-fluoropyridine   C-60085
[1480-64-4]
C_5H_3ClFN M 131.537
Liq. Bp_{100} 94-95°, Bp_{12} 59-60.5°.
Finger, G.C. *et al, J. Org. Chem.*, 1963, **28**, 1666 (*synth*)
Link, W.J. *et al, J. Heterocycl. Chem.*, 1967, **4**, 641 (*synth*)

4-Chloro-2-fluoropyridine   C-60086
[34941-92-9]
C_5H_3ClFN M 131.537
Ger. Pat., 2 128 540, (*1971*); *CA*, **76**, 59469 (*synth*)

4-Chloro-3-fluoropyridine   C-60087
[2546-56-7]
C_5H_3ClFN M 131.537
Bp 138°.
Picrate: [1799-38-8]. Cryst. (EtOH aq.). Mp 134°.
Talik, T. *et al, Rocz. Chem.*, 1964, **38**, 777; 1968, **42**, 1861
   (*synth*)

5-Chloro-2-fluoropyridine   C-60088
[1480-65-5]
C_5H_3ClFN M 131.537
Liq. Bp_{100} 81-83.5°.
Finger, G.C. *et al, J. Org. Chem.*, 1963, **28**, 1666 (*synth*)
Thomas, W.A. *et al, Org. Magn. Reson.*, 1970, **2**, 503 (*F nmr*)
Japan. Pat., 83 219 163, (*1983*); *CA*, **100**, 174855 (*synth*)

3-Chloro-4-hydroxybenzaldehyde, 9CI, 8CI   C-60089
Updated Entry replacing C-01246
[2420-16-8]
$C_7H_5ClO_2$ M 156.568
Cryst. (pet. ether). Mp 139°. Bp_{14} 150°.
Me ether: [4903-09-7]. *3-Chloro-4-*
   methoxybenzaldehyde.
   $C_8H_7ClO_2$ M 170.595
   Cryst. (pet. ether). Mp 57-58°.
Semicarbazone: Cryst. Mp 210°.
Buehler, C.A. *et al, J. Org. Chem.*, 1941, **6**, 902 (*synth*)
Denton, D.A. *et al, J. Chem. Soc.*, 1963, 4741 (*synth*)
Thaller, V. *et al, J. Chem. Soc., Perkin Trans. 1*, 1972, 2032
   (*synth*)
Riggs, R.M. *et al, J. Med. Chem.*, 1987, **30**, 1887 (*synth, deriv,
   pmr*)

4-Chloro-3-hydroxybenzaldehyde, 9CI, 8CI   C-60090
Updated Entry replacing C-01248
[56962-12-0]
$C_7H_5ClO_2$ M 156.568
Cryst. (AcOH aq.). Mp 121°.

Me ether: [13726-16-4]. *4-Chloro-3-methoxybenzaldehyde.*
$C_8H_7ClO_2$ M 170.595
Cryst. Mp 52°.
Semicarbazone: Cryst. Mp 238-239°.

Hodgson, H.H. *et al, J. Chem. Soc.*, 1926, 147 (*synth*)
Faith, H.E. *et al, J. Am. Chem. Soc.*, 1955, **77**, 543 (*synth*)
Riggs, R.M. *et al, J. Med. Chem.*, 1987, **30**, 1887 (*synth, deriv, pmr*)

3-Chloro-4-hydroxybenzoic acid, 9CI, 8CI   C-60091

Updated Entry replacing C-01255
[3964-58-7]
$C_7H_5ClO_3$ M 172.568
Cryst. (H$_2$O). Mp 169-171°.
Me ester: [3964-57-6].
  $C_8H_7ClO_3$ M 186.595
  Needles. Mp 106-107°.
Amide:
  $C_7H_6ClNO_2$ M 171.583
  Cryst. (EtOH). Mp 180-182°.
Me ether: [37908-96-6]. *3-Chloro-4-methoxybenzoic acid.*
  $C_8H_7ClO_3$ M 186.595
  Cryst. (EtOH aq.). Mp 213-215°.
Me ester, Me ether: [37908-98-8].
  $C_9H_9ClO_3$ M 200.621
  Cryst. (pet. ether). Mp 94-95°.
Nitrile: 2-Chloro-4-cyanophenol.
  C_7H_4ClNO M 153.568
  Mp 151-152° (148.5-149°).

Gray, G.W. *et al, J. Chem. Soc.*, 1954, 2556 (*synth*)
Simpson, H.N. *et al, J. Org. Chem.*, 1965, **30**, 2678 (*synth*)
Bergmann, J.J. *et al, Can. J. Chem.*, 1973, **51**, 162 (*synth*)
Yaegashi, T. *et al, Chem. Pharm. Bull.*, 1984, **32**, 4466 (*nitrile*)
Riggs, R.M. *et al, J. Med. Chem.*, 1987, **30**, 1887 (*synth, deriv, pmr*)

3-Chloro-5-hydroxybenzoic acid, 9CI   C-60092

[53984-36-4]
$C_7H_5ClO_3$ M 172.568
Me ether: [82477-67-6]. *3-Chloro-5-methoxybenzoic acid.*
  $C_8H_7ClO_3$ M 186.595
  Yellowish cryst. (CHCl$_3$). Mp 170-171°.

Hartmann, R.W. *et al, J. Med. Chem.*, 1984, **27**, 577.

4-Chloro-3-hydroxybenzoic acid, 9CI, 8CI   C-60093

Updated Entry replacing C-01257
$C_7H_5ClO_3$ M 172.568
Cryst. (H$_2$O). Mp 219-220°.
Me ether: 4-Chloro-3-methoxybenzoic acid.
  $C_8H_7ClO_3$ M 186.595
  Cryst. (EtOH). Mp 215-216°.

Beyer, P.H., *Recl. Trav. Chim. Pays-Bas*, 1921, **40**, 627 (*synth*)
Riggs, R.M. *et al, J. Med. Chem.*, 1987, **30**, 1887 (*synth, deriv, pmr*)

2-Chloro-3-hydroxy-7-chamigren-9-one   C-60094

8-Chloro-9-hydroxy-1,5,5,9-tetramethylspiro[5.5]-undec-1-en-3-one, 9CI
[108925-14-0]

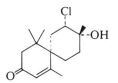

$C_{15}H_{23}ClO_2$ M 270.799
Constit. of *Laurencia obtusa*. Oil. $[\alpha]_D^{25}$ −46° (c, 0.22 in CHCl$_3$).

Brennan, M.R. *et al, Phytochemistry*, 1987, **26**, 1053.

2-Chloro-3-hydroxy-5-methylbenzoic acid   C-60095

6-Chloro-5-hydroxy-m-toluic acid

$C_8H_7ClO_3$ M 186.595
Me ether: [74510-43-3]. *2-Chloro-3-methoxy-5-methylbenzoic acid.*
  $C_9H_9ClO_3$ M 200.621
  Cryst. Mp 152°.

Ho, P.T., *Can. J. Chem.*, 1980, **58**, 861 (*synth, deriv*)

3-Chloro-4-hydroxy-5-methylbenzoic acid   C-60096

5-Chloro-4-hydroxy-m-toluic acid
$C_8H_7ClO_3$ M 186.595
Me ether: [62316-29-4]. *3-Chloro-4-methoxy-5-methylbenzoic acid.*
  $C_9H_9ClO_3$ M 200.621
  Cryst. (MeOH). Mp 204°.

Achenbach, H. *et al, Justus Liebigs Ann. Chem.*, 1977, 1 (*synth, deriv*)

5-Chloro-4-hydroxy-2-methylbenzoic acid   C-60097

5-Chloro-4-hydroxy-o-toluic acid
$C_8H_7ClO_3$ M 186.595
Me ether: [109803-47-6]. *5-Chloro-4-methoxy-2-methylbenzoic acid.*
  $C_9H_9ClO_3$ M 200.621
  Cryst. (C$_6$H$_6$). Mp 220-222°.
Me ether, Me ester: [109803-48-7].
  $C_{10}H_{11}ClO_3$ M 214.648
  Mp 88-90°.

Wyrick, S.D. *et al, J. Med. Chem.*, 1987, **30**, 1798 (*synth, deriv, pmr*)

3-Chloro-5-hydroxy-2-methyl-1,4-naphtho-quinone   C-60098

Updated Entry replacing C-01291
*3-Chloro-5-hydroxy-2-methyl-1,4-naphthalenedione,
9CI. 3-Chloroplumbagin*
[21890-57-3]

$C_{11}H_7ClO_3$ M 222.628
Constit. of *Drosera intermedia, D. angelica, Plumbago
zeylanica* and *Diospyros maritima*. Orange cryst. Mp
125°.

Bendz, G. *et al, Acta Chem. Scand.,* 1968, **22**, 2722 (*isol*)
Sidhu, G.S. *et al, Tetrahedron Lett.,* 1971, 2385 (*isol, struct*)
Sankaram, A.V.B. *et al, Phytochemistry,* 1976, **15**, 237 (*isol*)
Higa, M. *et al, Chem. Pharm. Bull.,* 1987, **35**, 4366 (*isol, uv, ir,
pmr, ms*)

2-Chloro-5-hydroxy-6-methylpyridine   C-60099

*6-Chloro-2-methyl-3-pyridinol. 6-Chloro-3-hydroxy-2-
picoline*
C_6H_6ClNO M 143.573
Plates (EtOAc). Mp 208°.

Parker, E.D. *et al, J. Am. Chem. Soc.,* 1947, **69**, 63 (*synth*)

2-Chloro-3-hydroxypyridine   C-60100

2-Chloro-3-pyridinol, 9CI
[6636-78-8]

C_5H_4ClNO M 129.546
Cryst. (H_2O). Mp 169-170°.
▷UU7707000.

Me ether: [52605-96-6]. *2-Chloro-3-methoxypyridine,
9CI.*
C_6H_6ClNO M 143.573
Needles (pet. ether). Mp 49°.

Den Hertog, H.J. *et al, Recl. Trav. Chim. Pays-Bas,* 1951, **70**,
182 (*synth*)
Heinert, D. *et al, Tetrahedron,* 1958, **3**, 49 (*synth*)
Lewicka, K. *et al, Recl. Trav. Chim. Pays-Bas,* 1959, **78**, 644
(*synth*)
Stogryn, E.L., *J. Heterocycl. Chem.,* 1974, **11**, 251 (*deriv*)
Iddon, B. *et al, J. Chem. Soc., Perkin Trans. 1,* 1980, 1370
(*cmr*)
Srivastava, S.L. *et al, Spectrochim. Acta, Part A,* 1984, **40**, 1101
(*uv*)

2-Chloro-5-hydroxypyridine   C-60101

6-Chloro-3-pyridinol, 9CI
[41288-96-4]
C_5H_4ClNO M 129.546
Cryst. (C_6H_6). Mp 152-159° dec.

Westland, R.D. *et al, J. Med. Chem.,* 1973, **16**, 319 (*synth*)

3-Chloro-5-hydroxypyridine   C-60102

5-Chloro-3-pyridinol, 9CI
[74115-12-1]

C_5H_4ClNO M 129.546
Mp 158°.
Me ether: [95881-83-7]. *3-Chloro-5-methoxypyridine,
9CI.*
C_6H_6ClNO M 143.573
Mp 40-41°.

Czuba, W., *Rocz. Chem.,* 1960, **34**, 905 (*synth*)
B.P., 2 025 953, (*1980*); *CA,* **93**, 114330 (*synth*)
Testaferri, L. *et al, Tetrahedron,* 1985, **41**, 1373 (*synth*)

5-Chloro-8-hydroxyquinoline   C-60103

Updated Entry replacing C-01330
5-Chloro-8-quinolinol, 9CI. Cloxyquin, USAN
[130-16-5]
C_9H_6ClNO M 179.606
Bactericide. Mp 129-130° (122-123°).
▷VC4590000.

B,HCl: [25395-13-5]. Yellow cryst. Mp 256-258°.
1-Oxide: [21168-34-3].
$C_9H_6ClNO_2$ M 195.605
Cryst. (ligroin). Mp 169-170°.

Das Gupta, S.J., *J. Indian Chem. Soc.,* 1952, **29**, 711 (*synth*)
Sukhina, L.F. *et al, Zh. Obshch. Khim.,* 1962, **32**, 1356.
Banerjee, T. *et al, Acta Crystallogr., Sect. C,* 1986, **42**, 1408
(*cryst struct*)

4-Chloro-1H-imidazo[4,5-c]pyridine, 9CI   C-60104

4-Chloro-1,3,5-triazaindene
[2770-01-6]

$C_6H_4ClN_3$ M 153.571
Cryst. (H_2O). Mp 217-218° dec.
Picrate: Cryst. (EtOH). Mp 179°.

Barlin, G.B., *J. Chem. Soc. (B),* 1966, 285 (*synth*)

6-Chloro-1H-imidazo[4,5-c]pyridine, 9CI   C-60105

6-Chloro-1,3,5-triazaindene
$C_6H_4ClN_3$ M 153.571
Pale-yellow cryst. (H_2O). Mp 236-238°.

Barlin, G.B., *J. Chem. Soc. (B),* 1966, 285 (*synth*)

1-Chloro-1-iodoethane, 9CI   C-60106

[594-00-3]

$$H_3CCHClI$$

C_2H_4ClI M 190.411
(±)-*form*
Bp 114-115°.

Kharasch, M.S. *et al, J. Am. Chem. Soc.,* 1934, **56**, 712 (*synth*)
Newman, R.C. *et al, J. Org. Chem.,* 1966, **31**, 1857 (*synth, pmr*)

2-Chloro-3-iodopyridine C-60107

[78607-36-0]

C_5H_3ClIN M 239.443
Mp 99°.

Magidson, O.Yu. *et al, CA*, 1929, **23**, 1640 (*synth*)
B.P., 2 053 189, (*1981*); *CA*, **95**, 80748 (*synth*)

2-Chloro-4-iodopyridine C-60108

C_5H_3ClIN M 239.443
Mp 43°.

Talik, T. *et al, Rocz. Chem.*, 1955, **29**, 1019; 1969, **43**, 489
(*synth*)

2-Chloro-5-iodopyridine C-60109

[69045-79-0]
C_5H_3ClIN M 239.443
Mp 98°.

Binz, A. *et al, Justus Liebigs Ann. Chem.*, 1931, **486**, 71 (*synth*)
Spiers, C.W.F. *et al, Recl. Trav. Chim. Pays-Bas*, 1937, **56**, 573.

3-Chloro-2-iodopyridine C-60110

[77332-89-9]
C_5H_3ClIN M 239.443
No phys. props. reported.

Gribble, G.W. *et al, Tetrahedron Lett.*, 1980, **21**, 4137 (*synth*)

3-Chloro-5-iodopyridine C-60111

[77332-90-2]
C_5H_3ClIN M 239.443
No phys. props. reported.

Gribble, G.W. *et al, Tetrahedron Lett.*, 1980, **21**, 4137 (*synth*)

4-Chloro-2-iodopyridine C-60112

[22918-03-2]
C_5H_3ClIN M 239.443
Mp 29°.

Talik, T. *et al, Rocz. Chem.*, 1969, **43**, 489 (*synth*)

4-Chloro-3-iodopyridine C-60113

C_5H_3ClIN M 239.443
Mp 79°.

Talik, T. *et al, Rocz. Chem.*, 1962, **36**, 417, 1049 (*synth*)

4-Chloro-5-methylfurazan C-60114

4-Chloro-5-methyl-1,2,5-oxadiazole

$C_3H_3ClN_2O$ M 118.523
2-Oxide: [65514-04-7]. *3-Chloro-4-methylfuroxan.*
$C_3H_3Cl_2N_2O_2$ M 169.975
Cryst. (pet. ether at low temp.). Mp 21-22°.
5-Oxide: [86988-90-1]. *4-Chloro-3-methylfuroxan.*
$C_3H_3ClN_2O_2$ M 134.522
Cryst. (pet. ether at low temp.). Mp 20.5-21.5°.

Calvino, R. *et al, J. Heterocycl. Chem.*, 1983, **20**, 783 (*oxides*)

2-Chloro-3-methyl-4-nitrophenol C-60115

Updated Entry replacing C-10159
*2-Chloro-4-nitro-*m-*cresol*

$C_7H_6ClNO_3$ M 187.582
Cryst. (C_6H_6). Mp 133.5-134.6°.
Me ether: 2-*Chloro-1-methoxy-3-methyl-4-nitroben-*
zene. 2-Chloro-3-methyl-4-nitroanisole. 2-Chloro-3-
methoxy-6-nitrotoluene.
$C_8H_8ClNO_3$ M 201.609
Needles (EtOAc/hexane). Mp 70-71°.

Cason, J. *et al, J. Org. Chem.*, 1948, **13**, 403.
Plattner, J.J. *et al, J. Heterocycl. Chem.*, 1983, **20**, 1059 (*synth*)
Stalder, H. *et al, Helv. Chim. Acta*, 1986, **69**, 1887 (*deriv*)

3-Chloro-4-methyl-5-nitrophenol C-60116

$C_7H_6ClNO_3$ M 187.582
Me ether: [102735-89-7]. *1-Chloro-5-methoxy-2-meth-*
yl-3-nitrobenzene. 2-Chloro-4-methoxy-6-
nitrotoluene.
$C_8H_8ClNO_3$ M 201.609
Cryst. (EtOH aq.). Mp 74-75°.

Büchi, G. *et al, J. Am. Chem. Soc.*, 1986, **108**, 4115 (*deriv,*
synth, uv, pmr, ms)

5-Chloro-3-methyl-1-pentene C-60117

[51174-45-9]

$H_2C{=}CHCH(CH_3)CH_2CH_2Cl$

$C_6H_{11}Cl$ M 118.606
(±)-*form* [104070-31-7]
Bp 118-120° (124-126°).

Beckwith, A.L.J. *et al, J. Chem. Soc., Perkin Trans. 2*, 1980,
1083 (*synth*)
Gadwood, R.C. *et al, J. Am. Chem. Soc.*, 1986, **108**, 6343
(*synth, ir, pmr*)

6-Chloro-1,2,3,4-naphthalenetetrol C-60118

6-Chloro-1,2,3,4-tetrahydroxynaphthalene

$C_{10}H_7ClO_4$ M 226.616
2,3-Di-Me ether, 1,4-di-Ac: [91431-42-4]. **Lonapalene**.
$C_{16}H_{15}ClO_6$ M 338.744
Antipsoriatic drug. Selective 5-lipoxygenase inhibitor.
Cryst. (Et_2O/hexane). Mp 91-92°.

Flynn, D.L. *et al, Tetrahedron Lett.*, 1986, **27**, 5075 (*synth*)
Perri, S.T. *et al, Tetrahedron Lett.*, 1987, **28**, 4507 (*synth*)

5-Chloro-2-nitrophenol C-60119

Updated Entry replacing C-01940
[611-07-4]

$C_6H_4ClNO_3$ M 173.556
Yellow needles (EtOH). Mp 42-43° subl. Steam-volatile.
▷Poss. hazardous synthesis

Me ether: [6627-53-8]. *4-Chloro-2-methoxy-1-nitrobenzene. 5-Chloro-2-nitroanisole.*
$C_7H_6ClNO_3$ M 187.582
Yellow needles (EtOH). Mp 71°.
Et ether: [29604-25-9]. *4-Chloro-2-ethoxy-1-nitrobenzene. 5-Chloro-2-nitrophenetole.*
$C_8H_8ClNO_3$ M 201.609
Needles (EtOH). Mp 63°.

Hodgson, H.H. *et al, J. Chem. Soc.,* 1925, **127**, 1599.
Fr. Pat., 1 581 400, (*1970*); *CA,* **73**, 35048
Robinson, N., *Chem. Br.,* 1987, 837 (*haz*)

2-Chloro-2-nitropropane, 9CI   C-60120

Updated Entry replacing C-10217
[594-71-8]

$$(H_3C)_2CClNO_2$$

$C_3H_6ClNO_2$ M 123.539
Acetone equivalent in cross-aldol reactions. d^{16} 1.179. Bp 134° sl. dec. Steam-volatile. Can be dist. if heated carefully.
▷Highly toxic. Explodes on rapid heating. TX5425000.

Barnes, M.W. *et al, J. Org. Chem.,* 1976, **41**, 733 (*synth*)
Russell, G.A. *et al, Synthesis,* 1981, 62 (*use*)
Amrollah-Madjdalbadi, A. *et al, Synthesis,* 1986, 828 (*synth*)
Sax, N.I., *Dangerous Properties of Industrial Materials,* 5th Ed., Van Nostrand-Reinhold, 1979, 496.

5-Chloro-1,3-pentadiene, 9CI   C-60121

[40596-30-3]

$$ClCH_2CH{=}CHCH{=}CH_2$$

C_5H_7Cl M 102.563
(*E*)-*form* [28070-18-0]
Bp_{240} 80-82°, Bp_{100} 57°.

Crombie, L. *et al, J. Chem. Soc.,* 1951, 2906 (*synth*)
Maruyama, K. *et al, J. Org. Chem.,* 1986, **51**, 5083 (*synth*)
Mayr, H. *et al, Tetrahedron,* 1986, **42**, 6657 (*synth*)

Chloropentamethylbenzene   C-60122

[5153-39-9]

$C_{11}H_{15}Cl$ M 182.693
Cryst. (EtOH). Mp 155-156°.

Baciocchi, E. *et al, J. Am. Chem. Soc.,* 1965, **87**, 3953 (*synth*)
Charbonneau, G. *et al, J. Chim. Phys.,* 1967, **64**, 273 (*cryst struct*)
Bugueno-Hoffmann, R., *Spectrochim. Acta, Part A,* 1981, **37**, 163; 1982, **38**, 223 (*ir*)
Kohl, F.X. *et al, Chem. Ber.,* 1987, **120**, 1539 (*pmr, cmr*)

1-Chloro-2,3-pentanedione, 9CI   C-60123

[91265-96-2]

$$ClCH_2COCOCH_2CH_3$$

$C_5H_7ClO_2$ M 134.562

2-Oxime:
$C_5H_8ClNO_2$ M 149.577
Cryst. (CCl_4). Mp 65-67°.

Hartman, G.J. *et al, J. Agric. Food Chem.,* 1984, **32**, 1015 (*ms*)
Gilchrist, T.L. *et al, J. Chem. Soc., Perkin Trans. 1,* 1985, 2769 (*deriv, synth, pmr*)

1-Chloro-10*H*-phenothiazine   C-60124

[1910-85-6]

$C_{12}H_8ClNS$ M 233.715
Leaflets (Me_2CO/MeOH). Mp 92-93°.

Massie, S.P. *et al, J. Org. Chem.,* 1956, **21**, 347 (*synth*)
Silberg, I.A. *et al, J. Prakt. Chem.,* 1976, **318**, 353 (*synth*)

2-Chloro-10*H*-phenothiazine   C-60125

[92-39-7]
$C_{12}H_8ClNS$ M 233.715
Cryst. (xylene). Mp 196-197°.
▷SN7580000.

Galbreath, R.J. *et al, J. Org. Chem.,* 1958, **23**, 1804 (*synth, ir*)

3-Chloro-10*H*-phenothiazine   C-60126

[1207-99-4]
$C_{12}H_8ClNS$ M 233.715
Cryst. (xylene). Mp 199-200°.
5-Oxide:
$C_{12}H_8ClNOS$ M 249.714
Cryst. (DMF aq.). Mp 280-281° dec.
5,5-Dioxide:
$C_{12}H_8ClNO_2S$ M 265.714
Cryst. (DMF aq.). Mp 295-297°.

Yale, H.L., *J. Am. Chem. Soc.,* 1955, **77**, 2270.

4-Chloro-10*H*-phenothiazine   C-60127

[7369-69-9]
$C_{15}H_8ClNS$ M 269.748
Mp 116°.
5-Oxide:
$C_{12}H_8ClNOS$ M 249.714
Fine cryst. (dioxan). Mp 241-243° dec.
5,5-Dioxide:
$C_{12}H_8ClNO_2S$ M 265.714
Plates (Me_2CO). Mp 283-284°.

Charpentier, P. *et al, C.R. Hebd. Seances Acad. Sci.,* 1952, **235**, 59 (*synth*)
Kano, H. *et al, Chem. Pharm. Bull.,* 1957, **5**, 389, 393 (*deriv, ir*)

2-Chloro-2-phenylethanol   C-60128

Updated Entry replacing C-20155
β-Chlorobenzeneethanol, 9CI
[1004-99-5]

$$Ph - \overset{\overset{\displaystyle Cl}{|}}{\underset{\underset{\displaystyle CH_2OH}{|}}{C}} \blacktriangleleft H \qquad (S)\text{-}form$$

C_8H_9ClO M 156.612

(S)-form [14252-67-6]

Ac: [33942-00-6].
    $C_{10}H_{11}ClO_2$ M 198.649
    Bp 72-74°. $[\alpha]_D^{28.5}$ −68.75° (neat).

(±)-form

Bp$_5$ 104-110°.

4-Nitrobenzoyl: Mp 109.5-110.5°.

Berti, G. *et al, J. Org. Chem.*, 1965, **30**, 4091 (*abs config*)
Nakai, H. *et al, Chem. Lett.*, 1977, 995 (*synth*)
Eisenbaumer, R.L. *et al, J. Org. Chem.*, 1979, **44**, 600 (*synth*)
Imuta, M. *et al, J. Am. Chem. Soc.*, 1979, **101**, 3990 (*synth*)
Loreto, M.A. *et al, Synth. Commun.*, 1981, **11**, 287 (*synth*)
Spawn, C.-L. *et al, Synthesis*, 1986, 315 (*synth*)

3-Chloro-2(1H)-pyridinethione   C-60129

3-Chloro-2-pyridinethiol. 3-Chloro-2-mercaptopyridine
[5897-94-9]
C_5H_4ClNS M 145.606
Yellow needles (MeOH). Mp 197-206°.

Me thioether: [98626-97-2]. *3-Chloro-2-(methylthio)-pyridine.*
    C_6H_6ClNS M 159.633
    Oil.

Me thioether, S-dioxide: [98626-98-3]. *3-Chloro-2-(methylsulfonyl)pyridine, 9CI.*
    $C_6H_6ClNO_2S$ M 191.632
    Mp 98-99°.

Bradsher, C.K. *et al, J. Heterocycl. Chem.*, 1966, **3**, 27 (*synth*)
Testaferri, L. *et al, Tetrahedron*, 1985, **41**, 1373 (*deriv*)

5-Chloro-2,4-(1H,3H)-pyrimidinedione, 9CI   C-60130

5-Chlorouracil
[1820-81-1]

$C_4H_3ClN_2O_2$ M 146.533
Incorp. into nucleic acids. Mp 324-325° (300-303°) dec.
▷YQ9410000.

Dornow, A. *et al, Arch. Pharm. (Weinheim, Ger.)*, 1953, **286**, 494 (*synth*)
West, R.A. *et al, J. Am. Chem. Soc.*, 1954, **76**, 3146 (*synth*)

8-Chloro-2(1H)-quinolinone, 9CI   C-60131

[23981-25-1]
C_9H_6ClNO M 179.606
Solid. Mp 208°.

Le Clerc, G. *et al, J. Med. Chem.*, 1986, **29**, 2433 (*synth*)

8-Chloro-4(1H)-quinolinone, 9CI   C-60132

[23833-96-7]
C_9H_6ClNO M 179.606
Mp 212-213°.

Heindel, N.D. *et al, J. Med. Chem.*, 1969, **12**, 797 (*synth*)

Chlororepdiolide   C-60133

[106566-98-7]

$C_{19}H_{23}ClO_7$ M 398.840
Constit. of *Centaurea repens*. Cryst. Mp 207-208°. $[\alpha]_D^{25}$ +80.2° (c, 0.109 in CHCl$_3$).

Stevens, K.L. *et al, J. Nat. Prod.*, 1986, **49**, 833.

2-Chloro-3,4,5,6-tetranitroaniline   C-60134

2-Chloro-3,4,5,6-tetranitrobenzenamine, 9CI
[102367-93-1]

$C_6H_2ClN_5O_8$ M 307.564
Yellow needles (CH$_2$Cl$_2$). Mp 163°.
▷Explosive

Atkins, R.L. *et al, J. Org. Chem.*, 1986, **51**, 2572 (*synth, ir, pmr*)

3-Chloro-1,2,4,5-tetranitrobenzene   C-60135

[102367-94-2]

$C_6HClN_4O_8$ M 292.549
Yellow cryst. (CH$_2$Cl$_2$/Et$_2$O). Mp 144-146°.
▷Explosive

Atkins, R.L. *et al, J. Org. Chem.*, 1986, **51**, 2572 (*synth, ir, pmr*)

3-Chlorothiophene, 9CI   C-60136

Updated Entry replacing C-02258
[17249-80-8]
C_4H_3ClS M 118.581
Liq. Bp 135-137°.

Profft, E. *et al, J. Prakt. Chem.*, 1964, **24**, 38 (*synth*)
Conde, S. *et al, Synthesis*, 1976, 412 (*synth*)
Dettmeier, U. *et al, Angew. Chem., Int. Ed. Engl.*, 1987, **26**, 468 (*synth*)

Chlorothricin, 9CI   C-60137

Updated Entry replacing C-02266
*K 818*A. *Antibiotic K 818*A
[34707-92-1]

$C_{50}H_{63}ClO_{16}$ M 955.491
Data applies to 5:1 mixt. with Deschlorothricin. Produced
by *Streptomyces antibioticus*. Shows antibiotic props.
Cryst. (CH_2Cl_2/MeOAc). Mp 206-207°. pK_{a1} 5.01,
pK_{a2} 7.91. Dibasic acid.

3^C-*Bromo-3^C-dechloro:* **Bromothricin**.
   $C_{50}H_{63}BrO_{16}$ M 999.942
   From *S. antibioticus* with KBr. Less biol. active than
   Chlorothricin. Powder.
Dechloro: [72656-14-5]. **Deschlorothricin**. From *S.
antibioticus*. pK_{a1} 5.14, pK_{a2} 7.72.
$2^A\beta$-*Hydroxy:* **2′′′-Hydroxychlorothricin**. *K 818*B. *Anti-
biotic K 818*B.
   $C_{50}H_{63}ClO_{17}$ M 971.491
   From *S* sp. K 818. Exhibits antitumour props. Powder.
   Mp 202°. $[\alpha]_D^{25}$ +2.4° (c, 0.5 in MeOH).

Keller-Schierlein, W. *et al, Experientia*, 1969, **25**, 786
   (*Bromothricin*)
Keller-Schierlein, W. *et al, Helv. Chim. Acta*, 1969, **52**, 127
   (*deriv*)
Muntwyler, R. *et al, Helv. Chim. Acta*, 1972, **55**, 2071 (*struct*)
Brufani, M. *et al, Helv. Chim. Acta*, 1972, **55**, 2094 (*struct*)
Hook, D. *et al, Biochemistry*, 1978, **17**, 556; 1981, **20**, 919
   (*biosynth, cmr*)
Floss, H.G. *et al, Antibiotics* (*N.Y.*), 1981, **4**, 193 (*rev*)
Hall, S.E. *et al, J. Org. Chem.*, 1982, **47**, 4611 (*synth*)
Snider, B.B. *et al, J. Org. Chem.*, 1983, **48**, 4370 (*synth*)
Ireland, R.E. *et al, J. Org. Chem.*, 1986, **51**, 635 (*synth*)
Lee, J.J. *et al, J. Antibiot.*, 1986, **39**, 1123 (*biosynth*)
Yamamoto, I. *et al, J. Antibiot.*, 1987, **40**, 1452 (*deriv*)

1-Chloro-1-(trichlorovinyl)cyclopropane   C-60138

1-Chloro-1-(trichloroethenyl)cyclopropane, 9CI
[82979-27-9]

$C_5H_4Cl_4$ M 205.899
Bp$_{22}$ 92°.

Liese, T. *et al, Chem. Ber.*, 1986, **119**, 2995 (*synth, ir, pmr, cmr,
ms*)

6-Chloro-2-(trifluoromethyl)-1*H*-imidazo[4,5-*b*]pyridine, 9CI, 8CI   C-60139

Fluromidine. Fluoromidine
[13577-71-4]

$C_7H_3ClF_3N_3$ M 221.569
Herbicide. Cryst. (EtOH). Mp 304-305° (293°).
4-Oxide: [19918-38-8].
   $C_7H_3ClF_3N_3O$ M 237.568
   Solid (EtOH). Mp 293°.

B.P., 1 114 199, (*1968*); *CA*, **69**, 67384 (*synth, deriv*)
Pesticide Manual, 7th Ed., No. 6650 (*use, rev*)

Chlorotrifluorooxirane, 9CI   C-60140

Chloroepoxytrifluoroethane, 8CI
[3935-49-7]

C_2ClF_3O M 132.470
(±)-*form*
Gas. Fp −130.2°. Bp −22°.

Chow, D. *et al, Can. J. Chem.*, 1969, **47**, 2491 (*synth, F nmr*)
Jolley, K.W. *et al, Spectrochim. Acta, Part A*, 1974, **30**, 1455 (*F
nmr*)
Jones, R.B. *et al, Biochem. Pharmacol.*, 1983, **32**, 2359 (*tox*)

2-Chloro-3,3,3-trifluoro-1-propene   C-60141

[2730-62-3]

$$F_3CCCl=CH_2$$

$C_3H_2ClF_3$ M 130.497
Mobile liq. or gas. Bp 15°.

Lang, R.W., *Helv. Chim. Acta*, 1986, **69**, 881.

3-Chloro-4,5,6-trifluoropyridazine, 9CI   C-60142

[88692-18-6]

$C_4ClF_3N_2$ M 168.506

Klauké, E. *et al, J. Fluorine Chem.*, 1983, **23**, 301 (*synth, F
nmr*)

4-Chloro-3,5,6-trifluoropyridazine, 9CI   C-60143

[26279-50-5]
$C_4ClF_3N_2$ M 168.506
Bp 155°. n_D^{20} 1.4562.

Klauké, E. *et al, J. Fluorine Chem.*, 1983, **23**, 301 (*synth, F
nmr*)

5-Chloro-2,4,6-trifluoropyrimidine, 9CI

C-60144

[697-83-6]

$C_4ClF_3N_2$ M 168.506

Dye intermediate. Bp 114-115°. n_D^{25} 1.4390.

Schroeder, H. et al, J. Org. Chem., 1962, **27**, 2580 (synth)
Banks, R.E. et al, J. Chem. Soc. (C), 1967, 1822 (F nmr)
Hitzke, J. et al, Org. Mass Spectrom., 1974, **9**, 435 (ms)

2-Chlorotryptophan

C-60145

Updated Entry replacing C-40143

2-Amino-3-(2-chloro-1H-indol-3-yl)propanoic acid

$C_{11}H_{11}ClN_2O_2$ M 238.673

(**S**)-form [89311-53-5]

L-form

Synthetic. Plates. Mp ca. 200° dec. $[\alpha]_D^{20}$ +0.9° (c, 0.55 in H_2O).

Phillips, R.S. et al, J. Am. Chem. Soc., 1986, **108**, 2023 (synth, ir, pmr, ms)

6-Chlorotryptophan, 9CI

C-60146

$C_{11}H_{11}ClN_2O_2$ M 238.673

Sweetening agent.

(**R**)-form [56632-86-1]

D-form

Mp 264° dec. $[\alpha]_D^{23}$ +28° (c, 1 in MeOH).

N^α-Formyl: [57233-89-3]. Cryst. + $1H_2O$. Mp 143-145°. $[\alpha]_D^{23}$ −47° (c, 1 in MeOH).

(±)-form [17808-21-8]

Off-white cryst. (AcOH). Mp 278° dec. (285-286°).

N^α-Formyl: [57233-85-9]. Cryst. (EtOAc). Mp 180-181°.

Fukuda, D.S. et al, Appl. Microbiol., 1971, **21**, 841 (manuf)
Yamada, S. et al, J. Agric. Food Chem., 1975, **23**, 653 (synth, resoln)
Perry, C.W. et al, Synthesis, 1977, 492 (synth)
Hengartner, U. et al, J. Org. Chem., 1979, **44**, 3748 (synth)

5-Cholestene-3,26-diol

C-60147

Updated Entry replacing C-02378

(3β,25R)-form

$C_{27}H_{46}O_2$ M 402.659

Constit. of human aortal atheroma placques and aortal tissue.

(3β,25R)-form [20380-11-4]

(25R)-26-Hydroxycholesterol

Needles (EtOAc/pet. ether). Mp 172-173°. $[\alpha]_D^{21}$ −33.5° (c, 1.5 in $CHCl_3$).

(3β,25S)-form [56845-83-1]

(25S)-26-Hydroxycholesterol

Cryst. Mp 171-172°. $[\alpha]_D^{25}$ −38.0° (c, 1.72 in $CHCl_3$).

Scheer, I. et al, J. Am. Chem. Soc., 1956, **78**, 4733 (synth)
Brooks, C.J.W. et al, Biochem. Biophys. Acta, 1966, **125**, 620 (isol)
Van Lier, J.E. et al, Biochemistry, 1967, **6**, 3269 (isol)
Varma, R.K. et al, J. Org. Chem., 1975, **40**, 3680 (synth, struct, stereochem)
Kirfel, A. et al, Acta Crystallogr., Sect. B, 1977, **33**, 895 (struct)
Kluge, A.F. et al, J. Org. Chem., 1985, **50**, 2359 (synth)
Uomori, A. et al, J. Chem. Soc., Perkin Trans. 1, 1987, 1713 (pmr)
Ferraboschi, P. et al, J. Chem. Soc., Perkin Trans. 1, 1987, 1749 (synth)

Chromophycadiol

C-60148

$C_{20}H_{34}O_2$ M 306.487

4-Ac: Chromophycadiol monoacetate.

$C_{22}H_{36}O_3$ M 348.525

Constit. of a *Dictyota* sp. Cryst. Mp 137-139°. $[\alpha]_D$ +26.5° (c, 0.65 in $CHCl_3$).

Clardy, J. et al, J. Chem. Soc., Chem. Commun., 1987, 767 (isol, cryst struct)

Citreomontanin

C-60149

Updated Entry replacing C-10273

4-Methoxy-5-methyl-6-(7,9,11-trimethyl-1,3,5,7,9,11-tridecahexaenyl)-2H-pyran-2-one, 9CI

[74474-66-1]

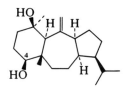

$C_{23}H_{28}O_3$ M 352.472

Metab. of *Penicillium pedomontanum*. Orange cryst. (MeOH). Mp 165-166°.

Rebuffat, S. et al, Phytochemistry, 1980, **19**, 427; 1981, **20**, 1279 (isol, struct, biosynth)
Brassy, C. et al, Acta Crystallogr., Sect. B, 1982, **38**, 1624 (cryst struct)
Patel, P. et al, Tetrahedron Lett., 1985, **26**, 4789 (synth)
Steyn, P.S. et al, J. Chem. Soc., Chem. Commun., 1985, 1531 (biosynth)
Venkataraman, H. et al, Tetrahedron Lett., 1987, **28**, 2455 (synth)
Cha, J.K. et al, Tetrahedron Lett., 1987, **28**, 2455 (synth)

Citreoviral   C-60150

Updated Entry replacing C-40159

$C_{11}H_{18}O_4$ M 214.261
Metab. of *Penicillium citreo-viride* B (IFO 6050). Oil.
$[\alpha]_D^{27}$ +2.7° (c, 7.0 in $CHCl_3$).

Shizuri, Y. *et al*, *Tetrahedron Lett.*, 1984, **25**, 4771 (*isol*)
Nishiyama, S. *et al*, *Tetrahedron Lett.*, 1985, **26**, 231 (*synth, abs config*)
Shizuri, Y. *et al*, *J. Chem. Soc., Chem. Commun.*, 1985, 292 (*synth*)
Williams, D.R. *et al*, *Tetrahedron Lett.*, 1985, **26**, 2529 (*synth*)
Trost, B.M. *et al*, *Tetrahedron Lett.*, 1987, **28**, 375 (*synth*)

Citrulline   C-60151

Updated Entry replacing C-02520
N^5-(*Aminocarbonyl*)*ornithine*, *9CI*. N^5-*Carbamoylornithine*, *8CI*. *2-Amino-5-ureidovaleric acid. 5-Ureidoornithine*

$C_6H_{13}N_3O_3$ M 175.187
(*S*)-*form* [372-75-8]
L-form
Occurs in the watermelon *Citrullus vulgaris* and the red alga *Grateloupia filicina*. Prisms (MeOH aq.). Mp 222°. $[\alpha]_D^{20}$ +3.7° (c, 2 in H_2O).
B,HCl: $[\alpha]_D^{22}$ +17.9° (c, 2 in H_2O). Dec. at 185°.
Cu salt: Blue prisms. Mp 257-258°.
α-*N-Benzoyl, Me ester:*
   $C_{14}H_{19}N_3O_4$ M 293.322
   Dec. at 120°.
α-*N-Benzoyl, amide:*
   $C_{13}H_{18}N_4O_3$ M 278.310
   Dec. at 140°.

Biochem. Prep., 1953, **3**, 100, 104 (*synth*)
Ashida, T. *et al*, *Acta Crystallogr., Sect. B*, 1972, **28**, 1367 (*cryst struct*)
Wakamiya, T. *et al*, *Tetrahedron*, 1984, **40**, 235 (*isol, bibl*)
Toffoli, P. *et al*, *Bull. Soc. Chim. Fr.*, 1986, 119 (*cryst struct*)

Citrusinol   C-60152

$C_{20}H_{16}O_5$ M 336.343
Constit. of *Citrus nobilis*. Yellow needles (Me_2CO). Mp 252-254°.

Wu, T.-S., *Phytochemistry*, 1987, **26**, 3094.

Clavularin *A*   C-60153

Updated Entry replacing C-30174
6-Methyl-7-(3-oxobutyl)-2-cyclohepten-1-one, *9CI*
[86582-92-5]

$C_{12}H_{18}O_2$ M 194.273
Constit. of *Clavularia koellikeri*. Cytotoxic. Oil.
7-Epimer: [86746-90-9]. **Clavularin B.**
   $C_{12}H_{18}O_2$ M 194.273
   Constit. of *C. koellikeri*. Cytotoxic. Oil.

Endo, M. *et al*, *J. Chem. Soc., Chem. Commun.*, 1983, 322 (*pmr, struct*)
Urech, R. *et al*, *J. Chem. Soc., Chem. Commun.*, 1984, 989 (*synth, struct*)
Urech, R. *et al*, *Aust. J. Chem.*, 1986, **39**, 433 (*synth*)
Still, I.W.J. *et al*, *Tetrahedron Lett.*, 1987, **28**, 2489 (*synth*)

8,11,13-Cleistanthatrien-19-ol   C-60154

[111179-61-4]

$C_{20}H_{30}O$ M 286.456
Constit. of *Vellozia flavicans*. Cryst. Mp 110-112°. $[\alpha]_D^{25}$ +33° (c, 0.2 in $CHCl_3$).
19-Aldehyde: [111150-44-8]. **8,11,13-Cleistanthatrien-19-al.**
   $C_{20}H_{28}O$ M 284.441
   From *V. flavicans*. Oil. $[\alpha]_D^{25}$ +66.5° (c, 1.94 in $CHCl_3$).
19-Carboxylic acid: [111150-42-6]. **8,11,13-Cleistanthatrien-19-oic acid.**
   $C_{20}H_{28}O_2$ M 300.440
   From *V. flavicans*.
19-Carboxylic acid, Me ester: Cryst. (EtOAc/hexane). Mp 92-95°. $[\alpha]_D^{25}$ +92° ($CHCl_3$).

Pinto, A. *et al*, *Phytochemistry*, 1987, **26**, 2409.

Clitocine   C-60155

N-(6-Amino-5-nitro-4-pyrimidinyl)-β-D-ribofuranosylamine, *9CI*. *6-Amino-5-nitro-4-imino-β-D-ribofuranosylpyrimidine*
[105798-74-1]

$C_9H_{13}N_5O_6$ M 287.232
Nucleoside isol. from the mushroom *Clitocybe inversa*. Insecticidal agent. Mp 228-230°.

Kubo, I. *et al*, *Tetrahedron Lett.*, 1986, **27**, 4277 (*isol, uv, ir, pmr, cmr, ms*)

Cochloxanthin
C-60156

6-Hydroxy-3-oxo-8'-apo-ε-caroten-8'-oic acid

$C_{30}H_{38}O_4$ M 462.628

Constit. of *Cochlospermum tinctorium*.

4,5-Dihydro: 4,5-Dihydro-6-hydroxy-3-oxo-8'-apo-ε-
caroten-8'-oic acid. **Dihydrocochloxanthin**.
$C_{30}H_{40}O_4$ M 464.644
From *C. tinctorium*.

Diallo, B. *et al, Phytochemistry*, 1987, **26**, 1491.

Coleon *C*
C-60157

Updated Entry replacing C-02684

*2,3,4,4a-Tetrahydro-5,6,8,10-tetrahydroxy-7-(2-
hydroxy-1-methylethyl)-1,1,4a-trimethyl-9(1H)-phen-
anthrenone, 9CI*

[35298-85-2]

$C_{20}H_{26}O_6$ M 362.422

Tautomeric with Coleon *D*, C-02685 . Constit. of yellow
glands of *Coleus aquaticus* and *Plectranthus* spp.
Cryst. (MeOH). Mp 210° dec. $[\alpha]_D^{25}$ +27° (CHCl$_3$).

Tetra-Ac: Cryst. (pentane/Et$_2$O/Me$_2$CO). Mp 207-
210°. $[\alpha]_D^{24}$ +31.2° (CHCl$_3$).

Rüedi, P. *et al, Helv. Chim. Acta*, 1971, **54**, 1606; 1975, **58**,
1899 (*isol, struct*)
Burnell, R.H. *et al, Can. J. Chem.*, 1985, **63**, 2769 (*synth*)
Matsumoto, T. *et al, Bull. Chem. Soc. Jpn.*, 1987, **60**, 2435
(*synth*)

Coleonol *B*
C-60158

Updated Entry replacing C-40166

[67921-05-5]

$C_{22}H_{34}O_7$ M 410.506

Constit. of *Coleus forskohlii*. Cryst. (Me$_2$CO/hexane).
Mp 210°. $[\alpha]_D$ +2.92° (c, 1 in CHCl$_3$).

13-Epimer: [67921-06-6]. **Coleonol C**.
$C_{22}H_{34}O_7$ M 410.506
From *C. forskohlii*. Cryst. (Me$_2$CO/hexane). Mp 205-
206°. $[\alpha]_D$ −6.7° (c, 1 in CHCl$_3$).

Tandon, J.S. *et al, Indian J. Chem., Sect. B*, 1978, **16**, 341 (*isol*)
Saksena, A.K. *et al, Tetrahedron Lett.*, 1985, **26**, 551 (*struct*)

Colletodiol
C-60159

Updated Entry replacing C-10294

*11,12-Dihydroxy-6,14-dimethyl-1,7-dioxacyclotetra-
deca-3,9-diene-2,8-dione, 9CI. 11,12-Dihydroxy-3,9-di-
methyl-2,8-dioxacyclotetradeca-5,13-diene-1,7-dione*

[21142-67-6]

$C_{14}H_{20}O_6$ M 284.308

Constit. of *Colletotrichum capsici*. Cryst. (Me$_2$CO/pet.
ether). Mp 163-164°. Biol. inactive.

Di-Ac: Rods. Mp 130-131°.

Dibenzoyl: Prisms (EtOAc/pet. ether). Mp 103-105°.

11-Ketone: [50376-39-1]. **Colletoketol**. *12-Hydroxy-
6,14-dimethyl-1,7-dioxacyclotetradeca-3,9-diene-
2,8,11-trione, 9CI*.
$C_{14}H_{18}O_6$ M 282.293
Metab. of *C. capsici*. Prisms (Me$_2$CO/pet. ether). Mp
138-139°.

9,10-Dihydro, 11,12-Diketone: [74838-13-4]. **Grahami-
mycin A$_1$**.
$C_{14}H_{18}O_6$ M 282.293
Prod. by *Cytospora* sp. Weakly active against gram-
positive and -negative bacteria and blue-green algae.
Yellow rhombs. Mp 91-92°. $[\alpha]_D^{22}$ −14.7° (c, 0.76 in
CHCl$_3$).

MacMillan, J. *et al, J. Chem. Soc., Perkin Trans. 1*, 1973, 1487
(*isol*)
Amstutz, R. *et al, Helv. Chim. Acta*, 1981, **64**, 1796 (*cryst
struct*)
Tsutsui, H. *et al, Tetrahedron Lett.*, 1984, **25**, 2159, 2163
(*synth*)
Schnurrenberger, P. *et al, Tetrahedron Lett.*, 1984, **25**, 2209
(*synth, bibl*)
Hillis, L.R. *et al, J. Org. Chem.*, 1985, **50**, 470 (*synth*)
Simpson, T.J. *et al, J. Chem. Soc., Chem. Commun.*, 1985, 1822
(*biosynth, nmr*)
Schnurrenberger, P. *et al, Justus Liebigs Ann. Chem.*, 1987, 733
(*synth, bibl*)

Columbin
C-60160

Updated Entry replacing C-02726

[546-97-4]

$C_{20}H_{22}O_6$ M 358.390

Constit. of Columbo root, *Jatrorrhiza palmata*. Cryst.
(MeOH). Mp 195-196°. $[\alpha]_D$ +52.7° (c, 1 in Py).

2β,3β-Epoxide: [23369-74-6]. **Jateorin**.
$C_{20}H_{22}O_7$ M 374.390

Constit. of *J. palmata*. Not isol. in pure state, readily isom. to 8-epimer.

2β,3β-Epoxide, 8-epimer: Isojateorin. 8-Epijateorin.
$C_{20}H_{22}O_7$ M 374.390
Cryst. (MeOH). Mp 165-167°. $[\alpha]_D$ +30° (c, 3.99 in Py).

8β-Hydroxy: 8β-Hydroxycolumbin.
$C_{20}H_{22}O_7$ M 374.390
Constit. of *Chasmanthera dependens*. Cryst. Mp 195-197°. $[\alpha]_D^{20}$ −24.6° (c, 0.6 in Py).

Barton, D.H.R. *et al*, *J. Chem. Soc.*, 1962, 4809 (*isol*)
Overton, K.H. *et al*, *J. Chem. Soc.* (*C*), 1966, 1482 (*struct*)
Ramstad, E. *et al*, *Phytochemistry*, 1975, **14**, 2719 (*isol*)
Oguakwa, J.U. *et al*, *Planta Med.*, 1986, 198 (*isol, deriv*)

Conocarpan   C-60161

Updated Entry replacing C-02760

4-[2,3-Dihydro-3-methyl-5-(1-propenyl)-2-benzofuranyl]phenol, 9CI. 2,3-Dihydro-2-(4-hydroxyphenyl)-3-methyl-5-(1-propenyl)benzofuran

[56319-02-9]

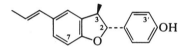

Absolute configuration

$C_{18}H_{18}O_2$ M 266.339
Constit. of the wood of *Conocarpus erectus* and roots of *Krameria cystisoides*. Noncryst.

Ac: Cryst. (pet. ether). Mp 94-95°.

3'-Methoxy: 2,3-Dihydro-2-(4-hydroxy-3-methoxyphenyl)-3-methyl-5-(1-propenyl)benzofuran.
$C_{19}H_{20}O_3$ M 296.365
Constit. of *K. cystisoides*. Cryst. Mp 105-107°. $[\alpha]_D^{21}$ +94° (c, 1.1 in MeOH).

7-Methoxy: 2,3-Dihydro-2-(4-hydroxyphenyl)-7-methoxy-3-methyl-5-(1-propenyl)benzofuran.
$C_{19}H_{20}O_3$ M 296.365
Constit. of *K. cystisoides*. Cryst. Mp 100-103°. $[\alpha]_D^{21}$ +57° (c, 0.67 in MeOH).

3',7-Dimethoxy: see Licarin A, L-20040

Hayashi, T., *Phytochemistry*, 1975, **14**, 1085 (*isol*)
Achenbach, H. *et al*, *Phytochemistry*, 1987, **26**, 1159 (*isol, pmr, cmr*)

Conphysodalic acid   C-60162

$C_{20}H_{18}O_{10}$ M 418.356
Constit. of *Flavoparmelia springtonensis*. Powder (Me₂CO/pet. ether). Mp >230° dec.

Elix, J.A. *et al*, *Aust. J. Chem.*, 1987, **40**, 417.

Constictic acid   C-60163

Updated Entry replacing C-02765

1,3-Dihydro-1,4-dihydroxy-5-(hydroxymethyl)-10-methoxy-8-methyl-3,7-dioxo-7H-isobenzofuro[4,5-b][1,4]benzodioxepin-11-carboxaldehyde, 9CI

[30287-05-9]

$C_{19}H_{14}O_{10}$ M 402.314
Constit. of the lichen *Usnea aciculifera*. Mp 195-200°.

α-Ac: α-Acetylconstictic acid.
$C_{21}H_{16}O_{11}$ M 444.351
Constit. of *Pseudocyphellaria faveolata*. Cryst. (Me₂CO/pet. ether). Mp >230° dec.

Yoshioka, I. *et al*, *Chem. Pharm. Bull.*, 1970, **18**, 2364 (*isol, struct*)
Elix, J.A. *et al*, *Aust. J. Chem.*, 1987, **40**, 417 (*deriv*)

Coralloidolide *A*   C-60164

[107748-88-9]

$C_{20}H_{24}O_4$ M 328.407
Constit. of *Alcyonium coralloides*. Cryst. Mp 112-113°. $[\alpha]_D^{20}$ −66.1° (c, 0.36 in EtOH).

D'Ambrosio, M. *et al*, *Helv. Chim. Acta*, 1987, **70**, 63.

Coralloidolide *B*   C-60165

$C_{20}H_{26}O_6$ M 362.422
Constit. of *Alcyonium coralloides*. Amorph. powder. $[\alpha]_D^{20}$ −85.4° (c, 0.78 in EtOH).

D'Ambrosio, M. *et al*, *Helv. Chim. Acta*, 1987, **70**, 63.

Cordatin   C-60166

C_{21}H_{26}O_6   M 374.433
Constit. of *Aparisthmium cordatum*. Cryst. (Me_2CO).
Mp 170-172°. $[\alpha]_D^{20}$ +140° (c, 0.1 in MeOH).

Dadoun, H. *et al*, *Phytochemistry*, 1987, **26**, 2108 (*cryst struct*)

Cornudentanone   C-60167

C_{22}H_{34}O_5   M 378.508
Constit. of *Ardisia cornudentata*. Yellow amorph. solid.
$[\alpha]_D^{22}$ −31.5° (c, 1.6 in CHCl_3).

Tian, Z. *et al*, *Phytochemistry*, 1987, **26**, 2361.

Coroglaucigenin   C-60168

Updated Entry replacing C-02814
3β,14β,19-Trihydroxy-5α-card-20(22)-enolide.
Cannogenol
[468-19-9]

C_{23}H_{34}O_5   M 390.519
Constit. of *Coronilla glauca* and *Peregularia tomentosa*.
Cryst. (EtOH). Mp 249°. $[\alpha]_D$ +23° (MeOH).
▷FH5386000.

3-O-D-Allomethyloside: [546-02-1]. **Frugoside**.
C_{29}H_{44}O_9   M 536.661
Constit. of *Gomphocarpus fructicosus*. Cryst. + 2H_2O
(MeOH aq.). Mp 169-170° and 237-242° (double
Mp). $[\alpha]_D^{21}$ −17.4° (MeOH aq.).
▷Toxic, LD_{50} 0.16 mg/Kg (cat). LS7200000.
3-O-Rhamnoside:
C_{29}H_{44}O_9   M 536.661
Constit. of *Maclotus philippinensis*. Cryst.
(Me_2CO/Et_2O). Mp 231-235°. $[\alpha]_D^{27}$ −38.1° (c, 1.04
in MeOH).
3-O-D-Cymaroside: **Maquiroside A**.
C_{30}H_{46}O_8   M 534.689
Constit. of *Maquira calophylla*. Powder. Mp 123-
125°. $[\alpha]_D^{25}$ +17.9° (c, 0.07 in CHCl_3).

Hunger, A. *et al*, *Helv. Chim. Acta*, 1952, **35**, 1073 (*struct*)
Roberts, K.D. *et al*, *Helv. Chim. Acta*, 1963, **46**, 2886 (*deriv*)
Idrissi, T.E. *et al*, *C. R. Hebd. Seances Acad. Sci.*, Ser. C, 1971,
    273, 598 (*isol*)

Fayez, M.B.E. *et al*, *J. Pharm. Sci.*, 1972, **61**, 765 (*ms*)
Yamauchi, T. *et al*, *Chem. Pharm. Bull.*, 1978, **26**, 2894 (*cmr*)
Rovinski, J.M. *et al*, *J. Nat. Prod.*, 1987, **50**, 211 (*Maquiroside
    A*)

7,13-Corymbidienolide   C-60169

C_{23}H_{34}O_4   M 374.519
Constit. of *Corymbium villosum*. Cryst. Mp 111°. $[\alpha]_D^{24}$
−161° (c, 0.54 in CHCl_3).

Zdero, C. *et al*, *Phytochemistry*, 1988, **27**, 227.

Corymbivillosol   C-60170

C_{43}H_{68}O_6   M 681.007
Constit. of *Corymbium villosum*. Gum. $[\alpha]_D^{24}$ −17° (c,
    0.18 in CHCl_3).
3-Ac:
C_{45}H_{70}O_7   M 723.044
From *C. villosum*. Gum. $[\alpha]_D^{24}$ −18° (c, 0.16 in
CHCl_3).

Zdero, C. *et al*, *Phytochemistry*, 1988, **27**, 227.

Crotalarin   C-60171

[109517-68-2]

C_{20}H_{18}O_5   M 338.359
Constit. of *Crotalaria madurensis*. Needles
(CHCl_3/MeOH). Mp 281-285°.

Chaturvedi, R. *et al*, *J. Nat. Prod.*, 1987, **50**, 266.

Crotarin   C-60172

[109517-69-3]

C_{20}H_{16}O_6   M 352.343
Constit. of *Crotalaria madurensis*. Cryst. (MeOH). Mp
260° dec.

Chaturvedi, R. *et al*, *J. Nat. Prod.*, 1987, **50**, 266.

Cryptotanshinone
C-60173

Updated Entry replacing C-02960
1,2,6,7,8,9-Hexahydro-1,6,6-trimethylphenanthro[1,2-b]furan-10,11-dione, 9CI

[4783-35-1]

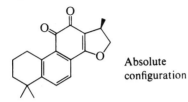

Absolute
configuration

$C_{19}H_{20}O_3$ M 296.365

(*R*)-form [35825-57-1]
Isol. from root of *Salvia miltiorrhiza* and from *Rosmarinus officinalis*. Red cryst. Mp 182°. $[\alpha]_D^{21}$ −91.4° (CHCl₃).

(±)-*form*
Orange-brown needles. Mp 174-175°.

Takiura, K. *et al*, *Chem. Pharm. Bull.*, 1962, **10**, 112 (*isol*)
Baillie, A.C. *et al*, *J. Chem. Soc.* (*C*), 1968, 48 (*synth*)
Inouye, Y. *et al*, *Bull. Chem. Soc. Jpn.*, 1969, **42**, 3318 (*synth*)
Tateishi, M. *et al*, *Tetrahedron*, 1971, **27**, 237 (*synth*)
Brieskorn, C.H. *et al*, *Planta Med.*, 1973, **24**, 190 (*isol*)
Romanova, A.S. *et al*, *Khim. Prir. Soedin.*, 1977, 414 (*isol*)
Tomita, Y. *et al*, *J. Chem. Soc., Chem. Commun.*, 1987, 1311 (*biosynth, abs config*)

Cubanecarboxylic acid
C-60174

Updated Entry replacing C-02967
Pentacyclo[4.2.0.0²,⁵.0³,⁸.0⁴,⁷]octane-1-carboxylic acid, 9CI, 8CI

[53578-15-7]

$C_9H_8O_2$ M 148.161
Cryst. (pentane). Mp 124-125°. pK_a 5.94 (25°, 50% EtOH aq.).

Eaton, P.E. *et al*, *J. Am. Chem. Soc.*, 1964, **86**, 3157 (*synth*)
Cole, T.W. *et al*, *J. Am. Chem. Soc.*, 1974, **96**, 4555.
Edward, J.T. *et al*, *J. Am. Chem. Soc.*, 1976, **98**, 3075 (*nmr*)
Della, E.W. *et al*, *Aust. J. Chem.*, 1986, **39**, 1061 (*synth, pmr*)

5,24-Cucurbitadiene-3,11,26-triol
C-60175

(3β,11α,24*E*)-form

$C_{30}H_{50}O_3$ M 458.723

(3β,11α,24*E*)-*form*
Carnosiflogenin C
Powder. $[\alpha]_D^{29}$ +41° (c, 0.39 in MeOH).
26-O-(β-D-Glucopyranosyl-(1→2)-β-D-glucopyranoside), 3-O-β-D-glucopyranoside: **Carnosifloside V**.
$C_{48}H_{80}O_{18}$ M 945.149
Constit. of *Hemsleya carnosiflora*. Sweet taste. Powder. $[\alpha]_D^{26}$ +4.0° (c, 0.77 in MeOH).
26-O-(β-D-Glucopyranosyl-(1→6)-β-D-glucopyranoside), 3-O-β-D-glucopyranoside: **Carnosifloside VI**.
$C_{48}H_{80}O_{18}$ M 945.149

Constit. of *H. carnosiflora*. Sweet taste. Powder. $[\alpha]_D^{26}$ +6.0° (c, 0.6 in MeOH).
11-Ketone: 3β,26-Dihydroxy-5,24E-cucurbitadien-11-one. Carnosiflogenin A.
$C_{30}H_{48}O_3$ M 456.707
Needles (CHCl₃/C₆H₆). Mp 170.5°. $[\alpha]_D^{16}$ +168.7° (c, 0.43 in MeOH).
11-Ketone, 26-O-(β-D-glucopyranosyl-(1→6)-glucopyranoside): Carnosifloside I.
$C_{42}H_{68}O_{13}$ M 780.991
Constit. of *H. carnosiflora*. Powder. $[\alpha]_D^{22}$ +59.6° (c, 1.42 in MeOH). Tasteless.
11-Ketone, 3-O-β-D-glucopyranoside, 26-O-(β-D-glucopyranosyl-(1→2)-glucopyranoside): Carnosifloside II.
$C_{48}H_{78}O_{18}$ M 943.133
Constit. of *H. carnosiflora*. Powder. $[\alpha]_D^{22}$ +55.9° (c, 0.95 in MeOH). Bitter taste.
11-Ketone, 3-O-β-D-glucopyranoside, 26-O-(β-D-glucopyranosyl-(1→6)-glucopyranoside): Carnosifloside III.
$C_{48}H_{78}O_{18}$ M 943.133
Constit. of *H. carnosiflora*. Powder. $[\alpha]_D^{14}$ +33.6° (c, 1.01 in MeOH). Bitter taste.

(3β,11α,24*Z*)-*form*
11-Ketone: 3β,27-Dihydroxy-5,24Z-cucurbitadien-11-one. Carnosiflogenin B.
$C_{30}H_{48}O_3$ M 456.707
Needles (MeOH aq.). Mp 155-156°. $[\alpha]_D^{18}$ +195.3° (c, 0.54 in MeOH).
3-O-β-D-Glucopyranoside, 26-O-(β-D-glucopyranosyl-(1→6)-glucopyranoside): Carnosifloside IV.
$C_{48}H_{78}O_{18}$ M 943.133
Constit. of *H. carnosiflora*. Powder. $[\alpha]_D^{18}$ +36.2° (c, 0.61 in MeOH). Bitter taste.

Kasai, R. *et al*, *Phytochemistry*, 1987, **26**, 1371.

Curculathyrane *A*
C-60176

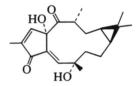

$C_{20}H_{28}O_4$ M 332.439
Constit. of *Jatropha curcus*. Cryst. (CHCl₃). Mp 168-170°. $[\alpha]_D^{22}$ −385° (c, 0.105 in CH₂Cl₂).

Naengchomnong, W. *et al*, *Tetrahedron Lett.*, 1986, **27**, 5675 (*cryst struct*)

Curculathyrane *B*
C-60177

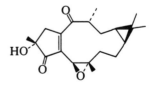

$C_{20}H_{28}O_4$ M 332.439
Constit. of *Jatropha aureus*. Cryst. (CHCl₃). Mp 178-179°. $[\alpha]_D^{22}$ 0° (c, 0.122 in CHCl₂).

Naengchomnong, W. *et al*, *Tetrahedron Lett.*, 1986, **27**, 5675 (*cryst struct*)

Curromycin *A*   C-60178

$C_{38}H_{55}N_3O_{10}$ M 713.867

Isol. from *Streptomyces hygroscopicus*. Active against
some kinds of gram-positive bacteria and shows cytotoxic props. It has a narrower antibacterial spectrum
than Oxozolomycin to which it is related. Yellow
amorph. powder. Mp 103-105°. $[\alpha]_D^{22}$ +39.0° (c, 0.115
in MeOH).

30-Demethoxy: 30-Demethoxycurromycin A. **Curramy-
cin B**. *IM 8443*T. *Antibiotic IM 8443*T.
$C_{37}H_{53}N_3O_9$ M 683.840
From *S. hygroscopicus*. Activity resembles that of
Curromycin *A*. Yellow amorph. powder. Mp 106-109°.
$[\alpha]_D^{22}$ +35° (c, 0.1 in MeOH).

Ogura, M. *et al*, *Agric. Biol. Chem.*, 1985, **49**, 1909; *J. Antibiot.*,
1985, **38**, 669 (*isol, uv, ir, cmr, pmr, ms*)
Okabe, T. *et al*, *J. Antibiot.*, 1985, **38**, 964 (*isol*)

Cyanuric acid   C-60179

Updated Entry replacing C-10334
1,3,5-Triazine-2,4,6(1H,3H,5H)-trione, 9CI. *Isocyanuric
acid*
[108-80-5]

$C_3H_3N_3O_3$ M 129.075

Predominant tautomer; see 1,3,5-Triazine-2,4,6-triol, T-
02371 . Used in rubber, resin manufacture. Cryst. +
$2H_2O$ (H_2O). Sol. hot EtOH, mod. sol. H_2O. d^0 1.768.
Mp >360°. pK_a 4.74 (H_2O, 25°). Dec. to Cyanic acid
on heating without melting. Reacts as mono-, di- or
tribasic acid.

▷Toxic, irritant to skin and eyes. Reacts violently with
EtOH. XZ1800000.

1,3-Di-Me: [6726-48-3].
$C_5H_7N_3O_3$ M 157.129
Mp 221-222°.
Tri-N-Me: [827-16-7]. *Trimethyl isocyanurate.*
$C_6H_9N_3O_3$ M 171.155
Prisms. Sol. EtOH. Mp 175-176°. Bp 274°.
Tri-N-Et: [715-63-9]. *Triethyl isocyanurate.*
$C_9H_{15}N_3O_3$ M 213.236
Prisms. Mp 95°. Bp760 276°.
*1,3-Dichloro: see 1,3-Dichloro-1,3,5-triazine-
2,4,6(1H,3H,5H)-trione, D-03204*

▷XZ1845000.

Smolin, E. *et al*, *s-Triazines and Derivatives*, *Chem. Heterocycl.
Compd.*, 1959, **13**, 20 (*bibl*)
Verschoor, G.C. *et al*, *Acta Crystallogr.*, *Sect. B*, 1971, **27**, 134
(*cryst struct*)
Sucharda-Sobezyk, A., *Rocz. Chem.*, 1976, **50**, 647 (*ir*)
Kolonko, K.J. *et al*, *J. Org. Chem.*, 1979, **44**, 3769 (*deriv*)
D'Silva, T.D.J. *et al*, *J. Org. Chem.*, 1986, **51**, 3781 (*deriv*)
Bretherick, L., *Handbook of Reactive Chemical Hazards*, 2nd
Ed., Butterworths, London and Boston, 1979, 425, 818.
Sax, N.I., *Dangerous Properties of Industrial Materials*, 5th
Ed., Van Nostrand-Reinhold, 1979, 528, 563.

Cyclic(*N*-methyl-L-alanyl-L-tyrosyl-D-tryptophyl-L-lysyl-L-valyl-L-phenyla-lanyl), 9CI   C-60180

[81377-02-8]

$$N\text{-Me-Ala-Tyr-D-Trp}$$
$$|\qquad\qquad\qquad|$$
$$Phe\text{-}Val\text{-}Lys$$

$C_{44}H_{56}N_8O_7$ M 808.976

Highly potent cyclic hexapeptide analogue of Somatosta-
tin.

B,AcOH: [99248-33-6]. Amorph. solid + $3H_2O$. $[\alpha]_D^{25}$
−65.5° (c, 1.07 in 1*M* AcOH).

Brady, S.F. *et al*, *J. Org. Chem.*, 1987, **52**, 764 (*synth, bibl*)

Cycloanticopalic acid   C-60181

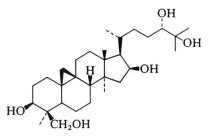

$C_{20}H_{30}O_2$ M 302.456

Constit. of *Pinus strobus*.

Me ester: $[\alpha]_D^{20}$ +87.5° (c, 1.4 in $CHCl_3$).

Zinkel, D.F. *et al*, *Phytochemistry*, 1987, **26**, 769.

3,16,24,25,30-Cycloartanepentol   C-60182

$C_{30}H_{52}O_5$ M 492.738

(3β,16β,24S)-form [107869-22-7]
Cyclofoetigenin B
Sapogenin from *Thalictrum foetidum*. Cryst. (Me_2CO).
Mp 240-242°. $[\alpha]_D^{22}$ +72° (c, 0.5 in MeOH).

Ganenko, T.V. *et al*, *Khim. Prir. Soedin.*, 1986, **22**, 288, 315.

1,2-Cyclobutanedione, 9CI   C-60183

Updated Entry replacing C-03159
[33689-28-0]

$C_4H_4O_2$ M 84.074
Yellow cryst. (pentane). Mp 65°.

Heine, H.-G., *Chem. Ber.*, 1971, **104**, 2869 (*synth*)
Barnier, J.P. *et al*, *Tetrahedron*, 1974, **30**, 1405 (*synth, spectra*)
Gleiter, R. *et al*, *Chem. Ber.*, 1985, **118**, 2127 (*uv, pe*)
Back, R.A. *et al*, *Can. J. Chem.*, 1986, **64**, 2152 (*spectra*)

1,3-Cyclobutanedithiol   C-60184

1,3-Dimercaptocyclobutane

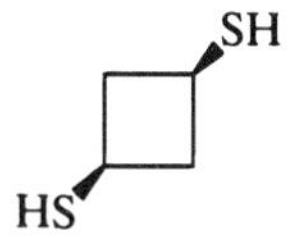

cis-form

$C_4H_8S_2$ M 120.227
cis-form [103562-76-1]
Oil.
1,3-Bis(2-propenyl): [103562-75-0]. *1,3-Bis(allylthio)-
cyclobutane.*
$C_{10}H_{16}S_2$ M 200.356
Oil.

Block, E. *et al, J. Org. Chem.*, 1986, **51**, 3428 (*synth, pmr, cmr,
ms*)

2,7-Cyclodecadien-1-one   C-60185

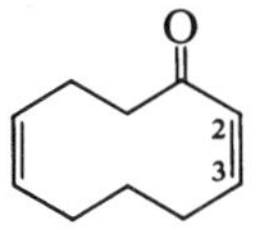

$C_{10}H_{14}O$ M 150.220
(Z,Z)-form [83298-49-1]
Cryst. (pentane). Mp 57-58°.

Gleiter, R. *et al, Helv. Chim. Acta*, 1986, **69**, 1872 (*synth, uv,
pmr, cmr*)

3,7-Cyclodecadien-1-one   C-60186

$C_{10}H_{14}O$ M 150.220
(Z,Z)-form [83298-50-4]
Oil. Purity >90%.

Gleiter, R. *et al, Helv. Chim. Acta*, 1986, **69**, 1872 (*synth, uv, ir,
pmr, cmr*)

Cyclodeca[1,2,3-*de*:6,7,8-*d'e'*]-
dinaphthalene, 8CI   C-60187

Updated Entry replacing C-03194
*1,2,3:6,7,8-Di(1',8'-naphth)[10]annulene. Din-
aphtho[1,8-ab,1',8'-fg]cyclodecene*
[20315-06-4]

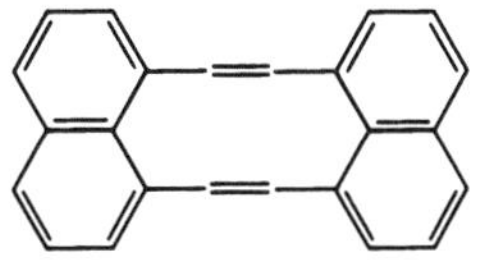

$C_{24}H_{16}$ M 304.390
A claimed synth. of this compd. was erroneous.

Gleiter, R. *et al, Helv. Chim. Acta*, 1987, **70**, 480.

Cyclodehydromyopyrone *A*   C-60188

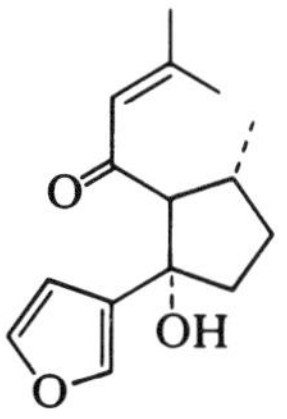

$C_{15}H_{20}O_3$ M 248.321
Constit. of *Eumorphia prostata.*

Hess, T. *et al, Tetrahedron Lett.*, 1987, **28**, 5643.

Cyclodehydromyopyrone *B*   C-60189

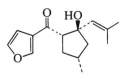

$C_{15}H_{20}O_3$ M 248.321
Constit. of *Eumorphia prostata.*

Hess, T. *et al, Tetrahedron Lett.*, 1987, **28**, 5643.

1,5,9-Cyclododecatrien-3-yne   C-60190

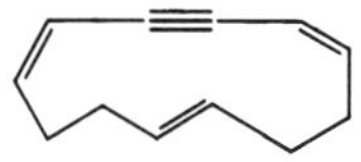

$C_{12}H_{14}$ M 158.243
(1Z,5E,9Z)-form [107170-30-9]
Liq.

Meier, H. *et al, Tetrahedron*, 1986, **42**, 1711 (*synth, pmr, cmr,
ms*)

1-Cyclododecenecarboxaldehyde, 9CI   C-60191

1-Formylcyclododecene

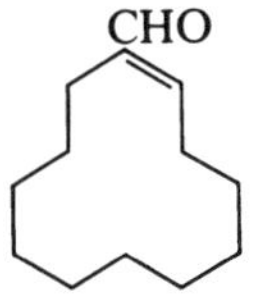

CHO

$C_{13}H_{22}O$ M 194.316
(E)-form [35721-53-0]
Bp$_{0.10}$ 120°. Chromatog. of *E/Z* mixt. gives entirely *E*.

Williams, D.R. *et al, Tetrahedron Lett.*, 1980, **21**, 4417 (*synth*)
Tamura, R. *et al, J. Org. Chem.*, 1986, **51**, 4368 (*synth*)

Cyclododecyne   C-60192

[1129-90-4]

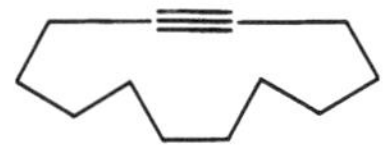

$C_{12}H_{20}$ M 164.290
Bp$_{11}$ 106-109°.

Prelog, V. *et al, Helv. Chim. Acta*, 1955, **38**, 1786 (*synth, ir*)

α-Cyclohallerin   C-60193

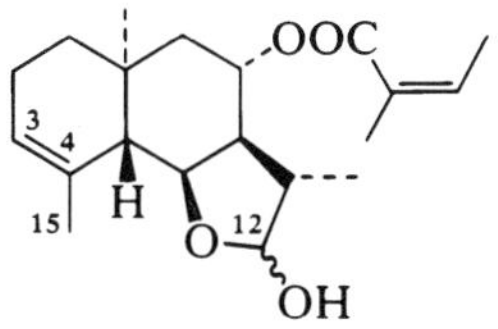

$C_{20}H_{30}O_4$ M 334.455
Constit. of *Laserpitium halleri* as a mixt. of 12-epimers.
Me ether (12β): Methyl-α-cyclohallerin. Gum. $[α]_D^{25}$
−140° (c, 0.51 in CHCl$_3$).
*Δ^{4,15}-Isomer: β-**Cyclohallerin**.*
$C_{20}H_{30}O_4$ M 334.455
Constit. of *L. halleri* as a mixt. of 12-epimers.
Δ^{4,15}-Isomer, Me ether (12β): Methyl-β-cyclohallerin.
Oil. $[α]_D^{25}$ +8° (c, 0.6 in CHCl$_3$).

Appendino, G. *et al*, *Phytochemistry*, 1987, **26**, 1755.

Cyclohept[*fg*]acenaphthylene-5,6-dione, 9CI   C-60194

Updated Entry replacing A-20008
5,6-Acepleiadylenedione
[85960-33-4]

C₁₆H₈O₂   M 232.238
Purple-red needles (C₆H₆/hexane). Mp 145° dec.
 Unstable.

Tsunetsugu, J. *et al*, *J. Chem. Soc., Perkin Trans. 1*, 1985, 785
 (*synth, uv, ir, pmr, cmr, ms*)

Cyclohept[*fg*]acenaphthylene-5,8-dione, 9CI   C-60195

Updated Entry replacing A-20009
5,8-Acepleiadylenedione
[85960-34-5]
C₁₆H₈O₂   M 232.238
Purple-red needles (C₆H₆/hexane). Mp 186° dec.

Tsunetsugu, J. *et al*, *J. Chem. Soc., Perkin Trans. 1*, 1985, 785
 (*synth, uv, ir, pmr, cmr, ms*)

2,6-Cycloheptadien-4-yn-1-one, 9CI   C-60196

Updated Entry replacing C-03250
4,5-Didehydrotropone
[74373-21-0]

C₇H₄O   M 104.108
Transient intermediate.

Nakazawa, T. *et al*, *Angew Chem., Int. Ed. Engl.*, 1975, **14**, 711
 (*synth*)
Nazakawa, T. *et al*, *Tetrahedron Lett.*, 1986, **27**, 3005 (*synth,
 props, bibl*)

Cyclohepta[*ef*]heptalene   C-60197

Pleiaheptalene

C₁₆H₁₂   M 204.271
Parent compd. currently unknown. Several substd. derivs.
 prepd., as stable red cryst. compds.

Hafner, K. *et al*, *Angew. Chem., Int. Ed. Engl.*, 1986, **25**, 633.

Cyclohepta[*de*]naphthalene, 9CI   C-60198

Updated Entry replacing C-03254
 peri-*Cycloheptanaphthalene. Pleiadene. Pleiadiene*
[208-20-8]

C₁₄H₁₀   M 178.233
The name Pleiadene is more usually associated with
 Pleiadene, P-02079 . Red bars (hexane). Mp 88-90°.

Kolc, J. *et al*, *J. Am. Chem. Soc.*, 1973, **95**, 7391 (*uv*)
Wendisch, D. *et al*, *Tetrahedron*, 1974, **30**, 295 (*nmr*)
Watson, C.R. *et al*, *J. Am. Chem. Soc.*, 1976, **98**, 2551 (*synth*)
Paquette, L.A. *et al*, *J. Am. Chem. Soc.*, 1986, **108**, 3739 (*synth,
 pmr, cmr*)

Cyclohepta[*a*]phenalene, 9CI   C-60199

[20542-67-0]

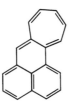

C₁₈H₁₂   M 228.293
Dark-green (C₆H₆/hexane). Mp 60.5° dec.

Sugihara, Y. *et al*, *Angew. Chem., Int. Ed. Engl.*, 1987, **26**, 1247
 (*synth, uv, pmr, cmr*)

5*H*-Cyclohepta[4,5]pyrrolo[2,3-*b*]indole, 9CI   C-60200

Indolo[2,3-b] [*1*]*azaazulene*
[108460-21-5]

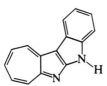

C₁₅H₁₀N₂   M 218.257
Red-brown needles (EtOH). Mp >300° dec.

Yamane, K. *et al*, *Bull. Chem. Soc. Jpn.*, 1986, **59**, 3326 (*synth,
 uv, ir, pmr, ms*)

Cyclohepta[*b*]pyrrol-2(1*H*)-one, 9CI   C-60201

1-Azaazulen-2-one
[2132-34-5]

C₉H₇NO   M 145.160
Yellow needles. Mp 163-164°.
N-*Ac*:
 C₁₁H₉NO₂   M 187.198
 Orange needles. Mp 116-117°.

Nozoe, T. *et al*, *Chem. Ind.* (*London*), 1954, 1356 (*synth*)

1,3,5-Cycloheptatriene-1,6-dicarboxylic acid   C-60202

Updated Entry replacing C-03272

HOOC⟶COOH

$C_9H_8O_4$ M 180.160
Needles (dioxan/C_6H_6). Mp 293-295° dec.
Mono-Me ester: [75538-94-2].
   $C_{10}H_{10}O_4$ M 194.187
   Needles (CH_2Cl_2/hexane). Mp 122-123°.
Di-Me ester:
   $C_{11}H_{12}O_4$ M 208.213
   Needles (Et_2O/hexane). Mp 53-54°.
Dichloride: [73875-02-2].
   $C_9H_6Cl_2O_2$ M 217.051
   Mp 63°.

Darms, R. *et al*, *Helv. Chim. Acta*, 1963, **46**, 2893 (*nmr, uv, synth*)
Vogel, E. *et al*, *Angew. Chem., Int. Ed. Engl.*, 1980, **19**, 4; 1986, **25**, 723 (*synth, derivs*)

2,4,6-Cycloheptatriene-1-thione, 9CI   C-60203

Updated Entry replacing C-03273
Troponethione
[30456-90-7]

C_7H_6S M 122.184
Red needles (CH_2Cl_2). Mp 20.5-21.5° (sinters at 19.5°).
pK_a −1.57 (21°). $t_{1/2}$ 9.1h at 25°, gives a mixt. of oligomers.

Machiguchi, T. *et al*, *Chem. Lett.*, 1974, 497 (*cmr*)
Dugger, H.A. *et al*, *Helv. Chim. Acta*, 1976, **59**, 747 (*synth, uv*)
Machiguchi, T. *et al*, *Tetrahedron Lett.*, 1987, **28**, 203 (*synth, uv, ir, pmr, cmr*)

2,4,6-Cycloheptatrien-1-one, 9CI   C-60204

Updated Entry replacing C-03277
Tropone
[539-80-0]

C_7H_6O M 106.124
Viscous hygroscopic oil. d_4^{22} 1.10. Mp −8° to −5°. Bp_{15} 113°. Forms a hydrochloride by protonation on oxygen.

Picrylsulfonate: Mp 266-267°.
Picrate: Mp 100-101°.

Dauben, H.J. *et al*, *J. Am. Chem. Soc.*, 1951, **73**, 876 (*synth, uv, ir*)
Yamaguchi, H. *et al*, *Tetrahedron*, 1968, **24**, 267 (*uv*)
McCullagh, L.N. *et al*, *Synthesis*, 1972, 422 (*synth*)
Barrow, M.J. *et al*, *J. Chem. Soc., Chem. Commun.*, 1973, 66 (*cryst struct*)
Bagli, J.F. *et al*, *Can. J. Chem.*, 1978, **56**, 578 (*cmr*)
Garfunkel, E. *et al*, *J. Org. Chem.*, 1979, **44**, 3725 (*synth*)
Reingold, I.D. *et al*, *J. Org. Chem.*, 1982, **47**, 3544 (*synth*)
Föhlisch, B. *et al*, *Justus Liebigs Ann. Chem.*, 1987, 1 (*synth*)

3-(1,3,6-Cycloheptatrien-1-yl-2,4,6-cyclo-heptatrien-1-ylidenemethyl)-1,3,5-cyclo-heptatriene, 9CI   C-60205

Di(1,3,5-cycloheptatrien-3-yl)(2,4,6-cycloheptatrienylidene)methane
[104422-32-4]

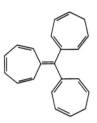

$C_{22}H_{20}$ M 284.400
Red oil, air- and acid-sensitive. Forms mono-, di- and tritropylium cations on deprotonation.

Mizumoto, K. *et al*, *Angew. Chem., Int. Ed. Engl.*, 1986, **25**, 916.

2,4,6-Cycloheptatrien-1-ylcyclopropyl-methanone, 9CI   C-60206

2,4,6-Cycloheptatrienyl cyclopropyl ketone
[97847-61-5]

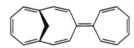

$C_{11}H_{12}O$ M 160.215
$Bp_{0.05}$ 46°.

Ritter, K. *et al*, *Chem. Ber.*, 1986, **119**, 3704 (*synth, ir, pmr, cmr*)

4-(2,4,6-Cycloheptatrien-1-ylidene)-bicyclo[5.4.1]dodeca-2,5,7,9,11-pentaene, 9CI   C-60207

[91511-06-7]

$C_{19}H_{16}$ M 244.335
Red-brown needles. Mp 104°.

Beck, A. *et al*, *Tetrahedron Lett.*, 1986, **27**, 485 (*synth, uv, cmr, pmr, cryst struct, props*)

Cycloheptatrienylidene(tetraphenylcyclo-pentadenylidene)ethylene   C-60208

7-[(2,3,4,5-Tetraphenyl-2,4-cyclopentadien-1-ylidene)-ethenylidene]-1,3,5-cycloheptatriene, 9CI
[73625-00-0]

$C_{38}H_{26}$ M 482.623
Mp 183-184° dec. No colour mentioned but λ_{max} 538 nm (ε 13200).

Toda, T. *et al*, *Angew. Chem., Int. Ed. Engl.*, 1987, **26**, 335 (*synth, uv, cmr*)

1,1-Cyclohexanedithiol   C-60209

1,1-Dimercaptocyclohexane

[3855-24-1]

HS SH

$C_6H_{12}S_2$ M 148.281

Liq. with unpleasant odour. Bp_{13} 70-72°, $Bp_{0.1}$ 43-44°.

▷GV0800000.

Cyclic trimer: [177-58-2]. *7,14,21-
Trithiatrispiro[5.1.5.1.5.1]heneicosane.*
$C_{18}H_{30}S_3$ M 342.615
Cryst. (MeOH). Mp 95-99°. $Bp_{0.1}$ 125°.

Fromm, E., *Ber.*, 1927, **60**, 2090 (*trimer*)
Djerassi, C. *et al, J. Org. Chem.*, 1962, **27**, 1041 (*synth, ir, pmr*)
Demuynck, M. *et al, Bull. Soc. Chim. Fr.*, 1967, 1213 (*synth, pmr, ir*)

Cyclohexanehexone, 9CI, 8CI   C-60210

Hexaketocyclohexane. Triquinoyl

[527-31-1]

C_6O_6 M 168.062

Has various industrial and pharmaceutical uses.
Hygroscopic, not obt. anhydrous, dec. in air.

Octahydrate: Cryst. (H_2O). Mp 95-96° dec.

Nietzki, R. *et al, Ber.*, 1885, **18**, 506 (*synth*)
Henle, F., *Justus Liebigs Ann. Chem.*, 1906, **350**, 330 (*synth*)
Bergel, F., *Ber.*, 1929, **62**, 490 (*props*)
Eistert, B. *et al, Angew. Chem.*, 1958, **70**, 595 (*synth*)
Fatiadi, A.J. *et al, CA*, 1963, **58**, 12450 (*synth*)
Japan. Pat., 64 19 498, (*1964*); *CA*, **66**, 104828e (*synth*)
Skujins, S. *et al, Tetrahedron*, 1968, **24**, 4805 (*synth, use*)
Kanakarajan, K. *et al, J. Org. Chem.*, 1986, **51**, 5241 (*synth*)

3-Cyclohexen-1-one, 9CI   C-60211

[4096-34-8]

C_6H_8O M 96.129

Bp_{10} 38-40°.

Oxime:
C_6H_9NO M 111.143
Mp 80.5-81.5°. Bp_8 115-118°.
2,4-Dinitrophenylhydrazone: Red prisms or hairlike nee-
dles (EtOH). Mp 170-171°.
Dimethyl ketal: [16831-48-4]. *4,4-
Dimethoxycyclohexene.*
$C_8H_{14}O_2$ M 142.197
Bp_{10} 48-50°.
Ethylene ketal: [7092-24-2]. *1,4-Dioxaspiro[4,5]dec-7-
ene.*
$C_8H_{12}O_2$ M 140.182
Bp_7 62-64°.

Shine, H.J. *et al, J. Am. Chem. Soc.*, 1958, **80**, 3064 (*synth, deriv*)

Hanack, M. *et al, Chem. Ber.*, 1963, **96**, 2937 (*synth, ir*)
Lambert, J.B. *et al, J. Am. Chem. Soc.*, 1986, **108**, 7575 (*deriv, synth, pe, pmr*)

2-Cyclohexyl-2-hydroxyacetic acid   C-60212

Updated Entry replacing C-03406

*α-Hydroxycyclohexaneacetic acid, 9CI. Cyclohexylgly-
collic acid. Hexahydromandelic acid*

[4442-94-8]

(R)-form

$C_8H_{14}O_3$ M 158.197

(R)-form [53585-93-6]

Intermed. in synth. of optically active polymers and
macrolides. Cryst. (C_6H_6). Mp 129°. $[\alpha]_D^{20}$ −25.5° (c,
1 in AcOH).

Me ester: [92587-21-8].
$C_9H_{16}O_3$ M 172.224
Bp_5 108°. $[\alpha]_D^{20}$ −31.3° (c, 2.36 in $CHCl_3$).

(S)-form

Prisms (Et_2O). Mp 129°. $[\alpha]_D^{25}$ +13.5° (EtOH).

Me ester: Bp_7 110°. $[\alpha]_D^{25}$ +23.11°.

(±)-form [10498-47-2]

Cryst. Mp 135°.

4-Bromophenacyl ester: Mp 70°.
Me ester: Bp_{30} 120°.
Me ether: [15540-18-8]. *2-Cyclohexyl-2-methoxyacetic
acid.*
$C_9H_{16}O_3$ M 172.224
Mp 67°.

Wood, C.E. *et al, J. Chem. Soc.*, 1924, 2630 (*synth, resoln*)
Newman, D.D.E. *et al, J. Chem. Soc.*, 1952, 4713 (*synth*)
Johnson, W.S. *et al, J. Am. Chem. Soc.*, 1953, **75**, 4995 (*synth*)
Stocker, J.H. *et al, J. Org. Chem.*, 1962, **27**, 2288 (*synth*)
Riley, D.P. *et al, J. Org. Chem.*, 1980, **45**, 5187 (*synth*)
Tsuboi, S. *et al, Tetrahedron Lett.*, 1986, **27**, 1915 (*synth, use,
bibl*)

Cyclo(hydroxyprolylleucyl)   C-60213

*Hexahydro-7-hydroxy-3-(2-methylpropyl)pyrrolo[1,2-
a]pyrazine-1,4-dione. Hydroxyproline leucine anhydride*

$C_{11}H_{18}N_2O_3$ M 226.275

L-L-form

Isol. from rabbit skin tissue. Plant growth regulator. Mp
178-179°. $[\alpha]_D^{28}$ −148.2° (c, 1 in H_2O).

Ienaga, K. *et al, Tetrahedron Lett.*, 1987, **28**, 1285 (*isol, pmr,
cmr, struct*)

Cyclolongipesin   C-60214

3-(3-Methyl-2-butenyl)furo[2,3,4-de]-1-benzopyran-2(5H)-one, 9CI
[110024-19-6]

$C_{15}H_{14}O_3$ M 242.274

Constit. of *Bothriocline longipes*. Cryst. Mp 124°.

9-Methyl: [110024-46-9]. **9-Methylcyclolongipesin**.
$C_{16}H_{16}O_3$ M 256.301
Constit. of *B. longipes*. Cryst. Mp 85°.

Jakupovic, J. *et al, Phytochemistry*, 1987, **26**, 1069.

3*H*-Cyclonona[*def*]biphenylene, 9CI   C-60215

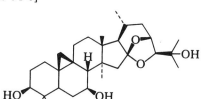

$C_{17}H_{12}$ M 216.282

(*Z,Z*)-**form** [104642-21-9]
Yellow solid (pentane or by subl.). Mp 111-117°.

Wilcox, C.F. *et al, J. Am. Chem. Soc.*, 1986, **108**, 7693 (*synth, uv, pmr, cmr, ms*)

Cyclooocta[1,2-*b*:5,6-*b'*]dinaphthalene   C-60216

Dinaphtho[2,3-a:2',3'-e]cyclooctene

$C_{24}H_{16}$ M 304.390
Cryst. (CHCl₃/heptane). Mp 249-250°.

Cava, M.P. *et al, J. Am. Chem. Soc.*, 1972, **94**, 6441 (*synth, uv*)
Shepherd, M.K., *J. Chem. Soc., Perkin Trans. 1*, 1985, 2689 (*synth*)

1-Cyclooctenecarboxaldehyde, 9CI   C-60217

1-Formylcyclooctene
[6038-12-6]

(*E*)-form

$C_9H_{14}O$ M 138.209
(*E*)-**form** [96308-48-4]
Liq. Bp₁₃ 150°.

2,4-Dinitrophenylhydrazone: Cryst. (EtOAc). Mp 186.8°.

Cope, A.C. *et al, J. Am. Chem. Soc.*, 1960, **82**, 5439 (*synth*)
Neumann, H. *et al, Chem. Ber.*, 1978, **111**, 2785 (*synth, ir, pmr*)
Tamura, R. *et al, J. Org. Chem.*, 1986, **51**, 4368 (*synth*)

Cycloorbigenin   C-60218

[106009-91-0]

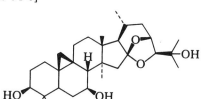

$C_{30}H_{48}O_5$ M 488.706

Sapogenin from *Astragalus orbiculatus*. Cryst. Mp 217-219°. $[\alpha]_D^{20}$ +28.3° (c, 1.19 in EtOH).

Agzamova, M.A. *et al, Khim. Prir. Soedin.*, 1986, **22**, 425.

Cyclopentadecanone, 9CI   C-60219

Updated Entry replacing C-03537
Exaltone. Oxocyclopentadecane. Normuscone
[502-72-7]

$C_{15}H_{28}O$ M 224.386
Used in perfumery. Mp 63°. Bp₀.₃ 120°.
▷GY0900000.

Oxime: [34341-05-4].
$C_{15}H_{29}NO$ M 239.400
Mp 75-76°.
Semicarbazone: Mp 187-188°.
2,4-Dinitrophenylhydrazone: Mp 105°.
Ethylene acetal:
$C_{17}H_{32}O_2$ M 268.439
Mp 27-28°.

Mathur, H.H. *et al, J. Chem. Soc.*, 1963, 3505 (*synth*)
Nozaki, H., *Tetrahedron Lett.*, 1967, 779 (*synth*)
Groth, P., *Acta Chem. Scand., Ser. A*, 1976, **30**, 294 (*cryst struct*)
Nokami, J. *et al, Chem. Lett.*, 1978, 1283 (*synth*)
Feldhues, M. *et al, Tetrahedron*, 1986, **42**, 1285 (*synth, deriv, ir, pmr, ms*)
Suginome, H. *et al, Tetrahedron Lett.*, 1987, **28**, 3963 (*synth*)

Cyclopentadecyne   C-60220

[6573-73-5]

$C_{15}H_{26}$ M 206.370
Bp₁₉ 168°, Bp₁ 106-108°.

Meier, H. *et al, Synthesis*, 1971, 215 (*synth*)
Mandeville, W.H. *et al, J. Org. Chem.*, 1986, **51**, 3257 (*synth*)

2-(2,4-Cyclopentadienylidene)-1,3-dithiole   C-60221

1,4-Dithiafulvalene

$C_8H_6S_2$ M 166.255

Orange cryst. (cyclohexane). Mp 100-102°.

4,5-Dihydro: [3357-53-7]. *2-(2,4-Cyclopentadien-1-ylidene)-1,3-dithiolane, 9CI. Ethylidenedimercaptofulvene.*
$C_8H_8S_2$ M 168.271
Orange cryst. (pet. ether). Mp 96-98°.

Gompper, R. *et al, Chem. Ber.*, 1965, **98**, 2825 (*synth, ir*)
Hartke, K. *et al, Chem. Ber.*, 1966, **99**, 3268 (*synth, ir, deriv*)

1,1-Cyclopentanedithiol   C-60222

1,1-Dithiocyclopentane. 1,1-Dimercaptocyclopentane
[1687-46-3]

$C_5H_{10}S_2$ M 134.254
Pinkish liq. Bp_{10} 63°.

Djerassi, C. *et al, J. Org. Chem.*, 1962, **27**, 1041 (*synth, ir*)
Jentzsch, J. *et al, Chem. Ber.*, 1962, **95**, 1764 (*synth, ir*)

Cyclopenta[*cd*]pyrene   C-60223

Updated Entry replacing C-20300
[27208-37-3]

$C_{18}H_{10}$ M 226.277
Component of carbon black and automobile exhaust. Mp 174-176°.

▷Carcinogenic to mice, bacterial mutagen. GY5850000.

Kanieczny, M. *et al, J. Org. Chem.*, 1979, **44**, 2158 (*synth, pmr, tox*)
Tintel, K. *et al, J. Chem. Soc., Chem. Commun.*, 1982, 185 (*synth*)
Veeraraghavan, S. *et al, J. Org. Chem.*, 1987, **52**, 1355 (*synth*)
Sax, N.I., *Dangerous Properties of Industrial Materials*, 5th Ed., Van Nostrand-Reinhold, 1979, 532.

Cyclopenta[*b*]thiapyran, 9CI   C-60224

Updated Entry replacing C-03631
Thialen
[271-17-0]

C_8H_6S M 134.195
Violet oil. Dec. in light and air.

1,3,5-Trinitrobenzene complex: Brown needles. Mp 128-130°.

Mayer, R. *et al, J. Prakt. Chem.*, 1965, **30**, 262 (*synth, uv*)
Klein, R.F.X. *et al, J. Org. Chem.*, 1986, **51**, 4644 (*synth, uv, pmr, cmr, ir, ms, props*)

Cyclopent[*b*]azepine   C-60225

Updated Entry replacing C-20304
4-Azaazulene
[275-61-6]

C_9H_7N M 129.161
Deep-blue oil.

N-Oxide: [99895-14-4].
C_9H_7NO M 145.160
Green cryst. Mod. stable. No Mp reported.

Hess, B.A. *et al, Tetrahedron*, 1975, **31**, 295 (*struct*)
Meth-Cohn, O. *et al, J. Chem. Soc., Chem. Commun.*, 1983, 1246 (*synth, uv, pmr, cmr*)
Meth-Cohn, O. *et al, J. Chem. Soc., Perkin Trans. 1*, 1985, 1793 (*oxide*)

2-Cyclopenten-1-one, 9CI   C-60226

Updated Entry replacing C-03664
3-Oxocyclopentene
[930-30-3]

C_5H_6O M 82.102
Liq. Mod. sol. H_2O. d^{15} 0.989. Bp 135-137°, Bp_{12} 40°.
n_D^{15} 1.4629.

Oxime:
C_5H_7NO M 97.116
Needles. Mp 52-53°.
2,4-Dinitrophenylhydrazone: [789-99-1]. Mp 169°.
Semicarbazone: Leaflets. Mp 214-215°.

Godchot, M. *et al, Bull. Soc. Chim. Fr.*, 1913, **13**, 548.
Kotz, A. *et al, Justus Liebigs Ann. Chem.*, 1913, **400**, 73 (*synth*)
Corbel, B. *et al, Can. J. Chem.*, 1978, **56**, 505 (*synth*)
Ito, Y. *et al, J. Org. Chem.*, 1978, **43**, 1011 (*synth*)
Salomon, R.G. *et al, J. Am. Chem. Soc.*, 1979, **101**, 3961 (*synth*)
Mihelich, E. *et al, J. Org. Chem.*, 1983, **48**, 4135 (*synth*)
Takano, S. *et al, Chem. Pharm. Bull.*, 1986, **34**, 3445 (*synth*)

2-Cyclopentylidenecyclopentanone, 9CI   C-60227

[825-25-2]

$C_{10}H_{14}O$ M 150.220
d 1.00. Bp_{20} 139-142°, Bp_5 102-103°. n_D^{20} 1.5231.
Oxime: [5834-10-6].
$C_{10}H_{15}NO$ M 165.235
Needles (EtOH). Mp 123-124°.
Semicarbazone: Needles. Mp 217°.

Wallach, O., *Ber.*, 1896, **29**, 2955 (*synth*)
Godchot, M. *et al, Bull. Soc. Chim. Fr.*, 1913, 12 (*synth*)
Goheen, G.E., *J. Am. Chem. Soc.*, 1941, **63**, 744 (*synth*)

[2.2.2](1,2,3)Cyclophane   C-60228
Updated Entry replacing C-20306
5,6,11,12-Tetrahydro-1,10-ethanodibenzo[a,e]-cyclooctene, 9CI

$C_{18}H_{18}$ M 234.340
Prisms (MeOH). Mp 171-172°.

Boekelheide, V., *CA*, 1979, **92**, 128465x.
Kleinschroth, J. *et al*, *Angew. Chem., Int. Ed. Engl.*, 1982, **21**, 469 (*rev*)
Yamato, T. *et al*, *J. Chem. Res. (S)*, 1987, 198 (*synth*)

[2.2.2](1,2,3)Cyclophane-1,9-diene   C-60229
1,10-Ethanodibenzo[a,e]cyclooctene

$C_{18}H_{14}$ M 230.309
Prisms (MeOH). Mp 72-73°.

Yamato, T. *et al*, *J. Chem. Res. (S)*, 1987, 198 (*synth, ir, pmr*)

13²,17³-Cyclopheophorbide enol   C-60230
6-Ethenyl-11-ethyl-15,19,20,21-tetrahydro-18-hydroxy-5,10,22,23-tetramethyl-4,7-imino-2,21-methano-14,16-metheno-9,12-nitrilo-17H-azuleno[1,8-bc][1,5]diazacyclooctadecin-17-one, 9CI
[103538-56-3]

$C_{33}H_{32}N_4O_2$ M 516.641
Chlorophyll A deriv. isol. from *Darwinella oxeata*, first porphyrin from a sponge. Lustrous black needles (hexane/CH_2Cl_2). Mp >360°. Opt. active but accurate$[\alpha]_D$ could not be determined.

Karuso, P. *et al*, *Tetrahedron Lett.*, 1986, **27**, 2177 (*isol, uv, ir, cryst struct*)

2-Cyclopropyl-2-oxoacetic acid   C-60231
α-Oxocyclopropaneacetic acid, 9CI. Cyclopropaneglyoxylic acid
[13885-13-7]

$C_5H_6O_3$ M 114.101
K salt: [56512-18-6]. Cryst. (Me_2CO). Mp 257-259° dec.
Me ester: [6395-79-5].
    $C_6H_8O_3$ M 128.127
    Bp_{20} 97-98°, Bp_{12} 85-90°.

Basnak, I. *et al*, *Collect. Czech. Chem. Commun.*, 1975, **40**, 1038 (*synth, ir, pmr*)
Laurie, D. *et al*, *Tetrahedron*, 1986, **42**, 1035 (*synth, pmr*)

Cyclopterospermol   C-60232
22-Methylene-3β-cycloartanol

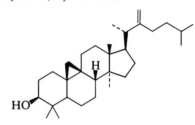

$C_{31}H_{52}O$ M 440.751
Constit. of *Pterospermum heyneanum*. Plates (MeOH). Mp 124-125°.

Anjaneyulu, A.S.R. *et al*, *Phytochemistry*, 1987, **26**, 2805.

Cymbodiacetal   C-60233
[111534-64-6]

$C_{20}H_{30}O_4$ M 334.455
Constit. of the essential oil of *Cymbopogon martinii*. Cryst. ($CHCl_3$). Mp 206-207°. $[\alpha]_D^{25}$ +26° (c, 0.12 in $CHCl_3$).

Bottini, A.T. *et al*, *Phytochemistry*, 1987, **26**, 2301 (*isol, cryst struct*)

Cystoketal   C-60234
Updated Entry replacing C-40243
[96253-60-0]

$C_{28}H_{38}O_4$ M 438.606
Metab. of *Cystoseira balearica*. Oil. $[\alpha]_D$ +11.5° (c, 1 in EtOH).
Z-Isomer: **Isocystoketal**.
    $C_{28}H_{38}O_4$ M 438.606

From *C. stricta*. Oil. $[\alpha]_D^{20}$ +24.9° (c, 0.54 in EtOH).

Amico, V. *et al*, *J. Nat. Prod.*, 1984, **47**, 947 (*Cystoketal*)
Amico, V. *et al*, *Phytochemistry*, 1987, **26**, 1719 (*Isocystoketal*)

Cytochalasin *P*   C-60236

$C_{30}H_{41}NO_6$ M 511.657

Prod. by *Phomopsis* sp. Cytotoxic mycotoxin. Cryst. (CHCl$_3$). Mp 117-118°. $[\alpha]_D$ −116° (MeOH).

Tomioka, T. *et al*, *Chem. Pharm. Bull.*, 1987, **35**, 902 (*isol, struct*)

Cytochalasin *O*   C-60235

$C_{28}H_{37}NO_4$ M 451.605

Prod. by *Phomopsis* sp. Cytotoxic mycotoxin. Needles (Me$_2$CO/hexane). Mp 187-188°. $[\alpha]_D$ +59.6° (MeOH).

*O*²¹*-Ac:* **Cytochalasin N**.

$C_{30}H_{39}NO_5$ M 493.642

From *P.* sp. Cytotoxic mycotoxin. Powder (Me$_2$CO). Mp 253-254°. $[\alpha]_D$ +85.4° (MeOH).

Tomioka, T. *et al*, *Chem. Pharm. Bull.*, 1987, **35**, 902 (*isol, struct*)

D

25-Dammarene-3,12,17,20,24-pentol　　　**D-60001**

$C_{30}H_{52}O_5$　　M 492.738

(3α,12β,17α,20S,24ξ)-form [105822-01-3]
Constit. of *Betula pendula* leaves. $[\alpha]_D^{20} -3°$ (c, 0.5 in CHCl₃).

Pokhilo, N.D. *et al, Khim. Prir. Soedin.*, 1986, **22**, 166.

Danshexinkun A　　　**D-60002**

[65907-75-7]

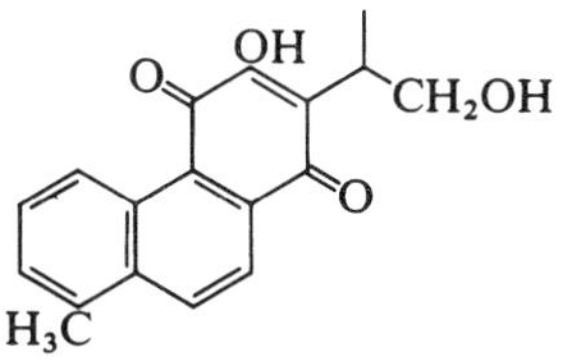

$C_{18}H_{16}O_4$　　M 296.322
Constit. of *Salvia miltiorrhiza*. Orange-red needles. Mp 184-186°.

Lee, A.-R. *et al, J. Nat. Prod.*, 1987, **50**, 157.

Daturilin　　　**D-60003**
(20S,22S)-21,24S-Epoxy-1-oxo-2,5,25-withatrienolide

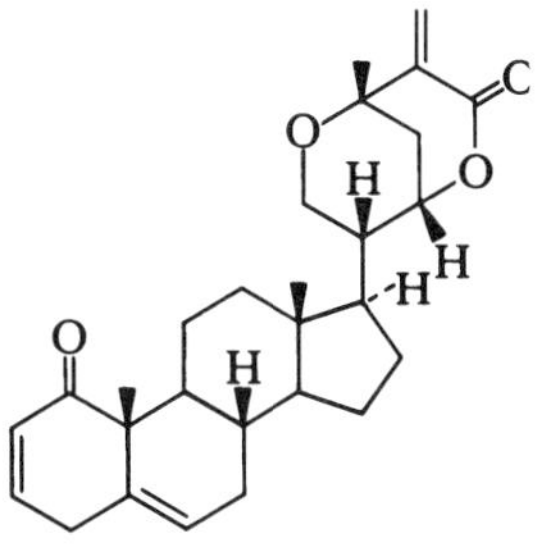

$C_{28}H_{36}O_4$　　M 436.590
Constit. of *Datura metel*. Cryst. (CHCl₃/pet. ether). Mp 206-207°. $[\alpha]_D^{23} -100.0°$ (c, 0.49 in CHCl₃).

Siddiqui, S. *et al, Phytochemistry*, 1987, **26**, 2641.

8-Daucene-3,6,14-triol　　　**D-60004**

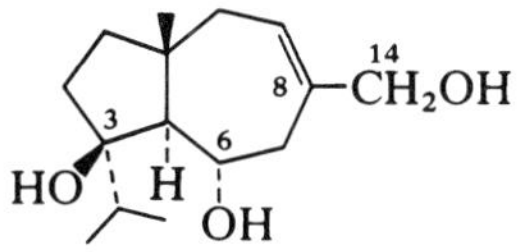

$C_{15}H_{26}O_3$　　M 254.369
(3β,6α)-form
Jaeschkeanadiol

O^6-(*4-Hydroxybenzyl*): **Feraginidin**.
$C_{22}H_{30}O_5$　　M 374.476
Constit. of *Ferula jaeachkeana*. Needles (MeOH). Mp 130°. $[\alpha]_D^{20} +6.0°$ (c, 1.5 in MeOH).

Garg, S.N. *et al, J. Nat. Prod.*, 1987, **50**, 253.

8(14)-Daucene-4,6,9-triol　　　**D-60005**

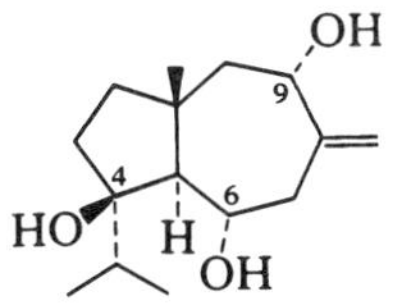

$C_{15}H_{26}O_3$　　M 254.369
(4β,6α,9α)-form
6-(*4-Hydroxybenzoyl*): [109517-72-8]. **Ferugin**.
$C_{22}H_{30}O_5$　　M 374.476
Constit. of *Ferula jaeschkeana*. Needles (MeOH). Mp 140°. $[\alpha]_D^{20} +15.0°$ (c, 1.5 in MeOH).

Garg, S.N. *et al, J. Nat. Prod.*, 1987, **50**, 253.

2,2,3,3,4,4,5,5,6,6-Decafluoropiperidine,　　　**D-60006**
9CI

[559-31-9]

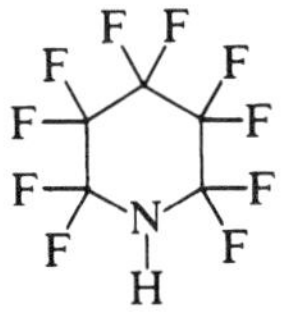

$C_5HF_{10}N$　　M 265.054
Bp 73.4°. Readily hydrolysed, moisture sensitive.
N-*Nitro*: [1840-07-9].
$C_5F_{10}N_2O_2$　　M 310.051
Liq. Bp 86.7°.

Banks, R.E. *et al, J. Chem. Soc.*, 1964, 2488 (*synth*)
Banks, R.E. *et al, J. Chem. Soc.* (*C*), 1967, 427 (*synth, F nmr*)
Banks, R.E. *et al, J. Fluorine Chem.*, 1978, **11**, 563 (*synth*)

1,1',2,2',3,4,5,5',6,6'-Decahydrobiphenyl　　　**D-60007**
4-Cyclohexylcyclohexene
[20247-93-2]

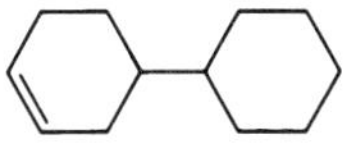

$C_{12}H_{20}$　　M 164.290
(±)-form
Bp 236-237°.

Schrauth, W. *et al, Ber.*, 1923, **56**, 1900 (*synth*)
Benkeser, R.A. *et al, J. Org. Chem.*, 1964, **29**, 1313 (*synth*)

1,1',2,3,4,4',5,5',6,6'-Decahydrobiphenyl　　　**D-60008**
3-Cyclohexylcyclohexene, 9CI
[1808-09-9]

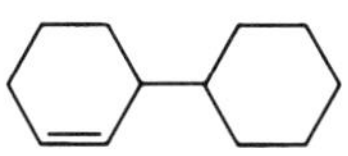

$C_{12}H_{20}$　　M 164.290

(±)-form
Bp 224°, Bp₁₅ 113°.

Hueckel, W. *et al*, *Chem. Ber.*, 1961, **94**, 1026 (*synth*)
Pehk, T. *et al*, *Org. Magn. Reson.*, 1971, **3**, 679 (*cmr*)
Salomon, R.G. *et al*, *J. Am. Chem. Soc.*, 1974, **96**, 1145 (*synth*)
Kropp, P.J. *et al*, *J. Org. Chem.*, 1980, **45**, 4471 (*synth*)
Goering, H.L. *et al*, *J. Org. Chem.*, 1986, **51**, 2884 (*synth, pmr*)

1,2,3,3′,4,4′,5,5′,6,6′-Decahydrobiphenyl   D-60009

1-Cyclohexylcyclohexene, *9CI*

[3282-54-0]

$C_{12}H_{20}$ M 164.290
Bp 235°, Bp₈ 127°.

Pines, H. *et al*, *J. Am. Chem. Soc.*, 1957, **79**, 1698 (*synth*)
Hueckel, W. *et al*, *Justus Liebigs Ann. Chem.*, 1959, **624**, 142 (*synth*)
Brooks, J.D. *et al*, *Aust. J. Chem.*, 1964, **17**, 55 (*synth*)
Pehk, T. *et al*, *Org. Magn. Reson.*, 1971, **3**, 679 (*cmr*)
Guyon, R. *et al*, *Bull. Soc. Chim. Fr.*, 1975, 2584 (*synth*)
Liu, H.-J. *et al*, *Chem. Lett.*, 1978, 923.

5,5a,6,6a,7,12,12a,13,13a,14-Decahydro-   D-60010
5,14:6,13:7,12-trimethanopentacene, *9CI*

[91790-20-4]

$C_{25}H_{24}$ M 324.465
Molecule used for the study of hyperconjugative effects. Needles (CH₂Cl₂/MeOH). Mp 166-168°. The analogue lacking the central methylene bridge was also synthesised.

Craig, D.C. *et al*, *Aust. J. Chem.*, 1986, **39**, 1587 (*synth, pmr, cryst struct*)

Decamethylcyclohexanone, *9CI*   D-60011

[92406-77-4]

$C_{16}H_{30}O$ M 238.412
Solid. Mp 245-250°.

Fitjer, L. *et al*, *Chem. Ber.*, 1986, **119**, 1162 (*synth, pmr, cmr*)

1,1,2,2,3,3,4,4,5,5-Decamethyl-6-methylen-   D-60012
ecyclohexane

[103457-86-9]

$C_{17}H_{32}$ M 236.440

Solid. Mp 272-276°.

Wehle, D. *et al*, *Chem. Ber.*, 1986, **119**, 3127 (*synth, pmr, cmr, ms*)

5,6-Decanediol   D-60013

$$H-\overset{(CH_2)_3CH_3}{\underset{}{C}}-OH$$
$$HO-\overset{}{\underset{(CH_2)_3CH_3}{C}}-H \qquad \textit{5S,6S-form}$$

$C_{10}H_{22}O_2$ M 174.283
(5S,6S)-form [99881-77-3]
Cryst. (EtOH). Mp 43.5-44°. $[\alpha]_{546}^{23}$ −40.2° (c, 0.4 in EtOH).
Benzyl ether: [99809-68-4].
$C_{17}H_{28}O_2$ M 264.407
Bp₀.₁ 100°. $[\alpha]_{546}^{22}$ +13.35° (c, 1.7 in CHCl₃).

Levy, G.C. *et al*, *Org. Magn. Reson.*, 1980, **14**, 214.
Matteson, D.S. *et al*, *J. Am. Chem. Soc.*, 1986, **108**, 810 (*synth, resoln, pmr*)

5-Decanone   D-60014

[820-29-1]

$$H_3C(CH_2)_4CO(CH_2)_3CH_3$$

$C_{10}H_{20}O$ M 156.267
Bp 204° (195-200°), Bp₁₃ 90°.

Franzen, V., *Chem. Ber.*, 1954, **87**, 1478 (*synth*)
Anderson, M.W. *et al*, *J. Chem. Soc., Perkin Trans. 1*, 1986, 1995 (*synth, ir, pmr*)

5-Decene-2,8-diyne   D-60015

$$H_3CC{\equiv}CCH_2CH{=}CHCH_2C{\equiv}CCH_3$$

$C_{10}H_{12}$ M 132.205
(E)-form [103722-81-2]
Cryst. Mp 52-54°. Bp₁₂ 92-94°.
(Z)-form [100758-30-3]
No phys. props. given.

Müller, P. *et al*, *Helv. Chim. Acta*, 1985, **68**, 975 (*synth, ir, pmr*)
Oppolzer, W. *et al*, *Tetrahedron*, 1985, **41**, 3497 (*synth, ir, pmr*)

2-Decen-4-ol   D-60016

$$H_3C(CH_2)_5CH(OH)CH{=}CHCH_3$$

$C_{10}H_{20}O$ M 156.267
(±)-(E)-form
Bp₁₃ 100°.

Boughdady, N.M. *et al*, *Aust. J. Chem.*, 1987, **40**, 767 (*synth, pmr, ms*)

Consult the *Dictionary of Organophosphorus Compounds* for *organic compounds containing phosphorus.*

Decumbeside *C*   D-60017
[111467-42-6]

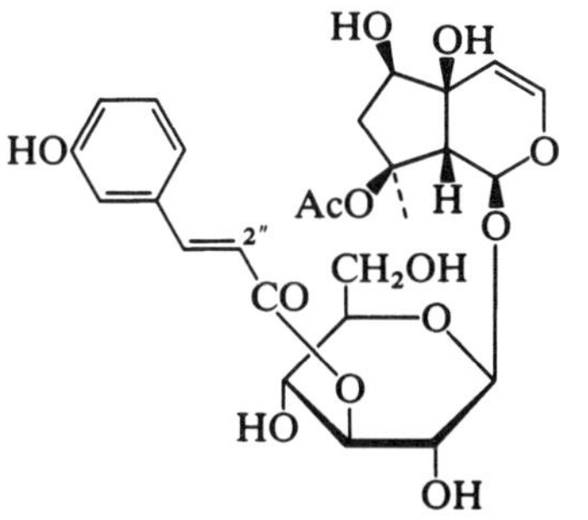

$C_{26}H_{32}O_{13}$ M 552.531
Constit. of *Ajuga decumbens*. Amorph. powder. $[\alpha]_D^{29}$
−108° (c, 0.5 in MeOH).

2''-Z-Isomer: [111394-30-0]. **Decumbeside D**.
$C_{26}H_{32}O_{13}$ M 552.531
Constit. of *A. decumbens*. Amorph. powder. $[\alpha]_D^{29}$
−158° (c, 0.53 in MeOH).

Takeda, Y. *et al*, *Phytochemistry*, 1987, **26**, 2303.

Dehydrochloroprepacifenol   D-60018

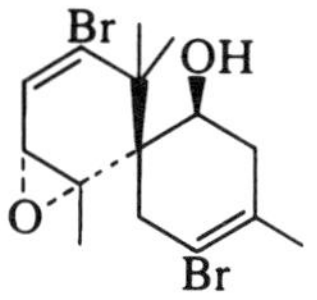

Absolute
configuration

$C_{15}H_{20}Br_2O_2$ M 392.130
Constit. of red alga *Laurencia majuscula*. Cryst.
(CH_2Cl_2/hexane). Mp 124-125°. $[\alpha]_D^{20}$ +51.7° (c, 1.1
in EtOH).

Caccamese, S. *et al*, *Tetrahedron*, 1987, **43**, 5893 (*cryst struct*)

Dehydromyoporone   D-60019
Updated Entry replacing D-00256
1-(3-Furanyl)-4,8-dimethyl-7-nonene-1,6-dione, 9CI
[38462-36-1]

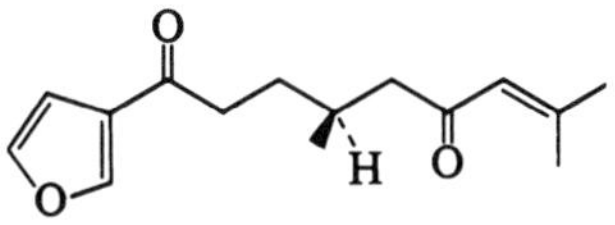

$C_{15}H_{20}O_3$ M 248.321
Constit. of *Myoporum* spp. and *Eremophila* spp. Cryst.
(Et_2O/hexane). Mp 32.5-33.5°. $[\alpha]_D$ −15°.

Blackburne, I.D. *et al*, *Aust. J. Chem.*, 1972, **25**, 1787 (*isol,
    struct*)
Hess, T. *et al*, *Tetrahedron Lett.*, 1987, **28**, 5643 (*synth, abs
    config*)

Dehydroosthol   D-60020
Updated Entry replacing D-40024
[38070-91-6]

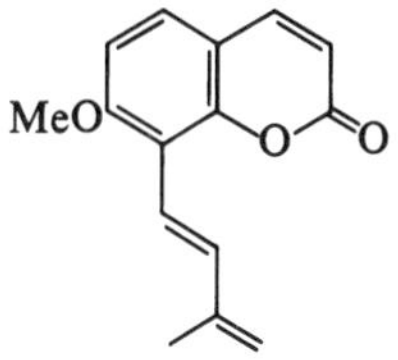

$C_{15}H_{14}O_3$ M 242.274
Constit. of *Choisya ternata*. Cryst. (Et_2O/pet. ether).
Mp 84°.

Z-Isomer: cis-*Dehydroosthol*.
$C_{15}H_{14}O_3$ M 242.274
Constit. of *Murraya exotica*. Oil.

Bohlmann, F. *et al*, *An. Quim.*, 1972, **68**, 765 (*isol*)
Ito, C. *et al*, *Heterocycles*, 1987, **26**, 1731 (*deriv*)

3',4'-Deoxypsorospermin   D-60021

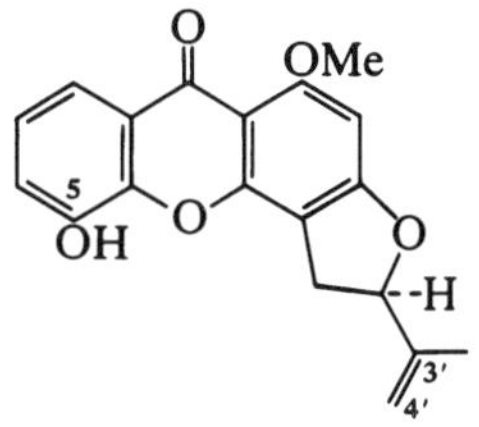

$C_{19}H_{16}O_5$ M 324.332
From *P. febrifugum*.

(*R*)-*form* [106212-04-8]
Constit. of *Psorospermum febrifugum*. Cryst. (MeOH).
Mp 228-230°. $[\alpha]_D^{20}$ −46° (c, 0.1 in MeOH).

3',4'-Dihydro, 3'-hydroxy, 5-Me ether: [106230-68-6].
O^5-*Methyl-3',4'-deoxypsorospermin-3'-ol*.
$C_{20}H_{20}O_6$ M 356.374
From *P. febrifugum*. Cryst. (MeOH). Mp 224-226°.
$[\alpha]_D^{20}$ −82° (c, 0.1 in MeOH).

3',4'-Dihydro, 3'-hydroxy, 4'-chloro: [74046-00-7]. *3',4'-
Deoxy-4'-chloropsorospermin-3'-ol*.
$C_{19}H_{17}ClO_6$ M 376.793
From *P. febrifugum*. Needles (Me_2CO/hexane). Mp
269-270°. $[\alpha]_D^{20}$ −114° (c, 0.1 in MeOH).

3',4'-Dihydro, 3',4'-dihydroxy: [106212-05-9]. *3',4'-
Deoxypsorospermin-3',4'-diol*.
$C_{19}H_{18}O_7$ M 358.347
Needles (MeOH/$CHCl_3$). Mp 278-279° dec. $[\alpha]_D^{20}$
−114° (c, 0.1 in MeOH).

3',4'-Epoxy: see Psorospermin, P-60189
Habib, A.M. *et al*, *J. Org. Chem.*, 1987, **52**, 412.

129

Desertorin *A*   D-60022

7,7'-Dihydroxy-4,4'-dimethoxy-5,5'-dimethyl[6,8'-bi-2H-1-benzopyran]-2,2'-dione, 9CI

[110325-63-8]

$C_{22}H_{18}O_8$ M 410.379

Constit. of *Emericella desertorum*. Cryst. (MeOH). Mp >300°.

7-Me ether: [110325-64-9]. **Desertorin B.**
$C_{23}H_{20}O_8$ M 424.406
Constit. of *E. desertorum*. Powder (CH_2Cl_2). Mp >300°.

7,7'-Di-Me ether: [110325-65-0]. **Desertorin C.**
$C_{24}H_{22}O_8$ M 438.433
Constit. of *E. desertorum*. Leaflets (MeOH). Mp 235-237°. $[\alpha]_D$ +16.8° (c, 1 in $CHCl_3$).

Nozawa, K. *et al, J. Chem. Soc., Perkin Trans. 1*, 1987, 1735.

1,3-Diacetylglycerol   D-60023

Updated Entry replacing D-00540

1,2,3-Propanetriol 1,3-diacetate. 1,3-Diacetin. Glycerol 1,3-diacetate

[105-70-4]

$$AcOCH_2CH(OH)CH_2OAc$$

$C_7H_{12}O_5$ M 176.169

Solvent. Hygroscopic liq. Sol. H_2O. d^{15} 1.179. Bp 259-261°, Bp_{12} 149°. n_D^{20} 1.4395. Component of coml. Diacetin (Diacetylglycerol), a mixt. with the 1,2-isomer.

Wegscheider, R. *et al, Monatsh. Chem.*, 1913, **34**, 1067, 1081 (*synth*)
Wahl, A., *Bull. Soc. Chim. Fr.*, 1925, **37**, 713 (*bibl*)
Lagenbeck, W. *et al, Naturwissenschaften*, 1955, **42**, 389 (*synth*)
Japan. Pat., 78 98 920, (*1977*); *CA*, **90**, 22339c (*synth*)
Suemune, H. *et al, Chem. Pharm. Bull.*, 1986, **34**, 3440 (*synth, pmr*)
Sax, N.I., *Dangerous Properties of Industrial Materials*, 5th Ed., Van Nostrand-Reinhold, 1979, 706.

1,3-Diacetyl-5-iodobenzene   D-60024

1,1'-(5-Iodo-1,3-phenylene)bisethanone, 9CI

[87533-53-7]

$C_{10}H_9IO_2$ M 288.084

Pale-orange needles (2-propanol/EtOAc). Mp 128-129°.

Ulrich, P. *et al, J. Med. Chem.*, 1984, **27**, 35 (*synth*)

3,4-Di-1-adamantyl-2,2,5,5-tetramethylhexane   D-60025

*1,2-Di-1-adamantyl-1,2-di-*tert-*butylethane*

(3RS,4RS)-(M)-form   *(3RS,4RS)-(P)-form*

R = 1-adamantyl

$C_{30}H_{50}$ M 410.725

The (*RS,RS*) or (±)-form represents the first simple aliphatic hydrocarbon to be obt. as two noninterconverting rotameric atropisomers.

(3RS,4RS)-(M)-form
Needles. Mp 203° dec.
(3RS,4RS)-(P)-form
Platelets. Mp 186° dec.
(3RS,4SR)-form
meso-*form*
Cryst. (Me_2CO). Mp 142-143° dec.

Rüchardt, C. *et al, Chem. Ber.*, 1986, **119**, 1492 (*synth, pmr, cmr, cryst struct*)

1,6-Diallyl-2,5-dithiobiurea, 8CI   D-60026

N,N'-Di-2-propenyl-1,2-hydrazinedicarbothioamide, 9CI. Diallyldithiocarbamidohydrazine. Dalzin

[539-97-9]

$$H_2C{=}CHCH_2NHCSNHNHCSNHCH_2CH{=}CH_2$$

$C_8H_{14}N_4S_2$ M 230.345

Used in anal. of heavy metals. Leaflets (EtOH). Mp 186°.

Busch, M. *et al, J. Prakt. Chem.*, 1914, **90**, 265 (*synth*)
Dutt, N.K. *et al, Anal. Chim. Acta*, 1956, **15**, 21, 102 (*use*)
Eberhardt, U. *et al, Pharmazie*, 1977, **32**, 458.
Nikolaeva, I.V. *et al, Zh. Prikl. Khim.* (*Leningrad*), 1985, **58**, 1189 (*synth*)

2,9-Diaminoacridine   D-60027

Updated Entry replacing D-00588

2,9-Acridinediamine, 9CI. 3,5-Diaminoacridine (*obsol.*)

[23043-62-1]

$C_{13}H_{11}N_3$ M 209.250

Orange cryst. (chlorobenzene). Mp 245° dec.

▷Potent mutagen

Albert, A. *et al, J. Chem. Soc.*, 1949, 1148 (*synth*)
Fukunaga, M. *et al, Chem. Pharm. Bull.*, 1987, **35**, 792 (*tox*)

3,6-Diaminoacridine   D-60028

Updated Entry replacing D-00590

3,6-Acridinediamine, 9CI. 2,8-Diaminoacridine (*obsol.*). *Proflavine*

[92-62-6]

$C_{13}H_{11}N_3$ M 209.250

Derivs. are antiseptics. Yellow cryst. Mp 284-286°. Green fluor. in EtOH, blue in H_2SO_4.

▷Potent mutagen. AR8670000.

B,H_2SO_4: [553-30-0]. Antiseptic. Red or red-brown cryst. + $1H_2O$. Spar. sol. H_2O.
▷AR9065000.
B,MeCl: [65589-70-0]. *Acriflavine.* Antiinfective. Commercial Acriflavine is normally a mixt. of B,MeCl.HCl and B,2HCl.

N^3,N^6-*Tetra-Me: see 3,6-Bis(dimethylamino)acridine,*
B-30172

Scherlin, S.M. *et al*, *Justus Liebigs Ann. Chem.*, 1935, **516**, 218
(*synth*)
Albert, A. *et al, J. Chem. Soc.*, 1936, 93 (*synth*)
Goldberg, A.A. *et al, J. Chem. Soc.*, 1946, 102 (*synth*)
Achari, A. *et al, Acta Crystallogr., Sect. B*, 1976, **32**, 2537
(*cryst struct*)
Fukunaga, M. *et al, Chem. Pharm. Bull.*, 1987, **35**, 792 (*tox*)
Merck Index, 9th Ed., Nos. 120 and 7568 (*derivs*)

3,9-Diaminoacridine   D-60029

Updated Entry replacing D-00591
3,9-Acridinediamine. 2,6-Diaminoacridine (obsol.)
[951-80-4]
C$_{13}$H$_{11}$N$_3$ M 209.250
Cryst. (EtOH/Et$_2$O). Mp 146°.

▷Potent mutagen. AR8690000.

Albert, A. *et al, J. Chem. Soc.*, 1936, 1618 (*synth*)
Albert, A. *et al, Chem. Ind. (London)*, 1942, **61**, 159 (*synth*)
Fukunaga, M. *et al, Chem. Pharm. Bull.*, 1987, **35**, 792 (*tox*)

1,4-Diamino-2,3-benzenedithiol   D-60030

*2,3-Dimercapto-p-phenylenediamine. 1,4-Diamino-2,3-
dimercaptobenzene*

C$_6$H$_8$N$_2$S$_2$ M 172.263
B,2HI: [107474-49-7]. Cryst. (CH$_2$Cl$_2$/MeOH/Et$_2$O).
Mp >150° dec.
Di-Me thioether: [107474-50-0]. *2,3-Bis(methylthio)-
1,4-benzenediamine.*
C$_8$H$_{12}$N$_2$S$_2$ M 200.316
Cryst. (as dihydriodide). Mp >200° dec.
(dihydriodide).

Lakshmikantham, M.V. *et al, J. Org. Chem.*, 1987, **52**, 1874
(*synth, deriv, ir, ms*)

1,4-Diamino-2,3,5,6-benzenetetrathiol   D-60031

*2,3,5,6-Tetramercapto-p-phenylenediamine. 1,4-Dia-
mino-2,3,5,6-tetramercaptobenzene*

HS ![structure] SH
HS ![structure] SH
(NH$_2$ top, NH$_2$ bottom)

C$_6$H$_8$N$_2$S$_4$ M 236.383
Tetra-Me thioether: [107474-53-3]. *2,3,5,6-Tetrakis-
(methylthio)-1,4-benzenediamine.*
C$_{10}$H$_{16}$N$_2$S$_4$ M 292.490
Yellow fluorescent prisms (Et$_2$O/hexane). Mp 115°.

Lakshmikantham, M.V. *et al, J. Org. Chem.*, 1987, **52**, 1874
(*synth, deriv, ir, ms*)

2,4-Diaminobenzoic acid   D-60032

Updated Entry replacing D-00645
[611-03-0]
C$_7$H$_8$N$_2$O$_2$ M 152.152

Mp ca. 140°. Unstable.
B,HCl: [61566-58-3]. Needles. Mp 270° dec.
2-N-Ac: [59156-43-3].
    C$_9$H$_{10}$N$_2$O$_3$ M 194.190
    Needles (EtOH). Mp 217°.
4-N-Ac: [43134-76-5].
    C$_9$H$_{10}$N$_2$O$_3$ M 194.190
    Needles (EtOH). Mp 207° dec.
2,4-N-Di-Ac: [73748-78-4].
    C$_{11}$H$_{12}$N$_2$O$_4$ M 236.227
    Needles. Mp 261° (248°).

Ullmann, F. *et al, Ber.*, 1903, **36**, 1803 (*synth*)
Lesiak, T. *et al, Rocz. Chem.*, 1957, **31**, 1033; *CA*, **52**, 8086
(*synth*)
Franc, J. *et al, J. Chromatogr.*, 1961, **6**, 396 (*glc*)
Stevenson, P.E., *J. Mol. Spectrosc.*, 1965, **15**, 220 (*uv*)
Mori, M. *et al, Chem. Pharm. Bull.*, 1986, **34**, 4859 (*synth, pmr,
ir, derivs*)

2,2′-Diaminobiphenyl   D-60033

Updated Entry replacing D-00669
[1,1′-Biphenyl]-2,2′-diamine, 9CI
[1454-80-4]

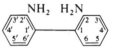

C$_{12}$H$_{12}$N$_2$ M 184.240
Cryst. (EtOH or pet. ether). Mp 80-81°. Bp$_{1.5}$ 155°. pK_a
3.81 (70% EtOH aq.).
2-N-Ac: [54147-74-9].
    C$_{14}$H$_{14}$N$_2$O   M 226.277
    Mp 89-90°.
2,2′-N-Di-Ac:
    C$_{16}$H$_{16}$N$_2$O$_2$ M 268.315
    Mp 161°.
2,2′-N-Di-Me: 2,2′-Bis(methylamino)biphenyl.
    C$_{14}$H$_{16}$N$_2$ M 212.294
    Mp 192-193°.
2,2′-N-Tetra-Me: [20627-78-5]. *2,2′-
Bis(dimethylamino)biphenyl.*
    C$_{16}$H$_{20}$N$_2$ M 240.347
    Plates (pet. ether). Mp 72-73°.

Stephenson, E.F.M., *J. Chem. Soc.*, 1949, 2620; 1954, 2354
(*synth*)
Moore, R.E. *et al, J. Org. Chem.*, 1958, **23**, 1504 (*synth*)
Lloyd, D. *et al, J. Chem. Soc.*, 1960, 4136 (*synth*)
Grantham, P.H. *et al, J. Org. Chem.*, 1961, **26**, 1008.
Dieteren, H.M.L. *et al, Recl. Trav. Chim. Pays-Bas*, 1963, **82**, 5
(*uv*)
Farnum, D.G. *et al, J. Org. Chem.*, 1977, **42**, 573 (*ir, pmr*)
Ottersen, T., *Acta Chem. Scand., Ser. A*, 1977, **31**, 480 (*cryst
struct*)
Smith, W.B., *J. Heterocycl. Chem.*, 1987, **24**, 745 (*synth, cmr,
pmr*)

2,6-Diamino-1,5-dihydro-4*H*-imidazo[4,5-   D-60034
c]pyridin-4-one, 9CI

8-Amino-3-deazaguanine
[103438-45-5]

![structure]

C$_6$H$_7$N$_5$O   M 165.154
Shows weak biological activity. Light-tan solid (H$_2$O).
Mp >300°.

Methanesulfonyl: [103438-47-7]. Cryst. (H$_2$O). Mp >300°.

Berry, D.A. *et al, J. Med. Chem.*, 1986, **29**, 2034 (*synth, ir, pmr, uv*)

2,8-Diamino-1,7-dihydro-6*H*-purin-6-one,　　　D-60035
9CI

2,8-Diaminohypoxanthine, 8CI. 2,8-Diamino-6-hydroxypurine. 8-Aminoguanine

[28128-41-8]

C$_5$H$_6$N$_6$O　　M 166.142

Potent inhibitor and substrate of purine nucleoside phosphorylase. Needles.

9-β-D-Ribofuranosyl: 8-Aminoguanosine.
C$_{10}$H$_{14}$N$_6$O$_5$　　M 298.258
Cryst. (H$_2$O). Mp >240°. Slowly darkens on exposure to air.

Fischer, H., *Hoppe - Seylers Z. Physiol. Chem.*, 1909, **60**, 69 (*synth*)
Holmes, R.E. *et al, J. Am. Chem. Soc.*, 1965, **87**, 1772 (*deriv, synth, uv*)
Stoeckler, J.D. *et al, Biochem. Pharm.*, 1982, **31**, 163 (*biochem*)

2,6-Diamino-3-heptenedioic acid　　　D-60036

HOOCCH(NH$_2$)CH$_2$CH=CHCH(NH$_2$)COOH

C$_7$H$_{12}$N$_2$O$_4$　　M 188.183

(*E*)-*form* [100910-16-5]
Solid. Presumably a mixt. of diastereoisomers.

Girodeau, J.-M. *et al, J. Med. Chem.*, 1986, **29**, 1023 (*synth, pmr*)

3,4-Diamino-1-methoxyisoquinoline　　　D-60037

1-Methoxy-3,4-isoquinolinediamine, 9CI

[110128-59-1]

C$_{10}$H$_{11}$N$_3$O　　M 189.216
Needles (pet. ether). Mp 119-120°.

Deady, L.W. *et al, Aust. J. Chem.*, 1986, **39**, 2089 (*synth, pmr*)

1,2-Diamino-4,5-methylenedioxybenzene　　　D-60038

C$_7$H$_8$N$_2$O$_2$　　M 152.152

Highly sensitive fluorogenic reagent for α-ketoacids.
B,2HCl: Needles (EtOH). Mp 176-179° dec.

Nakamura, M. *et al, Chem. Pharm. Bull.*, 1987, **35**, 687 (*synth, use*)

2,6-Diamino-4-methyleneheptanedioic acid,　　　D-60039
9CI

[100910-60-9]

C$_8$H$_{14}$N$_2$O$_4$　　M 202.210

(2*R**,6*R**)-*form* [100910-61-0]
B,2HCl: [100910-33-6]. [α]$_D^{20}$ +14° (c, 0.17 in H$_2$O).
(2*S**,6*S**)-*form* [100928-36-7]
B,2HCl: [100910-32-5]. [α]$_D^{20}$ −18° (c, 0.2 in H$_2$O).
(2*RS*,6*RS*)-*form*
(±)-*form*
Solid.
(2*RS*,6*SR*)-*form* [100910-34-7]
meso-*form*
Solid.

Girodeau, J.M. *et al, J. Med. Chem.*, 1986, **29**, 1023 (*synth, resoln, pmr*)

2,4-Diamino-6-methyl-1,3,5-triazine　　　D-60040

6-Methyl-1,3,5-triazine-2,4-diamine, 9CI.
Acetoguanamine

[542-02-9]

C$_4$H$_7$N$_5$　　M 125.133
Cryst. (H$_2$O). Mp 278-279°.
N^3-Oxide:
　C$_4$H$_7$N$_5$O　　M 141.132
　Cryst. (AcOH aq.). Mp >310°.
N^5-Oxide:
　C$_4$H$_7$N$_5$O　　M 141.132
　Cryst. (AcOH aq.). Mp >330°.
N-Ph:
　C$_{10}$H$_{11}$N$_5$　　M 201.230
　Mp 176-177°.
N,N-Di-Ph:
　C$_{16}$H$_{15}$N$_5$　　M 277.328
　Mp 188-189.5°.
N,N'-Di-Ph:
　C$_{16}$H$_{15}$N$_5$　　M 277.328
　Mp 233.5-235°.

Grundmann, C. *et al, Chem. Ber.*, 1950, **83**, 452 (*synth*)
Kreutzberger, A., *J. Am. Chem. Soc.*, 1957, **79**, 2629 (*synth*)
Shaw, J.T., *J. Org. Chem.*, 1962, **27**, 3890 (*deriv*)
Huffman, K.R. *et al, J. Org. Chem.*, 1963, **28**, 1812, 1816 (*synth, deriv*)
Yuki, Y. *et al, Bull. Chem. Soc. Jpn.*, 1970, **43**, 2123 (*deriv*)

2,4-Diamino-6-phenyl-1,3,5-triazine　　　D-60041

Updated Entry replacing D-01041
6-Phenyl-1,3,5-triazine-2,4-diamine, 9CI.
Benzoguanamine

[91-76-9]

C$_9$H$_9$N$_5$　　M 187.204
Used in manuf. of thermosetting resins, pesticides pharmaceuticals and dyestuffs. Mp 227-228°.
▷XY7000000.

Org. Synth., Coll. Vol., **4**, 78.
Stevens, M.F.G. et al, J. Chem. Soc. (C), 1970, 2298.
Alsofrom, D. et al, J. Heterocycl. Chem., 1976, **13**, 913.
Smyrl, N.R. et al, J. Heterocycl. Chem., 1982, **19**, 493.
Sax, N.I., Dangerous Properties of Industrial Materials, 6th Ed., Van Nostrand-Reinhold, 1984, 888.

2,4-Diaminopteridine, 8CI   D-60042

2,4-Pteridinediamine, 9CI. 2,4-Diaminopyrimido[4,5-b]-pyrazine

[1127-93-1]

$C_6H_6N_6$ M 162.154

Diuretic agent; parent of a number of antifolate drugs. Needles (HCOOH aq.). Dec. on heating.

5,8-Dioxide: [42346-97-4].
$C_6H_6N_6O_2$ M 194.152
Yellow cryst. powder. Mp >360°.

Mallette, M.F. et al, J. Am. Chem. Soc., 1947, **69**, 1814 (synth)
Albert, A. et al, J. Chem. Soc., Perkin Trans. 2, 1973, 1101 (uv)
Ewers, U. et al, Chem. Ber., 1973, **106**, 3951 (cmr)
Yamamoto, H. et al, Chem. Ber., 1973, **106**, 3175 (dioxide)
Schwalbe, C.H. et al, Acta Crystallogr., Sect. C, 1986, **42**, 1252 (cryst struct)

4,6-Diaminopteridine   D-60043

4,6-Pteridinediamine, 9CI

[19167-60-3]
$C_6H_6N_6$ M 162.154
Yellow cryst. (H_2O).

Albert, A. et al, J. Chem. Soc., 1956, 4621 (synth)
Kamiya, M., Bull. Chem. Soc. Jpn., 1971, **44**, 285 (uv)

4,7-Diaminopteridine, 8CI   D-60044

4,7-Pteridinediamine, 9CI

[771-41-5]
$C_6H_6N_6$ M 162.154
Pale-yellow needles (H_2O). Mp >300°.

Osdene, T.S. et al, J. Chem. Soc., 1955, 2036 (synth)
Kamiya, M., Bull. Chem. Soc. Jpn., 1971, **44**, 285 (uv)

6,7-Diaminopteridine   D-60045

6,7-Pteridinediamine, 9CI

[3747-69-1]
$C_6H_6N_6$ M 162.154
Flocculent needles (H_2O). Mp >250°.

Albert, A. et al, J. Chem. Soc., 1965, 27.

2,7-Diamino-4,6(3H,5H)-pteridinedione   D-60046

2,7-Diamino-1,5-dihydro-4,6-pteridinedione, 9CI. 7-Aminoxanthopterine

[71014-14-7]

$C_6H_6N_6O_2$ M 194.152

Artifact obt. from workup of biol. materials contg. pteridine. Beige microcryst. powder.

Netscher, T. et al, Justus Liebigs Ann. Chem., 1987, 259 (synth, uv, ms)

2,6-Diamino-4,7(3H,8H)-pteridinedione   D-60047

6-Aminoisoxanthopterine

[108120-13-4]

$C_6H_6N_6O_2$ M 194.152

CAS no. refers to (1H,8H)-tautomer. Artifact obt. in workup of biol. systems contg. Isoxanthopterin, I-60141 .

Netscher, T. et al, Justus Liebigs Ann. Chem., 1987, 259 (synth, uv, ms)

2,4-Diaminoquinazoline   D-60048

2,4-Quinazolinediamine, 9CI

[1899-48-5]

$C_8H_8N_4$ M 160.178

Light-yellow prisms by subl. Mp 250-252°.
B,HCl: Needles. Mp 308°.
Picrate: Mp 232°.

Kötz, A., J. Prakt. Chem., 1893, **47**, 303 (synth)
Vopicka, E. et al, J. Am. Chem. Soc., 1935, **57**, 1068 (synth)
Modest, E.J. et al, J. Org. Chem., 1965, **30**, 1837 (synth, uv)
Schwalbe, C.H. et al, Acta Crystallogr., Sect. C, 1986, **42**, 1101 (cryst struct)

2,5-Diamino-1,3,4-thiadiazole   D-60049

1,3,4-Thiadiazole-2,5-diamine, 9CI

[2937-81-7]

$C_2H_4N_4S$ M 116.140
Violet-brown cryst.

Senda, H. et al, Acta Crystallogr., Sect. C, 1987, **43**, 347 (synth, cryst struct)

Di-9-anthracenylmethanone   D-60050

Di-9-anthryl ketone

[102725-09-7]

$C_{29}H_{18}O$ M 382.461
Lemon-yellow needles (toluene). Mp 269-272°.

Becker, H.D. et al, J. Org. Chem., 1986, **51**, 2956 (synth, pmr)

Di-9-anthrylacetylene, 8CI   D-60051

9,9'-(1,2-Ethynediyl)bisanthracene, 9CI
[20199-19-3]

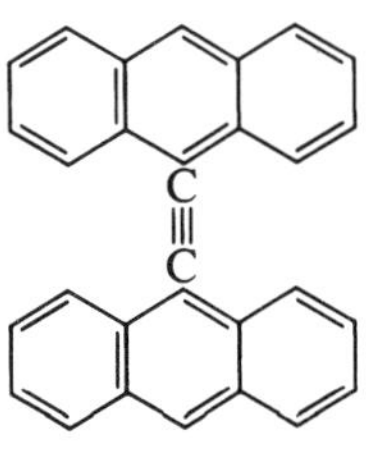

$C_{30}H_{18}$ M 378.472
Orange-red cryst. (C_6H_6). Mp 310° dec.

Akiyama, S. *et al, Bull. Chem. Soc. Jpn.*, 1971, **44**, 2231 (*synth*)

Diaporthin   D-60052

*8-Hydroxy-3-(2-hydroxypropyl)-6-methoxy-1H-2-ben-
zopyran-1-one. 8-Hydroxy-3-(2-hydroxypropyl)-6-
methoxyisocoumarin*

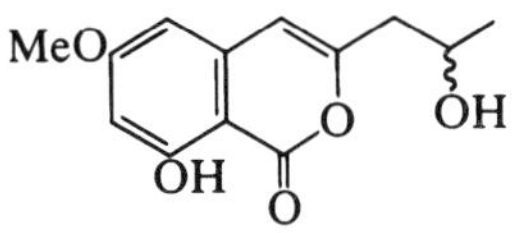

$C_{13}H_{14}O_5$ M 250.251
Isol. from cultures of *Endothia parasitica*. Phytotoxin.
Needles or plates. Mp 90-92° (83-85°). $[\alpha]_D$ +54° (c,
10.87 in $CHCl_3$).

Hardegger, E. *et al, Helv. Chim. Acta*, 1966, **49**, 1283 (*isol,
props*)

2,5-Diazabicyclo[4.1.0]heptane   D-60053

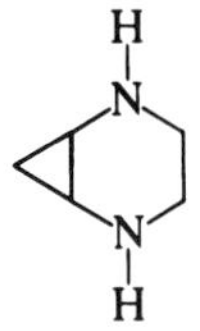

$C_5H_{10}N_2$ M 98.147
N,N'-*Bis(benzenesulfonyl):* Mp 169-171°.
N,N'-*Bis(4-methylbenzenesulfonyl):* Cryst. (MeOH).
Mp 185-187°.

Majchrzak, M.W. *et al, J. Heterocycl. Chem.*, 1983, **20**, 815.

1,4-Diazabicyclo[4.3.0]nonane   D-60054

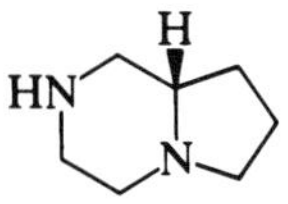

(S)-form

$C_7H_{14}N_2$ M 126.201
(**S**)-*form* [93643-24-4]
Bp_{10} 74-75°.

Botré, C. *et al, J. Med. Chem.*, 1986, **29**, 1814 (*synth*)

1,5-Diazabicyclo[5.2.2]undecane   D-60055

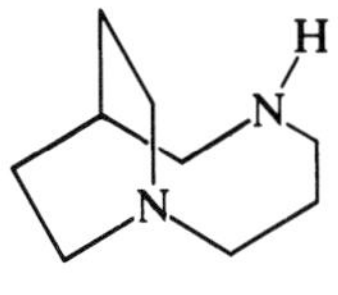

$C_9H_{18}N_2$ M 154.255

N-*Me:* [103094-97-9].
$C_{10}H_{20}N_2$ M 168.281
Oil.

Nelsen, S.F. *et al, J. Org. Chem.*, 1986, **51**, 3169 (*synth, pmr,
cmr*)

1,3-Diazaspiro[4.5]decane-2,4-dione, 9CI   D-60056

Spirohydantoin
[702-62-5]

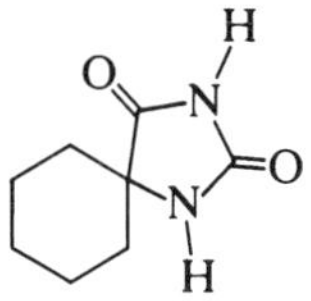

$C_8H_{12}N_2O_2$ M 168.195
Cryst. (EtOH aq.). Mp 218-220°.
▷HM2480000.

Tiffeneau, M. *et al, Bull. Soc. Chim. Fr.*, 1947, 445 (*synth*)
Oldfield, W. *et al, J. Med. Chem.*, 1965, **8**, 239 (*synth, biochem*)

1,3-Diazetidine-2,4-dione, 9CI   D-60057

2,4-Uretidinedione, 8CI. Dicyanic acid
[4455-27-0]

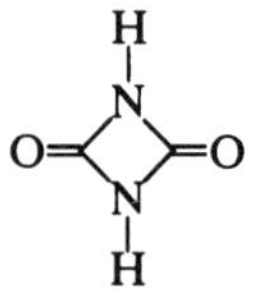

$C_2H_2N_2O_2$ M 86.050
Polymer intermed., derivs. have many industrial uses.
Only descr. in soln.
1,3-Di-Me: [36909-44-1].
$C_4H_6N_2O_2$ M 114.104
Plates (hexane). Mp 97-99°. Dimer of methyl
isocyanate.
1,3-Diisopropyl: [67463-80-3].
$C_8H_{14}N_2O_2$ M 170.211
Needles (pentane at −78°). Mp <15°. Bp_{14} 80-83°.
1,3-Di-Ph: [1025-36-1].
$C_{14}H_{10}N_2O_2$ M 238.245
Mp 176.2°. Dimer of phenyl isocyanate.

Brown, C.J., *J. Chem. Soc.*, 1955, 2931 (*cryst struct, deriv*)
Hayek, E. *et al, Monatsh. Chem.*, 1965, **96**, 516 (*struct*)
Bardour, J.-L. *et al, Acta Crystallogr., Sect. B*, 1974, **30**, 691
(*deriv, cryst struct*)
White, D.K. *et al, J. Org. Chem.*, 1978, **43**, 4530 (*deriv, synth,
ir, pmr, ms*)
D'Silva, T.D.J. *et al, J. Org. Chem.*, 1986, **51**, 3781 (*deriv,
synth, ir, pmr, cmr*)

2,2-Diazidopropane, 9CI   D-60058

[85620-95-7]

$(H_3C)_2C(N_3)_2$

$C_3H_6N_6$ M 126.121
Liq. $Bp_{0.5}$ 28-30°. Explosive.

Al-Khalil, S.I. *et al, J. Chem. Soc., Perkin Trans. 1*, 1986, 555
(*synth, ir, pmr*)

9-Diazo-9H-cyclopenta[1,2-b:4,3-b']-dipyridine, 9CI   D-60059
9-Diazo-1,8-diazafluorene
[1807-47-2]

$C_{11}H_6N_4$ M 194.195
Carbene precursor. Yellow-orange needles. Mp 94-95°.

Schönberg, A. *et al, Chem. Ber.*, 1962, **95**, 2137 (*synth*)
Li, Y.Z. *et al, J. Org. Chem.*, 1986, **51**, 3804 (*props*)

Diazocyclopropane, 9CI   D-60060
[54022-19-4]

$C_3H_4N_2$ M 68.078
Reactive intermed.

Kirmse, W. *et al, Chem. Ber.*, 1986, **119**, 1511.

1-(1-Diazoethyl)naphthalene, 9CI   D-60061
[102421-45-4]

$C_{12}H_{10}N_2$ M 182.224
Red oil.

Barcus, R.L. *et al, J. Am. Chem. Soc.*, 1986, **108**, 3928 (*synth, pmr*)

9-Diazo-9H-fluorene, 9CI   D-60062
Updated Entry replacing D-30051
[832-80-4]

$C_{13}H_8N_2$ M 192.220
Deoxygenating reagent for heterocyclic *N*-oxides. Needles (pentane). Mp 100-101° (99°).

Baltzly, R. *et al, J. Org. Chem.*, 1961, **26**, 3669 (*synth*)
Schweitzer, E.E. *et al, J. Org. Chem.*, 1964, **29**, 1744 (*use*)
Adamson, J.R. *et al, J. Chem. Soc., Perkin Trans. 1*, 1975, 2030 (*synth*)
Jonczyk, A. *et al, Bull. Soc. Chim. Belg.*, 1977, **86**, 739 (*synth*)
Handoo, K.L. *et al, Indian J. Chem., Sect. B*, 1981, **20**, 699 (*synth*)
Dudman, C.C. *et al, Synthesis*, 1982, 419 (*synth*)
Grieve, D.M.A. *et al, Helv. Chim. Acta*, 1985, **68**, 1427 (*synth*)
Bethell, D. *et al, J. Am. Chem. Soc.*, 1986, **108**, 895 (*props*)

4-Diazo-4H-imidazole, 9CI   D-60063
[89108-47-4]

$C_3H_2N_4$ M 94.076
Sensitive prod. kept in soln.

Amick, T.J. *et al, Tetrahedron Lett.*, 1986, **27**, 901 (*synth, use*)

1-(Diazomethyl)-4-methoxybenzene, 9CI   D-60064
4-Methoxyphenyldiazomethane
[23304-25-8]

$C_8H_8N_2O$ M 148.164
Reagent for synth. of 4-methoxybenzyl ethers and esters. Red oil. V. unstable. Stored in soln. at −80°.

▷Explosive when neat

Closs, G.L. *et al, J. Am. Chem. Soc.*, 1964, **86**, 4042 (*synth, uv*)
Kamaike, K. *et al, Tetrahedron*, 1986, **42**, 4701 (*synth*)

4-Diazo-2-nitrophenol   D-60065
4-Diazo-2-nitro-2,5-cyclohexadien-1-one, 9CI
[18396-83-3]

$C_6H_3N_3O_3$ M 165.108
Bronze plates (Me$_2$CO). Mp 170° (explodes).

▷Explosive

Atkins, R.L. *et al, J. Org. Chem.*, 1986, **51**, 2572 (*synth, ir, pmr*)

2-Diazo-1,1,3,3-tetramethylcyclopentane, 9CI   D-60066
[71690-98-7]

$C_9H_{16}N_2$ M 152.239
Red oil. Bp$_{0.5}$ 40°.

Guziec, F.S. *et al, J. Chem. Soc., Perkin Trans. 1*, 1985, 107 (*synth, ir, pmr*)

2-Diazo-1,1,3,3-tetramethylindane   D-60067

2-Diazo-2,3-dihydro-1,1,3,3-tetramethyl-1H-indene, 9CI

[81331-45-5]

C₁₃H₁₆N₂   M 200.283
Red cryst. Mp 67°. Bp₀.₅ 50°.

Guziec, F.S. *et al, J. Chem. Soc., Perkin Trans. 1*, 1985, 107 (*synth, ir, pmr*)

Dibenz[*b,g*]azocine-5,7(6*H*,12*H*)-dione   D-60068

C₁₅H₁₁NO₂   M 237.257
N-*Me:* [104014-40-6].
  C₁₆H₁₃NO₂   M 251.284
  Mp 127-127.5°.
N-*Benzyl:* [104014-34-8].
  C₂₂H₁₇NO₂   M 327.382
  Mp 156-156.5°.
N-*Ph:* [99233-89-3].
  C₂₁H₁₅NO₂   M 313.355
  Mp 181-182°.

Hellwinkel, D. *et al, Chem. Ber.*, 1986, **119**, 3165.

Dibenz[*b,h*]indeno[1,2,3-*de*][1,6]-naphthyridine, 9CI   D-60069

9,14-Diazabenz[a,e]*fluoranthene. 9,14-Diazadibenzo*[a,e]*aceanthrylene*

C₂₂H₁₂N₂   M 304.350
Bright-yellow needles (pet. ether). Mp 306-308°.
Picrate: Dark-red needles (toluene). Mp 270-273°.

Bloomfield, D.G. *et al, J. Chem. Soc., Perkin Trans. 1*, 1986, 857 (*synth*)
Upton, C., *J. Chem. Soc., Perkin Trans. 1*, 1986, 1225 (*synth*)

Dibenzo[*c,f*][1]benzopyrano[2,3,4-*ij*][2,7]-naphthyridine, 9CI   D-60070

10-Oxa-9,15-diazanaphtho[1,2,3-fg]*naphthacene*

[191-44-6]

C₂₂H₁₂N₂O   M 320.350
Yellow needles. Mp 238-239°.

▷Carcinogenic

Partridge, M.W. *et al, J. Chem. Soc.*, 1962, 632 (*synth, uv*)
Bloomfield, D.G. *et al, J. Chem. Soc., Perkin Trans. 1*, 1986, 857 (*synth*)

Dibenzo[*a,d*]cycloheptenylium, 9CI   D-60071

[33011-97-1]

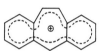

C₁₅H₁₁⊕   M 191.252 (ion)
Red ion in soln.
Perchlorate: [62114-58-3].
  C₁₅H₁₁ClO₄   M 290.703
  Red cryst. powder (Et₂O). Mp 134-138° dec.

Berti, G. *et al, Gazz. Chim. Ital.*, 1957, **87**, 293 (*synth*)
Berti, G. *et al, J. Org. Chem.*, 1957, **22**, 230 (*synth*)
Böhshar, M. *et al, Tetrahedron*, 1986, **42**, 1815 (*synth*)

Dibenzo[*jk,uv*]dinaphtho[2,1,8,7-*defg*:2′,1′,8′,7′-*opqr*]pentacene, 9CI   D-60072

Diphenanthro[5,4,3-abcd:5′,4′,3′-jklm]*perylene. Dinaphtho*[7′:1′-1:13] [1″:7″-6:8]*peropyrene*

[190-89-6]

C₃₈H₁₈   M 474.560
Shows v. high photoconductivity. Orange-red needles. Mp 338-339°. Green fluor.

Clar, E. *et al, J. Chem. Soc.*, 1956, 3878 (*synth, uv*)
Robertson, J.M. *et al, J. Chem. Soc.*, 1959, 2614 (*cryst struct*)
Oonishi, J. *et al, Bull. Chem. Soc. Jpn.*, 1978, **51**, 2256 (*cryst struct*)

2,3,11,12-Dibenzo-1,4,7,10,13,16-hex-athia-2,11-cyclooctadecadiene   D-60073

6,7,9,10,17,18,20,21-Octahydrodibenzo[b,k] [1,4,7,10,13,16]-hexathiacyclooctadecin, 9CI. Hexathiobenzo[18]crown-6
[105400-44-0]

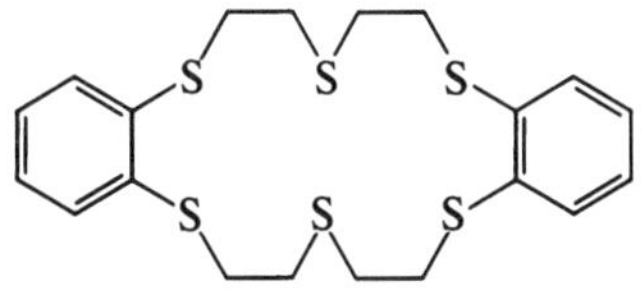

$C_{20}H_{24}S_6$ M 456.770
Cryst. Mp 161°.

Sellmann, D. *et al, Angew. Chem., Int. Ed. Engl.*, 1986, **25**, 1107 (*synth, pmr*)

Dibenzo[c,f]indeno[1,2,3-ij][2,7]-naphthyridine, 9CI   D-60074

9,14-Diazadibenz[a,e]*acephenanthrylene*
[193-40-8]

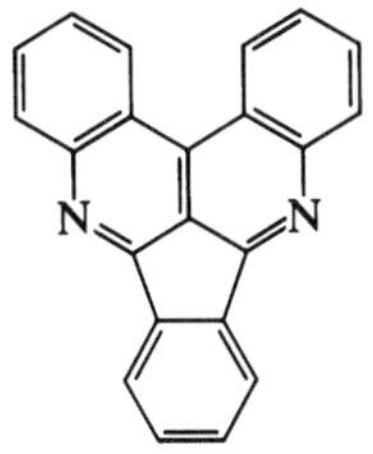

$C_{22}H_{12}N_2$ M 304.350
Pale-yellow prisms (xylene). Mp 270-272°.
▷Carcinogenic. HO4725000.
Picrate: Needles (AcOH). Mp 286-287°.

Partridge, M.W. *et al, J. Chem. Soc.*, 1962, 632 (*synth, uv*)
Bloomfield, D.G. *et al, J. Chem. Soc., Perkin Trans. 1*, 1986, 857 (*synth*)

Dibenzo[a,rst]naphtho[8,1,2-cde]-pentaphene, 9CI, 8CI   D-60075

Violanthrene B
[191-46-8]

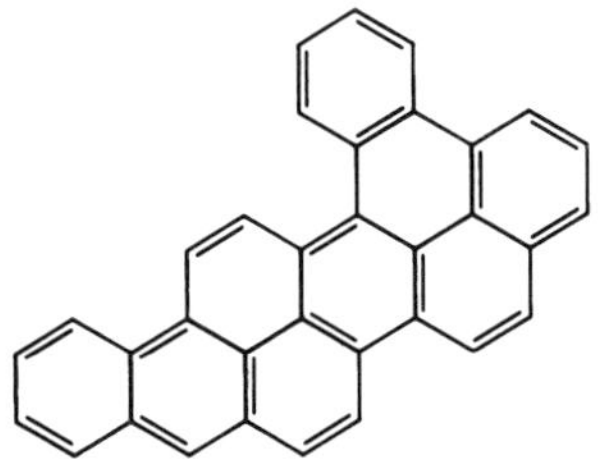

$C_{34}H_{18}$ M 426.516
Orange-red needles (chlorobenzene). Mp 339.4°.

Aoki, J. *et al, Bull. Chem. Soc. Jpn.*, 1977, **50**, 1017 (*synth*)
Aoki, J. *et al, J. Org. Chem.*, 1981, **46**, 3922 (*synth*)
Ueda, T. *et al, Org. Mass Spectrom.*, 1983, **18**, 105 (*ms*)

Dibenzo[cd:c′d′][1,2,4,5]tetrazino[1,6-a:4,3-a′]diindole, 9CI   D-60076

[1,2,4,5]Tetrazino[1,6-a:4,3-a′]bis[1]azaacenaphthylene
[107550-92-5]

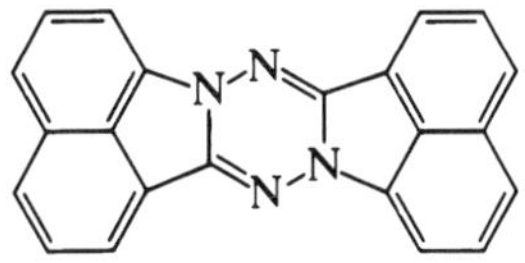

$C_{22}H_{12}N_4$ M 332.364
Dark cryst. with metallic sheen + ½H_2O. Mp >312° dec.

Eichenberger, T. *et al, Helv. Chim. Acta*, 1986, **69**, 1521 (*synth, ir, uv*)

Dibenzo[b,f]thiepin-3-carboxylic acid, 9CI   D-60077

[71474-60-7]

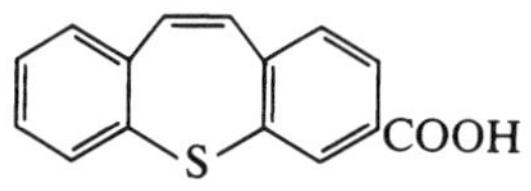

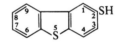

$C_{15}H_{10}O_2S$ M 254.303
Cryst. (DMF aq.). Mp 253-255°.
5,5-Dioxide: [71474-64-1].
   $C_{15}H_{10}O_4S$ M 286.302
   Cryst. (MeOH). Mp 266-268°.

Hands, D. *et al, J. Heterocycl. Chem.*, 1986, **23**, 1333 (*synth, ir, pmr*)

2-Dibenzothiophenethiol   D-60078

2-Mercaptodibenzothiophene

$C_{12}H_8S_2$ M 216.315
Mp 81-83°.
S-Me:
   $C_{13}H_{10}S_2$ M 230.342
   Solid. Mp 67-73°.
S-Benzyl:
   $C_{19}H_{14}S_2$ M 306.440
   Solid. Mp 56-58°.

Dunkerton, L.V. *et al, J. Heterocycl. Chem.*, 1987, **24**, 749 (*synth, pmr, cmr*)

4-Dibenzothiophenethiol   D-60079

4-Mercaptodibenzothiophene
$C_{12}H_8S_2$ M 216.315
Mp 86°.
S-Me:
   $C_{13}H_{10}S_2$ M 230.342
   Mp 68-70°.
S-Benzyl:
   $C_{19}H_{14}S_2$ M 306.440
   Mp 90-91°.

Dunkerton, L.V. *et al, J. Heterocycl. Chem.*, 1987, **24**, 749 (*synth, pmr, cmr*)

7H-Dibenzo[c,h]xanthen-7-one, 9CI   D-60080

Updated Entry replacing D-01349

[3264-24-2]

C$_{21}$H$_{12}$O$_2$ M 296.325

Yellow cryst. (EtOH). Mp 248-250° (240°). Green fluor. in conc. H$_2$SO$_4$.

Clar, E., *Ber.*, 1929, **62**, 350.
Kamel, M., *Helv. Chim. Acta*, 1959, **42**, 580 (*synth*)
Kimura, M. *et al*, *Chem. Pharm. Bull.*, 1987, **35**, 136 (*synth, pmr, cmr, ms*)

3,4-Dibenzoyl-2,5-hexanedione, 9CI   D-60081

1,2-Diacetyl-1,2-dibenzoylethane

[51439-47-5]

$$H_3CCOCH(COPh)CH(COPh)COCH_3$$

C$_{20}$H$_{18}$O$_4$ M 322.360

Prisms (C$_6$H$_6$). Mp 197.5-199.5° (173-175°).

Cannon, J.R. *et al*, *Aust. J. Chem.*, 1986, **39**, 1811 (*synth, pmr, cryst struct*)

Dibenzyl diselenide   D-60082

Bis(phenylmethyl)diselenide, 9CI

[1482-82-2]

$$PhCH_2-Se-Se-CH_2Ph$$

C$_{14}$H$_{14}$Se$_2$ M 340.185

Yellow cryst. (EtOH). Mp 92-94°.

Klayman, D.L. *et al*, *J. Am. Chem. Soc.*, 1973, **95**, 197 (*synth*)
Reich, H.J. *et al*, *J. Org. Chem.*, 1986, **51**, 2981 (*synth, pmr*)

1,2-Dibenzylidenecyclobutane   D-60083

1,1′-(1,2-Cyclobutanediylidenedimethylidyne)-bisbenzene, 9CI

[74725-44-3]

C$_{18}$H$_{16}$ M 232.324

(*E,E*)-*form* [94618-65-2]

Mp 140° (114-115°).

Davalian, D. *et al*, *J. Org. Chem.*, 1980, **45**, 4183 (*synth*)
Minami, T. *et al*, *J. Org. Chem.*, 1986, **51**, 3572 (*synth, ir, pmr, cmr*)

Dibenzyl pyrocarbonate   D-60084

Dicarbonic acid bis(phenylmethyl)ester, 9CI

[31139-36-3]

$$O(COOCH_2Ph)_2$$

C$_{16}$H$_{14}$O$_5$ M 286.284

Reagent for prepn. of *N*-benzyloxycarbonyl protected amino acids. Cryst. (hexane). Mp 28°.

Sennyey, G. *et al*, *Tetrahedron Lett.*, 1986, **27**, 5375 (*synth, pmr, ir, use*)

9,10-Dibromoanthracene, 9CI   D-60085

Updated Entry replacing D-01432

ms-Dibromoanthracene

[523-27-3]

C$_{14}$H$_8$Br$_2$ M 336.025

Yellow needles (xylene). Mp 226°.

1,3,5-Trinitrobenzene complex: Mp 179°.

Org. Synth., Coll. Vol., **1**, 201 (*synth*)
Mecke, R. *et al*, *Z. Naturforsch., A*, 1964, **19**, 41 (*ir*)
Stosser, R. *et al*, *J. Prakt. Chem.*, 1975, **317**, 591 (*pmr, cmr*)
Trotter, J., *Acta Crystallogr., Sect. C*, 1986, **42**, 862 (*cryst struct*)

2,3-Dibromobenzofuran, 9CI   D-60086

2,3-Dibromocoumarone

[64150-61-4]

C$_8$H$_4$Br$_2$O   M 275.927

Mp 27°. Bp 269-270°, Bp$_{0.5}$ 80°.

Robba, M. *et al*, *Bull. Soc. Chim. Fr.*, 1977, 139 (*synth*)

5,7-Dibromobenzofuran, 9CI   D-60087

5,7-Dibromocoumarone

[23145-08-6]

C$_8$H$_4$Br$_2$O   M 275.927

Needles (EtOH). Mp 70° (57.5°). Bp 278-280°.

Simonis, H. *et al*, *Ber.*, 1900, **33**, 1962 (*synth*)
Kurdukar, R. *et al*, *Proc. Indian Acad. Sci. (A)*, 1963, **58**, 336 (*synth*)
Fr. Pat., 1 537 206 (*1968*); *CA*, **71**, 61198 (*synth*)

2,6-Dibromo-1,4-benzoquinone   D-60088

Updated Entry replacing D-01491

2,6-Dibromo-2,5-cyclohexadiene-1,4-dione, 9CI

[19643-45-9]

C$_6$H$_2$Br$_2$O$_2$ M 265.889

Cryst. (EtOH). Mp 131°.

4-Oxime: 2,6-Dibromo-4-nitroso-1-naphthol.
  C$_6$H$_3$Br$_2$NO$_2$ M 280.903
  Yellow cryst. Mp 170°.

Hodgson, H.H. *et al*, *J. Chem. Soc.*, 1930, 1085 (*synth*)
Ramart-Lucas, P. *et al*, *Bull. Soc. Chim. Fr.*, 1949, 901 (*deriv*)
Flaig, W. *et al*, *Justus Liebigs Ann. Chem.*, 1958, **618**, 117 (*uv*)
Saá, J.M. *et al*, *Tetrahedron Lett.*, 1986, **27**, 5125 (*synth*)

2,2′-Dibromo-1,1′-binaphthyl   D-60089

2,2′-Dibromo-1,1′-binaphthalene, 9CI

[74866-28-7]

(*R*)-*form*

C$_{20}$H$_{12}$Br$_2$ M 412.123

(R)-form [86688-08-6]
Mp 157-157.5°. $[\alpha]_D^{25}$ +32.9° (c, 1 in Py).
(±)-form [76284-65-6]
Pale-yellow cryst. (EtOH). Mp 187.3-187.9°.

McKillop, A. *et al*, *J. Am. Chem. Soc.*, 1980, **102**, 6504 (*synth*)
Okamoto, Y. *et al*, *J. Am. Chem. Soc.*, 1981, **103**, 6971 (*resoln*)
Miyashita, A. *et al*, *Tetrahedron*, 1984, **40**, 1245 (*synth*)
Brown, K.J. *et al*, *J. Org. Chem.*, 1985, **50**, 4345 (*synth*)

4,4'-Dibromo-1,1'-binaphthyl   D-60090

4,4'-Dibromo-1,1'-binaphthalene, 9CI
[49610-35-7]
$C_{20}H_{12}Br_2$ M 412.123
(+)-form [73453-34-6]
Mp 215-217°. $[\alpha]_D^{23}$ +47° (c, 9.1 in $CHCl_3$).
(±)-form
Cryst. (toluene/pet. ether). Mp 217.5°.

Hutchins, L.G. *et al*, *J. Org. Chem.*, 1980, **45**, 2414 (*synth*)
McKillop, A. *et al*, *J. Am. Chem. Soc.*, 1980, **102**, 6504 (*synth*)

1,1-Dibromocycloheptane, 9CI   D-60091

[102450-37-3]

$C_7H_{12}Br_2$ M 255.980
$Bp_{2.5}$ 130°.

Napolitano, E. *et al*, *Synthesis*, 1986, 122 (*synth, ir, pmr*)

2,2-Dibromocyclopropanecarboxaldehyde   D-60092

[61782-63-6]

$C_4H_4Br_2O$ M 227.883
(±)-form
Bp_2 56-57°.
2,4-Dinitrophenylhydrazone: Cryst. (EtOH/C_6H_6). Mp 160°.

Marino, J.P. *et al*, *Tetrahedron Lett.*, 1976, 3241 (*synth*)
Holm, K.H. *et al*, *Acta Chem. Scand.*, Ser. B, 1978, **32**, 693 (*synth, ir, pmr*)

2,2-Dibromocyclopropanecarboxylic acid   D-60093

[5365-17-3]

$C_4H_4Br_2O_2$ M 243.882
(±)-form
Cryst. (hexane). Mp 94-95°.
Me ester: [71666-01-8].
$C_5H_6Br_2O_2$ M 257.909
Bp_9 90°.

Woodworth, R.C. *et al*, *J. Am. Chem. Soc.*, 1957, **79**, 2542 (*synth*)
Holm, K.H. *et al*, *Acta Chem. Scand.*, Ser. B, 1978, **32**, 693 (*synth*)
Kusuyama, Y., *Bull. Chem. Soc. Jpn.*, 1979, **52**, 1944 (*synth, deriv, pmr*)

1,2-Dibromocyclopropene   D-60094

[108186-05-6]

$C_3H_2Br_2$ M 197.857
Unstable, identified by pmr and readily trapped as adducts.

Dent, B.R. *et al*, *Aust. J. Chem.*, 1986, **39**, 1621.

1,2-Dibromo-1,4-dihydro-1,4-methanonaphthalene, 9CI   D-60095

1,2-Dibromobenzonorbornadiene

$C_{11}H_8Br_2$ M 299.992
(±)-form
Oil.

Paquette, L.A. *et al*, *J. Org. Chem.*, 1987, **52**, 2674 (*synth, ir, pmr, cmr*)

1,3-Dibromo-1,4-dihydro-1,4-methanonaphthalene   D-60096

1,3-Dibromobenzonorbornadiene
$C_{11}H_8Br_2$ M 299.992
(±)-form
Oil.

Paquette, L.A. *et al*, *J. Org. Chem.*, 1987, **52**, 2674 (*synth, ir, pmr, cmr*)

2,6-Dibromo-4,5-dihydroxy-2-cyclohexen-1-one   D-60097

$C_6H_6Br_2O_3$ M 285.920
(4S,5R,6S)-form
4-Ac: [110786-60-2]. *4-Acetoxy-2,6-dibromo-5-hydroxy-2-cyclohexen-1-one.*
$C_8H_8Br_2O_4$ M 327.957
Metabolite of acorn worm *Pytchodera* sp. Cryst. (EtOAc/hexane). Mp 136-138°. $[\alpha]_D^{25}$ +109.7° (c, 1.24 in $CHCl_3$).
4-Ac, 2,3-dihydro: *4-Acetoxy-2,6-dibromo-1,5-dihydroxy-2-cyclohexen-1-one.*
$C_8H_{10}Br_2O_4$ M 329.973
Metabolite of *P.* sp. Cryst. (EtOAc/hexane). Mp 150.5-151.5°.

(4S,5R,6R)-form
  4-Ac: [110850-10-7].
    C$_8$H$_8$Br$_2$O$_4$ M 327.957
    Metab. of *P.* sp. [α]$_D^{23}$ +130° (c, 0.1 in CH$_2$Cl$_2$).
  Higa, T. *et al*, *Tetrahedron*, 1987, **43**, 1603 (*cryst struct*)

3,5-Dibromo-1,6-dihydroxy-4-oxo-2-cyclo-   **D-60098**
hexene-1-acetonitrile, 9CI

(1R,5S,6S)-form

C$_8$H$_7$Br$_2$NO$_3$ M 324.956
(1R,5S,6S)-form [108526-92-7]
  Constit. of *Aplysina laevis*. Possesses antimicrobial activity.
(1R,5R,6S)-form [108590-99-4]
  Constit. of *A. laevis*. Possesses antimicrobial activity.
  Capon, R.J. *et al*, *Aust. J. Chem.*, 1987, **40**, 341.

2,4-Dibromo-2,4-dimethyl-3-pentanone, 9CI   **D-60099**
[17346-16-6]

$$(H_3C)_2CBrCOCBr(CH_3)_2$$

C$_7$H$_{12}$Br$_2$O   M 271.979
Liq. Bp$_{40}$ 116-118°, Bp$_{18}$ 92-94°.
  Faworski, A., *J. Prakt. Chem.*, 1913, **88**, 641 (*synth*)
  Bushby, R.J. *et al*, *J. Chem. Soc., Perkin Trans. 1*, 1975, 2513 (*synth*)
  Furuhata, T. *et al*, *Tetrahedron*, 1986, **42**, 5301 (*synth*)

4,5-Dibromo-1H-imidazole   **D-60100**
[2302-30-9]

C$_3$H$_2$Br$_2$N$_2$ M 225.870
Cryst. (EtOH aq.). Mp 225-226°.
B,HBr: Glistening plates (H$_2$O). Mp 250° dec.
  Balaban, I.E. *et al*, *J. Chem. Soc.*, 1922, 947 (*synth*)
  Iddon, B. *et al*, *Tetrahedron Lett.*, 1986, **27**, 1635 (*synth*)

4,5-Dibromo-1H-imidazole-2-carboxalde-   **D-60101**
hyde, 9CI
4,5-Dibromo-2-formyl-1H-imidazole
[106848-46-8]

C$_4$H$_2$Br$_2$N$_2$O   M 253.881
Cryst. (EtOAc). Mp 192-194°.
  Iddon, B. *et al*, *Tetrahedron Lett.*, 1986, **27**, 1635 (*synth*)

4,5-Dibromo-1H-imidazole-2-carboxylic   **D-60102**
acid
[74840-91-8]

C$_4$H$_2$Br$_2$N$_2$O$_2$ M 269.880
Cryst. (EtOH aq.). Mp 173-175° dec.
Et ester: [74840-99-6].
  C$_6$H$_6$Br$_2$N$_2$O$_2$ M 297.934
  Cryst. (AcOH aq.). Mp 154-156°.
  Dirlam, J.P. *et al*, *J. Heterocycl. Chem.*, 1980, **17**, 409 (*synth, deriv, ir, pmr*)
  Iddon, B. *et al*, *Tetrahedron Lett.*, 1986, **27**, 1635 (*synth*)

1,3-Dibromo-2-methylbenzene   **D-60103**
Updated Entry replacing D-01849
2,6-Dibromotoluene
[69321-60-4]
C$_7$H$_6$Br$_2$ M 249.932
d^{22} 1.81. Mp 5-6°. Bp 246°, Bp$_{23}$ 122°.
  Cohen, J.B. *et al*, *J. Chem. Soc.*, 1914, **105**, 505 (*synth*)
  Yamato, T. *et al*, *J. Chem. Soc., Perkin Trans. 1*, 1987, 1 (*synth*)

1,1-Dibromo-3-methyl-1-butene   **D-60104**
[32363-92-1]

$$Br_2C{=}CHCH(CH_3)_2$$

C$_5$H$_8$Br$_2$ M 227.926
Oil. d$_{20}^{20}$ 1.66. Bp 159-160°, Bp$_{12}$ 49-51°.
  Farrell, J.K. *et al*, *J. Am. Chem. Soc.*, 1935, **57**, 1281 (*synth*)
  Combret, J.C. *et al*, *Tetrahedron Lett.*, 1971, 1035 (*synth*)
  Eberbach, W. *et al*, *Tetrahedron*, 1986, **42**, 2221 (*synth, ir, pmr*)

1,2-Dibromopropane, 9CI   **D-60105**
Updated Entry replacing D-02101
Propylene dibromide
[78-75-1]

(S)-form

C$_3$H$_6$Br$_2$ M 201.888
▷TX8574000.
(S)-form [101916-65-8]
  Bp$_{55}$ 70°. [α]$_D^{20}$ −19.43° (neat).
(±)-form
  d$_4^{20}$ 1.933. Bp 141.6°, Bp$_{30}$ 52.3°.
  Mouneyrat, A., *C. R. Hebd. Seances Acad. Sci.*, 1898, **127**, 274 (*synth*)
  Kharasch, M.S. *et al*, *J. Am. Chem. Soc.*, 1935, **57**, 2463 (*synth*)
  Pirrung, M.C. *et al*, *J. Org. Chem.*, 1986, **51**, 2103 (*synth*)
  Sax, N.I., *Dangerous Properties of Industrial Materials*, 5th Ed., Van Nostrand-Reinhold, 1979, 942.

1,2-Dibromo-3,4,5,6-tetrafluorobenzene, 9CI   D-60106

[827-08-7]

C₆Br₂F₄   M 307.868

Liq. d₄²⁰ 2.26. Mp 12°. Bp 198°, Bp₂₀ 96-97°. n_D²⁵ 1.5151.

U.S.P., 2 967 894, (1961); *CA*, **55**, 8350 (*synth*)
Smithson, L.D. *et al, Org. Mass. Spectrom.*, 1970, **4**, 1 (*ms*)
Gerritsen, J. *et al, J. Magn. Reson.*, 1972, **8**, 20 (*F nmr*)
Green, J.H.S. *et al, Spectrochim. Acta, Part A*, 1977, **33**, 193 (*ir, raman*)

1,3-Dibromo-2,4,5,6-tetrafluorobenzene, 9CI   D-60107

[1559-87-1]

C₆Br₂F₄   M 307.868

Liq. Bp₁₈ 86-88°. n_D²³ 1.5178.

Fr. Pat., 1 360 917, (1964); *CA*, **61**, 13236 (*synth*)
Tilney-Basset, J.F., *Chem. Ind.* (*London*), 1965, 693 (*synth*)
Smithson, L.D. *et al, Org. Mass. Spectrom.*, 1970, **4**, 1 (*ms*)
Green, J.H.S. *et al, Spectrochim. Acta, Part A*, 1977, **33**, 193 (*ir, raman*)

1,4-Dibromo-2,3,5,6-tetrafluorobenzene, 9CI   D-60108

[344-03-6]

C₆Br₂F₄   M 307.868

Cryst. (MeOH aq.). Mp 76-77°.

Hellman, M. *et al, J. Am. Chem. Soc.*, 1953, **75**, 4590 (*synth*)
Bruce, M.I., *J. Chem. Soc.* (*A*), 1968, 1459 (*F nmr*)
Smithson, L.D. *et al, Org. Mass. Spectrom.*, 1970, **4**, 1 (*ms*)
Pawley, G.S. *et al, Acta Crystallogr., Sect. A*, 1977, **33**, 142 (*struct*)
Green, J.H.S. *et al, Spectrochim. Acta, Part A*, 1977, **33**, 193 (*ir, raman*)
Aldrich Library of FT-IR Spectra, 1st Ed., **1**, 1015B (*ir*)

2,4-Dibromothiazole   D-60109

[4175-77-3]

C₃HBr₂NS   M 242.916

Needles by subl. Mp 82°.

Reynaud, P. *et al, Bull. Soc. Chim. Fr.*, 1962, 1735 (*synth, ir*)

2,5-Dibromothiazole   D-60110

[4175-78-4]

C₃HBr₂NS   M 242.916

Cryst. (pet. ether). Mp 46-47°.

Erlenmeyer, H. *et al, Helv. Chim. Acta*, 1945, **28**, 985 (*synth*)

4,5-Dibromothiazole   D-60111

[67594-67-6]

C₃HBr₂NS   M 242.916

Cryst. by subl. Mp 75°.

Reynaud, P. *et al, Bull. Soc. Chim. Fr.*, 1962, 1735 (*synth, ir*)

1,1-Di-*tert*-butylcyclopropane   D-60112

1,1-Bis(1,1-dimethylethyl)cyclopropane, 9CI

[80816-10-0]

$$(H_3C)_3C \quad C(CH_3)_3$$

C₁₁H₂₂   M 154.295

Liq.

Fitjer, L. *et al, Chem. Ber.*, 1986, **119**, 1144 (*synth, ir, pmr, cmr*)

3,7-Di-*tert*-butyl-9,10-dimethyl-2,6-anthraquinone   D-60113

3,7-Bis(1,1-dimethylethyl)-9,10-dimethyl-2,6-anthracenedione

[105858-62-6]

C₂₄H₂₈O₂   M 348.484

Brown-orange cryst. (EtOAc). Mp 250° dec.

Boldt, P. *et al, Chem. Ber.*, 1987, **120**, 497 (*synth, ir, pmr, ms*)

Di-*tert*-butylthioketene   D-60114

2-(1,1-Dimethylethyl)-1-butene-1-thione, 9CI

[16797-75-4]

$$[(H_3C)_3C]_2C{=}C{=}S$$

C₁₀H₁₈S   M 170.312

Bp₆ 60-63°.

S-*Oxide:* [16797-76-5].
   C₁₀H₁₈OS   M 186.312
   Cryst. (pentane). Mp 34-35°.

Elam, E.U. *et al, J. Org. Chem.*, 1968, **33**, 2738 (*synth*)

5,6-Dichloroacenaphthene   D-60115

5,6-Dichloro-1,2-dihydroacenaphthylene, 9CI. *4,5-Dichloroacenaphthene* (*obsol.*)

[4208-97-3]

C₁₂H₈Cl₂   M 223.101

Needles (hexane/C₆H₆). Mp 165.8-167.5°.

Morgan, G.T. *et al, J. Soc. Chem. Ind.*, 1930, **49**, 413 (*synth*)
Miyamoto, H. *et al, Tetrahedron Lett.*, 1986, **27**, 2011 (*synth*)

9,10-Dichloroanthracene   D-60116

Updated Entry replacing D-02326
ms-Dichloroanthracene

[605-48-1]

C₁₄H₈Cl₂   M 247.123

Yellow needles (CCl₄). Sol. C₆H₆, spar. sol. Et₂O, EtOH. Mp 212-212.5°. Oxidn. → anthraquinone.

Ninaev, F., *CA*, 1931, **25**, 1252.
Karolina, P., *Rocz. Chem.*, 1974, **48**, 1453 (*synth*)
Sharpless, N.E. *et al, Org. Magn. Reson.*, 1974, **6**, 115 (*nmr*)
Stoesser, R. *et al, J. Prakt. Chem.*, 1975, **317**, 591 (*nmr*)

Trotter, J., *Acta Crystallogr., Sect. C*, 1986, **42**, 862 (*cryst struct*)

De Crauw, T., *Recl. Trav. Chim. Pays-Bas*, 1931, **50**, 26 (*synth*)
Schaefer, T. *et al*, *J. Magn. Reson.*, 1982, **46**, 325 (*deriv, pmr*)

2,3-Dichloro-1,4,9,10-anthracenetetrone, 9CI   D-60117

[109141-98-2]

$C_{14}H_4Cl_2O_4$ M 307.089
Cryst. (nitrobenzene/CS_2). Mp 224-226°.

Cano, P. *et al*, *J. Chem. Soc., Perkin Trans. 1*, 1986, 1923 (*synth, pmr*)

4,6-Dichloro-1,3-benzenedicarboxaldehyde, 9CI   D-60118

Updated Entry replacing D-02365
4,6-Dichloroisophthalaldehyde, 8CI
$C_8H_4Cl_2O_2$ M 203.024
Needles (2-propanol). Mp 163-164°.

Katritzky, A.R. *et al*, *J. Org. Chem.*, 1987, **52**, 2726 (*synth, pmr*)

2,3-Dichlorobenzenethiol, 9CI   D-60119

2,3-Dichlorothiophenol. 2,3-Dichlorophenyl mercaptan.
1,2-Dichloro-3-mercaptobenzene
[17231-95-7]

$C_6H_4Cl_2S$ M 179.064
Cryst. Mp 46.5°. Bp_7 108-110°.

Mahajanshetti, C.S. *et al*, *CA*, 1963, **59**, 8636d (*synth*)
Takikawa, Y. *et al*, *CA*, 1968, **68**, 59210a (*synth*)
Bártl, V. *et al*, *Collect. Czech. Chem. Commun.*, 1984, **49**, 2295.

2,4-Dichlorobenzenethiol, 9CI   D-60120

2,4-Dichlorothiophenol. 2,4-Dichlorophenyl mercaptan.
2,4-Dichloro-1-mercaptobenzene
[1122-41-4]
$C_6H_4Cl_2S$ M 179.064
Needles. Mp 20° (22°). Steam-volatile.

Baddeley, G. *et al*, *J. Chem. Soc.*, 1933, 46 (*synth*)
Sparke, M.B. *et al*, *J. Am. Chem. Soc.*, 1953, **75**, 4907 (*synth*)
Parke, D.V. *et al*, *Biochem. J.*, 1955, **59**, 415 (*synth, chromatog*)

2,5-Dichlorobenzenethiol, 9CI   D-60121

2,5-Dichlorothiophenol. 2,5-Dichlorophenyl mercaptan.
1,4-Dichloro-2-mercaptobenzene
[5858-18-4]
$C_6H_4Cl_2S$ M 179.064
Cryst. (pet. ether). Mp 27°. Bp_{50} 112-116°.
Me ether: [17733-24-3]. *1,4-Dichloro-2-(methylthio)-benzene. 2,5-Dichlorothioanisole.*
$C_7H_6Cl_2S$ M 193.090
Cryst. (EtOH). Mp 51°.

Gebauer-Fülnegg, E. *et al*, *Monatsh. Chem.*, 1927, **48**, 627 (*synth, deriv*)

2,6-Dichlorobenzenethiol, 9CI   D-60122

2,6-Dichlorothiophenol. 2,6-Dichlorophenyl mercaptan.
1,3-Dichloro-2-mercaptobenzene
[24966-39-0]
$C_6H_4Cl_2S$ M 179.064
Cryst. (pet. ether). Mp 48-50°. Bp_3 95-100°.

Kopylova, B.V. *et al*, *CA*, 1970, **73**, 14363q (*synth*)
Trani, A. *et al*, *J. Heterocycl. Chem.*, 1974, **11**, 257 (*synth, ir, pmr*)

3,4-Dichlorobenzenethiol, 9CI   D-60123

3,4-Dichlorothiophenol. 3,4-Dichlorophenyl mercaptan.
1,2-Dichloro-4-mercaptobenzene
[5858-17-3]
$C_6H_4Cl_2S$ M 179.064
Liq. Bp_{10} 119-120°.
Me ether: [17733-23-2]. *1,2-Dichloro-4-(methylthio)-benzene. 3,4-Dichlorothioanisole.*
$C_7H_6Cl_2S$ M 193.090
Liq.

Červená, I. *et al*, *Collect. Czech. Chem. Commun.*, 1976, **41**, 881 (*synth*)
Brunelle, D.J., *J. Org. Chem.*, 1984, **49**, 1309 (*synth, deriv*)

3,5-Dichlorobenzenethiol, 9CI   D-60124

3,5-Dichlorothiophenol. 3,5-Dichlorophenyl mercaptan.
1,3-Dichloro-5-mercaptobenzene
[17231-94-6]
$C_6H_4Cl_2S$ M 179.064
Cryst. Mp 63.5°.

Takikawa, Y. *et al*, *CA*, 1968, **68**, 59210a (*synth*)

2,3-Dichlorobenzofuran, 9CI   D-60125

2,3-Dichlorocoumarone
[69314-26-7]

$C_8H_4Cl_2O$ M 187.025
Leaflets (EtOH). Mp 25-26°. Bp 226-227°.

Stoermer, R., *Justus Liebigs Ann. Chem.*, 1899, **312**, 316 (*synth*)
Clavel, J.-M. *et al*, *Tetrahedron*, 1978, **34**, 1537 (*synth*)

5,7-Dichlorobenzofuran   D-60126

5,7-Dichlorocoumarone
[23145-06-4]
$C_8H_4Cl_2O$ M 187.025
Mp 60° (45°). $Bp_{0.15}$ 76°.

Fr. Pat., 1 537 206, (*1968*); *CA*, **71**, 61198 (*synth*)
Givens, E.N. *et al*, *J. Catalysis*, 1969, **15**, 319 (*synth*)
Givens, E.N. *et al*, *Tetrahedron*, 1969, **25**, 2407 (*ms*)
Kraus, G.A. *et al*, *Tetrahedron*, 1985, **41**, 2337 (*synth, ir, pmr*)

2,2-Dichlorocyclopropanecarboxylic acid   D-60127

[5365-14-0]

COOH structure with Cl, Cl on cyclopropane ring

$C_4H_4Cl_2O_2$ M 154.980

(±)-*form*

Cryst. (cyclohexane). Mp 76-77°.

Me ester: [3591-47-7].
   $C_5H_6Cl_2O_2$ M 169.007
   Bp$_{23}$ 79°.
4-Bromophenacyl ester: Cryst. (EtOH). Mp 122-123°.

Woodworth, R.C. *et al, J. Am. Chem. Soc.*, 1957, **79**, 2542 (*synth*)

Kusuyama, Y., *Bull. Chem. Soc. Jpn.*, 1979, **52**, 1944 (*synth, deriv*)

DeWeese, F.T. *et al, Tetrahedron*, 1986, **42**, 239 (*synth*)

2,2-Dichloro-3,3-difluorooxirane, 9CI   D-60128

1,1-Dichloroepoxy-2,2-difluoroethane, 8CI

[22940-91-6]

F, F, Cl, Cl oxirane structure

$C_2Cl_2F_2O$ M 148.924

Volatile liq. Fp −93.5°. Bp 20.2°.

Chow, D. *et al, Can. J. Chem.*, 1969, **47**, 2591 (*synth, ir*)

Jones, R.B. *et al, Biochem. Pharmacol.*, 1983, **32**, 2359 (*tox*)

1-(3,5-Dichloro-2,6-dihydroxy-4-methoxy-phenyl)-1-hexanone   D-60129

$CO(CH_2)_4CH_3$ with HO, OH, Cl, Cl, OMe on benzene ring

$C_{13}H_{16}Cl_2O_4$ M 307.173

Isol. from the amphibian *Dictyostelium disoideum.* Morphogen differentiation inducing factor. Representative of a new class of effector molecule.

Morris, H.R. *et al, Nature (London)*, 1987, **328**, 811 (*isol, ms, struct, synth*)

2,4-Dichloro-3,4-dimethyl-2-cyclobuten-1-one, 9CI   D-60130

[110655-89-5]

Cl, O, H_3C, Cl, CH_3 cyclobutenone structure

$C_6H_6Cl_2O$ M 165.019

Liq.

Ammann, A.A. *et al, Helv. Chim. Acta*, 1987, **70**, 321 (*synth, uv, ir, pmr*)

4,4-Dichloro-2,3-dimethyl-2-cyclobuten-1-one, 9CI   D-60131

[72284-72-1]

H_3C, O, H_3C, Cl, Cl cyclobutenone structure

$C_6H_6Cl_2O$ M 165.019

Oil, solidifying on standing. Mp 44.5°.

Ammann, A.A. *et al, Helv. Chim. Acta*, 1987, **70**, 321 (*synth, uv, ir, pmr*)

3,4-Dichloro-1-(4-fluorophenyl)-1H-pyrrole-2,5-dione, 9CI   D-60132

2,3-Dichloro-N-(4-fluorophenyl)maleimide.

Fluoromide

[41205-21-4]

$C_{10}H_4Cl_2FNO_2$ M 260.052

Fungicide. Pale-yellow cryst. Mp 240.5-241.8°.

Japan. Pat., 74 110 661, 74 110 662, (*1974*); *CA*, **82**, 139418, 139761 (*synth*)

Anon, *Jpn. Pestic. Inf.*, 1978, **34**, 26 (*rev, use, tox*)

Pesticide Manual, 7th Ed., No. 4250.

2,3-Dichloro-4-hydroxybenzoic acid   D-60133

COOH, Cl, Cl, OH benzene ring structure with numbering

$C_7H_4Cl_2O_3$ M 207.013

Me ether: [55901-80-9]. *2,3-Dichloro-4-methoxybenzoic acid.*
   $C_8H_6Cl_2O_3$ M 221.040
   Cryst. (MeOH/C$_6$H$_6$). Mp 221-224°.

Wyrick, S.D. *et al, J. Med. Chem.*, 1987, **30**, 1798 (*synth, deriv, pmr*)

4,5-Dichloro-1H-imidazole-2-carboxaldehyde, 9CI   D-60134

4,5-Dichloro-2-formylimidazole

[81293-97-2]

Cl, N, Cl, CHO, H imidazole structure

$C_4H_2Cl_2N_2O$ M 164.979

Cryst. (EtOH). Mp 190-191°.

Di-Et acetal: [81315-59-5].
   $C_8H_{12}Cl_2N_2O_2$ M 239.101
   Cryst. (H$_2$O). Mp 110-111°.

Dirlam, J.P. *et al, J. Org. Chem.*, 1982, **47**, 2196 (*synth, deriv, ir, pmr*)

Iddon, B. *et al, Tetrahedron Lett.*, 1986, **27**, 1635 (*synth*)

4,5-Dichloro-1*H*-imidazole-2-carboxylic acid   D-60135

[64736-53-4]

$C_4H_2Cl_2N_2O_2$ M 180.978
Cryst. (EtOH aq.). Mp 160-162°.

Dirlam, J.P. *et al*, *J. Heterocycl. Chem.*, 1980, **17**, 409 (*synth, ms*)
Iddon, B. *et al*, *Tetrahedron Lett.*, 1986, **27**, 1635 (*synth*)

6,7-Dichloroisoquinoline, 9CI   D-60136

$C_9H_5Cl_2N$ M 198.051
B,HCl: [73075-60-2]. Cryst. (EtOH). Mp 227-228°.
Kaiser, C. *et al*, *J. Med. Chem.*, 1986, **29**, 2381 (*synth*)

6,7-Dichloro-5,8-isoquinolinedione   D-60137

[84289-03-2]

$C_9H_3Cl_2NO_2$ M 228.034
Pale-green cryst. (MeOH/Et$_2$O). Mp 180-181°.

Shaikh, I.A. *et al*, *J. Med. Chem.*, 1986, **29**, 1329 (*synth, ir, pmr, ms*)

2,3-Dichloro-5-methylpyridine   D-60138

5,6-Dichloro-3-picoline
[59782-90-0]

$C_6H_5Cl_2N$ M 162.018
Cryst. (MeOH aq.). Mp 45-47°.

Setliff, F.L. *et al*, *J. Chem. Eng. Data*, 1976, **21**, 246 (*synth, ir, pmr*)

2,5-Dichloro-3-methylpyridine   D-60139

2,5-Dichloro-3-picoline
[59782-88-6]

$C_6H_5Cl_2N$ M 162.018
Cryst. (MeOH aq.). Mp 40-41°.

Setliff, F.L. *et al*, *J. Chem. Eng. Data*, 1976, **21**, 246 (*synth, pmr, ir*)
Martin, P. *et al*, *Tetrahedron*, 1985, **41**, 4057 (*synth, pmr, cmr*)

2,3-Dichloro-5-nitropyridine   D-60140

[22353-40-8]

$C_5H_2Cl_2N_2O_2$ M 192.989
Mp 52°. Bp 254°.

Takahashi, T. *et al*, *J. Pharm. Soc. Jpn.*, 1952, **72**, 378 (*synth*)
Batkowski, T., *Rocz. Chem.*, 1968, **42**, 2079 (*synth*)
Eur. Pat., 147 105, (*1985*); *CA*, **103**, 132311 (*synth*)

2,4-Dichloro-3-nitropyridine   D-60141

[5975-12-2]
$C_5H_2Cl_2N_2O_2$ M 192.989
Needles. Mp 61.5-62°.

den Hertog, H.J. *et al*, *Recl. Trav. Chim. Pays-Bas*, 1948, **67**, 29 (*synth*)
Mizuno, Y. *et al*, *Chem. Pharm. Bull.*, 1964, **12**, 866 (*synth*)
Montgomery, J.A. *et al*, *J. Med. Chem.*, 1966, **9**, 105 (*synth*)

2,4-Dichloro-5-nitropyridine   D-60142

[4487-56-3]
$C_5H_2Cl_2N_2O_2$ M 192.989
Light-yellow needles (pentane). Mp 43.5°.

Mizuno, Y. *et al*, *Chem. Pharm. Bull.*, 1964, **12**, 866 (*synth*)
Talik, T. *et al*, *CA*, 1966, **64**, 2046g (*synth*)
Kroon, C. *et al*, *Recl. Trav. Chim. Pays-Bas*, 1976, **95**, 127 (*synth*)

2,5-Dichloro-3-nitropyridine   D-60143

[21427-62-3]
$C_5H_2Cl_2N_2O_2$ M 192.989
Cryst. (EtOH). Mp 44°.

Batkowski, T., *Rocz. Chem.*, 1968, **42**, 2079 (*synth*)
Kroon, C. *et al*, *Recl. Trav. Chim. Pays-Bas*, 1976, **95**, 127 (*synth*)

2,6-Dichloro-3-nitropyridine   D-60144

[16013-85-7]
$C_5H_2Cl_2N_2O_2$ M 192.989
Cream needles (hexane). Mp 62.5-63.5°.

B.P., 2 059 947, (*1981*); *CA*, **95**, 132687 (*manuf*)
Colbry, N.L. *et al*, *J. Heterocycl. Chem.*, 1984, **21**, 1521 (*synth*)

2,6-Dichloro-4-nitropyridine   D-60145

[25194-01-8]
$C_5H_2Cl_2N_2O_2$ M 192.989
1-Oxide: [2587-01-1].
    $C_5H_2Cl_2N_2O_3$ M 208.988
    Cryst. (H$_2$O). Mp 177-178.5°.

Rousseau, R.J. *et al*, *J. Heterocycl. Chem.*, 1965, **2**, 196 (*oxide*)
Bagal, L.I. *et al*, *Zh. Org. Khim.*, 1969, **5**, 2016 (*synth*)
Eur. Pat., 53 306, (*1982*); *CA*, **97**, 162832 (*synth*)

3,4-Dichloro-5-nitropyridine   D-60146

[56809-84-8]
$C_5H_2Cl_2N_2O_2$ M 192.989
Mp 43-47° (39°).

Namirski, P.N., *CA*, 1962, **57**, 16554a (*synth*)

Delarge, J. et al, Pharm. Acta Helv., 1975, **50**, 188 (synth)

3,5-Dichloro-4-nitropyridine   D-60147

$C_5H_2Cl_2N_2O_2$ M 192.989

1-Oxide: [18344-58-6].
    $C_5H_2Cl_2N_2O_3$ M 208.988
    Needles (pet. ether). Mp 109-110°.

Johnson, C.D. et al, J. Chem. Soc. (B), 1967, 1235.

1,4-Dichlorophthalazine, 9CI   D-60148

[4752-10-7]

$C_8H_4Cl_2N_2$ M 199.039
Needles (C_6H_6 or Me_2CO). Mp 162-164°.

Drew, H.D.K. et al, J. Chem. Soc., 1937, 16 (synth)

6,7-Dichloro-5,8-quinolinedione, 9CI   D-60149

[6541-19-1]

$C_9H_3Cl_2NO_2$ M 228.034
Bright-yellow cryst. (CH_2Cl_2/MeOH). Mp 221-223°.
N-Oxide: [84289-01-0].
    $C_9H_3Cl_2NO_3$ M 244.034
    Bright-orange cryst.(CH_2Cl_2). Mp 209-210°.

Shaikh, I.A. et al, J. Med. Chem., 1986, **29**, 1329 (synth, ir, pmr, ms)

2,3-Dichloroquinoxaline, 9CI   D-60150

[2213-63-0]

$C_8H_4Cl_2N_2$ M 199.039
Cryst. (EtOH). Mp 150°.
▷VD1720000.
1-Oxide: [53870-24-9].
    $C_8H_4Cl_2N_2O$ M 215.038
    Mp 138-139°.

Hinsberg, O. et al, Ber., 1896, **29**, 784 (synth)
Cheeseman, G.W.H., J. Chem. Soc., 1955, 1804 (synth)
Mixan, C.E. et al, J. Org. Chem., 1977, **42**, 1869 (oxide)

1,2-Dichloro-3,4,5,6-tetrafluorobenzene, 9CI   D-60151

[1198-59-0]

$C_6Cl_2F_4$ M 218.966

Bladon, P. et al, Spectrochim. Acta, 1964, **20**, 1033 (F nmr)
Netherlands Pat., 6 411 449, (1965); CA, **63**, 11425 (synth)
Green, J.H.S. et al, Spectrochim. Acta, Part A, 1977, **33**, 193 (ir, raman)

1,3-Dichloro-2,4,5,6-tetrafluorobenzene, 9CI   D-60152

[1198-61-4]

$C_6Cl_2F_4$ M 218.966
Liq. Bp_{759} 156-157°. n_D^{20} 1.4678.

Boden, N. et al, Mol. Phys., 1964, **8**, 133 (F nmr)
Fuller, G., J. Chem. Soc., 1965, 6264 (synth)
Gage, J.C., Br. J. Ind. Med., 1970, **27**, 1 (tox)
Hitzke, J. et al, Org. Mass. Spectrom., 1972, **6**, 349 (ms)
Green, J.H.S. et al, Spectrochim. Acta, Part A, 1977, **33**, 193 (ir, raman)

1,4-Dichloro-2,3,5,6-tetrafluorobenzene, 9CI   D-60153

[1198-62-5]

$C_6Cl_2F_4$ M 218.966
Cryst. (EtOH). Mp 52-54°. Bp_{759} 157-158°.

Belf, L.J. et al, J. Chem. Soc., 1965, 3372 (synth)
Emsley, J.W. et al, Mol. Phys., 1966, **11**, 437 (F nmr)
Green, J.H.S. et al, Spectrochim. Acta, Part A, 1977, **33**, 193 (ir, raman)

2,4-Dichlorothiazole   D-60154

[4175-76-2]

C_3HCl_2NS M 154.014
Cryst. Mp 42-43°. Bp 184°.

Reynaud, P. et al, Bull. Soc. Chim. Fr., 1962, 1735.

2,5-Dichlorothiazole   D-60155

[16629-14-4]
C_3HCl_2NS M 154.014
Mobile, volatile liq. Bp 159-161°.

Reynaud, P. et al, Bull. Soc. Chim. Fr., 1962, 1735 (synth, ir)

4,5-Dichlorothiazole   D-60156

[16629-16-6]
C_3HCl_2NS M 154.014
Volatile needles with strong odour. Mp 38-39°. Bp 183°,
    Bp_{30} 104-105°.

Reynaud, P. et al, Bull. Soc. Chim. Fr., 1962, 1735 (synth, ir)

4,5-Dichloro-2-(trifluoromethyl)-1H-benz-imidazole, 9CI   D-60157

Chloroflurazole
[3615-21-2]

$C_8H_3Cl_2F_3N_2$ M 255.026

Herbicide. Mp 212-214°. pK_a 6.96 (7.4).

▷DD7350000.

Burton, D.E. *et al*, *Nature (London)*, 1965, **208**, 1166 (*rev, use, tox*)
Buechel, K.H. *et al*, *Z. Naturforsch., B*, 1970, **25**, 934 (*synth, props*)
Adamson, G.W. *et al*, *Pestic. Sci.*, 1984, **15**, 31 (*tox*)
Pesticide Manual, 7th Ed., No. 2370 (*bibl*)

4,6-Dichloro-2-(trifluoromethyl)-1*H*-benz-imidazole D-60158

[4228-88-0]

$C_8H_3Cl_2F_3N_2$ M 255.026
Herbicidal. Mp 190°. pK_a 6.78 (7.25).

▷DD7700000.

Buechel, K.H. *et al*, *Z. Naturforsch., B*, 1970, **25**, 934 (*synth*)
Adamson, G.W. *et al*, *Pestic. Sci.*, 1984, **15**, 31.

4,7-Dichloro-2-(trifluoromethyl)-1*H*-benz-imidazole D-60159

[4228-89-1]
$C_8H_3Cl_2F_3N_2$ M 255.026
Herbicidal. Mp 223-225°. pK_a 6.24 (6.85).

▷DD7875000.

Buechel, K.H. *et al*, *Z. Naturforsch., B*, 1970, **25**, 934 (*synth*)
Adamson, G.W. *et al*, *Pestic. Sci.*, 1984, **15**, 31 (*tox*)

5,6-Dichloro-2-(trifluoromethyl)-1*H*-benz-imidazole D-60160

[2338-25-2]
$C_8H_3Cl_2F_3N_2$ M 255.026
Cryst. Mp 240-242°. pK_a 7.4 (7.6).

▷DD8050000.

N-*Me:* [19517-16-9].
  $C_9H_5Cl_2F_3N_2$ M 269.053
  Solid. Mp 167.5°.
N-(*Phenyloxycarbonyl*): [14255-88-0]. *Phenyl 5,6-dich-loro-2-(trifluoromethyl)-1H-benzimidazole-1-car-boxylate, 9CI. Fenazaflor. Fenoflurazole.*
  $C_{15}H_7Cl_2F_3N_2O_2$ M 375.134
  Acaricide. Yellow cryst. Mp 103°.

  ▷DD6650000.

Buechel, K.H. *et al*, *Z. Naturforsch., B*, 1970, **25**, 934 (*synth*)
Adamson, G.W. *et al*, *Pestic. Sci.*, 1984, **15**, 31 (*tox*)
Pesticide Manual, 7th Ed., No. 6050 (*deriv*)
Agrochemicals Handbook, *Royal Society of Chemistry*, 1985, No. A194 (*deriv, rev*)

2,2-Dichloro-3,3,3-trifluoro-1-propanol, 9CI D-60161

[20411-84-1]

$$F_3CCCl_2CH_2OH$$

$C_3H_3Cl_2F_3O$ M 182.957
Synthetic equiv. of F_3CCOCH_2-. Readily obt. on large scale. Long needles with strong aromatic odour. Mp 61-62°. Bp 100-110°.

Ac:
  $C_5H_5Cl_2F_3O_2$ M 224.995

Oil.
4-Methylbenzenesulfonyl: Mp 39-41°.

Lang, R.W., *Helv. Chim. Acta*, 1986, **69**, 881 (*synth, ir, pmr, nmr, ms*)

Dictymal D-60162

$C_{20}H_{32}O$ M 288.472
Constit. of brown alga *Dictyota dichotoma*. Oil. $[\alpha]_D^{18}$ +16.4° (c, 0.88 in $CHCl_3$).

Segawa, M. *et al*, *Tetrahedron Lett.*, 1987, **28**, 3703.

9-(Dicyanomethyl)anthracene D-60163
9-Anthracenylpropanedinitrile, 9CI
[88015-28-5]

$C_{17}H_{10}N_2$ M 242.279
Pale-yellow needles (AcOH). Mp 190-192°.

Wellman, D.E. *et al*, *J. Am. Chem. Soc.*, 1984, **106**, 355 (*synth, ir, pmr*)

10-(Dicyanomethylene)anthrone D-60164
Updated Entry replacing D-50247
10-(Dicyanomethyl)-9(10H)-anthracenone. 10-Oxo-$\Delta^{9(10-H),a}$-anthracenemalononitrile, 8CI
[10395-02-5]

$C_{17}H_8N_2O$ M 256.263
Cryst. ($CHCl_3$). Mp 289-290°. Nonplanar config. Does not form complexes.

Takimoto, H. *et al*, *J. Org. Chem.*, 1962, **27**, 4688 (*synth*)
Silverman, J. *et al*, *J. Chem. Soc. (B)*, 1967, 194 (*synth, cryst struct*)

Dicyanotriselenide, 9CI D-60165
Selenium diselenocyanate
[2179-96-6]

$$Se(SeCN)_2$$

$C_2N_2Se_3$ M 288.915
Cryst. (C_6H_6). Mp 132° dec., Mp 150°.

Muthmann, W. *et al*, *Ber.*, 1900, **33**, 1765 (*synth*)
Kaufmann, H.P. *et al*, *Ber.*, 1926, **59**, 178 (*synth*)
Rheinboldt, H. *et al*, *J. Am. Chem. Soc.*, 1949, **71**, 1740 (*synth*)
Hange, S., *Acta Chem. Scand.*, 1971, **25**, 3081 (*synth, bibl*)

Di(2,4,6-cycloheptatrien-1-yl)ethanone, 9CI　　D-60166

Di(2,4,6-cycloheptatrien-1-yl) ketone

[42836-42-0]

C$_{15}$H$_{14}$O　　M 210.275

Pale-yellow liq. Bp$_{0.6}$ 130° (bath).

Kayama, Y. *et al, Synth. Commun.*, 1973, **3**, 53 (*synth, ir, pmr*)

Dicyclopenta[*ef,kl*]heptalene, 8CI　　D-60167

Updated Entry replacing D-03257

Pyraceheptylene. Azupyrene

[193-85-1]

C$_{16}$H$_{10}$　　M 202.255

Bronze plates (hexane or MeOH). Mp 257-259° (250-258°).

Anderson, A.G. *et al, J. Am. Chem. Soc.*, 1968, **90**, 2993 (*synth, nmr, ir, uv*)

Jutz, C. *et al, Angew. Chem., Int. Ed. Engl.*, 1971, **10**, 808 (*synth, uv*)

Anderson, A.G. *et al, J. Org. Chem.*, 1973, **38**, 1445 (*synth, nmr, uv, ms, ir*)

Jutz, C. *et al, Synthesis*, 1974, 193 (*synth, nmr, uv*)

Mullen, K., *Helv. Chim. Acta*, 1978, **61**, 2307 (*nmr*)

Anderson, A.G. *et al, J. Am. Chem. Soc.*, 1985, **107**, 1896 (*props*)

Anderson, A.G. *et al, J. Org. Chem.*, 1986, **51**, 2961 (*synth*)

2,3-Didehydro-1,2-dihydro-1,1-dimethyl-naphthalene, 9CI　　D-60168

2,3-Dehydro-1,2-dihydro-1,1-dimethylnaphthalene

[105282-83-5]

C$_{12}$H$_{12}$　　M 156.227

Reactive intermediate; generated in solution and trapped by cycloaddition. First example of a 1,2,4-cyclohexatriene deriv.

Miller, B. *et al, J. Am. Chem. Soc.*, 1987, **109**, 578.

5,6-Didehydro-1,4,7,10-tetramethyldibenzo[*ae*]cyclooctene, 9CI　　D-60169

[69459-93-4]

C$_{20}$H$_{18}$　　M 258.362

Stable cyclooctatrienyne. Yellow needles (pentane). Mp 88° dec. A previous report of this compd. was erroneous.

Chan, T.-L. *et al, Tetrahedron*, 1986, **42**, 655 (*synth, ir, pmr, cryst struct*)

1-[4-[6-(Diethylamino)-2-benzofuranyl]-phenyl]-1*H*-pyrrole-2,5-dione, 9CI　　D-60170

N-[4-[2-(6-Dimethylaminobenzofuranyl)phenyl]]-maleimide

[101046-20-2]

C$_{20}$H$_{16}$N$_2$O$_3$　　M 332.358

Sensitive fluorogenic reagent for biological thiols. Reddish-purple cryst. (Me$_2$CO). Mp 203-204°.

Akiyama, S. *et al, Bull. Chem. Soc. Jpn.*, 1985, **58**, 2192 (*synth, pmr*)

Nakashima, K. *et al, Chem. Pharm. Bull.*, 1986, **34**, 1678, 1684 (*use*)

2,2-Diethylbutanoic acid, 9CI　　D-60171

Triethylacetic acid

[813-58-1]

$$(H_3CCH_2)_3CCOOH$$

C$_8$H$_{16}$O$_2$　　M 144.213

d$_4^{40.3}$ 0.91. Mp 38°. Bp 220-222°, Bp$_{16}$ 121-122°.

Na salt: Glistening leaflets + 3H$_2$O.

Chloride: [35354-15-5].
　C$_8$H$_{15}$ClO　　M 162.659
　Liq. Bp$_{20}$ 65-66°.

Amide:
　C$_8$H$_{17}$NO　　M 143.228
　Needles (EtOH). Mp 108°. Bp$_{20}$ 148-149°.

Haller, A. *et al, C.R. Hebd. Seances Acad. Sci.*, 1909, **148**, 130 (*synth*)

Meerwein, H., *Justus Liebigs Ann. Chem.*, 1919, **419**, 121 (*synth*)

Eberson, L., *J. Org. Chem.*, 1962, **27**, 3706 (*synth*)

Hoffman, W.F. *et al, J. Med. Chem.*, 1986, **29**, 849 (*deriv, synth, pmr*)

1,3-Diethynyladamantane　　D-60172

1,3-Diethynyltricyclo[3.3.1.1^{3,7}]decane, 9CI

[106325-28-4]

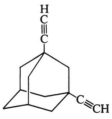

C$_{14}$H$_{16}$　　M 184.280

Needles (MeOH). Mp 45-46°.

Broxterman, A.B. *et al, Tetrahedron Lett.*, 1986, **27**, 1055 (*synth, ir, pmr, cmr*)

> The symbol ▷ in Entries highlights hazard or toxicity information

2,5-Difluorobenzenethiol, 9CI   D-60173

2,5-Difluorothiophenol. 2,5-Difluorophenyl mercaptan.
1,4-Difluoro-2-mercaptobenzene
[77380-28-0]

$C_6H_4F_2S$ M 146.154
Liq. Bp_2 60-62°.
Me ether: [54378-78-8]. *1,4-Difluoro-2-(methylthio)-*
  benzene. 2,5-Difluorothioanisole.
  $C_7H_6F_2S$ M 160.181
  Liq. Bp_{22} 84-86°.

Peach, M.E. *et al, J. Fluorine Chem.*, 1974, **4**, 399; 1977, **10**,
  319 (*synth, deriv, pmr, cmr, F nmr*)
Šindelář, K. *et al, Collect. Czech. Chem. Commun.*, 1982, **47**,
  3114 (*synth*)

2,6-Difluorobenzenethiol, 9CI   D-60174

2,6-Difluorothiophenol. 2,6-Difluorophenyl mercaptan.
1,3-Difluoro-2-mercaptobenzene
$C_6H_4F_2S$ M 146.154
Me ether: [91524-69-5]. *1,3-Difluoro-2-(methylthio)-*
  benzene. 2,6-Difluorothioanisole.
  $C_7H_6F_2S$ M 160.181
  Liq.

Baleja, J.D., *Synth. Commun.*, 1984, **14**, 215 (*synth, deriv*)
Schaefer, T. *et al, Can. J. Chem.*, 1986, **64**, 1376 (*pmr, F nmr*)

3,4-Difluorobenzenethiol, 9CI   D-60175

3,4-Difluorothiophenol. 3,4-Difluorophenyl mercaptan.
1,2-Difluoro-4-mercaptobenzene
[60811-24-7]
$C_6H_4F_2S$ M 146.154
Liq. Bp_{18} 135-140°.

Červená, I. *et al, Collect. Czech. Chem. Commun.*, 1976, **41**, 881
  (*synth, ir, pmr*)

3,5-Difluorobenzenethiol, 9CI   D-60176

3,5-Difluorothiophenol. 3,5-Difluorophenyl mercaptan.
1,3-Difluoro-5-mercaptobenzene
[99389-26-1]
$C_6H_4F_2S$ M 146.154
Me ether: [54378-77-7]. *1,3-Difluoro-5-(methylthio)-*
  benzene. 3,5-Difluorothioanisole.
  $C_7H_6F_2S$ M 160.181
  Liq. Bp_{22} 83-85°.

Peach, M.E. *et al, J. Fluorine Chem.*, 1974, **4**, 399 (*synth, deriv,*
  pmr, F nmr)

7,7-Difluorobicyclo[4.1.0]hepta-1,3,5-tri-   D-60177
ene, 9CI

7,7-Difluorobenzocyclopropene. 7,7-
Difluorocyclopropabenzene
[18238-55-6]

$C_7H_4F_2$ M 126.105
Liq. Bp_{25} 42°.

Gluck, C. *et al, Synthesis*, 1987, 260 (*synth, bibl*)

4,4'-Difluoro-[1,1'-biphenyl]-3,3'-   D-60178
dicarboxylic acid, 9CI

[27830-00-8]

$C_{14}H_8F_2O_4$ M 278.212
Cryst. (AcOH). Mp 365-370°.
Di-Me ester: [27830-01-9].
  $C_{16}H_{12}F_2O_4$ M 306.265
  Cryst. (EtOH). Mp 151.5-153°.

Martinez de Bertorello, M. *et al, J. Organomet. Chem.*, 1970,
  23, 285 (*synth*)

4,5-Difluoro-[1,1'-biphenyl]-2,3-   D-60179
dicarboxylic acid, 9CI

3,4-Difluoro-6-phenylphthalic acid
$C_{14}H_8F_2O_4$ M 278.212
Di-Me ester: [75599-97-2].
  $C_{16}H_{12}F_2O_4$ M 306.265
  Cryst. (hexane). Mp 92-93°.

England, D.C. *et al, J. Org. Chem.*, 1981, **46**, 144 (*synth, ir,*
  nmr)

6,6'-Difluoro-[1,1'-biphenyl]-2,2'-   D-60180
dicarboxylic acid, 9CI

6,6'-Difluorodiphenic acid
[567-85-1]
$C_{14}H_8F_2O_4$ M 278.212
Cryst. (EtOH). Mp 308-311°.
Di-Me ester: [111398-15-3].
  $C_{16}H_{12}F_2O_4$ M 306.265
  Prisms (MeOH). Mp 122-123°.

Stanley, W.M. *et al, J. Am. Chem. Soc.*, 1933, **55**, 706 (*synth,*
  resoln)
Cosmo, R. *et al, Aust. J. Chem.*, 1987, **40**, 35 (*synth, pmr*)

4,4-Difluoro-3,3-bis(trifluoromethyl)-1,2-   D-60181
oxathietane 2,2-dioxide

Bis-α-trifluoromethyldifluoroethane-β-sultone
[40325-85-7]

$C_4F_8O_3S$ M 280.089
Obt. by cycloaddition of SO_3 to 1,1,3,3,3-Pentafluoro-2-
  (trifluoromethyl)-1-propene, P-00373 . Liq. d_4^{20} 1.820.
  Bp 63°. n_D^{20} 1.3129.

Belaventsev, M.A. *et al, Izv. Akad. Nauk SSSR, Ser. Khim.*,
  1972, 2510 (*Engl. transl.* p. 2441) (*synth, F nmr*)

8,8-Difluoro-8,8-dihydro-4-methyl-2,2,6,6-tetrakis(trifluoromethyl)-2H,6H-[1,2]-iodoxolo[4,5,1-hi][1,2]benziodoxole, 9CI   D-60182

1,1-Difluoro-10-methyl-3,3,7,7-tetrakis(trifluoromethyl)-4,5,6-benzo-1-ioda-2,8-dioxabicyclo[3.3.1]-octane

[80360-39-0]

$C_{13}H_5F_{14}IO_2$ M 586.063

Strong oxidising agent. Cryst. by subl. Mp 205-210°.

Nguyen, T.T. *et al, J. Org. Chem.*, 1982, **47**, 1024 (*synth, pmr, F nmr, ms*)

1,1-Difluoro-2,2-diiodoethylene, 8CI   D-60183

[683-79-4]

$$F_2C{=}CI_2$$

$C_2F_2I_2$ M 315.828

Liq. d_4^{20} 2.961. Bp_{626} 133.5°. n_D^{20} 1.5869.

Park, J.D. *et al, J. Org. Chem.*, 1958, **23**, 1661 (*synth*)

Difluorodiphenylmethane   D-60184

1,1'-(Difluoromethylene)bisbenzene, 9CI

[360-11-2]

$$Ph_2CF_2$$

$C_{13}H_{10}F_2$ M 204.219

Liq. d_4^{20} 1.16. Mp −1.9 to −1.8°. Bp 260° dec., Bp_{10} 125°.

Henne, A.L. *et al, J. Am. Chem. Soc.*, 1938, **68**, 864 (*synth, props*)

Patrick, T.B. *et al, J. Org. Chem.*, 1981, **46**, 3917 (*synth, F nmr*)

(Difluoromethylene)-tetrafluorocyclopropane, 9CI   D-60185

Perfluoro(methylenecyclopropane).
Hexafluoro(methylenecyclopropane)

[51502-05-7]

C_4F_6 M 162.034

Gas. Bp 5-6°. Ir C=C 1818 cm⁻¹.

▷Extremely toxic

Smart, B.F., *J. Am. Chem. Soc.*, 1974, **96**, 927 (*synth, F nmr, tox, props*)

4,4-Difluoro-1,2-oxathietane 2,2-dioxide, 9CI   D-60186

2,2-Difluoroethane β-sultone

[41505-89-9]

$C_2H_2F_2O_3S$ M 144.093

Obt. by addition of SO_3 to 1,1-Difluoroethylene, D-03517. Liq. d_4^{20} 1.663. Bp_{15} 55°. n_D^{20} 1.3730.

Sokol'skii, G.A. *et al, Khim. Geterotsikl. Soedin.*, 1973, 178 (*Engl. transl.* p. 164); *CA*, **78**, 136135 (*synth, pmr, F nmr*)

2,3-Difluorooxirane, 9CI   D-60187

1,2-Difluoroethylene oxide

$C_2H_2F_2O$ M 80.034

Gillies, C.W. *et al, J. Am. Chem. Soc.*, 1975, **97**, 1276; 1977 **99**, 7239 (*synth, F nmr, pmr, ms, ir*)

Gillies, C.W. *et al, J. Mol. Spectrosc.*, 1978, **71**, 85 (*microwave*)

Agopovich, J.W. *et al, J. Am. Chem. Soc.*, 1982, **104**, 813 (*synth, pmr, F nmr, ir*)

Agopovich, J.W. *et al, J. Am. Chem. Soc.*, 1984, **106**, 2250 (*microwave*)

2,3-Difluoropyrazine, 9CI   D-60188

[52751-15-2]

$C_4H_2F_2N_2$ M 116.070

Van den Ham, D.M.W. *et al, J. Electron Spectrosc. Relat. Phenomena*, 1974, **3**, 479 (*synth, spectra*)

2,6-Difluoropyrazine, 9CI   D-60189

[33873-09-5]

$C_4H_2F_2N_2$ M 116.070

Liq. Bp 90-92°.

Cheeseman, G.W.H. *et al, J. Chem. Soc. (C)*, 1971, 2973 (*synth*)

Diformylacetic acid   D-60190

Updated Entry replacing D-03544

2-Formyl-3-hydroxy-2-propenoic acid, 9CI. 2-Formyl-malonaldehydic acid

$$HOOCCH(CHO)_2$$

$C_4H_4O_4$ M 116.073

Me ester: [39947-70-1]. *Methyl diformylacetate.*
Methoxycarbonylmalonaldehyde.
$C_5H_6O_4$ M 130.100

Reagent for photoannelation reactions. $Bp_{3.5}$ 54-55°.

Et ester: Ethyl diformylacetate.
Ethoxycarbonylmalonaldehyde.
$C_6H_8O_4$ M 144.127

Versatile synthetic reagent. Bp_3 115-117°, $Bp_{0.35}$ 30-32°.

Büchi, G. *et al*, *J. Am. Chem. Soc.*, 1970, **92**, 2165 (*synth, ir, uv, nmr*)
Bertz, S.H. *et al*, *J. Org. Chem.*, 1982, **47**, 2216 (*deriv, synth, pmr, ir, uv*)
Torii, S. *et al*, *Synthesis*, 1986, 400 (*deriv, synth, bibl, use*)
Fieser, M. *et al*, *Reagents for Organic Synthesis*, Wiley, 1967-84, **8**, 338 (*use*)

1,2-Di(2-furanyl)ethylene   D-60191

2,2'-(1,2-Ethenediyl)bisfuran, 9CI. 2,2'-Vinylenedifuran, 8CI

[5416-79-5]

(*E*)-form

$C_{10}H_8O_2$ M 160.172
(*E*)-**form** [1439-19-6]
 Plates (hexane). Mp 100-101°.
(*Z*)-**form** [18266-93-8]
 Cryst. by subl. Mp 38.5-42.5°. Obt. mixed with *ca.* 10% (*E*)-*form*.

Bruins, P.F., *J. Am. Chem. Soc.*, 1929, **51**, 1270 (*synth*)
Zimmerman, A.A. *et al*, *J. Org. Chem.*, 1969, **34**, 73 (*synth, ir, pmr*)
Hinz, W. *et al*, *Synthesis*, 1986, 620 (*synth*)

1,2-Dihydrazinoethane   D-60192

1,1'-(1,2-Ethanediyl)bishydrazine, 9CI. 1,2-Bis(hydrazino)ethane. Ethylenedihydrazine

[6068-98-0]

$$H_2{}^{\beta'}N^{\alpha'}NHCH_2CH_2{}^{\alpha}NH^{\beta}NH_2$$

$C_2H_{10}N_4$ M 90.128
B,2HCl: Cryst. (MeOH aq.). Mp 159°, Mp 167° dec.
$N^{\alpha},N^{\alpha'},N^{\beta},N^{\beta'}$-*Tetra-Ac:*
  $C_{10}H_{18}N_4O_4$ M 258.277
  Prisms (Me$_2$CO). Mp 163-165°.
$N^{\beta},N^{\beta},N^{\beta'},N^{\beta'}$-*Tetra-Me:* [2594-64-1].
  $C_6H_{18}N_4$ M 146.235
  Bp$_{16}$ 78-81°.
$N^{\alpha},N^{\alpha'},N^{\beta},N^{\beta},N^{\beta'},N^{\beta'}$-*Hexa-Me:* [49840-62-2].
  $C_8H_{22}N_4$ M 174.289
  Bp$_{16}$ 95-96°.

Daeniker, H.U. *et al*, *Helv. Chim. Acta*, 1957, **40**, 918 (*synth*)
Evans, R.F. *et al*, *J. Chem. Soc.*, 1963, 4023 (*synth*)
Nielsen, A.T., *J. Heterocycl. Chem.*, 1976, **13**, 101 (*deriv, synth, ir, pmr*)
Nelsen, S.F. *et al*, *J. Org. Chem.*, 1986, **51**, 2081 (*deriv, synth, pmr, cmr*)

2,3-Dihydro-6(1*H*)-azulenone, 9CI   D-60193

6-Oxo-1,2,3,6-tetrahydroazulene

[5291-91-8]

$C_{10}H_{10}O$ M 146.188
Needles (Et$_2$O or by subl.). Mp 107-108°.

Chapman, O.L. *et al*, *J. Org. Chem.*, 1966, **31**, 1042 (*synth, uv*)
Kende, A.S. *et al*, *Tetrahedron Lett.*, 1986, **27**, 6051 (*synth, ir, uv, pmr*)

1,5-Dihydro-2,4-benzodithiepin, 9CI   D-60194

[7216-19-5]

$C_9H_{10}S_2$ M 182.298
Cryst. by subl. Mp 155-156°.

Autenreith, W. *et al*, *Ber.*, 1901, **34**, 1772 (*synth*)
Kohrman, R. *et al*, *J. Org. Chem.*, 1971, **36**, 3971 (*synth, ir, uv, pmr, ms*)

3,4-Dihydro-2*H*-1,5-benzodithiepin, 9CI   D-60195

1,5-Benzodithiepan

[78269-38-2]

$C_9H_{10}S_2$ M 182.298
Cryst. (EtOH). Mp 59.5-60.5°.

Kyba, E.P. *et al*, *J. Am. Chem. Soc.*, 1981, **103**, 3868 (*synth, pmr*)
Ménard, D. *et al*, *Can. J. Chem.*, 1986, **64**, 2142 (*synth, pmr, cmr*)

2,2'-(4,8-Dihydrobenzo[1,2-*b*:5,4-*b'*]-dithiophene-4,8-diylidene)-bispropanedinitrile   D-60196

$C_{16}H_4N_4S_2$ M 316.354
Electron acceptor, forms highly conducting charge-transfer compd. with tetrathiafulvalene. Orange-red prisms (MeCN). Mp 325° dec. 2 other isomeric compds. prepd.

Kobayashi, K. *et al*, *J. Chem. Soc., Chem. Commun.*, 1986, 1779 (*synth, ir, pmr, use*)

2,3-Dihydro-2-benzofurancarboxylic acid, 9CI   D-60197

[1914-60-9]

$C_9H_8O_3$ M 164.160
(±)-*form*
 Cryst. (toluene). Mp 116-118°.

Edwards, C.R. *et al*, *J. Heterocycl. Chem.*, 1987, **24**, 495.

3,4-Dihydro-2*H*-1-benzopyran-2-ol, 9CI   D-60198
2-Chromanol. 2-Hydroxychroman

[32560-26-2]

$C_9H_{10}O_2$ M 150.177

(±)-*form*

Mp 28°. Bp$_{0.05}$ 88-89°. n_D^{21} 1.5620.

Ac:

   $C_{11}H_{12}O_3$ M 192.214

   Fragrant oil, solidifying on standing. Mp 36.5°. Bp$_{0.1}$ 88-89°. n_D^{25} 1.5268.

Parham, W.E. *et al, J. Am. Chem. Soc.*, 1962, **84**, 813 (*synth, ir*)

3,4-Dihydro-2*H*-1-benzopyran-3-ol, 9CI   D-60199
3-Chromanol, 8CI. 3-Hydroxychroman

[21834-60-6]

$C_9H_{10}O_2$ M 150.177

(±)-*form*

Needles (C_6H_6/hexane). Mp 80°.

Still, W.C. *et al, J. Org. Chem.*, 1970, **35**, 2282 (*synth, ir, ms*)
Clark-Lewis, J.W. *et al, Aust. J. Chem.*, 1973, **26**, 819 (*synth*)

3,4-Dihydro-2*H*-1-benzopyran-4-ol, 9CI   D-60200
4-Chromanol, 8CI. 4-Hydroxychroman

[1481-93-2]

$C_9H_{10}O_2$ M 150.177

(±)-*form*

Prisms (C_6H_6/hexane). Mp 41°.

Ac:

   $C_{11}H_{12}O_3$ M 192.214

   Bp$_{0.03}$ 81-82°. n_D^{27} 1.5296.

Parham, W.G. *et al, J. Am. Chem. Soc.*, 1962, **84**, 813 (*synth, ir*)
Clark-Lewis, J.W. *et al, Aust. J. Chem.*, 1973, **26**, 819 (*synth, pmr*)

3,4-Dihydro-2*H*-1-benzopyran-5-ol, 9CI   D-60201
5-Chromanol, 8CI. 5-Hydroxychroman

[13849-32-6]

$C_9H_{10}O_2$ M 150.177

Mp 68-70°.

Shiratsuchi, M. *et al, Chem. Pharm. Bull.*, 1987, **35**, 632 (*synth, pmr*)

3,4-Dihydro-2*H*-1-benzopyran-7-ol, 9CI   D-60202
7-Chromanol. 7-Hydroxychroman

[57052-72-9]

$C_9H_{10}O_2$ M 150.177

Cryst. Mp 77-78°. Bp$_{0.1}$ 130°.

Me ether:

   $C_{10}H_{12}O_2$ M 164.204

   Bp$_{0.05}$ 82-84°.

Graffe, B. *et al, J. Heterocycl. Chem.*, 1975, **12**, 247 (*synth, pmr*)

Shiratsuchi, M. *et al, Chem. Pharm. Bull.*, 1987, **35**, 632 (*synth, pmr*)

3,4-Dihydro-2*H*-1-benzopyran-8-ol, 9CI   D-60203
8-Chromanol. 8-Hydroxychroman

[1915-20-4]

$C_9H_{10}O_2$ M 150.177

Oil. Bp$_{1.2}$ 94-97°.

Me ether: 3,4-Dihydro-8-methoxy-2H-1-benzopyran. 8-Methoxychroman.

   $C_{10}H_{12}O_2$ M 164.204

   d_4^{20} 1.55. Bp$_{0.01}$ 76°. n_D^{20} 1.5532.

Willhalm, B. *et al, Tetrahedron*, 1964, **20**, 1185 (*synth, ms, deriv*)
Bramwell, P.S. *et al, J. Chem. Soc.*, 1965, 3882 (*synth*)
Shiratsuchi, M. *et al, Chem. Pharm. Bull.*, 1987, **35**, 632 (*synth, pmr*)

1,4-Dihydro-3*H*-2-benzoselenin-3-one, 9CI   D-60204
3,4-Dihydro-1H-2-benzoselenin-3-one. 3-Isoselenochromanone

[87216-47-5]

C_9H_8OSe M 211.122

Mp 111-114°.

Lemaire, C. *et al, J. Heterocycl. Chem.*, 1983, **20**, 811 (*synth, ms, pmr, cmr*)

1,4-Dihydro-3*H*-2-benzotellurin-3-one, 9CI   D-60205
3,4-Dihydro-1H-2-benzotellurin-3-one. 3-Isotellurochromanone

[87216-48-6]

C_9H_8OTe M 259.762

Mp 112-115°.

Lemaire, C. *et al, J. Heterocycl. Chem.*, 1983, **20**, 811 (*synth, ms, pmr, cmr*)

1,4-Dihydro-3*H*-2-benzothiopyran-3-one, 9CI   D-60206
3,4-Dihydro-1H-2-benzothiin-3-one. 3-Isothiochromanone

[87216-46-4]

C_9H_8OS M 164.222

Mp 105-106°.

Lemaire, C. *et al, J. Heterocycl. Chem.*, 1983, **20**, 811 (*synth, ms, pmr, cmr*)

3,4-Dihydro-2*H*-1,5-benzoxathiepin-3-one   D-60207

$C_9H_8O_2S$ M 180.221

Oil, cryst. on long standing. Mp 28-31°.

Sugihara, H. *et al, Chem. Pharm. Bull.*, 1987, **35**, 1919 (*synth, pmr*)

2,3-Dihydro-2,3-bis(methylene)furan   D-60208

[73567-98-3]

C_6H_6O M 94.113

Obt. in Ar matrix.

Münzel, N. *et al, Angew. Chem., Int. Ed. Engl.*, 1987, **26**, 471.

3,4-Dihydro-β-carboline   D-60209

Updated Entry replacing D-40161

*4,9-Dihydro-3*H-*pyrido[3,4-b]indole, 9CI. Norharmalan*

[4894-26-2]

$C_{11}H_{10}N_2$ M 170.213

Needles (Et₂O). Mp 175-176°. Forms Et₂O and C₆H₆ solvates with lower Mps than the pure cpd.

B,HClO₄: Cryst. (EtOH). Mp 216-217°.
Picrate: Cryst. (Et₂O). Mp 232-233°.

Szantay, C. *et al, Periodica Politechn.*, 1965, **9**, 231; *CA*, **65**, 3848.

Fleming, I. *et al, J. Chem. Soc. (C)*, 1966, 425 (*synth, uv*)
Kanaoka, Y. *et al, Chem. Pharm. Bull.*, 1967, **15**, 101 (*synth, uv*)
Whittaker, N., *J. Chem. Soc. (C)*, 1969, 85 (*synth, bibl*)

8-Dihydrocinnamoyl-5,7-dihydroxy-4-phenyl-2*H*-benzopyran-2-one   D-60210

$C_{24}H_{20}O_5$ M 388.419

Constit. of *Pityrogramma calomelanos*. Prisms (MeOH). Mp 65.5-66°, Mp 104° (double Mp).

Donnelly, D.M.X. *et al, Phytochemistry*, 1987, **26**, 1143 (*isol, cryst struct*)

Dihydroconfertin   D-60211

Updated Entry replacing D-40162

*Decahydro-5-hydroxy-4*a,*8-dimethyl-3-methyleneazuleno[6,5-b]furan-2(3*H)-*one, 9CI*

[68832-40-6]

$C_{15}H_{22}O_3$ M 250.337

Constit. of *Inula* spp. Cryst. Mp 147°. $[\alpha]_D^{24}$ +83.3° (c, 0.6 in CHCl₃).

4-Ketone: [19908-69-1]. **Confertin**. *Anhydrocumanin.*
$C_{15}H_{20}O_4$ M 264.321
Constit. of *Ambrosia* spp. Cryst. (Me₂CO/hexane). Mp 145-146°.

2α-Hydroxy, 11α,13-dihydro, 4-ketone: **2α-Hydroxy-11α,13-dihydroconfertin**.
$C_{15}H_{22}O_5$ M 282.336
Constit. of *Stevia isomeca*. Oil.

2α-Acetoxy, 11α,13-dihydro, 4-ketone: **2α-Acetoxy-11α,13-dihydroconfertin**.
$C_{17}H_{24}O_6$ M 324.373
From *S. isomeca*. Cryst. Mp 158°. $[\alpha]_D$ +141° (c, 1.3 in CHCl₃).

2,3-Didehydro, 11α-13-dihydro, 4-ketone: **2,3-Dehydro-11α,13-dihydroconfertin**.
$C_{15}H_{20}O_4$ M 264.321
From *S. isomeca*. Cryst. Mp 167°.

4-Ketone, 11,12-dihydro, 11β-hydroxy, 8-epimer: **11β-Hydroxy-11,13-dihydro-8-epi-confertin**.
$C_{15}H_{22}O_4$ M 266.336
Constit. of *Bedfordia arborescens*. Gum.

4-Ketone, 11β,13-Dihydro: **11β,13-Dihydroconfertin**.
$C_{15}H_{22}O_3$ M 250.337
Constit. of *Dittrichia graveolens*. Oil. $[\alpha]_D^{24}$ +17° (c, 0.26 in CHCl₃).

4-Ketone, 11β,13-Dihydro, 8-epimer: **11β,13-Dihydro-8-epi-confertin**.
$C_{15}H_{22}O_3$ M 250.337
From *D. graveolens*. Oil. $[\alpha]_D^{24}$ +62° (c, 0.4 in CHCl₃).

Romo, J. *et al, Can. J. Chem.*, 1968, **46**, 1535 (*isol*)
Marshall, J.A. *et al, J. Am. Chem. Soc.*, 1976, **98**, 4312 (*synth*)
Bohlmann, F. *et al, Phytochemistry*, 1978, **17**, 1165 (*isol, struct*)
Wender, P.A. *et al, J. Am. Chem. Soc.*, 1979, **101**, 2196 (*synth*)
Heathcock, C.H. *et al, J. Am. Chem. Soc.*, 1982, **104**, 1907 (*synth*)
Schultz, A.G. *et al, J. Am. Chem. Soc.*, 1982, **104**, 5800 (*synth*)
Ziegler, F.E. *et al, J. Am. Chem. Soc.*, 1982, **104**, 7174 (*synth*)
Bohlmann, F. *et al, Phytochemistry*, 1985, **24**, 1017 (*derivs*)
Zdero, C. *et al, Phytochemistry*, 1987, **26**, 1207 (*deriv*)
Rustaiyan, A. *et al, Phytochemistry*, 1987, **26**, 2603 (*derivs*)

4,5-Dihydrocyclobuta[*b*]furan   D-60212

[6681-01-2]

C_6H_6O M 94.113

Obt. in Ar matrix.

Münzel, N. *et al, Angew. Chem., Int. Ed. Engl.*, 1987, **26**, 471.

1,2-Dihydrocyclobuta[a]naphthalene-3,4-dione, 9CI   D-60213

[111170-38-8]

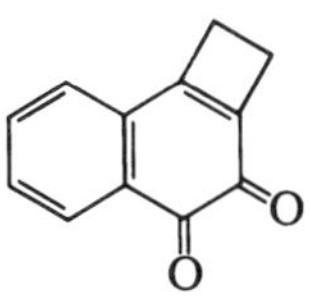

C₁₂H₈O₂   M 184.194
Mp 154°.

Ghera, E. *et al*, *Tetrahedron Lett.*, 1987, **28**, 709 (*synth, uv, ir, pmr*)

1,2-Dihydrocyclobuta[b]naphthalene-3,8-dione   D-60214

[26511-45-5]

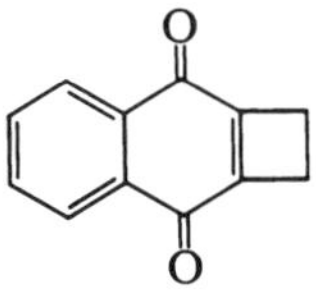

C₁₂H₈O₂   M 184.194
Yellow needles. Mp 195-200°, Mp 259-264° (double Mp).

Naito, T. *et al*, *Chem. Pharm. Bull.*, 1986, **34**, 1505 (*synth, uv, ir, pmr*)

1,2-Dihydrocyclobuta[a]naphthalen-3-ol   D-60215

3-Hydroxynaphtho[a]*cyclobutene*
[111170-37-7]

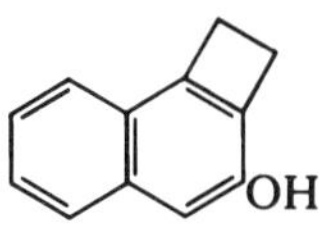

C₁₂H₁₀O   M 170.210
Mp 103-104°.

Ghera, E. *et al*, *Tetrahedron Lett.*, 1987, **28**, 709 (*synth, pmr*)

Dihydrocyclopenta[c,d]pentalene(2−)   D-60216

Dihydroacepentalenediide(2−)
[104984-21-6]

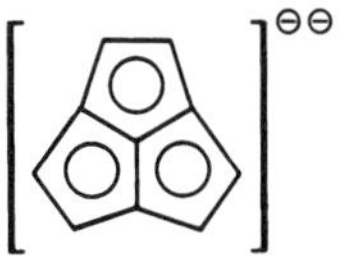

C₁₀H₄⊖⊖   M 124.142 (ion)
Di-K salt:
   C₁₀H₄K₂   M 202.338
   Reddish-brown powder. Obt. mixed with KH.

Lendvai, T. *et al*, *Angew. Chem., Int. Ed. Engl.*, 1986, **25**, 719 (*synth, pmr, cmr*)

10,11-Dihydro-5H-dibenzo[a,d]cycloheptene   D-60217

2,3:6,7-Dibenzosuberane
[833-48-7]

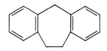

C₁₅H₁₄   M 194.276
Pale-yellow cryst. (EtOH aq.). Mp 73-74.5°.

Mychajlyszyn, V. *et al*, *Collect. Czech. Chem. Commun.*, 1959, **24**, 3955 (*synth*)
Moritani, I. *et al*, *Bull. Chem. Soc. Jpn.*, 1967, **40**, 1506 (*synth*)
Bordwell, F.G. *et al*, *J. Am. Chem. Soc.*, 1986, **108**, 7310 (*synth*)

10,11-Dihydro-5H-dibenzo[a,d]cyclohepten-5-one   D-60218

[1210-35-1]

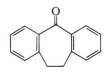

C₁₅H₁₂O   M 208.259
Oil. Bp₀.₀₅ 135-140°.
Hydrazone: [61047-38-9].
   C₁₅H₁₄N₂   M 222.289
   Cryst. (EtOH aq.). Mp 78-80°.

Campbell, T.W. *et al*, *Helv. Chim. Acta*, 1953, **49**, 1489 (*synth*)
Falshaw, C.P. *et al*, *J. Chem. Soc., Perkin Trans. 1*, 1985, 1837 (*deriv*)

9,10-Dihydrodicyclopenta[c,g]phenanthrene, 9CI   D-60219

1,12-Dihydrodicyclopenta[c,g]*phenanthrene*
[71634-96-3]

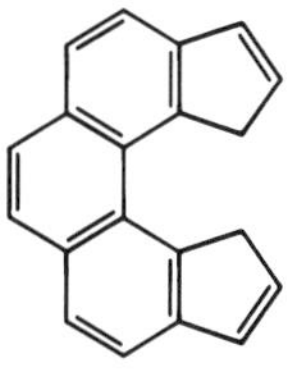

C₂₀H₁₄   M 254.331
Forms Fe complex. Mp 151-152°.

Katz, T.J. *et al*, *J. Am. Chem. Soc.*, 1979, **101**, 4259 (*synth, pmr, cmr, ir, ms*)

1,10-Dihydrodicyclopenta[a,h]naphthalene   D-60220

[71628-20-1]

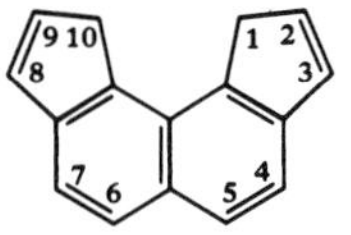

C₁₆H₁₂   M 204.271
Forms Fe and Co complex. Long, frail needles (CS₂/Et₂O at −78°). Mp 175-177°.

Katz, T.J. *et al*, *J. Am. Chem. Soc.*, 1979, **101**, 4259 (*synth, ir, pmr, cmr, ms*)

3,8-Dihydrodicyclopenta[*a,h*]naphthalene　　D-60221

[71628-26-7]

C$_{16}$H$_{12}$　　M 204.271
Mp 106-109°.

Katz, T.J. *et al, J. Am. Chem. Soc.*, 1979, **101**, 4259 (*synth, pmr, ir, ms*)

3,4-Dihydro-3,4-dihydroxy-2-(hydroxy-methyl)-2*H*-pyrrole　　D-60222

[108692-47-3]

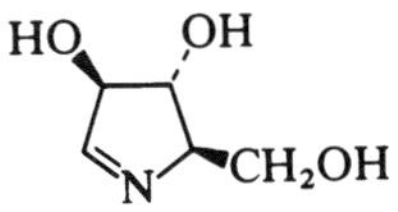

C$_5$H$_9$NO$_3$　　M 131.131
Prod. by *Nectria lucida*. Hypoglycemic and immunostimulant.

Japan. Pat., 87 36 355, (*1987*); *CA*, **107**, 5742 (*isol, props*)

3,8-Dihydro-6,6′;7,3′*a*-diliguetilide　　D-60223

[106533-38-4]

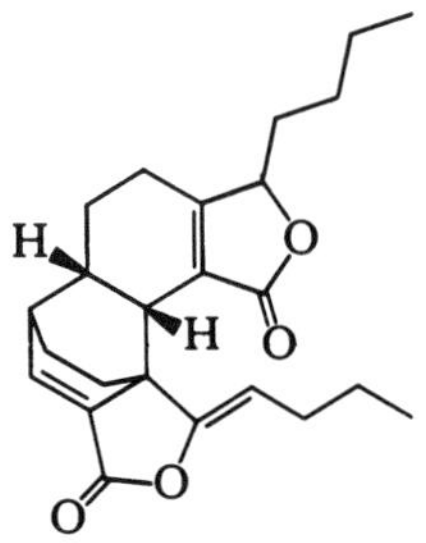

C$_{24}$H$_{30}$O$_4$　　M 382.499
Constit. of *Ligusticum wallichii*. Oil.

Kaouadji, M. *et al, J. Nat. Prod.*, 1986, **49**, 872.

9,10-Dihydro-9,9-dimethylanthracene　　D-60224

[42332-94-5]

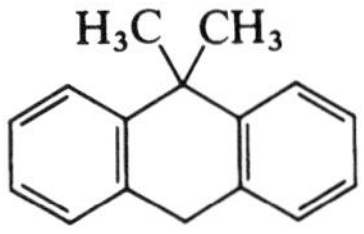

C$_{16}$H$_{16}$　　M 208.302
Cryst. (pet. ether). Mp 84-86°.

Davis, M.A. *et al, J. Med. Chem.*, 1964, **7**, 88 (*synth, uv*)
Falshaw, C.P. *et al, J. Chem. Soc., Perkin Trans. 1*, 1985, 1837 (*synth, pmr, ms*)

1,6-Dihydro-1,6-dimethyleneazulene　　D-60225

1,6-Dihydro-1,6-bis(methylene)azulene. 1,6-Azulylene

[104007-33-2]

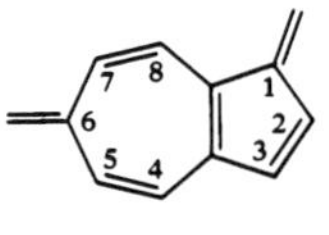

C$_{12}$H$_{10}$　　M 154.211
Generated by flash vapour pyrol. of [2.2](1,6)-Azulenophane, A-50394 . Reactive polyene.

Rudolf, K. *et al, J. Org. Chem.*, 1987, **52**, 641 (*synth, uv*)

2,6-Dihydro-2,6-dimethyleneazulene　　D-60226

2,6-Dihydro-2,6-bis(methylene)azulene. 2,6-Azulylene

[103322-24-3]

C$_{12}$H$_{10}$　　M 154.211
Generated by flash vacuum pyrol. of [2.2](2,6)-Azulenophane, A-60355 . Reactive polyene.

Koenig, T. *et al, J. Am. Chem. Soc.*, 1986, **108**, 5024 (*synth, uv, pmr*)

2,3-Dihydro-2,2-dimethyl-1*H*-inden-1-one, 9CI　　D-60227

2,2-Dimethyl-1-indanone

[10489-28-8]

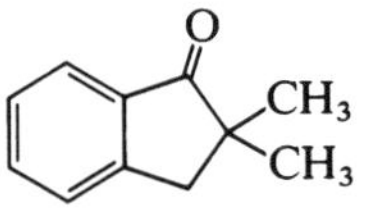

C$_{11}$H$_{12}$O　　M 160.215
Liq.

Orliac-LeMoing, A. *et al, Tetrahedron*, 1985, **41**, 4483 (*synth, ir, pmr*)

3,4-Dihydro-2,2-dimethyl-1(2*H*)-naphthalenone, 9CI　　D-60228

2,2-Dimethyltetralone

[2977-45-9]

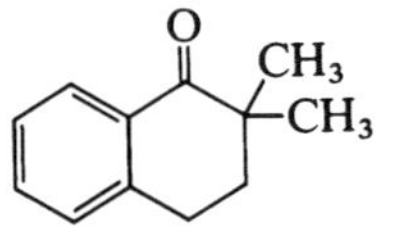

C$_{12}$H$_{14}$O　　M 174.242
Liq. Bp$_{12}$ 125-135°.

Orliac-Lemoing, A. *et al, Tetrahedron*, 1985, **41**, 4483 (*synth, ir, pmr*)
Lissel, M. *et al, Justus Liebigs Ann. Chem.*, 1987, 263 (*synth, ir, pmr*)

3,4-Dihydro-2,2-dimethyl-2*H*-pyrrole, 9CI　　D-60229

5,5-Dimethyl-1-pyrroline, 8CI

[2045-76-3]

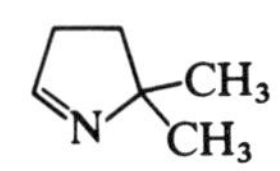

C$_6$H$_{11}$N　　M 97.160
Liq. Bp 104-108°.
N-*Oxide:* [3317-61-1].
　　C$_6$H$_{11}$NO　　M 113.159
　　Spintrap, scavenger of alkyl, hydroxyalkyl and alkoxy radicals. Hygroscopic solid. Bp$_{0.6}$ 66°, Bp$_{0.1}$ 53°.
B,HCl: [68373-50-2]. Hygroscopic solid.
Picrate: Yellow prisms (EtOH/Et$_2$O). Mp 120-124°.

Bonnett, R. *et al, J. Chem. Soc.*, 1959, 2087 (*synth, ir*)
Janzen, E.G. *et al, J. Magn. Reson.*, 1973, **9**, 510 (*deriv, use*)
Haire, D.L. *et al, J. Org. Chem.*, 1986, **51**, 4298 (*deriv, synth, pmr, bibl*)

1,2-Dihydro-2,2-dimethyl-3*H*-pyrrol-3-one, 9CI   D-60230

[106556-70-1]

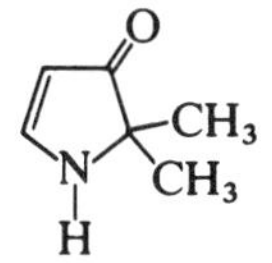

C₆H₉NO   M 111.143
Mp 125-127°.

Patens, J. *et al, Helv. Chim. Acta*, 1986, **69**, 905 (*synth, uv, pmr, cmr*)

3,4-Dihydro-3,3-dimethyl-2(1*H*)-quinolinone, 9CI   D-60231

[92367-59-4]

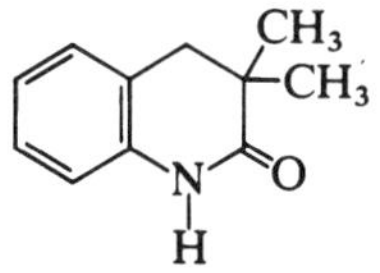

C₁₁H₁₃NO   M 175.230
Cryst. (THF/hexane). Mp 155-156°.

Robertson, D.W. *et al, J. Med. Chem.*, 1986, **29**, 1832 (*synth*)

2,3-Dihydro-2,2-dimethylthiophene   D-60232

[71690-19-2]

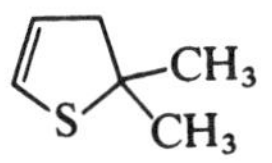

C₆H₁₀S   M 114.205
Oil.

Baldwin, J.E. *et al, J. Chem. Soc., Chem. Commun.*, 1979, 249 (*synth, pmr*)

2,5-Dihydro-2,5-dimethylthiophene   D-60233

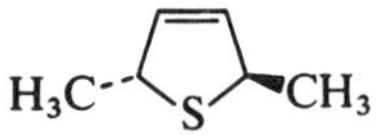

(2*RS*,5*RS*)-form

C₆H₁₀S   M 114.205
cis:trans ratio dep. on reaction condits.

1,1-Dioxide:
  C₆H₁₀O₂S   M 146.204
  Oil. No stereochem. given.

Trost, B.M. *et al, J. Am. Chem. Soc.*, 1971, **93**, 3825 (*synth, pmr*)
McIntosh, J.M. *et al, J. Org. Chem.*, 1975, **40**, 1294 (*synth*)
Chou, T.-S. *et al, J. Chem. Soc., Perkin Trans. 1*, 1986, 1039 (*deriv, synth, ir, pmr*)

3,5-Dihydro-4*H*-dinaphth[2,1-*c*:1′,2′-*e*]-azepine   D-60234

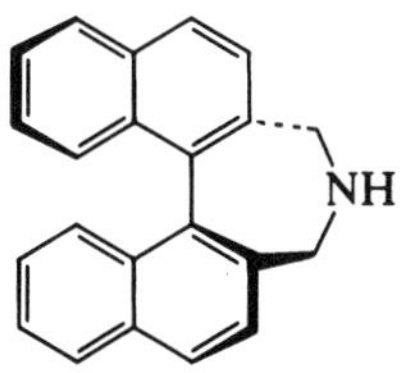

(*S*)-form

C₂₂H₁₇N   M 295.383

(*S*)-form [97551-09-2]
  Reagent for asymmetric synth. Solidified foam. Mp 73-84°. [α]$_D^{20}$ +620° (c, 0.78 in CHCl₃) (opt. pure).
(±)-*form* [102518-95-6]
  Foam. Mp 147-149°.

Maigrot, N. *et al, J. Org. Chem.*, 1985, **50**, 3916.
Hawkins, J.M. *et al, J. Org. Chem.*, 1986, **51**, 2820 (*synth, ir, pmr, cmr, resoln, use*)

1,2-Dihydro-4,6-diphenylpyrimidine, 9CI   D-60235

4,6-Diphenyl-1,2-dihydropyrimidine

[46898-00-4]

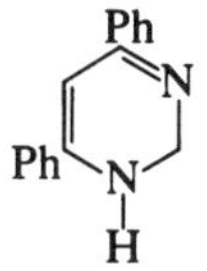

C₁₆H₁₄N₂   M 234.300
Yellow cryst. (hexane). Mp 110-111°. 2,5-Dihydro tautomer obs. in CHCl₃ soln.

Weis, A.L. *et al, J. Org. Chem.*, 1986, **51**, 4623 (*synth, uv*)

2,5-Dihydro-2,2-diphenyl-1,3,4-thiadiazole, 9CI   D-60236

2,2-Diphenyl-1,3,4-thiadiazoline

[79999-60-3]

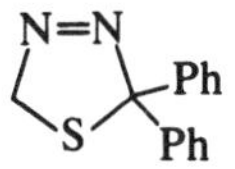

C₁₄H₁₂N₂S   M 240.322
Source of thiobenzophenone *S*-methylide. Cryst. at low temp. Deflagrates at −20°.

Kalwinsch, I. *et al, J. Am. Chem. Soc.*, 1981, **103**, 7032 (*synth*)
Huisgen, R. *et al, Tetrahedron Lett.*, 1986, **27**, 5475 (*use*)

2,5-Dihydro-1*H*-dipyrido[4,3-*b*:3′,4′-*d*]-pyrrol-1-one, 9CI   D-60237

*2,5-Dihydro-1*H*-pyrrolo[3,2-c:4,5-c′]dipyridin-1-one*

[106689-46-7]

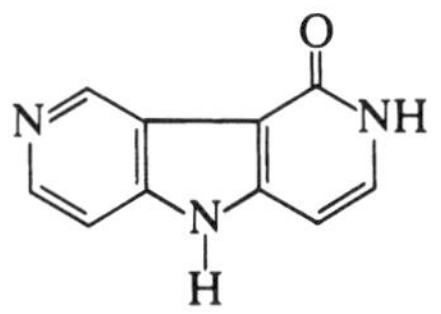

C₁₀H₇N₃O   M 185.185
Yellowish cryst. (DMF). Mp >280°.

Nguyen Chi Hung, *et al, Tetrahedron*, 1986, **42**, 2303 (*synth, pmr*)

2-(5,6-Dihydro-1,3-dithiolo[4,5-*b*][1,4]-dithiin-2-ylidene)-5,6-dihydro-1,3-dithiolo[4,5-*b*][1,4]dithiin, 9CI   D-60238

Bis(ethylenedithio)tetrathiafulvalene

[66946-48-3]

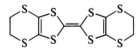

C₁₀H₈S₈   M 384.653

Forms salts with long linear anions which are superconductive "organic metals". Mp >250°.

Mizuno, M. *et al, J. Chem. Soc., Chem. Commun.*, 1978, 18 (*synth*)
Kobayashi, H. *et al, Mol. Cryst. Liq. Cryst.*, 1984, **107**, 33 (*deriv, cryst struct, props*)
Leung, P.C.W. *et al, Mol. Cryst. Liq. Cryst.*, 1985, **125**, 113 (*cryst struct, props, bibl*)
Williams, J.M. *et al, Acc. Chem. Res.*, 1985, **18**, 261 (*props*)
Emge, T.J. *et al, J. Am. Chem. Soc.*, 1986, **108**, 695 (*use, bibl*)

4,5-Dihydro-1,5-ethano-1*H*-1-benzazepin-2(3*H*)-one, 9CI   D-60239

2,3,4,5-Tetrahydro-2-oxo-1,5-ethanobenzazepine
[102586-88-9]

$C_{12}H_{13}NO$ M 187.241
Oil.

Somayaji, V. *et al, J. Org. Chem.*, 1986, **51**, 2676 (*synth, ir, pmr*)

4,7-Dihydro-4,7-ethanoisobenzofuran, 9CI   D-60240

[107826-05-1]

$C_{10}H_{10}O$ M 146.188
Diels-Alder diene. Solid. Mp 25°.

Stephan, D. *et al, Tetrahedron Lett.*, 1986, **27**, 4295 (*synth, pmr, use*)

1,4-Dihydrofuro[3,4-*d*]pyridazine, 9CI   D-60241

[105064-77-5]

$C_6H_6N_2O$ M 122.126
Source of the 3,4-dimethylenefuran biradical.

Stone, K.J. *et al, J. Am. Chem. Soc.*, 1986, **108**, 8088 (*synth, pmr, cmr, uv, use*)

Dihydro-4-hydroxy-2(3*H*)-furanone, 9CI   D-60242

3-Hydroxytetrahydrofuranone. 3-Hydroxy-4-butanolide. 3,4-Dihydroxybutanoic acid γ-lactone
[5469-16-9]

(*R*)-*form*
Absolute
configuration

$C_4H_6O_3$ M 102.090
(*R*)-*form* [58081-05-3]
Acid hydrol. prod. of Oscillatoxin *A*. Viscous oil. Bp$_{0.4}$ 103-105°, Bp$_{0.1}$ 90-95°. [α]$_D$ +94° (c, 1.5 in EtOH).

(*S*)-*form* [7331-52-4]
Metab. of alga *Lyngbya majuscula*. Bp$_{0.1}$ 120°. [α]$_D$ −80° (c, 0.5 in CHCl$_3$).

Mori, K. *et al, Tetrahedron*, 1979, **35**, 933 (*synth, ir, pmr*)
Moore, R.E. *et al, J. Org. Chem.*, 1984, **49**, 2484 (*synth, isol*)
Saito, S. *et al, Chem. Lett.*, 1984, 1389 (*synth*)
Ainslie, R.D. *et al, Phytochemistry*, 1986, **25**, 2654 (*isol, pmr*)
Henrot, S. *et al, Synth. Commun.*, 1986, **16**, 183 (*synth*)
Seebach, D. *et al, Synthesis*, 1986, 37 (*synth*)

2,3-Dihydro-2-hydroxy-1*H*-imidazole-4,5-dione   D-60243

2-Hydroxy-4,5-imidazolidinedione

$C_3H_4N_2O_3$ M 116.076
(±)-*form*
1,3-Di-Ph: [104716-68-9].
$C_{15}H_{12}N_2O_3$ M 268.271
Cryst. (MeOH aq.). Mp 178-185° dec.
1,3-Di-Ph, Me ether: [104716-69-0].
$C_{16}H_{14}N_2O_3$ M 282.298
Cryst. (toluene). Mp 150-151°.
1,3-Di-Ph, O-Ac: [104716-70-3].
$C_{17}H_{14}N_2O_4$ M 310.309
Cryst. (toluene). Mp 176-179°.

Barsa, E.A. *et al, J. Org. Chem.*, 1986, **51**, 4483 (*synth, pmr, cmr*)

2,3-Dihydro-2-(4-hydroxyphenyl)-5-(3-hydroxypropyl)-3-methylbenzofuran   D-60244

2,3-Dihydro-2-(4-hydroxyphenyl)-3-methyl-5-benzofuranpropanol, 9CI

Absolute
configuration

$C_{18}H_{20}O_3$ M 284.354
(*2R,3R*)-*form* [109194-67-4]
Constit. of *Krameria cystisoides*. Cryst. Mp 123-125°. [α]$_D^{21}$ +4.1° (c, 0.56 in MeOH).
3'-Methoxy: [109225-42-5]. *2,3-Dihydro-2-(4-hydroxy-3-methoxyphenyl)-5-(3-hydroxypropyl)-3-methylbenzofuran.*
$C_{19}H_{22}O_4$ M 314.380
Constit. of *K. cystisoides*. Oil. [α]$_D^{21}$ +37° (c, 0.76 in MeOH).

Achenbach, H. *et al, Phytochemistry*, 1987, **26**, 1159.

4,5-Dihydro-1*H*-imidazole, 9CI   D-60245

2-Imidazoline, 8CI
[504-75-6]

$C_3H_6N_2$ M 70.094
Waxy cryst. Mp 53-57°. Bp$_{14}$ 96-98°, Bp$_1$ 50°.
Picrate: Mp 202°.

Jentzsch, W. *et al, Chem. Ber.*, 1965, **98**, 1342 (*synth*)
Edge, P.J. *et al, Chem. Ind. (London)*, 1969, 203 (*synth*)
Ito, Y. *et al, J. Am. Chem. Soc.*, 1973, **95**, 4447 (*synth*)

Suzuki, H. *et al, Bull. Chem. Soc. Jpn.*, 1975, **48**, 1922 (*synth*)
Hitchcock, P.B. *et al, J. Chem. Soc., Dalton Trans.*, 1977, 2160 (*synth*)

1-(4,5-Dihydro-1*H*-imidazol-2-yl)-2-imida-   D-60246
zolidinethione, 9CI

Jaffe's base

[484-92-4]

$C_6H_{10}N_4S$ M 170.232
Cryst. (EtOH). Mp 235°.
B,2HCl: [54255-13-9]. Cryst. (MeOH). Mp 270°.
B,HI: [38631-03-7]. Cryst. (H_2O). Dec. at 296-299°.

Wittekind, R.R. *et al, J. Org. Chem.*, 1973, **38**, 1641 (*struct, bibl*)
Trani, A. *et al, J. Heterocycl. Chem.*, 1974, **11**, 257 (*synth*)
Marshall, W.D., *J. Agric. Food Chem.*, 1979, **27**, 295 (*synth*)
Herbstein, F.H. *et al, J. Am. Chem. Soc.*, 1986, **106**, 2367 (*struct*)

1,3-Dihydro-2*H*-imidazo[4,5-*b*]pyrazine-2-   D-60247
thione, 9CI

[79100-21-3]

$C_5H_4N_4S$ M 152.173
Various tautomers possible. 1*H*, 3*H*-form illus. Cryst. (H_2O). Mp 326.5-329°.

(1H,3H)-form

1-Me: [84996-46-3]. *1,3-Dihydro-1-methyl-2H-imidazo[4,5-b]pyrazine-2-thione, 9CI.*
$C_6H_6N_4S$ M 166.200
Cryst. Mp 251-253°.
1,3-Di-Me: [84996-50-9]. *1,3-Dihydro-1,3-dimethyl-2H-imidazo[4,5-b]pyrazine-2-thione, 9CI.*
$C_7H_8N_4S$ M 180.227
Cryst. (pet. ether). Mp 163-165°.

1H,5H-form

S-Me: [84996-48-5]. *2-(Methylthio)-1H-imidazo[4,5-b]pyrazine, 9CI.*
$C_6H_6N_4S$ M 166.200
Cryst. (H_2O). Mp 257-258°.
1,S-Di-Me: [84996-47-4]. *1-Methyl-2-methylthio-1H-imidazo[4,5-b]pyrazine.*
$C_7H_8N_4S$ M 180.227
Cryst. (pet. ether). Mp 122-123°.

(4H,5H)-form

4,S-Di-Me: [84996-49-6]. *4-Methyl-2-methylthio-4H-imidazo[4,5-b]pyrazine.*
$C_7H_8N_4S$ M 180.227
Cryst. (C_6H_6/pet. ether). Mp 164-165°.

Tong, Y.C., *J. Heterocycl. Chem.*, 1981, **18**, 751 (*synth, tautom*)
Barlin, G.B., *Aust. J. Chem.*, 1982, **35**, 2299 (*synth, deriv, pmr*)

1,5-Dihydro-6*H*-imidazo[4,5-*c*]pyridazin-   D-60248
6-one, 9CI

6-Hydroxyimidazo[4,5-c]pyridazine

[79690-94-1]

$C_5H_4N_4O$ M 136.113
Cryst. (H_2O). Mp >320°.

Barlin, G.B., *Aust. J. Chem.*, 1981, **34**, 1361 (*synth*)

1,3-Dihydro-2*H*-imidazo[4,5-*b*]pyridin-2-   D-60249
one, 9CI

2-Oxo-1H,3H-imidazo[4,5-b]pyridine. 2-Hydroxyimidazo[b]pyridine. 2-Hydroxy-4-azabenzimidazole

[16328-62-4]

$C_6H_5N_3O$ M 135.125
Cryst. Mp 270-272° (238-239°, 265-266°).

Petrow, V. *et al, J. Chem. Soc.*, 1948, 1389 (*synth*)
Vaughan, J.R. *et al, J. Am. Chem. Soc.*, 1949, **71**, 1885 (*synth*)
Dornow, A. *et al, Arch. Pharm. (Weinheim, Ger.)*, 1957, **290**, 20 (*synth*)
Harrison, D. *et al, J. Chem. Soc.*, 1959, 3157 (*synth*)

1,3-Dihydro-2*H*-imidazo[4,5-*c*]pyridin-2-   D-60250
one, 9CI

2-Hydroxy-1,3,5-triazaindene

[7397-68-4]

$C_6H_5N_3O$ M 135.125
Cryst. (H_2O). Mp 304-305° (315° dec.).
1-Benzyl: [61719-58-2].
   $C_{13}H_{11}N_3O$ M 225.249
   Cryst. Mp 214°.
1-Me: [40423-52-7].
   $C_7H_7N_3O$ M 149.152
   Prisms (dioxan). Mp 272°.

Barlin, G.B., *J. Chem. Soc. (B)*, 1966, 285 (*synth*)
Yutilov, Y.U., *Chem. Heterocycl. Compd. (Engl. Transl.)*, 1973, 129 (*deriv*)
Debeljak-Sustar, M. *et al, J. Org. Chem.*, 1978, **43**, 393 (*synth, pmr*)
Frankowski, A., *Tetrahedron*, 1986, **42**, 1511 (*deriv, synth, ir, pmr*)

2,3-Dihydro-5(1H)-indolizinone, 9CI D-60251

[101773-62-0]

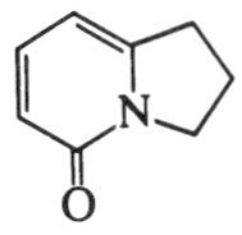

C_8H_9NO M 135.165

Low melting, hygroscopic solid. $Bp_{0.1}$ 79-90°.

Thomas, E.W. *et al, J. Org. Chem.,* 1986, **51**, 2184 (*synth, pmr, cmr, ir, ms*)

Dihydro-3-iodo-2(3H)-furanone, 9CI D-60252

[31167-92-7]

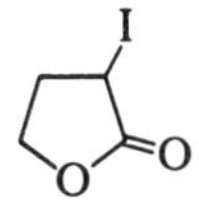

$C_4H_5IO_2$ M 211.987

(±)-*form*

Liq.

Evans, R.D. *et al, Synthesis,* 1986, 727 (*synth, ir, pmr*)

4,7-Dihydroisobenzofuran, 9CI D-60253

[56582-03-7]

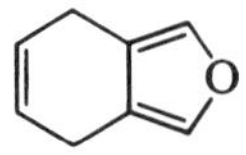

C_8H_8O M 120.151

Diels-Alder diene.

Roth, W.R. *et al, Chem. Ber.,* 1975, **108**, 1655 (*synth, ir, pmr, ms*)

Stephan, D. *et al, Tetrahedron Lett.,* 1986, **27**, 4295 (*synth, use*)

7,8-Dihydro-5(6H)-isoquinolinone, 9CI D-60254

5-Oxo-5,6,7,8-tetrahydroisoquinoline

[21917-86-2]

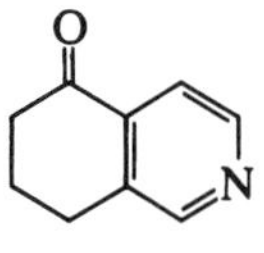

C_9H_9NO M 147.176

Cryst. (pet. ether). Mp 42-44°. Bp_4 113-115°.

B,HCl: Cryst. Mp 235-236°.

Sugimoto, N. *et al, J. Pharm. Soc. Jpn.,* 1956, **76**, 1308; *CA,* **51**, 5076e (*synth*)

Epsztajn, J. *et al, J. Chem. Soc., Perkin Trans. 1,* 1985, 213 (*synth, ir, pmr*)

7,12-Dihydro-5H-6,12-methanodibenz[c,f]azocine, 9CI D-60255

[67280-33-5]

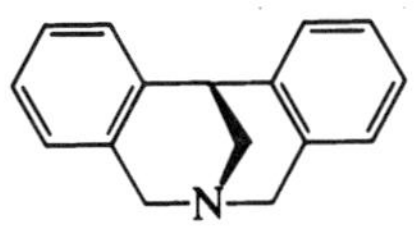

$C_{16}H_{15}N$ M 221.301

Cryst. (Me₂CO). Mp 133-134°.

Suzuki, T. *et al, Chem. Pharm. Bull.,* 1986, **34**, 1888 (*synth, pmr*)

1,4-Dihydro-1,4-methanonaphthalene, 9CI D-60256

2,3-Benzobicyclo[2.2.1]hepta-2,5-diene.
Benzonorbornadiene

[4453-90-1]

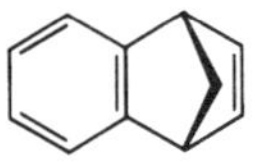

$C_{11}H_{10}$ M 142.200

Oil. Bp_{12} 82.5-83°, $Bp_{0.5}$ 55-56°.

Wittig, G. *et al, Chem. Ber.,* 1958, **91**, 895 (*synth, ir*)

Wilt, J.W. *et al, J. Org. Chem.,* 1967, **32**, 893 (*synth, uv*)

Mich, T.F. *et al, J. Chem. Educ.,* 1968, **45**, 272 (*synth, pmr*)

Howell, B.A. *et al, J. Magn. Reson.,* 1975, **20**, 141 (*cmr*)

3,9-Dihydro-9-methyl-1H-purine-2,6-dione, 9CI D-60257

Updated Entry replacing D-03929

9-Methylxanthine

[1198-33-0]

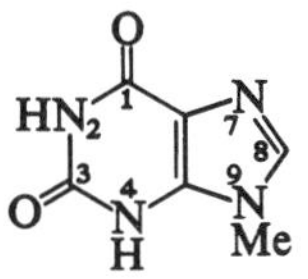

$C_6H_6N_4O_2$ M 166.139

Needles (H₂O). Mp 384° dec.

Koppel, H.C. *et al, J. Am. Chem. Soc.,* 1958, **80**, 2751 (*synth*)

Pfleiderer, W. *et al, Justus Liebigs Ann. Chem.,* 1960, **631**, 168 (*synth*)

11-Dihydro-12-norneoquassin D-60258

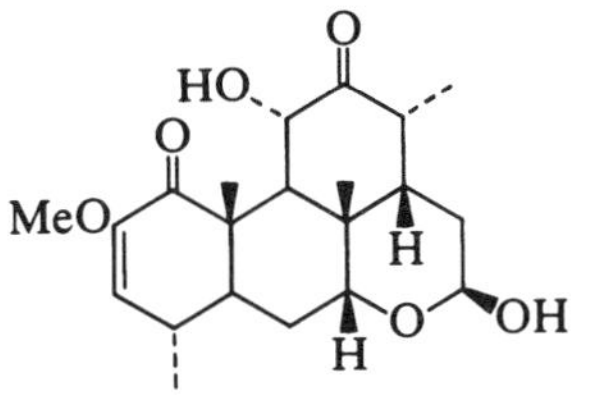

$C_{21}H_{30}O_6$ M 378.464

Constit. of *Quassia amara.* Prisms (MeOH). Mp 227-229°. $[\alpha]_D^{20}$ +21° (c, 0.19 in MeOH).

Grandolini, G. *et al, Phytochemistry,* 1987, **26**, 3085.

2,3-Dihydro-2-oxo-1H-indole-3-acetic acid, 9CI D-60259

2-(2-Oxo-3-indolinyl)acetic acid. Oxindole-3-acetic acid

[2971-31-5]

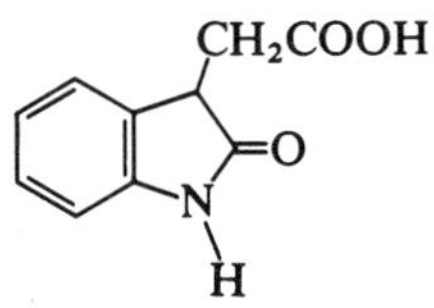

$C_{10}H_9NO_3$ M 191.186

(±)-*form*

Needles (Me₂CO/C₆H₆). Mp 146°.

Me ester:
   $C_{11}H_{11}NO_3$ M 205.213
   Needles (MeOH). Mp 164-167°.

Et ester: [40940-16-7].
   $C_{12}H_{13}NO_3$ M 219.240

Cryst. (pet. ether). Mp 94-98°.

Julian, P.L. *et al*, *J. Am. Chem. Soc.*, 1953, **75**, 5305 (*synth*)
Takase, S. *et al*, *Tetrahedron*, 1986, **42**, 5879 (*synth, ir, pmr*)

9,10-Dihydrophenanthrene, 9CI   D-60260

Updated Entry replacing D-03994
Phenanthrane
[776-35-2]
$C_{14}H_{12}$ M 180.249
Mp 34.5-35°. Bp$_{25}$ 183-184°.

Org. Synth., Coll. Vol., **4**, 313 (*synth*)
Maquestiau, A., *Org. Mass Spectrom.*, 1971, **5**, 1015 (*ms*)
Bowden, B.F. *et al*, *Aust. J. Chem.*, 1975, **28**, 65 (*synth*)
Shapiro, B.L., *J. Phys. Chem. Ref. Data*, 1977, **6**, 919 (*nmr*)
Cosmo, R. *et al*, *Aust. J. Chem.*, 1987, **40**, 35 (*synth, pmr*)

2,3-Dihydro-2-phenyl-1,2-benzisothiazole, 9CI   D-60261

[88841-60-5]

$C_{13}H_{11}NS$ M 213.297
Liq. Partial dec. on distillation.

Kanakarajan, K. *et al*, *Angew. Chem., Int. Ed. Engl.*, 1984, **23**, 244 (*synth*)

2,3-Dihydro-2-phenyl-2-benzofurancarboxylic acid   D-60262

[111080-51-4]

$C_{15}H_{12}O_3$ M 240.258
(±)-*form*
Mp 131-133°.

Edwards, C.R. *et al*, *J. Heterocycl. Chem.*, 1987, **24**, 495.

2,3-Dihydro-3-(phenylmethylene)-4*H*-1-benzopyran-4-one, 9CI   D-60263

3-Benzylidene-4-chromanone. Benzalchromanone
[30779-90-9]

(E)-form

$C_{16}H_{12}O_2$ M 236.270
(*E*)-*form* [24513-66-4]
   Yellow prisms. Mp 113°.
(*Z*)-*form* [24513-65-3]
   Yellow needles (pet. ether). Mp 67-68°.

Pfeiffer, P. *et al*, *Justus Liebigs Ann. Chem.*, 1949, **564**, 208 (*synth*)
Bennett, P. *et al*, *J. Chem. Soc., Perkin Trans. 1*, 1972, 1554 (*synth, pmr*)
Katrusiak, A. *et al*, *Acta Crystallogr., Sect. C*, 1987, **43**, 103 (*cryst struct*)

1,2-Dihydro-1-phenyl-1-naphthalenecarboxylic acid   D-60264

1,2-Dihydro-1-phenyl-1-naphthoic acid. Isatronic acid

$C_{17}H_{14}O_2$ M 250.296
(±)-*form*
   Cryst. (C_6H_6). Mp 159-162°.

Voigtländer, H.W. *et al*, *Arch. Pharm. (Weinheim, Ger.)*, 1959, **292**, 632 (*synth*)
Herz, W. *et al*, *J. Org. Chem.*, 1964, **29**, 1691 (*synth*)

3,4-Dihydro-2-phenyl-1(2*H*)-naphthalenone, 9CI   D-60265

1-Oxo-2-phenyl-1,2,3,4-tetrahydronaphthalene. 2-Phenyl-1-tetralone
[7498-87-5]

$C_{16}H_{14}O$ M 222.286
(±)-*form*
   Cryst. (Et$_2$O/pet. ether). Mp 78-79°.
Semicarbazone: Cryst. Mp 250-251.4° dec. (sinters at 245°).

Newman, M.S. *et al*, *J. Am. Chem. Soc.*, 1938, **60**, 2947 (*synth*)
Crawford, H.M., *J. Am. Chem. Soc.*, 1939, **61**, 608 (*synth*)
Johnson, W.S. *et al*, *J. Am. Chem. Soc.*, 1949, **71**, 1092 (*synth*)
McCague, R. *et al*, *J. Med. Chem.*, 1986, **29**, 2053 (*synth*)

3,4-Dihydro-3-phenyl-1(2*H*)-naphthalenone, 9CI   D-60266

1-Oxo-3-phenyl-1,2,3,4-tetrahydronaphthalene. 3-Phenyl-1-tetralone
[14944-26-4]
$C_{16}H_{14}O$ M 222.286
(±)-*form*
   Small needles (pet. ether). Mp 65°.
Semicarbazone: Prisms (MeOH). Mp 211°.

Spring, F.S., *J. Chem. Soc.*, 1934, 1332 (*synth*)
Crawford, H.M., *J. Am. Chem. Soc.*, 1939, **61**, 608 (*synth*)

3,4-Dihydro-4-phenyl-1(2*H*)-naphthalenone, 9CI   D-60267

1-Oxo-4-phenyl-1,2,3,4-tetrahydronaphthalene. 4-Phenyltetralone
[14578-68-8]
$C_{16}H_{14}O$ M 222.286
(±)-*form* [63646-27-5]
   Prisms (EtOH). Mp 75.5-76°.
Oxime: [50845-35-7].
   $C_{16}H_{15}NO$ M 237.301
   Cryst. (CHCl$_3$/hexane). Mp 114-115°.
2,4-Dinitrophenylhydrazone: Red-orange prisms (EtOH/EtOAc). Mp 221-223°.

Wawzonek, S. *et al*, *J. Am. Chem. Soc.*, 1954, **76**, 1641 (*synth*)
Sarges, R., *J. Org. Chem.*, 1975, **40**, 1216 (*synth*)

2,3-Dihydro-6-phenyl-2-thioxo-4(1*H*)-pyri-   **D-60268**
midinone, 9CI

5-Phenyl-2-thiouracil. 2-Thioxo-6-phenyl-4-pyrimidinone

[36822-11-4]

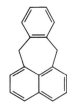

$C_{10}H_8N_2OS$ M 204.246
Solid. Mp 263-265°.

Anderson, G.W. *et al, J. Am. Chem. Soc.*, 1945, **67**, 2197 (*synth*)
Skulnick, H.I. *et al, J. Med. Chem.*, 1986, **29**, 1499 (*synth*)

7,12-Dihydropleiadene, 9CI, 8CI   **D-60269**
1,8-o-Xylylenenaphthalene

[4580-70-5]

$C_{18}H_{14}$ M 230.309
Needles (MeOH aq.). Mp 114.5-115°.

Rieche, A. *et al, Ber.*, 1932, **65**, 1380 (*synth*)
Lansbury, P.T. *et al, J. Am. Chem. Soc.*, 1966, **88**, 1477 (*synth*)

1,7-Dihydro-6*H*-purine-6-thione, 9CI   **D-60270**

Updated Entry replacing D-04036
Leukerin
[50-44-2]

1,9-Dihydro-*form*   1,7-Dihydro-*form*

$C_5H_4N_4S$ M 152.173
1,7-Dihydro-form predominant as cryst.; 1,9-dihydro-form predominant in soln.; 3,X-dihydro- and thiol forms minor tautomers. Yellow prisms + 1H_2O (H_2O). Mp 313-314° dec. (loses H_2O at 140°).

▷Exp. teratogen. UO9800000.

1,7-Dihydro-form
1-Me: [1006-22-0].
   $C_6H_6N_4S$ M 166.200
   Cryst. (H_2O). Mp >300°.
7-Me: [3324-79-6].
   $C_6H_6N_4S$ M 166.200
   Light-brown cryst. (H_2O). Mp 306-308°.

1,9-Dihydro-form [111079-50-6]
9-Me: [1006-20-8].
   $C_6H_6N_4S$ M 166.200
   Cryst. (H_2O). Mp 338-342°.
1,9-Di-Me: [5759-62-6].
   $C_7H_8N_4S$ M 180.227
   Yellow needles (AcOH). Mp 250°.

3,7-Dihydro-form
3-Me: [1006-12-8].
   $C_6H_6N_4S$ M 166.200
   Yellow prisms (H_2O). Mp >300° dec.
3,7-Di-Me: [5759-60-4].
   $C_7H_8N_4S$ M 180.227
   Yellow needles (H_2O). Mp 282-283°.

3,9-Dihydro-form
3,9-Di-Me:
   $C_7H_8N_4S$ M 180.227
   Yellow needles (AcOH/C_6H_6). Mp 276-278°.

6-Thiol-form
6-Mercaptopurine. 6-Purinethiol
S-*Me:* [50-66-8].
   $C_6H_6N_4S$ M 166.200
   Cryst. (H_2O). Mp 220° dec.
   ▷UO8976000.
S,3N-*Di-Me:* [1008-08-8].
   $C_7H_8N_4S$ M 180.227
   Cryst. (H_2O). Mp 220° dec.
S,7N-*Di-Me:* [1008-01-1].
   $C_7H_8N_4S$ M 180.227
   Needles (H_2O). Mp 212-213°.
S,9N-*Di-Me:* [1127-75-9].
   $C_7H_8N_4S$ M 180.227
   Needles (MeOH). Mp 171-172°.
5-Me, 9-β-D-ribofuranosyl: [342-69-8].
   $C_{11}H_{14}N_4O_4S$ M 298.316
   Needles. Mp 163-164°.
   ▷UO8985000.
9-β-D-Ribofuranosyl: [574-25-4].
   $C_{10}H_{12}N_4O_4S$ M 284.289
   Needles (AcOH aq.). Mp 208-210° dec. $[\alpha]_D^{25}$ −73° (c, 1 in 0.1M NaOH).
   ▷UP0710000.

Elion, G.B. *et al, J. Am. Chem. Soc.*, 1952, **74**, 411 (*synth*)
Johnson, J.A. *et al, J. Am. Chem. Soc.*, 1956, **78**, 3863 (*derivs*)
Fox, J.J. *et al, J. Am. Chem. Soc.*, 1958, **80**, 1669 (*derivs*)
Bergmann, F. *et al, J. Org. Chem.*, 1961, **26**, 1504.
Townsend, L.B. *et al, J. Org. Chem.*, 1962, **27**, 990.
Sletten, E. *et al, Acta Crystallogr.*, 1969, **325**, 1330 (*cryst struct*)
Lichtenberg, D. *et al, Isr. J. Chem.*, 1972, **10**, 805 (*nmr, uv, tautom*)
Benezra, S.A. *et al, Anal. Profiles Drug Subst.*, 1978, **7**, 343 (*rev*)
Nair, V. *et al, Synthesis*, 1986, 450 (*derivs, synth, bibl*)
Sax, N.I., *Dangerous Properties of Industrial Materials*, 5th Ed., Van Nostrand-Reinhold, 1979, 947.

3,8-Dihydro-9*H*-pyrazolo[4,3-*f*]quinazolin-   **D-60271**
9-one, 9CI

Pyrazolo[4,3-f]quinazolin-9(8H)-one. prox-*Benzoallopurinol*

[78416-44-1]

$C_9H_6N_4O$ M 186.173
Cryst. Mp >325°.

Cuny, E. *et al, Chem. Ber.*, 1981, **114**, 1624 (*synth, uv, pmr*)

4,5-Dihydro-3(2*H*)-pyridazinone, 9CI   D-60272

[61468-81-3]

C₄H₆N₂O   M 98.104
Mp 41-43° (35-41°). Bp₃ 94° (Bp₁ 100°).

Evans, R.C. *et al, J. Am. Chem. Soc.*, 1945, **67**, 60 (*synth*)
Reichelt, I. *et al, Synthesis*, 1984, 786 (*synth, ir, pmr*)

3,4-Dihydro-2(1*H*)-pyrimidinone, 9CI   D-60273

2-Oxo-1,2,3,4-tetrahydropyrimidine
[10167-11-0]

C₄H₆N₂O   M 98.104
Cryst. (EtOH). Mp 157-158°.

1-Me:
    C₅H₈N₂O   M 112.131
    Cryst. (EtOAc/hexane or by subl.). Mp 120-121°.
1-β-D-Ribofuranosyl: [102922-03-2].
    C₉H₁₄N₂O₅   M 230.220
3-β-D-Ribofuranosyl: [102922-02-1].
    C₉H₁₄N₂O₅   M 230.220
    Fluffy powder.

Škarić, V. *et al, Croat. Chim. Acta*, 1964, **36**, 87; 1966, **38**, 1
    (*synth, uv, pmr*)
Kim, C.-H. *et al, J. Med. Chem.*, 1986, **29**, 1374 (*deriv, synth, pmr*)

5,6-Dihydro-4(1*H*)-pyrimidinone, 9CI   D-60274

5,6-Dihydro-4-oxopyrimidine
[10167-06-3]

C₄H₆N₂O   M 98.104
Prisms (dioxane/hexane). Mp 137-138°.

3-Me:
    C₅H₈N₂O   M 112.131
    Cryst. (C₆H₆/hexane or by subl.). Mp 63-65°.
    Unstable to air and in protic solvents.

Škarić, V. *et al, Croat. Chim. Acta*, 1966, **38**, 1 (*synth, uv*)

6,7-Dihydro-5*H*-1-pyrindine, 9CI   D-60275

2,3-Cyclopentenopyridine. 2,3-Pyridane. 2,3-Trimethylenepyridine
[533-37-9]

C₈H₉N   M 119.166
Isol. from coal tar. Fp 67°. Bp 199.5°, Bp₁₁ 87-88°.
Picrate: Mp 181-182°.
N-*Oxide:* [90685-58-8].
    C₈H₉NO   M 135.165
    Mp 121-122°.

Eisch, J.J. *et al, J. Organomet. Chem.*, 1968, **14**, P13 (*synth*)
Breitmaier, E. *et al, Tetrahedron Lett.*, 1970, 3291 (*synth*)
Thummel, R.P. *et al, J. Org. Chem.*, 1977, **42**, 2742 (*synth, uv, pmr*)
Beschke, H., *Aldrichchimica Acta*, 1978, **11**, 13 (*rev*)
Irie, H. *et al, Heterocycles*, 1979, **12**, 771 (*synth*)
Abramovitch, R.A. *et al, J. Am. Chem. Soc.*, 1981, **103**, 1525 (*synth*)
Stefaniak, L. *et al, Org. Magn. Reson.*, 1984, **22**, 201 (¹⁵N nmr)
Neunhoeffer, H. *et al, Justus Liebigs Ann. Chem.*, 1985, 1732 (*synth*)

6,7-Dihydro-5*H*-2-pyrindine, 9CI   D-60276

3,4-Cyclopentenopyridine. 3,4-Pyridane. 3,4-Trimethylenepyridine
[533-35-7]

C₈H₉N   M 119.166
Bp₁₀ 78-80°, Bp₅ 52-53°.
Picrate: Cryst. (EtOH). Mp 144°.

Eisch, J.J. *et al, J. Organomet. Chem.*, 1968, **14**, P13 (*synth*)
Thummel, R.P. *et al, J. Org. Chem.*, 1977, **42**, 2742 (*synth, uv, pmr*)
Boger, D.L. *et al, J. Org. Chem.*, 1981, **46**, 2179 (*synth*)

2,3-Dihydro-1*H*-pyrrolizin-1-one, 9CI   D-60277

1-Azabicyclo[3.3.0]octa-5,7-dien-4-one. 1-Oxo-3H-1,2-dihydropyrrolo[1,2-a]pyrrole
[17266-64-7]

C₇H₇NO   M 121.138
V. pale-yellow needles (pet. ether). Mp 54°.
Semicarbazone: Long thin prisms (EtOH). Mp 211°.

Clemo, G.R. *et al, J. Chem. Soc.*, 1931, 49; 1942, 424 (*synth*)

3,4-Dihydro-2(1*H*)-quinazolinone, 9CI   D-60278

[66655-67-2]

C₈H₈N₂O   M 148.164
Cryst. (THF/hexane). Mp 241-242°.

Yoshida, T. *et al, J. Org. Chem.*, 1987, **52**, 1611 (*synth, pmr, ir*)

1,4-Dihydro-2,3-quinoxalinedione, 9CI   D-60279

2,3-Quinoxalinediol. 2,3-Dihydroxyquinoxaline

C₈H₆N₂O₂   M 162.148

Dione-form [15804-19-0]
Needles (H$_2$O). Mp >360°.
▷VD2100000.
1-Me: [20934-51-4].
C$_9$H$_8$N$_2$O$_2$ M 176.174
Needles (MeNO$_2$). Mp 287-288°.
1,4-Di-Me: [58175-07-8].
C$_{10}$H$_{10}$N$_2$O$_2$ M 190.201
Needles (EtOH). Mp 256-258°.

Monohydroxy-form

Et ether: 3-Ethoxy-2(1H)quinoxalinone.
C$_{10}$H$_{10}$N$_2$O$_2$ M 190.201
Fine needles (EtOH aq.). Mp 197-199°.

2,3-Dihydroxy-form

Mono-Et ether: 3-Ethoxy-2-quinoxalinol. 2-Ethoxy-3-hydroxyquinoxaline.
C$_{10}$H$_{10}$N$_2$O$_2$ M 190.201
Fine needles (EtOH aq.). Mp 197-199°.
Di-Me ether: [6333-43-3]. *2,3-Dimethoxyquinoxaline.*
C$_{10}$H$_{10}$N$_2$O$_2$ M 190.201
Fine needles (MeOH). Mp 92-94°.

Hinsberg, O., *Ber.*, 1908, **41**, 2031 (*synth*)
Newbold, G.T. *et al*, *J. Chem. Soc.*, 1948, 519 (*synth*)
Atkinson, C.H. *et al*, *J. Chem. Soc.*, 1956, 26 (*synth, deriv*)
Burakevich, J.V. *et al*, *J. Org. Chem.*, 1970, **35**, 2102 (*synth*)
Loev, B. *et al*, *J. Med. Chem.*, 1985, **28**, 363 (*synth*)

3,4-Dihydro-4-selenoxo-2(1*H*)-quinazolin-   **D-60280**
one, 9CI

[107135-39-7]

C$_8$H$_6$N$_2$OSe   M 225.108
Pale-yellow powder. Mp 233-234° dec.
Hydrate (+ $^1/_4$H$_2$O): Pale-yellow plates (THF/hexane).
Mp 233-234°.

Yoshida, T. *et al*, *J. Org. Chem.*, 1987, **52**, 1611 (*synth, cmr, pmr, ir*)

2,3-Dihydro-2-selenoxo-4(1*H*)-quinazolin-   **D-60281**
one, 9Ci

[107182-80-9]

C$_8$H$_6$N$_2$OSe   M 225.108
Cryst. (EtOH). Mp 245-247°.

Yun, L.M. *et al*, *Khim. Geterotsikl. Soedin.*, 1986, 417 (*synth, uv*)

Dihydro-3,3,4,4-tetramethyl-2(3*H*)-fur-   **D-60282**
anthione, 9CI

[103620-99-1]

C$_8$H$_{14}$OS   M 158.258
Needles by subl. Mp 121-122°.

Levine, J.A. *et al*, *J. Med. Chem.*, 1986, **29**, 1996 (*synth, pmr, ir, uv*)

4,5-Dihydro-3,3,5,5-tetramethyl-4-methy-   **D-60283**
lene-3*H*-pyrazole, 9CI

3,3,5,5-Tetramethyl-4-methylene-1-pyrazoline
[55790-78-8]

C$_8$H$_{14}$N$_2$ M 138.212
Cryst. by subl. Mp 35-36.2°.

Crawford, R.J. *et al*, *Can. J. Chem.*, 1974, **52**, 4033 (*synth, uv, pmr*)

2,5-Dihydro-2,2,5,5-tetramethyl-3-[[[(2-   **D-60284**
phenyl-3*H*-indol-3-ylidene)amino]oxy]-
carbonyl]-1*H*-pyrrol-1-yloxy, 9CI

3[(1-Oxyl-2,2,5,5-tetramethyl-2,5-dihydropyrrole-3-carbonyloxy)imino]-2-phenyl-3H-indole
[106367-37-7]

C$_{23}$H$_{22}$N$_3$O$_3$ M 388.445
Stable spin label. Cryst. (C$_6$H$_6$/pet. ether). Mp 205-207°.

Chiorboli, E. *et al*, *Synthesis*, 1986, 219 (*synth, ms, ir, esr, use*)

3,5-Dihydro-3,3,5,5-tetramethyl-4*H*-pyra-   **D-60285**
zole-4-thione, 9CI

3,3,5,5-Tetramethyl-4-pyrazolinethione
[65927-08-4]

C$_7$H$_{12}$N$_2$S   M 156.245
Deep-pink flakes. Mp 44-45.5°. Highly volatile.
S-Oxide: [65927-19-7].
C$_7$H$_{12}$N$_2$OS   M 172.245
Cryst. (pentane). Mp 84-85°.

Bushby, R.J. *et al*, *J. Chem. Soc., Perkin Trans. 1*, 1979, 2401 (*synth, uv, pmr, ms*)

3,5-Dihydro-3,3,5,5-tetramethyl-4*H*-pyrazol-4-one, 9CI   D-60286

3,3,5,5-Tetramethyl-1-pyrazolin-4-one

[30467-62-0]

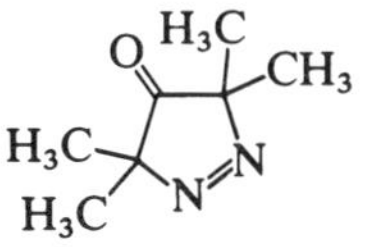

$C_7H_{12}N_2O$ M 140.185
Cryst. by subl. Mp 83.5-85°.

Crawford, R.J. *et al, Can. J. Chem.*, 1974, **52**, 4033 (*synth, pmr, uv*)
Bushby, R.J. *et al, J. Chem. Soc., Perkin Trans. 1*, 1979, 2401 (*synth, ir, pmr, cmr*)

4,5-Dihydro-3,3,4,4-tetramethyl-2(3*H*)-thiophenethione   D-60287

[103621-02-9]

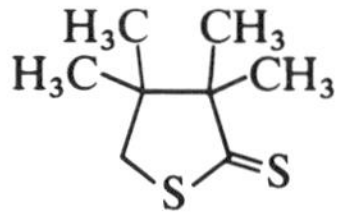

$C_8H_{14}S_2$ M 174.319
Yellow solid by subl. Mp 140-142°.

Levine, J.A. *et al, J. Med. Chem.*, 1986, **29**, 1996 (*synth, ir, pmr, uv*)

4,5-Dihydro-3,3,4,4-tetramethyl-2(3*H*)-thiophenone   D-60288

[103620-96-8]

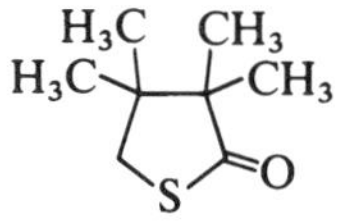

$C_8H_{14}OS$ M 158.258
Needles (EtOH aq. or by subl.). Mp 131-137°.

Levine, J.A. *et al, J. Med. Chem.*, 1986, **29**, 1996 (*synth, pmr, ir, uv*)

1,4-Dihydrothieno[3,4-*d*]pyridazine   D-60289

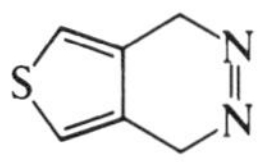

$C_6H_6N_2S$ M 138.187
Source of the 3,4-dimethylenethiophene biradical.

Stone, K.J. *et al, J. Am. Chem. Soc.*, 1986, **108**, 8088 (*synth, uv, pmr, use*)

2,3-Dihydrothiophene   D-60290

[1120-59-8]

C_4H_6S M 86.151
Liq.

Sosnovsky, G., *Tetrahedron*, 1962, **18**, 903 (*synth*)
Trahanovsky, W.S. *et al, J. Org. Chem.*, 1986, **51**, 113 (*synth, cmr, pmr*)

2,3-Dihydro-2-thioxopyrido[2,3-*d*]-pyrimidin-4(1*H*)-one, 9CI   D-60291

Updated Entry replacing D-30292
3,4-Dihydro-4-oxopyrido[2,3-d]pyrimidine-2(1H)-thione. Pyrido[3,2-e]-2-thiouracil
[37891-04-6]

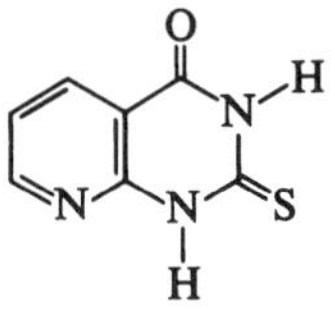

$C_7H_5N_3OS$ M 179.196
Cryst. (CHCl$_3$/pet. ether). Mp 286°.

Stanovnik, B. *et al, Synthesis*, 1972, 308 (*synth*)
Koščik, D. *et al, Collect. Czech. Chem. Commun.*, 1983, **48**, 3315 (*synth*)

2,3-Dihydro-2-thioxopyrido[3,2-*d*]-pyrimidin-4(1*H*)-one, 9CI   D-60292

[37891-05-7]

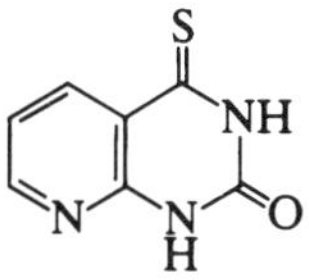

$C_7H_5N_3OS$ M 179.196
Cryst. (H$_2$O). Mp 300°.

Stanovnik, B. *et al, Synthesis*, 1972, 308 (*synth, pmr*)

2,3-Dihydro-2-thioxopyrido[3,4-*d*]-pyrimidin-4(1*H*)-one   D-60293

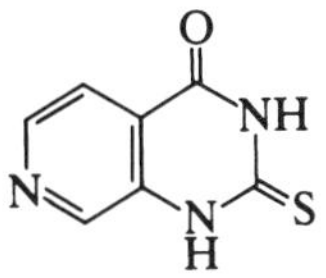

$C_7H_5N_3OS$ M 179.196
Tan powder. Mp >360° subl.

Fox, H.H. *et al, J. Org. Chem.*, 1952, **17**, 547 (*synth*)

2,3-Dihydro-2-thioxo-4*H*-pyrido[1,2-*a*]-1,3,5-triazin-4-one, 9CI   D-60294

4-Oxo-2-thionodihydro-2H-pyrido[1,2-a]-1,3,5-triazine
[37890-96-3]

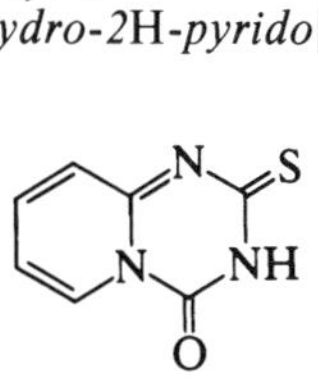

$C_7H_5N_3OS$ M 179.196
Cryst. (DMF/toluene). Mp 225-228°.

Stanovnik, B. *et al, Synthesis*, 1972, 308 (*synth, pmr*)

2,3-Dihydro-4-thioxo-2(1*H*)-quinazolinone   D-60295

*4-Thio-2,4(1*H,3H*)-quinazolinedione. 4-Thiono-2-oxo-tetrahydroquinazoline. 2-Hydroxy-4-quinazolinethione*

[17796-47-3]

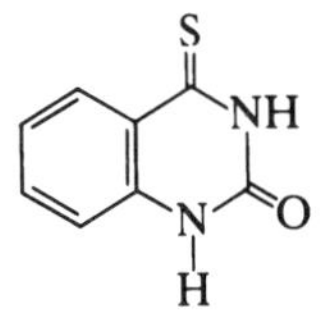

$C_8H_6N_2OS$ M 178.208
Mp 278-279°.

Trattner, R.B. *et al, J. Org. Chem.*, 1964, **29**, 2674 (*synth*)
Wagner, G. *et al, Z. Chem.*, 1968, **8**, 22 (*synth*)

2,3-Dihydro-2-thioxo-4(1*H*)-quinazolinone,   D-60296
9CI

*2-Thio-2,4(1*H,3H*)-quinazolinedione, 8CI. 2-Thiono-4-oxotetrahydroquinazoline*

[13906-09-7]

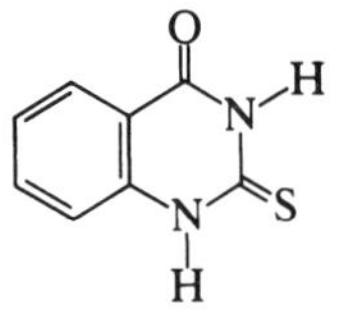

$C_8H_6N_2OS$ M 178.208
Mp 281-282°.

Rupe, H., *Ber.*, 1897, **30**, 1097 (*synth*)
Kappe, T. *et al, Monatsh. Chem.*, 1967, **98**, 214 (*synth*)

Dihydro-1,3,5-triazine-2,4(1*H,3H*)-dione,   D-60297
9CI

[27032-78-6]

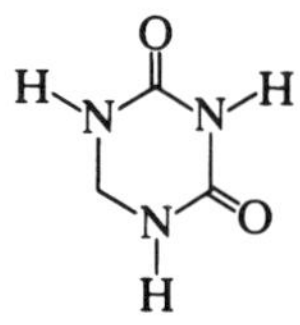

$C_3H_5N_3O_2$ M 115.091
Cryst. (H$_2$O). Mp 291-292° dec.

1,3,5-Tri-Me: [41221-01-6].
   $C_6H_{11}N_3O_2$ M 157.172
   Mp 95°.

Hartman, S.C. *et al, J. Am. Chem. Soc.*, 1955, **77**, 1051 (*synth*)
Piskala, A. *et al, Collect. Czech. Chem. Commun.*, 1961, **26**, 2519 (*synth*)
Etienne, A. *et al, Bull. Soc. Chim. Fr.*, 1975, 1419 (*deriv, synth, pmr*)

1,2-Dihydro-3*H*-1,2,4-triazole-3-thione,   D-60298
9CI

s-Triazole-3-thiol, 8CI. 3-Mercapto-1,2,4-thiazole

[3179-31-5]

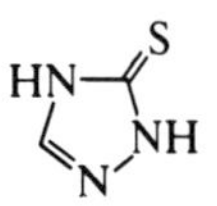

$C_2H_3N_3S$ M 101.126
Thione-form (illus.) is most probable. Mp 214-216°.
▷XZ5267500.

Freund, M., *Ber.*, 1896, **29**, 2483 (*synth*)
Goerdeler, J. *et al, Chem. Ber.*, 1957, **90**, 202 (*synth*)
Singh, H. *et al, Tetrahedron*, 1986, **42**, 1449 (*synth, ir, pmr*)

Dihydro-4-(trifluoromethyl)-2(3*H*)-furan-one, 9CI   D-60299

3-Trifluoromethyl-γ-butyrolactone

[110847-11-5]

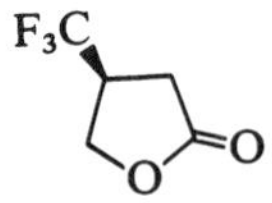

(*S*)-*form*

$C_5H_5F_3O_2$ M 154.089
Chiral reagent.
(*R*)-*form* [109719-07-5]
   $[\alpha]_D^{24}$ +16.8° (c, 2.24 in Et$_2$O).
(*S*)-*form* [109719-06-4]
   $[\alpha]_D^{19}$ −17.7° (c, 1.9 in Et$_2$O).

Taguchi, T. *et al, Tetrahedron Lett.*, 1986, **27**, 5117 (*synth, resoln, use*)

Dihydro-5-(trifluoromethyl)-2(3*H*)-furan-one, 9CI   D-60300

5,5,5-Trifluoro-γ-valerolactone

$C_5H_5F_3O_2$ M 154.089

(*R*)-*form* [108211-37-6]
Bp$_{11}$ 80-90°. $[\alpha]_D^{25}$ −15° (c, 1.6 in CHCl$_3$).

Seebach, D. *et al, Helv. Chim. Acta*, 1985, **68**, 2342 (*synth, ir, pmr, ms*)

3,4-Dihydro-3,3,8*a*-trimethyl-1,6(2*H*,8*aH*)-naphthalenedione, 9CI   D-60301

[75612-51-0]

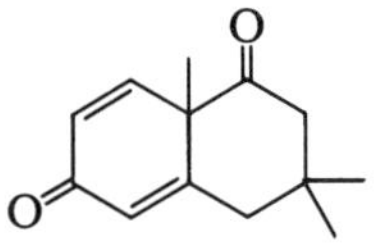

$C_{13}H_{16}O_2$ M 204.268
Mp 71-73°. Very sensitive to acids and bases.

Waring, A.J. *et al, J. Chem. Soc., Perkin Trans. 1*, 1985, 631 (*synth, uv, pmr*)

3,4-Dihydro-4,6,7-trimethyl-3-β-D-ribofuranosyl-9*H*-imidazo[1,2-*a*]purin-9-one, 9CI   D-60302

mimG

[108274-04-0]

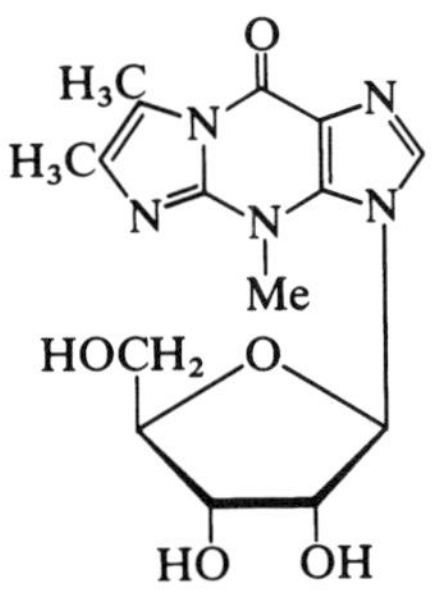

$C_{15}H_{19}N_5O_5$ M 349.346
Isol. from *Sulfolobus solfataricus*, *Thermoproteus neutrophilus* and *Pyrodictium occultum*. No phys. props. reported.

McCloskey, J.A. *et al, Nucleic Acids Res.*, 1987, **15**, 683 (*isol, struct*)

3,5-Dihydro-3,5,5-trimethyl-4*H*-triazol-4-one   D-60303

[104014-78-0]

$C_5H_9N_3O$ M 127.146

Oil. Unstable, obt. >95% pure. Other trialkyl-4*H*-triazol-4-ones also prepd.

Quast, H. *et al, Justus Liebigs Ann. Chem.*, 1986, 1891.

7,12-Dihydroxy-8,12-abietadiene-11,14-dione   D-60304

Updated Entry replacing D-04138
7-Hydroxyroyleanone

$C_{20}H_{28}O_4$ M 332.439

7α-form [21887-01-4]

Horminone

Constit. of *Plectranthus* spp. Golden cryst. (MeOH aq.). Mp 178°. [α]$_D$ −132° (c, 1.24 in CHCl$_3$).

Ac: 7α-Acetoxyroyleanone.
$C_{22}H_{30}O_5$ M 374.476
Isol. from whole plants of *Horminum pyrenaicum*, roots of *Salvia* spp. and leaf glands of *Coleus carnosus*.

7-Formyl ester: 7-O-Formylhorminone.
$C_{21}H_{28}O_5$ M 360.449
Constit. of *Plectranthus sanguineus*. Yellow-orange cryst. (hexane). Mp 152.5-154°.

7β-form [21764-41-0]

Taxoquinone

Isol. from *Taxodium distichum*. Mp 214°. [α]$_D^{25}$ +240° (c, 1.4 in CHCl$_3$).

7-Ac: [6812-88-0]. *7β-Acetoxyroyleanone*.
$C_{22}H_{30}O_5$ M 374.476
Constit. of *Inula royleana*. Orange-yellow cryst. (EtOAc). Mp 212-214.5°. [α]$_D$ −14° (c, 1 in CHCl$_3$).

Kupchan, S.M. *et al, J. Org. Chem.*, 1969, **34**, 3912 (*isol*)
Bhat, S.V. *et al, Tetrahedron*, 1975, **31**, 1001 (*isol, deriv*)
Hensch, M. *et al, Helv. Chim. Acta*, 1975, **58**, 1921 (*isol*)
Matsumoto, T. *et al, Bull. Chem. Soc. Jpn.*, 1979, **52**, 1459 (*synth*)
Hueso-Rodriguez, J.A. *et al, Phytochemistry*, 1983, **22**, 2005 (*isol, derivs*)
Matloubi-Moghadam, F. *et al, Helv. Chim. Acta*, 1987, **70**, 975 (*isol, deriv*)

11,12-Dihydroxy-6,8,11,13-abietatetraen-20-oic acid   D-60305

$C_{20}H_{26}O_4$ M 330.423

Di-Me ether, Me ester: Methyl 11,12-dimethoxy-6,8,11,13-abietatrien-20-oate.
$C_{23}H_{32}O_4$ M 372.503
Constit. of *Salvia canariensis*. Gum.

González, A.G. *et al, Phytochemistry*, 1987, **26**, 1471.

3,12-Dihydroxy-8,11,13-abietatrien-1-one   D-60306

$C_{20}H_{28}O_3$ M 316.439

3β-form

1-Oxohinokiol

Constit. of *Calocedrus formosana*. Cryst. Mp 228-229°. [α]$_D^{25}$ +130° (c, 0.68 in Me$_2$CO).

Fang, J.-M. *et al, Phytochemistry*, 1987, **26**, 853.

4,5-Dihydroxy-1,3-benzenedicarboxylic acid, 9CI   D-60307

4,5-Dihydroxyisophthalic acid, 8CI. 5-Carboxyprotocatechuic acid

[4707-77-1]

$C_8H_6O_6$ M 198.132

Prod. by a *Streptomyces* sp. Inhibitor of brain glutamate decarboxylase.

Di-Me ester: [33842-22-7].
$C_{10}H_{10}O_6$ M 226.185
Needles (MeOH aq.). Mp 141°.

Di-Me ether: [485-38-1]. *4,5-Dimethoxy-1,3-benzenedicarboxylic acid, 9CI. m-Hemipic acid. Isohemipic acid.*
$C_{10}H_{10}O_6$ M 226.185
Obt. from degradn. of lignins. Mp 261-262°.

Di-Me ether, di-Me ester: [17078-60-3].
$C_{12}H_{14}O_6$ M 254.239
Needles (MeOH aq.). Mp 55-56°.

Hunt, S.E. *et al, J. Chem. Soc.*, 1956, 3099 (*synth*)
Freudenberg, K. *et al, Chem. Ber.*, 1962, **95**, 2814 (*isol*)
Dallacker, F. *et al, Chem. Ber.*, 1971, **104**, 2347 (*synth*)
Endo, A. *et al, J. Antibiot.*, 1981, **34**, 1351 (*isol, props*)

2,2-Dihydroxy-1*H*-benz[*f*]indene-1,3(2*H*)-dione, 9CI   **D-60308**

Benzo[f]*ninhydrin*
[38627-57-5]

$C_{13}H_8O_4$ M 228.204

Colour reagent for α-amino acids, used in fingerprint detection. Plates (dioxane aq.). Mp 276-280°.

Jones, D.W. *et al*, *J. Chem. Soc., Perkin Trans. 1*, 1972, 2722 (*synth*)
Heffler, R. *et al*, *Tetrahedron Lett.*, 1987, 6539 (*synth, bibl*)

3,6-Dihydroxy-1,2-benzoquinone   **D-60309**

3,6-Dihydroxy-3,5-cyclohexadiene-1,2-dione, 9CI
$C_6H_4O_4$ M 140.095
Di-Me ether: [108213-73-6]. *3,6-Dimethoxy-1,2-benzoquinone.*
$C_8H_8O_4$ M 168.149
Dark-purple solid. Mp 120-122°.

Wriede, U. *et al*, *J. Org. Chem.*, 1987, **52**, 4485 (*synth, deriv, ir, pmr*)

1-[2,4-Dihydroxy-3,5-bis(3-methyl-2-butenyl)phenyl]-3-(4-hydroxyphenyl)-2-propen-1-one   **D-60310**

3′,5′-Diisopentenyl-2′,4,4′-trihydroxychalcone

$C_{25}H_{28}O_4$ M 392.494
(*E*)-form
Constit. of *Crotalaria medicaginea*. Yellow cryst. (C_6H_6). Mp 118-119°.

Rao, G.V.R. *et al*, *Phytochemistry*, 1987, **26**, 2866.

ent-6β,17-Dihydroxy-14,15-bisnor-7,11*E*-labdadien-13-one   **D-60311**

$C_{18}H_{28}O_3$ M 292.417
Di-Ac: ent-6β,17-Diacetoxy-14,15-bisnor-7,11E-labda-dien-13-one.
$C_{22}H_{32}O_5$ M 376.492
Constit. of *Rutidosis murchisonii*. Oil. $[\alpha]_D^{24}$ −138° (c, 0.1 in $CHCl_3$).

Zdero, C. *et al*, *Phytochemistry*, 1987, **26**, 1759.

2,3-Dihydroxybutanoic acid, 9CI   **D-60312**

Updated Entry replacing D-04339
2-Methylglyceric acid
[3413-97-6]

(2*R*,3*R*)-form
Absolute configuration

$C_4H_8O_4$ M 120.105
(2*R*,3*R*)-form [15851-57-7]
(−)-erythro-*form*
$[\alpha]_D^{25}$ −9.5° (c, 1.0 in H_2O).
(2*S*,3*S*)-form [23334-72-7]
(+)-erythro-*form*
$[\alpha]_D^{20}$ +10.3° (c, 3.2 in H_2O).
(2*RS*,3*RS*)-form [36294-90-3]
(±)-erythro-*form*
Cryst. (H_2O). Mp 81°.
Di-Ac:
$C_8H_{12}O_6$ M 204.179
Cryst. + H_2O (H_2O), oily liq. when anhyd. Mp 50°. Bp_4 127° (anhyd.).
Di-Ac, chloride:
$C_8H_{11}ClO_5$ M 222.625
Bp_3 79°.
Me ester:
$C_5H_{10}O_4$ M 134.132
Bp_{10} 109°.
Phenylhydrazide: Cryst. (EtOAc). Mp 123.5°.
(2*R*,3*S*)-form
(+)-threo-*form*
$[\alpha]_D^{20}$ +15.9° (c, 2.1 in H_2O).
(2*S*,3*R*)-form [15851-58-8]
(−)-threo-*form*
$[\alpha]_D^{25}$ −17.75° (c, 1.0 in H_2O).
(2*RS*,3*SR*)-form
(±)-threo-*form*
Prisms (H_2O). Mp 73-74°.

Izumi, Y. *et al*, *Bull. Chem. Soc. Jpn.*, 1966, **39**, 2223 (*synth*)
Bachelor, F.W. *et al*, *Can. J. Chem.*, 1967, **45**, 79; 1969, **47**, 4089 (*synth, abs config*)
Viscontini, M. *et al*, *Helv. Chim. Acta*, 1972, **55**, 570 (*synth*)
Sakai, T. *et al*, *Chem. Pharm. Bull.*, 1986, **59**, 3185 (*synth, ir, pmr*)

3,4-Dihydroxybutanoic acid   **D-60313**

[1518-61-2]

$C_4H_8O_4$ M 120.105
(*R*)-form
4-O-tert-Butyl, Et ester: [106058-89-3].
$C_{10}H_{20}O_4$ M 204.266
Chiral synthetic intermed. $Bp_{0.1}$ 80-82°. $[\alpha]_D$ +13.2° (c, 1.5 in EtOH).
4-O-tert-Butyl, Et ester, 3,5-dinitrobenzoyl: [106058-93-9]. Cryst. ($CHCl_3$/hexane). Mp 59-60°. $[\alpha]_D$ +19.0° (c, 1.0 in $CHCl_3$).

Moore, R.E. *et al*, *J. Org. Chem.*, 1984, **49**, 2484 (*synth*)
Seebach, D. *et al*, *Synthesis*, 1986, 37 (*deriv, synth, ms, ir, pmr, use*)

2,7-Dihydroxy-2,4,6-cycloheptatrien-1-one, 9CI   D-60314

2,7-Dihydroxytropone. 3-Hydroxytropolone

[33739-50-3]

$C_7H_6O_3$ M 138.123

Mp 143-144°.

Di-Ac: [107976-45-4].

  $C_{11}H_{10}O_5$ M 222.197

  Mp 90-91°.

Takeshita, H. *et al, Synthesis*, 1986, 578 (*synth, pmr, cmr*)

5,6-Dihydroxy-5-cyclohexene-1,2,3,4-tetrone, 9CI   D-60315

Dihydroxydiquinone. Rhodizonic acid

[118-76-3]

$C_6H_2O_6$ M 170.078

Salt solns. used as analytical indicators. Deep-orange needles. Mp 155-160°. Salt solns. have no carbonyl abs. in IR suggesting a delocalized struct.

Di-Na salt: [523-21-7]. Black-green solid.

Di-K salt: [13021-40-4]. Dark-purple prisms. Fairly sol. H_2O giving deep-yellow soln. Partially dec. on recryst. Solns. rapidly oxidised.

Feigl, F. *et al, Ind. Eng. Chem.* (*Anal. Ed.*), 1942, **14**, 840 (*use*)
Preisler, P.W. *et al, J. Am. Chem. Soc.*, 1942, **64**, 67 (*synth*)
Eistert, B. *et al, Angew. Chem.*, 1958, **70**, 595 (*synth*)
West, R. *et al, J. Am. Chem. Soc.*, 1960, **82**, 6204 (*struct*)
Chalmers, R.A. *et al, Mikrochim. Acta*, 1967, 1126 (*use*)
Moeckel, P. *et al, Z. Chem.*, 1967, **7**, 62 (*synth, bibl*)
Skujins, S. *et al, Tetrahedron*, 1968, **24**, 4805 (*ms*)
West, R. *et al, Chem. Carbonyl. Group*, 1970, **2**, 241 (*rev*)
Bailey, R.T., *J. Chem. Soc.* (*B*), 1971, 627 (*struct, ir, raman*)
Aihara, J., *Bull. Chem. Soc. Jpn.*, 1974, **47**, 2899 (*uv*)
Staedeli, W. *et al, Helv. Chim. Acta*, 1977, **60**, 948 (*cmr*)
Douglas, K.T. *et al, FEBS Lett.*, 1979, **106**, 393 (*props*)

3,10-Dihydroxy-5,11-dielmenthadiene-4,9-dione   D-60316

[106623-23-8]

$C_{20}H_{28}O_4$ M 332.439

Constit. of *Callitris macleayana*. Cryst. (heptane). Mp 149-150°. [α]$_D^{20}$ +55.1° (c, 0.55 in CHCl$_3$).

Carman, R.M. *et al, Aust. J. Chem.*, 1986, **39**, 1843.

5,8-Dihydroxy-2,3-dimethyl-1,4-naphthoquinone   D-60317

5,8-Dihydroxy-2,3-dimethyl-1,4-naphthalenedione, 9CI.
2,3-Dimethylnaphthazarin

[21418-10-0]

$C_{12}H_{10}O_4$ M 218.209

In the solid state this tautomer predominates over the 6,7-dimethyl tautomer. Mp 165° subl.

Di-Ac:

  $C_{16}H_{14}O_6$ M 302.283

  Cryst. (EtOH). Mp 184-186°. A 97:3 mixt. of 2,3-dimethyl and 6,7-dimethyl isomers.

Rodriguez, J.-G. *et al, Bull. Chem. Soc. Jpn.*, 1986, **59**, 3957 (*synth, pmr*)

1,7-Dihydroxy-3,7-dimethyl-2,5-octadien-4-one   D-60318

$C_{10}H_{16}O_3$ M 184.235

(2E,5E)-form

Constit. of *Artemisia aucheri*. Oil.

1-Ac: [111394-39-9]. *1-Acetoxy-7-hydroxy-3,7-dimethyl-2E,5E-octadien-4-one.*
  $C_{12}H_{18}O_4$ M 226.272
  From *A. aucheri*. Oil.

7-Hydroperoxide: [111480-78-5]. *7-Hydroxy-1-hydroxy-3,7-dimethyl-2E,5E-octadien-4-one.*
  $C_{10}H_{16}O_4$ M 200.234
  From *A. aucheri*. Oil.

7-Hydroxyperoxy, 1-Ac: [111394-41-3]. *1-Acetoxy-7-hydroperoxy-3,7-dimethyl-2E,5E-octadien-4-one.*
  $C_{12}H_{18}O_5$ M 242.271
  From *A. aucheri*. Oil.

Rustaiyan, A. *et al, Phytochemistry*, 1987, **26**, 2307.

12,19-Dihydroxy-5,8,10,14-eicosatetraenoic acid   D-60319

(5Z,8Z,10E,12S,14Z,19R)-form

$C_{20}H_{32}O_4$ M 336.470

(5Z,8Z,10E,12S,14Z,19R)-form [110066-02-9]

Probable metabolite of 12-Hydroxy-5,8,10,14-eicosatetraenoic acid, H-60131 .

(5Z,8Z,10E,12S,14Z,19S)-form [110066-03-0]

Probable metabolite of 12-Hydroxy-5,8,10,14-eicosatetraenoic acid, H-60131 .

Manna, S. *et al, Tetrahedron Lett.*, 1986, **27**, 2679 (*synth*)

167

1,8-Dihydroxy-3,7(11)-eremophiladien-12,8-olide   D-60320

$C_{15}H_{20}O_4$ M 264.321

(1β,8ξ,10β)-form
Istanbulin F
Constit. of *Smyrnium connatum*. Amorph.
Gören, N. *et al, Phytochemistry*, 1987, **26**, 2633.

1,5-Dihydroxyeriocephaloide   D-60321

$C_{15}H_{20}O_4$ M 264.321

(1β,5β)-form
Constit. of *Eriocephalus kingesii*. Oil.
Zdero, C. *et al, Phytochemistry*, 1987, **26**, 2763.

1,8-Dihydroxy-3,7(11)-eudesmadien-12,8-olide   D-60322

$C_{15}H_{20}O_4$ M 264.321

(1β,8β)-form
1-Ac: [108044-19-5]. *1β-Acetoxy-8β-hydroxy-3,7(11)-eudesmadien-12,8-olide.*
$C_{17}H_{22}O_5$ M 306.358
Constit. of *Smyrnium galaticum*. Amorph.
Ulubelen, A. *et al, J. Nat. Prod.*, 1986, **49**, 1104.

8,9-Dihydroxy-1(10),4,11(13)-germacra-trien-12,6-olide   D-60323

Updated Entry replacing D-04492

$C_{15}H_{20}O_4$ M 264.321
1(10)*E*,4*E*,6β,8α,9β-form illus.

(1(10)E,4E,6β,8α,9β)-form
9-(2R,3R-Epoxy-2-methylbutanoyl): [72638-72-3].
$C_{20}H_{26}O_5$ M 346.422
Constit. of *Montanoa hibiscifolia*. Oil.
9-Ac, 8-(2R,3R-epoxy-2-methylbutanoyl): [72690-77-8].
$C_{22}H_{28}O_6$ M 388.460

Constit. of *M. hibiscifolia*. Cryst. (CHCl₃/C₆H₆). Mp 193-194°. $[\alpha]_D$ −200° (c, 0.06 in CHCl₃).

(1(10)E,4E,6α,8β,9ξ)-form
8-(2-Methylbutanoyl): [73021-14-4].
$C_{20}H_{28}O_5$ M 348.438
Constit. of *Eupatorium mohrii*. $[\alpha]_D$ +50.3° (CHCl₃).

Herz, W. *et al, Phytochemistry*, 1979, **18**, 1337.
Herz, W. *et al, J. Org. Chem.*, 1980, **45**, 1113.
Chandra, A. *et al, Phytochemistry*, 1987, **26**, 1463.

9,15-Dihydroxy-1(10),4,11(13)-germacra-trien-12,6-olide   D-60324

Updated Entry replacing D-30347

$C_{15}H_{20}O_4$ M 264.321
(6α,7β,9α)-form illus.

(6α,7β,9α)-form
Sterophyllolide
Constit. of *Centaura aspera*. Cryst. Mp >200° dec. $[\alpha]_D^{20}$ +72.4° (MeOH).

(6α,7α,9β)-form
Di-Ac: 9β,15-Diacetoxy-1(10),4,11(13)-germacratrien-12,6α-olide. **Idomain**.
$C_{19}H_{24}O_6$ M 348.395
Constit. of *Gutenbergia cordifolia*. Cryst. (MeOH). Mp 161-162°. $[\alpha]_D^{25}$ +83° (c, 0.13 in MeOH).

Picher, M.-T. *et al, Phytochemistry*, 1984, **23**, 1995 (*isol*)
Amigo, J.-M. *et al, Phytochemistry*, 1984, **23**, 1999 (*cryst struct*)
Fujimoto, Y. *et al, Phytochemistry*, 1987, **26**, 2593 (*Idomain*)

3,5-Dihydroxy-4(15),10(14)-guaiadien-12,8-olide   D-60325

$C_{15}H_{20}O_4$ M 264.321

(1α,3β,5α,8α,11R)-form [106310-46-7]
Constit. of *Arctotis arctotoides*. Oil. $[\alpha]_D^{24}$ −74° (c, 0.5 in CHCl₃).

10α,14-Epoxide: [106310-47-8].
$C_{15}H_{20}O_5$ M 280.320
From *A. arctotoides*. Oil. $[\alpha]_D^{24}$ −29° (c, 0.09 in CHCl₃).

10,14-Dihydro, 10β-hydroxy,14-chloro, 3-Ac: [106310-45-6]. **Arctodecurrolide**.
$C_{17}H_{21}ClO_6$ M 356.802
Constit. of *A. arctotoides*. Oil. $[\alpha]_D^{24}$ +33° (c, 0.5 in CHCl₃).

Jakupovic, J. *et al, Planta Med.*, 1986, 365.

4,10-Dihydroxy-2,11(13)-guaiadien-12,6-olide   D-60326

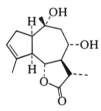

C₁₅H₂₀O₄   M 264.321

$C_{15}H_{20}O_4$ M 264.321

(1α,4α,5α,6α,10α)-form
Constit. of *Eriocephalus giessii*. Oil.
Zdero, C. *et al*, *Phytochemistry*, 1987, **26**, 2763.

2,8-Dihydroxy-3,10(14),11(13)-guaiatrien-12,6-olide   D-60327

$C_{15}H_{18}O_4$ M 262.305

(1α,2β,5α,6β,8β)-form [111576-37-5]
Constit. of *Hymenothrix wislizenii*. Oil.
Jakupovic, J. *et al*, *Phytochemistry*, 1987, **26**, 2543.

8,10-Dihydroxy-3-guaien-12,6-olide   D-60328

$C_{15}H_{22}O_4$ M 266.336

(1α,5α,6α,8α,10α,13S)-form
11β,13-Dihydroepiligustrin (*incorr.*)
Constit. of *Eriocephalus giessii*. Cryst. Mp 67°.
Zdero, C. *et al*, *Phytochemistry*, 1987, **26**, 2763.

5,6-Dihydroxyhexanoic acid   D-60329

[51460-81-2]

$$CH_2OH$$
$$HO{-}C{\blacktriangleleft}H$$ (S)-form
$$(CH_2)_3COOH$$

$C_6H_{12}O_4$ M 148.158

(S)-form
Me ester: [103367-32-4].
    $C_7H_{14}O_4$ M 162.185
    Oil. $[\alpha]_D^{20}$ −26.7° (c, 4.8 in EtOH).
Lactone: [89408-86-6]. 6-(*Hydroxymethyl*)*tetrahydro-
    2H-pyran-2-one*. 5-(*Hydroxymethyl*)-δ-*valerolactone*.
    $C_6H_{10}O_3$ M 130.143
    $[\alpha]_D^{20}$ +28.2° (c, 5.3 in EtOH).
Gerth, D.B. *et al*, *J. Org. Chem.*, 1986, **51**, 3726 (*synth, ir, pmr*)

2,3-Dihydroxy-6-iodobenzoic acid   D-60330

$C_7H_5IO_4$ M 280.019
Di-Me ether: [56221-41-1]. 6-*Iodo-2,3-dimethoxyben-
    zoic acid*. 6-*Iodo-o-veratric acid*.
    $C_9H_9IO_4$ M 308.072
    Cryst. (CHCl₃/pet. ether). Mp 137-138°.
Di-Me ether, Me ester: [56221-26-2].
    $C_{10}H_{11}IO_4$ M 322.099
    Cryst. Mp 57-59°.
Sugasawa, S. *et al*, *CA*, 1935, **29**, 161² (*synth*)
Dyke, S.F. *et al*, *Tetrahedron*, 1975, **31**, 561 (*synth, deriv, ir,
    pmr*)

2,4-Dihydroxy-3-iodobenzoic acid   D-60331

[86635-85-0]
$C_7H_5IO_4$ M 280.019
Prisms (C₆H₆). Mp 198-199°.

Di-Me ether: 3-*Iodo-2,4-dimethoxybenzoic acid*.
    $C_9H_9IO_4$ M 308.072
    Needles (Me₂CO). Mp 184-186°.

Miyazaki, T. *et al*, *Chem. Pharm. Bull.*, 1964, **12**, 1236 (*synth,
    deriv, ir*)
Ahluwalia, V.K. *et al*, *Tetrahedron*, 1982, **38**, 3673 (*synth, pmr*)

2,4-Dihydroxy-5-iodobenzoic acid   D-60332

$C_7H_5IO_4$ M 280.019
Di-Me ether: [3153-77-3]. 5-*Iodo-2,4-dimethoxybenzoic
    acid*.
    $C_9H_9IO_4$ M 308.072
    Needles (Me₂CO). Mp 209.5-210.5° dec.
Di-Me ether, Me ester: [3153-79-5].
    $C_{10}H_{11}IO_4$ M 322.099
    Needles (MeOH). Mp 123-123.5°.
Miyazaki, T. *et al*, *Chem. Pharm. Bull.*, 1964, **12**, 1236 (*synth,
    deriv, ir*)

2,5-Dihydroxy-4-iodobenzoic acid   D-60333

$C_7H_5IO_4$ M 280.019
Di-Me ether: 4-*Iodo-2,5-dimethoxybenzoic acid*.
    $C_9H_9IO_4$ M 308.072
    Cryst. (EtOH). Mp 175-177°.
Di-Me ether, Et ester:
    $C_{11}H_{13}IO_4$ M 336.126
    Cryst. (pet. ether). Mp 65-66°.
Sato, T., *Bull. Chem. Soc. Jpn.*, 1959, **32**, 1292 (*synth, deriv*)

2,6-Dihydroxy-3-iodobenzoic acid   D-60334

$C_7H_5IO_4$ M 280.019
Di-Me ether: [90347-70-9]. 3-*Iodo-2,6-dimethoxyben-
    zoic acid*.
    $C_9H_9IO_4$ M 308.072
    Needles (H₂O). Mp 162°.
Di-Me ether, Me ester:
    $C_{10}H_{11}IO_4$ M 322.099
    Prisms (MeOH). Mp 87°.
Doyle, F.P. *et al*, *J. Chem. Soc.*, 1963, 497 (*synth*)
ApSimon, J.W. *et al*, *J. Chem. Soc.*, 1965, 4156 (*synth*)
Cuyegkeng, M.A. *et al*, *Chem. Ber.*, 1987, **120**, 803 (*synth, pmr*)

3,4-Dihydroxy-5-iodobenzoic acid   D-60335
$C_7H_5IO_4$ M 280.019

Di-Me ether: 3-Iodo-4,5-dimethoxybenzoic acid. 5-Iodoveratric acid.
$C_9H_9IO_4$ M 308.072
Cryst. Mp 185°.
Di-Me ether, Me ester: [50772-81-1].
$C_{10}H_{11}IO_4$ M 322.099
Cryst. Mp 106°.

Erdtmann, H., *CA*, 1936, **30**, 449⁴ (*synth, deriv*)

3,5-Dihydroxy-2-iodobenzoic acid   D-60336
$C_7H_5IO_4$ M 280.019

Di-Me ether, Me ester: Methyl 2-iodo-3,5-dimethoxybenzoate.
$C_{10}H_{11}IO_4$ M 322.099
Plates (pet. ether). Mp 80-81°.

Sargent, M.V., *J. Chem. Soc., Perkin Trans. 1*, 1987, 2553 (*synth, deriv, pmr*)

3,5-Dihydroxy-4-iodobenzoic acid   D-60337
[76447-13-7]
$C_7H_5IO_4$ M 280.019
Cryst. (H_2O). Mp 231.5-234°.

Di-Me ether, Me ester: [65566-16-7]. Methyl 4-iodo-3,5-dimethoxybenzoate.
$C_{10}H_{11}IO_4$ M 322.099
Cryst. (heptane). Mp 113-114°.

Kompis, I. *et al, Helv. Chim. Acta*, 1977, **60**, 3025 (*synth, deriv, uv, pmr*)
Carey, F.A. *et al, J. Org. Chem.*, 1981, **46**, 1366 (*synth*)

4,5-Dihydroxy-2-iodobenzoic acid   D-60338
$C_7H_5IO_4$ M 280.019

Di-Me ether: [61203-48-3]. 2-Iodo-4,5-dimethoxybenzoic acid.
$C_9H_9IO_4$ M 308.072
Cryst. (EtOH aq.). Mp 159-160°.

Piatak, D.M. *et al, J. Org. Chem.*, 1977, **42**, 1068 (*synth, deriv, pmr*)

5,7-Dihydroxy-1(3H)-isobenzofuranone,   D-60339
9CI
Updated Entry replacing D-04536
5,7-Dihydroxyphthalide, 8CI
[27979-58-4]
$C_8H_6O_4$ M 166.133
Obt. from achenes of *Helichrysum arenarium* and from *Anaphalis contorta* whole plant. Cryst. (MeOH aq.). Mp 253-260° dec.
5-Me ether: [24953-77-3]. 7-Hydroxy-5-methoxyphthalide.
$C_9H_8O_4$ M 180.160
From *H. arenarium* and *H. polyphyllum*. Cryst. ($CHCl_3$). Mp 184-184.5°.
5-Me ether, glucoside:
$C_{15}H_{18}O_9$ M 342.302
From *H. arenarium*. Needles (H_2O). Mp 194-195°. $[\alpha]_D^{20}$ −49.3° (c, 0.4 in Py).
7-Me ether: [3465-55-2]. 5-Hydroxy-7-methoxyphthalide.
$C_9H_8O_4$ M 180.160
Prisms (MeOH). Mp 280-283° dec.

Di-Me ether: [3465-69-8]. 5,7-Dimethoxyphthalide.
$C_{10}H_{10}O_4$ M 194.187
Mp 151-152°.
O^7-(3-Methyl-2-butenyl): [76382-73-5]. **Anaphatol**.
$C_{13}H_{14}O_4$ M 234.251
Isol. from . *A. contorta*.
O^7-(3-Methyl-2-butenyl), O^5-Me: Mp 110°.

Allison, W.R. *et al, J. Chem. Soc.*, 1959, 3335 (*uv*)
Vrkoč, J. *et al, Collect. Czech. Chem. Commun.*, 1959, **24**, 3926.
Howells, E.M. *et al, J. Chem. Soc.*, 1965, 4592 (*deriv*)
Vrkoč, J. *et al, Phytochemistry*, 1972, **12**, 2062 (*isol, pmr*)
Talapatra, B. *et al, Indian J. Chem., Sect. B*, 1980, **19**, 927 (*isol, struct, derivs*)
Talapatra, B. *et al, J. Indian Chem. Soc.*, 1983, **40**, 1169 (*synth*)

4′,7-Dihydroxyisoflavone   D-60340
Updated Entry replacing D-30360
Daidzein. Dimethylbiochanin B. Daizeol
[486-66-8]
$C_{15}H_{10}O_4$ M 254.242
Isol. from roots of *Pueraria lobata* and from wood and bark of *Erythrina crista-galli*. Pale-yellow cryst. (EtOH aq.). Mp 330° (323°).
Di-Ac: [3682-01-7].
$C_{19}H_{14}O_6$ M 338.316
Cryst. (EtOH). Mp 188-189°.
7-O-β-D-Glucopyranosyl: [552-66-9]. **Daidzin**.
$C_{21}H_{20}O_9$ M 416.384
Isol. from soya bean meal, roots of *P. lobata* and shoots of *Piptanthus nepalensis*. Cryst. + $1H_2O$ (H_2O) becoming anhyd. at 120°. Mp 233-235°. $[\alpha]_D^{20}$ −36.4° (0.02N KOH).
4′,7-Di-O-β-D-glucopyranosyl: [53681-67-7].
$C_{27}H_{30}O_{14}$ M 578.526
Stress metab. of cell cultures of *Vigna angularis*. Isol. also from *P. nepalensis*. Needles. Mp 241°.
7-O-(Apiosyl-(1→6)-β-D-glucopyranoside): [108044-04-8]. **Ambonin**.
$C_{26}H_{28}O_{13}$ M 548.499
Constit. of *Neorautanenia amboensis*. Rosettes (Me_2CO aq.). Mp 225-227°. $[\alpha]_D$ −71° (c, 0.014 in H_2O).
4′-Apioside, 7-β-D-glucopyranoside: [108069-01-8]. **Neobanin**.
$C_{26}H_{28}O_{13}$ M 548.499
From *N. amboensis*. Rosettes ($CHCl_3$/MeOH). Mp 211-212°. $[\alpha]_D$ −72.6° (c, 0.018 in H_2O).
4′-Me ether: see 7-Hydroxy-4′-methoxyisoflavone, H-02182
7-Me ether: see 4′-Hydroxy-7-methoxyisoflavone, H-02181
Di-Me ether: [1157-39-7]. 4′,7-Dimethoxyisoflavone.
$C_{19}H_{14}O_4$ M 306.317
Cryst. (EtOH). Mp 162-164° (154-156°).

Farkas, L. *et al, Chem. Ber.*, 1959, **92**, 819 (*synth*)
Markham, K.R. *et al, Phytochemistry*, 1968, **7**, 791 (*isol*)
Gupta, S.R. *et al, Phytochemistry*, 1971, **10**, 877 (*synth*)
Dement, W.A. *et al, Phytochemistry*, 1972, **11**, 1089 (*isol*)
Inone, T. *et al, Chem. Pharm. Bull.*, 1974, **22**, 1422 (*biosynth*)
Deshpande, V.H. *et al, Indian J. Chem., Sect. B*, 1977, **15**, 201 (*isol*)
Nakayama, M. *et al, Bull. Chem. Soc. Jpn.*, 1978, **51**, 2398 (*synth*)
Jha, H.C. *et al, Can. J. Chem.*, 1980, **58**, 1211 (*cmr*)
Ayabe, S. *et al, J. Chem. Soc., Perkin Trans. 1*, 1982, 2725 (*biosynth, ms*)
Kobayashi, M. *et al, Phytochemistry*, 1983, **22**, 1257 (*isol*)
Breytenbach, J.C., *J. Nat. Prod.*, 1986, **49**, 1003 (*isol, deriv*)
Jain, A.C. *et al, J. Chem. Soc., Perkin Trans. 1*, 1986, 215 (*synth*)

ent-3β,19-Dihydroxy-15-kauren-17-oic acid   D-60341

$C_{20}H_{30}O_4$ M 334.455

3,19-Di-Ac: [110187-32-1]. *ent*-3β,19-Diacetoxy-15-kauren-17-oic acid.
$C_{24}H_{34}O_6$ M 418.529
Constit. of *Peteravenia malvaefolia*.

Ellmaurer, E. *et al*, *J. Nat. Prod.*, 1987, **50**, 221.

6,15-Dihydroxy-7-labden-17-oic acid   D-60342

$C_{20}H_{34}O_4$ M 338.486

(6α,13S)-form
Havardic acid C
Constit. of *Grindelia havardii*.
Me ester: Oil. $[\alpha]_D^{25}$ −2.4° (c, 3.1 in CHCl₃).

Jolad, S.D. *et al*, *Phytochemistry*, 1987, **26**, 483.

24,25-Dihydroxy-7,9(11)-lanostadien-3-one   D-60343

$C_{30}H_{48}O_3$ M 456.707

(25S)-form
Ganodermanondiol
Constit. of *Gonoderma lucidum*. Needles (MeOH). Mp 182-183°. $[\alpha]_D^{23}$ +45.8° (c, 0.5 in HCl₃).

Fujita, A. *et al*, *J. Nat. Prod.*, 1986, **49**, 1122.

3,15-Dihydroxy-7,9(11),24-lanostatrien-26-oic acid   D-60344

$C_{30}H_{46}O_4$ M 470.691

(3α,5α,15α,24E)-form
3-Ac: 3α-Acetoxy-15α-hydroxy-7,9(11),24E-lanostatrien-26-oic acid. **Ganoderic acid Mf**.
$C_{32}H_{48}O_5$ M 512.728
Metab. of *Ganoderma lucidum*. Syrup. $[\alpha]_D^{24}$ +42° (c, 0.2 in CHCl₃).
3,15-Di-Ac: 3α,15α-Diacetoxy-7,9(11),24E-lanostatrien-26-oic acid. **Ganoderic acid Me**.
$C_{34}H_{50}O_6$ M 554.765
Metab. of *G. lucidum*. Syrup. $[\alpha]_D^{24}$ +53° (c, 0.26 in MeOH).

Nishitoba, T. *et al*, *Agric. Biol. Chem.*, 1987, **51**, 619.

1,11-Dihydroxy-20(29)-lupen-3-one   D-60345

$C_{30}H_{48}O_3$ M 456.707

(1β,11α)-form
Constit. of *Salvia deserta*. Cryst. (MeOH). Mp 234-237°. $[\alpha]_D^{18}$ +108° (c, 0.084 in CHCl₃).

Savona, G. *et al*, *Phytochemistry*, 1987, **26**, 3305.

9,10-Dihydroxy-7-marasmen-5,13-olide   D-60346

$C_{15}H_{20}O_4$ M 264.321
Constit. of *Lactarius vellereus*. Oil. $[\alpha]_D^{20}$ +11.2° (c, 0.5 in CH₂Cl₂).

Daniewski, W.M. *et al*, *Phytochemistry*, 1988, **27**, 187.

2′,7-Dihydroxy-4′-methoxy-4-(2′,7-dihydroxy-4′-methoxyisoflavan-5′-yl)-isoflavan   D-60347

$C_{32}H_{30}O_8$ M 542.584

(−)-form
Constit. of *Dalbergia odorifera*. Pale-yellow powder. $[\alpha]_D$ −66.7° (MeOH).
3′-Hydroxy:
$C_{32}H_{30}O_9$ M 558.584
Constit. of *D. odorifera*. Needles (C₆H₆/Me₂CO). Mp 168-169°. $[\alpha]_D$ −130.8° (MeOH).
4′-Methoxy:
$C_{33}H_{32}O_9$ M 572.610
Constit. of *D. odorifera*. Pale-yellow powder. $[\alpha]_D$ −148.6° (MeOH).
3′-Hydroxy, 2′-Me ether:
$C_{33}H_{32}O_9$ M 572.610
Constit. of *D. odorifera*. Pale-yellow powder. $[\alpha]_D$ −111.3° (MeOH).

(+)-form
Constit. of *D. nitidula*.

Bezuidenhoudt, B.C.B. *et al*, *J. Chem. Soc., Perkin Trans. 1*, 1984, 2767 (*isol*)
Yahara, S. *et al*, *Chem. Pharm. Bull.*, 1985, **33**, 5130 (*cryst struct*)

4,7-Dihydroxy-5-methyl-2*H*-1-benzopyran-2-one, 9CI   D-60348

Updated Entry replacing D-04693
4,7-Dihydroxy-5-methylcoumarin
[23664-28-0]
$C_{10}H_8O_4$ M 192.171
Mp 265-267°.

Di-Me ether: [53377-54-1]. *4,7-Dimethoxy-5-methyl-2H-1-benzopyran-2-one. 4,7-Dimethoxy-5-methylcoumarin*. **Siderin**.
$C_{12}H_{12}O_4$ M 220.224
Found in *Sideritis* spp., *Cedrela toona, Clematis ligusticifolia* and other plants; and *Aspergillus variecolor*. Cryst. (MeOH). Mp 196-197°. Formerly assigned an isomeric struct.

Hay, J.V. *et al*, *J. Chem. Soc., Chem. Commun.*, 1972, 953.
Gonzalez, A.G. *et al*, *Chem. Ind.* (*London*), 1974, 166 (*struct, deriv*)
Lapper, R.D. *et al*, *Tetrahedron Lett.*, 1974, 4293 (*cmr, deriv*)
Venturella, P., *Tetrahedron Lett.*, 1974, 979 (*synth*)
Ayer, W.A. *et al*, *Phytochemistry*, 1975, **14**, 1457 (*deriv*)
Chexal, K.K. *et al*, *J. Chem. Soc., Perkin Trans. 1*, 1975, 554 (*deriv*)
Nagasampagi, B.A. *et al*, *Phytochemistry*, 1975, **14**, 1673 (*deriv*)
Ahluwalia, V.K. *et al*, *Indian J. Chem., Sect. B*, 1976, **14**, 589; 1977, **15**, 816 (*synth, ms*)
Joshi, B.S. *et al*, *Proc.-Indian Acad. Sci., Sect. A*, 1979, **88**, 185 (*synth*)
De Pascual, J. *et al*, *Phytochemistry*, 1981, **20**, 2778 (*deriv*)

5,7-Dihydroxy-6-methyl-1(3*H*)-isobenzo-furanone, 9CI   D-60349

Updated Entry replacing D-04719
5,7-Dihydroxy-6-methylphthalide
[55483-01-7]
$C_9H_8O_4$ M 180.160
Constit. of *Alectoria nigricans* and isol. from *Aspergillus duricaulis*. Needles (AcOH aq.). Mp 224-225° dec.
7-Me ether: [17811-38-0]. *5-Hydroxy-7-methoxy-6-methylphthalide.* **Nidulol.**
  $C_{10}H_{10}O_4$ M 194.187
  Prod. by *Aspergillus duricaulis*, *A. nidulans* and by *Emericella desertorum*. Cryst. (CHCl$_3$/MeOH). Mp 234°.
Di-Me ether: [17811-39-1]. *5,7-Dimethoxy-6-methylphthalide.*
  $C_{11}H_{12}O_4$ M 208.213
  Needles (MeOH). Mp 169.4°.

Solberg, Y., *Acta Chem. Scand., Ser. B*, 1975, **29**, 145 (*isol, struct*)
Fujita, M. *et al, Chem. Pharm. Bull.*, 1984, **32**, 2622 (*isol, struct*)
Achenbach, H. *et al, Justus Liebigs Ann. Chem.*, 1985, 1596 (*isol*)
Nozawa, K. *et al, J. Chem. Soc., Perkin Trans. 1*, 1987, 1735 (*isol*)

3,5-Dihydroxy-2-methyl-1,4-naphthoquin-one   D-60350

Updated Entry replacing D-50511
3,5-Dihydroxy-2-methyl-1,4-naphthalenedione, 9CI.
Droserone
[478-40-0]

$C_{11}H_8O_4$ M 204.182
Occurs in *Drosera* spp., *Plumbago zeylanica* and *Diospyros maritima*. Pale-yellow needles (EtOH or AcOH). Sol. hot H$_2$O. Mp 181°. Bp$_3$ 100-110° subl.
Di-Ac: [61836-58-6].
  $C_{15}H_{12}O_6$ M 288.256
  Mp 119°.
5-Me ether: [22267-00-1]. *3-Hydroxy-5-methoxy-2-methyl-1,4-naphthoquinone. Droserone 5-methyl ether.*
  $C_{12}H_{10}O_4$ M 218.209
  Isol. from heartwood of *D. melanoxylon*. Yellow needles. Mp 173-174° (171°).

Lugg, J.W.H. *et al, J. Chem. Soc.*, 1937, 1597.
Thomson, R.H., *J. Chem. Soc.*, 1949, 1277.
Sidhu, G.S. *et al, Indian J. Chem.*, 1968, **6**, 681 (*isol, uv, ir, deriv*)
Gunaherath, G.M.K.B. *et al, Phytochemistry*, 1983, **22**, 1245.
Higa, M. *et al, Chem. Pharm. Bull.*, 1987, **35**, 4366 (*isol, uv, ir, pmr*)

5,6-Dihydroxy-2-methyl-1,4-naphthoquin-one   D-60351

Updated Entry replacing D-40250
5,6-Dihydroxy-2-methyl-1,4-naphthalenedione, 9CI
[22273-47-8]

$C_{11}H_8O_4$ M 204.182
Deep red needles (Me$_2$CO/CCl$_4$). Mp 172°.
5-Me ether: [7539-90-4]. *6-Hydroxy-5-methoxy-2-methyl-1,4-naphthoquinone.* **Diomelquinone** A.
  $C_{12}H_{10}O_4$ M 218.209
  Constit. of heartwood of *Diospyros celebrica* and *D. melanoxylone*. Yellow needles (CHCl$_3$/pet. ether). Mp 152-153° (139-140°).
Di-Me ether: [4147-27-7]. *5,6-Dimethoxy-2-methyl-1,4-naphthoquinone.*
  $C_{13}H_{12}O_4$ M 232.235
  Yellow needles (MeOH). Mp 184°.

Sidhi, G.S. *et al, Justus Liebigs Ann. Chem.*, 1966, **691**, 172 (*isol, struct*)
Sidhi, G.S. *et al, Indian J. Chem.*, 1968, **6**, 681 (*synth, deriv*)
Laatsch, H., *Justus Liebigs Ann. Chem.*, 1985, 1847 (*synth, uv, ir, pmr*)
Musgrave, O.C. *et al, J. Chem. Soc., Perkin Trans. 1*, 1986, 675 (*uv, ir, pmr, ms*)

5,8-Dihydroxy-2-methyl-1,4-naphthoquin-one   D-60352

5,8-Dihydroxy-2-methyl-1,4-naphthalenedione, 9CI.
Methylnaphthazarin. Ramentone
[14554-09-7]
$C_{11}H_8O_4$ M 204.182
This tautomer preferred over the isomeric 6-methyl struct.

Fariña, F. *et al, Tetrahedron Lett.*, 1959, 9.
Rodriguez, J.-G. *et al, Bull. Chem. Soc. Jpn.*, 1986, **59**, 3957 (*pmr*)

2,5-Dihydroxy-3-methylpentanoic acid, 9CI   D-60353

Updated Entry replacing D-20289

$C_6H_{12}O_4$ M 148.158
(2S,3R)-form [53798-51-9]
Verrucarinic acid
Benzhydrylamide: Needles (Me$_2$CO/pet. ether). Mp 119-120°. [α]$_D^{23}$ −30° (c, 1.11 in Me$_2$CO).
Phenylhydrazide: Needles (Me$_2$CO/pet. ether). Mp 142-143°. [α]$_D^{22}$ −43° (c, 1.04 in Me$_2$CO).
Lactone: [1122-21-0]. *Tetrahydro-3-hydroxy-4-methyl-2H-pyran-2-one, 9CI. Verrucarinolactone.*
  $C_6H_{10}O_3$ M 130.143
  Needles (Et$_2$O). Mp 103-104°. [α]$_D^{22}$ −9° (c, 1.033 in CHCl$_3$). Enantiomer also prepd.
(2S,3S)-form
Lactone: Epiverrucarinolactone.
  $C_6H_{10}O_3$ M 130.143
  Cryst. Mp 88-89°. [α]$_D^{20}$ +103° (c, 1.9 in CHCl$_3$). Enantiomer also prepd.

Laloi, L. *et al, C. R. Hebd. Seances Acad. Sci.*, 1962, **255**, 2117 (*synth*)
Gutzwiller, J. *et al, Helv. Chim. Acta*, 1965, **48**, 157 (*synth, spectra*)
Herold, P. *et al, Helv. Chim. Acta*, 1983, **66**, 744 (*synth*)
Yamamato, Y. *et al, J. Chem. Soc., Chem. Commun.*, 1983, 774.
Grossen, P. *et al, Helv. Chim. Acta*, 1984, **67**, 1625 (*synth*)
Mulzer, J. *et al, Justus Liebigs Ann. Chem.*, 1986, 1172 (*synth, bibl*)

2,5-Dihydroxy-1,4-naphthoquinone, 8CI   D-60354

Updated Entry replacing D-04787
2,5-Dihydroxy-1,4-naphthalenedione, 9CI. 2-
Hydroxyjuglone
[4923-55-1]
$C_{10}H_6O_4$ M 190.155
Gelation inhibitor for uncured polyester resin. Orange-brown needles (AcOH). Mp 216-219° dec.

Di-Ac:
  $C_{14}H_{10}O_6$ M 274.229
  Yellow needles (pet. ether). Mp 152°.
5-Me ether: [71186-96-4]. *2-Hydroxy-5-methoxy-1,4-*
  naphthoquinone.
  $C_{11}H_8O_5$ M 220.181
  Yellow needles. Mp 174-176°.

MacLeod, J.W. *et al, J. Org. Chem.*, 1960, **25**, 36 (*synth*)
Bowie, J.H. *et al, J. Am. Chem. Soc.*, 1965, **87**, 5094 (*ms*)
Piette, L.H. *et al, J. Phys. Chem.*, 1967, **71**, 29 (*esr, nmr*)
Singh, I. *et al, Tetrahedron*, 1968, **24**, 6053 (*uv*)
Singh, H. *et al, Tetrahedron*, 1969, **25**, 5301 (*synth*)
U.S.P., 3 553 293, (*1970*); *CA*, **74**, 64869 (*use*)
Barre, G. *et al, Tetrahedron Lett.*, 1986, **27**, 6197 (*cmr, pmr*)
Kopanski, L. *et al, Justus Liebigs Ann. Chem.*, 1987, 793 (*synth, deriv, uv, ir, pmr, ms*)

2,8-Dihydroxy-1,4-naphthoquinone, 8CI   D-60355

Updated Entry replacing D-04790
2,8-Dihydroxy-1,4-naphthalenedione, 9CI. 3,5-Dihy-
droxy-1,4-naphthoquinone. 3-Hydroxyjuglone
[4923-58-4]
$C_{10}H_6O_4$ M 190.155
Uncured polyester resin gelation inhibitor. Ochre-brown needles (AcOH). Mp 215°, 220-221°.

Di-Ac: [61276-36-6].
  $C_{14}H_{10}O_6$ M 274.229
  Yellow needles (MeOH). Mp 137°.
8-Me ether: [13261-50-2]. *2-Hydroxy-8-methoxy-1,4-*
  naphthoquinone.
  $C_{11}H_8O_4$ M 204.182
  Yellow plates (pet. ether). Mp 211° dec.

MacLeod, J.W. *et al, J. Org. Chem.*, 1960, **25**, 36 (*synth*)
Bowie, J.H. *et al, J. Am. Chem. Soc.*, 1965, **87**, 5094 (*ms*)
Piette, L.H. *et al, J. Phys. Chem.*, 1967, **71**, 29 (*esr, nmr*)
Singh, I. *et al, Tetrahedron*, 1968, **24**, 6053 (*uv*)
Singh, H. *et al, Tetrahedron*, 1969, **25**, 5301 (*synth*)
Closse, A. *et al, Helv. Chim. Acta*, 1973, **56**, 619 (*synth, pmr, ir, uv, chromatog*)
Barre, G. *et al, Tetrahedron Lett.*, 1986, **27**, 6197 (*cmr, pmr*)

5,8-Dihydroxy-1,4-naphthoquinone, 8CI   D-60356

Updated Entry replacing D-04793
5,8-Dihydroxy-1,4-naphthalenedione, 9CI. Naphthazarin
[475-38-7]

$C_{10}H_6O_4$ M 190.155
Fungicide, nylon stabilizer, used in hair dyes. Red-brown needles (EtOH). Spar. sol. H_2O, sol. alkalis. Dec. on heating.
▷QL7970000.

Di-Ac:
  $C_{14}H_{10}O_6$ M 274.229
  Yellow prisms (CHCl$_3$). Mp 192-193°.
Monophenylsemicarbazone: Mp 218° dec.

Bowie, J.H. *et al, J. Am. Chem. Soc.*, 1965, **87**, 5094 (*ms*)
B.P., 1 003 600, (*1965*); *CA*, **63**, 12967 (*use*)
Brockmann, H. *et al, Chem. Ber.*, 1968, **101**, 4221 (*nmr*)
Cradwick, P.D. *et al, Acta Crystallogr., Sect. B*, 1971, **27**, 1990 (*cryst struct*)
Schmand, H.L.K. *et al, J. Am. Chem. Soc.*, 1975, **97**, 447 (*struct, ir, uv*)
Kobayashi, M. *et al, Tetrahedron Lett.*, 1976, 619 (*nmr*)
Rodriguez, J.-G. *et al, Bull. Chem. Soc. Jpn.*, 1986, **59**, 3957 (*struct, bibl, pmr*)

2′,3′-Dihydroxy-6′-nitroacetophenone   D-60357

1-(2,3-Dihydroxy-6-nitrophenyl)ethanone. 3-Acetyl-4-
nitrocatechol

$C_8H_7NO_5$ M 197.147
Di-Me ether: [98300-40-4]. *2′,3′-Dimethoxy-6′-*
  nitroacetophenone.
  $C_{10}H_{11}NO_5$ M 225.201
  Cryst. (2-Propanol). Mp 66-67°.

Bandurco, V.T. *et al, J. Med. Chem.*, 1987, **30**, 1421 (*synth, deriv*)

3′,6′-Dihydroxy-2′-nitroacetophenone   D-60358

1-(3,6-Dihydroxy-2-nitrophenyl)ethanone. 2-Acetyl-3-
nitroquinol
$C_8H_7NO_5$ M 197.147
Di-Me ether: 3′,6′-*Dimethoxy-2′-nitroacetophenone.*
  $C_{10}H_{11}NO_5$ M 225.201
  Yellow cryst. Mp 72-74°.

Howe, C.A. *et al, J. Org. Chem.*, 1960, **25**, 1245.
Bandurco, V.T. *et al, J. Med. Chem.*, 1987, **30**, 1421 (*synth, deriv*)

4′,5′-Dihydroxy-2′-nitroacetophenone   D-60359

1-(4,5-Dihydroxy-2-nitrophenyl)ethanone. 4-Acetyl-5-
nitrocatechol
$C_8H_7NO_5$ M 197.147
Di-Me ether: [4101-32-0]. *4′,5′-Dimethoxy-2′-*
  nitroacetophenone.
  $C_{10}H_{11}NO_5$ M 225.201
  Cryst. (EtOH). Mp 133-136°.

Fr. Pat., M3207, (*1965*); *CA*, **63**, 13287c (*synth, deriv*)

4,5-Dihydroxy-2-nitrobenzoic acid, 9CI   D-60360

Updated Entry replacing D-04843
6-Nitropyrocatechuic acid, 8CI
$C_7H_5NO_6$ M 199.120
4-Me ether: [31839-20-0]. *5-Hydroxy-4-methoxy-2-ni-*
  trobenzoic acid. 6-Nitroisovanillic acid.
  $C_8H_7NO_6$ M 213.146
  Yellow prisms (H_2O). Mp 181° dec.
4-Me ether, Me ester:
  $C_9H_9NO_6$ M 227.173

Mp 143°.
5-Me ether: 4-Hydroxy-5-methoxy-2-nitrobenzoic acid.
6-Nitrovanillic acid.
$C_8H_7NO_6$ M 213.146
Constit. of a *Cortinarius* sp. Pale-yellow needles
(EtOAc/pet. ether). Mp 181-182°.
Di-Me ether: [4998-07-6]. *4,5-Dimethoxy-2-nitroben-*
zoic acid.
$C_9H_9NO_6$ M 227.173
Yellow needles + ½H_2O. Mp 189-191°. pK_a 2.44
(25°).
Di-Me ether, chloride: [29568-78-3].
$C_9H_8ClNO_5$ M 245.619
Yellow needles (pet. ether). Mp 88-89°.
Di-Me ether, nitrile: [102714-71-6]. *1-Cyano-4,5-di-*
methoxy-2-nitrobenzene.
$C_9H_8N_2O_4$ M 208.173
Yellow needles (EtOH). Mp 168°.

Perkin, A.G. *et al, J. Chem. Soc.*, 1915, **107**, 206 (*synth*)
Greenwood, M. *et al, J. Chem. Soc.*, 1932, 1370 (*synth*)
Wittmer, F.B. *et al, J. Org. Chem.*, 1945, **10**, 527.
Carpenter, P.D. *et al, J. Chem. Soc., Perkin Trans. 1*, 1979, 103.
Sinha, S.K.P. *et al, J. Indian Chem. Soc.*, 1970, **47**, 925 (*synth*)
Gill, M. *et al, Phytochemistry*, 1987, **26**, 2815 (*isol*)

5,7-Dihydroxy-13-nor-8-marasmanone   D-60361

HOH₂C   H
HO
O   H

$C_{14}H_{22}O_3$ M 238.326
Constit. of *Lactarius vellereus*. Cryst. Mp 88-91°. $[\alpha]_D^{20}$
−125.9° (c, 1.6 in $CHCl_3$).

Daniewski, W.M. *et al, Phytochemistry*, 1988, **27**, 187.

5,8-Dihydroxy-13-nor-7-marasmanone   D-60362

HOH₂C   H
O
HO   H

$C_{14}H_{22}O_3$ M 238.326
Constit. of *Lactarius vellereus*. Oil. $[\alpha]_D^{20}$ −83.3° (c, 0.3
in $CHCl_3$).

Daniewski, W.M. *et al, Phytochemistry*, 1988, **27**, 187.

3,18-Dihydroxy-28-nor-12-oleanen-16-one,   D-60363
9CI

Updated Entry replacing D-10406
[96861-15-3]

HO   18   H
HO   3   O

$C_{29}H_{46}O_3$ M 442.681

(*3β,18β*)-*form*
Constit. of *Camellia japonica*. Cryst. ($CHCl_3$). Mp 215-
216.5°. $[\alpha]_D^{25}$ +30° (c, 0.22 in $CHCl_3$).
3-O-(β-D-Glucopyranosyl-(1→2)-β-D-galactopyrano-
syl-(1→4)-[α-D-galactopyranosyl-(1→2)]-β-D-glu-
curonopyranoside): [96827-23-5]. ***Camellidin II, 9CI.***
$C_{53}H_{84}O_{24}$ M 1105.232
From *C. japonica*. Antifungal agent. Cryst. Mp 211-
212°. $[\alpha]_D^{25}$ −6.0° (c, 0.5 in MeOH).
3-O-[β-D-Glucopyranosyl-(1→2)-β-D-
galactopyranosyl-(1→4)-[α-D-galactopyranosyl-
(1→2)]-β-D-glucuronopyranoside], 18-Ac: [96827-22-
4]. ***Camellidin I, 9CI.***
$C_{55}H_{86}O_{25}$ M 1147.269
Constit. of *C. japonica*. Antifungal agent. Cryst. Mp
208-209°. $[\alpha]_D^{25}$ +2.0° (c, 0.5 in MeOH).
3-Ketone: [81426-90-6]. *18β-Hydroxy-28-nor-3,16-*
oleanenedione.
$C_{29}H_{44}O_3$ M 440.665
Constit. of *C. japonica*. Cryst. Mp 232-233°. $[\alpha]_D^{26}$
+49° (c, 0.1 in $CHCl_3$).

Itokawa, H. *et al, Phytochemistry*, 1981, **20**, 2539 (*isol*)
Nagata, T. *et al, Agric. Biol. Chem.*, 1985, **49**, 1181 (*isol*)
Nishino, C. *et al, J. Chem. Soc., Chem. Commun.*, 1986, 720
(*struct*)

1,11-Dihydroxy-18-oleanen-3-one   D-60364

$C_{30}H_{48}O_3$ M 456.707

(*1β,11α*)-*form*
Constit. of *Salvia deserta*. Cryst. (MeOH). Mp 229-
233°. $[\alpha]_D^{20}$ +54° (c, 0.14 in $CHCl_3$).

Savona, G. *et al, Phytochemistry*, 1987, **26**, 3305.

6,17-Dihydroxy-15,17-oxido-16-spongian-   D-60365
one

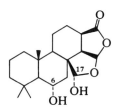

O
O
6   17   O
OH   OH

$C_{20}H_{30}O_5$ M 350.454
(*6α,17α*)-*form*
Di-Ac: [109152-40-1]. *6α,17α-Diacetoxy-15,17-oxido-*
16-spongianone.
$C_{24}H_{34}O_7$ M 434.528
Constit. of *Ceratosoma brevicaudatum*. Oil.
6-Butanoyl, 17-Ac: [109152-41-2]. *17α-Acetoxy-6α-bu-*
tanoyloxy-15,17-oxido-16-spongianone.
$C_{26}H_{38}O_7$ M 462.582
From *C. brevicaudatum*. Oil.
(*6α,17β*)-*form*
6-Butanoyl: [109181-97-7]. *6α-Butanoyloxy-17β-*
hydroxy-15,17-oxido-16-spongianone.
$C_{24}H_{36}O_6$ M 420.545
From *C. brevicaudatum*. Oil.
6-Ac: [109181-98-8]. *6α-Acetoxy-17β-hydroxy-15,17-*
oxido-16-spongianone.
$C_{22}H_{32}O_6$ M 392.491
From *C. brevicaudatum*. Oil.

Ksebati, M.B. *et al, J. Org. Chem.*, 1987, **52**, 3766.

2,9-Dihydroxy-8-oxo-1(10),4,11(13)-ger-macratrien-12,6-olide   D-60366

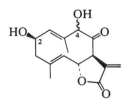

C$_{15}$H$_{18}$O$_5$ M 278.304
(*1(10)E,2β,4E,6α,9ξ*)-*form*
 Constit. of *Tanacetum vulgare*. Oil.

Chandra, A. *et al, Phytochemistry*, 1987, **26**, 1463.

11,15-Dihydroxy-9-oxo-5,13-prostadienoic acid, 9CI   D-60367

Updated Entry replacing D-20293

C$_{20}$H$_{32}$O$_5$ M 352.470
(*5Z,8R,11R,12R,13E,15S*)-*form* [363-24-6]
 Prostaglandin E$_2$. *PGE$_2$*. Dinoprostone
 Mp 62-64°. [α]$_D^{24}$ −52° (c, 1.15 in THF).
 ▷UK8000000.
(*5E,8R,11R,12R,13E,15S*)-*form* [36150-00-2]
 (*5E*)-*PGE$_2$*
 Plates (Et$_2$O/hexane). Mp 76-77°. [α]$_D$ −66° (c, 0.983
  in EtOH), −95° (c, 0.903 in CHCl$_3$).
 ▷Exp. teratogen
(*5Z,8R,11S,12R,13E,15S*)-*form* [38310-90-6]
 11-epi-*PGE$_2$*
 Oil. [α]$_D^{25}$ −26° (c, 0.0076 in EtOH).
(*5Z,8SR,11RS,12RS,13E,15SR*)-*form* [31660-17-0]
 (±)-*8*-iso-*PGE$_2$*
 Cryst. (EtOAc/heptane). Mp 90-92°.
(*5Z,8RS,11RS,12RS,13E,15RS*)-*form* [31660-13-6]
 (±)-*15*-epi-*PGE$_2$*

Bergström, S. *et al, Biochim. Biophys. Acta*, 1964, **90**, 207
  (*biosynth*)
Corey, E.J. *et al, J. Am. Chem. Soc.*, 1969, **91**, 5675;
  Tetrahedron Lett., 1970, 307 (*synth*)
Bundy, G.L. *et al, J. Am. Chem. Soc.*, 1972, **94**, 2124 (*synth*)
Floyd, D.M. *et al, Tetrahedron Lett.*, 1972, 3269 (*synth*)
Heather, J.B. *et al, Tetrahedron Lett.*, 1973, 2313 (*synth*)
Sih, C.J. *et al, J. Am. Chem. Soc.*, 1975, **97**, 865 (*synth*)
Andersen, N.H. *et al, Prostaglandins*, 1977, **14**, 61 (*synth*)
Schneider, W.P. *et al, J. Am. Chem. Soc.*, 1977, **99**, 1222
  (*synth, ms*)
Uekama, K. *et al, Chem. Lett.*, 1977, 1389 (*cd*)
Chen, S.-M.L. *et al, J. Org. Chem.*, 1978, **43**, 3450 (*synth, ir,
  pmr, cmr, ms*)
Nakamura, N. *et al, Tetrahedron Lett.*, 1978, 1549 (*synth, pmr*)
Newton, R.F. *et al, J. Chem. Soc., Perkin Trans. 1*, 1979, 2789
  (*synth*)
de Titta, G.T. *et al, Acta Crystallogr., Sect. B*, 1980, **36**, 638
  (*cryst struct, conformn*)
Donaldson, R.E. *et al, J. Org. Chem.*, 1983, **48**, 2167 (*synth*)
Tanaka, T. *et al, Tetrahedron Lett.*, 1985, **26**, 5575 (*synth*)
Johnson, C.R. *et al, J. Am. Chem. Soc.*, 1986, **108**, 5655 (*synth*)
Sax, N.I., *Dangerous Properties of Industrial Materials*, 5th
  Ed., Van Nostrand-Reinhold, 1979, 946.

3,5-Dihydroxy-4-oxo-4*H*-pyran-2-carboxylic acid   D-60368

Rubiginic acid

C$_6$H$_4$O$_6$ M 172.094
Mp 230°.

Aida, K. *et al, CA*, 1958, **52**, 5544 (*synth*)
Murooka, H. *et al, Agric. Biol. Chem.*, 1962, **26**, 135 (*synth*)

2,4-Dihydroxyphenylacetaldehyde   D-60369

2,4-Dihydroxybenzeneacetaldehyde

C$_8$H$_8$O$_3$ M 152.149
Dimethylhydrazone: [74321-33-8].
 C$_{10}$H$_{14}$N$_2$O$_2$ M 194.233
 Cryst. (CH$_2$Cl$_2$/MeOH). Mp 140° dec.
Di-Me ether, oxime:
 C$_9$H$_{11}$NO$_3$ M 181.191
 Prisms. Mp 115-116°.

Cocker, W. *et al, J. Chem. Soc.*, 1965, 1034.
Narasimhan, N. *et al, Indian J. Chem.*, 1969, **7**, 1004.
Severin, T. *et al, Chem. Ber.*, 1980, **113**, 970.

2,5-Dihydroxyphenylacetaldehyde   D-60370

2,5-Dihydroxybenzeneacetaldehyde, *9CI*
C$_8$H$_8$O$_3$ M 152.149
Di-Me ether: [33567-62-3]. *2,5-Dimethoxybenzeneace-
 taldehyde*, *9CI*. *2,5-Dimethoxyphenylacetaldehyde*.
 C$_{10}$H$_{12}$O$_3$ M 180.203
 Oil. Bp$_{0.7}$ 112-113°.

Leaf, G. *et al, Biochem. J.*, 1948, **43**, 606.
Billman, J.H. *et al, J. Pharm. Sci.*, 1971, **60**, 1188.

3,4-Dihydroxyphenylacetaldehyde   D-60371

3,4-Dihydroxybenzeneacetaldehyde, *9CI. Homoprotoca-
techualdehyde. Dopaldehyde. Dopal*
[5707-55-1]
C$_8$H$_8$O$_3$ M 152.149
Metab. of DOPA. Constit. of leaves of *Plectranthus
 caninus*. Oil.
2,4-Dinitrophenylhydrazone: Cryst. (EtOH). Mp 169-
 170°.
Semicarbazone: Cryst. (EtOH aq.). Mp 198-200°.
3-Me ether: [5703-24-2]. *4-Hydroxy-3-methoxypheny-
 lacetaldehyde. Homovanillin.*
 C$_9$H$_{10}$O$_3$ M 166.176
 Prisms (CCl$_4$). Mp 50-50.5°. Bp$_2$ 147-149°.
3-Me ether, oxime:
 C$_9$H$_{11}$NO$_3$ M 181.191
 Mp 115°.
3-Me ether, Ac:
 C$_{10}$H$_{10}$O$_4$ M 194.187
 Bp$_2$ 140-145°.
4-Me ether: 3-Hydroxy-4-methoxyphenylacetaldehyde.
 Homoisovanillin.
 C$_9$H$_{10}$O$_3$ M 166.176

Bp$_{0.2}$ 117-122°.
4-Me ether, semicarbazone: Mp 182-183°.
Di-Me ether: [5703-21-9]. *3,4-Dimethoxybenzeneacetal-dehyde, 9CI. 3,4-Dimethoxyphenylacetaldehyde.*
C$_{10}$H$_{12}$O$_3$ M 180.203
Oil. Bp$_2$ 125°, Bp$_{0.35}$ 124°.
Di-Me ether, phenylhydrazone: Cryst. (C$_6$H$_6$/hexane).
Mp 97-98°.
Di-Me ether, semicarbazone: Mp 158-160°.
Di-Me ether, di-Me acetal: [85452-72-8].
C$_{12}$H$_{18}$O$_4$ M 226.272
Liq. Bp$_{0.03}$ 115°.

Schöpf, C., *Justus Liebigs Ann. Chem.*, 1940, **544**, 30 (*deriv*)
Challis, A.A.L. *et al, J. Chem. Soc.*, 1947, 1692 (*deriv*)
Dornow, A. *et al, Arch. Pharm. (Weinheim, Ger.)*, 1951, **284**, 153 (*derivs*)
Fellman, J.H., *Nature (London)*, 1958, **182**, 311 (*synth*)
Kanaoka, Y. *et al, Chem. Pharm. Bull.*, 1966, **14**, 934 (*deriv*)
Arihara, S. *et al, Helv. Chim. Acta*, 1975, **58**, 447 (*isol*)
Ogura, K. *et al, Synthesis*, 1975, 385 (*deriv*)
Stamos, I.K., *Tetrahedron Lett.*, 1982, **23**, 459 (*deriv*)

2-(3,4-Dihydroxyphenyl)propanoic acid   D-60372

Updated Entry replacing D-04987
3,4-Dihydroxy-α-methylbenzeneacetic acid, 9CI. 3,4-Di-hydroxyhydratropic acid, 8CI. 2-(3,4-Dihydroxyphenyl)-propionic acid
[37697-50-0]

C$_9$H$_{10}$O$_4$ M 182.176
(±)-*form*
Mp 97°.
Di-Me ether: [50463-74-6].
C$_{11}$H$_{14}$O$_4$ M 210.229
Cryst. + 1H$_2$O (H$_2$O). Mp 60°.

Bougault, J., *Ann. Chim. (Paris)*, 1902, **25**, 483
Jeffrey, J., *J. Chem. Soc.*, 1955, 79 (*synth*)
Riggs, R.M. *et al, J. Med. Chem.*, 1987, **30**, 1914 (*synth, deriv, pmr*)

3,9-Dihydroxypterocarpan   D-60373

C$_{15}$H$_{12}$O$_4$ M 256.257
3-Me ether: [38822-00-3]. *9-Hydroxy-3-methoxyptero-carpan.* **Isomedicarpin**.
C$_{16}$H$_{14}$O$_4$ M 270.284
Constit. of *Psophocarpus tetragonolobus*. Cryst. Mp 106°. [α]$_D^{18}$ −201° (CHCl$_3$).
9-Me ether: see Medicarpin, M-30019

McMurry, T.B.H. *et al, Phytochemistry*, 1972, **11**, 3283 (*synth*)
Preston, N.W. *et al, Phytochemistry*, 1977, **16**, 2044 (*isol*)
Ingham, J.L. *et al, Phytochemistry*, 1980, **19**, 1203 (*isol*)
Prasad, A.V.K. *et al, J. Chem. Soc., Perkin Trans. 1*, 1986, 1561 (*synth*)

7,12-Dihydroxysterpurene   D-60374

C$_{15}$H$_{24}$O$_2$ M 236.353
Metabolite of fungus *Stereum purpureum*.

Abell, C. *et al, Tetrahedron Lett.*, 1987, **28**, 4887.

3′,5-Dihydroxy-4′,6,7,8-tetramethoxyfla-vone, 8CI   D-60375

5-Hydroxy-2-(3-hydroxy-4-methoxyphenyl)-6,7,8-tri-methoxy-4H-1-benzopyran-4-one, 9CI. **Gardenin D**
[29202-00-4]
C$_{19}$H$_{18}$O$_8$ M 374.346
Constit. of *Gardenia lucida* and aerial parts of *Sideritis* spp. Yellow cubes (C$_6$H$_6$). Mp 191-192°.

Rao, A.V.R. *et al, Indian J. Chem.*, 1970, **8**, 398 (*isol, uv*)
Krishnamurti, M. *et al, Indian J. Chem.*, 1970, **8**, 575 (*synth, uv, pmr*)
Rodriguez, B., *Phytochemistry*, 1977, **16**, 800 (*isol*)

5,7-Dihydroxy-3′,4′,6,8-tetramethoxyfla-vone   D-60376

2-(3,4-Dimethoxyphenyl)-5,7-dihydroxy-6,8-dimeth-oxy-4H-1-benzopyran-4-one, 9CI. **Hymenoxin**
[56003-01-1]
C$_{19}$H$_{18}$O$_8$ M 374.346
Constit. of *Helianthus simulans, H. angustifolius, Hymenoxys scaposa* and *H. linearifolia*. Yellow cryst. (MeOH). Mp 215-216°.

Thomas, M.B. *et al, J. Org. Chem.*, 1967, **32**, 3254 (*isol, synth, uv, ir, pmr*)
Waddell, T.G., *Phytochemistry*, 1973, **12**, 2061 (*isol, pmr, ms*)
Ohno, N. *et al, Phytochemistry*, 1981, **20**, 2393 (*isol*)
Herz, W. *et al, Phytochemistry*, 1983, **22**, 2021 (*isol*)

5,8-Dihydroxy-2,3,6,7-tetramethyl-1,4-naphthoquinone   D-60377

5,8-Dihydroxy-2,3,6,7-tetramethyl-1,4-naphthalene-dione, 9CI. **Tetramethylnaphthazarin**
[60583-16-6]

C$_{14}$H$_{14}$O$_4$ M 246.262
Mp 186-187° subl.
Di-Ac:
C$_{18}$H$_{18}$O$_6$ M 330.337
Yellow solid. Mp 219-220°.

Rodriguez, J.-G. *et al, Bull. Chem. Soc. Jpn.*, 1986, **59**, 3957 (*synth, pmr, cryst struct*)

7,12-Dihydroxy-3,11,15,23-tetraoxo-   **D-60378**
8,20(22)-lanostadien-26-oic acid

C$_{30}$H$_{40}$O$_8$ M 528.641

(7β,12β,20(22)E)-form
Ganoderenic acid E
Constit. of *Ganoderma lucidum*.
Me ester: Cryst. (EtOAc/MeOH). Mp 227-229°.

Nishitoba, T. *et al, Phytochemistry*, 1987, **26**, 1777.

7,20-Dihydroxy-3,11,15,23-tetraoxo-8-lan-   **D-60379**
osten-26-oic acid

C$_{30}$H$_{42}$O$_8$ M 530.657

(7β,20ξ)-form
Ganoderic acid N
Constit. of *Ganoderma lucidum*.
Me ester: Cryst. (EtOAc/cyclohexane). Mp 164-167°.
  [α]$_D^{24}$ +153° (c, 0.2 in MeOH).
7-Ketone: Ganoderic acid O. 20ξ-Hydroxy-3,7,11,15,23-
  pentaoxo-8-lanosten-26-oic acid.
  C$_{30}$H$_{40}$O$_8$ M 528.641
  Constit. of *G. lucidum*. Not the same as Granoderic
  acid *O* in 3,7,15,22-Tetrahydroxy-8,24-lanostadien-26-
  oic acid, T-60111.
7-Ketone, Me ester: Pale-yellow needles (Et$_2$O/hexane).
  Mp 168-171°.

Nishitoba, T. *et al, Phytochemistry*, 1987, **26**, 1777.

5,8-Dihydroxy-2,3,6-trimethyl-1,4-naphth-   **D-60380**
oquinone
5,8-Dihydroxy-2,3,6-trimethyl-1,4-naphthalenedione.
Trimethylnaphthazarin

C$_{13}$H$_{12}$O$_4$ M 232.235
Due to symmetry and tautomerism only one
  trimethylnaphthazarin exists which is tautomeric with
  the 2,6,7-trimethyl struct. Mp 165° subl.
Di-Ac:
  C$_{17}$H$_{16}$O$_6$ M 316.310
  Yellow solid. Mp 164-167°. A mixt. of 2,3,7- and
  2,6,7-trimethyl isomers in ratio 88:12.
Rodriguez, J.-G. *et al, Bull. Chem. Soc. Jpn.*, 1986, **59**, 3957
  (*synth, pmr*)

Handle all chemicals with care

2,3-Dihydroxy-12,20(30)-ursadien-28-oic   **D-60381**
acid

C$_{30}$H$_{46}$O$_4$ M 470.691

(2α,3α)-form
Constit. of *Prunella vulgaris*.
Me ester: Needles (MeOH). Mp 128-129°. [α]$_D^{23}$ +128°
  (c, 0.3 in CHCl$_3$).
24-Hydroxy: 2α,3α,24-Trihydroxy-12,20(30)-ursadien-
  28-oic acid.
  C$_{30}$H$_{46}$O$_5$ M 486.690
  Constit. of *P. vulgaris*.
24-Hydroxy, Me ester: Needles (MeOH). Mp 212-213°.
  [α]$_D^{19}$ +121.2° (c, 0.5 in CHCl$_3$).

Kojima, H. *et al, Phytochemistry*, 1987, **26**, 1107.

3,19-Dihydroxy-12-ursene-24,28-dioic acid   **D-60382**

C$_{30}$H$_{46}$O$_6$ M 502.690

(3β,19α)-form [108524-94-3]
Ilexgenin A
Constit. of the roots of *Ilex pubescens*. Powder. Mp
  >300°. [α]$_D^{20}$ +30.8° (c, 0.97 in Py).
Di-Me ester: Cryst. (MeOH). Mp 199-203°. [α]$_D^{20}$
  +78.2° (c, 1.01 in Py).
28-O-β-D-Glucopyranosyl ester: [108524-93-2]. **Ilexsa-**
  ponin A1.
  C$_{36}$H$_{56}$O$_{11}$ M 664.832
  From roots of *I. pubscens*. Amorph. powder. [α]$_D^{20}$
  +25.5° (c, 0.69 in Py).

Hidaka, K. *et al, Phytochemistry*, 1987, **26**, 2023.

1,3-Dihydroxy-12-ursen-28-oic acid   **D-60383**

C$_{30}$H$_{48}$O$_4$ M 472.707

(1β,3β)-form [107693-87-8]
Kaneric acid
Constit. of *Nerium oleander*. Plates (MeOH). Mp 122°.
  [α]$_D^{24}$ +16.66° (c, 0.6 in CHCl$_3$).
Siddiqui, S. *et al, J. Nat. Prod.*, 1986, **49**, 1086.

Diindolo[3,2-*a*:3′,2′-*c*]carbazole   **D-60384**
6,11-Dihydro-5H-diindolo[2,3-a:2′,3′-c]carbazole, 9CI

C$_{24}$H$_{15}$N$_3$ M 345.403

Mp >320°.

Bocchi, V. *et al*, *Tetrahedron*, 1986, **42**, 5019 (*synth, uv, pmr, cmr, ms*)

2,2'-Diiodo-1,1'-binaphthyl   D-60385

2,2'-Diiodo-1,1'-binaphthalene, 9CI

[76905-80-1]

(R)-form

$C_{20}H_{12}I_2$ M 506.124

(R)-form [86688-06-4]
Mp 215.2-216.8°. $[\alpha]_D^{23}$ +16.4° (c, 1.725 in Py).
(±)-form [86632-29-3]
Mp 225-227°.

Cava, M.P. *et al*, *J. Am. Chem. Soc.*, 1955, **77**, 6022 (*synth*)
Grachev, V.T. *et al*, *Zh. Strukt. Khim.*, 1980, **21**, 19 (*conformn*)
Brown, K.J. *et al*, *J. Org. Chem.*, 1985, **50**, 4345 (*synth*)

4,4'-Diiodo-1,1'-binaphthyl   D-60386

4,4'-Diiodo-1,1'-binaphthalene, 9CI

[62012-57-1]

$C_{20}H_{12}I_2$ M 506.124

(±)-form
Cryst. (toluene/pet. ether). Mp 228-233°.

McKillop, A. *et al*, *J. Am. Chem. Soc.*, 1980, **102**, 6504 (*synth*)

1,3-Diiodo-2,2-diazidopropane   D-60387

[102586-55-0]

$$I_2C(CH_2N_3)_2$$

$C_3H_4I_2N_6$ M 377.914
Needles (hexane). Mp 41-42°.

Hassner, A. *et al*, *J. Org. Chem.*, 1986, **51**, 2767 (*synth, cmr, pmr, ir, ms*)

2,4-Diiodothiazole   D-60388

[108306-55-4]

C_3HI_2NS M 336.917
Mp 110-112°.

Dondoni, A. *et al*, *Synthesis*, 1986, 757 (*synth, ir, pmr*)

2,5-Diiodothiazole   D-60389

[108306-62-3]
C_3HI_2NS M 336.917
Mp 103-105°.

Dondoni, A. *et al*, *Synthesis*, 1986, 757 (*synth, ir, pmr*)

1,3-Diisocyanatocyclopentane   D-60390

$C_7H_8N_2O_2$ M 152.152
Liq. $Bp_{0.05}$ 40-42°.

Ashcroft, P.L. *et al*, *J. Chem. Soc., Perkin Trans. 1*, 1986, 601 (*synth, ir, pmr, cmr*)

2,2'-Diisocyano-1,1'-binaphthyl   D-60391

2,2'-Diisocyano-1,1'-binaphthalene, 9CI

(R)-form

$C_{22}H_{12}N_2$ M 304.350
(R)-form [104144-79-8]
$[\alpha]_D^{23}$ −65° (C_6H_6).
(S)-form [104144-80-1]
$[\alpha]_D^{23}$ +69° (C_6H_6).

Yamamoto, Y. *et al*, *Inorg. Chim. Acta*, 1986, **115**, L35 (*synth, cryst struct*)

Diisopropylcyanamide   D-60392

Bis(1-methylethyl)cyanamide, 9CI

[3085-76-5]

$$[(H_3C)_2CH]_2NCN$$

$C_7H_{14}N_2$ M 126.201
Reagent for cyanation of organometallic compds. Bp_{15} 87-89°.

Mukaiyama, T. *et al*, *Bull. Chem. Soc. Jpn.*, 1954, **27**, 416 (*synth*)
Crossley, R. *et al*, *J. Chem. Soc., Perkin Trans. 1*, 1985, 2479 (*synth, ir, pmr, use*)

3,4-Diisopropyl-2,5-dimethylhexane   D-60393

2,5-Dimethyl-3,4-bis(1-methylethyl)hexane, 9CI.
1,1,2,2-Tetraisopropylethane

[102652-67-5]

$$[(H_3C)_2CH]_2CHCH[CH(CH_3)_2]_2$$

$C_{14}H_{30}$ M 198.391
Bruch, M. *et al*, *J. Org. Chem.*, 1986, **51**, 2969 (*synth, pmr, cmr*)

Dilophic acid   D-60394

9-(1,5-Dimethyl-4-hexenyl)-2,6-dimethyl-2,5-cyclonon-adiene-1-carboxylic acid, 9CI

[108864-15-9]

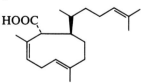

$C_{20}H_{32}O_2$ M 304.472
Constit. of *Dilophus guineensis*. Oil. $[\alpha]_D^{25}$ −116° (c, 2.35 in CHCl_3).

Schlenk, D. *et al*, *Phytochemistry*, 1987, **26**, 1081.

(Dimercaptomethylene)propanedioic acid, 9CI   D-60395

(Dimercaptomethylene)malonic acid

$$(HS)_2C=C(COOH)_2$$

$C_4H_4O_4S_2$ M 180.193

Di-S-Me, di-Me ester: [19607-08-0].
   $C_8H_{12}O_4S_2$ M 236.300
   Large rhombohedra (C_6H_6). Mp 76-78°.
Dinitrile: [4885-93-2]. *(Dimercaptomethylene)-malononitrile. 2,2-Dicyanoethylene-1,1-dithiol.*
   $C_4H_2N_2S_2$ M 142.193
   Useful reagent for cyclocondensations. V. small prisms + 3H_2O (EtOH aq.) (as di-Na salt).
Dinitrile, di-S-Me: [5147-80-8]. [*Bis(methylthio)-methylene*]*propanedinitrile*, 9CI.
   $C_6H_6N_2S_2$ M 170.247
   Useful reagent for cyclocondensations. Cryst. (MeOH). Mp 80-81°.

Söderbäck, E., *Acta Chem. Scand.*, 1963, **17**, 362 (*synth, derivs*)
Hummel, H.-U., *Acta Crystallogr.*, *Sect. C*, 1987, **43**, 41 (*cryst struct, nitrile*)

13,19:14,18-Dimethenoanthra[1,2-*a*]-benzo[*o*]pentaphene, 9CI   D-60396

Cyclo[d.e.d.e.e.d.e.d.e.e]*decakisbenzene*
[15123-45-2]

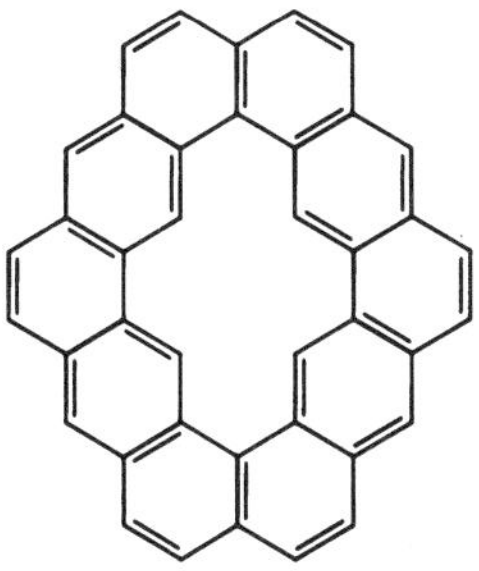

$C_{40}H_{20}$ M 500.598
Homologue of Kekulene, K-00069 . Yellow cryst. Mp >330° dec.

Funhoff, D.J.H. *et al*, *Angew. Chem., Int. Ed. Engl.*, 1986, **25**, 742 (*synth, ir, pmr, uv*)

5,6-Dimethoxy-1,3-cyclohexadiene   D-60397

[102698-33-9]

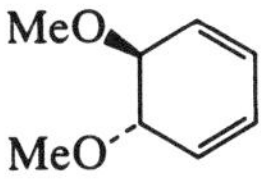

$C_8H_{12}O_2$ M 140.182
(5RS,6RS)-form
   (±)-trans-*form*
   Bp_{15} 65°.

Braun, H. *et al*, *Justus Liebigs Ann. Chem.*, 1986, 1360 (*synth, pmr, cmr*)

5,6-Dimethoxy-3-(4-methoxybenzyl)-phthalide   D-60398

5,6-Dimethoxy-3-[(4-methoxyphenyl)methyl]-1(3H)-isobenzofuranone, 9CI
[108907-02-4]

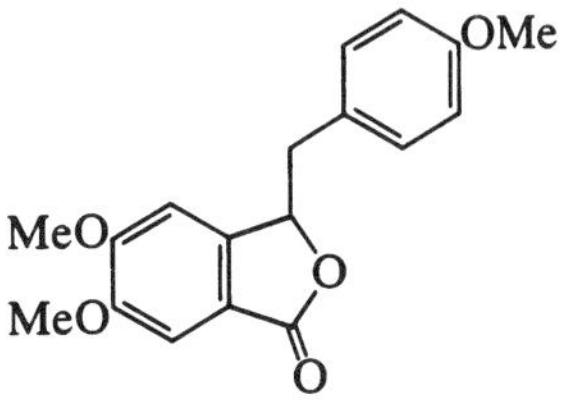

$C_{18}H_{18}O_5$ M 314.337
Constit. of *Frullania falciloba*. Cryst. Mp 78-80°.

Asakawa, Y. *et al*, *Phytochemistry*, 1987, **26**, 1023.

1-(Dimethoxymethyl)-2,3,5,6-tetrakis-(methylene)-7-oxabicyclo[2.2.1]heptane, 9CI   D-60399

[98202-88-1]

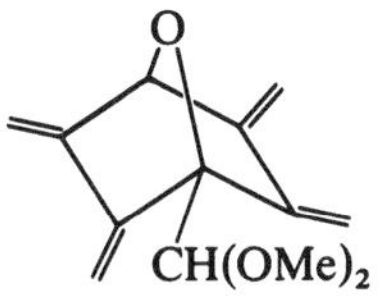

$C_{13}H_{16}O_3$ M 220.268
Versatile reagent for regioselective tandem cycloadditions. Needles. Mp 66-68°.

Métral, J.-L. *et al*, *Helv. Chim. Acta*, 1986, **69**, 1287 (*synth, uv, ir, pmr, cmr, ms*)

3,7-Dimethyladenine   D-60400

3,7-Dihydro-3,7-dimethyl-6H-purin-6-amine, 9CI

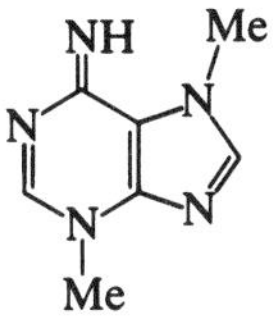

$C_7H_9N_5$ M 163.182
B,HI: [89885-39-2]. Prisms. Mp >300°. Other 3,7-dialkyladenines also prepd.

Broom, A.D. *et al*, *Biochemistry*, 1964, **3**, 494 (*synth*)
Fujii, T. *et al*, *Chem. Pharm. Bull.*, 1986, **34**, 1821 (*synth, uv, pmr*)

2-(Dimethylamino)benzaldehyde, 9CI   D-60401

Updated Entry replacing D-05464

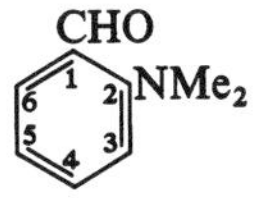

$C_9H_{11}NO$ M 149.192
Yellow oil. Bp 244°, $Bp_{4.5}$ 110-113°.
Oxime: [57678-40-7].
   $C_9H_{12}N_2O$ M 164.207
   Cryst. Mp 86-88°.
4-Nitrophenylhydrazone: Purple cryst. Mp 191°.
Di-Me acetal: [100656-07-3].
   $C_{11}H_{17}NO_7$ M 275.258
   $Bp_{0.09}$ 61°.

Cocker, W. *et al*, *J. Chem. Soc.*, 1938, 751 (*synth*)

Gale, D.J. *et al, Aust. J. Chem.*, 1975, **28**, 2447 (*synth*)
Wulff, G. *et al, Chem. Ber.*, 1986, **119**, 1876 (*synth, deriv*)

2-Dimethylamino-3,3-dimethylazirine, 9CI   D-60402

Updated Entry replacing D-10462
N,N,*2,2-Tetramethyl-2H-azirin-3-amine*, *9CI*
[54856-83-6]

$C_6H_{12}N_2$ M 112.174
Reagent for construction of peptides, e.g. these contg. the α-aminoisobutyric acid residue. Liq. Bp$_1$ 42°.

Rens, R. *et al, Tetrahedron Lett.*, 1970, 3765 (*synth*)
Vittorelli, P. *et al, Tet*, 1974, **30**, 3737 (*synth*)
Henriet, M. *et al, Tetrahedron Lett.*, 1980, **21**, 223 (*synth*)
Obrecht, D. *et al, Helv. Chim. Acta*, 1987, **70**, 329 (*use*)
Wipf, P. *et al, Helv. Chim. Acta*, 1987, **70**, 354 (*use*)

10,10-Dimethyl-9(10*H*)-anthracenone, 9CI   D-60403

10,10-Dimethylanthrone
[5447-86-9]

$C_{16}H_{14}O$ M 222.286
Cryst. (pet. ether). Mp 102-103°.
Hydrazone: [90624-29-6].
   $C_{16}H_{16}N_2$ M 236.316
   Gum.
Hydrazone, picrate: [101023-30-7]. Yellow cryst. (EtOH). Mp 193-197°.

Davis, M.A. *et al, J. Med. Chem.*, 1964, **7**, 88 (*synth*)
Falshaw, C.P. *et al, J. Chem. Soc., Perkin Trans. 1*, 1985, 1837 (*synth, ir, pmr, cmr, ms*)

1,6-Dimethylazulene, 9CI   D-60404

Updated Entry replacing D-05583
[52345-24-1]
$C_{12}H_{12}$ M 156.227
$λ_{max}$ 395 nm (log ε 2.51) (hexane).

Scott, L.T., *J. Chem. Soc., Chem. Commun.*, 1973, 882 (*synth*)
Rudolf, K. *et al, J. Org. Chem.*, 1987, **52**, 641 (*synth, pmr*)

7,12-Dimethylbenz[*a*]anthracene, 9CI   D-60405

Updated Entry replacing D-05623
9,10-Dimethyl-1,2-benzanthracene (*obsol.*)
[57-97-6]
$C_{20}H_{16}$ M 256.346
Leaflets (Me$_2$CO/EtOH). Mp 122-123°. Bluish-violet fluor. in uv light.
▷Highly carcinogenic. Teratogen. CW3850000.
Monopicrate: Black needles (EtOH). Mp 112-113°.
Dipicrate: Red needles (EtOH). Mp 102-106°.
1,2,3,4-Tetrahydro: [67242-54-0].
   $C_{20}H_{20}$ M 260.378
   Yellow cryst. Mp 89-90°. Carcinogenic.
▷CX3149000.

Bachmann, W.E. *et al, J. Am. Chem. Soc.*, 1938, **60**, 1024.
Newman, M.S., *J. Am. Chem. Soc.*, 1938, **60**, 1141 (*synth*)
Fieser, L.F. *et al, J. Am. Chem. Soc.*, 1940, **62**, 3103 (*synth*)
Crawford, R.J. *et al, J. Am. Chem. Soc.*, 1957, **79**, 3154 (*synth, tetrahydro*)
Grant, G.A. *et al, Nature* (*London*), 1969, **222**, 966
Newman, M.S. *et al, J. Org. Chem.*, 1971, **36**, 966 (*synth*)
Newman, M.S. *et al, Tetrahedron Lett.*, 1977, 2067 (*synth*)
Witiak, D.T. *et al, J. Org. Chem.*, 1986, **51**, 4499 (*bibl, tetrahydro*)
Sax, N.I., *Dangerous Properties of Industrial Materials*, 5th Ed., Van Nostrand-Reinhold, 1979, 599.

2,5-Dimethyl-1,4-benzenedithiol, 9CI   D-60406

Updated Entry replacing D-40310
2,5-Dimercapto-p-xylene
[104014-86-0]

$C_8H_{10}S_2$ M 170.287
Cryst. by subl. Mp 123-124°.
Di-Me ether: [34678-70-1]. *1,4-Dimethyl-2,5-bis(methylthio)benzene. 2,5-Bis(methylthio)-p-xylene.*
   $C_{10}H_{14}S_2$ M 198.341
   Cryst. (EtOH). Mp 83-85°.

Kloosterziel, H. *et al, J. Chem. Soc., Chem. Commun.*, 1971, 1365 (*synth*)
Beimling, P. *et al, Chem. Ber.*, 1986, **119**, 3198 (*synth, ir, pmr*)

3,6-Dimethyl-1,2,4,5-benzenetetracarboxylic acid   D-60407

$C_{12}H_{10}O_8$ M 282.206
Tetra-Me ester: [51082-19-0].
   $C_{16}H_{18}O_8$ M 338.313
   Cryst. (EtOH). Mp 142-143°.
Tetranitrile: [80717-49-3]. *1,2,4,5-Tetracyano-3,6-dimethylbenzene. Tetracyano-p-xylene.*
   $C_{12}H_6N_4$ M 206.206
   Needles (MeCN). Mp 335° dec.

Wong, H.N.C. *et al, Synthesis*, 1984, 787 (*synth, deriv, pmr*)
Staab, H.A. *et al, Chem. Ber.*, 1987, **120**, 541 (*synth, deriv*)

2,2-Dimethyl-2*H*-1-benzopyran-6-carboxylic acid   D-60408

$C_{12}H_{12}O_3$ M 204.225
Me ester: [34818-57-0].
   $C_{13}H_{14}O_3$ M 218.252
   Constit. of *Piper hostmannianum*. Oil.

Ahluwalia, V.K. *et al, Tetrahedron*, 1982, **38**, 3673 (*synth*)
Díaz, D.P.P. *et al, Phytochemistry*, 1987, **26**, 809 (*isol*)

7,7-Dimethylbicyclo[4.1.0]hept-3-ene   D-60409

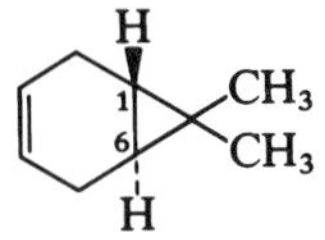

(1RS,6RS)-form

C_9H_{14} M 122.210

(**1RS,6RS**)-**form** [101934-24-1]

(±)-trans-form

Readily converts to cis-form photochemically or at 110°. Extremely acid sensitive.

(**1RS,6SR**)-**form** [36168-41-9]

cis-form

Liq.

Craig, M.D. et al, Org. Prep. Proc. Int., 1971, **3**, 275 (synth, ir, pmr)

Gassman, P.G. et al, J. Org. Chem., 1986, **51**, 2397 (synth, pmr, props)

4,4-Dimethylbicyclo[3.2.1]octane-2,3-dione, 9CI   D-60410

Carbocamphenilone

[27455-93-2]

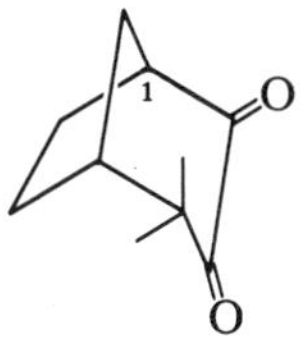

(1S)-form

$C_{10}H_{14}O_2$ M 166.219

(**1S**)-**form** [53549-12-5]

Mp 55-56° (49-52°). $[\alpha]_D$ +313° (c, 0.59 in cyclohexane). Rapidly hydrates in air.

Covalent hydrate: Mp 58-60°. Yellow melt.

Mono-2,4-dinitrophenylhydrazone: Yellow cryst. Mp 175° dec.

(±)-**form** [53537-91-0]

Cryst. (pentane). Mp 57-61°. $Bp_{1.5}$ 105°.

Jacob, G. et al, Bull. Soc. Chim. Fr., 1959, 1374 (synth, abs config)

Hug, W. et al, Helv. Chim. Acta, 1971, **54**, 633 (cd)

Noyori, R. et al, J. Org. Chem., 1975, **40**, 2681 (synth, bibl)

Ranganathan, S. et al, Tetrahedron, 1977, **33**, 2415 (synth)

Seymour, J.P. et al, Acta Crystallogr., Sect. B, 1977, **33**, 2667 (cryst struct)

3,3-Dimethylbicyclo[2.2.2]octan-2-one   D-60411

Homocamphenilone

[50682-96-7]

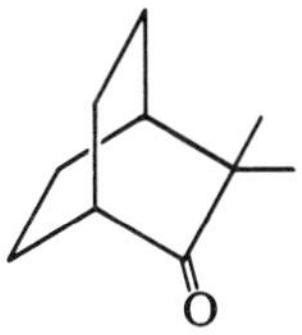

$C_{10}H_{16}O$ M 152.236

Cryst. by subl. Mp 97°.

Spreitzer, H. et al, Justus Liebigs Ann. Chem., 1986, 1578 (synth, ir, pmr, ms)

3,3-Dimethylbicyclo[2.2.2]oct-5-en-2-ene   D-60412

Dehydrohomocamphenilone

[53658-78-9]

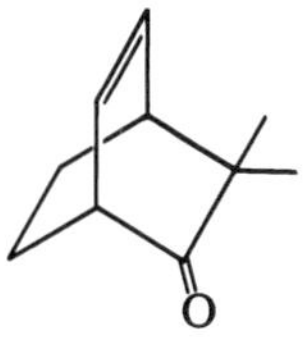

$C_{10}H_{14}O$ M 150.220

(±)-**form**

Cryst. by subl. Mp 99°.

Spreitzer, H. et al, Justus Liebigs Ann. Chem., 1986, 1578 (synth, ir, pmr, ms)

2,2′-Dimethyl-4,4′-biquinoline, 9CI   D-60413

4,4′-Biquinaldine

[52191-71-6]

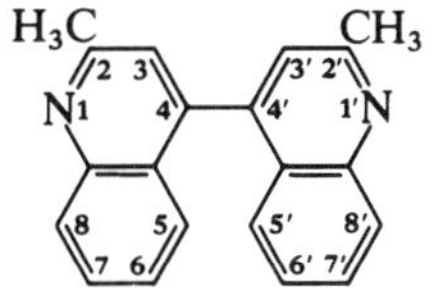

$C_{20}H_{16}N_2$ M 284.360

Mp 238°.

Vanderesse, R. et al, Tetrahedron Lett., 1986, **27**, 5483 (synth, pmr)

N,N-Dimethylcarbamohydroxamic acid   D-60414

$Me_2NCONHOH$

$C_3H_8N_2O_2$ M 104.108

Solid (EtOAc/EtOH). Mp 111-112° (107-109°).

Zinner, G. et al, Arch. Pharm. (Weinheim, Ger.), 1974, **307**, 7

Defoin, A. et al, Helv. Chim. Acta, 1987, **70**, 554 (synth)

2,3-Dimethyl-2-cyclobuten-1-one, 9CI   D-60415

[83897-48-7]

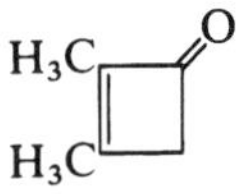

C_6H_8O M 96.129

Liq. Bp_{15} 50-60°.

Ammann, A.A. et al, Helv. Chim. Acta, 1987, **70**, 321 (synth, uv, pmr)

2,2-Dimethyl-5-cycloheptene-1,3-dione, 9CI   D-60416

[104091-52-3]

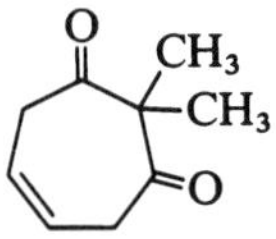

$C_9H_{12}O_2$ M 152.193

Yellow oil. $Bp_{0.01}$ 40-50°.

Föhlisch, B. et al, Justus Liebigs Ann. Chem., 1987, 1 (synth, pmr, cmr, ir, ms)

3,3-Dimethyl-1-cyclopropene-1-carboxylic acid, 9CI   D-60417

[108176-13-2]

$C_6H_8O_2$ M 112.128

Yellow oil. Unstable at r.t., stable at −20° in Et_2O soln.

Me ester:
$C_7H_{10}O_2$ M 126.155
Liq. $Bp_{0.1}$ 30°. Stored at −20° in $CHCl_3$ soln.

Baird, M.S. *et al*, *J. Chem. Soc., Perkin Trans. 1*, 1986, 1845 (*synth, pmr*)

3,3-Dimethyl-4,4-diphenyl-1-butene   D-60418

1,1'-(2,2-Dimethyl-3-butenylidene)bisbenzene, 9CI

[42842-41-1]

$$Ph_2CHC(CH_3)_2CH{=}CH_2$$

$C_{18}H_{20}$ M 236.356

Liq. which solidifies in refrigerator. $Bp_{0.076}$ 135-140°.

Mayr, H. *et al*, *Tetrahedron*, 1986, **42**, 4211 (*synth, pmr, ms*)

2,5-Dimethyl-3,4-diphenyl-2,4-hexadiene   D-60419

1,1'-[1,2-Bis(1-methylethylidene)-1,2-ethanediyl]-bisbenzene, 9CI

[102427-32-7]

$$(H_3C)_2C{=}CPhCPh{=}C(CH_3)_2$$

$C_{20}H_{22}$ M 262.394

Needles (MeOH). Mp 39.5-40°.

Kawamura, Y. *et al*, *Tetrahedron*, 1986, **42**, 6195 (*synth, pmr, ir, ms, uv*)

3,3-Dimethyl-5,5-diphenyl-4-pentenoic acid, 9CI   D-60420

[104394-22-1]

$$Ph_2C{=}CHC(CH_3)_2CH_2COOH$$

$C_{19}H_{20}O_2$ M 280.366

Light-yellow oil.

Me ester: [56405-97-1].
$C_{20}H_{22}O_2$ M 294.393
Oil.

Zimmerman, H.E. *et al*, *J. Org. Chem.*, 1978, **43**, 1997; 1986, **51**, 4604 (*synth, pmr, ir*)

2,5-Dimethyl-1,4-dithiane-2,5-diol, 9CI   D-60421

2,5-Dihydroxy-2,5-dimethyl-1,4-dithiane

[55704-78-4]

$C_6H_{12}O_2S_2$ M 180.280

Dimer of 1-Mercapto-2-propanone. Cryst. (C_6H_6). Mp 117-118°, Mp 136-138° (dimorph.). Cryst. form depends on method of prepn.

Haberl, R. *et al*, *Monatsh. Chem.*, 1955, **86**, 551 (*synth, ir, cryst struct*)
Brown, M.D. *et al*, *J. Chem. Soc., Perkin Trans. 1*, 1985, 1623 (*synth, ir, pmr, ms, use*)

Dimethylenebicyclo[1.1.1]pentanone   D-60422

Bis(methylene)bicyclo[1.1.1]pentanone

[85570-50-9]

C_7H_6O M 106.124

Unstable liq. $Bp_{0.000001}$ −60°. Dec. at −30°.

Dowd, P. *et al*, *Tetrahedron Lett.*, 1986, **27**, 2813 (*synth, pmr*)

3,4-Dimethyl-2(5H)-furanone   D-60423

[1575-46-8]

$C_6H_8O_2$ M 112.128

Mp 32-35°. Bp_{10} 99-102°, $Bp_{0.1}$ 75-85°.

Epstein, W.W. *et al*, *J. Org. Chem.*, 1967, **32**, 3390 (*synth, ir, pmr*)
Woo, E.P. *et al*, *J. Org. Chem.*, 1986, **51**, 3706 (*synth, pmr, cmr*)

2,5-Dimethyl-1,3,5-hexatriene, 9CI   D-60424

C_8H_{12} M 108.183

(E)-form [41233-74-3]

Bp_{20} 60°.

Vroegop, P.J. *et al*, *Tetrahedron*, 1973, **29**, 1393 (*synth, isom*)
Gielen, J.W.J. *et al*, *Tetrahedron Lett.*, 1976, 3751 (*isom*)
Ramadas, S.R. *et al*, *Org. Prep. Proced. Int.*, 1981, **13**, 9 (*synth, uv, ir, pmr*)
Brouwer, A.M. *et al*, *Tetrahedron Lett.*, 1986, **27**, 1395 (*props, bibl*)

2,2-Dimethyl-5-hexen-3-one, 9CI   D-60425

Allyl tert-butyl ketone

[55532-07-5]

$$(H_3C)_3CCOCH_2CH{=}CH_2$$

$C_8H_{14}O$ M 126.198

V. mobile liq. Bp_{50} 70-100°.

Bretsch, W. *et al*, *Justus Liebigs Ann. Chem.*, 1987, 175 (*synth, pmr*)

5,13-Dimethyl[2.2]metacyclophane   D-60426

6,13-Dimethyltricyclo[9.3.1.1^4,8]hexadeca-1(15),4,6,8(16),11,13-hexaene, 9CI

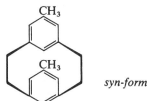

syn-form

$C_{18}H_{20}$ M 236.356

syn-form [107299-36-5]

Granules. Stable at r.t. in solid state, isomerises at 0° in THF to *anti*-form. Subl. at *ca.* 75° to *anti*-form.

anti-form [42053-19-0]

Prisms (EtOH). Mp 147-149°. Stable isomer.

Allinger, N.L. *et al, J. Org. Chem.*, 1967, **22**, 2272 (*synth*)
Fujise, Y. *et al, Tetrahedron Lett.*, 1986, **27**, 2907 (*synth*)

2,2-Dimethyl-3-methylenebicyclo[2.2.2]-octane   D-60427

Homocamphene
[102435-72-3]

$C_{11}H_{18}$ M 150.263

Cryst. with intense camphoraceous odour, by subl. Mp 123°.

Spreitzer, H. *et al, Justus Liebigs Ann. Chem.*, 1986, 1578 (*synth, ir, pmr, ms*)

5,5-Dimethyl-4-methylene-1,2-dioxolan-3-one, 9CI   D-60428

[104848-77-3]

$C_6H_8O_3$ M 128.127

Oil. Bp$_{0.1}$ 44-47°.

Adam, W. *et al, J. Org. Chem.*, 1986, **51**, 4479 (*synth, pmr, cmr, ir*)

4,4-Dimethyl-6-nitro-5-hexenoic acid   D-60429

$$O_2NCH{=}CHC(CH_3)_2CH_2CH_2COOH$$

$C_8H_{13}NO_4$ M 187.195

Nitrile: [5636-70-4]. *4,4-Dimethyl-6-nitro-5-hexenenitrile, 9CI. 5-Cyano-3,3-dimethyl-1-nitro-1-pentene. Ibanitrile.*
$C_8H_{12}N_2O_2$ M 168.195
Pale-yellow liq. Bp$_1$ 143-151°.

Hudak, N.J. *et al, J. Org. Chem.*, 1961, **26**, 1360 (*deriv*)
Poidevin, G. *et al, Bull. Soc. Chim. Fr.*, 1979, 196 (*deriv*)

3,7-Dimethyl-6-octenal   D-60430

Updated Entry replacing D-06677
Citronellal
[106-23-0]

$C_{10}H_{18}O$ M 154.252

(R)-form

Constit. of citronella oil. Bp 204-205°, Bp$_{14}$ 90°. [α]$_D^{15}$ +13.09°.

Arigoni, D. *et al, Helv. Chim. Acta*, 1954, **37**, 881 (*abs config*)
O'Donnell, G.W. *et al, Aust. J. Chem.*, 1966, **19**, 525.
Sokol'skii, D.V., *Dokl. Akad. Nauk SSSR*, 1978, **242**, 1126 (*synth*)
Akhila, A., *Phytochemistry*, 1986, **25**, 421 (*biosynth*)
Chimirri, A. *et al, Heterocycles*, 1987, **26**, 2469.

β,β-Dimethyloxiraneethanol   D-60431

2-Oxiranyl-2-methyl-1-propanol. 3,4-Epoxy-2,2-dimethyl-1-butanol

$C_6H_{12}O_2$ M 116.160

(R)-form

Chiral synthon. [α]$_D^{25}$ −15.2° (c, 1.8 in EtOAc).

(S)-form

Chiral synthon. Bp$_{1.3}$ 70°. [α]$_D^{25}$ +16.0° (c, 2 in EtOAc).

Lavallée, P. *et al, Tetrahedron Lett.*, 1986, **27**, 679 (*synth*)

4,4-Dimethyl-3-oxo-1-cyclopentene-1-carboxaldehyde, 9CI   D-60432

[109703-33-5]

$C_8H_{10}O_2$ M 138.166
Yellow oil.

Hudlicky, T. *et al, Synthesis*, 1986, 716 (*synth, ir, pmr, cmr, ms*)

3,3-Dimethyl-4-oxopentanoic acid, 9CI   D-60433

[23461-67-8]

$$H_3CCOC(CH_3)_2CH_2COOH$$

$C_7H_{12}O_3$ M 144.170
Liq. Bp$_{0.3}$ 105° (Bp$_2$ 102°).

Et ester: [64725-43-5].
$C_9H_{16}O_3$ M 172.224
Liq. Bp$_{20}$ 112-115°, Bp$_2$ 80-85°.

Cerfontain, H. *et al, Synthesis*, 1980, 490 (*synth, pmr*)
Guntrum, E. *et al, Synthesis*, 1986, 921 (*synth, ir, cmr*)

4,4-Dimethyl-5-oxopentanoic acid, 9CI   D-60434

4,4-Dimethylglutaraldehydic acid, 8CI. 4-Formyl-4,4-dimethylbutanoic acid. Ibaacid

[503-53-7]

$$OHCC(CH_3)_2CH_2CH_2COOH$$

$C_7H_{12}O_3$ M 144.170
Bp_1 103-105°.
Me ester: [4007-81-2].
   $C_8H_{14}O_3$ M 158.197
   Oil. $Bp_{0.5}$ 40-42°.
Oxime: [27579-06-2].
   $C_7H_{13}NO$ M 127.186
   Needles. Mp 90°.
2,4-Dinitrophenylhydrazone: Mp 151-151.5°.
Me ester, oxime: $Bp_{0.5}$ 98-100°.
Nitrile: [6140-61-0]. *4,4-Dimethylglutaraldehydonitrile,*
   8CI. 4-Cyano-2,2-dimethylbutyraldehyde.
   $C_7H_{11}NO$ M 125.170
   Bp_1 55-56°.
Nitrile, 2,4-dinitrophenylhydrazone: [25252-67-9]. Mp
   140-142°.

Brannock, K.C. *et al, J. Org. Chem.,* 1964, **29**, 801 (*synth*)
Cauquis, G. *et al, Bull. Soc. Chim. Fr.,* 1970, 183 (*synth*)

3,4-Dimethylpentanoic acid, 9CI   D-60435

[3302-06-5]

CH₂COOH
H─C─CH₃
CH(CH₃)₂   (*S*)-form

$C_7H_{14}O_2$ M 130.186
(*S*)-form [20180-65-8]
   Oil. Bp_{25} 140-150°. $[\alpha]_D^{21}$ −6.95° (c, 1.18 in C_6H_6).
Me ester: [34897-33-1].
   $C_8H_{16}O_2$ M 144.213
   Liq. Bp_{45} 125-135°.
(±)-form
   Bp_{14} 109°.

Levene, P.A. *et al, J. Biol. Chem.,* 1935, **111**, 299 (*synth*)
Giacomelli, G. *et al, J. Chem. Soc., Perkin Trans. 1,* 1975, 1795
   (*synth*)
Enders, D. *et al, Tetrahedron,* 1986, **42**, 2235 (*synth, pmr, ir, ms*)

3,4-Dimethyl-1-pentanol, 9CI   D-60436

[6570-87-2]

CH(CH₃)₂
H₃C─C─H
CH₂CH₂OH   (*S*)-form

$C_7H_{16}O$ M 116.203
(*R*)-form [58242-80-1]
   $[\alpha]_D$ +7.0° (c, 0.9 in hexane).
(*S*)-form [20180-66-9]
   Bp_{40} 105-110°. $[\alpha]_D^{22}$ −13.5° (neat).
(±)-form
   Bp 165°, Bp_{150} 118-119°.

Levene, P.A. *et al, J. Biol. Chem.,* 1935, **111**, 299 (*synth*)
Tsuda, K. *et al, J. Am. Chem. Soc.,* 1960, **82**, 3396 (*synth*)
Giacomelli, G. *et al, J. Chem. Soc., Perkin Trans. 1,* 1975, 1795
   (*synth*)
Wetter, H. *et al, Helv. Chim. Acta,* 1983, **66**, 118 (*synth, ir, pmr, ms*)
Enders, D. *et al, Tetrahedron,* 1986, **42**, 2235 (*synth, ir, pmr, cmr, ms*)

3,3-Dimethyl-4-pentenoic acid   D-60437

[7796-73-8]

$$H_2C{=}CHC(CH_3)_2CH_2COOH$$

$C_7H_{12}O_2$ M 128.171
Liq. Bp_4 83-85°.
Et ester: [7796-72-7].
   $C_9H_{16}O_2$ M 156.224
   Liq. Bp_{72} 94-98°, $Bp_{0.1}$ 35-45°.
Chloride: [88819-78-7].
   $C_7H_{11}ClO$ M 146.616
   Pale-yellow liq. Bp_{45} 71-72°.

Kleschick, W.A. *et al, J. Org. Chem.,* 1986, **51**, 5429 (*synth, ir, pmr*)
Nakada, Y. *et al, Bull. Chem. Soc. Jpn.,* 1979, **52**, 1511 (*deriv, synth, ir, pmr*)
Jöger, V. *et al, Tetrahedron Lett.,* 1977, 2543 (*synth*)

4,4-Dimethyl-2-pentyne, 9CI   D-60438

tert-*Butylmethylacetylene*
[999-78-0]

$$(H_3C)_3CC{\equiv}CCH_3$$

C_7H_{12} M 96.172
d_4^{15} 0.722. Bp 81-82°. n_D^{20} 1.4071.

De Graef, H., *Bull. Soc. Chim. Belg.,* 1925, **34**, 427 (*synth*)
Bock, H. *et al, Chem. Ber.,* 1986, **119**, 3766 (*synth, pmr*)

4,5-Dimethylphenanthrene, 9CI   D-60439

Updated Entry replacing D-06818
[3674-69-9]
$C_{16}H_{14}$ M 206.287
Prisms (MeOH). Mp 77-78°.
Picrate: Orange-red prisms (EtOH). Mp 109.5-110°.
9,10-Dihydro:
   $C_{16}H_{16}$ M 208.302
   Plates. Mp 106-107°.

Mosby, W.L., *J. Org. Chem.,* 1954, **19**, 294 (*ir*)
Friedel, R.A., *Appl. Spectrosc.,* 1957, **11**, 13 (*uv*)
Bestmann, H.J. *et al, Chem. Ber.,* 1966, **99**, 28 (*synth*)
Bartle, K.D. *et al, Spectrochim. Acta, Part A,* 1967, **23**, 1689
   (*pmr*)
Dougherty, R.C. *et al, Org. Mass Spectrom.,* 1971, **5**, 1321 (*ms*)
Cosmo, R. *et al, Aust. J. Chem.,* 1987, **40**, 35 (*deriv*)

2,2-Dimethyl-3-phenyl-2*H*-azirine, 9CI   D-60440

[14491-02-2]

$C_{10}H_{11}N$ M 145.204
Oil. Bp_{15} 85-87°.

Leonard, N.J. *et al, J. Am. Chem. Soc.,* 1967, **89**, 4456 (*synth, pmr*)
Anderson, W.K. *et al, J. Med. Chem.,* 1986, **29**, 2241 (*synth, uv, ir, pmr*)

3,3-Dimethyl-1-phenyl-1-butyne   D-60441

3,3-Dimethyl-1-butynylbenzene, 9CI. tert-
Butylphenylacetylene
[4250-82-2]

$(H_3C)_3CC{\equiv}CPh$

$C_{12}H_{14}$ M 158.243
Liq. Bp_{20} 100°, Bp_{10} 84°.

Kupin, B.S. *et al, J. Gen. Chem. USSR (Engl. Transl.)*, 1961, **31**, 2758 (*synth*)
Suzuki, A. *et al, J. Org. Chem.*, 1986, **51**, 4507 (*synth, pmr*)

2,5-Dimethyl-3-(2-phenylethenyl)pyrazine,   D-60442
9CI

Updated Entry replacing D-20402
2,5-Dimethyl-3-styrylpyrazine

$C_{14}H_{14}N_2$ M 210.278
(*E*)-*form* [54290-13-0]
Component of defence secretion of ant *Iridomyrmex humilis*. Oil. Bp_3 117-122°.
(*Z*)-*form*
Component of the defence secretion of *I. humilis*. Oil. Bp_3 105-118°.

Wheeler, J.W. *et al, Science*, 1973, **182**, 501 (*isol, synth*)
Cavill, G.W.K. *et al, Aust. J. Chem.*, 1974, **27**, 819 (*isol, synth, uv, pmr*)
Akita, Y. *et al, Chem. Pharm. Bull.*, 1986, **34**, 1447 (*synth, pmr*)

2,2-Dimethyl-1-phenyl-1-propanone, 9CI   D-60443

Updated Entry replacing D-06890
tert-*Butyl phenyl ketone. ω-Trimethylacetophenone. Pivalophenone. 2-Benzoyl-2-methylpropane*
[938-16-9]

$PhCOC(CH_3)_3$

$C_{11}H_{14}O$ M 162.231
Liq. Bp 219-221°, Bp_{15} 105-106°.
2,4-Dinitrophenylhydrazone: [59830-27-2]. Mp 171-172° (resolidifes with Mp 190-191°).

Grizzle, P.L. *et al, J. Org. Chem.*, 1975, **40**, 1902 (*synth*)
Org. Synth., 1976, **55**, 122 (*synth*)
Millard, A.A. *et al, J. Org. Chem.*, 1978, **43**, 1834 (*synth*)
Deuchert, K. *et al, Chem. Ber.*, 1979, **112**, 2045 (*synth*)
Lissel, M. *et al, Justus Liebigs Ann. Chem.*, 1987, 263 (*synth, ir, pmr*)

4,6-Dimethyl-2-phenylpyrimidine, 9CI   D-60444
[14164-34-2]

$C_{12}H_{12}N_2$ M 184.240
Mp 81-83°.

Haley, C.A.C. *et al, J. Chem. Soc.*, 1951, 3155 (*synth*)

(2,2-Dimethylpropyl)benzene   D-60445
2,2-Dimethyl-1-phenylpropane. Neopentylbenzene
[1007-26-7]

$PhCH_2C(CH_3)_3$

$C_{11}H_{16}$ M 148.247

Liq. Bp 185.5-185.8°, Bp_5 50°.

Bassindale, A.R. *et al, J. Chem. Soc. (C)*, 1969, 2505 (*synth, pmr*)
Bullpitt, M. *et al, J. Organomet. Chem.*, 1976, **116**, 187 (*cmr*)
Fry, A.J. *et al, J. Org. Chem.*, 1987, **52**, 2498 (*synth, pmr*)

1-(2,2-Dimethylpropyl)naphthalene   D-60446
2,2-Dimethyl-1-(1-naphthyl)propane. 1-Neopentylnaphthalene
[20411-45-4]

$C_{15}H_{18}$ M 198.307
Liq. Bp_1 115-116°.

Bullpitt, M. *et al, J. Organomet. Chem.*, 1976, **116**, 187; *Synthesis*, 1977, 316 (*synth, pmr, cmr*)

2-(2,2-Dimethylpropyl)naphthalene   D-60447
2,2-Dimethyl-1-(2-naphthyl)propane. 2-Neopentylnaphthalene
[61760-11-0]
$C_{15}H_{18}$ M 198.307
Cryst. (EtOH). Mp 65-66°.

Bullpitt, M. *et al, J. Organomet. Chem.*, 1976, **116**, 187; *Synthesis*, 1977, 316 (*synth, pmr, cmr*)

7,8-(2,2-Dimethylpyrano)-3,4′,5-trihydroxyflavan   D-60448

$C_{20}H_{20}O_5$ M 340.375
5-Me ether: 7,8-(2,2-Dimethylpyrano)-3,4′-dihydroxy-5-methoxyflavan.
$C_{21}H_{22}O_5$ M 354.402
Constit. of *Marshallia tenuifolia*. Gum. $[\alpha]_D^{20}$ −42.1° (c, 0.04 in $CHCl_3$).

Herz, W. *et al, Phytochemistry*, 1987, **26**, 1175.

7,6-(2,2-Dimethylpyrano)-3,4′,5-trihydroxyflavone   D-60449

$C_{20}H_{16}O_6$ M 352.343
Constit. of *Marshallia tenuifolia*. Cryst. (EtOAc/hexane). Mp 210°.

Herz, W. *et al, Phytochemistry*, 1987, **26**, 1175.

4,6-Dimethyl-2(1*H*)-pyrimidinethione, 9CI   D-60450

4,6-Dimethyl-2-thio-2,3-dihydropyrimidine. 2-Mercapto-4,6-dimethylpyrimidine. Acetylacetone-thiourea

[22325-27-5]

$C_6H_8N_2S$ M 140.203

Condensation prod. of acetylacetone and thiourea. Light-yellow prisms (EtOH). Mp 210°.

Na salt: [41840-27-1]. Cryst. (EtOH). Mp >300°.
1-Me:
   $C_7H_{10}N_2S$ M 154.229
   Small needles (C_6H_6). Mp 156.5°.

Hale, W.J. *et al, J. Am. Chem. Soc.,* 1915, **37**, 594 (*synth*)
Hunt, R.R. *et al, J. Chem. Soc.,* 1959, 525 (*synth*)
Voigt, H. *et al, Z. Chem.,* 1974, **14**, 436 (*deriv*)

3,3-Dimethyl-2,4(1*H*,3*H*)-quinolinedione, 9CI   D-60451

[106875-07-4]

$C_{11}H_{11}NO_2$ M 189.213
Cryst. (EtOH). Mp 150°.

Bergman, J. *et al, Tetrahedron,* 1986, **42**, 3689 (*synth, ir, cmr, pmr, ms*)

2,5-Dimethyl-3-thiophenecarboxaldehyde   D-60452

3-Formyl-2,5-dimethylthiophene

[26421-44-3]

C_7H_8OS M 140.200
Liq. $Bp_{0.7}$ 62-64°.

Glaze, A.P. *et al, J. Chem. Soc., Perkin Trans. 1,* 1985, 957 (*synth, pmr*)

2,7-Dimethylxanthone   D-60453

Updated Entry replacing D-07210
2,7-Dimethyl-9H-xanthen-9-one, 9CI
[7573-15-1]
$C_{15}H_{12}O_2$ M 224.259
Yellow needles (EtOH aq.). Mp 150-151° (143°). Sol. H_2SO_4 with blue-green fluor.

Schönberg, A. *et al, J. Chem. Soc.,* 1946, 609 (*synth*)
Köbrich, G., *Justus Liebigs Ann. Chem.,* 1963, **664**, 88.
Granoth, I. *et al, J. Org. Chem.,* 1975, **40**, 2088 (*synth, ms, nmr*)
Kimura, M. *et al, Chem. Pharm. Bull.,* 1987, **35**, 136 (*synth, ir, pmr, cmr, ms*)

1,3-Di-(1-naphthyl)benzene   D-60454

1,1'-(1,3-Phenylene)bisnaphthalene, 9CI. 1,1'-Bis(1,3-phenylene)naphthalene

[103068-16-2]

$C_{26}H_{18}$ M 330.428
Mp 130-132°.

Woods, G.F. *et al, J. Am. Chem. Soc.,* 1951, **73**, 3854 (*synth*)
Du, C.-J.F. *et al, J. Org. Chem.,* 1986, **51**, 3162 (*synth*)

1,3-Di(2-naphthyl)benzene   D-60455

2,2'-(1,3-Phenylene)bisnaphthalene, 9CI. 2,2'-Bis(1,3-phenylene)naphthalene

[103068-17-3]
$C_{26}H_{18}$ M 330.428
Mp 145-146°.

Woods, G.F. *et al, J. Am. Chem. Soc.,* 1951, **73**, 3854 (*synth*)
Du, C.-J.F. *et al, J. Org. Chem.,* 1986, **51**, 3162 (*synth*)

7,11:20,24-Dinitrilodibenzo[*b,m*][1,4,12,15]-tetraazacyclodocosine, 9CI   D-60456

4,5:15,16-Dibenzo-3,6,14,17,23,24-hexaazatricyclo[17.3.1.1^{8,12}]tetracosa-1(23),4,8(24),9,11,15,19,21-octaene

[31341-59-0]

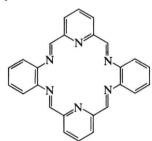

$C_{26}H_{18}N_6$ M 414.468
Forms numerous metal complexes. Yellow prisms (Py/C_6H_6). Spar. sol. most org. solvs.

Drew, M.G.B. *et al, J. Chem. Soc., Chem. Commun.,* 1979, 1033 (*use*)
Bell, T.W. *et al, J. Am. Chem. Soc.,* 1984, **106**, 6111 (*use*)
Bell, T.W. *et al, J. Chem. Soc., Chem. Commun.,* 1986, 769 (*synth, cryst struct, pmr*)

2,6-Dinitrobenzoic acid, 9CI, 8CI   D-60457

Updated Entry replacing D-07378
[603-12-3]
$C_7H_4N_2O_6$ M 212.118
Needles (H_2O). Mp 208°. pK_a 1.139 (25°, H_2O).
Me ester: [42087-82-1].
   $C_8H_6N_2O_6$ M 226.145
   Mp 147°.
Chloride: [59858-28-5].
   $C_7H_3ClN_2O_5$ M 230.564
   Mp 98°.
Nitrile: [35213-00-4]. *2-Cyano-1,3-dinitrobenzene.*
   $C_7H_3N_3O_4$ M 193.118
   Needles. Mp 58°.

Dippy, J.F.S. *et al*, *J. Chem. Soc.*, 1964, 154 (*synth*)
Köbrich, G. *et al*, *Angew. Chem., Int. Ed. Engl.*, 1966, **5**, 1044 (*synth*)
Mori, M. *et al*, *Chem. Pharm. Bull.*, 1986, **34**, 4859 (*synth, ir, pmr*)

3,5-Dinitrobenzoic acid, 9CI, 8CI   D-60458

Updated Entry replacing D-07380

[99-34-3]

$C_7H_4N_2O_6$ M 212.118

Corrosion inhibitor, also used in photography. Cryst. (EtOH aq.). Mp 206-207°. pK_a 2.824 (25°, H_2O).

Me ester: [2702-58-1].
$C_8H_6N_2O_6$ M 226.145
Mp 112°.

Et ester: [618-71-3].
$C_9H_8N_2O_6$ M 240.172
Mp 93°.

Chloride: [99-33-2].
$C_7H_3ClN_2O_5$ M 230.564
Reagent for isol. and characterisation of alcohols. Needles. Mp 74°. Bp_{11} 196°.

▷Highly toxic, irritant. DM6637000.

Amide: [121-81-3]. *3,5-Dinitrobenzamide.* **Nitromide, USAN**. *NSC 60719*.
$C_7H_5N_3O_5$ M 211.134
Coccidiostat for poultry. Leaflets. Mp 183°.

Anhydride: [40993-10-0].
$C_{14}H_6N_4O_{11}$ M 406.222
Mp 109°.

Nitrile: [4110-35-4]. *1-Cyano-3,5-dinitrobenzene.*
$C_7H_3N_3O_4$ M 193.118
Cryst. (diisopropyl ether). Mp 130-131°.

▷DI4365000.

Org. Synth., Coll. Vol., **3**, 337.
Bachman, G.B. *et al*, *J. Am. Chem. Soc.*, 1958, **80**, 5871 (*synth*)
Dippy, J.F.S. *et al*, *J. Chem. Soc.*, 1964, 154 (*synth*)
Mai, K. *et al*, *Tetrahedron Lett.*, 1986, **27**, 2203 (*nitrile*)
Fieser, M. *et al*, *Reagents for Organic Synthesis*, Wiley, 1967-84, **1**, 320.
Sax, N.I., *Dangerous Properties of Industrial Materials*, 5th Ed., Van Nostrand-Reinhold, 1979, 617.

2,6-Dinitrobenzyl alcohol   D-60459

Updated Entry replacing D-07392

2,6-Dinitrobenzenemethanol, 9CI. α-Hydroxy-2,6-dinitrotoluene

[96839-34-8]

$C_7H_6N_2O_5$ M 198.135

Pale-yellow plates (EtOH). Mp 96°.

Reich, S. *et al*, *Ber.*, 1912, **45**, 3055 (*synth*)
Mori, M. *et al*, *Chem. Pharm. Bull.*, 1986, **34**, 4859 (*synth, ir, pmr*)

2,2'-Dinitrobiphenyl   D-60460

Updated Entry replacing D-07395

[2436-96-6]

$C_{12}H_8N_2O_4$ M 244.206

Yellow needles (EtOH). Mp 123.5-124.5°.

Org. Synth., Coll. Vol., **3**, 339 (*synth*)

Kornblum, N. *et al*, *J. Am. Chem. Soc.*, 1952, **74**, 5782 (*synth*)
DeTar, D.F. *et al*, *J. Am. Chem. Soc.*, 1955, **77**, 3842 (*uv, ir*)
Thomas, C.B. *et al*, *J. Chem. Soc., Perkin Trans. 2*, 1972, 778 (*ms*)
Cofin, A., *J. Chromatogr.*, 1972, **68**, 89 (*ir, glc, tlc*)
Klemm, L.H. *et al*, *J. Chromatogr.*, 1978, **150**, 129 (*tlc*)
Cornforth, J. *et al*, *J. Chem. Soc., Perkin Trans. 1*, 1982, 2299 (*synth*)
Smith, W.B., *J. Heterocycl. Chem.*, 1987, **24**, 745 (*pmr, cmr*)

2,3-Dinitro-1*H*-pyrrole, 9CI   D-60461

$C_4H_3N_3O_4$ M 157.085

1-Me: [72795-78-9]. *1-Methyl-2,3-dinitro-1H-pyrrole.*
$C_5H_5N_3O_4$ M 171.112
Cryst. (hexane). Mp 98.5-99°.

Fournari, P. *et al*, *Bull. Soc. Chim. Fr.*, 1963, 488 (*synth*)
Grehn, L., *Chem. Scr.*, 1978, **13**, 67 (*synth, pmr, cmr*)

2,4-Dinitro-1*H*-pyrrole, 9CI   D-60462

[3130-54-9]

$C_4H_3N_3O_4$ M 157.085

Cryst. (H_2O). Mp 152° (149-150°).

1-Me: [2948-69-8]. *1-Methyl-2,4-dinitro-1H-pyrrole.*
$C_5H_5N_3O_4$ M 171.112
Cryst. (EtOH). Mp 95-95.5° (89-91°).

Rinkes, I.J., *Recl. Trav. Chim. Pays-Bas*, 1934, **53**, 1167 (*synth*)
Fournari, P., *Bull. Soc. Chim. Fr.*, 1963, 488 (*synth*)
Lippmaa, E. *et al*, *Org. Magn. Reson.*, 1972, **4**, 153 (*cmr, N nmr*)
Grehn, L., *Chem. Scr.*, 1978, **13**, 67 (*synth, pmr, cmr*)

2,5-Dinitro-1*H*-pyrrole, 9CI   D-60463

[32602-96-3]

$C_4H_3N_3O_4$ M 157.085

Needles (H_2O). Mp 173° (168-170°).

1-Me: [56350-95-9]. *1-Methyl-2,5-dinitro-1H-pyrrole.*
$C_5H_5N_3O_4$ M 171.112
Cryst. (hexane). Mp 99°.

Rinkes, I.J., *Recl. Trav. Chim. Pays-Bas*, 1934, **53**, 1167 (*synth*)
Fournari, P. *et al*, *Bull. Soc. Chim. Fr.*, 1963, 488 (*synth, deriv*)
Lippmaa, E. *et al*, *Org. Magn. Reson.*, 1972, **4**, 153 (*cmr, N nmr*)
Grehn, L., *Chem. Scr.*, 1978, **13**, 67 (*synth, pmr, cmr, deriv*)

3,4-Dinitro-1*H*-pyrrole, 9CI   D-60464

$C_4H_3N_3O_4$ M 157.085

1-Me: [68712-54-9]. *1-Methyl-3,4-dinitro-1H-pyrrole.*
$C_5H_5N_3O_4$ M 171.112
Cryst. (1,2-dichloroethane). Mp 169.5-170°.

Novikov, S.S. *et al*, *Izv. Akad. Nauk SSSR, Ser. Khim.*, 1959, 1098; *CA*, **54**, 1486 (*synth*)
Grehn, L., *Chem. Scr.*, 1978, **13**, 67 (*synth, pmr, cmr*)

1,7-Dioxa-4,10-diazacyclododecane   D-60465

[294-92-8]

$C_8H_{18}N_2O_2$ M 174.242
Cryst. Mp 80-83°.

Cram, D.J. et al, J. Am. Chem. Soc., 1986, **108**, 2989 (synth, pmr)

2H-1,5,2,4-Dioxadiazine-3,6(4H)dione   D-60466

$C_2H_2N_2O_4$ M 118.049
2,4-Di-tert-butyl: [103258-84-0].
  $C_{10}H_{18}N_2O_4$ M 230.263
  First example of 1,5,2,4-dioxadiazine ring system. Oil which slowly cryst. Mp 39.5-40.5°.

Stowell, J.C. et al, J. Org. Chem., 1986, **51**, 3355 (synth, ir, pmr, cmr)

8,13-Dioxapentacyclo[6.5.0.0²,⁶.0⁵,¹⁰.0³,¹¹]-tridecane-9,12-dione   D-60467

Octahydro-1,2,4-(epoxyethanylylidene)-5H-6-oxacyclobut[cd]indene-5,8-dione, 9CI

[80707-70-6]

$C_{11}H_{10}O_4$ M 206.198
Long needles (EtOH). Mp 348-352° dec.

Surapaneni, C.R. et al, J. Org. Chem., 1986, **51**, 2382 (synth, pmr, ir, cmr, ms, cryst struct)

2,6-Dioxaspiro[3.3]heptane, 9CI   D-60468

Updated Entry replacing D-07690
[174-79-8]

$C_5H_8O_2$ M 100.117
Cryst. Mp 89°. Bp 172°.
HgCl₂ complex: Cryst. Mp 130-132°.

Campbell, T.W., J. Org. Chem., 1957, **22**, 1029 (synth)
Ratz, L. et al, Z. Naturforsch. A, 1968, **23**, 2100 (ir)

1,2-Dioxetane   D-60469

Updated Entry replacing D-40361
[6788-84-7]

$C_2H_4O_2$ M 60.052

Obt. only in soln. Obt. as distillable CH_2Cl_2 soln. (Bp_{10} 20°). Dec. on heating. Unstable, $t_{1/2}$ 1.1 min. at 70° (extrapolated).
▷Explosive

O'Neal, H.E. et al, J. Am. Chem. Soc., 1970, **92**, 6553 (props)
Adam, W. et al, Angew. Chem., Int. Ed. Engl., 1984, **23**, 166.
Adam, W. et al, J. Am. Chem. Soc., 1985, **107**, 410.

3,6-Dioxo-4,7,11,15-cembratetraen-10,20-olide   D-60470

Updated Entry replacing D-10636

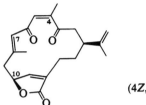

(4Z,7E,10β)-form

$C_{20}H_{24}O_4$ M 328.407
(4Z,7E,10β)-form [82443-67-2]
Lophodione
Metab. of the gorgonian coral *Lophogorgia alba*. Cryst. (EtOAc/2,2,3-trimethylpentane). Mp 172-174°. $[\alpha]_D^{23}$ −274.6° (c, 0.8 in $CHCl_3$).
4R,5S-*Epoxide:* **Epoxylaphodione**.
  $C_{20}H_{24}O_5$ M 344.407
  Metab. of *L. alba*. Solid. $[\alpha]_D^{23}$ −114.4° (c, 1.1 in $CHCl_3$).
(4E,7Z,10β)-form
Isolophodione
Metab. of *L. alba*. Cryst. Mp 172-175°. $[\alpha]_D^{18}$ −231.8° (c, 1.0 in $CHCl_3$).
(4Z,7E,10α)-form
Epilophodione
Constit. of *Gersemia rubiformis*. Needles (CH_2Cl_2/MeOH). Mp 153-155°. $[\alpha]_D^{25}$ +136.2° (c, 0.42 in $CHCl_3$).

Bandurraga, M.M. et al, Tetrahedron, 1982, **38**, 305 (cryst struct)
Williams, D. et al, J. Org. Chem., 1987, **52**, 332 (isol)

9,12-Dioxododecanoic acid, 9CI   D-60471

[51551-01-0]

$$OHCCH_2CH_2CO(CH_2)_7COOH$$

$C_{12}H_{20}O_4$ M 228.288
Cryst. (Et₂O/pentane). Mp 55-62°.
Me ester: [50266-44-9].
  $C_{13}H_{22}O_4$ M 242.314
  Oil. n_D^{25} 1.4788.
Me ester, ethylene acetal: [96202-77-6].
  $C_{15}H_{26}O_5$ M 286.367
  Solid (pentane). Mp 27-28°.

Brown, E. et al, Tetrahedron, 1975, **31**, 1047 (synth, ir, pmr)
Boga, C. et al, Synthesis, 1986, 212 (deriv, synth, pmr, ir)

3,23-Dioxo-7,24-lanostadien-26-oic acid   D-60472

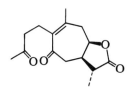

$C_{30}H_{44}O_4$ M 468.675

(*9β,24E*)-form [107584-83-8]
Firmanoic acid
Constit. of *Albies firma* and *A. sibirica*.
Me ester: Needles (CHCl₃/hexane). Mp 110-111°, Mp
210-212°. [α]_D +23° (c, 0.76 in CHCl₃).
$Δ^{25,27}$-*Isomer:* 3,23-*Dioxo-7,25(27)-lanostadien-26-oic*
acid. **Isofirmanoic acid.**
$C_{30}H_{44}O_4$ M 468.675
Constit. of *A. firma*.
$Δ^{25,27}$-*Isomer, Me ester:* Plates (hexane). Mp 163-164°.
[α]_D +24° (c, 0.48 in CHCl₃).

(*9β,24Z*)-form [107584-84-9]
From *A. sibirica*.
Me ester: Cryst. (MeCN). Mp 183-184°.

Roshchin, V.I. *et al, Khim. Prir. Soedin.*, 1986, **22**, 613.
Hasegawa, S. *et al, Phytochemistry*, 1987, **26**, 1095.

1,3-Dioxole   D-60473

[288-53-9]

$C_3H_4O_2$ M 72.063
Liq. Bp 51-52.5°. n_D^{26} 1.4032.

Field, N.D., *J. Am. Chem. Soc.*, 1961, **83**, 2304 (*synth, ir*)
Schaefer, T. *et al, Can. J. Chem.*, 1975, **53**, 2734 (*cmr*)

4,5-Dioxo-1(10)-xanthen-12,8-olide   D-60474

$C_{15}H_{20}O_4$ M 264.321

(*8β,11βH*)-form
Constit. of *Dittrichia graveolens*. Oil. [α]_D^{24} +15° (c, 0.1
in CHCl₃).

Rustaiyan, A. *et al, Phytochemistry*, 1987, **26**, 2603.

2,3-Diphenyl-2-butenal   D-60475

α-(1-Phenylethylidene)benzeneacetaldehyde, 9CI. 2,3-
Diphenylcrotonaldehyde

$C_{16}H_{14}O$ M 222.286

(*E*)-form [63904-67-6]
Cryst. (CHCl₃/hexane). Mp 105-106°.

(*Z*)-form [63904-66-5]
Yellow cryst. (CHCl₃/hexane). Mp 131-133°.
2,4-Dinitrophenylhydrazone: Orange needles. Mp 227-
228°.

Breslow, R. *et al, J. Am. Chem. Soc.*, 1962, **84**, 2793 (*synth*)
Padwa, A., *Tetrahedron Lett.*, 1965, 1049 (*synth, ir, uv*)
Friedrich, L.E. *et al, J. Org. Chem.*, 1978, **43**, 34 (*synth, ir, pmr,
ms*)

3,3-Diphenylcyclobutanone   D-60476

[54166-20-0]
$C_{16}H_{14}O$ M 222.286
Cryst. (EtOH). Mp 85.0-86.0°.

Michejda, C.J. *et al, J. Org. Chem.*, 1975, **40**, 1046 (*synth, ir,
pmr*)
Zimmerman, H.E. *et al, J. Am. Chem. Soc.*, 1986, **108**, 6276
(*synth*)

1,2-Diphenyl-1,4-cyclohexadiene   D-60477

[17351-29-0]

$C_{18}H_{16}$ M 232.324
Cryst. (EtOH). Mp 80-81°.

Carbanaro, A. *et al, J. Org. Chem.*, 1968, **33**, 3948 (*synth, uv, ir,
pmr*)

1,1-Diphenylcyclopropane, 8CI   D-60478

Updated Entry replacing D-07854
1,1'-Cyclopropylidenebisbenzene, 9CI
[3282-18-6]

$C_{15}H_{14}$ M 194.276
Liq. Bp₁₀ 132-134°. n_D^{20} 1.590.

Org. Synth., Coll. Vol., 5, 509 (*synth, bibl*)
Bloodworth, A.J. *et al, J. Org. Chem.*, 1986, **51**, 2110 (*synth*)

N,N'-Diphenylformamidine, 8CI   D-60479

Updated Entry replacing D-07904
N,N'-Diphenylmethanimidamide, 9CI
[622-15-1]

$$PhN{=}CHNHPh$$

$C_{13}H_{12}N_2$ M 196.251
Needles (EtOH). Mp 137° (142°).
B,HCl: Cryst. +3H₂O. Mp 255°.
Picrate: Mp 187°, 193°.
N-Me: [32189-59-6].
    $C_{14}H_{14}N_2$ M 210.278
    Bright-yellow oil. Bp₂₆ 218-219°.
N-Me; B,HCl: [40917-92-8]. Mp 228°.

Shoesmith, J.B. *et al, J. Chem. Soc.*, 1923, **123**, 2705 (*synth*)
Hinkel, L.E. *et al, J. Chem. Soc.*, 1930, 1834 (*synth*)
Bredereck, H. *et al, Chem. Ber.*, 1959, **92**, 837 (*deriv*)
Kiro, Z.B. *et al, Zh. Org. Khim.*, 1971, **7**, 2196; 1972, **8**, 2573
   (*deriv, ir, pmr, uv*)
Tritschler, W. *et al, Synthesis*, 1973, 423 (*synth*)
Bouillon, G. *et al, Chem. Ber.*, 1974, **112**, 2332 (*synth*)
Naulet, N. *et al, Org. Magn. Reson.*, 1975, **7**, 326 (*cmr*)
Kashima, C. *et al, Bull. Chem. Soc. Jpn.*, 1986, **59**, 3317 (*synth,
pmr*)

Fieser, M. *et al*, *Reagents for Organic Synthesis*, Wiley, 1967-84, **1**, 339.

1,7-Diphenyl-1,3,5-heptanetriol   D-60480

OH OH OH

Ph—1   3   5—Ph

C₁₉H₂₄O₃   M 300.397
(*1R,3R,5S*)-*form* [103654-25-7]
Yashabushitriol
Constit. of male flowers of *Alnus sieboldiana*. Cryst. Mp 89-90°. [α]_D +30.3° (CHCl₃).
3-Ketone: [103654-23-5]. *1,5-Dihydroxy-1,7-diphenyl-3-heptanone.* **Yashabushiketodiol A**.
C₁₉H₂₂O₄   M 314.380
Constit. of male flowers of *A. sieboldiana*. Cryst. Mp 62-63°. [α]_D +21.4° (MeOH).
3-Ketone, 1-Epimer: [103654-24-6]. **Yashabushiketodiol B**.
C₁₉H₂₂O₄   M 314.380
Constit. of *A. sieboldiana*. Cryst. Mp 60-61°. [α]_D –28.6° (CHCl₃), [α]_D –16.3° (MeOH).

Hashimoto, T. *et al*, *Chem. Pharm. Bull.*, 1986, **34**, 1846.

1,3-Diphenyl-1*H*-indene-2-carboxylic acid,   D-60481
9CI

[67845-24-3]

Ph

COOH

Ph

C₂₂H₁₆O₂   M 312.367
(±)-*form*
Microcryst. Mp 194°.
Me ester:
C₂₃H₁₈O₂   M 326.394
Prisms. Mp 75°.

Yamamura, K. *et al*, *Bull. Chem. Soc. Jpn.*, 1986, **59**, 3699 (*synth, pmr*)

Diphenylmethaneimine   D-60482
α-*Phenylbenzenemethanimine*, 9CI. *1,1-Diphenylmethylenimine*, 8CI. *Benzophenone imine*
[1013-88-3]

Ph₂C=NH

C₁₃H₁₁N   M 181.237
d₄¹⁹ 1.085. Bp₈ 151-152°, Bp₃.₅ 127-128°. n_D¹⁹ 1.162.
B,HCl: [5319-67-5]. Mp 295-300° dec.
N-*Me:* [22627-00-5]. Bp₂.₅ 126-128°, Bp₀.₆ 100-102°.
N-*Ac:* [22800-71-1]. N-(*Diphenylmethylene*)*acetamide*.
C₁₅H₁₃NO   M 223.274
Needles (pentane). Mp 43.5-44.5°.
N-*Nitro:* [4371-81-7]. *Benzophenone nitrimine.*
C₁₃H₁₀N₂O₂   M 226.234
Cryst. (MeOH or AcOH aq.). Mp 67-69°.

Org. Synth., Coll. Vol., **5**, 55 (*synth*)
Horner, L. *et al*, *Chem. Ber.*, 1961, **94**, 290 (*deriv*)
Perrier-Datin, A. *et al*, *Spectrochim. Acta, Part A*, 1969, **25**, 169 (*ir*)
van der Linde, R. *et al*, *Spectrochim. Acta, Part A*, 1969, **25**, 375 (*pmr*)

Allmann, R. *et al*, *Chem. Ber.*, 1984, **117**, 1604 (*synth, pmr, cmr, cryst struct, deriv*)
Brenner, D.G. *et al*, *J. Heterocycl. Chem.*, 1985, **22**, 805 (*synth*)

1,8-Diphenylnaphthalene, 9CI, 8CI   D-60483
Updated Entry replacing D-07979
[1038-67-1]
C₂₂H₁₆   M 280.368
Needles (hexane). Mp 149-150°.

House, H.O. *et al*, *J. Org. Chem.*, 1963, **28**, 2403 (*synth, uv, bibl*)
Clough, R.L. *et al*, *J. Org. Chem.*, 1976, **41**, 2252 (*synth, bibl*)
Gerson, F. *et al*, *Helv. Chim. Acta*, 1985, **68**, 1923 (*esr*)
Cosmo, R. *et al*, *Aust. J. Chem.*, 1987, **40**, 1107 (*synth, pmr, conformn*)

3,5-Diphenyl-1,2,4-oxadiazole, 9CI   D-60484
Updated Entry replacing D-07996
[888-71-1]

C₁₄H₁₀N₂O   M 222.246
Needles (MeOH or ligroin). Mp 108°. Bp₁₇ 210°. Sublimes. Steam-volatile.
4-Oxide: [20594-92-7].
C₁₄H₁₀N₂O₂   M 238.245
Mp 188-193° dec.

Morrocchi, S. *et al*, *Chim. Ind. (Milan)*, 1968, **50**, 558 (*synth, deriv*)
Bast, K., *Chem. Ber.*, 1972, **105**, 2825 (*synth*)
Goetz, N., *Synthesis*, 1976, 268 (*synth*)
Molina, P. *et al*, *Synthesis*, 1986, 843 (*synth, ir, pmr*)

2,4-Diphenyl-5(4*H*)-oxazolone, 9CI   D-60485
Updated Entry replacing D-10694
2,4-Diphenyl-2-oxazolin-5-one, 8CI. *Azlactone*
[28687-81-2]

Ph

O   N

Ph

C₁₅H₁₁NO₂   M 237.257
Needles (pet. ether). Mp 103-104°.

Gotthardt, H. *et al*, *J. Am. Chem. Soc.*, 1970, **92**, 4340 (*synth, ir, uv*)
Padwa, A. *et al*, *J. Am. Chem. Soc.*, 1974, **96**, 2414 (*pmr*)
Höfle, G. *et al*, *Chem. Ber.*, 1976, **109**, 2648 (*tautom*)
Sain, B. *et al*, *J. Chem. Soc., Perkin Trans. 1*, 1985, 773 (*synth, ir, pmr, ms*)

2,7-Diphenyloxepin   D-60486
[102342-18-7]

Ph   O   Ph

C₁₈H₁₄O   M 246.308
Orange-yellow cryst. (hexane). Mp 100.5-101.0°.

McManus, M.J. *et al*, *J. Org. Chem.*, 1986, **51**, 2784 (*synth, ir, uv, pmr, cmr, ms, cryst struct*)

1,5-Diphenyl-1,3-pentanediol, 9CI, 8CI   D-60487

Updated Entry replacing D-08015

$$\begin{array}{c} Ph \\ HO \blacktriangleright \overset{1}{C} \blacktriangleleft H \\ CH_2 \\ H \blacktriangleright \overset{3}{C} \blacktriangleleft OH \\ CH_2CH_2Ph \end{array} \quad (1S,3S)\text{-}form$$

$C_{17}H_{20}O_2$ M 256.344

(1S,3S)-form
(−)-erythro-*form*
Constit. of the wood of *Flindersia laevicarpa*. Mp 89-90°. $[\alpha]_D$ −19° (c, 3 in EtOH).

(1R,3S)-form
(+)-threo-*form*
Isol. from *F. laevicarpa* leaves. $[\alpha]_D$ +29.8° (c, 1 in MeOH).

(1S,3R)-form
(−)-threo-*form*
Cryst. (hexane/Et$_2$O). Mp 91-92°. $[\alpha]_D$ −23.3° (c, 1.0 in MeOH).

(1RS,3RS)-form
(±)-erythro-*form*
Needles (MeOH). Mp 86-87°.

(1RS,3SR)-form
(±)-threo-*form*
Needles (MeOH). Mp 79°.

Breen, G.J.W., *Aust. J. Chem.*, 1962, **15**, 819 (isol, uv, ir, synth)
Picker, K. *et al*, *Aust. J. Chem.*, 1976, **29**, 2023 (isol, ir, uv)
Niwa, M. *et al*, *Chem. Pharm. Bull.*, 1987, **35**, 108 (synth, abs config, pmr, cmr)

1,5-Diphenylpentanepentone   D-60488

Diphenyl pentaketone
[104779-80-8]

PhCOCOCOCOCOPh

$C_{17}H_{10}O_5$ M 294.263
Deep-coloured. λ_{max} 437 (ϵ 154), 547 (130) nm.
Covalent hydrate: [104779-82-0]. *3,3-Dihydroxy-5,5-diphenyl-1,2,4,5-pentanetetrone.*
$C_{17}H_{12}O_6$ M 312.278
Yellow. Mp 125-126°.

Gleiter, R. *et al*, *Angew. Chem., Int. Ed. Engl.*, 1986, **25**, 999 (synth, uv, ir, cmr)

1,5-Diphenyl-1,3,5-pentanetrione, 9CI   D-60489

Updated Entry replacing D-30566
1,3-Dibenzoylacetone
[1467-40-9]

PhCOCH$_2$COCH$_2$COPh

$C_{17}H_{14}O_3$ M 266.296
Dienol form prevails in cryst. state. Dark-brown cryst. (EtOH); bright-yellow cryst. (EtOH/HCl). Mp 106-109°.

Light, R.J. *et al*, *J. Org. Chem.*, 1960, **25**, 538 (synth)
Miles, M.L. *et al*, *J. Org. Chem.*, 1965, **30**, 1007 (synth)
Cea-Olivares, R. *et al*, *Aust. J. Chem.*, 1987, **40**, 1127 (cryst struct)

9,18-Diphenylphenanthro[9,10-b]-triphenylene, 9CI   D-60490

9,18-Diphenyltetrabenz[a,c,h,j]anthracene
[103692-62-2]

$C_{42}H_{26}$ M 530.667
Contains a twisted anthracene chromophore. Yellow cryst. Mp >350°.

Pascal, R.A. *et al*, *J. Am. Chem. Soc.*, 1986, **108**, 5652 (synth, pmr, ms, uv, cryst struct)

1,1-Diphenylpropene, 8CI   D-60491

Updated Entry replacing D-08085
1,1'-(1-Propenylidene)bisbenzene, 9CI. α-Ethylidenediphenylmethane
[778-66-5]

$H_3CCH{=}CPh_2$

$C_{15}H_{14}$ M 194.276
Plates (EtOH). Mp 52° (49°). Bp 280-281°, Bp$_1$ 98-100°.

Greenwald, R. *et al*, *J. Org. Chem.*, 1963, **28**, 1128 (synth)
Van der Linde, R. *et al*, *Spectrochim. Acta*, 1965, **21**, 1893 (nmr)
Corey, E.J. *et al*, *J. Am. Chem. Soc.*, 1966, **88**, 5652 (synth)
Bruson, H.A. *et al*, *J. Org. Chem.*, 1967, **32**, 3356 (synth)
Ohta, S. *et al*, *Chem. Pharm. Bull.*, 1986, **34**, 5145 (synth, pmr, ir)

2,6-Diphenyl-4H-thiopyran-4-one   D-60492

[1029-96-5]

$C_{17}H_{12}OS$ M 264.341
Cryst. (EtOAc). Mp 131-132°.
1,1-Dioxide: [41068-60-4].
$C_{17}H_{12}O_3S$ M 296.340
Cryst. (toluene). Mp 145-146°.

Arndt, F. *et al*, *Ber.*, 1925, **58**, 1633 (synth)
Chen, C.H. *et al*, *Heterocycles*, 1977, **7**, 231 (synth, pmr)
Chen, C.H. *et al*, *J. Org. Chem.*, 1977, **42**, 2777; 1986, **51**, 3282 (synth, deriv)

Suggestions for new DOC Entries are welcomed. Please write to the Editor, DOC 5, Chapman and Hall Ltd, 11 New Fetter Lane, London EC4P 4EE

Diphyllin   D-60493

Updated Entry replacing D-30577

9-(1,3-Benzodioxol-5-yl)-4-hydroxy-6,7-dimethoxyn-
aphtho[2,3-c]-furan-1(3H)-one, 9CI. 4-Hydroxy-6,7-di-
methoxy-9-(3,4-methylenedioxyphenyl)naphtho[2,3-c]-
furan-1(3H)-one

[22055-22-7]

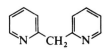

$C_{21}H_{16}O_7$ M 380.353

Lignan from roots of *Diphylleia grayi*, leaves of
Cleisthanthus collinus, Justicia procumbens and
Haplophyllum hispanicum. Needles (EtOH). Mp
291°.

▷Cytotoxin

Ac: Cryst. (C_6H_6). Mp 236-240°.

*O-(β-D-Glucopyranosyl(1→2)-3,4-di-O-methyl-D-
xylopyranosyl):* [86402-39-3]. **Cleisthanthoside** A.
$C_{34}H_{38}O_{16}$ M 702.665
Constit. of *C. patulus*. Cryst. Mp 238-241°. $[\alpha]_D$
+7.3° (c, 1.04 in dioxan).

O-(4-O-Methyl-β-D-xylopyranosyl): **Cleisthanthoside** B.
$C_{27}H_{26}O_{11}$ M 526.496
Constit. of *C. patulus*. Feathery needles (MeOH). Mp
178-179°. $[\alpha]_D$ −62.2° (c, 1 in dioxan).

Me ether: [25001-57-4]. **Justicidin** A.
$C_{22}H_{18}O_7$ M 394.380
Isol. from *J. hayati* and *J. procumbens*. Piscicidal
compd., cytotoxic. Mp 291°.

Dehydroxy: [17951-19-8]. **Justicidin** B.
Dehydrocollinusin.
$C_{21}H_{16}O_6$ M 364.354
Isol. from *J. hayati*. Piscicide. Mp 240°.

Horii, Z. et al, J. Chem. Soc., Chem. Commun., 1968, 653
(struct)
Govindachari, T.R. et al, Tetrahedron, 1969, 25, 2815 (isol)
Okigawa, M. et al, Tetrahedron, 1970, 26, 4301 (isol)
Stevenson, R. et al, J. Org. Chem., 1971, 36, 3450, 3453 (synth)
Arnold, B.J. et al, J. Chem. Soc., Perkin Trans. 1, 1973, 1266
(synth)
Horii, Z. et al, Chem. Pharm. Bull., 1977, 25, 1803 (synth)
Momose, T. et al, Chem. Pharm. Bull., 1978, 26, 3195 (synth)
Sastry, K.V. et al, Planta Med., 1983, 47, 227 (Cleisthanthoside)
Fukamiya, N. et al, J. Nat. Prod., 1986, 49, 348.
Sastry, K.V. et al, Phytochemistry, 1987, 26, 1153
(Cleisthanthoside B)

Dipyrido[1,2-b:1',2'-e][1,2,4,5]tetrazine,   D-60494
9CI

[881-64-1]

$C_{10}H_8N_4$ M 184.200

Yellow-green to dark-green powder (EtOAc). Mp 203-
203.5°.

Eichenberger, T. et al, Helv. Chim. Acta, 1986, 69, 1521 (synth,
ir, uv, cryst struct)

6H-Dipyrido[1,2-a:2',1'-d][1,3,5]triazin-5-   D-60495
ium, 9CI

N,N'-Methylene-2,2'-azapyridocyanine

$C_{11}H_{10}N_3^{\oplus}$ M 184.220 (ion)

Iodide: [22013-58-7].
$C_{11}H_{10}IN_3$ M 311.125
Golden flaky solid. Mp 258-260°.

Leubner, I.H., J. Org. Chem., 1973, 38, 1098 (synth)
Munavalli, S. et al, Synthesis, 1986, 402 (synth, pmr, uv)

Di-2-pyridylmethane   D-60496

Updated Entry replacing D-08290

2,2'-Methylenebispyridine, 9CI

[1132-37-2]

$C_{11}H_{10}N_2$ M 170.213

Bp_2 176-186°.

B,2HCl: [19087-65-1]. Mp 245°.

Dipicrate: Mp 196°.

Thiele, K., CA, 1968, 69, 51949y.
Canty, A.J. et al, Aust. J. Chem., 1986, 39, 1063 (synth, pmr)

Di-2-pyridyl sulfite   D-60497

2-Pyridinol sulfite (2:1) (ester), 9CI

[105125-43-7]

$C_{10}H_8N_2O_3S$ M 236.245

Reagent for prepn. of *N*-sulfinylamines, nitriles,
isocyanides and carbodiimides under neutral condns.
Yellow solid. Mp 73-74°. Fairly stable at 0°.

Kim, S. et al, Tetrahedron Lett., 1986, 27, 1925 (synth, pmr, ir,
use)

1,2-Diselenete, 9CI   D-60498

[16914-66-2]

$C_2H_2Se_2$ M 183.958

Diehl, F. et al, Angew. Chem., Int. Ed. Engl., 1987, 26, 343
(synth, pe)

Disidein   D-60499

Updated Entry replacing D-08307
17,17a-Dihydro-4,4,8,17-tetramethyl-3'H-in-
deno[1',2':17,17a]-D-homoandrost-17-ene-4',6',7'-triol,
9CI
[56012-79-4]

$C_{31}H_{46}O_3$ M 466.703
Constit. of *Disidea pallescens*. Cryst. (Et$_2$O). Mp 260°
dec. $[\alpha]_D$ +24° (c, 2.3 in dioxan).
Tri-Ac: Cryst. (MeOH). Mp 143-145°.
6'-Chloro: **6'-Chlorodisidein**.
   $C_{37}H_{51}ClO_6$ M 627.259
   Constit. of *D. pallescens*. Cryst. Mp 158-159°. $[\alpha]_D$
   +22.1° (c, 2 in CHCl$_3$).
6'-Bromo: **6'-Bromodisidein**.
   $C_{37}H_{51}BrO_6$ M 671.710
   Constit. of *D. pallescens*. Cryst. Mp 159-160°. $[\alpha]_D$
   +24.2° (c, 2.5 in CHCl$_3$).

Cimino, G. et al, Tetrahedron, 1975, **31**, 271 (isol, struct)
Cimino, G. et al, Tetrahedron, 1987, **43**, 4777 (cryst struct, abs
   config, derivs)

Dispiro[cyclopropane-1,5'-[3,8]-   D-60500
dioxatricyclo[5.1.0.0^{2,4}]octane-6',1''-cy-
clopropane], 9CI

7,8,9,10-Bisepoxydispiro[2.0.2.4]decane
[111192-70-2]

$C_{10}H_{12}O_2$ M 164.204
Cryst. Mp 112-113°.

de Meijere, A. et al, Tetrahedron, 1986, **42**, 6487 (synth, ir,
   pmr, cmr)

Dispiro[2.0.2.4]dec-8-ene-7,10-dione, 9CI   D-60501
[111192-77-9]

$C_{10}H_{10}O_2$ M 162.188
Mp 62-63°.

de Meijere, A. et al, Tetrahedron, 1986, **42**, 6487 (synth, ir,
   pmr, uv)

Distichol   D-60502

$C_{42}H_{32}O_9$ M 680.709
Constit. of the bark of *Shorea disticha*. Amorph. solid.
Mp 266-268°. $[\alpha]_D^{25}$ −44° (MeOH).

Sultanbawa, M.U.S. et al, Phytochemistry, 1987, **26**, 799.

Distyryl selenide   D-60503

1,1'-(Selenodi-2,1-ethenediyl)bisbenzene, 9CI. Bis(β-
styryl)selenide

$$PhCH{=}CHSeCH{=}CHPh$$

$C_{16}H_{14}Se$ M 285.247
(E,E)-form [67502-64-1]
   Mp 43-45°.
(E,Z)-form [105590-36-1]
   Mp 101-103°.
(Z,Z)-form [67502-78-7]
   Mp 38-40°.

Testaferri, L. et al, Tetrahedron, 1986, **42**, 63 (synth, pmr, cmr,
ms)

2-(1,3-Dithian-2-yl)pyridine, 9CI   D-60504

2-(2-Pyridyl)-1,3-dithiane
[80085-67-2]

$C_9H_{11}NS_2$ M 197.313
Brown oil.

Aloup, J.-C. et al, J. Med. Chem., 1987, **30**, 24 (synth)

2,11-Dithia[3.3]paracyclophane   D-60505

Updated Entry replacing D-08366
3,10-Dithiatricyclo[10.2.2.2^{5,8}]octadeca-5,7,12,14,15,17-
hexaene, 9CI
[28667-63-2]

$C_{16}H_{16}S_2$ M 272.422
Cryst. (CHCl$_3$). Mp 224°.

Vögtle, F. et al, Chem.-Ztg. Chem. Appar., 1970, **94**, 313
   (synth, pmr)
Brink, M., Synthesis, 1975, 807.
Chan, T. et al, Acta Crystallogr., Sect. C, 1986, **42**, 897 (cryst
   struct)

2,7-Dithiatricyclo[6.2.0.0³,⁶]deca-1(8),3(6)-diene-4,5,9,10-tetrone   D-60506

Dicyclobuta[1,4]dithiin-1,2,4,5-tetraone
[70597-76-1]

$C_8O_4S_2$ M 224.206
Orange cryst. (MeCN). Mp 174-176° (165°).

Seitz, G. *et al*, *Chem. Ber.*, 1979, **112**, 990 (*synth, ir, cmr*)
Schmidt, A.H. *et al*, *Synthesis*, 1984, 754 (*synth, ir*)
Bock, H. *et al*, *J. Am. Chem. Soc.*, 1986, **108**, 7844 (*synth*)

1,2-Dithiecane, 9CI   D-60507

1,2-Dithiacyclodecane
[6573-66-6]

$C_8H_{16}S_2$ M 176.334
Pale-yellow cryst. Mp 15-18°. Bp₂ 107-110°.

Schöberl, A. *et al*, *Justus Liebigs Ann. Chem.*, 1958, **614**, 66 (*synth*)
Harpp, D.N. *et al*, *Tetrahedron Lett.*, 1986, **27**, 441 (*synth*)

Dithiobisphthalimide   D-60508

[34251-41-7]

$C_{16}H_8N_2O_4S_2$ M 356.370
Sulfur transfer agent. Prisms (CHCl₃/MeOH). Mp 225°.

Huang, N.Z. *et al*, *J. Org. Chem.*, 1987, **52**, 169 (*synth, use*)

Dithiocarbazic acid, 8CI   D-60509

Updated Entry replacing D-08422
Hydrazinecarbodithioic acid, 9CI. Dithiocarbazinic acid. Hydrazinodithioformic acid. Aminodithiocarbamic acid. Hythizine
[471-32-9]

H₂NNHCSSH

$CH_4N_2S_2$ M 108.176
Unknown in free state.
Hydrazine salt: [53636-08-1]. Prisms. Mp 124° dec.
Me ester: [5397-03-5].
$C_2H_6N_2S_2$ M 122.203
Prisms (C₆H₆). Mp 79°.

Curtius, T. *et al*, *Ber.*, 1894, **27**, 55 (*salts*)
Sandstrom, J., *Ark. Kemi*, 1952, **4**, 297; *CA*, **47**, 9271 (*esters*)
Kubota, S. *et al*, *J. Med. Chem.*, 1978, **21**, 591 (*esters*)
Mattes, R. *et al*, *J. Chem. Soc., Dalton Trans.*, 1980, 423 (*cryst struct, bibl*)
Manogaran, S. *et al*, *Bull. Chem. Soc. Jpn.*, 1982, **55**, 2628 (*struct, bibl*)
Haines, R.A. *et al*, *Inorg. Chim. Acta*, 1983, **71**, 1 (*complexes*)
Gattow, G. *et al*, *Z. Anorg. Allg. Chem.*, 1985, **531**, 101 (*esters, bibl*)

1,4-Dithiocyano-2-butene   D-60510

2-Butene-1,4-diyl dithiocyanate

NCSCH₂CH=CHCH₂SCN

$C_6H_6N_2S_2$ M 170.247
(*E*)-*form*
Needles (CHCl₃). Mp 83-84°.
(*Z*)-*form*
Plates (CHCl₃/hexane). Mp 47-47.5°.

Huber, S. *et al*, *Helv. Chim. Acta*, 1986, **69**, 1898 (*synth, ir, pmr, cmr, ms*)

Dithiosilvatin   D-60511

$C_{18}H_{22}N_2O_3S_2$ M 378.503
Diketopiperazine antibiotic. Isol. from *Aspergillus silvaticus*. Plates (C₆H₆). Mp 100-102°. $[\alpha]_{435}^{25}$+35.9° (c, 1 in CHCl₃). Related to Silvathione.

Kawahara, N. *et al*, *J. Chem. Soc., Perkin Trans. 1*, 1987, 2099 (*isol, struct*)

2,6-Dithioxobenzo[1,2-d:4,5-d']bis[1,3]-dithiole-4,8-dione, 9CI   D-60512

[65160-04-5]

$C_8O_2S_6$ M 320.447
Electron acceptor, charge transfer complexes with electron donors show high electrical conductivity. Cryst. (C₆H₆). Mp 230° dec.

Demetriadis, N.G. *et al*, *Tetrahedron Lett.*, 1977, 2223 (*synth*)
Yamashita, Y. *et al*, *J. Chem. Soc., Chem. Commun.*, 1986, 1489 (*use*)

Dodecafluorobicyclobutylidene   D-60513

Hexafluoro(hexafluorocyclobutylidene)cyclobutane, 9CI. Perfluorobicyclobutylidene
[68252-05-1]

C_8F_{12} M 324.069
Bp 79-80°.

Bayliff, A.E. *et al*, *J. Chem. Soc., Perkin Trans. 1*, 1985, 1191 (*props, bibl*)
Chambers, R.D. *et al*, *J. Chem. Soc., Perkin Trans. 1*, 1980, 426 (*synth, ir, F nmr*)

Dodecahedrane   D-60514

Updated Entry replacing D-10729

Undecacyclo[9.9.0.0^{2,9}.0^{3,7}.0^{4,20}.0^{5,18}.0^{6,15}.0^{8,15}.0^{10,14}-.0^{12,19}.0^{13,17}]eicosane, 9CI

[4493-23-6]

$C_{20}H_{20}$ M 260.378

The structurally most complex symmetric member of the C_nH_n convex polyhedra. Cryst. Mp >450°.

C_6H_6 complex: [97350-56-6]. Cryst. powder (C_6H_6). Mp 280° dec.

Eaton, P.E., *Tetrahedron*, 1979, **35**, 2189 (*rev*)
Paquette, L.A. *et al, J. Am. Chem. Soc.*, 1982, **104**, 4503; 1986, **108**, 1716 (*synth, spectra*)
Fessner, W.-D. *et al, Angew. Chem., Int. Ed. Engl.*, 1987, **26**, 452 (*synth*)

Dodecahydro-3a,6,6,9a-tetramethyl-   D-60515
naphtho[2,1-*b*]furan, 9CI

Updated Entry replacing D-08536

1,1,4a,6-Tetramethyl-5-ethyl-6,5-oxidodecahydron-aphthalene. Bicyclofarnesyl epoxide

[6790-58-5]

(3aα,5aβ,9aα,9bβ)-*form*

$C_{16}H_{28}O$ M 236.397

Volatile constit. of Ambergris tincture. Important amber perfumery ingredient.

(3aα,5aβ,6α,9aα,9bβ)-form [6790-58-5]
Ambroxide. Ambrox. n-*Epoxide*
Cryst. (pet. ether). Mp 75-76°.

(3aα,5aα,9aβ,9bα)-form [68365-88-8]
Isoambrox. Isoepoxide
Mp 60°.

Stoll, M. *et al, Helv. Chim. Acta*, 1950, **33**, 1251; 1951, **34**, 1664 (*synth, config*)
Hinder, M. *et al, Helv. Chim. Acta*, 1953, **36**, 1995 (*synth, ir*)
Lucius, G. *et al, Arch. Pharm. (Weinheim, Ger.)*, 1958, **291**, 57; *CA*, **52**, 14635e (*config*)
Torii, S. *et al, J. Org. Chem.*, 1978, **43**, 4600 (*synth*)
Ohloff, G. *et al, Helv. Chim. Acta*, 1985, **68**, 2022.
Decorzaut, R. *et al, Tetrahedron*, 1987, **43**, 1871 (*synth*)

1,2,3,4,5,6,7,8,9,10,11,12-Dodecahydrotri-   D-60516
phenylene, 9CI

Tritetralin

[1610-39-5]

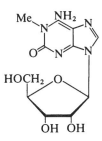

$C_{18}H_{24}$ M 240.388

Cryst. (EtOAc). Mp 232-233° subl.

Mannich, C., *Ber.*, 1907, **40**, 153 (*synth*)

Erickson, J.L.E. *et al, J. Org. Chem.*, 1966, **31**, 480 (*synth*)
Illuminati, G. *et al, J. Chem. Soc. (B)*, 1971, 2206 (*synth*)

Dolichol   D-60517

Updated Entry replacing D-30604

[2067-66-5]

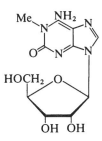

$C_{100}H_{164}O$ M 1382.395

Constit. of mammalian tissues, esp. endocrine organs. Of unknown biol. role. Viscous oil. Mp −10°.

Morton, R.A., *Biochem. J.*, 1972, **128**, 11P (*rev*)
Ekström, T. *et al, Acta Chem. Scand., Ser. B*, 1982, **36**, 411 (*biosynth*)
Suzuki, S. *et al, Tetrahedron Lett.*, 1983, **24**, 5103 (*synth*)
Eggers, I., *Acta Chem. Scand., Ser. B*, 1987, **41**, 67 (*biosynth*)
Appelkvist, E.-L., *Acta Chem. Scand., Ser. B*, 1987, **41**, 73 (*biosynth*)

Doridosine   D-60518

1,2-Dihydro-1-methyl-2-oxoadenosine, 9CI. N,6-Dide-hydro-1,2,3,6-tetrahydro-1-methyl-2-oxoadenosine, 9CI. 1-Methylisoguanosine

[70639-65-5]

$C_{11}H_{15}N_5O_5$ M 297.270

Prod. by the orange sponge *Tadania digitata* and the nudibranch *Anisdoris nobilis*. Cryst. (H_2O). Mp 262-263°. [α]$_D^{24}$ −65.4° (c, 1.0 in DMSO), [α]$_D^{22}$ −54.6° (c, 1.0 in H_2O).

Cook, A.F. *et al, J. Org. Chem.*, 1980, **45**, 4020 (*isol, uv, pmr, ms, cmr, synth*)
Fuhrman, F.A. *et al, Science*, 1980, **207**, 193 (*isol, pmr*)
Wong, R.L. *et al, Acta Crystallogr., Sect. C*, 1984, **40**, 1409 (*cryst struct*)
Davies, L.P., *Trends Pharmacol. Sci.*, 1985, **6**, 143 (*isol, rev*)

Drimenin   D-60519

Updated Entry replacing D-20485

[2326-89-8]

$C_{15}H_{22}O_2$ M 234.338

Constit. of *Drimys winteri*. Cryst. (MeOH). Mp 133°. [α]$_D$ −42° (c, 0.76 in C_6H_6).

3β-Acetoxy: 3β-Acetoxydrimenin.
$C_{17}H_{24}O_4$ M 292.374
Constit. of *D. winteri*. Cryst. (EtOAc/hexane). Mp 173-174°. [α]$_D^{20}$ −7° (c, 0.9 in CHCl$_3$).

Appel, H.H. *et al*, *J. Chem. Soc.*, 1960, 4685 (*isol*)
Kitahara, Y. *et al*, *J. Chem. Soc., Chem. Commun.*, 1969, 342 (*synth*)
Yanagawa, H. *et al*, *Synthesis*, 1970, 257 (*synth*)
Jalali-Naini, M. *et al*, *Tetrahedron*, 1983, **39**, 749 (*synth*)
Liapis, M. *et al*, *J. Chem. Soc., Perkin Trans. 1*, 1985, 815 (*synth*)
Sierra, J.R. *et al*, *Phytochemistry*, 1986, **25**, 253 (*Acetoxydrimenin*)
Ragoussis, V. *et al*, *J. Chem. Soc., Perkin Trans. 1*, 1987, 987 (*synth*)

9(11)-Drimen-8-ol　　　　　　　　D-60520

Updated Entry replacing D-20486
Decahydro-2,5,5,8a-tetramethyl-1-methylene-2-naphthalenol, 9CI

(5R,8R,10R)-form

$C_{15}H_{26}O$　　M 222.370

(5R,8R,10R)-form [86546-84-1]
Metab. of *Aspergillus oryzae*. Oil. $[\alpha]_D^{32}$ −132° (c, 0.08 in CHCl$_3$).

(5R,8S,10R)-form [86546-83-0]
Metab. of *A. oryzae*. Oil. $[\alpha]_D^{34}$ +4.4° (c, 0.14 in CHCl$_3$).

Wada, K. *et al.*, *Agric. Biol. Chem.*, 1983, **47**, 1075 (*isol*)
Leite, M.A.F. *et al.*, *J. Org. Chem.*, 1986, **51**, 5409 (*synth, abs config*)

Drosopterin　　　　　　　　D-60521

Updated Entry replacing D-08670
[33466-46-5]

Relative
configuration

$C_{15}H_{16}N_{10}O_2$　　M 368.357
Red eye pigment from *Drosophila melanogaster*. Red cryst. powder. Mp >350° dec. pK_{a1} 8.27, pK_{a2} 1.24, pK_{a3} 0.45. Opt. active.

Stereoisomer(?): [56711-41-2]. **Neodrosopterin**.
　$C_{15}H_{16}N_{10}O_2$　　M 368.357
　Pigment from *D. melanogaster*.
Enantiomer(?): [33466-47-6]. **Isodrosopterin**.
　$C_{15}H_{16}N_{10}O_2$　　M 368.357
　Pigment from *D. melanogaster*.

Schlobach, H. *et al*, *Helv. Chim. Acta.*, 1972, **55**, 2518, 2525, 2533, 2541 (*isol, uv, ord, cd, pmr, synth*)
Rokos, K. *et al*, *Chem. Ber.*, 1975, **108**, 2728 (*Neodrosopterim*)
Theobald, N. *et al*, *Chem. Ber.*, 1978, **111**, 3385 (*uv, pmr, cd, struct*)

Dukunolide A　　　　　　　　D-60522

Updated Entry replacing D-40437
[97804-04-1]

$C_{26}H_{26}O_9$　　M 482.486
Constit. of *Lansium domesticum*. Cryst. Mp 279-281°. $[\alpha]_D^{20}$ +166° (c, 0.9 in CHCl$_3$).

8α,9α-Epoxide: [99343-73-4]. **Dukunolide B**.
　$C_{26}H_{26}O_{10}$　　M 498.485
　From *L. domesticum*. Cryst. (CH$_2$Cl$_2$/hexane). Mp 248.5-251°. $[\alpha]_D^{20}$ +82.2° (c, 0.9 in CHCl$_3$).
30α-Acetoxy: [99343-74-5]. **Dukunolide C**.
　$C_{28}H_{28}O_{11}$　　M 540.523
　From *L. domesticum*. Cryst. (EtOAc/hexane). Mp 217-218°. $[\alpha]_D^{20}$ +206° (c, 0.89 in CHCl$_3$).

Nishizawa, M. *et al*, *J. Org. Chem.*, 1985, **50**, 5487.

Dukunolide D　　　　　　　　D-60523

$C_{26}H_{28}O_8$　　M 468.502
Constit. of *Lansium domesticum*. Cryst. (EtOAc/hexane). Mp 295.5-298°. $[\alpha]_D^{14.5}$ +175.3° (c, 0.57 in CHCl$_3$).

8α,9α-Epoxide: **Dukunolide E**.
　$C_{26}H_{28}O_9$　　M 484.502
　Constit. of *L. domesticum*. Cryst. Mp 270-272°. $[\alpha]_D^{13}$ +189° (c, 0.68 in CHCl$_3$).
8α,9α-Epoxide, 1,2-diepimer: **Dukunolide F**.
　$C_{26}H_{28}O_9$　　M 484.502
　Constit. of *L. domesticum*. Cryst. Mp 268-269°. $[\alpha]_D^{14}$ +167° (c, 0.73 in CHCl$_3$).

Nishizawa, M. *et al*, *Phytochemistry*, 1988, **27**, 237.

Dypnopinacol D-60524

(*2-Hydroxy-6-methyl-2,4,6-triphenyl-3-cyclohexen-1-yl)phenylmethanone, 9CI. Methyltriphenylhydroxycyclohexenophenone. Dypnopinacone*

[20511-09-5]

$C_{32}H_{28}O_2$ M 444.572

Needles (EtOH). Mp 164-165°.

Grimshaw, J. *et al, J. Chem. Soc.* (*C*), 1970, 817 (*synth, bibl*)
Declerq, J.P. *et al, Bull. Cl. Sci. Acad. R. Belg.*, 1978, **64**, 406; *CA*, **91**, 185154 (*cryst struct*)
Sosnovskikh, V.Y. *et al, CA*, 1981, **95**, 114979 (*synth*)

Dysoxylin D-60525

$C_{26}H_{30}O_8$ M 470.518

Constit. of *Dysoxylum richii*. Amorph. solid. Mp 247-251°. $[\alpha]_D^{25}$ +72.5° (c, 0.15 in $CHCl_3$).

Jogia, M.K. *et al, Phytochemistry*, 1987, **26**, 3309.

E

Ebuloside   E-60001

Updated Entry replacing E-50001

[103553-93-1]

$C_{21}H_{32}O_{10}$ M 444.478

Constit. of *Sambucus ebulus*. Amorph. powder. $[\alpha]_D^{20}$ −169.1° (c, 0.64 in MeOH).

6′-O-*Apiosyl:* [107882-65-5]. 6′-O-*Apiosylebuloside.*
$C_{26}H_{40}O_{14}$ M 576.594
From *S. ebulus*. Amorph. powder. $[\alpha]_D^{20}$ −215.5° (c, 0.51 in H_2O).

7β-*Alcohol:* 7,7-O-*Dihydroebuloside.*
$C_{21}H_{34}O_{10}$ M 446.494
From *S. ebulus*. Amorph. $[\alpha]_D^{20}$ −73.8° (c, 0.6 in MeOH).

7β-*Alcohol, penta-Ac:* Cryst. Mp 112°. $[\alpha]_D^{20}$ −65.2° (c, 0.45 in $CHCl_3$).

Gross, G.-A. *et al, Helv. Chim. Acta,* 1986, **69**, 156; 1987, **70**, 91 (*isol*)

Echinofuran   E-60002

Updated Entry replacing E-50006

[80348-64-7]

$C_{18}H_{18}O_5$ M 314.337

Isol. from callus cultures of *Echium lycopsis*. Orange oil. $[\alpha]_D^{25}$ −40.0° ($CHCl_3$).

Deacetoxy: [90685-55-5]. **Echinofuran B**.
$C_{16}H_{16}O_3$ M 256.301
Constit. of *Lithospermum erythrorhizon* callus cultures. Orange oil.

Inouye, H. *et al, Phytochemistry,* 1981, **20**, 1701 (*isol*)
Fukui, H. *et al, Phytochemistry,* 1984, **23**, 301 (*uv, ir, pmr, cmr*)

Ellagic acid   E-60003

Updated Entry replacing E-00091

2,3,7,8-Tetrahydroxy[1]benzopyrano[5,4,3-cde][1]-benzopyran-5,10-dione, 9CI

[476-66-4]

$C_{14}H_6O_8$ M 302.197

Occurs free and combined in galls. Isol. from the leaves of *Castanopsis* spp. and several other spp. Needles + 2Py (Py). Spar. sol. H_2O, EtOH, insol. Et_2O. Mp >360°.

▷DJ2620000.

Tetra-Ac: Mp 317-319°, 343-346°.

3,3′-Di-Me ether: **Nasutin C**.
$C_{16}H_{10}O_8$ M 330.250
Constit. of the haemolymph of the termite *Nasutitermes exitiosus*. Pale-cream needles (Me_2CO). Mp 336-338° dec.

3,3′-Di-Me ether, 4-glucoside:
$C_{22}H_{20}O_{13}$ M 492.392
Occurs in *Terminalia paniculata*. Prisms (MeOH aq.). Mp 214-215°. $[\alpha]_D^{30}$ +79° (c, 0.504 in MeOH).

3,3′,4-Tri-Me ether: **Nasutin B**.
$C_{17}H_{12}O_8$ M 344.277
Constit. of the haemolymph of *N. exitiosus*. Pale-cream platelets (EtOH). Mp 298° dec.

Tetra-Me ether: Mp 355° dec.

Perkin, A.G., *J. Chem. Soc.,* 1905, **87**, 1415 (*synth*)
Nierenstein, M., *Helv. Chim. Acta,* 1931, **14**, 912 (*isol*)
Zetzsche, F., *Helv. Chim. Acta,* 1931, **14**, 240 (*isol*)
Row, L.R., *Tetrahedron,* 1962, **18**, 357 (*isol*)
Moore, B.P., *Aust. J. Chem.,* 1964, **17**, 901 (*derivs*)
Arthur, H.R., *Aust. J. Chem.,* 1969, **22**, 597 (*isol*)
Sato, T., *Phytochemistry,* 1987, **26**, 2124 (*synth, cmr*)

Elvirol   E-60004

2-(1,5-Dimethyl-4-hexenyl)-4-methylphenol, 9CI. 2-(1,5-Dimethyl-4-hexenyl)-p-cresol, 8CI

[23479-73-4]

$C_{15}H_{22}O$ M 218.338

Isol. from the roots of *Elvira biflora*. Oil. $Bp_{0.06}$ 62°.

Ac: $Bp_{0.06}$ 59-60°.

Me ether: $Bp_{0.07}$ 62-63°.

Bohlmann, F. *et al, Tetrahedron Lett.,* 1969, 1005; *Chem. Ber.,* 1974, **107**, 1777 (*isol, synth*)
Dennison, N.R. *et al, Aust. J. Chem.,* 1975, **28**, 1339 (*synth*)
Pednekar, P.R. *et al, Indian J. Chem., Sect. B,* 1980, **19**, 1076 (*synth*)

Emindole *DA*   E-60005

Decahydro-5-(1H-indol-3-ylmethyl)-1,4a-dimethyl-6-methylene-1-(4-methyl-3-pentenyl)-2-naphthalenol, 9CI
[110883-36-8]

$C_{28}H_{39}NO$ M 405.622

Indolic diterpenoid antibiotic. Isol. from *Emericella desertorum* and *E. striata*. Prisms (C_6H_6/hexane). Mp 146-147°. [α]$_D$ −30.7° (c, 2.32 in MeOH).

Ac: Needles (hexane). Mp 142.5-143.5°.

5-Epimer: [110883-37-9]. **Emindole SA**.
$C_{28}H_{39}NO$ M 405.622
From *E. desertorum* and *E. striata*. Amorph. Mp 58-60°. [α]$_D$ +32° (c, 0.79 in MeOH).

Nozawa, K. *et al, J. Chem. Soc., Chem. Commun.*, 1987, 1157 (*isol, struct*)

Ephemeric acid   E-60006

$C_{20}H_{28}O_4$ M 332.439

Constit. of *Ephemerantha comata*. Cryst. ($CHCl_3$/hexane). Mp 183-185°. [α]$_D$ −75.0° (c, 0.56 in $CHCl_3$).

β-D-Glucopyranosyl ester: **Ephemeroside**.
$C_{26}H_{28}O_9$ M 484.502
Constit. of *E. comata*. Viscous liq. [α]$_D$ −43.2° (c, 1 in MeOH).

Niwa, M. *et al, Phytochemistry*, 1987, **26**, 3293.

3,6-Epidioxy-6-methoxy-4,16,18-eicosatrienoic acid   E-60007

$C_{21}H_{34}O_5$ M 366.497

Me ester: Methyl 3,6-epidioxy-6-methoxy-4,16,18-eicosatrienoate.
$C_{22}H_{36}O_5$ M 380.523
Constit. of sponge *Plakortis lita*. Antitumour agent. Solid. Mp 47.5°. [α]$_D^{20}$ +36.4° (c, 1.1 in MeOH).

Sakemi, S. *et al, Tetrahedron*, 1987, **43**, 263.

3,6-Epidioxy-6-methoxy-4-octadecenoic acid   E-60008

$C_{19}H_{34}O_5$ M 342.475

Me ester: Methyl 3,6-epidioxy-6-methoxy-4-octadecenoate.
$C_{20}H_{36}O_5$ M 356.501
Constit. of sponge *Plakortis lita*. Antitumour agent. Solid. Mp 49°. [α]$_D^{20}$ +38.1° (c, 5.3 in MeOH).

16,17-Didehydro, Me ester: Methyl 3,6-epidioxy-6-methoxy-4,16-octadecadienoate.
$C_{20}H_{34}O_5$ M 354.486
Constit. of *P. lita*. Antitumour agent. Solid. Mp 37.5°. [α]$_D^{20}$ +41.4° (c, 2.9 in MeOH).

14,15,16,17-Tetradehydro, Me ester: Methyl 3,6-epidioxy-6-methoxy-4,14,16-octadecatrienoate.
$C_{20}H_{32}O_5$ M 352.470
Constit. of *P. lita*. Antitumour agent. Solid. Mp 38-39°. [α]$_D^{20}$ +40.8° (c, 4.9 in MeOH).

Sakemi, S. *et al, Tetrahedron*, 1987, **43**, 263.

2,3-Epoxy-7,10-bisaboladiene   E-60009

4-(1,5-Dimethyl-1,4-hexadienyl)-1-methyl-7-oxabicyclo[4.1.0]heptane, 9CI. 2-(3,4-Epoxy-4-methylcyclohexyl)-6-methyl-2,5-heptadiene

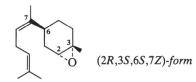

(2R,3S,6S,7Z)-form

$C_{15}H_{24}O$ M 220.354

(2R,3S,6S,7Z)-form

Sex pheromone of the Southern green stinkbug *Nezora viridula*. Oil. [α]$_D$ −14.7° (c, 1.2 in $CHCl_3$).

Baker, R. *et al, J. Chem. Soc., Chem. Commun.*, 1987, 414.

7,8-Epoxy-2,11-cembradiene-4,6-diol   E-60010

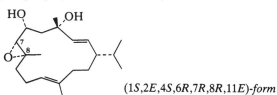

(1S,2E,4S,6R,7R,8R,11E)-form

$C_{20}H_{34}O_3$ M 322.487

(1S,2E,4S,6R,7R,8R,11E)-form

Constit. of green leaves and flowers of tobacco. Cryst. Mp 121-122°. [α]$_D$ +0.8° (c, 0.8 in $CHCl_3$).

(1S,2E,4S,6R,7S,8S,11E)-form

Constit. of green leaves and flowers of tobacco. Cryst. Mp 45-47°, Mp 67° (double Mp). [α]$_D$ +67° (c, 0.88 in $CHCl_3$).

Wahlberg, I. *et al, Acta Chem. Scand., Ser. B*, 1986, **40**, 855 (*isol, cryst struct*)

12,16-Epoxy-11,14-dihydroxy-5,8,11,13-abietatetraen-7-one   **E-60011**

[108572-36-7]

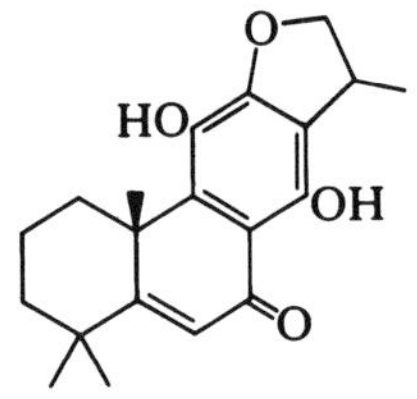

$C_{20}H_{24}O_4$ M 328.407

Constit. of *Premna integrifolia*. Yellow cryst.
(CHCl$_3$/MeOH). Mp 268°.

Rao, P.V.S. *et al*, *Indian J. Chem., Sect. B*, 1987, **26**, 191.

2,3-Epoxy-1,4-dihydroxy-7(11),8-eudesmadien-12,8-olide   **E-60012**

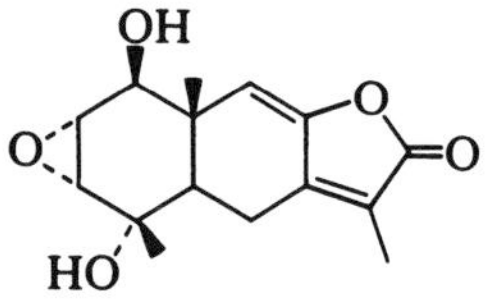

$C_{15}H_{18}O_5$ M 278.304

(*1β,2α,3α,4α*)-*form*

Constit. of *Smyrnium perfoliatum*. Amorph.

Gören, N. *et al*, *Phytochemistry*, 1987, **26**, 2585.

4,5-Epoxy-2,8-dihydroxy-1(10),11(13)-germacradien-12,6-olide   **E-60013**

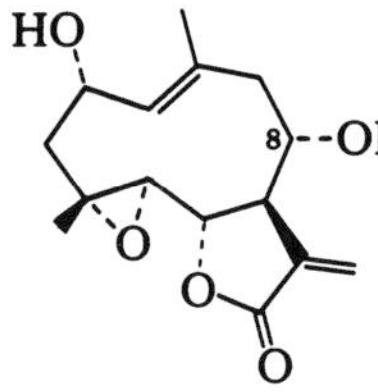

$C_{15}H_{20}O_5$ M 280.320

(*1(10)E,2α,4α,5α,6α,8α*)-*form*

8-(*3-Methyl-2-butenoyl*):
$C_{20}H_{26}O_6$ M 362.422
Constit. of *Elephantopus angustifolius*. Cryst. Mp 195°.

Jakupovic, J. *et al*, *Phytochemistry*, 1987, **26**, 1467.

6,11-Epoxy-6,12-dihydroxy-6,7-seco-8,11,13-abietatrien-7-al   **E-60014**

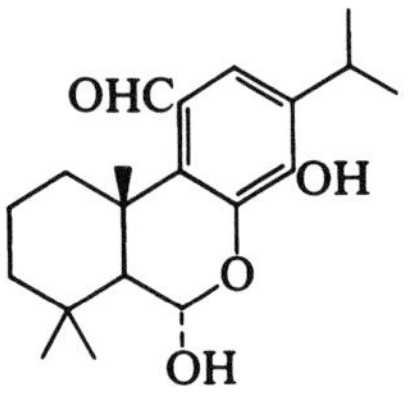

$C_{20}H_{28}O_4$ M 332.439

6α-form [111524-31-3]

Constit. of *Coleus barbatus*.

Di-Ac: Cryst. Mp 124-125°. $[\alpha]_D^{25}$ +64.6° (c, 0.73 in CHCl$_3$).

Kelecom, A. *et al*, *Phytochemistry*, 1987, **26**, 2337.

5,6-Epoxy-7,9,11,14-eicosatetraenoic acid   **E-60015**

Updated Entry replacing E-20026

3-(1,3,5,8-Tetradecatetraenyl)-oxiranebutanoic acid, 9CI

[71548-17-9]

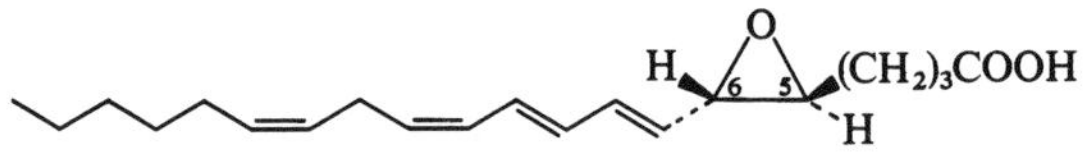

(*5S,6S,7E,9E,11Z,14Z*)-*form*

$C_{20}H_{30}O_3$ M 318.455

(*5S,6S,7E,9E,11Z,14Z*)-*form* [72059-45-1]

Leukotriene A$_4$. *LTA$_4$*
$[\alpha]_D^{25}$ −2.19° (c, 0.32 in cyclohexane).

(*5S,6S,7E,9Z,11Z,14Z*)-*form*
Metab. of arachidonic acid.

(*5RS,6RS,7Z,9E,11Z,14Z*)-*form*
Oil.

Corey, E.J. *et al*, *J. Am. Chem. Soc.*, 1980, **102**, 1436 (*synth*)
Atrache, V. *et al*, *Tetrahedron Lett.*, 1981, **22**, 3443 (*synth*)
Ernest, I. *et al*, *Tetrahedron Lett.*, 1982, **23**, 167 (*synth*)
Wang, Y. *et al*, *Tetrahedron Lett.*, 1986, **27**, 4583 (*synth*)

8,9-Epoxy-5,11,14-eicosatrienoic acid   **E-60016**

Updated Entry replacing E-20030

7-[3-(2,5-Undecadienyl)oxiranyl]-5-heptenoic acid, 9CI

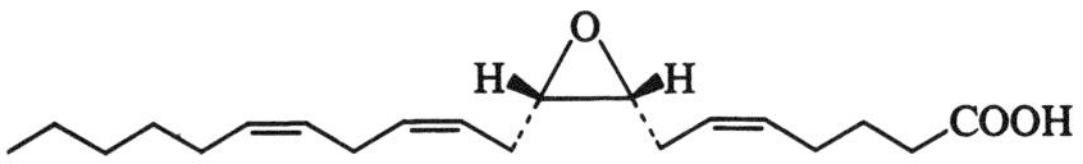

$C_{20}H_{32}O_3$ M 320.471

(*5Z,8R,*9R*,11Z,14Z*)-*form* [82864-43-5]

Arachidonic acid metab.

Falck, J.R. *et al*, *Tetrahedron Lett.*, 1982, **23**, 1755 (*synth, pmr*)
Mosset, P. *et al*, *Tetrahedron Lett.*, 1986, **27**, 6035 (*synth*)

11,12-Epoxy-5,8,14-eicosatrienoic acid   **E-60017**

Updated Entry replacing E-20031

10-[3-(2-Octenyl)oxiranyl]-5,8-decadienoic acid, 9CI

[81276-02-0]

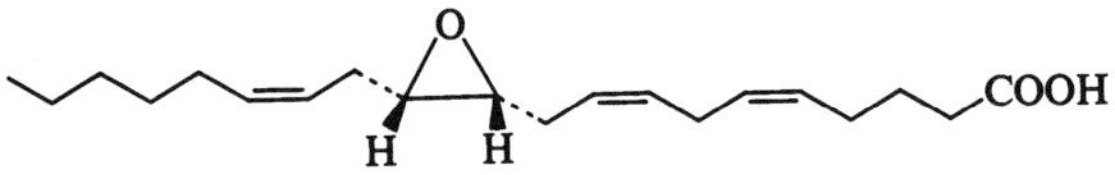

$C_{20}H_{32}O_3$ M 320.471

(*5Z,8Z,11RS,12SR,14Z*)-*form*
Oil.

Corey, E.J. *et al*, *J. Am. Chem. Soc.*, 1980, **102**, 1433 (*synth*)
Mosset, P. *et al*, *Tetrahedron Lett.*, 1986, **27**, 6035 (*synth*)

ent-5α,11-Epoxy-1β,4α,6α,8α,9β,14-　　　　　E-60018
eudesmanehexol

1α,4β,6β,8β,9α,14-Hexahydroxydihydro-β-agarofuran

C$_{15}$H$_{26}$O$_7$　　M 318.366
Related to Alatol, A-50077 .

1-Benzoyl, 9,14-Di-Ac: [107602-75-5]. *9α,14-Diace-
toxy-1α-benzoyloxy-4β,6β,8β-trihydroxydihydro-β-
agarofuran.*
C$_{26}$H$_{34}$O$_{10}$　　M 506.549
Constit. of *Rzedowskia tolantonguensis*. Cryst.
(EtOAc/hexane). Mp 278-281°.
1-Benzoyl, 6,9,14-Tri-Ac: [107602-77-7]. *6β,9α,14-
Triacetoxy-1α-benzoyloxy-4β,8β-dihydroxydihydro-
β-agarofuran.*
C$_{28}$H$_{36}$O$_{11}$　　M 548.586
From *R. tolantonguensis*. Cryst. (EtOAc/hexane). Mp
142-145°.
1-Benzoyl, 6,8,9,14-Tetra-Ac: [107602-74-4].
*6β,8β,9α,14-Tetraacetoxy-1α-benzoyloxy-4β-hydrox-
ydihydro-β-agarofuran.*
C$_{30}$H$_{38}$O$_{12}$　　M 590.623
From *R. tolantonguensis*. Cryst. (EtOAc/hexane). Mp
210-212°.
1,8-Dibenzoyl, 9,14-Di-Ac: [107602-78-8]. *9α,14-Diace-
toxy-1α,8β-dibenzoyloxy-4β,8β-dihydroxydihydro-β-
agarofuran.*
C$_{33}$H$_{38}$O$_{11}$　　M 610.657
From *R. tolantonguensis*. Cryst. (EtOAc/C$_6$H$_6$). Mp
202-204°.
8-Ketone, 1-benzoyl, 6,9,14-Tri-Ac: [107602-76-6].
*6β,9α,14-Triacetoxy-1α-benzoyloxy-4β-hydroxy-8-
oxodihydro-β-agarofuran.*
C$_{28}$H$_{34}$O$_{11}$　　M 546.570
From *R. tolantonguensis*. Cryst. (EtOAc/hexane). Mp
224-226°.

González, A.G. *et al*, *Heterocycles*, 1986, **24**, 3379.

1,4-Epoxy-6-eudesmanol　　　　　　　　　E-60019

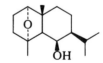

C$_{15}$H$_{26}$O$_2$　　M 238.369
(*1α,4α,6β*)-form
Ac: 6β-Acetoxy-1α,4α-epoxyeudesmane.
C$_{17}$H$_{28}$O$_3$　　M 280.406
Constit. of *Sideritis varoi*. Gum. [α]$_D^{20}$ −17.6° (c, 1 in
CHCl$_3$).

Cabrera, E. *et al*, *Phytochemistry*, 1988, **27**, 183.

8,12-Epoxy-3,7,11-eudesmatrien-1-ol　　　E-60020

C$_{15}$H$_{20}$O$_2$　　M 232.322
1β-form
Ac: [108044-18-4].
C$_{17}$H$_{22}$O$_3$　　M 274.359
Constit. of *Smyrnium galaticum*. Amorph.

Ulubelen, A. *et al*, *J. Nat. Prod.*, 1986, **49**, 1104.

3,4-Epoxy-11(13)-eudesmen-12-oic acid　　E-60021

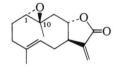

C$_{15}$H$_{22}$O$_3$　　M 250.337
(*3α,4α*)-form [108907-00-2]
Viscosic acid
Constit. of *Inula viscosa*. Oil.

Ulubelen, A. *et al*, *Phytochemistry*, 1987, **26**, 1223.

1,10-Epoxy-4,11(13)-germacradien-12,8-　　E-60022
olide

C$_{15}$H$_{20}$O$_3$　　M 248.321
(*1S*,4E,7R*,8S*,10R**)-form
Constit. of *Critonia quadrangularis*. Oil.

Jakupovic, J. *et al*, *Phytochemistry*, 1987, **26**, 451.

4,5-Epoxy-1(10),7(11)-germacradien-8-one　E-60023
Germacrone 4,5-epoxide

C$_{15}$H$_{22}$O$_2$　　M 234.338
(*4S,5S*)-form
Constit. of essential oil of Zedoariae Rhizoma. Biogenetic
precursor of sesquiterpenoid constits. of Zedoariae
Rhizoma. Prisms (hexane). Mp 59-60°. [α]$_D^{16}$ +399°
(c, 1.05 in CHCl$_3$).

Yoshihara, M. *et al*, *Chem. Pharm. Bull.*, 1984, **32**, 2059; 1986,
34, 434 (*isol, cryst struct*)

Epoxyisodihydrorhodophytin   E-60024

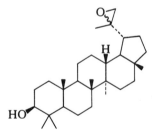

$C_{15}H_{20}BrClO_2$ M 347.679

Constit. of *Laurencia obtusa*. Needles (hexane). Mp 92-93°. $[\alpha]_D^{20}$ +23.7° (c, 0.815 in $CHCl_3$).

Imre, S. *et al, Z. Naturforsch., C*, 1987, **42**, 507.

14,15-Epoxy-8(17),12-labdadien-16-oic acid   E-60025

$C_{20}H_{30}O_3$ M 318.455

(12E,14ξ)-form

Me ester: Methyl 14ξ,15-epoxy-8(17),12E-labdadien-16-oate.
$C_{21}H_{32}O_3$ M 332.482
Constit. of *Aframomum daniellii*. Oil. $[\alpha]_D^{21}$ +10° (c, 0.6 in $CHCl_3$).

Kimbu, S.F. *et al, J. Nat. Prod.*, 1987, **50**, 230.

20(29)-Epoxy-3-lupanol   E-60026

$C_{30}H_{50}O_2$ M 442.724

3β-form

Metab. of *Pseudocyphellaria rubella*.

Corbett, R.E. *et al, Aust. J. Chem.*, 1987, **40**, 461.

5,6-Epoxyquinoline   E-60027

Quinoline 5,6-oxide

C_9H_7NO M 145.160

Liver microsomal metab. of quinoline.

Agarwal, S.K. *et al, Tetrahedron Lett.*, 1986, **27**, 4253 (*synth, pmr, use*)

7,8-Epoxyquinoline   E-60028

Quinoline 7,8-oxide

C_9H_7NO M 145.160

Agarwal, S.K. *et al, Tetrahedron Lett.*, 1986, **27**, 4253 (*synth, pmr*)

15,16-Epoxy-13,17-spatadiene-5,19-diol   E-60029

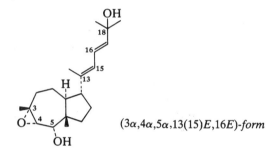

$C_{20}H_{30}O_3$ M 318.455

(5R)-form

19-Ac: [111576-31-9]. *19-Acetoxy-15,16-epoxy-13,17-spatadien-5α-ol*.
$C_{22}H_{32}O_4$ M 360.492
Constit. of *Stoechospermum marginatum*. Cryst. ($CHCl_3$/hexane). Mp 130-132°. $[\alpha]_D$ +48.5° (c, 0.8 in $CHCl_3$).

Rao, C.B. *et al, Indian J. Chem., Sect. B*, 1987, **26**, 79.

3,4-Epoxy-13(15),16-sphenolobadiene-5,18-diol   E-60030

$C_{20}H_{32}O_3$ M 320.471

(3α,4α,5α,13(15)E,16E)-form [108864-26-2]
Constit. of *Anastrophyllum minutum*. Oil. $[\alpha]_D^{22}$ +28.3° (c, 0.45 in EtOH).

5-Ac:
$C_{22}H_{34}O_5$ M 378.508
Constit. of *A. minutum*. Cryst. (EtOAc/hexane). Mp 46-48°.

(3α,4α,5α,13(15)Z,16E)-form
5-Ac: Constit. of *A. minutum*. Oil. $[\alpha]_D^{23}$ −11.5° (c, 0.59 in EtOH).

Beyer, J. *et al, Phytochemistry*, 1987, **26**, 1085.

3,4-Epoxy-13(15),16,18-sphenolobatrien-5-ol   E-60031

$C_{20}H_{30}O_2$ M 302.456

(3α,4α,5α,13(15)E,16E)-form

Ac:
$C_{22}H_{32}O_3$ M 344.493
Constit. of *Anastrophyllum minutum*. Oil. $[\alpha]_D^{25}$ +13° (c, 0.62 in $CHCl_3$).

(3α,4α,5α,13(15)Z,16E)-form
Constit. of *A. minutum*. Cryst. (EtOAc/hexane). Mp 77-79°.

Ac: Constit. of *A. minutum*. Oil. $[\alpha]_D^{24}$ +32° (c, 2.42 in $CHCl_3$).

Beyer, J. *et al, Phytochemistry*, 1987, **26**, 1085.

21,23-Epoxy-7,24-tirucalladien-3-ol   E-60032

$C_{30}H_{48}O_2$ M 440.708

3β-form

Cryst. (MeOH). Mp 144-146°. $[\alpha]_D$ −8° (CHCl₃).

Ac: 3β-Acetoxy-21,23-epoxy-7,24-tirucalladiene.
$C_{32}H_{50}O_3$ M 482.745
Constit. of *Cornus capitata*. Cryst. (MeOH). Mp 176°. $[\alpha]_D$ −6° (CHCl₃).

3-Ketone: 21,23-Epoxy-7,24-tirucalladien-3-one.
$C_{30}H_{46}O_2$ M 438.692
Constit. of *C. capitata*. Cryst. (hexane). Mp 188°. $[\alpha]_D$ −24° (CHCl₃).

Bhakuni, R.S. *et al*, *Phytochemistry*, 1987, **26**, 2607.

2,3-Epoxy-1,4,8-trihydroxy-7(11)-eudes-men-12,8-olide   E-60033

$C_{15}H_{20}O_6$ M 296.319

(1β,2α,3α,4α,8β)-form

8-Me ether: 2α,3α-Epoxy-1β,4α-dihydroxy-8β-methoxy-7(11)-eudesmen-12,8-olide.
$C_{16}H_{22}O_6$ M 310.346
Constit. of *Smyrnium perfoliatum*. Amorph. $[\alpha]_D^{24}$ −33.6°.

Gören, N. *et al*, *Phytochemistry*, 1987, **26**, 2585.

Eranthemoside   E-60034

$C_{15}H_{22}O_9$ M 346.333
Constit. of *Eranthemum pulchellum*. Amorph. powder. $[\alpha]_D^{20}$ −98° (c, 0.9 in EtOH).

Penta-Ac: Cryst. (EtOH). Mp 104-105°. $[\alpha]_D^{20}$ −109° (c, 0.4 in CHCl₃).

Jensen, H.F.W. *et al*, *Phytochemistry*, 1987, **26**, 3353.

Ergolide   E-60035

$C_{17}H_{22}O_5$ M 306.358
Constit. of *Erigeron khorassanicus*. Cryst. (EtOH). Mp 179-180°. $[\alpha]_D^{20}$ +123° (c, 4.88 in EtOH).

Ovezdurdyev, A. *et al*, *Khim. Prir. Soedin.*, 1986, **22**, 532.

Eriolin   E-60036

Updated Entry replacing E-20046
[27542-21-8]

$C_{15}H_{22}O_4$ M 266.336
Constit. of *Eriophyllum confertiflorum*. Cryst. Mp 238-240°. $[\alpha]_D$ −42°.

11,13-Didehydro: [87441-73-4]. *11,13-Dehydroeriolin.*
$C_{15}H_{20}O_4$ M 264.321
Constit. of *Schkuhria* spp. Cryst. (Me₂CO/Et₂O). Mp 173-174°. $[\alpha]_D$ −36° (CHCl₃).

Torrance, S.J. *et al*, *Phytochemistry*, 1969, **8**, 2381 (*isol*)
Romo de Vivar, A. *et al*, *Phytochemistry*, 1982, **21**, 2905 (*isol*)
Calderón, J.S. *et al*, *Phytochemistry*, 1987, **26**, 1747 (*cryst struct*)

Esperamicin   E-60037

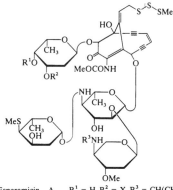

Esperamicin A₁   R¹ = H, R² = X, R³ = CH(CH₃)₂
             A₁ᵦ R¹ = H, R² = X, R³ = Et
             A₂   R¹ = X, R² = H, R³ = CH(CH₃)₂

Other components (A₃, A₄, B₁ and B₂) have been isol. but not characterized.

Esperamicin A₁ [99674-26-7]
$C_{59}H_{80}N_4O_{22}S_4$ M 1325.535

Isol. from *Actinomadura verrucosospora*. Potent antitumour agent. White or pale-yellow cryst. Mp 156-158° dec. $[\alpha]_D^{24}$ −207° (c, 0.035 in CHCl₃).

Esperamicin A₁ᵦ [88895-06-1]
FR 900405. WS 6049A. Antibiotic FR 900405. Antibiotic WS 6049A

$C_{58}H_{78}N_4O_{22}S_4$ M 1311.508
From *A. pulveracea* and *A. verrucosospora*. Active against gram-positive and gram-negative bacteria, tumours and some fungi. Powder. Mp 150° dec. $[\alpha]_D^{25}$ −208° (c, 1 in CHCl$_3$).

Esperamicin A₂
$C_{59}H_{80}N_4O_{22}S_4$ M 1325.535

From *A. verrucosospora*. Potent antitumour agent. Cryst. Mp 147-149°. $[\alpha]_D^{27}$ −179.4° (c, 0.5 in CHCl$_3$).

Antibiotic FR 900406
FR 900406. WS 6049B. Antibiotic WS 6049B
From *A. pulveracea*. Active against gram-positive and gram-negative bacteria, tumours and some fungi. Powder. Mp 145° dec. $[\alpha]_D^{25}$ −201° (c, 1 in CHCl$_3$).

Kiyoto, S. *et al, J. Antibiot.*, 1985, **38**, 835, 840, 955 (*isol, uv, ir, pmr, cmr, props*)
Konishi, M. *et al, J. Antibiot.*, 1985, **38**, 1605 (*isol, uv, ir, pmr, cmr*)
Golik, J. *et al, J. Am. Chem. Soc.*, 1987, **109**, 3462 (*struct*)

Esulone A   E-60038
Updated Entry replacing E-40035
[100215-74-5]

$R^1 = \alpha OH, H, R^2 = O$

$C_{38}H_{42}O_{12}$ M 690.743
Constit. of *Euphorbia esula*. Cryst. (MeOH). Mp 288-292°. $[\alpha]_D^{20}$ +10.6° (c, 10.4 in Me$_2$CO).
8-Ac: [100198-26-3]. **Esulone B.**
  $C_{40}H_{44}O_{13}$ M 732.780
  From *E. esula*. Cryst. (MeOH). Mp 274-276°. $[\alpha]_D^{20}$ −8.3° (c, 5.05 in Me$_2$CO).

Manners, G.D. *et al, J. Chem. Soc., Perkin Trans. 1*, 1985, 2075.

Esulone C   E-60039
As Esulone *A*, E-60038 with

$R^1 = O, R^2 = OH, H$

$C_{38}H_{42}O_{12}$ M 690.743
Constit. of *Euphorbia esula*. Cryst. Mp 210-213°.
Manners, G.D. *et al, Phytochemistry*, 1987, **26**, 727.

Ethanebis(dithioic)acid, 9CI   E-60040
Tetrathiooxalic acid
[82766-65-2]

HSCS-CSSH

$C_2H_2S_4$ M 154.278
Bis(tetraethylammonium) salt: [80733-35-3]. Yellow-orange cryst. Mp 186-189° dec.
Bis(tetraphenylphosphonium) salt: [82749-44-8]. Cryst. + 2H$_2$O (MeCN/Et$_2$O). Darkens at 130°, melts ca. 160°, resolidifies at ca. 190°.

Di-Me ester: [61485-47-0].
  $C_4H_6S_4$ M 182.331
  Cryst. (Et$_2$O). Mp 71-72°.

Hartke, K. *et al, Chem. Ber.*, 1980, **113**, 1898 (*ester*)
Jeroschewski, P. *et al, Z. Chem.*, 1981, **21**, 412; 1982, **22**, 223 (*synth*)
Lund, H. *et al, Acta Chem. Scand., Ser. B*, 1982, **36**, 207 (*synth*)

Ethanedithioamide, 9CI   E-60041
Dithiooxamide, 8CI. Rubeanic acid
[79-40-3]

$$H_2N-CS-CS-NH_2$$

$C_2H_4N_2S_2$ M 120.187
A reagent for anal. of Co, Cu and Ni. Also used as a stabilizer for ascorbic acid solns. Red cryst. Dec. at ca. 200°.

▷RP1575000.

Scott, T.A. *et al, J. Phys. Chem.*, 1959, **30**, 465 (*spectra, struct*)
Ray, P. *et al, J. Indian Chem. Soc.*, 1961, **38**, 535 (*rev*)
Wheatley, P.J., *J. Chem. Soc.*, 1965, 396 (*cryst struct*)
Burakevich, J.V. *et al, J. Org. Chem.*, 1970, **35**, 2102 (*synth*)
Hoppe, H. *et al, Arch. Pharm.* (*Weinheim*), 1975, **308**, 526 (*synth*)
Desseyn, H.O, *et al, Appl. Spectrosc.*, 1978, **32**, 101 (*ir, raman*)
Domnisse, R.A. *et al, Bull. Soc. Chim. Belg.*, 1979, **88**, 261 (*cmr*)

13,15-Ethano-3,17-diethyl-2,12,18-trimethylmonobenzo[*g*]porphyrin   E-60042
8,14-Diethyl-16,17-dihydro-9,13,24-trimethyl-5,22:12,15-diimino-20,18-metheno-7,10-nitrilobenzo[o]-cyclopent[b]azacyclononadecine, 9CI
[100813-35-2]
As 13,15-Ethano-17-ethyl-2,3,12,18-tetramethylmonobenzo[*g*]porphyrin, E-60043 with

$$R = CH_2CH_3$$

$C_{33}H_{32}N_4$ M 484.643
Monobenzoporphyrin from Venezuelan cretaceous crude oil.

Kaur, S. *et al, J. Am. Chem. Soc.*, 1986, **108**, 1347 (*isol, uv, pmr*)

13,15-Ethano-17-ethyl-2,3,12,18-tetramethylmonobenzo[*g*]porphyrin   E-60043
14-Ethyl-16,17-dihydro-8,9,13,24-tetramethyl-5,22:12,15-diimino-20,18-metheno-7,10-nitrilobenzo[o]-cyclopent[b]azacyclononadecine, 9CI
[100813-32-9]

$R = CH_3$

$C_{32}H_{30}N_4$ M 470.616
Monobenzoporphyrin from Venezuelan cretaceous crude oil.

Kaur, S. *et al, J. Am. Chem. Soc.*, 1986, **108**, 1347 (*isol, uv, pmr*)

1-Ethoxy-1-hydroperoxyethane   E-60044
(*1-Ethoxyethyl*) *hydroperoxide. Mozuku toxin* A

$$EtOCH(OOH)CH_3$$

$C_4H_{10}O_3$ M 106.121

Primary autoxidn. prod. present in unpurified Et_2O. Isol. from the brown alga *Sphaerotrichia divaricata*. Oil. The structs. of the Mozuku toxins and their relationship to ether oxidn. prods. require further investigation.

▷Highly toxic. LD_{50} 250 mg/kg (mice). Explosive

Anthoni, U. *et al, Acta Chem. Scand., Ser. B*, 1987, **41**, 216.

Ethuliacoumarin   E-60045
Updated Entry replacing E-00591
2-(3,3-Dimethyloxiranyl)-4-ethenyl-3,4-dihydro-2-hydroxy-4,10-dimethyl-2H,5H-pyrano[3,2-c] [1]-benzopyran-5-one, 9CI

[63893-02-7]

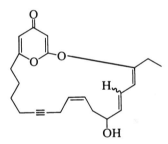

$C_{20}H_{22}O_5$ M 342.391

Constit. of *Ethulia conyzoides*. Cryst. (Et_2O/pet. ether). Mp 61°. $[\alpha]_D^{24}$ +28° (c, 0.1 in $CHCl_3$).

9-De-Me, 9-ethyl: 5-Methylethuliacoumarin.
$C_{21}H_{24}O_5$ M 356.418
Constit. of *Bothriocline ripensis*. Oil.

Bohlmann, F. *et al, Phytochemistry*, 1977, **16**, 1092 (*isol, struct*)
Jakupovic, J. *et al, Phytochemistry*, 1987, **26**, 1069 (*isol, deriv*)

1-Ethylcyclohexene   E-60046
[1453-24-3]

$$CH_2CH_3$$

C_8H_{14} M 110.199
d_{19} 0.82. Bp 134-136°, Bp_{30} 49°.

Wallach, O., *Justus Liebigs Ann. Chem.*, 1908, **360**, 26 (*synth*)
Bergmann, E. *et al, J. Am. Chem. Soc.*, 1937, **59**, 1443 (*synth*)
Benkeser, R.A. *et al, J. Am. Chem. Soc.*, 1955, **77**, 3230 (*synth*)

3-Ethylcyclohexene   E-60047
[2808-71-1]
C_8H_{14} M 110.199
(±)-*form*
Liq. Bp 160°, Bp_{30} 42-44°.

Bailey, B. *et al, J. Chem. Soc.*, 1954, 967 (*synth*)
Pearson, A.J. *et al, J. Org. Chem.*, 1986, **51**, 2505 (*synth, ir, pmr*)
Tseng, C.C. *et al, J. Org. Chem.*, 1986, **51**, 2884 (*synth, ir, pmr, cmr*)

4-Ethylcyclohexene   E-60048
[3742-42-5]

C_8H_{14} M 110.199
(±)-*form*
Bp 130-131°. n_D^{20} 1.4499.

Pines, H. *et al, J. Am. Chem. Soc.*, 1952, **74**, 4872 (*synth*)
Shelton, J.R. *et al, J. Org. Chem.*, 1970, **35**, 1576 (*synth*)

Ethylenetetracarboxylic acid   E-60049
Updated Entry replacing E-00758
Ethenetetracarboxylic acid, 9CI
[4363-44-4]

$$(HOOC)_2C{=}C(COOH)_2$$

$C_6H_4O_8$ M 204.093
Mp 155-165° (impure).
Tetra-Et ester:
$C_{14}H_{20}O_8$ M 316.307
Cryst. (EtOH). Mp 120-120.5°.
1,1-Dinitrile, Di-Me ester: [82849-49-8]. *DDED.*
$C_8H_6N_2O_4$ M 194.146
Liq. $Bp_{0.05}$ 75-76°.

Snyder, H.R. *et al, J. Am. Chem. Soc.*, 1958, **80**, 1942 (*synth*)
Hall, H.K. *et al, J. Org. Chem.*, 1982, **47**, 4572.
Cannon, J.R. *et al, Aust. J. Chem.*, 1986, **39**, 1811 (*cryst struct, ester*)

21-Ethyl-2,6-epoxy-17-hydroxy-1-oxacy-   E-60050
clohenicosa-2,5,14,18,20-pentaen-11-yn-
4-one

$C_{22}H_{26}O_4$ M 354.445
Metabolite of red alga *Phacelocarpus labillardieri*.
Shin, J. *et al, Tetrahedron Lett.*, 1986, **27**, 5189.

5-Ethyl-3-hydroxy-4-methyl-2(5H)-furan-   E-60051
one, 9CI
2,4-Dihydroxy-3-methyl-2-hexenoic acid γ-lactone. 5-Ethyl-4-methylisotetronic acid. Abhexone. 5-Ethylsotolone
[698-10-2]

$C_7H_{10}O_3$ M 142.154
Prod. in protein hydrol. by condensation of threoline residues. Hygroscopic yellow oil with intense odour of lovage.
(±)-*form*
Flavouring for foodstuffs. Mp 21°. $Bp_{0.2}$ 92-94°, $Bp_{0.003}$ 70-71°.
Ac:
$C_9H_{12}O_4$ M 184.191
$Bp_{0.05}$ 90-92°.

Monnin, J., *Helv. Chim. Acta*, 1957, **40**, 1983 (*synth*)

Sakan, T. *et al*, *Bull. Chem. Soc. Jpn.*, 1964, **37**, 1171 (*uv*, *ir*)
Ger. Pat., 1 955 390, (*1970*); *CA*, **73**, 77034 (*synth*, *use*)
Bonini, C.C. *et al*, *Org. Mass. Spectrom.*, 1980, **15**, 516 (*ms*)
Stach, H. *et al*, *Helv. Chim. Acta*, 1987, **70**, 369 (*synth*, *ir*, *pmr*, *cmr*, *bibl*)

7-Ethyl-5-methyl-6,8-dioxabicyclo[3.2.1]-oct-3-ene E-60052

Updated Entry replacing E-40055
[88525-42-2]

$C_9H_{14}O_2$ M 154.208
Constit. of male mouse (*Mus musculus*) urine. Phero-mone signalling male mouse aggression. Related to Brevicomin, B-40141 .

Wiesler, D.P. *et al*, *J. Org. Chem.*, 1984, **49**, 882 (*isol*, *synth*, *ir*, *pmr*, *ms*)
Mori, K. *et al*, *Tetrahedron*, 1986, **42**, 5901 (*synth*)

3-Ethyl-3-pentanethiol, 9CI E-60053

Triethylthiocarbinol. 1,1-Diethylpropanethiol. 3-Ethyl-3-mercaptopentane
[5827-80-5]

$$(H_3CCH_2)_3CSH$$

$C_7H_{16}S$ M 132.263
Oil. Bp_{95} 92-95°, Bp_{36} 72°.

Reinheckel, H. *et al*, *Chem. Ber.*, 1966, **99**, 23 (*synth*)
Barton, D.H.R. *et al*, *J. Chem. Soc.*, *Perkin Trans. 1*, 1986, 1603 (*synth*, *pmr*)

5-Ethyl-2,4(1H,3H)-pyrimidinedione, 9CI E-60054

5-Ethyluracil
[4212-49-1]

$C_6H_8N_2O_2$ M 140.141
Cryst. Mp 302-303°.
Mono-Hg salt: [25181-82-2]. Cryst. Mp >360°.
1-Ac:
  $C_8H_{10}N_2O_3$ M 182.179
  Cryst. Mp 142-145°.
1-β-D-Ribofuranosyl: [25110-76-3]. *5-Ethyluridine.*
  $C_{11}H_{16}N_2O_6$ M 272.257
  Cryst. (EtOH). Mp 184-186°.
1-(2-Deoxy-β-D-ribofuranosyl): [15176-29-1]. *2'-Deoxy-5-ethyluridine.*
  $C_{11}H_{16}N_2O_5$ M 256.258
  Antiviral against Herpes simplex and vaccinia. Cryst. (MeOH). Mp 154-155°.
▷YU7500000.

Shapira, J., *J. Org. Chem.*, 1962, **27**, 1918 (*synth*, *deriv*)
Gauri, K.K. *et al*, *Z. Naturforsch.*, *B*, 1969, **24**, 833 (*synth*, *deriv*, *uv*, *pmr*, *ord*)
Świerkowski, M. *et al*, *J. Med. Chem.*, 1969, **12**, 533 (*synth*, *deriv*, *pmr*, *use*)

Shealy, Y.F. *et al*, *J. Med. Chem.*, 1986, **29**, 79 (*deriv*, *props*, *bibl*)

3-(Ethylthio)-2-propenoic acid, 9CI E-60055

3-Ethylthioacrylic acid

$$EtSCH=CHCOOH$$

$C_5H_8O_2S$ M 132.177
Nitrile: [10568-83-9].
  C_5H_7NS M 113.177
  Bp_{18} 111-115°.
Nitrile, S-dioxide:
  $C_5H_7NO_2S$ M 145.176
  Cryst. (H_2O). Mp 44.5-45°.
(*E*)-*form* [101541-97-3]
Cryst. (Et_2O/pet. ether). Mp 82-83°.
Et ester:
  $C_7H_{12}O_2S$ M 160.231
  $Bp_{0.25}$ 54-55°.
(*Z*)-*form*
Et ester: $Bp_{0.25}$ 47-49°.

Fitger, P., *Ber.*, 1921, **54**, 2953 (*synth*)
U.S.P., 3 078 298, (*1963*); *CA*, **59**, 1493 (*deriv*, *synth*)
Barua, N.C. *et al*, *Tetrahedron*, 1986, **42**, 4471 (*synth*, *pmr*)

Ethynamine E-60056

Aminoacetylene
[52324-04-6]

$$HC\equiv CNH_2$$

C_2H_3N M 41.052
Characterised in the gas phase.

van Baar, B. *et al*, *Angew. Chem.*, *Int. Ed. Engl.*, 1986, **25**, 827.

2-Ethynylindole E-60057

2-Indolylacetylene
[75258-11-6]

$C_{10}H_7N$ M 141.172

Prikhod'ko, T.A. *et al*, *Izv. Akad. Nauk SSSR*, *Ser. Khim.*, 1980, 1690; *CA*, **93**, 186084d (*synth*)

3-Ethynylindole E-60058

3-Indolylacetylene
[62365-78-0]
$C_{10}H_7N$ M 141.172
Flakes (heptane). Mp 114-115° dec. (106-108°, 102° dec.).

Suvorov, N.N. *et al*, *J. Org. Chem. USSR* (*Engl. Transl.*), 1977, **13**, 181 (*synth*, *uv*)
Wentrup, C. *et al*, *Angew. Chem.*, *Int. Ed. Engl.*, 1978, **17**, 609 (*synth*)
Benzies, D.W.M. *et al*, *J. Chem. Soc.*, *Perkin Trans. 1*, 1986, 1651 (*synth*, *ir*, *pmr*, *cmr*)

3-Ethynyl-3-methyl-4-pentenoic acid, 9CI   E-60059
3-Methyl-3-vinyl-4-pentynoic acid

$$HC\equiv C-\underset{\underset{CH=CH_2}{|}}{\overset{\overset{CH_2COOH}{|}}{C}}-CH_3 \qquad (R)\text{-form}$$

$C_8H_{10}O_2$ M 138.166

(**R**)-*form* [100018-85-7]
$[\alpha]_D^{23}$ +23.08° (c, 10.84 in $CHCl_3$). ca. 95% opt. purity.
Me ester: [99947-54-3].
   $C_9H_{12}O_2$ M 152.193
   Liq. Bp_{25} 79-80°. $[\alpha]_D^{25}$ +20.9° (c, 10.9 in $CHCl_3$).
(**S**)-*form* [100018-84-6]
$Bp_{0.2}$ 73-75°. $[\alpha]_D^{25}$ −23.00° (c, 15.68 in $CHCl_3$). ca. 95%
opt. purity.
N,N-*Dimethylamide:* [99947-42-9].
   $C_{10}H_{15}NO$ M 165.235
   $Bp_{0.5}$ 69-70°. $[\alpha]_D^{25}$ −12.27° (c, 14.31 in $CHCl_3$).
(±)-*form* [99947-40-7]
$Bp_{0.2}$ 73-75°.
Et ester: [99947-39-4].
   $C_{10}H_{14}O_2$ M 166.219
   Bp_{30} 93-95°.

Stevens, R.V. *et al, J. Am. Chem. Soc.*, 1986, **108**, 1039 (*synth, resoln, ir, pmr, cmr*)

5-Ethynyl-2,4(1H,3H)-pyrimidinedione, 9CI   E-60060
5-Ethynyluracil
[59989-18-3]

$C_6H_4N_2O_2$ M 136.110
Mp >300° dec.
1-β-D-Ribofuranosyl: [69075-42-9]. *5-Ethynyluridine.*
   $C_{11}H_{12}N_2O_6$ M 268.226
   Cryst. (MeOH). No Mp given.
1-β-D-(2'-Deoxyribofuranosyl): [61135-33-9]. *2'-Deoxy-5-ethynyluridine, 9CI.*
   $C_{11}H_{12}N_2O_5$ M 252.226
   Mp 197-199° dec.

Barr, P.J. *et al, Nucleic Acids Res.*, 1976, **3**, 2845.
Perman, J. *et al, Tetrahedron Lett.*, 1976, 2427 (*synth, uv, pmr, ir*)
Barr, P.J. *et al, J. Chem. Soc., Perkin Trans. 1*, 1978, 1263 (*deriv, synth, uv, pmr*)

2-Ethynylthiazole, 9CI   E-60061
[111600-85-2]

C_5H_3NS M 109.145
Bp_{55} 118°.

Sakamoto, T. *et al, Chem. Pharm. Bull.*, 1987, **35**, 823 (*synth, pmr*)

4-Ethynylthiazole, 9CI   E-60062
[111600-89-6]
C_5H_3NS M 109.145
Bp_{50} 108°.

Sakamoto, T. *et al, Chem. Pharm. Bull.*, 1987, **35**, 823 (*synth, pmr*)

4,7(11)-Eudesmadiene-12,13-diol   E-60063
$C_{15}H_{24}O_2$ M 236.353
Di-Ac: **Coralloidin** D.
   $C_{19}H_{28}O_4$ M 320.428
   Constit. of *Alcyonium coralloides*. Oil. $[\alpha]_D^{20}$ −14.1°
   (c, 0.227 in EtOH).

D'Ambrosio, M. *et al, Helv. Chim. Acta*, 1987, **70**, 612.

3,5-Eudesmadien-1-ol   E-60064
1,2,5,7,8,8a-Hexahydro-4,8a-dimethyl-6-(1-methylethyl)-1-naphthalenol, 9CI. 1,2,6,7,8,8a-Hexahydro-6-isopropyl-4,8a-dimethyl-1-naphthol

$C_{15}H_{24}O$ M 220.354

1β-form [95457-17-3]
Constit. of *Sideritis varoi*. Cryst. Mp 48-50°. $[\alpha]_D^{20}$ −7°
(c, 1 in $CHCl_3$).

Garcia-Granados, A. *et al, Phytochemistry*, 1985, **24**, 97 (*synth*)
Cabrera, E. *et al, Phytochemistry*, 1988, **27**, 183 (*isol*)

5,7-Eudesmadien-11-ol   E-60065
$C_{15}H_{24}O$ M 220.354
(**4R*,10R***)-*form*
Ac: **Coralloidin** E.
   $C_{17}H_{26}O_2$ M 262.391
   Constit. of *Alcyonium coralloides*. Cryst. Mp 55°.
   $[\alpha]_D^{20}$ +305° (c, 0.09 in EtOH).

D'Ambrosio, M., *Helv. Chim. Acta*, 1987, **70**, 612.

5,7(11)-Eudesmadien-15-ol   E-60066

$C_{15}H_{24}O$ M 220.354
(**4R,10S**)-*form* [110299-97-3]
   Oil. $[\alpha]_D^{20}$ −172.9° (c, 0.14 in EtOH).
   Ac: [110299-91-7]. **Coralloidin** C.
   $C_{17}H_{26}O_2$ M 262.391
   Constit. of *Alcyonium coralloides*. Oil. $[\alpha]_D$ −168.7°
   (c, 0.34 in $CHCl_3$).

D'Ambrosio, M. *et al, Helv. Chim. Acta*, 1987, **70**, 612.

11-Eudesmene-1,5-diol
E-60067

$C_{15}H_{26}O_2$ M 238.369

(1α,5α)-form
α-Corymbolol
Constit. of *Cyperus articulatus*. Oil. $[\alpha]_D^{25}$ +43° (c, 2.9 in CHCl₃).

(1β,5α)-form
β-Corymbolol
Oil. $[\alpha]_D^{25}$ +25° (c, 0.35 in CHCl₃).

Nyasse, B. *et al*, *Phytochemistry*, 1988, **27**, 179.

4(15)-Eudesmene-1,5,6-triol
E-60068

$C_{15}H_{26}O_3$ M 254.369

(1β,5α,6β)-form
6-Ac: 6β-Acetoxy-4(15)-eudesmene-1β,5α-diol.
$C_{17}H_{28}O_4$ M 296.406
Constit. of *Sideritis varoi*.

Cabrera, E. *et al*, *Phytochemistry*, 1988, **27**, 183.

11-Eudesmen-5-ol
E-60069
5-Eudesmol

$C_{15}H_{26}O$ M 222.370

(4αH,5α)-form
Constit. of *Kleinia pendula*. Oil. $[\alpha]_D$ +32.8° (c, 0.7 in CHCl₃).

Elmi, A.H. *et al*, *Phytochemistry*, 1987, **26**, 3069.

Euglenapterin
E-60070
Updated Entry replacing E-10133
2-Dimethylamino-6-(trihydroxypropyl)-4(3H)-pteridinone
[73789-39-6]

Absolute configuration

$C_{11}H_{15}N_5O_4$ M 281.271
Occurs with 3'-o-phosphate and 2',3'-cyclic phosphate. All show yellow fluorescence. Found in the alga *Euglena gracilis*. Yellow cryst. (H₂O). Mp >200°.

Elstner, E. *et al*, *Arch. Biochem. Biophys.*, 1976, **173**, 614 (*isol*)
Jacobi, P.A. *et al*, *J. Org. Chem.*, 1981, **46**, 5416 (*synth*)
Böhme, M. *et al*, *Justus Liebigs Ann. Chem.*, 1986, 1705 (*isol, cryst struct, uv, cd, pmr, synth*)

Eumaitenin
E-60071

$C_{26}H_{34}O_{10}$ M 506.549
Constit. of *Maytenus boaria*. Cryst. (EtOAc/pet. ether). Mp 183-185°.

Becerra, J. *et al*, *Phytochemistry*, 1987, **26**, 3073.

Eumaitenol
E-60072

$C_{27}H_{32}O_{10}$ M 516.544
Constit. of *Maytenus boaria*. Cryst. (EtOAc/pet. ether). Mp 208-212°.

8α-Acetoxy: **Acetyleumaitenol**.
$C_{29}H_{34}O_{12}$ M 574.580
From *M. boaria*. Cryst. (EtOAc/pet. ether). Mp 95-97°.

Becerra, J. *et al*, *Phytochemistry*, 1987, **26**, 3073.

Eumorphistonol
E-60073
Updated Entry replacing E-01321
[68776-18-1]

$C_{15}H_{20}O_3$ M 248.321
Constit. of *Eumorphia* spp. Oil.

Bohlmann, F. *et al*, *Phytochemistry*, 1978, **17**, 1155 (*isol*)
Hess, T. *et al*, *Tetrahedron Lett.*, 1987, **28**, 5643 (*struct*)

Euphorianin
E-60074

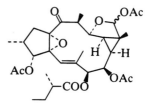

$C_{31}H_{42}O_{11}$ M 590.666
Constit. of *Euphorbia poisonii*. Cryst. (Et₂O/pet. ether). Mp 129-130°.

Fakunle, C.O. *et al*, *Tetrahedron Lett.*, 1978, 2119 (*isol, pmr*)
Okogun, J.I. *et al*, *Z. Naturforsch., B*, 1987, **42**, 243 (*cryst struct*)

Eupomatenoid 6   **E-60075**

Updated Entry replacing E-01361

*4-[3-Methyl-5-(1-propenyl)-2-benzofuranyl]phenol,
9CI. 2-(4-Hydroxyphenyl)-3-methyl-5-(1-propenyl)-
benzofuran. Ratanhiaphenol II*

[41744-26-7]

$C_{18}H_{16}O_2$ M 264.323

Constit. of *Eupomatia* spp. Needles (C_6H_6/pet. ether).
Mp 148-150°.

3'-Methoxy: [41744-28-9]. **Eupomatenoid 5**.
$C_{19}H_{18}O_3$ M 294.349
Constit. of *E.* spp. Needles (C_6H_6/pet. ether). Mp 114-
115°.

3',7-Dimethoxy: [41744-30-3]. **Eupamatenoid 7**.
$C_{20}H_{20}O_4$ M 324.376
Constit. of *E.* spp. Needles (pet. ether). Mp 105-106°.

3'-Methoxy, 4-Me ether: [41744-29-0]. **Eupomatenoid 4**.
$C_{20}H_{20}O_3$ M 308.376
Constit. of *E.* spp. Prisms (MeOH). Mp 96°.

3',7-Dimethoxy, 4-Me ether: [41365-37-1]. **Eupomaten-
oid 12**.
$C_{21}H_{22}O_4$ M 338.402
Constit. of *E.* spp. Needles (EtOH). Mp 115-116°.

O-Rhamnosyl: [41744-25-6]. **Eupomatenoid 2**. Constit.
of *E.* spp. Cryst. (EtOH). Mp 154°. $[\alpha]_D$ −140° (c, 1
in $CHCl_3$).

McCredie, R.S. *et al, Aust. J. Chem.*, 1969, **22**, 1011 (*isol,
struct*)
Bowden, B.F. *et al, Aust. J. Chem.*, 1972, **25**, 2659 (*isol, struct*)
Ahmed, R. *et al, Phytochemistry*, 1975, **14**, 2710 (*synth*)
Stahl, E. *et al, Planta Med.*, 1981, **42**, 144 (*isol*)
Achenbach, H. *et al, Phytochemistry*, 1987, **26**, 1159 (*isol, pmr,
cmr*)

F

Fercoperol
F-60001

[106533-43-1]

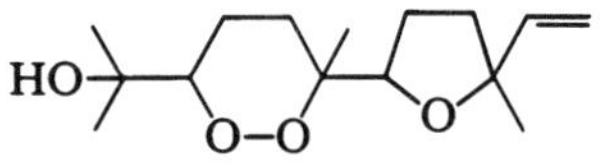

$C_{15}H_{26}O_4$ M 270.368

Constit. of *Ferula communis* subsp. *communis*. Gum.

Miski, M. *et al, J. Nat. Prod.*, 1986, **49**, 916.

9(11)-Fernene-3,7,19-triol
F-60002

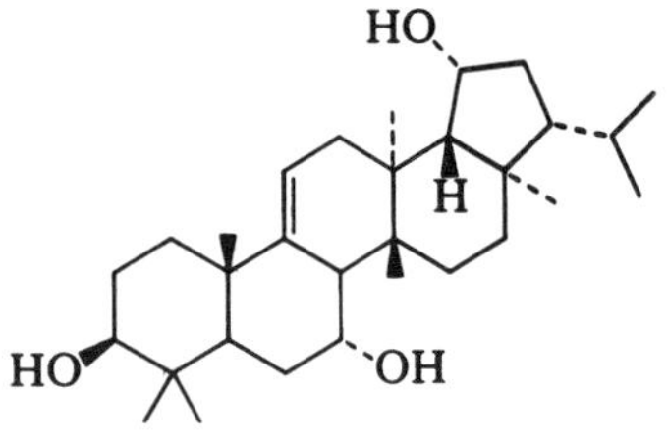

$C_{30}H_{50}O_3$ M 458.723

(3β,7α,19α)-form

Rubiatriol

Constit. of the Chinese drug "Qián Cáo Gén" from *Rubia cordifolia* roots. Mp 252-256°.

Arisawa, M. *et al, J. Nat. Prod.*, 1986, **49**, 1114.

Ferrioxamine *B*
F-60003

Updated Entry replacing F-00103

[14836-73-8]

As Ferrioxamine *A*, F-00102 with

n = 5

$C_{25}H_{45}FeN_6O_8$ M 613.513

Metab. of *Actinomyces* spp. Important Sideramine-type growth factor isol. from *Streptomyces pilosus*. Hygroscopic red-brown solid. pK_a 9.74 (methyl cellosolve).

N-*Ac:* **Ferrioxamine D_1.**

$C_{27}H_{47}FeN_6O_9$ M 655.550

Growth factor.

Bickel, H. *et al, Helv. Chim. Acta*, 1960, **43**, 2118, 2129 (*isol, ir*)

Keller-Schierlein, W. *et al, Helv. Chim. Acta*, 1961, **44**, 709 (*ir, struct*)

Prelog, V. *et al, Helv. Chim. Acta*, 1962, **45**, 631 (*ir, synth*)

Leong, J. *et al, J. Am. Chem. Soc.*, 1975, **97**, 293 (*struct, synth*)

Hossain, M.B. *et al, Acta Crystallogr., Sect. C*, 1986, **42**, 1275 (*cryst struct*)

Feruginin
F-60004

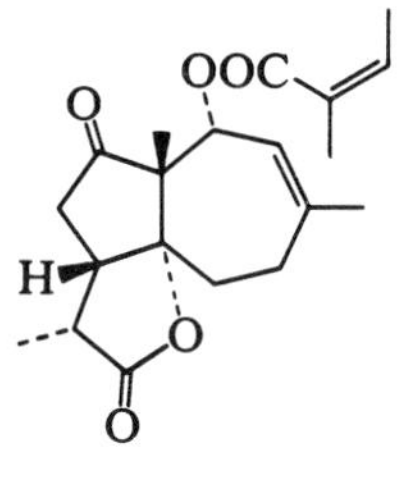

$C_{20}H_{26}O_5$ M 346.422

Constit. of *Ferula jaeschkeana*. Needles (Me₂CO). Mp 90°. $[\alpha]_D^{35}$ −466° (c, 0.6 in CHCl₃).

Garg, S.N. *et al, Phytochemistry*, 1987, **26**, 449.

Ferulenol
F-60005

Updated Entry replacing F-00136

4-Hydroxy-3-[(3,7,11-trimethyl-2,6,10-dodecatrienyl)]-2H-1-benzopyran-2-one, 9CI. 3-(1-Farnesyl)-4-hydroxycoumarin

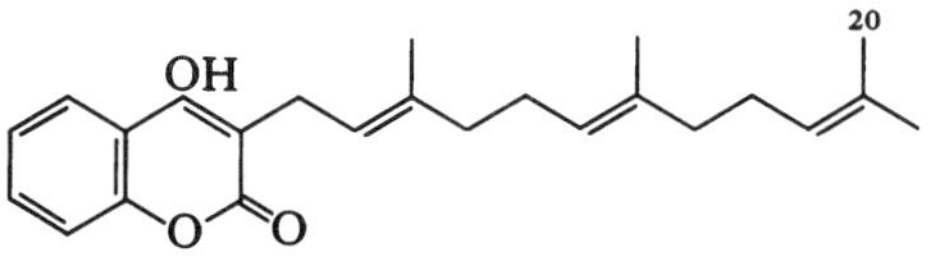

$C_{24}H_{30}O_3$ M 366.499

Constit. of the latex of *Ferula communis*. Shows haemorrhagic action. Mp 64-65°.

20-Hydroxy: ω-Hydroxyferulenol.

$C_{24}H_{30}O_4$ M 382.499

Constit. of *F. communis* var. *genuina*. Shows haemorrhagic action.

Carboni, S. *et al, Tetrahedron Lett.*, 1964, 2783 (*struct, pmr*)

Lamnaouer, D. *et al, Phytochemistry*, 1987, **26**, 1613 (*isol, pmr, cmr*)

Valle, M.G. *et al, Phytochemistry*, 1987, **26**, 253 (*deriv*)

Ferulinolone
F-60006

[107900-70-9]

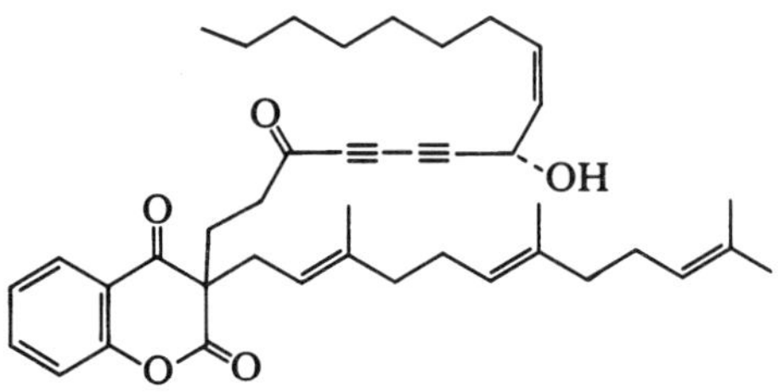

$C_{41}H_{52}O_5$ M 624.859

Constit. of *Ferula communis*. Oil. $[\alpha]_D$ +85.5° (c, 4 in CHCl₃).

Teresa, J.de.P. *et al, Planta Med.*, 1986, 458.

Fervanol
F-60007

$C_{15}H_{24}O$ M 220.354

Benzoyl: Fervanol benzoate.
  $C_{22}H_{28}O_2$ M 324.462
  Constit. of *Ferula haussknechtii*. Gum.
4-Hydroxybenzoyl: Fervanol p-hydroxybenzoate.
  $C_{22}H_{28}O_3$ M 340.461
  From *F. haussknechtii*. Gum.
4-Hydroxy-3-methoxybenzoyl: Fervanol vanillate.
  $C_{23}H_{30}O_4$ M 370.488
  From *F. haussknechtii*. Gum.

Miski, M. *et al*, *Phytochemistry*, 1987, **26**, 1733.

Fevicordin *A*
F-60008

$C_{31}H_{42}O_8$ M 542.668
Constit. of seeds of *Fevillea cordifolia*.

2-O-β-D-Glucopyranoside: Fevicordin A glucoside.
  $C_{37}H_{52}O_{13}$ M 704.810
  From *F. cordifolia* seeds.

Achenbach, H. *et al*, *J. Chem. Soc., Chem. Commun.*, 1987, 441.

Fexerol
F-60009
Updated Entry replacing C-40037

$C_{15}H_{26}O_3$ M 254.369
Constit. of *Ferula tschatkalensis*. Cryst. Mp 141-142°.

6-Tigloyl: [89803-99-6]. **Chatferin.**
  $C_{20}H_{32}O_4$ M 336.470
  Constit. of *F. tschatkalensis*. Oil. $[\alpha]_D^{20}$ +6.2° (c, 1.53 in EtOH).

Sagitdinova, G.V. *et al*, *Khim. Prir. Soedin.*, 1977, **13**, 665; 1983, **19**, 721 (*isol, struct, synth*)
Makmudov, M.K. *et al*, *Khim. Prir. Soedin.*, 1986, **22**, 39 (*cryst struct*)

Ficisterol
F-60010
Updated Entry replacing F-00147
23-Ethyl-24-methyl-27-nor-5,25-cholestadien-3β-ol
[74958-11-5]

$C_{29}H_{48}O$ M 412.698
Constit. of *Petrosia ficiformis*. Cryst. (MeOH). Mp 143-145°. $[\alpha]_D^{20}$ −10.6° (c, 0.85 in $CHCl_3$).
Ac: Cryst. Mp 102-103° (99-100°).

Khalil, M.W. *et al*, *J. Am. Chem. Soc.*, 1980, **102**, 2133 (*isol*)
Shu, A.Y.L. *et al*, *J. Chem. Soc., Perkin Trans. 1*, 1987, 1291 (*synth, struct*)

Ficulinic acid A
F-60011
2-Heptyl-10-oxo-11-nonadecenoic acid, 9CI

$$H_3C(CH_2)_6CH{=}CHCO(CH_2)_nCH(COOH)(CH_2)_6CH_3$$
$$(n = 7)$$

$C_{26}H_{48}O_3$ M 408.663
(E)-form [102791-30-0]
  Isol. from the sponge *Ficulina ficus*. Shows weak cytotoxic props. Mp 33-35°.

Guyot, M. *et al*, *J. Nat. Prod.*, 1986, **49**, 307 (*isol*)

Ficulinic acid B
F-60012
2-Heptyl-12-oxo-13-henicosenoic acid, 9CI
As Ficulinic acid A, F-60011 with

$$n = 9$$

$C_{28}H_{52}O_3$ M 436.717
(E)-form [102791-31-1]
  Isol. from the sponge *Ficulina ficus*. Shows weak cytotoxic props. Mp 31-32°.

Guyot, M. *et al*, *J. Nat. Prod.*, 1986, **49**, 307 (*isol*)

Firmanolide
F-60013

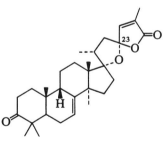

$C_{30}H_{42}O_4$ M 466.659
Constit. of *Abies firma*. Needles. Mp 193-195°. $[\alpha]_D$ −8° (c, 0.48 in $CHCl_3$).

23-Epimer: **23-epi-Firmanolide**.
  $C_{30}H_{42}O_4$ M 466.659
  Constit. of *A. firma*. Fine needles. Mp 193-195°. $[\alpha]_D$ +24° (c, 0.67 in $CHCl_3$).

Hasegawa, S. *et al*, *Phytochemistry*, 1987, **26**, 1095.

Flossonol   F-60014
3,4-Dihydro-4-hydroxy-6-methoxy-2,7-dimethyl-
1(2H)-naphthalenone

$C_{13}H_{16}O_3$ M 220.268
Constit. of *Pararistolochia flos-avis*. Cryst.
(Et$_2$O/hexane). Mp 93-95°.

Sun, N.-J. *et al, Phytochemistry*, 1987, **26**, 3051.

Fluorantheno[3,4-*cd*]-1,2-diselenole, 9CI   F-60015
[108079-35-2]

$C_{16}H_8Se_2$ M 358.159
Purplish-black needles (CHCl$_3$). Mp 238° dec.

Miyamoto, H. *et al, Tetrahedron Lett.*, 1986, **27**, 2011 (*synth,
pmr, uv, ms*)

Fluorantheno[3,4-*cd*]-1,2-ditellurole, 9CI   F-60016
[108079-36-3]

$C_{16}H_8Te_2$ M 455.439
Greenish-black needles (CS$_2$). Mp >300°.

Miyamoto, H. *et al, Tetrahedron Lett.*, 1986, **27**, 2011 (*synth,
pmr, uv, ms*)

Fluorantheno[3,4-*cd*]-1,2-dithiole, 9CI   F-60017
[108079-34-1]

$C_{16}H_8S_2$ M 264.359
Orange needles (hexane/C$_6$H$_6$). Mp 213-215°.

Miyamoto, H. *et al, Tetrahedron Lett.*, 1986, **27**, 2011 (*synth,
pmr, uv, ms*)

9*H*-Fluorene-9-selenone   F-60018
Selenofluorenone

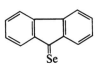

$C_{13}H_8Se$ M 243.166
Reactive intermediate, can be trapped by adduction with
dienes in soln.

Meinke, P.T. *et al, Tetrahedron Lett.*, 1987, **28**, 3887 (*synth,
props*)

9-Fluorenylmethyl pentafluorophenyl   F-60019
carbonate
[88744-04-1]

$C_{21}H_{11}F_5O_3$ M 406.308
Reagent for *N*-9-Fluorenylmethyloxycarbonyl protection
of amino acids and prepn. of their pentafluorophenyl
esters. Cryst. (hexane). Mp 84-86°.

Schön, I. *et al, Synthesis*, 1986, 303 (*synth, uv, ir, use*)

4-Fluoro-1,2-benzenediol, 9CI   F-60020
Updated Entry replacing F-20028
4-Fluoropyrocatechol, 8CI. *4-Fluorocatechol*
[367-32-8]
$C_6H_5FO_2$ M 128.103
Cryst. (Et$_2$O/pet. ether). Mp 90-91°.
Di-Me ether: [398-62-9]. *4-Fluoro-1,2-dimethoxyben-
zene. 4-Fluoroveratrole.*
$C_8H_9FO_2$ M 156.156
Liq. Bp$_{14}$ 98°.

Corse, J. *et al, J. Org. Chem.*, 1951, **16**, 1345 (*synth*)
Ahond, S.P. *et al, J. Org. Chem.*, 1975, **40**, 807 (*synth*)
Furlano, D.C. *et al, J. Org. Chem.*, 1986, **51**, 4073 (*synth*)
Kirk, K.L. *et al, J. Org. Chem.*, 1986, **51**, 4073 (*synth, deriv*)

2-Fluorocyclohexanone, 9CI   F-60021
Updated Entry replacing F-00367
[694-82-6]

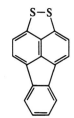

C_6H_9FO M 116.135
(±)-*form*
Bp$_{17}$ 78°. λ$_{max}$ 295 nm.

Cantacuzène, J. *et al, Bull. Soc. Chim. Fr.*, 1967, 1587 (*synth,
ir, uv*)
Pan, Y.-H. *et al, Can. J. Chem.*, 1967, **45**, 2943 (*nmr*)
Purrington, S.T. *et al, Tetrahedron Lett.*, 1986, **27**, 2715 (*synth,
nmr*)

2-Fluorocyclopentanone, 9CI   F-60022

Updated Entry replacing F-10050

[1755-12-0]

C_5H_7FO M 102.108

Bp_{16} 70°, Bp_{10} 57-59°.

Machleidt, H. *et al*, *Justus Liebigs Ann. Chem.*, 1964, **679**, 9 (*synth*)
Cantaeuzene, J. *et al*, *Bull. Soc. Chim. Fr.*, 1967, 1587 (*synth*)
Purrington, S.T. *et al*, *Tetrahedron Lett.*, 1986, **27**, 2715 (*synth, nmr*)

3-Fluoro-2,6-dihydroxybenzoic acid   F-60023

$C_7H_5FO_4$ M 172.113

Di-Me ether: [52189-67-0]. *3-Fluoro-2,6-dimethoxybenzoic acid.*
$C_9H_9FO_4$ M 200.166
Cryst. Mp 104-105°.

Durrani, A.A. *et al*, *J. Chem. Soc., Perkin Trans. 1*, 1980, 1658 (*synth, pmr*)

4-Fluoro-3,5-dihydroxybenzoic acid   F-60024

$C_7H_5FO_4$ M 172.113

Di-Me ether, Me ester: [65566-14-5]. *Methyl 4-fluoro-3,5-dimethoxybenzoate.*
$C_{10}H_{11}FO_4$ M 214.193
Cryst. (EtOH/heptane). Mp 84-85°.

Kompis, I. *et al*, *Helv. Chim. Acta*, 1977, **60**, 3025 (*synth, deriv, ir, pmr*)

5-Fluoro-2,3-dihydroxybenzoic acid   F-60025

$C_7H_5FO_4$ M 172.113

Di-Me ether: [91407-43-1]. *5-Fluoro-2,3-dimethoxybenzoic acid.*
$C_9H_9FO_4$ M 200.166
Cryst. (H_2O). Mp 137-138°.

Daukshas, V.K. *et al*, *J. Org. Chem. USSR* (*Engl. Transl.*), 1984, **20**, 465 (*synth, uv, pmr*)

2-Fluoro-1,3-dimethylbenzene, 9CI   F-60026

Updated Entry replacing F-00380

*2-Fluoro-*m-*xylene*

[443-88-9]

C_8H_9F M 124.158
Liq.

Hertz, E. *et al*, *Monatsh. Chem.*, 1947, **76**, 249 (*ir*)
Padhye, M.R. *et al*, *Curr. Sci.*, 1960, **29**, 129 (*uv*)
Yamato, T. *et al*, *J. Chem. Soc., Perkin Trans. 1*, 1987, 1 (*synth*)

2-Fluoro-3,5-dinitroaniline, 8CI   F-60027

2-Fluoro-3,5-dinitrobenzenamine, 9CI

[18646-02-1]

$C_6H_4FN_3O_4$ M 201.114
Yellow cryst. (C_6H_6/cyclohexane). Mp 105-106°.

Kirk, K.L. *et al*, *J. Org. Chem.*, 1969, **34**, 395 (*synth, uv*)

2-Fluoro-4,6-dinitroaniline, 8CI   F-60028

2-Fluoro-4,6-dinitrobenzenamine, 9CI

[367-78-2]

$C_6H_4FN_3O_4$ M 201.114
Yellow cryst. (EtOH). Mp 156.5-157°.

Deorha, D.S. *et al*, *J. Indian Chem. Soc.*, 1963, **40**, 975 (*synth*)
Kuznetsov, L.L. *et al*, *J. Org. Chem. USSR* (*Engl. Transl.*), 1974, **10**, 546 (*synth, props*)

4-Fluoro-2,6-dinitroaniline, 8CI   F-60029

4-Fluoro-2,6-dinitrobenzenamine, 9CI

[82366-44-7]

$C_6H_4FN_3O_4$ M 201.114
Cryst. Mp 133-134°.

Kuznetsov, L.L. *et al*, *J. Org. Chem. USSR* (*Engl. Transl.*), 1982, **18**, 595 (*synth, uv, props*)

4-Fluoro-3,5-dinitroaniline, 8CI   F-60030

4-Fluoro-3,5-dinitrobenzenamine, 9CI

[92367-07-2]

$C_6H_4FN_3O_4$ M 201.114
Yellow-orange cryst. (1,2-dichloroethane). Mp 149-150°.

Nielsen, A.T. *et al*, *J. Org. Chem.*, 1984, **49**, 4575 (*synth, pmr*)

5-Fluoro-2,4-dinitroaniline, 8CI   F-60031

5-Fluoro-2,4-dinitrobenzenamine, 9CI

[367-81-7]

$C_6H_4FN_3O_4$ M 201.114
Cryst. (EtOH). Mp 186-187°.

N-Ac:
$C_8H_4FN_3O_5$ M 241.135
Cryst. (EtOH). Mp 119°.

Bergmann, E.D. *et al*, *J. Org. Chem.*, 1961, **26**, 1480 (*synth*)
Matsui, K. *et al*, *CA*, 1966, **65**, 10699f (*synth*)

6-Fluoro-3,4-dinitroaniline, 8CI   F-60032

6-Fluoro-3,4-dinitrobenzenamine, 9CI

$C_6H_4FN_3O_4$ M 201.114
Yellow needles (H_2O). Mp 135° dec.

N-Ac:
$C_8H_6FN_3O_5$ M 243.151
Yellow needles (EtOH aq.). Mp 170-172°.

Camasava, M.J. *et al*, *J. Chem. Soc., Perkin Trans. 1*, 1987, 2317 (*synth, deriv, uv, pmr*)

2-Fluoro-2,2-diphenylacetaldehyde   F-60033

α-Fluoro-α-phenylbenzeneacetaldehyde, 9CI

[107365-22-0]

Ph₂CFCHO

C₁₄H₁₁FO   M 214.239
Liq. Dec. on dist. or on standing.
Purrington, S.T. *et al, Tetrahedron Lett.,* 1986, **27**, 2715 (*synth, nmr*)

2-Fluoroheptanal, 9CI   F-60034
[7740-66-1]

H₃C(CH₂)₄CHFCHO

C₇H₁₃FO   M 132.178
(±)-*form*
Liq. Bp₁₀₀ 92° dec., Bp₁₅ 42°. Dec. on standing.
Cantacuzene, J. *et al, Bull. Soc. Chim. Fr.,* 1967, 1587 (*synth, ir, uv, pmr*)
Purrington, S.T. *et al, Tetrahedron Lett.,* 1986, **27**, 2715 (*synth, nmr*)

2-Fluoro-3-hydroxybenzaldehyde   F-60035
[103438-86-4]

C₇H₅FO₂   M 140.114
Cryst. (Et₂O/pet. ether). Mp 112-113°.
Benzyl ether: [103438-90-0].
   C₁₄H₁₁FO₂   M 230.238
   Cryst. (cyclohexane). Mp 88-89°.
Kirk, K.L. *et al, J. Med. Chem.,* 1986, **29**, 1982 (*synth, pmr*)

2-Fluoro-5-hydroxybenzaldehyde   F-60036
[103438-84-2]
C₇H₅FO₂   M 140.114
Cryst. (Et₂O/pet. ether). Mp 74-75°.
Benzyl ether: [103438-92-2].
   C₁₄H₁₁FO₂   M 230.238
   Cryst. (cyclohexane). Mp 72-74°.
Kirk, K.L. *et al, J. Med. Chem.,* 1986, **29**, 1982 (*synth, pmr*)

4-Fluoro-3-hydroxybenzaldehyde   F-60037
[103438-85-3]
C₇H₅FO₂   M 140.114
Cryst. (Et₂O/pet. ether). Mp 109-110°.
Benzyl ether: [103438-91-1].
   C₁₄H₁₁FO₂   M 230.238
   Cryst. (cyclohexane). Mp 66-67°.
Kirk, K.L. *et al, J. Med. Chem.,* 1986, **29**, 1982 (*synth, pmr*)

2-Fluoro-5-hydroxybenzoic acid, 9CI   F-60038
[51446-30-1]

C₇H₅FO₃   M 156.113
Me ether: [367-83-9]. *2-Fluoro-5-methoxybenzoic acid, 9CI.*
   C₈H₇FO₃   M 170.140

Cryst. (dichloroethane). Mp 142-143°.
Hartmann, R.W. *et al, J. Med. Chem.,* 1984, **27**, 577.

2-Fluoro-6-hydroxybenzoic acid, 9CI   F-60039
6-Fluorosalicylic acid
[67531-86-6]
C₇H₅FO₃   M 156.113
Cryst. (C₆H₆). Mp 165-165.5°.
Hannah, J. *et al, J. Med. Chem.,* 1978, **21**, 1093 (*synth, props*)

3-Fluoro-2-hydroxybenzoic acid, 9CI   F-60040
3-Fluorosalicylic acid, 8CI
[341-27-5]
C₇H₅FO₃   M 156.113
Cryst. (C₆H₆). Mp 145-147°.
Me ester: [70163-98-3].
   C₈H₇FO₃   M 170.140
   Mp 33°.
Ferguson, L.N. *et al, J. Am. Chem. Soc.,* 1946, **68**, 2502; 1950, **72**, 5315 (*synth*)
Baine, O. *et al, J. Org. Chem.,* 1954, **19**, 510 (*synth*)

3-Fluoro-4-hydroxybenzoic acid, 9CI   F-60041
[350-29-8]
C₇H₅FO₃   M 156.113
Cryst. (H₂O). Mp 163-163.5°.
Me ester: [403-01-0].
   C₈H₇FO₃   M 170.140
   Mp 90-91.4°.
Me ether: [403-20-3]. *3-Fluoro-4-methoxybenzoic acid, 9CI.*
   C₈H₇FO₃   M 170.140
   Cryst. (H₂O). Mp 211-212°.
Minor, J.T. *et al, J. Org. Chem.,* 1952, **17**, 1425 (*synth*)
Gray, G.W. *et al, J. Chem. Soc.,* 1954, 2556 (*synth*)
Lock, G., *Monatsh. Chem.,* 1955, **86**, 511 (*synth*)
May, S.W. *et al, Biochemistry,* 1978, **17**, 1853 (*synth*)
Misaki, S., *J. Fluorine Chem.,* 1982, **21**, 191 (*synth*)

4-Fluoro-2-hydroxybenzoic acid   F-60042
4-Fluorosalicylic acid, 8CI
[345-29-9]
C₇H₅FO₃   M 156.113
Cryst. (EtOH aq.).
Me ester:
   C₈H₇FO₃   M 170.140
   Cryst. Mp 39-40°. Bp₁₀ 93°.
Amide:
   C₇H₆FNO₂   M 155.128
   Cryst. (H₂O). Mp 181°.
Zuber, F. *et al, Helv. Chim. Acta,* 1950, **33**, 1269 (*synth*)
Danek, O., *Collect. Czech. Chem. Commun.,* 1964, **29**, 730 (*synth*)

4-Fluoro-3-hydroxybenzoic acid, 9CI   F-60043
[51446-31-2]
C₇H₅FO₃   M 156.113
Cryst. (H₂O). Mp 215.5-216.5°.
May, S.W. *et al, Biochemistry,* 1978, **17**, 1853 (*synth*)

5-Fluoro-2-hydroxybenzoic acid, 9CI   F-60044
5-Fluorosalicylic acid, 8CI

[345-16-4]
C₇H₅FO₃   M 156.113
Cryst. (EtOH aq.). Mp 179-180°.
Me ester: [391-92-4].
  C₈H₇FO₃   M 170.140
  Mp 28°. Bp₃₈ 133-135°.
Me ether: [394-04-7]. *5-Fluoro-2-methoxybenzoic acid,*
  9CI.
  C₈H₇FO₃   M 170.140
  Cryst. (pet. ether). Mp 84-85°.

Kraft, K. *et al, Chem. Ber.,* 1952, **85**, 577 (*synth*)
Banie, O. *et al, J. Org. Chem.,* 1954, **19**, 510 (*synth*)
Danek, O., *Collect. Czech. Chem. Commun.,* 1964, **29**, 730
  (*synth*)
Durrani, A.A. *et al, J. Chem. Soc., Perkin Trans. 1,* 1979, 2079
  (*deriv*)
Misaki, S., *J. Fluorine Chem.,* 1982, **21**, 191 (*synth*)

2-Fluoro-4-iodopyridine   F-60045
[22282-70-8]

C₅H₃FIN   M 222.988
Mp 58°.

Talik, T. *et al, Rocz. Chem.,* 1968, **42**, 1861 (*synth*)

3-Fluoro-4-iodopyridine   F-60046
[22282-75-3]
C₅H₃FIN   M 222.988
Mp 87° (80-81°).
Picrate: [22282-76-4]. Mp 140°.

Talik, T. *et al, Rocz. Chem.,* 1968, **42**, 1861 (*synth*)
Gribble, G.W. *et al, Tetrahedron Lett.,* 1980, **21**, 4137 (*synth*)

1-Fluoro-2-isocyanatobenzene, 9CI   F-60047
*Isocyanic acid o-fluorophenyl ester, 8CI. 2-Fluorophenyl
isocyanate*
[16744-98-2]

C₇H₄FNO   M 137.113
Moisture-sensitive liq. d₄²⁰ 1.222. Bp₁₈ 65°. n_D²⁰ 1.5124.
▷Lachrymator

Malichenko, B.F. *et al, Zh. Obshch. Khim.,* 1967, **37**, 1796
  (*Engl. transl.* p. 1711); *CA,* **68**, 21618b (*synth*)
Aldrich Library of FT-IR Spectra, 1st Ed., **2**, 472A (*ir*)
Aldrich Library of NMR Spectra, 2nd Ed., **2**, 443A (*pmr*)
Sigma-Aldrich Library of Chemical Safety Data, 1st Ed., 970D
  (*haz*)

1-Fluoro-3-isocyanatobenzene, 9CI   F-60048
*Isocyanic acid m-fluorophenyl ester, 8CI. 3-Fluorophenyl
isocyanate*
[404-71-7]
C₇H₄FNO   M 137.113
Herbicide intermediate. Liq. d₄²⁰ 1.201. Bp₁₉ 69°. n_D²⁰
  1.5115. Moisture-sensitive.
▷Lachrymator; Flash pt. 41°

U.S.P., 2 689 861, (*1954*); *CA,* **49**, 11712 (*synth*)
Malichenko, B.F. *et al, Zh. Obsch. Khim.,* 1967, **37**, 1796 (*Engl.
  transl.* p. 1711); *CA,* **68**, 21618b (*synth*)
Katritzky, A.R. *et al, J. Am. Chem. Soc.,* 1970, **92**, 6855 (*ir*)
Aldrich Library of FT-IR Spectra, 1st Ed., **2**, 473B (*ir*)
Aldrich Library of NMR Spectra, 2nd Ed., **2**, 443C (*pmr*)
Sigma-Aldrich Library of Chemical Safety Data, 971A (*haz*)

1-Fluoro-4-isocyanatobenzene, 9CI   F-60049
*Isocyanic acid p-fluorophenyl ester, 8CI. 4-Fluorophenyl
isocyanate*
[1195-45-5]
C₇H₄FNO   M 137.113
Liq. d₄²⁰ 1.201. Bp₁₅ 68°. n_D²⁰ 1.5216.
▷Corrosive, lachrymator. Flash pt. 41°

Maciel, G.E. *et al, J. Chem. Phys.,* 1965, **42**, 2427 (*cmr*)
Malichenko, B.F. *et al, Zh. Obsch. Khim.,* 1967, **37**, 1796 (*Engl.
  transl.* p. 1711); *CA,* **68**, 21618 (*synth*)
Katritzky, A.R. *et al, J. Am. Chem. Soc.,* 1970, **92**, 6855 (*ir*)
Weigert, F.J. *et al, J. Org. Chem.,* 1976, **41**, 4006 (*F nmr*)
Aldrich Library of FT-IR Spectra, 1st Ed., **2**, 475D (*ir*)
Aldrich Library of NMR Spectra, 2nd Ed., **2**, 443C (*pmr*)
Sigma-Aldrich Library of Chemical Safety Data, 1st Ed., 971B.

3-Fluoro-3-methoxy-3H-diazirine, 9CI   F-60050
[4823-41-0]

C₂H₃FN₂O   M 90.057
Fluoromethoxycarbene precursor. Gas. Bp ca. 5°. Isol. in
  soln. at −196°, soln. fairly stable in dark at r.t.
▷Explosive

Mitsch, R.A. *et al, J. Heterocycl. Chem.,* 1965, **2**, 371 (*F nmr,
  pmr, ir, ms, uv, props*)
Zollinger, J.L. *et al, J. Org. Chem.,* 1973, **38**, 1065 (*synth*)
Moss, R.A. *et al, Tetrahedron Lett.,* 1986, **27**, 419 (*synth, uv, ir,
  F nmr*)

5-Fluoro-2-methyl-1,3-dinitrobenzene, 9CI   F-60051
4-Fluoro-2,6-dinitrotoluene
[102735-88-6]

C₇H₅FN₂O₄   M 200.126
Cryst. (hexane). Mp 44-47°.

Büchi, G. *et al, J. Am. Chem. Soc.,* 1986, **108**, 4115 (*synth, pmr,
  uv, ms*)

2-Fluoro-3-methylphenol, 9CI   F-60052
2-Fluoro-m-cresol. 2-Fluoro-3-hydroxytoluene
[77772-72-6]

C₇H₇FO   M 126.130
No phys. props. reported.
Misaki, S., *J. Fluorine Chem.,* 1981, **17**, 159 (*synth*)

2-Fluoro-4-methylphenol
F-60053

2-Fluoro-p-cresol. 3-Fluoro-4-hydroxytoluene
[452-81-3]
C_7H_7FO M 126.130
Bp 175°, Bp_{11} 65°.

Me ether: [399-55-3]. *2-Fluoro-1-methoxy-4-methyl-benzene, 9CI. 2-Fluoro-4-methylanisole. 3-Fluoro-4-methoxytoluene.*
C_8H_9FO M 140.157
Oil. Bp_{24} 81-82.5°, Bp_{12} 72°.

Bennett, E.L. *et al, J. Am. Chem. Soc.*, 1950, **72**, 1806 (*synth*)
Finger, G.C. *et al, J. Am. Chem. Soc.*, 1959, **81**, 94 (*synth*)
Misaki, S., *J. Fluorine Chem.*, 1981, **17**, 159 (*synth*)
Ng, J.S. *et al, J. Org. Chem.*, 1981, **46**, 2520 (*deriv*)

2-Fluoro-5-methylphenol
F-60054

6-Fluoro-m-cresol. 4-Fluoro-3-hydroxytoluene
[63762-79-8]
C_7H_7FO M 126.130
Bp_{30} 86-93°.

Me ether: [63762-78-7]. *1-Fluoro-2-methoxy-4-methyl-benzene. 2-Fluoro-5-methylanisole. 4-Fluoro-3-methoxytoluene.*
C_8H_9FO M 140.157
Bp_{30} 93-95°.

Flaugh, M.E. *et al, J. Med. Chem.*, 1979, **22**, 63 (*synth*)
Misaki, S., *J. Fluorine Chem.*, 1981, **17**, 159 (*synth*)

2-Fluoro-6-methylphenol
F-60055

6-Fluoro-o-cresol. 3-Fluoro-2-hydroxytoluene
[443-90-3]
C_7H_7FO M 126.130
Bp_{745} 158-162°.

U.S.P., 2 560 950, (*1951*); *CA*, **46**, 1040i (*synth*)
Misaki, S., *J. Fluorine Chem.*, 1981, **17**, 159 (*synth*)

3-Fluoro-2-methylphenol
F-60056

3-Fluoro-o-cresol. 2-Fluoro-6-hydroxytoluene
C_7H_7FO M 126.130
Bp_{748} 186-188°.

Ac: [38226-08-3].
$C_9H_9FO_2$ M 168.167
$Bp_{0.02}$ 33-35°.

Lock, G. *et al, Ber.*, 1936, **69**, 2253 (*synth*)
Krause, G.H. *et al, Z. Naturforsch., B*, 1972, **27**, 663 (*deriv*)

3-Fluoro-4-methylphenol
F-60057

3-Fluoro-p-cresol. 2-Fluoro-4-hydroxytoluene
[452-78-8]
C_7H_7FO M 126.130
Fp 30.1°. Bp 194-196°, Bp_{25} 105-110°.

Benzoyl:
$C_{14}H_{11}FO$ M 214.239
Plates (EtOH). Mp 77°.
Me ether: 1-Fluoro-3-methoxy-6-methylbenzene. 3-Fluoro-4-methylanisole. 2-Fluoro-4-methoxytoluene.
C_8H_9FO M 140.157
Oil. Bp_{10} 63.5-64°.

Brown, J.H. *et al, J. Chem. Soc., Suppl.*, 1949, **1**, 95 (*synth*)
Lock, G., *Monatsh. Chem.*, 1959, **90**, 680 (*synth*)
U.S.P., 2 950 325, (*1960*); *CA*, **55**, 7359 (*synth*)
Hoyer, H. *et al, Z. Naturforsch., B*, 1965, **20**, 617 (*deriv*)

4-Fluoro-2-methylphenol
F-60058

4-Fluoro-o-cresol. 5-Fluoro-2-hydroxytoluene
[452-72-2]
C_7H_7FO M 126.130
Mp 35°. Bp_{14} 87°.

Finger, G.C. *et al, J. Am. Chem. Soc.*, 1959, **81**, 94 (*synth*)
Misaki, S., *J. Fluorine Chem.*, 1981, **17**, 159 (*synth*)

4-Fluoro-3-methylphenol
F-60059

4-Fluoro-m-cresol, 8CI. 2-Fluoro-5-hydroxytoluene
[452-70-0]
C_7H_7FO M 126.130
Mp 32°. Bp_{12} 98-99°, Bp_5 76°.

Me ether: [2338-54-7]. *1-Fluoro-4-methoxy-3-methyl-benzene, 9CI. 4-Fluoro-3-methylanisole, 8CI. 2-Fluoro-5-methoxytoluene.*
C_8H_9FO M 140.157
Oil.

Finger, G.C. *et al, J. Am. Chem. Soc.*, 1959, **81**, 94 (*synth*)
Misaki, S., *J. Fluorine Chem.*, 1981, **17**, 159 (*synth*)
Hartmann, R.W. *et al, J. Med. Chem.*, 1984, **27**, 577 (*deriv*)

5-Fluoro-2-methylphenol
F-60060

5-Fluoro-o-cresol. 4-Fluoro-2-hydroxytoluene
[452-85-7]
C_7H_7FO M 126.130
Light-amber oil. Bp_4 66°.

p-*Nitrobenzoyl:* Mp 113°.

Allen, F.L. *et al, Tetrahedron*, 1959, **6**, 315 (*synth*)

5-Fluoro-2-nitrophenol
F-60061

Updated Entry replacing F-00552
[446-36-6]
$C_6H_4FNO_3$ M 157.101
Has fungicidal props. Yellow needles (pet. ether). Mp 32°. Steam-volatile.
▷Hazardous exothermic synth. on large scale
O-Benzoyl:
$C_{13}H_8FNO_4$ M 261.209
Mp 110-111°.
Me ether: 4-Fluoro-2-methoxy-1-nitrobenzene. 5-Fluoro-2-nitroanisole.
$C_7H_6FNO_3$ M 171.128
Cryst. (pet. ether). Mp 52°.

Hodgson, H.H. *et al, J. Chem. Soc.*, 1928, 1879 (*synth*)
Balasubramanian, A. *et al, Can. J. Chem.*, 1966, **44**, 961 (*uv*)
Robinson, N., *Chem. Br.*, 1987, 837 (*haz*)

2-Fluoro-2-phenylacetaldehyde
F-60062

α-Fluorobenzeneacetaldehyde, 9CI
[13344-76-8]

PhCHFCHO

C_8H_7FO M 138.141
(±)-*form*
Bp_2 46° dec., Bp_1 35-36°. Polymerises on standing.

Middleton, W.J. *et al, J. Am. Chem. Soc.*, 1980, **102**, 4845 (*synth, F nmr*)
Purrington, S.T. *et al, Tetrahedron Lett.*, 1986, **27**, 2715 (*synth, nmr*)

2-Fluoro-2-phenylpropanal　　　F-60063

α-Fluoro-α-methylbenzeneacetaldehyde, 9CI
[107365-21-9]

$$PhCF(CH_3)CHO$$

C_9H_9FO　　M 152.168

(±)-*form*
Liq. Dec. on dist. or on standing.

Purrington, S.T. *et al, Tetrahedron Lett.*, 1986, **27**, 2715 (*synth, nmr*)

2-Fluoro-3-phenylpropanal　　　F-60064

α-Fluorobenzenepropanal, 9CI
[107365-23-1]

$$PhCH_2CHFCHO$$

C_9H_9FO　　M 152.168

(±)-*form*
Liq. Unstable, dec. on dist. or on standing.

Purrington, S.T. *et al, Tetrahedron Lett.*, 1986, **27**, 2715 (*synth, nmr*)

2-Fluoro-1-phenyl-1-propanone, 9CI　　F-60065

2-Fluoropropiophenone
[21120-36-5]

$$PhCOCHFCH_3$$

C_9H_9FO　　M 152.168

(±)-*form*
Bp$_{12}$ 96°, Bp$_2$ 68-70°.

Olah, G.A. *et al, J. Org. Chem.*, 1979, **44**, 3872 (*synth*)
Purrington, S.T. *et al, Tetrahedron Lett.*, 1986, **27**, 2715 (*synth, pmr*)

1-Fluoropyridinium, 9CI　　　F-60066

$C_5H_5FN^{⊕}$　　M 98.100 (ion)
Trifluoromethanesulfonate: [107263-95-6].
　$C_6H_5F_4NO_3S$　　M 247.164
　Cryst. (MeCN/Et$_2$O). Mp 185-187°.
Tetrafluoroborate: [107264-09-5].
　$C_5H_5BF_5N$　　M 184.903
　Mp 90-91°.
Perchlorate: [107264-11-9].
　$C_5H_5ClFNO_4$　　M 197.550
　Mp 225-227.5° dec.
Hexafluoroantimonate: [107264-12-0].
　$C_5H_5F_7NSb$　　M 333.840
　Mp >300°.
Perfluorobutanesulfonate: [107264-13-1]. Mp 111-112°.

Umemoto, T. *et al, Tetrahedron Lett.*, 1986, **27**, 3271 (*synth, F nmr*)

Fontonamide　　　F-60067
[109217-15-4]

$C_{20}H_{22}ClNO_2$　　M 343.852
Isol. from blue-green alga *Hapalosiphon fontinalis*. Mp
　156-157°. [α]$_D$ −141° (c, 0.21 in CHCl$_3$). Related to
　Hapalindole *A*.

Moore, R.E. *et al, J. Org. Chem.*, 1987, **52**, 3773 (*isol, struct*)

11-Formamido-5-eudesmene　　　F-60068

$C_{16}H_{27}NO$　　M 249.395
7β*H*-form
Constit. of *Axinella cannabina*. Oil.
Isocyanide: **11-Isocyano-5-eudesmene**.
　$C_{16}H_{25}N$　　M 231.380
　Constit. of *A. cannabina*. Oil. [α]$_D$ −85.7° (c, 0.8 in
　CHCl$_3$). Has —NC replacing —NHCHO.
Isothiocyanate: **11-Isothiocyano-5-eudesmene**.
　$C_{16}H_{25}NS$　　M 263.440
　Constit. of *A. cannabina*. Oil. [α]$_D$ −89.7° (c, 0.8 in
　CHCl$_3$). Has —NCS replacing —NHCHO.

Ciminiello, P. *et al, Can. J. Chem.*, 1987, **65**, 518.

6-Formamido-4(15)-eudesmene　　　F-60069

$C_{16}H_{27}NO$　　M 249.395
(6α,10α)-*form* [110044-92-3]
Constit. of *Axinella cannabina*. Oil.
Isocyanide: [110044-90-1]. **6-Isocyano-4(15)-eudesmene**.
　$C_{16}H_{25}N$　　M 231.380
　Constit. of *A. cannabina* and *A. acuta*. Cryst. Mp 78-
　79°. [α]$_D$ +92.9° (c, 1.8 in CHCl$_3$). Has —NC
　replacing —NHCHO.
Isothiocyanate: **6-Isothiocyano-4(15)-eudesmene**.
　$C_{15}H_{25}NS$　　M 251.429
　Constit. of *A. cannabina* and *A. acuta*. Cryst. Mp 52-
　53°. [α]$_D$ +88.4° (c, 1 in CHCl$_3$). Has —NCS
　replacing —NHCHO.

Ciminiello, P. *et al, J. Nat. Prod.*, 1987, **50**, 217.

2-Formyl-3,4-dihydroxybenzoic acid　　　F-60070

3,4-Dihydroxyphthalaldehydic acid
[82177-66-0]

$C_8H_6O_5$　　M 182.132
No phys. props. reported.
Di-Me ether: [483-85-2]. *2-Formyl-3,4-dimethoxyben-
　zoic acid, 9CI. 3,4-Dimethoxyphthalaldehydic acid.
　Pseudoopianic acid.*
　$C_{10}H_{10}O_5$　　M 210.186

Long needles (H_2O or $CHCl_3$). Mp 121-122°.

Perkin, W.H., *J. Chem. Soc.*, 1890, **57**, 1064 (*deriv*)
Eaton, R.W. *et al*, *Arch. Biochem. Biophys.*, 1982, **216**, 289 (*synth*)

2-Formyl-3,5-dihydroxybenzoic acid, 9CI   F-60071

[92810-15-6]

$C_8H_6O_5$ M 182.132

Irregular prisms (H_2O). V. sol. hot H_2O. Mp 233° dec.

Me ester: [16849-78-8].
  $C_9H_8O_5$ M 196.159
  Cryst. (H_2O). Mp 156-157°.
Di-Me ether, Me ester: [52344-93-1].
  $C_{11}H_{12}O_5$ M 224.213
  Needles (MeOH). Mp 108°.

Birkinshaw, J.H. *et al*, *J. Chem. Soc.*, 1942, 368.
Hassall, C.H. *et al*, *J. Chem. Soc., Perkin Trans. 1*, 1973, 2853 (*deriv*)
Carey, F.A. *et al*, *J. Org. Chem.*, 1981, **46**, 1366 (*deriv*)

4-Formyl-2,5-dihydroxybenzoic acid   F-60072

$C_8H_6O_5$ M 182.132

Di-Me ether: [94930-47-9]. *4-Formyl-2,5-dimethoxy-benzoic acid, 9CI.*
  $C_{10}H_{10}O_5$ M 210.186
  Cryst. (EtOAc). Mp 196-197°.

Freskos, J.N. *et al*, *J. Org. Chem.*, 1985, **50**, 805 (*deriv*)

6-Formyl-2,3-dihydroxybenzoic acid   F-60073

5,6-Dihydroxyphthalaldehydic acid

$C_8H_6O_5$ M 182.132

Di-Me ether: [519-05-1]. *6-Formyl-2,3-dimethoxyben-zoic acid, 9CI. 5,6-Dimethoxyphthalaldehydic acid, 8CI. Opianic acid.*
  $C_{10}H_{10}O_5$ M 210.186
  Needles (H_2O). Mp 150°.

Wilson, J.W. *et al*, *J. Org. Chem.*, 1951, **16**, 792.
Blair, J. *et al*, *J. Chem. Soc.*, 1955, 708.
Paul, B. *et al*, *J. Heterocycl. Chem.*, 1976, **13**, 701 (*pmr*)
Napolitano, E. *et al*, *J. Org. Chem.*, 1983, **48**, 3653.

5-Formyl-4-phenanthrenecarboxylic acid   F-60074

[5684-15-1]

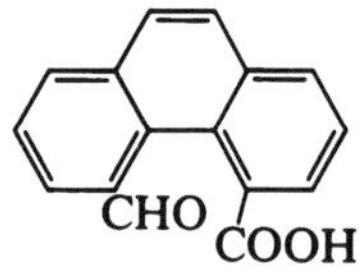

$C_{16}H_{10}O_3$ M 250.253
Cryst. (DMF/AcOH/H_2O). Mp 272-276°.

Org. Synth., Coll. Vol. 4, 484 (*synth*)
Rubin, Y. *et al*, *J. Org. Chem.*, 1986, **51**, 3270 (*synth*)

6-Formyl-1,2,4-triazine-3,5(2H,4H)-dione   F-60075

5-Formyl-6-azauracil

[97776-60-8]

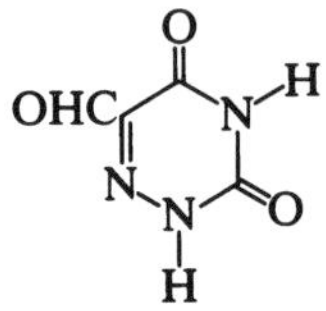

$C_4H_3N_3O_3$ M 141.086
Foam.

Mitchell, W.L. *et al*, *J. Med. Chem.*, 1986, **29**, 809 (*synth, ir, pmr, uv*)

Fruticulin A   F-60076

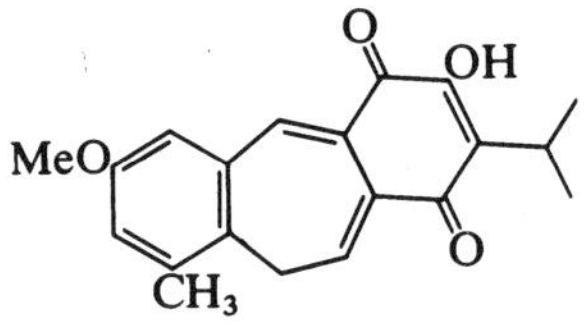

$C_{20}H_{20}O_4$ M 324.376
Constit. of *Salvia fruticulosa*. Cryst. Mp 190-193°.

O-De-Me: **Demethylfruticulin A**.
  $C_{19}H_{18}O_4$ M 310.349
  Constit. of *S. fruticulosa*. Cryst. Mp 200-203°.

Rodriguez-Hahn, L. *et al*, *Tetrahedron Lett.*, 1986, **27**, 5459 (*cryst struct*)

Fruticulin B   F-60077

11-Hydroxy-2-methoxy-19,20-dinor-1,3,5(10),7,9(11),13-abietahexaene-6,12-dione

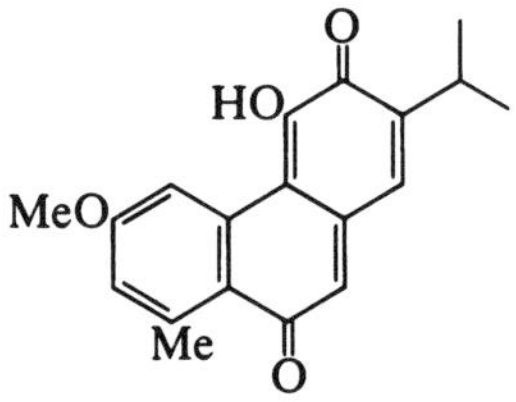

$C_{19}H_{18}O_6$ M 342.348
Constit. of *Salvia fruticulosa*. Cryst. Mp 173-174°.

Rodriguez-Hahn, L. *et al*, *Tetrahedron Lett.*, 1986, **27**, 5459.

FS-2   F-60078

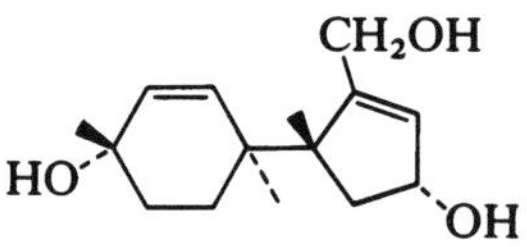

$C_{15}H_{24}O_3$ M 252.353
Isol. from *Fusarium sporotrichioides*. Mycotoxin. Oil.

Corley, D.G. *et al*, *J. Org. Chem.*, 1987, **26**, 4405.

Fuligorubin *A*   F-60079

[108343-55-1]

$C_{20}H_{23}NO_5$ M 357.405

Tetramic acid deriv. Isol. from the slime mold *Fuligo septica* (Myxomycetes). Red cryst. + 2H$_2$O. Mp 150° dec.

Casser, I. *et al*, *Angew. Chem., Int. Ed. Engl.*, 1987, **26**, 586 (*isol, struct*)

Fulvalene   F-60080

Updated Entry replacing F-00778

5-(2,4-Cyclopentadien-1-ylidene)-1,3-cyclopentadiene, 9CI. Bi-2,4-cyclopentadien-1-ylidene, 8CI. Pentafulvalene. 1,1'-Bis(cyclopentadienylidene)

[91-12-3]

$C_{10}H_8$ M 128.173

Deep-red semicryst. at −70°. λ_{max} 265, 278, 289, 299, 313 nm (log$_{10}\epsilon$ 3.54, 3.85, 4.18, 4.48, 4.57). Nonaromatic, very reactive in conc. solns. above −50°.

DeMore, W.B. *et al*, *J. Am. Chem. Soc.*, 1959, **81**, 5874.
Scherer, K.V. Jr., *et al*, *J. Am. Chem. Soc.*, 1963, **85**, 1550.
Escher, A. *et al*, *Helv. Chim. Acta*, 1986, **69**, 1644 (*synth, pmr, cmr, uv, props*)

Fulvic acid   F-60081

Updated Entry replacing F-10100

4,10-Dihydro-3,7,8-trihydroxy-3-methyl-10-oxo-1H,3H-pyrano[4,3-b][1]benzopyran-9-carboxylic acid, 9CI

[479-66-3]

$C_{14}H_{12}O_8$ M 308.244

Metab. from *Penicillium* spp. Yellow cryst. Mp 246° dec.

Oxford, A.E. *et al*, *Biochem. J.*, 1935, **29**, 1102 (*isol*)
Dean, F.M. *et al*, *Nature (London)*, 1957, **179**, 366 (*struct*)
Kurobane, I. *et al*, *Tetrahedron Lett.*, 1981, 493 (*biosynth*)
Yamauchi, M. *et al*, *J. Chem. Soc., Chem. Commun.*, 1983, 335; 1984, 1505, 1565 (*synth*)
Katayama, S. *et al*, *Heterocycles*, 1985, **23**, 227 (*synth*)
Yamauchi, M. *et al*, *J. Chem. Soc., Perkin Trans. 1*, 1987, 389 (*synth*)

Fumifungin   F-60082

4-Acetoxy-2-amino-3,5,14-trihydroxy-6-eicosenoic acid

[110231-33-9]

$H_3C(CH_2)_5CH(OH)(CH_2)_6CH=CHCH(OH)-$
$CH(OAc)CH(OH)CH(NH_2)COOH$

$C_{22}H_{41}NO_7$ M 431.568

Amino acid antibiotic. Prod. by *Aspergillus fumigatus*. Antifungal agent. Mp 108°.

Mukhopadhyay, T. *et al*, *J. Antibiot.*, 1987, **40**, 1050 (*isol, struct*)

3-Furanacetaldehyde   F-60083

3-Furfuraldehyde. 2-(3-Furyl)acetaldehyde

[99948-48-8]

$C_6H_6O_2$ M 110.112

Bp$_{0.4}$ 25°.

Baillargeon, V.P. *et al*, *J. Am. Chem. Soc.*, 1986, **108**, 452 (*synth, pmr, cmr, ir*)

2-(2-Furanyl)-4*H*-1-benzopyran-4-one   F-60084

2-(2-Furyl)chromone

[3034-14-8]

$C_{13}H_8O_3$ M 212.204

Needles (EtOH). Mp 135°.

Ollis, W.D. *et al*, *J. Chem. Soc.*, 1952, 3826 (*synth*)

2-(3-Furanyl)-4*H*-1-benzopyran-4-one   F-60085

2-(3-Furyl)chromone

[89002-85-7]

$C_{13}H_8O_3$ M 212.204

Plates (EtOAc). Mp 119-120°.

Costa, A.M. *et al*, *J. Chem. Soc., Perkin Trans. 1*, 1986, 1707 (*synth, ir, pmr*)

4-(2-Furanyl)-3-buten-2-one, 9CI   F-60086

2-Furfuralacetone. 2-(3-Oxo-1-butenyl)furan

[623-15-4]

$C_8H_8O_2$ M 136.150

Mp 37-39°. Bp$_{10}$ 114-116°.

Org. Synth., Coll. Vol., **1**, 283.
Petrini, M. *et al*, *Tetrahedron*, 1986, **42**, 51 (*synth, ir, pmr*)

1-(3-Furanyl)-4-methyl-1-pentanone, 9CI, 8CI   F-60087

Perilla ketone

[553-84-4]

$C_{10}H_{14}O_2$ M 166.219

Isol. from the essential oil of *Perilla frutescens*. Promotes intestinal propulsion in mice. Light-yellow liq. Bp$_{30}$ 130°.

▷SA8935000.

2,4-Dinitrophenylhydrazone: Red-orange needles (EtOH). Mp 149.5°.

Oxime: Needles (EtOH aq.). Mp 62-64°.

Kondo, K. *et al*, *Tetrahedron Lett.*, 1976, 4363 (*synth*)
Kitamura, T. *et al*, *Synth. Commun.*, 1977, **7**, 521 (*synth*)
Abdulla, R.F. *et al*, *J. Org. Chem.*, 1978, **43**, 4248 (*synth*)
Inomata, K. *et al*, *Chem. Lett.*, 1979, 709 (*synth*)
Garst, J.E. *et al*, *J. Agric. Food Chem.*, 1984, **22**, 1083 (*synth*)
Kaezuka, Y. *et al*, *Planta Med.*, 1985, 480 (*isol, props*)

1-(2-Furanyl)-2-(2-pyrrolyl)ethylene   F-60088
2-[2-(2-Furanyl)ethenyl]-1H-pyrrole, 9CI

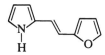

$C_{10}H_9NO$ M 159.187
(*E*)-**form** [107902-33-0]
Mp 104°.
N-*Me:* [99702-05-3].
    $C_{11}H_{11}NO$ M 173.214
    Mp 34-36°. Bp$_{0.15}$ 92-94°.

Hinz, W. *et al*, *Synthesis*, 1986, 620 (*synth, uv, pmr*)

2-(2-Furanyl)quinoxaline, 9CI   F-60089
2-(2-Furyl)quinoxaline, 8CI. 2-(2-Quinoxalinyl)furan.
Glucazidone
[494-21-3]

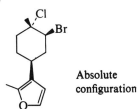

$C_{12}H_8N_2O$ M 196.208
Mp 101°.

Maurer, K. *et al*, *Ber.*, 1935, **68**, 1716; 1937, **70**, 1857 (*synth, derivs*)
Gomez-Sanchez, A. *et al*, *An. Quim., Ser. B*, 1954, **50**, 431 (*synth, struct*)
Sarkis, G.Y. *et al*, *J. Chem. Eng. Data*, 1973, **18**, 102 (*synth, uv, ir*)
Cimanis, A. *et al*, *Khim.-Farm. Zh.*, 1977, **11**, 48 (*synth, props*)
Galbis, P.J.A. *et al*, *An. Quim.*, 1977, **73**, 601 (*synth*)

Furocaespitane   F-60090
Updated Entry replacing F-00896
3-(3-Bromo-4-chloro-4-methylcyclohexyl)-2-
methylfuran
[51847-78-0]

Absolute
configuration

$C_{12}H_{16}BrClO$ M 291.615
Constit. of *Laurencia caespitosa.* Cryst. (C_6H_6). Mp 83-85°.

González, A.G. *et al*, *Tetrahedron Lett.*, 1973, 3625; 1979, 2719 (*isol, struct, cmr*)
Estrada, D.M. *et al*, *Tetrahedron Lett.*, 1987, **28**, 687 (*abs config*)

Furo[2,3-*d*]pyridazine   F-60091
[271-93-2]

$C_6H_4N_2O$ M 120.110
Cryst. (Me_2CO). Mp 108°.
B,HCl: Cryst. (EtOH). Mp 212°.
B,MeI: Cryst. (EtOH). Mp 212°.

Robba, M. *et al*, *Bull. Soc. Chim. Fr.*, 1968, 4959 (*synth, ir, pmr*)

Furo[3,4-*d*]pyridazine   F-60092
[270-80-4]

$C_6H_4N_2O$ M 120.110
Yellow cryst. Mp 161°.

Robba, M. *et al*, *C.R. Hebd. Seances Acad. Sci., Ser. C*, 1966, **263**, 301 (*synth, ir, pmr*)

Furo[2,3-*b*]pyridine-2-carboxaldehyde, 9CI   F-60093
2-Formylfuro[2,3-b]pyridine
[109274-92-2]

$C_8H_5NO_2$ M 147.133
Cryst. (EtOAc/hexane). Mp 136.5-137°.
Oxime:
    $C_8H_6N_2O_2$ M 162.148
    Cryst. (MeOH). Mixt. of (*E*)- and (*Z*)-isomers.

Morita, H. *et al*, *J. Heterocycl. Chem.*, 1986, **23**, 1465 (*synth, ir, pmr, ms*)

Furo[2,3-*b*]pyridine-3-carboxaldehyde, 9CI   F-60094
3-Formylfuro[2,3-b]pyridine
[109274-99-9]
$C_8H_5NO_2$ M 147.133
Cryst. (MeOH/Et_2O). Mp 121-124°.
Hydrazone: [109275-00-5].
    $C_8H_7N_3O$ M 161.163
    Mp 288-292°.
4-Methylbenzenesulfonylhydrazone: Mp 178-180°.

Morita, H. *et al*, *J. Heterocycl. Chem.*, 1986, **23**, 1465 (*synth, ir, pmr*)

Furo[3,2-*b*]pyridine-2-carboxaldehyde, 9CI   F-60095
2-Formylfuro[3,2-b]pyridine

$C_8H_5NO_2$ M 147.133
Cryst. (1,2-dimethoxyethane/hexane). Mp 151.5-152°.
Oxime:
    $C_8H_6N_2O_2$ M 162.148

Cryst. (MeOH). Mp 195-196°.

Morita, H. *et al, J. Heterocycl. Chem.*, 1987, **24**, 373 (*synth, pmr, ms*)

Furo[2,3-*c*]pyridine-2-carboxaldehyde, 9CI   F-60096
2-Formylfuro[2,3-c]pyridine

C$_8$H$_5$NO$_2$ M 147.133
Cryst. (Et$_2$O). Mp 120.5-121°.

Oxime:
   C$_8$H$_6$N$_2$O$_2$ M 162.148
   Cryst. (MeOH). Mp 210-211°.

Morita, H. *et al, J. Heterocycl. Chem.*, 1987, **24**, 373 (*synth, pmr, ms*)

Furo[3,2-*c*]pyridine-2-carboxaldehyde, 9CI   F-60097
2-Formylfuro[3,2-c]pyridine

C$_8$H$_5$NO$_2$ M 147.133
Cryst. (Et$_2$O). Mp 129.5-130°.

Oxime:
   C$_8$H$_6$N$_2$O$_2$ M 162.148
   Cryst. (MeOH). Mp 204-205°.

Morita, H. *et al, J. Heterocycl. Chem.*, 1987, **24**, 373 (*synth, pmr, ms*)

Furo[2,3-*b*]pyridine-2-carboxylic acid, 9CI   F-60098
[34668-26-3]

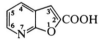

C$_8$H$_5$NO$_3$ M 163.132
Mp 281-282° dec.

Nitrile: [109274-95-5]. *2-Cyanofuro[2,3-b]pyridine.*
   C$_8$H$_4$N$_2$O   M 144.132
   Cryst. (Et$_2$O). Mp 74.5-76°.

Morita, H. *et al, J. Heterocycl. Chem.*, 1986, **23**, 1465 (*synth, ir, pmr, ms, bibl*)

Furo[2,3-*b*]pyridine-3-carboxylic acid   F-60099
C$_8$H$_5$NO$_3$ M 163.132

Amide: [109274-98-8].
   C$_8$H$_6$N$_2$O$_2$ M 162.148
   Cryst. (MeOH). Mp 204-205°.
Nitrile: [109274-96-6]. *3-Cyanofuro[2,3-b]pyridine.*
   C$_8$H$_4$N$_2$O   M 144.132
   Cryst. (Et$_2$O). Mp 127-129°. Basic hydrol. did not give
   the parent acid but led to ring opening.

Morita, H. *et al, J. Heterocycl. Chem.*, 1986, **23**, 1465 (*synth, ir, pmr*)

Furo[3,2-*b*]pyridine-2-carboxylic acid, 9CI   F-60100

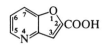

C$_8$H$_5$NO$_3$ M 163.132
Cryst. (H$_2$O). Mp >320°.

Nitrile: 2-Cyanofuro[3,2-b]pyridine.
   C$_8$H$_4$N$_2$O   M 144.132
   Cryst. (Et$_2$O). Mp 68.5-70.5°.

Morita, H. *et al, J. Heterocycl. Chem.*, 1987, **24**, 373 (*synth, pmr, ms*)

Furo[3,2-*b*]pyridine-3-carboxylic acid, 9CI   F-60101
C$_8$H$_5$NO$_3$ M 163.132

Amide:
   C$_8$H$_6$N$_2$O$_2$ M 162.148
   Cryst. (MeOH). Mp 160-161°.
Nitrile: 3-Cyanofuro[3,2-b]pyridine.
   C$_8$H$_4$N$_2$O   M 144.132
   Cryst. (Et$_2$O). Mp 146-147°. Alkaline hydrol. causes
   ring-opening.

Morita, H. *et al, J. Heterocycl. Chem.*, 1987, **24**, 373.

Furo[2,3-*c*]pyridine-2-carboxylic acid, 9CI   F-60102

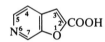

C$_8$H$_5$NO$_3$ M 163.132
Cryst. (H$_2$O). Mp >320°.

Nitrile: 2-Cyanofuro[2,3-c]pyridine.
   C$_8$H$_4$N$_2$O   M 144.132
   Mp 128-130°.

Morita, H. *et al, J. Heterocycl. Chem.*, 1987, **24**, 373 (*synth, pmr, ms*)

Furo[2,3-*c*]pyridine-3-carboxylic acid, 9CI   F-60103
C$_8$H$_5$NO$_3$ M 163.132

Amide:
   C$_8$H$_6$N$_2$O$_2$ M 162.148
   Cryst. (Me$_2$CO). Mp 227-229°.
Nitrile: 3-Cyanofuro[2,3-c]pyridine.
   C$_8$H$_4$N$_2$O   M 144.132
   Cryst. (Et$_2$O). Mp 121-122°. Alkaline hydrol. causes
   ring opening.

Morita, H. *et al, J. Heterocycl. Chem.*, 1987, **24**, 373.

Furo[3,2-*c*]pyridine-2-carboxylic acid, 9CI   F-60104

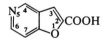

C$_8$H$_5$NO$_3$ M 163.132
Cryst. + $^1/_4$H$_2$O (EtOH aq.). Mp >320°.

Nitrile: 2-Cyanofuro[3,2-c]pyridine.
   C$_8$H$_4$N$_2$O   M 144.132
   Cryst. (H$_2$O). Mp 63.5-66°.

Morita, H. *et al, J. Heterocycl. Chem.*, 1987, **24**, 373 (*synth, pmr, ms*)

Furo[3,2-c]pyridine-3-carboxylic acid, 9CI F-60105

$C_8H_5NO_3$ M 163.132

Amide:
$C_8H_6N_2O_2$ M 162.148
Cryst. (MeOH). Mp 241-242°.
Nitrile: 3-Cyanofuro[3,2-c]pyridine.
$C_8H_4N_2O$ M 144.132
Cryst. (Et$_2$O). Mp 129-130°. Alkaline hydrol. causes
ring opening.

Morita, H. *et al, J. Heterocycl. Chem.*, 1987, **24**, 373.

Furo[2,3-d]pyrimidin-2(1H)-one, 9CI F-60106

5(7H)-*Furano*[2,3-d]*pyrimidin-6-one*
[62785-91-5]

$C_6H_4N_2O_2$ M 136.110
Cryst. (MeOH). Mp 260°.

Bleackley, R.C. *et al, Tetrahedron*, 1976, **32**, 2795 (*synth, pmr, uv*)

Fusarin A F-60107

[100079-50-3]

$C_{23}H_{29}NO_6$ M 415.485
Metab. of *Fusarium moniliforme*.

*17α-Hydroxy: **Fusarin D**.*
$C_{23}H_{29}NO_7$ M 431.485
From *F. moniliforme*.

Gelderblom, W.C.A. *et al, J. Chem. Soc., Chem. Commun.*, 1984, 122 (*isol*)
Steyn, P.S. *et al, J. Chem. Soc., Chem. Commun.*, 1985, 1189 (*biosynth*)

Fusarubin F-60108

Updated Entry replacing F-00937
5,8-*Dihydroxy*-2-(*hydroxymethyl*)-6-*methoxy*-3-(2-*ox-opropyl*)-1,4-*naphthalenedione*, 9CI. 3-*Acetonyl*-5,8-*di-hydroxy*-2-(*hydroxymethyl*)-6-*methoxy*-1,4-*naphtho-quinone*, 8CI. *Oxyjavanicin. Hydroxyjavanicin*
[1702-77-8]

$C_{15}H_{14}O_7$ M 306.271
Naphthoquinone-type antibiotic. Constit. of *Fusarium
solani* and other *F.* spp. Phytotoxic. Active against
gram-positive bacteria and fungi. Less active than
dihydro derivs. Red prisms (C$_6$H$_6$). Mp 218°.
Fusarubin may be an artifact formed nonenzymatically
during alkaline fermentation; also formed by some spp.
in acid conditions.

4aα,10aα-Dihydro: [67576-71-0]. **Dihydrofusarubin B**.
$C_{15}H_{16}O_7$ M 308.287
From *F. solanii*. Active against gram-positive bacteria,
fungi and tumours. Pale-yellow needles (MeOH). Mp
117-118°. $[\alpha]_D^{20}$ +23.3° (c, 0.31 in Me$_2$CO).

4aα,10aβ-Dihydro: [67533-03-3]. **Dihydrofusarubin A**.
$C_{15}H_{16}O_7$ M 308.287
From *F. solanii*. Active against gram-positive bacteria,
fungi and tumours. Reddish-orange needles (CH$_2$Cl$_2$).
Mp 153-154°. $[\alpha]_D^{20}$ +145.4° (c, 0.31 in Me$_2$CO).
During alk. fermentation → Fusarubin.

4aα,10aα-Dihydro, O³-Et: [76376-33-5]. **3-O-Ethyldihy-
drofusarubin B**. From *F. solani*.

4aα,10aβ-Dihydro, O³-Et: [76343-96-9]. **3-O-Ethyldihy-
drofusarubin A**.
$C_{17}H_{20}O_7$ M 336.341
From *F. marhi*. Shows moderate activity against gram-
positive bacteria, fungi and tumours. Reddish-orange
needles (CHCl$_3$/EtOH). Mp 138-139°. $[\alpha]_D^{25}$ +125.8°
(c, 0.1 in Me$_2$CO).

4a,10a-Dihydro-10a-hydroxy: From *F. solani*. Active
against gram-positive bacteria and fungi. Yellow
pigment.

*4aα,10aβ-Dihydro, O³-Me: **3-O-Methyldihydrofusarubin
A**.
$C_{16}H_{18}O_7$ M 322.314
From *F. martii*. Shows moderate activity against
gram-positive bacteria and tumours. Reddish-orange
needles (CHCl$_3$/EtOH). Mp 134-135°. $[\alpha]_D^{25}$ +128.5°
(c, 0.1 in Me$_2$CO).

O³-Et: [71724-91-9].
$C_{17}H_{18}O_7$ M 334.325
From *F. martii* and *F. solanii*. Artefact. Shows
moderate activity against gram-positive bacteria and
mouse leukaemia. Red. Mp 166-167° (197-198°).

O³-Me: [87667-46-7]. *Fusarubin methyl acetal.*
$C_{16}H_{16}O_7$ M 320.298
From *F. martii*. Artefact. Shows moderate activity
against gram-positive bacteria, fungi and mouse
leukaemia. Reddish-orange needles (CHCl$_3$/EtOH).
Mp 157-158° (188-190°).

O⁹-Me: [73618-70-9].
$C_{16}H_{16}O_7$ M 320.298
From *F. moniliforme* and *F. oxysporum*. Red-brown
needles (MeOH). Mp 138-139°.

3-Deoxy, 3,4-Didehydro, 9-Me ether: [96888-55-0]. **An-
hydrofusarubin 9-methyl ether**.
$C_{16}H_{14}O_6$ M 302.283
From *F. oxysporum*. Black needles (MeOH). Mp 175-
177°.

Ruelius, H.W. *et al, Justus Liebigs Ann. Chem.*, 1950, **569**, 38 (*isol*)
Birch, A.J. *et al, Chem. Ind.* (*London*), 1954, 1047 (*struct*)
Hardegger, E. *et al, Helv. Chim. Acta*, 1964, **47**, 2031 (*synth*)
Kern, H. *et al, Phytopathol. Z.*, 1967, **60**, 316 (*biochem*)
Kurobane, I. *et al, Can. J. Chem.*, 1978, **56**, 1593; 1980, **58**, 1380 (*derivs*)
Ammar, M.S. *et al, J. Antibiot.*, 1979, **32**, 679 (*isol*)
Gerber, N.N. *et al, J. Antibiot.*, 1979, **32**, 685 (*struct*)
Steyn, P.S. *et al, Tetrahedron*, 1979, **35**, 1551 (*isol*)
Kurobane, I. *et al, J. Antibiot.*, 1980, **33**, 1376; 1986, **39**, 205.
McCulloch, A.W. *et al, Can. J. Chem.*, 1982, **60**, 2943.
Tatum, J.H. *et al, Phytochemistry*, 1983, **22**, 543; 1985, **24**, 457 (*isol*)

G

Galipein G-60001

C$_{19}$H$_{20}$O$_4$ M 312.365
Constit. of *Galipea trifoliata*. Prisms (Me$_2$CO). Mp 88-90°.

Wirasutisna, K.R. *et al*, *Phytochemistry*, 1987, **26**, 3372.

Galiridoside G-60002

Updated Entry replacing G-00043
[30688-55-2]

C$_{15}$H$_{22}$O$_9$ M 346.333
Constit. of *Galeopsis tetrahit*. Cryst. Mp 189.5-191.5°. [α]$_D$ −78° (c, 1 in H$_2$O).
2′-(p-*Hydroxycinnamoyl*) (E-): **Decumbeside** A. C$_{24}$H$_{28}$O$_{11}$ M 492.479
Constit. of *Ajuga decumbens*. Amorph. powder. [α]$_D^{26}$ −60° (c, 0.52 in MeOH).
2′-(p-*Hydroxycinnamoyl*) (Z-): **Decumbeside** B. C$_{24}$H$_{28}$O$_{11}$ M 492.479
Constit. of *A. decumbens*. Amorph. powder. [α]$_D^{26}$ −13° (c, 0.59 in MeOH).

Sticher, O., *Helv. Chim. Acta*, 1970, **53**, 2010 (*isol, struct*)
Takeda, Y. *et al*, *Phytochemistry*, 1987, **26**, 2303 (*derivs*)

Garcinone E G-60003

1,3,6,7-Tetrahydroxy-2,5,8-triprenylxanthone

C$_{28}$H$_{32}$O$_6$ M 464.557
Constit. of *Garcinia mangostana*. Amorph.

Dutta, P.K. *et al*, *Indian J. Chem., Sect. B*, 1987, **26**, 281.

Garvalone *A* G-60004

C$_{27}$H$_{32}$O$_7$ M 468.546
Constit. of *Garveia annulata*. Pale-yellow oil.

Fahy, E. *et al*, *Can. J. Chem.*, 1987, **65**, 376.

Garvalone *B* G-60005

C$_{25}$H$_{26}$O$_7$ M 438.476
Mol. formula given in the paper as C$_{26}$H$_{28}$O$_6$, apparently incorrectly. Constit. of *Garveia annulata*. Light-yellow oil. [α]$_D$ +136.9° (c, 0.39 in CHCl$_3$).

Fahy, E. *et al*, *Can. J. Chem.*, 1987, **65**, 376.

Garveatin *A* quinone G-60006

C$_{20}$H$_{18}$O$_6$ M 354.359
Constit. of *Garveia annulata*. Red oil.

Fahy, E. *et al*, *Can. J. Chem.*, 1987, **65**, 376.

Garveatin *D* G-60007

C$_{21}$H$_{22}$O$_5$ M 354.402
Mol. formula given in the paper as C$_{21}$H$_{20}$O$_5$, apparently incorrectly. Constit. of *Garveia annulata*. Yellow oil.

Fahy, E. *et al*, *Can. J. Chem.*, 1987, **65**, 376.

Garvin *A*   **G-60008**

$C_{23}H_{26}O_6$ M 398.455
Constit. of *Garveia annulata*. Yellow solid.
Fahy, E. *et al*, *Can. J. Chem.*, 1987, **65**, 376.

Garvin *B*   **G-60009**

$C_{21}H_{20}O_6$ M 368.385
Enolised β-diketone. Constit. of *Garveia annulata*. Yellow solid.

2-Hydroxy: **2-Hydroxygarvin B**.
   $C_{21}H_{20}O_7$ M 384.385
   Constit. of *G. annulata*. Yellow solid.
Fahy, E. *et al*, *Can. J. Chem.*, 1987, **65**, 376.

Gelsemide   **G-60010**
[110309-27-8]

$C_{10}H_{12}O_5$ M 212.202
Constit. of *Gelsemium sempervirens*. Cryst. (EtOH). Mp
   179-180°. $[\alpha]_D^{20}$ −343° (c, 0.9 in MeOH).

7-β-D-Glucopyranoside: [110322-49-1]. *Gelsemide 7-glucoside*.
   $C_{16}H_{22}O_{10}$ M 374.344
   From *G. sempervirens*. Foam. $[\alpha]_D^{20}$ −199° (c, 0.6 in MeOH).

7-β-D-Glucopyranoside, tetra-Ac: Cryst.
   (toluene/EtOH). Mp 134-136°. $[\alpha]_D^{20}$ −179° (c, 0.5 in CHCl_3).
Jensen, S.R. *et al*, *Phytochemistry*, 1987, **26**, 1725 (*isol, cryst struct*)

Gelsemiol   **G-60011**
[110414-77-2]

$C_{10}H_{16}O_4$ M 200.234
Constit. of *Gelsemium sempervirens*. Cryst. Mp 91-93°.
   $[\alpha]_D^{20}$ +13° (c, 0.5 in MeOH).

1-O-β-D-Glucopyranoside: [110309-32-5]. *Gelsemiol 1-glucoside*.
   $C_{16}H_{26}O_9$ M 362.376

From *G. sempervirens*.
3-O-β-D-Glucopyranoside: [110309-33-6]. *Gelsemiol 3-glucoside*.
   $C_{16}H_{26}O_9$ M 362.376
   From *G. sempervirens*.
Jensen, S.R. *et al*, *Phytochemistry*, 1987, **26**, 1725.

Geodiamolide *A*   **G-60012**

$C_{28}H_{40}IN_3O_6$ M 641.545
Cyclic depsipeptide. Isol. from the marine sponge *Geodia*
sp. Exhibits some antifungal props. Prisms
(MeCN/CH_2Cl_2). Mp 217-218°. $[\alpha]_D^{25}$ +53° (c, 0.04
in CHCl_3).

Bromo analogue: **Geodiamolide B**.
   $C_{28}H_{40}BrN_3O_6$ M 594.545
   From *G.* sp. Exhibits some antifungal props. Cryst.
   (MeCN/CH_2Cl_2). Mp 203-204°. $[\alpha]_D^{22}$ +101° (c, 0.04
   in CHCl_3).
Chan, W.R. *et al*, *J. Org. Chem.*, 1987, **52**, 3091 (*isol, struct*)

Gerberinol 1   **G-60013**
3,3′-Methylenebis[4-hydroxy-5-methyl-2H-1-benzo-pyran-2-one], *9CI*. *Gerberinol*
[84153-78-6]

$C_{21}H_{16}O_6$ M 364.354
Constit. of *Gerbera lanuginosa*. Cryst. (EtOAc). Mp
   264-266°.
Chatterjea, J.N. *et al*, *Indian J. Chem., Sect. B*, 1986, **25**, 796.

1(10),4-Germacradiene-6,8-diol   **G-60014**
Updated Entry replacing G-40013

$C_{15}H_{26}O_2$ M 238.369
(1E,4E,6S,8S)-form [98941-66-3]
Tovarol
   Constit. of *Thapia villosa*, also as various esters. Cryst.
   (Et_2O/hexane). Mp 161-163°. $[\alpha]_D$ −64.5° (c, 0.78 in
   CHCl_3).
(1E,4E,6R,8R)-form [56283-44-4]
Angrendiol
   Cryst. (Et_2O). Mp 135-136°. $[\alpha]_D$ −86.1° (c, 0.9 in
   EtOH).
8-(4-Hydoxybenzoyl): [39380-12-6]. **Ferolin**.
   $C_{22}H_{30}O_4$ M 358.477
   Constit. of roots of *Ferula pallida*. Cryst. Mp 189-190.5°.
8-(4-Hydroxy-3-methoxybenzoyl): **Chimganidin**.
   $C_{23}H_{32}O_5$ M 388.503

From roots of *F. pallida*. Cryst. Mp 140-141°.

Saidkhodzhaev, A.I. *et al*, *Khim. Prir. Soedin.*, 1977, **13**, 434 (*isol*)
Terasa, J. de P. *et al*, *Phytochemistry*, 1985, **24**, 1779 (*isol*)
Makmudov, M.K. *et al*, *Khim. Prir. Soedin.*, 1986, **22**, 406 (*cryst struct*)

Gersemolide   G-60015

[106293-82-7]

$C_{20}H_{24}O_4$ M 328.407
Constit. of *Gersemia rubiformis*. Needles.

Williams, D. *et al*, *J. Org. Chem.*, 1987, **52**, 332 (*isol, cryst struct*)

Gersolide   G-60016

$C_{20}H_{24}O_4$ M 328.407
Constit. of soft coral *Gersemia rubiformis*. Needles
(MeOH). Mp 176-178°. $[\alpha]_D$ −51.9° (c, 0.31 in
CH_2Cl_2).

Williams, D.E. *et al*, *Tetrahedron Lett.*, 1987, **28**, 5079 (*cryst struct*)

Gibberellin A_2 G-60017

Updated Entry replacing G-00150
*2,4a,8-Trihydroxy-1,8-dimethylgibbane-1,10-
dicarboxylic acid 1,4a-lactone, 9CI. ent-3α,10β,15β-Tri-
hydroxy-20-norgibberellane-7,19-dioic acid 19,10-
lactone*

[561-68-2]

As Gibberellin A_1, G-20022 with

$$R^1 = H; R^2 = \alpha\text{-}OH, \beta\text{-}CH_3$$

$C_{19}H_{26}O_6$ M 350.411
See also Gibberellins, G-20021 . Produced by *Gibberella
fujikuroi*. Cryst. (EtOAc/pet. ether). Mp 235-237°
dec. or 256° dec. (dimorph.). $[\alpha]_D$ +11.7° (c, 7.2 in
MeOH).

Mori, K. *et al*, *Tetrahedron*, 1969, **25**, 1293 (*synth*)
Abouamer, K.M. *et al*, *J. Chem. Soc., Perkin Trans. 1*, 1987,
1991 (*cryst struct*)

Gibberellin A_9 G-60018

Updated Entry replacing G-30006
*4a-Hydroxy-1-methyl-8-methylenegibbane-1,10-
dicarboxylic acid 1,4a-lactone, 9CI. ent-10β-Hydroxy-
20-nor-16-gibberellene-7,19-dioic acid 19,10-lactone*

[427-77-0]

R = CH_2

$C_{19}H_{24}O_4$ M 316.396
See also Gibberellins, G-20021 . Metab. of *Gibberella
fujikuroi*. Cryst. (Me_2CO/pet. ether), gum. Mp 208-
211°. $[\alpha]_D^{17}$ −22° (c, 0.25 in EtOH).
2α-Hydroxy: [57672-81-8]. **Gibberellin A_{40}**.
    $C_{19}H_{24}O_5$ M 332.396
    From *G. fujikuroi*. Cryst. Mp 212-213°.
2β-Hydroxy: [56978-14-4]. **Gibberellin A_{51}**.
    $C_{19}H_{24}O_5$ M 332.396
    Found in immature seeds of *Pisum sativum*. Cryst.
    (EtOAc/hexane). Mp 190-193°.
15β-Hydroxy: [55812-47-0]. **Gibberellin A_{45}**.
    $C_{19}H_{24}O_5$ M 332.396
    Constit. of *Pyrus communis*.
3β,12α-Dihydroxy: [55035-85-3]. **Gibberellin A_{55}**.
    $C_{19}H_{24}O_6$ M 348.395
    Obt. from *Cucurbita maxima*. Gum.
3β,15β-Dihydroxy: [63351-80-4]. **Gibberellin A_{63}**.
    $C_{19}H_{24}O_6$ M 348.395
    From *P. communis*.

Hanson, J.R. *et al*, *J. Chem. Soc.*, 1965, 3550 (*struct*)
Mori, K. *et al*, *Tetrahedron*, 1969, **25**, 1293 (*synth*)
Bearder, J.R. *et al*, *Tetrahedron Lett.*, 1975, 669 (*deriv*)
Yamagushi, I. *et al*, *J. Chem. Soc., Perkin Trans. 1*, 1975, 996
    (*derivs*)
Beeley, L.J. *et al*, *J. Chem. Soc., Perkin Trans. 1*, 1976, 1022
    (*synth*)
Sponsel, V.M. *et al*, *Planta*, 1977, **135**, 129 (*deriv*)
Duri, Z.J. *et al*, *J. Chem. Soc., Perkin Trans. 1*, 1981, 161
    (*synth*)
Kamiya, Y. *et al*, *Phytochemistry*, 1983, **22**, 681 (*biosynth*)
Beale, M.H. *et al*, *Phytochemistry*, 1984, **23**, 565 (*deriv*)
Dolan, S.C. *et al*, *J. Chem. Soc., Perkin Trans. 1*, 1985, 651
    (*deriv*)
Abouamer, K.M. *et al*, *J. Chem. Soc., Perkin Trans. 1*, 1987,
    1991 (*cryst struct*)

Gibberellin A_{24} G-60019

Updated Entry replacing G-00171
ent-20-Oxo-16-gibberellene-7,19-dioic acid

[19427-32-8]

As Gibberellin A_{23}, G-00170 with

$$R^1 = CHO, R^2 = R^3 = H$$

$C_{20}H_{26}O_5$ M 346.422
See also Gibberellins, G-20021 . Metab. of *Gibberella
fujikuroi*. Cryst. (Me_2CO/pet. ether). Mp 199-203°.
$[\alpha]_D^{23}$ −88° (c, 0.6 in EtOH).
3β-Hydroxy: [38076-57-2]. **Gibberellin A_{36}**.
    $C_{20}H_{26}O_6$ M 362.422
    From *G. fujikuroi*. Cryst. (EtOAc/pet. ether). Mp
    205-208°.

Harrison, D.M. *et al*, *J. Chem. Soc. (C)*, 1971, 631 (*isol*)
Bearder, J.R. *et al*, *J. Chem. Soc., Perkin Trans. 1*, 1973, 2824
    (*deriv*)

Graebe, J.E. *et al*, *Phytochemistry*, 1974, **13**, 1433 (*biosynth*)
Kurogochi, S. *et al*, *Phytochemistry*, 1987, **26**, 2895 (*isol*)

Gibboside　　　　　　　　　　　　**G-60020**

$C_{16}H_{26}O_9$　　M 362.376
Constit. of *Patrinia gibbosa*. Amorph. powder. $[\alpha]_D$
−22.1° (c, 1 in MeOH).
Penta-Ac: Needles (EtOH). Mp 156.5-157.5°. $[\alpha]_D$
−3.87° (c, 1 in CHCl₃).

Uesato, S. *et al*, *Phytochemistry*, 1987, **26**, 561 (*isol, cryst struct*)

Glabone　　　　　　　　　　　　**G-60021**

$C_{18}H_{12}O_4$　　M 292.290
Constit. of *Pongamia glabra*. Cryst. (C₆H₆/pet. ether).
Mp 170-171°.

Das Kanungo, P. *et al*, *Phytochemistry*, 1987, **26**, 3373.

Glechomanolide　　　　　　　　　　**G-60022**
Updated Entry replacing G-00211
7,8,11,11a-Tetrahydro-3,6,10-trimethylcyclodeca[b]-
furan-2(4H)-one, 9CI
[38146-68-8]

$C_{15}H_{20}O_2$　　M 232.322
Constit. of *Glechoma hederacea*. Cryst. Mp 110°. $[\alpha]_D^{20}$
+120.5° (CHCl₃).
1β,10α; 4α,5β-Diepoxide, 8α-(2-methylpropanoyloxy):
1β,10α; 4α,5β-Diepoxy-8α-isobutoxyglechomanolide.
$C_{19}H_{28}O_5$　　M 336.427
Constit. of *Smyrnium perfoliatum*. Amorph.
1β,10α; 4α,5β-Diepoxide, 8β-(2-methylpropanoyloxy):
1β,10α; 4α,5β-Diepoxy-8β-isobutoxyglechomanolide.
$C_{19}H_{28}O_5$　　M 336.427
From *S. perfoliatum*. Cryst. Mp 155°.

Stahl, E. *et al*, *Justus Liebigs Ann. Chem.*, 1972, **757**, 23 (*isol*)
Gören, N. *et al*, *Phytochemistry*, 1987, **26**, 2585 (*derivs*)

The first digit of the DOC Number
defines the Supplement in which
the Entry is found. 0 indicates the
Main Work.

Gleinene　　　　　　　　　　　　**G-60023**

$C_{16}H_{18}O_4$　　M 274.316
Constit. of *Murraya gleinei*. Cryst. (CHCl₃/pet. ether).
Mp 176-178°.
*3′,4′-Didehydro: **Gleinadiene***.
$C_{16}H_{16}O_4$　　M 272.300
From *M. gleinei*. Pale-yellow cubes (EtOAc/pet.
ether). Mp 120-121°.

Kumar, V. *et al*, *Phytochemistry*, 1987, **26**, 511.

Gliotoxin *E*　　　　　　　　　　**G-60024**
[101623-21-6]

$C_{13}H_{14}N_2O_4S_3$　　M 358.445
Epithiodioxopiperazine antibiotic. Prod. by *Aspergillus
fumigatus*, *Penicillium terlikowskii* and *Thermoascus
crustaceus*. Cryst. (Et₂O). Mp 172-173°.

Waring, P. *et al*, *Aust. J. Chem.*, 1987, **40**, 991 (*isol, struct*)

Gloeosporone　　　　　　　　　　**G-60025**
Updated Entry replacing G-30009
*1-Hydroxy-6-pentyl-5,15-dioxabicyclo[10.2.1]-
pentadecane-4,13-dione*
[88936-02-1]

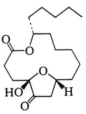

$C_{18}H_{30}O_5$　　M 326.432
Previously assigned struct. incorrect. Constit. of
Colletotrichum gloeosporioides. Germination
inhibitor. Cryst. (hexane). Mp 108-110°.

Meyer, W.L. *et al*, *Tetrahedron Lett.*, 1983, **24**, 5059.
Carling, R.W. *et al*, *Tetrahedron Lett.*, 1986, **27**, 6133 (*synth*)
Meyer, W.L. *et al*, *Helv. Chim. Acta*, 1987, **70**, 281 (*cryst
struct, bibl*)
Mortimore, M. *et al*, *Tetrahedron Lett.*, 1987, **28**, 3747 (*synth*)

γ-Glutamylmarasmine　　　　　　**G-60026**
N-γ-*Glutamyl-3-*[[*(methylthio)methyl]sulfinyl*]-
alanine, 9CI
[61481-40-1]

(2R,2′S,4S)-form

$C_{10}H_{18}N_2O_6S_2$　　M 326.382

(2R,2'S,4R)-form [106565-99-5]
Mp 169-172°. $[\alpha]_D^{25}$ −23° (c, 0.2 in H_2O).
(2R,2'S,4S)-form [106565-95-1]
Isol. from *Marasmius alliaceus*, *M. prasiosmus* and *M. scorodonius*. Mp 162-165°. $[\alpha]_D^{25}$ 0° (H_2O).

Gmelin, R. *et al*, *Phytochemistry*, 1976, **15**, 1717 (*isol*)
Cushman, M. *et al*, *J. Org. Chem.*, 1987, **52**, 1511 (*synth*)

γ-Glutamyltaurine G-60027

N-(*2-Sulfoethyl*)*glutamine*, *9CI*. *Glutaurine*. *Litoralon*

COOH
|
H_2N►C◄H   (*S*)-form
|
$CH_2CH_2CONHCH_2CH_2SO_3H$

$C_7H_{14}N_2O_6S$ M 254.257
(S)-form [56488-60-9]
L-form
Isol. from bovine parathyroids. Influences metab. of
Vitamin *A*, antagonizes the glucocorticoids and
triiodothyronine and possesses radioprotective props.
Cryst. (EtOH aq.). Mp 223.5-224°. $[\alpha]_D^{25}$ +23.3° (c,
2.06 in 1*M* HCl).

Gulyas, J. *et al*, *Org. Prep. Proced. Int.*, 1987, **19**, 64 (*synth, bibl*)

5(10)-Gluten-3-one G-60028

$C_{30}H_{48}O$ M 424.709
Constit. of *Andrachne cordifolia*. Cryst. (CHCl₃). Mp
250-254°.

Mukherjee, K.S. *et al*, *Phytochemistry*, 1987, **26**, 1539.

Glycerol 1-acetate G-60029

Updated Entry replacing G-00308
1,2,3-Propanetriol 1-acetate, *9CI*. *3-Acetoxy-1,2-propan-
ediol*. *α-Monoacetin*
[106-61-6]

CH_2OAc (*R*)-form
|
H►C◄OH   Absolute
| configuration
CH_2OH

$C_5H_{10}O_4$ M 134.132
▷AK3600000.
(R)-form [57416-04-3]
L-form
$[\alpha]_D$ −10.5° (dry Py).
(±)-form [93713-40-7]
Used in manuf. of smokeless powder and dynamite,
solvent for basic dyes and tanning leather. d_4^{20} 1.206.
Bp_4 134-135°, $Bp_{0.4}$ 103°. n_D^{20} 1.4157.
Di-Me ether:
$C_7H_{14}O_4$ M 162.185
Bp_{18} 100-110°.

Bis(4-nitrobenzoyl): Mp 129-130°. Bp 126-129°.

Baer, E. *et al*, *J. Am. Chem. Soc.*, 1945, **67**, 2031 (*synth, abs config*)
Williams, P.H. *et al*, *J. Am. Chem. Soc.*, 1960, **82**, 4883 (*synth*)
Ger. Pat., 1 810 568, (*1969*); *CA*, **71**, 90859 (*synth*)
Dahlhoff, W.V. *et al*, *Justus Liebigs Ann. Chem.*, 1975, 1914 (*synth*)
Gronowitz, S. *et al*, *Chem. Phys. Lipids*, 1976, **17**, 244 (*synth, cd*)
Suemune, H. *et al*, *Chem. Pharm. Bull.*, 1986, **34**, 3440 (*synth*)

Gnididione G-60030

Updated Entry replacing G-00628
[60498-89-7]

$C_{15}H_{16}O_3$ M 244.290
Constit. of *Gnidia latifolia*. Cryst. (MeOH). Mp 110-
111°. $[\alpha]_D^{29}$ +372° (c, 1 in CHCl₃).

Kupchan, S.M. *et al*, *J. Org. Chem.*, 1977, **42**, 348.
Jacobi, P.A. *et al*, *J. Am. Chem. Soc.*, 1984, **106**, 3041 (*synth*)
Dell, C.P. *et al*, *J. Chem. Soc., Chem. Commun.*, 1987, 349 (*synth*)

Goniothalamin G-60031

Updated Entry replacing G-40029
[17303-67-2]

Absolute
configuration

$C_{13}H_{12}O_2$ M 200.237
Constit. of bark of *Cryptocarya caloneura*, *Goniothala-
mus andersonii* and other *G*. spp. Cryst. (pet. ether).
Mp 85°. $[\alpha]_D$ +170.3° (c, 1.38 in CHCl₃).
▷UQ0590000.
7,8-Epoxide: **Goniothalamin oxide**.
$C_{13}H_{12}O_3$ M 216.236
Embryotoxic constit. of *Goniothalamus macrophyllus*.
Cryst. Mp 90-94°. $[\alpha]_D$ +100.7° (c, 0.70 in CHCl₃).

Hlubucek, J.R. *et al*, *Aust. J. Chem.*, 1967, **20**, 2199 (*struct, abs config*)
Jewers, K. *et al*, *Phytochemistry*, 1972, **11**, 2025 (*isol*)
El-Zayat, A.A.E. *et al*, *Tetrahedron Lett.*, 1985, **26**, 955 (*isol*)
O'Connor, B. *et al*, *Tetrahedron Lett.*, 1986, **27**, 5201 (*synth*)
Sam, T.W. *et al*, *Tetrahedron Lett.*, 1987, **28**, 2541 (*deriv*)

Gossypol G-60032

Updated Entry replacing G-10063
*1,1',6,6',7,7'-Hexahydroxy-3,3'-dimethyl-5,5'-bis(1-
methylethyl)-[2,2'-binaphthalene]-8,8'-dicarboxalde-
hyde*, *9CI*. *Thespesin*
[17273-29-9]

$C_{30}H_{30}O_8$ M 518.562

Atropisomeric compd., occurs naturally in both (+)- and (±)-forms.

▷Toxic

(+)-*form*

Constit. of *Thespesia populinea*. Pale-yellow needles (pet. ether), deep-yellow prisms + Me₂CO (Me₂CO), large elongated plates (Me₂CO aq.). Mp 181-183°. $[\alpha]_D^{19}$ +445° (c, 0.15 in CHCl₃).

Hexa-Me ether: [17273-30-2]. Colourless. Mp 242-244°. $[\alpha]_D$ +177° (CHCl₃).

(±)-*form* [303-45-7]

Toxic component of cotton boll cavities. Male antifertility agent, undergoing widespsread trials in the Peoples' Republic of China. Cryst. in three forms (Et₂O, CHCl₃, pet. ether). Mp 184°, 199°, 214°.

▷DU3100000.

Hexa-Ac: [30719-67-6]. Mp 276-279°.

▷DU3103000.

Bisphenylhydrazone: Yellow plates (C₆H₆). Mp 303°.

6-Me ether: [54302-42-0].
C₃₁H₃₂O₈   M 532.589
Constit. of the roots of *Gossypium* spp. Yellow cryst. (C₆H₆/hexane). Mp 146-149°.

6,6'-Di-Me ether: [1110-58-3].
C₃₂H₃₄O₈   M 546.616
From *G.* spp. Golden-yellow cryst. (C₆H₆/hexane). Mp 181-184°.

Hexa-Me ether: Cryst. in three forms, two colourless and one red. Mp 231-232°, 221° (colourless), 158-160° (red).

Adams, R. *et al*, *Chem. Rev.*, 1960, **60**, 555 (*rev*)
Bell, A., *Phytopathology*, 1967, **57**, 759.
Bhakuni, D.S. *et al*, *Experientia*, 1968, **24**, 109 (*isol*)
King, T.J. *et al*, *Tetrahedron Lett.*, 1968, 261 (*isol*)
Wood, A.B. *et al*, *Chem. Ind.* (*London*), 1969, 1738 (*conformn*)
Edwards, J.D., *J. Am. Oil Chem. Soc.*, 1970, **47**, 441 (*synth*)
Stipanovic, R.D. *et al*, *Phytochemistry*, 1975, **14**, 1077.
O'Brien, D.H. *et al*, *J. Org. Chem.*, 1978, **43**, 1105 (*nmr*)
Talipov, S.A. *et al*, *Khim. Prir. Soedin.*, 1985, **21**, 797; 1986, **22**, 108 (*cryst struct*)
Ibragimov, B.T. *et al*, *Khim. Prir. Soedin.*, 1985, **21**, 799; 1986, **22**, 110 (*cryst struct*)
Masciadri, R. *et al*, *J. Chem. Soc., Chem. Commun.*, 1985, 1573 (*biosynth*)
Stipanovic, R.D. *et al*, *J. Chem. Soc., Chem. Commun.*, 1986, 100 (*biosynth*)
Sampath, D.S. *et al*, *J. Chem. Soc., Chem. Commun.*, 1986, 649 (*cd*)
Lacombe, L. *et al*, *J. Nat. Prod.*, 1987, **50**, 277 (*cmr*)

Gracilin *A*   G-60033

Updated Entry replacing G-40031

C₂₃H₃₄O₅   M 390.519
Constit. of sponge *Spongionella gracilis*. Oil. $[\alpha]_D$ −60.5° (c, 1.3 in CHCl₃).

9α,11-Dihydro: [106231-25-8]. *9,11-Dihydrogracilin* A.
C₂₃H₃₆O₅   M 392.534
Constit. of *Dendrilla membranosa*. Oil. $[\alpha]_D$ −11° (c, 1.3 in CHCl₃).

Mayol, L. *et al*, *Tetrahedron Lett.*, 1985, **26**, 1357 (*isol*)
Molinski, T.F. *et al*, *J. Org. Chem.*, 1987, **52**, 296 (*deriv*)

Gracilin *B*   G-60034

Updated Entry replacing G-40032

[96313-95-0]

C₂₂H₂₈O₈   M 420.458
Constit. of sponge *Spongionella gracilis*. Cryst. Mp 167-168°. $[\alpha]_D$ +191.0° (c, 1.0 in CHCl₃).

12-Deacetyl, 12-propanoyl: [106849-36-9]. *Gracilin* D.
C₂₃H₃₀O₈   M 434.485
From *S. gracilis*. Oil. $[\alpha]_D^{25}$ +130.1° (c, 1.1 in CHCl₃).

8E-Isomer: [106621-85-6]. *Gracilin* C.
C₂₂H₂₈O₈   M 420.458
From *S. gracilis*. Cryst. (MeOH). Mp 240-241°. $[\alpha]_D^{25}$ +273.3° (c, 1.4 in CHCl₃).

Mayol, L. *et al*, *Tetrahedron Lett.*, 1985, **26**, 1253.
Mayol, L. *et al*, *J. Nat. Prod.*, 1986, **49**, 823 (*isol, struct*)

Gracilin *F*   G-60035

C₁₉H₃₀O₂   M 290.445
Constit. of *Spongionella gracilis*. Oil. $[\alpha]_D$ +0.5° (c, 1.2 in CHCl₂).

Ac: Gracilin E.
C₂₁H₃₂O₃   M 332.482
Constit. of *S. gracilis*. $[\alpha]_D$ −55.6° (c, 0.8 in CHCl₃).

Mayol, L. *et al*, *Tetrahedron*, 1986, **42**, 5369.

Graucin *A*   G-60036

C₂₆H₃₀O₁₀   M 502.517
Constit. of *Evodia grauca*. Cryst. (MeCN). Mp >310°. $[\alpha]_D^{15}$ −150° (c, 0.0005 in MeCN).

Nakatani, M. *et al*, *Bull. Chem. Soc. Jpn.*, 1987, **60**, 2503.

Gravelliferone   G-60037

Updated Entry replacing G-10072

3-(1,1-Dimethyl-2-propenyl)-7-hydroxy-6-(3-methyl-2-butenyl)-2H-1-benzopyran-2-one, 9CI. 3-(1,1-Dimethy-lallyl)-7-hydroxy-6-(3-methyl-2-butenyl)coumarin, 8CI

[21316-80-3]

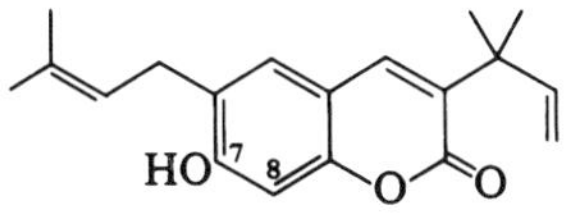

$C_{19}H_{22}O_3$ M 298.381

Isol. from *Ruta graveolens*. Mp 116-118°.

Me ether: [20958-62-7].
     $C_{20}H_{24}O_3$ M 312.408
     Found in *R.* spp. Cryst. (hexane or EtOH aq.). Mp 70-72° (67-70°).

8-Methoxy: [30430-91-2].
     $C_{20}H_{24}O_4$ M 328.407
     Constit. of *R. Graveolens*. Mp 131-133°.

Reisch, J. *et al, Experientia*, 1968, **24**, 992; *Tetrahedron Lett.*, 1968, 4395 (*isol, ir, uv*)
Reisch, J. *et al, Tetrahedron Lett.*, 1970, 4305 (*deriv*)
Gonzalez, A.G. *et al, An. Quim.*, 1972, **68**, 415; 1977, **73**, 1015 (*isol, uv, ir*)
Bergenthal, D. *et al, Arch. Pharm. (Weinheim, Ger.)*, 1978, **311**, 1026 (*cmr*)
Sharma, R.B. *et al, Indian J. Chem., Sect. B*, 1983, **22**, 538 (*synth*)
Cairns, N. *et al, J. Chem. Soc., Chem. Commun.*, 1987, 400 (*synth*)
Massanet, G.M. *et al, Heterocycles*, 1987, **26**, 1541 (*synth*)

Grevilline A   G-60038

Updated Entry replacing G-40039

3-Hydroxy-4-(4-hydroxyphenyl)-6-[(4-hydroxyphenyl-methylene)]-2H-pyran-2,5(6H)-dione, 9CI

[41744-32-5]

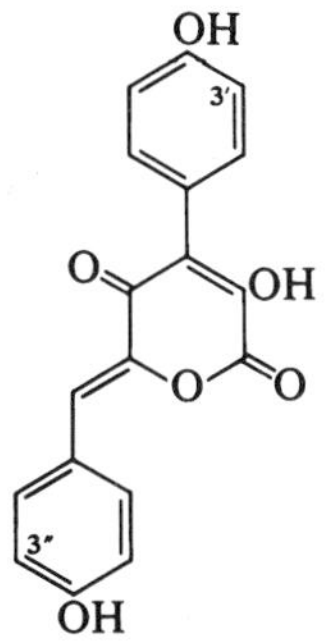

$C_{18}H_{12}O_6$ M 324.289

Pigment from *Suillus grevillei*.

3″-Hydroxy: [41744-33-6]. **Grevilline B**.
     $C_{18}H_{12}O_7$ M 340.289
     Pigment from *S. grevillei*.

3″-Hydroxy, tetra-Ac: Mp 180-181°.

3′,3″-Dihydroxy: [41744-34-7]. **Grevilline C**.
     $C_{18}H_{12}O_8$ M 356.288
     Pigment from *S. grevillei*.

3′,3″-Dihydroxy, penta-Ac: Mp 175-178°.

2′,3″,5′-Trihydroxy, 4′-deoxy: [54707-49-2]. **Grevilline D**.
     $C_{18}H_{12}O_8$ M 356.288
     Pigment from *S. grevillei*.

Steglich, W. *et al, Tetrahedron Lett.*, 1972, 4895 (*struct, pmr, uv*)
Lohrisch, H.J. *et al, Tetrahedron Lett.*, 1975, 2905 (*synth*)

Lohrisch, H.J. *et al, Justus Liebigs Ann. Chem.*, 1986, 177 (*synth*)
Pattenden, G. *et al, Tetrahedron Lett.*, 1987, **28**, 4749 (*synth*)

1,3,5,7(11),9-Guaiapentaen-14-al   G-60039

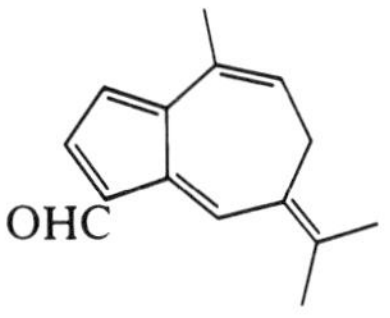

$C_{15}H_{16}O$ M 212.291

Constit. of *Lactarius sanguifluus*. Blood-red oil.

De Rosa, S. *et al, Phytochemistry*, 1987, **26**, 2007.

α-Guttiferin   G-60040

[11048-92-3]

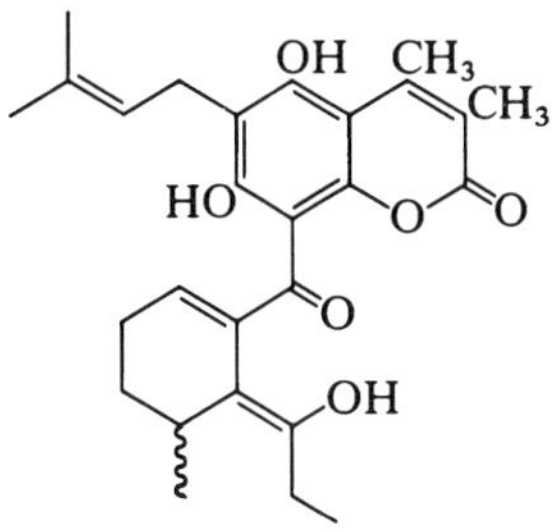

$C_{33}H_{38}O_2$ M 466.662

In *CA*, α-Guttiferin is confused with α₂-Guttiferin which is identical with Morellic acid. See M-04050. Constit. of Gamboge the resinous exodation of *Garcinia morella*. Mp 113-115°. $[\alpha]_D^{26}$ −475° (c, 1.5 in CHCl₃). CHCl₃).

▷MG1578700.

Py complex: Mp 115-117°. $[\alpha]_D^{26}$ −561.2° (c, 1.496 in CHCl₃).

Nageswara Rao, K.V. *et al, Experientia*, 1961, **17**, 213 (*struct*)

H

Haageanolide
H-60001

Updated Entry replacing H-10001

[68715-67-3]

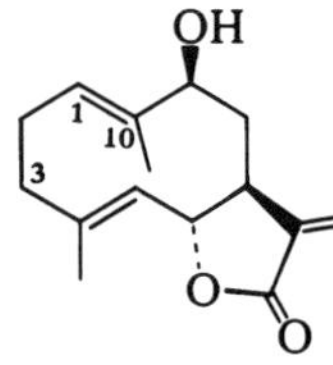

C₁₅H₂₀O₃   M 248.321

Constit. of *Zinnia haageana*. Amorph. powder.

Ac: Cryst. (EtOH). Mp 196-197°.

1β,10α-Epoxide: 1β,10α-Epoxyhaageanolide.
   $C_{15}H_{20}O_4$ M 264.321
   Constit. of *Inula heterolepsis*. Gum. $[\alpha]_D^{24}$ +7.5° (c, 0.24 in CHCl₃).

3β-Acetoxy, Ac: [83725-52-4]. *3β-Acetoxyhaageanolide acetate.*
   $C_{19}H_{24}O_6$ M 348.395
   Constit. of *Ursinia saxatilis*. Gum. $[\alpha]_D^{24}$ +53° (c, 0.2 in Et₂O).

Kisrel, W., *Phytochemistry*, 1978, **17**, 1059 (*isol*)
Bohlmann, F. *et al, Phytochemistry*, 1982, **21**, 1166, 1357 (*isol, deriv*)
Takahashi, T. *et al, Heterocycles*, 1987, **25**, 139 (*synth*)

Halenaquinol
H-60002

Updated Entry replacing H-40003

8,11-Dihydroxy-12b-methyl-1H-benzo[6,7]-phenanthro[10,1-bc]furan-3,6(2H,12bH)-dione, 9CI

[96603-02-0]

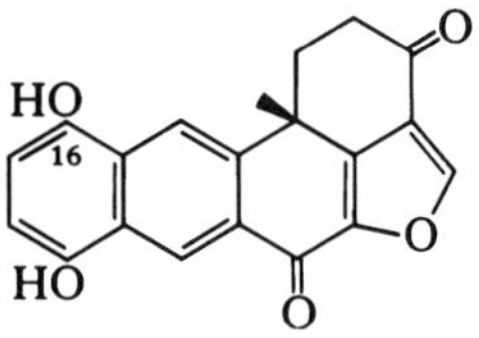

Absolute configuration

C₂₀H₁₄O₅   M 334.328

Constit. of sponge *Xestospongia sapra*. Yellow solid. $[\alpha]_{577}$+179° (Me₂CO). Unstable to heat and light.

O¹⁶-Sulfate: [99528-87-7].
   $C_{20}H_{14}O_8S$ M 414.386
   Constit. of *X. sapra*. Yellow solid. $[\alpha]_{577}$+106° (MeOH).

Quinone: [86690-14-4]. **Halenaquinone**.
   $C_{20}H_{12}O_5$ M 332.312
   Antibiotic constit. of *X. exigua*. Yellow solid. Mp >250° dec. $[\alpha]_D^{25}$ +22.2° (c, 0.124 in CH₂Cl₂).

Roll, D.M. *et al, J. Am. Chem. Soc.*, 1983, **105**, 6177 (*cryst struct*)
Kobayashi, M. *et al, Chem. Pharm. Bull.*, 1985, **33**, 1305 (*isol*)
Kobayashi, M. *et al, Tetrahedron Lett.*, 1985, **26**, 3833 (*abs config*)

Halichondramide
H-60003

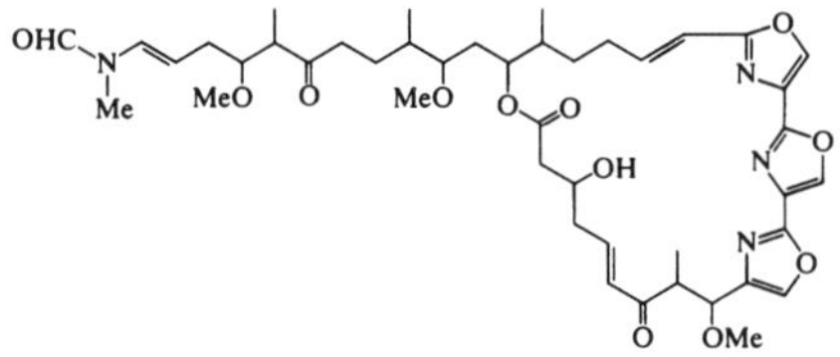

C₄₄H₆₀N₄O₁₂   M 836.978

Macrolide antibiotic. Prod. by sponge *Halichondria* sp. Antifungal agent. Mp 66-68°. Related to Kabiramide C.

Kernan, M.R. *et al, Tetrahedron Lett.*, 1987, **28**, 2809 (*isol, struct*)

Halleridone
H-60004

Updated Entry replacing H-30003

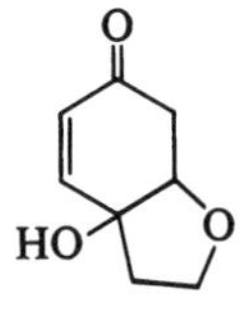

C₈H₁₀O₃   M 154.165

Constit. of *Halleria lucida*. Oil.

Messana, I. *et al, Phytochemistry*, 1984, **23**, 2617.
Breton, J.L. *et al, Tetrahedron*, 1987, **43**, 4447 (*synth*)

Hallerone
H-60005

Updated Entry replacing H-30004

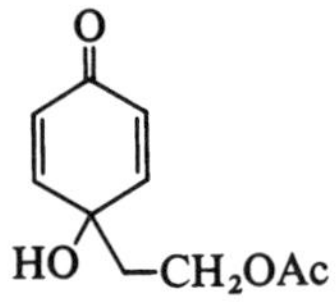

C₁₀H₁₂O₄   M 196.202

Oil. Constit. of *Halleria lucida*.

Messana, I. *et al, Phytochemistry*, 1984, **23**, 2617.
Breton, J.L. *et al, Tetrahedron*, 1987, **43**, 4447 (*synth*)

Hanphyllin
H-60006

Updated Entry replacing H-40006

3a,4,5,8,9,11a-Hexahydro-9-hydroxy-6,10-dimethyl-3-methylenecyclodeca[b]furan-2(3H)-one, 9CI. 3β-Hydroxy-1(10),4,11(13)-germacratrien-12,6α-olide

[60268-40-8]

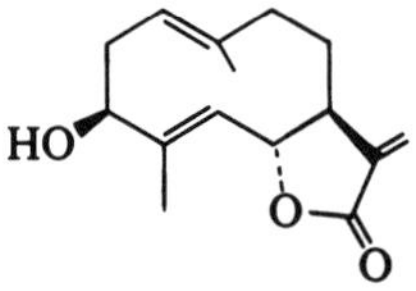

C₁₅H₂₀O₃   M 248.321

Constit. of *Handelia trichophylla* and *Artemisia klotzchiana*. Cryst. (Me₂CO/diisopropyl ether). Mp 180°.

2α-*Hydroxy, 3-Ac: 3β-Acetoxy-2α-hydroxy-1(10),4,11(13)-germacratrien-12,6α-olide. 2α-Hydroxyhanphyllin-3-O-acetate.*
$C_{17}H_{22}O_5$ M 306.358
Constit. of an *Eriocephalus* sp. Oil. $[\alpha]_D^{24}$ +131° (c, 0.37 in CHCl$_3$).

Tarasov, V.A. *et al, Khim. Prir. Soedin.*, 1976, 263; *CA*, **85**, 143313v (*isol*)
Tarasov, V.A. *et al, Khim. Prir. Soedin.*, 1978, 78; *CA*, **89**, 6438h (*struct*)
Mata, R. *et al, Phytochemistry*, 1985, **24**, 1515 (*cryst struct*)
Zdero, C. *et al, Phytochemistry*, 1987, **26**, 2763 (*deriv*)

Hardwickiic acid   H-60007

Updated Entry replacing H-50011

(−)-*form*
Absolute
configuration

$C_{20}H_{28}O_3$ M 316.439
(+)-*form* [24470-47-1]
Constit. of *Copaifera officinalis*. Mp 104-106°. $[\alpha]_D^{20}$ +125°.
(−)-*form* [1782-65-6]
Constit. of, *inter alia*, *Hardwickia pinnata*. Cryst. Mp 106-107°. $[\alpha]_D$ −114.7° (CHCl$_3$).
19-Hydroxy: [18411-75-1]. **Hautriwaic acid**.
$C_{20}H_{28}O_4$ M 332.439
Constit. of *Dodonnea viscosa* and *D. attenuata*. Also isol. as various esters from *Conyza scabrida*. Cryst. Mp 183-184°. $[\alpha]_D$ −105°.
19-Acetoxy: Hautriwaic acid acetate.
$C_{22}H_{30}O_5$ M 374.476
Constit. of *Baccharis macraei*. Gum. $[\alpha]_D^{25}$ −85° (c, 0.9 in CHCl$_3$).
19-Acetoxy, 1,2-didehydro:
$C_{22}H_{28}O_5$ M 372.460
Isol. from *C. scabrida* (as Me ester, after methylation of extract).
19-Oxo:
$C_{20}H_{26}O_4$ M 330.423
Isol. from *C. scabrida* as Me ester, after methylation.
19-Oxo, Me ester: $[\alpha]_D^{24}$ −67° (c, 0.15 in CHCl$_3$).
2β,19-Dihydroxy: 2-Hydroxyhautriwaic acid.
$C_{20}H_{28}O_5$ M 348.438
Constit. of *B. sarothroides*. Cryst. (C$_6$H$_6$). Mp 188-189°.

Cocker, W. *et al, Tetrahedron Lett.*, 1965, 1983 (*isol*)
Bohlmann, F. *et al, Chem. Ber.*, 1972, **105**, 3123 (*deriv*)
Payne, T.G. *et al, Tetrahedron*, 1973, **29**, 2575 (*struct*)
Ferguson, G. *et al, J. Chem. Soc., Chem. Commun.*, 1975, 299 (*stereochem*)
Misra, R. *et al, Tetrahedron*, 1979, **35**, 2301 (*abs config*)
Bohlmann, F. *et al, Justus Liebigs Ann. Chem.*, 1983, 2008 (*derivs*)
Arriaga-Giner, F.J. *et al, Phytochemistry*, 1986, **25**, 719 (*2-Hydroxyhautriwaic acid*)
Gambaro, V. *et al, Phytochemistry*, 1987, **26**, 475 (*isol, derivs*)

Havannahine   H-60008

[110201-59-7]

$C_{26}H_{34}O_{10}$ M 506.549
Constit. of *Xenia membranacea*. Cryst. Mp 181°. $[\alpha]_D$ +33° (c, 1.27 in CHCl$_3$).
11,19-Deepoxy: [110201-60-0]. **Deoxyhavannahine**.
$C_{26}H_{34}O_9$ M 490.549
From *X. membranaceae*. Amorph. $[\alpha]_D$ +37° (c, 1.04 in CHCl$_3$).

Lelong, H. *et al, J. Nat. Prod.*, 1987, **50**, 203 (*isol, cryst struct*)

Hebeclinolide   H-60009

Updated Entry replacing H-00060
6-(3-Furanyl)-5,6-dihydro-3-[2-(2,6,6-trimethyl-4-oxo-2-cyclohexen-1-yl)ethyl]-2H-pyran-2-one, 9CI
[63147-18-2]

$C_{20}H_{24}O_4$ M 328.407
Constit. of *Hebeclinium macrophyllum*. Oil. $[\alpha]_D^{24}$ −6° (c, 1.5 in CHCl$_3$). Isol. as a mixt. of 5-epimers.

5αH-*form*

*3β-Hydroxy: 3β-***Hydroxyhebeclinolide**.
$C_{20}H_{24}O_5$ M 344.407
Constit. of *H. macrophyllum*.
3β-Acetoxy: Cryst. Mp 112°.

5βH-*form*

3β-Hydroxy: From *H. macrophyllum*.

Bohlmann, F. *et al, Chem. Ber.*, 1977, **110**, 1321.
Warning, U. *et al, Phytochemistry*, 1987, **26**, 2331 (*deriv*)

Hebemacrophyllide   H-60010

$C_{20}H_{28}O_3$ M 316.439
6β-Acetoxy: [111508-89-5]. **6β-Acetoxyhebemacrophyllide**.
$C_{22}H_{30}O_5$ M 374.476
Constit. of *Hebeclinium macrophyllum*. Oil.

5-Epimer, 6β-Acetoxy: [111508-90-8]. **5-epi-6β-acetoxyhebemacrophyllide**.
$C_{22}H_{30}O_5$ M 374.476
From *H. macrophyllum*. Oil.
17-Hydroxy: **17-Hydroxyhebemacrophyllide**.
$C_{20}H_{28}O_4$ M 332.439
Constit. of *H. macrophyllum*. Oil. Isol. as a mixt. of 5-epimers.
17-Aldehyde: **17-Oxohebemacrophyllide**.
$C_{20}H_{26}O_4$ M 330.423
From *H. macrophyllide*. Oil. Isol. as a mixt. of 5-epimers.

Warning, U. *et al, Phytochemistry*, 1987, **26**, 2331.

Hebesterol   H-60011

$C_{29}H_{48}O$ M 412.698
Constit. of *Petrosia hebes*. Cryst. (MeCN). Mp 133-134°. $[\alpha]_D^{20}$ −2.1° (CHCl₃).

Cho, J.-H. *et al, J. Chem. Soc., Perkin Trans. 1*, 1987, 1307.

Hector's base   H-60012

Updated Entry replacing A-03001
4,5-Dihydro-5-imino-N,4-diphenyl-1,2,4-thiadiazol-3-amine, 9CI. 3-Amino-4-phenyl-1,2,4-thiadiazole-5(4H)-imine
[4115-26-8]

$C_{14}H_{12}N_4S$ M 268.336
Oxidn. prod. of phenylthiourea with H_2O_2. Cryst. (EtOH). Mp 239-240°.

Tsuchiya, T. *et al, Chem. Lett.*, 1976, 723.
Butler, A.R. *et al, Acta Crystallogr., Sect. B*, 1978, **34**, 3241 (*cryst struct*)
Butler, A.R. *et al, J. Chem. Res. (S)*, 1980, 114 (*cmr, N nmr*)
Kandror, I.I. *et al, Khim. Geterotsikl. Soedin.*, 1985, 205 (*cryst struct*)
Butler, A.R. *et al, Acta Chem. Scand., Ser. B*, 1986, **40**, 779 (*cmr*)

Heleniumlactone 1   H-60013

$C_{30}H_{42}O_3$ M 450.660

Constit. of *Helenium autumnale*. Cryst. Mp 80°.
5,6-Dihydro, 4,15-didehydro: **Heleniumlactone 2**.
$C_{30}H_{42}O_3$ M 450.660
Constit. of *H. autumnale*. Oil.

Matusch, R. *et al, Helv. Chim. Acta*, 1987, **70**, 342.

Heleniumlactone 3   H-60014

$C_{30}H_{42}O_3$ M 450.660
Constit. of *Helenium autumnale*. Oil.

Matusch, R. *et al, Helv. Chim. Acta*, 1987, **70**, 342.

Helicquinone   H-60015

$C_{15}H_{14}O_4$ M 258.273
Constit. of *Helicteres angustifolia*. Red cryst. Mp 360°.

Wang, M. *et al, Phytochemistry*, 1987, **26**, 578.

Helminthosporal   H-60016

Updated Entry replacing H-00094
1,7-Dimethyl-4-(1-methylethyl)bicyclo[3.2.1]oct-6-ene-6,8-dicarboxaldehyde, 9CI. 8-(Hydroxymethyl)-1,7-dimethyl-4-(1-methylethyl)bicyclo[3.2.1]oct-6-ene-6-carboxaldehyde, 9CI. 4-Isopropyl-1,7-dimethylbicyclo[3.2.1]oct-6-ene-6,8-dicarboxaldehyde
[723-61-5]

Absolute configuration

$C_{15}H_{22}O_2$ M 234.338
Prod. by *Helminthosporium sativum* and *Cochliobolus sativus*. Phytotoxin. Cryst. Mp 56-59°. $[\alpha]_D$ −49° (c, 1.2 in CHCl₃).

Corey, E.J. *et al, J. Am. Chem. Soc.*, 1965, **87**, 5728 (*synth*)
de Mayo, P. *et al, Can. J. Chem.*, 1965, **43**, 1357 (*isol*)
Piers, E. *et al, Can. J. Chem.*, 1977, **55**, 1039 (*synth*)
Sommereyns, G. *et al, Arch. Int. Physiol. Biochem.*, 1977, **85**, 431 (*isol, props*)
Yanagiya, M. *et al, Tetrahedron Lett.*, 1979, **20**, 1761 (*synth*)
Shizuri, Y. *et al, J. Chem. Soc., Chem. Commun.*, 1986, 63 (*synth*)
Gray, B.D. *et al, J. Chem. Soc., Chem. Commun.*, 1987, 1136 (*synth*)

Henricine   H-60017
[107783-46-0]

$C_{22}H_{26}O_6$ M 386.444

Constit. of *Schisandra henryi*. Cryst. Mp 73-75°. $[\alpha]_D^{21}$ +102° (CHCl$_3$).

Lian-niang, L. *et al*, *Planta Med.*, 1986, 493.

7,16-Heptacenedione   H-60018
7,16-Heptacenequinone
[13200-02-7]

$C_{30}H_{16}O_2$ M 408.455

Brown solid. Does not melt.

Clar, E., *Ber.*, 1942, **75**, 1330 (*synth*)
Smith, J.G. *et al*, *J. Org. Chem.*, 1986, **51**, 3762 (*synth, ir, uv*)

13-Heptacosene   H-60019
Updated Entry replacing H-00150
[61906-95-4]

$$H_3C(CH_2)_{12}CH_2CH{=}CH(CH_2)_{11}CH_3$$

$C_{27}H_{54}$ M 378.724

(*Z*)-*form* [54863-75-1]

Pheromone for *Musca autumnalis*.

Küpper, F. *et al*, *Z. Naturforsch., B*, 1976, **31**, 1256 (*synth*)
Subramanian, G.B.V. *et al*, *Tetrahedron*, 1986, **42**, 3967 (*synth, ir, pmr*)

2,4-Heptadienal, 9CI   H-60020
Updated Entry replacing H-00242
[4313-03-5]

$$H_3CCH_2CH{=}CHCH{=}CHCHO$$

$C_7H_{10}O$ M 110.155

(*E,E*)-*form*

Bp$_{12}$ 78°, Bp$_{0.06}$ 27-29°.

Di-Et acetal:
   $C_{11}H_{20}O_2$ M 184.278
   Bp$_{12}$ 96.5-97.5°.
2,4-Dinitrophenylhydrazone: Purple needles. Mp 157-159°.

Nishida, S. *et al*, *Chem. Lett.*, 1976, 1297 (*synth*)
Zimmermann, B. *et al*, *Chem. Ber.*, 1986, **119**, 2848 (*synth, ir, pmr*)

2,4-Heptadienoic acid, 9CI   H-60021
Updated Entry replacing H-00248
[17175-86-9]

$$CH_3CH_2CH{=}CHCH{=}CHCOOH$$

$C_7H_{10}O_2$ M 126.155

(*E,E*)-*form* [65518-46-9]
   Mp 44°. Bp$_{0.2}$ 105°.
(*2E,4Z*)-*form* [50915-66-7]
   Metabolite of *Sporobolomyces odorus*. Isol. as Me ester by glc.

Tahara, S. *et al*, *Agric. Biol. Chem.*, 1975, **39**, 71 (*synth*)
Markau, K. *et al*, *Chem. Ber.*, 1962, **95**, 889 (*synth*)
Devos, M.J. *et al*, *Tetrahedron Lett.*, 1976, 3911 (*synth*)
Zimmermann, B. *et al*, *Chem. Ber.*, 1986, **119**, 2848 (*synth, ir, pmr*)

3,4-Heptafulvalenedione   H-60022
5-(2,4-Cycloheptatrien-1-ylidene)-3,6-cyclohepta-diene-1,2-dione, 9CI
[109296-83-5]

$C_{14}H_{10}O_2$ M 210.232

Forms deep blue soln. in CH$_2$Cl$_2$. Stable only in soln.

Takahashi, K. *et al*, *Tetrahedron Lett.*, 1986, **27**, 5515 (*synth, pmr*)

2,3,5,14,20,22,25-Heptahydroxycholest-7-   H-60023
en-6-one, 9CI
Updated Entry replacing H-00268

$C_{27}H_{44}O_8$ M 496.640

(*2β,3β,5β,14α,20R,22R*)-*form* [18069-14-2]
   5β,20R-Dihydroxyecdysone. Polypodine B. 5β-Hydrox-ycrustecdysone. 5β-Hydroxyecdysterone. Ajugaster-one A
   Constit. of *Polypodium vulgare, Dacrydium intermedium* and *Vitex megapotamica*. Shows insect moulting hormone activity. Cryst. (Me$_2$CO). Mp 254-257°. $[\alpha]_D^{20}$ +93° (c, 0.18 in MeOH).
   2-Cinnamoyl: [38147-16-9].
     $C_{36}H_{50}O_9$ M 626.786
     Constit. of the bark of *D. intermedium*. Cryst. (Me$_2$CO). Mp 268-270°.

Jizba, J. *et al*, *Collect. Czech. Chem. Commun.*, 1967, **32**, 2867 (*isol*)
Rimpler, H., *Tetrahedron Lett.*, 1969, 329 (*isol, struct*)
Hikino, H. *et al*, *Tetrahedron Lett.*, 1969, 1417.
Russell, G.B. *et al*, *Aust. J. Chem.*, 1972, **25**, 1935 (*isol*)
Hardman, R. *et al*, *Phytochemistry*, 1976, **15**, 1515 (*isol*)
Nishimoto, N. *et al*, *Phytochemistry*, 1987, **26**, 2505 (*cmr*)

3,3',4',5,6,7,8-Heptahydroxyflavone   H-60024
Updated Entry replacing H-40021
2-(3,4-Dihydroxyphenyl)-3,5,6,7,8-pentahydroxy-4H-1-benzopyran-4-one, 9CI

$C_{15}H_{10}O_9$ M 334.239

3',6,7,8-Tetra-Me ether: 3,4',5-Trihydroxy-3',6,7,8-tetramethoxyflavone.
   $C_{19}H_{18}O_9$ M 390.346
   Constit. of *Gymnosperma glutinosum*.
3',4',5,6,7,8-Hexa-Me ether: [35154-55-3]. *3-Hydroxy-3',4',5,6,7,8-hexamethoxyflavone.* **Natsudaidan**.
   $C_{21}H_{22}O_9$ M 418.399
   Constit. of peel oil of *C. natsudaidai*. Yellow cryst. (EtOH). Mp 154-156° (141-143°).
3',4',5,6,7,8-Hexa-Me ether, Ac: [35154-56-4]. Mp 125°.
3,3',4',6,7,8-Hexa-Me ether: [1176-88-1]. *5-Hydroxy-3,3',4',6,7,8-hexamethoxyflavone.*
   $C_{21}H_{22}O_9$ M 418.399

Isol. from peel of *Citrus sinensis*. Yellow cryst. (MeOH). Mp 110-111°.

Sastry, G.P. *et al*, *Tetrahedron*, 1961, **15**, 111 (*synth*)
Row, L.R. *et al*, *Indian J. Chem.*, 1963, **1**, 207 (*synth*)
Gentili, B. *et al*, *Tetrahedron*, 1964, **20**, 2313 (*synth*)
Kinoshita, K. *et al*, *Yakugaku Zasshi*, 1971, **91**, 1105 (*isol*)
Tatum, J.H. *et al*, *Phytochemistry*, 1972, **11**, 2283 (*isol, uv*)
Bittner, M. *et al*, *Phytochemistry*, 1983, **22**, 1523.
Yu, S. *et al*, *Phytochemistry*, 1988, **27**, 171 (*isol*)

6-(1-Heptenyl)-5,6-dihydro-2*H*-pyran-2-one, 9CI   H-60025

Updated Entry replacing H-10021
2,6-Dodecadien-5-olide

$$H_3C(CH_2)_4$$

$C_{12}H_{18}O_2$ M 194.273

(*R,Z*)-form [64543-31-3]
   Argentilactone
   Constit. of *Aristolochia argentina*. $Bp_{0.6}$ 139-143°. $[\alpha]_D$ −21.1° (c, 2.25 in EtOH).

Priestap, H.A. *et al*, *Phytochemistry*, 1977, **16**, 1579 (*isol, uv, ir, pmr, ms*)
Fehr, C. *et al*, *Helv. Chim. Acta*, 1981, **64**, 1247 (*synth*)
O'Connor, B. *et al*, *Tetrahedron Lett.*, 1986, **27**, 5201 (*synth*)

6-Heptyne-2,5-diamine, 9CI   H-60026
3,5-Diamino-1-heptyne
[81645-70-7]

$C_7H_{14}N_2$ M 126.201

(*2R,5R*)-form [88192-22-7]
   Most active isomer as ornithinecarboxylase inhibitor.
   B,2HCl: [106539-45-1]. Cryst. Mp 236°. $[\alpha]_D$ −13.6° (c, 1.51 in H_2O).

(*2R,5S*)-form [106539-42-8]
   B,2HCl: [106539-46-2]. Mp 230°. $[\alpha]_D$ +28° (c, 0.515 in H_2O).

(±)-*form*
   Ornithine decarboxylase inhibitor.
   B,2HCl: Needles (Et_2O aq.). Mp 230°. Approx. equal mixt. of the two racemates.

Casara, P. *et al*, *J. Chem. Soc., Perkin Trans. 1*, 1985, 2201 (*synth, cryst struct, pmr, ir, props*)

Herbacin   H-60027
4,5,5a,6,7,9a-Hexahydro-5a,6,9a-trimethylnaphtho[1,2-b]furan

$C_{15}H_{20}O$ M 216.322

The name Herbacin is also used for a flavonoid. Isol. from marine sponge *Dysidea herbaceae*. Needles (C_6H_6). Mp 40-41°. $[\alpha]_D$ 9.2° (c, 0.2 in $CHCl_3$).

Sarma, N.S. *et al*, *Indian J. Chem., Sect. B*, 1986, **25**, 1001 (*isol, struct*)

Heritol   H-60028
[108295-47-2]

$C_{15}H_{16}O_3$ M 244.290
Constit. of *Heritiera littoralis*. Ichthyotoxin. Needles (MeOH). Mp 271-272°. $[\alpha]_D^{25}$ +261.3°.

Miles, D.H. *et al*, *J. Org. Chem.*, 1987, **52**, 2930 (*isol, cryst struct*)

Hermosillol   H-60029
4-[2-[2-Methoxy-5-(1-propenyl)phenyl]-2-propenyl]-phenol, 9CI
[110241-11-7]

$C_{19}H_{20}O_2$ M 280.366
Constit. of *Krameria sonorae*. Oil.

Dominguez, X.A. *et al*, *Phytochemistry*, 1987, **26**, 1821.

Heteronemin   H-60030
Updated Entry replacing H-00419
[62008-04-2]

$C_{29}H_{44}O_6$ M 488.663
Constit. of *Heteronema erecta*. Cryst. (pet. ether). Mp 182°.
12-Ac: [107748-93-6].
   $C_{31}H_{46}O_7$ M 530.700
   Constit. of *Hyrtios erecta*. Cryst. Mp 211-212°. $[\alpha]_D$ −30° (c, 0.1 in $CHCl_3$).
12-Epimer, 12-Ac: [107748-94-7].
   $C_{31}H_{46}O_7$ M 530.700
   Constit. of *Hyrtios erecta*. Oil.

Kashman, Y. *et al*, *Tetrahedron*, 1977, **33**, 2997.
Crews, P. *et al*, *J. Nat. Prod.*, 1986, **40**, 1041 (*isol, deriv*)

Hexabenzo[*bc,ef,hi,kl,no,qr*]coronene   H-60031

Updated Entry replacing H-00433
Hexaperibenzocoronene
[190-24-9]

C$_{42}$H$_{18}$ M 522.604
Long flat needles (pyrene). Mp >700°.

Clar, E. *et al, J. Chem. Soc.*, 1959, 142 (*synth, uv*)
Hendel, W. *et al, Tetrahedron*, 1986, **42**, 1127 (*synth, uv, bibl, props*)

Hexabutylbenzene   H-60032

[106821-88-9]

C$_{30}$H$_{54}$ M 414.757
Cryst. (heptane). Mp 59-60°.

Lutz, E.F., *J. Am. Chem. Soc.*, 1961, **83**, 2551 (*synth*)
Jhingan, A.K. *et al, J. Org. Chem.*, 1987, **52**, 1161 (*synth, pmr, cmr*)

6,15-Hexacenedione, 9CI   H-60033

Updated Entry replacing H-00456
2,3-Benzo-6,7-naphthaanthraquinone
[13214-71-6]

C$_{26}$H$_{14}$O$_2$ M 358.395
Orange-yellow leaflets. Mp 295-310°.

Clar, E., *Ber.*, 1942, **75**, 1283 (*synth*)
Verine, A. *et al, Bull. Soc. Chim. Fr.*, 1973, 1154 (*synth, ir, uv*)
Smith, J.G. *et al, J. Org. Chem.*, 1986, **51**, 3762 (*synth, pmr, ir, uv*)

2,2',4,4',5,5'-Hexachlorobiphenyl   H-60034

[35065-27-1]

C$_{12}$H$_4$Cl$_6$ M 360.882
Main component (ca. 22%) of polychlorobiphenyls detected in human tissues.
▷DV5350000.

McKinney, J.D. *et al, Toxicol. Appl. Pharmacol.*, 1976, **36**, 64 (*synth*)

Singh, P. *et al, Acta Crystallogr., Sect. C*, 1986, **42**, 1172 (*cryst struct*)
Geise, H.J. *et al, Acta Crystallogr., Sect. C*, 1986, **42**, 1176 (*cryst struct*)

Hexacyclo[5.4.0.0^{2,6}.0^{3,10}.0^{5,9}.0^{8,11}]-undecan-4-one   H-60035

Hexahydro-1,2,3,5-ethanediylidene-1H-cyclobuta[cd]-pentalen-4(1aH)one, 9CI. Homopentaprismanone
[30417-04-0]

C$_{11}$H$_{10}$O   M 158.199
Cryst. (Et$_2$O). Mp 154.8-155.2°.

Ward, J.S. *et al, J. Am. Chem. Soc.*, 1971, **93**, 262 (*synth*)
Eaton, P.E. *et al, Tetrahedron*, 1986, **42**, 1621 (*synth, ir, pmr, cmr, ms*)

2-Hexadecenal   H-60036

$$H_3C(CH_2)_{12}CH=CHCHO$$

C$_{16}$H$_{30}$O   M 238.412
(*E*)-*form* [22644-96-8]
Oil. Bp$_1$ 135°.

Hino, T. *et al, J. Chem. Soc., Perkin Trans. 1*, 1986, 1687 (*synth, ir, pmr, ms*)

5-Hexadecyne   H-60037

[71899-37-1]

$$H_3C(CH_2)_9C\equiv C(CH_2)_3CH_3$$

C$_{16}$H$_{30}$ M 222.413
Liq. Bp$_{0.04}$ 71-73°.

Sikorski, J.A. *et al, J. Org. Chem.*, 1986, **51**, 4521 (*synth, pmr, cmr*)

1,2,9,10,17,18-Hexadehydro[2.2.2]-paracyclophane   H-60038

Updated Entry replacing H-20050
Tetracyclo[14.2.2.2^{4,7}.2^{10,13}]tetracosa-2,4,6,8,10,12,14,16,18,19,21,23-dodecaene, 9CI. [2$_3$]-Paracyclophanetriene. [2.2.2]Paracyclophane-1,9,17-triene. [2.2.2]Paracyclophene
[16337-16-9]

C$_{24}$H$_{18}$ M 306.406
(*Z,Z,Z*)-*form* [84180-01-8]
Cryst. (EtOH aq.). Mp 136-136.8°.

Cram, D.J. *et al, J. Am. Chem. Soc.*, 1959, **81**, 5963 (*synth*)
Trueblood, K. *et al, Acta Crystallogr., Sect. B*, 1982, **38**, 2428 (*cryst struct*)
Tanner, D. *et al, Tetrahedron*, 1986, **42**, 4499 (*synth, pmr, ms*)

1,5-Hexadiene-3,4-dione   H-60039
Divinylglyoxal
[104910-78-3]

$$H_2C\!=\!CHCOCOCH\!=\!CH_2$$

$C_6H_6O_2$ M 110.112
Kramme, R. *et al, Angew. Chem., Int. Ed. Engl.,* 1986, **25**, 1116
   (*synth, pmr, cmr, ir, uv, pe*)

1,5-Hexadiyn-3-one, 9CI   H-60040
Ethynyl propargyl ketone
[66737-76-6]

$$HC\!\equiv\!CCOCH_2C\!\equiv\!CH$$

C_6H_4O M 92.097
No phys. props. reported.

Dowd, P. *et al, Synth. Commun.,* 1978, **8**, 205 (*synth*)

3,5-Hexadiyn-2-one, 9CI   H-60041
Acetylbutadiyne
[31097-80-0]

$$H_3CCOC\!\equiv\!CC\!\equiv\!CH$$

C_6H_4O M 92.097
Mp 3-6°. Darkens in air at r.t.

Mestres, R., *J. Chem. Soc., Perkin Trans. 1,* 1972, 805 (*synth*)

Hexaethylbenzene, 9CI, 8CI   H-60042
Updated Entry replacing H-00610
[604-88-6]

$C_{18}H_{30}$ M 246.435
Cryst. (EtOH), needles (heptane). Mp 130°. Bp 298°.
▷DA3000000.

Wertyporoch, E. *et al, Justus Liebigs Ann. Chem.,* 1933, **500**,
   287 (*synth*)
Smith, L. *et al, J. Am. Chem. Soc.,* 1940, **62**, 2631 (*synth*)
Richards, R. *et al, Proc. R. Soc. London, Ser. A,* 1948, **195**, 1
   (*ir*)
Jhingan, A.K. *et al, J. Org. Chem.,* 1987, **52**, 1161 (*synth, pmr,
   cmr, ms*)

1,1,1,4,4,4-Hexafluoro-2,3-butanedione,   H-60043
9CI
Perfluoro-2,3-butanedione. Perfluorobiacetyl.
Hexafluorobiacetyl
[685-24-5]

$$F_3CCOCOCF_3$$

$C_4F_6O_2$ M 194.033
Deep-yellow liq. Mp −20°. Bp 20°. Reacts
   exothermically with H_2O to give a hydrate.

Moore, L.O. *et al, J. Org. Chem.,* 1965, **30**, 2472 (*synth, uv, ir,
   props*)
Earl, B.L. *et al, Inorg. Chem.,* 1966, **5**, 2184 (*synth*)
Moore, L.O., *J. Org. Chem.,* 1970, **35**, 3999 (*synth*)
Krespan, C.G. *et al, J. Am. Chem. Soc.,* 1977, **99**, 1214 (*synth*)

1,2,3,4,5,5-Hexafluoro-1,3-cyclopenta-   H-60044
diene, 9CI
Perfluorocyclopentadiene
[699-39-8]

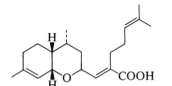

C_5F_6 M 174.045
Liq. Bp 27-32°.

Soelch, R. *et al, J. Org. Chem.,* 1985, **50**, 5845 (*synth, F nmr,
   ms, bibl*)

2,3,4,4a,5,6-Hexahydro-7H-1-benzo-   H-60045
pyran-7-one, 9CI
4a,5,6,7-Tetrahydrochroman-7-one
[17422-92-3]

$C_9H_{12}O_2$ M 152.193
Mp 57-58°. $Bp_{1.5}$ 120-123°.

Blickenstaft, R.T. *et al, Tetrahedron,* 1968, **24**, 2495 (*synth*)
Forchiassin, M. *et al, J. Heterocycl. Chem.,* 1983, **20**, 493
   (*synth, pmr*)

2-[(3,4,4a,5,6,8a-Hexahydro-2H-1-benzo-   H-60046
pyran-2-yl)ethylidene]-6-methyl-5-hep-
tenoic acid

Absolute
configuration

$C_{20}H_{30}O_3$ M 318.455
Compd. not named in the lit. Constit. of *Eremophila fo-
liosissimum.*

Me ester: Oil. $Bp_{0.3}$ 220° (bath). $[\alpha]_D$ −94.1° (c, 0.5 in
   $CHCl_3$).

Forster, P.G. *et al, Tetrahedron,* 1987, **43**, 2999.

5,6,7,8,9,10-Hexahydro-4H-cyclonona[c]-   H-60047
furan, 9CI
3,4-Heptamethylenefuran. Cyclonona[c]furan
[102652-07-3]

$C_{11}H_{16}O$ M 164.247
Bp_{15} 124-127°. Unstable.

Hunger, J. *et al, Chem. Ber.,* 1986, **119**, 2698 (*synth, pmr, cmr*)

Hexahydrocyclopenta[*cd*]pentalene-1,3,5(2*H*)-trione, 9CI H-60048

Tricyclo[5.2.1.0^{4,10}]decane-2,5,8-trione

[78168-48-6]

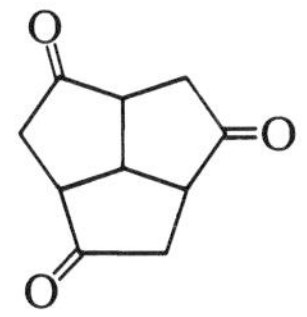

$C_{10}H_{10}O_3$ M 178.187

Cryst. (Me$_2$CO). Mp 170-171°.

Osborn, M.E. *et al*, *J. Org. Chem.*, 1981, **46**, 3379 (*synth, pmr, ir*)

Carceller, E. *et al*, *Tetrahedron*, 1986, **42**, 1831 (*synth, ir, pmr, cmr, ms*)

Hexahydro-2*H*-cyclopenta[*b*]thiophene, 9CI H-60049

2-Thiabicyclo[3.3.0]octane

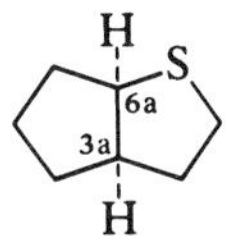

$C_7H_{12}S$ M 128.232

(**3aRS,6aRS**)-*form* [104808-82-4]

(±)-cis-*form*

No phys. props. reported.

Ceré, V. *et al*, *J. Org. Chem.*, 1986, **51**, 4880 (*synth, pmr, cmr*)

2-[(2,3,3*a*,4,5,7*a*-Hexahydro-3,6-dimethyl-2-benzofuranyl)ethylidene]-6-methyl-5-heptenoic acid H-60050

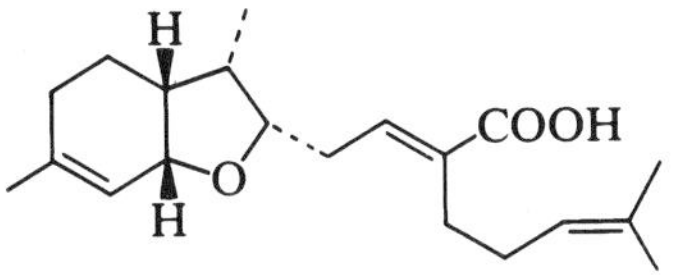

$C_{20}H_{30}O_3$ M 318.455

Compd. not named in the reference. Constit. of *Eremophila foliosissima*.

Me ester: Oil. Bp$_{0.25}$ 175° (bath). [α]$_D$ −49.5° (c, 0.6 in CHCl$_3$).

Forster, P.G. *et al*, *Tetrahedron*, 1987, **43**, 2999.

Hexahydro-1,4-dithiino[2,3-*b*]-1,4-dithiin, 9CI H-60051

2,5,7,10-Tetrathiabicyclo[4.4.0]decane. 1,4,5,8-Tetrathiadecalin

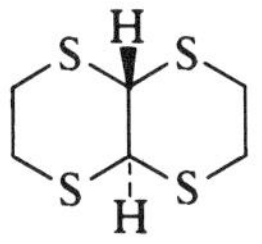

$C_6H_{10}S_4$ M 210.385

trans-form [111727-91-4]

Mp 134.2°.

Pericas, M.A. *et al*, *Tetrahedron*, 1986, **42**, 2717 (*synth, ir, pmr, cmr, ms, conformn*)

Hexahydro-3-(1-methylpropyl)pyrrolo[1,2-*a*]pyrazine-1,4-dione, 9CI H-60052

*3-sec-Butylhexahydropyrrolo[1,2-*a]*pyrazine-1,4-dione, 8CI. Cyclo(prolylisoleucyl)*

[58917-56-9]

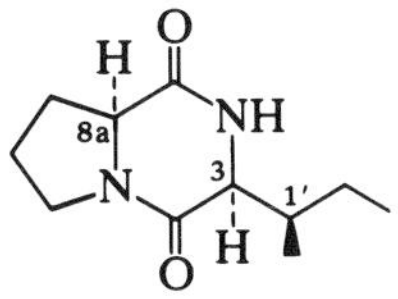

(3*R*,8a*S*,1′*R*)-*form*

$C_{11}H_{18}N_2O_2$ M 210.275

(**3R,8aS,1′R**)-*form* [61117-55-3]

L-D-allo-*form*

Cryst. (hexane/Et$_2$O/Me$_2$CO). Mp 107-108°.

(**3R,8aS,1′S**)-*form* [61117-54-2]

L-D-*form*

Cryst. (Me$_2$CO). Mp 143-145°.

Bycroft, B.W. *et al*, *J. Chem. Soc., Chem. Commun.*, 1975, 988 (*synth*)

Young, P.E. *et al*, *J. Am. Chem. Soc.*, 1976, **98**, 5358, 5365 (*synth, conformn*)

Suzuki, K. *et al*, *Chem. Pharm. Bull.*, 1981, **29**, 233 (*synth*)

1,2,3,5,8,8*a*-Hexahydronaphthalene, 9CI H-60053

[62690-65-7]

$C_{10}H_{14}$ M 134.221

Christl, M. *et al*, *Angew. Chem., Int. Ed. Engl.*, 1987, **26**, 449 (*synth, pmr, cmr*)

1,2,3,7,8,8*a*-Hexahydronaphthalene, 9CI H-60054

[62690-66-8]

$C_{10}H_{14}$ M 134.221

Christl, M. *et al*, *Angew. Chem., Int. Ed. Engl.*, 1987, **26**, 449 (*synth, pmr, cmr*)

3,4,4*a*,5,6,7-Hexahydro-2*H*-pyrano[2,3-*b*]pyrilium, 9CI H-60055

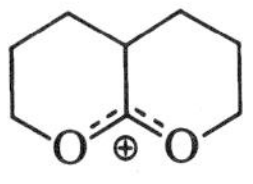

$C_8H_{13}O_2^{\oplus}$ M 141.190 (ion)

Tetrafluoroborate: [75967-78-1].

$C_8H_{13}BF_4O_2$ M 227.993

Solid.

Khouri, F.F. *et al*, *J. Am. Chem. Soc.*, 1986, **108**, 6683 (*synth, pmr*)

Hexahydropyrimidine H-60056

Updated Entry replacing H-40058

1,3-Diazacyclohexane. Piperimidine

[505-19-1]

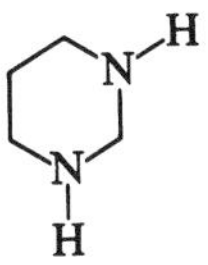

$C_4H_{10}N_2$ M 86.136

Earlier refs. claimed a tautomeric open-chain struct.

Fuming liq. Bp$_{20}$ 58-60°. Monomeric in C_6H_6 soln. at r.t.

N,N′-*Dinitroso:* [15973-99-6].
C$_4$H$_8$N$_4$O$_2$ M 144.133
Mp 61-63°.

▷Exp. carcinogen
N,N′-*Dinitro:* [5754-89-2].
C$_4$H$_8$N$_4$O$_4$ M 176.132
Mp 84-86°.

▷Potentially explosive
1,3-Di-Me: [10556-96-4].
C$_6$H$_{14}$N$_2$ M 114.190
Bp 125-126° (132-136°).
1,3-Di-Et: [18468-57-0].
C$_8$H$_{18}$N$_2$ M 142.244
Bp$_{25}$ 84°, Bp$_{17}$ 60-4°.
1,3-Diisopropyl: [28286-13-7].
C$_{10}$H$_{22}$N$_2$ M 170.297
Oil. Bp$_{0.04}$ 44°.
1,3-Di-tert-butyl: [16077-41-1].
C$_{12}$H$_{26}$N$_2$ M 198.351
Oil. Bp$_{20}$ 118°, Bp$_{0.2}$ 53°.

Krässig, H. *et al, Makromol. Chem.,* 1955, **17**, 77 (*deriv synth*)
Bell, J.A. *et al, J. Chem. Soc. (C),* 1966, 870.
Evans, R.F., *Aust. J. Chem.,* 1967, **20**, 1643 (*synth, ir, pmr, props*)
Riddell, F.G., *J. Chem. Soc. (B),* 1967, 560 (*deriv, synth, pmr, conformn*)
Böhme, H. *et al, Justus Liebigs Ann. Chem.,* 1969, **723**, 41 (*deriv, synth*)
Jones, R.A.Y. *et al, J. Chem. Soc. (B),* 1970, 131 (*deriv, synth, ir, pmr, conformn*)
Willer, R.L. *et al, J. Org. Chem.,* 1984, **49**, 5147 (*synth*)

Hexahydro-1*H*-pyrrolizin-1-one, 9CI   H-60057

1-Oxopyrrolizidine. 1-Ketopyrrolizidine. 1-Pyrrolizidin-one. 1-Azabicyclo[3.3.0]octan-4-one
[14174-83-5]

(*S*)-*form*

C$_7$H$_{11}$NO   M 125.170
(*S*)-**form** [18881-07-7]
[α]$_D^0$ −59.2° (c, 2.3 in dioxan). Est. opt. purity >75%, optically unstable.
(±)-**form**
Mobile liq. d$_4^{20}$ 1.08. Bp$_3$ 55-56°. Rapidly darkens in air.
Picrate: [14671-09-1]. Yellow needles (EtOH). Mp 176.5-179° (162-164°).
Oxime: [15912-22-8].
C$_7$H$_{12}$N$_2$O   M 140.185
Prisms. Mp 158-160°.
Oxime; B,HCl: [15912-21-7]. Prisms (EtOH). Mp 172-173°.

Adams, R. *et al, J. Am. Chem. Soc.,* 1960, **82**, 1466 (*synth*)
Kochetkov, N.K. *et al, J. Gen. Chem. USSR (Engl. Transl.),* 1961, **31**, 3225 (*synth*)
Aaron, H.S. *et al, J. Org. Chem.,* 1966, **31**, 3502 (*ir, deriv*)
Kunieda, T. *et al, Chem. Pharm. Bull.,* 1967, **15**, 337, 490 (*synth, deriv, abs config*)

Hexahydro-3*H*-pyrrolizin-3-one, 9CI   H-60058

3-Pyrrolizidinone. 1-Azabicyclo[3.3.0]octan-2-one. 3-Oxopyrrolizidine. 3-Ketopyrrolizidine
[32548-24-6]

C$_7$H$_{11}$NO   M 125.170
(±)-*form*
Liq. Bp$_{12}$ 107-108°.

Galinovsky, F. *et al, Ber.,* 1944, **77**, 138 (*synth*)
Edwards, O.E. *et al, Can. J. Chem.,* 1971, **49**, 1648 (*synth*)

3,4,6,7,8,9-Hexahydro-2*H*-quinolizin-2-one, 9CI   H-60059

1,10-Didehydro-2-quinolizidone
[98959-58-1]

C$_9$H$_{13}$NO   M 151.208
Cryst. (Et$_2$O). Mp 82-83°.

Bohlmann, F. *et al, Chem. Ber.,* 1961, **94**, 1767 (*synth, ir, uv*)
Brandi, A. *et al, J. Chem. Soc., Chem. Commun.,* 1986, 813 (*synth*)

Hexahydro-1,3,5-triazine   H-60060

1,3,5-Triazacyclohexane. Trimethylenetriamine
[110-90-7]

C$_3$H$_9$N$_3$ M 87.124
Intermed. present in soln. in reaction of HCHO and NH$_3$, not isol.
1,3,5-Tri-Ac: [26028-46-6].
C$_9$H$_{15}$N$_3$O$_8$ M 293.233
Hygroscopic cryst. Mp 96-98° (hydrate), Mp 71.5-73.5°.
1,3,5-Tri-Me: [108-74-7]. *Hexahydro-1,3,5-trimethyl-1,3,5-triazine, 9CI.*
C$_6$H$_{15}$N$_3$ M 129.205
Bp 166°, Bp$_{1.5}$ 50-52°.
▷XY9300000.
1,3,5-Tri-Et: [7779-27-3].
C$_9$H$_{21}$N$_3$ M 171.285
Bp 190-196°.
▷XY9275000.
1,3,5-Tri-Et; B,HCl: Long needles.
1,3,5-Triisopropyl: [10556-98-6].
C$_{12}$H$_{27}$N$_3$ M 213.365
Bp$_{16}$ 104-106°, Bp$_1$ 81°.
1,3,5-Tri-tert-butyl
: [10560-39-1].
C$_{15}$H$_{33}$N$_3$ M 255.446
Bp$_{20}$ 125-130°, Bp$_1$ 145°.
1,3,5-Trinitroso: [13980-04-6].
C$_3$H$_6$N$_6$O$_3$ M 174.119
Cryst. Mp 105°.
▷Carcinogen. XY9470000.
1,3,5-Trihydroxy: see Hexahydro-1,3,5-trihydroxy-1,3,5-triazine, H-00709
1,3,5-Trinitro: see Hexahydro-1,3,5-trinitro-1,3,5-triazine, H-00710

Graymore, J., *J. Chem. Soc.*, 1924, 2284 (*deriv, synth*)
Gradsten, M.A. *et al*, *J. Am. Chem. Soc.*, 1948, **70**, 3079 (*deriv, synth*)
Richmond, H.H. *et al*, *J. Am. Chem. Soc.*, 1948, **70**, 3659 (*synth*)
Lehn, J.M. *et al*, *J. Chem. Soc. (B)*, 1967, 387 (*deriv, synth, conformn, pmr*)
Jones, R.A.Y. *et al*, *J. Chem. Soc. (B)*, 1970, 135 (*deriv, synth, conformn*)

2,3,6,7,10,11-Hexahydrotrisimidazo[1,2-a;1′,2′-c;1″,2″-e][1,3,5]-triazine   H-60061

[28584-89-6]

$C_9H_{12}N_6$ M 204.234
Cryst. (H_2O). Mp 320-325°.

Schaefer, F.C., *J. Am. Chem. Soc.*, 1955, **77**, 5922 (*synth*)
Hansen, G.R. *et al*, *J. Heterocycl. Chem.*, 1970, **7**, 997 (*synth, ir, pmr*)

3,3′,4,4′,5,5′-Hexahydroxybibenzyl   H-60062

5,5′-(1,2-Ethanediyl)bis[1,2,3-benzenetriol]. 1,2-Bis(3,4,5-trihydroxyphenyl)ethane

$C_{14}H_{14}O_6$ M 278.261
3,4;3′,4′-Dimethylene ether: [80357-98-8]. *6,6′-(1,2-Ethanediyl)bis-1,3-benzodioxol-4-ol, 9CI. 3,3′-Dihydroxy-4,5;4′,5′-dimethylenedioxybibenzyl.*
$C_{16}H_{14}O_6$ M 302.283
Constit. of *Frullania* spp.

Asakawa, Y. *et al*, *Phytochemistry*, 1987, **26**, 1117.

2,3,14,20,22,25-Hexahydroxy-7-cholesten-6-one, 9CI   H-60063

Updated Entry replacing H-30053

[5289-74-7]

$C_{27}H_{44}O_7$ M 480.640
▷FZ8060000.

(2β,3β,5β,20R,22R)-form

Crustecdysone. *β-Ecdysone. Ecdysterone. 20-Hydroxyecdysone. Polypodine A. Viticosterone. Commisterone*
Isol. from sea water crayfish, saturniid oak-silk moth pupae, the silkworm, the wood of *Podocarpus elatus* and in *Polypodium vulgare*. Crustacean moulting hormone. Plates (EtOAc/THF). Mp 237.5-239.5° (243°). [α]$_D$ +61.8° ($CHCl_3$).

25-Ac: [22033-96-1]. **Viticosterone E**.
$C_{29}H_{46}O_8$ M 522.678
From *Vitex megapotamica*. Mp 198-199°.
2-Cinnamoyl: [38147-15-8].
$C_{36}H_{50}O_8$ M 610.786
From the bark of *Dacrydium intermedium*. Mp 254-256°.
25-Me ether: [52677-91-5]. **Polypodaurein**.
$C_{28}H_{46}O_7$ M 494.667
Constit. of *Polypodium aureum*. Cryst. (MeOH). Mp 251-253°.
2-Cinnamoyl: [38147-15-8]. *β-Ecdysone 2-cinnamate*.
$C_{36}H_{50}O_8$ M 610.786
3-(4-Hydroxycinnamoyl): [38147-16-9]. *β-Ecdysone 3-p-coumarate*.
$C_{36}H_{50}O_9$ M 626.786
From *D. intermedium*. Cryst. Mp 265-267°.
3-O-α-D-Galactopyranoside: **Sileneoside**.
$C_{33}H_{54}O_{12}$ M 642.782
Constit. of *Silene brahuica*. Cryst. (MeOH/Me_2CO). Mp 240-242°. [α]$_D^{20}$ +91.2° (c, 1.01 in MeOH).
22-Benzoyl: *Ecdysterone 22-O-benzoate*.
$C_{34}H_{48}O_8$ M 584.748
Constit. of *S. scabrifolia*. Cryst. (MeOH aq.). Mp 202-205°. [α]$_D^{20}$ +45° (c, 1.1 in EtOH).

(2α,3β,5β,20R,22R)-form [84580-28-9]

Paristerone
Constit. of tubers of *Paris polyphylla*. Cryst. Mp 216-220°. [α]$_D$ +41.9°.

(2β,3α,5β,20R,22R)-form [54053-93-9]

3-epi-20-Hydroxyecdysone
$C_{27}H_{44}O_7$ M 480.640
Constit. of the tobacco hornworm *Manduca sexta*.

Galbraith, M.N. *et al*, *J. Chem. Soc., Chem. Commun.*, 1966, 905 (*isol*)
Hüppi, G. *et al*, *J. Am. Chem. Soc.*, 1967, **89**, 6790 (*synth*)
Horn, D.H.S. *et al*, *Biochem. J.*, 1968, **109**, 399 (*isol*)
Kerb, U. *et al*, *Tetrahedron Lett.*, 1968, 4277 (*synth*)
Dammeier, B. *et al*, *Chem. Ber.*, 1971, **104**, 1660 (*struct, abs config*)
Russell, G.B. *et al*, *Aust. J. Chem.*, 1972, **25**, 1935 (*deriv*)
Jizba, J. *et al*, *Phytochemistry*, 1974, **13**, 1915 (*Polypodaurein*)
Zatsny, I. *et al*, *Khim. Prir. Soedin.*, 1975, **11**, 155 (*ms*)
Kaplanis, J.N. *et al*, *Steroids*, 1979, **34**, 333 (*3-epi-20-Hydroxyecdysone*)
Kametani, T. *et al*, *Tetrahedron Lett.*, 1980, 4855 (*synth*)
Singh, S.B. *et al*, *Tetrahedron*, 1982, **38**, 2189 (*Paristerone*)
Nishimoto, N. *et al*, *Phytochemistry*, 1983, **26**, 2505 (*cmr*)
Saatov, Z. *et al*, *Khim. Prir. Soedin.*, 1984, **20**, 741 (*isol*)
Kubo, I. *et al*, *Agric. Biol. Chem.*, 1985, **49**, 243 (*pmr*)
Saatov, Z. *et al*, *Khim. Prir. Soedin.*, 1986, **22**, 71 (*Ecdysterone 22-O-benzoate*)

2′,3,5,5′,6,7-Hexahydroxyflavone   H-60064

$C_{15}H_{10}O_8$ M 318.239
3,5,7-Tri-Me ether: 2′,5′,6-Trihydroxy-3,5,7-trimethoxyflavone.
$C_{18}H_{16}O_6$ M 328.321
Constit. of *Blumea malcomii*. Cryst. (Me_2CO). Mp 252°.
2′,3,5,7-Tetra-Me ether: 5′,6-Dihydroxy-2′,3,5,7-trimethoxyflavone.
$C_{19}H_{18}O_8$ M 374.346
Constit. of *B. malcomii*. Cryst. Mp 189-191°.
2′,3,5,5′,7-Penta-Me ether: 6-Hydroxy-2′,3,5,5′,7-pentamethoxyflavone.
$C_{20}H_{20}O_8$ M 388.373
Constit. of *B. malcomii*. Cryst. Mp 166-168°.

Kulkarni, M.M. *et al*, *Phytochemistry*, 1987, **26**, 2079.

Hexahydroxy-1,4-naphthoquinone H-60065

Updated Entry replacing H-00728
2,3,5,6,7,8-Hexahydroxy-1,4-naphthalenedione, 9CI.
Spinochrome E
[476-37-9]

$C_{10}H_6O_8$ M 254.153
Isol. from sea urchins *Paracentrantus lividus*,
Psammechinus milioris and *Strongylocentrotus
droebachiensis*. Brownish-red needles (MeOH).
Sublimes at 300-20° without melting.
Hexa-Ac:
  $C_{22}H_{18}O_{14}$ M 506.376
  Yellow cryst. Mp 192°.
2-Me ether: [15308-24-4]. *2,3,5,6,8-Pentahydroxy-7-
  methoxy-1,4-naphthoquinone. 2,5,6,7,8-Pentahy-
  droxy-3-methoxy-1,4-naphthoquinone.*
  Namakochrome.
  $C_{11}H_8O_8$ M 268.179
  Isol. from *Polycheira rufescens*. Cryst. (AcOH). Mp
  218°. Tautomeric to give the 7-Me ether struct.
2,6-Di-Me ether: [15308-22-2]. *2,5,6,8-Tetrahydroxy-
  3,7-dimethoxy-1,4-naphthoquinone, 8CI.*
  $C_{12}H_{10}O_8$ M 282.206
  Isol. from spines of *Acanthaster planci*.
2,7-Di-Me ether: [14090-99-4]. *2,5,7,8-Tetrahydroxy-
  3,6-dimethoxy-1,4-naphthoquinone, 8CI.*
  $C_{12}H_{10}O_8$ M 282.206
  Isol. from spines of *A. planci*.
Tetra-Me ether:
  $C_{14}H_{14}O_8$ M 310.260
  Brown cryst. Mp 185°.

Lederer, E., *Biochim. Biophys. Acta*, 1952, **9**, 92 (*isol*)
Mukai, T. *et al*, *Bull. Chem. Soc. Jpn.*, 1960, **33**, 453, 1234
  (*deriv*)
Smith, J. *et al*, *J. Chem. Soc.*, 1961, 1008 (*isol*)
Singh, I. *et al*, *J. Am. Chem. Soc.*, 1965, **87**, 4023 (*synth*)
Anderson, H.A. *et al*, *J. Chem. Soc. (C)*, 1966, 426 (*synth*)
Singh, H. *et al*, *Experientia*, 1967, **23**, 624; *Tetrahedron*, 1967,
  23, 3271 (*derivs*)
Kol'tsova, E.A., *Khim. Pir. Soedin.*, 1977, 202 (*isol*)

1,2,3,5,6,7-Hexahydroxyphenanthrene H-60066

1,2,3,5,6,7-Phenanthrenehexol

$C_{14}H_{10}O_6$ M 274.229
1,2,5,6-Tetra-Me ether: [108925-82-1]. *3,4,7,8-Tetra-
  methoxy-2,6-phenanthrenediol, 9CI. 3,7-Dihydroxy-
  1,2,5,6-tetramethoxyphenanthrene.* **Confusaridin**.
  $C_{18}H_{18}O_6$ M 330.337
  Constit. of *Eria confusa*. Cryst. (EtOAc/pet. ether).
  Mp 192°.

Majumder, P.L. *et al*, *Phytochemistry*, 1987, **26**, 1127.

Hexakis(dichloromethyl)benzene H-60067
[33624-92-9]

$C_{12}H_6Cl_{12}$ M 575.615
Cryst. (tetrachloroethane). Mp >300°.

Kahr, B. *et al*, *J. Org. Chem.*, 1987, **52**, 3713 (*synth, pmr, cryst
  struct*)

2,2,3,3,4,4-Hexamethylcyclobutanol, 9CI H-60068
[103547-83-7]

$C_{10}H_{20}O$ M 156.267
Solid. Mp 58°.

Fitjer, L. *et al*, *Chem. Ber.*, 1986, **119**, 1162 (*synth, ir, pmr,
  cmr*)

Hexamethylcyclobutanone, 9CI H-60069
[1703-85-1]

$C_{10}H_{18}O$ M 154.252
Bp_{10} 80° (bath).
2,4-Dinitrophenylhydrazone: Mp 192-193°.

Cookson, R.C. *et al*, *J. Chem. Soc., Chem. Commun.*, 1965, 98.
Braun, M. *et al*, *Chem. Ber.*, 1975, **108**, 2368 (*synth, ir, pmr*)
Fitjer, L. *et al*, *Chem. Ber.*, 1986, **119**, 1162 (*synth, cmr*)

1,1,2,2,3,3-Hexamethylcyclohexane, 9CI H-60070
[103495-83-6]

$C_{12}H_{24}$ M 168.322
Solid. Mp 77°.

Fitjer, L. *et al*, *Chem. Ber.*, 1986, **119**, 1144 (*synth, pmr, cmr*)

2,2,3,3,4,4-Hexamethylpentane, 9CI H-60071
[60302-27-4]

$$(H_3C)_3CC(CH_3)_2C(CH_3)_3$$

$C_{11}H_{24}$ M 156.311
Glassy solid. Mp 64-66°.

Hellmann, S. *et al*, *Chem. Ber.*, 1983, **116**, 2219 (*pmr*)
Fitjer, L. *et al*, *Chem. Ber.*, 1986, **119**, 1144 (*synth, cmr*)

1,1,3,3,5,5-Hexamethyl-2,4,6-tris(methylene)cyclohexane, 9CI   H-60072

[103495-77-8]

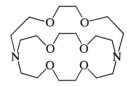

$C_{15}H_{24}$ M 204.355
Oil, cryst. on cooling. Mp 32-35°. Bp$_{0.01}$ 60-65°.

Fitjer, L. *et al*, *Chem. Ber.*, 1986, **119**, 1144 (*synth, pmr, cmr*)

4,7,13,16,21,24-Hexaoxa-1,10-diazabicyclo[8.8.8]hexacosane, 9CI   H-60073

Cryptand 222. Kriptofix 222

[23978-09-8]

$C_{18}H_{36}N_2O_6$ M 376.492
Metal chelating agent. Cryst. (hexane). Mp 68-69°.

▷MP4750000.

Dietrich, B. *et al*, *Tetrahedron*, 1973, **29**, 1629 (*synth, pmr*)
Metz, B. *et al*, *J. Chem. Soc., Perkin Trans. 2*, 1976, 423 (*cryst struct*)
Foerster, H.G. *et al*, *J. Am. Chem. Soc.*, 1980, **102**, 6984 (^{15}N nmr)
Kulstad, S. *et al*, *Tetrahedron Lett.*, 1980, **21**, 643 (*synth*)
Anelli, P.L. *et al*, *J. Org. Chem.*, 1985, **50**, 3453 (*synth*)

Hexaphenylbenzene   H-60074

Updated Entry replacing H-00826
3′,4′,5′,6′-Tetraphenyl-1,1′:2′,1″-terphenyl, 9CI.
2′,4′,5′,6′-Tetraphenyl-m-terphenyl, 8CI

[992-04-1]

$$\underset{Ph}{\overset{Ph}{\underset{Ph}{\overset{Ph}{\bigcirc}}}}$$

Ph Ph
Ph Ph
Ph

$C_{42}H_{30}$ M 534.699
Plates (diphenyl ether), cryst. (C$_6$H$_6$/CHCl$_3$). Mp 454-456° (sealed capillary), Mp 478°.

Beynon, J.H. *et al*, *J. Chem. Soc., Chem. Commun.*, 1965, 7052 (*ms*)
Org. Synth., 1966, **46**, 44.
Storek, W. *et al*, *Z. Naturforsch., A*, 1979, **34**, 1334 (*cmr*)
Jhingan, A.K. *et al*, *J. Org. Chem.*, 1987, **52**, 1161 (*synth, pmr*)

Consult the *Dictionary of Antibiotics and Related Substances* for a fuller treatment of antibiotics and related compounds.

1,4,7,10,13,16-Hexathiacyclooctadecane, 9CI   H-60075

S_6-*Ethano-18*

[296-41-3]

$C_{12}H_{24}S_6$ M 360.682
Complexing agent. Cryst. (CH$_2$Cl$_2$/Et$_2$O). Mp 91-93°.

Ochrymowycz, L.A. *et al*, *J. Org. Chem.*, 1974, **39**, 2079 (*synth*)
Blake, A.J. *et al*, *Angew. Chem., Int. Ed. Engl.*, 1986, **25**, 274 (*synth, ir, use*)

2-Hexenal, 9CI   H-60076

Updated Entry replacing H-00853
3-Propylacrolein

[505-57-7]

$$H_3CCH_2CH_2CH{=}CHCHO$$

$C_6H_{10}O$ M 98.144
Constit. of many foods, hornbeam leaves, and of scent gland of many bugs, e.g., *Pternistria bispina*; alarm pheromone of, e.g., *Cimex lectularius*.

▷MP5880000.

(*E*)-*form* [6728-26-3]
Leaf aldehyde
Used in perfumery and flavourings. Bp 146-147°, Bp$_{12}$ 43°.

▷MP5900000.

2,4-Dinitrophenylhydrazone: [2122-02-3]. Red needles. Mp 147°.

Semicarbazone: Cryst. (MeOH). Mp 179°.

Curtius, T. *et al*, *Justus Liebigs Ann. Chem.*, 1913, **404**, 93 (*isol*)
Elkik, E., *Bull. Soc. Chim. Fr.*, 1968, 283 (*synth*)
Hatanaka, A. *et al*, *Phytochemistry*, 1976, **15**, 1125 (*biosynth*)
Ito, Y. *et al*, *J. Org. Chem.*, 1978, **43**, 1011 (*synth*)
Zimmermann, B. *et al*, *Chem. Ber.*, 1986, **119**, 2848 (*synth, ir, pmr*)

4-Hexen-3-ol, 9CI   H-60077

Updated Entry replacing H-00874

[4798-58-7]

$$H_3CCH_2CH(OH)CH{=}CHCH_3$$

$C_6H_{12}O$ M 100.160
(+)-(*E*)-*form*
$[\alpha]_D^{15}$ +13.5°.
Hydrogen phthalate: Mp 70.5°. $[\alpha]_{546}^{15}$+15° (CHCl$_3$).
Brucine salt: Mp 168°.

(−)-(*E*)-*form*
Hydrogen phthalate: Mp 70.5°. $[\alpha]_{546}^{15}$−15° (CHCl$_3$).
Brucine salt: Mp 125-126°.

(±)-(*E*)-*form* [29478-27-1]
Liq. with strong odour. Bp 135-135.5°, Bp$_{13}$ 44-45°.
Ac: [19393-86-3].
$C_8H_{14}O_2$ M 142.197
Bp 154-157°.

Me ether: 4-Methoxy-2-hexene.
$C_7H_{14}O$ M 114.187
Bp 110-113°.

(±)-(Z)-form [29478-30-6]
Bp_{14} 44-46°.
1-Naphthylurethane: Mp 74.5°.

v. Auwers, K. *et al*, *Ber.*, 1921, **54**, 2993 (*props*)
Airs, R.S. *et al*, *J. Chem. Soc.*, 1942, 18 (*synth*)
Kieffer, R., *Bull. Soc. Chim. Fr.*, 1967, 3026 (*synth*)
Boughdady, N.M. *et al*, *Aust. J. Chem.*, 1987, **40**, 767 (*synth, pmr, ir*)

5-Hexyldihydro-2(3H)-furanone, 9CI   H-60078

Updated Entry replacing H-00922
4-Hydroxydecanoic acid lactone, 8CI. γ-*Hexylbutyrolactone.* γ-*Decalactone.* 4-Decanolide
[706-14-9]

$H_3C(CH_2)_5$

$C_{10}H_{18}O_2$ M 170.251
Formed by *Sporobolomyces odorus*, also found in oil from *Pityosporum* spp. Flavorant and perfume ingredient. Component of apricot, plum and strawberry aromas. Oil with peach-like odour. Bp 281°, $Bp_{7.7}$ 140-145°.

▷LU4600000.

Honkanen, E. *et al*, *Acta Chem. Scand.*, 1965, **19**, 370 (*ms*)
Heiba, E.I. *et al*, *J. Am. Chem. Soc.*, 1974, **96**, 7977 (*synth*)
Pyysalo, H. *et al*, *Finn. Chem. Lett.*, 1975, 136; *CA*, **84**, 69088m (*synth, pmr, cmr, ms*)
Opdyke, D.L.J., *Food Cosmet. Toxicol.*, 1976, **14** (*Suppl.*), 741; *CA*, **91**, 198665r (*rev*)
Grieco, P.A. *et al*, *Tetrahedron Lett.*, 1978, 419 (*synth*)
Labows, J.N. *et al*, *Appl. Envirom. Microbiol.*, 1979, **38**, 412; *CA*, **91**, 189438s (*isol*)
Bravo, P. *et al*, *Tetrahedron*, 1987, **43**, 4635 (*synth*)

2-Hexyl-5-methyl-3(2H)furanone, 9CI   H-60079

2-Hexyl-5-methyl-3-oxo-2H-furan. Norcepanone
[33922-66-6]

H_3C $(CH_2)_5CH_3$

$C_{11}H_{18}O_2$ M 182.262
Constit. of onions, leeks and shallots.

(±)-form
Bp_3 79-81° ($Bp_{0.15}$ 78°).
S-Oxide (exo-): [104808-81-3].
$C_7H_{12}OS$ M 144.231
$Bp_{0.4}$ 113°.
S-Oxide (endo-): [104872-75-5]. $Bp_{0.4}$ 98°.

Boelens, M. *et al*, *J. Agric. Food Chem.*, 1971, **19**, 984 (*isol, ms, pmr, ir*)
de Rijke, D. *et al*, *Recl. Trav. Chim. Pays-Bas*, 1973, **92**, 731 (*synth, ir, pmr*)
Galetto, W.G. *et al*, *J. Agric. Food Chem.*, 1976, **24**, 854 (*synth*)
Thomas, A.F. *et al*, *Tetrahedron Lett.*, 1986, **27**, 505 (*synth*)

5-Hexyne-1,4-diamine   H-60080

3,6-Diamino-1-hexyne
[69355-11-9]

$$CH_2CH_2CH_2NH_2$$
$$H_2N\!\blacktriangleright\!\overset{|}{\underset{|}{C}}\!\blacktriangleleft H$$
$$C\!\equiv\!CH$$

(R)-form

$C_6H_{12}N_2$ M 112.174
(R)-form [66640-91-3]
Ornithine decarboxylase inhibitor. *S*-form is biol. inactive.
B,2HCl: Cryst. (EtOH/Et$_2$O). Mp 165°. $[\alpha]_D^{26}$ −20.2° (c, 0.41 in H$_2$O).

Metcalf, B.W. *et al*, *J. Am. Chem. Soc.*, 1978, **100**, 2551 (*biochem*)
Casara, P. *et al*, *J. Chem. Soc., Chem. Commun.*, 1982, 1190 (*synth, biochem*)

Hildecarpidin   H-60081

$C_{21}H_{18}O_7$ M 382.369
Constit. of *Tephrosia hildebrandtii*. $[\alpha]_D$ −237° (c, 1.45 in MeOH).

Lwande, W. *et al*, *Phytochemistry*, 1987, **26**, 2425.

Hinokiflavone   H-60082

Updated Entry replacing H-00972
6-[4-(5,7-Dihydroxy-4-oxo-4H-1-benzopyran-2-yl)-phenoxy]-5,7-dihydroxy-2-(4-hydroxyphenyl)-4H-1-benzopyran-4-one, 9CI
[19202-36-9]

$C_{30}H_{18}O_{10}$ M 538.466
Obt. from leaves of *Chamaecyparis obtusa*. Cryst. (hydrate). Mp 353-355° dec.
Penta-Ac: Mp 239-240° dec.
4‴-Me ether: [22012-97-1]. **Cryptomerin A**.
$C_{31}H_{20}O_{10}$ M 552.493
Obt. from leaves of *Cryptomeria japonica*. Yellow prisms (MeOH/Py). Mp 308-310° dec.
7-Me ether: [20931-36-6]. **Neocryptomerin**.
$C_{31}H_{20}O_{10}$ M 552.493
From the leaves of *Podocarpus macrophylla*.
7″-Me ether: [20931-58-2]. **Isocryptomerin**.
$C_{31}H_{20}O_{10}$ M 552.493
From *C. obtusa* and *C. pisafera*. Yellow prisms (MeOH/Py). Mp 310° dec.
4‴,7″-Di-Me ether: [22012-98-2]. **Cryptomerin B**.
$C_{32}H_{22}O_{10}$ M 566.520

From *C. japonica*. Yellow prisms (Py). Mp 302-303°
dec.
7,7″ Di-Me ether: [20931-35-5]. **Chamaecyparin**.
$C_{32}H_{22}O_{10}$ M 566.520
From *C. pisifera* and *C. obtusa*. Amorph.

Miura, H. *et al*, *Chem. Pharm. Bull.*, 1966, **14**, 1404; 1968, **16**, 1838 (*deriv*)
Nakazawa, K., *Tetrahedron Lett.*, 1967, 5223 (*struct, synth*)
Miura, H. *et al*, *Tetrahedron Lett.*, 1968, 2339 (*derivs*)
Nakazawa, K., *Chem. Pharm. Bull.*, 1968, **16**, 2503 (*synth*)
Markham, K.R. *et al*, *Phytochemistry*, 1987, **26**, 3335 (*cmr*)

Histidine trimethylbetaine   H-60083

α-Carboxy-N,N,N-trimethyl-1H-imidazole-4-ethana-
minium hydroxide inner salt, 9CI. (1-Carboxy-2-imida-
zol-4-ylethyl)trimethylammonium hydroxide inner salt,
8CI. Hercycin. Erzinine. Herzynine
[507-29-9]

$C_9H_{15}N_3O_2$ M 197.236
(S)-form [534-30-5]
L-form
Occurs in rubber latex, prod. by fungi, eg. *Boletus edulis*,
and microorganisms. Intermed. in biosynth. of
ergothioneine from histidine. Mp 237-238° dec. $[\alpha]_D^{22}$
+44.5° (c, 1 in 5M HCl).

Barger, G. *et al*, *J. Chem. Soc.*, 1911, 2336 (*synth*)
Reinhold, V.N. *et al*, *J. Med. Chem.*, 1968, **11**, 258 (*synth*)
Tan, C.H. *et al*, *Phytochemistry*, 1968, **7**, 109 (*isol*)
Espersen, W.G. *et al*, *J. Phys. Chem.*, 1976, **80**, 741 (*cmr*)

Homoalethine   H-60084

N,N′-(Dithiodi-3,1-propanediyl)bis[3-aminopropana-
mide], 9CI. N,N′-[Dithiobis(trimethylene)]bis[3-amino-
propionamide], 8CI. [N-(β-Alanyl)-3-aminopropyl]-
disulfide. N-(β-Alanyl)homocystamine
[4759-45-9]

$[H_2NCH_2CH_2CONHCH_2CH_2CH_2S]_2$

$C_{12}H_{26}N_4O_2S_2$ M 322.483
B,2HCl: Cryst. (EtOH). Mp 159-160° dec.

Felder, E. *et al*, *Helv. Chim. Acta*, 1963, **46**, 752 (*synth*)

Homoazulene-1,5-quinone   H-60085

Bicyclo[5.3.1]undeca-1,4,6,9-tetraene-3,8-dione, 9CI
[102795-79-9]

$C_{11}H_8O_2$ M 172.183

Scott, L.T. *et al*, *Tetrahedron Lett.*, 1986, **27**, 779 (*synth, uv*)

Homoazulene-1,7-quinone   H-60086

Bicyclo[5.3.1]undeca-1,4,6,8-tetraene-3,10-dione, 9CI

[102795-80-2]
$C_{11}H_8O_2$ M 172.183
Scott, L.T. *et al*, *Tetrahedron Lett.*, 1986, **27**, 779 (*synth, uv*)

Homoazulene-4,7-quinone   H-60087

Bicyclo[5.3.1]undeca-3,6,8,10-tetraene-2,5-dione, 9CI
[102795-81-3]
$C_{11}H_8O_2$ M 172.183
Unstable, stored in soln.

Scott, L.T. *et al*, *Tetrahedron Lett.*, 1986, **27**, 779 (*synth, uv*)

Homocyclolongipesin   H-60088

[110024-20-9]

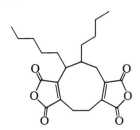

$C_{16}H_{16}O_4$ M 272.300
Ac: [110024-47-0].
$C_{18}H_{18}O_5$ M 314.337
Constit. of *Bothriocline longipes*. Oil.
Propanoyl: [110024-48-1].
$C_{19}H_{20}O_5$ M 328.364
Constit. of *B. longipes*. Oil.

Jakupovic, J. *et al*, *Phytochemistry*, 1987, **26**, 1069.

Homoheveadride   H-60089

$C_{22}H_{28}O_6$ M 388.460
Constit. of *Cladonia polycarpoides*. Gel. $[\alpha]_D^{20}$ +118° (c,
0.5 in CH_2Cl_2).

Archer, A.W. *et al*, *Phytochemistry*, 1987, **26**, 2117.

Homopantetheine   H-60090

2,4-Dihydroxy-N-[3-[(3-mercaptopropyl)amino]-3-ox-
opropyl]-3,3-dimethylbutanamide, 9CI
[4488-18-0]

$C_{12}H_{24}N_2O_4S$ M 292.393
(R)-form [19901-58-7]
D-form
Oil. $[\alpha]_D^{22}$ +26.3° (c, 1 in MeOH).
S-Benzoyl: $[\alpha]_D^{22}$ +18.9° (c, 2 in dioxane).
S-(2-Naphthoyl): Cryst. (EtOAc). Mp 82°. $[\alpha]_D^{23}$
+24.3° (c, 1 in dioxane).

Disulfide: [7061-35-0]. **Homopantethine**.
$C_{24}H_{46}N_4O_8S_2$ M 582.769
Glass. $[\alpha]_D^{22}$ +27.9° (c, 1 in MeOH).

Felder, E. *et al, Helv. Chim. Acta*, 1963, **46**, 752 (*derivs*)
Nagase, O. *et al, Chem. Pharm. Bull.*, 1968, **16**, 977 (*synth*)

2-[(2-Hydrazinoethyl)amino]ethanol, 9CI   H-60091

[88303-65-5]

$$H_2NNHCH_2CH_2NHCH_2CH_2OH$$

$C_4H_{13}N_3O$ M 119.166
Fp ∼5°. Bp$_{0.05}$ 110°. Extremely hygroscopic, dec. on prolonged heating or in soln. at r.t.

Showalter, H.D.H. *et al, J. Heterocycl. Chem.*, 1986, **23**, 1491 (*synth, ir, p mr*)

1-Hydrazinophthalazine   H-60092

Updated Entry replacing H-01133
Hydralazine, BAN, INN
[86-54-4]

NHNH$_2$

$C_8H_8N_4$ M 160.178
Antihypertensive agent. Sensitive reagent for nitrite and formaldehyde. Yellow cryst. (MeOH). Mp 171-173°.
▷TH8925000.

B,HCl: [304-20-1]. *Apresoline. Lopress.* Yellow cryst. Mp 273° dec.
▷TH9000000.

Druey, J. *et al, Helv. Chim. Acta*, 1951, **34**, 195 (*synth*)
Oishi, E., *J. Pharm. Soc. Jpn.*, 1969, **89**, 959; *CA*, **72**, 83624 (*synth*)
Ogiso, T. *et al, Chem. Pharm. Bull.*, 1984, **32**, 3155 (*metab, bibl*)
Noda, H. *et al, Chem. Pharm. Bull.*, 1986, **34**, 3499 (*use, bibl*)

4-Hydrazino-2(1H)-pyridinone, 9CI   H-60093

[106689-41-2]

$C_5H_7N_3O$ M 125.130
Microcryst. + ¼H$_2$O (EtOH). Mp 245° (subl. at 170°).

Nguyen Chi Hung, *et al, Tetrahedron*, 1986, **42**, 2303 (*synth*)

as-Hydrindacene, 8CI   H-60094

1,2,3,6,7,8-Hexahydro-as-indacene, 9CI. Benzo[1,2:3,4]-dicyclopentene
[1076-17-1]

$C_{12}H_{14}$ M 158.243
Needles (MeOH). Mp 40-41°. Bp 252-253°.

Rapoport, H. *et al, J. Am. Chem. Soc.*, 1960, **82**, 1171 (*synth, uv*)
Le Guillanton, G., *Bull. Soc. Chim. Fr.*, 1963, 611 (*synth*)
Wolinsky, J. *et al, J. Org. Chem.*, 1972, **37**, 121 (*synth*)
Buchanan, G.W. *et al, Can. J. Chem.*, 1973, **51**, 2357 (*cmr*)
Isabelle, M.E. *et al, Can. J. Chem.*, 1977, **55**, 3268 (*synth*)
Sepiol, J. *et al, Synthesis*, 1977, 701 (*synth*)
Thummel, R.P. *et al, J. Org. Chem.*, 1977, **42**, 300 (*synth, pmr, ir, uv*)
Allen, F.H., *Acta Crystallogr., Sect. B*, 1981, **37**, 900 (*struct*)

s-Hydrindacene, 8CI   H-60095

1,2,3,5,6,7-Hexahydro-s-indacene, 9CI. Benzo[1,2:4,5]-dicyclopentene
[495-52-3]

$C_{12}H_{14}$ M 158.243
Prisms (EtOH). Mp 53-54°. Bp$_{18}$ 122-125°.

Arnold, R.T. *et al, J. Am. Chem. Soc.*, 1944, **66**, 960 (*synth*)
Buchanan, G.W. *et al, Can. J. Chem.*, 1973, **51**, 2357 (*cmr*)
Thummel, R.P. *et al, J. Org. Chem.*, 1977, **42**, 300 (*synth, uv, pmr*)
Vejdelek, Z.J. *et al, Collect. Czech. Chem. Commun.*, 1977, **42**, 3094 (*synth*)
Allen, F.H., *Acta Crystallogr., Sect. B*, 1981, **37**, 900 (*struct*)

7-Hydro-8-methylpteroylglutamylglutamic acid   H-60096

[108402-49-9]

$C_{25}H_{30}N_8O_9$ M 586.560
Isol. from actinomycete strain SK 2049. Antifolate. Yellowish powder + 5H$_2$O. Mp >196° dec. $[\alpha]_D^{20}$ +24° (c, 0.25 in H$_2$O).

Murata, M. *et al, J. Antibiot.*, 1987, **40**, 251 (*isol, struct, props*)

2-Hydroperoxy-2-methyl-6-methylene-3,7-octadiene   H-60097

β-Myrcene hydroperoxide

$C_{10}H_{16}O_2$ M 168.235
Constit. of *Artemisia annua*. Unstable oil.

Rücker, G. *et al, J. Nat. Prod.*, 1987, **50**, 287.

3-Hydroperoxy-2-methyl-6-methylene-1,7-octadiene   H-60098

α-Myrcene hydroperoxide
$C_{10}H_{16}O_2$ M 168.235
Constit. of *Artemisia annua*. Unstable oil.

Rücker, G. *et al, J. Nat. Prod.*, 1987, **50**, 287.

4-Hydroxyacenaphthene   H-60099

1,2-Dihydro-4-acenaphthylenol, 9CI. 4-Acenaphthenol, 8CI

[6296-98-6]

$C_{12}H_{10}O$ M 170.210
Cryst. by subl. Mp 148-150°.

Ac: [111013-17-3].
   $C_{14}H_{12}O_2$ M 212.248
   Cryst. Mp 86-87°.
Me ether: [111013-13-9]. *4-Methoxyacenaphthene.*
   $C_{13}H_{12}O$ M 184.237
   Mp 89-90°.

Brown, R.F.C. *et al, Aust. J. Chem.*, 1987, **40**, 107 (*synth, ir, pmr, cmr, ms*)

8-Hydroxyachillin   H-60100

Updated Entry replacing H-01173
3,3a,4,5,8a,8b-Hexahydro-4-hydroxy-3,6,9-trimethylazuleno[4,5-b]furan-2,7-dione, 9CI

$C_{15}H_{18}O_4$ M 262.305
8α-form [35879-92-6]
Constit. of *Achillea lanulosa*. Cryst. Mp 161-162°. $[\alpha]_D^{22}$ +110° (c, 1.6 in MeOH).

Ac: [35866-60-5].
   $C_{17}H_{20}O_5$ M 304.342
   Constit. of *A. canulosa*. Cryst. Mp 193-194°. $[\alpha]_D$ +116° (c, 1.6 in CHCl₃).
1β,10β-Epoxide: **1β,10β-Epoxy-8α-hydroxyachillin**.
   $C_{15}H_{18}O_4$ M 262.305
   Constit. of *A. lanata* and *A. frigida*. Cryst. (EtOAc/pet. ether). Mp 273° dec.

White, E.H. *et al, Tetrahedron Lett.*, 1963, 137 (*isol*)
Bachelor, F.W. *et al, Can. J. Chem.*, 1972, **50**, 333 (*struct*)
González, A.G. *et al, An. Quim.*, 1976, **72**, 695 (*isol*)
Yong-Long *et al, J. Nat. Prod.*, 1981, **44**, 722 (*isol*)
Collado, I.G. *et al, J. Chem. Soc., Perkin Trans. 1*, 1987, 1641 (*struct*)

2-Hydroxy-1,3-benzenedicarboxylic acid,   H-60101
9CI

Updated Entry replacing H-01277
2-Hydroxyisophthalic acid, 8CI
[606-19-9]

$C_8H_6O_5$ M 182.132
Needles + 1H₂O (H₂O). Mp 244.5-245° (hydrate), 250-250.5° (anhyd.).
Mono-Me ester: [101670-85-3].
   $C_9H_8O_5$ M 196.159
   Needles. Mp 135°.
Di-Me ester: [36669-06-4].
   $C_{10}H_{10}O_5$ M 210.186
   Mp 72°.
Di-Et ester: [88544-98-3].
   $C_{12}H_{14}O_5$ M 238.240

Mp 112°.
Ac: [90772-05-7]. *2-Acetoxy-1,3-benzenedicarboxylic acid. 3-Carboxyaspirin.*
   $C_{10}H_8O_6$ M 224.170
   Cryst. (Et₂O/hexane). Mp 160-161°.
Me ether: [36727-13-6]. *2-Methoxy-1,3-benzenedicarboxylic acid.*
   $C_9H_8O_5$ M 196.159
   Prisms (H₂O). Mp 216-218°.

Wohl, A., *Ber.*, 1910, **43**, 3474.
Benica, W.S. *et al, J. Am. Pharm. Assoc.*, 1945, **34**, 42.
Moshfegh, A. *et al, Helv. Chim. Acta*, 1957, **40**, 1157.
Suh, J. *et al, J. Am. Chem. Soc.*, 1986, **108**, 3057 (*deriv*)

1-Hydroxy-1,2-benziodoxol-3(1H)-one,   H-60102
9CI, 8CI

Updated Entry replacing H-01290
1,3-Dihydro-1-hydroxy-3-oxo-1,2-benziodoxole
[131-62-4]

$C_7H_5IO_3$ M 264.019
Major tautomeric form of 2-Iodosylbenzoic acid, I-30055
. Enzyme inhibitor. Reagent for determination of xanthates. Catalyst for cleavage of reactive phosphates in cationic surfactant soln. Cryst. (H₂O). Mp 231-232° dec. pK_a 6.22.
▷DE3850000.
Ac:
   $C_9H_7IO_4$ M 306.056
   Cryst. (Ac₂O). Mp 167-169°.
Me ether: 1-Methoxy-1,2-benziodoxol-3(1H)one.
   $C_8H_7IO_3$ M 278.046
   Cryst. (MeOH). Mp 166-168° and ca. 190° (double Mp).

Shefter, E. *et al, Nature (London)*, 1964, **203**, 512 (*cryst struct*)
Baker, G.P. *et al, J. Chem. Soc.*, 1965, 3721 (*synth, struct, ir, bibl*)
Siebert, H. *et al, Z. Anorg. Allg. Chem.*, 1976, **426**, 173 (*ir, raman, struct*)
Verma, K.K. *et al, Fresenius' Z. Anal. Chem.*, 1977, **285**, 263 (*use*)
Moss, R.A. *et al, J. Am. Chem. Soc.*, 1983, **105**, 681 (*props, bibl*)

4-Hydroxy-5-benzofurancarboxylic acid,   H-60103
9CI

Karanjic acid
[487-56-9]

$C_9H_6O_4$ M 178.144
Prisms (EtOH). Mp 219-220° dec.
Ac:
   $C_{11}H_8O_5$ M 220.181
   Prisms (AcOH aq.). Mp 173°.
Me ester: [60077-57-8].
   $C_{10}H_8O_4$ M 192.171
   Mp 105-106°.
Me ether: 4-Methoxy-5-benzofurancarboxylic acid.
   $C_{10}H_8O_4$ M 192.171

Mp 148°.
Me ether, chloride:
   C$_{10}$H$_7$ClO$_3$ M 210.617
   Cryst. (CCl$_4$). Mp 72°.

Foster, R.T. *et al*, *J. Chem. Soc.*, 1948, 115 (*synth*)
Seshandri, T.R. *et al*, *J. Chem. Soc.*, 1954, 1871; 1955, 2048
   (*synth, deriv*)

ent-7α-Hydroxy-15-beyeren-19-oic acid   H-60104

C$_{20}$H$_{30}$O$_3$ M 318.455
Constit. of *Stevia aristata*. Cryst. Mp 240°. [α]$_D^{24}$ +10°
   (c, 0.4 in CHCl$_3$).

Zdero, C. *et al*, *Phytochemistry*, 1987, **26**, 463.

ent-12β-Hydroxy-15-beyeren-19-oic acid   H-60105
C$_{20}$H$_{30}$O$_3$ M 318.455
Constit. of *Stevia aristata*. Oil.

Zdero, C. *et al*, *Phytochemistry*, 1987, **26**, 463.

ent-18-Hydroxy-15-beyeren-19-oic acid   H-60106

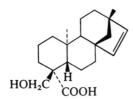

C$_{20}$H$_{30}$O$_3$ M 318.455
Ac: [110187-33-2]. ent-*18-Acetoxy-15-beyeren-19-oic*
   acid. 19-(Acetyloxy)-13-methoxy-17-norkaur-15-en-
   18-oic acid, 9CI.
   C$_{22}$H$_{32}$O$_4$ M 360.492
   Constit. of *Petervenia malvaefolia*.
Ac, Me ester: Oil. [α]$_D^{24}$ −9° (c, 0.77 in CHCl$_3$).

Ellmauerer, E. *et al*, *J. Nat. Prod.*, 1987, **50**, 221.

6-Hydroxy-2,7,10-bisabolatrien-9-one   H-60107
Atlanton-6-ol

C$_{15}$H$_{22}$O$_2$ M 234.338
(*E*)-*form*
   Constit. of *Cedrus libanotica*. Oil. [α]$_D^{25}$ +0.15° (c, 1.2 in
      CHCl$_3$).

Avcibasi, H. *et al*, *Phytochemistry*, 1987, **26**, 2852.

12-Hydroxybromosphaerol   H-60108
Updated Entry replacing H-10108

C$_{20}$H$_{32}$Br$_2$O$_2$ M 464.280
12β-form illus.
12α-form [108907-23-9]
   Constit. of *Sphaerococcus coronopifolius*. Cryst. Mp 89-
      92°. [α]$_D$ −34° (CHCl$_3$).
12β-form [81509-31-1]
   Constit. of *S. coronopifolius*. Cryst. Mp 88-90°. [α]$_D$
      +39° (CHCl$_3$). Formerly descr. as oil.

   Cafieri, F. *et al*, *Phytochemistry*, 1987, **26**, 471.

2-Hydroxycarbazole   H-60109
Updated Entry replacing H-01430
9H-*Carbazol-2-ol, 9CI*
[86-79-3]

C$_{12}$H$_9$NO   M 183.209
Mp 276° (259-262°).
Ac:
   C$_{14}$H$_{11}$NO$_2$ M 225.246
   Mp 188°.
N-*Me:* [51846-67-4]. *9-Methyl-9H-carbazol-2-ol. 2-*
   Hydroxy-9-methylcarbazole.
   C$_{13}$H$_{11}$NO   M 197.236
   Cryst. (Me$_2$CO/pet. ether). Mp 165°.
Et ether: 2-Ethoxycarbazole.
   C$_{14}$H$_{13}$NO   M 211.263
   Mp 217°.

Haglid, F. *et al*, *Acta Chem. Scand.*, 1961, **15**, 1761 (*synth*)
Oikawa, Y. *et al*, *J. Org. Chem.*, 1976, **41**, 1118 (*synth, ir, uv,*
   pmr)
Flo, C. *et al*, *Justus Liebigs Ann. Chem.*, 1987, 509 (*synth, ir,*
   pmr, ms, deriv)

7-Hydroxy-8(17),13-corymbidienolide   H-60110

C$_{22}$H$_{32}$O$_5$ M 376.492
(3β,13E)-*form*
   Constit. of *Salvia villosum*. Gum.

   Zdero, C. *et al*, *Phytochemistry*, 1988, **27**, 227.

6-Hydroxycyclonerolidol   H-60111

C$_{15}$H$_{26}$O$_2$ M 238.369

6α-form

Constit. of *Asteriscus sericeus*. Oil. [α]$_D$ +16.5° (c, 0.27 in CHCl$_3$).

6-Ketone: 6-Oxocyclonerolidol.
C$_{15}$H$_{24}$O$_2$ M 236.353
Constit. of *A. sericeus*. Oil.

Jakupovic, J. *et al*, *Phytochemistry*, 1987, **26**, 2854.

2-Hydroxycyclononanone, 9CI   H-60112

Azeloin. Azelaoin
[496-83-3]

C$_9$H$_{16}$O$_2$ M 156.224

(±)-form

Cryst. (Et$_2$O/pentane). Mp 43°. Bp$_{17}$ 124-139°, Bp$_{0.2}$ 60-70°.

Oxime:
C$_{19}$H$_{17}$NO$_2$ M 291.349
Mp 117-118°.

Prelog, V. *et al*, *Helv. Chim. Acta*, 1947, **30**, 1741 (*synth*)
Kashima, C. *et al*, *Bull. Chem. Soc. Jpn.*, 1965, **38**, 255 (*synth*)
Org. Synth., 1977, **57**, 1 (*synth*)
Friedrich, E. *et al*, *Chem. Ber.*, 1980, **113**, 1245 (*synth, ir, pmr*)

3-Hydroxycyclononanone, 9CI   H-60113

[85814-86-4]
C$_9$H$_{16}$O$_2$ M 156.224

(±)-form

Oil.

Urbina, E. *et al*, *Synthesis*, 1983, 113 (*synth*)

2-Hydroxycyclopentanone, 9CI   H-60114

Glutaroin
[473-84-7]

C$_5$H$_8$O$_2$ M 100.117

(±)-form [99493-88-6]

Bp$_2$ 50-53°. (*R*)-form known but not well characterised.

Ac: [52789-75-0].
C$_7$H$_{10}$O$_3$ M 142.154
Bp$_2$ 85-88°.
Benzoyl: [59058-16-1].
C$_{12}$H$_{12}$O$_3$ M 204.225
Cryst. (EtOH aq.). Mp 90-91°.
Benzoyl, 2,4-dinitrophenylhydrazone: Cryst. (EtOH aq.). Mp 128-129°.
Oxime: [14352-55-7].
C$_5$H$_9$NO$_2$ M 115.132
Mp 74-75°. Bp$_4$ 118-119°.
Phenylhydrazone: Mp 144°.
Me ether: [35394-09-3]. *2-Methoxycyclopentanone, 9CI*.
C$_6$H$_{10}$O$_2$ M 114.144
Bp$_{15}$ 68-70°.

Schraepler, U. *et al*, *Chem. Ber.*, 1964, **97**, 1383 (*synth*)
Barco, A. *et al*, *Synthesis*, 1972, 626 (*deriv*)

David, S., *C.R. Hebd. Seances Acad. Sci.*, *Ser. C*, 1974, **278**, 1051 (*synth*)
Friedrich, E. *et al*, *Chem. Ber.*, 1980, **113**, 1245 (*synth*)
Moriarty, R.M. *et al*, *Tetrahedron Lett.*, 1984, **25**, 691 (*synth*)
Lee, L.G. *et al*, *J. Org. Chem.*, 1986, **51**, 25 (*synth*)

3-Hydroxycyclopentanone, 9CI   H-60115

[26831-63-0]
C$_5$H$_8$O$_2$ M 100.117

(±)-form

Oil. Bp$_{0.1}$ 61-63°.

Crispin, D.J. *et al*, *J. Chem. Soc. (C)*, 1970, 10 (*synth*)
Salomon, R.G. *et al*, *J. Am. Chem. Soc.*, 1984, **106**, 6049.

4-Hydroxy-2-cyclopenten-1-one, 9CI   H-60116

Updated Entry replacing H-20125
[61305-27-9]

C$_5$H$_6$O$_2$ M 98.101
Intermediate in prostaglandin synthesis.

(*R*)-form [59995-47-0]

Bp$_{0.9}$ 88-91°. [α]$_D^{20}$ +81° (CHCl$_3$).

Ac: [59995-48-1].
C$_7$H$_9$O$_3$ M 141.146
[α]$_D^{20}$ +76° (c, 0.017 in CCl$_4$).

(*S*)-form [59995-49-2]

Bp$_{0.7}$ 83°. [α]$_D^{20}$ −94.1° (c, 3.4 in CHCl$_3$).

(±)-form [61740-29-2]

Oil. Bp$_2$ 90-92°.

Benzoyl: [29555-14-4].
C$_{12}$H$_{10}$O$_3$ M 202.209
Plates (Et$_2$O/hexane). Mp 87.5-88.5°.

Ogura, K. *et al*, *Tetrahedron Lett.*, 1976, 759 (*abs config*)
Tanaka, T. *et al*, *Tetrahedron*, 1976, **32**, 1713 (*synth*)
Nara, M. *et al*, *Tetrahedron*, 1980, **36**, 3161 (*synth*)
Gill, M. *et al*, *Aust. J. Chem.*, 1981, **34**, 2587 (*synth*)
Baraldi, P.G. *et al*, *Synthesis*, 1986, 781 (*synth, pmr, ir*)
Laumen, K. *et al*, *J. Chem. Soc., Chem. Commun.*, 1986, 1298 (*synth*)

1-Hydroxycyclopropanecarboxylic acid   H-60117

C$_4$H$_6$O$_3$ M 102.090

Me ether: [100683-08-7]. *1-Methoxycyclopropanecarboxylic acid*. Viscous oil solidifying on cooling. Mp ~20°.

Johnston, L.J. *et al*, *J. Am. Chem. Soc.*, 1986, **108**, 2343 (*deriv, synth, pmr*)

7-Hydroxy-6,7-dihydro-5*H*-pyrrolo[1,2-*a*]-imidazole   H-60118

C$_6$H$_8$N$_2$O   M 124.142

(±)-form

Prod. of reacn. of imidazole with acrolein. Cryst. (Me₂CO). Mp 138-140°.

Weintraub, P.M. *et al, J. Heterocycl. Chem.,* 1987, **24**, 561 (*synth, cryst struct, ir, pmr*)

7-Hydroxy-3,4-dimethyl-2*H*-1-benzo-   H-60119
pyran-2-one

7-Hydroxy-3,4-dimethylcoumarin

[2107-78-0]

$C_{11}H_{10}O_3$ M 190.198
Mp 256°.

Et ether: 7-Ethoxy-3,4-dimethylcoumarin.
$C_{13}H_{14}O_3$ M 218.252
Constit. of *Edgeworthia gardneri.* Cryst. (C₆H₆). Mp 121-122°.

Trividi, K.N., *J. Indian Chem. Soc.,* 1965, **42**, 273 (*synth*)
Chatterjee, A. *et al, Indian J. Chem., Sect. B,* 1987, **26**, 81.

4-Hydroxy-2,5-dimethyl-3(2*H*)-furanone,   H-60120
9CI

Updated Entry replacing H-10130

Furaneol

[3658-77-3]

$C_6H_8O_3$ M 128.127
Constit. of pineapple and strawberry. Solid with "fruity caramel" or "burnt pineapple" aroma. Mp 75-78° (71°).

Ac:
$C_8H_{10}O_4$ M 170.165
Oil.

4-Me ether: [4077-47-8]. *4-Methoxy-2,5-dimethyl-3(2H)-furanone.*
$C_7H_{10}O_3$ M 142.154
Oil.

Rodin, J.O. *et al, J. Food Sci.,* 1965, **30**, 280 (*isol, ir, uv, pmr*)
Henry, D.W. *et al, J. Org. Chem.,* 1966, **31**, 2391 (*synth, ir, pmr, uv*)
Büchi, G. *et al, J. Org. Chem.,* 1973, **38**, 123 (*synth, pmr*)
Re, L. *et al, Helv. Chim. Acta,* 1973, **56**, 1882 (*synth*)
Briggs, M.A. *et al, J. Chem. Soc., Perkin Trans. 1,* 1985, 795 (*synth*)

6-Hydroxy-2,4-dimethylheptanoic acid   H-60121

(2S,4R,6R)-form

$C_9H_{18}O_3$ M 174.239
(2S,4R,6R)-form [110454-83-6]
Needles (heptane/Et₂O). Mp 83-85°. [α]²³_D +4.3° (c, 8.50 in CHCl₃).

(2R,4S,6S)-form [110397-95-0]
Needles (hexane/Et₂O). Mp 83.5-85°. [α]²³_D −4.2° (c, 8.69 in CHCl₃).

Mori, K. *et al, Tetrahedron,* 1986, **42**, 5539 (*synth, ir, pmr*)

3-(2-Hydroxy-4,8-dimethyl-3,7-   H-60122
nonadienyl)benzaldehyde

$C_{18}H_{24}O_2$ M 272.386
Ac:
$C_{20}H_{26}O_3$ M 314.424
Constit. of green algae *Halimeda scabra, H. maero-loba* and *H. discoidea.* Oil. [α]²⁵_D +2.4° (c, 0.5 in CHCl₃).

Paul, V.J. *et al, Tetrahedron,* 1984, **40**, 3053.

2-(7-Hydroxy-3,7-dimethyl-2-octenyl)-6-   H-60123
methoxy-1,4-benzoquinone

(E)-form

$C_{17}H_{24}O_4$ M 292.374
(E)-form [109954-47-4]
Verapliquinone C
Constit. of an *Aplidium* sp.
(Z)-form [109954-46-3]
Verapliquinone D
Constit. of an *A.* sp.

Guella, G. *et al, Helv. Chim. Acta,* 1987, **26**, 621.

4-Hydroxy-3,3-dimethyl-2-oxobutanoic   H-60124
acid, 9CI, 8CI

Ketopantoic acid

[470-30-4]

$$HOCH_2C(CH_3)_2COCOOH$$

$C_6H_{10}O_4$ M 146.143
Intermed. in biosynth. of pantothenic acids. No phys. props. reported.

Wieland, T., *Chem. Ber.,* 1948, **81**, 323 (*oxime*)
Maas, W.K. *et al, J. Bacteriol.,* 1953, **65**, 388 (*biosynth*)
McIntosh, E.N. *et al, J. Biol. Chem,* 1957, **228**, 499 (*biosynth*)
Powers, S.G. *et al, J. Biol. Chem,* 1976, **251**, 3780, 3786 (*biosynth*)
Japan. Pat., 83 198 480, (*1983*); *CA,* **100**, 138939 (*synth*)
Aberhart, D.J. *et al, J. Am. Chem. Soc.,* 1984, **106**, 4902, 4907 (*biosynth*)

10-Hydroxy-11-dodecenoic acid, 9CI   H-60125

[85288-97-7]

$$H_2C{=}CHCH(OH)(CH_2)_8COOH$$

$C_{12}H_{22}O_3$ M 214.304
(±)-form
Cryst. (Et₂O/pet. ether). Mp 43-44°.
Me ester: [106753-87-1].
$C_{13}H_{24}O_3$ M 228.331

Oil.

Cameron, A.G. *et al, J. Chem. Soc., Perkin Trans. 1,* 1986, 161 (*synth, pmr, ir*)

12-Hydroxy-10-dodecenoic acid, 9CI   H-60126
[85288-99-9]

$$HOCH_2CH{=}CH(CH_2)_8COOH$$

$C_{12}H_{22}O_3$ M 214.304

(*E*)-form
Cryst. (Et$_2$O/pet. ether). Mp 39-40°.
Me ester: [106753-90-6].
  $C_{13}H_{24}O_3$ M 228.331
  Oil. Bp$_{0.1}$ 175° (oven).
Lactone: 10-Dodecen-12-olide.
  $C_{12}H_{20}O_2$ M 196.289
  Bp$_{0.1}$ 85° (oven).

Cameron, A.G. *et al, J. Chem. Soc., Perkin Trans. 1,* 1986, 161 (*synth, pmr, ir*)

5-Hydroxy-6,8,11,14-eicosatetraenoic acid,   H-60127
9CI
Updated Entry replacing H-20167
5-HETE
[71030-39-2]

$$H_3C(CH_2)_3(CH_2CH{=}CH)_3CH{=}CHCH(OH)(CH_2)_3{-}$$
$$COOH$$

$C_{20}H_{32}O_3$ M 320.471

(*5S,6E,8Z,11Z,14Z*)-form [70608-72-9]
Metab. of arachidonic acid.
Me ester: [78037-99-7].
  $C_{21}H_{34}O_3$ M 334.498
  $[\alpha]_{436}^{23}{+}12.42°$, $[\alpha]_D^{23}{+}4.73°$ (c, 0.99 in EtOH).

Porter, N.A. *et al, J. Org. Chem.,* 1979, **44**, 3177 (*ms*)
Boeynaems, J.M. *et al, Anal. Biochem.,* 1980, **104**, 259 (*synth*)
Corey, E.J. *et al, J. Am. Chem. Soc.,* 1980, **102**, 1435 (*synth*)
Rabinovitch, H. *et al, Lipids,* 1981, **16**, 518 (*metab*)
Rokach, J. *et al, Tetrahedron Lett.,* 1983, **24**, 5185 (*synth*)
Gunn, B.P., *Tetrahedron Lett.,* 1985, **26**, 2869 (*synth*)
Nicolaou, K.C. *et al, Synthesis,* 1986, 344 (*synth*)

8-Hydroxy-5,9,11,14-eicosatetraenoic acid,   H-60128
9CI
Updated Entry replacing H-20169
8-HETE

$$H_3C(CH_2)_4CH{=}CHCH_2CH{=}CHCH{=}CHCH(OH){-}$$
$$CH_2CH{=}CH(CH_2)_3COOH$$

$C_{20}H_{32}O_3$ M 320.471

(*8S,5Z,9E,11Z,14Z*)-form [98462-03-4]
Metab. of arachidonic acid.
Me ester:
  $C_{21}H_{34}O_3$ M 334.498
  Oil. $[\alpha]_D^{22}{-}4.75°$ (c, 0.4 in CHCl$_3$).

Porter, N.A. *et al, J. Org. Chem.,* 1979, **44**, 3177 (*synth, ms*)
Boeynaems, J.M. *et al, Anal. Biochem.,* 1980, **104**, 259 (*synth*)
Rabinovitch, H. *et al, Lipids,* 1981, **16**, 518 (*metab*)
Adams, J. *et al, Tetrahedron Lett.,* 1984, **25**, 35 (*synth*)
Just, G. *et al, J. Org. Chem.,* 1986, **51**, 4796 (*synth, ms, pmr*)
Yadagiri, P. *et al, Tetrahedron Lett.,* 1986, **27**, 6039 (*synth*)

9-Hydroxy-5,7,11,14-icosatetraenoic acid   H-60129
Updated Entry replacing H-20170
9-HETE

$$H_3C(CH_2)_4(CH{=}CHCH_2)_2CH(OH)(CH{=}CH)_2{-}$$
$$(CH_2)_3COOH$$

$C_{20}H_{32}O_3$ M 320.471

(*5Z,7E,9S,11Z,14Z*)-form
Metab. of arachidonic acid (poss. as the 9*R*-enantiomer).
Me ester:
  $C_{21}H_{34}O_3$ M 334.498
  $[\alpha]_D^{22}{-}7.1°$ (c, 2.05 in CHCl$_3$).

Porter, N.A. *et al, J. Org. Chem.,* 1977, **44**, 3177 (*synth, ms*)
Boeynaems, J.M. *et al, Anal. Biochem.,* 1980, **104**, 259 (*synth*)
Rabinovitch, M. *et al, Lipids,* 1981, **16**, 518 (*metab*)
Capdevila, J. *et al, Proc. Natl. Acad. Sci. USA,* 1982, **79**, 767 (*synth, ms*)
Adams, J. *et al, Tetrahedron Lett.,* 1984, **25**, 35 (*synth*)
Just, G. *et al, J. Org. Chem.,* 1986, **51**, 4796 (*deriv, synth, pmr, uv, ms*)

11-Hydroxy-5,8,12,14-icosatetraenoic acid   H-60130
Updated Entry replacing H-20171
11-HETE

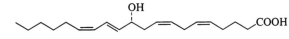

(5Z,8Z,11R,12E,14Z)-form

$C_{20}H_{32}O_3$ M 320.471
Metab. of arachidonic acid.

(*5Z,8Z,11R,12E,14Z*)-form [73347-43-0]
  $[\alpha]_D^{21}{+}11.2°$ (CH$_2$Cl$_2$).
Me ester: [79083-18-4].
  $C_{21}H_{34}O_3$ M 334.498
  $[\alpha]_D^{23}{+}10.17°$ (c, 1.0 in CHCl$_3$).

(*5Z,8Z,11S,12E,14Z*)-form
  $[\alpha]_D^{21}{-}11.3°$ (CH$_2$Cl$_2$).
Me ester: $[\alpha]_D^{22}{-}10.3°$ (c, 1.2 in CHCl$_3$).

Porter, N.A. *et al, J. Org. Chem.,* 1979, **44**, 3177 (*synth, ms*)
Boeynaems, J.M. *et al, Anal. Biochem.,* 1980, **104**, 259 (*synth*)
Corey, E.J. *et al, J. Am. Chem. Soc.,* 1980, **102**, 1433 (*synth*)
Corey, E.J. *et al, J. Am. Chem. Soc.,* 1981, **103**, 4618 (*synth, pmr*)
Rabinovitch, H., *Lipids,* 1981, **16**, 518 (*metab*)
Capdevila, J. *et al, Proc. Natl. Acad. Sci. USA,* 1982, **79**, 767 (*synth*)
Just, G. *et al, Tetrahedron Lett.,* 1982, **23**, 1331, 2285 (*synth*)
Just, G. *et al, J. Org. Chem.,* 1986, **51**, 4796 (*deriv, synth, uv, pmr, ms*)

12-Hydroxy-5,8,10,14-eicosatetraenoic   H-60131
acid, 9CI
Updated Entry replacing H-20172
12-HETE
[59985-28-3]

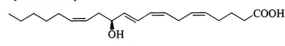

(5Z,8Z,10E,12S,14Z)-form

$C_{20}H_{32}O_3$ M 320.471

(*5Z,8Z,10E,12S,14Z*)-form [54397-83-0]
Metab. of arachidonic acid.
Me ester: [57872-14-7].
  $C_{21}H_{34}O_3$ M 334.498
  $[\alpha]_D^{22}{+}1.3°$ (c, 0.3 in CHCl$_3$).

(5Z,8Z,10Z,12S,14Z)-form

$[\alpha]_D^{21}$ −1.87° (c, 6.1 in CHCl$_3$).

Me ester: [81623-72-5]. $[\alpha]_D^{23}$ −2.06° (c, 5.4 in CHCl$_3$).

Corey, E.J. et al, J. Am. Chem. Soc., 1978, **100**, 1942 (synth, pmr)
McGuire, J.C. et al, Prep. Biochem., 1978, **8**, 147 (metab)
Porter, N.A. et al, J. Org. Chem., 1979, **44**, 3177 (synth, ms)
Boeynaems, J.M. et al, Anal. Biochem., 1980, **104**, 259 (synth)
Corey, E.J. et al, J. Am. Chem. Soc., 1980, **102**, 1433 (synth)
Rabinovitch, H. et al, Lipids, 1981, **16**, 518 (metab)
Russell, S.W. et al, J. Chem. Soc., Perkin Trans. 1, 1982, 545 (synth)
Corey, E.J. et al, Tetrahedron Lett., 1984, **25**, 5115 (synth)
Just, G. et al, J. Org. Chem., 1986, **51**, 4796 (deriv, synth, pmr, ir, ms)
Nicolaou, K.C. et al, Synthesis, 1986, 344 (synth)
Yadagiri, P. et al, Tetrahedron Lett., 1986, **27**, 6039 (synth)

20-Hydroxyelemajurinelloide   H-60132

C$_{20}$H$_{26}$O$_7$ M 378.421

Constit. of *Jurinella moschus*. Oil. $[\alpha]_D$ +38° (c, 0.2 in CHCl$_3$).

Rustaiyan, A. et al, Phytochemistry, 1987, **26**, 2857.

1-(2-Hydroxyethyl)-1,4-cyclohexanediol,   H-60133
9CI

Updated Entry replacing H-40120

Rengyol

[93675-85-5]

C$_8$H$_{16}$O$_3$ M 160.213

Constit. of crude drug "rengyo", the fruits of *Forsythia suspensa*. Cryst. (MeOH). Mp 123-124°.

4-Epimer: [101489-38-7]. **Isorengyol**.
C$_8$H$_{16}$O$_3$ M 160.213
Constit. of *F. suspensa*. Powder.

Endo, K. et al, Can. J. Chem., 1984, **62**, 2011 (isol)
Abdallahi, H. et al, Phytochemistry, 1986, **25**, 2821 (isol)
Endo, K. et al, Tetrahedron, 1987, **43**, 2681 (rel config, isomer)
Breton, J.L. et al, Tetrahedron, 1987, **43**, 4447 (synth)

O-(2-Hydroxyethyl)hydroxylamine   H-60134
2-(Aminooxy)ethanol, 9CI

[3279-95-6]

HOCH$_2$CH$_2$ONH$_2$

C$_2$H$_7$NO$_2$ M 77.083
Liq. Bp$_1$ 61-62°, Bp$_{0.07}$ 55°.

B,HCl: [23156-68-5]. Cryst. (2-propanol/Et$_2$O). Mp 62-65°. Hygroscopic.
Picrate: Yellow needles (EtOH). Mp 144-145°.

Bruno, I. et al, Helv. Chim. Acta, 1962, **45**, 358 (synth)
Schumann, E.L. et al, J. Med. Chem., 1962, **5**, 464 (deriv)
Dhanak, D. et al, J. Chem. Soc., Chem. Commun., 1986, 903 (synth)

9-Hydroxy-4,11(13)-eudesmadien-12-oic   H-60135
acid

C$_{15}$H$_{22}$O$_3$ M 250.337

9β-form

Ac: [111394-37-7]. *9β-Acetoxy-4,11(13)-eudesmadien-12-oic acid. 9β-Acetoxy-4,5-dehydro-4(15)-dihydrocostic acid.*
C$_{17}$H$_{24}$O$_4$ M 292.374
Constit. of *Artemisia tournefortiana*. Oil. $[\alpha]_D^{24}$ +48° (c, 0.3 in CHCl$_3$).
9-Ketone: [111394-35-5]. *9-Oxo-4,11(13)-eudesmadien-12-oic acid. 9-Oxo-4,5-dehydro-4(15)-dihydrocostic acid.*
C$_{15}$H$_{20}$O$_3$ M 248.321
From *A. tournefortiana*.

Rustaiyan, A. et al, Phytochemistry, 1987, **26**, 2307.

3-Hydroxy-4,11(13)-eudesmadien-12,8-   H-60136
olide

Updated Entry replacing H-40124

C$_{15}$H$_{20}$O$_3$ M 248.321
(3α,8β)-form illus.

(3α,8β)-form

Constit. of *Artemisia iwayomogi*. Oil.

3-Peroxide: 3α-Hydroperoxy-4,11(13)-eudesmadien-12,8β-olide. 3α-Peroxy-4,11(13)-eudesmadien-12,8β-olide.
C$_{15}$H$_{20}$O$_4$ M 264.321
From *A. iwayamogi*. Oil.
3-Ketone: 3-Oxo-4,11(13)-eudesmadien-12,8β-olide.
C$_{15}$H$_{18}$O$_3$ M 246.305
From *A. iwayomogi*. Cryst. Mp 170°. $[\alpha]_D$ +142° (c, 0.05 in CHCl$_3$).
4α,5α-Epoxide: 4α,5α-Epoxy-3α-hydroxy-11(13)-eudesmen-12,8β-olide.
C$_{15}$H$_{20}$O$_4$ M 264.321
Constit. of *A. iwayomogi*. Oil. $[\alpha]_D$ +140° (c, 0.18 in CHCl$_3$).

(3β,8β)-form

Ac: 3β-Acetoxy-4,11(13)-eudesmadien-12,8β-olide.
C$_{17}$H$_{22}$O$_4$ M 290.358
Constit. of *Calea szyszylowiczii*. Oil. $[\alpha]_D$ −25° (c, 0.11 in CHCl$_3$).
4α,5α-Epoxide: 4α,5α-Epoxy-3β-hydroxy-11(13)-eudesmen-12,8β-olide.
C$_{15}$H$_{20}$O$_4$ M 264.321
From *A. iwayomogi*. Oil. $[\alpha]_D$ +250° (c, 0.05 in CDCl$_3$).

Bohlmann, F. et al, Justus Liebigs Ann. Chem., 1983, 2227 (isol)

Greger, H. *et al, Phytochemistry*, 1986, **25**, 891 (*isol*)

9-Hydroxy-4,11(13)-eudesmadien-12,6-olide   H-60137

$C_{15}H_{20}O_3$ M 248.321

(*6β,9β*)-*form* [111394-33-3]
9β-Hydroxytournefortiolide
Constit. of *Artemisia tournefortiana*. Oil. $[\alpha]_D^{24}$ −12° (c, 0.6 in CHCl₃).

Ac: [111394-34-4]. *9β-Acetoxytournefortiolide*.
  $C_{17}H_{22}O_4$ M 290.358
  From *A. tournefortiana*. Oil. $[\alpha]_D^{24}$ +14° (c, 0.2 in CHCl₃).
9-Ketone: [111420-61-2]. *9-Oxo-4,11(13)-eudesmadien-12,16β-olide*. *9-Oxotournefortiolide*.
  $C_{15}H_{18}O_3$ M 246.305
  From *A. tournefortiana*. Cryst. Mp 165°. $[\alpha]_D^{24}$ +44° (c, 2.2 in CHCl₃).

Rustaiyan, A. *et al, Phytochemistry*, 1987, **26**, 2307.

15-Hydroxy-4,11(13)-eudesmadien-12,8-olide   H-60138

$C_{15}H_{20}O_3$ M 248.321

8β-form
  15-O-(6-Acetyl-β-D-glucopyranoside): *Absinthifolide*.
  $C_{23}H_{32}O_9$ M 452.500
  Constit. of *Bahia absinthifolia*. Yellow gum. $[\alpha]_D$ −18.27° (c, 0.208 in CHCl₃).

Pérez, C.A.L. *et al, Phytochemistry*, 1987, **26**, 765.

6-Hydroxy-1,4-eudesmadien-3-one   H-60139

$C_{15}H_{22}O_2$ M 234.338

6β-form
Constit. of *Sideritis varoi*. Gum. $[\alpha]_D^{20}$ +98.4° (c, 1.7 in CHCl₃).

Cabrera, E. *et al, Phytochemistry*, 1988, **27**, 183.

1-Hydroxy-3,7(11),8-eudesmatrien-12,8-olide   H-60140

$C_{15}H_{18}O_3$ M 246.305

1β-form
  Ac: *1β-Acetoxy-3,7(11),8-eudesmatrien-12,8-olide*.
  $C_{17}H_{20}O_4$ M 288.343
  Constit. of *Smyrnium perfoliatum*. Amorph.

Gören, N. *et al, Phytochemistry*, 1987, **26**, 2585.

1-Hydroxy-4(15),7(11),8-eudesmatrien-12,8-olide   H-60141

$C_{15}H_{18}O_3$ M 246.305

1β-form
  Ac: *1β-Acetoxy-4(15),7(11),8-eudesmatrien-12,8-olide*.
  $C_{17}H_{20}O_3$ M 272.343
  Constit. of *Smyrnium perfoliatum*. Amorph.

Gören, N. *et al, Phytochemistry*, 1987, **26**, 2585.

4-Hydroxy-11(13)-eudesmen-12,8-olide   H-60142

$C_{15}H_{22}O_3$ M 250.337

(*4β,8β*)-*form*
  Septuplinolide
  Constit. of *Calea septuplinervia*. Cryst. Mp 171-172°.

Ober, A.G. *et al, Phytochemistry*, 1987, **26**, 848.

6-Hydroxyflavone, 8CI   H-60143

Updated Entry replacing H-10159
6-Hydroxy-2-phenyl-4H-1-benzopyran-4-one, 9CI. 6-Hydroxy-2-phenylchromone
[6665-83-4]
$C_{15}H_{10}O_3$ M 238.242
Yellow needles (EtOH aq.). Mp 235.5-236.5° (231-232°).

Me ether: [26964-24-9]. *6-Methoxyflavone*.
  $C_{16}H_{12}O_3$ M 252.269
  Mp 163-164°.

Kostanecki, S. *et al, Ber.*, 1899, **32**, 326 (*synth*)
Simonis, H. *et al, Ber.*, 1926, **59**, 2914 (*deriv*)
Looker, J.H. *et al, J. Org. Chem.*, 1962, **27**, 381 (*synth, ir*)
Fozdar, B.I. *et al, Chem. Ind.* (*London*), 1986, 586 (*synth*)

7-Hydroxyflavone   H-60144

Updated Entry replacing H-10160
7-Hydroxy-2-phenyl-4H-1-benzopyran-4-one, 9CI. 7-Hydroxy-2-phenylchromone
[6665-86-7]
$C_{15}H_{10}O_3$ M 238.242
Needles (EtOH aq.). Mp 244° (240°).

Me ether: [22395-22-8]. *7-Methoxyflavone*.
  $C_{16}H_{12}O_3$ M 252.269
  Mp 110-111°.

Emilewicz, T. *et al, Ber.*, 1899, **32**, 309 (*deriv*)
Looker, J.H. *et al, J. Org. Chem.*, 1962, **27**, 381 (*synth, ir*)
Audier, H., *Bull. Soc. Chim. Fr.*, 1966, 2892 (*ms*)
Naik, G.N. *et al, Indian J. Chem.*, 1966, **4**, 273 (*pmr*)
Fozdar, B.I. *et al, Chem. Ind.* (*London*), 1986, 586 (*synth*)

8-Hydroxyflavone H-60145

Updated Entry replacing H-10161

[77298-64-7]

$C_{15}H_{10}O_3$ M 238.242

Mp 250-252° (249-250°).

▷LK8650100.

Me ether: [26964-26-1]. *8-Methoxyflavone.*

   $C_{16}H_{12}O_3$ M 252.269

   Mp 200-201°.

Gupta, D.S. *et al, CA,* 1955, **49**, 1713 *(synth)*

Looker, J.H. *et al, J. Org. Chem.,* 1962, **27**, 381 *(synth, ir)*

Fozdar, B.I. *et al, Chem. Ind. (London),* 1986, 586 *(synth)*

4-Hydroxyfuro[2,3-*d*]pyridazine H-60146

*Furo[2,3-*d]*pyridazin-4(5H)-one, 8CI. 4-Oxo-4,5-dihydrofuro[2,3-*d]*pyridazine*

[14757-77-8]

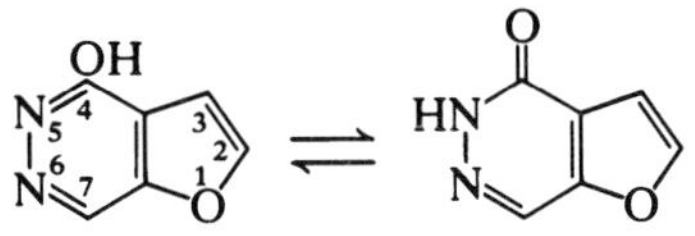

$C_6H_4N_2O_2$ M 136.110

OH-form said to predominate in soln. Cryst. (EtOH). Mp 212°.

Robba, M. *et al, Bull. Soc. Chim. Fr.,* 1968, 4959 *(synth, ir, pmr)*

7-Hydroxyfuro[2,3-*d*]pyridazine H-60147

*Furo[2,3-*d]*pyridazin-7(6H)-one, 8CI. 7-Oxo-6,7-dihydrofuro[2,3-*d]*pyridazine*

[13177-73-6]

$C_6H_4N_2O_2$ M 136.110

OH-form predominates in soln. Cryst. (EtOH). Mp 198°.

Robba, M. *et al, Bull. Soc. Chim. Fr.,* 1968, 4959 *(synth, ir, pmr)*

2-Hydroxygarveatin *B* H-60148

7-Ethyl-2,8,9-trihydroxy-2,4,4,6-tetramethyl-1,3(2H,4H)anthracenedione, 9CI

[109894-11-3]

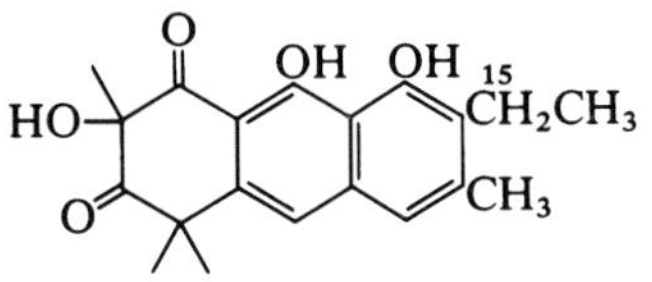

$C_{20}H_{22}O_5$ M 342.391

Constit. of *Garveia annulata.* Yellow solid.

15-Oxo: [109894-10-2]. *2-Hydroxygarveatin* A.

   $C_{20}H_{20}O_6$ M 356.374

   Constit. of *G. annulata.* Orange solid.

Fahy, E. *et al, Can. J. Chem.,* 1987, **65**, 376.

11-Hydroxy-4-guaien-3-one H-60149

[20482-28-4]

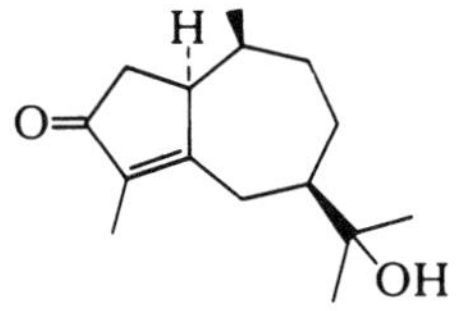

$C_{15}H_{24}O_2$ M 236.353

Constit. of *Euryops pedunculatus.* Oil. $[\alpha]_D^{24}$ +18° (c, 0.2 in CHCl₃).

Jakupovic, J. *et al, Phytochemistry,* 1987, **26**, 1049.

7-Hydroxy-5-heptynoic acid H-60150

[41300-59-8]

$$HOCH_2C{\equiv}C(CH_2)_3COOH$$

$C_7H_{10}O_3$ M 142.154

Intermed. in prostagladin synth.

Me ester: [50781-91-4].

   $C_8H_{12}O_3$ M 156.181

   Oil. Bp₀.₀₅ 150°.

Corey, E.J. *et al, J. Am. Chem. Soc.,* 1973, **95**, 8483 *(synth)*

Ger. Pat., 2 313 868, *(1974); CA,* **80**, 59566u

Casy, G. *et al, Tetrahedron,* 1986, **42**, 5849 *(synth, ir, pmr)*

3-Hydroxy-22,23,24,25,26,27-hexanor-20-dammaranone H-60151

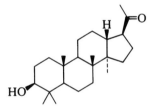

$C_{24}H_{40}O_2$ M 360.579

3β-form

Constit. of *Euphorbia supina.* Cryst. (CHCl₃/CHCl₃).

   Mp 195-197°. $[\alpha]_D$ +55° (c, 1.02 in CHCl₃).

Mills, J.S. *et al, J. Chem. Soc.,* 1956, 2196 *(synth)*

Tanaka, R. *et al, Phytochemistry,* 1987, **26**, 3365 *(isol)*

7-Hydroxy-3-(4-hydroxybenzylidene)-4-chromanone H-60152

2,3-Dihydro-7-hydroxy-3-[(4-hydroxyphenyl)-methylene]-4H-1-benzopyran-4-one, 9CI

[110064-50-1]

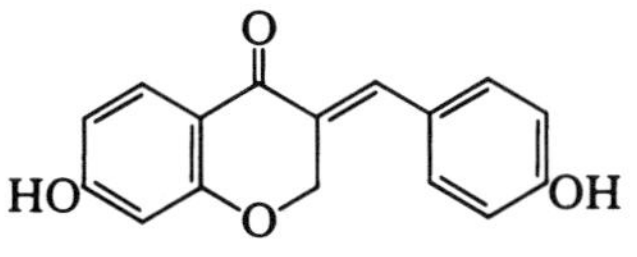

$C_{16}H_{12}O_4$ M 268.268

Constit. of *Caesalpinia sappan.* Yellow needles (Me₂CO/hexane). Mp 248-249°.

Namikoshi, M. *et al, Phytochemistry,* 1987, **26**, 1831.

4-Hydroxy-4-(2-hydroxyethyl)-cyclohexanone, 9CI   H-60153

[107389-91-3]

$C_8H_{14}O_3$ M 158.197

Constit. of *Isoplexis canariensis* var. *tomentosa*. Oil.

Llera, L.D. *et al*, *J. Nat. Prod.*, 1987, **50**, 251.

4-Hydroxy-3-(2-hydroxy-3-methyl-3-butenyl)benzoic acid   H-60154

$C_{12}H_{14}O_4$ M 222.240

Me ester:

  $C_{13}H_{16}O_4$ M 236.267

  Constit. of *Piper hostmannianum*. Cryst. Mp 120°.

  $[\alpha]_D^{18}$ −6.7° (c, 0.006 in CHCl₃).

Díaz, D.P.P. *et al*, *Phytochemistry*, 1987, **26**, 809.

5-Hydroxy-6-(hydroxymethyl)-7-methoxy-2-methyl-4H-1-benzopyran-4-one, 9CI   H-60155

Updated Entry replacing H-01923

5-Hydroxy-6-hydroxymethyl-7-methoxy-2-methylchromone. 6-Hydroxymethyleugenin

[37042-21-0]

$C_{12}H_{12}O_5$ M 236.224

Constit. of *Roccella fuciformis* and metab. of *Chaetomium minutum*. Yellow cryst. (MeOH/CH₂Cl₂). Mp 198-199°.

Hauser, D. *et al*, *Experientia*, 1972, **28**, 1114 (*isol*)
Huneck, S. *et al*, *Phytochemistry*, 1972, **11**, 1489 (*struct*)
Hanson, J.R. *et al*, *J. Chem. Res. (S)*, 1987, 8 (*cryst struct, biosynth*)

5-Hydroxy-2,4-imidazolidinedione   H-60156

5-Hydroxyhydantoin

[29410-13-7]

$C_3H_4N_2O_3$ M 116.076

Cryst. (AcOH). Mp 140-142°. Compd. descr. in early ref. had different props.

Biltz, H. *et al*, *Ber.*, 1921, **54**, 1802 (*synth*)
Abblard, J. *et al*, *Bull. Soc. Chim. Fr.*, 1971, 942 (*synth, pmr*)

4-Hydroxyimidazo[4,5-b]pyridine   H-60157

1H-Imidazo[4,5-c]pyridin-4(5H)-one, 9CI. 6-Hydroxy-3-deazapurine. 3-Deazahypoxanthine

[3243-24-1]

$C_6H_5N_3O$ M 135.125

Needles (H₂O). Mp >320° dec.

B,HCl: Mp 302° dec.

Salemink, C.A. *et al*, *Recl. Trav. Chim. Pays-Bas*, 1949, **68**, 1013.

6-Hydroxyindole   H-60158

Updated Entry replacing H-01972

1H-Indol-6-ol, 9CI

[2380-86-1]

C_8H_7NO M 133.149

Cryst. Mp 124-126°.

Picrate: Red needles (C₆H₆/pet. ether). Mp 137°.

Me ether: [3189-13-7]. *6-Methoxyindole.*

  C_9H_9NO M 147.176

  Plates (pet. ether). Mp 91-92°.

Benzyl ether: [15903-94-3].

  $C_{15}H_{13}NO$ M 223.274

  Mp 118°.

Kermack, W.O. *et al*, *J. Chem. Soc.*, 1922, **121**, 1879.
Beer, R.J.S. *et al*, *J. Chem. Soc.*, 1948, 1605.
Stoll, A. *et al*, *Helv. Chim. Acta*, 1955, **38**, 1452.
Eich, E. *et al*, *Pharm. Acta Helv.*, 1966, **41**, 109.
Daly, J.W. *et al*, *J. Am. Chem. Soc.*, 1967, **89**, 1032 (*pmr*)
Gerecs, A. *et al*, *Acta Chim. Acad. Sci. Hung.*, 1968, **56**, 311; *CA*, **69**, 96381.
Lloyd, D.H. *et al*, *Tetrahedron Lett.*, 1983, **24**, 4561 (*deriv, synth*)
Feldman, P.L. *et al*, *Synthesis*, 1986, 735 (*deriv, synth, ir, pmr*)

3-Hydroxy-2-iodo-6-methylpyridine   H-60159

2-Iodo-6-methyl-3-pyridinol, 9CI, 8CI. 5-Hydroxy-6-iodo-2-picoline

[23003-30-7]

C_6H_6INO M 235.024

Cryst. (EtOH). Mp 179-183° dec. (174° subl.).

O-Ac:

  $C_8H_8INO_2$ M 277.061

  Prisms (pet. ether). Mp 66-67°.

Undheim, K. *et al*, *Acta Chem. Scand., Ser. B*, 1969, **23**, 1704 (*synth*)
Norton, S.J. *et al*, *J. Heterocycl. Chem.*, 1970, **7**, 699 (*synth*)

3-Hydroxy-2-iodopyridine   H-60160

2-Iodo-3-pyridinol, 9CI

[40263-57-8]

C_5H_4INO M 220.997

Mp 198-201°.

Urbanski, T. *et al*, *CA*, 1954, **48**, 1337 (*synth*)
Takahashi, T. *et al*, *Chem. Pharm. Bull.*, 1958, **6**, 611 (*synth*)
Blank, B. *et al*, *J. Med. Chem.*, 1974, **17**, 1065 (*synth*)

4-Hydroxyisobacchasmacranone   H-60161

C$_{20}$H$_{24}$O$_5$ M 344.407
Isobacchasmacranone not known.

4β-form

Constit. of *Baccharis macraei*. Cryst. (EtOAc/pet. ether). Mp 178-179°. [α]$_D^{25}$ −68.5° (c, 0.8 in CHCl$_3$).

Gambaro, V. *et al*, *Phytochemistry*, 1987, **26**, 475.

18-Hydroxy-8,15-isopimaradien-7-one   H-60162

[110201-62-2]

C$_{20}$H$_{30}$O$_2$ M 302.456
Constit. of *Nepeta tuberosa* subsp. *reticulata*. Oil. [α]$_D$ +85.4° (c, 1 in CHCl$_3$).

Teresa, J.de.P. *et al*, *Phytochemistry*, 1987, **26**, 1481.

8-Hydroxyisopimar-15-ene   H-60163

Updated Entry replacing H-02039
7-Ethenyldodecahydro-1,1,4a,7-tetramethyl-8a(2H)-phenanthrenol, 9CI. 8-Hydroxy-15-sandaracopimarene
C$_{20}$H$_{34}$O   M 290.488

8β-form [14699-32-2]

Nezukol

Constit. of *Dacrydium colensoi*, *Thuja standishi* and *Osteospermum* spp. Cryst. (MeOH aq.). Mp 40-41°. [α]$_D^{20}$ −6.8° (c, 2.3 in CHCl$_3$).

Corbett, R.E. *et al*, *J. Chem. Soc. (C)*, 1967, 300 (*isol*)
Bohlmann, F. *et al*, *Chem. Ber.*, 1973, **106**, 826 (*isol*)

4-Hydroxy-2-isopropyl-5-benzofurancar-boxylic acid   H-60164

4-Hydroxy-2-(1-methylethyl)-5-benzofurancarboxylic acid, 9CI. Rotenic acid. Isotubaic acid
[526-49-8]

C$_{12}$H$_{12}$O$_4$ M 220.224
Degradn. prod. of rotenone. Fine cryst. (C$_6$H$_6$). Mp 185°.
Me ester: [71243-11-3].
    C$_{13}$H$_{14}$O$_4$ M 234.251

Needles (MeOH). Mp 39-39.5°.

Batu, G. *et al*, *J. Org. Chem.*, 1979, **44**, 3948 (*synth, bibl*)

2-Hydroxy-2-isopropylbutanedioic acid   H-60165

2-Hydroxy-2-(1-methylethyl)butanedioic acid, 9CI. 2-Isopropylmalic acid, 8CI. 2-Hydroxy-2-isopropylsuccinic acid. β-Hydroxy-β-carboxyisocaproic acid
[3237-44-3]

C$_7$H$_{12}$O$_5$ M 176.169
Conflicting assignments of abs. config. Cryst. struct. determination gives (S) for the natural (−)-form but other work gives (R).

(−)-form [43119-99-9]

Prod. by various bacteria. Intermed. in leucine biosynth. Mp 171-173°. [α]$_D^{20}$ −20.0° (c, 2.5 in MeOH).

(±)-form

Cryst. (EtOAc/pet. ether). Mp 145-147°. Bp$_{10}$ 118-120°.

1-Amide: 2-Hydroxy-2-isopropylsuccinamic acid.
    C$_7$H$_{13}$NO$_4$ M 175.184
    Cryst. (EtOAc). Mp 156-157°.
Diamide: 2-Hydroxy-2-isopropylsuccinamide.
    C$_7$H$_{14}$N$_2$O$_3$ M 174.199
    Cryst. (H$_2$O). Mp 195-197°.

Yamashita, M., *J. Org. Chem.*, 1958, **23**, 835 (*synth, derivs*)
Calvo, J.M. *et al*, *Biochemistry*, 1962, **1**, 1157 (*synth, props*)
Sai, T., *Agric. Biol. Chem.*, 1968, **32**, 522 (*isol*)
Cole, F.E. *et al*, *Biochemistry*, 1973, **12**, 3346 (*abs config, cryst struct*)
Brandange, S. *et al*, *Acta Chem. Scand., Ser. B*, 1974, **28**, 153 (*abs config*)
Edgar, J.A. *et al*, *Tetrahedron Lett.*, 1980, **21**, 2657 (*synth*)

2-Hydroxy-3-isopropylbutanedioic acid   H-60166

2-Hydroxy-3-(1-methylethyl)butanedioic acid, 9CI. 3-Isopropylmalic acid, 8CI. 2-Hydroxy-3-isopropylsuccinic acid. α-Hydroxy-β-carboxyisocaproic acid
[16048-89-8]

C$_7$H$_{12}$O$_5$ M 176.169
Intermed. in leucine biosynth.

(2R,3S)-form

D-erythro-*form*
Isol. from *Neurospora crassa*. Long needles. Mp 146-147°. [α]$_D^{24}$ −5.2° (H$_2$O).

(2RS,3RS)-form [111408-27-6]

(±)-threo-*form*
Cryst. (C$_6$H$_6$). Mp 122-122.3°.

(2RS,3SR)-form

(±)-erythro-*form*
Cryst. (Et$_2$O/ligroin). Mp 119-119.5°.

Calvo, J.M. *et al*, *Biochemistry*, 1962, **1**, 1157; 1964, **3**, 2024 (*synth, abs config*)

Winterfeldt, E. *et al, Chem. Ber.*, 1969, **102**, 2336 (*ester*)
Calvo, J.M. *et al, Methods Enzymol.*, 1970, **17A**, 791 (*isol*)
Yamada, T. *et al, Chem. Lett.*, 1987, 1745 (*synth*)

ent-3β-Hydroxy-15-kauren-17-oic acid　　　H-60167

$C_{20}H_{30}O_3$　　M 318.455

3-Ac: [110187-31-0]. ent-*3β-Acetoxy-15-kauren-17-oic acid.*
$C_{22}H_{32}O_4$　　M 360.492
Constit. of *Peteravenia malvaefolia*.

3-Ac, Me ester: [110187-35-4]. Oil. $[\alpha]_D^{24}$ −43° (c, 1 in CHCl₃).

Ellmauerer, E. *et al, J. Nat. Prod.*, 1987, **50**, 221.

15-Hydroxy-7-labden-17-oic acid　　　H-60168

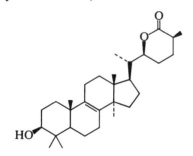

$C_{20}H_{34}O_3$　　M 322.487

Constit. of *Halimium viscosum*. Oil. $[\alpha]_D^{22}$ −24.1° (c, 1.31 in CHCl₃).

15-Ac: 15-Acetoxy-7-labden-17-oic acid.
$C_{22}H_{36}O_4$　　M 364.524
From *H. viscosum*. Oil. $[\alpha]_D^{22}$ −36.6° (c, 1.1 in CHCl₃).

15-Ac, Me ester:
$C_{23}H_{38}O_4$　　M 378.551
From *H. viscosum*. Oil. $[\alpha]_D^{22}$ −84.5° (c, 1.18 in CHCl₃).

(13S)-form
Havardic acid B
Constit. of *Grindelia havardii*.

Me ester: Oil. $[\alpha]_D^{25}$ −65° (c, 4.2 in CHCl₃).

Jolad, S.D. *et al, Phytochemistry*, 1987, **26**, 483.
Urones, J.G. *et al, Phytochemistry*, 1987, **26**, 3037.

3-Hydroxy-8-lanosten-26,22-olide　　　H-60169

$C_{30}H_{48}O_3$　　M 456.707

(3α,22S,25S)-form
3-epi-Astrahygrol
Metab. of *Astraeus hygrometricus*. Needles (MeOH). Mp 193-194°. $[\alpha]_D^{20}$ +101.0° (c, 0.5 in CHCl₃).

3-Ketone: 3-Oxo-8-lanosten-26,22-olide. **Astrahygrone**.
$C_{30}H_{46}O_3$　　M 454.692
Metab. of *A. hygrometricus*. Needles (MeOH). Mp 168-169°. $[\alpha]_D^{20}$ +58° (c, 0.5 in CHCl₃).

(3β,22S,25S)-form
Astrahygrol
Metab. from *A. hydrometricus*. Needles (MeOH). Mp 186-187°. $[\alpha]_D^{20}$ +18.0° (c, 0.5 in CHCl₃).

Takaishi, Y. *et al, Phytochemistry*, 1987, **26**, 2341.

1-Hydroxy-7-methylanthraquinone　　　H-60170

Updated Entry replacing H-50254
1-Hydroxy-7-methyl-9,10-anthracenedione. Barleriaquinone

[68963-23-5]

$C_{15}H_{10}O_3$　　M 238.242
Constit. of roots of *Barleria buxifolia*. Reddish-yellow needles (CHCl₃). Mp 183-184° (171-172°).

Gupta, R.C. *et al, J. Chem. Soc., Chem. Commun.*, 1982, 929 (*synth*)
Gopalakrishnan, S. *et al, Chem. Pharm. Bull.*, 1984, **32**, 4137 (*isol, uv, ir, pmr, cmr, ms*)

2-(Hydroxymethyl)benzaldehyde, 9CI　　　H-60171

Updated Entry replacing H-30153
α-Hydroxy-o-tolualdehyde, 8CI. o-*Formylbenzyl alcohol. 1,3-Dihydro-1-isobenzofuranol. 1-Hydroxy-1,3-dihydrobenzofuran*

[55479-94-2]

$C_8H_8O_2$　　M 136.150
Tautomeric, with acetal predominating in soln., open-chain form in cryst. state. Cryst. (EtOAc/hexane). Mp 35-38°. Bp₀.₆ 104-105°.

2,4-Dinitrophenylhydrazone: Cryst. (EtOH). Mp 207°.
Di-Me acetal: [87656-32-4]. *2-(Dimethoxymethyl)-benzenemethanol, 9CI. 2-(Dimethoxymethyl)benzyl alcohol.*
$C_{10}H_{14}O_3$　　M 182.219
Isobenzofuran precursor. Liq. Bp₁.₂₅ 98-106°. Can decompose when distilled.

Davey, W. *et al, J. Org. Chem.*, 1961, **26**, 3699.
Smith, J.G. *et al, J. Org. Chem.*, 1983, **48**, 5361 (*deriv, synth, pmr, ir, use*)
Wulff, G. *et al, Chem. Ber.*, 1986, **119**, 1876.

2-(Hydroxymethyl)-1,4-benzodioxan　　　H-60172

2,3-Dihydro-1,4-benzodioxin-2-methanol, 9CI. 1,4-Benzodioxan-2-methanol, 8CI

[3663-82-9]

$C_9H_{10}O_3$　　M 166.176
Mp 89-90°.

Carbamate: [15567-77-8]. Cryst. (MeOH aq.). Mp 82-84°.

Koo, J. *et al, J. Am. Chem. Soc.*, 1955, **77**, 5373 (*synth*)
Mndzhoyen, A.L. *et al, CA*, 1968, **68**, 105117 (*synth*)

5-Hydroxy-4-methyl-2*H*-1-benzopyran-2-one　　　H-60173

5-Hydroxy-4-methylcoumarin

[2373-34-4]
$C_{10}H_8O_3$ M 176.171
Cryst. (MeOH). Mp 263°.

Me ether: 5-*Methoxy-4-methylcoumarin.*
   $C_{11}H_{10}O_3$ M 190.198
   Constit. of *Edgeworthia gardneri.* Cryst. (C_6H_6 or MeOH). Mp 141-142° (137-138°).

Okogun, J.I. *et al, Tetrahedron*, 1978, **34**, 1221 (*synth, uv, pmr*)
Chatterjee, A. *et al, Indian J. Chem., Sect. B*, 1987, **26**, 81 (*isol, deriv*)

2-Hydroxy-3-methyl-2-cyclopenten-1-one, H-60174
9CI

Updated Entry replacing H-20213
Cyclotene
[80-71-7]

$C_6H_8O_2$ M 112.128
Flavour constit. of coffee. Used in tobacco and food flavouring. Cryst. (H_2O). Mp 102-103°.
▷GY7298000.

Me ether: [14189-85-6]. 2-*Methoxy-3-methyl-2-cyclo-penten-1-one.*
   $C_7H_{10}O_2$ M 126.155
   Pale-yellow oil. Bp_1 45-48°.

Gianturco, M.A. *et al, Tetrahedron*, 1963, **19**, 2051 (*isol*)
Leir, C.M., *J. Org. Chem.*, 1970, **35**, 3203 (*synth*)
Sato, K. *et al, J. Org. Chem.*, 1973, **38**, 551 (*synth*)
Naoshima, Y. *et al, Agric. Biol. Chem.*, 1974, **38**, 2273 (*synth*)
Forsskahl, I. *et al, Carbohydr. Res.*, 1976, **48**, 13 (*synth*)
Shono, T. *et al, J. Chem. Soc., Chem. Commun.*, 1977, 712 (*synth*)
Strunz, G.M. *et al, Can. J. Chem.*, 1982, **60**, 572 (*synth, bibl*)
Schow, S.R. *et al, J. Am. Chem. Soc.*, 1986, **108**, 2662 (*deriv, synth, pmr*)

4-Hydroxy-2-methyl-2-cyclopenten-1-one, H-60175
9CI

[23535-17-3]

$C_6H_8O_2$ M 112.128
(±)-*form* [107708-98-5]
Oil.

Scettri, A. *et al, Tetrahedron*, 1979, **35**, 135 (*synth, ir, pmr*)
Baraldi, P.G. *et al, Synthesis*, 1986, 781 (*synth, pmr, ir*)

3-Hydroxymethyldibenzo[b,f]thiepin H-60176
Dibenzo[b,f]*thiepin-3-methanol, 9CI*
[77167-91-0]

$C_{15}H_{12}OS$ M 240.319
Cryst. (cyclohexane). Mp 110-112°.
5,5-*Dioxide:* [77167-93-2].
   $C_{15}H_{12}O_3S$ M 272.318

Antagonist of lung prostanoids. Cryst. (EtOH). Mp 161-163°.

Hands, D. *et al, J. Heterocycl. Chem.*, 1986, **23**, 1333 (*synth, pmr, bibl*)

2-Hydroxy-3,4-methylenedioxybenzoic H-60177
acid, 8CI
4-*Hydroxy-1,3-benzodioxole-5-carboxylic acid, 9CI.* 2-*Hydroxypiperonylic acid*

$C_8H_6O_5$ M 182.132
Needles (EtOH aq.). Mp 235° dec.

Me ether: [484-32-2]. 2-*Methoxy-3,4-methylenedioxy-benzoic acid.* 2-*Methoxypiperonylic acid. Croweacic acid.*
   $C_9H_8O_5$ M 196.159
   Oxidation product of croweacin from *Eriostemon crowei.* Needles (EtOH aq.). Mp 156°.
Me ether, Me ester: [23731-78-4].
   $C_{10}H_{10}O_5$ M 210.186
   Needles (hexane). Mp 25-26°. Bp_{2-3} 114-118°.

Baker, W. *et al, J. Chem. Soc.*, 1938, 1602; 1939, 439 (*synth, deriv*)
Dallacker, F. *et al, Chem. Ber.*, 1969, **102**, 2663; *Z. Naturforsch., C*, 1978, **33**, 465 (*synth*)
McKittrick, B.A. *et al, J. Chem. Soc., Perkin Trans. 1*, 1984, 709 (*synth, pmr*)

3-Hydroxy-4,5-methylenedioxybenzoic H-60178
acid

Updated Entry replacing M-00601
7-*Hydroxy-1,3-benzodioxole-5-carboxylic acid*
$C_8H_6O_5$ M 182.132

Me ether: [526-34-1]. 3-*Methoxy-4,5-methylenedioxy-benzoic acid.* 5-*Methoxypiperonylic acid. Myristicic acid. Myristicin acid.*
   $C_9H_8O_5$ M 196.159
   Isol. from seeds of *Apium graveolens* and from Chinese Gaoben. Needles (MeOH). Mp 212°.
Me ether, Me ester: [22934-58-3].
   $C_{10}H_{10}O_5$ M 210.186
   Cryst. (EtOAc/2-propanol). Mp 91°. Bp_2 138°.
Me ether, chloride: [76015-47-9].
   $C_9H_7ClO_4$ M 214.605
   Needles (C_6H_6/pet. ether). Mp 105°. Bp_{20} 189-190°.
Me ether, amide:
   $C_9H_9NO_4$ M 195.174
   Needles + $1H_2O$ (H_2O). Mp 184°.
Me ether, nitrile: [6443-68-1]. 3-*Methoxy-4,5-methylne-dioxybenzonitrile, 8CI.*
   $C_9H_7NO_3$ M 177.159
   Cryst. (MeOH). Mp 154-155°.

Salway, A.H., *J. Chem. Soc.*, 1909, **95**, 1161; 1911, **99**, 268 (*synth*)
Baker, W. *et al, J. Chem. Soc.*, 1932, 1283 (*synth*)
Seshadri, T.R., *Proc. Indian Acad. Sci., Sect. A*, 1950, **32**, 25 (*synth*)
Dallacker, F. *et al, Monatsh. Chem.*, 1960, **91**, 1089, 1103; 1969, **100**, 560 (*synth*)
Dallacker, F. *et al, Z. Naturforsch., C*, 1978, **33**, 465 (*synth*)
Garg, G.P. *et al, Indian J. Chem., Sect. B*, 1978, **16**, 658; 1979, **18**, 352 (*isol*)
Baba, K. *et al, CA*, 1984, **101**, 97724 (*isol*)

4-Hydroxy-2,3-methylenedioxybenzoic acid   H-60179

7-Hydroxy-1,3-benzodioxole-4-carboxylic acid, 9CI

$C_8H_6O_5$ M 182.132

Me ether: [23724-57-4]. *7-Methoxy-1,3-benzodioxole-4-carboxylic acid, 9CI. 4-Methoxy-2,3-methylenedioxybenzoic acid.*

$C_9H_8O_5$ M 196.159

Needles (EtOH). Mp 257.5°.

Me ether, Me ester: [23812-55-7].

$C_{10}H_{10}O_5$ M 210.186

Needles (MeOH/AcOH). Mp 123.5°.

Dallacker, F. *et al, Chem. Ber.*, 1969, **102**, 2663; *Z. Naturforsch., C*, 1978, **33**, 465 (*synth*)

6-Hydroxy-2,3-methylenedioxybenzoic acid   H-60180

5-Hydroxy-1,3-benzodioxole-4-carboxylic acid, 9CI

$C_8H_6O_5$ M 182.132

Me ether: [68803-47-4]. *5-Methoxy-1,3-benzodioxole-4-carboxylic acid, 9CI. 6-Methoxy-2,3-methylenedioxybenzoic acid.*

$C_9H_8O_5$ M 196.159

Cryst. (EtOAc/cyclohexane). Mp 142°.

Doyle, F.P. *et al, J. Chem. Soc.*, 1962, 1453 (*synth*)
Dallacker, F. *et al, Z. Naturforsch., C*, 1978, **33**, 465 (*synth*)

6-Hydroxy-3,4-methylenedioxybenzoic acid   H-60181

6-Hydroxy-1,3-benzodioxole-5-carboxylic acid, 9CI. 6-Hydroxypiperonylic acid, 8CI

[4890-01-1]

$C_8H_6O_5$ M 182.132

Cryst. (CHCl$_3$). Mp 238°.

Me ether: [7168-93-6]. *6-Methoxy-1,3-benzodioxole-5-carboxylic acid, 9CI. 6-Methoxy-3,4-methylenedioxybenzoic acid.*

$C_9H_8O_5$ M 196.159

Needles (CHCl$_3$). Mp 153-153.2°.

Me ether, chloride:

$C_9H_7ClO_4$ M 214.605

Mp 118-121°.

Dallacker, F. *et al, Justus Liebigs Ann. Chem.*, 1966, **694**, 98 (*deriv*)
Moron, J. *et al, Bull. Soc. Chim. Fr.*, 1967, 130 (*synth*)
Stout, G.H. *et al, Tetrahedron*, 1969, **25**, 5295 (*deriv*)
Rall, G.J.H. *et al, Tetrahedron*, 1970, **26**, 5007 (*deriv*)
Dallacker, F. *et al, Z. Naturforsch., C*, 1978, **33**, 465 (*deriv*)

5-Hydroxymethyl-2(5H)-furanone, 9CI   H-60182

Updated Entry replacing H-20218
5-Hydroxy-2-penten-4-olide. 5-Hydroxymethyl-2-butenolide

$C_5H_6O_3$ M 114.101

(S)-form [78508-96-0]

Oil which solidifies on standing. Mp 42-43°. Bp$_{0.9}$ 130°. $[\alpha]_D^{20}$ −155.4° (c, 1.15 in H$_2$O), $[\alpha]_D^{24}$ −136.3° (c, 0.25 in H$_2$O).

β-D-Glucopyranoside: [644-69-9]. **Ranunculin.**

$C_{11}H_{16}O_8$ M 276.243

Common in Ranunculaceae. Cryst. (MeOH). Mp 140-141°. $[\alpha]_D^{20}$ −81° (c, 2 in H$_2$O).

Ac: [85846-83-9].

$C_7H_8O_4$ M 156.138

Liq. $[\alpha]_D^{20}$ −150.9° (c, 3.41 in CHCl$_3$).

(±)-form

Bp$_1$ 140°.

Benn, M.H. *et al, Can. J. Chem.*, 1968, **46**, 729 (*abs config*)
Boll, M., ACS, 1968, **22**, 3245 (*synth*)
Camps, P. *et al, Tetrahedron*, 1982, **38**, 2395 (*synth*)
Cardellach, J. *et al, Tetrahedron*, 1982, **38**, 2377 (*synth*)
Takano, S. *et al, Synthesis*, 1986, 403 (*synth*)
Häfele, B. *et al, Justus Liebigs Ann. Chem.*, 1987, 85 (*synth, ir, pmr, cmr*)

5-Hydroxy-4-methyl-3-heptanone   H-60183

[79314-57-1]

(4R,5R)-form

$C_8H_{16}O_2$ M 144.213

(4R,5R)-form

Minor component (<0.5%) of sitophilure. Bp$_{17}$ 93-95°. $[\alpha]_D^{23}$ −37.8° (c, 1.20 in Et$_2$O).

(4R,5S)-form

Sitophilure

Aggregation pheromone of rice and maize weevils, *Sitophilus* spp. Bp$_6$ 80-82°. $[\alpha]_D^{20}$ −26.7° (c, 1.52 in Et$_2$O).

(4S,5S)-form [100483-73-6]

Bp$_3$ 69-74° (bath). $[\alpha]_D^{22}$ +36.8° (c, 1.25 in Et$_2$O).

(4S,5R)-form

Bp$_5$ 90-105° (bath). $[\alpha]_D^{20}$ +27.0° (c, 1.24 in Et$_2$O).

(±)-form

Bp$_{0.2}$ 55°. Mixt. of diastereoisomers.

Heathcock, C.H. *et al, J. Org. Chem.*, 1979, **44**, 4294 (*cmr*)
Smith, A.B. *et al, Synthesis*, 1981, 567 (*synth*)
Schmuff, N.R. *et al, Tetrahedron Lett.*, 1984, **25**, 1533 (*isol, ms, pmr*)
Mori, K. *et al, Tetrahedron*, 1986, **42**, 4421 (*synth, ir, pmr, cmr, ms, abs config*)

2-Hydroxy-2-methylhexanoic acid, 9CI   H-60184

(R)-form

$C_7H_{14}O_3$ M 146.186

(R)-form [70954-68-6]

Cryst. (hexane). Mp 68-70°. $[\alpha]_D$ −8.2° (c, 1 in H$_2$O).

(±)-form [70908-63-3]

Bp$_1$ 110-120°.

Cason, J. *et al, J. Org. Chem.*, 1954, **19**, 1947 (*synth*)
Meyers, A.I. *et al, J. Org. Chem.*, 1980, **45**, 2785 (*synth*)
Frater, G. *et al, Tetrahedron Lett.*, 1981, **22**, 4221 (*synth*)

2-Hydroxy-3-methylhexanoic acid, 9CI   H-60185

(2R,3S)-form

$C_7H_{14}O_3$ M 146.186

(2R,3S)-form [76713-40-1]

Bp$_{0.1}$ 91-95° (bath). $[\alpha]_D^{23}$ −15.8° (c, 0.378 in EtOH).

(2S,3R)-form [76713-39-8]

Bp$_{0.1}$ 87-91° (bath). $[\alpha]_D^{22.5}$ +15.8° (c, 0.621 in EtOH).

(±)-form

Bp$_{0.05}$ 110°. n_D^{20} 1.4467. Mixt. of diasteroisomers.

Achmatowicz, O. *et al*, *Rocz. Chem.*, 1962, **36**, 1791; *CA*, **59**, 8610g.

Mori, K. *et al*, *Tetrahedron*, 1980, **36**, 2209 (*synth*)

2-(6-Hydroxy-4-methyl-4-hexenylidene)-   H-60186
6,10-dimethyl-7-oxo-9-undecenal, 9CI

10-Formyl-16-hydroxy-2,6,14-trimethyl-2,10,14-hexa-decatrien-5-one

$C_{20}H_{32}O_3$ M 320.471

(10E,14Z)-form [110601-42-8]

(6E)-*10,11-Dihydro-12,19-dioxogeranylnerol*

Constit. of *Milleria quinqueflora*. Oil.

Δ³,⁴-*Isomer, 2-hydroxy:* [110601-43-9]. *10-Hydroxy-2-(6-hydroxy-4-methyl-4-hexylidene)-6,10-dimethyl-7-oxo-8-undecenal, 9CI. 15-Hydroxy-12,19-dioxo-13,14-dehydro-10,11,14,15-tetrahydrogeranylnerol. 10-Formyl-2,16-dihydroxy-2,6,14-trimethyl-3,10,14-hexadecatrien-5-one.*

$C_{20}H_{32}O_4$ M 336.470

Constit. of *M. quinqueflora*. Oil.

Jakupovic, J. *et al*, *Phytochemistry*, 1987, **26**, 2011.

5-Hydroxymethyl-4-methyl-2(5H)-furan-   H-60187
one, 9CI

Updated Entry replacing H-20221

5-Hydroxy-3-methyl-2-penten-4-olide. 4-Hydroxy-methyl-3-methyl-2-buten-1-olide. **Umbelactone**

(R)-form

$C_6H_8O_3$ M 128.127

(R)-form [69534-86-7]

Constit. of *Memycelon umbelatum*. Solid. Mp 65°. $[\alpha]_D$ +52°.

(±)-form [84412-93-1]

Solid. Mp 60-62°.

Agarwal, S.K. *et al*, *Phytochemistry*, 1978, **17**, 1663 (*isol*)

Caine, D. *et al*, *J. Org. Chem.*, 1983, **48**, 740 (*synth*)

Ortuño, R.M. *et al*, *Tetrahedron*, 1987, **43**, 2199 (*synth, abs config*)

3-(Hydroxymethyl)-2-naphthalenecarboxal-   H-60188
dehyde

3-(Hydroxymethyl)-2-naphthaldehyde. 1,3-Dihydro-1-hydroxynaphtho[2,3-c]furan

[65539-69-7]

$C_{12}H_{10}O_2$ M 186.210

Conts. 20% of hemiacetal tautomer. Cryst. (toluene/hexane). Mp 106-107°.

Dao, L.H. *et al*, *Can. J. Chem.*, 1977, **55**, 3791 (*synth*)

Smith, J.G. *et al*, *J. Org. Chem.*, 1986, **51**, 3762 (*synth, ir, pmr, cmr*)

5-Hydroxy-7-methyl-1,2-naphthoquinone   H-60189

5-Hydroxy-7-methyl-1,2-naphthalenedione, 9CI

$C_{11}H_8O_3$ M 188.182

Me ether: [1936-10-3]. *5-Methoxy-7-methyl-1,2-naphthoquinone.*

$C_{12}H_{10}O_3$ M 202.209

Red needles (pet. ether). Mp 207-208° dec.

Brown, A.G. *et al*, *J. Chem. Soc.*, 1965, 2355 (*synth, uv, ir*)

8-Hydroxy-3-methyl-1,2-naphthoquinone   H-60190

8-Hydroxy-3-methyl-1,2-naphthalenedione, 9CI

$C_{11}H_8O_3$ M 188.182

Me ether: [22267-03-4]. *8-Methoxy-3-methyl-1,2-naphthoquinone.*

$C_{12}H_{10}O_3$ M 202.209

Constit. of heartwood of a *Diospyros* sp. Red cryst. (CH$_2$Cl$_2$/pet. ether). Mp 155°.

Sidhu, G.S. *et al*, *Indian J. Chem.*, 1968, **6**, 681 (*uv, ir, pmr*)

12-Hydroxy-24-methyl-24-oxo-16-scalar-   H-60191
ene-22,25-dial

$C_{26}H_{38}O_4$ M 414.584

12α-form

O¹²-Ac:

$C_{28}H_{40}O_5$ M 456.621

Constit. of sponges *Dictyoceratida* and *Halichondria* spp. Antimicrobial. Amorph.

22-Carboxylic acid: 12α-Hydroxy-24-methyl-24,25-dioxo-14-scalaren-22-oic acid.

$C_{26}H_{38}O_5$ M 430.583

Constit. of *D.* and *H.* spp. Amorph. solid.

Nakagawa, M. *et al*, *Tetrahedron Lett.*, 1987, **28**, 431.

3-Hydroxy-2-methylpentanoic acid, 9CI   H-60192

Updated Entry replacing H-02544
3-Hydroxy-2-methylvaleric acid, *8CI*
[28892-73-1]

(2R,3S)-*form*
Absolute
configuration

$C_6H_{12}O_3$ M 132.159

(2R,3S)-*form* [77405-43-7]
Degradn. prod. of Mycobactin. Oil. $Bp_{0.1}$ 90-100°. $[\alpha]_D^{25}$
−14.8° (c, 4.3 in MeOH).

4-Bromophenacyl ester: Mp 89.5-90°. $[\alpha]_D^{18}$ −15° (c, 2.5
in MeOH).

(2S,3S)-*form*
Me ester: Bp_5 69-70°. $[\alpha]_D^{23}$ +12.2° (c, 1.12 in $CHCl_3$).

(2RS,3SR)-*form*
Me ester: [67498-21-9].
$C_7H_{14}O_3$ M 146.186
Oil. Bp_4 68-70°.

Snow, G.A., *J. Chem. Soc.*, 1954, 4080 (*isol, synth*)
Snow, G.A., *Biochem. J.*, 1965, **94**, 160 (*abs config*)
Kirmse, W. *et al*, *Justus Liebigs Ann. Chem.*, 1976, 1333
  (*synth*)
Aten, R.W. *et al*, *Synthesis*, 1978, 400 (*synth*)
Heathcock, C.H. *et al*, *J. Org. Chem.*, 1979, **44**, 4294 (*cmr*)
Mori, K. *et al*, *Tetrahedron*, 1986, **42**, 4685 (*synth, ir, pmr*)

3-Hydroxy-5-methyl-2,4,6-trinitrobenzoic   H-60193
acid, 9CI

2,4,6-Trinitro-3,5-cresotic acid. Nitrococussic acid
[602-14-2]

$C_8H_5N_3O_9$ M 287.142
Cryst. (C_6H_6/pet. ether). Mp 179-180° (sealed tube).
Me ether, Me ester:
  $C_{10}H_9N_3O_9$ M 315.196
  Cryst. (MeOH aq.). Mp 135-136°.

Barton, D.H.R. *et al*, *Tetrahedron*, 1959, **6**, 48 (*synth*)
Huang, L.-S. *et al*, *CA*, 1974, **81**, 3655.

4-Hydroxy-1,2-naphthoquinone   H-60194

4-Hydroxy-1,2-naphthalenedione, *9CI*

$C_{10}H_6O_3$ M 174.156
Parent compd. is unfavoured tautomer of 2-Hydroxy-1,4-
  naphthoquinone, H-50272 .

Me ether: [18916-57-9]. *4-Methoxy-1,2-*
naphthoquinone.
  $C_{11}H_8O_3$ M 188.182

Orange-yellow needles (EtOH). Mp 190-191°.
Et ether: [7473-19-0]. *4-Ethoxy-1,2-naphthoquinone.*
  $C_{12}H_{10}O_3$ M 202.209
Orange needles (C_6H_6/hexane). Mp 121-123°.

Teuber, H.-J. *et al*, *Chem. Ber.*, 1954, **87**, 1236 (*synth, uv*)
Takuwa, A. *et al*, *Chem. Pharm. Bull.*, 1986, **59**, 2959 (*synth, ir,
  pmr*)

2-Hydroxy-3-nitro-1,4-naphthoquinone   H-60195

2-Hydroxy-3-nitro-1,4-naphthalenedione, *9CI*
[54808-30-9]

$C_{10}H_5NO_5$ M 219.153
Shows antiallergic activity. Light-yellow cryst. (HCl aq.).
  Mp 162-163° dec.

Inoue, A. *et al*, *CA*, 1960, **54**, 4504g (*synth*)
Buckle, D.R. *et al*, *J. Med. Chem.*, 1977, **20**, 1059 (*synth*)

N-Hydroxy-N-nitrosoaniline   H-60196

N-*Hydroxy-N-nitrosobenzenamine*, *9CI*. N-*Nitroso-N-*
phenylhydroxylamine, *8CI*

PhN(OH)NO

$C_6H_6N_2O_2$ M 138.126

NH₄ salt: [135-20-6]. *Cupferron.* Reagent for separating
  Cu and Fe from other metals and quantitative analysis.
  Cryst. Mp 163-164°.

▷Highly toxic, exp. carcinogen. Dec. on htg. to NH_3 and
  NO_x. NC4725000.

Org. Synth., Coll. Vol., **1**, 177 (*synth*)
U.S.P., 3 413 349, (*1968*); *CA*, **70**, 47070 (*synth*)
Abou el Ela, A.H. *et al*, *Z. Naturforsch., B*, 1973, **28**, 610 (*uv,
  ir*)
Iida, H. *et al*, *CA*, 1978, **89**, 179627 (*synth*)

4-Hydroxy-3-nitrosobenzaldehyde, 9CI   H-60197

Deferriviridomycin A
[57350-38-6]

$C_7H_5NO_3$ M 151.121
Isol. from *Streptomyces viridans*. Antibacterial.

Fe complex: **Viridomycin A**. Isol. from *S. viridans*.
  Antibacterial agent. Mp >300°.

Blinova, I.N. *et al*, *Khim. Prir. Soedin.*, 1975, **11**, 490 (*isol*)
Yang, C.C. *et al*, *Antimicrob. Agents Chemother.*, 1981, **20**, 558
  (*props*)
Kurobane, I. *et al*, *J. Antibiot.*, 1987, **40**, 1131 (*isol, struct, bibl*)

15-Hydroxy-17-nor-8-labden-7-one H-60198

$C_{19}H_{32}O_2$ M 292.461

(13S)-form
Havardiol
Constit. of *Grindelia havardii*. Oil. $[\alpha]_D^{25}$ +22.4° (c, 0.3 in CHCl₃).

15-Carboxylic acid: 17-Nor-7-oxo-8-labden-15-oic acid. **Havardic acid F.**
$C_{19}H_{30}O_3$ M 306.444
From *G. harvardii*.
15-Carboxylic acid, Me ester: Cryst. Mp 82-83°. $[\alpha]_D^{25}$ +31.3° (c, 0.5 in CHCl₃).

Jolad, S.D. *et al*, *Phytochemistry*, 1987, **26**, 483.

3-Hydroxy-30-nor-12,20(29)-oleanadien-28-oic acid H-60199

Updated Entry replacing H-50275

3α-form

$C_{29}H_{44}O_3$ M 440.665

3α-form
3α-Akebonoic acid
Constit. of tissue cultures of *Akibia quinata*.
Me ester: Cryst. (CHCl₃/MeOH). Mp 200-202°. $[\alpha]_D^{24}$ +118.1° (c, 0.166 in CHCl₃).

3β-form [104777-60-8]
Akebonoic acid
Constit. of tissue cultures of *A. quinata*.
Me ester: Cryst. (CHCl₃/MeOH). Mp 152-155°. $[\alpha]_D^{24}$ +127.7° (c, 0.658 in CHCl₃).
3-O-[α-L-Rhamnopyranosyl(1→3)-α-L-rhamnopyranosyl](1→2)[β-D-glucopyranosyl(1→3)]-α-L-arabinopyranoside: **Guaianin D.**
$C_{52}H_{82}O_{20}$ M 1027.208
Constit. of bark of *Guaiacum officinale*.
3-O-[α-L-Rhamnopyranosyl(1→3)-α-L-rhamnopyranosyl(1→2)[β-D-glucopyranosyl(1→3)]-α-L-arabinopyranoside, 28-O-β-D-glucopyranoside: **Guaianin E.**
$C_{58}H_{92}O_{52}$ M 1621.334
Constit. of *G. officinale*. $[\alpha]_D^{20}$ +3.84° (c, 0.26 in MeOH).

Ikuta, A. *et al*, *Phytochemistry*, 1986, **25**, 1625.
Ahmad, V.U. *et al*, *Tetrahedron*, 1988, **43**, 247 (*derivs*)

3-Hydroxy-28,13-oleananolide H-60200

$C_{30}H_{48}O_3$ M 456.707
Constit. of *Salvia lanigera*.
Ac: Needles (CHCl₃). Mp 282°.

Al-Hazimi, H.M.G. *et al*, *Phytochemistry*, 1987, **26**, 1091.

17-Hydroxy-15,17-oxido-16-spongianone H-60201

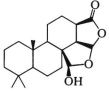

$C_{20}H_{30}O_4$ M 334.455
17β-form [106009-81-8]
Constit. of *Ceratosoma brevicaudatum*. Cryst. (CHCl₃).
Ksebati, M.B. *et al*, *J. Org. Chem.*, 1987, **52**, 3766.

9-Hydroxy-3-oxo-1,4(15),11(13)-eudesma-trien-12,6-olide H-60202

$C_{15}H_{16}O_4$ M 260.289
(6α,9β)-form
Ac: 9β-Acetoxy-3-oxo-1,4(15),11(13)-eudesmatrien-12,6-olide. **Gutenbergin**.
$C_{17}H_{18}O_5$ M 302.326
Constit. of *Gutenbergia cordifolia*. Cryst. (MeOH). Mp 189-191°. $[\alpha]_D^{25}$ +87° (c, 0.23 in MeOH).
Fujimoto, Y. *et al*, *Phytochemistry*, 1987, **26**, 2593.

6-Hydroxy-5-oxo-7,15-fusicoccadien-15-al H-60203

[108605-66-9]

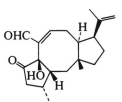

$C_{20}H_{28}O_3$ M 316.439
Metab. of *Cercospora traversiana*. Cryst. (EtOAc). Mp 234-235°.
Stoessl, A. *et al*, *J. Chem. Soc., Chem. Commun.*, 1987, 508 (*isol, struct, biosynth*)

11-Hydroxy-3-oxo-13-nor-7(11)-eudesmen-12,6-olide H-60204

$C_{14}H_{18}O_4$ M 250.294
Constit. of *Crepis pygmaea*. Cubes (Et₂O). Mp 179-180°. $[\alpha]_D$ +25.7° (c, 2.9 in CHCl₃).

Rossi, C. *et al*, *J. Chem. Res. (S)*, 1985, 160 (*synth*)
Rossi, C. *et al*, *Phytochemistry*, 1985, **24**, 603 (*isol*)

2-Hydroxy-6-(8,11-pentadecadienyl)-benzoic acid   H-60205

Updated Entry replacing H-10249
6-Pentadecadienylsalicylic acid
[18654-17-6]

$C_{22}H_{32}O_3$ M 344.493
(***all-Z***)-*form* [16611-84-0]
Anacardic acid
Found in cashew nut shell. Mp 25-26°.

11′,12′-Dihydro: 2-*Hydroxy*-6-(8-*pentadecenyl*)*benzoic acid.* 6-(8-*Pentadecenyl*)*salicylic acid.* **Ginkgoic acid.**
$C_{22}H_{34}O_3$ M 346.509
Constit. of *Ginkgo biloba*. Needles by subl. Mp 45.3-48°.

8′,9′-Dihydro, Δ$^{10′}$ *isomer:* 2-*Hydroxy*-6-(10-*pentadecenyl*)*benzoic acid.* 6-(10-*Pentadecenyl*)-*salicylic acid.*
$C_{22}H_{34}O_3$ M 346.509
Constit. of *Ozoroa mucronata*. Inhibits prostaglandin synthetase. Needles. Mp 45.8-46.2°.

Tetrahydro: 2-*Hydroxy*-6-*pentadecylbenzoic acid.* 6-*Pentadecylsalicyclic acid.*
$C_{22}H_{36}O_3$ M 348.525
Constit. of root of *O. mucronata*. Inhibits prostaglandin synthetase. Needles (hexane). Mp 90.2-91.5°.

Tyman, J.H.P., *Chem. Soc. Rev.*, 1979, **8**, 499 (*rev*)
Kubo, I. *et al*, *Chem. Lett.*, 1987, 1101 (*struct*)
Yamigawa, Y. *et al*, *Tetrahedron*, 1987, **43**, 3387 (*synth*)

2-Hydroxy-3*H*-phenoxazin-3-one, 9CI   H-60206

[1915-49-7]

$C_{12}H_7NO_3$ M 213.192
Red cryst. Mp 264-265° dec.

Me ether: 2-*Methoxy*-3H-*phenoxazin*-3-*one.*
$C_{13}H_9NO_3$ M 227.219
Pale-yellow cryst. Mp 255° dec.

Cavill, G.W.K. *et al*, *Tetrahedron*, 1961, **12**, 139 (*synth*)
Bolognese, A. *et al*, *J. Heterocycl. Chem.*, 1986, **23**, 1003 (*synth, uv, pmr*)

2-Hydroxyphenylacetaldehyde   H-60207

2-*Hydroxybenzeneacetaldehyde*, 9CI
[7451-95-8]

$C_8H_8O_2$ M 136.150
Bp$_{0.01}$ 75-80° (bath).

2,4-Dinitrophenylhydrazone: [24007-54-3]. Orange cryst. Mp 165-168°.
Me ether: [33567-59-8]. 2-*Methoxybenzeneacetaldehyde*, 9CI. (2-*Methoxyphenyl*)*acetaldehyde.*
$C_9H_{10}O_2$ M 150.177
Bp$_{0.3}$ 65-67°.
Me ether, oxime: [82204-36-2].
$C_9H_{11}NO_2$ M 165.191

Mp 87-90°.
Me ether, semicarbazone: Cryst. (MeOH/EtOH). Mp 161-162°.

Bruce, J.M. *et al*, *J. Chem. Soc. (C)*, 1970, 649 (*synth*)
Billman, J.H. *et al*, *J. Pharm. Sci.*, 1971, **60**, 1188 (*deriv*)
Beugelmans, R. *et al*, *J. Chem. Soc., Chem. Commun.*, 1980, 508 (*deriv*)
Stamos, I.K., *Tetrahedron Lett.*, 1982, **23**, 459 (*deriv*)

3-Hydroxyphenylacetaldehyde   H-60208

3-*Hydroxybenzeneacetaldehyde*, 9CI
[81104-39-4]
$C_8H_8O_2$ M 136.150
Bp$_{0.85}$ 96-98°.

Me ether, 2,4-dinitrophenylhydrazone: Red. cryst. (EtOH/EtOAc). Mp 134-135°.
Me ether, oxime:
$C_9H_{11}NO_2$ M 165.191
Mp 92°.
Me ether, semicarbazone: Needles. Mp 131.5-132.7°.

Lin, K.H. *et al*, *J. Chem. Soc.*, 1938, 2005 (*deriv*)
Auterhoff, H. *et al*, *Arch. Pharm. (Weinheim, Ger.)*, 1956, **289**, 470 (*synth*)
Nelson, N.A. *et al*, *J. Am. Chem. Soc.*, 1958, **80**, 6626 (*synth*)
Gupta, A.K.D. *et al*, *J. Sci. Ind. Res., Sect. B*, 1961, **20**, 394 (*deriv*)

4-Hydroxyphenylacetaldehyde   H-60209

4-*Hydroxybenzeneacetaldehyde*, 9CI
[7339-87-9]
$C_8H_8O_2$ M 136.150
No phys. props. reported.

Oxime: [23745-82-6]. 4-*Hydroxyphenylacetaldoxime.*
$C_8H_9NO_2$ M 151.165
Isol. from *Streptomyces nigellus*. Inhibits β-galactosidase. Needles (CHCl$_3$). Mp 110-111°.
Semicarbazone: Cryst. (EtOH aq.). Mp 187.5-189° dec. Subl. at 186°.
Ac, oxime:
$C_{10}H_{11}NO_3$ M 193.202
Needles (C$_6$H$_6$/pet. ether). Mp 92-94°.
Me ether: [5703-26-4]. 4-*Methoxyphenylacetaldehyde.*
$C_9H_{10}O_2$ M 150.177
Liq. d$_4^{20}$ 1.096. Bp$_{12}$ 123-124°.
Me ether, oxime:
$C_9H_{11}NO_2$ M 165.191
Mp 121°.
Me ether, semicarbazone: Mp 181-182°.

Harries, C. *et al*, *Ber.*, 1916, **49**, 1029 (*deriv*)
Robbins, J.H., *Arch. Biochem. Biophys.*, 1966, **114**, 576 (*synth*)
Zielinski, W., *CA*, 1968, **69**, 2679b (*deriv*)
Veeraswamy, M. *et al*, *Biochem. J.*, 1976, **159**, 807 (*synth*)
Hazato, T. *et al*, *J. Antibiot.*, 1979, **32**, 91, 212 (*oxime*)
Severin, T. *et al*, *Chem. Ber.*, 1980, **113**, 970 (*deriv*)

2-Hydroxy-2-phenylacetaldehyde   H-60210

α-*Hydroxybenzeneacetaldehyde*, 9CI. Mandelaldehyde, 8CI. Phenylglycolaldehyde
[34025-29-1]

$C_8H_8O_2$ M 136.150

Exists as dimer.

(S)-form

Me ether: [86544-37-8]. $[\alpha]_D$ +132° (CHCl$_3$).

(±)-form

Cryst. (pet. ether). Mp 137-138°.

Oxime: Mp 93-94°.

Phenylhydrazone: Cryst. (benzaldehyde). Mp 103°.

Ac: [22094-21-9].
C$_{10}$H$_{10}$O$_3$ M 178.187
Bp$_{0.8}$ 92-93°.

Ac, 2,4-dinitrophenylhydrazone: Orange-yellow needles (EtOH). Mp 134-136.5° (121°).

Me ether: [19190-53-5]. *α-Methoxybenzeneacetalde-hyde, 9CI. 2-Methoxy-2-phenylacetaldehyde.*
C$_9$H$_{10}$O$_2$ M 150.177
Bp$_{30}$ 130-133°.

Me ether, 2,4-dinitrophenylhydrazone: [4881-00-9]. Mp 148°.

Di-Me acetal: [21504-23-4]. *α-(Dimethoxymethyl)-benzenemethanol, 9CI. 2,2-Dimethoxy-1-phenylethanol.*
C$_{10}$H$_{14}$O$_3$ M 182.219
Pale-yellow oil. Bp$_{0.25}$ 75-78°.

Riehl, J.-J. *et al, Bull. Soc. Chim. Fr.*, 1968, 4083 (*derivs*)
Russell, G.A. *et al, J. Org. Chem.*, 1969, **34**, 3618 (*synth*)
Kauffmann, T. *et al, Chem. Ber.*, 1977, **110**, 3034 (*deriv*)
Shono, T. *et al, Tetrahedron Lett.*, 1982, **23**, 4801 (*deriv*)
Guanti, G. *et al, Tetrahedron Lett.*, 1983, **24**, 817 (*deriv*)
Annunziata, R. *et al, J. Chem. Soc., Perkin Trans. 1*, 1985, 255 (*deriv*)

(4-Hydroxyphenyl)ethylene   H-60211

Updated Entry replacing H-10259
4-Ethenylphenol, 9CI. p-Vinylphenol, 8CI. 4-Hydroxystyrene
[2628-17-3]
C$_8$H$_8$O   M 120.151
Mp 68-69°.
▷SN3800000.

Ac: [2628-16-2].
C$_{10}$H$_{10}$O$_2$ M 162.188
Fp 7.4° and 8.2° (dimorph.). Bp$_3$ 99-100°, Bp$_{0.6}$ 73-75°.

Benzoyl: [32568-59-5].
C$_{15}$H$_{12}$O$_2$ M 224.259
Needles (MeOH). Mp 75.5-76.5°.

4-O-(β-D-Xylopyranosyl(1→6)-β-D-glucopyranoside): [90899-20-0]. **Ptelatoside A.**
C$_{19}$H$_{26}$O$_{10}$ M 414.408
Constit. of bracken fern *Pteridium aquilinum*. Cryst. (Me$_2$CO aq.). Mp 183-185°. $[\alpha]_D^{22}$ −104° (c, 0.68 in H$_2$O).

4-O-(α-L-Rhamnopyranosyl(1→2)-β-D-glucopyranoside): [90852-99-6]. **Ptelatoside B.**
C$_{20}$H$_{28}$O$_{10}$ M 428.435
Constit. of *P. aquilinum*. Amorph. $[\alpha]_D^{23}$ −94.8° (c, 1 in H$_2$O).

Corson, B.B. *et al, J. Org. Chem.*, 1958, **23**, 544 (*synth*)
Arshady, R. *et al, J. Polym. Sci., Polym. Chem. Ed.*, 1974, **12**, 2017 (*synth*)
Hatakeyama, H. *et al, Polymer*, 1978, **19**, 593 (*synth*)
Ojika, M. *et al, Chem. Lett.*, 1984, 397.
Ojika, M. *et al, Tetrahedron*, 1987, **43**, 5275 (*deriv, synth*)

2-(4-Hydroxyphenyl)-7-methoxy-5-(1-pro-   H-60212
penyl)-3-benzofurancarboxaldehyde, 9CI

3-Formyl-2-(4-hydroxyphenyl)-7-methoxy-5-(1-propenyl)benzofuran
[109194-68-5]

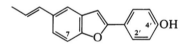

C$_{19}$H$_{16}$O$_4$ M 308.333
Constit. of *Krameria cystisoides*. Yellow solid. Mp 258-261°.

Achenbach, H. *et al, Phytochemistry*, 1987, **26**, 1159.

3-(4-Hydroxyphenyl)-2-propenal, 9CI   H-60213

Updated Entry replacing H-10270
p-Hydroxycinnamaldehyde, 8CI. p-Coumaraldehyde
[2538-87-6]
C$_9$H$_8$O$_2$ M 148.161
Constit. of *Alpinia galanga* rhizomes. Cryst. (EtOAc). Mp 136-138°.

Semicarbazone: Mp 224°.

Me ether: [1963-36-0]. *3-(4-Methoxyphenyl)-2-propenal. p-Methoxycinnamaldehyde.*
C$_{10}$H$_{10}$O$_2$ M 162.188
Mp 58-59°. Bp$_{15}$ 171°.

Pauly, H. *et al, Ber.*, 1923, **56**, 603.
Friedrich, K. *et al, Chem. Ber.*, 1961, **94**, 838.
Aulin-Erdtman, G. *et al, Acta Chem. Scand.*, 1968, **22**, 1187 (*uv*)
Sadler, I.H. *et al, J. Chem. Soc., Chem. Commun.*, 1969, 773.
Barik, B.R. *et al, Phytochemistry*, 1987, **26**, 2126 (*isol*)

2-(4-Hydroxyphenyl)-5-(1-propenyl)-   H-60214
benzofuran

4-[5-(1-Propenyl)-2-benzofuranyl]phenol, 9CI
[109194-69-6]

C$_{17}$H$_{14}$O$_2$ M 250.296
Constit. of *Krameria cystisoides*. Cryst. Mp 208-211°.

2′-Hydroxy: [109194-71-0]. *4-[5-(1-Propenyl)-2-benzo-furanyl]-2,3-benzenediol, 9CI. 2-(2,4-Dihydroxy-phenyl)-5-(1-propenyl)benzofuran.*
C$_{17}$H$_{14}$O$_3$ M 266.296
Constit. of *K. cystisoides*. Cryst. Mp 181-184°.

2′-Hydroxy,4′-Me ether: [79214-54-3]. *2-(2-Hydroxy-4-methoxyphenyl)-5-(1-propenyl)benzofuran.* **Ratan-hiaphenol I.**
C$_{18}$H$_{16}$O$_3$ M 280.323
Constit. of *K. triandra* and *K. cystisoides*. Cryst. Mp 182-184°.

7-Methoxy: [109194-70-9]. *2-(4-Hydroxyphenyl)-7-methoxy-5-(1-propenyl)benzofuran.*
C$_{18}$H$_{16}$O$_3$ M 280.323
Constit. of *K. cystisoides*. Cryst. Mp 177-179°.

2′-Hydroxy, 7-Methoxy: [109194-72-1]. *2-(2,4-Dihy-droxyphenyl)-7-methoxy-5-(1-propenyl)benzofuran.*
C$_{18}$H$_{16}$O$_4$ M 296.322
Constit. of *K. cystisoides*. Cryst. Mp 172-174°.

2′-Hydroxy, 4′-Me ether, 7-Methoxy: [109145-64-4]. *2-(2-Hydroxy-4-methoxyphenyl)-7-methoxy-5-(1-propenyl)benzofuran.* **Toltecol.**
C$_{19}$H$_{18}$O$_4$ M 310.349

Constit. of *K. cystisoides*. Cryst. Mp 101-102°.

Stahl, E. *et al, Planta Med.*, 1981, **42**, 144 (*isol*)
Achenbach, H. *et al, Phytochemistry*, 1987, **26**, 1159 (*isol, pmr, cmr*)

1-(4-Hydroxyphenyl)-2-(4-propenylphen-   H-60215
oxy)-1-propanol

[109194-73-2]

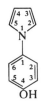

C$_{18}$H$_{20}$O$_3$ M 284.354

Constit. of *Krameria cystisoides*. Oil. [α]$_D^{21}$ −6.4° (c, 0.44 in MeOH).

2-Methoxy: [109225-43-6]. *1-(4-Hydroxyphenyl)-2-(2-methoxy-4-propenylphenoxy)-1-propanol*.
C$_{19}$H$_{22}$O$_4$ M 314.380
Constit. of *K. cystisoides*. [α]$_D^{21}$ −11.6° (c, 0.79 in MeOH).

Achenbach, H. *et al, Phytochemistry*, 1987, **26**, 1159.

1-(4-Hydroxyphenyl)pyrrole   H-60216
4-(1H-Pyrrol-1-yl)phenol, 9CI
[23351-09-9]

C$_{10}$H$_9$NO   M 159.187
Mp 119-121°.

Me ether: [5145-71-1]. *1-(4-Methoxyphenyl)pyrrole*.
C$_{11}$H$_{11}$NO   M 173.214
Mp 111-113°.

Gross, H. *et al, Chem. Ber.*, 1962, **95**, 2270 (*synth*)
Matsuo, T. *et al, Bull. Chem. Soc. Jpn.*, 1968, **41**, 2849 (*spectra*)

2-(2-Hydroxyphenyl)pyrrole   H-60217
2-(1H-Pyrrol-2-yl)phenol, 9CI. Pseudilin
[42041-50-9]
C$_{10}$H$_9$NO   M 159.187

Me ether: [69640-32-0]. *2-(2-Methoxyphenyl)pyrrole*.
C$_{11}$H$_{11}$NO   M 173.214
Oil.

ApSimon, J.W. *et al, J. Chem. Soc., Perkin Trans. 1*, 1978, 1588 (*deriv*)
Ger. Pat., 2 835 439, (*1980*); *CA*, **93**, 132363 (*synth*)

8-Hydroxy-7-phenylquinoline   H-60218
7-Phenyl-8-quinolinol, 9CI
[52793-99-4]
C$_{15}$H$_{11}$NO   M 221.258
Plates (hexane). Mp 146-149°.

Barton, D.H.R. *et al, J. Chem. Soc., Perkin Trans. 1*, 1985, 2657 (*synth*)

3-(4-Hydroxyphenyl)-1-(2,4,6-trihydroxy-   H-60219
phenyl)-1-propanone

Updated Entry replacing H-10278
2,4,4′,6-Tetrahydroxydihydrochalcone. Dihydronari-genin. Asebogenol. Phloretin
[60-82-2]

HO⟨4′⟩COCH$_2$CH$_2$⟨4⟩OH (with OH, OH substituents)

C$_{15}$H$_{14}$O$_5$ M 274.273
Needles (EtOH aq.). Insol. Et$_2$O. Mp ca. 262-264° dec.

4′-Me ether: [520-42-3]. *1-(2,6-Dihydroxy-4-methoxy-phenyl)-3-(4-hydroxyphenyl)-1-propanone*. **Asebo-genin**. *Asebotol*.
C$_{16}$H$_{16}$O$_5$ M 288.299
Found in *Piperis japonica*, *Rhododendron* spp. and *Pitryogramma calomelanos*. Cryst. (EtOH). Mp 168°.

2′-Glucosyl: [60-81-1]. **Phloridzin**. *Phlorrhizin. Phlorhizin.*
C$_{21}$H$_{24}$O$_6$ M 372.417
Occurs in *Micromelum teprocarpum*, in apple (*Malus*), *R.* spp., *Kalmia latifolia* and *Piperis japonica*. Produces glucosuria in man. Herbivore antifeedant. Needles. [α]$_D^{18}$ −52.1° (EtOH aq.). Various Mp's, some double, recorded between 108° and 170°.

▷UC2080000.

4′-Me ether, 2′-Glucoside: [11075-15-3]. **Asebotin**. *Asebotoside*.
C$_{22}$H$_{26}$O$_6$ M 386.444
Constit. of *Andromeda japonica*, *R.* spp., *K. latifolia* and *P. japonica*. Mp 148° (135-136°). [α]$_D^{25}$ −46.2° (EtOH).

Eykman, J.F., *Ber.*, 1883, **16**, 2769 (*isol*)
Tamura, K., *J. Chem. Soc. Jpn.*, 1936, **57**, 1141 (*synth, Asebogenin*)
Zemplén, R. *et al, Ber.*, 1942, **75**, 645, 1298 (*synth*)
Murakami, S., *J. Pharm. Soc. Jpn.*, 1955, **75**, 573, 603 (*isol*)
Batterham, T.J. *et al, Aust. J. Chem.*, 1964, **17**, 428 (*pmr*)
Williams, A.-H. *et al, Nature (London)*, 1964, **202**, 824 (*isol*)
Rice, G.L., *Allelopathy*, Academic Press, N.Y., 1974 (*deriv*)
King, B. *et al, Phytochemistry*, 1975, **14**, 1448 (*deriv*)
Hutz, C. *et al, Z. Naturforsch. C*, 1982, **37**, 337 (*isol*)
Mancini, S.D. *et al, J. Nat. Prod.*, 1979, **42**, 483 (*isol*)

3-(4-Hydroxyphenyl)-1-(2,4,6-trihydroxy-   H-60220
phenyl)-2-propen-1-one, 9CI

Updated Entry replacing H-20235
2′,4,4′,6′-Tetrahydroxychalcone, 8CI. 4-Hydroxystyryl 2,4,6-trihydroxyphenyl ketone. **Chalconaringenin**
[73692-50-9]
C$_{15}$H$_{12}$O$_5$ M 272.257

2′-(O-Rhamnosyl(1→4)xyloside): [82344-84-1].
C$_{26}$H$_{30}$O$_{13}$ M 550.515
Pigment from *Acacia dealbata*.

2′-Me ether, 4′-glucoside: [61826-89-9]. **Helichrysin**. *Dehydro-p-asebotin*.
C$_{22}$H$_{24}$O$_{10}$ M 448.426
Constit. of the flowers of *Helichrysum* spp. and of *Gnaphalium affine*. Yellow cryst. (MeOH). Mp 246° dec. (199-200°). [α]$_D^{30}$ −85.7° (MeOH).

4′,6′-Di-Me ether: [56798-34-6]. *2′,4-Dihydroxy-4′,6′-dimethoxychalcone*.
C$_{17}$H$_{16}$O$_5$ M 300.310
Orange-red needles (C$_6$H$_6$). Mp 188°.

4′,6′-Di-Me, 2′,4-di-Ac:
C$_{21}$H$_{20}$O$_7$ M 384.385

Pale-yellow needles (MeOH). Mp 147°.

2',4',6'-Tri-Me ether: 4-Hydroxy-2',4',6'-trimethoxychalcone.
$C_{18}H_{18}O_5$ M 314.337
Golden-yellow cryst. (MeOH). Mp 195-196°.

2',4',6'-Tri-Me, 4-Ac:
$C_{20}H_{20}O_6$ M 356.374
Yellow cryst. (MeOH). Mp 108°.

4,4',6'-Tri-Me ether: [3420-72-2]. 2'-Hydroxy-4,4',6'-trimethoxychalcone.
$C_{18}H_{18}O_5$ M 314.337
Constit. of *Piper methysticum* and *Dahlia tenuicaulis*. Yellow needles (EtOH). Mp 113°.

4,4',6'-Tri-Me, 2-Ac:
$C_{20}H_{20}O_6$ M 356.374
Yellowish leaflets (EtOH). Mp 120°, 108-109° (dimorph.).

Tetra-Me ether: [25163-67-1].
$C_{19}H_{20}O_5$ M 328.364
Pale-yellow cryst. (EtOH aq.). Mp 119-121°.

4'-(3-Methyl-2-butenyl):
$C_{20}H_{20}O_5$ M 340.375
Constit. of *H. athrisciifolium*. Gum.

Dihydro, 6'-Me ether: 3-(4-Hydroxyphenyl)-1-(2,4-dihydroxy-6-methoxyphenyl)-1-propanone, 9CI. 2',4,4'-Trihydroxy-6'-methoxydihydrochalcone.
$C_{16}H_{16}O_5$ M 288.299
Constit. of *Coptis japonica* var. *dissecta*. Needles (EtOH). Mp 195°.

Kostanecki, S.v. *et al, Ber.*, 1904, **37**, 792 (*synth*)
Mosimann, W. *et al, Ber.*, 1916, **49**, 1701 (*synth*)
B.P., 914 248, (*1962*); *CA*, **58**, 12472e (*synth*)
Guise, G.B. *et al, Aust. J. Chem.*, 1962, **15**, 314 (*isol, deriv*)
Mahanthy, P., *Indian J. Chem.*, 1965, **3**, 121 (*ir*)
Rakosi-David, E. *et al, Acta Phys. Chim.*, 1968, **14**, 145 (*uv*)
Ramakrishnan, V.T. *et al, J. Org. Chem.*, 1970, **35**, 2901 (*synth*)
Aritomi, M. *et al, Chem. Pharm. Bull.*, 1974, **22**, 1800 (*Dehydroasebotin*)
Lam, J. *et al, Phytochemistry*, 1975, **14**, 1621 (*isol, deriv*)
Wright, W.G., *J. Chem. Soc., Perkin Trans. 1*, 1976, 1819 (*Helichrysin*)
Duddeck, H. *et al, Phytochemistry*, 1978, **17**, 1369 (*pmr*)
Imperato, F., *Phytochemistry*, 1982, **21**, 480 (*rhamnosylxyloside*)
Bohlmann, F. *et al, Phytochemistry*, 1984, **23**, 1338 (*isol*)
Mizuno, M. *et al, Phytochemistry*, 1987, **26**, 2071 (*deriv*)

1-Hydroxypinoresinol   **H-60221**

Updated Entry replacing H-50311

[81426-17-7]

$C_{20}H_{22}O_7$ M 374.390
Constit. of bark of *Olea europaea*. Cryst. powder. Mp 183-185°. $[\alpha]_D^{15}$ +39.0° (c, 0.65 in EtOH).

1-Ac: [81426-14-4]. 1-Acetoxypinoresinol.
$C_{22}H_{24}O_8$ M 416.427
Constit. of *O. europaea*. Amorph. powder. $[\alpha]_D^{21}$ +31.4° (c, 0.99 in EtOH).

1-O-β-D-Glucopyranoside:
$C_{26}H_{32}O_{12}$ M 536.532
Constit. of *O. europaea* and *O. africana*. Amorph. powder. Mp 179-183°. $[\alpha]_D^{23}$ −17.5° (c, 0.38 in EtOH).

4-O-β-D-Glucopyranoside:
$C_{26}H_{32}O_{12}$ M 536.532
Constit. of *O. europaea*, *O. africana* and *Fraxinus mandshurica*. Mp 127-129°. $[\alpha]_D^{23}$ −9.3° (c, 0.42 in MeOH).

1-Ac, 4'-O-β-D-glucopyranoside: 1-Acetoxypinoresinol 4'-O-β-D-glucopyranoside.
$C_{28}H_{34}O_{13}$ M 578.569
Constit. of bark of *O. europaea* and *O. africana*. Needles. Mp 183.5-185°. $[\alpha]_D^{22}$ +7.9° (c, 1.0 in EtOH).

4"-Me ether: [81426-18-8].
$C_{21}H_{24}O_7$ M 388.416
Constit. of *O. europaea*. Amorph. powder. $[\alpha]_D^{23}$ +37.9° (c, 0.81 in CHCl$_3$).

4"-Me ether, 1-Ac:
$C_{23}H_{26}O_8$ M 430.454
Constit. of *O. europaea*. Amorph. powder. $[\alpha]_D^{23}$ +29.9° (c, 1.14 in CHCl$_3$).

4"-Me ether, 1-Ac, 4'-O-β-D-glucopyranoside:
$C_{29}H_{36}O_{13}$ M 592.596
Constit. of *O. europaea* and *O. africana*. Amorph. powder. $[\alpha]_D^{20}$ +9.1° (c, 1.6 in EtOH).

5',5"-Dimethoxy: [89199-95-1]. **1-Hydroxysyringaresinol**.
$C_{22}H_{26}O_9$ M 434.442
Constit. of *F. mandshurica* var. *japonica* and *F. japonica*. Cryst. powder. Mp 95-97°. $[\alpha]_D^{26}$ +24.6° (c, 0.12 in CHCl$_3$).

Tsukamoto, H. *et al, Chem. Pharm. Bull.*, 1984, **32**, 2730, 4482; 1985, **33**, 1232.

3-Hydroxy-2(1*H*)-pyridinethione   **H-60222**
3-Hydroxy-2-mercaptopyridine
[23003-22-7]

C_5H_5NOS M 127.161
Cryst. (EtOH). Mp 144-145°.

Udheim, K. *et al, Acta Chem. Scand.*, 1969, **23**, 1704 (*synth, uv*)
Davies, J.S. *et al, Tetrahedron Lett.*, 1980, **21**, 2191 (*synth*)

3-Hydroxy-2(1*H*)-pyridinone, 9CI   **H-60223**
Updated Entry replacing H-03343
[16867-04-2]

$C_5H_5NO_2$ M 111.100
Mp 248°.

N-Me: [19365-01-6].
$C_6H_7NO_2$ M 125.127
Mp 130-131°.

3-Ac:
$C_7H_7NO_3$ M 153.137
Mp 155°.

Diol-form

2,3-Pyridinediol. 2,3-Dihydroxypyridine

Di-Me ether: [52605-97-7]. 2,3-Dimethoxypyridine.
$C_7H_9NO_2$ M 139.154
Bp$_{17}$ 100°.

Bain, B.M. *et al*, *J. Chem. Soc.*, 1961, 5216 (*synth*)
Moehrle, H. *et al*, *Tetrahedron*, 1970, **26**, 3779 (*synth*)
Stogryn, E.C. *et al*, *J. Heterocycl. Chem.*, 1974, **11**, 251 (*deriv*)
Elguero, J. *et al*, *Tautomerism of Heterocycles*, 1976, Academic Press, London, 111 (*tautom*)

4-Hydroxy-2(1*H*)-pyridinone, 9CI   H-60224

Updated Entry replacing H-03344
3-Deazauracil
$C_5H_5NO_2$ M 111.100
2-One-form predominates. Cryst. (H_2O or EtOH). Mp 260-265° dec.

N-*Me*: [40357-87-7].
  $C_6H_7NO_2$ M 125.127
  Mp 171-172°.
O^4,N-*Di-Me*: [41759-19-7]. *4-Methoxy-1-methyl-2(1H)-pyridinone.*
  $C_7H_9NO_2$ M 139.154
  Mp 113-114°.

4-One-form
2-Hydroxy-4(1H)-pyridinone
O^2,N-*Di-Me*: *2-Methoxy-1-methyl-4(1H)-pyridinone.*
  $C_7H_9NO_2$ M 139.154
  Mp 141-143°.

Diol-form
2,4-Pyridinediol. 2,4-Dihydroxypyridine
Di-Me ether: [18677-43-5]. *2,4-Dimethoxypyridine.*
  $C_7H_9NO_2$ M 139.154
  Bp 200-201°.
Di-Me ether, 1-oxide:
  $C_7H_9NO_3$ M 155.153
  Cryst. (C_6H_6). Mp 85°.
Di-Et ether: [52311-30-5]. *2,4-Diethoxypyridine.*
  $C_9H_{13}NO_2$ M 167.207
  Oil.
Di-Et ether, picrate: Mp 136-138°.

Talik, Z., *Rocz. Chem.*, 1961, **35**, 475; *CA*, **57**, 15066 (*derivs*)
Pieterse, M.J. *et al*, *Recl. Trav. Chim. Pays-Bas*, 1962, **81**, 855 (*derivs*)
Tiecklemann, H., *Chem. Heterocycl. Compd.*, 1974, **14**, (*Suppl. 3*), 597
Shone, R.C. *et al*, *J. Heterocycl. Chem.*, 1975, **12**, 389 (*synth*)
Elguero, J. *et al*, *Tautomerism of Heterocycles*, 1976, Academic Press, London, 108.
Law, J.N. *et al*, *Acta Crystallogr.*, Sect. C, 1983, **39**, 1688 (*cryst struct*)

5-Hydroxy-2(1*H*)-pyridinone, 9CI   H-60225

Updated Entry replacing H-03345
[5154-01-8]
$C_5H_5NO_2$ M 111.100
Pyridone-form predominates. Cryst. (H_2O). Mp 248°.
B,HCl: Mp 154°.
5-Ac:
  $C_7H_7NO_3$ M 153.137
  Mp 156°.
N-*Me*: [29094-75-5].
  $C_6H_7NO_2$ M 125.127
  Cryst. (C_6H_6/Et_2O). Mp 150-153°.

Diol-form
2,5-Pyridinediol. 2,5-Dihydroxypyridine
Di-Et ether: 2,5-Diethoxypyridine.
  $C_9H_{13}NO_2$ M 167.207

Bp 216-217°.
Di-Et ether, picrate: Mp 118-119°.

Den Hertog, H.J. *et al*, *Recl. Trav. Chim. Pays-Bas*, 1950, **69**, 700 (*derivs*)
Moehrle, H. *et al*, *Tetrahedron*, 1970, **26**, 3779 (*synth*)
Nantka-Namirski, P. *et al*, *Acta Pol. Pharm.*, 1974, **31**, 433; *CA*, **82**, 139899 (*synth*)
Elguero, J. *et al*, *Tautomerism of Heterocycles*, 1976, Academic Press, London, 107.

6-Hydroxy-2(1*H*)-pyridinone, 9CI   H-60226

Updated Entry replacing H-03346
Glutaconimide
[626-06-2]
$C_5H_5NO_2$ M 111.100
OH-form predominates in H_2O, *CH*-form in dioxan. Mp 195°.
N-*Me*: [6231-17-0].
  $C_6H_7NO_2$ M 125.127
  Cryst. (EtOAc/pet. ether). Mp 162-163°.
Me ether: 6-Methoxy-2(1H)-pyridinone.
  $C_6H_7NO_2$ M 125.127
  Solid. Subl. readily.
Me ether, N-Me: [6231-16-9].
  $C_7H_9NO_2$ M 139.154
  Cryst. (EtOAc/pet. ether). Mp 52-54°.

Diol-form
2,6-Pyridinediol. 2,6-Dihydroxypyridine
Di-Me ether: [6231-18-1]. *2,6-Dimethoxypyridine.*
  $C_7H_9NO_2$ M 139.154
  Oil.
Di-Et ether: [13472-57-6]. *2,6-Diethoxypyridine.*
  $C_9H_{13}NO_2$ M 167.207
  Mp 21.5°.

Bickel, A.F. *et al*, *Recl. Trav. Chim. Pays-Bas*, 1946, **65**, 65.
Katritzky, A.R. *et al*, *J. Chem. Soc. (B)*, 1966, 562 (*tautom*)
Spinner, E. *et al*, *Aust. J. Chem.*, 1971, **24**, 2557 (*tautom*)
Kaneko, C. *et al*, *Chem. Pharm. Bull.*, 1986, **34**, 3658 (*deriv*)

3-Hydroxy-4(1*H*)-pyridinone, 9CI   H-60227

Updated Entry replacing H-03348
[10182-48-6]
$C_5H_5NO_2$ M 111.100
Pyridone-form predominates. Cryst. + 1H_2O (H_2O). Mp 239.5-240° dec., Mp >250° dec.
3-Me ether: [50700-60-2]. *3-Methoxy-4(1H)-pyridinone.*
  $C_6H_7NO_2$ M 125.127
  Cryst. + 3H_2O (H_2O). Mp 173°, 180.5-181.5.
3-Et ether: 3-Ethoxy-4(1H)-pyridinone.
  $C_7H_9NO_2$ M 139.154
  Cryst. +1 H_2O (H_2O). Mp 112-113°, Mp 135-136° (anhyd.).
N-*Me:*
  $C_6H_7NO_2$ M 125.127
  Mp 227-228° dec.
O^3,N-*Di-Me: 3-Methoxy-1-methyl-4(1H)-pyridinone.*
  $C_7H_9NO_2$ M 139.154
  Mp 91-92°.

Diol-form
3,4-Dihydroxypyridine. 3,4-Pyridinediol
Di-Me ether: 3,4-Dimethoxypyridine.
  $C_7H_9NO_2$ M 139.154
  Mp 172-174.5° (as picrate).
Di-Me ether, 1-oxide: [769-71-1].
  $C_7H_9NO_3$ M 155.153

Cryst. (C₆H₆). Mp 105°.
Di-Et ether: 3,4-Diethoxypyridine.
C₉H₁₃NO₂   M 167.207
Bp₂₀ 120°.
Di-Et ether, picrate: Mp 169-170°.

Bickel, A.F., *J. Am. Chem. Soc.*, 1947, **69**, 1801.
Kleipool, R.J.C. *et al*, *Recl. Trav. Chim. Pays-Bas*, 1950, **69**, 37 (*derivs*)
Den Hertog, H.J. *et al*, *Recl. Trav. Chim. Pays-Bas*, 1950, **69**, 700 (*derivs*)
Elguero, J. *et al*, *Tautomerism of Heterocycles*, 1976, Academic Press, London, 110.

3-Hydroxy-2-quinolinecarboxylic acid, 9CI   H-60228

3-Hydroxyquinaldic acid, 8CI
[15462-45-0]

C₁₀H₇NO₃   M 189.170
Isol. from *Streptomyces griseoflavus*. Yellow needles (Me₂CO). Mp 200° dec. (196-198°).
Ca salt: [110429-27-1]. Mp 170°.
Amide: [15462-44-9].
C₁₀H₈N₂O₂   M 188.185
Pale-yellow needles (Me₂CO). Mp 215-216°.
Nitrile: [15462-43-8]. *2-Cyano-3-hydroxyquinoline.*
C₁₀H₆N₂O   M 170.170
Needles (MeOH). Mp >300° dec.
Me ether, Me ester: [110429-26-0].
C₁₂H₁₁NO₃   M 217.224
Mp 78-80°.

Kaneko, C. *et al*, *Chem. Pharm. Bull.*, 1967, **15**, 663 (*synth*)
Breiding-Mack, S. *et al*, *J. Antibiot.*, 1987, **40**, 953 (*isol, props*)

12-Hydroxy-13-tetradecenoic acid, 9CI   H-60229

[85288-98-8]

$$H_2C{=}CHCH(OH)(CH_2)_{10}COOH$$

C₁₄H₂₆O₃   M 242.358
(±)-*form*
Mp 51-52°.
Me ester: [106753-88-2].
C₁₅H₂₈O₃   M 256.384
Cryst. Mp 36-38°.

Cameron, A.G. *et al*, *J. Chem. Soc., Perkin Trans. 1*, 1986, 161 (*synth, ir, pmr*)

14-Hydroxy-12-tetradecenoic acid, 9CI   H-60230

[85289-00-5]

$$HOCH_2CH{=}CH(CH_2)_{10}COOH$$

C₁₄H₂₆O₃   M 242.358
(*E*)-*form* [106753-93-9]
Cryst. (Et₂O/pet. ether). Mp 47-48°.
Lactone: 12-Tetradecen-14-olide.
C₁₄H₂₄O₂   M 224.342
Bp₀.₀₁ 110° (oven).

(*Z*)-*form* [106541-96-2]
Cryst. (Et₂O/pet. ether). Mp 42.5-43.0°.
Lactone: Bp₀.₀₃ 110° (oven).

Ames, D.E. *et al*, *J. Chem. Soc.*, 1963, 5889 (*synth*)
Cameron, A.G. *et al*, *J. Chem. Soc., Perkin Trans. 1*, 1986, 161 (*synth, ir, pmr*)

ent-12-Hydroxy-13,14,15,16-tetranor-1(10)-halimen-18-oic acid   H-60231

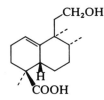

C₁₆H₂₆O₃   M 266.380
Ac: [109163-64-6].
C₁₈H₂₈O₄   M 308.417
Constit. of *Halimium viscosum*.
Ac, Me ester: [109163-57-7]. Oil. [α]$_D^{22}$ 34.5° (c, 1.2 in CHCl₃).

Urones, J.G. *et al*, *Phytochemistry*, 1987, **26**, 1077.

3-Hydroxy-25,26,27-trisnor-24-cycloartanal   H-60232

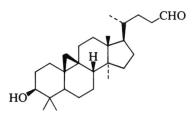

C₂₇H₄₄O₂   M 400.643
3β-*form*
Di-Me acetal: 24,24-Dimethoxy-25,26,27-trisnor-3-cycloartanol.
C₂₉H₅₀O₃   M 446.712
Constit. of *Euphorbia broteri*.
Di-Me acetal, Ac: Cryst. (MeOH). Mp 98-100°. [α]$_D^{25}$ +50.3° (c, 0.68 in CHCl₃).

Teresa, J.de.P. *et al*, *Phytochemistry*, 1987, **26**, 1767.

3-Hydroxy-28,20-ursanolide   H-60233

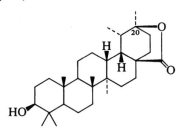

C₃₀H₄₈O₃   M 456.707
(3β,20β)-*form*
Cryst. (CH₂Cl₂/hexane). Mp 277-281°.
Ac: 3β-Acetoxy-28,20β-ursanolide.
C₃₂H₅₀O₄   M 498.745
Constit. of *Opilia celtidifolia*. Cryst. (CH₂Cl₂/hexane). Mp 314-321°.

Druet, D. *et al*, *Can. J. Chem.*, 1987, **65**, 851 (*cryst struct*)

11-Hydroxy-1(10)-valencen-2-one   H-60234

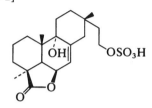

C$_{15}$H$_{24}$O$_2$ M 236.353
Constit. of *Teucrium carolipaui*. Thick oil. [α]$_D^{19}$
+131.2° (c, 0.188 in EtOH).

Savona, G. *et al, Phytochemistry*, 1987, **26**, 571.

Hymatoxin *A*   H-60235
[109621-33-2]

C$_{20}$H$_{30}$O$_7$S   M 414.513
Major phytotoxin from fungus *Hypoxylon mammatum*,
a parasite of aspen. Noncryst. solid. [α]$_D^{22}$ −49° (c,
0.50 in MeOH) (presumably as salt).

Bodo, B. *et al, Tetrahedron Lett*., 1987, **28**, 2355.

Hyperlatolic acid   H-60236
[90332-20-0]

C$_{27}$H$_{36}$O$_7$ M 472.577
Metab. of *Fuscidea viridis*. Cryst. (cyclohexane). Mp
113°.

Culberson, C.F. *et al, Mycologia*, 1984, **76**, 148 (*isol*)
Elix, J.A. *et al, Aust. J. Chem*., 1987, **40**, 425 (*synth*)

I

Illudin *M*   I-60001

Updated Entry replacing I-20008

2′,3′-Dihydro-3′,6′-dihydroxy-2′,2′,4′,6′-tetramethyl-spiro[cyclopropane-1,5′-[5H]inden]-7′(6′H)-one, 9CI

[1146-04-9]

R=CH₃

C₁₅H₂₀O₃   M 248.321

Sesquiterpene antibiotic. Metab. of *Clitocybe illudens*.
Cryst. (EtOH aq.). Mp 130-131°. [α]_D −126°.

▷WH0204350.

6′-Deoxy: **6-Deoxyilludin M**.
    C₁₅H₂₀O₂   M 232.322
    Prod. by *Pleurotus japonicus*. Antitumour agent. Pale-
    yellow solid. [α]_D^{23} −11° (c, 1 in MeOH).

McMorris, T.C. *et al, J. Am. Chem. Soc.*, 1965, **87**, 1594 (*isol,
    struct*)
Matsumoto, T. *et al, Tetrahedron Lett.*, 1970, 1171 (*synth*)
Hanson, J.R. *et al, J. Chem. Soc., Perkin Trans. 1*, 1976, 876
    (*biosynth*)
Bradshaw, A.P.W. *et al, J. Chem. Soc., Perkin Trans. 1*, 1982,
    2445 (*biosynth*)
Hara, M. *et al, J. Antibiot.*, 1987, **40**, 1643 (*deriv*)

Illudin *S*   I-60002

Updated Entry replacing I-20009

Lampterol

[1149-99-1]

As Illudin *M*, I-60001 with

R = CH₂OH

C₁₅H₂₀O₄   M 264.321

Sesquiterpene. Metab. of *Clitocybe illudens* and
Lampteromyces japonicus. Cryst. (EtOAc). Mp 137-
138°.

▷WH0204300.

6′-Deoxy: **6-Deoxyilludin S**.
    C₁₅H₂₀O₃   M 248.321
    Prod. by *Pleurotus japonicus*. Antitumour agent. Pale-
    yellow solid. [α]_D^{23} −13° (c, 1 in MeOH).

Harada, N. *et al, J. Chem. Soc., Chem. Commun.*, 1970, 310
    (*abs config*)
Matsumoto, T. *et al, Tetrahedron Lett.*, 1971, 2049 (*synth*)
Bradshaw, A.P.W. *et al, J. Chem. Soc., Perkin Trans. 1*, 1982,
    2445 (*biosynth*)
Hara, M. *et al, J. Antibiot.*, 1987, **40**, 1643 (*deriv*)

Iloprost   I-60003

5-[Hexahydro-5-hydroxy-4-(3-hydroxy-4-methyl-1-oc-ten-6-ynyl)-2(1H)-pentalenylidene]pentanoic acid, 9CI.

ZK 36374

[78919-13-8]

C₂₂H₃₂O₄   M 360.492

Analogue of prostacyclin with similar vasodilator activity
but increased stability.

(5Z)-Isomer: Isoiloprost. ZK 36375.
    C₂₂H₃₂O₄   M 360.492
    Shows lower biol. activity than Iloprost.

Schenker, K.V. *et al, Helv. Chim. Acta*, 1986, **69**, 1718 (*cmr,
    pmr, config, bibl*)

1*H*-Imidazole-4-carboxylic acid, 9CI   I-60004

Updated Entry replacing I-00114

[1072-84-0]

C₄H₄N₂O₂   M 112.088

Mp 284°.

Me ester: [17325-26-7].
    C₅H₆N₂O₂   M 126.115
    Mp 154-156°.
Et ester: [23785-21-9].
    C₆H₈N₂O₂   M 140.141
    Mp 160-162°.
Amide: [26832-08-6].
    C₄H₅N₃O   M 111.103
    Cryst. + 1H₂O. Mp 214°.
Methylamide:
    C₅H₇N₃O   M 125.130
    Mp 145°.
Nitrile: [57090-88-7]. *4(5)-Cyanoimidazole.*
    C₄H₃N₃   M 93.088
    Light-yellow powder. Mp 141-144°.
*3-Me: see 1-Methyl-1H-imidazole-5-carboxylic acid,
M-02077*

Balaban, I.E., *J. Chem. Soc.*, 1930, 270.
Heubner, C.F. *et al, J. Am. Chem. Soc.*, 1949, **71**, 2801 (*synth*)
Onishchuk, A.E., *J. Gen. Chem. USSR (Engl. Transl.)*, 1955,
    25, 949; *CA*, **50**, 3412
Mitsuhashi, K. *et al, J. Heterocycl. Chem.*, 1983, **20**, 1103
    (*nitrile, synth, pmr, ir*)
Matthews, D.P. *et al, J. Org. Chem.*, 1986, **51**, 3228 (*nitrile, ir,
    pmr*)

Imidazo[1,2-*a*]pyrazine, 9CI   I-60005

Updated Entry replacing I-00131

[274-79-3]

$C_6H_5N_3$ M 119.126

Cryst. (C_6H_6/hexane). Mp 83-85°.

de Pompei, M.F. *et al, J. Heterocycl. Chem.*, 1975, **12**, 861.
Bonnet, P.A. *et al, Aust. J. Chem.*, 1984, **37**, 1357 (*cmr*)
Pugmire, R.J. *et al, J. Heterocycl. Chem.*, 1987, **24**, 805 (*cmr*)

2*H*-Imidazo[4,5-*b*]pyrazin-2-one, 9CI   I-60006

2-Hydroxyimidazo[b]*pyrazine*

[16328-63-5]

$C_5H_4N_4O$ M 136.113

Pale-yellow cryst. (EtOH). Mp 336°.

1,3-Dihydro-form

1-Me: [84996-45-2]. *1,3-Dihydro-1-methyl-2H-imi-*
dazo[4,5-b]pyrazin-2-one, 9CI.
$C_6H_6N_4O$ M 150.140
Cryst. (EtOH). Mp 241°.

1,3-Di-Me: [84996-53-2]. *1,3-Dihydro-1,3-dimethyl-*
2H-imidazo[4,5-b]pyrazin-2-one, 9CI.
$C_7H_8N_4O$ M 164.166
Cryst. (cyclohexane). Mp 142-143.5°.

1,4-Dihydro-form

1,4-Di-Me: [84996-55-4]. *1,4-Dihydro-1,4-dimethyl-*
2H-imidazo[4,5-b]pyrazin-2-one, 9CI.
$C_7H_8N_4O$ M 164.166
Cryst. (C_6H_6). Mp 233-235°.

OH-form

4-Me, Me ether: [84996-54-3]. *2-Methoxy-4-methyl-*
4H-imidazo[4,5-b]pyrazine, 9CI.
$C_7H_8N_4O$ M 164.166
Cryst. (cyclohexane). Mp 147-149°.

Muehlmann, F.L. *et al, J. Am. Chem. Soc.*, 1956, **78**, 242
(*synth*)
Mason, S.F. *et al, J. Chem. Soc.*, 1957, 4874 (*tautom*)
Hershenson, F.M. *et al, J. Org. Chem.*, 1968, **33**, 2543 (*synth, ir,*
pmr, ms)
Tong, Y.C., *J. Heterocycl. Chem.*, 1981, **18**, 751 (*synth, tautom*)
Barlin, G.B., *Aust. J. Chem.*, 1982, **35**, 2299 (*deriv, synth, pmr*)

Imidazo[1,2-*b*]pyridazine, 9CI   I-60007

Updated Entry replacing I-00133

[766-55-2]

$C_6H_5N_3$ M 119.126

Many subsd. derivs. show CNS activity. Cryst. (EtOAc-
/pet. ether). Mp 53-55°. pK_a 4.4 (20°, H_2O). λ_{max} 222
(ϵ 19 480) and 334 nm (3990) (EtOH).

B,HCl: [18087-70-2]. Mp 240°.
B,MeI: Mp 285-286°. Site of CH_3 not known.

Kobe, J. *et al, Tetrahedron*, 1968, **24**, 234 (*synth, pmr, uv*)
Pugmire, R.J. *et al, J. Heterocycl. Chem.*, 1976, **13**, 1057; 1987,
24, 805 (*cmr*)

Imidazo[4,5-*c*]pyridine   I-60008

Updated Entry replacing I-10015

1,3,5-Triazaindene

[272-97-9]

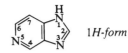

1H-form

$C_6H_5N_3$ M 119.126

Prisms (EtOAc). Mp 170-171°, Mp 210-215° dec. Bp$_{0.1}$
180° subl. pK_{a1} 6.10, pK_{a2} 10.88.

▷NJ5108000.

1-β-D-Ribofuranosyl: [6703-46-4].
$C_{11}H_{13}N_3O_4$ M 251.241
Solid. Mp 200-204°. $[\alpha]_D^{20}$ −49.2° (c, 0.5 in DMF).

Albert, A., *J. Chem. Soc.*, 1956, 4683 (*synth*)
Barlin, G.B., *J. Chem. Soc. (B)*, 1966, 285 (*uv*)
Stanovnik, B. *et al, Synthesis*, 1974, 120 (*synth, pmr*)
May, J.A. *et al, J. Chem. Soc., Perkin Trans. 1*, 1975, 125
(*deriv, synth, uv, bibl*)
Barlin, G.B. *et al, Aust. J. Chem.*, 1981, **34**, 1341 (*cmr*)
Krenitzky, T.A. *et al, J. Med. Chem.*, 1986, **29**, 138 (*deriv,*
synth, uv, pmr)

Imidazo[4,5-*g*]quinazoline-6,8(5*H*,7*H*)-   I-60009
dione

lin-*Benzoxanthine*

[60189-64-2]

$C_9H_6N_4O_2$ M 202.172

Powder. Mp >300°.

5,7-Di-Me: [76822-71-4]. lin-*Benzotheophylline.*
$C_{11}H_{10}N_4O_2$ M 230.226
Powder (DMF). Mp 289-292°.

1,5,7-Tri-Me: [76832-42-3]. lin-*Benzocaffeine.*
$C_{12}H_{12}N_4O_2$ M 244.252
Plates (DMSO). Mp 355-356°. Purple fluorescent
soln.

3,5,7-Tri-Me: [76822-74-7].
$C_{12}H_{12}N_4O_2$ M 244.252
Needles (DMSO). Mp 341-343°.

Keyser, G.E. *et al, J. Org. Chem.*, 1979, **44**, 2989 (*synth, pmr*)
Schneller, S.W. *et al, J. Org. Chem.*, 1981, **46**, 1699 (*deriv,*
synth, pmr, ir)

Imidazo[4,5-*g*]quinazoline-4,8,9(3*H*,7*H*)-   I-60010
trione

[105664-72-0]

$C_9H_4N_4O_3$ M 216.156

CAS no. refers to a different tautomer. Hypoxanthine analogue, substrate for xanthine oxidase. Bright-yellow
solid (AcOH/conc. HCl). Mp >300° dec.

Lee, C.-H. *et al, J. Org. Chem.*, 1986, **51**, 4784 (*synth, pmr,*
cmr, ms)

Imidazo[4,5-*f*]quinazolin-9(8*H*)-one   I-60011
prox-*Benzohypoxanthine*
[53449-52-8]

C$_9$H$_6$N$_4$O   M 186.173
Needles (H$_2$O). Mp >320° (>400° dec.).

Morrice, A.G. *et al*, *J. Org. Chem.*, 1975, **40**, 363 (*synth*, *pmr*)
Schneller, S.W. *et al*, *J. Org. Chem.*, 1986, **51**, 4067 (*synth*, *pmr*, *ir*)

Imidazo[4,5-*g*]quinazolin-8-one   I-60012
lin-*Benzohypoxanthine*
[53449-18-6]

C$_9$H$_6$N$_4$O   M 186.173
Cryst. (H$_2$O). Mp >320°.

Leonard, N.J. *et al*, *J. Org. Chem.*, 1975, **40**, 356 (*synth*, *pmr*)
Leonard, N.J. *et al*, *J. Am. Chem. Soc.*, 1976, **98**, 3987 (*synth*, *ms*)

Imidazo[4,5-*h*]quinazolin-6-one   I-60013
dist-*Benzohypoxanthine*
[53449-49-3]

C$_9$H$_6$N$_4$O   M 186.173
Cryst. (H$_2$O). Mp >320°.

Morrice, A.G. *et al*, *J. Org. Chem.*, 1975, **40**, 363 (*synth*, *pmr*)

2,2′-Iminodibenzoic acid, 8CI   I-60014
2,2′-Iminobisbenzoic acid, 9CI. *Diphenylamine-2,2′-dicarboxylic acid*
[579-92-0]

C$_{14}$H$_{11}$NO$_4$ M 257.245
Cryst. (EtOH). Mp 314-316° (295°).
▷DH2960000.
Di-Me ester: [34069-89-1].
     C$_{16}$H$_{15}$NO$_4$ M 285.299
     Cryst. (MeOH). Mp 102-103°.
Dichloride: [32621-46-8].
     C$_{14}$H$_9$Cl$_2$NO$_2$ M 294.137
     Yellow prisms. Mp 161-163° (155°).
Diamide: [32615-84-2]. *2,2′-Iminobisbenzamide*, 9CI.
     2,2′-Dicarbamoyldiphenylamine.
     C$_{14}$H$_{13}$N$_3$O$_2$ M 255.276
     Pale-yellow needles (DMF/MeOH). Mp 219-220°.

Hellwinkel, D. *et al*, *Chem. Ber.*, 1971, **104**, 1001 (*ester*)

Banerji, A. *et al*, *J. Chem. Soc., Perkin Trans. 1*, 1977, 1162 (*synth*)
Ozaki, K. *et al*, *J. Org. Chem.*, 1981, **46**, 1571 (*diamide*)

2,3′-Iminodibenzoic acid, 8CI   I-60015
2,3′-Iminobisbenzoic acid, 9CI. *Diphenylamine-2,3′-dicarboxylic acid*
[27693-67-0]
C$_{14}$H$_{11}$NO$_4$ M 257.245
Needles (EtOH). Mp 296°.

Ullmann, F. *et al*, *Justus Liebigs Ann. Chem.*, 1907, **355**, 355 (*synth*)

2,4′-Iminodibenzoic acid, 8CI   I-60016
2,4′-Iminobisbenzoic acid, 9CI. *Diphenylamine-2,4′-dicarboxylic acid*
[17332-57-9]
C$_{14}$H$_{11}$NO$_4$ M 257.245
Needles (EtOH). Mp 290° (278-279°).

Di-Me ester:
     C$_{16}$H$_{15}$NO$_4$ M 285.299
     Mp 180-181°.

Ullmann, F. *et al*, *Justus Liebigs Ann. Chem.*, 1907, **355**, 356 (*synth*)
Picciola, G. *et al*, *Farmaco, Ed. Sci.*, 1968, **23**, 502 (*synth*, *props*)

3,3′-Iminodibenzoic acid, 8CI   I-60017
3,3′-Iminobisbenzoic acid, 9CI. *Diphenylamine-3,3′-dicarboxylic acid*
[19039-48-6]
C$_{14}$H$_{11}$NO$_4$ M 257.245
Green solid. Dec. >300°.

Michels, J.G. *et al*, *J. Am. Chem. Soc.*, 1950, **72**, 888 (*synth*)

4,4′-Iminodibenzoic acid, 8CI   I-60018
4,4′-Iminobisbenzoic acid, 9CI. *Diphenylamine-4,4′-dicarboxylic acid*
[20800-00-4]
C$_{14}$H$_{11}$NO$_4$ M 257.245
Plates. Mp 322-324°.

Di-Me ester: [17104-81-3].
     C$_{16}$H$_{15}$NO$_4$ M 285.299
     Cryst. (EtOH aq.). Mp 177°.
Dinitrile: [36602-05-8]. *4,4′-Iminobisbenzonitrile*, 9CI.
     4,4′-Dicyanodiphenylamine.
     C$_{14}$H$_9$N$_3$ M 219.245
     Cryst. (EtOAc). Mp 265-266°.

Kizber, A.I. *et al*, *Zh. Obshch. Khim.*, 1954, **24**, 2195; *CA*, **50**, 207e (*synth*)
Bergmann, E.D. *et al*, *Tetrahedron*, 1968, **24**, 6449 (*deriv*)
Yoshida, K. *et al*, *J. Org. Chem.*, 1972, **37**, 4145 (*dinitrile*)
Julliard, M. *et al*, *Helv. Chim. Acta*, 1980, **63**, 456.

The symbol ▷ in Entries highlights hazard or toxicity information

Indamine
I-60019

N-(*4-Imino-2,5-cyclohexadien-1-ylidene)-1,4-benzene-diamine, 9CI*. N-(*4-Imino-2,5-cyclohexadien-1-ylidene*)-*p-phenylenediamine*. N-(*4-Aminophenyl*)-*p-benzoquin-one diimine*. *Phenylene blue*

[101-78-0]

$$H_2N-\!\!\!\bigcirc\!\!\!-N=\!\!\!\bigcirc\!\!\!=NH$$

$C_{12}H_{11}N_3$ M 197.239

Parent compd. of a family of dyes for which the term indamine dyes is used. No phys. props. reported.

N-*Tetra-Me: see Bindschedler's green, B-60110*

Corbett, J.F. *et al, J. Chem. Soc., Perkin Trans. 2*, 1972, 1531 (*synth, bibl*)

Indazolo[3,2-*a*]isoquinoline
I-60020

$C_{15}H_{10}N_2$ M 218.257

Yellow needles (toluene/hexane). Mp 92-93°.

Stanforth, S.P., *J. Heterocycl. Chem.*, 1987, **24**, 531 (*synth, ir, pmr*)

[2.2](4,7)(7,4)Indenophane
I-60021

1,5,6,8,12,13-Hexahydro-4,14:7,11-diethenodicyclopenta[a,g]*cyclododecene*

[103665-30-1]

$C_{22}H_{20}$ M 284.400

Plates. Mp 235° dec.

Hopf, H. *et al, Tetrahedron*, 1986, **42**, 1655 (*synth, ir, uv, pmr, cmr, ms*)

Indeno[1,2,3-*cd*]pyren-1-ol
I-60022

1-Hydroxyindeno[*1,2,3-*cd]*pyrene*

[99520-65-7]

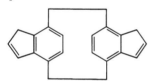

$C_{22}H_{12}O$ M 292.336

Yellow needles (toluene). Mp 236-238°.

▷ Prob. mutagenic

Ac: [102420-57-5].
    $C_{24}H_{14}O_2$ M 334.373
    Fine yellow needles (C_6H_6). Mp 205-206°.

Rice, J.E. *et al, J. Org. Chem.*, 1986, **51**, 2428 (*synth, ir, uv, pmr, ms*)

Indeno[1,2,3-*cd*]pyren-2-ol
I-60023

2-Hydroxyindeno[*1,2,3-*cd]*pyrene*

[99520-66-8]

$C_{22}H_{12}O$ M 292.336

Yellow-tan cryst. (C_6H_6/EtOH). Mp 248-251°.

▷ Prob. mutagenic

Ac: [102420-58-6].
    $C_{24}H_{14}O_2$ M 334.373
    Yellow prisms (xylene). Mp 236-238°.

Rice, J.E. *et al, J. Org. Chem.*, 1986, **51**, 2428 (*synth, pmr, uv, ir, ms*)

Indeno[1,2,3-*cd*]pyren-6-ol
I-60024

6-Hydroxyindeno[*1,2,3-*cd]*pyrene*

[99520-67-9]

$C_{22}H_{12}O$ M 292.336

Yellow needles (MeOH aq.). Mp 213-214°.

▷ Prob. mutagenic

Ac: [102420-62-2].
    $C_{24}H_{14}O_2$ M 334.373
    Yellow-orange needles. Mp 254-256°.

Rice, J.E. *et al, J. Org. Chem.*, 1986, **51**, 2428 (*synth, pmr, ir, uv, ms*)

Indeno[1,2,3-*cd*]pyren-7-ol
I-60025

7-Hydroxyindeno[*1,2,3-*cd]*pyrene*

[102420-63-3]

$C_{22}H_{12}O$ M 292.336

Yellow needles (C_6H_6). Mp 226-228° dec.

▷ Prob. mutagenic

Me ether: [102420-67-7]. *7-Methoxyindeno*[*1,2,3-*cd]-*pyrene*.
    $C_{23}H_{14}O$ M 306.363
    Yellow needles (hexane). Mp 209-211°.

Rice, J.E. *et al, J. Org. Chem.*, 1986, **51**, 2428 (*synth, ir, pmr, uv, ms*)

Indeno[1,2,3-*cd*]pyren-8-ol
I-60026

8-Hydroxyindeno[*1,2,3-*cd]*pyrene*

[99520-58-8]

$C_{22}H_{12}O$ M 292.336

Major metabolite of Indeno[1,2,3-*cd*]pyrene, I-00240 .
    Fine yellow cryst. (C_6H_6). Mp 221-222.5° dec.

▷ Potent mutagen

Me ether: [102420-65-5]. *8-Methoxyindeno*[*1,2,3-*cd]-*pyrene*.
    $C_{23}H_{14}O$ M 306.363
    Orange cryst. (C_6H_6/EtOAc). Mp 197-201°.

Rice, J.E. *et al, J. Org. Chem.*, 1986, **51**, 2428 (*synth, ir, pmr, uv, ms*)

Indicoside *A* I-60027

[109575-71-5]

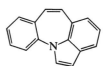

C$_{35}$H$_{62}$O$_{12}$ M 674.868

Metabolite of starfish *Astropecten indicus*. Glass. [α]$_D$ −69.4° (c, 1 in MeOH).

Riccio, R. *et al, Tetrahedron Lett.*, 1987, **28**, 2291.

Indisocin I-60028

3-(Acetyloxy)-3-(2-chloro-3-isocyanoethenyl)-2-oxoin-dole. 3-Acetoxy-3-(2-chloro-2-isocyanovinyl)-2-oxoindole

[90632-49-8]

C$_{13}$H$_9$ClN$_2$O$_3$ M 276.679

Indole antibiotic. Isol. from *Nocardia blackwellii*. Active against gram-positive and -negative bacteria and fungi. [α]$_D^{25}$ +20° (c, 0.03 in MeOH). Stable in org. solvents. Related to Antibiotic B371.

N-*Me:* N-*Methylindisocin.*
C$_{14}$H$_{11}$ClN$_2$O$_3$ M 290.706
Semisynthetic. Active against gram-positive and -negative bacteria and fungi. [α]$_D^{25}$ +22° ±5° (c, 0.04 in MeOH).

Isshiki, K. *et al, J. Antibiot.*, 1987, **40**, 1195, 1199, 1202 (*isol, struct, synth, props*)

1*H*-Indole-5-carboxaldehyde I-60029

5-Formylindole

[1196-69-6]

C$_9$H$_7$NO M 145.160
Mp 99-101°.

Fr. Pat., 1 344 579, (*1963*); *CA*, **60**, 11988h (*synth*)
Fr. Pat., 1 373 316, (*1964*); *CA*, **62**, 4008d (*synth*)
Moyer, M.P. *et al, J. Org. Chem.*, 1986, **51**, 5106 (*synth, pmr*)

1*H*-Indole-6-carboxaldehyde I-60030

6-Formylindole

[1196-70-9]

C$_9$H$_7$NO M 145.160
Mp 127-129°.

Fr. Pat., 1 344 579, (*1963*); *CA*, **60**, 11988h (*synth*)
Fr. Pat., 1 373 316, (*1964*); *CA*, **62**, 4008d
Moyer, M.P. *et al, J. Org. Chem.*, 1986, **51**, 5106.
Somei, M., *Chem. Pharm. Bull.*, 1986, **34**, 4109 (*synth, ir, pmr, bibl*)

1*H*-Indole-7-carboxaldehyde I-60031

7-Formylindole

[1074-88-0]

C$_9$H$_7$NO M 145.160
Mp 87-89°.

Fr. Pat., 1 344 579, (*1963*); *CA*, **60**, 11988h (*synth*)
Fr. Pat., 1 373 316, (*1964*); *CA*, **62**, 4008d (*synth*)
Moyer, M.P. *et al, J. Org. Chem.*, 1986, **51**, 5106 (*synth, pmr*)

1*H*-Indole-6-methanol I-60032

6-Hydroxymethylindole

[1075-26-9]

C$_9$H$_9$NO M 147.176
Prisms (C$_6$H$_6$). Mp 62-63° (52-53°).

Somei, M., *Chem. Pharm. Bull.*, 1986, **34**, 4105 (*synth, pmr, bibl*)

Indolizine, 9CI I-60033

Updated Entry replacing I-00279
Pyrrocoline. Pyrrolo[1,2-a]pyridine

[274-40-8]

C$_8$H$_7$N M 117.150
Mp 75°. Bp 205°. Steam-volatile. Dec. by acids.
Picrate: Mp 101°.

Black, P.J. *et al, Aust. J. Chem.*, 1964, **17**, 1128 (*pmr*)
Flitsch, W. *et al, Chem. Ber.*, 1969, **102**, 1309 (*synth*)
Swinbourne, F.J. *et al, Adv. Heterocycl. Chem.*, 1978, **23**, 103 (*rev*)
Pugmire, R.J. *et al, J. Am. Chem. Soc.*, 1971, **93**, 1887 (*cmr, pmr*)
Pugmire, R.J. *et al, J. Heterocycl. Chem.*, 1987, **24**, 805 (*cmr*)

Indolo[1,7-*ab*][1]benzazepine, 9CI I-60034

[202-01-7]

C$_{16}$H$_{11}$N M 217.270
Yellow cryst. (hexane). Mp 113-114°.

Hallberg, A. *et al, Heterocycles*, 1982, **19**, 75 (*synth*)
Hallberg, A. *et al, J. Heterocycl. Chem.*, 1983, **20**, 37 (*pmr*)

3*H*-Indol-3-one, 9CI, 8CI I-60035

3-Oxoindole. 3-Pseudoindolone

[67285-12-5]

C$_8$H$_5$NO M 131.134
Disproportionates spontaneously to indoxyl and dehydroindigo.

Oxime: [76983-82-9]. *3-Hydroxyiminoindole. 3-Nitrosoindole.*
C$_8$H$_5$N$_2$O M 145.140
Brown needles (C$_6$H$_6$). Mp 180-182° dec. Tautomeric.
1-Oxide: [5814-98-2]. *Isatogen.*
C$_8$H$_5$NO$_2$ M 147.133

1-Oxide, oxime: [69111-91-7].
  $C_8H_6N_2O_2$ M 162.148
  Yellow-green needles (H_2O). Mp 167-169°. Subl. at
  ca. 150°.

Neunhoeffer, O. *et al, Chem. Ber.*, 1961, **94**, 2965.
Stamm, H., *Method. Chim.*, 1975, **6**, 329 (*rev*)
Acheson, R.M. *et al, J. Chem. Soc., Perkin Trans. 1*, 1978, 1117
  (*deriv*)
Hiremath, S.P. *et al, Adv. Heterocycl. Chem.*, 1978, **22**, 123
  (*rev*)
Hiremath, S.P. *et al, Indian J. Chem., Sect. B*, 1980, **19**, 767
  (*deriv*)

5*H*-Indolo[2,3-*b*]quinoxaline, 9CI   I-60036

Indophenazine

[243-59-4]

                                                      6H-form

$C_{14}H_9N_3$ M 219.245
Exists predominantly as 6*H*-form; *CA* refers to 5*H*-form
  which is not a favoured tautomer. Yellow needles
  (C_6H_6/MeOH or AcOH). Mp 296-298°.
5,11-Dioxide: [32861-73-7].
  $C_{14}H_9N_3O_2$ M 251.244
  Bright-yellow rosettes (AcOH). Mp 284° dec.

Badger, G.M. *et al, J. Chem. Soc.*, 1962, 3926 (*synth, struct*)
Abu El-Haj, M.J. *et al, J. Org. Chem.*, 1972, **37**, 589 (*synth,
  deriv*)
Carter, S.D. *et al, Tetrahedron*, 1978, **34**, 981 (*synth*)
Sarkis, G.Y. *et al, J. Heterocycl. Chem.*, 1980, **17**, 813 (*synth,
  uv, ir, pmr*)
Niume, K. *et al, Bull. Chem. Soc. Jpn.*, 1982, **55**, 2293 (*synth*)

Inuviscolide   I-60037

Updated Entry replacing I-20024
[63109-30-8]

$C_{15}H_{20}O_3$ M 248.321
Constit. of *Inula viscosa*. Oil. $[\alpha]_D^{24}$ −18.6° (c, 0.35 in
  $CHCl_3$).
11β,13-Dihydro: [84093-50-5]. **11β,13-
  Dihydroinuviscolide**.
  $C_{15}H_{22}O_3$ M 250.337
  Constit. of *Geigeria aspera*. Gum. $[\alpha]_D^{24}$ +28° (c, 0.13
  in $CHCl_3$).
1-Epimer: **1-epi-Inuviscolide**.
  $C_{15}H_{20}O_3$ M 248.321
  Constit. of *Dittrichia graveolens*. Oil.
8-Epimer: [108885-57-0]. **8-epi-Inuviscolide**.
  $C_{15}H_{20}O_3$ M 248.321
  Constit. of *Bedfordia arborescens*. Oil.
4,8-Diepimer: [108885-58-1]. **4,8-Bis-epi-inuviscolide**.
  $C_{15}H_{20}O_3$ M 248.321
  Constit. of *B. arborescens*. Oil.
1-Epimer, 11β,13-Dihydro: **11β,13-Dihydro-1-epi-
  inuviscolide**.
  $C_{15}H_{22}O_3$ M 250.337
  Constit. of *D. graveolens*. Oil.

Bohlmann, F. *et al, Chem. Ber.*, 1977, **110**, 1330 (*isol, struct*)
Bohlmann, F. *et al, Phytochemistry*, 1982, **21**, 1679 (*isol*)
Zdero, C. *et al, Phytochemistry*, 1987, **26**, 1207 (*isol, struct*)
Rustaiyan, A. *et al, Phytochemistry*, 1987, **26**, 2603 (*deriv*)

Invictolide   I-60038

Updated Entry replacing I-20025
*Tetrahydro-3,4-dimethyl-6-(1-methylbutyl)-2H-pyran-
2-one*

$C_{12}H_{22}O_2$ M 198.305
Queen recognition pheromone of red fire ant *Solenopsis
  invicta*. Oil.

Rocca, J.R. *et al, Tetrahedron Lett.*, 1983, **24**, 1893 (*synth*)
Mori, K. *et al, Tetrahedron*, 1986, **42**, 6459 (*synth*)
Senda, S. *et al, Agric. Biol. Chem.*, 1987, **51**, 1379 (*synth*)
Wakamatsu, T. *et al, Heterocycles*, 1987, **26**, 1761 (*synth*)

2-Iodoacetophenone   I-60039

Updated Entry replacing I-50047
*2-Iodo-1-phenylethanone, 9CI. Phenacyl iodide. Iodo-
methyl phenyl ketone. ω-Iodoacetophenone*

[4636-16-2]

$$PhCOCH_2I$$

C_8H_7IO M 246.047
Cryst. (EtOH aq.). Mp 34.4°.
Oxime (Z-): [67155-45-7].
  C_8H_8INO M 261.062
  Mp 112°.

Rheinboldt, H. *et al, J. Am. Chem. Soc.*, 1947, **69**, 3148 (*synth*)
Tronov, B.V. *et al, Zh. Obshch. Khim.*, 1956, **26**, 1393 (*synth*)
Jones, R.N. *et al, Can. J. Chem.*, 1958, **36**, 1020 (*ir*)
Spinner, E. *et al, Spectrochim. Acta*, 1961, **17**, 558 (*uv*)
Molin, Y.N., *Dokl. Akad. Nauk. SSSR*, 1965, **163**, 402 (*pmr*)
Pannell, K.H., *Org. Mass Spectrom.*, 1975, **10**, 550 (*ms*)
Barluenga, J. *et al, Synthesis*, 1986, 678 (*synth, ir, pmr, cmr*)
Schauble, J.H. *et al, Synthesis*, 1986, 727 (*synth, ir, pmr*)

2-Iodobutanal, 9CI   I-60040

$$H_3CCH_2CHICHO$$

C_4H_7IO M 198.003
(±)-*form* [20175-17-1]
  Bp_{35} 68-69°, Bp_{15} 50-52°.

Barluenga, J. *et al, Synthesis*, 1986, 678 (*synth, ir, pmr, cmr*)

3-Iodo-2-butanone, 9CI   I-60041

$$H_3CCHICOCH_3$$

C_4H_7IO M 198.003
(±)-*form* [30719-18-7]
  Bp_{667} 148-150° dec., Bp_{15} 62-64°. Dec. at r.t. in air.

Modarai, B. *et al, J. Org. Chem.*, 1977, **42**, 3527 (*synth, ir, pmr*)
Barluenga, J. *et al, Synthesis*, 1986, 678 (*synth, ir, pmr, cmr*)

2-Iodocyclohexanone, 9CI   I-60042

[35365-19-6]

C_6H_9IO M 224.041
Liq.

(±)-*form*
Bp$_1$ 55°.

Evans, R.D. *et al*, *Synthesis*, 1986, 727 (*synth, ir, pmr*)

2-Iodocyclopentanone, 9CI   I-60043

[69381-32-4]

C_5H_7IO M 210.014

(±)-*form*
Bp$_1$ 51°.

Evans, R.D. *et al*, *Synthesis*, 1986, 727 (*synth, ir, pmr*)

1-Iodo-3,3-dimethyl-2-butanone, 9CI   I-60044

tert-*Butyl iodomethyl ketone*
[99714-13-3]

$$(H_3C)_3CCOCH_2I$$

$C_6H_{11}IO$ M 226.057
Bp$_{15}$ 90-92°.

Barluenga, J. *et al*, *Synthesis*, 1986, 678 (*synth, ir, pmr, cmr*)

2-Iododiphenylmethane   I-60045

*1-Iodo-2-(phenylmethyl)benzene, 9CI. (2-Iodophenyl)-
phenylmethane*
[35444-93-0]

$C_{13}H_{11}I$ M 294.134
Bp$_{11}$ 182°.

Collette, J. *et al, J. Am. Chem. Soc.*, 1956, **78**, 3819 (*synth*)
Ogata, Y. *et al, J. Chem. Soc., Perkin Trans. 1*, 1972, 180
    (*synth*)

4-Iododiphenylmethane   I-60046

*1-Iodo-4-(phenylmethyl)benzene, 9CI. (4-Iodophenyl)-
phenylmethane*
[35444-94-1]
$C_{13}H_{11}I$ M 294.134
Mp 41.0-41.5°.

Ogata, Y. *et al, J. Chem. Soc., Perkin Trans. 1*, 1972, 180
    (*synth*)
Wilson, S.R. *et al, J. Org. Chem.*, 1986, **51**, 4833 (*synth, pmr,
    ms*)

3-Iodo-2-(iodomethyl)-1-propene   I-60047

1,3-Diiodo-2-methylenepropane
[17616-43-2]

$$H_2C{=}C(CH_2I)_2$$

$C_4H_6I_2$ M 307.900
Cryst. (pet. ether). Mp 32-33°. Bp$_5$ 83-85°.
▷Lachrymator and vesicant

Skell, P.S. *et al, J. Am. Chem. Soc.*, 1967, **89**, 4688 (*synth*)
Looker, B.E. *et al, Tetrahedron*, 1971, **27**, 2567 (*synth*)
Schultze, V.K. *et al, J. Prakt. Chem.*, 1977, **319**, 463 (*synth*)

Iodomethanesulfonic acid   I-60048

Updated Entry replacing I-00525

$$ICH_2SO_3H$$

CH_3IO_3S M 221.997
Free acid little studied and apparently not descr.

Na salt: [126-31-8]. *Methiodal sodium, BAN.*
    Radioopaque medium. Cryst. Sol. H_2O, spar. sol. org.
    solvs.
    ▷PB2250000.
Bromide: [91586-90-2].
    CH_2BrIO_2S M 284.894
    Pale-violet liq. Bp$_{0.06}$ 78-80°.

D.R. Pat., 532 766, (*1929*); *CA*, **26**, 480 (*synth*)
Lauer, W.M. *et al, J. Am. Chem. Soc.*, 1935, **57**, 2361 (*synth*)
Block, E. *et al, J. Am. Chem. Soc.*, 1986, **108**, 4568 (*bromide*)

2-Iodo-3-methylbutanal, 9CI   I-60049

[74067-77-9]

$$(H_3C)_2CHCHICHO$$

C_5H_9IO M 212.030

(±)-*form*
Bp$_{15}$ 68-70° part dec.

Barluenga, J. *et al, Synthesis*, 1986, 678 (*synth, ir, pmr, cmr*)

4-Iodo-3-methylphenol   I-60050

4-Iodo-m-cresol
C_7H_7IO M 234.036

Me ether: [63452-69-7]. *1-Iodo-4-methoxy-2-methyl-
    benzene. 4-Iodo-3-methylanisole. 2-Iodo-5-
    methoxytoluene.* Cryst. Mp 43-45°. Bp$_{12}$ 129-130°.

Tashiro, M. *et al, Org. Prep. Proced. Int.*, 1976, **8**, 249 (*deriv*)
Cornforth, J. *et al, J. Chem. Soc., Perkin Trans. 1*, 1987, 871
    (*synth, deriv*)

2-Iodo-2-methylpropanal, 9CI   I-60051

[20175-18-2]

$$(H_3C)_2CICHO$$

C_4H_7IO M 198.003
Bp$_{15}$ 44-46° part dec.

Barluenga, J. *et al, Synthesis*, 1986, 678 (*synth, ir, pmr, cmr*)

5-Iodo-2-nitrobenzaldehyde   I-60052

[105728-31-2]
$C_7H_4INO_3$ M 277.018
Cryst. (Et$_2$O/hexane). Mp 64-65°.

Venuti, M.C. *et al, J. Med. Chem.*, 1987, **30**, 303 (*synth*)

3-Iodo-2-octanone, 9CI   I-60053
[73746-49-3]

$$H_3C(CH_2)_4CHICOCH_3$$

$C_8H_{15}IO$ M 254.110

(±)-*form*
Bp$_3$ 61-63°, Bp$_{0.1}$ 60-62°.

Barluenga, J. *et al*, *Synthesis*, 1986, 678 (*synth, ir, pmr, cmr*)

2-Iodopentanal, 9CI   I-60054
[108350-37-4]

$$H_3CCH_2CH_2CHICHO$$

C_5H_9IO M 212.030

(±)-*form*
Bp$_{15}$ 72-74°.

Barluenga, J. *et al*, *Synthesis*, 1986, 678 (*synth, ir, pmr, cmr*)

1-Iodo-3-pentanol   I-60055
[108161-72-4]

$$ICH_2CH_2CH(OH)CH_2CH_3$$

$C_5H_{11}IO$ M 214.046

Hamilton, R.J. *et al*, *Tetrahedron*, 1986, **42**, 2881 (*synth, ir, pmr, ms*)

5-Iodo-2-pentanol   I-60056
[90397-87-8]

$$H_3CCH(OH)CH_2CH_2CH_2I$$

$C_5H_{11}IO$ M 214.046

(±)-*form*
Ac: [82131-06-4].
   $C_7H_{13}IO_2$ M 256.083
   Liq.

Node, M. *et al*, *Tetrahedron Lett.*, 1984, **25**, 219 (*synth*)
Yadav, V.K. *et al*, *J. Org. Chem.*, 1986, **51**, 3372 (*synth, ir, pmr*)

5-Iodo-2-pentanone   I-60057
[3695-29-2]

$$ICH_2CH_2CH_2COCH_3$$

C_5H_9IO M 212.030
Light-yellow liq. Bp$_{17-20}$ 101-106°.
Ethylene acetal: [3695-28-1]. 2-(3-*Iodopropyl*)-2-*methyl-1,3-dioxolane*, 9CI.
   $C_7H_{13}IO_2$ M 256.083
   Dark-brown liq. Bp$_{0.2}$ 105°, Bp$_{0.05}$ 76°.

Findlay, J.A. *et al*, *J. Chem. Soc.* (*C*), 1970, 2631 (*synth, ir, pmr*)
Sterczycki, R., *Synthesis*, 1979, 724 (*synth*)
Cornish, C.A. *et al*, *J. Chem. Soc., Perkin Trans. 1*, 1985, 2585 (*synth, ir, pmr*)

5-Iodo-4-penten-1-ol   I-60058

$$ICH{=}CHCH_2CH_2CH_2OH$$

C_5H_9IO M 212.030

(*E*)-*form* [93782-93-5]
Bp$_{0.7}$ 75-85°.

Reich, H.J. *et al*, *J. Am. Chem. Soc.*, 1986, **108**, 7791 (*synth, pmr, ir, cmr*)

3-Iodo-2-phenyl-4*H*-1-benzopyran-4-one, 9CI   I-60059
3-Iodoflavone
[98153-12-9]

$C_{15}H_9IO_2$ M 348.139
Small cubes (EtOAc). Mp 128°.

Costa, A.M. *et al*, *J. Chem. Soc., Perkin Trans. 1*, 1985, 799 (*synth, ir, pmr*)

2-Iodo-1-phenyl-1-butanone, 9CI   I-60060
[108350-39-6]

$$PhCOCHICH_2CH_3$$

$C_{10}H_{11}IO$ M 274.101

(±)-*form*
Bp$_{0.1}$ 86-88°.

Barluenga, J. *et al*, *Synthesis*, 1986, 678 (*synth, ir, pmr, cmr*)

2-Iodo-1-phenyl-2-(phenylsulfonyl)-ethanone, 9CI   I-60061

$$PhSO_2CHICOPh$$

$C_{14}H_{11}IO_3S$ M 386.204

(±)-*form*
Mp 149-151°.

Kokkou, S.C. *et al*, *Acta Crystallogr., Sect. C*, 1986, **42**, 1074 (*cryst struct*)

5-Iodo-2,4(1*H*,3*H*)-pyrimidinedione, 9CI   I-60062
5-Iodouracil
[696-07-1]

$C_4H_3IN_2O_2$ M 237.984
Antimetabolite incorp. into nucleic acids. Cryst. (H$_2$O). Mp 272° dec.
▷YR0525000.

Johnson, T.B. *et al*, *J. Biol. Chem.*, 1905-6, **1**, 305 (*synth*)
Sternglanz, H. *et al*, *Acta Cryst. Sect. B*, 1975, **31**, 1393 (*cryst struct*)

2-Iodosophenylacetic acid   I-60063

2-Iodosobenzeneacetic acid

[89942-33-6]

$C_8H_7IO_3$ M 278.046

Believed to exist as cyclic tautomer. Catalyst for cleavage of reactive phosphates in cationic surfactant soln. Cryst. Mp 130° (126-128°).

Leffler, J.E. *et al, J. Am. Chem. Soc.*, 1963, **85**, 3443 (*synth, ir, props*)

2-Iodothiazole, 9CI   I-60064

Updated Entry replacing I-00784

[3034-54-6]

C_3H_2INS M 211.020

Yellow oil. Bp_{40} 118°, Bp_{10} 85.5°.

Tavagli, G., *Gazz. Chim. Ital.*, 1955, **85**, 926.
Vincent, E.J. *et al, Bull. Soc. Chim. Fr.*, 1966, 3524 (*pmr*)
Iversen, P.E., *Acta Chem. Scand.*, 1968, **22**, 1690.
Dondoni, A. *et al, Synthesis*, 1986, 757 (*synth*)

4-Iodothiazole   I-60065

[108306-60-1]

C_3H_2INS M 211.020

Oil.

Dondoni, A. *et al, Synthesis*, 1986, 757 (*synth, ir, pmr*)

5-Iodothiazole   I-60066

[108306-61-2]

C_3H_2INS M 211.020

Mp 79-81°.

Dondoni, A. *et al, Synthesis*, 1986, 757 (*synth, ir, pmr*)

3-Iodothiophene, 9CI   I-60067

Updated Entry replacing I-00786

[10486-61-0]

C_4H_3IS M 210.032

Fp −13.4°. Bp_{11} 77°.

Steinkopf, W. *et al, Justus Liebigs Ann. Chem.*, 1937, **527**, 237.
Gronowitz, S. *et al, Ark. Kemi*, 1963, **21**, 191 (*synth*)
Gronowitz, S. *et al, Chem. Scr.*, 1975, **7**, 76.
Robien, W. *et al, Monatsh. Chem.*, 1985, **116**, 685 (*cmr*)
Dettmeier, U. *et al, Angew. Chem., Int. Ed. Engl.*, 1987, **26**, 468 (*synth*)

2-Iodo-3-thiophenecarboxaldehyde   I-60068

2-Iodo-3-formylthiophene

[18812-40-3]

C_5H_3IOS M 238.043

Cryst. (Et_2O/hexane). Mp 69-70°. Bp_1 90°.

Oxime: [18799-97-8].
  C_5H_4INOS M 253.057
  Mp 115°.
Thiosemicarbazone: Mp 234°.

Guilard, R. *et al, Bull. Soc. Chim. Fr.*, 1967, 4121 (*synth, pmr*)

3-Iodo-2-thiophenecarboxaldehyde   I-60069

3-Iodo-2-formylthiophene

[930-97-2]

C_5H_3IOS M 238.043

Mp 80°.

Oxime: [18799-94-5].
  C_5H_4INOS M 253.057
  Mp 149°.
Thiosemicarbazone: Mp 256°.

Guilard, R. *et al, Bull. Soc. Chim. Fr.*, 1967, 4121 (*synth, pmr*)
Antonioletti, R. *et al, J. Chem. Soc., Perkin Trans. 1*, 1986, 1755 (*pmr, ir, ms*)

4-Iodo-2-thiophenecarboxaldehyde   I-60070

4-Iodo-2-formylthiophene

[18812-38-9]

C_5H_3IOS M 238.043

Cryst. by subl. Mp 76°. Bp_{14} 140-144°.

Oxime: [18799-95-6].
  C_5H_4INOS M 253.057
  Mp 186°.
Thiosemicarbazone: Mp 238°.

Guilard, R. *et al, Bull. Soc. Chim. Fr.*, 1967, 4121 (*synth, pmr*)

4-Iodo-3-thiophenecarboxaldehyde   I-60071

3-Formyl-4-iodothiophene

[18799-84-3]

C_5H_3IOS M 238.043

Cryst. (Et_2O/pentane). Mp 73°.

Oxime: [18799-98-9].
  C_5H_4INOS M 253.057
  Mp 163°.
Thiosemicarbazone: Mp 233°.

Guilard, R. *et al, Bull. Soc. Chim. Fr.*, 1967, 4121 (*synth, pmr*)

5-Iodo-2-thiophenecarboxaldehyde   I-60072

2-Formyl-5-iodothiophene

[5370-19-4]

C_5H_3IOS M 238.043

Cryst. by subl. Mp 49°.

Oxime: [18799-96-7].
  C_5H_4INOS M 253.057
  Mp 162°.
Thiosemicarbazone: Mp 238°.

Guilard, R. *et al, Bull. Soc. Chim. Fr.*, 1967, 4121 (*synth, pmr*)

5-Iodo-3-thiophenecarboxaldehyde   I-60073

2-Iodo-4-formylthiophene

[18799-85-4]

C_5H_3IOS M 238.043

Needles (pentane). Mp 74°.

Oxime: [18799-99-0].
C_5H_4INOS M 253.057
Mp 125°.
Thiosemicarbazone: Mp 188°.

Guilard, R. *et al, Bull. Soc. Chim. Fr.*, 1967, 4121 (*synth, pmr*)

2-Iodo-3-thiophenecarboxylic acid   I-60074

[18895-00-6]

$C_5H_3IO_2S$ M 254.042
Mp 178°.
Amide: [18800-06-1].
C_5H_4INOS M 253.057
Mp 160°.
Nitrile: [18800-01-6]. *3-Cyano-2-iodothiophene.*
C_5H_2INS M 235.042
Mp 57°.

Guilard, R. *et al, Bull. Soc. Chim. Fr.*, 1967, 4121 (*synth*)

3-Iodo-2-thiophenecarboxylic acid   I-60075

$C_5H_3IO_2S$ M 254.042
Mp 200° (193-195°).
Amide: [18800-03-8].
C_5H_4INOS M 253.057
Mp 140°.
Nitrile: [18800-00-5]. *2-Cyano-3-iodothiophene.*
C_5H_2INS M 235.042
Cryst. (pentane). Mp 54°.

Steinkopf, W. *et al, Justus Liebigs Ann. Chem.*, 1937, **527**, 237 (*synth*)
Guilard, R. *et al, Bull. Soc. Chim. Fr.*, 1967, 4121 (*synth*)

4-Iodo-2-thiophenecarboxylic acid   I-60076

$C_5H_3IO_2S$ M 254.042
Mp 116°.
Amide: [18800-04-9].
C_5H_4INOS M 253.057
Mp 170°.
Nitrile: [18894-98-9]. *2-Cyano-4-iodothiophene.*
C_5H_2INS M 235.042
Cryst. (hexane/Et_2O). Mp 74°.

Gronowitz, S. *et al, Ark. Kemi*, 1963, **21**, 191 (*synth*)
Guilard, R. *et al, Bull. Soc. Chim. Fr.*, 1967, 4121 (*synth*)

4-Iodo-3-thiophenecarboxylic acid   I-60077

$C_5H_3IO_2S$ M 254.042
Mp 170°.
Amide: [18800-07-2].
C_5H_4INOS M 253.057
Mp 184°.
Nitrile: [18894-99-0]. *3-Cyano-4-iodothiophene.*
C_5H_2INS M 235.042
Cryst. (pentane). Mp 54°.

Steinkopf, W. *et al, Justus Liebigs Ann. Chem.*, 1937, **527**, 237 (*synth*)
Guilard, R. *et al, Bull. Soc. Chim. Fr.*, 1967, 4121 (*synth*)

5-Iodo-2-thiophenecarboxylic acid   I-60078

$C_5H_3IO_2S$ M 254.042
Mp 135°.
Amide: [18800-05-0].
C_5H_4INOS M 253.057
Mp 182°.
Nitrile: [18945-81-8]. *2-Cyano-5-iodothiophene.*
C_5H_2INS M 235.042
Cryst. (hexane). Mp 56°.

Gatterman, L. *et al, Ber.*, 1886, **19**, 690 (*synth*)
Schick, J.W. *et al, J. Am. Chem. Soc.*, 1948, **70**, 286 (*synth*)
Guilard, R. *et al, Bull. Soc. Chim. Fr.*, 1967, 4121 (*synth*)

5-Iodo-3-thiophenecarboxylic acid   I-60079

[18895-01-7]
$C_5H_3IO_2S$ M 254.042
Mp 145°.
Amide: [18800-08-3].
C_5H_4INOS M 253.057
Mp 162°.
Nitrile: [18800-02-7]. *4-Cyano-2-iodothiophene.*
C_5H_2INS M 235.042
Cryst. (pentane). Mp 65°.

Guilard, R. *et al, Bull. Soc. Chim. Fr.*, 1967, 4121 (*synth*)

γ-Ionone   I-60080

Updated Entry replacing I-00807
4-(2,2-Dimethyl-6-methylenecyclohexyl)-3-buten-2-one, 9CI
[24190-32-7]

$C_{13}H_{20}O$ M 192.300
Constit. of *Tamarindus indica.* Oil.
Semicarbazone: Cryst. Mp 144-144.5°.

Buchecker, R. *et al, Helv. Chim. Acta*, 1973, **56**, 2548 (*abs config*)
Mukaiyama, T. *et al, Chem. Lett.*, 1976, 1033 (*synth*)
Leyendecker, F. *et al, Tetrahedron*, 1987, **43**, 85 (*synth*)

Ircinianin   I-60081

Updated Entry replacing I-00827
[63555-48-6]

$C_{25}H_{32}O_4$ M 396.525
Constit. of *Ircinia* spp. Cryst. (Me_2CO/toluene). Mp
165-167°. $[\alpha]_D^{25}$ −232° (c, 0.5 in CHCl_3).

Hofheinz, W. *et al, Helv. Chim. Acta*, 1977, **60**, 1367 (*isol, struct*)
Takeda, K. *et al, Tetrahedron Lett.*, 1986, **27**, 3903 (*synth*)

Ircinic acid   I-60082

[106534-60-5]

$C_{25}H_{32}O_4$ M 396.525
Constit. of an *Ircinia* sp. Oil.

Manes, L.V. *et al, J. Nat. Prod.*, 1986, **49**, 787.

Ircinin 1   I-60083

Updated Entry replacing I-00828

[35731-89-6]

$C_{25}H_{30}O_5$ M 410.509
Constit. of marine sponge *Ircinia oros*. Oil. $[\alpha]_D^{19.5}$
$-34.12°$ (MeOH).

Δ^{11} *isomer* (Z-): [35761-52-5]. *Ircinin 2*.
$C_{25}H_{30}O_5$ M 410.509
From *I.* spp. Oil. $[\alpha]_D^{19.5}$ $-40.2°$ (MeOH).

Cimino, G. *et al, Tetrahedron*, 1972, **28**, 333 (*isol*)
Manes, L.V. *et al, J. Nat. Prod.*, 1986, **49**, 787 (*isol, struct*)

Iridodial   I-60084

Updated Entry replacing I-00833
2-Formyl-α,3-dimethylcyclopentaneacetaldehyde, 9CI
[550-45-8]

Absolute
configuration

$C_{10}H_{16}O_2$ M 168.235
From *Iridomyrmex* spp. Defensive secretion of Rove
beetles (*Staphylinus olens*). Oil. $Bp_{1.0}$ 90-92°.

Bis-2,4-dinitrophenylhydrazone: Cryst. (EtOH). Mp
224-225° dec.

Cavill, G.W. *et al, Chem. Ind.* (*London*), 1956, 465 (*struct, isol*)
Clark, K.J. *et al, Tetrahedron*, 1959, **6**, 217 (*abs config*)
Achmad, S. *et al, Proc. Chem. Soc., London*, 1963, 166 (*struct*)
Achmad, S.A. *et al, Aust. J. Chem.*, 1965, **18**, 1989 (*synth*)
Ritterskamp, P. *et al, J. Org. Chem.*, 1984, **49**, 1155 (*synth*)
Uesato, S. *et al, Tetrahedron Lett.*, 1986, **27**, 2896 (*biosynth*)
Uesato, S. *et al, J. Chem. Soc., Chem. Commun.*, 1987, 1020
(*synth*)
Uesato, S. *et al, Tetrahedron Lett.*, 1987, **28**, 4431 (*biosynth*)

12-Isoagathen-15-oic acid   I-60085

$C_{20}H_{32}O_2$ M 304.472
2,3-Dihydroxypropyl ester:
$C_{23}H_{38}O_4$ M 378.551
Constit. of nudibranch *Archidoris montereyensis*.
Cryst. (Et_2O/hexane). Mp 125-126°. $[\alpha]_D$ $-12.5°$
(CHCl_3).

Gustafson, K. *et al, Tetrahedron Lett.*, 1984, **25**, 11 (*cryst
struct*)

Isoambrettolide   I-60086

*Oxacycloheptadec-10-en-2-one, 9CI, 8CI. 16-Hydroxy-9-
hexadecenoic acid o-lactone. Δ^9-Isoambrettolide*

[28645-51-4]

(Z)-form

$C_{16}H_{28}O_2$ M 252.396
Used as musk odour in perfume industry.

(**E**)-*form* [63286-42-0]
$Bp_{0.2}$ 115-116°.
(**Z**)-*form* [93635-21-3]
$Bp_{0.4}$ 150-160° (bath).

Mookherjee, B.D. *et al, J. Org. Chem.*, 1972, **37**, 3846 (*synth*)
Venkataraman, K. *et al, Tetrahedron Lett.*, 1980, **21**, 1893
(*synth*)
Chatterjea, J.N. *et al, Chem. Ind.* (*London*), 1983, 43; *Indian J.
Chem., Sect. B*, 1984, **23**, 733 (*synth*)

Isoamericanin A   I-60087

[109063-85-6]

$C_{18}H_{16}O_6$ M 328.321
Constit. of seeds of *Phytolacca americana*. Prostaglandin
I_2 inducer. Cryst. Mp 177-178°. $[\alpha]_D$ ±0°.

Hasegawa, T. *et al, Chem. Lett.*, 1987, 329.

Isobicyclogermacrenal   I-60088

Updated Entry replacing I-30068
*7,11,11-Trimethylbicyclo[8.1.0]undeca-2,6-diene-3-car-
boxaldehyde, 9CI*

OHC

$C_{15}H_{22}O$ M 218.338
(−)-form illus.

(+)-*form* [110268-36-5]
Constit. of *Aristolochia manshuriensis*. Cryst. Mp 54-
56°. $[\alpha]_D^{20}$ $+341°$ (c, 0.7 in CHCl_3).
(−)-*form* [73256-82-3]
Constit. of *Lepidozia vitrea*. Oil. $[\alpha]_D$ $-168°$ (c, 1 in
CHCl_3).

Matsuo, A. *et al, J. Chem. Soc., Perkin Trans. 1*, 1984, 203
(*isol, struct*)
Magari, H. *et al, J. Chem. Soc., Chem. Commun.*, 1987, 1196
(*synth*)
Rücker, G. *et al, Phytochemistry*, 1987, **26**, 1529 (*isol, cryst
struct*)

7-Isocyanato-2,10-bisaboladiene I-60089

7-Isocyanato-7,8-dihydro-α-bisabolene

$C_{16}H_{25}NO$ M 247.380

(6R,7R)-form [105281-35-4]
Constit. of a *Ciocalypta* sp. Oil. $[\alpha]_D^{25}$ −24.3° (c, 0.094 in hexane).

Gulavita, N.K. *et al*, *J. Org. Chem.*, 1986, **51**, 5136 (*isol, cryst struct*)

8-Isocyano-10,14-amphilectadiene I-60090

[108695-81-4]

$C_{21}H_{31}N$ M 297.483
Constit. of a *Halichondria* sp. Oil. $[\alpha]_D$ −79.8° (c, 2 in $CHCl_3$).

Molinski, T.F. *et al*, *J. Org. Chem.*, 1987, **52**, 3334.

3-Isocyano-7,9-bisaboladiene I-60091

4-(1,5-Dimethyl-1,3-hexadienyl)-1-isocyano-1-methyl-cyclohexane, 9CI. 3-Isocyanotheonellin

$C_{16}H_{25}N$ M 231.380
(7E,9E)-form [105281-40-1]
Constit. of a *Phyllidia* sp. Oil.

Gulavita, N.K. *et al*, *J. Org. Chem.*, 1986, **51**, 5136.

7-Isocyano-3,10-bisaboladiene I-60092

7-Isocyano-7,8-dihydro-α-bisabolene

$C_{16}H_{25}N$ M 231.380
(6R,7R)-form [105281-36-5]
Constit. of a *Ciocalypta* sp. Oil. $[\alpha]_D^{22}$ −49.9° (c, 0.033 in hexane).

Gulavita, N.K. *et al*, *J. Org. Chem.*, 1986, **51**, 5136.

7-Isocyano-1-cycloamphilectene I-60093

[108695-79-0]

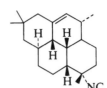

$C_{21}H_{31}N$ M 297.483
Constit. of a *Halichondria* sp. Needles (hexane). Mp 182-183°. $[\alpha]_D$ −14° (c, 0.41 in $CHCl_3$).

Molinski, T.F. *et al*, *J. Org. Chem.*, 1987, **52**, 3334 (*isol, cryst struct*)

7-Isocyano-11-cycloamphilectene I-60094

[108695-80-3]

$C_{21}H_{31}N$ M 297.483
Constit. of a *Halichondria* sp. Prisms (hexane). Mp 134°.

Molinski, T.F. *et al*, *J. Org. Chem.*, 1987, **52**, 3334 (*isol, cryst struct*)

8-Isocyano-1(12)-cycloamphilectrene I-60095

Updated Entry replacing I-00980
[108695-82-5]
$C_{21}H_{31}N$ M 297.483
Constit. of *Adocia* spp. and a *Halicondria* sp. Gum. $[\alpha]_D^{20}$ +39.6° (c, 0.6 in $CHCl_3$).

Koslauskas, R. *et al*, *Tetrahedron Lett.*, 1980, 315 (*isol*)
Molinski, T.F. *et al*, *J. Org. Chem.*, 1987, **52**, 3334 (*isol, cryst struct*)

Isocyclocalamin I-60096

[111004-32-1]

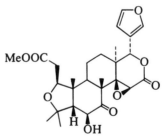

$C_{27}H_{34}O_9$ M 502.560
Constit. of *Citrus reticulata* var. *austera* × *Fortunella* sp.

Herman, Z. *et al*, *Phytochemistry*, 1987, **26**, 2247.

Isodalbergin   I-60097

7-Hydroxy-6-methoxy-4-phenyl-2H-1-benzopyran-2-one, 9CI. 7-Hydroxy-6-methoxy-4-phenylcoumarin, 8CI. 6-Methoxy-4-phenylumbelliferone

[605-09-4]

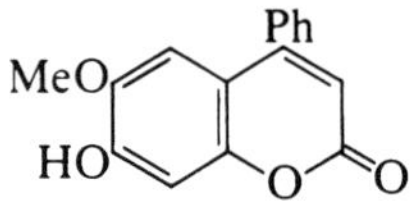

$C_{16}H_{12}O_4$ M 268.268

Isol. from *Dalbergia sissoo*. Needles (EtOAc). Mp 195.5-197°.

De Graw, J.I. *et al*, *J. Med. Chem.*, 1968, **11**, 375 (*synth*)
Mukerjee, S.K. *et al*, *Tetrahedron*, 1971, **27**, 799 (*isol*)

Isodidymic acid   I-60098

3-Hydroxy-7-methoxy-9-pentyl-1-propyldibenzofuran-2-carboxylic acid, 9CI

[106533-72-6]

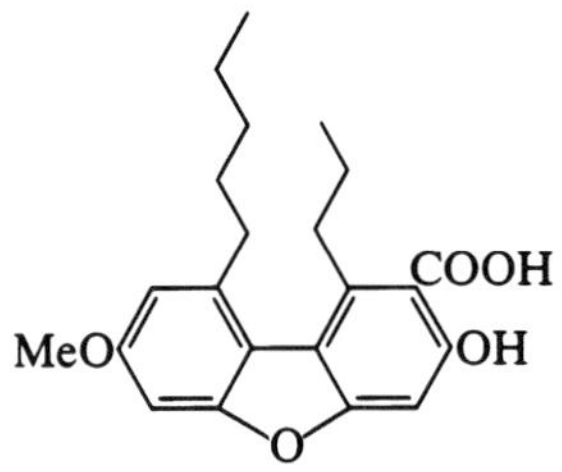

$C_{22}H_{26}O_5$ M 370.444

Constit. of *Cladonia didyma*. Cryst. (CH_2Cl_2/pet. ether). Mp 148-150°.

Chester, D.O. *et al*, *Aust. J. Chem.*, 1986, **39**, 1759 (*synth*)
Carvalho, C.F. *et al*, *Aust. J. Chem.*, 1986, **39**, 1765 (*synth*)

Isoelephantopin   I-60099

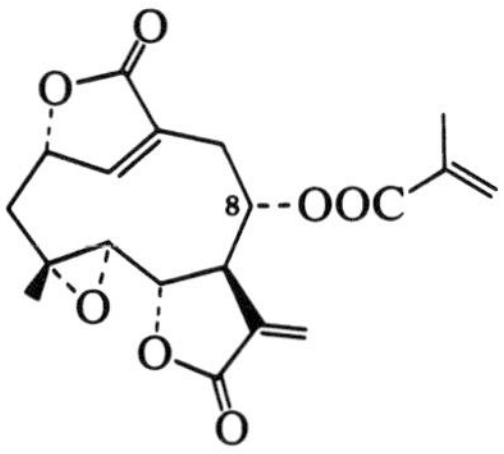

$C_{19}H_{20}O_7$ M 360.363

Constit. of *Elephantopus angustifolius*. Gum.

8-Deacyl, 8-(3-Methyl-2-butenoyl): Desacylisoelephantopin senecioate.
$C_{20}H_{22}O_7$ M 374.390
From *E. angustifolius*. Oil.
8-Deacyl, 8-(2-Methyl-2E-butenoyl): Desacylisoelephantopin tiglate.
$C_{20}H_{22}O_7$ M 374.390
From *E. angustifolius*. Oil.

Zhang, D. *et al*, *Phytochemistry*, 1986, **25**, 899 (*synth*)
Jakupovic, J. *et al*, *Phytochemistry*, 1987, **26**, 1467 (*isol*)

Isohallerin   I-60100

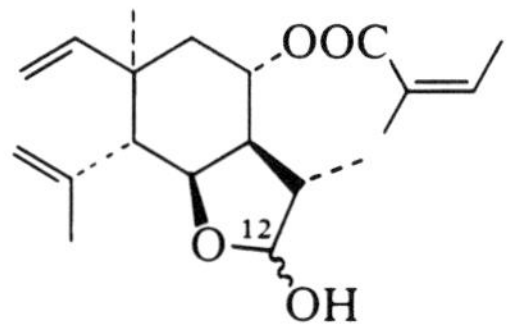

$C_{20}H_{30}O_4$ M 334.455

Constit. of *Laserpitium halleri* as a mixt. of 12-epimers. *Me ether (12β): Methylisohallerin*. Oil. $[\alpha]_D^{25}$ +30° (c, 1.83 in $CHCl_3$).

Appendino, G. *et al*, *Phytochemistry*, 1987, **26**, 1755.

Isohyperlatolic acid   I-60101

[90332-19-7]

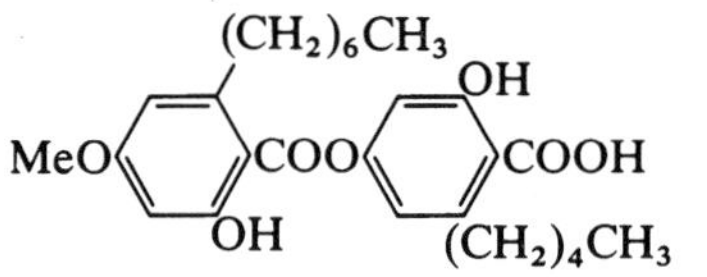

$C_{27}H_{36}O_7$ M 472.577

Metab. of *Fuscidea viridis*. Cryst. (cyclohexane). Mp 103°.

Culberson, C.F. *et al*, *Mycologia*, 1984, **76**, 148 (*isol*)
Elix, J.A. *et al*, *Aust. J. Chem.*, 1987, **40**, 425 (*synth*)

1H-Isoindolin-1-one-3-carboxylic acid   I-60102

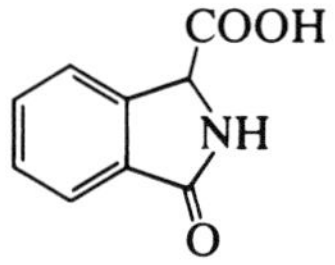

$C_9H_7NO_3$ M 177.159

(±)-*form*

Pale-yellow cryst. (H_2O). Mp 153-154°.

Lowe, J.A. *et al*, *J. Heterocycl. Chem.*, 1987, **24**, 877 (*synth, ir, pmr, bibl*)

Isolecanoric acid   I-60103

[110064-65-8]

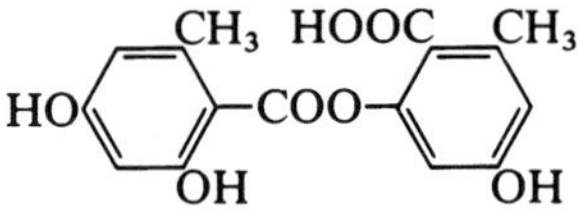

$C_{16}H_{14}O_7$ M 318.282

Metab. of *Parmelia tinctorum*. Powder. Mp >300°.

Sakurai, A. *et al*, *Bull. Chem. Soc. Jpn.*, 1987, **60**, 1917.

Isoligustroside I-60104

[108789-18-0]

$C_{25}H_{32}O_{12}$ M 524.521

Constit. of *Syringa vulgaris*.

Penta-Ac: [108789-15-7]. Cryst. Mp 45-50°. $[\alpha]_D^{30}$ −122.7° (c, 1.1 in CHCl$_3$).

Kikuchi, M. *et al, Yakugaku Zasshi*, 1987, **107**, 245.

Isolimbolide I-60105

$C_{30}H_{38}O_9$ M 542.625

Constit. of *Azadirachta indica*. Plates (CHCl$_3$). Mp 92-94°. $[\alpha]_D^{22}$ +33.3° (c, 0.03 in CHCl$_3$).

Siddiqui, S. *et al, Heterocycles*, 1987, **26**, 1827.

Isolobophytolide I-60106

3,4-Epoxy-7,11,15(17)-cembratrien-16,2-olide
[66275-29-4]

$C_{20}H_{28}O_3$ M 316.439

Constit. of soft coral *Lobophytum crassum*. Oil. $[\alpha]_D$ −103° (c, 0.04 in CHCl$_3$).

Bowden, B.F. *et al, Tetrahedron Lett.*, 1977, 3661 (*isol, struct*)
Marshall, J.A. *et al, Tetrahedron Lett.*, 1986, **27**, 5197 (*synth*)
Marshall, J.A. *et al, J. Org. Chem.*, 1987, **52**, 2378 (*struct, synth*)

Suggestions for new DOC Entries are welcomed. Please write to the Editor, DOC 5, Chapman and Hall Ltd, 11 New Fetter Lane, London EC4P 4EE

Isomarchantin *C* I-60107

$C_{28}H_{24}O_4$ M 424.495

Constit. of *Marchantia polymorpha*. Cryst. Mp 216-218°.

Asakawa, Y. *et al, Phytochemistry*, 1987, **26**, 1811.

Isomaresiic acid *C* I-60108

3,23-Dioxo-8(14→13R)-abeo-17,13-friedo-9β-lanosta-7,14,24E-trien-26-oic acid

$C_{30}H_{42}O_4$ M 466.659

Constit. of seeds of *Abies mariesii*.

Me ester: Gum. $[\alpha]_D$ −154° (c, 3.18 in CHCl$_3$).

Hasegawa, S. *et al, Tetrahedron*, 1987, **43**, 1775.

Isonimolicinolide I-60109

$C_{30}H_{36}O_9$ M 540.609

Constit. of *Azadiracta indica*. Cryst. (EtOAc). Mp 100-102°. $[\alpha]_D$ +20° (c, 0.2 in CHCl$_3$).

Siddiqui, S. *et al, J. Chem. Soc., Perkin Trans. 1*, 1987, 1429.

Isonimolide I-60110

$C_{29}H_{38}O_7$ M 498.615

Constit. of *Azadirachta indica*. Cryst. (MeOH). Mp 145-148°. $[\alpha]_D^{22}$ +50° (c, 0.04 in CHCl$_3$).

Siddiqui, S. *et al*, *Heterocycles*, 1987, **26**, 1827.

10-Isopentenylemodinanthran-10-ol   I-60111

$C_{20}H_{20}O_5$ M 340.375

Constit. of *Psorospermum tenuifolium*. Cryst. (MeOH). Mp 190-192°.

Delle Monache, G. *et al*, *Phytochemistry*, 1987, **26**, 2611.

Isopicropolin   I-60112

[27968-82-7]

$C_{22}H_{26}O_8$ M 418.443

Constit. of *Teucrium polium*.

Brieskorn, C.H. *et al*, *Chem. Ber.*, 1967, **100**, 1995 (*synth*)
Piozzi, F., *Heterocycles*, 1981, **15**, 1489 (*isol, struct*)

8(14),15-Isopimaradiene-2,18-diol   I-60113

8(14),15-Sandaracopimaradiene-2,18-diol

$C_{20}H_{32}O_2$ M 304.472

2α-form

Constit. of *Tetradenia riparia*. Cryst. (cyclohexane). Mp 203-204°. [α]$_D^{20}$ −21.7° (c, 0.35 in CHCl₃).

Van Puyvelde, L. *et al*, *Phytochemistry*, 1987, **26**, 493.

Isoporphobilinogen   I-60114

2-(Aminomethyl)-4-(carboxymethyl)-1H-pyrrole-3-propanoic acid, 9CI

[526-51-2]

$C_{10}H_{14}N_2O_4$ M 226.232

Needles. Mp 192-195° dec. (190° sealed tube).

Di-Et ester; B,HCl: [68541-99-1]. Needles (EtOH/Et₂O). Mp 99-100°.

Valasinas, A. *et al*, *J. Org. Chem.*, 1974, **39**, 2872 (*synth*)
Ufer, G. *et al*, *Can. J. Chem.*, 1978, **56**, 2437 (*deriv*)
Battersby, A.R. *et al*, *J. Chem. Soc., Perkin Trans. 1*, 1981, 2771.

1-Isopropenylcyclohexene   I-60115

1-(1-Methylethenyl)cyclohexene, 9CI. 1-Propen-2-ylcyclohexene

[6252-18-2]

C_9H_{14} M 122.210

Bp₂₀ 65-67°, Bp₉ 54-55°.

Wharton, P.S. *et al*, *J. Org. Chem.*, 1966, **31**, 3787 (*synth*)
House, H.O. *et al*, *J. Org. Chem.*, 1975, **40**, 1460 (*synth*)
Meyers, A.I. *et al*, *J. Org. Chem.*, 1976, **41**, 1735 (*synth*)
Herz, W. *et al*, *J. Org. Chem.*, 1985, **50**, 618.

3-Isopropenylcyclohexene   I-60116

3-(1-Methylethenyl)cyclohexene, 9CI. 3-Propen-2-ylcyclohexene

[56814-40-5]

C_9H_{14} M 122.210

(±)-form

Bp₇₁ 73-75°, Bp₂₀ 60°.

Bellassoued, M. *et al*, *Synthesis*, 1977, 205 (*synth, pmr*)
Goering, H.L. *et al*, *J. Org. Chem.*, 1986, **51**, 2884 (*synth, pmr, cmr*)

4-Isopropenylcyclohexene   I-60117

4-(1-Methylethenyl)cyclohexene, 9CI. 4-Propen-2-ylcyclohexene

[26325-89-3]

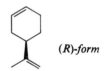

(R)-form

C_9H_{14} M 122.210

(R)-form [62393-64-0]

Bp₂ ca. 50°. [α]$_D$ +67.7° (c, 4 in CHCl₃).

(±)-form

Bp 157°.

Johnstone, R.A.W. *et al*, *J. Chem. Soc.*, 1963, 935 (*synth*)
Ceder, O. *et al*, *Acta Chem. Scand., Ser. B*, 1976, **30**, 908 (*synth*)
Uijttwaal, A.P. *et al*, *J. Org. Chem.*, 1979, **44**, 3157 (*synth*)
Barber, J.J. *et al*, *J. Org. Chem.*, 1979, **44**, 3603 (*synth*)

1-Isopropylcyclohexene, 8CI   I-60118

1-(1-Methylethyl)cyclohexene, 9CI

[4292-04-0]

CH(CH₃)₂

C_9H_{16} M 124.225

Liq. d₂₀ 0.83. Bp 155°.

Wallach, O., *Justus Liebigs Ann. Chem.*, 1908, **360**, 26 (*synth*)
Benkeser, R.A. *et al*, *J. Org. Chem.*, 1964, **29**, 1313 (*synth*)
Roach, L.C. *et al*, *J. Chem. Soc., Chem. Commun.*, 1970, 606 (*synth*)
Servis, K.L. *et al*, *J. Am. Chem. Soc.*, 1975, **97**, 73 (*synth*)
Barillier, D. *et al*, *Tetrahedron*, 1983, **39**, 767 (*cmr*)
Benkeser, R.A. *et al*, *J. Org. Chem.*, 1983, **48**, 2796 (*synth*)
Apparu, M. *et al*, *J. Org. Chem.*, 1984, **49**, 2125 (*synth*)

3-Isopropylcyclohexene, 8CI   I-60119

3-(1-Methylethyl)cyclohexene, 9CI

[3983-08-2]

C_9H_{16} M 124.225

(±)-*form*

d_{28} 0.82. Bp 150-152°, Bp_{15} 47°.

Berlaude, M.A. *et al, Bull. Soc. Chim. Fr.*, 1942, 644 (*synth*)
Biggerstaff, W.R. *et al, J. Org. Chem.*, 1954, **19**, 934 (*synth*)
Hueckel, W. *et al, Justus Liebigs Ann. Chem.*, 1959, **624**, 142 (*synth*)
Barillier, D. *et al, Tetrahedron*, 1983, **39**, 767 (*cmr*)
Benkeser, R.A. *et al, J. Org. Chem.*, 1983, **48**, 2796 (*synth*)
Goering, H.L. *et al, J. Org. Chem.*, 1986, **51**, 2884 (*synth, pmr*)
Tseng, C.C. *et al, J. Org. Chem.*, 1986, **51**, 2884 (*synth, ir, pmr, cmr*)

4-Isopropylcyclohexene, 8CI   I-60120

4-(1-Methylethyl)cyclohexene, 9CI

[14072-82-3]

C_9H_{16} M 124.225

(±)-*form*

Bp 155°. n_D^{20} 1.456.

Pines, H. *et al, J. Am. Chem. Soc.*, 1952, **74**, 4872 (*synth*)
Stork, G. *et al, J. Am. Chem. Soc.*, 1956, **78**, 4604 (*synth*)
Benkeser, R.A. *et al, J. Org. Chem.*, 1964, **29**, 1313 (*synth*)
Nelson, S.J. *et al, Tetrahedron Lett.*, 1973, 447 (*synth*)
Barillier, D. *et al, Tetrahedron*, 1983, **39**, 767 (*cmr*)

2-Isopropylideneadamantane   I-60121

2-(1-Methylethylidene)tricyclo[3.3.1.1³,⁷]decane, 9CI

[20441-18-3]

$C_{13}H_{20}$ M 176.301
Mp 41.6-42.2°.

Burhard, J. *et al, Z. Chem.*, 1969, **9**, 29 (*synth*)
Fry, J.L. *et al, J. Am. Chem. Soc.*, 1972, **94**, 4628 (*synth*)

2-Isopropyl-5-methyl-1,3-benzenediol   I-60122

5-Methyl-2-(1-methylethyl)-1,3-benzenediol, 9CI. p-*Cymene-3,5-diol. Isocymorcin*

[4389-63-3]

$C_{10}H_{14}O_2$ M 166.219
Leaflets (toluene). Mp 89.5-90°.

Treibs, W. *et al, J. Prakt. Chem.*, 1961, **13**, 291 (*synth*)

2-Isopropyl-6-methyl-1,4-benzenediol   I-60123

2-Methyl-6-(1-methylethyl)-1,4-benzenediol. m-*Cymene-2,5-diol*

$C_{10}H_{14}O_2$ M 166.219

Boscott, R.J., *Chem. Ind.* (*London*), 1955, 201

3-Isopropyl-6-methyl-1,2-benzenediol   I-60124

3-Methyl-6-(1-methylethyl)-1,2-benzenediol, 9CI. p-*Cymene-2,3-diol, 8CI. 3-Isopropyl-6-methylcatechol. Cymopyrocatechol*

[490-06-2]

$C_{10}H_{14}O_2$ M 166.219
Cryst. (pet. ether). Mp 47-47.5°. Bp 264°, Bp_{15} 143-145°.

Di-Ac:
    $C_{14}H_{18}O_4$ M 250.294
    Cryst. (EtOH). Mp 71.5-72°.
Bis(4-nitrobenzoyl): Cryst. (EtOH). Mp 156-156.5°.

Treibs, W. *et al, J. Prakt. Chem.*, 1959, **8**, 123 (*synth*)
Ansell, M.F. *et al, J. Chem. Soc.* (*C*), 1971, 1401 (*synth*)
Dallacker, F. *et al, Z. Naturforsch., B*, 1983, **38**, 1243.

5-Isopropyl-2-methyl-1,3-benzenediol   I-60125

2-Methyl-5-(1-methylethyl)-1,3-benzenediol, 9CI. p-*Cymene-2,6-diol. Cymorcin. 5-Hydroxycarvacrol*

[4389-62-2]

$C_{10}H_{14}O_2$ M 166.219
Isol. from essential oil of *Pseudocaryophyllus guili*. Mp 130-132°. Bp 294°.

Dibenzoyl: Mp 80°.
Di-Me ether:
    $C_{12}H_{18}O_2$ M 194.273
    Bp_{20} 142°.

Treibs, W. *et al, Ber.*, 1931, **64**, 2184; *J. Prakt. Chem.*, 1933, **138**, 284 (*synth*)
De Fenik, I.J.S. *et al, CA*, 1975, **83**, 103138 (*isol*)

5-Isopropyl-3-methyl-1,2-benzenediol   I-60126

3-Methyl-5-(1-methylethyl)-1,2-benzenediol, 9CI. m-*Cymene-5,6-diol*

[95509-70-9]

$C_{10}H_{14}O_2$ M 166.219
Cryst. (hexane). Mp 90.5-91.5°.

Carman, R.M. *et al, Aust. J. Chem.*, 1984, **37**, 2607 (*synth*)

5-Isopropyl-4-methyl-1,3-benzenediol   I-60127

4-Methyl-5-(1-methylethyl)-1,3-benzenediol, 9CI. o-*Cymene-4,6-diol*

[81633-96-7]

$C_{10}H_{14}O_2$ M 166.219
No phys. props. reported.

Gesson, J.P. *et al, Nouv. J. Chim.*, 1982, **6**, 477 (*synth*)

2-Isopropyl-5-methyl-4-cyclohexen-1-one   I-60128

p-*Menth-1(6)-en-3-one*

$C_{10}H_{16}O$ M 152.236

Constit. of *Pulicaria undulata*. Oil.

Sacco, T. *et al, Planta Med.*, 1983, **47**, 49 (*isol*)
Metwally, M. *et al, Chem. Pharm. Bull.*, 1986, **34**, 378 (*isol*)

8-Isopropyl-5-methyl-2-naphthalenecarboxaldehyde   I-60129

5-Methyl-8-(1-methylethyl)-2-naphthalenecarboxaldehyde

$C_{15}H_{16}O$ M 212.291

Constit. of *Calocedrus formosana*.

3,4-Dihydro: 3,4-Dihydro-8-isopropyl-5-methyl-2-naphthalenecarboxaldehyde.
$C_{15}H_{18}O$ M 214.307
From *C. formosana*.
Carboxylic acid, Me ester: Methyl 8-isopropyl-5-methyl-2-naphthalenecarboxylate.
$C_{16}H_{18}O_2$ M 242.317
Constit. of *C. formosana*.
Carboxylic acid, 3,4-dihydro, Me ester: Methyl 3,4-dihydro-8-isopropyl-5-methyl-2-naphalenecarboxylate.
$C_{16}H_{20}O_2$ M 244.333
From *C. formosana*.

Fang, J.-M. *et al, Phytochemistry*, 1987, **26**, 853.

2-Isopropyl-1,3,5-trimethylbenzene   I-60130

2-(1-Methylethyl)-1,3,5-trimethylbenzene, 9CI
[5980-96-1]

$C_{12}H_{18}$ M 162.274

Liq. Bp$_{0.1}$ 110°.

Reetz, M.T. *et al, Chem. Ber.*, 1987, **120**, 123 (*synth, pmr*)

5,8-Isoquinolinequinone   I-60131

Updated Entry replacing I-20084
5,8-Isoquinolinedione
[50-46-4]

$C_9H_5NO_2$ M 159.144

Yellow powder by subl. or cryst. Mp 135-138° dec.
▷NX0505000.
8-Oxime: [25132-36-9].
$C_9H_6N_2O_2$ M 174.159
Olive cryst. (DMF). Mp 235° dec.

Joseph, P.K. *et al, J. Med. Chem.*, 1964, **7**, 801 (*synth*)
Joullié, M.M. *et al, J. Heterocycl. Chem.*, 1969, **6**, 697 (*synth, pmr*)
Cameron, D.W. *et al, Aust. J. Chem.*, 1982, **35**, 1439 (*synth*)
Potts, K.T. *et al, J. Org. Chem.*, 1986, **51**, 2011 (*synth*)

1(2H)-Isoquinolinethione   I-60132

1-Mercaptoisoquinoline
[4702-25-4]

C_9H_7NS M 161.221

Thione tautomer predominates. Orange-brown cryst. (EtOH). Mp 171°.

S-Me: [42088-41-5]. *1-Methylthioisoquinoline*.
$C_{10}H_9NS$ M 175.248
Bp$_{0.08}$ 100°.

Albert, A. *et al, J. Chem. Soc.*, 1959, 2384 (*synth, uv, tautom*)

3(2H)-Isoquinolinethione   I-60133

3-Mercaptoisoquinoline. 3-Isoquinolinethiol

C_9H_7NS M 161.221

Thione tautomer predominates. Orange-red cryst. (C_6H_6). Mp 217°.

S-Me; B,HCl:
$C_{10}H_9NS$ M 175.248
Pale-yellow cryst. by subl. Mp 197-199°.

Albert, A. *et al, J. Chem. Soc.*, 1959, 2384 (*synth, tautom, uv*)

1-(1-Isoquinolinyl)-1-(2-pyridinyl)ethanol   I-60134

α-Methyl-α-2-pyridinylisoquinolinemethanol, 9CI

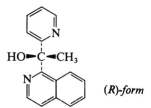

(R)-form

$C_{16}H_{14}N_2O$ M 250.299

Prob. abs. config. shown.

(R)-form [106623-38-5]
Alkylation or attachment to a resin gives chiral chelating ligands for transition metals. Cryst. (EtOH aq.). Mp 93-94°. $[\alpha]_D^{26}$ −62° (c, 0.0119 in CH_2Cl_2).
Me ether: [106623-40-9]. *1-[1-Methoxy-1-(2-pyridinyl)ethyl]isoquinoline, 9CI.*
$C_{17}H_{16}N_2O$ M 264.326
Chiral ligand for transition metals. Cryst. (pet. ether). Mp 84-85°. $[\alpha]_D^{26}$ +172° (c, 0.0129 in CH_2Cl_2).
(S)-form [106624-74-2]
Cryst. (EtOH aq.). Mp 93-94°. $[\alpha]_D^{26}$ +62° (c, 0.0119 in CH_2Cl_2).
Me ether: [106623-39-6]. Chiral chelating ligand for transition metals. Cryst. (pet. ether). Mp 84-85°. $[\alpha]_D^{26}$ +172° (c, 0.0129 in CH_2Cl_2).
(±)-form
Cryst. (EtOH aq.). Mp 78-79°.
Me ether: [106551-87-5]. Cryst. (pet. ether). Mp 115-116°.

Elman, B. *et al*, *Tetrahedron*, 1986, **42**, 223 (*synth, ir, pmr*)

Isoriccardin *C*   I-60135

$C_{28}H_{24}O_4$ M 424.495

Constit. of *Marchantia polymorpha*. Powder. Mp 218-219°.

Asakawa, Y. *et al*, *Phytochemistry*, 1987, **26**, 1811.

Isosphaeric acid   I-60136

[90332-22-2]

$C_{23}H_{28}O_7$ M 416.470

Metab. of *Dimelaena oreina*. Cryst. (cyclohexane). Mp 142°.

Culberson, C.F. *et al*, *Mycologia*, 1984, **76**, 148 (*isol*)
Elix, J.A. *et al*, *Aust. J. Chem.*, 1987, **40**, 425 (*synth*)

Isospongiadiol   I-60137

2α,19-Dihydroxy-13(16),14-spongiadien-3-one

$C_{20}H_{28}O_4$ M 332.439

Constit. of sponge *Spongia* sp. Cytotoxic and antiviral. Cryst. (MeOH aq.). Mp 181-183°. $[\alpha]_D^{20}$ −50° (c, 3.0 in CH_2Cl_2).

Kohmoto, S. *et al*, *Chem. Lett.*, 1987, 1687.

Isotanshinone IIB   I-60138

12,16-Epoxy-18-hydroxy-20-nor-5(10),6,8,12,15-abietapentaene-11,14-dione

[109664-01-9]

$C_{19}H_{18}O_4$ M 310.349

Constit. of *Salvia miltiorrhiza*. Red platelets. Mp 206-209°.

Lee, A.-R. *et al*, *J. Nat. Prod.*, 1987, **50**, 157.

Isotaxiresinol   I-60139

Updated Entry replacing I-01574
1-(3,4-Dihydroxyphenyl)-1,2,3,4-tetrahydro-7-hydroxy-6-methoxy-2,3-naphthalenedimethanol, 9CI, 8CI

[26194-57-0]

$C_{19}H_{22}O_6$ M 346.379

Constit. of *Taxus baccata*, *T. cuspida* and *Fitzroya cupressoides*. Cryst. (AcOH aq.). Mp 171°. Optical activity not recorded.

3'-Me ether: [548-29-8]. **Isolariciresinol**. α-Conidendryl alcohol.
$C_{20}H_{24}O_6$ M 360.406
Constit. of *F. cupressoides*. Needles ($CHCl_3$/MeOH). Mp 157-158°. $[\alpha]_D^{22}$ +69.5° (c, 1.19 in Me_2CO).

3'-Me ether, enantiomer: ent-*Isolariciresinol*.
$C_{20}H_{24}O_6$ M 360.406
Constit. of *Reseda suffruticosa*. Cryst. ($CHCl_3$/MeOH). Mp 152-154°. $[\alpha]_D^{25}$ −53.8° (c, 1.47 in MeOH).

7-Me ether: [23141-17-5].
$C_{20}H_{24}O_6$ M 360.406
Constit. of *F. cupressoides*. Needles ($CHCl_3$/MeOH). Mp 181-182°. $[\alpha]_D^{24}$ +34.9° (c, 1.49 in Me_2CO).

3',4'-Di-Me ether: Isolariciresinol 4'-methyl ether.
$C_{21}H_{26}O_6$ M 374.433
Constit. of *Araucaria angustifolia*. Cryst. (MeOH/Me_2CO). Mp 188-190°. $[\alpha]_D^{25}$ +16° (c, 1 in MeOH).

3',4',7-Tri-Me ether: Mp 161-164°. $[\alpha]_D$ +15.8° (c, 1.79 in $CHCl_3$).

King, F.E. *et al*, *J. Chem. Soc.*, 1952, 17 (*isol*)
Erdtman, H. *et al*, *Acta Chem. Scand.*, 1969, **23**, 2021 (*isol*)
Mujumdar, R.B. *et al*, *Indian J. Chem.*, 1972, **10**, 677 (*isol*)
Ahmed, R. *et al*, *Tetrahedron*, 1976, **32**, 1339 (*synth*)
Fonseca, S.F. *et al*, *Phytochemistry*, 1978, **17**, 499 (*cmr*)
Charlton, J.L. *et al*, *J. Org. Chem.*, 1986, **51**, 3490 (*synth*)
Urones, J.G. *et al*, *Phytochemistry*, 1987, **26**, 1540 (*isol, deriv*)

1-Isothiocyanato-5-(methylthio)pentane, 9CI   I-60140

Updated Entry replacing I-01596
5-(Methylthio)pentyl isothiocyanate

[4430-42-6]

$$MeS(CH_2)_5NCS$$

$C_7H_{13}NS_2$ M 175.306

Enzymic hydrol. prod. from the glucoside, Glucoberterain, occurring in *Berteroa incana* and also *Erysimum rhaeticum*. Oil. Bp$_{10}$ 155°.

S-Oxide: [646-23-1]. *1-Isothiocyanato-5-(methylsulfinyl)pentene*, 9CI. **Alyssin**.
$C_7H_{13}NOS_2$ M 191.306
Isol. from seeds of *Alyssum argentum*. Oil.

Kjaer, A. *et al*, *Acta Chem. Scand.*, 1955, **9**, 1311; 1956, **10**, 1100; *Phytochemistry*, 1973, **12**, 929 (*isol, deriv*)
Cole, R.A. *et al*, *Phytochemistry*, 1976, **15**, 759 (*isol*)

Isoxanthopterin   I-60141

2-Amino-4,7(1H,8H)pteridinedione, 9CI
[529-69-1]

$C_6H_5N_5O_2$ M 179.138

Widespread insect pigment found in amphibian and fish skin; normal constit. of urine. Mp >300°. pK_{a1} 7.34, pK_{a2} 10.06 (H_2O, 20°).

▷UO3425000.

Pfleiderer, W. *et al, Chem. Ber.*, 1961, **94**, 1 (*struct, bibl, uv*)
Konrad, G. *et al, Chem. Ber.*, 1970, **103**, 735 (*uv, props*)
Taylor, E.C. *et al, J. Org. Chem.*, 1975, **40**, 2341 (*synth, bibl*)

5-Isoxazolecarboxylic acid   I-60142

Updated Entry replacing I-01636
[21169-71-1]
$C_4H_3NO_3$ M 113.073
Pale-yellow cryst. Mp 149° subl.

Me ester: [15055-81-9].
   $C_5H_5NO_3$ M 127.099
   Cryst. (EtOH). Mp 49-50°. Bp_{12} 100-101°.
Et ester:
   $C_6H_7NO_3$ M 141.126
   Oil. Bp_1 110°.
Amide:
   $C_4H_4N_2O_2$ M 112.088
   Mp 173-174°.
Nitrile: [68776-59-0]. *5-Cyanoisoxazole.*
   $C_4H_2N_2O$ M 94.073
   Oil. Bp 168°.
Chloride: [62348-13-4].
   $C_4H_2ClNO_2$ M 131.518
   Liq. Bp_{20} 74-76°.

Quilico, A. *et al, Gazz. Chim. Ital.*, 1949, **79**, 654 (*synth*)
Spiegler, W. *et al, Synthesis*, 1986, 69 (*synth, deriv, ir, pmr*)

Ixocarpanolide   I-60143

6,7-Epoxy-5,20,22-trihydroxy-1-oxoergost-2-en-26-oic acid δ-lactone, 9CI. 6α,7α-Epoxy-5α,20R-dihydroxy-1-oxo-22R,24S,25R-with-2-enolide
[108157-59-1]

$C_{28}H_{40}O_6$ M 472.620

Constit. of *Physalis ixocarpa*. Cryst. (MeOH). Mp 252-253°. $[\alpha]_D$ +27° (c, 0.83 in $CHCl_3$).

14α-Hydroxy: [107221-65-8]. *6α,7α-Epoxy-5α,14α,20R-trihydroxy-1-oxo-22R,24S,25R-with-2-enolide. 14α-Hydroxyixocarpanolide.*
$C_{28}H_{40}O_7$ M 488.620
From *P. angulata*. Cryst. (MeOH). Mp 245-250°. $[\alpha]_D^{20}$ +29° (c, 1.18 in $CHCl_3$).

Abdullaev, N.D. *et al, Khim. Prir. Soedin.*, 1986, **22**, 300 (*isol*)
Vasina, O.E. *et al, Khim. Prir. Soedin.*, 1986, **22**, 560 (*deriv*)

J

Jaborol
J-60001

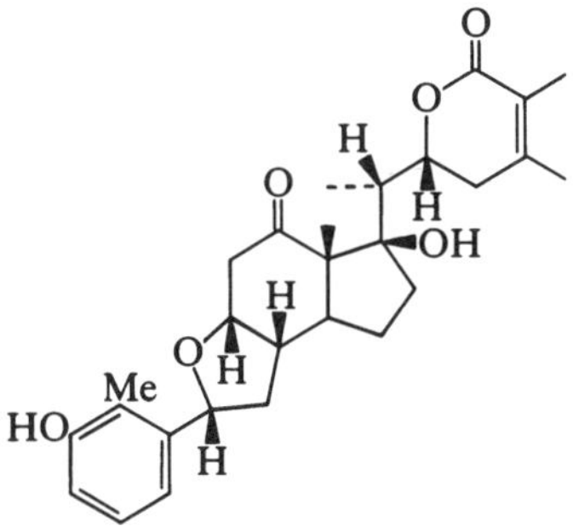

$C_{28}H_{36}O_6$ M 468.589
Constit. of *Jaborosa magellanica*. Cryst. (MeOH). Mp
135°. $[\alpha]_D$ +77.2° (c, 0.40 in MeOH).

Fajardo, V. *et al*, *Tetrahedron*, 1987, **43**, 3875 (*cryst struct*)

Jasmonic acid
J-60002

Updated Entry replacing J-50010
3-Oxo-2-(2-pentenyl)-cyclopentaneacetic acid, *9CI, 8CI*
[6894-38-8]

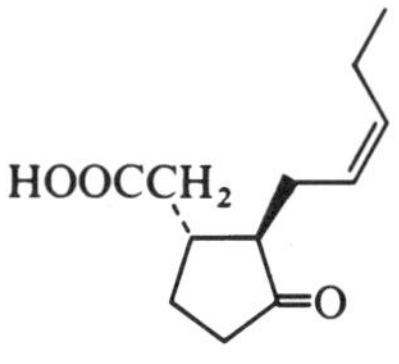

Absolute
configuration

$C_{12}H_{18}O_3$ M 210.272
Esters are present in *Jasminum grandiflorum* and are
  responsible for its odour. Viscous oil. Bp$_{0.001}$ 125°. $[\alpha]_D$
  −83.5° (c, 0.97 in CHCl$_3$).
2,4-Dinitrophenylhydrazone: Pale citron-yellow cryst.
  (Et$_2$O/ligroin). Mp 152-154° and 157-159° (double
  Mp).
Me ester: [20073-13-6]. *Methyl jasmonate.*
  $C_{13}H_{20}O_3$ M 224.299
  From *J. grandiflorum*. Has characteristic odour of
  jasmine; used in perfumery. Oil. Bp$_{0.001}$ 81-84°.
Me ester, 2,4-dinitrophenylhydrazone: Mp 50-54°.
  Mixture of geom. isomers.
2-Epimer: **7-Isojasmonic acid**.
  $C_{12}H_{18}O_3$ M 210.272
  Metab. of *Botryodiplodia theobromae*. Plant growth
  regulator. Oil. $[\alpha]_D$ +64° (c, 0.1 in MeOH).

Hill, R.K. *et al*, *Tetrahedron*, 1965, **21**, 1501 (*abs config*)
Fukui, H. *et al*, *Agric. Biol. Chem.*, 1977, **41**, 189 (*synth*)
Dubs, P. *et al*, *Helv. Chim. Acta*, 1978, **61**, 990.
Gerlach, H. *et al*, *Helv. Chim. Acta*, 1978, **61**, 2503.
Näf, F. *et al*, *Helv. Chim. Acta*, 1978, **61**, 2524.
Sato, T. *et al*, *Bull. Chem. Soc. Jpn.*, 1981, **54**, 505 (*synth*)
Johnson, F. *et al*, *J. Org. Chem.*, 1982, **47**, 4254 (*synth*)
Kitahara, T. *et al*, *Agric. Biol. Chem.*, 1982, **46**, 1369 (*synth*)
Oppolzer, W. *et al*, *Helv. Chim. Acta*, 1983, **66**, 2140 (*synth*)
Nishiyama, H. *et al*, *Tetrahedron Lett.*, 1984, **25**, 2487 (*synth*)
Posner, G.H. *et al*, *J. Org. Chem.*, 1985, **50**, 2589 (*synth*)
Smith, A.B. *et al*, *J. Org. Chem.*, 1985, **50**, 3239 (*synth*)
Miersch, O. *et al*, *Phytochemistry*, 1986, **26**, 1037 (*isol, deriv*)
Kataoka, H. *et al*, *Tetrahedron*, 1987, **43**, 4107 (*synth*)
Kitahara, T. *et al*, *Agric. Biol. Chem.*, 1987, **51**, 1129 (*synth*)

Jurinelloide
J-60003

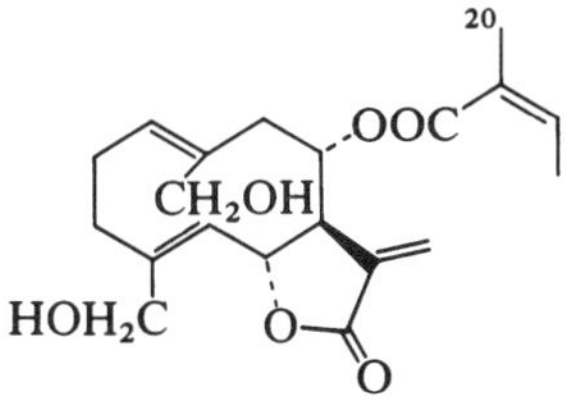

$C_{20}H_{26}O_6$ M 362.422
Constit. of *Jurinella moschus*. Oil. $[\alpha]_D$ +135° (c, 0.2 in
CHCl$_3$).
20-Hydroxy: *20-Hydroxyjurinelloide.*
  $C_{20}H_{26}O_7$ M 378.421
  Constit. of *J. moschus*. Oil.

Rustaiyan, A. *et al*, *Phytochemistry*, 1987, **26**, 2857.

K

Kachirachirol *B*　　　　　　　　　　K-60001

Updated Entry replacing K-30002

2,3-Dihydro-2-(3,4-dihydroxyphenyl)-7-methoxy-3-methyl-5-propenylbenzofuran

[94513-61-8]

$C_{19}H_{20}O_4$　　M 312.365

Constit. of *Magnolia kachirachirai*. Needles (C_6H_6). Mp 64-66°. $[\alpha]_D$ −60° (c, 0.65 in $CHCl_3$).

2,3-Didehydro, 3′-deoxy: [73027-11-9]. **Kachirachirol A**. 2-(4-Hydroxyphenyl)-7-methoxy-3-methyl-5-propenylbenzofuran. *Eupomatenoid 13*.
$C_{19}H_{18}O_3$　　M 294.349
Constit. of *M. kachirachirai* and *Krameria cystisoides*. Prisms ($CHCl_3$). Mp 193-196°.

Kaouadji, M. *et al*, *Phytochemistry*, 1978, **17**, 2134 (*isol*)
Ito, K. *et al*, *Phytochemistry*, 1984, **23**, 2643 (*isol*)
Achenbach, H. *et al*, *Phytochemistry*, 1987, **26**, 1159 (*isol, pmr, cmr*)

Kahweol　　　　　　　　　　K-60002

Updated Entry replacing K-00006

[6894-43-5]

$C_{20}H_{26}O_3$　　M 314.424

Occurs in unsaponifiable fraction of coffee bean oil. Cryst. Mp 88-90°. $[\alpha]_D^{20}$ −270°.

Di-Ac: Cryst. (MeOH). Mp 115°. $[\alpha]_D^{20}$ −245°.

Kaufmann, H.P. *et al*, *Chem. Ber.*, 1963, **96**, 2489 (*isol, struct*)
Corey, E.J. *et al*, *Tetrahedron Lett.*, 1987, **28**, 5403 (*synth*)

Kainic acid　　　　　　　　　　K-60003

Updated Entry replacing C-00301

2-Carboxy-4-(1-methylethenyl)-3-pyrrolidineacetic acid, 9CI. 3-Carboxymethyl-4-isopropenylproline. α-Kaininic acid. Digenic acid

[487-79-6]

$C_{10}H_{15}NO_4$　　M 213.233

Constit. of red algae *Digenea simplex* and *Centroceras clavulatum*. Neurotoxin, anthelmintic agent. Cryst. + 1H_2O (EtOH aq.). Mp 253-254° dec. $[\alpha]_D^{24}$ −14.8° (c, 1 in H_2O).

▷Highly neurotoxic

N-Ac: [59845-92-0]. Mp 161-162°. $[\alpha]_D$ −53° (H_2O).
Di-Me ester: [4071-37-8]. Bp_4 145°. $[\alpha]_D^{20}$ +23°.
4-Epimer: [4071-39-0]. α-**Allokainic acid**. *α-Allokaininic acid*.
$C_{10}H_{15}NO_4$　　M 213.233
Constit. of *D. simplex*. Mp 238-242° dec. $[\alpha]_D^{20}$ +7.7° (c, 1.3 in H_2O).

Nitta, I. *et al*, *Nature* (*London*), 1958, **181**, 761 (*struct*)
Impellizzeri, G. *et al*, *Phytochemistry*, 1975, **14**, 1549 (*isol*)
Oppolzer, W. *et al*, *Tetrahedron Lett.*, 1978, 3397; *Helv. Chim. Acta*, 1979, **62**, 2282 (*synth, bibl*)
Kraus, G.A. *et al*, *Tetrahedron Lett.*, 1983, **24**, 3427 (*synth*)
DeShong, P. *et al*, *Tetrahedron Lett.*, 1986, **27**, 3979 (*synth*)

ent-2α,16β,17-Kauranetriol　　　　　　　　　　K-60004

$C_{20}H_{34}O_3$　　M 322.487

O^2,O^{17}-*Bis-β-D-glucopyranoside:*
$C_{32}H_{54}O_{13}$　　M 646.771
Constit. of seeds of *Turbina corymbosa*. Plates (EtOH aq.). Mp 239-241°. $[\alpha]_D$ −40.6° (c, 2 in Py).

Nair, M.G. *et al*, *J. Chem. Res.* (*S*), 1987, 318.

ent-3β,16β,17-Kauranetriol　　　　　　　　　　K-60005

$C_{20}H_{34}O_3$　　M 322.487

Constit. of *Croton lacciferus*. Cryst. Mp 218-220°. $[\alpha]_D$ −39° (c, 0.59 in MeOH).

3-Ac: ent-*3β-Acetoxy-16β,17-kauranediol*.
$C_{22}H_{36}O_4$　　M 364.524
Constit. of *Peteravenia malvaefolia*. Oil.

3,17-Di-Ac: ent-*3β,17-Diacetoxy-16β-kauranol*.
$C_{24}H_{38}O_5$　　M 406.561
Constit. of *P. malvaefolia*. Oil. $[\alpha]_D^{24}$ −19° (c, 0.25 in $CHCl_3$).

Hanson, J.R. *et al*, *Tetrahedron*, 1970, **26**, 4839 (*synth*)
Ellmaurer, E. *et al*, *J. Nat. Prod.*, 1987, **50**, 221 (*isol*)
Bandara, B.M.R. *et al*, *Phytochemistry*, 1988, **27**, 225.

ent-11-Kaurene-16β,18-diol　　　　　　　　　　K-60006
Sidendrodiol

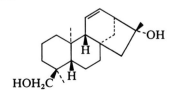

$C_{20}H_{32}O$　　M 288.472

Constit. of *Sideritis dendrochahorra*. Cryst. Mp 116-118°. The mol. formula is incorr. given as $C_{20}H_{30}O$ and the ms determination agrees with this figure, which throws doubt on the struct. assignment.

Fraga, B.M. *et al*, *Phytochemistry*, 1987, **26**, 775.

ent-15-Kaurene-3β,17-diol   K-60007

$C_{20}H_{32}O_2$ M 304.472
Cryst. Mp 174-175°. $[\alpha]_D$ −22.7° (c, 0.44 in $CHCl_3$).
3-Ac: ent-*3β-Acetoxy-15-kauren-17-ol.*
  $C_{22}H_{34}O_3$ M 346.509
  Constit. of *Croton lacciferus*. Cryst. Mp 130-131°.
  $[\alpha]_D$ −50° (c, 0.4 in $CHCl_3$).

Bandara, B.M.R. *et al*, *Phytochemistry*, 1988, **27**, 225.

ent-15-Kauren-17-oic acid   K-60008

[14696-36-7]
$C_{20}H_{30}O_2$ M 302.456
Constit. of *Peteravenia malvaefolia*.
Me ester: [110187-34-3]. Cryst. Mp 138°. $[\alpha]_D^{24}$ −29° (c, 1.6 in $CHCl_3$).

Ellmauerer, E. *et al*, *J. Nat. Prod.*, 1987, **50**, 221.

Kayamycin   K-60009

12-Deoxy-10,11-dihydro-4′-hydroxypicromycin, 9CI.
10,11-Dihydro-5-O-mycaminosylnarbonolide. Al R6-4.
Antibiotic Al R6-4
[102907-96-0]

$C_{28}H_{49}NO_8$ M 527.697
Macrolide antibiotic. Prod. by *Nocardiopsis* sp. Weakly active against gram-positive bacteria.

Japan. Pat., 85 155 189, (*1985*); *CA*, **105**, 23094 (*isol, struct, props*)
Rengaraju, N.S. *et al*, *CA*, 1987, **106**, 152654 (*isol*)

Khellin   K-60010

Updated Entry replacing K-00097
4,9-Dimethoxy-7-methyl-5H-furo[3,2-g] [1]-benzopyran-5-one, 9CI
[82-02-0]

$C_{14}H_{12}O_5$ M 260.246
Constit. of the fruit of *Ammi visnaga*. Exhibits spasmolytic activity, used in treatment of heart disease. Needles (MeOH aq. or Et_2O). Mp 150.3° and 154-155° (dimorph.).
▷LV1050000.

Baxter, R.A. *et al*, *J. Chem. Soc.*, 1949, Suppl., 30 (*synth*)
Clarke, J.R. *et al*, *J. Chem. Soc.*, 1949, 302 (*synth*)
Gardner, T.S. *et al*, *J. Org. Chem.*, 1950, **15**, 841 (*synth*)
Geissman, T.A. *et al*, *J. Am. Chem. Soc.*, 1951, **73**, 1280 (*synth*)
Beale, J.P. *et al*, *Cryst. Struct. Commun.*, 1973, **2**, 125 (*cryst struct*)
Carpy, A. *et al*, *Cryst. Struct. Commun.*, 1979, **8**, 835 (*cryst struct*)
Elgamal, M.H.A. *et al*, *Phytochemistry*, 1979, **18**, 139 (*cmr*)
Hassan, M.A. *et al*, *Anal. Profiles Drug Subst.*, 1980, **9**, 371 (*rev*)
Gammill, R.B. *et al*, *J. Org. Chem.*, 1983, **48**, 3863 (*synth*)
Yamashita, A., *J. Am. Chem. Soc.*, 1985, **107**, 5823 (*synth, bibl*)

Khellinone   K-60011

Updated Entry replacing K-50022
1-(6-Hydroxy-4,7-dimethoxy-5-benzofuranyl)ethanone, 9CI. 6-Hydroxy-4,7-dimethoxy-5-acetylcoumarone
[484-51-5]

$C_{12}H_{12}O_5$ M 236.224
Degradn. prod. of Khellin, K-60010 . Cryst. (MeOH aq.). Mp 99-101°.
Ac: Cryst. (Et_2O/ligroin). Mp 73.5-74° (*in vacuo*).
Oxime: [52631-78-4]. Mp 145°.

Baxter, R.A. *et al*, *J. Chem. Soc.*, 1949, Suppl., 30 (*synth*)
Clarke, J.R. *et al*, *J. Chem. Soc.*, 1949, 302 (*synth*)
Gardner, T.S. *et al*, *J. Org. Chem.*, 1950, **15**, 841 (*synth*)
Geissman, T.A. *et al*, *J. Am. Chem. Soc.*, 1951, **73**, 1280 (*synth*)
Hishmat, O.H. *et al*, *Z. Naturforsch., B*, 1978, **33**, 1491 (*synth*)
Gammill, R.B. *et al*, *Tetrahedron Lett.*, 1985, **26**, 1385 (*synth*)
Reed, M.W. *et al*, *J. Org. Chem.*, 1987, **52**, 3491 (*synth*)

Kielcorin *B*   K-60012

[110784-15-1]

$C_{24}H_{20}O_8$ M 436.417

Constit. of *Kielmeyera coriacea*.
Pinto, M.M.deM. *et al, Phytochemistry*, 1987, **26**, 2045.

Kolavenic acid K-60013

Updated Entry replacing K-10011
ent-*3,13-Clerodadien-15-oic acid*
[25436-90-2]

$R^1 = CH_3$,
$R^2 = COOH$
Absolute
configuration

$C_{20}H_{32}O_2$ M 304.472
Constit. of *Hardwickia pinnata* and *Solidago altissima*.
Me ester: [23527-24-4].
  $C_{21}H_{34}O_2$ M 318.498
  Constit. of roots of *Solidago elongata*. Liq. $Bp_{0.4}$ 179-
  180°. $[\alpha]_D$ −65.6° ($CHCl_3$).
2-Oxo: [83725-59-1]. *2-Oxokolavenic acid*.
  $C_{20}H_{30}O_3$ M 318.455
  Constit. of *Xylopia aethiopica*. Cubes (EtOAc/pet.
  ether). Mp 187-189°. $[\alpha]_D^{21}$ −56.5° (c, 0.52 in $CHCl_3$).
2-Oxo, Me ester: [38076-35-6]. Prisms (Et_2O). Mp 85-
  93°.

Misra, R. *et al, Tetrahedron Lett.*, 1964, 3751; 1968, 2681 (*isol,
  struct, abs config*)
Anthonsen, T. *et al, Acta Chem. Scand.*, 1969, **23**, 1068 (*isol*)
Ferguson, G. *et al, J. Chem. Soc., Chem. Commun.*, 1975, 299
  (*stereochem*)
Hasan, C.M. *et al, Phytochemistry*, 1982, **21**, 1365 (*deriv*)
Iio, H. *et al, J. Chem. Soc., Chem. Commun.*, 1987, 358 (*synth*)
Lopes, L.M.X. *et al, Phytochemistry*, 1987, **26**, 2781 (*cmr*)

Kurubasch aldehyde K-60014

$C_{15}H_{24}O_2$ M 236.353
Benzoyl: Kurubasch aldehyde benzoate.
  $C_{22}H_{28}O_3$ M 340.461
  Constit. of *Ferula haussknechtii*. Gum.
*4-Hydroxy-3-methoxybenzoyl: Kurubasch aldehyde
  vanillate*.
  $C_{23}H_{30}O_5$ M 386.487
  From *F. hausaknechtii*. Gum.
*15-Carboxylic acid, 2-methyl-2-butenoyl: Kurubashic
  acid angelate*.
  $C_{20}H_{30}O_4$ M 334.455
  From *F. haussknechtii*. Gum.
15-Carboxylic acid, benzoyl: Kurubashic acid benzoate.
  $C_{22}H_{28}O_4$ M 356.461
  From *F. haussknechtii*. Gum.
*15-Carboxylic acid, 1β,10α-epoxide, 2-methyl-2-
  butenoyl: 1β,10α-Epoxykurubashic acid angelate*.
  $C_{20}H_{30}O_5$ M 350.454
  From *F. haussknechtii*. Gum.
*15-Carboxylic acid, 1β,10α-epoxide, benzoyl: 1β,10α-
  Epoxykurubashic acid benzoate*.
  $C_{22}H_{28}O_5$ M 372.460
  From *F. haussknechtii*. Gum.

*15-Carboxylic acid, 1α,10β-epoxide, benzoyl: 1α,10β-
  Epoxykurubaschic acid benzoate*.
  $C_{22}H_{28}O_5$ M 372.460
  From *F. haussknechtii*. Gum.

Miski, M. *et al, Phytochemistry*, 1987, **26**, 1733.

Kuwanon J K-60015

Updated Entry replacing K-30022
[83709-26-6]

$C_{40}H_{38}O_{10}$ M 678.734
Metab. of tissue cultures of *Morus alba*. Yellow amorph.
  powder. $[\alpha]_D^{17}$ +85° (c, 0.04 in MeOH).
16″-Deoxy: [89803-86-1]. **Kuwanon Q**.
  $C_{40}H_{38}O_9$ M 662.735
  Pigment from culture cells of *M. alba*. Yellow amorph.
  powder. $[\alpha]_D^{16}$ +160° (c, 0.081 in Me_2CO).
2-Deoxy: [89803-85-0]. **Kuwanon R**.
  $C_{40}H_{38}O_9$ M 662.735
  Pigment from culture cells of *M. alba*. Yellow amorph.
  powder. $[\alpha]_D^{17}$ +56° (c, 0.165 in Me_2CO).
2,16″-Dideoxy: [89803-84-9]. **Kuwanon V**.
  $C_{40}H_{38}O_8$ M 646.735
  Pigment from culture cells of *M. alba*. Yellow amorph.
  powder. $[\alpha]_D^{23}$ +145° (c, 0.11 in Me_2CO).

Ueda, S. *et al, Chem. Pharm. Bull.*, 1982, **30**, 3042 (*isol*)
Ikuta, J. *et al, Chem. Pharm. Bull.*, 1986, **34**, 2471 (*isol, derivs*)

Kwakhurin K-60016

4′,6′,7-Trihydroxy-3′-methoxy-2′-prenylisoflavone

$C_{21}H_{20}O_6$ M 368.385
Constit. of *Pueraria mirifica*.

Tahara, S. *et al, Z. Naturforsch., C*, 1987, **42**, 510.

L

---

ent-7,13E-Labdadiene-3β,15-diol   L-60001

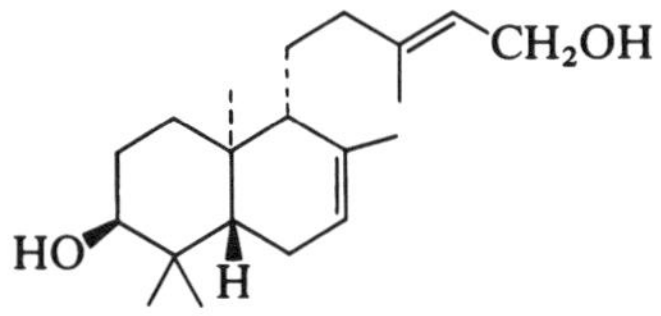

$C_{20}H_{34}O_2$ M 306.487

Constit. of *Corymbium villosum* also present as various esters. Gum. $[\alpha]_D^{24}$ −9° (c, 2.81 in CHCl₃).

Zdero, C. *et al, Phytochemistry*, 1988, **27**, 227.

8,13-Labdadiene-2,6,7,15-tetrol   L-60002

$C_{20}H_{34}O_4$ M 338.486

(2α,6β,7α,13E)-form

2,7-Di-Ac, O⁶-(3-methylbutanoyl): *2α,7α-Diacetoxy-6β-isovaleroyloxy-8,13E-labdadien-15-ol.*
$C_{29}H_{46}O_7$ M 506.678
Constit. of marine pulmonate *Trimusculus reticulatus.* Oil. $[\alpha]_D$ +41.2° (c, 0.35 in CHCl₃).

Manker, D.C. *et al, Tetrahedron*, 1987, **43**, 3677.

8,13-Labdadiene-6,7,15-triol   L-60003

$C_{20}H_{34}O_3$ M 322.487

(6β,7α,13E)-form

O⁶-(3-Methylbutanoyl): *6β-Isovaleroyloxy-8,13E-labdadiene-7α,15-diol.*
$C_{25}H_{42}O_4$ M 406.604
Constit. of marine pulmonate *Trimusculus reticulatus* and its defensive mucus. Oil. $[\alpha]_D$ +38.7° (c, 0.83 in CHCl₃).

Manker, D.C. *et al, Tetrahedron*, 1987, **43**, 3677.

---

ent-7,12E,14-Labdatriene-6β,17-diol   L-60004

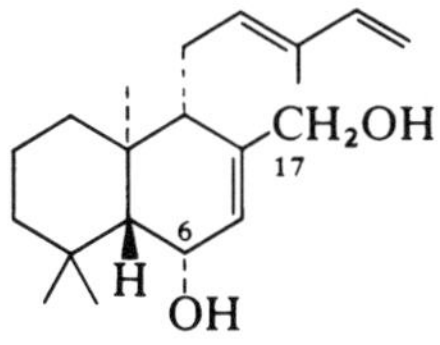

$C_{20}H_{32}O_2$ M 304.472
Cryst. Mp 70°. $[\alpha]_D^{24}$ −16° (c, 0.13 in CHCl₃).
6-Ac: *ent-6β-Acetoxy-7,12E,14-labdatrien-17-ol.*
  $C_{22}H_{34}O_3$ M 346.509
  Constit. of *Rutidosis murchisonii.* Oil.
6,17-Di-Ac: *ent-6β,17-Diacetoxy-7,12E,14-labdatriene.*
  $C_{24}H_{36}O_4$ M 388.546
  From *R. murchisonii.* Oil.
17-Carboxylic acid: *ent-6β-Hydroxy-7,12E,14-labdatrien-17-oic acid.* Oil.
17-Carboxylic acid, 6-Ac: *ent-6β-Acetoxy-7,12E,14-labdatrien-17-oic acid.*
  $C_{22}H_{32}O_4$ M 360.492
  Constit. of *R. murchisonii.* Oil.

Zdero, C. *et al, Phytochemistry*, 1987, **26**, 1759.

---

ent-7,11E,14-Labdatriene-6β,13ξ,17-triol   L-60005

$C_{20}H_{32}O_3$ M 320.471

6,17-Di-Ac: *ent-6β,17-Diacetoxy-7,11E,14-labdatrien-13ξ-ol.*
$C_{24}H_{36}O_5$ M 404.545
Constit. of *Rutidosis murchisonii.* Oil.

Zdero, C. *et al, Phytochemistry*, 1987, **26**, 1759.

7-Labdene-15,17-dioic acid   L-60006

$C_{20}H_{32}O_4$ M 336.470

(13S)-form

Havardic acid A
Constit. of *Grindelia havardii.*
Di-Me ester: Oil. $[\alpha]_D^{25}$ −56° (c, 4.3 in CHCl₃).

Jolad, S.D. *et al, Phytochemistry*, 1987, **26**, 483.

7-Labdene-15,18-dioic acid   L-60007

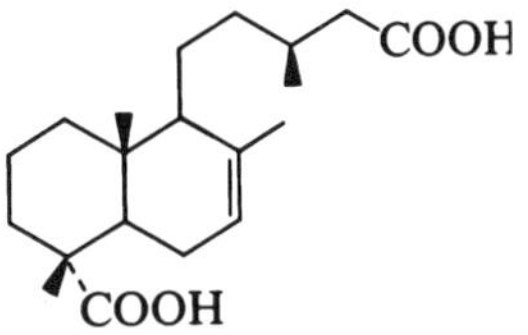

$C_{20}H_{32}O_4$ M 336.470

(13S)-form

8,17H-7,8-Dehydropinifolic acid
Constit. of *Ericameria linearifolia.*
Di-Me ester: Oil. $[\alpha]_D^{24}$ +16.44° (c, 4.34 in CHCl₃).

Dentali, S.J. *et al, Phytochemistry*, 1987, **26**, 3025.

Ladibranolide   L-60008

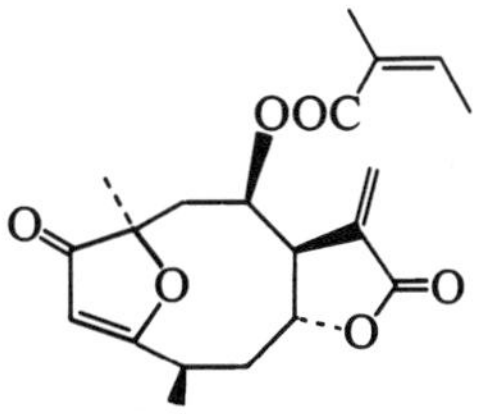

$C_{20}H_{24}O_6$ M 360.406
Constit. of *Viguiera ladibractate.* Prisms (EtOAc).

Gao, F. *et al, Phytochemistry*, 1987, **26**, 779.

Laferin
L-60009

Updated Entry replacing L-50021

4-Acetoxy-10-puteninone

[58561-94-7]

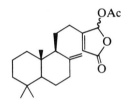

C$_{22}$H$_{26}$O$_7$ M 402.443

Constit. of *Ferula olgae*. Cryst. (Et$_2$O/pet. ether). Mp 142-144°. [α]$_D^{20}$ −3.1° (c, 2.97 in CHCl$_3$).

Konovalova, O.A. *et al*, *Khim. Prir. Soedin.*, 1975, 590 (*isol*)
Rychlewska, U. *et al*, *Collect. Czech. Chem. Commun.*, 1985, **50**, 2607 (*cryst struct*)

Lagerstronolide
L-60010

C$_{22}$H$_{32}$O$_4$ M 360.492

Constit. of *Lagerstroemia lancasteri*. Cryst. (CHCl$_3$/MeOH). Mp 162°. [α]$_D$ −7.1° (c, 0.2 in CHCl$_3$).

Chaudhuri, P.K., *Phytochemistry*, 1987, **26**, 3361.

Lambertianic acid
L-60011

Updated Entry replacing L-00098

5-[2-(3-Furanyl)ethyl]decahydro-1,4a-dimethyl-6-methylene-1-naphthalenecarboxylic acid, 9CI. 15,16-Epoxy-8(17),13(16),14-labdatrien-19-oic acid

C$_{20}$H$_{28}$O$_3$ M 316.439

(+)-*form* [4966-13-6]

Constit. of *Pinus lambertiana*. Cryst. (hexane). Mp 126.5-127.5°. [α]$_D^{22}$ +55° (c, 3 in 95% EtOH).

12-Ketone: 15,16-Epoxy-12-oxo-8(17),13(16),14-labdatrien-19-oic acid. **12-Oxolambertianic acid**.
C$_{20}$H$_{26}$O$_4$ M 330.423
Constit. of *Brickellia glomerata*. Gum.

(−)-*form* [1235-77-4]

ent-*form. Daniellic acid. Illurinic acid*
Constit. of *Daniellia oliveri*. Cryst. (MeOH). Mp 129-130.5°. [α]$_D$ −58° (c, 1 in MeOH).

Haeuser, J. *et al*, *Tetrahedron*, 1961, **12**, 205 (*isol, struct*)
Dauben, W.G. *et al*, *Tetrahedron*, 1966, **22**, 679 (*isol*)
Mills, J.S., *Phytochemistry*, 1973, **12**, 2479 (*isol*)
Bell, R.A. *et al*, *Can. J. Chem.*, 1976, **54**, 141 (*synth*)
Calderón, J.S. *et al*, *Phytochemistry*, 1987, **26**, 2639 (*deriv*)

7,9(11)-Lanostadiene-3,24,25,26-tetrol
L-60012

C$_{30}$H$_{50}$O$_4$ M 474.723

3β-form

Ganoderiol A

Constit. of *Ganoderma lucidum*. Needles (CHCl$_3$). Mp 232-234°. [α]$_D^{23}$ +20° (c, 0.1 in EtOH).

3-Ketone: 24,25,26-Trihydroxy-7,9(11)-lanostadien-3-one. **Ganodermanontriol**.
C$_{30}$H$_{48}$O$_4$ M 472.707
From *G. lucidum*. Cryst. Mp 168-170°. [α]$_D^{24}$ +41° (c, 0.2 in MeOH).

Sato, H. *et al*, *Agric. Biol. Chem.*, 1986, **50**, 2887.
Fujita, A. *et al*, *J. Nat. Prod.*, 1986, **49**, 1122.

Lansilactone
L-60013

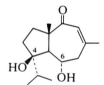

C$_{27}$H$_{44}$O$_3$ M 416.643

Constit. of *Lansium anamallayum*. Cryst. (C$_6$H$_6$/hexane). Mp 239°. [α]$_D^{30}$ +20° (c, 1 in CHCl$_3$).

Purushothaman, K.K. *et al*, *Can. J. Chem.*, 1987, **65**, 150.

Lapidol
L-60014

Updated Entry replacing L-40009

4β,6α-Dihydroxy-8-daucen-10-one

[79863-23-3]

C$_{15}$H$_{24}$O$_3$ M 252.353

Several numbering systems are in use for daucanes.
Yellow oil. [α]$_D^{20}$ +123° (c, 1.2 in CHCl$_3$).

6-Angeloyl: [79863-24-4]. **Lapidin**.
C$_{20}$H$_{30}$O$_4$ M 334.455
Constit. of *Ferula lapidosa*. Cryst. (hexane). Mp 80-81°. [α]$_D^{20}$ +166° (c, 1.5 in CHCl$_2$).

6-(3,4,5-Trimethoxybenzoyl): [80535-89-3]. **Palliferin**.
C$_{24}$H$_{32}$O$_7$ M 432.513
From *F. pallida*.

6-(3,4-Methylenedioxy-5-methoxybenzoyl): [80535-88-2]. **Palliferinin**.
C$_{23}$H$_{28}$O$_7$ M 416.470
From *F. pallida*.

Golovina, L.A. *et al*, *Khim. Prir. Soedin.*, 1981, **17**, 318 (*isol, struct*)
Kushmuradov, A.Y. *et al*, *Khim. Prir. Soedin.*, 1981, **17**, 523 (*isol*)
Moiseeva, G.P. *et al*, *Khim. Prir. Soedin.*, 1984, **20**, 45 (*cd*)

Lapidolinol   L-60015

Updated Entry replacing L-40010

8α,9α-Epoxy-2α,4β,6α,10α-daucanetetrol

[85202-12-6]

$C_{15}H_{26}O_5$ M 286.367

Several numbering systems are in use for daucanes.
Cryst. Mp 260-261° dec. $[\alpha]_D^{20}$ +58° (c, 1.1 in EtOH).

2,10-Di-Ac, 6-angeloyl: [85179-08-4]. **Lapidolin**.
$C_{24}H_{36}O_8$ M 452.544
Constit. of *Ferula lapidosa*. Cryst. (EtOH). Mp 188-189°. $[\alpha]_D^{20}$ +27.7° (c, 1.3 in CHCl₃).

2,10-Di-Ac, 6-(3,4-dimethoxybenzoyl): [85179-07-3].
Lapidolinin.
$C_{28}H_{38}O_{10}$ M 534.602
From *F. lapidosa*. Cryst. (EtOH). Mp 182-183°. $[\alpha]_D^{20}$ +54.5° (c, 1.1 in CHCl₃).

Kemoklidze, Z.S. *et al*, *Khim. Prir. Soedin.*, 1982, **18**, 788 (*isol*)
Moiseeva, G.P. *et al*, *Khim. Prir. Soedin.*, 1984, **20**, 45 (*cd*)
Tashkhodzhaev, B. *et al*, *Khim. Prir. Soedin.*, 1984, **20**, 309 (*cryst struct*)

Laserpitine   L-60016

Updated Entry replacing L-00154

[7067-12-1]

$C_{25}H_{38}O_7$ M 450.571

Constit. of *Laserpitum latifolium*. Cryst. Mp 117°. $[\alpha]_D^{20}$ +119° (EtOH).

8-Deangeloyl: **Lascrol**.
$C_{20}H_{32}O_6$ M 368.469
Constit. of *L. latifolium*. Cryst. Mp 148°. $[\alpha]_D^{20}$ +108.0°.

5-Alcohol: [16836-38-7]. **Laserpitinol**.
$C_{25}H_{24}O_7$ M 436.460
Constit. of *L. latifolium*. Cryst. Mp 139°. $[\alpha]_D^{20}$ ±0°.

Holub, M. *et al*, *Monatsh. Chem.*, 1967, **98**, 1138.
Holub, M. *et al*, *Collect. Czech. Chem. Commun.*, 1970, **35**, 3597.

Lasianthin   L-60017

$C_{22}H_{30}O_5$ M 374.476

Constit. of *Salvia lasiantha*. Cryst. (Me₂CO/hexane). Mp 174-176°. $[\alpha]_D^{20}$ –93.46° (c, 0.49 in CHCl₃).

Sánchez, A.-A. *et al*, *Phytochemistry*, 1987, **26**, 479 (*isol, cryst struct*)

Laurenene   L-60018

Updated Entry replacing L-00191

1,2,2a,3,4,4a,5,7,8,9,10,10a-Dodecahydro-3,3,4a,7,10a-pentamethylpentaleno[1,6-cd]azulene, 9CI

[72779-23-8]

$C_{20}H_{32}$ M 272.473

Constit. of *Dacrydium cupressinum*. Oil. Bp₀.₄ 58°. $[\alpha]_D^{20}$ –7.1° (c, 0.78 in CHCl₃).

Corbett, R.E. *et al*, *J. Chem. Soc., Perkin Trans. 1*, 1979, 1774, 1791 (*isol, cryst struct*)
Tsumoda, T. *et al*, *Tetrahedron Lett.*, 1987, **28**, 2537 (*synth*)

Lecithin   L-60019

Phosphatidylcholine

$$^1CH_2OOCR'$$
$$R^2COO\!-\!\!{}^2\overset{}{C}\!\!\leftarrow\!H \quad O$$
$$^3CH_2OPOCH_2CH_2\overset{\oplus}{N}Me_3$$
$$\underset{O_\ominus}{\|}$$

Lecithins, now called phosphatidylcholines, are compds. with the general struct. shown. Natural lecithins are mixtures of compds. with different acyl groups and have the stereochem. indicated. Individual members are listed under their systematic names, e.g., Glycerol 1-octadecanoate 2-octadec-9-enoate 3-phosphocholine.

Lepidopterin   L-60020

3-[2-Amino-4-hydroxy-7(3H)-pteridinylidene]-2-iminopropanoic acid, 9CI. 2-Amino-4-hydroxy-α-imino-$\Delta^{7(3H)\beta}$-pteridinepropionic acid, 8CI

[29067-92-3]

$C_9H_8N_6O_3$ M 248.201

Constit. of flour moth *Ephestia kühnella*.

Viscontini, M. *et al*, *Helv. Chim. Acta*, 1961, **44**, 1783; 1962, **45**, 2479; 1963, **46**, 51 (*isol, synth, uv*)

Lespedazaflavone *B*   L-60021

4′,5,7-Trihydroxy-3′,8-diprenylflavanone

$C_{25}H_{28}O_5$ M 408.493

Constit. of *Lespedeza davidii*. Needles (C₆H₆). Mp 141-142°. $[\alpha]_D^{16}$ –29.13° (c, 0.515 in MeOH).

Wang, M. _et al_, _Phytochemistry_, 1987, **26**, 1218.

Lespedezaflavanone _A_   L-60022

2',5,7-Trihydroxy-4'-methoxy-6,8-diprenylflavanone

$C_{26}H_{30}O_6$ M 438.519

Constit. of _Lespedeza davidii_. Yellow needles
(EtOAc/pet. ether). Mp 157-158°. $[\alpha]_D^{11.5}$ −60° (c,
0.25 in CHCl₃).

Wang, M. _et al_, _Phytochemistry_, 1987, **26**, 1218.

Leucettidine   L-60023

[79121-29-2]

$C_{10}H_{12}N_4O_3$ M 236.230

Isol. from the calcareous sponge _Leucetta microraphis_.
$[\alpha]_D^{21}$ −35.9° (c, 1.26 in MeOH).

Cardellina, J.H. _et al_, _J. Org. Chem._, 1981, **46**, 4782 (isol, uv, ir,
pmr, ms, struct)

Licarin _C_   L-60024

_2,3-Dihydro-7-methoxy-3-methyl-5-(1-propenyl)-2-
(3,4,5-trimethoxyphenyl)benzofuran_, 9CI
[60297-83-8]

$C_{22}H_{26}O_5$ M 370.444

Isol. from nutmeg _Nectandra miranda_ and _Licaria_ sp.
Cryst. (MeOH). Mp 100-101°.

Braz, R. _et al_, _Tetrahedron Lett._, 1976, 1157 (pmr)
Gottlieb, O.R. _et al_, _Phytochemistry_, 1977, **16**, 745, 1003 (isol,
abs config)
Davis, D.V. _et al_, _J. Agric. Food Chem._, 1982, **30**, 495 (ms)

Licoricidin   L-60025

[30508-27-1]

$C_{26}H_{32}O_5$ M 424.536

Constit. of _Glycyrrhiza glabra_ and _G. uralensis_. Cryst.
(CHCl₃/Et₂O). Mp 154-156°. $[\alpha]_D^{22.5}$ +20° (c, 1 in
MeOH).

5-Me ether: 5-_O-Methyllicoricidin._
$C_{27}H_{34}O_5$ M 438.563
Constit. of _G. uralensis_.

Shibata, S. _et al_, _Chem. Pharm. Bull._, 1968, **16**, 1932 (isol)
Kinoshita, T. _et al_, _Chem. Pharm. Bull._, 1978, **26**, 141 (isol)
Shih, T.L. _et al_, _J. Org. Chem._, 1987, **52**, 2029 (synth)

Ligustilide   L-60026

Updated Entry replacing L-50045
3-Butylidene-4,5-dihydro-1(3H)-isobenzofuranone, 9CI.
3-Butylidene-4,5-dihydrophthalide, 8CI
[4431-01-0]

$C_{12}H_{14}O_2$ M 190.241
Constit. of _Ligusticum and Angelica_ spp. Oil. Bp₆ 168-
169°.

9-Hydroxy: [94530-84-4]. **Senkyunolide F**.
$C_{12}H_{14}O_3$ M 206.241
Constit. of _C. officinale_.
6,7-Dihydro, cis-6,7-Dihydroxy: [94596-27-7]. **Senkyun-
olide H**.
$C_{12}H_{16}O_4$ M 224.256
Constit. of _C. officinale_.
3,8-Dihydro: see Senkyunolide A, S-50058
6,7-Epoxide: (**Z**)-**6,7-Epoxyligustilide**.
$C_{12}H_{14}O_3$ M 206.241
From _L. wallichii_. Oil.

Stahl, E. _et al_, _Naturwissenschaften_, 1967, **54**, 118 (isol)
Mitsuhashi, H., _CA_, 1969, **71**, 88456 (isol)
Nikonov, G.K. _et al_, _Khim. Prir. Soedin._, 1971, **7**, 387 (isol)
Yamagishi, T. _et al_, _CA_, 1975, **83**, 84751; 1976, **84**, 132662;
Yakugaku Zasshi, 1977, **97**, 237 (isol)
Kobayashi, M. _et al_, _Chem. Pharm. Bull._, 1985, **32**, 3770.
Kaouadjii, M. _et al_, _J. Nat. Prod._, 1986, **49**, 872 (deriv)

Lineatin   L-60027

Updated Entry replacing L-20046
3,3,7-Trimethyl-2,9-dioxatricyclo[3.3.1.0^{4,7}]nonane, 9CI

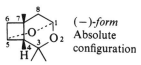

(−)-_form_
Absolute
configuration

$C_{10}H_{16}O_2$ M 168.235

(+)-**_form_** [65035-34-9]
Pheromone from _Tryptodendron lineatum_. Oil. $[\alpha]_D^{22}$
+36° (pentane).
(±)-**_form_** [71899-16-6]
Oil. Bp₂₀ 110°.

Mori, K. _et al_, _Tetrahedron_, 1980, **36**, 2197; 1983, **39**, 1735
(synth, cryst struct, abs config)
McKay, W.R. _et al_, _Can. J. Chem._, 1982, **60**, 872 (synth,
spectra)
White, J.D. _et al_, _J. Am. Chem. Soc._, 1982, **104**, 5486 (synth)
Johnston, B.D. _et al_, _J. Org. Chem._, 1985, **50**, 114 (synth, bibl)
Kandil, A.A. _et al_, _J. Org. Chem._, 1985, **50**, 5649 (synth)
Skattebøl, L. _et al_, _Acta Chem. Scand., Ser. B_, 1985, **39**, 291
(synth)
Ailjancic-Solaja, I. _et al_, _Helv. Chim. Acta_, 1987, **70**, 1302
(synth)

Lipiarmycin   L-60028

Updated Entry replacing L-00356

[56645-60-4]

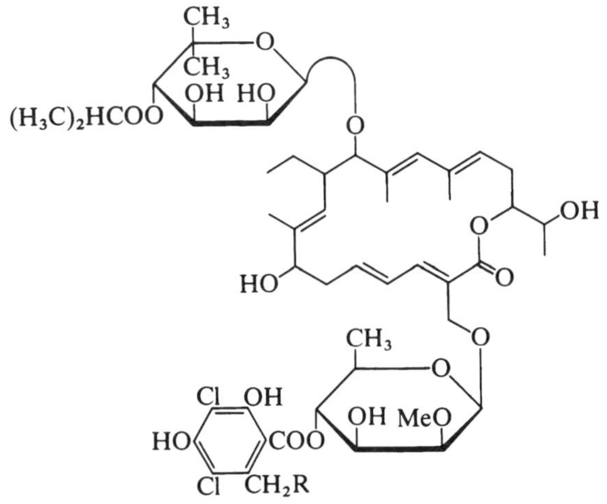

Lipiarmycin A_3 R = CH$_3$
               A_4 R = H

$C_{52}H_{74}Cl_2O_{18}$ M 1058.052

Macrolide antibiotic complex. Isol. from *Actinoplanes deccanensis*. Active against gram-positive bacteria. Plates. Mp 173-175°. $[\alpha]_D^{20}$ −5.5° (c, 1.98 in MeOH). pK_a 6.1. Factors A_3 and A_4 are major components.

Lipiarmycin A_3

Clpstp,ocom B$_1$. *Tiacumicin* B
   $C_{52}H_{74}Cl_2O_{18}$ M 1058.052
   Also isol. from *Dactylosporangium aurantiacum* ssp. *hamdenensis* and *Micromonospora echinospora* ssp. *armeniaca*. Cryst. (EtOAc/hexane). Mp 161-165°.
   $[\alpha]_D$ −6.2° (c, 2 in MeOH).

Lipiarmycin A_4
   $C_{51}H_{72}Cl_2O_{18}$ M 1044.025

Cryst. (EtOAc/hexane). Mp 138-140°. $[\alpha]_D$ −9.4° (c, 0.15 in MeOH).

Parenti, F. *et al*, *J. Antibiot.*, 1975, **28**, 247 (*isol*)
Coronelli, C. *et al*, *J. Antibiot.*, 1975, **28**, 253.
Talpaert, M. *et al*, *Biochem. Biophys. Res. Commun.*, 1975, **63**, 328.
Martinelli, F. *et al*, *J. Antibiot.*, 1983, **36**, 1312 (*struct, pmr, cmr*)
Omura, S. *et al*, *J. Antibiot.*, 1986, **39**, 1407 (*isol, props*)
Takahashi, Y. *et al*, *J. Antibiot.*, 1986, **39**, 1413 (*isol*)
Arnone, A. *et al*, *J. Chem. Soc., Perkin Trans. 1*, 1987, 1353 (*pmr, cmr, struct*)
Theriault, R.J. *et al*, *J. Antibiot.*, 1987, **40**, 567 (*Tiacumicin*)
Hochlowski, J.E. *et al*, *J. Antibiot.*, 1987, **40**, 575 (*Tiacumicin*)

Consult the Dictionary of Alkaloids for a comprehensive treatment of alkaloid chemistry.

Liquidambin   L-60029

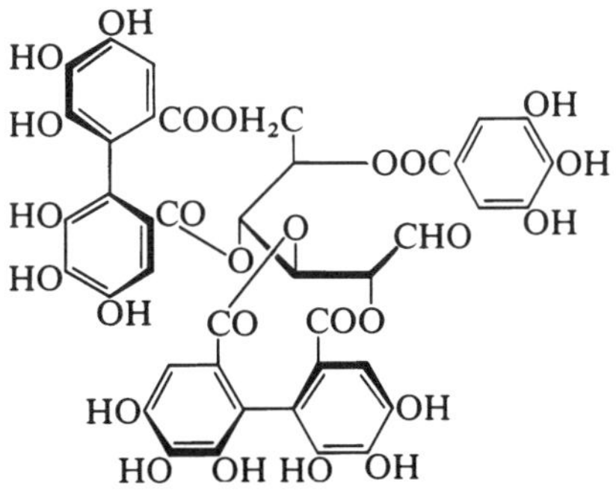

$C_{41}H_{28}O_{26}$ M 936.657

Constit. of *Liquidambar formosana*. Amorph. powder.
[α]$_D$ +69° (c, 0.5 in Me$_2$CO).

Okuda, T. *et al*, *Phytochemistry*, 1987, **26**, 2053.

Lonchocarpol *A*   L-60030

2,3-Dihydro-5,7-dihydroxy-2-(4-hydroxyphenyl)-6,8-bis(3-methyl-2-butenyl)-4H-1-benzopyran-4-one, 9CI.
4′,5,7-Trihydroxy-6,8-diprenylflavanone

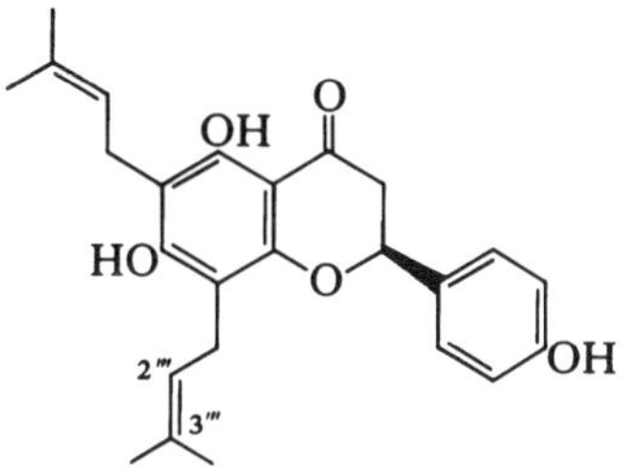

$C_{25}H_{28}O_5$ M 408.493

(**S**)-*form* [68236-11-3]

   Constit. of *Lonchocarpus minimiflorus*. Green oil. [α]$_D^{34}$ −4.7°.

   2‴,3‴-Dihydro, 2‴,3‴-dihydroxy: [111545-12-1]. **Lonchocarpol B**.
   $C_{25}H_{30}O_7$ M 442.508
   Constit. of *L. minimiflorus*. Greenish-yellow oil.

Roussis, V. *et al*, *Phytochemistry*, 1987, **26**, 2371.

Lonchocarpol *C*   L-60031

[111545-13-2]

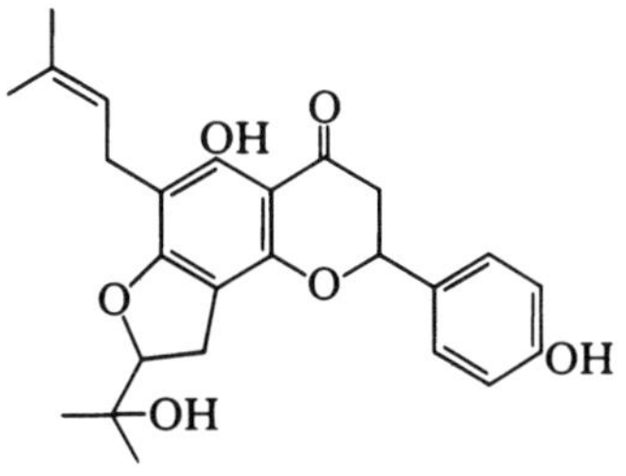

$C_{25}H_{28}O_6$ M 424.493

Constit. of *Lonchocarpus minimiflorus*. Yellow oil. [α]$_D^{34}$ −7.75°.

Roussis, V. *et al*, *Phytochemistry*, 1987, **26**, 2371.

Lonchocarpol *D*　　　　　　　　　　L-60032

[111545-14-3]

C$_{25}$H$_{28}$O$_6$　　M 424.493

Constit. of *Lonchocarpus minimiflorus*. Yellow oil.

Roussis, V. *et al*, *Phytochemistry*, 1987, **26**, 2371.

Lonchocarpol *E*　　　　　　　　　　L-60033

[111567-20-5]

C$_{25}$H$_{28}$O$_7$　　M 440.492

Constit. of *Lonchocarpus minimiflorus*. Orange-yellow oil.

Roussis, V. *et al*, *Phytochemistry*, 1987, **26**, 2371.

Longipesin　　　　　　　　　　　　L-60034

4-Hydroxy-5-(hydroxymethyl)-3-(3-methyl-2-butenyl)-2H-1-benzopyran-2-one. 4-Hydroxy-5-hydroxymethyl-3-prenylcoumarin

C$_{15}$H$_{16}$O$_4$　　M 260.289

9-Ac: [110024-35-6].
　C$_{17}$H$_{18}$O$_5$　　M 302.326
　Constit. of *Bothriocline longipes*.
9-Ac, 4-Me ether: [110042-12-1]. Cryst. Mp 33°.

Jakupovic, J. *et al*, *Phytochemistry*, 1987, **26**, 1069.

Lophirone *A*　　　　　　　　　　　L-60035

C$_{30}$H$_{22}$O$_8$　　M 510.499

Constit. of *Lophira lanceolata*. Noncryst. solid. [α]$_D^{23}$ +65° (c, 0.56 in MeOH).

Ghogomu, R. *et al*, *Tetrahedron Lett.*, 1987, **28**, 2967.

Lucidenic acid *H*　　　　　　　　　L-60036

3β,7β-Dihydroxy-4α-hydroxymethyl-4β,14α-dimethyl-11,15-dioxo-5α-chol-8-en-24-oic acid, 9CI. 3β,7β,28-Trihydroxy-11,15-dioxo-25,26,27-trisnor-8-lanosten-24-oic acid

C$_{27}$H$_{40}$O$_7$　　M 476.609

Constit. of *Ganoderma lucidum*.

Me ester: Cryst. (EtOAc/hexane). Mp 190-192°. [α]$_D^{23}$ +136° (c, 0.2 in MeOH).

7-Ketone: *3β-Hydroxy-4α-hydroxymethyl-4β,14α-dimethyl-7,11,15-trioxo-5α-chol-8-en-24-oic acid, 9CI. 3β,28-Dihydroxy-7,11,15-trioxo-25,26,27-trisnor-8-lanosten-24-oic acid*. **Lucidenic acid I**.
　C$_{27}$H$_{38}$O$_7$　　M 474.593
　Constit. of *G. lucidum*.

7-Ketone, Me ester: Pale-yellow syrup. [α]$_D^{23}$ +118° (c, 0.1 in MeOH).

7-Ketone, 12β-Hydroxy: *3β,12β-Dihydroxy-4α-hydroxymethyl-4β,14α-dimethyl-7,11,15-trioxo-5α-chol-8-en-24-oic acid, 9CI. 3β,12β,28-Trihydroxy-7,11,15-trioxo-25,26,27-trisnor-8-lanosten-24-oic acid*. **Lucidenic acid J**.
　C$_{27}$H$_{38}$O$_8$　　M 490.592
　Constit. of *G. lucidum*.

7-Ketone, 12β-hydroxy, Me ester: Pale-yellow syrup. [α]$_D^{25}$ +78° (c, 0.1 in MeOH).

Nishitoba, T. *et al*, *Phytochemistry*, 1987, **26**, 1777.

Lucidenic acid *M*　　　　　　　　　L-60037

3β,7α,15α-Trihydroxy-4,4,14α-trimethyl-11-oxo-5α-chol-8-en-24-oic acid, 9CI. 3β,7α,15α-Trihydroxy-11-oxo-25,26,27-trisnor-8-lanosten-24-oic acid

C$_{27}$H$_{42}$O$_6$　　M 462.625

Constit. of *Ganoderma lucidum*.

7,15-Diketone, 12β-hydroxy: *3β,12β-Dihydroxy-4,4,14α-trimethyl-7,11,15-trioxo-5α-chol-8-en-24-oic acid, 9CI. 3β,12β-Dihydroxy-7,11,15-trioxo-25,26,27-trisnor-8-lanosten-24-oic acid*. **Lucidenic acid L**.
　C$_{27}$H$_{38}$O$_7$　　M 474.593
　Constit. of *G. lucidum*.

3,7,15-Triketone, 12α-hydroxy: *12α-Hydroxy-4,4,14α-trimethyl-3,7,11,15-tetraoxo-5α-chol-8-en-24-oic acid, 9CI. 12α-Hydroxy-3,7,11,15-tetraoxo-25,26,27-trisnor-8-lanosten-24-oic acid*. **Lucidenic acid K**.
　C$_{27}$H$_{36}$O$_7$　　M 472.577
　Constit. of *G. lucidum*.

Nishitoba, T. *et al*, *Phytochemistry*, 1987, **26**, 1777.

Luminamicin   L-60038
Coloradocin
[99820-21-0]

$C_{32}H_{38}O_{12}$ M 614.645

Macrolide-related antibiotic. Isol. from *Nocardioides* sp.
and *Actinoplanes coloradoensis*. Selectively active
against anaerobic bacteria especially *Clostridium* sp.
Needles (CHCl₃). Mp 245°. $[\alpha]_D^{29}$ +6.3° (c, 1 in
MeCN).

Omura, S. *et al, J. Antibiot.*, 1985, **38**, 1322 (*isol, props*)
Jackson, M. *et al, J. Antibiot.*, 1987, **40**, 1375 (*isol, props*)
Rasmussen, R.R. *et al, J. Antibiot.*, 1987, **40**, 1383 (*struct*)

20(29)-Lupene-3,16-diol, 9CI   L-60039
Updated Entry replacing L-00492

$C_{30}H_{50}O_2$ M 442.724

(3β,16β)-form [10070-48-1]
Thurberin. Calenduladiol. Beyeriadiol
Constit. of *Beyeria leschenaultii* and *Lemairocereus
thurberi*. Cryst. (Me₂CO). Mp 218-219°. $[\alpha]_D$ +23°
(c, 1.2 in CHCl₃).
Di-Ac: [65043-59-6]. Mp 200-201°. $[\alpha]_D$ +42° (c, 1.14
in CHCl₃).
3-Tetradecanoyl: [108906-99-6].
$C_{44}H_{76}O_3$ M 653.083
From *Inula britannica*.
3-Hexadecanoyl: [110268-39-8].
$C_{46}H_{80}O_3$ M 681.136
Constit. of *I. britannica*.

Baddeley, G.V. *et al, Aust. J. Chem.*, 1964, **17**, 908 (*isol*)
Jolad, S.D. *et al, J. Org. Chem.*, 1969, **34**, 1367 (*isol*)
Protiva, J. *et al, Collect. Czech. Chem. Commun.*, 1977, **42**, 140
  (*struct*)
Wenkert, E. *et al, Org. Magn. Reson.*, 1978, **11**, 337 (*cmr*)
Öksüz, S. *et al, Phytochemistry*, 1987, **26**, 3082 (*deriv*)

Lycopadiene   L-60040

$C_{40}H_{78}$ M 559.056
Constit. of green alga *Botryococcus braunii*.

Metzger, P. *et al, Tetrahedron Lett.*, 1987, **28**, 3931.

Lycopene   L-60041
Updated Entry replacing L-30042
ψ,ψ-Carotene, 9CI. Lycopine
[502-65-8]

$C_{40}H_{56}$ M 536.882
Constit. of tomatoes and many other fruits. Cryst. (pet.
ether). Mp 175°.
1,2-Epoxide: [51599-09-8]. *1,2-Epoxy-1,2-
dihydrolycopene.*
$C_{40}H_{56}O$ M 552.882

Isol. from tomatoes.
5,6-Epoxide: [51599-10-1]. *5,6-Epoxy-5,6-
dihydrolycopene.*
$C_{40}H_{56}O$ M 552.882
Constit. of tomatoes.

(7Z,9Z,7′Z,9′Z)-form
Constit. of *Lycopersicon esculentum*.

Karrer, P. *et al, Helv. Chim. Acta*, 1950, **33**, 1349 (*synth*)
Isler, O. *et al, Helv. Chim. Acta*, 1956, **39**, 463 (*synth*)
Ben-Aziz, A. *et al, Phytochemistry*, 1973, **12**, 2759 (*isol*)
Berset, D. *et al, Helv. Chim. Acta*, 1984, **67**, 964 (*isol, struct,
  synth*)
Pattenden, G. *et al, Tetrahedron Lett.*, 1987, **28**, 5751 (*synth*)

Lycopersiconolide   L-60042
*3β,16β,20S-Trihydroxy-5α-pregnane-20-carboxylic acid
22,16-lactone*

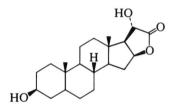

$C_{21}H_{32}O_4$ M 348.481
Constit. of tomato roots. Needles (CHCl₃/MeOH). Mp
295-297°. $[\alpha]_D^{20}$ −48° (c, 0.088 in MeOH).

Nagaoka, T. *et al, Phytochemistry*, 1987, **26**, 2113.

M

Maackiasin M-60001

[103147-87-1]

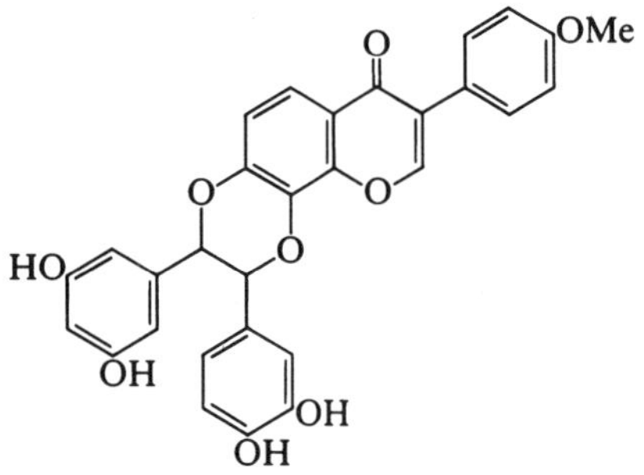

C$_{30}$H$_{22}$O$_9$ M 526.498
Constit. of *Maackia amurensis*. Cryst. (dioxane). Mp
243.5°.

Maksimov, O.B. *et al, Khim. Prir. Soedin.*, 1985, **21**, 735 (*isol*)
Krivoshchekova, O.E. *et al, Khim. Prir. Soedin.*, 1986, **22**, 35
(*struct*)

Machilin *B* M-60002

[110209-90-0]

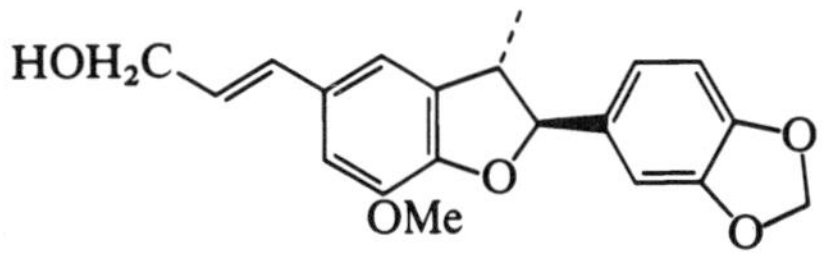

C$_{20}$H$_{20}$O$_5$ M 340.375
Constit. of *Machilus thunbergii*. Oil. [α]$_D^{25}$ −40.1° (c,
0.11 in CHCl$_3$).

Shimomura, H. *et al, Phytochemistry*, 1987, **26**, 1513.

Machilin *C* M-60003

[110269-51-7]

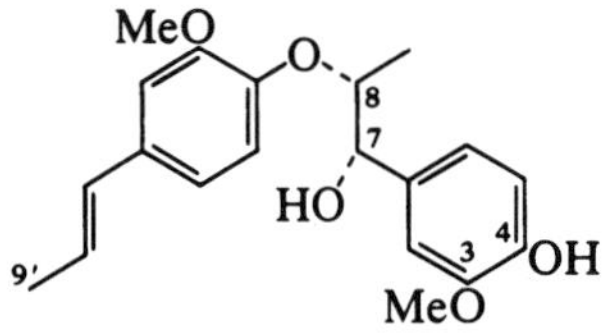

C$_{20}$H$_{24}$O$_5$ M 344.407
Compds. appear to be wrongly named in *CA*. Constit. of
Machilus thunbergii. Oil. [α]$_D^{25}$ −16.5° (c, 0.27 in
CHCl$_3$).
7-Epimer: [110269-53-9]. **Machilin *D*.**
  C$_{20}$H$_{24}$O$_5$ M 344.407
  From *M. thunbergii*. Oil. [α]$_D^{25}$ +38.1° (c, 0.07 in
  CHCl$_3$).
9′-Hydroxy, 7-Ac, 3,4-methylene analogue: **Machilin** E.
  C$_{22}$H$_{24}$O$_7$ M 400.427
  From *M. thunbergii*. Yellow oil. [α]$_D^{25}$ +29.2° (c, 0.11
  in CHCl$_3$).

Shimomura, H. *et al, Phytochemistry*, 1987, **26**, 1513.

Maesanin M-60004

Updated Entry replacing M-30003
*2-Hydroxy-5-methoxy-3-(10-pentadecenyl)-1,4-
benzoquinone*

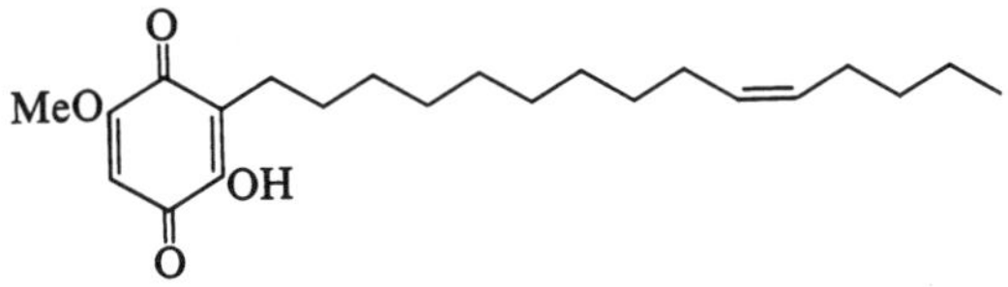

C$_{22}$H$_{34}$O$_4$ M 362.508
Constit. of fruits of *Maesa lanceolata*. Host defence
stimulant. Yellow needles. Mp 77°.

Kubo, I. *et al, Tetrahedron*, 1987, **43**, 2653 (*struct, synth*)

Magireol *A* M-60005

[109063-83-4]

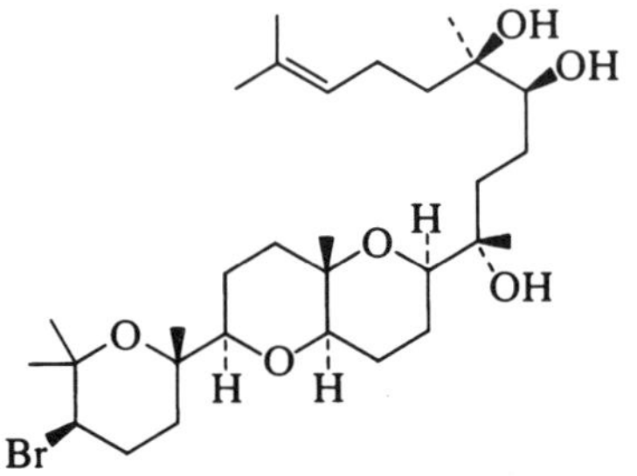

C$_{30}$H$_{53}$BrO$_6$ M 589.649
Constit. of red alga *Laurencia obtusa*. Cytotoxic. Cryst.
Mp 98.5-100°. [α]$_D$ ±0°.

Suzuki, T. *et al, Chem. Lett.*, 1987, 361.

Magireol *B* M-60006

[109063-84-5]

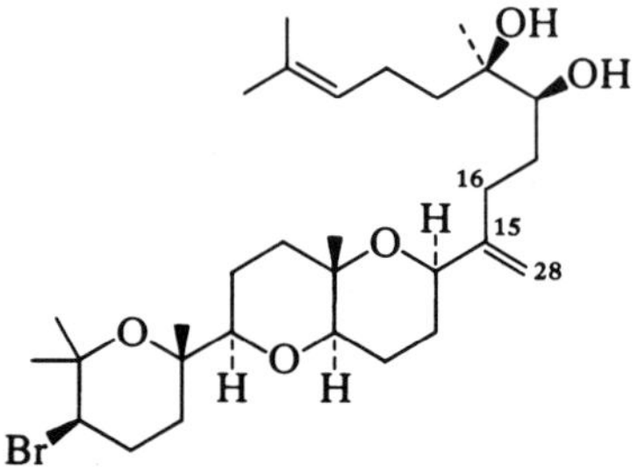

C$_{30}$H$_{51}$BrO$_5$ M 571.634
Constit. of red alga *Laurencia obtusa*. Cytotoxic. Cryst.
Mp 64.5-66°. [α]$_D$ +7.8° (c, 1.00 in CHCl$_3$).
Δ^{15}-*Isomer:* [109028-21-9]. **Magireol** C.
  C$_{30}$H$_{51}$BrO$_5$ M 571.634
  Constit. of *L. obtusa*. Cytotoxic. Mp 67-69°. [α]$_D$
  +6.4° (c, 1.00 in CHCl$_3$).

Suzuki, T. *et al, Chem. Lett.*, 1987, 361.

298

Magnoshinin   M-60007

Updated Entry replacing M-30006
[86702-02-5]

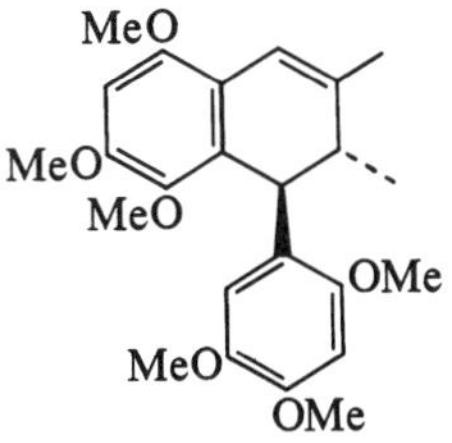

C$_{24}$H$_{30}$O$_6$ M 414.497
Constit. of _Magnolia salicifolia_. Cryst. (Et$_2$O). Mp
113.5-115°. [α]$_D$ 0° (CHCl$_3$).

Kikuchi, T. _et al, Chem. Pharm. Bull._, 1983, **31**, 1112.
Kadota, S. _et al, Tetrahedron Lett._, 1987, **28**, 2857 (_synth_)

14(26),17,21-Malabaricatrien-3-ol   M-60008

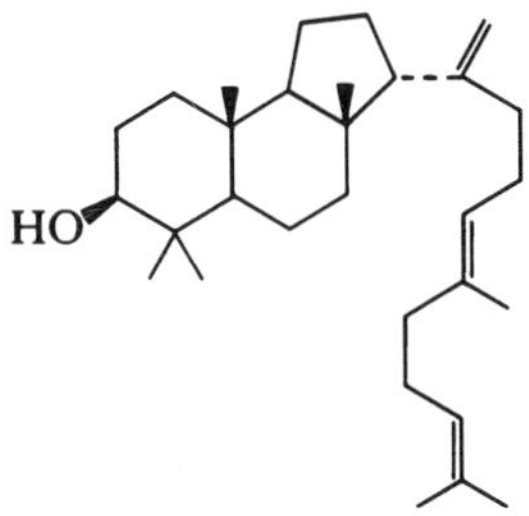

C$_{30}$H$_{50}$O   M 426.724
(3β,17E)-_form_

Ac: _3β-Acetoxy-14(26),17E,21-malabaricatriene._
C$_{32}$H$_{52}$O$_2$ M 468.762
Constit. of _Pyrethrum santolinoides_. Cryst. Mp 56°.
[α]$_D^{24}$ +14° (c, 1 in CHCl$_3$).
3-Ketone: _14(26),17E,21-Malabaricatrien-3-one._
C$_{30}$H$_{48}$O   M 424.709
From _P. santolinoides_. Oil.

Jakupovic, J. _et al, Phytochemistry_, 1987, **26**, 1536.

Malyngamide _C_   M-60009

[70622-52-5]

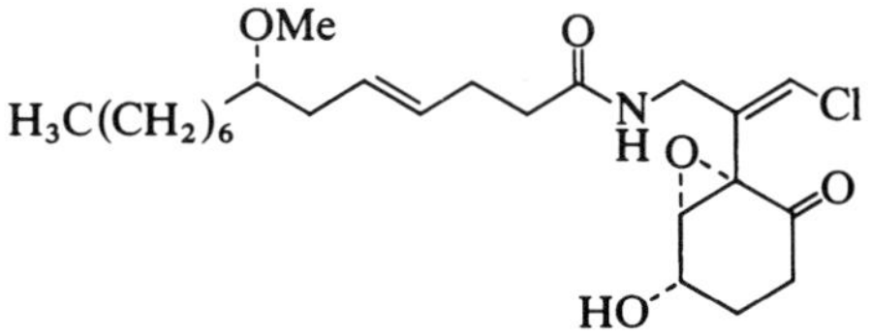

C$_{24}$H$_{38}$ClNO$_5$ M 456.021
Isol. from _Lyngbya majuscula_. Oil. [α]$_D$ −27.4° (c, 5.8
in EtOH).

Ainslie, R.D. _et al, J. Org. Chem._, 1985, **50**, 2859 (_isol, struct_)

Mariesiic acid _B_   M-60010

_3α,23R-Dihydroxy-17,13-friedo-9β-lanosta-7,12,24E-
trien-26-oic acid_

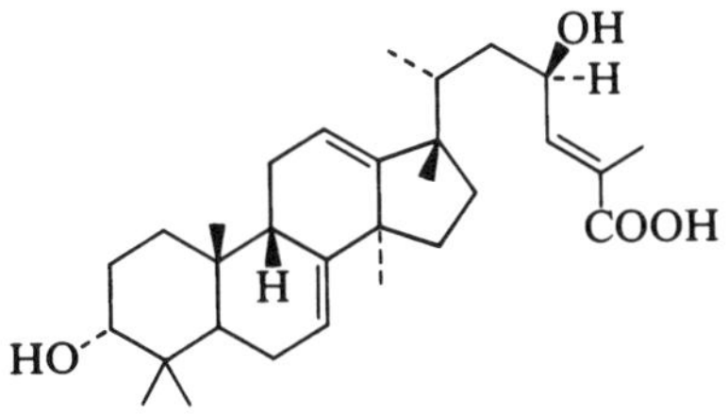

C$_{30}$H$_{46}$O$_4$ M 470.691
Constit. of seeds of _Abies mariesii_. Prisms (C$_6$H$_6$). Mp
200-202°. [α]$_D$ 98.7° (c, 0.85 in Me$_2$CO).

Me ester: Gum.
23-Ketone: **23-Oxomariesiic acid B**.
C$_{30}$H$_{44}$O$_4$ M 468.675
Constit. of _A. mariesii_ and _A. firma_.
22-Ketone, _Me ester:_ Gum. [α]$_D$ −157.4° (c, 1.98 in
CHCl$_3$).

Hasegawa, S. _et al, Tetrahedron_, 1987, **43**, 1775.

Mariesiic acid _C_   M-60011

_3,23-Dioxo-8(14→13R)-abeo-17,13-friedo-9β-lanosta-
7,14(30),24E-trien-26-oic acid_

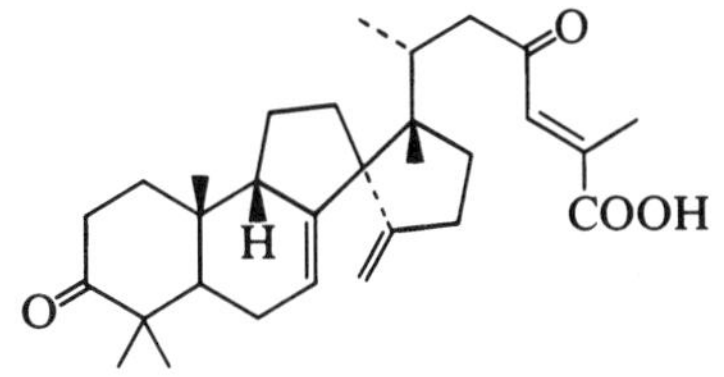

C$_{30}$H$_{42}$O$_4$ M 466.659
Constit. of seeds of _Abies mariesii_.

Me ester: Gum. [α]$_D$ −33.6° (c, 6.17 in CHCl$_3$).

Hasegawa, S. _et al, Tetrahedron_, 1987, **43**, 1775 (_cryst struct_)

7-Megastigmene-5,6,9-triol   M-60012

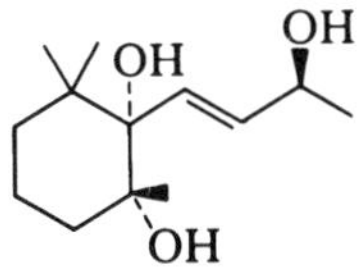

Absolute
configuration

C$_{13}$H$_{24}$O$_3$ M 228.331
Constit. of Greek tobacco. Cryst. Mp 118.5-119°. [α]$_D$
−5.0° (c, 0.12 in CHCl$_3$).

Wahlberg, I. _et al, Acta Chem. Scand., Ser. B_, 1987, **41**, 455
(_isol, cryst struct_)

Melampodin _D_   M-60013

[68612-45-3]

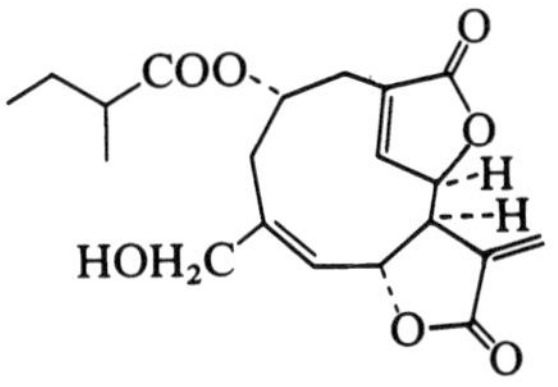

C$_{20}$H$_{24}$O$_7$ M 376.405
Constit. of _Melampodium argophyllum_ and _Polymnia
maculata_.

Malcolm, A.J. _et al, J. Nat. Prod._, 1987, **50**, 167.

Melcanthin *F*   M-60014

Updated Entry replacing M-20032

[79405-85-9]

As Melcanthins, M-30023 with

$$R^1 = Ac, R^2 = H_3CCH_2CH(CH_3)CO—$$

$C_{25}H_{32}O_{11}$ M 508.521

Constit. of *Melampodium leucanthum*.

9-Deacetoxy: **9-Desacetoxymelcanthin** F.
$C_{21}H_{28}O_7$ M 392.448
Constit. of *Polymnia maculata*. Gum.

Klimash, J.W. *et al, Phytochemistry*, 1981, **20**, 840 (*isol*)
Olivier, E.J. *et al, Phytochemistry*, 1983, **22**, 1453 (*struct*)
Malcolm, A.J. *et al, J. Nat. Prod.*, 1987, **50**, 167 (*deriv*)

Melianolone   M-60015

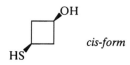

$C_{26}H_{36}O_{10}$ M 508.564

O^1-*Cinnamoyl:* **1-Cinnamoylmelianone**.
$C_{35}H_{42}O_{11}$ M 638.710
Constit. of *Melia azedarach*. Shows insecticidal props.

Lee, S.M. *et al, Tetrahedron Lett.*, 1987, **28**, 3543.

Membranolide   M-60016

[106231-26-9]

$C_{21}H_{28}O_4$ M 344.450
Constit. of *Dendrilla membranosa*. Oil. [α]$_D$ −28.8° (c, 2.25 in CHCl$_3$).

Molinski, T.F. *et al, J. Org. Chem.*, 1987, **52**, 296.

4′-Mercaptoacetophenone   M-60017

Updated Entry replacing M-00324
1-(4-Mercaptophenyl)ethanone, 9CI. p-*Acetylbenzenethiol*

[3814-20-8]

C_8H_8OS M 152.211
Liq. Bp$_{11}$ 140.5-141°, Bp$_7$ 135-136°.

Ac:
$C_{10}H_{10}O_2S$ M 194.248
Cryst. (EtOH). Mp 62.5-63.5°.
Me thioether: [1778-09-2]. *1-[4-(Methylthio)phenyl]-ethanone, 9CI.*
$C_9H_{10}OS$ M 166.237
Prisms (EtOH). Mp 80°.

Burton, H. *et al, J. Chem. Soc.*, 1948, 601 (*deriv, synth*)
Overberger, C.L. *et al, J. Am. Chem. Soc.*, 1956, **78**, 4792 (*synth*)
Walker, D. *et al, J. Org. Chem.*, 1963, **28**, 3077 (*synth*)
Loveridge, E.L. *et al, J. Org. Chem.*, 1971, **36**, 221 (*ir, nmr*)

4-Mercapto-2-butenoic acid   M-60018

$$HSCH_2CH=CHCOOH$$

$C_4H_6O_2S$ M 118.150
Parent compd. not known.

(*E*)-form
Me ester: [95968-93-7].
$C_5H_8O_2S$ M 132.177
Bp$_{0.01}$ 150-160°.
tert-*Butyl ester:* [89936-87-8].
$C_8H_{14}O_2S$ M 174.257
Bp$_2$ 90-100°.
Benzyl ester: [109202-01-9].
$C_{11}H_{12}O_2S$ M 208.275
Bp$_{0.025}$ 108-120°.
Nitrile: [109202-04-2].
C_4H_5NS M 99.150
Bp$_3$ 32-35°.

Anklam, E. *et al, Angew. Chem., Int. Ed. Engl.*, 1984, **23**, 364 (*synth, uv, pmr*)
Anklam, E. *et al, Helv. Chim. Acta*, 1984, **67**, 2198 (*ir, cmr, pmr*)
Bunce, R.A. *et al, Tetrahedron Lett.*, 1986, **27**, 5583 (*deriv, synth, ir, pmr, cmr*)

3-Mercaptocyclobutanol, 9CI   M-60019

cis-form

C_4H_8OS M 104.167

cis-form [103562-48-7]
Yellowish liq.
S-Ac: [103562-67-0]. *3-(Acetylthio)cyclobutanol*.
$C_6H_{10}O_2S$ M 146.204
Liq.
O-(*4-Methylbenzenesulfonyl*): [103562-50-1]. Yellowish oil.
Disulfide: [103562-70-5]. *Bis(3-hydroxycyclobutyl)-disulfide. 3,3′-Dithiobiscyclobutanol*.
$C_8H_{14}O_2S_2$ M 206.317
Cryst. (CHCl$_3$). Mp 114-115°. Config. not detd.

trans-form [103562-47-6]
Obt. as a mixt. with *cis*-form.
O-(*4-Methylbenzenesulfonyl*): [103562-49-8]. Yellowish oil.

Block, E. *et al, J. Org. Chem.*, 1986, **51**, 3428 (*synth, ir, pmr, cmr*)

4-Mercaptocyclohexanecarboxylic acid, 9CI   M-60020

$C_7H_{12}O_2S$ M 160.231

cis-form [105676-06-0]
Cryst. (H$_2$O). Mp 58-60°.
S-Triphenylmethyl: [105676-10-6]. *4-[(Triphenylmethyl)thio]cyclohexanecarboxylic acid, 9CI.*
$C_{26}H_{26}O_2S$ M 402.550
Cryst. (EtOAc/hexane). Mp 172-174°.
Amide: [105676-08-2].
$C_7H_{13}NOS$ M 159.246

Cryst. (2-propanol/hexane). Mp 198-200°.
Nitrile: [105675-99-8]. *4-Mercaptocyclohexanecarbonitrile, 9CI. 4-Cyanocyclohexanethiol.*
$C_7H_{11}NS$ M 141.231
Cryst. Mp 42-44°.

Ueda, Y. *et al, Can. J. Chem.*, 1986, **64**, 2184 (*synth, ir, pmr*)

3-Mercaptocyclopentanecarboxylic acid,   M-60021
9CI

[105676-05-9]

$C_6H_{10}O_2S$ M 146.204
(1RS,3SR)-form
cis-*form*
Oil.
Amide: [105676-07-1].
$C_6H_{11}NOS$ M 145.219
Cryst. (2-propanol/hexane). Mp 138-140°.
S-*Triphenylmethyl:* [105676-09-3]. *3-[(Triphenylmethyl)thio]cyclopentanecarboxylic acid, 9CI.*
$C_{25}H_{24}O_2S$ M 388.523
Cryst. (Et$_2$O/hexane). Mp 153-154°.

Ueda, Y. *et al, Can. J. Chem.*, 1986, **64**, 2184 (*synth, ir, pmr*)

2-Mercaptoethanesulfinic acid, 9CI   M-60022
[105456-49-3]

$$HSCH_2CH_2SO_2H$$

$C_2H_6O_2S_2$ M 126.188
Li salt: [105456-38-0]. Solid.

Harman, J.P. *et al, J. Org. Chem.*, 1986, **51**, 5235 (*synth*)

2-Mercapto-3-methylbutanoic acid   M-60023

$C_5H_{10}O_2S$ M 134.193
(R)-form [39801-53-1]
Cryst. Mp 35°. $[\alpha]_D^{20}$ +23.3° (c, 0.68 in Et$_2$O). Opt. pure.
S-*Benzoyl:* [103499-61-2].
$C_{12}H_{14}O_3S$ M 238.301
Cryst. (hexane). Mp 91.8-92.4°. $[\alpha]_D^{21}$ +83.3° (c, 1 in CHCl$_3$).

Strijtveen, B. *et al, J. Org. Chem.*, 1986, **51**, 3664 (*synth, pmr, cmr*)

2-Mercapto-2-phenylacetic acid   M-60024
Updated Entry replacing M-00368
α-Mercaptobenzeneacetic acid, 9CI. α-Mercaptophenylacetic acid. Thiomandelic acid
[4695-09-4]

$C_8H_8O_2S$ M 168.210

(R)-form [16201-51-7]
Mp 88-88.5°. $[\alpha]_D^{25}$ −106.2° (c, 0.5 in EtOH).
Me ester: [16201-52-8].
$C_9H_{10}O_2S$ M 182.237
$[\alpha]_D^{25}$ −101.2° (c, 1.4 in EtOH).
S-*Benzyl:* [13136-52-2].
$C_{15}H_{12}O_3S$ M 272.318
Mp 91.5-92°. $[\alpha]_D$ −161° (c, 1.05 in EtOH).
(S)-form [103616-08-6]
Needles (C_6H_6/hexane). Mp 88.5-90°. $[\alpha]_D^{25}$ +112.2° (c, 2 in 95% EtOH). Optical purity 85%.
S-*Benzyl:* [13136-51-1]. Mp 91-91.5°. $[\alpha]_D^{25}$ +161° (c, 0.89 in EtOH).
(±)-form [16201-50-6]
Vile-smelling oil or cryst. (C_6H_6). Mp 64-65° (60.5°). Bp$_{0.5}$ 147°. n_D^{24} 1.5665.
S-*Benzyl:* Mp 86°.

Parravano, N. *et al, Gazz. Chim. Ital.*, 1909, **39**, 62 (*synth*)
Janczewski, M. *et al, CA*, 1964, **60**, 52989 (*synth*)
Bonner, W.A., *J. Org. Chem.*, 1967, **32**, 2495; 1968, **33**, 1831 (*synth, resoln, abs config, ir, nmr*)
Strijtveen, B. *et al, J. Org. Chem.*, 1986, **51**, 3664 (*synth*)

3-Mercapto-2-phenyl-4*H*-1-benzopyran-4-   M-60025
one, 9CI
3-Mercaptoflavone
[98153-13-0]

$C_{15}H_{10}O_2S$ M 254.303
Prisms (EtOAc). Mp 115°.
S-*Me:* [56986-80-2]. *3-Methylthio-2-phenyl-4H-1-benzopyran-4-one.*
$C_{16}H_{12}O_2S$ M 268.330
Yellow oil.

Costa, A.M. *et al, J. Chem. Soc., Perkin Trans. 1*, 1985, 799 (*synth, ir, pmr*)

2-Mercapto-3-phenylpropanoic acid   M-60026
Updated Entry replacing M-00369
α-Mercaptobenzenepropanoic acid, 9CI. α-Mercaptohydrocinnamic acid
[90536-15-5]

$C_9H_{10}O_2S$ M 182.237
(R)-form [84800-12-4]
$[\alpha]_D^{20}$ −9.5° (c, 1 in MeOH). Optical purity 93%.
Me ester: [103499-59-8].
$C_{10}H_{12}O_2S$ M 196.264
Bp$_2$ 100-105°.
(±)-form
Mp 48-49°. Bp$_{4-5}$ 146-150°.

Fischer, E. *et al, Ber.*, 1914, **47**, 2469 (*synth*)
Schöberl, A. *et al, Justus Liebigs Ann. Chem.*, 1936, **522**, 97 (*synth*)
Strijtveen, B. *et al, J. Org. Chem.*, 1986, **51**, 3664 (*synth*)

2-Mercaptopropanoic acid, 9CI   M-60027

Updated Entry replacing M-00379

Thiolactic acid

[79-42-5]

$$COOH$$
$$H \blacktriangleright C \blacktriangleleft SH$$
$$CH_3$$

(*R*)-*form*
Absolute
configuration

$C_3H_6O_2S$ M 106.139

▷UF5250000.

(***R***)-*form* [33178-96-0]

Oil. Bp_{16} 95-100°. $[\alpha]_D^{23}$ +56.4° (c, 5 in EtOAc).
Optically pure.

Et ester: [103616-07-5].
$C_5H_{10}O_2S$ M 134.193
Liq. Bp_{15} 75°. $[\alpha]_D^{22}$ +60.5° (c, 3 in $CHCl_3$). >98% e.e.

S-*Benzoyl:* [33179-02-1].
$C_{10}H_{10}O_3S$ M 210.247
Needles (cyclohexane). Mp 62.5-63°. $[\alpha]_D^{23}$ +102.0°
(c, 3.5 in $CHCl_3$).

Et ester, S-Ac: [78560-77-7].
$C_7H_{14}O_3S$ M 178.246
Oil. Bp_7 50°. $[\alpha]_{578}^{20}$+137.5° (c, 3 in $CHCl_3$). Optically
pure.

(***S***)-*form* [57965-30-7]
Oil. $d^{19.2}$ 1.19. $[\alpha]_D^{15}$ −45.47°.

(±)-*form* [71563-86-5]
Used in depilatory and hair-wave preparations. Oil with
disagreeable odour. Sol. H_2O, EtOH, Et_2O. d_4^{15} 1.22.
Mp ca. 10°. Bp_{20} 110-115°, Bp_{14} 98.5-99°.

Me ester, S-Me ether: [61366-76-5].
$C_5H_{10}O_2S$ M 134.193
Mp 272°.

Et ester: [66707-26-4]. Liq. Spar. sol. H_2O.

S-*Ac:* [6431-92-1].
$C_5H_8O_3S$ M 148.176
Bp_3 55°.

Amide:
C_3H_7NOS M 105.154
Mp 123-124°.

S-*Me:* [58809-73-7]. *2-(Methylthio)propanoic acid, 9CI.*
$C_4H_8O_2S$ M 120.166
Mp 17.3°. Bp_9 106.5°. Known also in opt. active forms.

Billmann, E., *Justus Liebigs Ann. Chem.*, 1906, **348**, 120 (*synth*)
Mellander, A., *Ark. Kemi, Sect. B*, 1936, **12**, 27; *Ark. Kemi,
Sect. A*, 1937, **12**, 16 (*deriv*)
Eugster, C.H. *et al, Helv. Chim. Acta*, 1962, **45**, 1750 (*synth*)
Solladié-Cavallo, A. *et al, Bull. Soc. Chim. Fr.*, 1967, 517
(*synth, ir*)
Scopes, P.M. *et al, J. Chem. Soc. (C)*, 1971, 1671 (*cd, abs
config*)
Skiles, J.W. *et al, J. Med. Chem.*, 1986, **29**, 784 (*derivs*)
Strijtveen, B. *et al, J. Org. Chem.*, 1986, **51**, 3664 (*synth*)

1-Mercapto-2-propanone, 9CI   M-60028

*Acetonylmercaptan. Mercaptoacetone. 2-
Oxopropanethiol*

[24653-75-6]

$$H_3CCOCH_2SH$$

C_3H_6OS M 90.140
Malodorous solid. Mp 112-114° (109-111°). Unstable
above 0°.

Hromatka, O. *et al, Monatsh. Chem.*, 1948, **78**, 29 (*synth*)
Asinger, F. *et al, Justus Liebigs Ann. Chem.*, 1963, **661**, 95
(*synth*)
Klein, R.F.X. *et al, J. Org. Chem.*, 1986, **51**, 4644 (*synth*)

[4]Metacyclophane   M-60029

Bicyclo[4.3.1]deca-1(10),6,8-triene, 9CI

[107575-48-4]

$C_{10}H_{12}$ M 132.205
Reactive intermediate, generated by thermolysis of its
Dewar isomer; it gives a mixt. of [2+2] and [4+2]
dimers.

Kostermans, G.B.M. *et al, J. Am. Chem. Soc.*, 1987, **109**, 7887.

[3^{4,10}][7]Metacyclophane   M-60030

$C_{16}H_{22}$ M 214.350

in-form

Waxy solid.

Pascal, R.A. *et al, J. Am. Chem. Soc.*, 1987, **109**, 6878 (*synth,
uv, ir, pmr*)

[2.0.2.0]Metacyclophane   M-60031

Updated Entry replacing M-00426

*Pentacyclo[19.3.1.1^{2,6}.1^{9,13}.1^{14,18}]octacosa-
1(25),2,4,6(28),9,11,13(27),14,16,18(26),21,23-dode-
caene, 9CI. [2.2](3,3')Biphenylophane*

[24656-54-0]

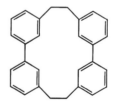

$C_{28}H_{24}$ M 360.498
Cryst. (EtOH/C_6H_6). Mp 184-185° (174-176°).

Vögtle, F., *Justus Liebigs Ann. Chem.*, 1969, **728**, 17 (*synth*)
Leach, D.N. *et al, J. Org. Chem.*, 1978, **43**, 2484 (*synth, ms,
pmr*)
Olsson, T. *et al, Tetrahedron*, 1981, **37**, 3473 (*pmr*)
Bates, R.B. *et al, Acta Crystallogr., Sect. C*, 1986, **42**, 1199
(*cryst struct*)

Methanesulfenyl thiocyanate   M-60032

Cyano methyl disulfide, 9CI

[104157-40-6]

$$MeSSCN$$

$C_2H_3NS_2$ M 105.172
Yellow liq. $Bp_{0.2}$ 33°.

Brintzinger, H. *et al, Chem. Ber.*, 1953, **86**, 557 (*synth*)
Morel, G. *et al, J. Org. Chem.*, 1986, **51**, 4043 (*synth, pmr, cmr,
ir*)

Methanesulfonyl azide, 9CI   M-60033

Mesyl azide

[1516-70-7]

$$MeSO_2N_3$$

$CH_3N_3O_2S$ M 121.114

Diazo transfer reagent. d_{20} 1.40. $Bp_{0.5}$ 56°. Unstable above 100°.

▷Potentially explosive

Boyer, J.H. *et al, J. Org. Chem.*, 1958, **23**, 1051 (*synth, ir*)
Taber, D.F. *et al, J. Org. Chem.*, 1986, **51**, 4077 (*use*)

1,4-Methano-1,2,3,4-tetrahydronaphthalene, 9CI   M-60034

Benzonorbornene. 2,3-Benzobicyclo[2.2.1]hept-2-ene

[4486-29-7]

$C_{11}H_{12}$ M 144.216

Oil. Mp −5.5° to −4.5°. Bp_{12} 78-79°.

Wittig, G. *et al, Chem. Ber.*, 1958, **91**, 895 (*synth*)
Tari, K. *et al, Can. J. Chem.*, 1964, **42**, 926 (*pmr*)
Wilt, J.W. *et al, J. Org. Chem.*, 1967, **32**, 893 (*synth, uv*)
Burn, P.K. *et al, Org. Magn. Reson.*, 1978, **11**, 370 (*cmr*)

N-Methionylalanine   M-60035

$$MeSCH_2CH_2CH(NH_2)CONHCH(COOH)CH_3$$

$C_8H_{16}N_2O_3S$ M 220.286

L-L-form [3061-96-9]

Me ester; B,HCl: Hygroscopic cryst. (MeOH/Et$_2$O). Mp 75-80°. $[\alpha]_D$ +3.8° (c, 1 in MeOH).
Boc-Met-Ala-OMe: [18670-99-0]. Cryst. (EtOAc/pet. ether). Mp 87-88°. $[\alpha]_D$ −31.6° (c, 1 in MeOH).
Boc-Met-Ala-OPh: [98149-05-4]. Cryst. (EtOAc). Mp 85-86°. $[\alpha]_D^{20}$ −66.7° (c, 1 in MeOH).
p-*Nitrophenyl ester; B,HCl:* Mp 138-141°. $[\alpha]_D^{21}$ −29.5° (HMPT).

Wieland, T. *et al, Chem. Ber.*, 1965, **98**, 504 (*conformn*)
Shattenerk, C. *et al, Recl. Trav. Chim. Pays-Bas*, 1973, **92**, 92 (*synth*)
Hollitzer, O. *et al, Angew. Chem., Int. Ed. Engl.*, 1976, **15**, 444.
Galpin, I.J. *et al, Tetrahedron*, 1985, **41**, 895.

12-Methoxy-8,11,13-abietatrien-20-oic acid   M-60036

Updated Entry replacing M-00514
O-Methylpisiferic acid
[76235-94-4]

$C_{21}H_{30}O_3$ M 330.466

Constit. of *Chamaecyparis pisifera*. Cryst. Mp 80-82°. $[\alpha]_D^{25}$ +127° (c, 0.3 in MeOH).

Yatagai, M. *et al, Phytochemistry*, 1980, **19**, 1149 (*isol*)
Mori, K. *et al, Tetrahedron*, 1986, **42**, 5531 (*synth*)

1-Methoxycyclohexene   M-60037

[931-57-7]

$C_7H_{12}O$ M 112.171

Bp 137-140° (140-145°).

Lindsay, D.G. *et al, Tetrahedron*, 1965, **21**, 1673 (*synth, uv, pmr*)
Wohl, R.A., *Synthesis*, 1974, 38 (*synth, ir, pmr*)
Martinez, A.G. *et al, J. Chem. Soc., Perkin Trans. 1*, 1986, 1595 (*synth, ir, ms*)
Shishido, K. *et al, J. Chem. Soc., Perkin Trans. 1*, 1986, 829 (*synth, ir, pmr*)

6-Methoxy-2-(3,7-dimethyl-2,6-octadienyl)-1,4-benzoquinone   M-60038

(E)-form

$C_{17}H_{22}O_3$ M 274.359

(E)-form [31415-00-6]
Verapliquinone A
Constit. of an *Aplidium* sp.
(Z)-form [109954-48-5]
Verapliquinone B
Constit. of an *A.* sp.

Guella, G. *et al, Helv. Chim. Acta*, 1987, **26**, 621.

6-Methoxy-2-[2-(4-methoxyphenyl)ethyl]-4H-1-benzopyran-4-one, 9CI   M-60039

6-Methoxy-2-[2-(4-methoxyphenyl)ethyl]chromone
[111286-05-6]

$C_{19}H_{18}O_4$ M 310.349

Constit. of *Aquilaria agallocha*. Cryst. Mp 84-85°.

Nakanishi, T. *et al, J. Nat. Prod.*, 1986, **49**, 1106.

1-Methyladamantanone   M-60040

1-Methyltricyclo[3.3.1.1^{3,7}]decan-2-one
[26832-19-9]

$C_{11}H_{16}O$ M 164.247

Waxy solid by subl. Mp 106.5-108.5°.

Lenoir, D. *et al, J. Org. Chem.*, 1971, **36**, 1821 (*synth, ir, pmr, ms*)
Raber, D.J. *et al, Tetrahedron*, 1986, **42**, 4347 (*pmr*)

5-Methyladamantanone   M-60041
5-Methyltricyclo[3.3.1.1³,⁷]decan-2-one
[21365-84-4]
$C_{11}H_{16}O$ M 164.247

(±)-*form*
Mp 124-124.5°.

McKervey, M.A. *et al*, *Tetrahedron Lett.*, 1968, 5165 (*synth*)
Bone, J.A. *et al*, *J. Chem. Soc., Perkin Trans. 1*, 1972, 2644
   (*synth*)
Raber, D.J. *et al*, *Tetrahedron*, 1986, **42**, 4347 (*synth, pmr*)

4-Methyl-1-azulenecarboxylic acid, 9CI   M-60042
Updated Entry replacing M-00897
[10527-10-3]
$C_{12}H_{10}O_2$ M 186.210
Cryst. (Et₂O/hexane). Mp 185-188°, 192-193° dec.

Me ester:
   $C_{13}H_{12}O_2$ M 200.237
   Constit. of *Helichrysum acuminatum*. Cryst. or
   amorph. violet powder. Mp 48-49°.

Meuche, D. *et al*, *Chem. Ber.*, 1966, **99**, 2669 (*synth*)
McDonald, R.N. *et al*, *J. Org. Chem.*, 1976, **41**, 1822 (*nmr,
   synth*)
Jakupovic, J. *et al*, *Phytochemistry*, 1987, **26**, 803 (*isol*)

1-Methylazupyrene   M-60043
1-Methyldicyclopenta[ef,kl]heptalene, 9CI
[102830-05-7]

$C_{17}H_{12}$ M 216.282
Gold-green leaflets. Mp 134-135°.

Anderson, A.G. *et al*, *J. Org. Chem.*, 1986, **51**, 2961 (*synth, uv,
   pmr, ms*)

9*b*-Methyl-9*b*H-benz[*cd*]azulene   M-60044
[101998-46-3]

$C_{14}H_{12}$ M 180.249
First tricyclic[12]annulene. Blue. Stable in soln. up to
80°, rapidly polymerises in solid state. Highly antiaro-
matic.

Hafner, K. *et al*, *Angew. Chem., Int. Ed. Engl.*, 1986, **25**, 632
   (*synth, uv, pmr*)

N-Methylbenzenecarboximidic acid, 9CI   M-60045
N-*Methylbenzimidic acid, 8CI*
[88070-48-8]

$$PhC(OH){=}NMe$$

C_8H_9NO M 135.165

(*E*)-*form*
   Me ester: [55504-08-0].
      $C_9H_{11}NO$ M 149.192
      Bp₁₅ 71°.

Wolfgang, W. *et al*, *Chem. Ber.*, 1977, **110**, 2463 (*synth, pmr,
   isom*)
Challis, B.C. *et al*, *J. Chem. Soc., Perkin Trans. 2*, 1978, 192
   (*synth*)
Crist, D.R. *et al*, *J. Org. Chem.*, 1986, **51**, 3266 (*synth*)

2-Methyl-1,3-benzenedicarboxaldehyde,   M-60046
9CI
2-Methylisophthalaldehyde, 8CI
[51689-50-0]

CHO
CH₃
CHO

$C_9H_8O_2$ M 148.161
Needles (MeOH). Mp 101-102°.

Mitchell, R.H. *et al*, *J. Am. Chem. Soc.*, 1974, **96**, 1547 (*synth*)

2-Methyl-1,4-benzenedicarboxaldehyde,   M-60047
9CI
Methylterephthalaldehyde, 8CI
[27587-17-3]
$C_9H_8O_2$ M 148.161
Isol. from oil of *Stewartiella baluchistanica*. Leaflets
   (H₂O). Mp 70-71°.

Glotzmann, C. *et al*, *Monatsh. Chem.*, 1975, **106**, 763 (*synth*)
Ashraf, M. *et al*, *Pak. J. Sci. Ind. Res.*, 1980, **23**, 70 (*isol*)

4-Methyl-1,2-benzenedicarboxaldehyde   M-60048
4-Methylphthalaldehyde, 8CI
[15158-36-8]
$C_9H_8O_2$ M 148.161
Cryst. (hexane). Mp 37-38°. Bp₁ 100-105°.

Pappas, J.J. *et al*, *J. Org. Chem.*, 1968, **33**, 787 (*synth*)

4-Methyl-1,3-benzenedicarboxaldehyde,   M-60049
9CI
4-Methylisophthalaldehyde, 8CI
[23038-58-6]
$C_9H_8O_2$ M 148.161
Cryst. (H₂O). Mp 48-49°.

Glotzmann, C. *et al*, *Monatsh. Chem.*, 1975, **106**, 763 (*synth*)

5-Methyl-1,3-benzenedicarboxaldehyde,   M-60050
9CI
5-Methylisophthalaldehyde. Uvitaldehyde
[1805-67-0]
$C_9H_8O_2$ M 148.161
Cryst. (hexane). Mp 97°.

Bis(tosylhydrazone): Prisms + 2C₆H₆ (C₆H₆/MeOH).
   Mp 117-119° dec.

Chan, T.-L. *et al*, *J. Chem. Soc., Perkin Trans. 2*, 1980, 672
   (*synth, cryst struct*)
Sreenivasulu, M. *et al*, *Indian J. Chem., Sect. B*, 1987, **26**, 173
   (*synth*)

2-Methylbenzo[*gh*]perimidine, 9CI   M-60051

2-Methyl-1,3-diazapyrene

[96012-97-4]

$C_{15}H_{10}N_2$ M 218.257
Cryst. (EtOH aq.). Mp 180-180.5°.

Edel, A. *et al*, *Tetrahedron Lett.*, 1985, 727 (*synth, pmr*)

1-Methyl-8*H*-benzo[*cd*]triazirino[*a*]-indazole, 9CI   M-60052

$C_{11}H_9N_3$ M 183.212
Obt. in soln.

Kaupp, G. *et al*, *Angew. Chem., Int. Ed. Engl.*, 1986, **25**, 828.

3-Methylbiphenyl, 8CI   M-60053

Updated Entry replacing M-01146
m-*Phenyltoluene*

[643-93-6]

$C_{13}H_{12}$ M 168.238
Oil. Mp 4.53°. Bp 272.7°, Bp$_{20}$ 148-150°.

Sherwood, I.R. *et al*, *J. Chem. Soc.*, 1932, 1832 (*synth*)
Skvarchenko, V.R. *et al*, *J. Gen. Chem. USSR* (*Engl. Transl.*),
1962, **32**, 1712 (*synth*)
Mostecky, J. *et al*, *Anal. Chem.*, 1970, **42**, 1132 (*glc*)
Korzeniowski, S.H. *et al*, *Tetrahedron Lett.*, 1977, 1871 (*synth*)
Rao, M.S.C. *et al*, *Synthesis*, 1987, 231 (*synth, ir, pmr*)

Methyl 3-(3-bromo-4-chloro-4-methylcy-clohexyl)-4-oxo-2-pentenoate   M-60054

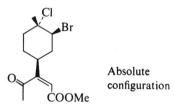

Absolute configuration

$C_{13}H_{18}BrClO_3$ M 337.640
Metabolite of red alga *Laurencia caespitosa*. Cryst. Mp
52-53°. [α]$_D$ +37° (c, 0.14 in CHCl$_3$).

Estrada, D.M. *et al*, *Tetrahedron Lett.*, 1987, **28**, 687 (*cryst struct*)

3-[(3-Methylbutylnitrosoamino]-2-butan-one, 9CI   M-60055

N-*Nitroso*-N-(*1-methyl-2-oxopropyl*)-*3-methyl-1-bu-tanamine*. N-*Nitroso*-N-(*1-methylacetonyl*)-*3-methylbutylamine*

[71016-15-4]

(H$_3$C)$_2$CHCH$_2$CH$_2$N(NO)CH(CH$_3$)COCH$_3$

$C_9H_{18}N_2O_2$ M 186.253
Isol. from moldy corn of Linxian, China. Pale-yellow oil.
▷Carcinogen

Singer, G.M. *et al*, *J. Agric. Food Chem.*, 1987, **35**, 130 (*synth, bibl*)

9-Methyl-9*H*-carbazole, 9CI   M-60056

Updated Entry replacing M-01252
N-*Methylcarbazole*

[1484-12-4]

$C_{13}H_{11}N$ M 181.237
Plates (EtOH). Mp 87°.

Graebe, C. *et al*, *Justus Liebigs Ann. Chem.*, 1880, **202**, 23.
Ehrenreich, F., *Monatsh. Chem.*, 1911, **32**, 1103.
Burton, H. *et al*, *J. Chem. Soc.*, 1924, **125**, 2501.
Flo, C. *et al*, *Justus Liebigs Ann. Chem.*, 1987, 509 (*synth, ms, pmr*)

24-Methyl-5,22,25-cholestatrien-3-ol   M-60057

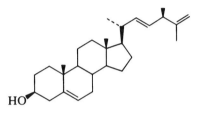

$C_{28}H_{44}O$ M 396.655
(3β,22E,24R)-*form*
Constit. of *Clerodendrum fragans*. Cryst. Mp 133-134°.

Akihisa, T. *et al*, *Phytochemistry*, 1988, **27**, 241.

Methylcyclononane, 9CI   M-60058

[874-99-7]

$C_{10}H_{20}$ M 140.268
Oil. d$_4^{20}$ 0.842. n$_D^{20}$ 1.4620.

Cope, A.C. *et al*, *J. Am. Chem. Soc.*, 1961, **82**, 4663 (*synth*)
Furrer, J. *et al*, *Helv. Chim. Acta*, 1987, **70**, 862 (*synth*)

2-Methylcyclooctanone, 9CI   M-60059

[10363-27-6]

$C_9H_{16}O$ M 140.225
(±)-*form*
Oil. Bp$_{12}$ 100-110°.

Milenkov, B. *et al*, *Helv. Chim. Acta*, 1986, **69**, 1323 (*synth, ir, pmr, cmr, ms*)

11*b*-Methyl-11*bH*-Cyclooct[*cd*]azulene,   M-60060
9CI

15-Methyltricyclo[6.5.2¹³,¹⁴]pentadeca-1,3,5,7,9,11,13-heptene

[102521-04-0]

C₁₆H₁₄   M 206.287

Orange oil. Stable in soln. at r.t., polymerises slowly when pure.

Neumann, G. *et al, J. Am. Chem. Soc.*, 1986, **108**, 4105 (synth, pmr, uv, ms)

2-Methyl-1,3-dinitrobenzene, 9CI   M-60061

Updated Entry replacing M-01580

2,6-Dinitrotoluene, 8CI

[606-20-2]

C₇H₆N₂O₄   M 182.135

Intermed. in polyurethane and elastomer manuf. Needles (EtOH). Mp 66°. p*K*ₐ 3.53 (ethylenediamine, 25°).

▷Toxic. Carcinogen and/or mutagen. XT1925000.

Gibson, W.H. *et al, J. Chem. Soc.*, 1922, **121**, 278 (synth)
Conduit, C.P., *J. Chem. Soc.*, 1959, 3273 (ir, uv)
Beynon, J.H. *et al, Ind. Chim. Belg.*, 1964, **29**, 311 (ms)
Mathias, A. *et al, Anal. Chim. Acta*, 1966, **35**, 376 (pmr)
Coon, C.L. *et al, J. Org. Chem.*, 1973, **38**, 4243 (synth)
Mori, M. *et al, Chem. Pharm. Bull.*, 1984, **32**, 4070; 1986, **34**, 4859 (metab, tox)
Bretherick, L., *Handbook of Reactive Chemical Hazards*, 2nd Ed., Butterworths, London and Boston, 1979, 162.
Sax, N.I., *Dangerous Properties of Industrial Materials*, 5th Ed., Van Nostrand-Reinhold, 1979, 619.

2-Methyl-1,6-dioxaspiro[4.5]decane, 9CI   M-60062

Updated Entry replacing M-01611

[68108-92-9]

C₉H₁₆O₂   M 156.224

Minor component of the odour of the common wasp, *Paravespula vulgaris* which appears to act as aggression inhibitor. Ident. by glc/ms.

Erdmann, H., *Justus Liebigs Ann. Chem.*, 1885, **228**, 176 (synth)
Francke, W. *et al, Angew. Chem., Int. Ed. Engl.*, 1978, **17**, 862 (isol, synth)
Iwata, C. *et al, Tetrahedron Lett.*, 1987, **28**, 2255 (synth)

2-Methyl-1,3-dioxolane, 9CI   M-60063
Glycol acetal

[497-26-7]

C₄H₈O₂   M 88.106
Bp 82-83°.

▷JI3509000.

Hibbert, H. *et al, J. Am. Chem. Soc.*, 1924, **46**, 1283 (synth)
Egyed, J. *et al, Bull. Soc. Chim. Fr.*, 1972, 2287 (synth)

Van de Sande, C.C. *et al, J. Am. Chem. Soc.*, 1975, **97**, 4617 (ms)
Kantlehner, W. *et al, Justus Liebigs Ann. Chem.*, 1979, 1362 (synth)
Maitte, P. *et al, Bull. Soc. Chim. Fr.*, 1979, 264 (synth)
Musavirov, R.S. *et al, Zh. Obshch. Khim.*, 1982, **52**, 1394 (synth)
Trofimov, B.A. *et al, Izv. Akad. Nauk SSSR, Ser. Khim.*, 1984, 1670 (synth)
Grindley, T.B. *et al, J. Carbohydr. Chem.*, 1985, **4**, 171 (pmr, cmr, struct)

5-Methyl-3,6-dioxo-2-piperazineacetic acid   M-60064
Cyclo(alanylaspartyl)

    (2*S*,5*S*)-*form*

C₇H₁₀N₂O₄   M 186.167
(2*S*,5*S*)-*form* [110954-19-3]
L-L-form
Benzyl ester: [77935-54-7]. Mp 189-192°. [α]²⁰_D −20.0° (c, 1 in MeOH).

Suzuki, K. *et al, Chem. Pharm. Bull.*, 1981, **29**, 233 (ester)
Gorbitz, C.H., *Acta Chem. Scand., Ser. B*, 1987, **41**, 83 (cryst struct)

1-Methyl-4,4-diphenylcyclohexene   M-60065
[50592-48-8]

C₁₉H₂₀   M 248.367
Cryst. (Et₂O/hexane). Mp 49.5-50.5°.

Fortin, C.J. *et al, Can. J. Chem.*, 1973, **51**, 3445 (synth, ms)

3-Methyl-4,4-diphenylcyclohexene   M-60066

C₁₉H₂₀   M 248.367
(±)-*form* [103367-60-8]
Solid (EtOH). Mp 60.0-62.0°.

Zimmerman, H.E. *et al, J. Am. Chem. Soc.*, 1986, **108**, 6276 (synth, ir, pmr)

5-Methyl-1,3-dithiane   M-60067
[38761-25-0]

C₅H₁₀S₂   M 134.254
Liq. Bp₀.₃ 35-43°.

Juaristi, E. *et al, J. Am. Chem. Soc.*, 1986, **108**, 2000 (synth, pmr)

Methyleneadamantane   M-60068

Updated Entry replacing M-30090
Methylenetricyclo[3.3.1.1³,⁷]decane, 9CI
[875-72-9]

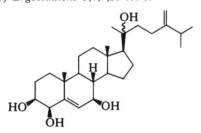

$C_{11}H_{16}$ M 148.247
Waxy solid by subl. Mp 135.8-136.5°.

Schleyer, P.v.R. *et al*, *J. Am. Chem. Soc.*, 1961, **83**, 182 (*synth*)
Dekkers, A.W.J.D. *et al*, *Tetrahedron*, 1973, **29**, 1691 (*uv*)
Fry, J.L. *et al*, *Collect. Czech. Chem. Commun.*, 1975, **40**, 2173 (*synth*)
Lapalme, R. *et al*, *Can. J. Chem.*, 1979, **57**, 3272 (*synth, pmr*)
Bishop, R. *et al*, *Aust. J. Chem.*, 1987, **40**, 249 (*synth, cmr*)

4-Methylenebicyclo[3.1.0]hex-2-ene   M-60069

Homofulvene
[19405-19-7]

C_7H_8 M 92.140
Polymerises fairly readily. For a cryst. struct. detn. on a deriv., see 4-Methylene-1,2,3,5-tetraphenylbicyclo[3.1.0]hex-2-ene, M-60079 .

Rey, M. *et al*, *Tetrahedron Lett.*, 1968, 3583 (*synth, uv, pmr*)

24-Methylene-5-cholestene-3,4,7,20-tetrol   M-60070

5,24(28)-Ergostadiene-3,4,7,20-tetrol

$C_{28}H_{46}O_4$ M 446.669
(*3β,4β,7β,20ξ*)-form
Lansisterol A
Constit. of *Lansium anamallayanum*. Cryst. (Me$_2$CO aq.). Mp 130°. $[\alpha]_D^{30} -107°$ (c, 2 in CHCl$_3$).

Purushothaman, K.K. *et al*, *Can. J. Chem.*, 1987, **65**, 150.

4-Methylene-2,5-cyclohexadien-1-one, 9CI   M-60071

p-*Benzoquinonemethide. Quinomethane*. p-*Quinonemethide*
[502-87-4]

C_7H_6O M 106.124

Reactive lignin intermed. V. unstable.

Pospisek, J. *et al*, *Collect. Czech. Chem. Commun.*, 1974, **40**, 142 (*synth, uv*)
Leary, G. *et al*, *J. Chem. Soc., Perkin Trans. 2*, 1977, 1737.
Lasne, M.C. *et al*, *Tetrahedron*, 1981, **37**, 503 (*synth, bibl*)

6-Methylene-2,4-cyclohexadien-1-one, 9CI   M-60072

o-*Quinomethane*. o-*Quinonemethide*
[27890-67-1]

C_7H_6O M 106.124
Unstable. Forms mixt. of dimer and trimer on isol.
Trimer: [49664-20-2].
  $C_{21}H_{18}O_3$ M 318.371
  Plates (Et$_2$O/pet. ether). Mp 190-192°.

Cavitt, S.B. *et al*, *J. Org. Chem.*, 1962, **27**, 1211 (*trimer*)
Wentrup, C. *et al*, *Tetrahedron Lett.*, 1973, 2915 (*synth*)
Mao, Y.-L. *et al*, *Proc. Natl. Acad. Sci. U.S.A.*, 1980, **77**, 1732 (*synth*)
Katada, T. *et al*, *J. Chem. Soc., Perkin Trans. 1*, 1984, 2649 (*synth, bibl*)

9-Methylene-1,3,5,7-cyclononatetraene, 9CI   M-60073

Updated Entry replacing M-50136
Nonafulvene
[13538-66-4]

$C_{10}H_{10}$ M 130.189
Isol. at −40° to −25°. Cryst. with camporaceous odour (pentane). Mp −21° to −19°. V. reactive, undergoes rapid valence isom. to indenes at r.t.

Otter, A. *et al*, *Helv. Chim. Acta*, 1986, **69**, 124 (*pmr, cmr*)
Furrer, J. *et al*, *Helv. Chim. Acta*, 1987, **70**, 862 (*synth, pmr, cmr, uv*)

7-Methylene-1,3,5-cyclooctatriene   M-60074

[2570-13-0]

C_9H_{10} M 118.178
Unstable.

Osborn, C.L. *et al*, *J. Am. Chem. Soc.*, 1965, **87**, 3158 (*ir, pmr, uv*)
Ferber, P.H. *et al*, *Tetrahedron Lett.*, 1980, 2447 (*synth*)

2-Methylene-1,3-dioxolane, 9CI

M-60075

Updated Entry replacing M-10119

[4362-23-6]

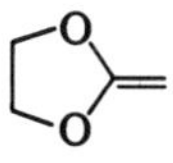

$C_4H_6O_2$ M 86.090

Liq. which rapidly polymerises. Bp_{50} 50-55°, Bp_{15} 44°.

McElvain, S.M. *et al, J. Am. Chem. Soc.*, 1948, **70**, 3781 (*synth*)

Taskinen, E. *et al, Tetrahedron*, 1978, **34**, 2365 (*synth, pmr, cmr*)

Capon, B. *et al, J. Am. Chem. Soc.*, 1981, **103**, 1765 (*synth*)

Fukuda, H. *et al, Tetrahedron Lett.*, 1986, **27**, 1587 (*synth, ir, pmr*)

4-Methylene-1,3-dioxolane, 9CI

M-60076

[4362-24-7]

$C_4H_6O_2$ M 86.090

Liq. d_4^{20} 1.05. Bp_{758} 93-95°, Bp_{45-50} 24-26°. n_D^{20} 1.4348.

Fischer, H.O.L. *et al, Ber.*, 1930, **63**, 1732 (*synth*)

Kankaanperä, A. *et al, Acta Chem. Scand.*, 1966, **20**, 2622 (*synth*)

Salomaa, P. *et al, Acta Chem. Scand.*, 1967, **21**, 2479 (*props*)

1-(3,4-Methylenedioxyphenyl)-1-tetradecene

M-60077

5-(1-Tetradecenyl)-1,3-benzodioxole

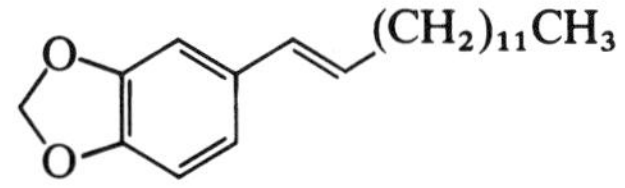

$C_{21}H_{32}O_2$ M 316.483

(*E*)-*form*

Constit. of fruit of *Piper sarmentosum*. Solid. Mp 33.5-36.5°.

Likhitwitayawiud, K. *et al, Tetrahedron*, 1987, **43**, 3689 (*isol, synth*)

1-Methyleneindane

M-60078

2,3-Dihydro-1-methylene-1H-indene, 9CI

[1194-56-5]

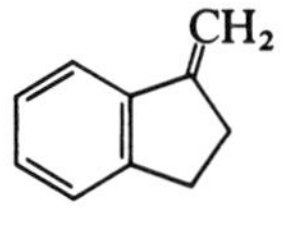

$C_{10}H_{10}$ M 130.189

Bp_{29} 99.5-101.5°. n_D^{23} 1.5759.

Goodman, A.L. *et al, J. Am. Chem. Soc.*, 1964, **86**, 908 (*synth, ir*)

Buchanan, G.W. *et al, Can. J. Chem.*, 1979, **57**, 3028 (*synth, ir, cmr*)

4-Methylene-1,2,3,5-tetraphenylbicyclo[3.1.0]hex-2-ene, 9CI

M-60079

Tetraphenylhomofulvene

[106868-03-5]

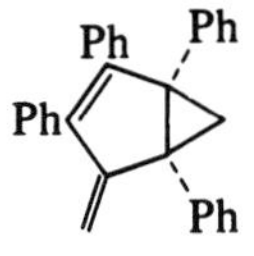

$C_{31}H_{24}$ M 396.531

Needles.

Debaerdemaeker, T. *et al, Acta Crystallogr., Sect. C*, 1987, **43**, 101 (*cryst struct*)

11-Methylene-1,5,9-triazabicyclo[7.3.3]-pentadecane, 9CI

M-60080

[104875-18-5]

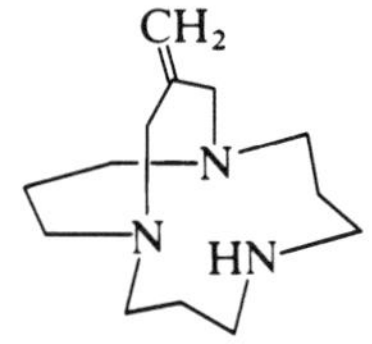

$C_{13}H_{25}N_3$ M 223.361

Unexpectedly strong base, rapidly protonated.

B,HI: [104875-17-4]. Prisms ($CHCl_3$). Mp 199-200°.

Bell, T.W. *et al, J. Am. Chem. Soc.*, 1986, **108**, 7427 (*synth, pmr, cryst struct*)

Methylene-1,3,5-trioxane

M-60081

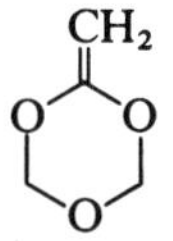

$C_4H_6O_3$ M 102.090

Cyclic ketene acetal. Bp_{15} 67-68°.

Fukuda, H. *et al, Tetrahedron Lett.*, 1986, **27**, 1587 (*synth, pmr, cmr*)

13¹-Methyl-13,15-ethano-13²,17-prop-13²(15²)-enoporphyrin

M-60082

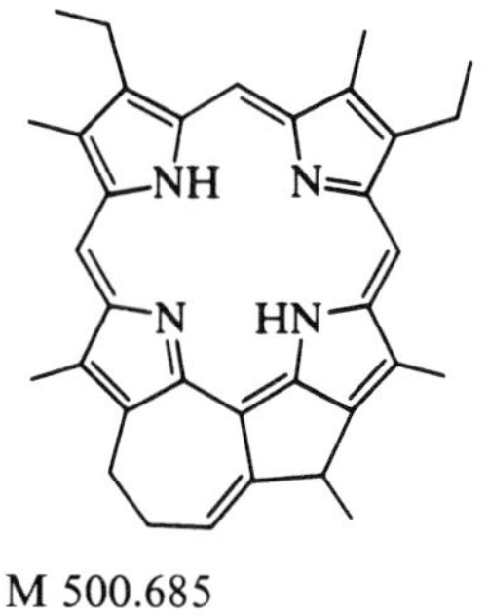

$C_{34}H_{36}N_4$ M 500.685

Isol. from marine Serpiano oil shale as Vanadium complex.

Chicarelli, M.I. *et al, Tetrahedron Lett.*, 1986, **27**, 4653 (*isol, pmr, uv*)

5-Methyl-2(5H)-furanone, 9CI

M-60083

Updated Entry replacing M-01848

4-Hydroxy-2-pentenoic acid γ-lactone. β-Angelica lactone. Δ¹-Angelica lactone

[591-11-7]

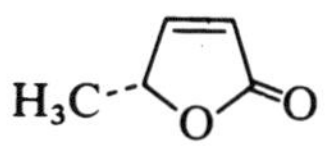

(*R*)-*form*

$C_5H_6O_2$ M 98.101

▷LU5100000.

(*R*)-*form* [62322-48-9]

Bp_{14} 100°. $[\alpha]_D^{20}$ −95.89° (c, 0.73 in $CHCl_3$).

(*S*)-*form* [92694-51-4]
Bp$_{15}$ 98-100°. [α]$_D^{20}$ +93.8° (c, 0.5 in CHCl$_3$).
(±)-*form* [70428-45-4]
Liq. d$_{20}^4$ 1.081. Fp <−80°. Bp 208-209°, Bp$_{10}$ 86.5°.
Forms a dimer.

Wineburg, J.P. *et al*, *J. Heterocycl. Chem.*, 1975, **12**, 749.
Kiyoshi, I., *Bull. Chem. Soc. Jpn.*, 1977, **50**, 242.
Camps, P. *et al*, *Tetrahedron*, 1983, **39**, 395 (*synth, pmr, ir*)
Ortuño, R.M. *et al*, *Tetrahedron Lett.*, 1986, **27**, 1079 (*synth*)

2-Methylfuro[2,3-*b*]pyridine, 9CI　　　M-60084

[75332-26-2]

C$_8$H$_7$NO　　M 133.149
Bp$_{25}$ 145°.

B,MeI: Sl. yellow needles (MeOH/Me$_2$CO). Mp 175-176°.

Morita, H. *et al*, *J. Heterocycl. Chem.*, 1986, **23**, 1465 (*synth, ir, pmr, ms*)

3-Methylfuro[2,3-*b*]pyridine, 9CI　　　M-60085

[109274-90-0]
C$_8$H$_7$NO　　M 133.149
Oil. Bp$_{15}$ 120-130° (bath).

Morita, H. *et al*, *J. Heterocycl. Chem.*, 1986, **23**, 1465 (*synth, ir, pmr, ms*)

3-Methyl-2-heptenoic acid　　　M-60086

C$_8$H$_{14}$O$_2$　　M 142.197
(*E*)-*form* [30801-90-2]

Dicyclohexylammonium salt: Cryst. (pet. ether). Mp 82-83°.

Armstrong, F.B. *et al*, *J. Chem. Soc., Perkin Trans. 1*, 1985, 691 (*synth, pmr*)

4-Methyl-3-hepten-2-one　　　M-60087

H$_3$CCOCH=C(CH$_3$)CH$_2$CH$_2$CH$_3$

C$_8$H$_{14}$O　　M 126.198
(*E*)-*form* [23732-22-1]
Bp$_1$ 30°.
(*Z*)-*form* [23732-23-2]
Liq.

Faulk, D.D. *et al*, *J. Org. Chem.*, 1970, **35**, 364 (*synth, pmr, ir, ms, uv*)
Dieter, R.K. *et al*, *J. Org. Chem.*, 1986, **51**, 4687 (*synth, pmr*)

6-Methyl-1*H*-indole, 9CI　　　M-60088

Updated Entry replacing M-02135
[3420-02-8]
C$_9$H$_9$N　　M 131.177
Mp 26-27°, Mp 58-59°. Bp$_{9.5}$ 126-126.5°, Bp$_1$ 75-78°.
Picrate: Mp 160-161°.
N-*Me:* [5621-15-8]. *1,6-Dimethyl-1*H-*indole.*
　C$_{10}$H$_{11}$N　　M 145.204

Liq. Bp$_{0.2}$ 68-73°.
N-*Me, picrate:* Mp 148°.

Marion, L. *et al*, *Can. J. Res., Sect. B*, 1947, **25**, 1.
Süs, O. *et al*, *Justus Liebigs Ann. Chem.*, 1955, **593**, 91.
Andrisano, R. *et al*, *Gazz. Chim. Ital.*, 1957, **87**, 949.
Sasse, W., *J. Chem. Soc.*, 1960, 526.
Safe, S. *et al*, *Org. Mass Spectrom.*, 1972, **6**, 33 (*ms*)
Ito, Y. *et al*, *Bull. Chem. Soc. Jpn.*, 1984, **57**, 73 (*synth*)
Tischler, A.N. *et al*, *Tetrahedron Lett.*, 1986, **27**, 1653 (*synth*)

2-Methyl-6-methylene-1,3,7-octatriene　　　M-60089

[105553-45-5]

C$_{10}$H$_{14}$　　M 134.221
(*E*)-*form* [107841-93-0]
C$_{10}$ terpene synthon. Oil. Bp$_9$ 54°.

Mignani, G. *et al*, *Tetrahedron Lett.*, 1986, **27**, 2591 (*synth, pmr, ir, ms, uv, use*)

5-Methyl-2-(1-methyl-1-phenylethyl)-cyclohexanol, 9CI　　　M-60090

(1*R*,2*S*,5*R*)-*form*

C$_{16}$H$_{24}$O　　M 232.365
(*1R,2S,5R*)-*form* [65253-04-5]
(−)-*8-Phenylmenthol. Phenmenthol*
Chiral auxiliary reagent for asymm. induction.
(*1S,2R,5R*)-*form* [100101-42-6]
8-Phenylisomenthol. Epientphenmenthol
Present as an impurity in samples of phenmenthol prepd.
from pulegone. Chiral auxiliary reagent. It has been
shown that the chiral centres at C-1 and C-2 are
responsible for the asymm. induction, therefore in
chiral inductions with phenmenthol, the presence of
epientphenmenthol as impurity is equivalent to using
phenmenthol of enantiomeric purity less than 100%.
(*1S,2S,5R*)-*form* [104870-75-9]
(+)-*8-Phenylneomenthol*
Chiral auxiliary reagent. [α]$_D$ +36.1° (c, 0.75 in CHCl$_3$).
(*1S,2R,5S*)-*form* [57707-91-2]
(+)-*8-Phenylmenthol*
Oil. [α]$_D^{22}$ +26.3° (c, 2.30 in EtOH).

Corey, E.J. *et al*, *J. Am. Chem. Soc.*, 1975, **97**, 6908.
Ensley, H.E. *et al*, *J. Org. Chem.*, 1978, **43**, 1610 (*synth, ir, pmr*)
Quinkert, G. *et al*, *Angew. Chem., Int. Ed. Engl.*, 1986, **25**, 992 (*synth, cryst struct, use, bibl*)
Whitesell, J.K. *et al*, *J. Org. Chem.*, 1986, **51**, 551.

2-Methyl-1,3-naphthalenediol, 9CI   **M-60091**

1,3-Dihydroxy-2-methylnaphthalene

[20034-31-5]

$C_{11}H_{10}O_2$ M 174.199

Needles (H_2O). Mp 139-140°.

Di-Ac: [14958-06-6].

$C_{15}H_{14}O_4$ M 258.273

Prisms (MeOH). Mp 118°.

Soliman, G. *et al, J. Chem. Soc.*, 1944, 53 (*synth*)
Bisanz, T. *et al, CA*, 1967, **66**, 28560u; 1968, **69**, 96313c.

2-Methyl-1,5-naphthalenediol, 9CI   **M-60092**

1,5-Dihydroxy-2-methylnaphthalene

[79786-99-5]

$C_{11}H_{10}O_2$ M 174.199

Needles (C_6H_6). Mp 179-180°.

Wurm, G. *et al, Arch. Pharm.* (*Weinheim, Ger.*), 1981, **314**, 861 (*synth*)
Möhrle, H. *et al, Z. Naturforsch. B*, 1987, **42**, 1181 (*synth, ir, pmr, ms*)

3-Methyl-1,2-naphthalenediol, 9CI   **M-60093**

1,2-Dihydroxy-3-methylnaphthalene

[61978-16-3]

$C_{11}H_{10}O_2$ M 174.199

Cryst. Mp 96-97°.

Maruyama, K. *et al, J. Chem. Soc., Perkin Trans. 2*, 1979, 255 (*synth, ir, pmr*)

4-Methyl-1,3-naphthalenediol, 9CI   **M-60094**

2,4-Dihydroxy-1-methylnaphthalene

[2089-76-1]

$C_{11}H_{10}O_2$ M 174.199

Cryst. (H_2O). Mp 108-110°.

Di-Ac: [49583-10-0].

$C_{15}H_{14}O_4$ M 258.273

Cryst. Mp 87°.

Budzikiewicz, H. *et al, Monatsh. Chem.*, 1973, **104**, 876 (*deriv*)
Buckle, D.R. *et al, J. Med. Chem.*, 1977, **20**, 1059 (*synth*)

6-Methyl-1,2-naphthalenediol, 9CI   **M-60095**

1,2-Dihydroxy-6-methylnaphthalene

[61978-33-4]

$C_{11}H_{10}O_2$ M 174.199

Needles. Mp 74-74.5°.

Maruyama, K. *et al, J. Chem. Soc., Perkin Trans. 2*, 1979, 255 (*synth, ir, pmr*)

1-Methylnaphtho[2,1-*b*]thiophene   **M-60096**

[69736-21-6]

$C_{13}H_{10}S$ M 198.282

Prisms (MeOH). Mp 60° (58-60°).

Dann, O. *et al, Chem. Ber.*, 1958, **91**, 172 (*synth*)
Tominaga, Y. *et al, J. Heterocycl. Chem.*, 1981, **18**, 977 (*synth, pmr*)

2-Methylnaphtho[2,1-*b*]thiophene   **M-60097**

[16587-35-2]

$C_{13}H_{10}S$ M 198.282

Needles (MeOH). Mp 88° (86°).

Picrate: [86474-81-9]. Orange needles. Mp 138°.

Clarke, K. *et al, J. Chem. Soc. (C)*, 1969, 1274 (*synth*)
Tominaga, Y. *et al, J. Heterocycl. Chem.*, 1983, **20**, 487 (*synth, pmr*)

4-Methylnaphtho[2,1-*b*]thiophene   **M-60098**

[67388-22-1]

$C_{13}H_{10}S$ M 198.282

Needles (MeOH). Mp 75°.

Picrate: [86474-80-8]. Orange needles. Mp 147°.

Tominaga, Y. *et al, J. Heterocycl. Chem.*, 1983, **20**, 487 (*synth, pmr*)

5-Methylnaphtho[2,1-*b*]thiophene   **M-60099**

[86474-77-3]

$C_{13}H_{10}S$ M 198.282

Needles (MeOH). Mp 79°.

Picrate: [86474-78-4]. Orange needles. Mp 150°.

Tominaga, Y. *et al, J. Heterocycl. Chem.*, 1983, **20**, 487 (*synth, pmr*)

6-Methylnaphtho[2,1-*b*]thiophene   **M-60100**

[4567-37-7]

$C_{13}H_{10}S$ M 198.282

Needles (MeOH). Mp 120°.

Picrate: Orange-red needles. Mp 142°.

Carruthers, W. *et al, J. Chem. Soc.*, 1965, 6221.
Tominaga, Y. *et al, J. Heterocycl. Chem.*, 1983, **20**, 487 (*synth, pmr*)

7-Methylnaphtho[2,1-*b*]thiophene   **M-60101**

[4567-39-9]

$C_{13}H_{10}S$ M 198.282

Needles (MeOH). Mp 98°.

Picrate: Orange-red needles. Mp 133°.

Tominaga, Y. *et al, J. Heterocycl. Chem.*, 1983, **20**, 487 (*synth, pmr*)

8-Methylnaphtho[2,1-*b*]thiophene   **M-60102**

[4567-36-6]

$C_{13}H_{10}S$ M 198.282

Needles (MeOH). Mp 66°.

Picrate: Orange-red needles. Mp 135°.

Carruthers, W. *et al, J. Chem. Soc.*, 1965, 6221.
Hunt, D.F. *et al, Anal. Chem.*, 1982, **54**, 574.
Tominaga, Y. *et al, J. Heterocycl. Chem.*, 1983, **20**, 487 (*synth, pmr*)

9-Methylnaphtho[2,1-*b*]thiophene   **M-60103**

[4567-34-4]

$C_{13}H_{10}S$ M 198.282

Needles (MeOH). Mp 77°.

Picrate: Orange-red needles. Mp 157° (152°).

Carruthers, W. *et al, J. Chem. Soc.*, 1965, 6221.
Tominaga, Y. *et al, J. Heterocycl. Chem.*, 1983, **20**, 487 (*synth, pmr*)

4-Methyl-5-nitropyrimidine, 9CI   M-60104

[99593-51-8]

$C_5H_5N_3O_2$ M 139.113

Liq.

Marcelis, A.T.M. *et al, J. Org. Chem.*, 1986, **51**, 67 (*synth, pmr*)

24-Methyl-25-nor-12,24-dioxo-16-sca-laren-22-oic acid   M-60105

$C_{25}H_{36}O_4$ M 400.557

Constit. of sponges *Dictyoceratida* and *Halichondria* spp. Antimicrobial. Amorph. solid.

Nakagawa, M. *et al, Tetrahedron Lett.*, 1987, **28**, 431.

2-Methyl-2-octenal   M-60106

$C_9H_{16}O$ M 140.225

(*E*)-*form* [49576-57-0]

Bp_{14} 125° (Bp_{20} 98-101°).

Chamberlin, A.R. *et al, Synthesis*, 1979, 44 (*synth, pmr*)
Tamura, R. *et al, J. Org. Chem.*, 1986, **51**, 4368 (*synth*)

5-Methyl-2-oxazolidinone, 9CI   M-60107

[1072-70-4]

$C_4H_7NO_2$ M 101.105

Bp_{2-3} 143-145°.

3-Me: [15833-10-0]. *3,5-Dimethyl-2-oxazolidinone.*
$C_5H_9NO_2$ M 115.132
Bp_5 94-96°.

3-Ph: [708-57-6]. *5-Methyl-3-phenyl-2-oxazolidinone.*
$C_{10}H_{11}NO_2$ M 177.202
Cryst. (C_6H_6/hexane). Mp 78-80°.

Najer, H. *et al, Bull. Soc. Chim. Fr.*, 1959, 1841 (*synth*)
Dyer, M.E. *et al, Chem. Rev.*, 1967, **67**, 197 (*rev*)
Shibata, I. *et al, J. Org. Chem.*, 1986, **51**, 2177 (*deriv, synth, ir, pmr*)

5-Methyl-2-oxo-4-oxazolidinecarboxylic acid, 9CI   M-60108

[1195-19-3]

(*4S,5R*)-*form*

$C_5H_7NO_4$ M 145.115

(*4S,5R*)-*form* [1195-20-6]
Mp 136-138°. No opt. rotn. reported.

(*4RS,5RS*)-*form* [50896-26-9]
(±)-cis-*form*
Mp 196-197° dec.

(*4RS,5SR*)-*form* [37791-36-9]
(±)-trans-*form*
Cryst. (EtOAc). Mp 127-128°.

N-*Benzoyl:* [37791-33-6]. Cryst. (H_2O). Mp 154-155°.
N-*Benzoyl, Me ester:* [37791-34-7]. Cryst. (EtOAc/pet. ether). Mp 116-117°.

Inui, T. *et al, Bull. Chem. Soc. Jpn.*, 1972, **45**, 1254; 1973, **46**, 3308 (*synth, pmr*)
Shanzer, A. *et al, J. Org. Chem.*, 1979, **44**, 3967 (*synth, pmr*)

4-Methyl-2-pentanone, 9CI   M-60109

Updated Entry replacing M-02911

Isobutyl methyl ketone. Isopropylacetone. Hexone

[108-10-1]

$(H_3C)_2CHCH_2COCH_3$

$C_6H_{12}O$ M 100.160

Solv. for cellulose esters and other coating system. Used in adhesives. d_4^{20} 0.801. Bp 116.8°. n_D^{17} 1.3969.

▷Irritant, TLV 410. Highly flammable, flash pt. 17°. SA9275000.

Semicarbazone: Mp 134°.
2,4-Dinitrophenylhydrazone: Mp 95°.

Law, H.O., *J. Chem. Soc.*, 1912, **101**, 1547 (*synth*)
Grignard, V. *et al, Ann. Chim. (Paris)*, 1928, **9**, 13 (*synth*)
Fuge, E.T.J. *et al, J. Phys. Chem.*, 1952, **56**, 1013 (*synth*)
Ejchart, A., *Org. Magn. Reson.*, 1980, **13**, 368 (*cmr*)
Bretherick, L., *Handbook of Reactive Chemical Hazards*, 2nd Ed., Butterworths, London and Boston, 1979, 601.
Sax, N.I., *Dangerous Properties of Industrial Materials*, 5th Ed., Van Nostrand-Reinhold, 1979, 750.
Hazards in the Chemical Laboratory, (Bretherick, L., Ed.), 3rd Ed., Royal Society of Chemistry, London, 1981, 401.

5-Methyl-3-phenyl-1,2,4-oxadiazole, 9CI   M-60110

Updated Entry replacing M-03096

[1198-98-7]

$C_9H_8N_2O$ M 160.175

Cryst. (MeOH aq.). Mp 41°. $Bp_{0.1}$ 84-86°.

Morocchi, S. *et al, Tetrahedron Lett.*, 1967, 331 (*synth, uv*)
Micetich, R.G., *Can. J. Chem.*, 1970, **48**, 2006 (*nmr*)
Fétizon, M. *et al, Tetrahedron*, 1975, **31**, 165 (*synth, ir*)
Molina, P. *et al, Synthesis*, 1986, 843 (*synth, ir, pmr*)

4-Methyl-5-phenyl-2-oxazolidinethione, 9CI   M-60111
[91794-28-4]

$C_{10}H_{11}NOS$ M 193.263

(4R,5S)-form
Chiral reagent. Prisms (EtOAc/hexane). Mp 81-82°.
$[\alpha]_D^{20}$ +219.2° (c, 0.44 in CHCl₃).

Nagao, Y. *et al, J. Chem. Soc., Perkin Trans. 1*, 1985, 2361
(*synth, ir, uv, pmr, use*)

4-(4-Methylphenyl)-2-pentanone, 9CI   M-60112
4-p-Tolyl-2-pentanone. Curcumone
[451-25-2]

$C_{12}H_{16}O$ M 176.258

(S)-form [69657-27-8]
Bp₁₀ 120°. $[\alpha]_D^{30}$ +48.21° (neat).
(±)-form [65075-32-3]
Bp₁₂ 130°.
Semicarbazone: Mp 144-145°.
Oxime: Bp₁₀ 160-161°.

Gandhi, R.P. *et al, Tetrahedron*, 1959, **7**, 236 (*synth*)
Banno, K., *Bull. Chem. Soc. Jpn.*, 1976, **49**, 2284 (*synth*)
Hansson, A.T. *et al, Acta Chem. Scand., Ser. B*, 1978, **32**, 483
(*synth*)
Ho, T.L., *Synth. Commun.*, 1981, **11**, 579 (*synth*)
John, T.K. *et al, Indian J. Chem., Sect. B*, 1985, **24**, 35 (*synth*)

2-(2-Methylphenyl)propanoic acid   M-60113
α,2-Dimethylbenzeneacetic acid, 9CI. α-o-Tolylpropionic acid
[62835-95-4]

$C_{10}H_{12}O_2$ M 164.204

(±)-form
Cryst. (heptane). Mp 93-94°.

Tanner, D.D. *et al, J. Org. Chem.*, 1987, **52**, 4689 (*synth, pmr*)

2-(4-Methylphenyl)propanoic acid   M-60114
Updated Entry replacing M-03149
α,4-Dimethylbenzeneacetic acid, 9CI. p-Methylhydratro-pic acid, 8CI. α-p-Tolylpropionic acid. Methyl-p-tolyla-cetic acid
[938-94-3]
$C_{10}H_{12}O_2$ M 164.204

(±)-form
Mp 40-41°. Bp 280°, Bp₁₂₅ 161-161.5°.
Et ester:
$C_{12}H_{16}O_2$ M 192.257

Oil. Bp₁₁ 123.5°.
Amide:
$C_{10}H_{13}NO$ M 163.219
Cryst. (C₆H₆). Mp 195°.
Nitrile:
$C_{10}H_{11}N$ M 145.204
Bp 246.5-247.5°, Bp₁₂₋₁₅ 123°.

Rupe, H. *et al, Helv. Chim. Acta*, 1924, **7**, 654.
Tanner, D.D. *et al, J. Org. Chem.*, 1987, **52**, 4689 (*synth, pmr*)

2-Methyl-5-phenyl-1H-pyrrole, 9CI   M-60115
Updated Entry replacing M-03243
[3042-21-5]
$C_{11}H_{11}N$ M 157.215
Leaves or flakes (MeOH aq.), cryst. (pet. ether). Mp
103° (98-99°). Subl. with part. dec.
▷UX9641550.

1-Ph: [3771-57-1]. *2-Methyl-1,5-diphenyl-1H-pyrrole,*
9CI.
$C_{17}H_{15}N$ M 233.312
Mp 81-82°.

Tedder, J.M. *et al, J. Chem. Soc.*, 1960, 3270 (*synth*)
Severin, T. *et al, Chem. Ber.*, 1965, **98**, 3847 (*synth*)
Thompson, T.W., *J. Chem. Soc., Chem. Commun.*, 1968, 532
(*synth, nmr*)
Gotthardt, H. *et al, J. Am. Chem. Soc.*, 1970, **92**, 4340 (*synth*)
Benages, I.A. *et al, J. Org. Chem.*, 1978, **43**, 4273 (*synth, ir*)
Hinz, W. *et al, Tetrahedron*, 1986, **42**, 3753 (*synth, pmr, cmr*)

4-Methyl-5-phenyl-1,2,3-thiadiazole   M-60116
[18212-30-1]

$C_9H_8N_2S$ M 176.236
Bp₀.₀₀₁ 101°.

Seybold, G. *et al, Chem. Ber.*, 1977, **110**, 1225 (*synth*)

5-Methyl-4-phenyl-1,2,3-thiadiazole   M-60117
[64273-28-5]
$C_9H_8N_2S$ M 176.236
Cryst. Mp 41.5°. Bp₀.₀₀₁ 94°.

Schaumann, E. *et al, Chem. Ber.*, 1979, **112**, 1769 (*synth, pmr*)
Caron, M., *J. Org. Chem.*, 1986, **51**, 4075 (*synth*)

6-Methyl-3-phenyl-1,2,4-triazine, 9CI   M-60118
Updated Entry replacing M-50199
[37967-82-1]
$C_{10}H_9N_3$ M 171.201
Pale-yellow needles (hexane/Et₂O). Mp 68-69°.

Konno, S. *et al, Chem. Pharm. Bull.*, 1987, **35**, 1378 (*synth, pmr, props*)

2-Methyl-2-piperidinecarboxylic acid, 9CI   M-60119
α-Methylpipecolic acid
[72518-41-3]

(*R*)-*form*

$C_7H_{13}NO_2$ M 143.185
(*R*)-*form* [105141-61-5]
Mp >330°. $[\alpha]_D^{20}$ +7.7° (c, 1 in 6*M* HCl). (*S*)-form also known.
B,HBr: Mp 261-263°. $[\alpha]_D$ +5.1° (c, 2 in MeOH).
(±)-*form*
Monohydrate. Mp 340° subl.
Me ester:
  $C_8H_{15}NO_2$ M 157.212
  Bp_8 67°.
Et ester:
  $C_9H_{17}NO_2$ M 171.239
  Bp_{11} 78°. n_D^{21} 1.4497.

Overberger, C.G. *et al, Eur. Polym. J.*, 1983, **19**, 1055 (synth, resoln, pmr, ir)
Schöllkopf, U. *et al, Angew. Chem., Int. Ed. Engl.*, 1987, **26**, 143 (synth, pmr)

2-Methyl-1,3-propanediol   M-60120
[2163-42-0]

$$H_3CCH(CH_2OH)_2$$

$C_4H_{10}O_2$ M 90.122
Bp_3 83.5-84°, $Bp_{0.025}$ 61-67°.
Bis(4-methylbenzenesulfonyl): [24330-53-8]. Prisms (MeOH aq.). Mp 83-84°.

Adkins, H. *et al, J. Am. Chem. Soc.*, 1948, **70**, 3121 (synth)
Eliel, E.L. *et al, J. Am. Chem. Soc.*, 1969, **91**, 2703 (synth)

2-Methyl-1,3-propanedithiol   M-60121
[5337-95-1]

$$H_3CCH(CH_2SH)_2$$

$C_4H_{10}S_2$ M 122.243
Bp_{20} 77°, $Bp_{2.5}$ 37-39°.
S,S′-Bis(2,4-dinitrophenyl): Yellow cryst. (EtOAc). Mp 134.5-135°.

Eliel, E.L. *et al, J. Am. Chem. Soc.*, 1969, **91**, 2703; 1976, **98**, 3583 (synth)

2-Methyl-4-propyl-1,3-oxathiane, 9CI   M-60122
Updated Entry replacing M-03419
[67715-80-4]

(2*R*,4*S*)-*form*

$C_8H_{16}OS$ M 160.274
Isol. from juice of passion fruit *Passiflora edulis*. Important organoleptic compd. Bp_{12} 85-86°. The isolate from passion fruit consisted of a mixt. of *cis*- and *trans*-isomers, ca. 10:1, of so-far undetd. chirality.

(2*R*,4*S*)-*form* [90243-46-2]
  (−)-*cis-form*
  Synthetic. Less powerful odorant than the (2*S*,4*R*)-form, with fresher, more ester-type odour and taste. $[\alpha]_D$ −35.0°.
(2*S*,4*R*)-*form* [90243-47-3]
  (+)-*cis-form*
  Synthetic. Powerful odorant and flavourant characteristic of passion fruit and similar fruits. $[\alpha]_D^{20}$ +36.5° (c, 1.15 in CHCl₃).

Winter, M. *et al, Helv. Chim. Acta*, 1976, **59**, 1613 (isol, synth, pmr, ms)
Ger. Pat., 2 534 162, (1976); CA, **85**, 37096s
Pickenhagen, W. *et al, Helv. Chim. Acta*, 1984, **67**, 947 (synth)

N-Methylprotoporphyrin IX   M-60123
Updated Entry replacing M-03437
Green Pigment
[84192-00-7]

$C_{35}H_{36}N_4O_4$ M 576.694
Hepatic pigment probably formed from cytochrome-P450. Ferrochelatase inhibitor. Mp >300°. The 4-isomeric *N*-methyl compds. are formed *in vivo* as their Fe complexes. Analogous *N*-substituted pigments are also formed from unsaturated drugs containing allyl, ethenyl or ethinyl groups.
Di-Me ester: [77553-75-4]. Greenish-red cryst. All four isomeric *N*-methyl di-Me esters have been synthesised.

de Matteis, F. *et al, Biochem. J.*, 1980, **188**, 145, 241.
de Matteis, F. *et al, FEBS Lett.*, 1980, **119**, 109.
Ortiz de Montellano, P. *et al, J. Am. Chem. Soc.*, 1981, **103**, 4225 (synth)
Ortiz de Montellano, P. *et al, J. Biol. Chem.*, 1981, **256**, 6708.
Smith, K.M. *et al, Tetrahedron Lett.*, 1986, **27**, 2717 (synth)

5-Methyl-1*H*-pyrrole-2-carboxylic acid, 9CI   M-60124
Updated Entry replacing M-03531
[3757-53-7]
$C_6H_7NO_2$ M 125.127
Cryst. (ligroin). Mp 137-138°.
Me ester:
  $C_7H_9NO_2$ M 139.154
  Mp 100°.
Benzyl ester: [87462-15-5].
  $C_{13}H_{13}NO_2$ M 215.251
  Intermed. in porphyrin synth. Cream needles (MeOH aq.). Mp 96-97°.

Nicolaus, R.A. *et al, Ann. Chim. (Rome)*, 1956, **46**, 847
Kresze, G. *et al, Angew. Chem.*, 1964, **76**, 439.
Smith, K.M. *et al, J. Org. Chem.*, 1986, **51**, 4667 (deriv, synth, ir, pmr, ms)

5-Methyl-1*H*-pyrrole-3-carboxylic acid, 9CI   M-60125

Updated Entry replacing M-03532
$C_6H_7NO_2$ M 125.127
Cryst. (EtOH). Mp 212-214°.
Me ester: [40611-76-5].
    $C_7H_9NO_2$ M 139.154
    Cryst. (pet. ether). Mp 120-121°.
Et ester:
    $C_8H_{11}NO_2$ M 153.180
    Cryst. (pet. ether). Mp 70-71°.
Nitrile: [42046-60-6]. *2-Methyl-4-cyanopyrrole.*
    $C_6H_6N_2$ M 106.127
    Cryst. Mp 61-62°.

Jones, R.G., *J. Am. Chem. Soc.*, 1955, **77**, 4069.
Nicolaus, R.A. *et al, Ann. Chim.* (*Rome*), 1956, **46**, 847
Padwa, A. *et al, J. Am. Chem. Soc.*, 1986, **108**, 6739 (*nitrile*)

2-Methyl-2-pyrrolidinecarboxylic acid   M-60126

Updated Entry replacing M-03347
2-Methylproline, 9CI
[42856-71-3]

(*R*)-*form*

$C_6H_{11}NO_2$ M 129.158
(*R*)-*form* [63399-77-9]
    Ppt. (MeOH/Et$_2$O). Mp 330° dec. $[\alpha]_D^{20}$ +75.3° (c, 1 in MeOH).
(±)-*form*
    Cryst. (MeOH/Et$_2$O). Mp 263-264.5°.

Ellington, J.J. *et al, J. Org. Chem.*, 1974, **39**, 104 (*synth*)
Schöllkopf, U. *et al, Angew. Chem., Int. Ed. Engl.*, 1987, **26**, 143 (*synth, pmr*)

Methyl selenocyanate   M-60127

Selenocyanic acid methyl ester, 9CI
[2179-80-8]

MeSeCN

C_2H_3NSe M 120.012
Colourless or pale-yellow liq. Bp$_{15}$ 57°.

Franklin, W.J. *et al, Tetrahedron Lett.*, 1965, 3003 (*synth*)
Sakaizumi, T. *et al, Bull. Chem. Soc. Jpn.*, 1978, **51**, 3411; 1986, **59**, 3791 (*synth, ir*)

4-Methyl-2,2,6,6-tetrakis(trifluoromethyl)-2*H*,6*H*-[1,2]bromoxolo[4,5,1-*hi*][1,2]-benzobromoxole, 9CI   M-60128

10-Methyl-3,3,7,7-tetrakis(trifluoromethyl)-4,5,6-benzo-1-bromo-2,8-dioxabicyclo[3.3.1]octane
[76220-92-3]

$C_{13}H_5BrF_{12}O_2$ M 501.066
First example of organobromine(III) species. Strong oxidizing agent. Mp 153-154°. 10-*tert*-Butyl analogue also prepd.

Nguyen, T.T. *et al, J. Am. Chem. Soc.*, 1986, **108**, 3803 (*synth, ir, pmr, F nmr, ms, cryst struct*)

2-Methyl-3,4,5,6-tetranitroaniline   M-60129

2-Methyl-3,4,5,6-tetranitrobenzenamine, 9CI
[84432-57-5]

$C_7H_5N_5O_8$ M 287.145
Yellow cryst. (CH$_2$Cl$_2$). Mp 183-185°.
▷Explosive

Atkins, R.L. *et al, J. Org. Chem.*, 1986, **51**, 3261 (*synth*)

3-Methyl-2,4,5,6-tetranitroaniline   M-60130

3-Methyl-2,4,5,6-tetranitrobenzenamine, 9CI
[84432-56-4]
$C_7H_5N_5O_8$ M 287.145
Yellow needles (CHCl$_3$). Mp 192-193°.
▷Explosive

Atkins, R.L. *et al, J. Org. Chem.*, 1986, **51**, 3261 (*synth, ir, pmr*)

4-Methyl-2,3,5,6-tetranitroaniline   M-60131

4-Methyl-2,3,5,6-tetranitrobenzeneamine, 9CI
[84432-53-1]
$C_7H_5N_5O_8$ M 287.145
Fine yellow needles (CH$_2$Cl$_2$).
▷Explosive

Atkins, R.L. *et al, J. Org. Chem.*, 1986, **51**, 3261 (*synth*)

6-Methyl-1,3,5-triazine-2,4(1*H*,3*H*)-dione   M-60132

6-Methyl-1,3,5-triazine-2,4-diol. 6-Methyl-5-azauracil
[933-19-7]

$C_4H_5N_3O_2$ M 127.102
Several tautomers possible. Cryst. (H$_2$O). Mp 301-302°.
1-Me: [40265-80-3]. *1,6-Dimethyl-5-azauracil.*
    $C_5H_7N_3O_2$ M 141.129
    Cryst. (EtOH). Mp 221-222°.
3-Me: [69032-72-0]. *3,6-Dimethyl-5-azauracil.*
    $C_5H_7N_3O_2$ M 141.129
    Cryst. (EtOH). Mp 226°.
1,3-Di-Me: [40265-81-4]. *1,3,6-Trimethyl-5-azauracil.*
    $C_6H_9N_3O_2$ M 155.156
    Cryst. (EtOAc). Mp 98-99°.
1-Me, 4-Me ether: 4-Methoxy-1,6-dimethyl-1,3,5-triazin-2(1H)-one.
    $C_6H_9N_3O_2$ M 155.156
    Cryst. (EtOAc/pet. ether). Mp 109-110°.
2,4-Di-Me ether: [4000-78-6]. *2,4-Dimethoxy-6-methyl-1,3,5-triazine.*
    $C_6H_9N_3O_2$ M 155.156

Cryst. (pet. ether). Mp 68-69°.

Piškala, A. *et al*, *Collect. Czech. Chem. Commun.*, 1963, **28**, 1681, 2365 (*synth, derivs*)

1'-Methylzeatin   M-60133

2-Methyl-4-(1H-purin-6-ylamino)-2-penten-1-ol, 9CI.
6-(4-Hydroxy-1,3-dimethyl-2-butenylamino)purine
[101512-26-9]

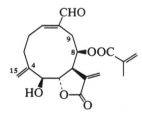

$C_{11}H_{15}N_5O$ M 233.272

Cytokinin from the culture filtrate of *Pseudomonas syringae* p.v. *savastanoi*. Mp 201-202°. $[\alpha]_D^{26}$ −109° (c, 0.153 in EtOH).

9-β-D-Ribofuranosyl: [98211-30-4].
  $C_{16}H_{23}N_5O_5$ M 365.388
  Isol. from *P. syringae* p.v. *savastanoi*. Cryst. + ½H_2O. Mp 130-132°. $[\alpha]_D^{14}$ −117° (c, 0.102).

Surico, G. *et al*, *Phytochemistry*, 1985, **24**, 1499 (*isol, pmr, cmr, ms, uv*)
Evidente, A. *et al*, *Phytochemistry*, 1986, **25**, 525 (*isol, pmr, cmr, uv, ms*)
Itaya, T. *et al*, *Tetrahedron Lett.*, 1986, **27**, 6349 (*synth, cd, abs config*)

Microglossic acid   M-60134

6-[3-(3-Furanyl)propyl]-2-(4-methyl-3-pentenyl)-2-heptenedioic acid 1-methyl ester, 9CI
[111150-39-1]

$C_{21}H_{30}O_5$ M 362.465

Constit. of *Microglossa zeylanica*. Oil.

10,11-Dihydro: [111150-38-0]. **10,11-Dihydromicroglossic acid**.
  $C_{21}H_{32}O_5$ M 364.481
  From *M. zeylanica*. Oil.

Gunatilaka, A.A.L. *et al*, *Phytochemistry*, 1987, **26**, 2408.

Mikrolin   M-60135

Updated Entry replacing M-03934
Microline
[60958-71-6]

$C_{14}H_{15}ClO_5$ M 298.723

Exists in soln. as equilib. mixt. with tetracyclic form predominant. Metab. of *Gilmaniella humicola*. Shows antifungal activity. Cryst. (EtOAc/hexane or Me_2CO/hexane). Mp 113-114°. $[\alpha]_D^{20}$ +135.6° (c, 0.648 in $CHCl_3$). Related to Gilmicolin and Mycorrhizin *A*.

Dechloro: [60958-72-7]. **Dechloromikrolin**.
Dechloromicroline.
  $C_{14}H_{16}O_5$ M 264.277
  Metab. of *G. humicola*. Shows antifungal activity. Oil.

Bollinger, P. *et al*, *Helv. Chim. Acta*, 1976, **59**, 1809 (*isol, struct, ir, uv, ms, nmr*)
Weber, H.P. *et al*, *Helv. Chim. Acta*, 1976, **59**, 1821 (*cryst struct*)
Chexal, K.K. *et al*, *Helv. Chim. Acta*, 1978, **61**, 2002 (*biosynth, pmr, cmr, ms*)
Koft, E.R. *et al*, *J. Am. Chem. Soc.*, 1982, **104**, 2659 (*struct*)
Smith, A.B. *et al*, *Tetrahedron Lett.*, 1987, **28**, 3659, 3663 (*synth*)

1(10)-Millerenolide   M-60136

$C_{19}H_{22}O_6$ M 346.379

(1(10)E)-form [110601-24-6]
Constit. of *Milleria quinqueflora*. Oil. $[\alpha]_D^{24}$ −255° (c, 0.26 in $CHCl_3$).

8,9-Didehydro (E): **1(10)E,8E-Millerdienolide**.
  $C_{19}H_{20}O_6$ M 344.363
  From *M. quinqueflora*. Oil. $[\alpha]_D^{24}$ −260° (c, 0.43 in $CHCl_3$).

8,9-Didehydro, 4β,15-Epoxide: **4β,15-Epoxy-1(10)E,8E-millerdienolide**.
  $C_{19}H_{20}O_7$ M 360.363
  From *M. quinqueflora*. Oil.

Jakupovic, J. *et al*, *Phytochemistry*, 1987, **26**, 2011.

9-Millerenolide   M-60137

$C_{19}H_{22}O_6$ M 346.379

(9E)-form [110601-29-1]

Constit. of *Milleria quinqueflora*. Oil. $[\alpha]_D^{24}$ −5° (c, 0.6 in CHCl₃).

4β,15-Epoxide: **4β,15-Epoxy-9E-millerenolide**.

$C_{19}H_{22}O_7$ M 362.379

From *M. quinqueflora*. Cryst. Mp 196-198°. $[\alpha]_D^{24}$ −131° (c, 1.8 in CHCl₃).

(9Z)-form

Constit. of *M. quinqueflora*. Oil.

1β-Ethoxy: **1-Ethoxy-9Z-millerenolide**.

$C_{21}H_{26}O_7$ M 390.432

From *M. quinqueflora*.

4β,15-Epoxide: **4β,15-Epoxy-9Z-millerenolide**.

$C_{19}H_{22}O_7$ M 362.379

From *M. quinqueflora*. Oil. $[\alpha]_D$ −27° (c, 0.35 in CHCl₃).

1β-Ethoxy, 4β,15-Epoxide: **4β,15-Epoxy-1β-ethoxy-4Z-millerenolide**.

$C_{21}H_{26}O_8$ M 406.432

From *M. quinqueflora*. Oil.

1β-Methoxy, 4β,15-epoxide: **4β,15-Epoxy-1β-methoxy-9Z-millerenolide**.

$C_{20}H_{24}O_8$ M 392.405

From *M. quinqueflora*. Oil.

Jakupovic, J. *et al*, *Phytochemistry*, 1987, **26**, 2011.

Misakinolide *A*   M-60138

Bistheonellide A

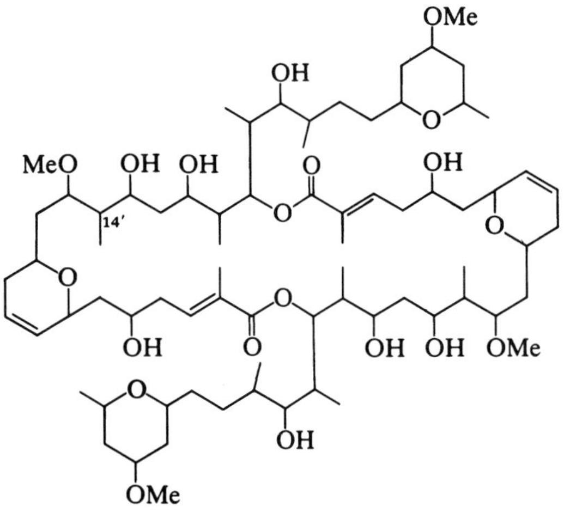

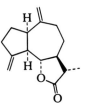

$C_{74}H_{128}O_{20}$ M 1337.813

Constit. of sponge *Theonella* sp. Antitumour agent. Oil. $[\alpha]_D^{20}$ −21.4° (c, 5.6 in CHCl₃).

14′-Desmethyl: **Bistheonellide B**.

$C_{73}H_{126}O_{20}$ M 1323.786

Constit. of sponge *Theonella* sp. Cytotoxic.

Sakai, R. *et al*, *Chem. Lett.*, 1986, 1499 (*isol*)
Kato, Y. *et al*, *Tetrahedron Lett.*, 1987, **28**, 6225 (*struct*)

Miyabenol *A*   M-60139

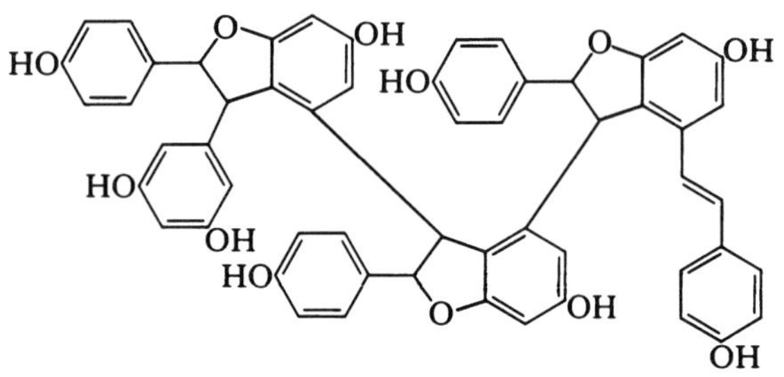

$C_{56}H_{42}O_{12}$ M 906.941

Constit. of *Carex fedia* var. *miyabei*. Pale-yellow solid. $[\alpha]_D^{23}$ −91° (c, 0.18 in MeOH).

Suzuki, K. *et al*, *Agric. Biol. Chem.*, 1987, **51**, 1003.

Miyabenol *B*   M-60140

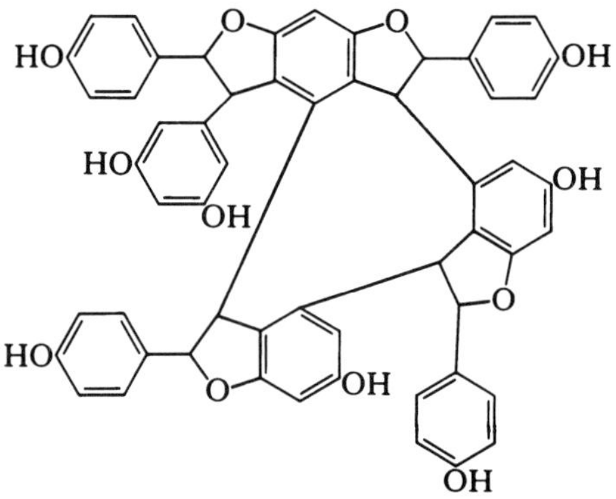

$C_{56}H_{40}O_{12}$ M 904.925

Constit. of *Carex fedia* var. *miyabei*. Pale-yellow solid. $[\alpha]_D^{23}$ −126° (c, 0.23 in MeOH).

Suzuki, K. *et al*, *Agric. Biol. Chem.*, 1987, **51**, 1003.

Miyabenol *C*   M-60141

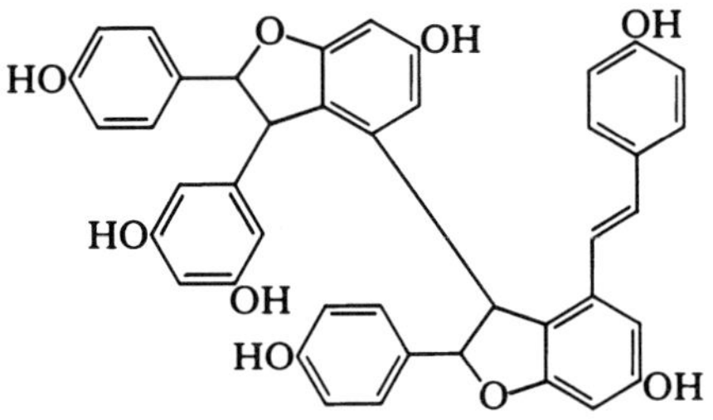

$C_{42}H_{32}O_9$ M 680.709

Constit. of *Carex fedia* var. *miyabei*. Pale-brown solid. $[\alpha]_D^{23}$ +69° (c, 0.07 in MeOH).

Suzuki, K. *et al*, *Agric. Biol. Chem.*, 1987, **51**, 1003.

Mokkolactone   M-60142

Dehydrodihydrocostus lactone

[4955-03-7]

$C_{15}H_{20}O_2$ M 232.322

Constit. of the Chinese crude drug Sen-mokko and *Saussurea lappa*. Cryst. Mp 35-37°. [α]$_D$ +18.2° (CHCl$_3$).

Hikino, H. *et al*, *Chem. Pharm. Bull.*, 1964, **12**, 632 (*isol*)
Bankar, N.S. *et al*, *Indian J. Chem.*, 1972, **10**, 952 (*isol*)

Moracin *D* M-60143

*7-(6-Hydroxy-2-benzofuranyl)-2,2-dimethyl-2*H-*1-benzopyran-5-ol*, 9CI

[69120-07-6]

$C_{19}H_{16}O_4$ M 308.333
Isol. from *Morus alba* infected with *Fusarium solani*. Phytoalexin. Mp 130-131°.

Di-Ac: Mp 125-126°.

Takasugi, M. *et al*, *Chem. Lett.*, 1978, 1239 (*isol, struct*)

Moracin *E* M-60144

*5-(6-Hydroxy-2-benzofuranyl)-2,2-dimethyl-2*H-*1-benzopyran-7-ol*, 9CI

[73338-84-8]

$C_{19}H_{16}O_4$ M 308.333
Isol. from *Morus alba* infected with *Fusarium solani*. Phytoalexin. Mp 184-185°.

Di-Ac: Mp 87-88°.

Takasugi, M. *et al*, *Tetrahedron Lett.*, 1979, 4675 (*isol, struct, props*)

Moracin *G* M-60145

5-(7,10-Dihydro-8-methylfuro[2,3-g][1]benzoxepin-2-yl)-1,3-benzenediol, 9CI

[73338-86-0]

$C_{19}H_{16}O_4$ M 308.333
Isol. from *Morus alba* infected with *Fusarium solani*. Phytoalexin. Mp 198-199°.

4′-Methoxy: [73338-87-1]. **Moracin H.**
$C_{20}H_{18}O_5$ M 338.359
From *M. alba* infected with *F. solani*. Phytoalexin. Mp 191-192°.

4-Methoxy, Di-Ac: Mp 140-141°.

Takasugi, M. *et al*, *Tetrahedron Lett.*, 1979, 4675 (*isol, struct, props*)

Moracin *M* M-60146

5-(6-Hydroxy-2-benzofuranyl)-1,3-benzenediol, 9CI. *2-(3,5-Dihydroxyphenyl)-6-hydroxybenzofuran*

[56317-21-6]

$C_{14}H_{10}O_4$ M 242.231
Isol. from *Morus alba* infected with *Fusarium solani*. Phytoalexin.

5-Hydroxy, O^6-Me: [73338-89-3]. **Moracin J.** *5-(5-Hydroxy-6-methoxy-2-benzofuranyl)-1,3-benzenediol*, 9CI. *2-(3,5-Dihydroxyphenyl)-5-hydroxy-6-methoxybenzofuran.*
$C_{20}H_{20}O_5$ M 340.375
From *M. alba* infected with *F. solani*. Phytoalexin.

5′-Methoxy, O^6-Me: [73338-85-9]. **Moracin F.** *5-(5,6-Dimethoxy-2-benzofuranyl)-1,3-benzenediol*, 9CI. *2-(3,5-Dihydroxyphenyl)-5,6-dimethoxybenzofuran.*
$C_{16}H_{14}O_5$ M 286.284
Isol. from *M. alba* infected with *F. solani*. Phytoalexin.

4′-(3-Methyl-2-butenyl): [69120-06-5]. **Moracin C.** *5-(6-Hydroxy-2-benzofuranyl)-2-(3-methyl-2-butenyl)-1,3-benzenediol*, 9CI. *2-(3,5-Dihydroxy-4-prenylphenyl)-6-hydroxybenzofuran.*
$C_{19}H_{18}O_4$ M 310.349
Isol. from *M. alba* infected with *E. solani*. Phytoalexin. Mp 198-199°.

4′-(3-Methyl-2-butenyl), tri-Ac: Mp 156-157°.

2′-(3-Methyl-2-butenyl), O$^{3′}$-Me: [73338-88-2]. **Moracin I.** *6-Hydroxy-2-(5-hydroxy-3-methoxy-2-prenylphenyl)benzofuran.*
$C_{20}H_{20}O_4$ M 324.376
From *M. alba* infected with *F. solani*. Phytoalexin.

Deshpande, V.H. *et al*, *Indian J. Chem.*, 1975, **13**, 453 (*isol*)
Takasugi, M. *et al*, *Tetrahedron Lett.*, 1979, 4675 (*Moracin F, Moracin J*)
Takasugi, M. *et al*, *CA*, 1980, **92**, 160540 (*isol, struct*)

3-Morpholinone, 9CI, 8CI M-60147

3-Oxomorpholine. Morpholone

[109-11-5]

$C_4H_7NO_2$ M 101.105
Cryst. (Et$_2$O/hexane). Mp 106°. Bp$_7$ 142°.

4-Benzoyl: [61883-62-3].
$C_{11}H_{11}NO_3$ M 205.213
Cryst. (Et$_2$O/pet. ether). Mp 50-52°.

4-Ph: [29518-11-4]. *4-Phenyl-3-morpholinone*, 9CI.
$C_{10}H_{11}NO_2$ M 177.202
Mp 113-114°.

Lehn, J.-M. *et al*, *Helv. Chim. Acta*, 1976, **59**, 1566 (*synth*)
Perrone, R. *et al*, *Synthesis*, 1976, 598 (*deriv*)
Arya, V.P. *et al*, *Indian J. Chem., Sect. B*, 1977, **15**, 720 (*synth*)

Mugineic acid   M-60148

2-Carboxy-α-[(3-carboxy-3-hydroxypropyl)amino]-β-hydroxy-1-azetidinebutanoic acid, 9CI

[69199-37-7]

$C_{12}H_{20}N_2O_8$ M 320.299

Excreted from the roots of barley *Hordeum vulgare*. Phytosiderophore. Mp 210-212° dec. $[\alpha]_D$ −70.7° (c, 0.97 in H_2O).

2′-Epimer: [74281-81-5]. **Isomugeneic acid.**
$C_{12}H_{20}N_2O_8$ M 320.299
From roots of barley *H. vulgare*. $[\alpha]_D$ −97° (c, 1.24 in H_2O).

2′-Deoxy: 2′-*Deoxymugeneic acid.*
$C_{12}H_{20}N_2O_7$ M 304.299
From wheat *Triticum aestivum*. Cryst. Mp 198.5-200.5°. $[\alpha]_D$ −70.5° (H_2O).

3-Hydroxy: [74235-23-7]. *3-Hydroxymugeneic acid.*
$C_{12}H_{20}N_2O_9$ M 336.298
From roots of barley *H. vulgare*.

Nomoto, K. *et al, Chimia*, 1981, **35**, 249 (*deriv*)
Ohfune, Y. *et al, J. Am. Chem. Soc.*, 1981, **103**, 2409 (*deriv*)
Sugiura, Y. *et al, Biochemistry*, 1983, **22**, 4842 (*pmr, cmr, cd*)
Sugiura, Y. *et al, J. Am. Chem. Soc.*, 1985, **107**, 4667 (*pmr, cmr, cd*)
Hamada, Y. *et al, J. Org. Chem.*, 1986, **51**, 5489 (*synth, bibl*)

Mulberrofuran *R*   M-60149

$C_{27}H_{20}O_7$ M 456.451
Constit. of *Morus lhou*. Amorph. powder.
Kohno, H. *et al, Heterocycles*, 1987, **26**, 759.

Murraol   M-60150

$C_{15}H_{16}O_4$ M 260.289
Constit. of *Murraya exotica*. Prisms. Mp 105-107°.
Ito, C. *et al, Heterocycles*, 1987, **26**, 1731.

Murraxocin   M-60151

$C_{17}H_{20}O_5$ M 304.342

Constit. of *Murraya exotica*.
Ac: Cryst. (Et_2O). Mp 138-140°.
Barik, B.R. *et al, Phytochemistry*, 1987, **26**, 3319.

Murraxonin   M-60152

$C_{12}H_{10}O_5$ M 234.208
Constit. of *Murraya exotica*. Cryst. (C_6H_6/Me_2CO). Mp 212-215°.
Barik, B.R. *et al, Phytochemistry*, 1987, **26**, 3319.

Muscarine   M-60153

Updated Entry replacing M-04106
Tetrahydro-4-hydroxy-N,N,N,5-tetramethyl-2-furanmethanaminium, 9CI

[7619-12-7]

$C_9H_{20}NO_2^{\oplus}$ M 174.263 (ion)

(2S,4R,5S)-form [300-54-9]
Main toxic constit. of the fly fungus *Amanita muscaria*.
▷QG3325000.
Chloride: [2303-35-7].
$C_9H_{20}ClNO_2$ M 209.716
Needles. Mp 181.5-182° (179-180°). $[\alpha]_D$ +7.4° (c, 3.1 in H_2O).
▷QG3500000.
Iodide: [24570-49-8].
$C_9H_{20}INO_2$ M 301.167
Needles. Mp 149-149.5° (138-142°). $[\alpha]_D$ +6.3° (c, 1.1 in H_2O). Turns yellow within a few hours.
Tetraphenylborate: [104487-60-7].
$C_{33}H_{40}BNO_2$ M 493.495
Mp 193°. $[\alpha]_D$ +9.7° (c, 1.34 in Me_2CO).

(2R,4S,5R)-form
Chloride: Mp 179-180°. $[\alpha]_D$ −8.4° (EtOH).

(2RS,4SR,5RS)-form
Reineckate: Mp 174-175°.
Tetrachloroaurate: [6032-87-7]. Mp 119-120°.

Hardegger, E. *et al, Helv. Chim. Acta*, 1957, **40**, 2383.
Eugster, C.H. *et al, Helv. Chim. Acta*, 1969, **52**, 708.
Matsumoto, T. *et al, Tetrahedron*, 1969, **25**, 5889 (*synth*)
Whiting, J. *et al, Can. J. Chem.*, 1972, **50**, 3322 (*synth, pmr*)
Nitta, K. *et al, Helv. Chim. Acta*, 1977, **60**, 1747.
Mulzer, J. *et al, Justus Liebigs Ann. Chem.*, 1987, 7 (*synth, pmr, ir*)

Mycosinol   M-60154

7-(2,4-Hexadiynylidene)-1,6-dioxaspiro[4.4]nona-2,8-dien-4-ol

(E)-form

$C_{13}H_{10}O_3$ M 214.220

(E)-form

Constit. of *Coleostephus myconis* infected with *Botrytis cinerea*. Antifungal phytoalexin. Oil.

(Z)-form

Constit. of *C. myconis* and formed by photochemical action on (*E*)-mycosinol.

Marshall, P.S. *et al, Phytochemistry*, 1987, **26**, 2493.

Myomontanone   M-60155

Updated Entry replacing M-20249

3-Furanyl[4-methyl-2-(2-methylpropyl)-1-cyclopenten-1-yl]methanone, 9CI

[86989-09-5]

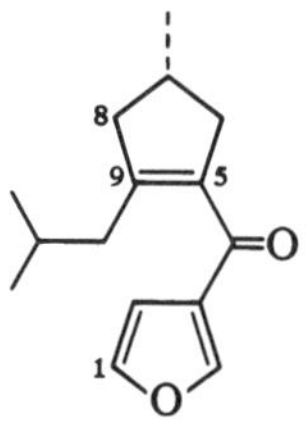

$C_{15}H_{20}O_2$ M 232.322

Constit. of leaves of *Myoporum montenum*. Cryst. (2,2,3-trimethylpentane). Mp 45°. $[\alpha]_D$ +26° (c, 3.5 in CHCl$_3$).

Δ^8-*Isomer:* [86989-08-4]. ***Isomyomontanone***.
$C_{15}H_{20}O_2$ M 232.322
From *M. montanum*. Oil.

Métra, P.L. *et al, Tetrahedron Lett.*, 1983, **24**, 1749 (*isol, struct*)
Hess, T. *et al, Tetrahedron Lett.*, 1987, **28**, 5643 (*synth, abs config*)

Myoporone   M-60156

Updated Entry replacing M-40168

1-(3-Furanyl)-4,8-dimethyl-1,6-nonanedione, 9CI

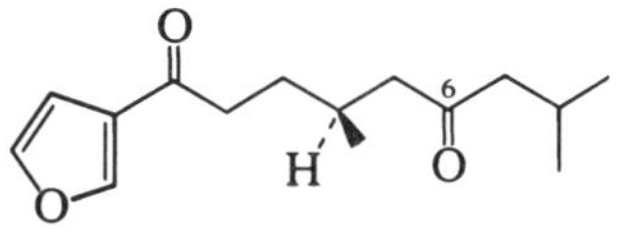

$C_{15}H_{22}O_3$ M 250.337

(S)-form [19479-15-3]

Constit. of *Myoporum* and *Eremophila* spp. Cryst. (MeOH). Mp 15.5-16.5°. $[\alpha]_D$ −5.0° (c, 5.7 in CHCl$_3$).

(6R)-*Alcohol:* [72145-16-5]. ***Dihydromyoporone***.
$C_{15}H_{24}O_3$ M 252.353
Stress metab. of sweet potato (*Ipomoea batatis*). Oil. $[\alpha]_D^{20}$ −3.6° (c, 1.95 in MeOH) (synthetic).

Blackburne, I.D. *et al, Aust. J. Chem.*, 1972, **25**, 1787 (*isol*)
Burka, L. *et al, Phytochemistry*, 1979, **18**, 873 (*isol*)
Still, W.C. *et al, J. Am. Chem. Soc.*, 1980, **102**, 7385 (*synth*)
Johnson, W.S. *et al, Tetrahedron Lett.*, 1984, **25**, 3951 (*synth, abs config*)
Hess, T. *et al, Tetrahedron Lett.*, 1987, **28**, 5643 (*synth, abs config*)

N

1,2-Naphthalenedicarboxaldehyde — N-60001

[74057-36-6]

$C_{12}H_8O_2$ M 184.194

2-Di-Me acetal: [103668-60-6]. *2-(Dimethoxymethyl)-1-naphthalenecarboxaldehyde. 2-(Dimethoxymethyl)-1-naphthaldehyde.*
$C_{14}H_{14}O_3$ M 230.263
Oil.

Japan. Pat., 80 8 007 235, (*1980*); *CA*, **93**, 26167e
Smith, J.G. *et al*, *J. Org. Chem.*, 1986, **51**, 3762 (*deriv, synth, ir, pmr, ms*)

1,3-Naphthalenedicarboxaldehyde, 9CI — N-60002

Updated Entry replacing N-00047
Naphthalene-1,3-dialdehyde. 1,3-Diformylnaphthalene
$C_{12}H_8O_2$ M 184.194
Mp 124°.

Bis-2,4-dinitrophenylhydrazone: Mp 327°.

Ried, W. *et al*, *Chem. Ber.*, 1958, **91**, 2479.
Raju, B. *et al*, *Synthesis*, 1987, 197 (*synth, ir, pmr*)

2,3-Naphthalenedicarboxylic acid, 9CI — N-60003

Updated Entry replacing N-00063

[2169-87-1]

$C_{12}H_8O_4$ M 216.193
Prisms (AcOH or by subl.). Sol. hot EtOH, spar. sol. Et$_2$O. Mp 246° (239-241°).

Di-Me ester: [13728-34-2].
$C_{14}H_{12}O_2$ M 212.248
Large plates (Et$_2$O/pet. ether). Mp 47°. Bp$_1$ 141-145°.

Di-Et ester: [50919-54-5].
$C_{16}H_{16}O_2$ M 240.301
Cryst. (Me$_2$CO). Mp 88.5-89.5°.

Mononitrile: *3-Cyano-2-naphthoic acid.*
$C_{12}H_7NO_2$ M 197.193
Yellow cryst. Mp 273-274°.

Dinitrile: [22856-30-0]. *2,3-Dicyanonaphthalene.*
$C_{12}H_6N_2$ M 178.193
Needles (EtOH). Mp 251°.

Imide:
$C_{12}H_7NO_2$ M 197.193
Microneedles (CHCl$_3$/EtOH). Mp 275°. Softens at 250°.

Freund, M. *et al*, *Justus Liebigs Ann. Chem.*, 1913, **402**, 68.
Waldmann, H. *et al*, *Ber.*, 1931, **64**, 1713.
Bradbrook, E.F. *et al*, *J. Chem. Soc.*, 1936, 1739.
Patton, J.W. *et al*, *J. Org. Chem.*, 1965, **30**, 3869.
Org. Synth., Coll. Vol., **5**, 810.
Davies, C., *Fuel*, 1973, **52**, 270; 1974, **53**, 105 (*ir, uv*)
Carlson, R.G. *et al*, *J. Org. Chem.*, 1986, **51**, 3978 (*synth*)

2-Naphthalenemethanol, 9CI, 8CI — N-60004

2-(Hydroxymethyl)naphthalene

[1592-38-7]

$C_{11}H_{10}O$ M 158.199
Cryst. (EtOH or hexane). Mp 80°.

Hauser, C.R. *et al*, *J. Org. Chem.*, 1958, **23**, 354 (*synth*)
Ouellette, R.J. *et al*, *Tetrahedron*, 1969, **25**, 819 (*pmr, conformn*)
Eisch, J.J. *et al*, *J. Org. Chem.*, 1982, **47**, 5051 (*synth*)
Gellert, E. *et al*, *Aust. J. Chem.*, 1983, **36**, 157 (*synth*)
Lau, C.K. *et al*, *J. Org. Chem.*, 1986, **51**, 3038 (*synth*)

1,4,5-Naphthalenetriol, 9CI — N-60005

Updated Entry replacing N-00140
1,4,5-Trihydroxynaphthalene

[481-40-3]

$C_{10}H_8O_3$ M 176.171
Occurs in green shells of unripe walnuts. Leaflets or needles (H$_2$O). Insol. C$_6$H$_6$, pet. ether, CHCl$_3$. Mp 168-170°.

Tri-Ac: [24308-04-1].
$C_{16}H_{14}O_6$ M 302.283
Prisms (EtOH). Mp 129-130°.
5-Me ether, 1-Ac:
$C_{13}H_{12}O_4$ M 232.235
Needles (EtOAc/pet. ether). Mp 82-83°.
1,5-Di-Me ether: [3843-55-8]. *4,8-Dimethoxy-1-naphthol.*
$C_{12}H_{12}O_3$ M 204.225
Cryst. (pet. ether). Mp 160° (155.5-156.5°).
1,5-Di-Me ether, Ac:
$C_{14}H_{14}O_4$ M 246.262
Cryst. (H$_2$O). Mp 119-120°.
4,5-Di-Me ether: [61836-40-6]. *4,5-Dimethoxy-1-naphthol.*
$C_{12}H_{12}O_3$ M 204.225
Needles. Mp 109°.
4,5-Di-Me ether, Ac:
$C_{14}H_{14}O_4$ M 246.262
Cryst. (CH$_2$Cl$_2$/pet. ether). Mp 62°.
Tri-Me ether: [64636-39-1]. *1,4,5-Trimethoxynaphthalene.*
$C_{13}H_{14}O_3$ M 218.252
Cryst. (C$_6$H$_6$). Mp 119°.

Willstätter, R. *et al*, *Ber.*, 1914, **47**, 2796 (*synth*)
Daglish, C., *J. Am. Chem. Soc.*, 1950, **72**, 4859 (*uv*)
Elsevier's Encycl. Org. Chem., 1953, Ser. III, **12B**, 2155 (*bibl*)
Corey, E.J. *et al*, *Tetrahedron Lett.*, 1975, 2389 (*deriv*)
Hannan, R.L. *et al*, *J. Org. Chem.*, 1979, **44**, 2153 (*deriv, synth*)
Laatsch, H., *Justus Liebigs Ann. Chem.*, 1980, 140 (*deriv, synth, pmr*)
Chorn, T.A. *et al*, *J. Chem. Soc., Perkin Trans. 1*, 1984, 1339 (*deriv, synth, ir, pmr*)
Malesani, G. *et al*, *J. Heterocycl. Chem.*, 1987, **24**, 513 (*synth, pmr, derivs*)

Naphth[2,3-*a*]azulene-5,12-dione, 9CI N-60006
[76319-75-0]

C_{18}H_{10}O_2 M 258.276
Cryst. (EtOAc/C_6H_6). Mp 279-280°.

Bindl, J. *et al, Chem. Ber.*, 1983, **116**, 2408 (*synth, ms, ir, uv, pmr*)

Naphtho[2,1-*b*:6,5-*b'*]bis[1]benzothiophene, N-60007
9CI
[91538-70-4]

C_{22}H_{12}S_2 M 340.457
Prisms. Mp 293°.

Tedjamulia, M.L. *et al, J. Heterocycl. Chem.*, 1984, **21**, 321
(*synth, pmr*)

Naphtho[2,3-*c*]furan N-60008
Benzo[f]*isobenzofuran*
[2586-62-1]

C_{12}H_8O M 168.195
Unstable intermediate which can be trapped by a variety
of dienophiles.

Mir-Mohamad-Sadeghy, B. *et al, J. Org. Chem.*, 1983, **48**, 2237
(*synth*)
Smith, J.G. *et al, J. Org. Chem.*, 1986, **51**, 3762 (*synth*)

Naphtho[1,2,3,4-*ghi*]perylene, 9CI N-60009
1:12-o-Phenyleneperylene
[190-84-1]

C_{26}H_{14} M 326.397
Long yellow needles (C_6H_6). Mp 268-270°.

Clar, E. *et al, Tetrahedron*, 1959, **6**, 358 (*synth, uv*)
Bunte, R. *et al, Chem. Ber.*, 1986, **119**, 3521 (*synth, pmr*)

4*H*-Naphtho[1,2-*b*]pyran-4-one, 9CI N-60010
Updated Entry replacing N-00210
Benzo[h]*chromone. 7,8-Benzochroman-4-one*
[3528-23-2]

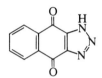

C_{13}H_8O_2 M 196.205
Needles (EtOH aq.). Mp 125°.

Pfeiffer, P. *et al, Ber.*, 1917, **50**, 922 (*synth*)
v. Strandtmann, M. *et al, J. Heterocycl. Chem.*, 1972, **9**, 171
(*synth*)
Ravikumar, K. *et al, Acta Crystallogr., Sect. C*, 1986, **42**, 1043
(*cryst struct*)

1*H*-Naphtho[2,3-*d*]triazole-4,9-dione, 9CI N-60011
4,9-Dihydro-4,9-dioxo-1H-naphtho[2,3-d]-v-triazole
[3915-98-8]

C_{10}H_5N_3O_2 M 199.168
Cryst. + 1H_2O (Me_2CO aq.). Mp 250° dec.

Fieser, L.F. *et al, J. Am. Chem. Soc.*, 1935, **57**, 1844 (*synth*)
Buckle, D.R. *et al, J. Med. Chem.*, 1983, **26**, 714 (*synth,
biochem, ir, uv, pmr*)

Naphthvalene N-60012
Updated Entry replacing N-40007
*2,3-Dihydro-1,2,3-metheno-1H-indene, 9CI.
Benzobenzvalene*
[34305-47-0]

C_{10}H_8 M 128.173
Liq. Stable at r.t., dec. on heating at 175° to
benzofulvene. λ_{max} 278 (ε 494), 271 (465), 264 (375),
235 (975). Obt. contaminated with 9% naphthalene.

Katz, T.J. *et al, J. Am. Chem. Soc.*, 1971, **93**, 3782 (*synth*)
Gleiter, R. *et al, Helv. Chim. Acta*, 1981, **64**, 1312 (*pe, struct*)
Kjell, D.P. *et al, Tetrahedron Lett.*, 1985, **26**, 5731 (*props, bibl*)
Kjell, D.P. *et al, J. Am. Chem. Soc.*, 1986, **108**, 4111 (*synth*)

1-(1-Naphthyl)piperazine N-60013

1-(1-Naphthalenyl)piperazine, 9CI

[57536-86-4]

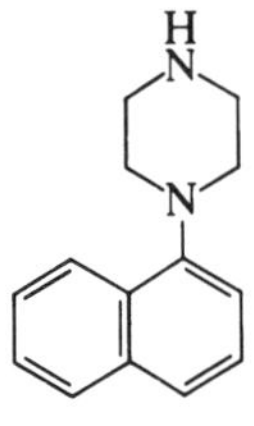

$C_{14}H_{16}N_2$ M 212.294

Pale-yellow oil. Bp$_{0.1}$ 148-156°.

B,HCl: [104113-71-5]. Cryst. (EtOH). Mp 313-315°.

B,HBr: [79144-85-7]. Cryst. (EtOH). Mp 293-295°.

Prelog, V. *et al, Collect. Czech. Chem. Commun.*, 1934, **6**, 211 (*synth*)

Červená, I. *et al, Collect. Czech. Chem. Commun.*, 1975, **40**, 1612 (*synth*)

Glennon, R.A. *et al, J. Med. Chem.*, 1986, **29**, 2375 (*synth*)

1-(2-Naphthyl)piperazine N-60014

1-(2-Naphthalenyl)piperazine, 9CI

[57536-91-1]

$C_{14}H_{16}N_2$ M 212.294

Pale-yellow oil. Bp$_{0.25}$ 138-140°.

B,HCl: [104090-87-1]. Cryst. (EtOH). Mp 265-266°.

B,HBr: Cryst. (EtOH). Mp 278-280°.

Prelog, V. *et al, Collect. Czech. Chem. Commun.*, 1934, **6**, 211 (*synth*)

Červená, I. *et al, Collect. Czech. Chem. Commun.*, 1975, **40**, 1612 (*synth*)

Glennon, R.A. *et al, J. Med. Chem.*, 1986, **29**, 2375 (*synth*)

1-(1-Naphthyl)-2-propanol N-60015

α-Methyl-1-naphthaleneethanol, 9CI

[27653-13-0]

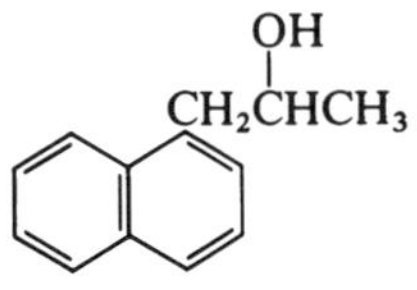

$C_{13}H_{14}O$ M 186.253

(±)-*form*

Bp$_5$ 162°.

Barcus, R.L. *et al, J. Am. Chem. Soc.*, 1986, **108**, 3928 (*synth, ir, pmr*)

Nectriafurone N-60016

Updated Entry replacing N-20029

5,8-Dihydroxy-1-(1-hydroxyethyl)-7-methoxyn-aphtho[2,3-c]furan-4,9-dione, 9CI

[87596-55-2]

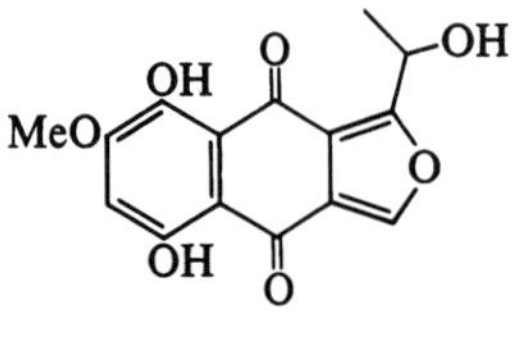

$C_{15}H_{12}O_7$ M 304.256

Metabolite of fungi *Nectria haematococca* and *Fusarium oxysporum*. Yellow-brown cryst. (MeOH). Mp 230°.

5-Me ether: [111660-59-4]. Yellow needles (MeOH). Mp 202-203°.

8-Me ether: [111660-60-7].

$C_{16}H_{14}O_7$ M 318.282

Metabolite of *F. oxysporum*. Yellow-brown needles (MeOH). Mp 214-222° dec.

Parisot, D. *et al, Phytochemistry*, 1983, **22**, 1301 (*isol*)
Tatum, J.H. *et al, Phytochemistry*, 1987, **26**, 2499 (*isol*)

Neocryptotanshinone N-60017

5,6,7,8-Tetrahydro-3-hydroxy-2-(2-hydroxy-1-methy-lethyl)-8,8-dimethyl-1,4-phenanthrenedione, 9CI. *12,16-Dihydroxy-20-nor-5(10),6,8,12-abietatetraene-11,14-dione*

[109664-02-0]

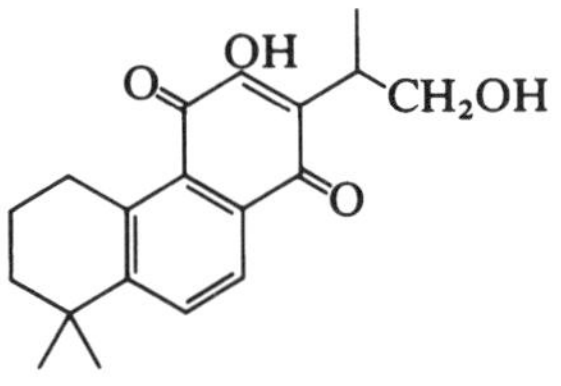

$C_{19}H_{22}O_4$ M 314.380

Constit. of *Salvia miltiorrhiza*. Orange-red needles. Mp 165-167°. $[\alpha]_D^{23}$ +29.8° (c, 0.84 in CHCl$_3$).

Lee, A.-R. *et al, J. Nat. Prod.*, 1982, **50**, 157.

Neoleuropein N-60018

[108789-16-8]

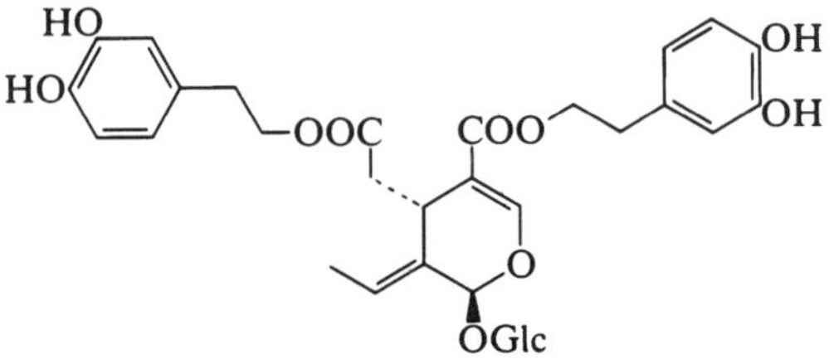

$C_{32}H_{38}O_{15}$ M 662.643

Constit. of *Syringa vulgaris*. Amorph. $[\alpha]_D^{30}$ −73.5° (c, 0.5 in CHCl$_3$).

Kikuchi, M. *et al, Yakugaka Zasshi*, 1987, **107**, 245.

Neovasinone N-60019

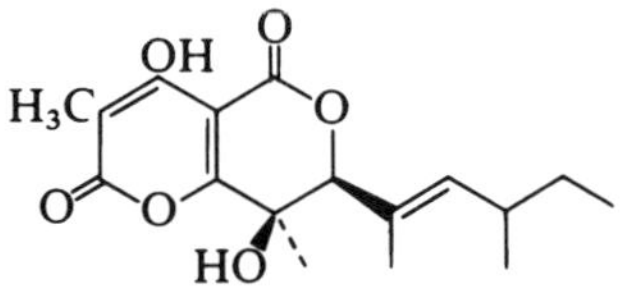

$C_{17}H_{22}O_6$ M 322.357

Metab. of *Neocosmospora vasinfecta*. Plates (EtOAc/hexane). Mp 193-195°. $[\alpha]_D^{25}$ −90° (c, 0.2 in MeOH).

Nakajima, H. *et al, Agric. Biol. Chem.*, 1987, **51**, 1221.

Nepetalactone N-60020

Updated Entry replacing N-00529

5,6,7,7a-Tetrahydro-4,7-dimethylcyclopenta[c]pyran-1(4aH)-one, 9CI

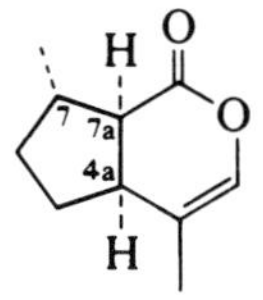

(4aS,7S,7aR)-*form*

$C_{10}H_{14}O_2$ M 166.219

(4aS,7S,7aR)-form [21651-62-7]

Constit. of *Nepeta cataria*. Oil. $[\alpha]_D^{21}$ +37° (c, 27 in CHCl$_3$).

(4aS,7S,7aS)-form [17257-15-7]

Constit. of *N. cataria*. Oil. $[\alpha]_D^{21}$ −24.4° (c, 6.2 in CHCl$_3$).

(4aR,7S,7aS)-form [21651-53-6]

Constit. of *N. mussini*. Oil. Bp$_{0.1}$ 60°. $[\alpha]_D^{23}$ +81° (c, 0.04 in CHCl$_3$).

(4aR,7R,7aS)-form [105660-81-9]

Constit. of *N. elliptica*. Oil. $[\alpha]_D^{25}$ −222° (c, 0.84 in CHCl$_3$).

(4aR,7S,7aR)-form

Constit. of *Nepeta nuda*.

Achmad, S.A. *et al*, *Proc. Chem. Soc.*, London, 1963, 166 (*synth*)

Eisenbraun, E.J. *et al*, *J. Org. Chem.*, 1980, **45**, 3811 (*struct, bibl*)

Bellesia, F. *et al*, *Phytochemistry*, 1984, **23**, 83 (*biosynth*)

Bottini, A.T. *et al*, *Phytochemistry*, 1987, **26**, 1200 (*isol*)

De Pooter, H.L. *et al*, *Phytochemistry*, 1987, **26**, 2311 (*isol*)

Nimbidiol   N-60021

C$_{17}$H$_{22}$O$_3$ M 274.359

Constit. of *Azadirachta indica*. Cryst. (EtOAc/C$_6$H$_6$/pet. ether). Mp 226°. $[\alpha]_D$ +3.4° (CHCl$_3$).

Majumder, P.L. *et al*, *Phytochemistry*, 1987, **26**, 3021.

Nimolicinoic acid   N-60022

C$_{26}$H$_{34}$O$_6$ M 442.551

Constit. of *Azadiracta indica*. Cryst. (CHCl$_3$). Mp 92-94°. $[\alpha]_D$ −14.28° (c, 0.07 in CHCl$_3$).

Siddiqui, S. *et al*, *J. Chem. Soc.*, *Perkin Trans. 1*, 1987, 1429.

1-Nitro-1,3-butadiene   N-60023

[927-66-2]

$$H_2C{=}CHCH{=}CHNO_2$$

C$_4$H$_5$NO$_2$ M 99.089

(E)-form

Liq. Bp$_{0.5}$ 40°. Stable at −20°, acid and base sensitive.

Bloom, A.J. *et al*, *Tetrahedron Lett.*, 1986, **27**, 873 (*synth, ir, pmr*)

Nitrocyclobutane, 9CI   N-60024

Updated Entry replacing N-10054

[2625-41-4]

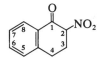

C$_4$H$_7$NO$_2$ M 101.105

d^{20} 1.096. Bp$_{40}$ 78-79°. pK_a 9.53. n$_D^{20}$ 1.4429.

Iffland, I.O.C. *et al*, *J. Am. Chem. Soc.*, 1953, **75**, 4044 (*synth*)

Hawthorne, M.F., *J. Am. Chem. Soc.*, 1957, **79**, 2510 (*synth*)

Amrollah-Madjdabadi, A. *et al*, *Synthesis*, 1986, 826 (*synth, ir, pmr*)

Nitrocycloheptane, 9CI   N-60025

Updated Entry replacing N-10055

[2562-40-5]

C$_7$H$_{13}$NO$_2$ M 143.185

d^{20} 1.063. Bp$_{15}$ 101°. n$_D^{20}$ 1.4721.

Emmons, W.D. *et al*, *J. Am. Chem. Soc.*, 1955, **77**, 4557 (*synth*)

Kornblum, N. *et al*, *J. Am. Chem. Soc.*, 1956, **78**, 1497 (*synth*)

Corey, E.J. *et al*, *Tetrahedron Lett.*, 1980, **21**, 1117 (*synth*)

Amrollah-Madjdabadi, A. *et al*, *Synthesis*, 1986, 826 (*synth, ir, pmr*)

2-Nitro-3,4-dihydro-1(2H)naphthalenone, N-60026
9CI

2-Nitro-1-oxo-1,2,3,4-tetrahydronaphthalene. 2-Nitro-1-tetralone

C$_{10}$H$_9$NO$_3$ M 191.186

(±)-form

Cryst. (Et$_2$O/hexane). Mp 70-71°. A soln. in CHCl$_3$ contains ~2% of the enol form.

Feuer, H. *et al*, *J. Org. Chem.*, 1966, **31**, 3152 (*synth, uv, ir, pmr*)

Benkert, E. *et al*, *Helv. Chim. Acta*, 1987, **70**, 2166 (*synth, pmr, cmr*)

4-Nitroheptane, 9CI   N-60027

[2625-37-8]

$$H_3CCH_2CH_2CH(NO_2)CH_2CH_2CH_3$$

C$_7$H$_{15}$NO$_2$ M 145.201

Liq. Bp$_3$ 58-60°, Bp$_{0.25}$ 44-46°.

Emmons, W.D. *et al*, *J. Am. Chem. Soc.*, 1955, **77**, 4557 (*synth, bibl*)

Turner, M.J. *et al*, *J. Med. Chem.*, 1986, **29**, 2439 (*synth, ir, pmr*)

5-Nitrohistidine, 9CI N-60028

α-Amino-5-nitro-1H-imidazole-4-propanoic acid

$C_6H_8N_4O_4$ M 200.154

(S)-form [41934-74-1]

L-form

Cryst. + 1H$_2$O. $[\alpha]_D^{25}$ +26.8° (1.5M HCl). Dec. at 197-198°.

B,HCl: [65092-35-5]. Cryst. (MeOH). Mp 253°.

N$^\alpha$-Ac: [41367-00-4].

  $C_8H_{10}N_4O_5$ M 242.191

  Cryst. (H$_2$O). Mp 237-238.5°. $[\alpha]_D^{25}$ −35.9° (1.5M HCl).

Me ester; B,HCl: [62013-43-8]. Cryst. + MeOH (MeOH/Et$_2$O). Mp 198-200° dec.

Me ester, N$^\alpha$-Ac: [41429-88-3].

  $C_9H_{12}N_4O_5$ M 256.218

  Cryst. (H$_2$O). Mp 202-205°.

Tautz, W. *et al, J. Med. Chem.*, 1973, **16**, 705 (*synth, props*)
Kelley, J.L. *et al, J. Med. Chem.*, 1977, **20**, 506.
Nagarajan, K. *et al, Indian J. Chem., Sect. B*, 1977, **15**, 629 (*synth*)
Giralt, E. *et al, An. Quim.*, 1979, **75**, 331 (*synth*)
Solans, X. *et al, Acta Crystallogr., Sect. B*, 1981, **37**, 2111 (*cryst struct*)
Pedroso, E. *et al, J. Heterocycl. Chem.*, 1986, **23**, 921 (*tautom*)

1-Nitroisoquinoline, 9CI, 8CI N-60029

[19658-76-5]

$C_9H_6N_2O_2$ M 174.159

Pale-yellow cryst. (C$_6$H$_6$/pet. ether). Mp 65-66°.

Hayashi, B. *et al, Yakugaku Zasshi*, 1967, **87**, 1342; *CA*, **69**, 2847 (*synth*)
Taylor, E.C. *et al, J. Org. Chem.*, 1982, **47**, 552 (*synth*)

4-Nitroisoquinoline N-60030

[36073-93-5]

$C_9H_6N_2O_2$ M 174.159

Cryst. (EtOH). Mp 64.5°.

Bryson, A., *J. Am. Chem. Soc.*, 1960, **82**, 4871 (*synth*)
Ochiai, E. *et al, Chem. Pharm. Bull.*, 1960, **8**, 24 (*synth*)
Bunting, J.W. *et al, Org. Prep. Proced. Int.*, 1972, **4**, 9 (*synth*)

5-Nitroisoquinoline N-60031

[607-32-9]

$C_9H_6N_2O_2$ M 174.159

Long yellow needles (EtOH). Mp 110°.

2-Oxide: [57554-78-6].

  $C_9H_6N_2O_3$ M 190.158

  Yellow needles (Me$_2$CO). Mp 220-222°.

Ochiai, E. *et al, J. Pharm. Soc. Jpn.*, 1953, **73**, 666; *CA*, **48**, 7014b.
Dewar, M.J.S. *et al, J. Chem. Soc.*, 1957, 2521 (*synth*)
Henry, R.A. *et al, J. Org. Chem.*, 1972, **37**, 3206 (*synth, pmr*)
Potts, K.T. *et al, J. Org. Chem.*, 1986, **51**, 2011 (*synth*)

6-Nitroisoquinoline N-60032

[70538-57-7]

$C_9H_6N_2O_2$ M 174.159

Mp 122.5-123°.

Miller, R.B. *et al, J. Org. Chem.*, 1980, **45**, 5312 (*synth*)

7-Nitroisoquinoline N-60033

[13058-73-6]

$C_9H_6N_2O_2$ M 174.159

B,HCl: Mp 177-178°.

Potter, M.D. *et al, J. Chem. Soc.*, 1953, 1320 (*synth*)
Miller, R.B. *et al, J. Org. Chem.*, 1980, **45**, 5312 (*synth*)
Saczewski, F., *Synthesis*, 1984, 170 (*synth*)

8-Nitroisoquinoline N-60034

$C_9H_6N_2O_2$ M 174.159

Yellow needles (EtOH aq.). Mp 87-88°.

Picrate: Mp 183-184°.

2-Oxide: [65464-36-0].

  $C_9H_6N_2O_3$ M 190.158

  Cryst. (Me$_2$CO). Mp 188-189°.

Dewar, M.J.S. *et al, J. Chem. Soc.*, 1957, 2521 (*synth*)
Popp, F.D. *et al, J. Org. Chem.*, 1961, **26**, 956 (*synth*)
Hamana, M. *et al, Heterocycles*, 1977, **8**, 403 (*synth*)

6-Nitro-1,4-naphthoquinone N-60035

6-Nitro-1,4-naphthalenedione

[58200-82-1]

$C_{10}H_5NO_4$ M 203.154

Cryst. Mp 147-148°.

Babu Rao, K. *et al, CA*, 1961, **55**, 22690i (*synth*)
Cameron, D.W. *et al, Aust. J. Chem.*, 1979, **32**, 575; 1980, **33**, 1805 (*synth, uv, ir, pmr*)

7-Nitro-1,2-naphthoquinone N-60036

7-Nitro-1,2-naphthalenedione

[18398-35-1]

$C_{10}H_5NO_4$ M 203.154

Needles (EtOH). Mp >230°.

Babu, B.H. *et al, J. Indian Chem. Soc.*, 1934, **11**, 411 (*synth*)

3-Nitropropanal, 9CI N-60037

[58657-26-4]

$$O_2NCH_2CH_2CHO$$

$C_3H_5NO_3$ M 103.077

Bp$_{0.05}$ ca. 60°. Stable at 0°.

▷Potentially explosive

Di-Me acetal: [72447-81-5]. *1,1-Dimethoxy-3-nitropropane.*

  $C_5H_{11}NO_4$ M 149.146

  Bp$_{15}$ 96°, Bp$_{0.05}$ 40-50°.

Di-Et acetal: [107833-73-8]. *1,1-Diethoxy-3-nitropropane.*

  $C_7H_{15}NO_4$ M 177.200

  Pale-yellow liq. Bp$_{0.05}$ 45-54°.

Öhrlein, R. *et al, Synthesis*, 1986, 535 (*synth, ir, pmr*)

2-Nitro-1,3-propanediol, 9CI   N-60038

Updated Entry replacing N-30059

Bis(hydroxymethyl)nitromethane

[1794-90-7]

$$O_2NCH(CH_2OH)_2$$

$C_3H_7NO_4$ M 121.093

Cryst. Mp 53-55°.

▷TZ0815000.

Seebach, D. *et al, Helv. Chim. Acta*, 1984, **67**, 261 (*synth, pmr*)
Amrollah-Madjdabadi, A. *et al, Synthesis*, 1986, 826 (*synth, ir, pmr*)
Ndibwami, A. *et al, Can. J. Chem.*, 1986, **64**, 1788 (*synth*)

3-Nitro-1-propanol, 9CI, 8CI   N-60039

Updated Entry replacing N-01341

[25182-84-7]

$$O_2NCH_2CH_2CH_2OH$$

$C_3H_7NO_3$ M 105.093

Found in *Astragalus* spp. Thick oil with weakly stinging odour and taste. Sol. H_2O, EtOH, Et_2O. d^{13} 1.173. $Bp_{0.8}$ 75°.

▷UB8752200.

Ac: [21461-49-4].
    $C_5H_9NO_4$ M 147.130
    Liq. $Bp_{2.5}$ 93°.

1-O-β-D-Glucoside: see Miserotoxin, M-03978

tert-*Butyl ether:*
    $C_7H_{14}NO_3$ M 160.193
    Oil. $Bp_{0.01}$ 26-30°.

Henry, L., *Chem. Zentralbl.*, 1897, **2**, 337 (*synth*)
Zee-Chung, K.-Y. *et al, J. Med. Chem.*, 1969, **12**, 157 (*synth, ir*)
Kim, H.K. *et al, J. Med. Chem.*, 1971, 301 (*synth, ir, pmr*)
Williams, M.C. *et al, Phytochemistry*, 1975, **14**, 2306 (*isol*)
Bordwell, F.G. *et al, J. Org. Chem.*, 1978, **43**, 3101 (*synth, pmr*)
Öhrlein, R. *et al, Synthesis*, 1986, 535 (*synth, deriv, ir, pmr*)

1-Nitropyrene, 9CI   N-60040

[5522-43-0]

$C_{16}H_9NO_2$ M 247.253

Yellow needles (MeCN). Mp 151-152°.

▷UR2480000.

Paputa-Peck, M.C. *et al, Anal. Chem.*, 1983, **55**, 1946 (*synth, uv, pmr, glc*)
van den Braken-van Leersum, A.M. *et al, Recl. Trav. Chim. Pays-Bas*, 1987, **106**, 120 (*synth, uv, pmr, cmr, ms*)

2-Nitropyrene, 9CI   N-60041

[789-07-1]

$C_{16}H_9NO_2$ M 247.253

Yellow needles (Me_3CN). Mp 198.5-199.5°.

Bolton, R., *J. Chem. Soc.*, 1964, 4637 (*synth*)
Paputa-Peck, M.C. *et al, Anal. Chem.*, 1983, **55**, 1946 (*synth, uv, pmr, glc*)
van den Braken-van Leersum, A.M. *et al, Recl. Trav. Chim. Pays-Bas*, 1987, **106**, 120 (*synth, uv, pmr, cmr, ms*)

4-Nitropyrene, 9CI   N-60042

[57835-92-4]

$C_{16}H_9NO_2$ M 247.253

Orange needles (Me_2CO/MeOH). Mp 196-197.5° (185-186°).

Bavin, P.M.G., *Can. J. Chem.*, 1959, **37**, 1614 (*synth*)
Paputa-Peck, M.C. *et al, Anal. Chem.*, 1983, **55**, 1946 (*synth, uv, pmr, glc*)
van der Braken-van Leersum, A.M. *et al, Recl. Trav. Chim. Pays-Bas*, 1987, **106**, 120 (*synth, uv, pmr, cmr, ms*)

2-Nitropyrimidine   N-60043

[79917-54-7]

$C_4H_3N_3O_2$ M 125.087

Mp 57-58°.

Taylor, E.C. *et al, J. Org. Chem.*, 1982, **47**, 552 (*synth, pmr*)

3-Nitropyrrole   N-60044

Updated Entry replacing N-01367

[5930-94-9]

$C_4H_4N_2O_2$ M 112.088

Formerly confused with the 2-isomer. Cryst. (pet. ether). Mp 99-101°.

Morgan, K.J. *et al, Tetrahedron*, 1966, **22**, 57.
Lippmaa, E. *et al, Org. Magn. Reson.*, 1972, **4**, 153 (*synth, cmr, N nmr*)
Anderson, H.J. *et al, Can. J. Chem.*, 1985, **63**, 896 (*synth*)

Nitrosin   N-60045

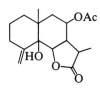

$C_{17}H_{24}O_5$ M 308.374

Constit. of *Artemesia nitrosa*. Cryst. (Et_2O/hexane). Mp 187-188°. $[\alpha]_D^{19}$ +37.7° (c, 1.2 in EtOH).

Adekenov, S.M. *et al, Khim. Prir. Soedin.*, 1986, **22**, 608.

4-Nitrosobenzaldehyde, 9CI   N-60046

Updated Entry replacing N-01396

[74663-99-3]

$C_7H_5NO_2$ M 135.122

Yellow needles (AcOH). Mp 137-138°. Steam-volatile. Readily polymerised.

Dimer: [105333-29-7]. Cryst. Mp 185-187°.

Alway, F.J., *Ber.*, 1903, **36**, 2303 (*synth*)
Lüttke, W., *Z. Elektrochem.*, 1957, **61**, 976 (*ir*)
Wan, P. *et al, Can. J. Chem.*, 1986, **64**, 2076 (*synth, ir, pmr*)

2-Nitro-3-(trifluoromethyl)phenol, 9CI   N-60047

*α,α,α-Trifluoro-2-nitro-*m-*cresol. 3-Hydroxy-2-nitrobenzotrifluoride*

[386-72-1]

$C_7H_4F_3NO_3$ M 207.109
Mp 71-73°. pK_a 5.6.

4-Nitrobenzoyl: Mp 130°.

Smith, M.A. *et al, J. Chem. Eng. Data*, 1961, **6**, 607 (*uv, ir, props*)
Balasubramanian, A. *et al, Can. J. Chem.*, 1966, **44**, 961 (*uv*)

2-Nitro-4-(trifluoromethyl)phenol, 9CI   N-60048

*α,α,α-Trifluoro-2-nitro-*p-*cresol, 8CI. 4-Hydroxy-3-nitrobenzotrifluoride*

[400-99-7]

$C_7H_4F_3NO_3$ M 207.109
Pale-yellow oil. d$_4^{20}$ 1.473. Bp$_{12}$ 92-94°, Bp$_5$ 79°. n$_D^{25}$ 1.5024.

▷GP2984950.

Me ether: [394-25-2]. *1-Methoxy-2-nitro-4-(trifluoromethyl)benzene, 9CI. 2-Nitro-4-(trifluoromethyl)anisole, 8CI. α,α,α-Trifluoro-3-nitro-4-methoxytoluene.*
$C_8H_6F_3NO_3$ M 221.136
Cryst. (pet. ether). Mp 46-47° (49°).
4-Methylbenzenesulfonyl: Mp 58-59°.
4-Nitrobenzoyl: Mp 98-99°.

Yaguipolskii, L.M. *et al, Ukr. Khim. Zh.*, 1955, **21**, 81; *CA*, **49**, 8867 (*synth*)
Cox, R. *et al, J. Mol. Spectrosc.*, 1970, **33**, 172 (*deriv, pmr, F nmr*)
Jacobs, R.L. *et al, J. Org. Chem.*, 1971, **36**, 242 (*synth, bibl*)
Lavagnino, E.R. *et al, Org. Prep. Proceed. Int.*, 1977, **9**, 96 (*synth, pmr, ir*)
Aldrich Library of FT-IR Spectra, **1**, 1362B (*ir*)
Aldrich Library of NMR Spectra, **2**, 1165C (*pmr*)

2-Nitro-5-(trifluoromethyl)phenol   N-60049

*α,α,α-Trifluoro-6-nitro-*m-*cresol. 3-Hydroxy-4-nitrobenzotrifluoride*

$C_7H_4F_3NO_3$ M 207.109
Liq. Bp$_{34}$ 108-111°.

4-Nitrobenzoyl: Solid (EtOH). Mp 106° (103°).

Belcher, R. *et al, J. Chem. Soc.*, 1954, 3846 (*synth*)

2-Nitro-6-(trifluoromethyl)phenol, 9CI   N-60050

*α,α,α-Trifluoro-6-nitro-*o-*cresol, 8CI. 2-Hydroxy-3-nitrobenzotrifluoride*

[1548-62-5]

$C_7H_4F_3NO_3$ M 207.109
Yellow cryst. (EtOH aq.). Mp 71.2-72.0°.

Filler, R. *et al, J. Org. Chem.*, 1962, **27**, 4660 (*synth*)

3-Nitro-4-(trifluoromethyl)phenol   N-60051

*α,α,α-Trifluoro-3-nitro-*p-*cresol, 8CI. 4-Hydroxy-2-nitrobenzotrifluoride*

[25889-36-5]

$C_7H_4F_3NO_3$ M 207.109
Cox, R., *J. Mol. Spectrosc.*, 1970, **33**, 172 (*pmr, F nmr*)

3-Nitro-5-(trifluoromethyl)phenol   N-60052

*α,α,α-Trifluoro-5-nitro-*m-*cresol. 3-Nitro-5-hydroxybenzotrifluoride*

$C_7H_4F_3NO_3$ M 207.109
Pale-yellow plates (C_6H_6/pet. ether). Mp 92°.

Me ether: 1-Methoxy-3-nitro-5-(trifluoromethyl)-benzene. 3-Nitro-5-(trifluoromethyl)anisole.
$C_8H_6F_3NO_3$ M 221.136
Cryst. (pet. ether). Mp 37.5°.

Whalley, W.B., *J. Chem. Soc.*, 1949, 3016 (*synth*)
Benkeser, R.A. *et al, J. Am. Chem. Soc.*, 1952, **74**, 3011 (*synth*)
Gitis, S.S. *et al, Zh. Obshch. Khim.*, 1964, **34**, 2250 (*Engl. transl.* p. 2262) (*deriv*)

4-Nitro-2-(trifluoromethyl)phenol, 9CI   N-60053

*α,α,α-Trifluoro-4-nitro-*o-*cresol, 8CI. 2-Hydroxy-5-nitrobenzotrifluoride*

[1548-61-4]

$C_7H_4F_3NO_3$ M 207.109
Solid (C_6H_6/pet. ether). Mp 105-106°, Mp 134-135°.

Me ether: [654-76-2]. *1-Methoxy-4-nitro-2-(trifluoromethyl)benzene, 9CI. 4-Nitro-2-(trifluoromethyl)anisole, 8CI. 2-Methoxy-5-nitrobenzotrifluoride.*
$C_8H_6F_3NO_3$ M 221.136
Solid (EtOH). Mp 79-79.5°.

Filler, R. *et al, J. Org. Chem.*, 1962, **27**, 4660 (*synth, deriv*)
Blakitnyi, A.N. *et al, Zh. Org. Khim.*, 1977, **13**, 2149 (*Engl. transl.* p. 2000) (*synth*)

4-Nitro-3-(trifluoromethyl)phenol, 9CI   N-60054

*α,α,α-Trifluoro-4-nitro-*m-*cresol, 8CI. 2-Nitro-5-hydroxybenzotrifluoride*

[88-30-2]

$C_7H_4F_3NO_3$ M 207.109
Lampricide. Solid (C_6H_6/pet. ether). Mp 75-76°. Bp$_{0.01}$ 135-138°. pK_a 6.07.

▷GP3520000.

Me ether: [344-39-8]. *4-Methoxy-1-nitro-2-(trifluoromethyl)benzene, 9CI. 4-Nitro-3-(trifluoromethyl)anisole, 8CI.*
$C_8H_6F_3NO_3$ M 221.136
Prisms (pet. ether). Mp 39°.

Smith, M.A. *et al, Anal. Chem.*, 1960, **32**, 1670 (*anal*)
Smith, M.A. *et al, J. Chem. Eng. Data*, 1961, **6**, 607 (*uv, ir, props*)
Canadian Pat., 666 608, (*1963*); *CA*, **60**, 2826g (*synth*)

Nonachlorophenalenyl   N-60055
Perchlorophenalenyl
[106680-46-0]

$C_{13}Cl_9$ M 475.220

Thermally-stable free radical. Hexachloroantimonate salt
of the corresponding cation also described.

Haddon, R.C. *et al, J. Org. Chem.*, 1987, **52**, 711 (*synth, esr*)

Nonafluoromorpholine, 9CI, 8CI   N-60056
Perfluoromorpholine
[378-94-9]

C_4F_9NO M 249.036

Mild fluorinating agent. Liq. Bp 34.5°.

Banks, R.E. *et al, J. Chem. Soc.*, 1965, 6077 (*synth*)
Lee, J. *et al, Trans. Faraday Soc.*, 1967, **63**, 16 (*F nmr*)

Nonamethylcyclopentanol, 9CI   N-60057
[103457-83-6]

$C_{14}H_{28}O$ M 212.375

Mp 108°. Dehydrates readily.

Wehle, D. *et al, Chem. Ber.*, 1986, **119**, 3127 (*synth, pmr, cmr*)

1-Nonen-3-ol, 9CI   N-60058
Updated Entry replacing N-30071
Hexylvinylcarbinol
[21964-44-3]

$$H_3C(CH_2)_5CH(OH)CH{=}CH_2$$

$C_9H_{18}O$ M 142.241

Isol. from *Petasites japonicus*.

(±)-*form* [79605-61-1]
Bp 193-194°, Bp$_{20}$ 92.5°.

4-Nitrobenzoyl: Plates (EtOH). Mp 36-37°.

Dumont, W. *et al, Angew. Chem., Int. Ed. Engl.*, 1974, **13**, 804
 (*synth*)
Ueno, Y. *et al, Synthesis*, 1980, 1011 (*synth, pmr*)
Halazy, S. *et al, Tetrahedron Lett.*, 1981, 2135 (*synth*)
Martin, V.S. *et al, J. Am. Chem. Soc.*, 1981, **103**, 6237 (*resoln*)
Boughdady, N.M. *et al, Aust. J. Chem.*, 1987, **40**, 767 (*synth,
 pmr, ir*)

30-Norcyclopterospermol   N-60059
22-Methylene-30-nor-3β-cycloartanol

$C_{30}H_{50}O_2$ M 442.724

Constit. of *Pterospermum heyneanum*. Needles
(MeOH). Mp 144-145°. [α]$_D$ +62.0°.

3-Ketone: **30-Norcyclopterospermone**. *22-Methylene-30-
nor-3-cycloartanone*.
 $C_{30}H_{48}O_2$ M 440.708
 Cryst. (MeOH). Mp 68-69°. [α]$_D$ +35°.

Anjaneyulu, A.S.R. *et al, Phytochemistry*, 1987, **26**, 2805.

Nordinone   N-60060
[109872-63-1]

$C_{18}H_{24}O_5$ M 320.385

Metab. of *Monocillium nordinii*. Powder. Mp 130-134°.

4,5-Dihydroxy: [109872-64-2]. **Nordinonediol**.
 $C_{18}H_{24}O_7$ M 352.383
 Metab. of *M. nordinii*. Cryst. Mp 194-196°.
4,5-Didehydro: [75207-14-6]. **Monocillin IV**.
 $C_{18}H_{22}O_5$ M 318.369
 Metab. of *M. nordinii*. Cryst. (Et$_2$O). Mp 139-140°.
4,5,8,9-Tetradehydro: [75207-15-7]. **Monocillin II**.
 $C_{18}H_{20}O_5$ M 316.353
 Metab. of *M. nordinii*. Cryst. (Et$_2$O). Mp 198-200°.
4R,5R-Epoxide: [75207-12-4]. **Monocillin V**.
 $C_{18}H_{22}O_6$ M 334.368
 Metab. of *M. nordinii*. Liq.
8,9-Didehydro, 4R,5R-epoxide: [75207-11-3]. **Monocil-
lin III**.
 $C_{18}H_{20}O_6$ M 332.352
 Metab. of *M. nordinii*. Cryst. (Et$_2$O). Mp 204-205°.
6,7,8,9-Tetradehydro, 4R,5R-epoxide: [75207-13-5].
 Monocillin I.
 $C_{18}H_{18}O_6$ M 330.337
 Metab. of *M. nordinii*. Oil.

Ayer, W.A. *et al, Can. J. Microbiol.*, 1980, **26**, 766 (*isol, struct*)
Ayer, W.A. *et al, Phytochemistry*, 1987, **26**, 1353 (*isol, struct*)

Norjuslimdiolone N-60061

ent-*15,16-Dihydroxy-19-nor-4-rosen-3-one*

[42715-03-7]

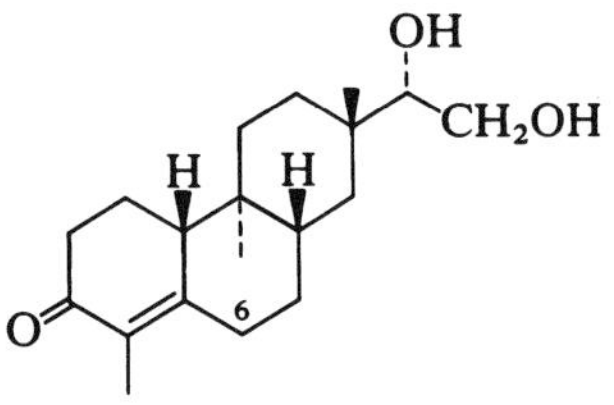

$C_{19}H_{30}O_3$ M 306.444

Constit. of *Hymenothrix wislizenii* and *Croptilon
divaricatum*. Oil.

6β-Hydroxy: [111576-39-7]. ent-*6α,15,16-Trihydroxy-
19-nor-4-rosen-3-one. 6β-Hydroxynorjuslimdiolone.*
$C_{19}H_{30}O_4$ M 322.444
From *H. wislizenii*. Oil.

Dominguez, X.A. *et al, Rev. Latinoamer. Quim.*, 1973, **3**, 177
(*isol*)
Jakupovic, J. *et al, Phytochemistry*, 1987, **26**, 2543 (*isol*)

Nuatigenin N-60062

Updated Entry replacing N-20091

22S,25S-Epoxy-5-furostene-3β,26-diol

[6811-35-4]

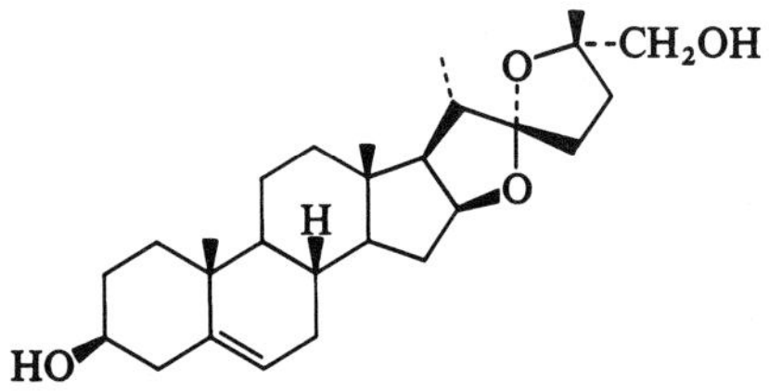

$C_{27}H_{42}O_4$ M 430.626

Saponin from *Solanum sisymbriifolium*. Cryst. Mp 227-
228°. $[\alpha]_D^{20}$ −86° (c, 0.1 in CHCl₃).

*3-O-[α-L-Rhamnosyl-(1→4)-[β-D-glycopyranosyl-
(1→2)]-β-D-glucopyranoside], 26-O-β-D-
glucopyranoside:* [24915-65-9]. **Avenacoside A**.
$C_{51}H_{82}O_{23}$ M 1063.195
Constit. of *Avena sativa*. $[\alpha]_D^{24}$ +52° (c, 1 in H₂O).

*3-O-[α-L-Rhamnosyl-(1→4)-[β-D-glucopyranosyl-
(1→3)-β-D-glucopyranosyl-(1→3)-β-D-
glucopyranosyl-(1→2)-β-D-glucopyranoside], 26-O-
β-D-glucopyranoside:* [35920-91-3]. **Avenacoside B**.
$C_{57}H_{92}O_{28}$ M 1225.337
Constit. of *A. sativa*. $[\alpha]_D^{24}$ +52° (c, 1 in H₂O).

*3-O-[α-L-Rhamnopyranosyl-(1→2)-α-L-
rhamnopyranosyl-(1→4)-β-D-glucopyranoside, 26-O-
β-D-glucopyranoside:* **Aculeatiside A**.
$C_{51}H_{82}O_{22}$ M 1047.196
Constit. of *S. aculeatissimum*. Needles (MeOH aq.).
Mp 196-204° dec. $[\alpha]_D^{22}$ −96.7° (c, 1.08 in Py).

*3-O-[α-L-Rhamnopyranosyl-(1→2)-β-D-
glucopyranosyl-(1→3)-β-D-galactopyranoside], 26-
O-β-D-glucopyranoside:* **Aculeatiside B**.
$C_{51}H_{82}O_{23}$ M 1063.195
Constit. of *S. aculeatissimum*. Amorph. powder. $[\alpha]_D^{21}$
−82° (c, 1 in Py).

3-O-α-L-Rhamnopyranosyl(1→2)-β-D-glucopyranoside:
$C_{39}H_{62}O_{13}$ M 738.911
Constit. of *Allium vineale*.

Tschesche, R. *et al, Chem. Ber.*, 1969, **102**, 2072; 1971, **104**,
3549; 1978, **111**, 3300 (*isol*)
Saijo, R. *et al, Phytochemistry*, 1983, **22**, 733 (*isol*)
Chen, S. *et al, Tetrahedron Lett.*, 1987, **28**, 5603 (*deriv*)

O

Oblongolide O-60001

Updated Entry replacing O-40001

[97344-05-3]

Absolute
configuration

$C_{14}H_{20}O_2$ M 220.311

Metab. of *Phomopsis oblonga*. Cryst. (pet. ether). Mp 105-106°.

Begley, M.J. *et al*, *J. Chem. Soc.*, *Perkin Trans. 1*, 1985, 861 (*isol*)
Shing, T.K.M., *J. Chem. Soc.*, *Chem. Commun.*, 1986, 49 (*synth, abs config*)

Obscuronatin O-60002

Updated Entry replacing O-50003

[74176-06-0]

Absolute
configuration

$C_{20}H_{34}O$ M 290.488

Constit. of coral *Xenia obscuronata* and brown alga *Pachydictyon coriaceum*. Oil. $[\alpha]_D$ −112° (c, 0.49 in $CHCl_3$).

Kashman, Y. *et al*, *J. Org. Chem.*, 1980, **45**, 3814 (*struct*)
Kodama, M. *et al*, *Tetrahedron Lett.*, 1984, **25**, 5781 (*synth, rel config*)
Ishitsuka, M. *et al*, *Tetrahedron Lett.*, 1986, **27**, 2639 (*abs config*)

Ochromycinone O-60003

Updated Entry replacing O-50005

3,4-Dihydro-8-hydroxy-3-methylbenz[a]*anthracene-1,7,12(2H)-trione*, *9CI*

[28882-53-3]

$C_{19}H_{14}O_4$ M 306.317

Benzanthraquinone antibiotic.

(S)-form

Prod. by several *Streptomyces* spp. Active against gram-positive bacteria. Yellow cryst. Mp 152-153°. $[\alpha]_D^{25}$ +204.5° ($CHCl_3$). λ_{max} 265 (log ϵ 4.42), 405 nm (3.55).

Ac: Mp 175-176°.
Me ether: [85202-11-5]. **Ochromycinone methyl ether**. *X 14881C. Antibiotic X 14881C. 8-O-Methylochramycinone.*
$C_{20}H_{16}O_4$ M 320.344

Isol. from cultures of an actinomycete X-14881. Shows little biol. activity. Yellow prisms. Mp >235° dec.

(±)-form [111540-00-2]

Mp 168-169°.

Ac: [111540-01-3]. Mp 175-176°.

Bowie, J.H., *Tetrahedron Lett.*, 1967, 1449 (*nmr, ms*)
Maehr, H. *et al*, *J. Antibiot.*, 1982, **35**, 1627 (*struct, deriv*)
Guingant, A. *et al*, *Tetrahedron Lett.*, 1987, **28**, 3107 (*synth*)
Katsuura, K. *et al*, *Can. J. Chem.*, 1987, **65**, 124 (*synth*)

2,4-Octadecadien-1-ol O-60004

$$H_3C(CH_2)_{12}CH{=}CHCH{=}CHCH_2OH$$

$C_{18}H_{34}O$ M 266.466

(E,E)-form [93255-83-5]

Solid. Mp 52-54°.

Cardillo, G. *et al*, *Tetrahedron*, 1986, **42**, 917 (*synth, ir, pmr, cmr, ms*)

3-Octadecene-1,2-diol O-60005

$C_{18}H_{36}O_2$ M 284.481

(R,Z)-form

Cryst. (hexane). Mp 56-57°. $[\alpha]_D^{20}$ −7.5° (c, 2.0 in $CHCl_3$).

Dibenzoyl:
$C_{32}H_{44}O_4$ M 492.697
$[\alpha]_D^{20}$ +38.0° (c, 3.25 in $CHCl_3$).

(S,Z)-form

Platelets (hexane). Mp 56-57°. $[\alpha]_D^{20}$ +8.1° (c, 3.23 in $CHCl_3$).

1-Benzoyl:
$C_{25}H_{40}O_3$ M 388.589
$[\alpha]_D^{20}$ +11.6° (c, 2.2 in Py).
Dibenzoyl: $[\alpha]_D^{20}$ −38.2° (c, 3.25 in $CHCl_3$).

Mulzer, J. *et al*, *Tetrahedron*, 1986, **42**, 5961 (*synth, ir, pmr, cmr*)

9-Octadecene-1,12-diol O-60006

$C_{18}H_{36}O_2$ M 284.481

(R,E)-form [35732-93-5]

(+)-trans-*form*. Ricinelaidyl alcohol
Mp 51.2-52.4°. $[\alpha]_D^{27}$ +6.0° (c, 1.51 in MeOH).
12-Ac: Yellow oil. $Bp_{0.5}$ 178°.

(*R,Z*)-form [540-11-4]

(+)-cis-*form. Ricinoleyl alcohol*
Bp$_{0.5}$ 175°. [α]$_D^{27}$ +6.7° (c, 1.65 in MeOH).

12-Ac: [2581-27-3]. *Gyplure.*
  C$_{20}$H$_{38}$O$_3$ M 326.518
  Insect sex attractant. Oil. Bp$_{0.5}$ 182°. [α]$_D^{30}$ +7.4° (c, 1 in CHCl$_3$).

Di-Ac:
  C$_{22}$H$_{40}$O$_4$ M 368.556
  Liq. Bp$_{1.3}$ 180°. [α]$_D^{30}$ +8.7° (c, 1 in CHCl$_3$).

Jacobson, M. *et al, J. Org. Chem.*, 1962, **27**, 2523 (*synth*)
Applewhite, T.H. *et al, J. Org. Chem.*, 1967, **32**, 1173 (*synth, ord*)
Eiter, K. *et al, Justus Liebigs Ann. Chem.*, 1967, **709**, 29 (*synth*)

2-Octadecen-1-ol   O-60007

[22104-84-3]

$$H_3C(CH_2)_{14}CH{=}CHCH_2OH$$

C$_{18}$H$_{36}$O   M 268.482

(*E*)-form [41207-34-5]
Oil.

(*Z*)-form [2831-86-9]
Low-melting solid.

Borgini, A. *et al, J. Chem. Soc., Perkin Trans. 1*, 1986, 1339, 1345 (*synth, ir, pmr, cmr*)

Octaethyltetramethyl[26]porphyrin[3.3.3.3]   O-60008

C$_{48}$H$_{64}$N$_4$$^{\oplus\oplus}$ M 697.060 (ion)
Stretched porphyrin showing claimed 'superaromatic' character.

Dibromide: [105162-50-3].
  C$_{48}$H$_{64}$Br$_2$N$_4$ M 856.868
  λ_{max} 547 nm (ϵ 909600).

Gosmann, M. *et al, Angew. Chem., Int. Ed. Engl.*, 1986, **25**, 1100 (*synth, ms, pmr, uv*)

2,2,3,3,5,5,6,6-Octafluoromorpholine, 9CI   O-60009

[13580-54-6]

C$_4$HF$_8$NO   M 231.045
Liq.

Banks, R.E. *et al, J. Chem. Soc. (C)*, 1967, 427 (*synth, F nmr*)

Banks, R.E. *et al, J. Fluorine Chem.*, 1978, **11**, 563 (*synth*)

4,5,7,8,12,13,15,16-Octafluoro[2.2]-paracyclophane   O-60010

5,6,11,12,13,14,15,16-Tricyclo[8.2.2.2^{4,7}]hexadeca-4,6,10,12,13,15-hexaene, 9CI

[1785-64-4]

C$_{16}$H$_8$F$_8$ M 352.226
Cryst. (EtOH). Mp >180° subl.

Filler, R. *et al, Chem. Ind. (London)*, 1965, 767; *J. Am. Chem. Soc.*, 1969, **91**, 1862; *J. Org. Chem.*, 1987, **52**, 511 (*synth, uv, ir, pmr*)

Octahydro-7*H*-1-benzopyran-7-one, 9CI   O-60011

Hexahydrochroman-7-one

C$_9$H$_{14}$O$_2$ M 154.208

(*4aRS,8aRS*)-form [86379-88-6]
(±)-trans-*form*
Oil. Bp$_{2.5}$ 91-93°.

2,4-Dinitrophenylhydrazone: Yellow plates (EtOH). Mp 146-148°.

Forchiassin, M. *et al, J. Heterocycl. Chem.*, 1983, **20**, 493 (*synth, pmr*)

Octahydrobenzo[*b*]thiophene, 9CI   O-60012

7-Thiabicyclo[4.3.0]nonane. 1-Thiahydrindane. Octahydrothianaphthene

[5745-52-8]

C$_8$H$_{14}$S   M 142.259
Component of Middle Eastern kerosine.

(*3aRS,7aRS*)-form [19516-14-4]
(±)-cis-*form*
d^{20} 1.04. Mp −44°. Bp$_{20}$ 102-102.5°.

1,1-Dioxide:
  C$_8$H$_{14}$O$_2$S   M 174.257
  Cryst. (Et$_2$O). Mp 38.24°. Bp$_{1.4}$ 139-141°.

(*3aRS,7aSR*)-form
(±)-trans-*form*
d^{20} 1.02. Mp −11.5°. Bp$_{21}$ 102.7-103.7°.

1,1-Dioxide: Cryst. (Et$_2$O). Mp 92.8-93.5°.

Birch, S.F. *et al, Ind. Eng. Chem.*, 1955, **47**, 240 (*isol*)
Birch, S.F. *et al, J. Org. Chem.*, 1955, **20**, 1178 (*synth, ir*)

Octahydro-2,7-dihydroxy-5*H*,10*H*-dipyrrolo[1,2-*a*:1′,2′-*d*]pyrazine-5,10-dione, 9CI   O-60013

Octahydro-2,7-dihydroxypyrocoll.
Cyclo(hydroxyprolylhydroxyprolyl)

(2*R*,5a*R*,7*R*,10a*R*)-*form*

C$_{10}$H$_{14}$N$_2$O$_4$ M 226.232

(2*R*,5a*R*,7*R*,10a*R*)-form [53990-68-4]
 Cryst. (EtOH). Mp 220-221°. [α]$_D^{28}$ +65.5° (c, 1 in H$_2$O).
(2*R*,5a*R*,7*R*,10a*S*)-form [53990-67-3]
 Cryst. (EtOH/Et$_2$O). Mp 210-211°. [α]$_D^{28}$ −18.5° (c, 1 in H$_2$O).
(2*R*,5a*S*,7*R*,10a*S*)-form [52363-44-7]
 Mp 248-249°. [α]$_D^{19}$ −153.4° (c, 2.05 in H$_2$O).
(2*R*,5a*S*,7*S*,10a*R*)-form [54002-75-4]
 Mp 275-276°. [α]$_D$ 0°.

 Eguchi, C. *et al, Bull. Chem. Soc. Jpn.,* 1974, **47**, 2277 (*synth*)
 Shirota, F.N. *et al, J. Med. Chem.,* 1977, **20**, 1176 (*synth*)

Octahydro-5a,10a-dihydroxy-5*H*,10*H*-dipyrrolo[1,2-*a*:1′,2′-*d*]pyrazine-5,10-dione, 9CI   O-60014

Octahydro-5a,10a-dihydroxypyrocoll

(±)-*cis-form*

C$_{10}$H$_{14}$N$_2$O$_4$ M 226.232

(5a*RS*,10a*RS*)-form [40030-54-4]
 (±)-cis-*form*
 Mp 168-173°.
(5a*RS*,10a*SR*)-form [40030-53-3]
 trans-*form*
 Cryst. (MeOH). Mp 190-195° dec.

 Oehler, E. *et al, Chem. Ber.,* 1973, **106**, 165, 396 (*synth*)

1,2,3,4,4a,5,6,7-Octahydro-4a,5-dimethyl-2-(1-methylethenyl)naphthalene   O-60015

C$_{15}$H$_{24}$ M 204.355

(2*R*,4a*R*,5*R*)-form
 Constit. of *Helichrysum davyi.* Oil. [α]$_D^{24}$ −20° (c, 0.05 in CHCl$_3$).

 Jakupovic, J. *et al, Phytochemistry,* 1987, **26**, 1841.

Octahydro-5*H*,10*H*-dipyrrolo[1,2-*a*:1′,2′-*d*]pyrazine-5,10-dione, 9CI   O-60016

Cyclo(prolylprolyl). Prolylproline anhydride.
Octahydropyrocoll
[6708-06-1]

(5a*S*,10a*S*)-*form*

C$_{10}$H$_{14}$N$_2$O$_2$ M 194.233

(5a*R*,10a*R*)-form [53990-71-9]
 D-D-form
 Mp 141-143°. [α]$_D^{30}$ +147.5° (c, 1 in H$_2$O).
(5a*S*,10a*S*)-form [19943-27-2]
 L-L-form
 Mp 143-144°. [α]$_D^{30}$ −149.5° (c, 1 in H$_2$O).
(5a*R*,10a*S*)-form [53990-72-0]
 L-D-form
 Mp 193-195°. Opt. inactive (*meso-*).
(5a*RS*,10a*RS*)-form
 (±)-*form*
 Cryst. (EtOH). Mp 149-150°.

 Rothe, M. *et al, Angew. Chem., Int. Ed. Engl.,* 1972, **11**, 293 (*synth*)
 Eguchi, C. *et al, Bull. Chem. Soc. Jpn.,* 1974, **47**, 2277 (*synth*)
 Benedetti, E. *et al, Cryst. Struct. Commun.,* 1975, **4**, 641 (*cryst struct*)
 Kralj, B. *et al, Biomed. Mass Spectrom.,* 1975, **2**, 215 (*synth, ms*)
 Kricheldorf, H.R. *et al, Org. Magn. Reson.,* 1980, **13**, 52 (*synth, cmr, N nmr*)
 Ueda, T. *et al, Bull. Chem. Soc. Jpn.,* 1983, **56**, 568 (*synth*)
 Valentine, B. *et al, Int. J. Pept. Protein Res.,* 1985, **25**, 56 (^{17}O nmr)

Octahydro-2-hydroxy-5*H*,10*H*-dipyrrolo[1,2-*a*:1′,2′-*d*]pyrazine-5,10-dione, 9CI   O-60017

Hydroxyprolylproline anhydride.
Cyclo(hydroxyprolylprolyl)

(2*R*,5a*S*,10a*S*)-*form*

C$_{10}$H$_{14}$N$_2$O$_3$ M 210.232

(2*R*,5a*R*,10a*R*)-form
 Mp 174-175°. [α]$_D^{28}$ +90.8° (c, 1 in H$_2$O).
(2*R*,5a*S*,10a*S*)-form [36099-80-6]
 Isol. from rabbit skin tissue extract. Plant growth regulator. Mp 141-142°. [α]$_D^{28}$ −134.7° (c, 1 in H$_2$O).
(2*S*,5a*R*,10a*S*)-form [55903-97-4]
 Cryst. (MeOH/Et$_2$O). Mp 195-198°. [α]$_D$ −10.2° (c, 0.48 in DMF).

 Justova, V. *et al, Collect. Czech. Chem. Commun.,* 1975, **40**, 662 (*synth*)
 Anteunis, M.J.O. *et al, Bull. Soc. Chim. Belg.,* 1978, **87**, 41 (*pmr, conformn*)
 Ienaga, K. *et al, Tetrahedron Lett.,* 1987, **28**, 1285 (*isol, synth, props*)

1,2,3,4,5,6,7,8-Octahydropyridazino[4,5-d]pyridazine, 9CI   O-60018

3,4,8,9-Tetraazabicyclo[4.4.0]dec-1(6)-ene
[104835-53-2]

$C_6H_{12}N_4$ M 140.188
Oxygen-sensitive.

Dowd, P. *et al, J. Am. Chem. Soc.*, 1986, **108**, 7416 (*synth*)

Octahydro-4H-quinolizin-4-one, 9CI   O-60019

4-Oxoquinolizidine. 4-Quinolizidinone. Norlupinone
[491-40-7]

$C_9H_{15}NO$ M 153.224

(±)-*form*

Bp$_{4.5}$ 113-114°, Bp$_{0.05}$ 83°.

B,HCl: [38910-53-1]. Mp 144-145° (139-141°).

Goldberg, S.I. *et al, J. Org. Chem.*, 1970, **35**, 242 (*synth*)
Watson, T.R. *et al, Aust. J. Chem.*, 1970, **23**, 1057 (*synth, ms*)
Cahill, R. *et al, Org. Magn. Reson.*, 1972, **4**, 283; 1973, **5**, 295 (*pmr, config*)
Bohlmann, F. *et al, Chem. Ber.*, 1975, **108**, 1043 (*cmr*)
Perkowska, A. *et al, J. Mol. Struct.*, 1983, **101**, 147 (*cryst struct*)
Edwards, P.D. *et al, Tetrahedron Lett.*, 1984, **25**, 939 (*synth*)

Octahydro-1,2,5,6-tetrazocine   O-60020

Perhydro-1,2,5,6-tetrazocine. 1,2,5,6-Tetraazacyclooctane
[6577-17-9]

$C_4H_{12}N_4$ M 116.166
Cryst. (EtOH aq.). Mp 149-165° dec. Becomes dark-brown and oily in air at r.t.

B,2HCl: [59416-98-7]. Cryst. (conc. HCl). Mp 208-210° dec.

B,H₂SO₄: [59416-99-8]. Amorph. solid. Mp 205-210° dec.

Dipicrate: [59417-01-5]. Flat yellow prisms (EtOH aq.). Mp 177-180°, Mp 235-240° (double Mp).

Tetra-Ac: [59416-97-6].
$C_{12}H_{20}N_4O_4$ M 284.314
Cryst. (Me₂CO). Mp 192-193°.

Nielsen, A.T., *J. Heterocycl. Chem.*, 1976, **13**, 101 (*synth, ir, pmr*)

2,3,4,5,6,7,8,9-Octahydro-1H-triindene, 9CI   O-60021

Trindan, 8CI. Tricyclotrimethylenebenzene.
2,3,4,5,6,7,8,9-Octahydro-1H-cyclopent[e]-as-indacene.
Tristrimethylenebenzene
[1206-79-7]

$C_{15}H_{18}$ M 198.307
Shiny needles (MeOH or by subl.). Mp 96-96.5°. Bp 238-239°. Slowly subl. in air.

Wallach, O., *Ber.*, 1897, **30**, 1094 (*synth*)
Mayer, R., *Chem. Ber.*, 1956, **89**, 1443 (*synth*)
Boyko, E.R. *et al, Acta Cryst.*, 1964, **17**, 152 (*cryst struct*)
Illuminati, G. *et al, J. Chem. Soc.* (*B*), 1971, 2206 (*synth*)

2′,3,4′,5,5′,6,7,8-Octahydroxyflavone   O-60022

Updated Entry replacing O-40022
$C_{15}H_{10}O_{10}$ M 350.238

3,4′,6,8-Tetra-Me ether: 2′,5,5′,7-Tetrahydroxy-3,4′,6,8-tetramethoxyflavone.
$C_{19}H_{18}O_{10}$ M 406.345
Constit. of *Gutierrezia grandis.*

3,5′,6,8-Tetra-Me ether: 2′,4′,5,7-Tetrahydroxy-3,5′,6,8-tetramethoxyflavone.
$C_{19}H_{18}O_{10}$ M 406.345
From *G. grandis* and *G. microcephala.* Yellow cryst. (MeOH). Mp 198-199.5°.

2′,3,6,8-Tetra-Me ether:
$C_{19}H_{18}O_{10}$ M 406.345
From *Gymnosperma glutinosum.*

3,4′,5′,6,8-Penta-Me ether: 4′,5,5′,7-Tetrahydroxy-2′,3,6,8-tetramethoxyflavone.
$C_{20}H_{20}O_{10}$ M 420.372
Constit. of *Gutierrezia microcephala.* Yellow prisms (C₆H₆). Mp 184-185°.

2′,3,5′,6,8-Penta-Me ether: 4′,5,7-Trihydroxy-2′,3,5′,6,8-pentamethoxyflavone.
$C_{20}H_{20}O_{10}$ M 420.372
From *Gymnosperma glutinosum.*

2′,3,6,7,8-Penta-Me ether: 4′,5,5′-Trihydroxy-2′,3,6,7,8-pentamethoxyflavone.
$C_{20}H_{20}O_{10}$ M 420.372
From *Gymnosperma glutinosum.*

2′,3,4′,5′,6,8-Hexa-Me ether: 5,7-Dihydroxy-2′,3,4′,5′,6,8-hexamethoxyflavone.
$C_{21}H_{22}O_{10}$ M 434.399
From *Gymnospermum glutinosum.2′,5,7-Trihydroxy-3,4′,5′,6,8-pentamethoxyflavone*

Roitman, J.N. *et al, Phytochemistry*, 1985, **24**, 835.
Fang, N. *et al, Phytochemistry*, 1985, **24**, 2693.
Iinuma, M. *et al, Chem. Pharm. Bull.*, 1986, **34**, 2228 (*synth*)
Yu, S. *et al, Phytochemistry*, 1988, **27**, 171 (*isol*)

3,3′,4′,5,5′,6,7,8-Octahydroxyflavone   O-60023

Updated Entry replacing O-40024
$C_{15}H_{10}O_{10}$ M 350.238

3,6,8-Tri-Me ether: 3′,4′,5,5′,7-Pentahydroxy-3,6,8-trimethoxyflavone.
$C_{18}H_{16}O_{10}$ M 392.318

Constit. of *Gutierrezia grandis*.
3,3',5',6,8-Penta-Me ether: 4',5,7-*Trihydroxy-3,3',5',6,8-pentamethoxyflavone*.
$C_{20}H_{20}O_{10}$ M 420.372
From *G. grandis* and *G. microcephala*. Yellow needles (MeOH aq.). Mp 176-177°.
3,4',5',6,8-Penta-Me ether: 3',5,7-*Trihydroxy-3,4',5',6,8-pentamethoxyflavone*.
$C_{20}H_{20}O_{10}$ M 420.372
From *G. microcephala*. Yellow needles (MeOH). Mp 207-208°.
3,3',4',5',6,8-Hexa-Me ether: 5,7-*Dihydroxy-3,3',4',5',6,8-hexamethoxyflavone*.
$C_{21}H_{22}O_{10}$ M 434.399
From *G. grandis* and *G. microcephala*. Yellow needles (MeOH). Mp 178-180°.

Roitman, J.N. *et al, Phytochemistry*, 1985, **24**, 835.
Fang, N. *et al, Phytochemistry*, 1985, **24**, 2693.

2,2,3,3,4,4,5,5-Octamethylcyclopentanol, O-60024
9CI

[103547-87-1]

$C_{13}H_{26}O$ M 198.348
Solid. Mp 96-98°.

Fitjer, L. *et al, Chem. Ber.*, 1986, **119**, 1162 (*synth, pmr, cmr*)

Octamethylcyclopentanone, 9CI   O-60025
[92406-82-1]

$C_{13}H_{24}O$ M 196.332
Solid. Mp 191-193°.

Fitjer, L. *et al, Chem. Ber.*, 1986, **119**, 1162 (*synth, ir, pmr, cmr*)

Octamethylcyclopentene, 9CI   O-60026
[79816-98-1]

$C_{13}H_{24}$ M 180.333
Bp_3 62-71°.

Klein, H. *et al, Angew. Chem., Int. Ed. Engl.*, 1981, **20**, 1027 (*synth, cmr*)
Fitjer, L. *et al, Chem. Ber.*, 1986, **119**, 1162 (*synth, pmr*)

1,1,2,2,3,3,4,4-Octamethyl-5-methylenecy-   O-60027
clopentane

[103457-85-8]

$C_{14}H_{26}$ M 194.359

Solid. Mp 72-74°.

Wehle, D. *et al, Chem. Ber.*, 1986, **119**, 3127 (*synth, pmr, cmr, ms*)

4-Octene-1,7-diyne   O-60028
[57488-06-9]

$$HC{\equiv}CCH_2CH{=}CHCH_2C{\equiv}CH$$

C_8H_8 M 104.151
(**E**)-*form* [20060-07-5]
Large plates (pentane). Mp 15°. Bp_{25} 59-60°.
(**Z**)-*form* [100758-27-8]
Bp_{20} 50°.

Gaoni, Y. *et al, J. Am. Chem. Soc.*, 1968, **90**, 4940 (*synth, ir, pmr*)
Müller, P. *et al, Helv. Chim. Acta*, 1985, **68**, 975 (*synth, ir, pmr*)

2-Octen-1-ol, 9CI   O-60029
Updated Entry replacing O-00409
[22104-78-5]

$$H_3C(CH_2)_4CH{=}CHCH_2OH$$

$C_8H_{16}O$ M 128.214
(**E**)-*form* [18409-17-1]
Found in grapes, ripe bananas and mushroom volatiles.
Bp_{11} 87-89°.
Ac: [3913-80-2].
$C_{10}H_{18}O_2$ M 170.251
Bp_{13} 82-84°.
(**Z**)-*form* [26001-58-1]
Bp_{10} 85-87°.
Dinitrobenzoyl: Mp 60°.

Colonge, J., *Bull. Soc. Chim. Fr.*, 1955, 953 (*synth*)
Osbond, J., *J. Chem. Soc.*, 1961, 5270 (*synth*)
Schlosser, M. *et al, Synthesis*, 1972, 575 (*synth*)
Sood, R. *et al, Tetrahedron Lett.*, 1974, 423 (*synth*)
Pyysalo, H., *Acta Chem. Scand., Ser. B*, 1976, **30**, 235 (*isol*)
Tsuji, J. *et al, Bull. Chem. Soc. Jpn.*, 1976, **49**, 1701.
Tsuji, J. *et al, Tetrahedron Lett.*, 1979, 39 (*synth, spectra*)
Porter, N.A. *et al, J. Org. Chem.*, 1985, **50**, 2252 (*synth, pmr, cmr*)
Nicolaou, K.C. *et al, Synthesis*, 1986, 453 (*synth, ir, pmr*)
Binns, M.R. *et al, Aust. J. Chem.*, 1987, **40**, 281 (*synth*)

4-Octen-2-one   O-60030
[33665-27-9]

$$H_3CCH_2CH_2CH{=}CHCH_2COCH_3$$

$C_8H_{14}O$ M 126.198
(**E**)-*form* [37720-70-0]
Liq. Bp_{50} 55-58°.
2,4-Dinitrophenylhydrazone: Mp 71°.

Heilman, R. *et al, Bull. Soc. Chim. Fr.*, 1957, 119 (*synth, uv, ir*)

2-Octylamine
O-60031

Updated Entry replacing O-00433

2-Octanamine, 9CI. 2-Aminooctane

[693-16-3]

$$CH_3$$
$$H_2N{-}C{-}H$$
$$(CH_2)_5CH_3$$

(*R*)-*form*
Absolute
configuration

$C_8H_{19}N$ M 129.245

(***R***)-*form* [34566-05-7]

Bp$_{21}$ 65.5°. [M]$_D$ −8.93°.

(***S***)-*form* [34566-04-6]

Bp$_{25}$ 70°. Opt. rotn. affected markedly by solv. [M]$_D$ +8.62° (pure liq.).

B,HCl: Mp 90-91°. [M]$_D^{21.5}$ +6.75° (0.078 in H$_2$O).

N-Benzoyl:

$C_{15}H_{13}NO$ M 223.274

Mp 101-102°.

(±)-*form* [44855-57-4]

Bp 163-165°, Bp$_{13}$ 58-59°.

N-Benzoyl: Mp 77-78°.

Mann, F.G. *et al, J. Chem. Soc.*, 1944, 456; 1950, 3384 (*synth, props*)

Streitwieser, A. *et al, J. Am. Chem. Soc.*, 1956, **78**, 5597 (*abs config*)

Eggert, H. *et al, J. Am. Chem. Soc.*, 1973, **95**, 3710 (*cmr*)

Inoue, Y. *et al, Synthesis*, 1986, 332 (*synth*)

9(11),12-Oleanadien-3-one
O-60032

[80113-53-7]

$C_{30}H_{46}O$ M 422.693

Constit. of *Vellozia compacta.* Cryst. (hexane). Mp 206-208°. [α]$_D^{25}$ +373° (c, 1.07 in CHCl$_3$).

Barnes, R.A. *et al, Chem. Pharm. Bull.*, 1984, **32**, 3674.

13(18)-Oleanene-2,3-diol
O-60033

Updated Entry replacing O-00502

$C_{30}H_{50}O_2$ M 442.724

(*2α,3α*)-*form*

Hirsudiol

Constit. of *Cocculus hirsutus.* Needles (MeOH). Mp 238°. [α]$_D$ −25° (c, 0.25 in CHCl$_3$).

(*2β,3β*)-*form* [60828-09-3]

Constit. of *Salvia horminum.* Cryst. (EtOH). Mp 228°.

Ulubelen, A. *et al, Phytochemistry*, 1977, **16**, 790 (*isol*)

Ahmad, V.U. *et al, Phytochemistry*, 1987, **26**, 793 (*isol*)

12-Oleanene-1,2,3,11-tetrol
O-60034

$C_{30}H_{50}O_4$ M 474.723

(*1β,2α,3β,11α*)-*form*

3-Ac: 3β-Acetoxy-12-oleanene-1β,2α,11α-triol.

$C_{32}H_{52}O_5$ M 516.760

Constit. of *Salvia argentea.* Cryst. (EtOAc/hexane). Mp 236-238°. [α]$_D^{20}$ +39.3° (c, 0.178 in CHCl$_3$).

Bruno, M. *et al, Phytochemistry*, 1987, **26**, 497.

12-Oleanene-3,16,23,28-tetrol
O-60035

$C_{30}H_{50}O_4$ M 474.723

(*3β,16β*)-*form*

23-Hydroxylongispinogenin

Cryst. (MeOH/CHCl$_3$). Mp 253-255°. [α]$_D$ +46° (c, 0.5 in Py).

3-O-β-D-Galactopyranoside: Corchorusin C.

$C_{36}H_{60}O_9$ M 636.865

Constit. of *Corchorus acutangulus.* Cryst. (MeOH). Mp 220-222°. [α]$_D$ +25.6° (c, 0.6 in MeOH).

Mahato, S.B. *et al, J. Chem. Soc., Perkin Trans. 1*, 1987, 629.

12-Oleanene-1,3,11-triol
O-60036

$C_{30}H_{50}O_3$ M 458.723

(*1β,3β,11α*)-*form*

Constit. of *Maytenus horrida.* Gum.

3-Ac: Cryst. Mp 222-226°.

González, A.G. *et al, Phytochemistry*, 1987, **26**, 2785.

12-Oleanene-2,3,11-triol
O-60037

$C_{30}H_{50}O_3$ M 458.723

(*2α,3β,11α*)-*form*

3-Ac: 3β-Acetoxy-12-oleanene-2α,11α-diol.

$C_{32}H_{52}O_4$ M 500.760

Constit. of *Salvia argentea.* Cryst. (EtOAc/hexane). Mp 243-246°. [α]$_D^{20}$ +40° (c, 0.162 in CHCl$_3$).

Bruno, M. *et al, Phytochemistry*, 1987, **26**, 497.

12-Oleanene-3,16,28-triol
O-60038

Updated Entry replacing O-00518

$C_{30}H_{50}O_3$ M 458.723

(*3β,16α*)-*form* [465-95-2]

Primulagenin A

Constit. of *Primula officinalis, Aegiceras corniculatum* and *Jacquinia* spp. Cryst. (CHCl$_3$/MeOH). Mp 249.5-250°. [α]$_D$ +58° (c, 0.7 in CHCl$_3$).

(*3β,16β*)-*form* [465-94-1]

Longispinogenin

Constit. of *Lemaireocereus longispinus.* Cryst. (Me$_2$CO). Mp 247-249°. [α]$_D^{25}$ +51° (CHCl$_3$).

3-O-β-D-Galactopyranoside: Corchorusin A.

$C_{36}H_{60}O_8$ M 620.865

Constit. of *Corchorus acutangulus.* Cryst. (MeOH/CHCl$_3$). Mp 282-284°. [α]$_D$ +22.5° (c, 0.25 in MeOH).

Djerassi, C. *et al, J. Am. Chem. Soc.*, 1954, **76**, 4089 (*struct*)

Itô, S. *et al, Tetrahedron Lett.*, 1969, 2905 (*pmr*)

Allen, J. *et al, J. Chem. Soc., Perkin Trans. 1*, 1972, 2994 (*synth*)

Tori, K. *et al, Tetrahedron Lett.*, 1976, 4163 (*cmr*)

Baigert, D.R. *et al, Aust. J. Chem.*, 1978, **31**, 1375 (*isol*)

Mahato, S.B. *et al, J. Chem. Soc., Perkin Trans. 1*, 1987, 629 (*Corchorusin A*)

Oliveric acid   O-60039

Updated Entry replacing O-00541

5-Carboxydecahydro-β-5,8a-trimethyl-2-methylene-1-naphthalenepentanoic acid, 9CI. 13S-Labd-8(17)-ene-15,19-dioic acid

$C_{20}H_{32}O_4$ M 336.470

(−)-(*ent*-Labdane) form shown.

(+)-*form* [41787-69-3]

Constit. of *Brickellia glomerata*. Gum. $[\alpha]_D^{25}$ +30.5° (c, 0.59 in EtOH).

(−)-*form* [29455-26-3]

ent-*form*

Constit. of *Daniella oliveri*. Cryst. (MeOH). Mp 142°. $[\alpha]_D$ −2° (c, 1 in EtOH).

Haeuser, J. *et al*, *Tetrahedron*, 1970, **26**, 3461 (isol)
Calderón, J.S. *et al*, *Phytochemistry*, 1987, **26**, 2639 (isol)

Olmecol   O-60040

2-(4-Hydroxyphenyl)-3-methyl-5-benzofuranpropanol, 9CI. 2-(4-Hydroxyphenyl)-5-(3-hydroxypropyl)-3-methylbenzofuran

[109145-65-5]

$C_{18}H_{18}O_3$ M 282.338

Constit. of *Krameria cystisoides*. Cryst. Mp 161-162°.

Achenbach, H. *et al*, *Phytochemistry*, 1987, **26**, 1159.

N-Ornithyl-β-alanine   O-60041

$H_2N(CH_2)_3CH(NH_2)CONHCH_2CH_2COOH$

$C_8H_{17}N_3O_3$ M 203.241

(*S*)-*form*

L-*form*

B,HCl: [90970-63-1]. Hygroscopic cryst. + 0.8H₂O. $[\alpha]_D$ +3.3° (c, 1 in H₂O).

Tada, M. *et al*, *J. Agric. Food Chem.*, 1984, **32**, 992 (synth)
Huynh-ba, T. *et al*, *J. Agric. Food Chem.*, 1987, **35**, 165 (synth, cmr)

N-Ornithyltaurine   O-60042

2-[(2,5-Diamino-1-oxopentyl)amino]ethanesulfonic acid, 9CI

$H_2N(CH_2)_3CH(NH_2)CONHCH_2CH_2SO_3H$

$C_7H_{17}N_3O_4S$ M 239.289

(*S*)-*form*

L-*form*

B,HCl: [90970-64-2]. Cryst. + 1H₂O. Mp 125-130°. $[\alpha]_D^{25}$ +8.5° (c, 1 in H₂O).

N^α,N^δ-*Bis(benzyloxycarbonyl)*: [90990-60-6]. Cryst. + 2.5H₂O. Mp 85-88°. $[\alpha]_D^{25}$ −13.5° (c, 1 in H₂O).

Huynh-ba, T. *et al*, *J. Agric. Food Chem.*, 1987, **35**, 165 (synth)

Orotinichalcone   O-60043

$C_{26}H_{28}O_6$ M 436.504

Constit. of *Lonchocarpus orotinus*. Red needles (Me₂CO). Mp 166°.

Waterman, P.G. *et al*, *Phytochemistry*, 1987, **26**, 1189.

Orotinin   O-60044

$C_{25}H_{26}O_6$ M 422.477

Constit. of *Lonchocarpus orotinus*. Yellow needles (EtOAc/pet. ether). Mp 186°. $[\alpha]_D^{21}$ −220° (c, 1 in CHCl₃).

5-Me ether:

$C_{26}H_{28}O_6$ M 436.504

Constit. of *L. orotinus*.

Waterman, P.G. *et al*, *Phytochemistry*, 1987, **26**, 1189.

Orthopappolide   O-60045

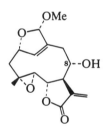

$C_{16}H_{20}O_6$ M 308.330

Parent compd. not yet isolated.

8-(Methylpropenoyl): Orthopappolide methacrylate.

$C_{20}H_{24}O_7$ M 376.405

Constit. of *Elephantopus angustifolius*. Cryst. Mp 85°.

8-(3-Methyl-2-butenoyl): Orthopappolide senecioate.

$C_{21}H_{26}O_7$ M 390.432

From *E. angustifolius*. Cryst. Mp 89°.

8-(2-Methyl-2E-butenoyl): Orthopappolide tiglate.

$C_{21}H_{26}O_7$ M 390.432

From *E. angustifolius*. Oil.

Jakupovic, J. *et al*, *Phytochemistry*, 1987, **26**, 1467.

Ovalene, 9CI, 8CI   O-60046
[190-26-1]

C₃₂H₁₄   M 398.463
Light-orange needles. Mp 475°. Semiconductor.

Clar, E., *Nature* (*London*), 1948, **161**, 238 (*synth*)
Clar, E. *et al*, *Chem. Ber.*, 1949, **82**, 46 (*synth*)
Lang, K.F. *et al*, *Chem. Ber.*, 1961, **94**, 1075 (*synth*)
Bursey, M.M. *et al*, *Org. Mass Spectrom. Suppl.*, 1970, **4**, 615 (*ms*)
Hazell, R.G. *et al*, *Z. Kristallogr., Kristallgeom., Kristallphys., Kristallchem.*, 1973, **137**, 159 (*cryst struct*)

7-Oxabicyclo[4.1.0]hept-3-ene   O-60047
4,5-Epoxycyclohexene
[6253-27-6]

C₆H₈O   M 96.129
Liq. Bp 152°, Bp₁₁ 39-40°.

Ali, M.E. *et al*, *J. Chem. Soc.*, 1958, 1066 (*synth*)
Davies, S.G. *et al*, *J. Chem. Soc., Perkin Trans. 1*, 1986, 1277 (*synth, pmr, ir*)

Oxalohydroxamic acid, 8CI   O-60048
N,N′-Dihydroxyethanediamide, 9CI. *N,N′-Dihydroxyoxamide. Oxalodihydroxamic acid. Oxalylhydroxamic acid. Dihydroxyglyoxime. Ethanedihydroxamic acid*
[1687-60-1]

$$HON{=}C(OH)C(OH){=}NOH \rightleftharpoons$$
$$HONHCOCONHOH$$

C₂H₄N₂O₄   M 120.065
Diketo tautomer predominates in cryst. state. Propellant. Forms metal complexes, used in anal. for metals. Cryst. (H₂O). V. spar. sol. cold H₂O. Mp 165° (deflagrates). pK_{a1} 6.68, pK_{a2} 8.49, pK_{a3} 10.97.
▷Salts are explosive

Lossen, W. *et al*, *Ber.*, 1894, **27**, 1105 (*synth*)
Ponzio, G., *Gazz. Chim. Ital.*, 1926, **56**, 709; *CA*, **21**, 1097 (*synth, haz*)
Przyborowski, L. *et al*, *CA*, 1975, **83**, 125675u (*synth, props*)
Lowe-Ma, C.K. *et al*, *Acta Crystallogr., Sect. C*, 1986, **42**, 1648 (*cryst struct*)

2-(1,3-Oxathian-2-yl)pyridine   O-60049
[82081-51-4]

C₉H₁₁NOS   M 181.252
(±)-**form**
Brown oil.

Aloup, J.-C. *et al*, *J. Med. Chem.*, 1987, **30**, 24 (*synth*)

[1,4]Oxathiino[3,2-*b*:5,6-*c′*]dipyridine, 9CI   O-60050
1,8-Diazaphenoxathiin. [*1,4*]*Oxathiino*[*3,2-c:5,6-b′*]-*dipyridine* (*incorr.*)

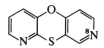

C₁₀H₆N₂OS   M 202.230
Cryst. (Et₂O). Mp 146-147°.
8-Oxide:
  C₁₀H₆N₂O₂S   M 218.230
  Mp 91-91.5°.

Lindsy, C.M. *et al*, *J. Heterocycl. Chem.*, 1987, **24**, 211 (*synth, pmr, cmr, ms*)

2-Oxatricyclo[3.3.1.1³,⁷]decane, 9CI   O-60051
Updated Entry replacing O-00721
2-Oxaadamantane
[281-24-3]

C₉H₁₄O   M 138.209
Mp 232.5°.

Averina, N.V. *et al*, *J. Chem. Soc., Chem. Commun.*, 1973, 197 (*synth*)
Wishnok, J.S. *et al*, *J. Org. Chem.*, 1973, **38**, 539 (*synth, bibl*)
Duddeck, H. *et al*, *Org. Magn. Reson.*, 1976, **8**, 593 (*cmr*)
Suginome, H. *et al*, *Synthesis*, 1986, 741 (*synth*)

2-Oxatricyclo[4.1.0¹,⁶.0³,⁵]heptane   O-60052

C₆H₈O   M 96.129
Bp 120°.

Müller, E. *et al*, *Tetrahedron Lett.*, 1963, 1047 (*synth, pmr*)

7-Oxatricyclo[4.1.1.0²,⁵]octane, 9CI   O-60053

C₇H₁₀O   M 110.155
(*1α,2β,5β,6α*)-**form** [107784-07-6]
Cryst. (pentane). Mp 42-43.5°.

Kas'jan, L.A. *et al*, *Tetrahedron Lett.*, 1986, **27**, 2921 (*synth, cmr*)

8-Oxatricyclo[3.3.0.0²,⁷]octane, 9CI   O-60054

[20145-38-4]

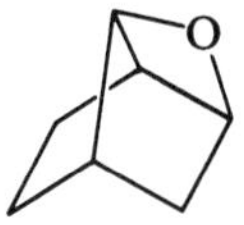

$C_7H_{10}O$ M 110.155

Cryst. (pentane). Mp 84-85° dec. Unstable on storage.

Kas'jan, L.A. *et al, Tetrahedron Lett.*, 1986, **27**, 2921 (*synth, cmr*)

4-Oxatricyclo[4.3.1.1³,⁸]undecan-5-one, 9CI   O-60055

[21898-84-0]

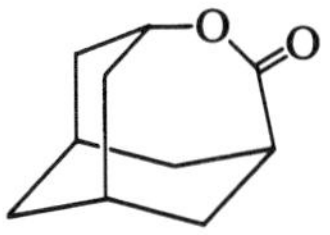

$C_{10}H_{14}O_2$ M 166.219

Mp 288-290°.

Udding, A.C. *et al, Tetrahedron Lett.*, 1968, 5719 (*synth, pmr*)

3,3-Oxetanedimethanol, 9CI, 8CI   O-60056

3,3-Bis(hydroxymethyl)oxetane

[2754-18-9]

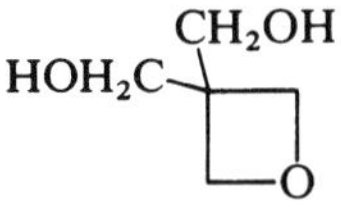

$C_5H_{10}O_3$ M 118.132

Mp 84°. Bp$_{0.35}$ 123°.

O,O-Dinitrate: 3,3-Bis(nitratomethyl)oxetane.
$C_5H_8N_2O_7$ M 208.127
Monomer for prodn. of energetic polymers for explosive propellants.

Ratz, L. *et al, Z. Naturforsch., A*, 1968, **23**, 2100 (*synth, ir*)
George, C. *et al, Acta Crystallogr., Sect. C*, 1986, **42**, 1161 (*cryst struct, dinitrate*)

2-Oxo-2H-benzopyran-4-acetic acid   O-60057

Coumarin-4-acetic acid

[24526-73-6]

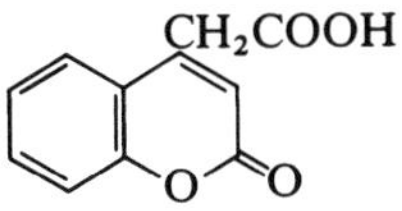

$C_{11}H_8O_4$ M 204.182

Cryst. (EtOH). Mp 182-185°.

Et ester: [24526-71-4].
$C_{13}H_{12}O_4$ M 232.235
Cryst. (H$_2$O). Mp 118-120°.
Amide:
$C_{11}H_9NO_3$ M 203.197
Cryst. (EtOH). Mp 225°.
Chloride: [24526-70-3].
$C_{11}H_7ClO_3$ M 222.628
Cryst. (cyclohexane). Mp 116-118°.
Nitrile: [24526-74-7]. *4-(Cyanomethyl)coumarin, 8CI.*
$C_{11}H_7NO_2$ M 185.182

Cryst. (EtOH). Mp 158-160°.

Checchi, S. *et al, Gazz. Chim. Ital.*, 1969, **99**, 501 (*synth, deriv, ir*)
Brubaker, A.N. *et al, J. Med. Chem.*, 1986, **29**, 1094 (*synth*)

1-Oxo-1H-2-benzopyran-3-acetic acid, 9CI   O-60058

Isocoumarin-3-acetic acid

[39153-95-2]

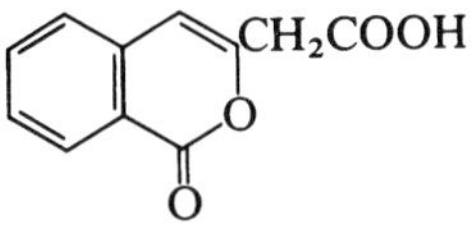

$C_{11}H_8O_4$ M 204.182

Cream needles (AcOH). Mp 220°.

Me ester: [39153-96-3].
$C_{12}H_{10}O_4$ M 218.209
Light-yellow needles (C$_6$H$_6$/pet. ether). Mp 115-117°.

Chatterjea, J.N. *et al, J. Indian Chem. Soc.*, 1972, **49**, 895 (*synth, ir*)

4-Oxo-4H-1-benzopyran-3-acetic acid   O-60059

3-Chromoneacetic acid

[50878-09-6]

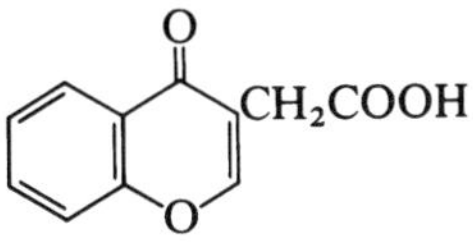

$C_{11}H_8O_4$ M 204.182

Cryst. (2-propanol). Mp 220-222°.

Amide: [50878-10-9].
$C_{11}H_9NO_3$ M 203.197
Cryst. (EtOH). Mp 210-212°.
Nitrile: [50878-08-5].
$C_{11}H_7NO_2$ M 185.182
Cryst. (EtOAc). Mp 137-139°.

Klutchko, S. *et al, J. Heterocycl. Chem.*, 1974, **11**, 183 (*synth, ir*)

3-Oxo-1-cyclohexene-1-carboxaldehyde, 9CI   O-60060

3-Formyl-2-cyclohexen-1-one

[62952-40-3]

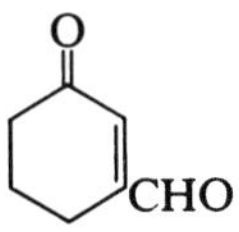

$C_7H_8O_2$ M 124.139

Quesada, M.L. *et al, Synth. Commun.*, 1976, **6**, 555 (*synth*)
Baillargeon, V.P. *et al, J. Am. Chem. Soc.*, 1986, **108**, 452 (*synth, ir, pmr, cmr*)
Albeck, A. *et al, J. Am. Chem. Soc.*, 1986, **108**, 4614 (*uv, pmr, ms*)

α-Oxo-3-cyclopentene-1-acetaldehyde, 9CI   O-60061
3-Cyclopentene-1-glyoxal
[80344-66-7]

$C_7H_8O_2$ M 124.139
Bp$_{0.5}$ 48-50°. Exists as a mixt. of hydrated forms in soln.
2,4-Dinitrophenylhydrazone: Mp 248-250°.

Deshpande, M.N. *et al, J. Org. Chem.*, 1986, **51**, 2436 (*synth, ir, pmr, ms*)

3-Oxo-1-cyclopentenecarboxaldehyde   O-60062
3-Formyl-2-cyclopentenone
[102574-14-1]

$C_6H_6O_2$ M 110.112

Albeck, A. *et al, J. Am. Chem. Soc.*, 1986, **108**, 4614 (*synth, pmr*)

10-Oxodecanoic acid, 9CI   O-60063
9-Formylnonanoic acid. 9-Aldehydononanoic acid. Sebacic semialdehyde
[5578-80-3]

$$OHC(CH_2)_8COOH$$

$C_{10}H_{18}O_3$ M 186.250
Glistening plates (Me$_2$CO/pet. ether). Mp 56-57°. Bp$_{0.1}$ 150°.
Semicarbazone: Nodules (EtOH). Mp 170°.
Oxime:
  $C_{10}H_{19}NO_3$ M 201.265
  Prisms (MeOH). Mp 111°.
Me ester: [14811-73-5].
  $C_{11}H_{20}O_3$ M 200.277
  Pale-yellow oil. Bp$_2$ 119-121°, Bp$_{0.6}$ 108-110°.
Me ester, semicarbazone: Mp 98-100°.
Me ester, 2,4-Dinitrophenylhydrazone: Orange prisms (MeOH). Mp 86-87°.
Me ester, Di-Me acetal: [65157-90-6].
  $C_{13}H_{26}O_4$ M 246.346
  Bp$_{0.25}$ 137-138°.

Noller, C.R. *et al, J. Am. Chem. Soc.*, 1926, **48**, 1074 (*deriv, synth*)
Hargreaves, G.H. *et al, J. Chem. Soc.*, 1947, 753 (*synth*)
Gokhale, P.D. *et al, Synthesis*, 1974, 718 (*synth*)
Adlof, R.O. *et al, J. Amer. Oil Chem. Soc.*, 1977, **54**, 414 (*synth*)
Cameron, A.G. *et al, J. Chem. Soc., Perkin Trans. 1*, 1986, 161 (*deriv, synth, pmr*)

7-Oxodihydrogmelinol   O-60064
[108044-10-6]

$C_{22}H_{26}O_8$ M 418.443
Constit. of *Gmelina arborea*. Cryst. (C$_6$H$_6$/pet. ether). Mp 70°. $[\alpha]_D^{28}$ +63.05° (c, 1.134 in CHCl$_3$).

Satyanarayana, P. *et al, J. Nat. Prod.*, 1986, **49**, 1061.

μ-Oxodiphenylbis(trifluoroacetato-*O*)-diiodine, 9CI   O-60065
μ-Oxobis[trifluoroacetato(phenyl)iodine]
[91879-79-7]

$$F_3CCOI(Ph)\text{—}O\text{—}I(Ph)COCF_3$$

$C_{16}H_{10}F_6I_2O_5$ M 650.051
Oxidising agent. Cryst. (hexane). Mp 110-112°.

Gallos, J. *et al, J. Chem. Soc., Perkin Trans. 1*, 1985, 757 (*synth, props, cryst struct*)

12-Oxododecanoic acid, 9CI   O-60066
10-Formylundecanoic acid
[3956-80-7]

$$OHC(CH_2)_{10}COOH$$

$C_{12}H_{22}O_3$ M 214.304
Mp 65-66°. Bp$_{0.05}$ 151-153°.
Me ester: [2009-59-8].
  $C_{13}H_{24}O_3$ M 228.331
  Pale-yellow oil, solidifies on refrigeration. Bp$_5$ 154°, Bp$_{0.1}$ 109-110°.
Me ester, Di-Me acetal: [1931-67-5].
  $C_{15}H_{30}O_4$ M 274.400
  Liq. Bp$_{0.05}$ 129-130°.

Tomecko, C.G. *et al, J. Am. Chem. Soc.*, 1927, **49**, 522 (*deriv, synth*)
Adlof, R.O. *et al, J. Amer. Oil Chem. Soc.*, 1977, **54**, 414 (*synth*)
Cameron, A.G. *et al, J. Chem. Soc., Perkin Trans. 1*, 1986, 161 (*deriv, synth, pmr*)

1-Oxo-7(11)-eudesmen-12,8-olide   O-60067

$C_{15}H_{20}O_3$ M 248.321
8α-form [108044-20-8]
Constit. of *Smyrnium galaticum*. Amorph.

Ulubelen, A. *et al, J. Nat. Prod.*, 1986, **49**, 1104.

1-Oxo-4-germacren-12,6-olide　　　　　O-60068

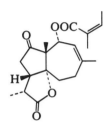

$C_{15}H_{22}O_3$　　M 250.337

(*4E,6α,11S,10β*)-*form*

Constit. of *Artemisia maritima*. Oil.

10,14-Didehydro: 1-Oxo-4,10(14)-germacradien-12,6α-olide.
$C_{15}H_{20}O_3$　　M 248.321
From *A. maritima*. Cryst. Mp 132°.

Pathak, V.P. *et al*, *Phytochemistry*, 1987, **26**, 2103.

ent-15-Oxo-1(10),13-halimadien-18-oic　　　O-60069
acid

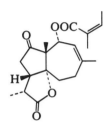

$C_{20}H_{30}O_3$　　M 318.455

(*13E*)-*form* [109181-77-3]

Constit. of *Halimium viscosum*.

Me ester: [109163-59-9]. Oil. $[\alpha]_D^{22}$ +64.12° (c, 1.05 in $CHCl_3$).

15-Carboxylic acid: [109163-62-4]. **ent-1(10)13E-Halimadiene-15,18-dioic acid**.
$C_{20}H_{30}O_4$　　M 334.455
Constit. of *H. viscosum*.

15-Carboxylic acid, di-Me ester: Oil. $[\alpha]_D^{22}$ +73.35° (c, 1.2 in $CHCl_3$).

(*13Z*)-*form* [109163-65-7]

Constit. of *H. viscosum*.

Me ester: [109163-58-8]. Oil. $[\alpha]_D^{22}$ +61.54° (c, 1.31 in $CHCl_3$).

15-Carboxylic acid: [109163-61-3]. **ent-1(10),13Z-Halimadiene-15,18-dioic acid**.
$C_{20}H_{30}O_4$　　M 334.455
Constit. of *H. viscosum*.

15-Carboxylic acid, di-Me ester: Oil. $[\alpha]_D^{22}$ +68.43° (c, 0.8 in $CHCl_3$).

Urones, J.G. *et al*, *Phytochemistry*, 1987, **26**, 1077.

4-Oxoheptanal, 9CI　　　　　　　　　O-60070

[74327-28-9]

$$H_3CCH_2CH_2COCH_2CH_2CHO$$

$C_7H_{12}O_2$　　M 128.171
Bp_{16} 94-95°.

Larcheveque, M. *et al*, *Tetrahedron*, 1979, **35**, 1745 (*synth, ir, pmr*)
Kulinkovich, O.G. *et al*, *Synthesis*, 1984, 886 (*synth*)
Miyakoshi, T., *Synthesis*, 1986, 766 (*synth*)

4-Oxoheptanedioic acid, 9CI　　　　　O-60071

Updated Entry replacing O-00830

Acetonediacetic acid. Hydrochelidonic acid. γ-Oxopimelic acid

[502-50-1]

$$CO(CH_2CH_2COOH)_2$$

$C_7H_{10}O_5$　　M 174.153
Plates (H_2O). Mp 139°. Heat at Mp → anhydride.

Oxime:
$C_7H_{11}NO_5$　　M 189.168
Prisms (H_2O). Mp 129° dec.

Semicarbazone: Mp 183-184° dec.

Di-Me ester: [22634-92-0].
$C_9H_{14}O_5$　　M 202.207
Needles (EtOH). Mp 56°. Bp 276-277° dec.

Di-Me ester, oxime:
$C_9H_{15}NO_5$　　M 217.221
Needles (CS_2). Mp 52°.

Di-Et ester: [40420-22-2].
$C_{11}H_{18}O_5$　　M 230.260
$Bp_{0.3}$ 116-121°. n_D^{25} 1.4397.

Me ester, mononitrile:
$C_8H_{11}NO_3$　　M 169.180
Bp_{30} 230-235°.

Dinitrile: [66619-32-7]. *4-Oxoheptanedinitrile, 9CI. 1,5-Dicyano-3-pentanone.*
$C_7H_8N_2O$　　M 136.153
Mp 37°.

Org. Synth., Coll. Vol., **4**, 302 (*deriv*)
Colonge, J. *et al*, *Bull. Soc. Chim. Fr.*, 1962, 832 (*synth*)
Ger. Pat., 2 520 981, (*1976*); *CA*, **86**, 120792 (*synth*)
Baldwin, J.E. *et al*, *Tetrahedron*, 1986, **42**, 4247 (*deriv, synth, ir, pmr, ms, cmr*)

4-Oxo-2-heptenedioic acid　　　　　　O-60072

3-Oxo-1-pentene-1,5-dicarboxylic acid. Furonic acid
[502-51-2]

$$HOOCCH\!\!=\!\!CHCOCH_2CH_2COOH$$

$C_7H_8O_5$　　M 172.137
Platelets (H_2O). Mp 186°.

Di-Me ester: [22503-69-1].
$C_9H_{12}O_5$　　M 200.191
Cryst. (Et_2O/pet. ether). Mp 105°. Bp_{18} 175°.

2,4-Dinitrophenylhydrazone: [22856-34-4]. Mp 182°.

Di-Me ester, 2,4-Dinitrophenylhydrazone: [22503-68-0]. Mp 112°.

Kaplan, L.A. *et al*, *J. Org. Chem.*, 1962, **27**, 780 (*synth*)
Simura, T. *et al*, *Nippon Kagaku Zasshi*, 1968, **89**, 695; *CA*, **70**, 11072 (*synth*)
Wegmann, H. *et al*, *Justus Liebigs Ann. Chem.*, 1980, 1736 (*ester*)

4-Oxohexanal, 9CI　　　　　　　　　O-60073

[25346-59-2]

$$H_3CCH_2COCH_2CH_2CHO$$

$C_6H_{10}O_2$　　M 114.144
Bp_{16} 83-84°, $Bp_{4.5}$ 60-61°.

Cavill, G.W.K. *et al*, *Aust. J. Chem.*, 1970, **23**, 83 (*synth*)
Barnier, J.P. *et al*, *Bull. Soc. Chim. Fr.*, 1975, 1659 (*synth, ir, pmr, ms*)
Miyakoshi, T., *Synthesis*, 1986, 766 (*synth*)

2-Oxo-4-imidazolidinecarboxylic acid, 9CI, 8CI   O-60074

Updated Entry replacing O-00859
2-Imidazolidone-4-carboxylic acid
[21277-16-7]

$C_4H_6N_2O_3$ M 130.103
(S)-form [41371-53-3]
L-form
Mp 245-248°.
(±)-form
Cryst. (H_2O). Mp 190-201°.

Dittmer, K. *et al, J. Biol. Chem.,* 1946, **164**, 19 (*synth*)
Williamson, J.M. *et al, J. Biol. Chem,* 1982, **257**, 12039 (*synth, props*)

ent-15-Oxo-16-kauren-18-oic acid   O-60075

ent-*Kauren-15-one-18-oic acid*
$C_{20}H_{28}O_3$ M 316.439
Constit. of *Porella densifolia* subsp. *appendiculata.*
  Cryst. Mp 227-228°. $[\alpha]_D$ −110° (c, 0.12 in $CHCl_3$).

Asakawa, Y. *et al, Phytochemistry,* 1987, **26**, 1019.

ent-6-Oxo-7,12E,14-labdatrien-17,11α-olide   O-60076

[110201-83-7]

$C_{20}H_{26}O_3$ M 314.424
Constit. of *Rutidosis murchisonii.* Oil.

Zdero, C. *et al, Phytochemistry,* 1987, **26**, 1759.

6-Oxo-7-labdene-15,17-dioic acid   O-60077

$C_{20}H_{30}O_5$ M 350.454
(13S)-form
  Havardic acid D
  Constit. of *Grindelia havardii.*

Jolad, S.D. *et al, Phytochemistry,* 1987, **26**, 483.

α-Oxo-1-naphthaleneacetic acid, 9CI   O-60078

1-Naphthylglyoxylic acid, 8CI. α-Naphthoylformic acid
[26153-26-4]

$C_{12}H_8O_3$ M 200.193
Cryst. Mp 111-114°.
Et ester: [33656-65-4].
  $C_{14}H_{12}O_3$ M 228.247

Liq. $Bp_{0.6}$ 135-141°.

Leermakers, P.A. *et al, J. Am. Chem. Soc.,* 1964, **86**, 1768 (*synth, deriv*)
Middleton, W.J. *et al, J. Org. Chem.,* 1980, **45**, 2883 (*synth, deriv*)
Miura, M. *et al, J. Org. Chem.,* 1987, **52**, 2623 (*synth*)

α-Oxo-2-naphthaleneacetic acid, 9CI   O-60079

2-Naphthylglyoxylic acid, 8CI. β-Naphthoylformic acid
[14289-45-3]
$C_{12}H_8O_3$ M 200.193
Yellow cubes (pet. ether). Mp 91-92°.
Et ester: [73790-09-7].
  $C_{14}H_{12}O_3$ M 228.247
  Liq. $Bp_{1.2}$ 130-133°.
2,4-Dinitrophenylhydrazone: Orange needles (EtOH).
  Mp 219°.

Cymerman-Craig, J. *et al, Aust. J. Chem.,* 1956, **9**, 222 (*synth*)
Middleton, W.J. *et al, J. Org. Chem.,* 1980, **45**, 2883 (*synth, deriv*)

7-Oxo-11-nordrim-8-en-12-oic acid   O-60080

$C_{14}H_{20}O_3$ M 236.310
Me ester:
  $C_{15}H_{22}O_3$ M 250.337
  Constit. of fungus *Lepista glaucocana.* Oil. $[\alpha]_D^{20}$ +28.2° (c, 0.90 in $CHCl_3$).

Errington, S.G. *et al, J. Chem. Res. (S),* 1987, 47.

4-Oxooctanal, 9CI   O-60081

[66662-22-4]

$$H_3C(CH_2)_3COCH_2CH_2CHO$$

$C_8H_{14}O_2$ M 142.197
Bp_2 60-61°.

Kulinkovich, O.G. *et al, Synthesis,* 1984, 886 (*synth, pmr*)
Miyakoshi, T. *et al, Synthesis,* 1986, 766 (*synth, ir, pmr*)

2-Oxo-4-oxazolidinecarboxylic acid, 9CI   O-60082

4,5-Dihydro-2-oxo-3H-oxazole-4-carboxylic acid
[89033-27-2]

(S)-form

$C_4H_5NO_4$ M 131.088
(S)-form [19525-95-2]
L-form
Mp 118.5-120°. $[\alpha]_D^{22.5}$ −17.7° (c, 3.44 in H_2O).
(±)-form [104975-37-3]
Mp 125-125.8°.

Kaneko, T. *et al, Bull. Chem. Soc. Jpn.,* 1968, **41**, 974 (*synth*)

4-Oxopentanal, 9CI   O-60083

Updated Entry replacing O-00920
Levulinaldehyde, 8CI. Levulinic aldehyde

[626-96-0]

$$H_3CCOCH_2CH_2CHO$$

$C_5H_8O_2$ M 100.117
Oil. Misc. H_2O, EtOH, Et_2O. $Bp_{8.5}$ 66°. Steam-volatile.
Dioxime:
  $C_5H_{10}N_2O_2$ M 130.146
  Prisms (Et_2O). Mp 73-74°.
Disemicarbazone: Cryst. (MeOH). Mp 180-182°.
Bis-4-nitrophenylhydrazone: Cryst. Mp 284-285°.
Di-Me acetal: [3209-78-7]. *5,5-Dimethoxy-2-pentanone.*
  $C_7H_{14}O_3$ M 146.186
  Oil. Misc. EtOH, Et_2O, sol. 6 parts H_2O. Bp_{13} 79-80°.

Harries, C. *et al, Justus Liebigs Ann. Chem.,* 1910, **374**, 338 (*synth*)
Fischer, F.G. *et al, Ber.,* 1932, **65**, 1467 (*synth*)
Kirrmann, A. *et al, Bull. Soc. Chim. Fr.,* 1935, 2143 (*acetal*)
Mondon, A., *Angew. Chem.,* 1952, **64**, 224 (*synth*)
Salomon, R.G. *et al, J. Am. Chem. Soc.,* 1977, **99**, 3501 (*synth*)
Miyakoshi, T., *Synthesis,* 1986, 766 (*synth, ir, pmr*)

3-Oxopentanedioic acid, 9CI   O-60084

Updated Entry replacing O-00922
3-Oxoglutaric acid. Acetonedicarboxylic acid
[542-05-2]

$$HOOCCH_2COCH_2COOH$$

$C_5H_6O_5$ M 146.099
Intermed. in tropinone synth. Needles (EtOAc). Sol.
  H_2O, EtOH, insol. $CHCl_3$, C_6H_6. Mp 135° dec. Dec.
  by hot H_2O, acids or alkalis.
Di-Me ester: [1830-54-2].
  $C_7H_{10}O_5$ M 174.153
  Bp_{25} 150°.
Amide:
  $C_5H_7NO_4$ M 145.115
  Needles. Spar. sol. H_2O, EtOH. Mp 86°.
Dianilide: [40315-17-1].
  $C_{17}H_{16}N_2O_3$ M 296.325
  Needles (EtOH). Spar. sol. Et_2O, $CHCl_3$, C_6H_6. Mp
  155°.
Oxime:
  $C_5H_7NO_5$ M 161.114
  Mp 53-54°.
Ethylene ketal: [5694-91-7]. *1,3-Dioxolane-2,2-diacetic*
  acid. 3,3-(Ethylenedioxy)pentanedioic acid.
  $C_7H_{10}O_6$ M 190.152
  Cryst. (EtOAc). Mp 91-93°.
Ethylene ketal, anhydride: [32296-88-1]. *1,4,8-*
  Trioxaspiro[4.5]decan-8-one.
  $C_7H_8O_5$ M 172.137
  Cryst. ($CHCl_3$/pet. ether). Mp 115.5-116.5°.

Org. Synth., Coll. Vol., **1**, 9 (*synth*)
Jerdan, J., *J. Chem. Soc.,* 1899, **75**, 809 (*synth*)
Ger. Pat., 2 429 627 (*synth*)
U.S.P., 3 963 775, (*1976*) (*use*)
Brutcher, F.V. *et al, J. Org. Chem.,* 1972, **37**, 297 (*deriv*)
Kita, Y. *et al, J. Org. Chem.,* 1986, **51**, 4150 (*deriv, ir, pmr*)
Fieser, M. *et al, Reagents for Organic Synthesis,* Wiley, 1967-
  84, **1**, 6.

4-Oxo-2-phenyl-4H-1-benzopyran-3-car-   O-60085
boxylic acid

3-Flavonecarboxylic acid
[77037-45-7]

$C_{16}H_{10}O_4$ M 266.253
Long needles (Me_2CO/Et_2O). Mp 177-179°.
Et ester: [77037-46-8].
  $C_{18}H_{14}O_4$ M 294.306
  Long needles (Et_2O/pentane). Mp 89-90°.

Costa, A.M.B.S.R.C.S., *J. Chem. Soc., Perkin Trans. 1,* 1985,
  799 (*synth, ir, pmr*)

4-Oxo-3-phenyl-4H-1-benzopyran-2-car-   O-60086
boxylic acid

Isoflavone-2-carboxylic acid
[7622-78-8]

$C_{16}H_{10}O_4$ M 266.253
Needles ($CHCl_3$/hexane). Mp 212°.

Baker, W. *et al, J. Chem. Soc.,* 1953, 1852 (*synth, deriv*)
Costa, A.M.B.S.R.C.S., *J. Chem. Soc., Perkin Trans. 1,* 1985,
  799 (*synth, ir, pmr*)

4-Oxo-2-phenyl-4H-1-benzothiopyran-3-   O-60087
carboxylic acid, 9CI

[98153-25-4]

$C_{16}H_{10}O_3S$ M 282.313
Powder. Mp 220°.

Costa, A.M. *et al, J. Chem. Soc., Perkin Trans. 1,* 1985, 799
  (*synth, ir, pmr*)

2-Oxo-4-phenylthio-3-butenoic acid, 9CI   O-60088

$$PhSCH{=}CHCOCOOH$$

$C_{10}H_8O_3S$ M 208.231
(*Z*)-*form* [105633-41-8]
  Mp 113°.
  Et ester: [105633-54-3].
    $C_{12}H_{12}O_3S$ M 236.285
    Mp 83°.

Hojo, M. *et al, Synthesis,* 1986, 137 (*synth, ir, pmr*)

4-Oxo-2-piperidinecarboxylic acid O-60089

Updated Entry replacing O-00943
4-Piperidone-2-carboxylic acid. 4-Oxopipecolic acid. 4-Ketopipecolic acid

(*S*)-form

$C_6H_9NO_3$ M 143.142

(*S*)-form [65060-18-6]
L-form
Amino acid present in hydrolysates of the virginiamycin/ostreogrycin family of cyclodepsipeptide antibiotics. Prisms. $[\alpha]_D^{20}$ −14° (c, 1 in H_2O).
B,*HCl*: Plates (2-propanol). Mp 175-180° dec. $[\alpha]_D^{20}$ +3.8° (c, 2 in H_2O).

(±)-*form*

N-*Me, Me ester:*
  $C_8H_{13}NO_3$ M 171.196
  Oil.

Eastwood, F.W. *et al, J. Chem. Soc.*, 1960, 2286 (*isol*)
Crooy, P. *et al, J. Antibiot.*, 1972, **25**, 371.
Essawy, M.Y. *et al, J. Heterocycl. Chem.*, 1983, **20**, 478 (*synth, pmr*)

3-Oxopropanoic acid, 9CI O-60090

Updated Entry replacing O-00950
Malonaldehydic acid, 8CI. Formylacetic acid. Aldehydoacetic acid. Formylethanoic acid
[926-61-4]

$$OHCCH_2COOH \rightleftharpoons HOCH=CHCOOH$$

$C_3H_4O_3$ M 88.063

Oxo-form
Pale-yellow oil. Dec. above 50°.
Oxime (E-): *Isonitrosopropanoic acid.*
  $C_3H_5NO_3$ M 103.077
  Mp 117-118°.
Semicarbazone: Cryst. Mp 116° dec.
2,4-Dinitrophenylhydrazone: [23130-58-7]. Cryst. (EtOH). Mp 160° dec.
Me ester: [63857-17-0].
  $C_4H_6O_3$ M 102.090
  Unstable, appears to be known only as derivs.
Me ester, semicarbazone: Cryst. (MeOH). Mp 163°.
Me ester, 2,4-dinitrophenylhydrazone: [1928-98-9]. Cryst. (EtOH). Mp 117°.
Me ester, di-Et acetal: Methyl 3,3-diethoxypropionate.
  $C_8H_{16}O_4$ M 176.212
  Oily liq. Spar. sol. H_2O. Bp 193°, Bp_{24} 95°.
Et ester, di-Et acetal: [10601-80-6]. *Ethyl 3,3-diethoxypropionate.*
  $C_9H_{18}O_4$ M 190.239
  Liq. Bp_{15} 91-98°.
Et ester, semicarbazone: Cryst. Mp 147-148°.
Et ester, 2,4-dinitrophenylhydrazone: Cryst. (ligroin/EtOH). Mp 158°.
Isopropyl ester:
  $C_6H_{10}O_3$ M 130.143
  Bp_8 30-50°.
Isopropyl ester, semicarbazone: Cryst. (EtOAc). Mp 138-141°.
tert-*Butyl ester:*
  $C_7H_{12}O_3$ M 144.170

Bp_{11} 35-60°.
tert-*Butyl ester, semicarbazone:* Cryst. (hexane/Me_2CO). Mp 183-184°.
Benzyl ester:
  $C_{10}H_{10}O_3$ M 178.187
  $Bp_{0.05}$ 75-85°.
Benzyl ester, semicarbazone: Cryst. (hexane/Me_2CO). Mp 147-149°.
Nitrile: see Cyanoacetaldehyde, C-03037

Enol-form
3-Hydroxy-2-propenoic acid. 3-Hydroxyacrylic acid
Me ester, O-benzoyl: Methyl 3-benzoyloxyacrylate.
  $C_{11}H_{10}O_4$ M 206.198
  Cryst. Mp 38-40°.
Et ester, O-Ac: Ethyl 3-acetoxyacrylate.
  $C_7H_{10}O_4$ M 158.154
  Oil. Bp_{18} 94-97°.

Wohl, A. *et al, Ber.*, 1900, **33**, 2760 (*synth*)
Straus, F. *et al, Ber.*, 1926, **59**, 1681 (*synth*)
Rinkes, J., *Recl. Trav. Chim. Pays-Bas*, 1927, **46**, 268 (*synth*)
Thompson, J.E. *et al, Can. J. Biochem.*, 1967, **45**, 563 (*isol*)
Lovett, E.G. *et al, J. Org. Chem.*, 1977, **42**, 2574 (*synth*)
Syatkovskii, A.I. *et al, Zh. Org. Khim.*, 1977, **13**, 1569 (*struct*)
Sato, M. *et al, Synthesis*, 1986, 672 (*derivs*)

2-(3-Oxo-1-propenyl)benzaldehyde, 9CI O-60091

2-Formylcinnamaldehyde

(*E*)-form

$C_{10}H_8O_2$ M 160.172

(*E*)-form [61650-52-0]
  Yellow needles (EtOAc/cyclohexane). Mp 62-63°.
(*Z*)-form [93273-84-8]
  Readily isom. to (*E*)-form on attempted purification.

Darby, N. *et al, J. Org. Chem.*, 1977, **42**, 1960 (*synth, ir, pmr*)
Zadok, E. *et al, Tetrahedron Lett.*, 1984, **25**, 4175 (*synth, pmr*)
Larson, R.A. *et al, Tetrahedron Lett.*, 1986, **27**, 3987 (*synth*)

3-Oxopyrazolo[1,2-*a*]pyrazol-8-ylium-1-olate O-60092

3-Hydroxy-1-oxo-1H-pyrazolo[1,2-a]pyrazol-4-ium hydroxide inner salt, 9CI
[97938-47-1]

$C_6H_4N_2O_2$ M 136.110
Cross-conjugated mesomeric betaine. Yellow irreg. prisms (THF/hexane). Dec. on attempted chemical characterization. Dec. slowly >55-65°.

Potts, K.T. *et al, J. Chem. Soc., Chem. Commun.*, 1986, 144 (*synth, ir, pmr, cmr*)

4-Oxo-3,4-secoambrosan-12,6-olid-3-oic acid   O-60093

$C_{15}H_{22}O_5$ M 282.336

Constit. of *Ambrosia artemisiifolia*. Cryst. (Me₂CO/pet. ether). Mp 171.5-176°. $[\alpha]_D^{20}$ −0.88° (c, 0.91 in Me₂CO).

Stefanović, M. *et al*, *Phytochemistry*, 1987, **26**, 850.

16-Oxo-17-spongianal   O-60094

4,4-Dimethyl-17-oxo-18-nor-16-oxaandrostane-8-carboxaldehyde, 9CI. 17-Spongianal-16-one

[109152-42-3]

$C_{20}H_{30}O_3$ M 318.455

Constit. of *Ceratosoma brevicaudatum*. Oil.

Ksebati, M.B. *et al*, *J. Org. Chem.*, 1987, **52**, 3766.

18-Oxo-3-virgene   O-60095

$C_{20}H_{32}O$ M 288.472

Constit. of *Nicotiana tabacum*. Oil. $[\alpha]_D$ −69.1°.

Uegaki, R. *et al*, *Phytochemistry*, 1987, **26**, 3029.

1,1'-Oxybis[2-iodoethane], 9CI   O-60096

Bis(2-iodoethyl) ether. β,β'-Diiododiethyl ether

[34270-90-1]

$$ICH_2CH_2OCH_2CH_2I$$

$C_4H_8I_2O$ M 325.916

Intermed. for crown ether synth. Liq. Bp₁₀ 123.5-124°.

Gibson, C.S. *et al*, *J. Chem. Soc.*, 1930, 2525 (*synth*)
Kulstad, S. *et al*, *Acta Chem. Scand., Ser. B*, 1979, **33**, 469 (*synth, use*)

3,3'-Oxybis-1-propanol, 9CI   O-60097

3-Oxa-1,7-heptanediol. Bis(3-hydroxypropyl) ether

[2396-61-4]

$$HOCH_2CH_2CH_2OCH_2CH_2CH_2OH$$

$C_6H_{14}O_3$ M 134.175

Intermed. in prepn. of macrocyclic ethers. Oil. Bp₀.₀₁ 98-102°.

Samat, A. *et al*, *J. Chem. Soc., Perkin Trans. 1*, 1985, 1717 (*synth, pmr*)

2,2'-Oxybispyridine, 9CI   O-60098

Updated Entry replacing O-10085
2,2'-Dipyridyl ether. 2,2'-Dipyridyl oxide

[53258-94-9]

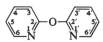

$C_{10}H_8N_2O$ M 172.186
Mp 49°.

Picrate: Mp 124-125°.

de Villiers, P.A. *et al*, *Recl. Trav. Chim. Pays-Bas*, 1957, **76**, 647 (*synth*)
Trovato, G. *et al*, *Gazz. Chim. Ital.*, 1973, **103**, 709 (*uv*)
Summers, L.A., *J. Heterocycl. Chem.*, 1987, **24**, 533 (*rev, uv, pmr*)

2,3'-Oxybispyridine   O-60099

Updated Entry replacing O-10086
2-(3-Pyridinyloxy)pyridine, 9CI. 2,3'-Dipyridyl oxide. 2,3'-Dipyridyl ether

[10168-50-0]

$C_{10}H_8N_2O$ M 172.186
Needles. Mp 34-35°. Bp₀.₁ 95-115°.

Picrate: Mp 163-164°.

de Villiers, P.A. *et al*, *Recl. Trav. Chim. Pays-Bas*, 1957, **76**, 647 (*synth*)
Yoneda, F., *Yakugaku Zasshi*, 1957, **77**, 944; *CA*, **52**, 2855b (*synth*)
Kajihara, S., *Nippon Kagaku Zasshi*, 1965, **86**, 1060; *CA*, **65**, 16936f (*synth*)
Trovato, G. *et al*, *Gazz. Chim. Ital.*, 1973, **103**, 709 (*uv*)
Summers, L.A., *J. Heterocycl. Chem.*, 1987, **24**, 533 (*rev, uv, pmr*)

3,3'-Oxybispyridine, 9CI   O-60100

Updated Entry replacing O-10087
3,3'-Dipyridyl oxide. 3,3'-Dipyridyl ether

[53258-95-0]

$C_{10}H_8N_2O$ M 172.186
Bp₁₄ 145-147°.

N-Oxide: [76167-53-8].
   $C_{10}H_8N_2O_2$ M 188.185
   Cryst. (2-propanol/Et₂O). Mp 116-118°.
N,N'-Dioxide: [76167-54-9].
   $C_{10}H_8N_2O_3$ M 204.185
   Mp 220-222°.

Trovato, G. *et al*, *Gazz. Chim. Ital.*, 1973, **103**, 709 (*uv*)
Butler, D.E. *et al*, *J. Med. Chem.*, 1981, **24**, 346 (*synth*)
Summers, L.A., *J. Heterocycl. Chem.*, 1987, **24**, 533 (*rev, uv, pmr*)

4,4'-Oxybispyridine, 9CI   O-60101

Updated Entry replacing O-10088
4,4'-Dipyridyl ether. 4,4'-Dipyridyl oxide

[53258-96-1]

$C_{10}H_8N_2O$ M 172.186
Needles. Mp 61-62°. Formerly incorrectly descr. as having Mp 177-8°.

Rockley, J.E. *et al*, *Chem. Ind. (London)*, 1981, 97 (*synth*)
Summers, L.A., *J. Heterocycl. Chem.*, 1987, **24**, 533 (*rev*)

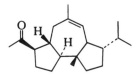

Oxyphencyclimine, BAN   O-60102

Updated Entry replacing O-01035

α-Cyclohexyl-α-hydroxybenzeneacetic acid (1,4,5,6-te-trahydro-1-methyl-2-pyrimidinyl)methyl ester, 9CI. α-Phenylcyclohexaneglycolic acid (1,4,5,6-tetrahydro-1-methyl-2-pyridmidinyl)methyl ester, 8CI. 1-Methyl-1,4,5,6-tetrahydro-2-pyrimidylmethyl α-cyclohexyl-α-phenylglycolate

[125-53-1]

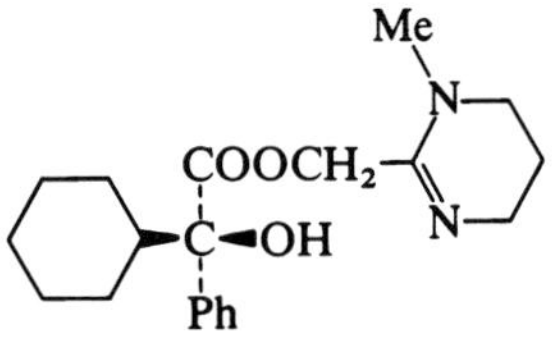

(*R*)-*form*

$C_{20}H_{28}N_2O_3$ M 344.453

Used for treatment of ulcers.

(*R*)-*form*

B,HCl: Mp 234-236° dec. $[\alpha]_D^{20}$ +9.69° (c, 2.1 in MeOH).

(*S*)-*form*

B,HCl: Cryst. (EtOH). Mp 229-230° dec. $[\alpha]_D^{20}$ −9.51° (c, 2.2 in MeOH).

(±)-*form*

B,HCl: [125-52-0]. *Daricon. Naridan.* Anticholinergic, antispasmodic. Spar. sol. H_2O. Mp 231-232° dec.

▷GV3030000.

Faust, J.A. *et al*, *J. Am. Chem. Soc.*, 1959, **81**, 2214 (*synth*)

Kuhnert-Brandstaetter, M. *et al*, *Sci. Pharm.*, 1974, **42**, 150 (*uv, props*)

Schjelderup, L. *et al*, *Acta Chem. Scand.*, *Ser. B*, 1986, **40**, 601 (*synth, abs config, pmr, cmr, ir*)

P

Palauolide
P-60001

Updated Entry replacing P-10003

[82205-22-9]

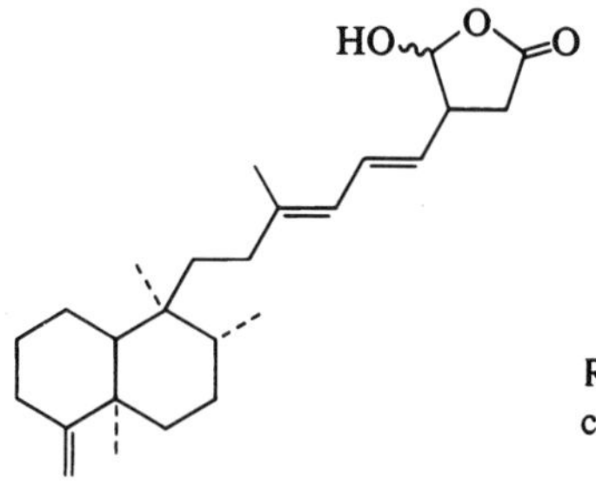

Relative configuration

$C_{25}H_{38}O_3$ M 386.573

Antimicrobial constit. of an unidentified Palauan sponge. Active against *Bacillus subtilis* and *Staphylococcus aureus*. Yellow oil. $[\alpha]_D$ +1.5° (c, 0.2 in CHCl$_3$).

Sullivan, B. *et al*, *Tetrahedron Lett.*, 1982, **23**, 907 (*isol, struct*)
Piers, E. *et al*, *J. Chem. Soc., Chem. Commun.*, 1987, 1342 (*synth*)

Pandoxide
P-60002

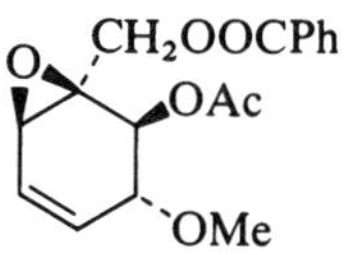

$C_{17}H_{18}O_6$ M 318.326

Constit. of *Uvaria pandensis*. Oil. $[\alpha]_D^{28}$ +35.5° (c, 1.13 in CHCl$_3$).

Nkunya, M.H.H. *et al*, *Phytochemistry*, 1987, **26**, 2563.

Panduratin *B*
P-60003

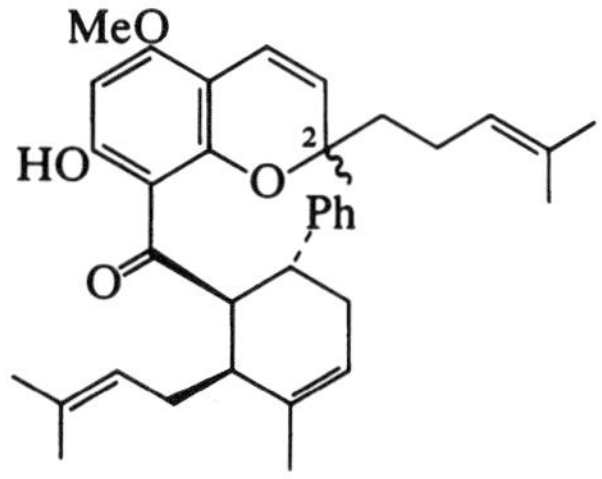

$C_{36}H_{44}O_4$ M 540.741

Constit. of *Boesenbergia pandurata* red rhizomes as a mixt. of 2-epimers. Oil.

Pancharven, O. *et al*, *Aust. J. Chem.*, 1987, **40**, 455.

Papakusterol
P-60004

Updated Entry replacing P-20012

23-(2-Methylcyclopropyl)-24-norchola-5,22-dien-3-ol, 9CI. 22-Dehydro-24,26-cyclocholesterol. 24,26-Cyclo-5,22-cholestadien-3β-ol. Glaucasterol

[84871-06-7]

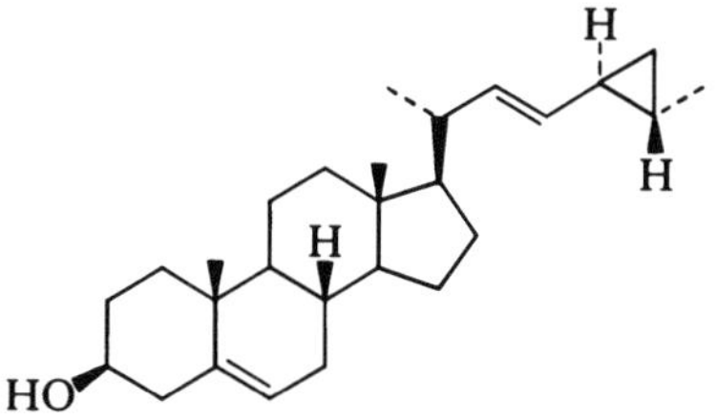

$C_{27}H_{42}O$ M 382.628

Mixt. of 24*R*,25*R*- and 24*S*,25*S*-forms. Constit. of six gorgonians (as yet unidentified). Cryst. Mp 112-113°.

5α,6-Dihydro: (24S,25S)-24,26-*Cyclo-5α-cholest-22E-en-3β-ol. 5α,6-Dihydroglaucasterol.*

$C_{27}H_{44}O$ M 384.644

Constit. of soft coral *Sarcophyton glaucum*. Cryst. Mp 120-124°.

Bonini, C. *et al*, *Tetrahedron Lett.*, 1983, **24**, 277 (*isol, synth*)
Catalan, C.A.N. *et al*, *Tetrahedron Lett.*, 1983, **24**, 3461 (*abs config*)
Kobayashi, M. *et al*, *Chem. Pharm. Bull.*, 1983, **31**, 1803 (*deriv*)
Fujimoto, Y. *et al*, *Tetrahedron Lett.*, 1984, **25**, 1805 (*synth, abs config*)

[2.2]Paracyclo(4,8)[2.2]-metaparacyclophane
P-60005

Updated Entry replacing P-00102

Pentacyclo[18.2.2.2^{8,11}.0^{4,14}.0^{5,17}]hexacosa-4,8,10,14,16,20,22,23,25-nonaene, 9CI

[52903-50-1]

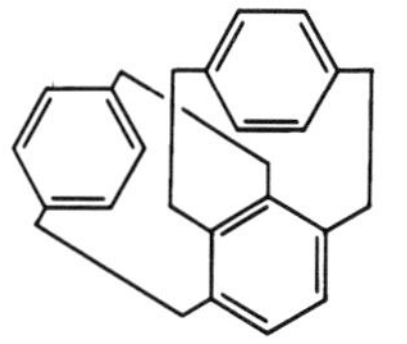

$C_{26}H_{26}$ M 338.491

Prisms (C$_6$H$_6$/hexane). Mp 209-210°. Cryst. struct. detn. is on 6-bromo deriv.

Misumi, S. *et al*, *Bull. Chem. Soc. Jpn.*, 1978, **51**, 2668.
Koizumi, Y. *et al*, *Bull. Chem. Soc. Jpn.*, 1986, **59**, 3511 (*cryst struct*)

[4]Paracyclophane   P-60006

Bicyclo[4.2.2]deca-6,8,9-triene, 9CI

[7124-96-1]

$C_{10}H_{12}$ M 132.205

Reactive intermediate generated photochemically at −20°.

Kostermans, G.B.M. *et al, J. Am. Chem. Soc.*, 1987, **109**, 2471.

[2.2][2.2]Paracyclophane-5,8-quinone   P-60007

Pentacyclo[18.2.2.2^{9,12}.0^{4,15}.0^{6,17}]hexacosa-4(15),6(17),9,11,20,22,23,25-octaene-5,16-dione, 9CI

[57468-59-4]

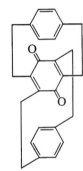

$C_{26}H_{24}O_2$ M 368.474

Yellow prismatic cryst. Mp >250° subl. dec.

Machida, H. *et al, Bull. Chem. Soc. Jpn.*, 1980, **53**, 2943 (*synth, pmr, uv*)
Toyoda, T. *et al, Bull. Chem. Soc. Jpn.*, 1986, **59**, 3994 (*cryst struct*)

[2.2][2.2]Paracyclophane-12,15-quinone   P-60008

Pentacyclo[18.2.2^{9,12}.0^{4,15}.0^{6,17}]hexacosa-4,6(17),9(26),11,15,20,22,23-octaene-10,25-dione, 9CI

[57468-58-3]

$C_{26}H_{24}O_2$ M 368.474

Reddish-orange cryst. (C$_6$H$_6$). Mp 225° dec.

Machida, H. *et al, Bull. Chem. Soc. Jpn.*, 1980, **53**, 2943 (*synth, uv, pmr*)

Paralycolin *A*   P-60009

$C_{22}H_{22}O_4$ M 350.413

Constit. of roots of *Clusia paralycola*. Cytotoxic. Light-brown solid (CH$_2$Cl$_2$/heptane). Mp 95-98°. $[\alpha]_D^{20}$ −88° (c, 0.66 in CHCl$_3$).

Monache, F.D. *et al, Tetrahedron Lett.*, 1987, **28**, 563.

β-Patchoulene   P-60010

Updated Entry replacing P-00166

1,2,3,4,5,6,7,8-Octahydro-1,4,9,9-tetramethyl-4,7-methanoazulene, 9CI

[514-51-2]

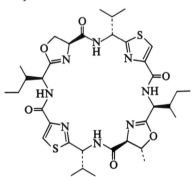

$C_{15}H_{24}$ M 204.355

Constit. of guaiac wood oil (*Bulnesia sarmienti*) and patchouli oil (*Pogostemon patchouli*). Oil. Bp$_{0.6}$ 66.8°. $[\alpha]_D^{30}$ −42.6° (c, 10.5 in CHCl$_3$).

Büchi, G. *et al, J. Am. Chem. Soc.*, 1961, **83**, 927 (*isol, struct*)
Bates, R.B. *et al, J. Am. Chem. Soc.*, 1962, **84**, 1307 (*synth*)
Akhita, A. *et al, Phytochemistry*, 1987, **26**, 2705 (*biosynth*)

Patellamide *A*   P-60011

Updated Entry replacing P-30017

[81120-73-2]

$C_{35}H_{50}N_8O_6S_2$ M 742.950

Cyclopeptide from the marine tunicate *Lissoclinum patella*. Cytotoxic, shows antineoplastic props. Cryst. (C$_6$H$_6$). Mp 228-229°. $[\alpha]_D^{24}$ +140.7° (+113.9°) (c, 0.27 in CHCl$_3$).

Ireland, C.M. *et al, J. Org. Chem.*, 1982, **47**, 1807 (*isol, ir, pmr, cmr*)
Biskupiak, J.E. *et al, J. Org. Chem.*, 1983, **48**, 2302 (*abs config*)
Hamada, Y. *et al, Tetrahedron Lett.*, 1985, **26**, 6501 (*synth*)

Patellamide *B*   P-60012

Updated Entry replacing P-30018

[81098-23-9]

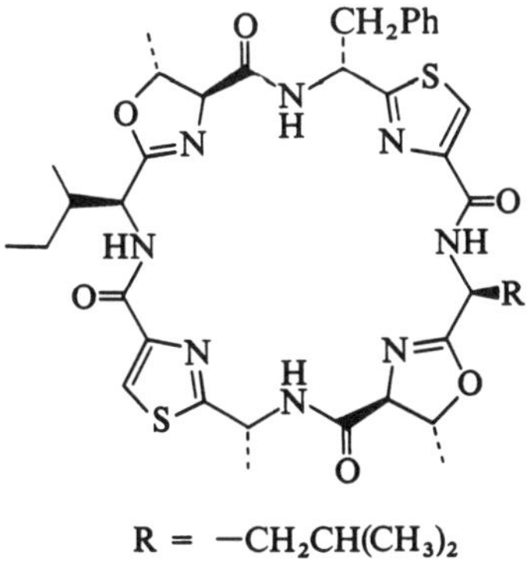

R = —CH₂CH(CH₃)₂

R = $-CH_2CH(CH_3)_2$

$C_{38}H_{48}N_8O_6S_2$ M 776.967

Cyclopeptide from the marine tunicate *Lissoclinum pa-tella*. Cytotoxic, shows antineoplastic props. $[\alpha]_D$ +29.4° (c, 0.34 in CH_2Cl_2).

Ireland, C.M. *et al*, *J. Org. Chem.*, 1982, **47**, 1807 (*isol, ir, pmr, cmr*)
Biskupiak, J.E. *et al*, *J. Org. Chem.*, 1983, **48**, 2302 (*abs config*)
Hamada, Y. *et al*, *Tetrahedron Lett.*, 1985, **26**, 5155, 5159 (*synth, struct*)
Schmidt, U. *et al*, *Tetrahedron Lett.*, 1986, **27**, 163, 179 (*synth, struct*)

Patellamide *C*   P-60013

Updated Entry replacing P-30019

[81120-74-3]

As Patellamide *B*, P-60012 with

R = —CH(CH₃)₂

R = $-CH(CH_3)_2$

$C_{37}H_{46}N_8O_6S_2$ M 762.940

Cyclopeptide from the marine tunicate *Lissoclinum pa-tella*. Antineoplastic agent. $[\alpha]_D$ +19° (c, 0.21 in CH_2Cl_2).

Ireland, C.M. *et al*, *J. Org. Chem.*, 1982, **47**, 1807 (*isol, ir, pmr, cmr*)
Biskupiak, J.E. *et al*, *J. Org. Chem.*, 1983, **48**, 2302 (*abs config*)
Hamada, Y. *et al*, *Tetrahedron Lett.*, 1985, **26**, 5155, 5159 (*synth, struct*)

Pedonin   P-60014

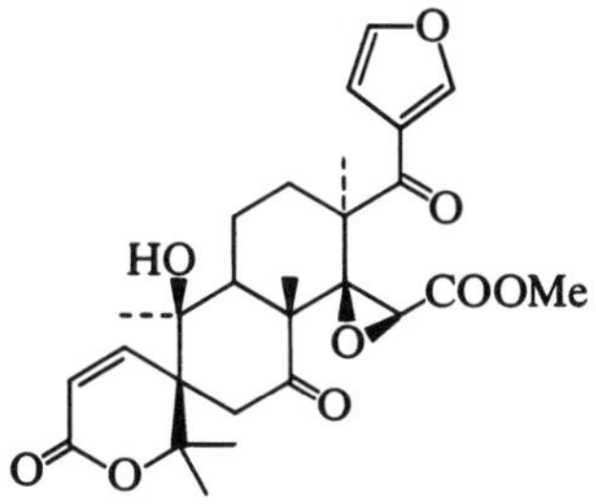

$C_{27}H_{32}O_9$ M 500.544

Constit. of *Harrisonia abyssinica*. Needles (Me₂CO). Mp 259-261°.

Hassanali, A. *et al*, *Phytochemistry*, 1987, **26**, 573 (*isol, cryst struct*)

Penlanpallescensin   P-60015

Updated Entry replacing P-40016

*3-[2-(2,2-Dimethyl-6-methylenecyclohexyl)ethyl]furan,
9Cl. Dihydropallescensin 2*

[83631-17-8]

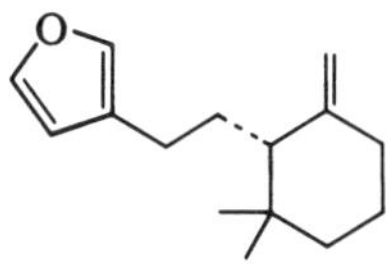

$C_{15}H_{22}O$ M 218.338

(*S*)-form [83631-17-8]

Constit. of *Dysidea fragilis*. Oil. $[\alpha]_D^{20}$ +6° (c, 0.3 in CHCl₃).

Guella, G. *et al*, *Helv. Chim. Acta*, 1985, **68**, 39.
Kurth, M.J. *et al*, *Tetrahedron Lett.*, 1987, **28**, 1031 (*synth*)

Penstebioside   P-60016

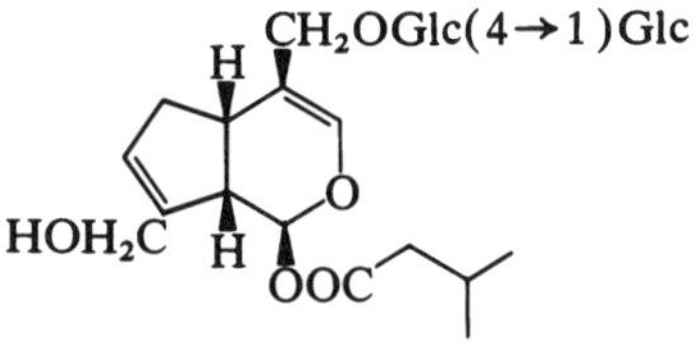

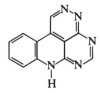

$C_{27}H_{42}O_{15}$ M 606.620

Constit. of *Penstemon richardsonii*. Yellow amorph. powder (CHCl₃/MeOH). Mp 84-86°.

Gering, B. *et al*, *Phytochemistry*, 1987, **26**, 753.

7*H*-2,3,4,6,7-Pentaazabenz[*de*]anthracene,   P-60017
9Cl

[109178-63-4]

$C_{12}H_7N_5$ M 221.221

CAS number refers to 3*H*-tautomer. Cryst. (MeOH/DMF). Mp >300°.

Stanovnik, B. *et al*, *Synthesis*, 1986, 807 (*synth, pmr*)

1,5,9,13,17-Pentaazacycloeicosane, 9Cl   P-60018

[84030-65-9]

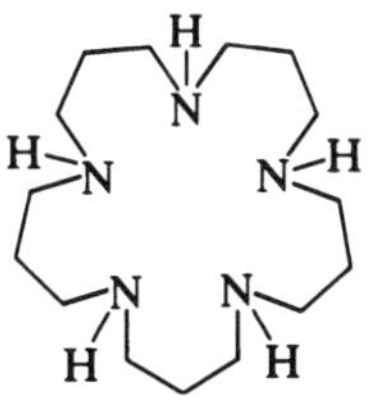

$C_{15}H_{35}N_5$ M 285.475

Cryst. (pet. ether). Mp 79-81°.

B,5HBr: [108577-49-7]. Cryst. + 1½H₂O (2-propanol aq.). Mp 251-254° dec.

Pentakis(*4-methylbenzenesulfonyl*): Cryst. (CHCl₃/MeOH). Mp 211-212°.

Osvath, P. *et al*, *Aust. J. Chem.*, 1987, **40**, 347 (*synth, pmr, cmr*)

1,4,7,10,14-Pentaazacycloheptadecane, 9CI   P-60019
[108577-38-4]

$C_{12}H_{29}N_5$ M 243.395

Hygroscopic low-melting cryst. $Bp_{0.5}$ 155-165°.

B,5HBr: [108577-41-9]. Cryst. + $\frac{1}{2}H_2O$ (HBr aq-./MeOH). Mp 250-254° dec.

Pentakis(4-methylbenzenesulfonyl): Cryst. (CHCl$_3$/EtOH). Mp 214-215°.

Osvath, P. *et al, Aust. J. Chem.,* 1987, **40**, 347 (*synth, pmr, cmr*)

1,4,7,11,14-Pentaazacycloheptadecane, 9CI   P-60020
[66802-95-7]

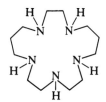

$C_{12}H_{29}N_5$ M 243.395

Extremely hygroscopic cryst.

B,5HBr: [108577-42-0]. Cryst. (HBr aq.). Mp 251-254° dec.

Pentakis(4-methylbenzenesulfonyl): Cryst. (CHCl$_3$/EtOH). Mp 219-220°.

Osvath, P. *et al, Aust. J. Chem.,* 1987, **40**, 347 (*synth, pmr, cmr*)

1,4,7,10,13-Pentaazacyclohexadecane, 9CI   P-60021
Updated Entry replacing P-30023

[29783-72-0]

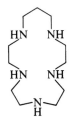

$C_{11}H_{27}N_5$ M 229.368

Cryst. + 1 H_2O (pet. ether). Mp 97-104° (94°).

B,5HBr: [108577-40-8]. Dihydrate. Mp 250-252° dec.

Pentakis(4-methylbenzenesulfonyl): [77320-30-0]. Cryst. (CHCl$_3$/MeOH). Mp 222-223°.

Org. Synth., 1978, **58**, 86 (*synth*)
Bencini, A. *et al, Inorg. Chem.,* 1981, **20**, 2544 (*synth*)
Bambieri, G. *et al, Inorg. Chim. Acta,* 1982, **61**, 43 (*synth*)
Osvath, P. *et al, Aust. J. Chem.,* 1987, **40**, 347 (*synth, pmr, cmr*)

1,4,8,12,16-Pentaazacyclononadecane, 9CI   P-60022
[108577-36-2]

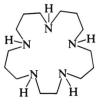

$C_{14}H_{33}N_5$ M 271.448

V. hygroscopic oil or cryst. Mp 45-53°. $Bp_{0.5}$ 170-175°.

B,5HBr: [108577-47-5]. Cryst. + $1\frac{1}{2}H_2O$ (2-propanol aq.). Mp 247.5-249° dec.

Pentakis(4-methylbenzenesulfonyl): Cryst. (Me_2CO/EtOH aq.). Mp 194-195°.

Osvath, P. *et al, Aust. J. Chem.,* 1987, **40**, 347 (*synth, pmr, cmr*)

1,4,7,11,15-Pentaazacyclooctadecane, 9CI   P-60023
[108577-39-5]

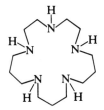

$C_{13}H_{31}N_5$ M 257.421

Extremely hygroscopic oil.

B,5HBr: [108577-44-2]. Cryst. + $1\frac{1}{2}H_2O$ (HBr/2-propanol).

Pentakis(4-methylbenzenesulfonyl): Cryst. (EtOH). Mp 101-103°.

Osvath, P. *et al, Aust. J. Chem.,* 1987, **40**, 347 (*synth, pmr, cmr*)

1,4,8,11,15-Pentaazacyclooctadecane, 9CI   P-60024
[108577-37-3]

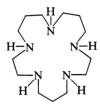

$C_{13}H_{31}N_5$ M 257.421

V. hygroscopic oil.

B,5HBr: [108603-02-7]. Cryst. + $\frac{1}{2}H_2O$ (HBr aq-./MeOH). Mp 258-262° dec.

Pentakis(4-methylbenzenesulfonyl): Cryst. (CHCl$_3$/EtOH). Mp 165-167°.

Osvath, P. *et al, Aust. J. Chem.,* 1987, **40**, 347 (*synth, pmr, cmr*)

1,4,7,10,13-Pentaazacyclopentadecane, 9CI   P-60025

Updated Entry replacing P-30024

[295-64-7]

$C_{10}H_{25}N_5$ M 215.341

Mp 97-100° (92°). Bp$_{0.5}$ 110-120°.

B,5HCl: [58066-59-4]. Mp 245-260° dec.

B,5HBr: [108603-00-5]. Cryst. + 3 H$_2$O (HBr aq.). Mp 251-254° dec.

Pentakis(4-methylbenzenesulfonyl): [52601-74-8]. Mp 278-280°.

Org. Synth., 1978, **58**, 86 (synth)

Bencini, A. *et al, Inorg. Chem.*, 1981, **20**, 2544 (synth)

Hay, R.W. *et al, J. Chem. Soc., Dalton Trans.*, 1982, 2131 (synth)

Osvath, P. *et al, Aust. J. Chem.*, 1987, **40**, 347 (synth, pmr, cmr)

1,5,9,13,17-Pentaazaheptadecane   P-60026

N-(*3-Aminopropyl*)-N′-[*3-[(3-aminopropyl)amino]-propyl*]-*1,3-propanediamine, 9CI*

[13274-42-5]

$$H_2N(CH_2)_3NH(CH_2)_3NH(CH_2)_3NH(CH_2)_3NH_2$$

$C_{12}H_{31}N_5$ M 245.410

Bp$_{0.2}$ 150-154°.

Osvath, P. *et al, Aust. J. Chem.*, 1987, **40**, 347 (synth, pmr, cmr, ms)

Pentacyclo[5.3.0.0^{2,5}.0^{3,9}.0^{4,8}]decan-6-one   P-60027

C$_2$-Bishomocuban-6-one

$C_{10}H_{10}O$ M 146.188

(+)-*form* [62928-74-9]

Cryst. by subl. Mp 123-124°. [α]$_D^{15}$ +11.0° (c, 0.519 in CHCl$_3$).

(±)-*form*

Mp 124-126°.

Cookson, R.C. *et al, Tetrahedron Lett.*, 1960, **no. 22**, 99 (synth, ir)

Nakasaki, M. *et al, J. Org. Chem.*, 1977, **42**, 2985 (synth, cd)

Klunder, A.J.H. *et al, Tetrahedron Lett.*, 1986, **27**, 2543 (synth)

Pentacyclo[6.4.0.0^{2,7}.0^{3,12}.0^{6,9}]dodeca-4,10-diene   P-60028

2a,2b,4a,4b,4c,4d,4e,4f-Octahydrodicyclobuta[def,jkl]-biphenylene, 9CI

[104286-26-2]

$C_{12}H_{12}$ M 156.227

Benzene dimer. Cryst. (MeOH) with camphene odour. Mp 73-75°.

Yang, N.C. *et al, Tetrahedron Lett.*, 1986, **27**, 543 (synth, ms, pmr, cmr)

Pentacyclo[12.2.2.2^{2,5}.2^{6,9}.2^{10,13}]tetracosa-1,5,9,13-tetraene, 9CI   P-60029

[99922-00-6]

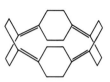

$C_{24}H_{32}$ M 320.517

Forms a highly symmetrical Ag complex. Mp >300°.

McMurry, J.E. *et al, J. Am. Chem. Soc.*, 1986, **108**, 515 (synth, props)

Pentacyclo[5,4.0.0^{2,6}.0^{3,10}.0^{5,9}]undecane-1,11-dione   P-60030

[2958-72-7]

$C_{11}H_{10}O_2$ M 174.199

Cookson, R.C. *et al, J. Chem. Soc.*, 1964, 3062.

Marchand, A.P. *et al, J. Org. Chem.*, 1974, **39**, 1596.

Sasaki, T. *et al, Tetrahedron*, 1974, **30**, 2707.

Pentacyclo[6.3.0.0^{2,6}.0^{5,9}]undecane-4,7,11-trione   P-60031

Hexahydro-1,3,5-methenocyclopenta[cd]pentalene-2,4,6(1H)-trione, 9CI. D$_3$-Trishomocubanetrione

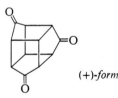

(+)-*form*

$C_{11}H_8O_3$ M 188.182

(+)-*form*

Cryst. by subl. Mp 290-291°. [α]$_D^{20}$ +949° (c, 0.407 in CHCl$_3$).

(−)-*form*

Mp 290-291°. [α]$_D^{20}$ −923° (0.374 in CHCl$_3$).

(±)-*form* [108635-94-5]

Needles by subl. Mp 290-291°.

Tris(ethylenethioketal): Long needles (CH$_2$Cl$_2$/EtOH). Mp >340°.

Fessner, W.-D. *et al, Tetrahedron*, 1986, **42**, 1797 (synth, resoln, ir, uv, pmr, cmr, ms)

3,3,4,4-Pentafluoro-1-butyne, 9CI, 8CI   P-60032

(Pentafluoroethyl)acetylene

[7096-51-7]

$$F_3CCF_2C{\equiv}CH$$

C_4HF_5 M 144.044
Gas. Bp −12°.

Haszeldine, R.N. *et al*, *J. Chem. Soc.*, 1952, 3483 (*synth*)
Cullen, W.R. *et al*, *Can. J. Chem.*, 1969, **47**, 3093 (*F nmr*)

1,3,3,4,4-Pentafluorocyclobutene   P-60033
[374-31-2]

C_4HF_5 M 144.044
Bp 25-26°.

Buxton, M.W. *et al*, *J. Chem. Soc.*, 1954, 1177 (*synth*)

Pentafluoroiodosobenzene, 9CI   P-60034
Pentafluoroiodosylbenzene. PFIB
[14353-90-3]

$(C_6F_5)IO$

C_6F_5IO M 309.962
Epoxidising agent. Yellow solid. Mp 88-89° dec.

Schmeisser, M. *et al*, *Chem. Ber.*, 1967, **100**, 1633 (*synth*)

Pentafluoroisocyanatobenzene, 9CI   P-60035
Isocyanic acid pentafluorophenyl ester, 8CI
[1591-95-3]

C_6F_5NCO

C_7H_5NO M 119.123
Bp 164-170°. Readily hydrolyses in air.

Malichenko, B.F. *et al*, *Zh. Obsch. Khim.*, 1968, **38**, 2497 (*Engl. transl.* p. 2413) (*synth*)
De Pasquale, R.J. *et al*, *J. Fluorine Chem.*, 1976, **8**, 311 (*trimer, use*)

Pentafluoroisocyanobenzene, 9CI   P-60036
Pentafluorophenyl isocyanide
[58751-09-0]

$(C_6F_5)NC$

C_7F_5N M 193.076
Pink foul-smelling liq.

Banks, R.E. *et al*, *J. Chem. Soc., Perkin Trans. 1*, 1975, 2451 (*synth, F nmr, ms*)
Lentz, D., *Ann. Chim. (Paris)*, 1984, **9**, 665 (*synth*)

Pentafluoroisothiocyanatobenzene, 9CI   P-60037
Pentafluorophenylisothiocyanate
[35923-79-6]

C_6F_5NCS

C_7F_5NS M 225.136
Liq. Bp_{10} 71°.

Herkes, F.E. *et al*, *J. Fluorine Chem.*, 1979, **13**, 1 (*synth, F nmr, props*)

Pentafluoromethylbenzene, 9CI   P-60038
2,3,4,5,6-Pentafluorotoluene, 8CI
[771-56-2]

$C_7H_3F_5$ M 182.093
Bp 115-118°.

Barbour, A.K. *et al*, *J. Chem. Soc.*, 1961, 808 (*synth, ir*)

Pentafluoronitrosobenzene, 9CI   P-60039
[1423-13-8]

$(C_6F_5)NO$

C_6F_5NO M 197.064
Monomeric in soln., dimeric in cryst. state. Green cryst. Mp 44-45°.

Brooke, G.M. *et al*, *Chem. Ind. (London)*, 1961, 832 (*synth*)
Birchall, J.M. *et al*, *J. Chem. Soc.*, 1962, 4977 (*synth*)
Pushkina, L.N. *et al*, *Org. Magn. Reson.*, 1972, **4**, 607 (*F nmr*)
Prout, C.K. *et al*, *Cryst. Struct. Commun.*, 1974, **3**, 39 (*cryst struct, dimer*)
Furin, G.G. *et al*, *J. Fluorine Chem.*, 1983, **22**, 231, 345 (*N nmr, F nmr*)

1-(Pentafluorophenyl)ethanol   P-60040
2,3,4,5,6-Pentafluoro-α-methylbenzenemethanol, 9CI.
2,3,4,5,6-Pentafluoro-α-methylbenzyl alcohol, 8CI
[830-50-2]

$C_8H_5F_5O$ M 212.119
(**R**)-*form* [104371-21-3]
Cryst. (pentane). Mp 42.5-43°. $[\alpha]_D^{29}$ +7.0° (c, 1.1 in pentane).
(**S**)-*form* [104371-20-2]
Cryst. (pentane or by subl.). Mp 42°. $[\alpha]_D^{28}$ −7.1° (c, 1.0 in pentane).
(**±**)-*form* [75853-08-6]
Bp_{30} 102-103°. n_D^{17} 1.4394.

Vorozhtsov, N.N. *et al*, *Dokl. Akad. Nauk SSSR, Ser. Sci. Khim.*, 1964, **159**, 125; *CA*, **62**, 4045c (*synth*)
Meese, C.O., *Justus Liebigs Ann. Chem.*, 1986, 2004 (*synth, pmr*)

Pentafluoro(trifluoromethyl)sulfur, 9CI, 8CI   P-60041
(Trifluoromethyl)sulfur pentafluoride
[373-80-8]

F_3CSF_5

CF_8S M 196.058
Electrical insulator. Gas. Mp −86.9°. Bp −20.4°. Stable to alkali.

Silvey, G.A., *J. Am. Chem. Soc.*, 1950, **72**, 3624 (*synth*)
Clifford, A.F. *et al*, *J. Chem. Soc.*, 1953, 2372 (*synth*)
Merrill, C.I. *et al*, *Inorg. Chem.*, 1962, **1**, 215 (*F nmr*)
Griffiths, J.E., *Spectrochim. Acta, Part A*, 1967, **23**, 2145 (*ir, raman*)

Marsden, C.J. *et al*, *J. Mol. Struct.*, 1985, **131**, 299 (*struct, ed*)

Pentahydroxybenzaldehyde   P-60042

$C_7H_6O_6$ M 186.121

Penta-Me ether: [86475-22-1].
Pentamethoxybenzaldehyde.
$C_{12}H_{16}O_6$ M 256.255
Oil. Bp_1 122-125°.

Dallacker, F., *Justus Liebigs Ann. Chem.*, 1963, **665**, 78 (*synth*)
Brossi, A. *et al*, *Helv. Chim. Acta*, 1983, **66**, 795 (*synth, ir, pmr*)

Pentahydroxybenzoic acid   P-60043

$C_7H_6O_7$ M 202.120

Penta-Me ether: Pentamethoxybenzoic acid.
$C_{12}H_{16}O_7$ M 272.254
Cryst. (EtOAc). Mp 95.5°.
Penta-Me ether, Me ester:
$C_{13}H_{18}O_7$ M 286.281
Leaflets (MeOH aq.). Mp 48.5°. Bp_2 150-153°.

Dallacker, F., *Justus Liebigs Ann. Chem.*, 1963, **665**, 78 (*synth, deriv, ir*)
Parker, K.A. *et al*, *J. Org. Chem.*, 1987, **52**, 674 (*synth, deriv, ir, pmr*)

3,3',4,4',5-Pentahydroxybibenzyl   P-60044

5-[2-(3,4-Dihydroxyphenyl)ethyl]-1,2,3-benzenetriol. 1-(3,4-Dihydroxyphenyl)-2-(3,4,5-trihydroxyphenyl)-ethane

$C_{14}H_{14}O_5$ M 262.262

3,4-Methylene ether, 3',5-Di-Me ether: [80784-20-9]. *4'-Hydroxy-3',5-dimethoxy-3,4-methylenedioxybibenzyl.*
$C_{17}H_{18}O_5$ M 302.326
Constit. of *Frullania* spp. Cryst. Mp 62-63°.

3,3',5-Tri-Me ether: 4,4'-Dihydroxy-3,3',5-trimethoxybibenzyl. **Aloifol II. Moscatilin.**
$C_{17}H_{20}O_5$ M 304.342
Constit. of *Cymbidium aloifolium* and of *Dendrobium moscatum*. Viscous compound or cryst. (EtOAc/pet. ether). Mp 84°.

Asakawa, Y. *et al*, *Phytochemistry*, 1987, **26**, 1117 (*isol*)
Juneja, R.K. *et al*, *Phytochemistry*, 1987, **26**, 1123 (*isol*)
Mujumder, P.L. *et al*, *Phytochemistry*, 1987, **26**, 2121 (*isol*)

2',3,5,7,8-Pentahydroxyflavanone   P-60045

$C_{15}H_{12}O_7$ M 304.256

7,8-Di-Me ether: 2',3,5-Trihydroxy-7,8-dimethoxyflavanone.
$C_{17}H_{16}O_7$ M 332.309

Constit. of *Notholaena neglecta*. Cryst. Mp 212°.
7,8-Di-Me ether, 2'-Ac: 2'-Acetoxy-3,5-dihydroxy-7,8-dimethoxyflavone.
$C_{19}H_{18}O_8$ M 374.346
Constit. of *N. neglecta*.

Scheele, C. *et al*, *J. Nat. Prod.*, 1987, **50**, 181.

3',4',5,6,7-Pentahydroxyflavanone   P-60046

Updated Entry replacing P-30041
$C_{15}H_{12}O_7$ M 304.256

7-Me ether: 3',4',5,6-Tetrahydroxy-7-methoxyflavanone.
$C_{18}H_{18}O_7$ M 346.336
Constit. of *Holocarpha obconica*.
4',6,7-Tri-Me ether: 3',5-Dihydroxy-4',6,7-trimethoxyflavanone.
$C_{16}H_{14}O_7$ M 318.282
Constit. of *Vitex negundo*. Yellow cryst. (CHCl₃/pet. ether). Mp 136-138°.
6-Me ether: 3',4',5,7-Tetrahydroxy-6-methoxyflavanone.
$C_{16}H_{14}O_7$ M 318.282
Constit. of *Eupatorium subhastatum*. Cryst. (MeOH). Mp 218-220°.

Achari, B. *et al*, *Phytochemistry*, 1984, **23**, 703.
Chandra, S. *et al*, *Indian J. Chem., Sect. B*, 1987, **26**, 82 (*synth*)
Crins, W.J. *et al*, *Phytochemistry*, 1987, **26**, 2128.
Ferraro, G. *et al*, *Phytochemistry*, 1987, **26**, 3092 (*deriv*)

3,4',5,6,7-Pentahydroxyflavone   P-60047

Updated Entry replacing P-20044
[4324-55-4]
$C_{15}H_{10}O_7$ M 302.240
Yellow cryst. (EtOH/Me₂CO). Mp 328-330°.

5,6-Di-Me ether: 3',4',7-Trihydroxy-5,6-dimethoxyflavone.
$C_{17}H_{14}O_7$ M 330.293
Constit. of *Adenostoma sparsifolium*.
4',6,7-Tri-Me ether: [4324-53-2]. *3,5-Dihydroxy-4',6,7-trimethoxyflavone.* **Mikanin**.
$C_{18}H_{16}O_7$ M 344.320
Constit. of *Mikania cordata*. Bright-yellow cryst. (C₆H₆ or CHCl₃/MeOH). Mp 222-224°.
5,6,7-Tri-Me ether: 3,4'-Dihydroxy-5,6,7-trimethoxyflavone. **Candidol**.
$C_{18}H_{16}O_7$ M 344.320
Constit. of *Tephrosia candida*. Cryst. Mp 253-254°.
3,6,7-Tri-Me ether: 4',5-Dihydroxy-3,6,7-trimethoxyflavone. **Penduletin**.
$C_{18}H_{16}O_7$ M 344.320
Constit. of *Tephrosia candida*. Yellow cryst. (MeOH). Mp 222°.
3,4',5,7-Tetra-Me ether: 6-Hydroxy-3,4',5,7-tetramethoxyflavone.
$C_{19}H_{18}O_7$ M 358.347
Constit. of *Blumea malcomii*. Cryst. Mp 178-180°.
3,4',6,7-Tetra-Me ether: [14787-34-9]. *5-Hydroxy-3,4',6,7-tetramethoxyflavone.*
$C_{19}H_{18}O_7$ M 358.347
Isol. from *Dodonaea lobulata* and *D. viscosa*. Cryst. (Et₂O), yellow needles (Me₂CO/hexane). Mp 151.5-153°, 176°.
Penta-Me ether: 3,4',5,6,7-Pentamethoxyflavone.
$C_{20}H_{20}O_7$ M 372.374
Needles. Mp 157-158° (151-153°).

Kiang, A.K. *et al*, *J. Chem. Soc.*, 1965, 6371 (*synth*)

Wagner, H. *et al*, *Tetrahedron Lett.*, 1965, 3849 (*synth*)
Dawson, R.M. *et al*, *Aust. J. Chem.*, 1966, **19**, 2133 (*isol*)
Sim, K.Y., *J. Chem. Soc.* (*C*), 1967, 976 (*synth*)
Wagner, H. *et al*, *Chem. Ber.*, 1967, **100**, 1768 (*synth*)
Rodriguez, E. *et al*, *Phytochemistry*, 1972, **11**, 3509 (*isol, struct*)
Southwick, L. *et al*, *Phytochemistry*, 1972, **11**, 2351 (*isol, struct*)
Proksch, M. *et al*, *Phytochemistry*, 1982, **21**, 2893 (*isol*)
Dutt, S.K. *et al*, *Phytochemistry*, 1983, **22**, 325 (*isol*)
Sachdev, K. *et al*, *Phytochemistry*, 1983, **22**, 1253 (*isol*)
Kulkarni, M.M. *et al*, *Phytochemistry*, 1987, **26**, 2079.
Parmar, V.S. *et al*, *Tetrahedron*, 1987, **43**, 4241 (*cryst struct*)

3,4',5,6,8-Pentahydroxyflavone   P-60048

Updated Entry replacing P-40035

$C_{15}H_{10}O_7$ M 302.240

3,4',8-Tri-Me ether: 5,6-Dihydroxy-3,4',8-
trimethoxyflavone.
$C_{18}H_{16}O_7$ M 344.320
Orange-yellow needles (MeOH). Mp 188.5-189.5°.
Formerly thought to occur in *Conyza stricta*; however,
this compd. was shown to be the 5,7-dihydroxy isomer.
3,6,8-Tri-Me ether: 4',5-Dihydroxy-3,6,8-trimethoxy-
flavone. **Candirone.**
$C_{18}H_{16}O_7$ M 344.320
Constit. of seeds of *Tephrosia candida*. Needles
(MeOH). Mp 232-234°.

Horie, T. *et al*, *Bull. Chem. Soc. Jpn.*, 1982, **55**, 2933 (*synth, struct*)
Parmar, V.S. *et al*, *Tetrahedron*, 1987, **43**, 4241 (*Candirone*)

3',4',5,6,7-Pentahydroxyisoflavone   P-60049

$C_{15}H_{10}O_7$ M 302.240

6,7-Di-Me ether: 3',4',5-Trihydroxy-6,7-dimethoxyiso-
flavone. 6-Methoxyorobol 7-methyl ether.
$C_{17}H_{14}O_7$ M 330.293
Constit. of *Wyethia reticulata*.

McCormick, S.P. *et al*, *Phytochemistry*, 1987, **26**, 2421.

1,2,5,6,7-Pentahydroxyphenanthrene   P-60050

1,2,5,6,7-Phenanthrenepentol

$C_{14}H_{10}O_5$ M 258.230

1,5,6-Tri-Me ether: [108909-02-0]. *1,5,6-Trimethoxy-
2,7-phenanthrenediol, 9CI. 2,7-Dihydroxy-1,5,6-tri-
methoxyphenanthrene.* **Confusarin.**
$C_{17}H_{16}O_5$ M 300.310
Constit. of *Eria confusa*. Cryst. (EtOAc/pet. ether).
Mp 185°.

Majumder, P.L. *et al*, *Phytochemistry*, 1987, **26**, 1127.

1,2,3,4,8-Pentahydroxyxanthone   P-60051

$C_{13}H_8O_7$ M 276.202

1,2,4-Tri-Me ether: [110187-38-7]. *3,8-Dihydroxy-
1,2,4-trimethoxyxanthone.*
$C_{16}H_{14}O_7$ M 318.282
Constit. of *Psorospermum febrifugum*. Yellow needles
(MeOH/Et₂O). Mp 182-184°.
Penta-Me ether: [22804-63-3]. *1,2,3,4,8-
Pentamethoxyxanthone.*
$C_{18}H_{18}O_7$ M 346.336

Cryst. Mp 110-112°.

Habib, A.M. *et al*, *J. Nat. Prod.*, 1987, **50**, 141.

1,2,5,6,8-Pentahydroxyxanthone   P-60052

Updated Entry replacing P-20049

1,2,5,6,8-Pentahydroxy-9H-xanthen-9-one, 9CI.
1,3,4,7,8-Pentahydroxyxanthone (incorrect). Bellidin
$C_{15}H_{12}O_7$ M 304.256

2,5-Di-Me ether: [5041-99-6]. *1,3,8-Trihydroxy-4,7-di-
methoxyxanthone. 4,7-Di-O-methylbellidin.*
$C_{17}H_{16}O_7$ M 332.309
Isol. from roots of *Gentiana bellidifolia*. Yellow cryst.
(C₆H₆). Mp 220-221°.
2,6-Di Me ester: [64181-95-9]. *1,5,8-Trihydroxy-2,6-di-
methoxyxanthone.* **Lanceolin.**
$C_{15}H_{12}O_7$ M 304.256
Constit. of *Triptospermum lanceolatum*. Yellow
needles (MeOH). Mp 233-234°.
2,6-Di Me ether, 5-glucoside: [81992-00-9]. **Lanceoside.**
$C_{21}H_{22}O_{12}$ M 466.398
Isol. from *T. lanceolatum*. Yellow needles (MeOH).
Mp 238-242°.
1,2,5-Tri-Me ether: [111509-25-2]. *6,8-Dihydroxy-
1,2,5-trimethoxyxanthone.* **Methyllanceolin.**
$C_{16}H_{14}O_7$ M 318.282
From *T. lanceolatum*. Yellow needles (MeOH). Mp
218-220°.

Markham, K.R., *Tetrahedron*, 1965, **21**, 3687 (*isol*)
Lin, C.-N. *et al*, *Phytochemistry*, 1982, **21**, 205 (*Lanceoside*)
Lin, C.-N. *et al*, *Phytochemistry*, 1987, **26**, 2381 (*Lanceolin, Methyllanceolin*)

Pentaisopropylidenecyclopentane   P-60053

Pentakis(1-methylethylidene)cyclopentane, 9CI.
Decamethyl[5]radialene
[108404-40-6]

$C_{20}H_{30}$ M 270.457
Prisms. Mp 100-101°.

Iyoda, M. *et al*, *J. Chem. Soc., Chem. Commun.*, 1986, 1794
(*synth, pmr, cmr, ms, uv, cryst struct*)

Pentalenic acid   P-60054

Updated Entry replacing P-00429
[69394-19-0]

$C_{15}H_{22}O_3$ M 250.337
Metab. of *Streptomyces* spp.

Seto, H. *et al*, *Tetrahedron Lett.*, 1978, 4411.
Crimmins, M.T. *et al*, *J. Org. Chem.*, 1984, **49**, 2076 (*synth*)
Ihara, M. *et al*, *J. Chem. Soc., Chem. Commun.*, 1987, 721
(*synth*)

Pentalenolactone *E*   P-60055

Updated Entry replacing P-20051
[72715-03-8]

$C_{15}H_{18}O_4$ M 262.305
Sesquiterpene antibiotic. Constit. of *Streptomyces* spp.

Cane, D.E. *et al, Tetrahedron Lett.*, 1979, 2973 (*isol*)
Paquette, L.A. *et al, J. Am. Chem. Soc.*, 1981, **103**, 6526
   (*synth*)
Ohtsuka, T. *et al, Tetrahedron Lett.*, 1983, **24**, 3851 (*synth*)
Thomas, P.J. *et al, J. Am. Chem. Soc.*, 1984, **106**, 5295 (*synth*)
Taber, D.F. *et al, J. Am. Chem. Soc.*, 1985, **107**, 5289 (*synth*)
Hua, D.H. *et al, Tetrahedron Lett.*, 1987, **28**, 5465 (*synth*)
Marino, J.P. *et al, J. Org. Chem.*, 1987, **52**, 4139 (*synth*)

2,3,5,6,7-Pentamethylenebicyclo[2.2.2]-   P-60056
octane

2,3,5,6,7-Pentamethylidenebicyclo[2.2.2]octane

$C_{13}H_{14}$ M 170.254
Cryst. (pentane). Mp 65-66°.

Burnier, G. *et al, Helv. Chim. Acta*, 1986, **69**, 1310 (*synth, ir,
   uv, pmr, cmr*)

1,2,3,4,5-Pentamethyl-6-nitrobenzene   P-60057

[13171-59-0]

$C_{11}H_{15}NO_2$ M 193.245
Mp 158-159°.

Susuki, H. *et al, Bull. Chem. Soc. Jpn.*, 1970, **43**, 473 (*synth, ir*)

Pentamethylphenol, 9CI, 8CI   P-60058

Updated Entry replacing P-00462
Hydroxypentamethylbenzene
[2819-86-5]

$C_{11}H_{16}O$ M 164.247
Needles (EtOH or pet. ether). Mp 126°. Bp 267°.
Ac: [73396-43-7].
   $C_{13}H_{18}O_2$ M 206.284
   Mp 70-71°.
Benzoyl:
   $C_{18}H_{20}O_2$ M 268.355

Plates (EtOH). Mp 127°.
Me ether: [14804-37-6]. *Methoxypentamethylbenzene,
   9CI. Pentamethylanisole, 8CI.*
   $C_{12}H_{18}O$ M 178.274
   Plates (Et$_2$O). Mp 119°. Bp 155-156°.

Hofmann, A.W., *Ber.*, 1885, **18**, 1826 (*synth*)
Kolka, A.J. *et al, J. Am. Chem. Soc.*, 1939, **61**, 1463 (*synth*)
Burawoy, A. *et al, J. Chem. Soc.*, 1949, 624 (*synth*)
Keumi, T. *et al, J. Org. Chem.*, 1986, **51**, 3439 (*deriv, synth,
   pmr*)

2,4-Pentanedione, 9CI, 8CI   P-60059

Updated Entry replacing P-00481
Acetylacetone. Diacetylmethane
[123-54-6]

$$H_3CCOCH_2COCH_3 \rightleftharpoons H_3CC(OH){=}CHCOCH_3$$

$C_5H_8O_2$ M 100.117
Tautomeric. Enol-form predominates in pure liq., vapour
and most solvs., oxo-form predominates (88%) in aq.
soln. Used in metal extraction. Protecting reagent for
aminoacids. Cleaves phenylhydrazones and
semicarbazones, intermed. in heterocyclic synth. Forms
many metal complexes.

▷Mod. toxic, flammable. SA1925000.

Oxo-form

Liq. Sol. dil. HCl, spar. sol. hot H$_2$O, misc. org. solvs. d_4^{25}
0.927. Mp −23°. Bp 134-136°, (Bp$_{746}$ 139°).
Imine: [870-74-6]. *4-Imino-2-pentanone, 9CI. 4-Amino-
3-penten-2-one.*
   C_5H_9NO M 99.132
   Cryst. (EtOAc). Mp 40°, 43°.
Monoanil: [880-12-6]. *4-(Phenylimino)-2-pentanone,
   9CI. 4-Anilino-3-penten-2-one.*
   $C_{11}H_{13}NO$ M 175.230
   Cryst. (pet. ether). Mp 51-53°. Bp 285-286°.
Dioxime: [2157-56-4].
   $C_5H_{10}N_2O_2$ M 130.146
   Prisms (Et$_2$O). Mp 151-152°.

Enol-form [1522-20-9]

4-Hydroxy-3-penten-2-one
Mp −9°. pK_a 8.93 (25°). Forms salts with many metals.
Exhibits intramolecular hydrogen bonding.
Me ether: [2845-83-2]. *4-Methoxy-3-penten-2-one.*
   $C_6H_{10}O_2$ M 114.144
   Bp$_{10}$ 58-59°. Most prepns. give mixt. of (*E*)- and (*Z*)-
   isomers.
Ac: [41002-50-0].
   $C_7H_{10}O_3$ M 142.154
   Bp$_{20}$ 95-96°.

Org. Synth., Coll. Vol., **3**, 16 (*synth*)
Koshimura, H. *et al, Bull. Chem. Soc. Jpn.*, 1973, **46**, 632 (*pmr*)
Grens, E. *et al, Spectrochim. Acta, Part A*, 1975, **31**, 555 (*ir*)
Shapet'ko, N.N. *et al, Org. Magn. Reson.*, 1975, **7**, 237 (*cmr*)
Soendergaard, N.C. *et al, Acta Chem. Scand., Ser. A*, 1975, **29**,
   709 (*tautom*)
Hush, N.S. *et al, Aust. J. Chem.*, 1987, **40**, 599 (*pe, tautom*)

> *The symbol ▷ in Entries highlights
> hazard or toxicity information*

4,7,13,16,21-Pentaoxa-1,10-diazabicyclo[8.8.5]tricosane, 9CI   P-60060

Cryptand 2.2.1

[31364-42-8]

$C_{16}H_{32}N_2O_5$ M 332.439

Metal chelating agent. Oil. $Bp_{0.001}$ 175°.

Dietrich, B. *et al, Tetrahedron*, 1973, **29**, 1629 (*synth*)
Foerster, H.G. *et al, J. Am. Chem. Soc.*, 1980, **102**, 6984 (^{15}N *nmr*)
Anelli, P.L. *et al, J. Org. Chem.*, 1985, **50**, 3453 (*synth*)

4-Pentene-2,3-dione   P-60061

Methylvinylglyoxal

[91238-45-8]

$$H_2C=CHCOCOCH_3$$

$C_5H_6O_2$ M 98.101

Dienophile.

Kramme, R. *et al, Angew. Chem., Int. Ed. Engl.*, 1986, **25**, 1116 (*synth, pmr, cmr, uv, pe*)

1-Penten-3-one, 9CI   P-60062

Updated Entry replacing P-00536

Ethyl vinyl ketone. Propionylethylene

[1629-58-9]

$$H_2C=CHCOCH_2CH_3$$

C_5H_8O M 84.118

Important synthetic intermed., esp. in annellation reactions for terpene ring construction. Flavour ingredient. Bp_{740} 102°, Bp_{90} 44°.

▷SB3800000.

2,4-Dinitrophenylhydrazone: [21454-46-6]. Mp 125-127°.

Woodward, R.B. *et al, J. Am. Chem. Soc.*, 1952, **74**, 4239 (*synth*)
Archer, A. *et al, J. Org. Chem.*, 1957, **22**, 92 (*synth*)
Grieco, P.A. *et al, J. Chem. Soc., Chem. Commun.*, 1974, 497 (*synth*)
Floyd, J.C., *Tetrahedron Lett.*, 1974, 2877 (*synth*)
Byrne, B. *et al, Synthesis*, 1986, 870 (*synth*)
Fieser, M. *et al, Reagents for Organic Synthesis*, Wiley, 1967-84, **1**, 388.

Perforenone   P-60063

[66113-28-8]

$C_{15}H_{22}O$ M 218.338

Not the same as Perforenone *A*, P-00618 . Constit. of *Laurencia perforata*. Oil. $[\alpha]_D$ −120°.

Gonzalez, A.G. *et al, Tetrahedron Lett.*, 1977, 3375; 1978, 481 (*isol, struct, synth*)
Majetich, G. *et al, Heterocycles*, 1987, **25**, 271 (*synth*)

Perrottetianal *A*   P-60064

Updated Entry replacing P-00654

4,4a,5,6,7,8-Hexahydro-2,5-dimethyl-5-(4-methyl-3-pentenyl)-1,8a(3H)-naphthalenedicarboxaldehyde, 9CI

[73483-87-1]

R = H

$C_{20}H_{30}O_2$ M 302.456

Constit. of *Porella perrottetiana*. Cryst. Mp 68-69°. $[\alpha]_D$ +282° (c, 2 in CHCl₃).

Asakawa, Y. *et al, Phytochemistry*, 1979, **18**, 1681 (*isol, struct*)
Hagiwara, H. *et al, J. Chem. Soc., Chem. Commun.*, 1987, 1351 (*synth*)

Persicaxanthin   P-60065

Updated Entry replacing P-10055

5,6-Epoxy-5,6-dihydro-12'-apo-β-carotene-3,12'-diol

[80952-82-5]

$C_{25}H_{36}O_3$ M 384.558

Constit. of *Prunus domistica*. Yellow plates (C_6H_6/pet. ether). Mp 92°.

12'-Aldehyde: **Apo-12'-violaxanthal**.
$C_{25}H_{34}O_3$ M 382.542
Constit. of *P. domestica*. Yellow pigment.

Gross, J. *et al, Phytochemistry*, 1981, **20**, 2267 (*isol*)
Molnár, P. *et al, Phytochemistry*, 1987, **26**, 1493 (*abs config*)

Petrostanol   P-60066

$C_{29}H_{50}O$ M 414.713

Constit. of *Petrosia hebes*. Cryst. (MeOH). Mp 122-123°.

Cho, J.-H. *et al, J. Chem. Soc., Perkin Trans. 1*, 1987, 1307.

Phaseolinic acid   P-60067

Tetrahydro-4-methyl-5-oxo-2-pentyl-3-furancarboxylic acid, 9CI
[109667-12-1]

$C_{11}H_{18}O_4$ M 214.261
Metab. of *Macrophomina phaseolina*. Needles (EtOAc).
Mp 139-140°. $[\alpha]_D$ −150° (c, 0.2 in $CHCl_3$).

Mahato, S.B. *et al, J. Nat. Prod.*, 1987, **50**, 245 (*isol, cryst struct*)

3,4-Phenanthrenedicarboxylic acid   P-60068

Updated Entry replacing P-00755
$C_{16}H_{10}O_4$ M 266.253
Di-Me ester:
 $C_{18}H_{14}O_4$ M 294.306
 Prisms (Et_2O/pet. ether). Mp 114.5-114.8°.
Anhydride: [5723-54-6].
 $C_{16}H_8O_3$ M 248.237
 Lemon-yellow needles (xylene). Mp 253.5-254°.

Fieser, L.F. *et al, J. Am. Chem. Soc.*, 1936, **58**, 2322.
Bunte, R. *et al, Chem. Ber.*, 1986, **119**, 3521 (*synth, uv, pmr, anhydride*)

3,6-Phenanthrenedimethanol, 9CI   P-60069

3,6-Bis(hydroxymethyl)phenanthrene
[66888-85-5]

$C_{16}H_{14}O_2$ M 238.285
Needles ($CHCl_3$). Mp 159-160°.

DuVernet, R.B. *et al, J. Am. Chem. Soc.*, 1978, **100**, 2457 (*synth, pmr*)

4,5-Phenanthrenedimethanol, 9CI   P-60070

4,5-Bis(hydroxymethyl)phenanthrene
[71628-27-8]
$C_{16}H_{14}O_2$ M 238.285
Needles + EtOH (EtOH). Mp 171-172°.
Di-Ac:
 $C_{20}H_{18}O_4$ M 322.360
 Prisms (pet. ether). Mp 105°.

Badger, G.M. *et al, J. Chem. Soc.*, 1950, 2326 (*synth*)
Katz, T.J. *et al, J. Am. Chem. Soc.*, 1979, **101**, 4259 (*synth, pmr*)
Rubin, Y. *et al, J. Org. Chem.*, 1986, **51**, 3270 (*synth, pmr*)

2,4,5-Phenanthrenetriol   P-60071

2,4,5-Trihydroxyphenanthrene
[70205-58-2]
$C_{14}H_{10}O_3$ M 226.231
4-Me ether: [108335-06-4]. *4-Methoxy-2,5-phenanthrenediol. 2,5-Dihydroxy-4-methoxyphenanthrene.*
Moscatin.
 $C_{15}H_{12}O_3$ M 240.258
 Constit. of *Dendrobium moscatum*. Cryst.
 (EtOAc/pet. ether). Mp 163-164°.

Majumder, P.L. *et al, Indian J. Chem., Sect. B*, 1987, **26**, 18.

Phenanthro[1,10-*bc*:8,9-*b'*,*c'*]bisthiopyran,   P-60072
9CI

3,10-Dithiaperylene
[102283-96-5]

$C_{18}H_{10}S_2$ M 290.397
Yellow leaflets (cyclohexane). Mp 228-230° dec.

Nakasuji, K. *et al, J. Am. Chem. Soc.*, 1986, **108**, 3460 (*synth, uv, pmr*)

Phenanthro[9,10-*b*]furan, 9CI   P-60073

Updated Entry replacing P-00818
[235-98-3]

$C_{16}H_{10}O$ M 218.254
Cryst. (pet. ether). Mp 117-118°.

Padwa, A. *et al, J. Am. Chem. Soc.*, 1966, **88**, 3759.
Müller, P. *et al, Helv. Chim. Acta*, 1986, **69**, 855 (*synth, ir, pmr, cmr, ms*)

Phenanthro[9,10-*c*]furan   P-60074

[235-94-9]

$C_{16}H_{10}O$ M 218.254
Cryst. (pet. ether). Mp 99-102° (102-103° dec.).

Stringer, M.B. *et al, Tetrahedron Lett.*, 1980, **21**, 3831 (*synth, pmr, uv*)
Litinas, K.E. *et al, J. Chem. Soc., Perkin Trans. 1*, 1985, 429 (*synth, uv, ms*)

Phenanthro[9,10-*g*]isoquinoline   P-60075

[110520-17-7]

$C_{21}H_{13}N$ M 279.340
Mp 139-150° dec.

Tanga, M.J. *et al, J. Heterocycl. Chem.*, 1987, **24**, 39 (*synth, pmr, cmr, ir, uv*)

Phenanthro[3,4,5,6-*bcdef*]ovalene, 9CI   P-60076

Circumanthracene

[190-28-3]

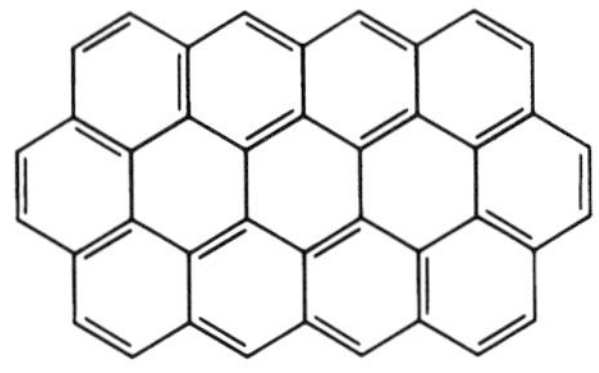

$C_{40}H_{16}$ M 496.566

Unknown. A reported synthesis (1956) gave a different product and a reputed x-ray cryst. struct. detn. was erroneous.

Clar, E. *et al, J. Am. Chem. Soc.*, 1981, **103**, 1320.

Phenanthro[4,5-*bcd*]thiophene, 9CI   P-60077

Updated Entry replacing P-00843

[30796-92-0]

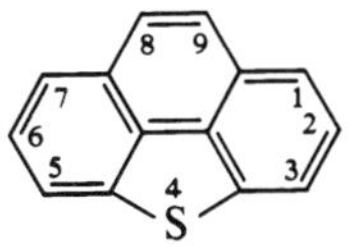

$C_{14}H_8S$ M 208.277

Obt. from coal and wood tar. Needles (EtOH). Mp 139-140°.

4-Oxide: [30796-93-1].
$C_{14}H_8OS$ M 224.277
Cryst. Mp 200.5-202°.

4,4-Dioxide: [30796-94-2].
$C_{14}H_8O_2S$ M 240.276
Cryst. Mp 271.5-273° dec.

Klemm, L.H. *et al, J. Heterocycl. Chem.*, 1970, **7**, 1347 (*synth, ms, pmr, uv*)
Borwitzky, H. *et al, CA*, 1977, **87**, 204080 (*isol*)
Klemm, L.H. *et al, J. Heterocycl. Chem.*, 1987, **24**, 357 (*synth*)

1*H*-Phenothiazin-1-one   P-60078

[54819-17-9]

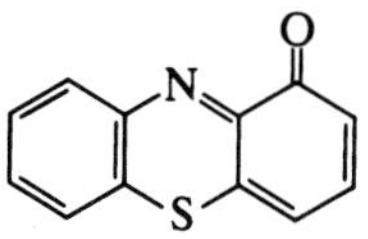

$C_{12}H_7NOS$ M 213.253

Dark blue-violet cryst. Mp 110-111°. Unstable, readily dimerises.

Silberg, I.A. *et al, Tetrahedron Lett.*, 1974, 3801 (*synth, uv*)

3-(Phenylazo)-2-butenenitrile   P-60079

3-(Phenylazo)crotononitrile, 8CI. Azipyrazole

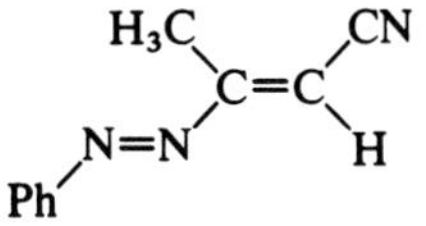

(*E,E*)-*form*

$C_{10}H_9N_3$ M 171.201

(*E,E*)-*form* [24514-98-5]
Red. Mp 80-81°.

(*Z,E*)-*form* [24515-82-0]
Mp 109°.

Searles, S. *et al, J. Am. Chem. Soc.*, 1957, **79**, 3175 (*synth, struct*)
Smith, P.A.S. *et al, J. Org. Chem.*, 1970, **35**, 2215 (*synth*)

2-Phenyl-4*H*-3,1,2-benzooxathiazine, 9CI   P-60080

[108288-67-1]

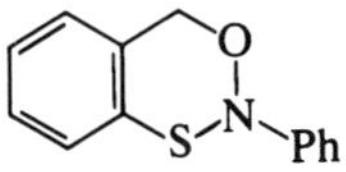

$C_{13}H_{11}NOS$ M 229.296

Jacob, D. *et al, Tetrahedron Lett.*, 1986, **27**, 5703 (*synth, ir, pmr, cmr, ms*)

2-Phenyl-1,5-benzothiazepin-4(5*H*)-one, 9CI   P-60081

[5667-03-8]

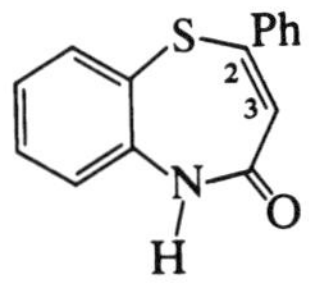

$C_{15}H_{11}NOS$ M 253.318

Main prod. of condensation of PhC≡CCOOH with 2-aminothiophenol. Yellow cryst. (MeCN). Mp 222-223°.

▷DL0675000.

N-Ac: [104505-68-2].
$C_{17}H_{13}NO_2S$ M 295.355
Cryst. (Ac₂O). Mp 157°.

N-Me: [30752-15-9]. *5-Methyl-2-phenyl-1,5-benzothiazepin-4(5H)-one.*
$C_{16}H_{13}NOS$ M 267.345
Mp 72-74° subl.

2,3-Dihydro: [29476-22-0].
$C_{15}H_{13}NOS$ M 255.334
Main prod. of condensation of PhCH=CHCOOH with 2-aminothiophenol. Cryst. (EtOH). Mp 177°.

2,3-Dihydro, N-Ac: [104505-67-1].
$C_{17}H_{15}NO_2S$ M 297.371
Cryst. (Ac₂O). Mp 161° (155-156°).

2,3-Dihydro, N-Me: [97038-01-2].
$C_{16}H_{15}NOS$ M 269.361
Cryst. (MeOH). Mp 95°.

Kaupp, G. *et al, Chem. Ber.*, 1986, **119**, 3109 (*synth, uv, pmr, ms, bibl*)

2-Phenyl-4*H*-1-benzothiopyran-4-one, 9CI   P-60082

Updated Entry replacing P-50115

*2-Phenyl-4H-benzo[b]thiin-4-one. 1-Thioflavone, 8CI.
Thioflavone*

[784-62-3]

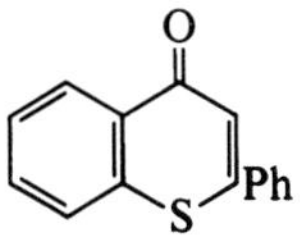

Needles (EtOH). Mp 129-130°. Yellow col. with H_2SO_4. Forms salts which are readily hyd. by H_2O.

S-*Oxide:* [65373-82-2].
$C_{15}H_{10}O_2S$ M 254.303
Mp 133-135°.
S,S-*Dioxide:* [22810-28-2].
$C_{15}H_{10}O_3S$ M 270.302
Mp 136.5-137°.

Ruhemann, S., *Ber.*, 1913, **46**, 2197.
Arndt, F. *et al, Ber.*, 1925, **58**, 1620.
Bossett, F., *Justus Liebigs Ann. Chem.*, 1964, **680**, 40; *Tetrahedron Lett.*, 1968, 4377.
Schumann, D. *et al, Chem. Ber.*, 1969, **102**, 3192.
Still, I.W.J. *et al, Can. J. Chem.*, 1976, **54**, 280 (*cmr*)
Nakazumi, H. *et al, J. Heterocycl. Chem.*, 1984, **21**, 193 (*oxides*)
Chen, C.H. *et al, J. Org. Chem.*, 1986, **51**, 3282 (*dioxide*)

2-Phenyl-2*H*-benzotriazole   P-60083

[1916-72-9]

$C_{12}H_9N_3$ M 195.223
Prisms (C_6H_6 or pet. ether). Steam-volatile. Mp 109-110°.

1-N-*Oxide:* [51750-18-6].
$C_{12}H_9N_3O$ M 211.223
Needles (pet. ether). Mp 88.5°.

Werner, A. *et al, Ber.*, 1899, **32**, 3256 (*synth, deriv*)
Bamberger, E. *et al, Ber.*, 1903, **36**, 3822 (*synth, deriv*)
Houghton, P.G. *et al, J. Chem. Soc., Perkin Trans. 1*, 1985, 1471 (*synth*)

2-Phenyl-1,3-benzoxathiol-1-ium, 9CI   P-60084

$C_{13}H_9OS^\oplus$ M 213.274 (ion)
Tetrafluoroborate: [58948-35-9].
$C_{13}H_9BF_4OS$ M 300.077
Mp 204-205°.
Perchlorate: [53755-85-4].
$C_{13}H_9ClO_5S$ M 312.724
Mp 232-233° dec.

Degani, I. *et al, J. Heterocycl. Chem.*, 1974, **11**, 507 (*synth, uv, props*)
Degani, I. *et al, J. Chem. Soc., Perkin Trans. 1*, 1976, 323 (*synth*)
Barbero, M. *et al, Synthesis*, 1986, 1074 (*synth, pmr, cmr*)

1-Phenyl-3-buten-1-one, 9CI   P-60085

3-Butenophenone. 3-Benzoylpropene
[6249-80-5]

$$PhCOCH_2CH{=}CH_2$$

$C_{10}H_{10}O$ M 146.188
Oil.

Hegedus, L.S. *et al, J. Am. Chem. Soc.*, 1975, **97**, 5448 (*synth, ir, pmr*)
Curran, D.P. *et al, Synthesis*, 1986, 312 (*synth, ir, pmr*)

2-Phenylcyclopentylamine, 8CI   P-60086

Updated Entry replacing P-01131
2-Phenylcyclopentanamine, 9CI. Cypenamine, BAN
[15301-54-9]

$C_{11}H_{15}N$ M 161.246
CNS stimulant, antidepressant.

(*1R,2R*)-*form*
(−)-cis-*form*
B,HCl: [102778-41-6]. Mp 184-185°. $[\alpha]_D$ −97.4°.
(*1S,2S*)-*form*
(+)-cis-*form*
B,HCl: [102778-42-7]. Mp 185-187°. $[\alpha]_D^{20}$ +97°.
(*1RS,2RS*)-*form* [40264-04-8]
(±)-cis-*form*
Liq. $Bp_{0.7}$ 80-82°.
N-*Benzoyl:*
$C_{18}H_{19}NO$ M 265.354
Needles (H_2O). Mp 162-162.5° (154°).
(*1RS,2SR*)-*form* [6604-06-4]
(±)-*trans*-form
Liq. Bp_{750} 148°, $Bp_{0.8}$ 70-83°. n_D^{20} 1.4504.
B,HCl: Cryst. (MeOH/EtOAc). Mp 143.5-145°.
Benzoyl:
$C_{18}H_{19}NO$ M 265.354
Mp 151.5°.

Govindachari, T.R. *et al, J. Chem. Soc.*, 1956, 4280 (*synth*)
Kaiser, C. *et al, J. Med. Chem.*, 1962, **5**, 1243 (*synth*)
Rathke, M.W. *et al, J. Am. Chem. Soc.*, 1966, **88**, 2870 (*synth*)
McGrath, W.R. *et al, Arch. Int. Pharmacodyn. Ther.*, 1968, **172**, 405 (*pharmacol*)
Wiehl, W. *et al, Chem. Ber.*, 1986, **119**, 2668 (*synth*)
Wiehl, W. *et al, Acta Crystallogr., Sect. C*, 1987, **43**, 83 (*cryst struct, abs config*)

2-Phenylcyclopropanecarboxylic acid, 9CI   P-60087

Updated Entry replacing P-01136
[5685-38-1]

$C_{10}H_{10}O_2$ M 162.188
Reported resolutions of both racemates are said to be troublesome and nonreproducible.

(*1R,2R*)-*form* [3471-10-1]
(−)-trans-*form*
$[\alpha]_D^{24}$ −368° (c, 0.931 in $CHCl_2$).
(*1S,2S*)-*form* [23020-15-7]
(+)-trans-*form*
$Bp_{0.18}$ 128°. $[\alpha]_D^{14}$ +381° (c, 0.960 in $CHCl_3$).
Me ester: [16205-72-4].
$C_{11}H_{12}O_2$ M 176.215
$[\alpha]_D^{20}$ +324.7° (c, 1.24 in $CHCl_3$).
(*1R,2S*)-*form* [48126-51-8]
(−)-cis-*form*
$[\alpha]_D^{20}$ −6.49° (c, 2.03 in $CHCl_3$).
Me ester: [67528-62-5]. $[\alpha]_D^{25}$ −10.0° (c, 2.80 in $CHCl_3$).

(1S,2R)-form [23020-18-0]
(+)-cis-form
$[\alpha]_D^{20}$ +22.9° (c, 2.03 in CHCl₃).
Me ester: [67528-63-6]. $[\alpha]_D^{20}$ +32.8° (c, 1.99 in CHCl₃).
(1RS,2RS)-form [42916-14-3]
(±)-trans-form
Needles (C₆H₆/pentane). Mp 105-106°.
(1RS,2SR)-form [67528-68-1]
(±)-cis-form
Cryst. (C₆H₆/pentane). Mp 104-106°.

Org. Synth., 1970, **50**, 94 (*synth*)
Aratani, T. *et al*, *Tetrahedron*, 1970, **26**, 1675 (*abs config*)
Yoshiaki, I. *et al*, *Bull. Soc. Chem. Jpn.*, 1977, **50**, 1784 (*cmr*)
Krieger, P.E. *et al*, *J. Org. Chem.*, 1978, **43**, 4447 (*synth*)
Scholl, B. *et al*, *Helv. Chim. Acta*, 1986, **69**, 1936 (*synth, resoln, bibl*)

3-Phenyl-1,4,2-dithiazole-5-thione, 9CI   P-60088

[14730-25-7]

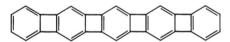

C₈H₅NS₃   M 211.314
Yellow cryst. (EtOH). Mp 118°.

Noel, D. *et al*, *Bull. Soc. Chim. Fr.*, 1967, 2239 (*synth*)
Greig, D.J. *et al*, *J. Chem. Soc., Perkin Trans. 1*, 1985, 1205 (*synth, cmr, ms*)

[4]Phenylene   P-60089

C₂₄H₁₂   M 300.359
Currently known as 2,3-bistrimethylsilyl deriv.

Hirthammer, M. *et al*, *J. Am. Chem. Soc.*, 1986, **108**, 2481 (*synth, uv, pmr, cmr*)

[5]Phenylene   P-60090

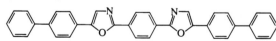

C₃₀H₁₄   M 374.441
Known as tetrakis(trimethylsilyl) deriv. See 2,3,9,10-Tetrakis(trimethylsilyl)[5]phenylene, T-60130 .

Blanco, L. *et al*, *Angew. Chem., Int. Ed. Engl.*, 1987, **26**, 1246.

2,2'-(1,4-Phenylene)bis[5,1-[1,1'-biphenyl]-4-yl]oxazole, 9CI   P-60091

2,2'-p-Phenylenebis[5-(4-biphenylyl)]oxazole, 8CI. 1,4-Bis[5-(4-biphenylyl)-2-oxazolyl]benzene. BOPOB
[494-67-7]

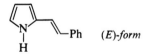

C₃₆H₂₄N₂O₂   M 516.598
Scintillation solute. Mp 292-294°.

Hayes, F.N. *et al*, *J. Am. Chem. Soc.*, 1955, **77**, 1850 (*synth*)
Leggate, P. *et al*, *Mol. Cryst.*, 1968, **4**, 357 (*uv, props*)
Paul, S.D. *et al*, *J. Indian Chem. Soc.*, 1972, **49**, 579 (*synth*)

1-Phenylethenol   P-60092

Updated Entry replacing P-01172
α-Methylenebenzenemethanol, 9CI. 1-Phenylvinyl alcohol. α-Hydroxystyrene

[4383-15-7]

$$H_2C=C(Ph)OH$$

C₈H₈O   M 120.151
Enol form of Acetophenone, A-60017 . Photohydration prod. of phenylacetylene, detected by flash photolysis.
Me ether: 1-Methoxy-1-phenylethylene. α-Methoxystyrene.
C₉H₁₀O   M 134.177
Oil with aromatic odour. d₀²¹ 1.003. Bp 197°.
Et ether: 1-Ethoxy-1-phenylethylene. α-Ethoxystyrene.
C₁₀H₁₂O   M 148.204
Oil. d₄²⁰ 0.97. Bp 209-210°, Bp₁₁ 88-89°.
Ph ether: 1-Phenoxy-1-phenylethylene. α-Phenoxystyrene.
C₁₄H₁₂O₂   M 212.248
Viscousoil.

Moureu, C., *C. R. Hebd. Seances Acad. Sci.*, 1903, **137**, 259 (*synth*)
Tiffeneau, M., *C. R. Hebd. Seances Acad. Sci.*, 1907, **145**, 811 (*synth*)
v. Auwers, K., *Ber.*, 1911, **44**, 3514 (*synth*)
Henne, A., *Angew. Chem.*, 1976, **88**, 4451 (*synth, nmr*)
Oku, A. *et al*, *Tetrahedron Lett.*, 1983, **24**, 4699 (*synth, pmr, ir, ms*)
Tureček, F., *Tetrahedron Lett.*, 1986, **27**, 4219 (*synth, props*)
Chiang, Y. *et al*, *Helv. Chim. Acta*, 1986, **69**, 1331 (*synth*)

2-(2-Phenylethenyl)benzothiazole, 9CI   P-60093

2-Styrylbenzothiazole
[1483-30-3]

C₁₅H₁₁NS   M 237.319
(E)-form [59066-61-4]
Light-yellow cryst. (EtOH). Mp 115° (107-109°).
(Z)-form [104505-71-7]
Labile. uv λ_max 246 sh (log ε 3.99), 247 (3.94), 264 (3.94), 290 sh (4.09), 301 (4.13), 330 sh nm (3.87).

Ried, W. *et al*, *Justus Liebigs Ann. Chem.*, 1956, **600**, 47 (*synth*)
Wilhelm, M. *et al*, *Helv. Chim. Acta*, 1970, **53**, 1697 (*synth*)
Kaupp, G. *et al*, *Chem. Ber.*, 1986, **119**, 3109 (*synth, ir, uv, pmr, cmr, ms*)

2-(2-Phenylethenyl)-1H-pyrrole, 9CI   P-60094

1-Phenyl-2-(2-pyrrolyl)ethylene. 2-Styrylpyrrole

C₁₂H₁₁N   M 169.226
(E)-form [2761-76-4]
Mp 141-142°.
N-Me: [2761-79-7].
C₁₃H₁₃N   M 183.252
Mp 71-72°.
(Z)-form [2433-58-1]
Bp₁ 140°.
N-Me: [2433-77-4].
C₁₃H₁₃N   M 183.252

Bp₁ 136°.

Jones, R.A. *et al*, *Aust. J. Chem.*, 1965, **18**, 875 (*synth, ir*)
Hinz, W. *et al*, *Synthesis*, 1986, 620 (*synth, uv, pmr*)

2-(2-Phenylethyl)-1,4-benzenediol, 9CI   P-60095

Phenethylhydroquinone, 8CI. 2,5-Dihydroxybibenzyl

[19312-07-3]

C₁₄H₁₄O₂   M 214.263
Mp 120-120.5°.

Manecke, V.G. *et al*, *Makromol. Chem.*, 1967, **108**, 198 (*synth*)

4-(2-Phenylethyl)-1,2-benzenediol, 9CI   P-60096

4-Phenethylpyrocatechol. 2,3-Dihydroxybibenzyl

[53515-94-9]
C₁₄H₁₄O₂   M 214.263
Mp 58-61°.

Murakami, M. *et al*, *CA*, 1957, **51**, 13840 (*synth*)
Scheline, R.R. *et al*, *Experientia*, 1974, **30**, 880 (*isol*)

4-(2-Phenylethyl)-1,3-benzenediol, 9CI   P-60097

2,4-Dihydroxybibenzyl. 4-Phenethylresorcinol

[94-77-9]
C₁₄H₁₄O₂   M 214.263
Plant growth inhibitor. Cryst. (EtOH). Mp 133-134°.

Bhumgara, K.S. *et al*, *CA*, 1948, **42**, 891h (*synth*)
U.S.P., 3 971 651, (*1976*); *CA*, **85**, 155074 (*synth, props*)

5-(2-Phenylethyl)-1,3-benzenediol, 9CI   P-60098

*5-Phenethylresorcinol, 8CI. 3,5-Dihydroxybibenzyl.
Dihydropinosylvin*

[14531-52-3]
C₁₄H₁₄O₂   M 214.263
Constit. of *Pinus* spp.; also prod. by *Dioscorea batatus* inoculated with *Pseudomonas cichorii*. Phytoalexin; active against various bacteria and fungi. Oil.

Mono-Me ether: Mp 49-50°.

Birch, A.J. *et al*, *J. Chem. Soc.*, 1960, 4395 (*synth*)
Erdtman, H. *et al*, *Phytochemistry*, 1966, **5**, 927 (*isol*)
Mitscher, L.A. *et al*, *Phytochemistry*, 1981, **20**, 781.
Takasugi, M. *et al*, *Phytochemistry*, 1987, **26**, 371 (*isol, props*)

2-(Phenylethynyl)thiazole, 9CI   P-60099

[35070-01-0]

C₁₁H₇NS   M 185.243
Mp 43-45°. Bp₄ 147°.

Picrate: Mp 135-136° dec.

Sakamoto, T. *et al*, *Chem. Pharm. Bull.*, 1987, **35**, 823 (*synth, pmr*)

4-(Phenylethynyl)thiazole, 9CI   P-60100

[111600-88-5]
C₁₁H₇NS   M 185.243
Bp₅ 150°.

Sakamoto, T. *et al*, *Chem. Pharm. Bull.*, 1987, **35**, 823 (*synth, pmr*)

N-Phenylformamidine, 8CI   P-60101

N-*Phenylmethanimidamide, 9CI*

[13484-76-9]

$$PhN{=}CHNH_2$$

C₇H₈N₂   M 120.154
Mp 143°.

Picrate: Mp 198-200° (191°).

Ali, M.U. *et al*, *J. Indian Chem. Soc.*, 1985, **62**, 666 (*synth*)
Kashima, C. *et al*, *Bull. Chem. Soc. Jpn.*, 1986, **59**, 3317 (*synth, pmr*)

3-Phenyl-2(5H)furanone, 9CI   P-60102

[57200-23-4]

C₁₀H₈O₂   M 160.172
Cryst. (C₆H₆/pet. ether). Mp 88°.

Kurihara, T. *et al*, *Chem. Pharm. Bull.*, 1986, **34**, 4620 (*synth, ir, pmr*)

4-Phenyl-2(5H)-furanone   P-60103

2-Oxo-4-phenyl-2,5-dihydrofuran. β-Phenyl-Δ^{α,β}-butenolide

[1575-47-9]

C₁₀H₈O₂   M 160.172
Cryst. (Me₂CO/hexane). Mp 92-93.5°.

Rubin, M. *et al*, *J. Org. Chem.*, 1941, **6**, 260 (*synth*)
Krauser, S.F. *et al*, *J. Org. Chem.*, 1978, **43**, 3400 (*synth*)
Ciattini, P.G. *et al*, *Synthesis*, 1986, 70 (*synth, ir, pmr*)

2-Phenylfuro[2,3-b]pyridine   P-60104

C₁₃H₉NO   M 195.220
Needles. Mp 91-91.5°.

Ogata, M. *et al*, *Tetrahedron*, 1969, **25**, 5217.
Sakamoto, T. *et al*, *Chem. Pharm. Bull.*, 1986, **34**, 2719 (*synth, pmr*)

2-Phenylfuro[3,2-c]pyridine   P-60105

C₁₃H₉NO   M 195.220
Needles. Mp 120-121°.

Sakamoto, T. *et al*, *Chem. Pharm. Bull.*, 1986, **34**, 2719 (*synth, pmr*)

7-Phenyl-5-heptynoic acid   P-60106

[88255-07-6]

$$PhCH_2C{\equiv}C(CH_2)_3COOH$$

$C_{13}H_{14}O_2$ M 202.252
Semisolid.

Me ester: [88255-19-0].
  $C_{14}H_{16}O_2$ M 216.279
  Bp$_{0.2}$ 105°.

Friary, R. *et al, J. Org. Chem.*, 1986, **51**, 3214 (*synth, ir, pmr, ms*)

7-Phenyl-3-heptyn-2-ol   P-60107

[106575-40-0]

$$Ph(CH_2)_3C{\equiv}CCH(OH)CH_3$$

$C_{13}H_{16}O$ M 188.269

(±)-*form*
Oil.

Nishida, A. *et al, Chem. Pharm. Bull.*, 1986, **34**, 1423 (*synth, ir, pmr, ms*)

1-Phenyl-1,4-hexadiyn-3-one, 9CI   P-60108

[16272-83-6]

$$PhC{\equiv}CCOC{\equiv}CCH_3$$

$C_{12}H_8O$ M 168.195
Possesses antifungal and insecticidal props. Cryst. (EtOH). Mp 71°.

Semicarbazone: Cryst. (EtOH). Mp 171°.

Chauvelier, J. *et al, Bull. Soc. Chim. Fr.*, 1950, 272 (*synth*)
Fontaine, M. *et al, Bull. Soc. Chim. Fr.*, 1962, 2145 (*synth, ir*)
Fr. Pat., 1 453 839, (*1966*); *CA*, **67**, 99860 (*synth, use*)
Mueller, E. *et al, Z. Naturforsch., B*, 1971, **26**, 1003 (*ms*)

1-Phenyl-2,4-hexadiyn-1-one, 9CI   P-60109

2,4-Hexadiynophenone, 8CI. 1-Benzoyl-1,3-pentadiyne.
Capillin

[495-74-9]

$$PhCOC{\equiv}CC{\equiv}CCH_3$$

$C_{12}H_8O$ M 168.195
Constit. of essential oil from *Artemisia capillaris, A. dracunculus, Chrysantheum* spp. and *Santolina rosmarinifolia.* Antifungal agent. Pale-cream needles (hexane). Mp 82-83°.

▷MM3622000.

2,4-Dinitrophenylhydrazone: Orange needles (EtOH). Mp 225° dec. (214°).

Bohlmann, F. *et al, Chem. Ber.*, 1963, **96**, 226; 1964, **97**, 1179 (*isol*)
Reisch, J. *et al, Arch. Pharm.* (*Weinheim, Ger.*), 1964, **297**, 628 (*synth*)
Nash, B.W. *et al, J. Chem. Soc.*, 1965, 2983 (*synth, bibl*)
Hearn, M.T.W., *J. Magn. Reson.*, 1975, **19**, 401 (*cmr*)
De Pascual, T.J. *et al, CA*, 1982, **96**, 109953 (*isol*)
Jones, G.E. *et al, Tetrahedron Lett.*, 1982, **23**, 3203 (*synth, bibl*)

6-Phenyl-3,5-hexadiyn-2-one, 9CI   P-60110

[1595-73-9]

$$PhC{\equiv}CC{\equiv}CCOCH_3$$

$C_{12}H_8O$ M 168.195
Pale-yellow needles (pet. ether). Mp 47°.

2,4-Dinitrophenylhydrazone: Orange needles (EtOH). Mp 170° dec.

Nash, B.W. *et al, J. Chem. Soc.*, 1965, 2983 (*synth*)
Mueller, E. *et al, Synthesis*, 1970, 147 (*synth*)
Gorgues, A. *et al, Bull. Soc. Chim. Fr.*, 1976, 125 (*synth*)

2-Phenyl-1*H*-imidazole-4(5)-carboxaldehyde   P-60111

[68282-47-3]

$C_{10}H_8N_2O$ M 172.186
Cryst. (MeCN/THF). Mp 169-170°.

1-Me: [94938-02-0]. *1-Methyl-2-phenyl-1H-imidazole-4-carboxaldehyde.*
  $C_{11}H_{10}N_2O$ M 186.213
  Cryst. (EtOAc/cyclohexane). Mp 109-110°.
3-Me: [94938-03-1]. *1-Methyl-2-phenyl-1H-imidazole-5-carboxaldehyde.*
  $C_{11}H_{10}N_2O$ M 186.213
  Cryst. (diisopropyl ether). Mp 94-95°.

Sircar, I. *et al, J. Med. Chem.*, 1986, **29**, 261 (*synth, uv, ir, pmr*)

3-(Phenylimino)-1(3*H*)-isobenzofuranone, 9CI   P-60112

3-(Phenylimino)phthalide, 8CI. N-*Phenylisophthali-mide.* N-*Phenylphthalisoimide. Isophthalanil*

[487-42-3]

$C_{14}H_9NO_2$ M 223.231
Pale-yellow cryst. Mp 120-121.5°.

B,HClO₄: [24259-39-0]. Mp 157° dec.

Roderick, W.R. *et al, J. Org. Chem.*, 1963, **28**, 2018 (*synth*)
Howe, R.K., *J. Org. Chem.*, 1973, **38**, 4164 (*synth*)
Boyd, G.V. *et al, J. Chem. Soc., Perkin Trans. 1*, 1978, 1338 (*synth*)
Ganin, E.V. *et al, Khim. Geterotsikl. Soedin.*, 1984, 1280; *CA*, **101**, 230277 (*synth*)

6-Phenyl-1*H*-indole, 9CI   P-60113

Updated Entry replacing P-01294
[106851-31-4]
$C_{14}H_{11}N$ M 193.248
Platelets (Me$_2$CO). Mp 160-161°.

Süs, O. *et al, Justus Liebigs Ann. Chem.*, 1955, **593**, 91 (*synth*)
Tischler, A.N. *et al, Tetrahedron Lett.*, 1986, **27**, 1653 (*synth*)

2-Phenyl-3*H*-indol-3-one, 9CI P-60114
3-Oxo-2-phenylindolenine
[2989-63-1]

$C_{14}H_9NO$ M 207.231
Red cryst. (Et$_2$O). Mp 102°. Unstable in water.
N-*Oxide:* [1969-74-0]. *2-Phenylisatogen.*
  $C_{14}H_9NO_2$ M 223.231
  Red-orange cryst. (MeOH). Mp 188-189°.
Oxime: [4676-99-7]. *3-Nitroso-2-phenylindole.*
  $C_{14}H_{10}N_2O$ M 222.246
  Yellow-orange cryst. (AcOH). Mp 278-280°.
  Tautomeric. The nitroso form may be preferred.
N-*Oxide, oxime:* [31917-92-7]. *2-Phenylisatogen oxime.*
  $C_{14}H_{10}N_2O_2$ M 238.245
  Orange plates (EtOH). Mp 242° dec.
N-*Oxide, oxime, Et ether:*
  $C_{16}H_{14}N_2O_2$ M 266.299
  Orange cryst. (pet. ether). Mp 96°.

Angeli, A. *et al, Atti R. Accad. Lincei,* 1907, **15II**, 761 (*deriv*)
Pfeiffer, P., *Justus Liebigs Ann. Chem.,* 1916, **411**, 72 (*deriv*)
Richman, R.J. *et al, J. Org. Chem.,* 1968, **33**, 2548 (*synth, uv, ir, props, bibl*)
Hiremath, S.P. *et al, Adv. Heterocycl. Chem.,* 1978, **22**, 123 (*bibl*)
Chiorboli, E. *et al, Synthesis,* 1986, 219 (*deriv, synth*)

1-Phenyl-2-naphthol, 8CI P-60115
1-Phenyl-2-naphthalenol, 9CI
[4919-96-4]

$C_{16}H_{12}O$ M 220.270
Cryst. (MeOH). Mp 90-91°.
Me ether: [75907-52-7]. *2-Methoxy-1-phenylnaphthalene.*
  $C_{17}H_{14}O$ M 234.297
  Cryst. (MeOH). Mp 70-72°. Bp$_1$ 170°.

Huisgen, R. *et al, Chem. Ber.,* 1969, **102**, 3405 (*uv, ir, pmr*)
Repinskaya, I.B. *et al, J. Org. Chem. USSR, Engl. transl.,* 1980, **16**, 1463 (*synth, deriv, uv, ir, pmr*)
Barton, D.H.R. *et al, J. Chem. Soc., Perkin Trans. 1,* 1987, 241 (*synth*)

3-Phenyl-2-naphthol, 8CI P-60116
3-Phenyl-2-naphthalenol, 9CI
[30889-48-6]
$C_{16}H_{12}O$ M 220.270
Cryst. (methylcyclohexane). Mp 117-118°.
Fields, D.L., *J. Org. Chem.,* 1971, **36**, 3002 (*synth*)

4-Phenyl-2-naphthol, 8CI P-60117
4-Phenyl-2-naphthalenol, 9CI
[36159-74-7]
$C_{16}H_{12}O$ M 220.270
Cryst. (pet. ether). Mp 75-76°.

Koptgug, V.A. *et al, J. Org. Chem. USSR, Engl. transl.,* 1971, **7**, 2490 (*synth, uv, ir, pmr*)

5-Phenyl-2-naphthol, 8CI P-60118
5-Phenyl-2-naphthalenol, 9CI
$C_{16}H_{12}O$ M 220.270
Me ether: [27331-47-1]. *2-Methoxy-5-phenylnaphthalene.*
  $C_{17}H_{14}O$ M 234.297
  Liq. Bp$_{0.1}$ 156-158°.

Bergmann, F. *et al, J. Am. Chem. Soc.,* 1947, **69**, 1773 (*synth, deriv*)
Leznoff, C.C. *et al, Can. J. Chem.,* 1970, **48**, 1842 (*synth, deriv, uv, pmr*)

8-Phenyl-2-naphthol, 8CI P-60119
8-Phenyl-2-naphthalenol, 9CI
[29430-70-4]
$C_{16}H_{12}O$ M 220.270
Cryst. Mp 48-50°. Bp$_{19}$ 214-215°.

Me ether: [27331-38-0]. *7-Methoxy-1-phenylnaphthalene.*
  $C_{17}H_{14}O$ M 234.297
  Oil.

Howell, W.N. *et al, J. Chem. Soc.,* 1936, 587 (*synth*)
Snykers, F. *et al, Helv. Chim. Acta,* 1970, **53**, 1294 (*synth, deriv, ir, pmr*)

3-Phenyl-1,2-naphthoquinone P-60120
3-Phenyl-1,2-naphthalenedione, 9CI
[51670-51-0]

$C_{16}H_{10}O_2$ M 234.254
Red needles (MeOH). Mp 161-162°.

Fieser, L.F. *et al, J. Am. Chem. Soc.,* 1951, **73**, 681 (*synth*)
Mackenzie, N.E. *et al, J. Chem. Soc., Perkin Trans. 1,* 1986, 2233 (*synth, ir, pmr, ms*)

4-Phenyl-1,2-naphthoquinone P-60121
4-Phenyl-1,2-naphthalenedione, 9CI
[73671-07-5]
$C_{16}H_{10}O_2$ M 234.254
Orange needles (MeOH). Mp 118-120°.

Cassebaum, H. *et al, Chem. Ber.,* 1957, **90**, 339 (*synth*)
Mackenzie, N.E. *et al, J. Chem. Soc., Perkin Trans. 1,* 1986, 2233 (*synth, ir, pmr, ms*)

5-Phenyl-1,2,4-oxadiazole-3-carboxaldehyde, 9CI P-60122
[73217-79-5]

$C_9H_6N_2O_2$ M 174.159
Cryst. Mp 67-69°. Bp$_{15}$ 136-139°.
Covalent hydrate: [73217-63-7].
  $C_9H_8N_2O_3$ M 192.174
  Needles (H$_2$O). Mp 80-82° (90-91°).
Oxime: [103499-08-7].
  $C_9H_7N_3O_2$ M 189.173
  Plates (EtOH). Mp 162-163°.

Mnatsakanova, T.R. *et al, Khim. Geterotsikl. Soedin.,* 1969, **5**, 212 (*Engl. transl.* p. 160) (*synth*)

Palazzo, G. *et al*, *J. Heterocycl. Chem.*, 1979, **16**, 1469 (*synth, deriv*)
Bedford, C.D. *et al*, *J. Med. Chem.*, 1986, **29**, 2174 (*deriv, pmr*)

2-Phenyl-1,3-oxazepine, 9CI   P-60123

[49679-57-4]

C$_{11}$H$_9$NO   M 171.198
Yellow oil. Grad. resinifies on standing.

Mukai, T. *et al*, *Tetrahedron Lett.*, 1973, 1835 (*synth, uv, pmr, ms*)

5-Phenyl-1,4-oxazepine   P-60124

C$_{11}$H$_9$NO   M 171.198
Orange oil.

Kurita, J. *et al*, *J. Chem. Soc., Chem. Commun.*, 1986, 1188 (*synth, ir, uv, pmr*)

5-Phenyloxazole, 9CI   P-60125

Updated Entry replacing P-01393
[1006-68-4]
C$_9$H$_7$NO   M 145.160
Cryst. Mp 41-42°.

Bachstez, M., *Ber.*, 1914, **47**, 3163.
Theilig, G., *Chem. Ber.*, 1953, **86**, 96.
Demchenko, N.P. *et al*, *Zh. Obshch. Khim.*, 1962, **32**, 1219.
Brown, D.J. *et al*, *J. Chem. Soc. (B)*, 1969, 270 (*pmr*)
Maquestiau, A. *et al*, *Tetrahedron Lett.*, 1986, **27**, 4023 (*synth, pmr*)

4-Phenyl-2-oxazolidinone, 9CI   P-60126

4-Phenyl-1,3-oxazolidine-2-one
[7480-32-2]

C$_9$H$_9$NO$_2$ M 163.176
(±)-*form*
Mp 136.8-137.8°.
3-Ph: [13606-71-8]. *3,4-Diphenyl-2-oxazolidinone, 9CI.*
3,4-Diphenyl-1,3-oxazolidin-2-one.
C$_{15}$H$_{13}$NO$_2$ M 239.273
Prisms (CHCl$_3$/pet. ether). Mp 128-129°.

Newman, M.S. *et al*, *J. Am. Chem. Soc.*, 1954, **76**, 1840 (*synth*)
Saettone, M.F. *et al*, *Tetrahedron Lett.*, 1966, 6009 (*deriv, synth*)
Shibata, I. *et al*, *J. Org. Chem.*, 1986, **51**, 2177 (*deriv, synth, pmr, ir*)

1-Phenyl-3-penten-1-one, 9CI   P-60127

1-Benzoyl-2-butene

$$PhCOCH_2CH=CHCH_3$$

C$_{11}$H$_{12}$O   M 160.215

(*E*)-*form* [74157-93-0]
Bp$_{0.7}$ 81-83°.
(*Z*)-*form* [61752-45-2]
Oil. Chromatog. gives *E*-form.

van der Weerdt, A.J. *et al*, *J. Chem. Soc., Perkin Trans. 2*, 1980, 592 (*synth, pmr, uv, ir*)
Grignon-Dubois, M. *et al*, *Can. J. Chem.*, 1981, **59**, 802 (*synth, ir, pmr*)
Curran, D.P. *et al*, *Synthesis*, 1986, 312 (*synth, ir, pmr*)

2-Phenyl-4-(phenylmethylene)-5(4*H*)-oxa-   P-60128
zolone, 9CI

4-Benzylidene-2-phenyloxazolin-5-one
[842-74-0]

(*Z*)-*form*

C$_{16}$H$_{11}$NO$_2$ M 249.268
(*E*)-*form* [15732-43-1]
Mp 149.5-151.5°. Labile isomer. Also called *cis*-form.
(*Z*)-*form* [17606-70-1]
Mp 165-166°. Stable isomer. Also called *trans*-form.

Brocklehurst, K. *et al*, *Tetrahedron*, 1974, **30**, 351 (*isom, spectra*)
Rao, Y.S. *et al*, *Synthesis*, 1975, 749 (*synth, rev, bibl*)
Kumar, K. *et al*, *Can. J. Chem.*, 1978, **56**, 232 (*spectra*)

3-Phenyl-2-propyn-1-amine, 9CI   P-60129

Phenylpropargylamine. 3-Amino-1-phenylpropyne
[78168-74-8]

$$PhC{\equiv}CCH_2NH_2$$

C$_9$H$_9$N   M 131.177
B,HCl: [30011-36-0]. Platelets (EtOH/Et$_2$O). Mp 216-217°.

Simon, D.Z. *et al*, *J. Med. Chem.*, 1970, **13**, 1249 (*synth*)
Klemm, L.M. *et al*, *J. Org. Chem.*, 1976, **41**, 2571 (*synth, ir, pmr*)

3-Phenyl-2-propyn-1-ol, 9CI   P-60130

Updated Entry replacing P-10180
Phenylpropargyl alcohol. Phenylpropiolic alcohol.
Hydroxymethylphenylacetylene
[1504-58-1]

$$PhC{\equiv}CCH_2OH$$

C$_9$H$_8$O   M 132.162
Bp$_5$ 100-103°, Bp$_{0.8}$ 89-95°.
3,5-Dinitrobenzoyl: Mp 131-132°.

Bates, E.B. *et al*, *J. Chem. Soc.*, 1954, 1854 (*synth*)
Jorgenson, M.J., *Tetrahedron Lett.*, 1962, 559 (*synth*)
Tomita, K. *et al*, *Chem. Pharm. Bull.*, 1968, **16**, 914 (*synth*)
Cassar, L., *J. Organomet. Chem.*, 1975, **93**, 253 (*synth*)
Hagihara, N. *et al*, *Tetrahedron Lett.*, 1975, 4467 (*synth*)
Org. Synth., Coll. Vol., **5**, 880 (*synth*)
Denis, J.-N. *et al*, *J. Org. Chem.*, 1986, **51**, 46 (*synth, ir, pmr*)

4-Phenyl-2(1*H*)-quinolinone   P-60131

Updated Entry replacing P-01642
2-Hydroxy-4-phenylquinoline. 4-Phenyl-2-quinolinol.
4-Phenylcarbostyril

[5855-57-2]
$C_{15}H_{11}NO$ M 221.258
Needles (AcOH). Mp 260-261°.
Me ether: [41443-48-5]. *2-Methoxy-4-phenylquinoline.*
$C_{16}H_{13}NO$ M 235.285
Mp 80-81°.

Boherg, F. *et al, Justus Liebigs Ann. Chem.,* 1973, 256 (*synth, deriv*)
Merault, G. *et al, Bull. Soc. Chim. Fr.,* 1974, 1949 (*synth*)
Singh, P.N. *et al, Indian J. Chem.,* 1974, **12**, 1016 (*synth*)
Kaupp, G. *et al, Chem. Ber.,* 1986, **119**, 3109 (*synth, uv, pmr, ms*)

2-Phenyl-4(1H)-quinolinone   P-60132

Updated Entry replacing P-01643
[14802-18-7]

$C_{15}H_{11}NO$ M 221.258
Oxo-form
  Plates (EtOH). Mp 255-257°. Major tautomer.
  N-*Me: 1-Methyl-2-phenyl-4(1*H)-quinolinone*.
  $C_{16}H_{13}NO$ M 235.285
  Alkaloid from *Balfourodendron riedelianum* (Rutaceae). Needles (MeOH). Mp 143-144°.
OH-form [1144-20-3]
  2-Phenyl-4-quinolinol. 4-Hydroxy-2-phenylquinoline
  Me ether: [22680-62-2]. *4-Methoxy-2-phenylquinoline.*
  $C_{16}H_{13}NO$ M 235.285
  Alkaloid from the leaves of *Lunasia amara* (Rutaceae). Needles (pentane). Mp 69-70° (66-67°).
  Me ether; B,HClO4: Rods (EtOH). Mp 215-218°.
  Me ether; B,MeI: Bright-yellow needles (MeOH). Mp 148-151°.

Goodwin, S. *et al, J. Am. Chem. Soc.,* 1957, **79**, 2239 (*isol, uv, deriv*)
Price, J.R. *et al, Aust. J. Chem.,* 1959, **12**, 589 (*ir*)
Rapoport, H. *et al, J. Am. Chem. Soc.,* 1960, **82**, 4395 (*isol, deriv*)
Ogata, Y. *et al, Tetrahedron,* 1971, **27**, 2765 (*synth, uv*)

9-(Phenylseleno)phenanthrene, 9CI   P-60133

[65490-23-5]

$C_{20}H_{14}Se$ M 333.291
Cryst. Mp 158.5-160.5°.

Pierini, A.B. *et al, J. Org. Chem.,* 1979, **44**, 4667.
Dent, B.R. *et al, Aust. J. Chem.,* 1986, **39**, 1789 (*synth, cmr*)

3-Phenyltetrazolo[1,5-a]pyridinium   P-60134

$C_{11}H_9N_4^{\oplus}$ M 197.219 (ion)

Bromide: [94971-81-0].
  $C_{11}H_9BrN_4$ M 277.123
  Cryst. (MeOH/Et2O). Mp 300°.
Tetrafluoroborate:
  $C_{11}H_9BF_4N_4$ M 284.023
  Mp 246°.

Messmer, A. *et al, Tetrahedron,* 1986, **42**, 4827 (*synth, pmr*)

4-Phenylthieno[3,4-b]furan   P-60135

$C_{12}H_8OS$ M 200.255
First example of thieno[3,4-b]furan ring system. Cryst. (Et2O). Mp 75-77°. Se analogue also prepd.

Shafiee, A. *et al, J. Heterocycl. Chem.,* 1982, **19**, 227 (*synth, ms, pmr*)

2-Phenylthieno[2,3-b]pyridine   P-60136

$C_{13}H_9NS$ M 211.281
Pale-yellow needles (hexane). Mp 96-97°.

Sakamoto, T. *et al, Chem. Pharm. Bull.,* 1986, **34**, 2719 (*synth, pmr*)

(Phenylthio)nitromethane   P-60137

Updated Entry replacing P-40119
(Nitromethyl)thiobenzene. Nitromethyl phenyl sulfide
[60595-16-6]

$$PhSCH_2NO_2$$

$C_7H_7NO_2S$ M 169.198
Synthetic intermed. Homologation reagent. Yellow oil. $Bp_{0.05}$ 85-95°.
S-Dioxide: [21272-85-5]. *(Nitromethyl)sulfonylbenzene. Nitromethyl phenyl sulfone.*
  $C_7H_7NO_4S$ M 201.197
  Needles (hexane). Mp 78.0-78.5°.

Bordwell, F.G. *et al, J. Org. Chem.,* 1978, **43**, 3101 (*synth*)
Miyashita, M. *et al, J. Chem. Soc., Chem. Commun.,* 1978, 362 (*synth*)
Banks, B.J. *et al, J. Chem. Soc., Chem. Commun.,* 1984, 670 (*use*)
Barrett, A.G.M. *et al, J. Org. Chem.,* 1987, **52**, 4693 (*synth, use*)

1-Phenyl-1H-1,2,3-triazole-4-carboxylic acid, 9CI   P-60138

[4600-04-8]

$C_9H_7N_3O_2$ M 189.173
Prisms + 1MeOH (MeOH), needles (H2O). Mp 151° dec.
Me ester: [2055-52-9].
  $C_{10}H_9N_3O_2$ M 203.200

Prisms (MeOH). Mp 121°.
Et ester: [4915-97-3].
  $C_{11}H_{11}N_3O_2$ M 217.227
  Needles (EtOH aq.). Mp 88°.
Amide: [2055-53-0].
  $C_9H_8N_4O$ M 188.188
  Mp 233°.
Nitrile: [61456-87-9]. *4-Cyano-1-phenyl-1,2,3-triazole.*
  $C_9H_6N_4$ M 170.173
  Needles (CCl_4). Mp 123°.

Dimroth, O., *Ber.*, 1902, **35**, 1029 (*synth, derivs*)

1-Phenyl-1*H*-1,2,3-triazole-5-carboxylic acid, 9CI   P-60139

3-Phenyl-3H-1,2,3-triazole-4-carboxylic acid
[15966-72-0]
$C_9H_7N_3O_2$ M 189.173
Needles (EtOH). Mp 176° dec.
Me ester:
  $C_{10}H_9N_3O_2$ M 203.200
  Needles (MeOH aq.). Mp 101°.
Et ester:
  $C_{11}H_{11}N_3O_2$ M 217.227
  Needles (pet. ether/Et_2O). Mp 54-55°.
Amide:
  $C_9H_8N_4O$ M 188.188
  Prisms (H_2O). Mp 146°.

Dimroth, O., *Ber.*, 1902, **35**, 1029 (*synth, derivs*)

2-Phenyl-2*H*-1,2,3-triazole-4-carboxylic acid, 9CI   P-60140

[13306-99-5]

$C_9H_7N_3O_2$ M 189.173
Needles (H_2O). Mp 191-192°.
Me ester: [62289-79-6].
  $C_{10}H_9N_3O_2$ M 203.200
  Needles (MeOH). Mp 89-90°.
Et ester:
  $C_{11}H_{11}N_3O_2$ M 217.227
  Needles. Mp 59°.
Amide:
  $C_9H_8N_4O$ M 188.188
  Needles (H_2O). Mp 143.5°.
Nitrile: [36386-83-1]. *4-Cyano-2-phenyl-1,2,3-triazole.*
  $C_9H_6N_4$ M 170.173
  Plates (MeOH). Mp 94.5°.

v. Pechmann, H. *et al*, *Ber.*, 1888, **21**, 2751 (*synth, derivs*)
v. Pechmann, H. *et al*, *Justus Liebigs Ann. Chem.*, 1891, **262**, 277 (*synth, derivs*)

Phillygenin   P-60141

Updated Entry replacing P-50175
4-[4-(3,4-Dimethoxyphenyl)-1H,3H-furo[3,4-c]furan-1-yl]-2-methoxyphenol, 9CI

$C_{21}H_{24}O_6$ M 372.417
(+)-*form* [487-39-8]
  Isol. from fruit of *Forsythia suspensa*. Mp 133-134°.
  $[\alpha]_D^{22}$ +120° (c, 0.04 in MeOH).
  Ac: Mp 122.5-123°. $[\alpha]_D^{25}$ +98.3° (c, 0.29 in MeOH).
  β-D-Glucoside: [487-41-2]. **Phillyrin.**
    $C_{27}H_{34}O_{11}$ M 534.559
    From *F. suspensa*. Mp 146-148°. $[\alpha]_D^{21}$ +46.9° (c, 0.25 in MeOH).

(−)-*form*
  Constit. of *Pararistolochia flos-avis*. Cryst. (EtOAc/hexane). Mp 128-130°. $[\alpha]_D^{20}$ −114° (c, 0.5 in $CHCl_3$).

Chiba, M. *et al*, *Chem. Pharm. Bull.*, 1977, **25**, 3435 (*isol, struct*)
Chiba, M. *et al*, *Phytochemistry*, 1980, **19**, 335 (*isol*)
Sun, N.-J. *et al*, *Phytochemistry*, 1987, **26**, 3051 (*isol*)

Phrymarolin II   P-60142

Updated Entry replacing P-01789
1-Acetoxy-6-(2-methoxy-4,5-methylenedioxyphenyl)-2-(3,4-methylenedioxyphenoxy)-3,7-dioxabicyclo[3.3.0]-octane
[23720-86-7]

$C_{23}H_{22}O_{10}$ M 458.421
Isol. from root of *Phyrma leptostachya*. Cryst. (Et_2O and EtOAc). Mp 161-162°. $[\alpha]_D$ +117.6° (c, 2.72 in dioxan).
6′-Methoxy: [38303-95-6]. **Phrymarolin I**.
  $C_{24}H_{24}O_{11}$ M 488.447
  From *P. leptostachya*. Mp 155-157°.

Taniguchi, E. *et al*, *Agric. Biol. Chem.*, 1969, **33**, 466; 1972, **36**, 1489 (*isol*)
Ishibachi, F. *et al*, *Agric. Biol. Chem.*, 1986, **50**, 3119; *Chem. Lett.*, 1771 (*synth, abs config*)

Phthalidochromene   P-60143

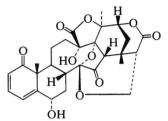

C₁₄H₁₄O₄   M 246.262

$C_{14}H_{14}O_4$ M 246.262

Constit. of *Anaphalis araneosa*. Gum.

Jakupovic, J. *et al, Phytochemistry*, 1987, **26**, 580.

Physalactone   P-60144

5β,6β-Epoxy-4β,17α,20R-trihydroxy-3β-methoxy-1-oxo-8(14),24-withadienolide

[64766-16-1]

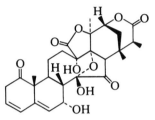

$C_{29}H_{40}O_8$ M 516.630

Constit. of *Physalis viscosa*. Amorph. $[\alpha]_D$ −4.3° (c, 2.9 in MeOH).

Maslennikova, V.A. *et al, Khim. Prir. Soedin.*, 1977, **13**, 443 (*isol*)
Abdullaev, N.D. *et al, Khim. Prir. Soedin.*, 1984, **20**, 182 (*struct*)

Physalin D   P-60145

[54980-22-2]

$C_{28}H_{32}O_{11}$ M 544.554

Constit. of *Physalis minima* and *P. angulata*. Cryst. (Me₂CO/MeOH). Mp 286-287°. $[\alpha]_D$ −68° (MeOH).

5-Me ether: **Physalin I.**
$C_{29}H_{34}O_{11}$ M 558.581
From *P. angulata*. Cryst. (MeOH). Mp 305-307°. $[\alpha]_D$ +12° (c, 0.5 in Me₂CO).

Row, L.R. *et al, Phytochemistry*, 1980, **19**, 1175.

Physalin G   P-60146

[76045-38-0]

$C_{28}H_{30}O_{10}$ M 526.539

Constit. of *Physalis angulata*. Cryst. (Me₂CO). Mp 295-296°. $[\alpha]_D$ +17° (c, 0.5 in Me₂CO).

Row, L.R. *et al, Phytochemistry*, 1980, **19**, 1175.

Physalin L   P-60147

$C_{28}H_{32}O_{10}$ M 528.555

Constit. of *Physalis alkekengi*. Cryst. (2-propanol). Mp 248-249°. $[\alpha]_D^{24}$ −118° (c, 0.3 in Me₂CO).

Kawai, M. *et al, Phytochemistry*, 1987, **26**, 3313.

Physalolactone   P-60148

6α-Chloro-4β,5β,14α,17β,20S-pentahydroxy-1-oxo-22R-witha-2,24-dienolide

[71339-25-8]

$C_{28}H_{39}ClO_8$ M 539.064

Constit. of *Physalis peruviana*. Cryst. (MeOH). Mp 227-228°. $[\alpha]_D$ +29.4° (c, 0.2 in Py).

Ray, A.B. *et al, J. Indian Chem. Soc.*, 1978, **55**, 1175.

Physanolide   P-60149

14α,17β,20R-Trihydroxy-1,4-dioxo-22R-witha-5,24-dienolide

[79199-28-3]

$C_{28}H_{38}O_7$ M 486.604

Constit. of *Physalis viscosa*. Cryst. (MeOH). Mp 170-175°.

Tursunova, R.N. *et al, Khim. Prir. Soedin.*, 1981, **17**, 145.

Physarochrome *A*   P-60150

$C_{24}H_{27}N_3O_6$ M 453.494

Pigment from the slime mold *Physarum polycephalum*. Yellow powder. $[\alpha]_D$ +7.2° (c, 0.042 in MeOH).

Steffan, B. *et al, Tetrahedron Lett.*, 1987, **28**, 3667 (*isol, struct*)

Physoperuvine   P-60151

8-Methyl-8-azabicyclo[3.2.1]octan-1-ol, 9CI. 4-(Methylamino)cyclohexanone, 9CI. 1-Hydroxytropane. 1-Tropanol

[60723-27-5]

Absolute configuration

$C_8H_{15}NO$ M 141.213

Free base tautomeric. Alkaloid from *Physalis peruviana* (Solanaceae). Mp 47-48°, Mp 68-70°. $[\alpha]_D$ −0.8° (c, 1 in MeOH). May be a partial racemate.

▷Mod. toxic

B,HCl: Mp 153°. $[\alpha]_D$ −0.8° (c, 1 in MeOH). Tropane bicyclic struct.

N-Benzoyl: Mp 136-137°. Monocyclic struct.

Ray, A.B. *et al, Heterocycles*, 1982, **19**, 1233 (*cryst struct, cmr, pmr, cd, abs config*)
McPhail, A.T. *et al, Tetrahedron*, 1984, **40**, 1661 (*cryst struct, synth, resoln*)

Phytol   P-60152

Updated Entry replacing P-10203

3,7,11,15-Tetramethyl-2-hexadecen-1-ol

$C_{20}H_{40}O$ M 296.535

(*2E,7R,11R*)-form [150-86-7]

Constit. of nettles and other plants. Oil. $Bp_{0.02}$ 132°. $[\alpha]_D^{18}$ +0.2°.

▷TJ3490000.

1-Aldehyde: [13754-69-3]. *3,7,11,15-Tetramethyl-2-hexadecenal*. **Phytal**.
$C_{20}H_{38}O$ M 294.520
Constit. of *Tetragonia tetragonoides*. Oil.

(*2Z,7R,11R*)-form [5492-30-8]

Constit. of *Gracilaria andersoniana*. Oil.

Burrell, J.W.K. *et al, J. Chem. Soc.* (*C*), 1966, 2144 (*synth, abs config*)
Shilleter, D.N. *et al, Phytochemistry*, 1970, **9**, 153 (*biosynth*)
Goodman, R.A. *et al, J. Am. Chem. Soc.*, 1973, **95**, 7553 (*cmr*)
Sims, J.J. *et al, Phytochemistry*, 1976, **15**, 1076 (*isol*)
Aoki, T. *et al, Phytochemistry*, 1982, **21**, 1361 (*isol*)
Schmid, M. *et al, Helv. Chim. Acta*, 1982, **65**, 684 (*synth*)
Gramatica, P. *et al, Tetrahedron*, 1987, **43**, 4481 (*synth*)

Pilosanone *A*   P-60153

$C_{20}H_{32}O_3$ M 320.471

Constit. of *Portulaca pilosa*. Cryst. Mp 98.5-99°. $[\alpha]_D^{30.5}$ −51.1° (c, 0.56 in EtOH).

18-Hydroxy: **Pilosanone B**.
$C_{20}H_{32}O_4$ M 336.470
Constit. of *P. pilosa*. Oil. $[\alpha]_D^{30.5}$ −52.2° (c, 1.27 in EtOH).

Ohsaki, A. *et al, J. Chem. Soc., Chem. Commun.*, 1987, 151.

ent-8(14)-Pimarene-2α,3α,15*R*,16-tetrol   P-60154

[110601-38-2]

$C_{20}H_{34}O_4$ M 338.486

Constit. of *Milleria quinqueflora*. Oil. $[\alpha]_D^{24}$ −28° (c, 0.86 in CHCl₃).

15-Ketone: [110601-39-3]. ent-*2α,3α,16-Trihydroxy-8(14)-pimaren-15-one*.
$C_{20}H_{32}O_4$ M 336.470
Constit. of *M. quinqueflora*. Oil.

Jakupovic, J. *et al, Phytochemistry*, 1987, **26**, 2011.

Pinguisanin   P-60155

Updated Entry replacing P-01912
4a,5,6,7,7a,8-Hexahydro-4,4a,7,7a-tetramethyl-5,8-ep-oxy-4H-indeno[5,6-b]furan, 9CI

[73020-90-3]

$C_{15}H_{20}O_2$ M 232.322
Constit. of *Porella platyphylla*. Cryst. Mp 50-51°. $[\alpha]_D$ +18° (c, 1.7 in MeOH).

Asakawa, Y. *et al, Phytochemistry*, 1979, **18**, 1349 (*isol*)
Asakawa, Y. *et al, J. Chem. Res. (S)*, 1987, 82 (*struct*)

Pinusolide   P-60156

Updated Entry replacing P-01927
[31685-80-0]

$C_{21}H_{30}O_4$ M 346.466
Constit. of the oleoresin of *Pinus sibirica* and *P. koraiensis*.

Parent acid: **Demethylpinusolide**.
  $C_{20}H_{28}O_4$ M 332.439
  Constit. of *Brickellia glomerata*. Gum.

Raldugin, V.A. *et al, Khim. Pir. Soedin.*, 1970, **6**, 541; *CA*, **74**, 84003s (*isol, struct*)
Calderón, J.S. *et al, Phytochemistry*, 1987, **26**, 2639 (*deriv*)

2,5-Piperazinedicarboxylic acid   P-60157

Updated Entry replacing P-40162
Hexahydro-2,5-pyrazinedicarboxylic acid

(2RS,5RS)-form

$C_6H_{10}N_2O_4$ M 174.156
(2RS,5RS)-form [96705-91-8]
  (±)-*cis-form*
  Cryst. Mp >280°.
  Di-Me ester: [96705-93-0]. Cryst. Mp 65-67°.
  Diamide: [96705-94-1]. Solid. Mp 201-203°.
(2RS,5SR)-form [96705-92-9]
  (±)-*trans-form*
  Cryst. Mp >280°.
  Di-Me ester: [96728-88-0].
    $C_8H_{14}N_2O_4$ M 202.210
    Needles (Me₂CO/hexane). Mp 116-118°.
  Diamide: [96705-95-2].
    $C_6H_{12}N_4O_2$ M 172.186
    Solid. Mp >260° (discolours >230°).

Felder, E. *et al, Helv. Chim. Acta*, 1960, **43**, 888 (*synth*)
Witiak, D.T. *et al, J. Med. Chem.*, 1985, **28**, 1228 (*synth, ir, pmr*)

α-Pipitzol   P-60158

Updated Entry replacing P-01993
[2211-20-3]

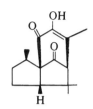

$C_{15}H_{20}O_3$ M 248.321
Constit. of *Perezia cuernavacana* and obt. by pyrolysis of Perezone. Cryst. (Me₂CO/hexane). Mp 146-147°. $[\alpha]_D^{20}$ +192° (CHCl₃).

Walls, F. *et al, Tetrahedron*, 1966, **22**, 2387 (*isol, struct*)
Taira, Z. *et al, Cryst. Struct. Commun.*, 1977, **6**, 23.
Joseph-Nathan, P. *et al, Tetrahedron*, 1980, **36**, 731 (*stereochem*)
Joseph-Nathan, P. *et al, J. Nat. Prod.*, 1986, **49**, 79 (*cmr*)
Joseph-Nathan, P. *et al, J. Org. Chem.*, 1987, **52**, 759 (*synth*)

Pipoxide   P-60159

Updated Entry replacing P-20133
1-[(Benzoyloxy)methyl]-7-oxabicyclo[4.1.0]hept-4-ene-2,3-diol 3-benzoate, 9CI

Absolute configuration

$C_{21}H_{18}O_6$ M 366.370
(*S*)-form illus.
(R)-form [71481-01-1]
  Constit. of *Uvaria pandensis*. Mp 148-150°. $[\alpha]_D^{28}$ −36.5° (c, 0.17 in CHCl₃).
(S)-form [29399-87-9]
  Constit. of the leaves of *Piper hookeri* and *U. purpurea*. Cryst. (C₆H₆). Mp 152-154°. $[\alpha]_D$ +36.3° (c, 0.67 in CHCl₃).

Singh, J. *et al, Tetrahedron*, 1970, **26**, 4403 (*isol*)
Holbert, G. *et al, Tetrahedron Lett.*, 1979, 715 (*isol*)
Schlessinger, R.H. *et al, J. Org. Chem.*, 1981, **46**, 5252 (*synth*)
Schulte, G.R. *et al, Tetrahedron Lett.*, 1982, **23**, 4299 (*abs config*)
Ogawa, S. *et al, J. Org. Chem.*, 1985, **50**, 2356 (*synth*)
Nkunya, M.H.H. *et al, Phytochemistry*, 1987, **26**, 2563 (*isol*)

Pityol   P-60160

Tetrahydro-α,α,5-trimethyl-2-furanmethanol, 9CI. 2-(1-Hydroxy-1-methylethyl)-5-methyltetrahydrofuran

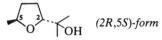

(2R,5S)-form

$C_8H_{16}O_2$ M 144.213
Male-specific attractant of the bark beetle *Pityophthorus pityographus*.
(2R,5S)-form [105814-94-6]
  (+)-*trans-form*
  Bp₄₀ 95°. $[\alpha]_D^{24}$ +18.2° (c, 1.57 in CHCl₃). n_D^{26} 1.4364.
(2S,5R)-form [105814-93-5]
  (−)-*trans-form*
  Bp₄₈ 89-91°. $[\alpha]_D^{23.5}$ −17.5° (c, 1.12 in CHCl₃). n_D^{25} 1.4360.

Mori, K. *et al*, *Justus Liebigs Ann. Chem.*, 1987, 271 (*synth, ir, pmr, cmr, ms, ord*)

Platypterophthalide   P-60161

$C_{13}H_{12}O_4$ M 232.235
Constit. of *Helichrysum platypterum*. Gum.
Jakupovic, J. *et al*, *Phytochemistry*, 1987, **26**, 580.

Pluramycin *A*   P-60162

Updated Entry replacing P-20145
  MA 321A₃. Antibiotic MA 321A₃
  [11016-27-6]

$C_{43}H_{52}N_2O_{11}$ M 772.891
Anthracycline-type antibiotic. Produced by *Streptomyces pluricolorescens*. Shows antitumour properties. Orange needles or prisms. Mp 177° (darkens)(needles), 200-215° (darkens)(prisms). $[\alpha]_D$ +38.5°. Similar to Hedamycin.
▷CB4584250.

De-O-*Ac*: [77849-09-3]. **Rubiflavin A**.
  $C_{41}H_{50}N_2O_{10}$ M 730.853
  From *S*. sp. SC 3728. Antitumour agent. Amorph.

Kondo, S. *et al*, *J. Antibiot.*, 1977, **30**, 1143.
Nadig, H. *et al*, *Helv. Chim. Acta*, 1980, **63**, 2446 (*Rubiflavin A*)
Ceroni, M. *et al*, *Helv. Chim. Acta*, 1982, **65**, 302 (*struct, config*)

Podocarpusflavanone   P-60163

$C_{33}H_{28}O_{10}$ M 584.578
Constit. of *Podocarpus taxifolia*.
Roy, S.K. *et al*, *Phytochemistry*, 1987, **26**, 1985.

Podoverine *A*   P-60164

3′,4′,5,7-Tetrahydroxy-3-methoxy-2′-prenylflavone
[107882-43-9]

$C_{21}H_{20}O_7$ M 384.385
Constit. of *Podophyllum versipelle*. Cryst. Mp 82-84°.
Arens, H. *et al*, *Planta Med.*, 1986, 468.

Podoverine *C*   P-60165

[107882-40-6]

$C_{36}H_{30}O_{14}$ M 686.625
Constit. of *Podophyllum versipelle*. Cryst. Mp 171-173°.
3′-Hydroxy: [107882-41-7]. **Podoverine B**.
  $C_{36}H_{30}O_{15}$ M 702.624
  From *P. versipelle*. Cryst. Mp 180-182°.
Arens, H. *et al*, *Planta Med.*, 1986, 468.

Polemannone   P-60166

$C_{20}H_{22}O_4$ M 326.391
Parent compd. unknown.
4,5-Dimethoxy, 4′,5′-(methylenedioxy): 4,5-Dimethoxy-4′,5′-methylenedioxypolemannone.
  $C_{23}H_{26}O_8$ M 430.454
  Constit. of *Polemannia montana*. Cryst. Mp 197°.
4,5;4′,5′-Bis(methylenedioxy): 4,5;4′,5′-Bismethylenedioxypolemannone.
  $C_{22}H_{22}O_8$ M 414.411
  From *P. montana*. Cryst. Mp 170°.
4,4′,5,5′-Tetramethoxy: 4,4′,5,5′-Tetramethoxypolemannone.
  $C_{24}H_{30}O_8$ M 446.496
  From *P. montana*. Cryst. Mp 140°.
Jakupovic, J. *et al*, *Phytochemistry*, 1987, **26**, 2427.

Polivione   P-60167

Updated Entry replacing P-40175

3,4-Dihydro-6,7-dihydroxy-3-(1-hydroxy-3-oxo-1-bu-tenyl)-4-oxo-2H-1-benzopyran-5-carboxylic acid

[98599-90-7]

$C_{14}H_{12}O_8$ M 308.244

Enolised triketone, mixt. of tautomers. Metab. of *Penicillium frequentans*. Converts readily to Citromycetin, C-02518 .

Demetriadou, A.K. *et al, J. Chem. Soc., Chem. Commun.*, 1985, 762, 764 (*isol, biosynth*)

Laurencia Polyketal   P-60168

$C_{15}H_{20}O_8$ M 328.318

Not named in the paper. Constit. of red alga *Laurencia chilensis*. Cryst. (EtOAc). Mp 182-185°. Racemate.

Bittner, M. *et al, Tetrahedron Lett.*, 1987, 28, 4031 (*cryst struct*)

Praeruptorin *A*   P-60169

[73069-27-9]

$C_{21}H_{22}O_7$ M 386.401

Constit. of *Peucedanum praeruptorum*. Shows calcium antagonistic acitivity. Cryst. Mp 145-147.5°.

Okuyama, T. *et al, Planta Med.*, 1981, 42, 89.

Prangeline   P-60170

Updated Entry replacing P-40182

Pabulenol. Gosferol

[37551-62-5]

$C_{16}H_{14}O_5$ M 286.284

(*R*)-*form* [33889-70-2]

Constit. of *Angelica prancicii, Heracleum sosnowskyi, Hippomarathrum caspicium, Ruta graveolens* and *R. pinnata*. Cryst. Mp 136.5-138.5° (123-125°). $[\alpha]_D^{18}$ +8.6° (c, 0.46 in EtOH).

2′-*Angeloyl:* [33884-53-6]. ***Angeloylprangeline***.

$C_{21}H_{20}O_6$ M 368.385

From *A. prancicii*. Cryst. Mp 87-88°. $[\alpha]_D$ −27°.

2′-*Ketone:* [5058-15-1]. ***Pabulenone***.

$C_{16}H_{12}O_5$ M 284.268

From *Peucedanum ostruthium*. Cryst. Mp 119-121°. Achiral.

(*S*)-*form* [33889-70-2]

From *A. apaensis, Cachrys sicula, Peucedanum ostruthum* and *Prangos pabularia*. Mp 134-135°. $[\alpha]_D^{25}$ −3.8° (EtOH).

(±)-*form*

From *Prangos ferulacea, P. lophoptera* and *Peucedanum stenocarpum*. Cryst. (EtOH). Mp 136.5-138.5°.

Ognyanov, I. *et al, Doklady, Bolg. Akad. Nauk.*, 1971, 24, 315 (*isol*)
Basa, S.C. *et al, Tetrahedron Lett.*, 1971, 1977 (*isol*)
Chatterjea, A. *et al, Tetrahedron*, 1972, 28, 5175 (*isol*)
Abyshev, A.Z. *et al, Khim. Prir. Soedin.*, 1972, 8, 49 (*isol*)
Reisch, J. *et al, Phytochemistry*, 1975, 14, 1889 (*isol*)
Babu, K. *et al, Chem. Pharm. Bull.*, 1981, 29, 2565 (*isol*)
Mendez, J. *et al, Phytochemistry*, 1983, 22, 2599 (*isol*)
Grande, M. *et al, Phytochemistry*, 1986, 25, 505 (*isol*)

Precapnelladiene   P-60171

Updated Entry replacing P-02210

[72715-05-0]

$C_{15}H_{24}$ M 204.355

Constit. of *Capnella imbricata*. Oil.

Ayanoglu, E. *et al, Tetrahedron*, 1979, 35, 1035.
Mehta, G. *et al, J. Org. Chem.*, 1987, 52, 2875 (*synth*)

Pringleine   P-60172

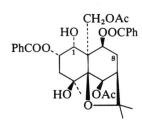

$C_{33}H_{38}O_{11}$ M 610.657

Constit. of *Celastrus pringlei*. Amorph. powder. Mp 128°.

1-Ac: ***Acetylpringleine***.

$C_{35}H_{40}O_{12}$ M 652.694

From *C. pringlei*. Amorph. powder. Mp 118°.

1-Ac, 8α-benzoyloxy: ***8α-Benzoyloxyacetylpringleine***.

$C_{42}H_{44}O_{14}$ M 772.801

From *C. pringlei*. Amorph. powder. Mp 98°.

Sánchez, A.-A. *et al, Phytochemistry*, 1987, 26, 2631.

N-Prolylserine   P-60173

$C_8H_{14}N_2O_4$ M 202.210

L-L-form [71835-80-8]

Mp 215°. $[\alpha]_D^{20}$ −47.9° (H_2O).

Z-Pro-Ser-OMe: [2481-26-7]. Cryst. (EtOAc/pet. ether). Mp 103-104°. $[\alpha]_D^{22}$ −56.8° (c, 1 in AcOH).

Z-Pro-Ser-NHNH₂: [2951-15-7]. Mp 185-186°. $[\alpha]_D^{22}$ −76.8° (c, 1 in AcOH).

Abderhalden, E. *et al*, *CA*, 1936, **30**, 7592 (*synth*)
Luebke, K. *et al*, *Justus Liebigs Ann. Chem.*, 1964, **679**, 195 (*deriv*)
Okada, Y. *et al*, *Chem. Pharm. Bull.*, 1980, **28**, 2254 (*deriv*)
Aubry, A. *et al*, *Int. J. Pept. Protein Res.*, 1984, **23**, 113, 123 (*conformn*)

2-(1,2-Propadienyl)benzothiazole, 9CI   P-60174

2-Allenylbenzothiazole

[109948-61-0]

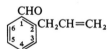

$C_{10}H_7NS$ M 173.232
Oil.

Babudri, F. *et al*, *Synthesis*, 1986, 638 (*synth, ir, pmr*)

2,2-Propanedithiol   P-60175

2,2-Dimercaptopropane

[1687-47-4]

$$(H_3C)_2C(SH)_2$$

$C_3H_8S_2$ M 108.216
Mp 4-6° (6-8°). Bp_{100} 57°, Bp_{70} 40°.

Jentzsch, J. *et al*, *Chem. Ber.*, 1962, **95**, 1764 (*synth, ir*)
Magnusson, B., *Acta Chem. Scand.*, 1963, **17**, 273 (*synth*)
Demuynck, M. *et al*, *Bull. Soc. Chim. Fr.*, 1967, 1213 (*synth, pmr*)

2-(2-Propenyl)benzaldehyde   P-60176

Updated Entry replacing P-30180

o-*Allylbenzaldehyde*

[62708-42-3]

$C_{10}H_{10}O$ M 146.188
Oil. $Bp_{0.35}$ 51-52°.

Semmelhack, M.F. *et al*, *J. Am. Chem. Soc.*, 1983, **105**, 2034 (*synth, spectra*)
Kampmeier, J.A. *et al*, *J. Org. Chem.*, 1984, **49**, 621 (*synth, ir, pmr, ms*)
Hickey, D.M.B. *et al*, *J. Chem. Soc., Perkin Trans. 1*, 1986, 1113 (*synth, pmr*)
Ashby, E.C. *et al*, *J. Org. Chem.*, 1987, **52**, 4079 (*synth*)

4-(2-Propenyl)benzaldehyde   P-60177

p-*Allylbenzaldehyde*

[77785-94-5]

$C_{10}H_{10}O$ M 146.188
Liq. Bp_{12} 113°.

Semicarbazone: Leaflets (EtOH). Mp 197°.

Quelet, R., *Bull. Soc. Chim. Fr.*, 1929, **45**, 255 (*synth*)

1-(1-Propenyl)cyclohexene, 9CI   P-60178

[20457-80-1]

C_9H_{14} M 122.210
Bp_{15} 56-58°.

(*E*)-*form* [54354-35-7]
No phys. props. reported.

(*Z*)-*form* [5680-41-1]
No phys. props. reported.

Marvel, E.N. *et al*, *J. Org. Chem.*, 1965, **30**, 3991 (*synth*)
Roumestant, M.L. *et al*, *Bull. Soc. Chim. Fr.*, 1972, 591 (*synth*)
McIntosh, J.M. *et al*, *J. Org. Chem.*, 1975, **40**, 1294 (*synth*)
Pasto, D.J. *et al*, *J. Org. Chem.*, 1978, **43**, 1382 (*synth*)

1-(2-Propenyl)cyclohexene, 9CI   P-60179

1-Allylcyclohexene, 8CI

[13511-13-2]

C_9H_{14} M 122.210
Bp 154-158°.

Barluenga, J. *et al*, *J. Chem. Soc., Chem. Commun.*, 1978, 847 (*synth*)
Duboudin, J.G. *et al*, *J. Organomet. Chem.*, 1979, **172**, 1 (*synth*)
Barluenga, J. *et al*, *J. Chem. Res. (S)*, 1980, 41 (*synth*)

3-(2-Propenyl)cyclohexene, 9CI   P-60180

3-Allylcyclohexene, 8CI

[15232-95-8]

C_9H_{14} M 122.210

Bartlett, P.D. *et al*, *Tetrahedron, Suppl. 8*, 1966, 399 (*synth*)
Goering, H.L. *et al*, *J. Org. Chem.*, 1986, **51**, 2884 (*synth*)

4-(1-Propenyl)cyclohexene, 9CI   P-60181

[14033-65-9]

C_9H_{14} M 122.210

(±)-*form*
Bp_{100} 91.5-93°.

Petrov, A.A. *et al*, *Zh. Obshch. Khim.*, 1957, **27**, 1795 (*synth*)
Corey, E.J. *et al*, *J. Am. Chem. Soc.*, 1966, **88**, 5652 (*synth*)
Warwel, S. *et al*, *Angew. Chem., Int. Ed. Engl.*, 1982, **21**, 700 (*synth*)

4-(2-Propenyl)cyclohexene, 9CI   P-60182

4-Allylcyclohexene

[15414-36-5]

C_9H_{14} M 122.210

(±)-*form*
Bp 147-149°, Bp_{25} 25°.

Butler, G.B. *et al*, *J. Macromol. Chem., Sci. Chem.*, 1974, **A8**, 1139.

3-(2-Propenyl)indole
P-60183

3-Allylindole

[16886-09-2]

CH₂CH=CH₂ (structure)

$C_{11}H_{11}N$ M 157.215

$Bp_{0.03}$ 94-98°.

Brown, J.B. *et al*, *J. Chem. Soc.*, 1952, 3172 (*synth, ir*)
Wenkert, E. *et al*, *J. Org. Chem.*, 1986, **51**, 2343 (*synth, uv, ir, pmr, cmr, ms*)

2-Propyn-1-amine, 9CI
P-60184

Updated Entry replacing P-30192

3-Aminopropyne. Propargylamine

[2450-71-7]

$$HC{\equiv}CCH_2NH_2$$

C_3H_5N M 55.079

Bp 81-83°.

▷UK5250000.

B,HCl: [15430-52-1]. Sol. H_2O, EtOH. Unstable.

N-Me; B,HI:
  C_4H_7N M 69.106
  Mp 83°.

N-Nitro:
  $C_3H_4N_2O_2$ M 100.077
  Liquid propellant for rockets, etc. $Bp_{0.1}$ 59-60°. n_D^{25}
  1.4982.

N-Di-Me: [7223-38-3].
  C_5H_9N M 83.133
  Bp 82°.

N-Di-Et:
  C_7H_9N M 107.155
  Bp 120°. n_D^{20} 1.4320.

Paal, C. *et al*, *Ber.*, 1889, **22**, 3080; 1891, **24**, 3040 (*synth*)
Campbell, K.N. *et al*, *J. Org. Chem.*, 1952, **17**, 1141 (*deriv*)
Fegley, M.F. *et al*, *J. Am. Chem. Soc.*, 1957, **79**, 4140 (*deriv*)
U.S.P., 3 120 566, (*1964*); *CA*, **60**, 9148 (*synth, deriv*)
Ger. Pat., 2 405 370, (*1974*); *CA*, **81**, 135434 (*synth*)
Verkruijsse, H.D. *et al*, *Recl. Trav. Chim. Pays-Bas*, 1981, **100**, 244 (*derivs*)
Riggs, N.V., *Aust. J. Chem.*, 1987, **40**, 435 (*conformn, bibl*)

4-(2-Propynyl)-2-azetidinone, 9CI
P-60185

4-Propargyl-2-azetidinone

[81355-47-7]

HC≡CCH₂ — NH / O (structure)

C_6H_7NO M 109.127

Synthon for carbapenem synth.

(±)-*form*

Cryst. (Et_2O). Mp 45-46°.

Nishida, A. *et al*, *Chem. Pharm. Bull.*, 1986, **34**, 1423, 1434 (*synth, ir, pmr, use*)

Protosappanin A
P-60186

3,10,11-Trihydroxy-7,8-dihydro-6H-dibenz[b,d]*oxcin-7-one*

[102036-28-2]

(structure)

$C_{15}H_{12}O_5$ M 272.257

Constit. of heartwood of *Caesalpinia sappan* (Sappan Lignum). Prob. precursor of #50173-9. Mp 250-251°.

Nagai, M. *et al*, *Chem. Pharm. Bull.*, 1986, **34**, 1 (*cryst struct*)

Pseudopterosin A
P-60187

[104855-20-1]

(structure)

$C_{25}H_{36}O_6$ M 432.556

Constit. of *Pseudopterogorgia elisabethae*. Shows high antiinflammatory and analgesic activities. Amorph. solid. $[\alpha]_D^{20}$ −85° (c, 0.69 in $CHCl_3$).

2'-Ac: [104855-21-2]. *Pseudopterosin B*.
  $C_{27}H_{38}O_7$ M 474.593
  From *P. elisabethae*. Oil. $[\alpha]_D^{20}$ −55.2° (c, 2.1 in $CHCl_3$).

3'-Ac: [104881-78-9]. *Pseudopterosin C*.
  $C_{27}H_{38}O_7$ M 474.593
  From *P. elisabethae*. Cryst. (EtOAc/EtOH). Mp 113.5-115°. $[\alpha]_D^{20}$ −77° (c, 1.09 in $CHCl_3$).

4'-Ac: Pseudopterosin D.
  $C_{27}H_{38}O_7$ M 474.593
  From *P. elisabethae*. Oil. $[\alpha]_D^{20}$ −107.3° (c, 0.55 in $CHCl_3$).

Look, S.A. *et al*, *J. Org. Chem.*, 1986, **51**, 5140 (*isol, cryst struct*)

Psorolactone
P-60188

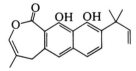

$C_{20}H_{20}O_4$ M 324.376

Constit. of *Psorospermum glaberrimum*.

Botta, B. *et al*, *Tetrahedron Lett.*, 1987, **28**, 567.

Psorospermin   P-60189
[74045-97-9]

C$_{19}$H$_{16}$O$_6$ M 340.332
Constit. of *Psorospermum febrifugum*. Needles (Me-$_2$CO/hexane). Mp 229-230°.

Kupchan, S.M. *et al, J. Nat. Prod.*, 1980, **43**, 296 (*isol*)
Ho, D.K. *et al, J. Org. Chem.*, 1987, **52**, 342 (*synth*)
Habib, A.M. *et al, J. Org. Chem.*, 1987, **52**, 412 (*isol, struct*)

Ptaquiloside   P-60190
Updated Entry replacing P-30200
[87625-62-5]

Absolute
configuration

C$_{20}$H$_{30}$O$_8$ M 398.452
Constit. of *Pteridium aquilinum* var *latiusculum*.
  Amorph. powder. [α]$^{22}_D$ −188° (c, 1.00 in MeOH).
▷Potent carcinogen

Tetra-Ac: [87701-33-5]. Cryst. (MeOH). Mp 173-174°
  dec.

Ojika, M. *et al, Tetrahedron*, 1987, **43**, 5261.

2,4(1*H*,3*H*)-Pteridinedione, 9CI   P-60191
Updated Entry replacing P-02635
Lumazine, 8CI
[487-21-8]

C$_6$H$_4$N$_4$O$_2$ M 164.123
Yellow cryst. (H$_2$O). Mp >350°. Bluish-green fluor. in
  aq. soln., green in alkali, blue in acid. Subl. 145°/0.05
  mm.

1-Me: [50256-18-3].
  C$_7$H$_6$N$_4$O$_2$ M 178.150
  Mp 290-291° (271°).
3-Me: [50256-19-4].
  C$_7$H$_6$N$_4$O$_2$ M 178.150
  Cryst. (H$_2$O). Mp 332°.
1,3-Di-Me: [13401-18-8].
  C$_8$H$_8$N$_4$O$_2$ M 192.177
  Cryst. (H$_2$O). Mp 200°.

Pfleiderer, W., *Chem. Ber.*, 1957, **90**, 2582 (*derivs*)
Brown, D.J. *et al, J. Chem. Soc.*, 1961, 4413 (*synth*)
Pfleiderer, W. *et al, Chem. Ber.*, 1973, **106**, 3149 (*synth*)
Ewers, J. *et al, Chem. Ber.*, 1974, **107**, 3275 (*nmr*)
Tsuzuki, K. *et al, J. Heterocycl. Chem.*, 1986, **23**, 1299 (*deriv*)

Pterosterone   P-60192
Updated Entry replacing P-02660
2β,3β,14α,20ξ,22ξ,24ξ-Hexahydroxy-5β-cholest-7-en-6-one
[18089-44-6]

C$_{27}$H$_{44}$O$_7$ M 480.640
Isol. from *Lastrea thelypteris, Onoclea sensibilis* and
  Vitex megapotamica. Insect moulting hormone. Cryst.
  + H$_2$O. Mp 229-230°. [α]$_D$ +7.4° (MeOH).
Tetra-Ac: Mp 116-117°. [α]$_D$ −9.8° (CHCl$_3$).

Takemoto, T. *et al, Chem. Pharm. Bull.*, 1967, **15**, 1816 (*isol*)
Takemoto, T. *et al, Tetrahedron Lett.*, 1968, 375 (*struct*)
Rimpler, H., *Tetrahedron Lett.*, 1969, 329 (*isol*)
Rimpler, H., *Arch. Pharm. (Weinheim, Ger.)*, 1972, **305**, 746
  (*isol*)
Blunt, J.W. *et al, Aust. J. Chem.*, 1979, **32**, 779 (*struct*)
Nishimoto, N. *et al, Phytochemistry*, 1987, **26**, 2505 (*cmr*)

12-Ptilosarcenol   P-60193

C$_{24}$H$_{31}$ClO$_8$ M 482.957
Constit. of sea fern *Ptilosarcus gurneyi*. Shows insecticid-
  al props. [α]$^{26}_D$ −62.9° (c, 0.56 in CH$_2$Cl$_2$).
12-Ac:
  C$_{26}$H$_{33}$ClO$_9$ M 524.994
  Constit. of *P. gurneyi*. [α]$^{26}_D$ −82.4° (c, 0.40 in
  CH$_2$Cl$_2$).
12-Propanoyl:
  C$_{27}$H$_{35}$ClO$_9$ M 539.021
  Constit. of *P. gurneyi*. [α]$^{26}_D$ −117.0° (c, 0.90 in
  CH$_2$Cl$_2$).

Hendrickson, R.L. *et al, Tetrahedron*, 1986, **42**, 6565.

Ptilosarcenone   P-60194

C$_{24}$H$_{29}$ClO$_8$ M 480.941
Constit. of sea fern *Ptilosarcus gurneyi*. Shows insecticid-
  al props. Cryst. Mp 153-155°. [α]$^{26}_D$ −72.4° (c, 1.01 in
  CH$_2$Cl$_2$).

11α-Hydroxy: **11-Hydroxyptilosarcenone**.
  C$_{24}$H$_{29}$ClO$_9$ M 496.941

Constit. of *P. gurneyi*. $[\alpha]_D^{26}$ −62.5° (c, 0.74 in CH$_2$Cl$_2$).

Hendrickson, R.L. *et al*, *Tetrahedron*, 1986, **42**, 6565.

Ptilosarcol P-60195

C$_{28}$H$_{39}$ClO$_{10}$ M 571.063
Constit. of sea fern *Ptilosarcus gurneyi*. Shows insecticidal props. $[\alpha]_D^{26}$ −55.8° (c, 1.46 in CH$_2$Cl$_2$).

Hendrickson, R.L. *et al*, *Tetrahedron*, 1986, **42**, 6565.

Ptilosarcone P-60196

Updated Entry replacing P-02666
[64597-86-0]

C$_{28}$H$_{37}$ClO$_{10}$ M 569.047
Constit. of *Ptilosarcus gurneyi*. Insecticidal agent. $[\alpha]_D^{26}$ −76.7° (c, 0.64 in CH$_2$Cl$_2$).

Wratten, S.J. *et al*, *Tetrahedron Lett.*, 1977, 1559 (*isol*)
Hendrickson, R.L. *et al*, *Tetrahedron*, 1986, **42**, 6565 (*struct*)

Punctaporonin *A* P-60197

Updated Entry replacing P-30210
Punctatin A. M 95464. Antibiotic M 95464
[91161-74-9]

C$_{15}$H$_{24}$O$_3$ M 252.353
Sesquiterpene antibiotic. Metab. of *Poronia punctata*. Cryst. (EtOAc). Mp 187-192°. $[\alpha]_D^{20}$ −26° (c, 1 in MeOH). Not to be confused with homoisoflavone Punctatins.

9-Epimer: **Punctaporonin D**. *Punctatin D. M 167906. Antibiotic M 167906.*
C$_{15}$H$_{24}$O$_3$ M 252.353
From *P. punctata*. Plates (EtOH). Mp 199-201°. $[\alpha]_D^{20}$ +125° (c, 1 in MeOH).

Anderson, J.R. *et al*, *J. Chem. Soc., Chem. Commun.*, 1984, 405, 917; 1986, 984.
Paquette, L.A. *et al*, *J. Am. Chem. Soc.*, 1986, **108**, 3841 (*synth, abs config*)
Poyser, J.P. *et al*, *J. Antibiot.*, 1986, **39**, 167 (*isol, struct*)
Sugimura, T. *et al*, *J. Am. Chem. Soc.*, 1987, **109**, 3017 (*synth*)

Pyrazino[2′,3′:3,4]cyclobuta[1,2-*g*]-quinoxaline, 9CI P-60198

[107427-60-1]

C$_{12}$H$_6$N$_4$ M 206.206
Yellow cryst. Mp >150° dec. Air-sensitive. Subl. >130°.

Shepherd, M.K., *J. Chem. Soc., Perkin Trans. 1*, 1986, 1495 (*synth, pmr, cmr, ms*)

Pyrazino[2,3-*b*]pyrido[3′,2′-*e*][1,4]thiazine, 9CI P-60199

9H-10-Thia-1,4,5,9-tetraazaanthracene. 1,4,6-Triazaphenothiazine
[80127-31-7]

C$_9$H$_6$N$_4$S M 202.233
Greenish-yellow powder. Mp 213-214°.

Okafor, C.O., *J. Org. Chem.*, 1982, **47**, 592 (*synth, ir, pmr, ms*)

Pyrazino[2,3-*g*]quinazoline-2,4-(1*H*,3*H*)-dione, 9CI P-60200

lin-*Benzolumazine*
[78795-11-6]

C$_{10}$H$_6$N$_4$O$_2$ M 214.183
Light-tan amorph. solid (DMSO aq.). Mp >400° dec.

Schneller, S.W. *et al*, *J. Heterocycl. Chem.*, 1981, **18**, 539 (*synth, ir, pmr*)

Pyrazino[2,3-*f*]quinazoline-8,10-(7*H*,9*H*)-dione, 9CI P-60201

prox-*Benzolumazine*
[78754-84-4]

C$_{10}$H$_6$N$_4$O$_2$ M 214.183
Mp >350° dec.

Schneller, S.W. *et al*, *J. Heterocycl. Chem.*, 1981, **18**, 653 (*synth, pmr*)

1*H*-Pyrazolo[3,4-*d*]pyrimidine-4,6(5*H*,7*H*)-dione, 9CI   P-60202

Updated Entry replacing P-02776
Oxypurinol. BW 55-5. Oxoallopurinol
[2465-59-0]

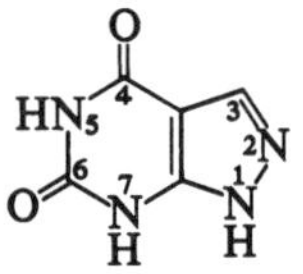

1*H*-form

$C_5H_4N_4O_2$ M 152.112
Purine numbering may also be used. 1*H*-Form predominates. Specific inhibitor of the reduced form of xanthine oxidase. Cryst. (H₂O). Mp >300°.

1*H*-form
1,5-*Di-Me:* [7254-33-3].
   $C_7H_8N_4O_2$ M 180.166
   Granules. Mp 297-298°.
5,7-*Di-Me:* [4680-51-7].
   $C_7H_8N_4O_2$ M 180.166
   Cryst. (EtOH). Mp 280-281°.
1,5,7-*Tri-Me:* [4318-52-9].
   $C_8H_{10}N_4O_2$ M 194.193
   Cryst. (EtOH). Mp 202-204°.

2*H*-form
2,5,7-*Tri-Me:* [10505-26-7].
   $C_8H_{10}N_4O_2$ M 194.193
   Cryst. Mp 267-269°.

Robins, R.K., *J. Am. Chem. Soc.*, 1956, **78**, 784 (synth)
Seela, F. *et al, Justus Liebigs Ann. Chem.*, 1986, 1213 (cmr, bibl, derivs)

1*H*-Pyrazolo[4,3-*g*]quinazoline-5,7(6*H*,8*H*)-dione, 9CI   P-60203

*1,8-Dihydro-5*H*-pyrazolo[4,3-g]quinazolin-5,7(6H)-dione*
[73908-01-7]

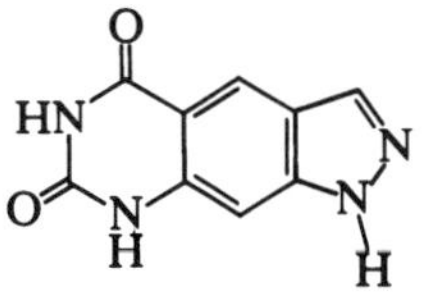

$C_9H_6N_4O_2$ M 202.172
Mp >330°.

Cuny, E. *et al, Chem. Ber.*, 1981, **114**, 1624 (synth, uv, pmr)

Pyrazolo[3,4-*f*]quinazolin-9(8*H*)-one   P-60204

prox-*Benzoisoallopurinol*
[73907-90-1]

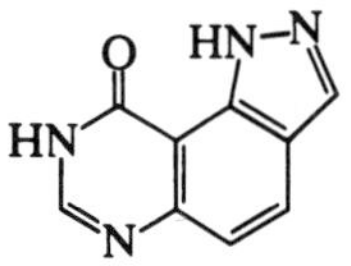

$C_9H_6N_4O$ M 186.173
Tan needles. Mp >300°.

Cuny, E. *et al, Tetrahedron Lett.*, 1980, **21**, 3029 (synth, pmr, uv)
Foster, R.H. *et al, J. Org. Chem.*, 1980, **45**, 3072 (synth, pmr, uv)

Pyrazolo[4,3-*g*]quinazolin-5(6*H*)-one   P-60205

*1,6-Dihydro-5*H*-pyrazolo[4,3-g]quinazolin-5-one. lin-Benzoallopurinol*
[71785-46-1]

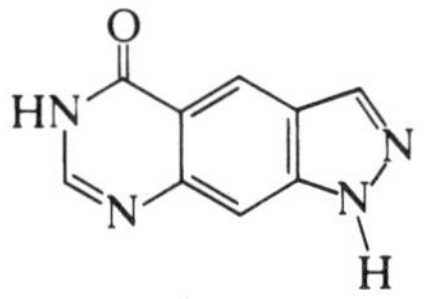

$C_9H_6N_4O$ M 186.173
Powder (DMF). Mp >300°. Fluorescent.

Foster, R.H. *et al, J. Org. Chem.*, 1979, **44**, 4609 (synth, uv, pmr)
Cuny, E. *et al, Chem. Ber.*, 1981, **114**, 1624 (synth, cmr, pmr, uv)

Pyrazolo[3,4-*c*]quinoline, 9CI   P-60206

3H-2,3,5-Triazabenz[e]indene. 1,2,6-Triaza-4,5-benzindene (obsol.)

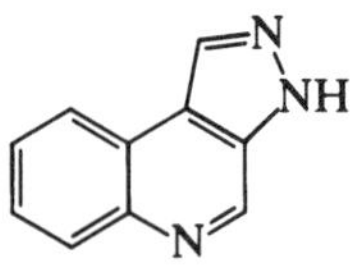

$C_{10}H_7N_3$ M 169.185
Fawn cryst. (C₆H₆/Py). Mp 224°.

Ockenden, D.W. *et al, J. Chem. Soc.*, 1953, 1915.

Pyrazolo[1,5-*b*][1,2,4]triazine, 9CI   P-60207

[89212-61-3]

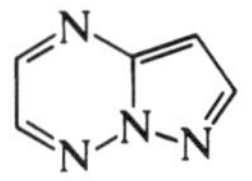

$C_5H_4N_4$ M 120.113
Cryst. (cyclohexane). Mp 118-119°.

Sliskovic, D.R. *et al, Synthesis*, 1986, 71 (synth, ir, uv, pmr, ms)

1-Pyreneacetic acid, 9CI   P-60208

[64709-55-3]

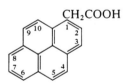

$C_{18}H_{12}O_2$ M 260.292
Powder. Mp 214-219°.

Me ester: [73654-19-0].
   $C_{19}H_{14}O_2$ M 274.318
   Plates (hexane). Mp 90-91.5°.

Shozda, R.J. *et al, J. Am. Chem. Soc.*, 1956, **78**, 1716 (synth, deriv)
Deck, L.M. *et al, J. Org. Chem.*, 1983, **48**, 3577 (synth, deriv)

4-Pyreneacetic acid, 9CI   P-60209

[22245-55-2]
$C_{18}H_{12}O_2$ M 260.292
Cryst. (chlorobenzene). Mp 242-243°.
Me ester: [22245-56-3].
   $C_{19}H_{14}O_2$ M 274.318
   Cryst. (MeOH). Mp 130.5-132°.

Gerasimenko, Y.E. *et al*, *J. Org. Chem. USSR* (*Engl. Transl.*), 1968, **4**, 2120 (*synth, deriv*)
Konieczny, M. *et al*, *J. Org. Chem.*, 1979, **44**, 2158 (*synth, pmr*)
Veerarghavan, S. *et al*, *J. Org. Chem.*, 1987, **52**, 1355 (*synth, pmr*)

1-Pyrenesulfonic acid   P-60210

[26651-23-0]

$C_{16}H_{10}O_3S$ M 282.313

Cryst. + $2H_2O$ (H_2SO_4 aq.). Mp 128° (182-184° anhydrous).

Chloride: [61494-52-8].
$C_{16}H_9ClO_2S$ M 300.759
Mp 175-176°.

Colter, A.K. *et al*, *Can. J. Chem.*, 1978, **56**, 585 (*synth, deriv, ir, pmr*)
Cerfontain, H. *et al*, *Recl. Trav. Chim. Pays-Bas*, 1983, **102**, 210 (*pmr*)
Menger, F.M. *et al*, *J. Org. Chem.*, 1987, **52**, 3793 (*synth*)

2-Pyrenesulfonic acid   P-60211

[64350-83-0]
$C_{16}H_{10}O_3S$ M 282.313

Vollmann, H. *et al*, *Justus Liebigs Ann. Chem.*, 1937, **531**, 2 (*synth*)
Kachurin, O.I. *et al*, *CA*, 1977, **87**, 151884p (*synth*)

4-Pyrenesulfonic acid   P-60212

$C_{16}H_{10}O_3S$ M 282.313

Kachurin, O.I. *et al*, *CA*, 1977, **87**, 151884p (*synth*)
Cerfontain, H. *et al*, *Recl. Trav. Chim. Pays-Bas*, 1983, **102**, 210 (*pmr*)

Pyrenocin A   P-60213

4-Methoxy-6-methyl-5-(1-oxo-2-butenyl)-2H-pyran-2-one, 9CI. Citreopyrone

[76868-97-8]

$C_{11}H_{12}O_4$ M 208.213

Metab. of fungi *Pyrenochaeta terrestris* and *P. citreoviride* B. Phytotoxin. Cryst. (Et_2O/hexane). Mp 107.3-108.3°.

2′,3′-Dihydro, 3′-hydroxy: [72674-29-4]. **Pyrenocin** B.
$C_{11}H_{14}O_5$ M 226.229
Metab. of *P. terrestris*. Phytotoxin. Cryst.
(Et_2O/hexane). Mp 103-103.5°.

Sato, H. *et al*, *Agric. Biol. Chem.*, 1979, **43**, 2409 (*isol, struct*)
Niwa, M. *et al*, *Tetrahedron Lett.*, 1980, 4481 (*isol, struct*)
Ichihara, A. *et al*, *Tetrahedron*, 1987, **43**, 5245 (*synth*)

Pyreno[4,5-b]furan, 9CI   P-60214

[104690-12-2]

$C_{18}H_{10}O$ M 242.276
Microcryst. Mp 196°.

Demerseman, P. *et al*, *J. Heterocycl. Chem.*, 1985, **22**, 1337 (*synth, ms, pmr, uv*)

Pyrichalasin H   P-60215

[111631-97-1]

$C_{31}H_{41}NO_6$ M 523.668
Isol. from *Pyricularia grisea*. Phytotoxin. Plates (EtOAc). Mp 207-209°. $[\alpha]_D^{23}$ −18.4° (c, 1.05 in $CHCl_3$), $[\alpha]_D^{23}$ +59° (c, 1.05 in MeOH). Similar to Cytochalasin *H*.

Nukina, M., *Agric. Biol. Chem.*, 1987, **51**, 2625 (*isol, struct, props*)

Pyriculol   P-60216

Updated Entry replacing P-50288
2-(3,4-Dihydroxy-1,5-heptadienyl)-6-hydroxybenzaldehyde, 9CI

[24868-59-5]

$C_{14}H_{16}O_4$ M 248.278
Isol. from *Pyricularia oryzae*. Phytotoxin. Yellow cryst. Mp 96-97°. $[\alpha]_D$ −54.3° ($CHCl_3$). All 4 stereoisomers are known.

Iwasaki, S. *et al*, *Tetrahedron Lett.*, 1969, 3977 (*isol*)
Bousquet, J.F. *et al*, *Ann. Phytopathol.*, 1973, **5**, 285; *CA*, **81**, 87897s (*isol*)
Nukina, M. *et al*, *Agric. Biol. Chem.*, 1981, **45**, 2161 (*isol*)
Suzuki, M. *et al*, *Agric. Biol. Chem.*, 1986, **60**, 2159; 1987, **51**, 1121 (*synth, abs config*)

3,5-Pyridinedipropanoic acid   P-60217

3,5-Pyridinebis(propanoic acid)

$$HOOCH_2CH_2C \quad CH_2CH_2COOH$$

$C_{11}H_{13}NO_4$ M 223.228
B,HCl: [98329-63-6]. Mp 160°.
Di-Et ester:
$C_{15}H_{21}NO_4$ M 279.335
Bp_2 184°.

Momenteau, M. *et al*, *J. Chem. Soc., Perkin Trans. 1*, 1985, 61, 221 (*synth, pmr*)

2,3-Pyridinedithiol   P-60218
3-Mercapto-2(1H)-pyridinethione, 9CI. 2,3-Dimercaptopyridine
[69212-29-9]

$C_5H_5NS_2$ M 143.221

Thione form predominates. This series given under the dimercapto form for convenient presentation. Shows antifungal props. Yellow needles (H_2O). Mp 216-220° effervsc. Evolves H_2S >130°.

Di-Me thioether: [69212-36-8]. *2,3-Bis(methylthio)-pyridine, 9CI.*
$C_7H_9NS_2$ M 171.275
Oil. Bp$_{0.3}$ 85°.
Di-Me thioether, S-tetraoxide: [85330-79-6]. *2,3-Bis(methylsulfonyl)pyridine, 9CI.*
$C_7H_9NO_4S_2$ M 235.272
Mp 176-177°.

Krowicki, K., *Pol. J. Chem.*, 1978, **52**, 2039 (*synth, uv*)
B.P., 2 053 189, (*1981*); *CA*, **95**, 80748 (*synth*)
Woods, S.G. *et al*, *J. Heterocycl. Chem.*, 1984, **21**, 97 (*deriv*)
Testaferri, L. *et al*, *Tetrahedron*, 1985, **41**, 1373 (*deriv*)

2,4-Pyridinedithiol   P-60219
4-Mercapto-2(1H)-pyridinethione, 9CI. 2,4-Dimercaptopyridine
[28508-52-3]
$C_5H_5NS_2$ M 143.221
Yellow solid. Mp 135-137°.

4-S-Me: [71506-84-8]. *4-(Methylthio)-2(1H)-pyridinethione.*
$C_6H_7NS_2$ M 157.248
Yellow solid. Mp 220-223° (rapid htg.).
N,S-Di-Me: [71506-86-0]. *1-Methyl-4-(methylthio)-2(1H)-pyridinethione.*
$C_7H_9NS_2$ M 171.275
Pale-yellow solid. Mp 155-156°.
Di-S-Me: [71506-85-9]. *2,4-Bis(methylthio)pyridine.*
$C_7H_9NS_2$ M 171.275
Bp$_2$ 130°.

Krowicki, K., *Pol. J. Chem.*, 1979, **53**, 701 (*synth, pmr*)

2,5-Pyridinedithiol   P-60220
5-Mercapto-2(1H)-pyridinethione, 9CI. 2,5-Dimercaptopyridine
[71554-62-6]
$C_5H_5NS_2$ M 143.221
Yellow solid. Mp 210-212° (evolves H_2S).

Di-Me thioether: [85330-62-7]. *2,5-Bis(methylthio)-pyridine, 9CI.*
$C_7H_9NS_2$ M 171.275
Oil. Bp$_{0.1}$ 90°.
Di-Me thioether, S-tetraoxide: [85330-63-8]. *2,5-Bis(methylsulfonyl)pyridine, 9CI.*
$C_7H_9NO_4S_2$ M 235.272
Mp 207-208°.

Krowicki, K., *Pol. J. Chem.*, 1979, **53**, 889 (*synth*)
Woods, S.G. *et al*, *J. Heterocycl. Chem.*, 1984, **21**, 97 (*deriv*)
Testaferri, L. *et al*, *Tetrahedron*, 1985, **41**, 1373 (*deriv*)

3,4-Pyridinedithiol, 9CI   P-60221
3-Mercapto-4(1H)-pyridinethione. 3,4-Dimercaptopyridine
[66242-97-5]
$C_5H_5NS_2$ M 143.221
Yellow solid. No defined Mp, dimerises on htg.

Krowicki, K. *et al*, *Rocz. Chem.*, 1977, **51**, 2435 (*synth*)

3,5-Pyridinedithiol, 9CI   P-60222
3,5-Dimercaptopyridine
[70999-07-4]
$C_5H_5NS_2$ M 143.221
Cryst. (H_2O). Mp 122-124°.

Di-Me thioether: [70999-08-5]. *3,5-Bis(methylthio)-pyridine.*
$C_7H_9NS_2$ M 171.275
Oil. Bp$_{0.05}$ 90°.
Di-Me thioether, S-tetraoxide: [91164-64-6]. *3,5-Bis(methylsulfonyl)pyridine, 9CI.*
$C_7H_9NO_4S_2$ M 235.272
Mp 239-241° (229-231°).

Krowicki, K., *Pol. J. Chem.*, 1979, **53**, 503 (*synth, uv, pmr*)
Testaferri, L. *et al*, *Tetrahedron*, 1985, **41**, 1373 (*deriv*)

2-Pyridinethiol, 9CI   P-60223
Updated Entry replacing P-02838
2-Mercaptopyridine. 2-Pyridyl mercaptan

C_5H_5NS M 111.161
Minor tautomeric form of P-60224; predominates in vapour phase.

S-Me: [18438-38-5].
C_6H_7NS M 125.188
Liq. Bp$_{23}$ 100-104°.
▷UT6207000.
S-Me; B,MeI: Pale-yellow needles (Me_2CO). Mp 155-157°.
S-Me, N-Oxide:
C_6H_7NOS M 141.187
Needles (EtOAc/hexane). Mp 81°.

Albert, A. *et al*, *J. Chem. Soc.*, 1959, 2384 (*synth*)
Kwiatkowski, J.S., *J. Mol. Struct.*, 1971, **10**, 245 (*struct*)
Barlin, G.B. *et al*, *J. Chem. Soc., Perkin Trans. 2*, 1974, 790 (*uv, pmr*)
Beak, P. *et al*, *J. Am. Chem. Soc.*, 1976, **98**, 171; *J. Org. Chem.*, 1980, **45**, 1347, 1354 (*struct*)
Still, I.W.J. *et al*, *Can. J. Chem.*, 1976, **54**, 280 (*cmr*)
Martin, A., *Indian J. Chem., Sect. A*, 1977, **15**, 947 (*struct*)
Barton, D.H.R. *et al*, *J. Chem. Soc., Perkin Trans. 1*, 1986, 39 (*deriv*)

2(1H)-Pyridinethione, 9CI   P-60224
Updated Entry replacing P-02841
2-Thiopyridone
[2637-34-5]

C_5H_5NS M 111.161
Major tautomeric form of 2-Pyridinethiol, P-60223.
Yellowish prisms (C_6H_6). Mp 130-132° (128°).
▷UT8575000.

N-*Me:* [2044-27-1].
   C₆H₇NS   M 125.188
   Cryst. (MeOH aq.). Mp 85-87°.
▷UT9830000.
1-Hydroxy: [1121-30-8].
   C₅H₅NOS   M 127.161
   Cryst. (EtOH aq.). Mp 63-64°.
▷UT9625000.
1-Acetoxy:
   C₇H₇NO₂S   M 169.198
   Yellow-green semisolid.

Jones, R.A. *et al, J. Chem. Soc.*, 1958, 3610 (*synth, struct*)
Albert, A. *et al, J. Chem. Soc.*, 1959, 2384 (*synth*)
Jones, R.A. *et al, J. Chem. Soc.*, 1960, 2937 (*deriv, synth, tautom*)
Cook, M.J. *et al, J. Chem. Soc., Perkin Trans. 2*, 1972, 1295 (*struct*)
Beak, P. *et al, J. Am. Chem. Soc.*, 1976, **98**, 171; *J. Org. Chem.*, 1980, **45**, 1347, 1354 (*struct*)
Pauls, H. *et al, Chem. Ber.*, 1976, **109**, 3653 (*deriv*)
Stefaniak, L., *Org. Magn. Reson.*, 1978, **11**, 385; 1979, **12**, 379 (*nmr*)
Ohms, U. *et al, Acta Crystallogr., Sect. B*, 1982, **38**, 831 (*cryst struct*)
Barton, D.H.R. *et al, J. Chem. Soc., Perkin Trans. 1*, 1986, 39 (*deriv, synth, pmr*)

4(1*H*)-Pyridinethione, 9CI   P-60225

Updated Entry replacing P-02842
   4-Thiopyridone
   [19829-29-9]

C₅H₅NS   M 111.161
Major tautomeric form of 4-Pyridinethiol, P-02840 .
   Yellow needles (EtOH). Sol. H₂O. Mp 177°, 186°.
▷UT9610000.
N-*Me:* [6887-59-8].
   C₆H₇NS   M 125.188
   Yellow plates (EtOH). Mp 161-163°.
1-Hydroxy:
   C₅H₅NOS   M 127.161
   Cryst. (EtOH aq.). Mp 140° dec.

King, H. *et al, J. Chem. Soc.*, 1939, 873 (*synth*)
Jones, R. A. *et al, J. Chem. Soc.*, 1958, 3610 (*synth, struct*)
Jones, R.A. *et al, J. Chem. Soc.*, 1960, 2937 (*deriv, synth*)
Still, I.W.J. *et al, Can. J. Chem.*, 1976, **54**, 280 (*cmr*)
Stefaniak, L, *Org. Magn. Reson.*, 1979, **12**, 379 (*cmr, nmr*)
Beak, P. *et al, J. Org. Chem.*, 1980, **45**, 1347 (*struct*)
Barton, D.H.R. *et al, J. Chem. Soc., Perkin Trans. 1*, 1986, 39 (*deriv, synth, pmr*)

Consult the <u>Dictionary</u> <u>of</u>
<u>Organophosphorus</u> <u>Compounds</u> for
organic compounds containing
phosphorus.

[3.3][2.6]Pyridinophane   P-60226
[100994-29-4]

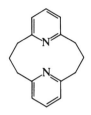

C₁₆H₁₈N₂   M 238.332
Cryst. by subl. Mp 169-170°.

Shinmyozu, T. *et al, J. Org. Chem.*, 1986, **51**, 1551 (*synth, pmr*)

[3](2.2)[3](5.5)Pyridinophane   P-60227
6,14-Diazatricyclo[10.2.2.2⁵,⁸]octadeca-5,7,12,14,15,17-
hexaene, 9CI

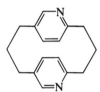

C₁₆H₁₈N₂   M 238.332
Two stereoisomers known.
Low-melting-form [101053-45-6]
   Cryst. by subl. Mp 115.5-117°.
High-melting-form [100994-20-5]
   Cryst. by subl. Mp 172-173.5°.

Shinmyozu, T. *et al, J. Org. Chem.*, 1986, **51**, 1551 (*synth, pmr*)

[3](2.5)[3](5.2)Pyridinophane   P-60228
6,13-Diazatricyclo[10.2.2.2⁵,⁸]octadeca-5,7,12,14,15,17-
hexaene, 9CI

C₁₆H₁₈N₂   M 238.332
Two stereoisomers known.
High-melting-form [101053-47-8]
   Cryst. by subl. Mp 156-158°.
Low-melting-form [100994-22-7]
   Cryst. by subl. Mp 83-85°.

Shinmyozu, T. *et al, J. Org. Chem.*, 1986, **51**, 1551 (*synth, pmr*)

1-(2-Pyridinyl)-2-propanone   P-60229
2-(2-Oxopropyl)pyridine. 2-Acetonylpyridine. 2-
Pyridylacetone
[6302-02-9]

C₈H₉NO   M 135.165
Bp₁.₅ 74-75°, Bp₀.₄ 60-65°.

Goldberg, N.N. *et al, J. Am. Chem. Soc.*, 1951, **73**, 4301 (*synth*)

Cassity, R.P. *et al, J. Org. Chem.*, 1978, **43**, 2286 (*synth, pmr*)

1-(3-Pyridinyl)-2-propanone   P-60230

3-(2-Oxopropyl)pyridine. 3-Acetonylpyridine. 3-Pyridylacetone

[6302-03-0]

C_8H_9NO M 135.165

Liq. Bp_1 119-123°, $Bp_{0.5}$ 79°.

Oxime:
  $C_8H_{10}N_2O$ M 150.180
  Cryst. Mp 117.5-119°.
Semicarbazone: Cryst. (AcOH aq.). Mp 184.5-185°.
Picrate: [53873-12-4]. Cryst. (EtOH aq.). Mp 154-155°.

Burger, A. *et al, J. Am. Chem. Soc.*, 1950, **72**, 1988 (*synth*)
Reynolds, S. *et al, J. Am. Chem. Soc.*, 1960, **82**, 472 (*synth*)
Krottinger, D.L. *et al, J. Chem. Eng. Data*, 1974, **19**, 392 (*synth*)

1-(4-Pyridinyl)-2-propanone   P-60231

4-(2-Oxopropyl)pyridine. 4-Acetonylpyridine. 4-Pyridylacetone

[6304-16-1]

C_8H_9NO M 135.165

Cryst. (EtOH). Mp 155-155.4°. Bp_{3-4} 110-115°, $Bp_{0.3}$ 78-80°.

Picrate: Cryst. (EtOH). Mp 154°.
4-Nitrophenylhydrazone: Brown-yellow cryst. (EtOH). Mp 199°.

Hey, J.W. *et al, Recl. Trav. Chim. Pays-Bas*, 1953, **72**, 522 (*synth*)
Osuch, C. *et al, J. Org. Chem.*, 1957, **22**, 939 (*synth*)

Pyrido[2′,3′:3,4]cyclobuta[1,2-*g*]quinoline, 9CI   P-60232

[107427-57-6]

$C_{14}H_8N_2$ M 204.231
Pale-yellow cryst. Mp 172-174° dec.

Shepherd, M.K., *J. Chem. Soc., Perkin Trans. 1*, 1986, 1495 (*synth, pmr, cmr, ms*)

Pyrido[3′,2′:3,4]cyclobuta[1,2-*g*]quinoline, 9CI   P-60233

[107427-58-7]
$C_{14}H_8N_2$ M 204.231
Pale-yellow cryst. Mp 184-186°. Subl. >120°.

Shepherd, M.K., *J. Chem. Soc., Perkin Trans. 1*, 1986, 1495 (*synth, pmr, cmr, ms*)

Pyrido[2″,1″:2′,3′]imidazo[4′,5′:4,5]-imidazo[1,2-*a*]pyridine   P-60234

Dipyrido[1,2-a:1′,2′-e]-1,3,4,6-tetraazapentalene. 5,5b,10,10b-Tetraazadibenzo[a,e]pentalene

[104716-50-9]

$C_{12}H_8N_4$ M 208.222

Forms 1:1 charge-transfer complexes with π-deficient aromatic cpds. Pale-yellow solid (Me_2CO). Mp 292-294°. Highly fluorescent.

Picrate: [104716-55-4]. Mp 278-281° dec.
B,EtBr: [104716-56-5]. Mp 262-263°.
2,4,7-Trinitrofluorenone complex: [104716-60-1]. Black needles. Mp 240-242° dec.
Tetracyano-1,4-quinodimethane complex: [104716-12-3]. Green-black powder. Mp 290-293° dec.

Groziak, M.P. *et al, J. Am. Chem. Soc.*, 1986, **108**, 8002 (*synth, deriv, pmr, cmr, ir, uv, ms, cryst struct*)

Pyrido[2′,1′:2,3]imidazo[4,5-*c*]isoquinoline   P-60235

$C_{14}H_9N_3$ M 219.245
Cryst. + $\frac{1}{2}H_2O$. Mp 162-164°.

Paolini, J.P. *et al, J. Heterocycl. Chem.*, 1987, **24**, 549 (*synth, uv, cryst struct*)

Pyrido[2,3-*d*]pyrimidine-2,4(1*H*,3*H*)-dione   P-60236

1,2,3,4-Tetrahydro-2,4-dioxopyrido[2,3-d]pyrimidine. 2,4-Dihydroxypyrido[2,3-d]pyrimidine

[21038-66-4]

$C_7H_5N_3O_2$ M 163.135
Cryst. (AcOH) or by subl. Mp 365°.

Robins, R.K. *et al, J. Am. Chem. Soc.*, 1955, **77**, 2256 (*synth*)
Beckwith, A.L.J. *et al, J. Chem. Soc.* (*C*), 1968, 2756 (*synth, ir*)

Pyrido[3,4-*d*]pyrimidine-2,4(1*H*,3*H*)-dione, 9CI   P-60237

1,2,3,4-Tetrahydro-2,4-dioxopyrido[3,4-d]pyrimidine. Copazoline-2,4(1H,3H)-dione

[21038-67-5]

$C_7H_5N_3O_2$ M 163.135

Cryst. by subl. Mp 365° (>320°).

Gabriel, S. *et al*, *Ber.*, 1902, **35**, 2832 (*synth*)
Beckwith, A.L.J. *et al*, *J. Chem. Soc. (C)*, 1968, 2756 (*synth, ir*)

Pyrido[3,4-*d*]-1,2,3-triazin-4(3*H*)-one, 9CI   P-60238
[64188-96-1]

$C_6H_4N_4O$ M 148.124
CAS No. refers to 1*H*-form. Mp 251° dec.

Debeljak-Šuštar, M. *et al*, *J. Org. Chem.*, 1978, **43**, 393 (*synth, pmr*)

2-(2-Pyridyl)ethanethiol   P-60239
2-Pyridineethanethiol, 9CI. 2-(*2-Mercaptoethyl*)*pyridine*
[2044-28-2]

C_7H_9NS M 139.215
Oil. Bp$_{46}$ 137-138°, Bp$_{0.15}$ 57-58°. Dec. slowly at r.t.
B,HCl: Cryst. (2-propanol). Mp 98-99°. Hygroscopic.
Picrate: Yellow needles (EtOH). Mp 89°.

Bauer, L. *et al*, *J. Org. Chem.*, 1961, **26**, 82 (*synth*)

2-(4-Pyridyl)ethanethiol   P-60240
4-Pyridineethanethiol, 9CI. 4-(*2-Mercaptoethyl*)*pyridine*
[2127-05-1]
C_7H_9NS M 139.215
Oil. Bp$_{0.2}$ 92°. Slowly dec. at r.t.
B,HCl: [6298-11-9]. Cryst. (2-propanol). Mp 189°, Mp 250° dec. (double Mp).

Bauer, L. *et al*, *J. Org. Chem.*, 1961, **26**, 82 (*synth*)

2-(2-Pyridyl)-1,3-indanedione, 8CI   P-60241
2-(*2-Pyridinyl*)-*1*H-*indene-1,3(2*H*)-dione, 9CI*.
Pyrophthalone
[641-63-4]

$C_{14}H_9NO_2$ M 223.231
Exists mainly as hydrogen-bonded enol-form. Mp 295-296° (292° dec.). Sinters at 285°.

Lombardino, J.G., *J. Org. Chem.*, 1967, **32**, 1988 (*synth, struct*)
Mosher, W.A. *et al*, *J. Org. Chem.*, 1971, **36**, 1561 (*synth*)
Dainis, I., *Aust. J. Chem.*, 1972, **25**, 1549 (*ms*)
Wolfe, J.F. *et al*, *J. Org. Chem.*, 1974, **39**, 2006 (*synth*)

2-(3-Pyridyl)-1,3-indanedione, 8CI   P-60242
2-(*3-Pyridinyl*)-*1*H-*indene-1,3(2*H*)-dione, 9CI*. 2-
(*3(4*H)-Pyridinylidene*)-*1*H-*indene-1,3(2*H*)-dione, 9CI*.
β-*Pyrophthalone*

[10478-89-4]
$C_{14}H_9NO_2$ M 223.231
Tautomeric. CA gives both names. Cryst. (EtOH). Mp 306°.
B,$^1/_2$HCl: [10478-90-7]. Mp 333°.
Hemipicrate: [10478-94-1]. Mp 240° dec.

Lombardino, J.G., *J. Org. Chem.*, 1967, **32**, 1988 (*synth, struct*)

2-(4-Pyridyl)-1,3-indanedione, 8CI   P-60243
2-(*4-Pyridinyl*)-*1*H-*indene-1,3(2*H*)-dione, 9CI*. 2-
(*4(1*H)-Pyridinylidene*)-*1*H-*indene-1,3(2*H*)-dione, 9CI*.
γ-*Pyrophthalone*
[10478-99-6]
$C_{14}H_9NO_2$ M 223.231
Tautomeric. CA gives both names. Cryst. (EtOH). Mp 325° dec.

Lombardino, J.G., *J. Org. Chem.*, 1967, **32**, 1988 (*synth, struct*)
Mosher, W.A. *et al*, *J. Org. Chem.*, 1971, **36**, 1561 (*synth*)
Wolfe, J.F. *et al*, *J. Org. Chem.*, 1974, **39**, 2006 (*synth*)
Ploquin, J. *et al*, *J. Heterocycl. Chem.*, 1980, **17**, 961.
Rehse, K. *et al*, *Arch. Pharm. (Weinheim, Ger.)*, 1984, **317**, 54 (*synth, props*)

Pyrimido[4,5-*i*]imidazo[4,5-*g*]cinnoline   P-60244
[96245-29-3]

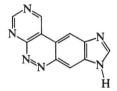

$C_{11}H_6N_6$ M 222.209
Mp >300°.

d'Alarcao, M. *et al*, *J. Org. Chem.*, 1985, **50**, 2456 (*synth, pmr, ms*)

Pyrimido[4,5-*b*]quinoline-2,4(3*H*,10*H*)-dione, 9CI, 8CI   P-60245
2,4-*Dioxo-1,2,3,4-tetrahydropyrimido[4,5-b]quinoline*.
2,4-*Dihydroxypyrimido[4,5-b]quinoline*. 5-*Deazaflavin*
[26908-38-3]

$C_{11}H_7N_3O_2$ M 213.195
CAS reg. no. refers to (1*H*,3*H*)-form. Flavin-shaped nicotinamide analogue; chemistry resembles that of NAD rather than that of flavin. Cryst. (AcOH) or granular solid. Mp >360°.
N^{10}-(ribo-*2,3,4,5-tetrahydroxypentyl*): [19342-73-5].
10-*Deazariboflavin*.
$C_{18}H_{21}N_3O_6$ M 375.380
Cryst. (MeOH). Mp 286-288° dec. Different numbering system is used for riboflavin compared with its parent heterocycle.

Taylor, E.C. *et al*, *J. Am. Chem. Soc.*, 1956, **78**, 5108 (*synth*)
O'Brien, D.E. *et al*, *J. Heterocycl. Chem.*, 1970, **7**, 99 (*synth, uv*)

Tanaka, K. *et al*, *Chem. Pharm. Bull.*, 1987, **35**, 1397 (*bibl*)

4*H*,6*H*-Pyrrolo[1,2-*a*][4,1]benzoxazepine   P-60246
[79100-18-8]

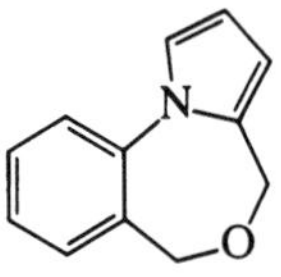

C₁₂H₁₁NO   M 185.225

$C_{12}H_{11}NO$ M 185.225
Oil.

Cheeseman, G.W.H. *et al*, *J. Heterocycl. Chem.*, 1985, **22**, 809 (*synth, pmr*)

9*H*-Pyrrolo[1,2-*a*]indol-9-one, 9CI   P-60247
Fluorazone
[525-24-6]

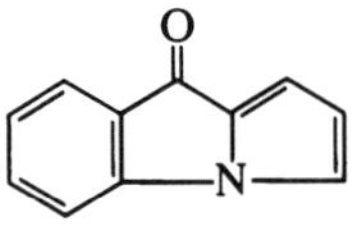

$C_{11}H_7NO$ M 169.182
Yellow prisms. Mp 123-124°. Bp$_{0.28}$ 143-146°.
Hydrazone: [94785-96-3].
    $C_{11}H_9N_3$ M 183.212
    Pale-yellow needles (EtOH aq.). Mp 171.4-172.2°.
Semicarbazone: Mp 212-214°.

Flitsch, W. *et al*, *Chem. Ber.*, 1978, **111**, 2407 (*pmr*)
Bailey, A.S. *et al*, *J. Chem. Soc., Perkin Trans. 1*, 1980, 97 (*synth, bibl*)
Cartoon, M.E.K. *et al*, *J. Organomet. Chem.*, 1981, **212**, 1 (*synth*)
Rault, S. *et al*, *Tetrahedron Lett.*, 1985, **26**, 2305 (*synth, bibl*)

5*H*-Pyrrolo[2,3-*b*]pyrazine, 9CI   P-60248
Updated Entry replacing P-03020
[4745-93-1]

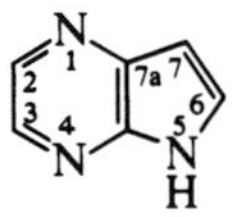

$C_6H_5N_3$ M 119.126
Cryst. (C₆H₆). Mp 155-156°.
5-Me:
    $C_7H_7N_3$ M 133.152
    Mp 111°.

Azimov, V.A. *et al*, *Khim. Geterotsikl. Soedin.*, 1973, **6**, 858; *CA*, **79**, 105191 (*synth*)
Clark, B.A.J. *et al*, *J. Chem. Soc., Perkin Trans. 1*, 1976, 1361.
Clark, B.A.J. *et al*, *Org. Mass Spectrom.*, 1977, **12**, 421 (*ms*)
Marcelis, A.T.M. *et al*, *J. Heterocycl. Chem.*, 1987, **24**, 545 (*synth, pmr, deriv*)

1*H*-Pyrrolo[3,2-*b*]pyridine, 9CI   P-60249
Updated Entry replacing P-03024
1,4-Diazaindene. 4-Azaindole
[272-49-1]

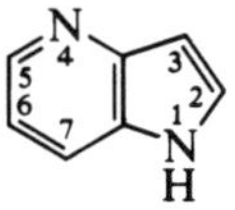

$C_7H_6N_2$ M 118.138
Prisms (C₆H₆). Mp 127-128°.
▷UY8715000.
   N-*Ac:* [24509-73-7].
     $C_9H_8N_2O$ M 160.175
     Mp 77-78°. Bp$_{10}$ 128°.
   N-*Benzoyl:*
     $C_{14}H_{10}N_2O$ M 222.246
     Needles. Mp 92-96°.

Clemo, G.R. *et al*, *J. Chem. Soc.*, 1948, 198.
Frydman, B. *et al*, *J. Org. Chem.*, 1968, **33**, 3762.
Willette, R.E., *Adv. Heterocycl. Chem.*, 1968, **9**, 27 (*rev*)
Yakhontov, L.N. *et al*, *Tetrahedron Lett.*, 1969, 1909.
Yakhontov, L.N., *Khim. Geterotsikl. Soedin.*, 1977, **10**, 1425; *CA*, **88**, 37665 (*synth*)
Sakamoto, T. *et al*, *Chem. Pharm. Bull.*, 1986, **34**, 2362 (*synth, pmr*)

1-(2-Pyrrolyl)-2-(2-thienyl)ethylene   P-60250
2-[2-(2-Thienyl)ethenyl]-1H-pyrrole, 9CI

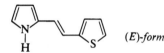
(*E*)-*form*

$C_{10}H_9NS$ M 175.248
(***E***)-*form* [107902-35-2]
Mp 141-144°.
   N-*Me:* [99702-07-5].
     $C_{11}H_{11}NS$ M 189.275
     Mp 65-66°. Bp$_{1.0}$ 65-66°.
(***Z***)-*form* [107902-36-3]
Mp 72-73°.

Hinz, W. *et al*, *Synthesis*, 1986, 620 (*synth, uv, pmr*)

Q

Quadrangolide Q-60001

$C_{15}H_{20}O_3$ M 248.321
Constit. of *Eupatorium quadrangularae*. Cryst.
(MeOH). Mp 118-120°. $[\alpha]_D^{22}$ +227°.

Hubert, T.D. *et al*, *Phytochemistry*, 1987, **26**, 1751.

5,8-Quinazolinediol, 9CI, 8CI Q-60002
5,8-Dihydroxyquinazoline
[24271-83-8]

$C_8H_6N_2O_2$ M 162.148
Yellow needles (EtOAc). Mp 253°.

Di-Me ether: [17944-05-7]. *5,8-Dimethoxyquinoxaline.*
$C_{10}H_{10}N_2O_2$ M 190.201
Light-yellow cryst. (Et$_2$O). Mp 119°.

Malesano, G. *et al*, *J. Med. Chem.*, 1970, **13**, 161 (*synth, ir, uv*)
Shaikh, I.A. *et al*, *J. Med. Chem.*, 1986, **29**, 1329 (*deriv, synth, pmr, ms*)

3-Quinolinethiol, 9CI Q-60003
3-Mercaptoquinoline
[76076-35-2]

C_9H_7NS M 161.221
Interconvertible pale-pink or bright-red cryst. by subl.
Mp 58°.

B,MeI: Cryst. (EtOH/MeOH). Mp 229-231°.
S-Me: [51934-46-4].
$C_{10}H_9NS$ M 175.248
Bp$_{0.2}$ 118-119°.
S-Me; B,HCl: Cryst. (butanol or by subl.). Mp 205-209°.
S-Me; B,MeI: Yellow needles (EtOH). Mp 245°.

Albert, A. *et al*, *J. Chem. Soc.*, 1959, 2384 (*synth, uv, tautom*)
Barlin, G.B. *et al*, *J. Chem. Soc., Perkin Trans. 2*, 1975, 298
(*deriv, synth*)

5-Quinolinethiol, 9CI Q-60004
5-Mercaptoquinoline
[3056-03-9]

C_9H_7NS M 161.221
Pale-pink solid.

Monohydrate: Red cryst. (EtOH aq.). Mp 87.5-89°.
S-Me: [26114-57-8]. *5-Methylthioquinoline.*
$C_{10}H_9NS$ M 175.248

Bp$_{0.1}$ 104°.
S-Me; B,HCl: Yellow cryst. (butanol or by subl.). Mp
241-243.5°.
S-Me; B,MeI: Cryst. (EtOH). Mp 189°.

Albert, A. *et al*, *J. Chem. Soc.*, 1959, 2384 (*synth, uv, tautom*)

6-Quinolinethiol, 9CI Q-60005
6-Mercaptoquinoline
C_9H_7NS M 161.221
Red oil. Bp$_{0.1}$ 114°.

B,MeCl: Cryst. (EtOH). Mp 219-221.5°.
B,MeI: Cryst. Mp 225-227°.
S-Me: [73420-44-7]. *6-Methylthioquinoline.*
$C_{10}H_9NS$ M 175.248
Cryst. (pet. ether). Mp 44-46°.
S-Me; B,MeI: Yellow cryst. (EtOH). Mp 237-238.5°.

Albert, A. *et al*, *J. Chem. Soc.*, 1959, 2384 (*synth, uv, tautom*)

4(1H)-Quinolinethione, 9CI Q-60006
4-Mercaptoquinoline. 4-Quinolinethiol
[76076-27-2]

C_9H_7NS M 161.221
Thione tautomer predominates. Yellow solid (toluene) or
red cryst. Mp 158-162° dec. The yellow form subl. at
125-135°/0.005 mm to give the red form.

S-Me: [46000-25-3]. *4-Methylthioquinoline.*
$C_{10}H_9NS$ M 175.248
Cryst. (pet. ether). Mp 70-72°.
S-Me; B,MeI: Yellow cryst. (EtOH). Mp 247-248°.
N-Me:
$C_{10}H_9NS$ M 175.248
Yellow needles (toluene or EtOH). Mp 209-211°.

Albert, A. *et al*, *J. Chem. Soc.*, 1959, 2384 (*synth, derivs, uv, tautom*)
Barlin, G.B. *et al*, *J. Chem. Soc., Perkin Trans. 2*, 1975, 298
(*deriv*)
Maslankiewicz, A., *Pol. J. Chem.*, 1980, **54**, 2069 (*deriv, synth*)

Quino[7,8-h]quinoline-4,9(1H,12H)dione, 9CI Q-60007
[107798-57-2]

$C_{16}H_{10}N_2O_2$ M 262.267
Mp 375-377° dec.

Zirnstein, M.A. *et al*, *Angew. Chem., Int. Ed. Engl.*, 1987, **26**,
460 (*synth*)

5,8-Quinoxalinediol, 9CI, 8CI   **Q-60008**
5,8-Dihydroxyquinoxaline
[19506-18-4]
$C_8H_6N_2O_2$ M 162.148
Tan solid. Mp 238-239°.
Di-Me ether: [19506-19-5].
    $C_{10}H_{10}N_2O_2$ M 190.201
    Greenish-yellow cryst. (CH$_2$Cl$_2$). Mp 147-148°.

Warren, J.D. *et al*, *J. Heterocycl. Chem.*, 1979, **16**, 1616 (*synth, ir, pmr, ms*)
Shaikh, I.A. *et al*, *J. Med. Chem.*, 1986, **29**, 1329 (*deriv, synth, ir, pmr, ms*)

6,7-Quinoxalinediol, 9CI   **Q-60009**
6,7-Dihydroxyquinoxaline
[19506-20-8]
$C_8H_6N_2O_2$ M 162.148
Yellow cryst. solid by subl. Mp >260° dec.
B,HCl: Violet cryst.
Mono-Me ether: 6-Hydroxy-7-methoxyquinoxaline.
    $C_9H_8N_2O_2$ M 176.174
    Silvery plates (EtOH aq.). Mp 238-239°.
Di-Me ether: [6295-29-0]. *6,7-Dimethoxyquinoxaline.*
    $C_{10}H_{10}N_2O_2$ M 190.201
    Long needles (EtOH). Mp 150-151°.

Erlich, J. *et al*, *J. Org. Chem.*, 1947, **12**, 522 (*synth, deriv*)

5,8-Quinoxalinedione, 9CI   **Q-60010**
Quinoxaline 5,8-quinone
[15250-38-1]

$C_8H_4N_2O_2$ M 160.132
Cryst. (Me$_2$CO). Mp 172-173°.

Warren, J.D. *et al*, *J. Heterocycl. Chem.*, 1979, **16**, 1617 (*synth, ir, pmr, ms*)

R

Ramosissin   R-60001

2-(Pentamethoxyphenyl)-5-(1-propenyl)benzofuran, 9CI
[110784-16-2]

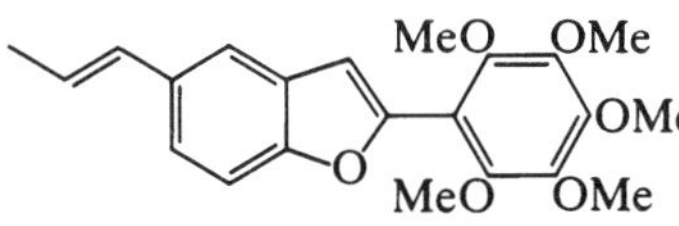

$C_{22}H_{24}O_6$ M 384.428
Constit. of *Krameria ramosissima*. Cryst. Mp 92-94°.

Achenbach, H. *et al, Phytochemistry*, 1987, **26**, 2041.

RA-V   R-60002

Updated Entry replacing R-20007
5-(N-*Methyl*-L-*tyrosine*)*bouvardin*, 9CI.
Deoxybouvardin
[64725-24-2]

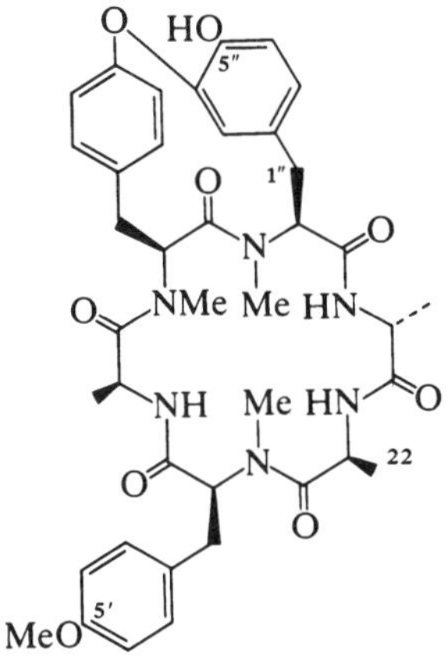

$C_{40}H_{48}N_6O_9$ M 756.854
Cyclic hexapeptide antibiotic. Isol. from the roots of
Rubia cordifolia and *R. akane*. Exhibits antitumour
activities. Powder (MeOH). Mp >300°. $[\alpha]_D^{21}$ −225°
(c, 0.3 in CHCl₃).
Me ether: [86229-97-2]. *RA-VII*.
   $C_{41}H_{50}N_6O_9$ M 770.881
   From *R. radix*. Needles (MeOH). Mp >300°. $[\alpha]_D^{21}$
   −229° (c, 0.1 in CHCl₃).
Me ether, 1'-hydroxy: [70840-66-3]. *RA-III*.
   $C_{41}H_{50}N_6O_{10}$ M 786.880
   Isol. from *R. radix*. Needles (MeOH). Mp >300°.
   $[\alpha]_D^{28}$ −199° (c, 0.1 in CHCl₃).
Me ether, 1"S-hydroxy: [86849-13-0]. *RA-IV*.
   $C_{41}H_{50}N_6O_{10}$ M 786.880
   From *R. radix* and *R. cordifolia*. Antitumour agent.
   Powder (MeOH). Mp 247-255°. $[\alpha]_D^{28}$ −126° (c, 0.07
   in CHCl₃).
22-Hydroxy: RA-I.
   $C_{40}H_{48}N_6O_{10}$ M 772.853
   From *R. cordifolia*. Antitumour agent. Powder
   (MeOH). Mp 284° dec. $[\alpha]_D^{21}$ −216° (c, 0.08 in
   CHCl₃/MeOH).
O⁵'-De-Me, O⁵"-Me: RA-II.
   $C_{40}H_{48}N_6O_9$ M 756.854
   From *R. cordifolia*. Antitumour agent. Needles
   (MeOH). Mp 261° dec. $[\alpha]_D^{28}$ −201° (c, 0.1 in
   CHCl₃).

Itokawa, H. *et al, Chem. Pharm. Bull.*, 1983, **31**, 1424; 1986, **34**,
3762 (*isol, struct, props*)

Rehmaglutin C   R-60003

[103744-81-6]

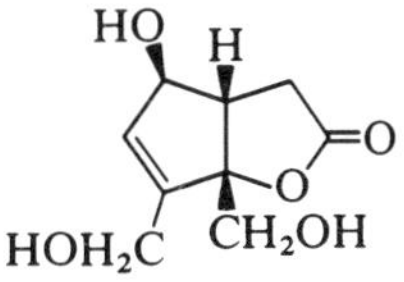

$C_9H_{12}O_5$ M 200.191
Constit. of dried root of *Rehmannia glutinosa* (Rehman-
niae Radix). Oil. $[\alpha]_D^{24}$ −51.4° (MeOH).

Yoshikawa, M. *et al, Chem. Pharm. Bull.*, 1986, **34**, 1403.

Reiswigin A   R-60004

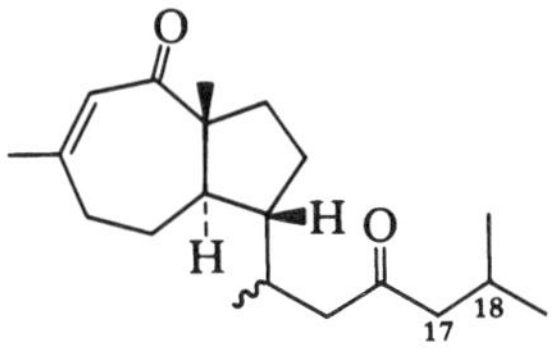

$C_{20}H_{32}O_2$ M 304.472
Constit. of sponge *Epipolasis reiswigi*. Antiviral agent.
Tan oil. $[\alpha]_D^{20}$ −10° (c, 0.1 in CHCl₃).
17,18-Didehydro: **Reiswigin B**.
   $C_{20}H_{30}O_2$ M 302.456
   Constit. of *E. reiswigi*. Antiviral agent. Tan oil. $[\alpha]_D^{20}$
   −20° (c, 0.1 in CHCl₃).

Kashman, Y. *et al, Tetrahedron Lett.*, 1987, **28**, 5461.

Repensolide   R-60005

[106310-35-4]

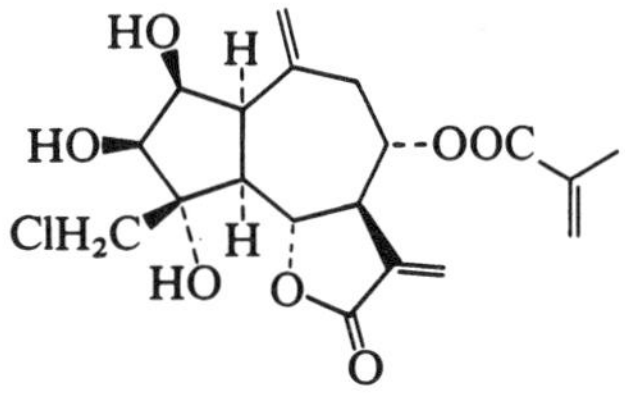

$C_{19}H_{23}ClO_7$ M 398.840
Constit. of *Centaurea* spp. Oil.

Jakupovic, J. *et al, Planta Med.*, 1986, 398.

Resinoside   R-60006

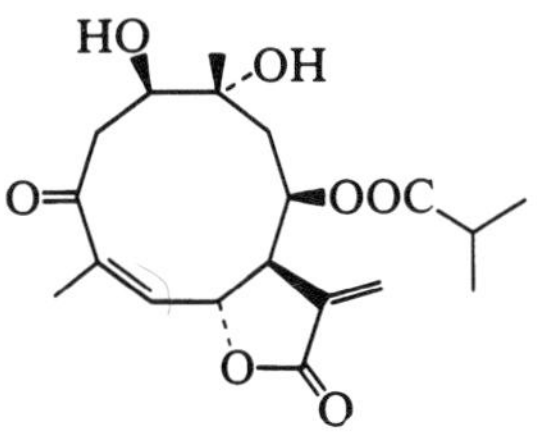

$C_{19}H_{26}O_7$ M 366.410

Constit. of *Greenmaniella resinosa*. Gum.

Zdero, C. *et al*, *Phytochemistry*, 1987, **26**, 1999.

Rhynchosperin *A*   R-60007

$C_{20}H_{20}O_6$ M 356.374

Constit. of *Rhynchospermum verticillatum*. Needles (MeOH). Mp 236-237°. $[\alpha]_D^{25}$ −195° (c, 0.37 in $CHCl_3$).

6β-Hydroxy: **Rhynchosperin B**.
  $C_{20}H_{20}O_7$ M 372.374
  Constit. of *R. verticillatum*. Needles (MeOH). Mp 277-279°. $[\alpha]_D^{25}$ −202° (c, 0.35 in $CHCl_3$).

6β-Tigloyloxy: **Rhynchosperin C**.
  $C_{25}H_{26}O_8$ M 454.476
  Constit. of *R. verticillatum*. Amorph. powder. $[\alpha]_D^{25}$ −153° (c, 1.76 in $CHCl_3$).

Seto, M. *et al*, *Phytochemistry*, 1987, **26**, 3289.

Rhynchospermoside *A*   R-60008

$C_{26}H_{34}O_{10}$ M 506.549

Constit. of *Rhynchospermum verticillatum*. Amorph. powder. $[\alpha]_D^{25}$ −162° (c, 1.93 in MeOH).

12-Epimer: **Rhynchospermoside B**.
  $C_{26}H_{34}O_{10}$ M 506.549
  Constit. of *R. verticillatum*. Amorph. powder. $[\alpha]_D^{25}$ −165° (c, 0.98 in MeOH).

Seto, M. *et al*, *Phytochemistry*, 1987, **26**, 3289.

Riligustilide   R-60009

6,8′;7,3′-Diligustide

[89354-45-0]

$C_{24}H_{28}O_4$ M 380.483

Constit. of *Ligusticum acutilobum*. Oil.

Kaouadji, M. *et al*, *J. Nat. Prod.*, 1986, **49**, 872.

Rosaniline   R-60010

4-[(4-Aminophenyl)(4-imino-2,5-cyclohexadien-1-ylidene)methyl]-2-methylbenzenamine, 9CI. *C.I. Basic Violet 14*

[3248-93-9]

$C_{20}H_{19}N_3$ M 301.390

Used as a dye or in the manuf. of dyestuffs. Possesses antifungal props. Brown-red cryst. Dec. at 186°.

▷ZE9705000.

B,HCl: [632-99-5]. *Fuchsine. Magenta I. C.I. Basic Violet 14*. Metallic green cryst. Dec. >200°.
  ▷CX9855000.

Fischer, E. *et al*, *Ber.*, 1880, **13**, 2204 (*synth*)

Michaelis, L. *et al*, *J. Am. Chem. Soc.*, 1943, **67**, 1212 (*struct*)

ent-5-Rosene-3α,15,16,19-tetrol   R-60011
Juslimtetrol

$C_{20}H_{34}O_4$ M 338.486

Constit. of *Croptilon divaricatum*. Oil.

4-Epimer: *ent*-5-Rosene-3α,15,16,18-tetrol. *4-epi-Juslimtetrol*.
  $C_{20}H_{32}O_3$ M 320.471
  Constit. of *Hymenothrix wislizenii*. Oil.

Dominguez, X.A. *et al*, *Rev. Latinoamer. Quim.*, 1973, **3**, 177 (*isol*)

Jakupovic, J. *et al*, *Phytochemistry*, 1987, **26**, 2543 (*isol*)

ent-5-Rosene-15,16,19-triol   R-60012

3-Desoxyjuslimtetrol

$C_{20}H_{34}O_3$ M 322.487

Constit. of *Hymenothrix wislizenii*.

4-Epimer: *ent*-5-Rosene-15,16,18-triol.
  $C_{20}H_{34}O_3$ M 322.487
  Constit. of *H. wislizenii*.

Jakupovic, J. *et al*, *Phytochemistry*, 1987, **26**, 2543.

Rosmanoyl carnosate   R-60013

$C_{40}H_{52}O_8$ M 660.846
Constit. of *Salvia canariensis*.

González, A.G. *et al*, *Phytochemistry*, 1987, **26**, 1471.

Dalbergia Rotenolone   R-60014

$C_{22}H_{22}O_8$ M 414.411
Constit. of *Dalbergia volubilis*. Brown semisolid.

Chawla, H.M. *et al*, *J. Chem. Res. (S)*, 1987, 168.

Royleanone   R-60015

Updated Entry replacing R-20028
12-Hydroxy-8,12-abietadiene-11,14-dione
[6812-87-9]

$C_{20}H_{28}O_3$ M 316.439
Constit. of *Inula royleana*. Orange-yellow cryst.
(AcOH). Mp 181.5-183°. $[\alpha]_D$ +134° (c, 1.03 in
$CHCl_3$).

6,7-Didehydro: [6855-99-8]. *6,7-Dehydroroyleanone.*
12-Hydroxy-6,8,12-abietatriene-11,14-dione.
$C_{20}H_{26}O_3$ M 314.424
Constit. of *I. royleana* and *Plectranthus* spp. Cryst.
(MeOH). Mp 167°. $[\alpha]_D$ −620° (c, 0.2 in $CHCl_3$).

7α-Acetoxy, 20-hydroxy: 7α-Acetoxy-12,20-dihydroxy-
8,12-abietadiene-11,14-dione.
$C_{22}H_{30}O_6$ M 390.475
Constit. of *Salvia lanata*. Brown plates
($CHCl_3$/hexane). Mp 160-162°.

2β-Hydroxy: 2β,12-Dihydroxy-8,12-abietadiene-11,14-
dione. 2β-Hydroxyroyleanone.
$C_{20}H_{28}O_4$ M 332.439
Constit. of *S. cryptantha*. Oil.

Hersch, M. *et al*, *Helv. Chim. Acta*, 1975, **58**, 1921 (*isol*)
Matsumoto, T. *et al*, *Bull. Chem. Soc. Jpn.*, 1977, **50**, 266
(*synth*)
Mukherjee, K.S. *et al*, *Phytochemistry*, 1983, **22**, 1296 (*deriv*)
Ulubelen, A. *et al*, *Phytochemistry*, 1987, **26**, 1534 (*2β-*
Hydroxyroyleanone)

Rubifolide   R-60016

[106231-29-2]

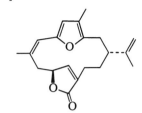

$C_{20}H_{24}O_3$ M 312.408
Constit. of *Gersemia rubiformis*. Cryst.
(CH_2Cl_2/MeOH). Mp 159-160°. $[\alpha]_D^{25}$ +31.7° (c,
0.39 in $CHCl_3$).

Williams, D. *et al*, *J. Org. Chem.*, 1987, **52**, 332.

S

Saikogenin F — S-60001

Updated Entry replacing S-00013

13β,28-Epoxy-11-oleanene-3β,16β,23-triol

[14356-59-3]

$C_{30}H_{48}O_4$ M 472.707

Genuine sapogenin from *Bupleurum falcatum* root.
Cryst. + ½CHCl$_3$ (CHCl$_3$). Mp 265-273°. [α]$_D$ +107.8°.

3-O-[β-D-Glucopyranosyl-(1→3)]-β-D-fucopyranoside:
[20736-09-8]. **Saikosaponin A**.
$C_{42}H_{68}O_{13}$ M 780.991
Constit. of *B. falcatum*. Cryst. Mp 225-232°. [α]$_D$ +46° (EtOH).

3-O-β-D-Galactopyranoside: **Corchorusin B**.
$C_{36}H_{58}O_9$ M 634.849
Constit. of *Corchorus acutangulus*. Cryst. (MeOH).
Mp 278-280°. [α]$_D$ +74.3° (c, 0.6 in MeOH).

Kubota, T. *et al, Tetrahedron Lett.*, 1966, 5045; 1968, 303 (*isol, struct*)
Tori, K. *et al, Tetrahedron Lett.*, 1976, 4163 (*cmr*)
Mahato, S.B. *et al, J. Chem. Soc., Perkin Trans. 1*, 1987, 629 (*deriv*)

Sainfuran — S-60002

2-(2-Hydroxy-4-methoxyphenyl)-5-hydroxy-6-methoxybenzofuran

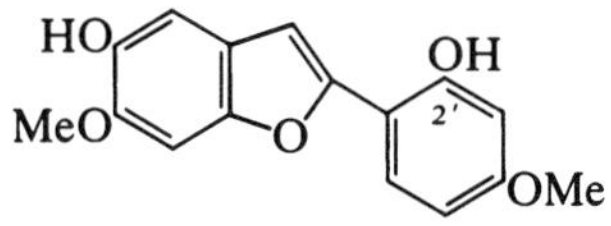

$C_{16}H_{14}O_5$ M 286.284

Constit. of *Onobrychis viciifolia*. Antifungal and insect antifeedant. Cryst. Mp 150-152°.

2'-Me ether: **Methylsainfuran**. *2-(2,4-Dimethylphenyl)-5-hydroxy-6-methoxybenzofuran*.
$C_{17}H_{16}O_5$ M 300.310
Constit. of *O. viciifolia*. Antifungal and insect antifeedant. Cryst. Mp 147-148°.

Russell, G.B. *et al, Phytochemistry*, 1984, **23**, 1417 (*isol, struct*)
Burke, J.M. *et al, J. Chem. Res. (S)*, 1987, 179 (*synth*)

Salvicin — S-60003

ent-6β,15-Dihydroxy-3,13E-clerodadien-19-oic acid
[104700-94-9]

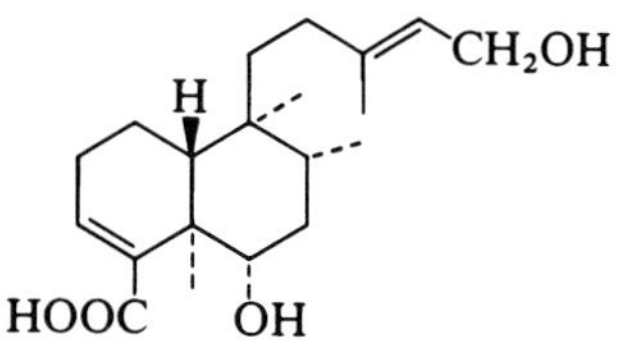

$C_{20}H_{32}O_4$ M 336.470

Constit. of *Pulicaria salviifolia*. [α]$_D^{20}$ −75.4° (c, 0.26 in MeOH).

Nurmukhamedova, M.R. *et al, Khim. Prir. Soedin.*, 1986, **22**, 277.

Salvisyriacolide — S-60004

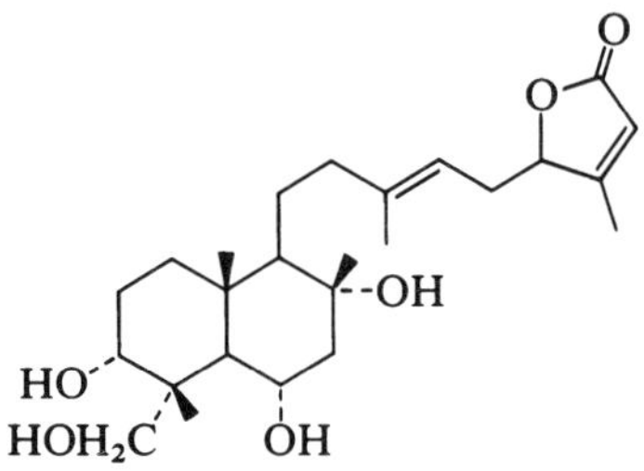

$C_{25}H_{40}O_6$ M 436.587

Constit. of *Salvia syriaca*. Oil. [α]$_D^{28}$ −18° (c, 0.2 in CHCl$_3$).

Rustaiyan, A. *et al, Phytochemistry*, 1987, **26**, 3078.

Sanadaol — S-60005

Updated Entry replacing S-20013

[83643-92-9]

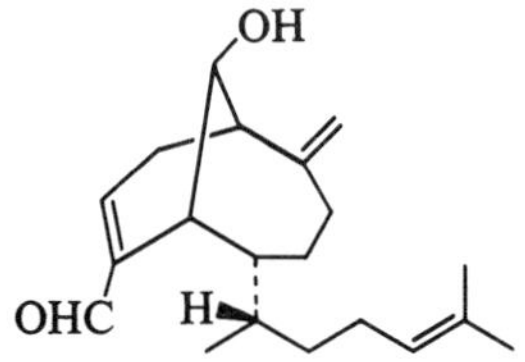

$C_{20}H_{30}O_2$ M 302.456

Constit. of brown alga *Pachydictyon coriaceum*. Oil.
[α]$_D$ +74.8° (c, 1.33 in CHCl$_3$). May be artifact produced from dictyodial.

Ac: [83643-93-0]. *Acetylsanadaol*.
$C_{22}H_{32}O_3$ M 344.493
Constit. of *P. coriaceum*. Oil. [α]$_D$ +42.5° (c, 0.89 in CHCl$_3$).

Ishitsuka, M. *et al, Tetrahedron Lett.*, 1982, **23**, 3179.
Nagaoka, H. *et al, Tetrahedron Lett.*, 1987, **28**, 2021 (*synth*)

Sanguinone A — S-60006

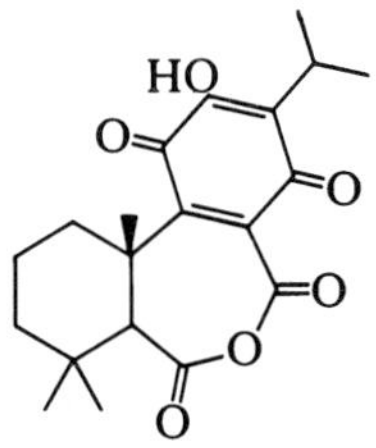

$C_{20}H_{24}O_6$ M 360.406

Constit. of *Plectranthus sanguineus*. Yellow cryst. (Et$_2$O). Mp 174.2-174.6°.

Matloubi-Moghadam, F. *et al, Helv. Chim. Acta*, 1987, **70**, 975.

Santamarine   S-60007

Updated Entry replacing S-20019
1β-Hydroxy-3,11(13)-eudesmadien-12,6α-olide.
Balchanin
[4290-13-5]

$C_{15}H_{20}O_3$ M 248.321

Constit. of *Artemisia balchanorum* and *Chrysanthemum*
parthenium. Cryst. (isopropyl ether/Me$_2$CO). Mp
142° (134-136°). [α]$_D^{20}$ +96.6° (CHCl$_3$).

8α-Hydroxy: [84305-05-5]. *8α-Hydroxybalchanin.*
$C_{15}H_{20}O_4$ M 264.321
Constit. of *Leucanthemella serotina.* Cryst.
(CHCl$_3$/Et$_2$O). Mp 80-82°. [α]$_D^{20}$ +150.6° (c, 0.29 in
CHCl$_3$).

11β,13-Dihydro:
$C_{15}H_{22}O_3$ M 250.337
Constit. of *A. canariensis.* Cryst. Mp 132-133°. [α]$_D$
+71° (c, 1.0 in CHCl$_3$).

9α-Hydroxy, 11β,13-dihydro: 9α-Hydroxy-11β,13-
dihyrosantamarine.
$C_{15}H_{22}O_4$ M 266.336
From *Pluchea dioscoridis.* Oil. [α]$_D^{24}$ +20° (c, 0.5 in
CHCl$_3$).

9α-Hydroxy: 9α-Hydroxysantamarine.
$C_{15}H_{20}O_4$ M 264.321
Constit. of *P. dioscoridis.* Cryst. Mp 149°. [α]$_D^{24}$
+56.7° (c, 0.2 in CHCl$_3$).

Suchý, M. *et al, Collect. Czech. Chem. Commun.,* 1962, **27,**
2925 (*isol*)
Romo de Vivar, A. *et al, Tetrahedron,* 1965, **21,** 1741 (*isol,*
struct)
Pathak, S.P. *et al, Chem. Ind.* (*London*), 1970, 1147 (*struct*)
Ando, M. *et al, Tetrahedron,* 1977, **33,** 2785 (*synth*)
Yamakawa, K. *et al, Heterocycles,* 1977, **8,** 103 (*synth*)
Rodrigues, A.A.S. *et al, Phytochemistry,* 1978, **17,** 953 (*synth*)
Holub, M. *et al, Collect. Czech. Chem. Commun.,* 1982, **47,**
2927 (*deriv*)
Gonzalez, A.G. *et al, Phytochemistry,* 1983, **22,** 1509 (*isol*)
Bohlmann, F. *et al, Phytochemistry,* 1984, **23,** 1975 (*derivs*)

Saponaceolide *A*   S-60008

$C_{30}H_{46}O_7$ M 518.689
Constit. of mushroom *Tricholoma saponaceum.* Needles
(Me$_2$CO/hexane). Mp 145-146°. [α]$_D^{20}$ +78.14° (c,
1.1 in CHCl$_3$).

▷Cytotoxic

De Bernardi, M. *et al, Tetrahedron,* 1988, **44,** 235 (*cryst struct*)

Sargahydroquinoic acid   S-60009

$C_{27}H_{38}O_4$ M 426.595
Plastoquinone constit. of brown alga *Sargassum saga-*
mianum var. *yezoense.* Oil.

Segawa, M. *et al, Chem. Lett.,* 1987, 1365.

Sargaquinone   S-60010

2-Methyl-6-(3,7,11,15-tetramethyl-2,6,10,14-hexadeca-
tetraenyl)-1,4-benzoquinone. 2-Geranylgeranyl-6-
methylbenzoquinone
[57576-82-6]

$C_{27}H_{38}O_2$ M 394.596
Constit. of *Sargassum tortile.* Oil.

9′-Methoxy: [72239-42-0]. *9′-Methoxysargaquinone. 2-*
(9-Methoxygeranylgeranyl)-6-methyl-1,4-
benzoquinone.
$C_{28}H_{40}O_3$ M 424.622
From *S. tortile.* Yellow oil. [α]$_D$ +2.68° (CHCl$_3$).

Δ$^{9'}$-Isomer, 11′-methoxy: [72239-43-1]. *11′-Methoxy-*
sargaquinone. 2-(11-Methoxygeranylgeranyl)-6-
methyl-1,4-benzoquinone.
$C_{28}H_{40}O_3$ M 424.622
From *S. tortile.* Yellow oil. [α]$_D$ 0°. Postive opt. rotn.
at short wavelengths.

8′,9′-Dihydroxy: [72239-44-2]. *8′,9′-Dihydroxysarga-*
quinone. 2-(8,9-Dihydroxygeranylgeranyl)-6-methyl-
1,4-benzoquinone.
$C_{27}H_{38}O_4$ M 426.595
From *S. tortile.* Yellow oil. Stereochem. unknown.

8′,9′-Dihydroxy, 5-methyl: [72239-45-3]. *8′,9′-Dihy-*
droxy-5-methylsargaquinone. 2-(8,9-Dihydroxyger-
anylgeranyl)-5,6-dimethyl-1,4-benzoquinone.
$C_{28}H_{40}O_4$ M 440.622
From *S. tortile.* Yellow oil.

Kumarireng, A.S. *et al, Chem. Lett.,* 1973, 1045 (*synth*)
Ishitsuka, M. *et al, Chem. Lett.,* 1979, 1269 (*isol, struct*)

Sarothralen *B*   S-60011

[105239-72-3]

$C_{33}H_{42}O_8$ M 566.690
Constit. of *Hypericum japonicum.* Cryst. Mp 116-119°.
Opt. inactive.

Ishigura, K. *et al, Planta Med.,* 1986, 288.

Schkuhridin *B*   S-60012

C₁₅H₂₀O₄   M 264.321
Constit. of *Schkuhria schkuhrioides*. Pale-yellow oil.
$[\alpha]_D^{25}$ +60.54° (c, 0.18 in MeOH).
14-(2-Hydroxy-3-methylbutanoyl): **Schkuhridin** A.
C₂₀H₂₈O₆   M 364.438
From *S. schkuhrioides*. Pale-yellow oil. $[\alpha]_D^{25}$
+104.58° (c, 0.1 in CHCl₃).

Delgado, G. *et al*, *Phytochemistry*, 1987, **26**, 755.

Scleroderolide   S-60013

C₁₈H₁₆O₆   M 328.321
Metab. of *Gremmeniella abietina*. Yellow cryst. (EtOH-
/pet. ether). Mp 232-233°. $[\alpha]_D$ −116° (c, 0.3 in
CHCl₃).

Ayer, W.A. *et al*, *Can. J. Chem.*, 1987, **65**, 748, 760 (*isol, cryst
struct, biosynth*)

Scleroderris green   S-60014

C₃₈H₃₅NO₁₀   M 665.695
Metab. of *Gremmeniella abietina*. Amorph. green-yellow
powder. Mp >340°.

Ayer, W.A. *et al*, *Can. J. Chem.*, 1987, **65**, 754, 760 (*isol, struct,
biosynth*)

Scopadulcic acid *A*   S-60015

Absolute
configuration

C₂₇H₃₄O₆   M 454.562
Constit. of *Scoparia dulcis*. Prisms (MeOH). Mp 172-
174°. $[\alpha]_D^{27}$ −5.7° (MeOH).

Hayashi, T. *et al*, *Tetrahedron Lett.*, 1987, **28**, 3693.

Scopadulcic acid *B*   S-60016

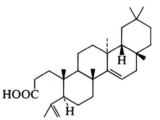

C₂₇H₃₄O₅   M 438.563
Constit. of *Scoparia dulcis*. Prisms (MeOH). Mp 228-
232°. $[\alpha]_D^{27}$ −49.6° (MeOH).

Hayashi, T. *et al*, *Tetrahedron Lett.*, 1987, **28**, 3693.

Scutellone *A*   S-60017
Scuterivulactone C₁

C₂₉H₃₈O₉   M 530.614
Constit. of *Scutellaria rivularis*. Prisms (Me₂CO), cryst.
(EtOAc). Mp 290-292° (268-272°). $[\alpha]_D$ −7.0°
(AcOH), $[\alpha]_D$ +1.26° (c, 1.0 in MeOH).
13-Epimer: **Scuterivulactone** C₂.
C₂₉H₃₈O₉   M 530.614
Constit. of *S. rivularis*.
13-epimer, Ac: Amorph. $[\alpha]_D$ −31° (CHCl₃).

Kikuchi, T. *et al*, *Chem. Lett.*, 1987, 987 (*isol*)
Lin, Y.-L. *et al*, *J. Chem. Res. (S)*, 1987, 320 (*isol, cryst struct*)

3,4-Seco-4(23),14-taraxeradien-3-oic acid   S-60018

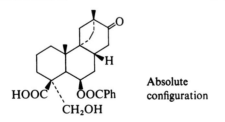

C₃₀H₄₈O₂   M 440.708
Constit. of *Euphorbia broteri*.

Teresa, J.de.P. *et al*, *Phytochemistry*, 1987, **26**, 1767.

Secotrinervitane   S-60019

Updated Entry replacing S-00221
3α-Acetoxy-15β-hydroxy-7,16-seco-7,11-trinervitadiene

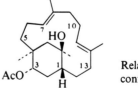

Relative
configuration

C₂₂H₃₆O₃   M 348.525
Constit. of *Nasutitermes princeps*. Cryst. (pentane). Mp
117°. $[\alpha]_D$ +61° (c, 0.52 in CHCl₃).

Braekman, J.C. *et al*, *Tetrahedron Lett.*, 1980, 2761 (*cryst struct*)
Kato, T. *et al*, *Tetrahedron Lett.*, 1987, **28**, 1439 (*synth*)

Semioxamazide, 8CI   S-60020
Aminooxoacetic acid hydrazide, 9CI. Aminooxamide. Oxamic hydrazide
[515-96-8]

$$H_2NCOCONHNH_2$$

$C_2H_5N_3O_2$ M 103.080
Reagent for characteristion of aldehydes and ketones. Leaflets (H_2O). Mp 220-221° dec.

▷AF2620000.

Kerp, W. *et al*, *Ber.*, 1897, **30**, 586 (*synth*)
Leonard, N.J. *et al*, *J. Org. Chem.*, 1950, **15**, 42 (*use*)
Tripathi, G.N.R. *et al*, *J. Mol. Struct.*, 1979, **54**, 19 (*ir, raman, struct*)

Semperoside   S-60021
[110344-57-5]

$C_{16}H_{24}O_9$ M 360.360
Constit. of *Gelsemium sempervirens*. Cryst. (EtOH). Mp 179-181°. $[\alpha]_D^{20}$ +52° (c, 0.3 in MeOH).
9β-Hydroxy: [110309-28-9]. **9-Hydroxysemperoside**.
   $C_{16}H_{24}O_{10}$ M 376.360
   From *G. sempervirens*. Cryst. (EtOH). Mp 132-135°. $[\alpha]_D^{20}$ +58° (c, 1.1 in MeOH).

Jensen, S.R. *et al*, *Phytochemistry*, 1987, **26**, 1725.

Senecioodontol   S-60022
[69868-02-6]

$C_{20}H_{26}O_5$ M 346.422
2-Me ether: [69905-06-2]. **3-O-*Methylsenecioodontol**.
   $C_{21}H_{28}O_5$ M 360.449
   Constit. of *Senecio oxyodontus*. Oil.
2,4-Di-Me ether: [69905-08-4]. **3,5-Di-O-methylsenecioodontol**.
   $C_{22}H_{30}O_5$ M 374.476
   From *S. oxyodontus*. Oil.
2-Angeloyl: [69905-11-9]. **3-O-*Angeloylsenecioodontol**.
   $C_{25}H_{32}O_6$ M 428.524
   From *S. oxyodontus*. Oil.
2-Me ether, 1,4-quinone: [69905-05-1]. **3-O-*Methyl-2,5-dehydrosenecioodontol. 2-O-Methyl-1,4-dehydrosenecioodontol**.
   $C_{21}H_{26}O_5$ M 358.433
   From *S. oxyodontus*. Yellow oil. $[\alpha]_D^{24}$ +12.4° (c, 0.6 in $CHCl_3$).

Bohlmann, F. *et al*, *Phytochemistry*, 1978, **17**, 1591.

Senepoxide   S-60023
Updated Entry replacing S-20045
1-Benzoyloxymethyl-7-oxabicyclo[4.1.0]hept-4-ene-2,3-diol diacetate, 9CI

(−)-*form*
Absolute
configuration

$C_{18}H_{18}O_7$ M 346.336
(−)-*form* [17550-38-8]
   Constit. of the fruits of *Uvaria catocarpa*. Displays an interesting spectrum of biological activity. Cryst. (MeOH). Mp 85°. $[\alpha]_D$ −197° (c, 1.19 in $CHCl_3$).
   1,6-Diepimer: [86747-02-6]. **β-Senepoxide**.
      $C_{18}H_{18}O_7$ M 346.336
      Constit. of *U. ferruginea*. Mp 72-73°. $[\alpha]_D^{25}$ +62° (c, 0.55 in $CHCl_3$).
(±)-*form* [55304-68-2]
   Mp 97-98°.

Hollands, R. *et al*, *Tetrahedron*, 1968, **24**, 1633 (*isol, struct*)
Ichihara, A. *et al*, *Tetrahedron Lett.*, 1974, 4235 (*synth*)
Ichihara, A., *Nippon Nogei Kagaku Kaishi*, 1975, **49**, 27 (*rev*)
Ducruix, A. *et al*, *Acta Crystallogr., Sect. B*, 1976, **32**, 1589 (*struct*)
Schlessinger, R.H. *et al*, *J. Org. Chem.*, 1981, **46**, 5252 (*synth*)
Ogawa, S. *et al*, *J. Org. Chem.*, 1985, **50**, 2356 (*synth*)
Nkunya, M.H.H. *et al*, *Phytochemistry*, 1987, **26**, 2563 (*isol, cryst struct*)

N-Serylmethionine   S-60024
Updated Entry replacing S-00330

$$HOCH_2CH(NH_2)CONHCH(COOH)CH_2CH_2SMe$$

$C_8H_{16}N_2O_4S$ M 236.285
L-L-form [3227-09-6]
   Cryst. (EtOH aq.). Mp 215-216°. $[\alpha]_D^{27}$ −11.4° (c, 5.6 in 1M HCl).
   Me ester: [10233-78-0].
      $C_9H_{18}N_2O_4S$ M 250.312
      Mp 84-85°.

Hofmann, K. *et al*, *J. Am. Chem. Soc.*, 1957, **79**, 1636 (*synth*)

N-Serylproline, 9CI, 8CI   S-60025
Updated Entry replacing S-00332
[23827-93-2]

$C_8H_{14}N_2O_4$ M 202.210
L-L-form
   Mp 196-198°. $[\alpha]_D^{24}$ −109.3° (c, 2.83 in AcOH).

Camble, R. *et al*, *J. Am. Chem. Soc.*, 1972, **94**, 2091 (*synth*)

Sesebrinic acid   S-60026

[110241-17-3]

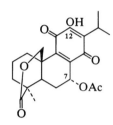

C$_{20}$H$_{26}$O$_6$ M 362.422

Constit. of *Seseli sibiricum*. Cryst. (Me$_2$CO/hexane).
Mp 155-156°.

Banerjee, S.K. *et al*, *Phytochemistry*, 1987, **26**, 1817.

Sessein   S-60027

C$_{22}$H$_{26}$O$_7$ M 402.443

Constit. of *Salvia sessei* and *S. regla*. Cryst. (Me$_2$CO-
/diisopropyl ether). Mp 187-189°.

O-De-Ac: Deacetylsessein.
    C$_{20}$H$_{24}$O$_6$ M 360.406
    Constit. of *S. regla*. Cryst. (Me$_2$CO/diisopropyl
    ether). Mp 105-107°.

Hernández, M. *et al*, *Phytochemistry*, 1987, **26**, 3297.

Shanzhiside   S-60028

Updated Entry replacing S-30036
*(1-β-D-Glucopyranosyloxy)-1,4a,5,6,7,7a-hexahydro-
5,7-dihydroxy-7-methylcyclopenta[c]pyran-4-carboxyl-
ic acid, 9CI*

[29836-27-9]

C$_{16}$H$_{24}$O$_{11}$ M 392.359

Constit. of *Gardenia jasminoides*. Cryst. Mp 82-90°.
[α]$_D$ −81.7° (EtOH).

Penta-Ac: Cryst. Mp 111-112°.

Me ester: [64421-28-9].
    C$_{17}$H$_{26}$O$_{11}$ M 406.386
    Constit. of *Mussaenda parviflora* and *Salvia
    digitaloides*. Powder. [α]$_D^{30}$ −110.8° (c, 0.42 in
    MeOH).

*Me ester, 6′-O-β-D-glucosyl: Shanzhisin methyl ester
gentiobioside*.
    C$_{22}$H$_{36}$O$_{16}$ M 556.517
    Constit. of *Canthium subcordatum*. Cryst.
    (MeOH/CHCl$_3$). Mp >150° dec. [α]$_D^{20}$ −56°
    (MeOH).

6-(4-Hydroxy-3,5-dimethoxybenzoyl), 8-Ac, Me ester:
[86450-76-2].
    C$_{28}$H$_{36}$O$_{16}$ M 628.583
    Constit. of Chinese drug Bai-Yun-Shen (*S.
    digitaloides*). Powder. [α]$_D^{24}$ −71.0° (c, 1.01 in
    MeOH).

8-Ac, Me ester: [57420-46-9]. *Barlerin*.
    C$_{19}$H$_{28}$O$_{12}$ M 448.423
    Constit. of *Barleria prionitis*. Cryst. (EtOAc). Mp
    180°. [α]$_D$ −8.5° (c, 0.8 in MeOH).

6,8-Di-O-Ac, Me ester: [57420-45-8]. *Acetylbarlerin*.
    C$_{21}$H$_{30}$O$_{13}$ M 490.460
    From *B. prionitis*. Hygroscopic powder. [α]$_D$ −99° (c,
    2.0 in MeOH).

6-Ac, Me ester: [110186-13-5]. *6-O-Acetylshanghiside
methyl ester*.
    C$_{19}$H$_{28}$O$_{12}$ M 448.423
    Constit. of *B. lupulina*. Needles (MeOH/EtOAc). Mp
    227-228°. [α]$_D^{22}$ −118.7° (c, 0.17 in MeOH).

Inouye, H. *et al*, *Tetrahedron Lett.*, 1970, 3581 (*isol*)
Taneja, S.C. *et al*, *Tetrahedron Lett.*, 1975, 1995 (*Barlerin*)
Takeda, Y. *et al*, *Phytochemistry*, 1977, **16**, 1401 (*isol*)
Achenbach, H. *et al*, *Tetrahedron Lett.*, 1980, 3677 (*isol*)
Damtoft, S. *et al*, *Phytochemistry*, 1981, **20**, 2717 (*cmr*)
Damtoft, S. *et al*, *Tetrahedron Lett.*, 1982, **23**, 4155 (*isol*)
Tanaka, T. *et al*, *Chem. Pharm. Bull.*, 1983, **31**, 780 (*isol*)
Byrne, L.T. *et al*, *Aust. J. Chem.*, 1987, **40**, 785 (*isol, cryst
    struct*)

Siccanin   S-60029

Updated Entry replacing S-00405
*1,2,3,4,4a,5,6,6a,11b,13b-Decahydro-4,4,6a,9-tetra-
methyl-13H-benzo[a]furo[2,3,4-mn]xanthen-11-ol, 9CI*

[22733-60-4]

C$_{22}$H$_{30}$O$_3$ M 342.477

Isol. from cultures of *Helminthosporium siccans*, a
parasitic organism of ryegrass. Antifungal agent.
Cryst. Mp 139-140°. [α]$_D^{20}$ −136° (c, 2 in CHCl$_3$). pK$_a$
10.9.

4-Bromobenzenesulfonyl: Orthorhombic cryst. Mp 156°.

Ishibashi, K., *J. Antibiot., Ser. A*, 1962, **15**, 161 (*isol*)
Hirai, K. *et al*, *Acta Crystallogr., Sect. B*, 1969, **25**, 2630 (*cryst
    struct*)
Hirai, K. *et al*, *Tetrahedron*, 1971, **27**, 6057 (*struct, uv, ir, ms,
    pmr*)
Suzuki, K.T. *et al*, *Bioorg. Chem.*, 1974, **3**, 72 (*biosynth*)
Bellotti, M.G. *et al*, *Chemoterapia*, 1985, **4**, 431 (*props*)
Kato, M. *et al*, *Tetrahedron*, 1987, **43**, 711 (*synth*)

Siccanochromene *E*   S-60030

Updated Entry replacing S-00409
*1,2,3,4,4a,5,6,6a-Octahydro-11-hydroxy-4,4,6a,9-tetra-
methyl-12bH-benzo[a]xanthene-12b-methanol, 9CI*

[24211-27-6]

C$_{22}$H$_{30}$O$_3$ M 342.477

Minor metab. of *Helminthosporium siccans*. Cryst. Mp
191-192°. [α]$_D^{20}$ −86° (EtOH).

Hirai, K. *et al*, *Tetrahedron*, 1971, **27**, 6057 (*isol*)
Nozoe, S. *et al*, *Tetrahedron*, 1974, **30**, 2773 (*isol, struct*)

Kato, M. *et al*, *Tetrahedron*, 1987, **43**, 711 (*synth*)

Sigmoidin *D*   S-60031

2,2′,3,3′-Tetrahydro-3′,5,7,8′-tetrahydroxy-2′,2′-di-methyl[2,6′-bi-4H-1-benzopyran]-4-one, 9CI

[106533-44-2]

$C_{20}H_{20}O_7$ M 372.374
Constit. of *Erythrina sigmoidea*. Cryst. Mp 220°. $[\alpha]_D^{24}$ −18.5° (MeOH).

Promsattha, R. *et al*, *J. Nat. Prod.*, 1986, **49**, 932.

Sikkimotoxin   S-60032

[18651-67-7]

$C_{23}H_{26}O_8$ M 430.454
Constit. of *Podophyllum sikkimensis*. Cryst. Mp 125-130°.

Takano, S. *et al*, *Heterocycles*, 1987, **25**, 69.

3-Silphinenone   S-60033

Updated Entry replacing S-30060
3-Oxosilphinene

$C_{15}H_{22}O$ M 218.338
Constit. of *Dugaldi hoopesii*. Oil. Bp$_{0.1}$ 85°. $[\alpha]_D$ +7.1° (c, 0.4 in CHCl$_3$).

Bohlmann, F. *et al*, *J. Nat. Prod.*, 1984, **47**, 658.
Ihara, M. *et al*, *J. Chem. Soc., Perkin Trans. 1*, 1987, 1331 (*synth*)

Sinularene   S-60034

Updated Entry replacing S-00437
[64845-75-6]

$C_{15}H_{24}$ M 204.355
Constit. of *Sinularia mayi*. Oil. $[\alpha]_D^{20}$ −142° (c, 0.55 in CCl$_4$).

Beechan, C.M. *et al*, *Tetrahedron Lett.*, 1977, 2395.

Collins, P.A. *et al*, *Aust. J. Chem.*, 1979, **32**, 1819 (*synth*)
Oppolzer, W. *et al*, *Tetrahedron Lett.*, 1982, **23**, 4673 (*synth*)
Antczak, K. *et al*, *Can. J. Chem.*, 1985, **58**, 993 (*synth*)
Piers, E. *et al*, *Can. J. Chem.*, 1985, **58**, 996 (*synth*)
Antczak, K. *et al*, *Can. J. Chem.*, 1987, **65**, 114 (*synth*)

Skimminin   S-60035

$C_{16}H_{18}O_5$ M 290.315
Constit. of *Skimmia laureola*. Cryst. Mp 150-152°.

Razdan, T.K. *et al*, *Phytochemistry*, 1987, **26**, 2063.

Smenospongine   S-60036

$C_{21}H_{29}NO_3$ M 343.465
Quinoterpenoid antibiotic. Prod. by the sponge *Smenospongia* sp. Cytotoxic and antimicrobial agent. Red cryst. Mp 153-155°.

Kondracki, M.-L. *et al*, *Tetrahedron Lett.*, 1987, **28**, 5815 (*isol, struct, props*)

Sophorocarpan *A*   S-60037

3-Hydroxy 6,8,9-trimethoxypterocarpan

R = R′ = Me

$C_{17}H_{16}O_5$ M 300.310
Constit. of *Sophora tomentosa*. Needles (MeOH aq.).
Mp 163-165°. $[\alpha]_D^{25}$ −110° (c, 0.33 in MeOH).

Kinoshita, T. *et al*, *Chem. Pharm. Bull.*, 1986, **34**, 3067.

Sophorocarpan *B*   S-60038

3-Hydroxy-6-methoxy-8,9-methylenedioxypterocarpan.
6β-Methoxymaackiain

As Sophorocarpan *A*, S-60037 with

RR′ = −CH$_2$−

$C_{17}H_{14}O_6$ M 314.294
Constit. of *Sophora tomentosa*. Amorph. solid. $[\alpha]_D^{25}$ −135° (c, 0.15 in MeOH).

Kinoshita, T. *et al*, *Chem. Pharm. Bull.*, 1986, **34**, 3067.

Specionin   S-60039

Updated Entry replacing S-50099

[96944-53-5]

C$_{20}$H$_{26}$O$_8$ M 394.421

Isol. from leaves of *Catalpa speciosa*. Antifeedant against Eastern spruce budworm.

Van der Eycken, E. *et al, Tetrahedron Lett.*, 1985, **26**, 367; 1987, **28**, 3519 (*isol, synth*)
Van der Eycken, E. *et al, Tetrahedron*, 1986, **19**, 5385 (*synth, struct*)
Hussain, N. *et al, Tetrahedron Lett.*, 1987, **28**, 4871 (*synth*)

Spirobi[9*H*-fluorene], 9CI   S-60040

[159-66-0]

C$_{25}$H$_{16}$ M 316.401

Forms stable inclusion complexes. Cryst. (EtOH or C$_6$H$_6$). Mp 206° (197-198°).

Clarkson, R.G. *et al, J. Am. Chem. Soc.*, 1930, **52**, 2884 (*synth*)
Kenner, G.W. *et al, J. Chem. Soc.*, 1962, 1756 (*synth*)
Schweig, A. *et al, Angew. Chem., Int. Ed. Engl.*, 1973, **12**, 310 (*pe*)
Weber, E. *et al, Angew. Chem., Int. Ed. Engl.*, 1986, **25**, 747 (*use*)

Spirobrassinin   S-60041

C$_{11}$H$_{10}$N$_2$OS$_2$ M 250.333

Phytoalexin of *Raphanus sativus* var. *hortensis*. Cryst. Mp 158-159°. [α]$_D$ −69.5° (c, 1.14 in CHCl$_3$).

Takasugi, M. *et al, Chem. Lett.*, 1987, 1631.

Spiro[4.4]non-2-ene-1,4-dione, 9CI   S-60042

[105365-16-0]

C$_9$H$_{10}$O$_2$ M 150.177

Dienophile. Cryst. (pentane).

Bach, R.D. *et al, Tetrahedron Lett.*, 1986, **27**, 1983 (*synth, pmr, ir*)

1,2,3-Spirostanetriol   S-60043

Updated Entry replacing S-10084

(1β,2β,3α,5β,20S, 22R,25R)-*form*

C$_{27}$H$_{44}$O$_5$ M 448.642

(1β,2β,3α,5β,20S,22R,25R)-form [547-01-3]
  Tokorigenin
  Sapogenin from *Dioscorea tokoro*. Cryst. (MeOH). Mp 266-268°. [α]$_D^{22}$ −49.6°.
  1-O-α-L-Arabinopyranoside: **Tokoronin**.
  C$_{32}$H$_{52}$O$_9$ M 580.757
  Constit. of *D. tokoro*. Needles (MeOH aq.). Mp 275-277°. [α]$_D$ −13.0°.

(1β,2β,3α,5β,20S,22R,25S)-form
  Neotokorigenin
  1-O-α-L-Arabinopyranoside: [20312-78-1].
  Neotokoronin.
  C$_{32}$H$_{52}$O$_9$ M 580.757
  Constit. of *Rhodea japonica*. Needles (MeOH aq.). Mp 284-280°. [α]$_D$ −17.2°.

(1β,2β,3α,5β,20S,22S,25S)-form
  22-Epitokorogenin
  1-O-α-L-Arabinopyranoside:
  C$_{32}$H$_{52}$O$_9$ M 580.757
  Constit. of *R. japonica*. Prisms (MeOH aq.). Mp 287-290°. [α]$_D$ +19.4° (Py).

Akakori, A. *et al, Chem. Pharm. Bull.*, 1968, **16**, 1994 (*isol, struct*)
Tomita, Y. *et al, J. Chem. Soc., Chem. Commun.*, 1971, 284 (*biosynth*)
Seo, S. *et al, J. Chem. Soc., Chem. Commun.*, 1981, 895 (*biosynth*)
Kudo, K. *et al, Chem. Pharm. Bull.*, 1984, **32**, 4229 (*isol*)

5-Spirostene-3,25-diol   S-60044

Updated Entry replacing S-30078

C$_{27}$H$_{42}$O$_4$ M 430.626

(3β,20S,22S,25R)-form
  Isonuatigenin
  Constit. of *Vestia lycioides*. Cryst. (MeOH/Me$_2$CO). Mp 260-262°.
  3-O-α-L-Rhamnopyranosyl(1→2)-β-D-glucopyranoside:
  C$_{39}$H$_{62}$O$_{13}$ M 738.911
  Constit. of *Allium vineale*.

Faini, F. *et al, Phytochemistry*, 1984, **23**, 1301 (*isol*)
Chen, S. *et al, Tetrahedron Lett.*, 1987, **28**, 5603 (*saponin*)

5-Spirosten-3-ol   S-60045

Updated Entry replacing S-10086

C$_{27}$H$_{42}$O$_3$ M 414.627

(3α,25R)-form [66289-52-9]
  3-epi-Diosgenin
  Isol. from the crude drug Sensokushichikon. Cryst. Mp 244-246°. [α]$_D$ −145°.
  3-O-β-D-Glucopyranoside: [66289-51-8].
  C$_{33}$H$_{52}$O$_8$ M 576.769
  Constit. of *Gynura japonica*. Cryst. Mp 218-221°. [α]$_D$ −122°.

(3β,25R)-form [512-04-9]
Diosgenin. *Nitrogenin*

Sapogenin from *Dioscorea* spp., *Solarium* spp. and *Trillium erectum*. Cryst. (Me$_2$CO). Mp 204-207°. $[\alpha]_D^{25}$ −129° (c, 1.4 in CHCl$_3$).

▷WH1322870.

Ac: Mp 195-198°.
Benzoyl: Mp 236-241°. $[\alpha]_D^{17}$ −45° (Py).
3-O-[6-Deoxy-α-L-mannopyranosyl-(1→2)-O-6-deoxy-α-L-mannopyranosyl-(1→4)-β-D-glucopyranoside]: [19057-60-4]. **Dioscin**.
C$_{45}$H$_{72}$O$_{16}$ M 869.054
Constit. of *D.* spp. and *T. terrestris*. Cryst. Mp 275-277° dec. $[\alpha]_D^{13}$ −115° (c, 0.4 in EtOH).
3-O-β-D-Glucopyranosyl: [14144-06-0]. **Trillin**. *Disogluside, INN. Alliumoside* A. *Funkioside*.
C$_{33}$H$_{52}$O$_8$ M 576.769
Constit. of *D.* spp., *Yucca* spp., *T. terrestris*, and *T. kamschaticum*. Drug used for treatment of menorrhagia. Cryst. + $\frac{1}{2}$H$_2$O (MeOH). Mp 275-280°. $[\alpha]_D$ −103° (dioxan).
3-O-(α-L-Rhamnosyl-β-D-glucosyl-β-D-glucoside): [19083-00-2]. **Gracillin**.
C$_{45}$H$_{72}$O$_{17}$ M 885.054
Constit. of *D. gracillima*. Cryst. Mp 290-293°.
3-O-[4-O-β-D-Glucopyranosyl-β-D-galactopyranoside]: [60454-77-5]. **Funkioside C**.
C$_{39}$H$_{62}$O$_{13}$ M 738.911
Constit. of *Funkia ovata*.

▷LW5440600.
3-O-[β-D-Glucopyranosyl-(1→2)-β-D-glucopyranosyl-(1→4)-β-D-galactopyranoside: [60454-78-6]. **Funkioside D**.
C$_{45}$H$_{72}$O$_{18}$ M 901.053
From *F. ovata*.

▷LW5440650.
3-O-[α-L-Rhamnopyranosyl-(1→4)-β-D-glucopyranosyl-(1→2)-glucopyranosyl-(1→4)-β-D-galactopyranoside]: [60454-79-7]. **Funkioside E**.
C$_{51}$H$_{82}$O$_{22}$ M 1047.196
From *F. ovata*.
3-O-[β-D-Glucopyranosyl-(1→2)-[D-xylopyranosyl-(1→3)]-β-D-glucopyranosyl-(1→4)-β-D-galactopyranoside]: [60454-80-0]. **Funkioside F**.
C$_{50}$H$_{80}$O$_{22}$ M 1033.169
From *F. ovata*.

▷LW5440800.
3-O-[α-L-Rhamnopyranosyl-(1→4)-β-D-glucopyranosyl-(1→2)-β-D-xylopyranosyl-(1→3)-β-D-glucopyranosyl-(1→4)-β-D-galactopyranoside]: [60454-81-1]. **Funkioside G**.
C$_{59}$H$_{90}$O$_{26}$ M 1215.344
From *F. ovata*.

▷LW5440400.
3-O-(α-L-Rhamnopyranosyl-(1→2)-[α-L-arabinopyranosyl-1→4)]-β-D-glucopyranoside: [76296-72-5]. **Polyphyllin D**.
C$_{44}$H$_{70}$O$_{16}$ M 855.027
Constit. of *Paris polyphylia*. Haemostatic agent. Cryst. (MeOH). Mp 275-280°. $[\alpha]_D$ −113° (c, 0.53 in MeOH).
3-O-α-L-Rhamnopyranosyl(1→2)-β-D-glucopyranoside:
C$_{39}$H$_{62}$O$_{12}$ M 722.912
Constit. of *Ophiogon planiscapus* and of *Allium vineale*.
3-O-[α-L-Rhamnopyranosyl(1→2)] [β-D-glucopyranosyl(1→4)]-β-D-glucopyranoside:
C$_{45}$H$_{72}$O$_{17}$ M 885.054

Constit. of *O. planiscapus* and *A. vineale*. Molluscicidal.
3-O-β-D-Glucopyranosyl(1→4)-α-L-rhamnopyranosyl(1→4)-β-D-glucopyranoside:
C$_{45}$H$_{72}$O$_{17}$ M 885.054
Constit. of *A. vineale*. Molluscicidal.
3-O-[α-L-Rhamnopyranosyl(1→2)] [β-D-glucopyranosyl(1→4)-α-L-rhamnopyranosyl(1→4)]-β-D-glucopyranoside:
C$_{45}$H$_{72}$O$_{16}$ M 869.054
Constit. of *A. vineale*. Molluscicidal.

(3β,25S)-form
Yamogenin

Sapogenin from rhizomes of Mexican *Dioscorea* spp. Cryst. (Me$_2$CO). Mp 201°. $[\alpha]_D^{25}$ −123°.

O-α-L-Rhamnopyranosyl-(1→2)-β-D-glucopyranosyl-(1→4)-[α-L-rhamnopyranosyl-(1→2)]-β-D-glucopyranoside: [79975-21-6]. **Balanitin 1**.
C$_{51}$H$_{82}$O$_{22}$ M 1047.196
Constit. of *Balanites aegyptiaca*. Molluscicide.
O-β-D-Xylopyranosyl-(1→6)-β-D-glucopyranosyl-(1→3)-[α-L-rhamnopyranosyl-(1→2)]-β-D-glucopyranoside: [79975-20-5]. **Balanitin 2**.
C$_{50}$H$_{80}$O$_{21}$ M 1017.169
Constit. of *B. aegyptiaca*. Molluscicide.
O-α-L-Rhamnopyranosyl-(1→2)-β-D-glucopyranosyl-(1→4)-β-D-glucopyranoside: [82358-27-8]. **Balanitin 3**.
C$_{45}$H$_{72}$O$_{17}$ M 885.054
Constit. of *B. aegyptiaca*. Molluscicide.

Marker, R.E. *et al, J. Am. Chem. Soc.*, 1947, **69**, 2167 (isol, struct)
Kawasaki, T. *et al, Chem. Pharm. Bull.*, 1962, **10**, 703; 1963, **11**, 1546 (struct, synth)
Bite, P. *et al, Acta Chim. Acad. Sci. Hung.*, 1967, **52**, 79 (isol)
Kessar, S.V. *et al, Tetrahedron*, 1968, **24**, 899 (synth)
Hardman, R. *et al, Planta Med.*, 1971, **20**, 193 (biosynth)
Miyahara, K. *et al, Chem. Pharm. Bull.*, 1972, **20**, 2506; 1974, **22**, 1407 (struct, pmr)
Eggert, H. *et al, Tetrahedron Lett.*, 1975, 3635 (cmr)
Lazur'evzkii, G.V. *et al, Dokl. Akad. Nauk SSSR, Ser. Sci. Khim.*, 1976, **230**, 476 (Funkiosides)
Takahira, N. *et al, Tetrahedron Lett.*, 1977, 3647.
Takahira, M. *et al, J. Pharm. Soc. Jpn.*, 1979, **99**, 264 (isol)
Seo, S. *et al, J. Chem. Soc., Chem. Commun.*, 1981, 895 (biosynth)
Liu, H.-W. *et al, Tetrahedron*, 1982, **38**, 513 (isol)
Watanabe, Y. *et al, Chem. Pharm. Bull.*, 1983, **31**, 3486 (saponins)
Ma, J.C.N. *et al, Phytochemistry*, 1985, **24**, 1561 (Polyphyllin D)
Chen, S. *et al, Tetrahedron Lett.*, 1987, **28**, 5603 (saponins)

Spiro[5.5]undeca-1,3-dien-7-one   **S-60046**
[82390-19-0]

C$_{11}$H$_{14}$O   M 162.231
Oil.

Fuchs, B. *et al, J. Org. Chem.*, 1982, **47**, 3474 (synth, ir, uv, pmr)

Spongiolactone   S-60047

C$_{25}$H$_{38}$O$_4$ M 402.573

Constit. of *Spongionella gracilis*. Oil. [α]$_D$ +67.6° (c, 1.7 in CHCl$_3$).

Mayol, L. *et al, Tetrahedron Lett.*, 1987, **28**, 3601.

Spongionellin   S-60048

C$_{21}$H$_{30}$O$_5$ M 362.465

Constit. of sponge *Spongionella gracilis*. Amorph. solid. [α]$_D$ +1.2° (c, 1.0 in CHCl$_3$).

Mayol, L. *et al, Tetrahedron*, 1986, **42**, 5369.

Squamolone   S-60049

2-Oxo-1-pyrrolidinecarboxamide, 9CI. 1-Carbamoyl-2-pyrrolidone

[40451-67-0]

C$_5$H$_8$N$_2$O$_2$ M 128.130

An incorrect diazepinedione struct. was originally assigned. Constit. of *Anona squamosa* and *Hexalobus crispiflorus* (Annonaceae). Prisms (C$_6$H$_6$). Mp 146-147° (142-143°).

Yang, T.H. *et al, J. Chin. Chem. Soc. (Taipei)*, 1972, **19**, 149; *CA*, **78**, 16151w (isol, uv, pmr, ms)
Marquez, V.E. *et al, J. Org. Chem.*, 1980, **45**, 5308 (struct, synth, ir, pmr, ms)
Achenbach, H. *et al, Justus Liebigs Ann. Chem.*, 1982, 1623 (isol, cmr)

Stachydrine   S-60050

Updated Entry replacing S-00721

2-Carboxy-1,1-dimethylpyrrolidinium hydroxide inner salt, 9CI. Hygric acid methylbetaine, 8CI. N-Methylproline methylbetaine. Cadabine. Chrysanthemine

[471-87-4]

C$_7$H$_{13}$NO$_2$ M 143.185

Chrysanthemine was a mixt. of Stachydrine and choline. Systolic depressant. Stachydrine-contg. spp. e.g. *Capparis* spp. are widely used against rheumatism and many other diseases.

(S)-form

L-form

Alkaloid from *Aspergillus oryzae, Betonica officinalis, Capparis tomentosa, Chrysanthemum* spp., *Citrus* spp., *Galeopsis grandiflora, Lagochilus hirtus, Medicago sativa* and *Stachys* spp. (Labiatae, Capparidaceae, Compositae, Rutaceae, Leguminosae). Mp 116-118° (monohydrate), 235° dec. (anhyd.). [α]$_D$ −40.25° (c, 4 in H$_2$O). Fairly readily racemised, (±)-form often obt.

B,HCl: Cryst. (EtOH). Mp 222°. [α]$_D^{20}$ −28.1° (c, 4.83 in H$_2$O).

Et ester: **Stachydrine ethyl ester**.
C$_9$H$_{17}$NO$_2$ M 171.239
Present in roots of *C. virgata* (isol. as periodide) (Capparidaceae). [α]$_D$ −4° (H$_2$O).

Et ester, picrate: Mp 101°.

(±)-form [50298-93-6]

Mp 235°.

B,HCl: Mp 195-196°.
Picrate: Cryst. (EtOH). Mp 195-196°.

King, H. *et al, J. Chem. Soc.*, 1950, 2866, (isol, struct)
Cornforth, J.W. *et al, J. Chem. Soc.*, 1952, 597, 601 (isol, props)
Marian, L. *et al, Phytochemistry*, 1962, **1**, 209 (biosynth)
Paudler, W.W. *et al, Chem. Ind. (London)*, 1963, 1693
Mandava, N. *et al, Justus Liebigs Ann. Chem.*, 1970, **741**, 167 (pmr)
Musich, J.A. *et al, J. Org. Chem.*, 1977, **42**, 139 (synth)
Massiot, G. *et al, Alkaloids (N.Y.)*, 1986, **27**, 310 (bibl, pharmacol)

Stacopin P1   S-60051

Valylvalylvalyl-N-(1-formyl-2-phenylethyl)valinamide, 9CI

[107191-87-7]

C$_{29}$H$_{47}$N$_5$O$_5$ M 545.721

Oligopeptide antibiotic. Prod. by *Staphylococcus tanabensis*. Cysteine protease inhibitor.

4-Hydroxy: [107191-86-6]. **Stacopin P2**.
C$_{29}$H$_{47}$N$_5$O$_6$ M 561.720
From *S. tanabensis*. Cysteine protease inhibitor.

Japan. Pat., 86 106 600, (1986); *CA*, **106**, 118150 (isol)

Sterebin *D*   S-60052

7β,8α-Dihydroxy-14,15-dinor-11-labden-13-one

C$_{18}$H$_{30}$O$_3$ M 294.433

Constit. of *Stevia rebaudiana*. Powder. [α]$_D$ +22.8° (c, 0.05 in MeOH).

6α-Hydroxy: **Sterebin A**. *6α,7β,8α-Trihydroxy-14,15-dinor-11-labden-13-one.*
C$_{18}$H$_{30}$O$_4$ M 310.433

Isol. from leaves of *S. rebaudiana*. Needles (Me$_2$CO). Mp 157-158°. [α]$_D$ +39.6° (c, 0.63 in Me$_2$CO).
6α-Acetoxy: **Sterebin B**.
C$_{20}$H$_{32}$O$_5$ M 352.470
Constit. of *S. rebaudiana*. Powder. [α]$_D$ +15.9° (c, 0.11 in MeOH).
6α-Hydroxy, 7-Ac: **Sterebin C**.
C$_{20}$H$_{32}$O$_5$ M 352.470
Constit. of *S. rebaudiana*. Powder. [α]$_D$ 34.8° (c, 0.07 in MeOH).

Oshima, Y. *et al*, *Tetrahedron*, 1986, **42**, 6443.

Sterpuric acid
S-60053

Updated Entry replacing S-00781
[79367-59-2]

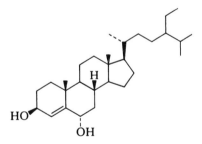

Relative configuration

C$_{15}$H$_{22}$O$_3$ M 250.337
Constit. of *Stereum purpureum*. Phytotoxin; causative agent of silver leaf disease. Cryst. (EtOAc). Mp 203-207°. [α]$_D^{25}$ +72° (c, 0.03 in MeOH).
13-Hydroxy: [79367-60-5]. **13-Hydroxysterpuric acid**.
C$_{15}$H$_{22}$O$_4$ M 266.336
Constit. of *S. purpureum*. Phytotoxin. Isol. as Me ester.
13-Hydroxy, 3,13-ethylidene acetal: [79367-61-6].
C$_{17}$H$_{24}$O$_4$ M 292.374
Constit. of *S. purpureum*. Isol. as Me ester.

Ayer, W.A. *et al*, *Tetrahedron, Suppl.*, 1981, No. 1, 379 (*cryst struct*)
Ayer, W.A. *et al*, *Can. J. Chem.*, 1984, **62**, 531 (*biosynth*)
Paquette, L.A. *et al*, *Tetrahedron Lett.*, 1987, **28**, 5017 (*synth*)

4-Stigmastene-3,6-diol
S-60054

C$_{29}$H$_{50}$O$_2$ M 430.713
(3β,6α)-form
Constit. of *Lagerstroemia lancasteri*. Cryst. (MeOH). Mp 220°. [α]$_D$ +21.5° (c, 0.5 in CHCl$_3$).

Chaudhuri, P.K. *et al*, *Phytochemistry*, 1987, **26**, 3361.

Stolonidiol
S-60055

Absolute configuration

C$_{20}$H$_{32}$O$_4$ M 336.470
Constit. of soft coral *Clavularia* sp. Cytotoxic and ichthyotoxic. Viscous oil. [α]$_D$ −31.0° (c, 1.4 in CHCl$_3$).
17-Ac: **Stolonidiol acetate**.
C$_{22}$H$_{34}$O$_5$ M 378.508
Constit. of *C.* sp. Cytotoxic and ichthyotoxic. Viscous oil. [α]$_D$ −26.8° (c, 0.38 in CHCl$_3$).

Mori, K. *et al*, *Tetrahedron Lett.*, 1987, **28**, 5673 (*cryst struct*)

Strictaepoxide
S-60056

C$_{28}$H$_{40}$O$_5$ M 456.621
Constit. of brown alga *Cystoseria stricta*. Oil. [α]$_D$ −5.1° (c, 5.6 in EtOH).

Amico, V. *et al*, *Tetrahedron*, 1986, **42**, 6015.

Strigol
S-60057

Updated Entry replacing S-30091
[11017-56-4]

C$_{19}$H$_{22}$O$_6$ M 346.379
Constit. of the root of *Gossypium hirsutum*. Potent seed germination stimulant for *Striga lutea*. Cryst. (C$_6$H$_6$/hexane). Mp 200-202° dec.

Cook, C.E. *et al*, *J. Am. Chem. Soc.*, 1972, **94**, 6198 (*isol*)
Coggon, P. *et al*, *J. Chem. Soc., Perkin Trans. 2*, 1973, 455 (*cryst struct*)
Heather, J.B. *et al*, *J. Am. Chem. Soc.*, 1976, **98**, 3661 (*synth*)
MacAlpine, G.A. *et al*, *J. Chem. Soc., Perkin Trans. 1*, 1976, 410 (*synth*)
Johnson, A.W. *et al*, *J. Chem. Soc., Perkin Trans. 1*, 1981, 1734 (*bibl*)
Brooks, D.W. *et al*, *J. Org. Chem.*, 1985, **50**, 628, 3779 (*synth, cryst struct*)
Berlage, U. *et al*, *Tetrahedron Lett.*, 1987, **28**, 3091, 3095 (*synth*)
Dailey, O.D., *J. Org. Chem.*, 1987, **52**, 1984 (*synth*)

Subexpinnatin *C*   S-60058

Updated Entry replacing S-40062

[99305-07-4]

$C_{17}H_{22}O_4$ M 290.358

Constit. of *Centaura canariensis*. Cryst. Mp 186-188°.

8-(2-Hydroxymethyl-2-propenoyl): **Subexpinnatin B**.
$C_{21}H_{26}O_6$ M 374.433
From *C. canariensis*. Non-cryst.

Collado, I.G. *et al*, *Phytochemistry*, 1985, **24**, 2107.
Collado, I.G. *et al*, *J. Org. Chem.*, 1987, **52**, 3323 (*synth*)

Substance *P*, 9CI   S-60059

Updated Entry replacing S-20090

[33507-63-0]

Arg-Pro-Lys⁴-Pro-Gln-Gln-Phe-Phe-Gly-Leu¹¹-MetNH₂

$C_{63}H_{98}N_{18}O_{13}S$ M 1347.640

Present in the brain of vertebrate species, in spinal ganglia and in the intestines, especially the duodenum and jejunum. Substance with a wide variety of pharmacological properties including the ability to cause transient hypotension and stimulate salivary secretion on intravenous injection and also to bring about contraction of smooth muscle preparations.$[\alpha]_D^{27}$ −76.0 (5% AcOH aq.). Fragment 4-11 is reported to be as active as Substance *P* itself.

Zuber, H. *et al*, *Angew. Chem., Int. Ed. Engl.*, 1962, **1**, 160 (*isol*)
Chang, M.M. *et al*, *J. Biol. Chem.*, 1970, **245**, 4748; *Nature New Biol.*, 1971, **232**, 86 (*isol, struct, synth*)
Studer, R.O. *et al*, *Helv. Chim. Acta*, 1973, **56**, 860 (*isol, struct*)
Yajima, H. *et al*, *Chem. Pharm. Bull.*, 1973, **21**, 682 (*synth*)
Bayer, E. *et al*, *Chem. Ber.*, 1974, **107**, 1344 (*synth*)
Leeman, S. *et al*, *Pept. Neurobiol.*, (Gainer, H., Ed.), 1977, Plenum, N.Y., 99 (*rev*)
Kitagawa, K. *et al*, *Chem. Pharm. Bull.*, 1978, **26**, 1604 (*synth*)
Iversen, L.L., *Br. Med. Bull.*, 1982, **38**, 277 (*rev*)
Chassaing, G. *et al*, *J. Org. Chem.*, 1983, **48**, 1757 (*synth*)
Schwyzer, R. *et al*, *Helv. Chim. Acta*, 1986, **69**, 1789, 1798, 1807 (*cd, struct, props*)

Substictic acid   S-60060

$C_{18}H_{12}O_9$ M 372.287

Constit. of *Aspicilia mashiginensis*. Cryst. (Me₂CO/pet. ether). Mp >270° dec.

Elix, J.A. *et al*, *Aust. J. Chem.*, 1987, **40**, 417.

Sulcatine   S-60061

[110219-85-7]

$C_{14}H_{22}O_2$ M 222.327

Metab. of *Laurilia sulcata*. Cryst. (Et₂O/hexane). Mp 155-160°. $[\alpha]_D^{20}$ −61.3° (c, 2.4 in CHCl₃).

Arnone, A. *et al*, *Phytochemistry*, 1987, **26**, 1739.

Sulfuryl chloride isocyanate, 9CI   S-60062

Updated Entry replacing S-00963

Chlorosulfonyl isocyanate

[1189-71-5]

ClSO₂NCO

$CClNO_3S$ M 141.529

Versatile synthetic reagent. Liq. which fumes in moist air. d_4^{20} 1.626. Bp₁₀₀ 54-56°. n_D^{27} 1.4435.

▷Reacts violently with H₂O. V. strong irritant

Org. Synth., 1966, **46**, 23 (*synth*)
Graf, R., *Angew. Chem., Int. Ed. Engl.*, 1968, **7**, 172 (*rev*)
Dhar, D.N. *et al*, *Synthesis*, 1986, 437 (*rev*)
Fieser, M. *et al*, *Reagents for Organic Synthesis*, Wiley, 1967-84, **7**, 65.
Bretherick, L., *Handbook of Reactive Chemical Hazards*, 2nd Ed., Butterworths, London and Boston, 1979, 279.
Sax, N.I., *Dangerous Properties of Industrial Materials*, 5th Ed., Van Nostrand-Reinhold, 1979, 501.

Superlatolic acid   S-60063

Prasinic acid

[90332-21-1]

$C_{29}H_{40}O_7$ M 500.631

Metab. of *Fuscidea viridis*.

Culberson, C.F. *et al*, *Mycologia*, 1984, **76**, 148.

Swertialactone *D*   S-60064

$C_{30}H_{46}O_3$ M 454.692

Constit. of *Swertia petiolata*. Cryst. (MeOH). Mp 304.5°.

$\Delta^{17(21)}$-*Isomer*: **Swertialactone C**.
$C_{30}H_{46}O_3$ M 454.692
From *S. petiolata*. Cryst. (MeOH). Mp 308°.

Bhan, S. *et al*, *Phytochemistry*, 1987, **26**, 3363.

Sydonic acid   S-60065
3-Hydroxy-4-(1-hydroxy-1,5-dimethylhexyl)benzoic acid
[65967-73-9]

$C_{15}H_{22}O_4$ M 266.336
Metab. of *Aspergillus sydowi*. Cryst. Mp 157-159°.

Hamisaki, T. *et al, Agric. Biol. Chem.*, 1978, **42**, 37 (*isol*)
Murali, D. *et al, Indian J. Chem., Sect. B*, 1987, **26**, 156 (*synth*)

Synrotolide   S-60066
[110300-20-4]

$C_{16}H_{22}O_8$ M 342.345
Constit. of *Syncolostemon rotundifolius*. Cryst. (EtOAc). Mp 168-170°. $[\alpha]_D^{24}$ −29° (c, 0.06 in MeOH).

Coleman, M.T.D. *et al, Phytochemistry*, 1987, **26**, 1497.

Syringopicrogenin *A*   S-60067

$C_{18}H_{20}O_6$ M 332.352
Constit. of *Syringa reticulata*.
Di-Ac: $[\alpha]_D^{30}$ −75° (c, 1 in CHCl₃).
3″-Hydroxy: **Syringopicrogenin B**.
  $C_{18}H_{20}O_7$ M 348.352
  From *S. reticulata*.
3″-Hydroxy, tri-Ac: $[\alpha]_D^{30}$ −65.2° (c, 0.5 in CHCl₃).
3′-Methoxy, 3″-hydroxy: **Syringopicrogenin C**.
  $C_{19}H_{22}O_8$ M 378.378
  From *S. reticulata*.
3′-Methoxy, 3″-hydroxy, tri-Ac: $[\alpha]_D^{30}$ −36.3° (c, 0.8 in CHCl₃).
3″-Hydroxy, 8-O-β-D-glucopyranoside: **Syringopicroside B**.
  $C_{24}H_{30}O_{12}$ M 510.494
  From *S. reticulata*.
3″-Hydroxy, 8-O-β-D-glucopyranoside, hexa-Ac: Cryst. Mp 76-78°. $[\alpha]_D^{30}$ −81.1° (c, 2.1 in CHCl₃).
3′-Methoxy, 3″-hydroxy, 8-O-β-D-glucopyranoside: **Syringopicroside C**.
  $C_{25}H_{32}O_{13}$ M 540.520
  From *S. reticulata*.
3′-Methoxy, 3″-hydroxy, 8-O-β-D-glucopyranoside, hexa-Ac: Cryst. Mp 74-76°. $[\alpha]_D^{30}$ −86.9° (c, 1.7 in CHCl₃).

Kikuchi, M. *et al, Yakugaku Zasshi*, 1987, **107**, 23.

T

Taccalonolide *B*   T-60001

[108885-69-4]

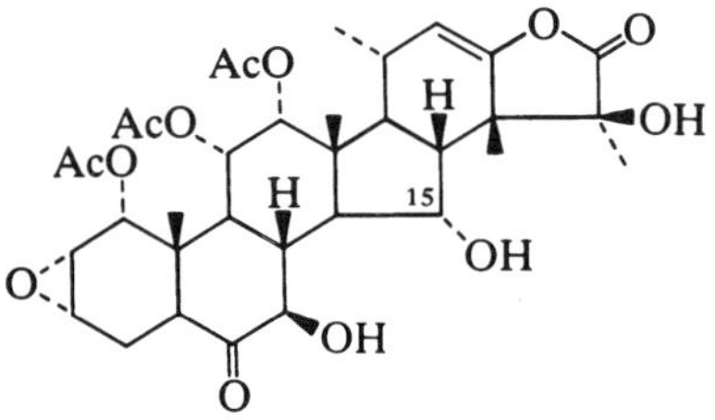

C$_{34}$H$_{44}$O$_{13}$ M 660.714

Constit. of *Tacca plantaginea*. Cryst. Mp 266°. [α]$_D^{14}$ +15.9° (c, 0.019 in CHCl$_3$).

15-Ac: [108885-68-3]. **Taccalonolide A**.
C$_{36}$H$_{46}$O$_{14}$ M 702.751
Constit. of *T. plantaginea*. Mp 215°. [α]$_D^{14}$ +39.5° (c, 0.025 in CHCl$_3$).

Chen, Z. *et al, Tetrahedron Lett.*, 1987, **28**, 1673 (*cryst struct*)

Tagitinin *A*   T-60002

Updated Entry replacing T-00015
[59979-61-2]

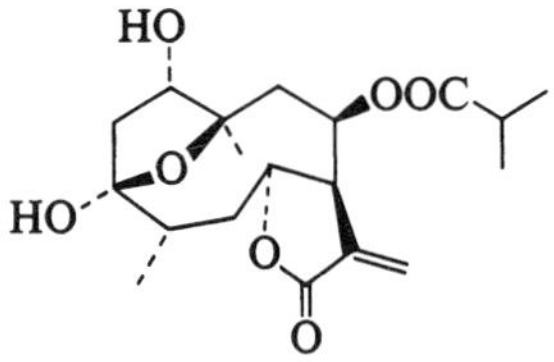

C$_{19}$H$_{28}$O$_7$ M 368.426

Constit. of *Tithonia diversifolia*. Possesses insecticidal props. Cryst. (CHCl$_3$/pet. ether). Mp 168-170°. [α]$_D$ −154° (c, 1 in EtOH).

Pal, R. *et al, Indian J. Chem., Sect. B*, 1976, **14**, 259 (*isol*)
Barauh, N.C. *et al, J. Org. Chem.*, 1979, **44**, 1831 (*struct*)
Sarma, J.C. *et al, Phytochemistry*, 1987, **26**, 2406 (*cryst struct*)

Talaromycin *A*   T-60003

Updated Entry replacing T-30001
9-Ethyl-4-hydroxy-1,7-dioxaspiro[5.5]undecane-3-methanol, 9CI
[83720-10-9]

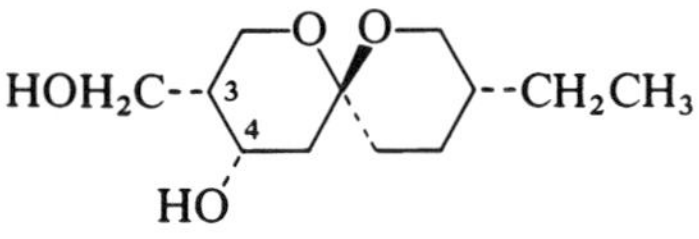

C$_{12}$H$_{22}$O$_4$ M 230.303

Spiroketal antibiotic. Metab. of *Talaromyces stipitatus*. Mycotoxin. Oil.

3-Epimer: [83780-27-2]. **Talaromycin B**.
C$_{12}$H$_{22}$O$_4$ M 230.303
From *T. stipitatus*. Oil.

4-Epimer: [89885-86-9]. **Talaromycin C**.
C$_{12}$H$_{22}$O$_4$ M 230.303
From *T. stipitatus*. Mycotoxin. No phys. props. reported.

3,4-Diepimer: [111465-42-0]. **Talaromycin E**.
C$_{12}$H$_{22}$O$_4$ M 230.303
From *T. stipitatus*. Mycotoxin. No phys. props. reported.

Lynn, D.G. *et al, J. Am. Chem. Soc.*, 1982, **104**, 7319 (*isol, struct*)
Kay, I.T. *et al, Tetrahedron Lett.*, 1984, **25**, 2035 (*synth*)
Kozikowski, A.P. *et al, J. Am. Chem. Soc.*, 1984, **106**, 353 (*synth*)
Smith, A.B. *et al, J. Org. Chem.*, 1984, **49**, 1469 (*synth*)
Midland, M.M. *et al, J. Org. Chem.*, 1985, **50**, 1143 (*synth*)
Schreiber, S.L. *et al, Tetrahedron Lett.*, 1985, **26**, 17 (*synth*)
Iwata, C. *et al, Tetrahedron Lett.*, 1987, **28**, 3135 (*synth*)
Mori, K. *et al, Tetrahedron*, 1987, **43**, 45 (*synth*)
Phillips, N.J. *et al, Tetrahedron Lett.*, 1987, **28**, 1619 (*epimers*)
Whitby, R. *et al, J. Chem. Soc., Chem. Commun.*, 1987, 906 (*synth*)

Tanavulgarol   T-60004

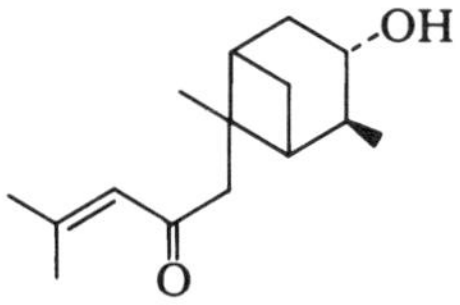

C$_{15}$H$_{24}$O$_2$ M 236.353

Constit. of *Tanacetum vulgare*. Oil. [α]$_D^{26}$ +80° (c, 0.3 in CHCl$_3$).

Chandra, A. *et al, Phytochemistry*, 1987, **26**, 3077.

Tannunolide *A*   T-60005

[110209-99-9]

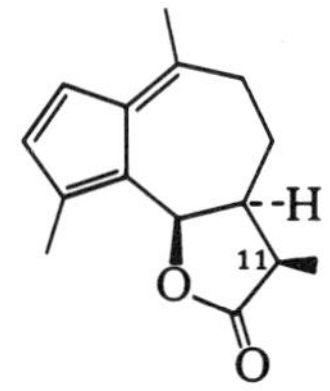

C$_{15}$H$_{18}$O$_2$ M 230.306

Constit. of *Tanacetum annuum*. Yellow cryst. (Et$_2$O). Mp 108-109°. [α]$_D$ −100° (c, 1 in CHCl$_3$).

11-Epimer: [110210-00-9]. **Tannunolide B**.
C$_{15}$H$_{18}$O$_2$ M 230.306
From *T. annuum*. Yellow cryst. (Et$_2$O). Mp 101-102°. [α]$_D$ −29.8° (c, 1.52 in CHCl$_3$).

Barrero, A.F. *et al, Phytochemistry*, 1987, **26**, 1531.

Taonianone   **T-60006**

Updated Entry replacing T-50007

[79203-37-5]

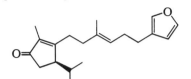

Absolute
configuration

$C_{20}H_{28}O_2$ M 300.440

Constit. of *Taonia australasica*. Oil. $[\alpha]_D^{21}$ +12.4° (c, 0.5
in $CHCl_3$).

Murphy, P.T. *et al, Tetrahedron Lett.*, 1981, 1555 (*isol*)
Kido, F. *et al, J. Chem. Soc., Chem. Commun.*, 1986, 590
     (*synth, abs config*)
Huckestein, M. *et al, Helv. Chim. Acta*, 1987, **70**, 445 (*synth,
     abs config*)

Taraktophyllin   **T-60007**

*1-(β-D-Glucopyranosyloxy)-4-hydroxy-2-cyclopentene-
1-carbonitrile, 9CI*

[110115-55-4]

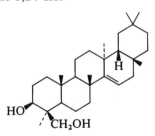

$C_{12}H_{17}NO_7$ M 287.269

Constit. of leaves of *Taraktogenos heterophylla* and
     Hydrocarpus anthelmintica. Syrup. $[\alpha]_D^{23}$ −75° (c, 1 in
     MeOH).

1,4-Diepimer: [109905-56-8]. **Epivolkenin**.
     $C_{12}H_{17}NO_7$ M 287.269
     Constit. of *T. heterophylla* and *H. anthelmintica*.
     Syrup. $[\alpha]_D^{23}$ +43° (c, 1 in MeOH).

Jaroszewski, J.W. *et al, Tetrahedron*, 1987, **43**, 2349.

14-Taraxerene-3,24-diol   **T-60008**

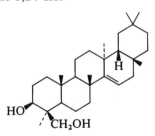

$C_{30}H_{50}O_2$ M 442.724

3β-form

Constit. of *Parsonia laevigata*. Cryst. ($CHCl_3$). Mp
     >320°.

Di-Ac: Cryst. (EtOH). Mp 234-235°. $[\alpha]_D^{25}$ +10.1° (c,
     0.99 in $CHCl_3$).

Ogihara, K. *et al, Phytochemistry*, 1987, **26**, 783.

The symbol ▷ in Entries highlights
hazard or toxicity information

Tarchonanthus lactone   **T-60009**

[71939-67-8]

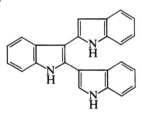

Absolute
configuration

$C_{17}H_{20}O_6$ M 320.341

Constit. of *Tarchonanthus trilobus*. Oil. $[\alpha]_D$ 67.2° (c,
     2.3 in $CHCl_3$).

Bohlmann, F. *et al, Phytochemistry*, 1979, **18**, 677 (*isol*)
Nakata, T. *et al, Tetrahedron Lett.*, 1987, **28**, 5661 (*synth*)

Tecomaquinone I   **T-60010**

Updated Entry replacing T-30006
Dehydrotectol

[89355-02-2]

$C_{30}H_{24}O_4$ M 448.517

Constit. of *Tecoma pentaphylla, T. grandis, Tabebuia
     rosea, T. chryantha* and *Phyllarthron comorense*.
     Green plates (EtOAc/pet. ether). Mp 219-221°.

Sandermann, W. *et al, Chem. Ber.*, 1964, **97**, 588 (*isol*)
Burnett, A.R. *et al, J. Chem. Soc. (C)*, 1968, 850 (*isol*)
Joshi, K.C. *et al, Phytochemistry*, 1974, **13**, 663 (*isol*)
Rohatgi, B.K. *et al, Indian J. Chem., Sect. B*, 1983, **22**, 886
     (*isol*)
Khanna, R.N. *et al, J. Chem. Soc., Perkin Trans. 1*, 1987, 1821
     (*struct*)

2,3′:2′,3″-Ter-1H-indole, 9CI   **T-60011**

2,3′-2′,3″-Triindolyl

[88919-86-2]

$C_{24}H_{17}N_3$ M 347.418

Mp 239-243°.

Bocchi, V. *et al, Tetrahedron*, 1986, **42**, 5019 (*synth, ir, pmr, uv,
     cmr, ms*)

1,2,3,5-Tetraaminobenzene   **T-60012**

Updated Entry replacing T-00182
1,2,3,5-Benzenetetramine, 9CI
$C_6H_{10}N_4$ M 138.172
B,3HCl: Needles + H_2O.

B,2H₂SO₄: Plates.
N-*Tetra-Ac:*
  $C_{14}H_{18}N_4O_4$ M 306.321
  Needles (AcOH). Mp 245°.
N-*Octa-Me:* [104779-70-6]. *1,2,4,5-*
  Tetrakis(dimethylamino)benzene.
  $C_{14}H_{26}N_4$ M 250.386
  V. strong electron donor. Cryst. Mp 95°. Readily
  oxidised in soln.

Nietzki, R. *et al, Ber.*, 1897, **30**, 539 (*synth*)
Borsche, W., *Ber.*, 1923, **56**, 1939 (*synth*)
Elbl, K. *et al, Angew. Chem., Int. Ed. Engl.*, 1986, **25**, 1023
  (*deriv*)

1,2,4,5-Tetraaminobenzene   T-60013

Updated Entry replacing T-00183
1,2,4,5-Benzenetetramine, 9CI
[3204-61-3]
$C_6H_{10}N_4$ M 138.172
Free base oxidises very readily in air.
B,4HCl: [4506-66-5]. Prisms, needles (H₂O). Sol. H₂O,
  spar. sol. HCl.
B,H₂SO₄: Needles. Mod. sol. H₂O.
B₂,3H₂SO₄: Plates. Spar. sol. H₂O.
1,2,4,5-N-Tetra-Ac:
  $C_{14}H_{18}N_4O_4$ M 306.321
  Needles (AcOH). Mp 285°.
N-*Octa-Me:* *1,2,4,5-Tetrakis(dimethylamino)benzene.*
  $C_{14}H_{26}N_4$ M 250.386
  Forms a diamagnetic dication, v. strong electron donor.
  Cryst. by subl. Mp 95°.
N-*Octa-Me; 2,HBr:* Cryst. + 2H₂O (MeOH). Mp 226-
  227°.
N-*Octa-Me, Bis(tetrafluoroborate):* Cryst. (MeOH).
  Mp 290-291°.

Nietzki, R. *et al, Ber.*, 1887, **20**, 328; 1889, **22**, 440 (*synth*)
Hewitt, J.T. *et al, Ber.*, 1898, **31**, 1789 (*synth*)
Knobloch, W. *et al, Chem. Ber.*, 1958, **91**, 2562 (*synth*)
Cheeseman, G.W.H., *J. Chem. Soc.*, 1962, 1170 (*synth*)
Elbl, K. *et al, Angew. Chem., Int. Ed. Engl.*, 1986, **25**, 1023
  (*deriv, synth, pmr, ms, ir, uv, props*)
Staab, H.A. *et al, Tetrahedron Lett.*, 1986, **27**, 5119 (*deriv,
  props, cryst struct*)

3,3′,5,5′-Tetraamino-2,2′,4,4′,6,6′-hexanitrobiphenyl   T-60014

2,2′,4,4′,6,6′-Hexanitro-[1,1′-biphenyl]-3,3′,5,5′-tetramine, 9CI
[21985-91-1]

$$
\begin{array}{c}
\text{H}_2\text{N} \quad \text{NO}_2 \quad \text{O}_2\text{N} \quad \text{NH}_2 \\
\text{O}_2\text{N} \qquad\qquad\qquad \text{NO}_2 \\
\text{H}_2\text{N} \quad \text{NO}_2 \quad \text{O}_2\text{N} \quad \text{NH}_2
\end{array}
$$

$C_{12}H_8N_{10}O_{12}$ M 484.255
Bronze-yellow plates (THF/MeOH). Mp 266-271° dec.
  An earlier mention of this compd. in *CA* was
  erroneous.

Bell, A.J. *et al, Aust. J. Chem.*, 1987, **40**, 175 (*synth, ir, uv, pmr,
  cmr, cryst struct*)

1,4,7,10-Tetraazacyclododecane-1,4,7,10-tetraacetic acid, 9CI   T-60015

dota
[60239-18-1]

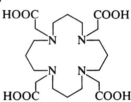

$C_{16}H_{28}N_4O_8$ M 404.419
Complexing agent. pK_{a1} 4.41, pK_{a2} 4.54, pK_{a3} 9.73, pK_{a4}
  11.36 (20°, 0.1M KCl aq.). Ir 1715 cm⁻¹.

Stetter, H. *et al, Tetrahedron*, 1981, **37**, 767 (*synth*)
Riesen, A. *et al, Helv. Chim. Acta*, 1986, **69**, 2067, 2074
  (*complexes*)

1,5,9,13-Tetraazacyclohexadecane-1,5,9,13-tetraacetic acid, 9CI   T-60016

heta
[107321-16-4]

$C_{20}H_{36}N_4O_8$ M 460.526
Solid + 1H₂O. Mp 246° dec.

Riesen, A. *et al, Helv. Chim. Acta*, 1986, **69**, 2074 (*synth, ir,
  pmr, use*)

1,4,8,11-Tetraazacyclotetradecane-1,4,8,11-tetraacetic acid, 9CI   T-60017

teta
[60239-22-7]

$C_{18}H_{32}N_4O_8$ M 432.473
Complexing agent. pK_{a1} 3.66, pK_{a2} 4.41, pK_{a3} 10.31, pK_{a4}
  10.95 (20°, 0.1M KCl aq.).

Stetter, H. *et al, Tetrahedron*, 1981, **37**, 767 (*synth*)
Riesen, A. *et al, Helv. Chim. Acta*, 1986, **69**, 2067, 2074
  (*complexes*)

5,6,8,9-Tetraaza[3.3]paracyclophane　　　T-60018

6,7,17,18-Tetraazatricyclo[10.2.2.2^{5,8}]octadeca-5,7,12,14,15,17-hexaene, 9CI

[109305-25-1]

$C_{14}H_{16}N_4$　　M 240.307

Red cryst. (pet. ether). Mp 140-141°.

Neugebauer, F.A. *et al*, *Tetrahedron Lett.*, 1986, **27**, 5367 (*synth, pmr, uv*)

Tetrabenzotetracyclo[5.5.1.0^{4,13}.0^{10,13}]-tridecane　　　T-60019

Fenestrindane

$C_{29}H_{20}$　　M 368.477

Cryst. (EtOH/C$_6$H$_6$). Mp 325-330° dec. Achiral.

Kuck, D. *et al*, *J. Am. Chem. Soc.*, 1986, **108**, 8107 (*synth, pmr, cmr, ir, ms, uv, cryst struct*)

1,1,2,2-Tetrabenzoylethane　　　T-60020

2,3-Dibenzoyl-1,4-diphenyl-1,4-butanedione, 9CI

[4440-93-1]

$$(PhCO)_2CHCH(COPh)_2$$

$C_{30}H_{22}O_4$　　M 446.501

Tetraoxo form predominates. Prisms. Mp 213-215° (204-208°).

v.Wolf, L. *et al*, *J. Prakt. Chem.*, 1962, **17**, 69 (*synth*)
Cava, M.P. *et al*, *J. Am. Chem. Soc.*, 1973, **95**, 2561 (*synth*)
VandenBorn, H.W. *et al*, *J. Am. Chem. Soc.*, 1974, **96**, 4296 (*pmr, uv, tautom*)
Cannon, J.R. *et al*, *Aust. J. Chem.*, 1986, **39**, 1811 (*cryst struct, pmr*)

Tetrabenzoylethylene　　　T-60021

2,3-Dibenzoyl-1,4-diphenyl-2-butene-1,4-dione, 9CI

[5860-38-8]

$$(PhCO)_2C=C(COPh)_2$$

$C_{30}H_{20}O_4$　　M 444.486

Fluorescent cryst. Mp 184°. Light-sensitive, turning yellow. Dimorphic; only one cryst. modification shows photochromism.

Schmid, H. *et al*, *Helv. Chim. Acta*, 1948, **31**, 1899 (*ir*)
VandenBorn, H.W. *et al*, *J. Am. Chem. Soc.*, 1974, **96**, 4296 (*synth*)

1,2,4,5-Tetrabromo-3-chloro-6-methylbenzene, 9CI　　　T-60022

2,3,5,6-Tetrabromo-4-chlorotoluene

$C_7H_3Br_4Cl$　　M 442.170

Cryst. (CHCl$_3$). Mp 264-265°.

Hart, H. *et al*, *Tetrahedron*, 1987, **43**, 5203 (*synth, pmr, ms*)

Tetrabromofuran　　　T-60023

[32460-09-6]

C_4Br_4O　　M 383.659

Needles (EtOH) with characteristic smell. Mp 64-65°. Bp$_4$ 118°. Unstable in soln. at r.t.

Hill, H.B. *et al*, *Justus Liebigs Ann. Chem.*, 1885, **232**, 42 (*synth*)
Sornay, R. *et al*, *Bull. Soc. Chim. Fr.*, 1971, 990 (*synth*)
Shoppee, C.W., *J. Chem. Soc., Perkin Trans. 1*, 1985, 45 (*synth, cmr*)

1,1,1′,1′-Tetra-*tert*-butylazomethane　　　T-60024

Bis[1-(1,1-dimethylethyl)-2,2-dimethylpropyl]diazene, 9CI

$$[(H_3C)_3C]_2CHN=NCH[C(CH_3)_3]_2$$

$C_{18}H_{38}N_2$　　M 282.512

(**E**)-*form* [100515-68-2]
Large light-yellow plates. Mp 61-63°.
(**Z**)-*form* [100515-67-1]
Intensely yellow solid.

Bernlöhr, W. *et al*, *Chem. Ber.*, 1986, **119**, 1911 (*synth, uv, pmr, ms*)

1,3,5,7-Tetra-*tert*-butyl-*s*-indacene　　　T-60025

1,3,5,7-Tetrakis(1,1-dimethylethyl)-s-indacene, 9CI

[101998-70-3]

$C_{28}H_{40}$　　M 376.624

Red needles. Mp 190° dec. Stable to air and heat, solns. v. sensitive to acid and O$_2$.

Hafner, K. *et al*, *Angew. Chem., Int. Ed. Engl.*, 1986, **25**, 630 (*synth, pmr, cmr, uv, ir, cryst struct*)

1,4,5,8-Tetrachloro-9,10-anthraquinodimethane
T-60026

1,4,5,8-Tetrachloro-9,10-dihydro-9,10-bis(methylene)-anthracene, 9CI

[105020-57-3]

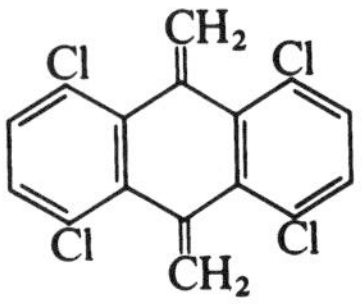

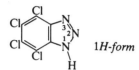

$C_{16}H_8Cl_4$ M 342.051

Stable, non-planar *p*-quinodimethane. Prisms (CCl₄). Mp 238-239°.

Staab, H.A. *et al, Chem. Ber.*, 1987, **120**, 93 (*synth, pmr, cryst struct*)

4,5,6,7-Tetrachlorobenzotriazole, 9CI
T-60027

[2338-10-5]

1H-form

$C_6HCl_4N_3$ M 256.906

1H-form

Cryst. (MeOH or MeNO₂). Mp 256-260°.

1-Me:
$C_7H_3Cl_4N_3$ M 270.933
Cryst. (CCl₄). Mp 193.5-196.5°.

2H-form

2-Me:
$C_7H_3Cl_4N_3$ M 270.933
Needles (MeOH or HNO₃). Mp 181-184°.

Wiley, R.H. *et al, J. Am. Chem. Soc.*, 1955, **77**, 5105 (*synth, uv, derivs*)

2,3,4,5-Tetrachloro-1,1-dihydro-1-[(methoxycarbonyl)imino]thiophene, 9CI
T-60028

N-*Methoxycarbonyl-(2,3,4,5-tetrachloro-1-thiophenio)-amide*

[106550-54-3]

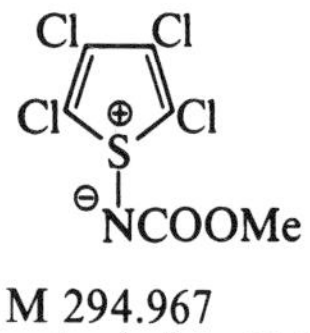

$C_6H_3Cl_4NO_2S$ M 294.967
Cryst. (EtOAc/pet. ether). Mp 134-135°.

Meth-Cohn, O. *et al, J. Chem. Soc., Perkin Trans. 1*, 1986, 233, 245 (*synth, ir, pmr, cmr, use*)

1,1,2,3-Tetrachloro-1-propene, 9CI
T-60029

[10436-39-2]

$$Cl_2C{=}CClCH_2Cl$$

$C_3H_2Cl_4$ M 179.861
Bp₇₅₇ 164°. ca. 95% pure.
▷UD1925000.

Bauer, H. *et al, Chem. Ber.*, 1986, **119**, 1890 (*synth, ir, pmr, cmr*)

29,29,30,30-Tetracyanobianthraquinodimethane
T-60030

[10-[(10-*Dicyanomethylene*)-9(10H)-*anthracenylidene*]-9(10H)-*anthracenylidene*]*propanedinitrile, 9CI*

[105754-75-4]

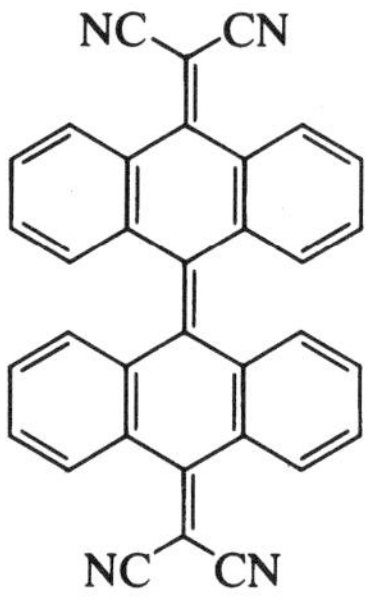

$C_{34}H_{16}N_4$ M 480.527
Orange prisms (DMSO). Mp >300°.

Yamaguchi, S. *et al, Tetrahedron Lett.*, 1986, **27**, 2411 (*synth, ir, pmr, uv, cryst struct*)

Tetracyanoethylene
T-60031

Updated Entry replacing T-00490
Ethenetetracarbonitrile, 9CI. Tetracyanoethene

[670-54-2]

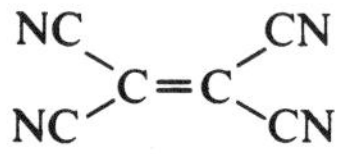

C_6N_4 M 128.093
Very reactive dienophile. Forms π-complexes with aromatic compds. Cryst. (chlorobenzene). Mp 198-200°. Bp 223°.

▷Highly toxic. KM7300000.

Org. Synth., Coll. Vol., **4**, 877 (*synth*)
Cairns, T.L. *et al, Angew. Chem.*, 1961, **73**, 520.
Fatiadi, A.J., *Synthesis*, 1986, 249 (*rev*)
Fieser, M. *et al, Reagents for Organic Synthesis*, Wiley, 1967-84, **6**, 567.
Sax, N.I., *Dangerous Properties of Industrial Materials*, 5th Ed., Van Nostrand-Reinhold, 1979, 1013.

Tetracyclo[6.2.1.1³,⁶.0²,⁷]dodeca-2(7),4,9-triene
T-60032

1,4,5,8-Tetrahydro-1,4:5,8-dimethanonaphthalene, 9CI. Sesquinorbornatriene

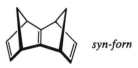

syn-forn

$C_{12}H_{12}$ M 156.227
syn-form [97253-51-5]
Oil. V. oxygen-sensitive, quickly yellows at r.t.
anti-form [97806-44-5]
Cryst. solid. Air-sensitive.

Paquette, L.A. *et al, J. Am. Chem. Soc.*, 1986, **108**, 3453 (*synth, ir, pmr*)

Tetracyclo[6.2.1.1^{3,6}.0^{2,7}]dodec-2(7)-ene   T-60033

Updated Entry replacing T-20035

1,2,3,4,5,6,7,8-Octahydro-1,4:5,8-dimethanonaphthalene, 9CI. Sesquinorbornene

$C_{12}H_{16}$ M 160.258

(*1α,3β,6β,8α*)*-form* [73679-39-7]

anti-*form*

Mp 64-65°. Bp_{0.6} 40°. *Syn*-form also known but not fully descr.

Paquette, L.A. *et al*, *J. Am. Chem. Soc.*, 1980, **102**, 1186 (*synth*)
Bartlett, P.D. *et al*, *J. Am. Chem. Soc.*, 1980, **102**, 1383 (*synth*, *pmr*, *cmr*)
DeLucchi, O. *et al*, *Tetrahedron Lett.*, 1986, **27**, 4347 (*synth*)

1,1,2,2-Tetracyclopropylethane   T-60034

1,1′,1″,1‴-1,2-Ethanediylidenetetrakiscyclopropane

[102652-65-3]

$C_{14}H_{22}$ M 190.328

Liq.

Bruch, M. *et al*, *J. Org. Chem.*, 1986, **51**, 2969 (*synth*, *ir*, *pmr*, *cmr*)

Tetracyclo[5.5.1.0^{4,13}.0^{10,13}]tridecane   T-60035

Updated Entry replacing T-50054

Dodecahydropentaleno[1,6-cd]pentalene, 9CI. [5.5.5.5]-*Fenestrane. Staurane*

[67490-05-5]

$C_{13}H_{20}$ M 176.301

Luyten, M. *et al*, *Angew. Chem., Int. Ed. Engl.*, 1984, **23**, 390 (*synth*, *ir*, *pmr*, *cmr*)
Luyten, M. *et al*, *Helv. Chim. Acta*, 1984, **67**, 2242 (*synth*)
Venkatachalam, M. *et al*, *Tetrahedron Lett.*, 1985, **26**, 4863 (*synth*)
Luef, W. *et al*, *Helv. Chim. Acta*, 1987, **70**, 543 (*struct*)

Tetracyclo[5.5.1.0^{4,13}.0^{10,13}]trideca-2,5,8,11-tetraene   T-60036

2,5,8,11-Stauranetetraene. [5.5.5.5]*Fenestratetraene*

$C_{13}H_{12}$ M 168.238

Solid by subl. Mp 90°.

Venkatachalam, M. *et al*, *Tetrahedron*, 1986, **42**, 1597 (*synth*, *ir*, *pmr*, *cmr*, *ms*)

Tetracyclo[4.4.1.0^{3,11}.0^{9,11}]undecane   T-60037

[4.4.5.5]*Fenestrane*

$C_{11}H_{16}$ M 148.247

Rao, V.B. *et al*, *Tetrahedron*, 1986, **42**, 1549 (*synth*, *ir*, *pmr*, *cmr*)

Tetradecahydro-4,6-dihydroxy-9,9,12a-tri-   T-60038
methyl-6*H*-phenanthro[1,10a-c]furan-3-carboxylic acid, 9CI

15,16-Dideoxy-15,17-dihydroxy-15,17-oxido-16-spongianoic acid

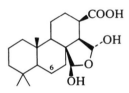

$C_{20}H_{32}O_5$ M 352.470

(*15α,17β*)*-form*

Di-Ac, Me ester: [109152-44-5]. *Methyl 15α,17β-diacetoxy-15,16-dideoxy-15,17-oxido-16-spongianoate.*
$C_{25}H_{38}O_7$ M 450.571
Constit. of *Ceratosoma brevicaudatum*. Oil.
6α-Acetoxy, 15,17-di-Ac, Me ester: [109181-99-9].
Methyl 6α,15α,17β-triacetoxy-15,16-dideoxy-15,17-oxido-16-spongianoate.
$C_{27}H_{40}O_9$ M 508.608
From *C. brevicaudatum*. Oil.
6α-Butanoyloxy, 15,17-Di-Ac, Me ester: [109152-43-4].
Methyl 15α,17β-diacetoxy-6α-butanoyloxy-15,16-dideoxy-15,17-oxido-16-spongianate.
$C_{29}H_{44}O_9$ M 536.661
From *C. brevicaudatum*. Oil.

Ksebati, M.B. *et al*, *J. Org. Chem.*, 1987, **52**, 3766.

5-Tetradecanone   T-60039

[31857-89-3]

$$H_3C(CH_2)_8CO(CH_2)_3CH_3$$

$C_{14}H_{28}O$ M 212.375

Mp 23°. Bp_{16} 145-146°, Bp_{0.7} 85-88°.

Carrington, R.A.G. *et al*, *J. Chem. Soc.*, 1957, 1701 (*synth*)
Anderson, M.W. *et al*, *J. Chem. Soc., Perkin Trans. 1*, 1986, 1995 (*synth*, *ir*, *pmr*)

1-Tetradecylcyclopropanecarboxylic acid   T-60040

[70858-10-5]

$$H_3C(CH_2)_{13} \quad COOH$$

$C_{18}H_{34}O_2$ M 282.465
Cryst. (MeOH). Mp 53-55°.

Ho, W. *et al, J. Med. Chem.*, 1986, **29**, 2184 (*synth*)

2-Tetradecyloxiranecarboxylic acid   T-60041

[68170-97-8]

$$(CH_2)_{13}CH_3 \\ —COOH$$

$C_{17}H_{32}O_3$ M 284.438

(±)-*form*

Cryst. (Me₂CO). Mp 77-79°.
Na salt: [85216-79-1]. Cryst. + 2H₂O (MeOH aq.). Mp
94-136°.
Me ester: [69207-52-9].
   $C_{18}H_{34}O_3$ M 298.465
   Cryst. (MeOH). Mp 43-45°.

Ho, W. *et al, J. Med. Chem.*, 1986, **29**, 2184 (*synth, ir, uv, pmr*)

7-Tetradecyn-1-ol   T-60042

[37011-94-2]

$$H_3C(CH_2)_5C{\equiv}C(CH_2)_5CH_2OH$$

$C_{14}H_{26}O$ M 210.359
Bp₀.₀₁ 117-118°.

Voaden, D.J. *et al, J. Med. Chem.*, 1972, **15**, 619 (*synth*)
Brown, H.C. *et al, J. Org. Chem.*, 1986, **51**, 4518 (*synth, ir, pmr, cmr*)

3,3,4,4-Tetrafluorocyclobutene, 9CI   T-60043

[2714-38-7]

$C_4H_2F_4$ M 126.053
Liq. d₄²⁰ 1.358. Bp 54-56°. n_D^{25} 1.3086.

Anderson, J.L. *et al, J. Am. Chem. Soc.*, 1961, **83**, 382 (*synth*)
Park, J.D. *et al, J. Org. Chem.*, 1963, **28**, 1008 (*synth*)
Harris, R.K. *et al, J. Magn. Reson.*, 1969, **1**, 362 (*F nmr*)

1,2,3,4-Tetrafluoro-5,6-diiodobenzene, 9CI   T-60044

[2708-97-6]

$C_6F_4I_2$ M 401.869
Cryst. (EtOH aq.). Mp 50.5-51.8°.

Plummer, W.J. *et al, J. Res. Nat. Bur. Stand.*, 1959, **62**, 113
   (*synth*)

Cooper, M.A., *Org. Magn. Reson.*, 1969, **1**, 363 (*F nmr*)
Cvitas, T. *et al, J. Chem. Soc., Perkin Trans. 2*, 1977, 962 (*pe*)
Green, J.H.S. *et al, Spectrochim. Acta, Part A*, 1977, **33**, 193
   (*ir, raman*)
Yadav, R. *et al, J. Raman Spectrosc.*, 1983, **14**, 353 (*ir, raman*)

1,2,3,5-Tetrafluoro-4,6-diiodobenzene, 9CI   T-60045

[67815-57-0]

$C_6F_4I_2$ M 401.869

Deacon, G.B. *et al, J. Fluorine Chem.*, 1980, **15**, 85 (*synth*)
Glidewell, C. *et al, Chem. Scr.*, 1985, **25**, 142 (*struct*)

1,2,4,5-Tetrafluoro-3,6-diiodobenzene, 9CI   T-60046

[392-57-4]

$C_6F_4I_2$ M 401.869
Cryst. (MeOH aq.). Mp 109-111°.

▷Irritant

Hellmann, M. *et al, J. Am. Chem. Soc.*, 1955, **77**, 3650 (*synth*)
Cvitas, T. *et al, J. Chem. Soc., Perkin Trans. 2*, 1977, 962 (*pe*)
Green, J.H.S. *et al, Spectrochim. Acta, Part A*, 1977, **33**, 193
   (*ir, raman*)
Pawley, G.S. *et al, Acta. Crystallogr., Sect. A*, 1977, **33**, 142
   (*struct*)
Chaplot, S.L. *et al, Acta Crystallogr., Sect. B*, 1981, **37**, 2210
   (*cryst struct*)

1,1,2,2-Tetrafluoro-3,4-dimethylcyclobu-tane   T-60047

$C_6H_8F_4$ M 156.123
Cis- and *trans-*isomers sepd. by glc.

Dobier, W.R. *et al, Tetrahedron*, 1986, **42**, 3763 (*synth, pmr, F nmr, cmr*)

2,2,4,4-Tetrafluoro-1,3-dithietane, 9CI   T-60048

[1717-50-6]

$C_2F_4S_2$ M 164.136
Source of Carbanothioic difluoride, C-00206 by pyroly-
sis. Liq. d₄²⁰ 1.604. Mp −6°. Bp 48°. n_D^{25} 1.3908. δ_F
+1.17 (CFCl₂CFCl₂), singlet.
1-Oxide: [85963-75-3].
   $C_2F_4OS_2$ M 180.135
   Bp 117°. n_D^{20} 1.4415.
1,1-Dioxide: [767-45-3].
   $C_2F_4O_2S_2$ M 196.134
   Liq. Bp 105°. n_D^{25} 1.4108.
1,1,3-Trioxide: [85963-76-4].
   $C_2F_4O_3S_2$ M 212.134
   Mp 81°.
1,1,3,3-Tetroxide: [73090-23-0].
   $C_2F_4O_4S_2$ M 228.133

Cryst. solid. Mp 134° subl. Sensitive to hydrolysis.

Middleton, W.J. *et al*, *J. Org. Chem.*, 1965, **30**, 1375 (*synth*)
Long, R.C. *et al*, *J. Chem. Phys.*, 1971, **54**, 1563 (*pmr, cmr*)
Smith, Z. *et al*, *Acta Chem. Scand., Ser. A*, 1976, **30**, 759 (*ed*)
Chiang, J.F. *et al*, *J. Phys. Chem.*, 1977, **81**, 1682 (*ed*)
Klaboe, P. *et al*, *Spectrochim. Acta, Part A*, 1978, **34**, 789 (*ir, raman*)
Balbach, B. *et al*, *Justus Liebigs Ann. Chem.*, 1980, 1981 (*tetraoxide, cryst struct*)
Schomburg, D. *et al*, *J. Am. Chem. Soc.*, 1981, **103**, 406 (*dioxide, cryst struct*)
Eschwey, M. *et al*, *Chem. Ber.*, 1983, **116**, 1623 (*oxides, synth, ir, ms, F nmr*)

Tetrafluorooxirane, 9CI   T-60049

Epoxytetrafluoroethane, 8CI. Tetrafluoroethylene oxide

[694-17-7]

C_2F_4O M 116.015

Monomer, synth. intermed. Gas. Mp −118°. Bp −63.5°. Rearranges to CF_3COF above Bp.

▷Addition of C_2F_4 and O_2 can be explosive

Caglioti, V. *et al*, *J. Chem. Soc.*, 1964, 5430 (*synth, ir, F nmr*)
Gozzo, F. *et al*, *Tetrahedron*, 1966, **22**, 1765 (*synth, bibl*)
Eleuterio, H.S., *J. Macromol. Sci. Chem.*, 1972, **6**, 1027 (*rev, haz*)
Craig, N.C. *et al*, *Spectrochim. Acta, Part A*, 1972, **28**, 1195 (*ir*)
Resnick, P.R., *Kirk-Othmer. Encycl. Chem. Technol.*, 3rd Ed., 1978, **10**, 956 (*bibl*)
Vilenchik, Y.M. *et al*, *Zh. Org. Khim.*, 1978, **14**, 1587 (*Engl. transl.* p. 1483) (*synth*)
Agopovich, J.W. *et al*, *J. Am. Chem. Soc.*, 1984, **106**, 2250 (*struct*)

4,5,7,8-Tetrafluoro[2.2]paracyclophane   T-60050

5,6,15,16-Tetrafluorotricyclo[8.2.2.2^{4,7}]hexadeca-4,6,10,12,13,15-hexaene, 9CI

[22557-12-6]

$C_{16}H_{12}F_4$ M 280.264

Cryst. Mp 188.5-190°.

Filler, R. *et al*, *J. Am. Chem. Soc.*, 1969, **91**, 1862 (*synth, uv, pmr, reactions*)
Filler, R. *et al*, *J. Org. Chem.*, 1987, **52**, 511 (*synth, uv, pmr, props*)

3,3,4,4-Tetrafluoro-2-(trifluoromethyl)-1,2-   T-60051
oxazetidine, 9CI

Perfluoro(2-methyl-1,2-oxazetidine)

[515-85-5]

C_3F_7NO M 199.028

Source of $F_3CN{=}CF_2$. Gas. Bp −6.8° (extrap).

Haszeldine, R.N. *et al*, *J. Chem. Soc.*, 1955, 1891 (*synth*)
Lee, J. *et al*, *Trans. Faraday Soc.*, 1965, **61**, 2342 (*F nmr*)

1,3,4,10-Tetrahydro-9(2H)-acridinone, 9CI   T-60052

Updated Entry replacing T-00673

1,2,3,4-Tetrahydroacridone

[13161-85-8]

$C_{13}H_{13}NO$ M 199.252

Needles (EtOH). Mp 358°. Green fluor. in EtOH.

N-*Me*: [5464-89-1].
     $C_{14}H_{15}NO$ M 213.279
     Prisms (Py or C_6H_6). Mp 172-174°.

Tiedtke, H., *Ber.*, 1909, **42**, 621 (*synth*)
Reed, R.A., *J. Chem. Soc.*, 1944, 425 (*synth, derivs*)
Chong, R.J. *et al*, *Tetrahedron Lett.*, 1986, **27**, 5323 (*synth*)

1,2,3,4-Tetrahydroanthracene, 8CI   T-60053

Updated Entry replacing T-20053

Tethracene

[2141-42-6]

$C_{14}H_{14}$ M 182.265

Leaflets (EtOH). Mp 103-105°. Bp_{14} 170-173°.

Schroeter, G., *Ber.*, 1924, **57**, 2003 (*synth*)
Garlock, E.A. *et al*, *J. Am. Chem. Soc.*, 1945, **67**, 2255 (*synth*)
Davies, D.I. *et al*, *J. Chem. Soc. (C)*, 1968, 1865 (*synth*)
Ishikawa, T. *et al*, *Chem. Pharm. Bull.*, 1982, **30**, 1594 (*synth, pmr, ms*)
Tius, M.A., *Tetrahedron Lett.*, 1986, **27**, 2571 (*synth, ir, pmr, cmr, ms*)

1,2,3,4-Tetrahydroanthraquinone   T-60054

Updated Entry replacing T-00676

1,2,3,4-Tetrahydro-9,10-anthracenedione, 9CI.
Tethracenequinone

[4923-66-4]

$C_{14}H_{12}O_2$ M 212.248

Golden-yellow needles (AcOH or EtOAc). Mp 158-159°.

Schroeter, G., *Ber.*, 1924, **57**, 2003 (*synth*)
Skita, A., *Ber.*, 1925, **58**, 2685 (*synth*)
Matsuura, R. *et al*, *CA*, 1976, **85**, 32701 (*synth*)
Negishi, E. *et al*, *Tetrahedron Lett.*, 1986, **27**, 4869 (*synth*)

6,7,8,9-Tetrahydro-5H-benzocycloheptene-6-carboxaldehyde   T-60055

[101955-05-9]

$C_{12}H_{14}O$ M 174.242

(±)-*form*

Bp$_{0.01}$ 83-86°.

Jenneskens, L.W. *et al*, *J. Org. Chem.*, 1986, **51**, 2162 (*synth, pmr, ms*)

6,7,8,9-Tetrahydro-5H-benzocycloheptene-7-carboxaldehyde   T-60056

[101955-04-8]

$C_{12}H_{14}O$ M 174.242

Bp$_{0.05}$ 70°.

Jenneskens, L.W. *et al*, *J. Org. Chem.*, 1986, **51**, 2162 (*synth, pmr, ms*)

1,2,3,4-Tetrahydro-7H-benzocyclohepten-7-one, 9CI   T-60057

[108185-85-9]

$C_{11}H_{12}O$ M 160.215

Mp 96-97°.

Kende, A.S. *et al*, *Tetrahedron Lett.*, 1986, **27**, 6051 (*synth, ir, pmr*)

1,3,4,5-Tetrahydro-2H-1,3-benzodiazepin-2-one   T-60058

[41921-63-5]

$C_9H_{10}N_2O$ M 162.191

Mp 170.5-171°.

Jen, T. *et al*, *J. Med. Chem.*, 1973, **16**, 407 (*synth*)
Yoshida, T. *et al*, *Tetrahedron Lett.*, 1986, **27**, 3037 (*synth*)

5a,6,11a,12-Tetrahydro[1,4]-benzoxazino[3,2-b][1,4]benzoxazine, 9CI   T-60059

Calcium red

[104462-81-9]

$C_{14}H_{12}N_2O_2$ M 240.261

Two incorrect structs. have formerly been proposed, Glyoxal bis 2-hydroxyanil and 2,2′-Bibenzoxazoline. Prod. of condensation of 2-aminophenol and glyoxal. Forms chelate complexes with divalent metal ions. Cryst. Mp 231° (212°).

Tauer, E. *et al*, *Chem. Ber.*, 1986, **119**, 3316 (*synth, struct, cryst struct, pmr, uv*)

4a,4b,8a,8b-Tetrahydrobiphenylene, 9CI   T-60060

Updated Entry replacing T-00687

Tricyclo[6.4.0.0^{2,7}]dodeca-3,5,9,11-tetraene. Benzene dimer

$C_{12}H_{12}$ M 156.227

(4aα,4bβ,8aβ,8bα)-form illus.

(**4aα,4bβ,8aβ,8bα**)-*form* [21657-71-6]

Cryst. Mp 15°. Thermolabile, dec. → C_6H_6, $t^{1/2}$, 3$^{1/2}$hr. in CCl_4 at 40°.

(**4aα,4bα,8aα,8bα**)-*form* [91279-91-3]

Cryst. (pentane). Mp 45-46°.

▷Thermally more stable than the (α,β,β,α)-form

Schroder, G. *et al*, *Angew. Chem., Int. Ed. Engl.*, 1969, **8**, 69.
Yang, N.C. *et al*, *J. Am. Chem. Soc.*, 1987, **109**, 3158 (*synth, uv, ir, pmr, cmr*)

Tetrahydrobis(4-hydroxyphenyl)-3,4-dimethylfuran   T-60061

4,4′-(Tetrahydro-3,4-dimethyl-2,5-furandiyl)bisphenol, 9CI. 2,5-Bis(4-hydroxyphenyl)-3,4-dimethyltetrahydrofuran

(2R*,3R*,4S*,5R*)-*form*

$C_{18}H_{20}O_3$ M 284.354

(**2R*,3R*,4S*,5R***)-*form* [109194-74-3]

(2α,3β,4β,5β)-*form*

Constit. of *Krameria cystisoides*. Cryst. Mp 158-162°. $[\alpha]_D^{21}$ +78° (c, 0.142 in MeOH).

3′-*Methoxy*: [109194-75-4]. *Tetrahydro-2-(4-hydroxyphenyl)-5-(4-hydroxy-3-methoxyphenyl)-3,4-dimethylfuran. 5-(4-Hydroxyphenyl)-2-(4-hydroxy-3-methoxyphenyl)-3,4-dimethyltetrahydrofuran.*
$C_{19}H_{22}O_4$ M 314.380
Constit. of *K. cystisoides*. Oil. $[\alpha]_D^{21}$ +56° (c, 0.358 in MeOH).

(**2R*,3R*,4S*,5S***)-*form*

(2α,3β,4β,5α)-*form*

3′-*Methoxy*: [109280-45-7].
$C_{19}H_{22}O_4$ M 314.380
Constit. of *K. cystisoides*. Oil. $[\alpha]_D^{21}$ +3.7° (c, 0.54 in MeOH).

Auchenbach, H. *et al*, *Phytochemistry*, 1987, **26**, 1159.

3,4,5,6-Tetrahydro-4,5-bis(methylene)-pyridazine, 9CI   T-60062

3,4,5,6-Tetrahydro-4,5-dimethylenepyridazine
[104835-54-3]

$C_6H_8N_2$ M 108.143

Source of tetramethyleneethane diradical. Unstable. Obt. only in soln.

Dowd, P. *et al, J. Am. Chem. Soc.*, 1986, **108**, 7416 (synth, pmr, cmr, use)

5a,5b,11a,11b-Tetrahydrocyclobuta[1,2-b:4,3-b']benzothiopyran-11,12-dione, 9CI   T-60063

(5aα,5bα,11aα,11bα)-form

$C_{18}H_{12}O_2S_2$ M 324.412

Photodimer of Thiochrome, T-02030 . Incorrect structs. were originally assigned. A fourth dimer, Mp 138-139°, was also prepd. but could not be obt. as suitable cryst. for structural analysis. It could be head-to-tail.

(5aα,5bα,11aα,11bα)-form [78738-44-0]
Mp 250-251°. Originally considered to be the *all-cis* head-to-tail dimer and because of crystal disorder this cannot be ruled out despite cryst. struct. determination.
(5aα,5bα,11aα,11bβ)-form [104113-47-5]
Mp 245-246°. Originally considered to be the (5aα-,5bα,11aα,11bα)-isomer.
(5aα,5bβ,11aβ,11bα)-form [78680-95-2]
Mp 180-182°. Originally considered to be the *trans*-head-to-tail dimer.

Still, I.W.J. *et al, Tetrahedron Lett.*, 1981, **22**, 1183 (synth, pmr, uv, cmr)
Nyberg, S.C. *et al, Acta Crystallogr., Sect. C*, 1986, **42**, 816 (cryst struct)

1,2,3,4-Tetrahydro-2,4-dioxo-5-pyrimidin-ecarboxaldehyde, 9CI   T-60064

5-Formyluracil
[1195-08-0]

$C_5H_4N_2O_3$ M 140.098
Cryst. Mp 302-303°.

Wiley, R.H. *et al, J. Org. Chem.*, 1960, **25**, 1906 (synth)
Brossmer, R. *et al, Tetrahedron Lett.*, 1966, 5253 (synth, uv)

4-(Tetrahydro-2-furanyl)morpholine, 9CI   T-60065

Tetrahydro-2-(4-morpholinyl)furan. 2-Morpholinotetrahydrofuran
[20024-89-9]

$C_8H_{15}NO_2$ M 157.212
Bp$_4$ 76-77°.

Fuchigami, T. *et al, J. Org. Chem.*, 1986, **51**, 366 (synth, pmr)

1,2,3,4-Tetrahydro-4-hydroxyisoquinoline   T-60066

Updated Entry replacing T-40043
1,2,3,4-Tetrahydro-4-isoquinolinol, 9CI, 8CI. 4-Hydroxy-1,2,3,4-tetrahydroisoquinoline
[51641-23-7]

(R)-form

$C_9H_{11}NO$ M 149.192
(R)-form [105181-85-0]
Mp 108.5°. $[\alpha]_{405}^{26}$ −7.2° (c, 0.78 in EtOH).
(S)-form [105181-84-8]
Mp 108.5°. $[\alpha]_{405}^{27}$ +8.7° (c, 1.2 in EtOH).
B,HCl: [105181-86-0]. Mp 194-196°. $[\alpha]_D^{22}$ +19.4° (c, 1.3 in H_2O).
(±)-form [75240-38-9]
Needles. Mp 80.5°.
B,HCl: [105121-97-9]. Mp 203° dec.
N-Ac: [105121-98-0].
$C_{11}H_{13}NO_2$ M 191.229
Cryst (EtOAc/hexane). Mp 97-98°.

Snatzke, G. *et al, Justus Liebigs Ann. Chem.*, 1987, 81 (synth, uv, ir, cd, pmr, ms, resoln)

3,4,5,6-Tetrahydro-4-hydroxy-6-methyl-2H-pyran-2-one, 9CI   T-60067

3-Hydroxy-5-hexanolide

$C_6H_{10}O_3$ M 130.143
(4S,6S)-form [33275-54-6]
Constit. of *Osmunda japonica*. Prisms (Et$_2$O). Mp 67-70°.

O-β-D-*Glucopyranoside:* [33276-04-9]. **Parasorboside**.
$C_{12}H_{20}O_8$ M 292.285
Constit. of *Sorbus aucuparia*. Needles (Me$_2$CO). Mp 68.9°, Mp 143.4° (double Mp). $[\alpha]_D^{20}$ −22° (c, 0.95 in H_2O).

Tschesche, R. *et al, Chem. Ber.*, 1971, **104**, 1420.
Numata, A. *et al, Chem. Pharm. Bull.*, 1984, **32**, 2815.

1,2,3,6-Tetrahydro-3-hydroxy-2-pyridine-   T-60068
carboxylic acid

*1,2,3,6-Tetrahydro-3-hydroxypicolinic acid. 3-
Hydroxybaikiain*

$C_6H_9NO_3$ M 143.142

(2S,3R)-form

Isol. from mushroom *Russula subnigricans*. Needles
(MeOH aq.). Mp 300-302°. $[\alpha]_D^{20} -332.7°$ (c, 0.3 in
H_2O).

Kusano, G. *et al, Chem. Pharm. Bull.*, 1987, **35**, 3482 *(isol)*

Tetrahydro-3-iodo-2*H*-pyran-2-one, 9CI   T-60069
[63641-53-2]

$C_5H_7IO_2$ M 226.014

(±)-form

Liq. Bp_1 78°.

Evans, R.D. *et al, Synthesis*, 1986, 727 *(synth, ir, pmr)*

1,2,3,4-Tetrahydro-1-isoquinolinecarboxy-   T-60070
lic acid, 9CI

Updated Entry replacing T-00895
[41034-52-0]

$C_{10}H_{11}NO_2$ M 177.202

(±)-form

Mp 299° dec.

Et ester:
   $C_{12}H_{15}NO_2$ M 205.256
   Oil. $Bp_{0.5}$ 120°.
Et ester; B, HCl: [103733-33-1]. Cryst. (EtOAc/Et_2O).
   Mp 132-134°.

Solomon, W., *J. Chem. Soc.*, 1947, 129 *(synth)*
Klutchko, S. *et al, J. Med. Chem.*, 1986, **29**, 1953 *(synth, ir,
pmr)*

4,5,6,7-Tetrahydroisoxazolo[4,5-*c*]pyridin-   T-60071
3-ol

4,5,6,7-Tetrahydroisoxazolo[4,5-c]pyridin-3(2H)-one
[53602-00-9]

$C_6H_8N_2O_2$ M 140.141
Zwitterionic. GABA agonist. Cryst. + $1H_2O$ (EtOH
aq.). Mp 254-256° dec.

B, HBr: [53601-99-3]. Cryst. (MeOH/Et_2O).
Me ether: [82988-65-6]. *4,5,6,7-Tetrahydro-3-methox-
yisoxazolo[4,5-c]pyridine.*
   $C_7H_{10}N_2O_2$ M 154.168
   Cryst. (MeOH/EtOAc) (as hydrochloride). Mp 191°
   dec. (hydrochloride).

Me ether; B, MeCl: [95597-35-6]. Cryst. (MeOH/Et_2O).
   Mp 210°.
Et ether: [95597-32-3]. *3-Ethoxy-4,5,6,7-tetrahydroi-
soxazolo[4,5-c]pyridine.*
   $C_8H_{12}N_2O_2$ M 168.195
   Cryst. (MeCN). Mp 190-193° (as hydrochloride).

Krogsgaard-Larsen, P. *et al, Acta Chem. Scand., Ser. B*, 1974,
   28, 533 *(synth, ir, pmr, conformn, uv)*
Sauerberg, P. *et al, J. Med. Chem.*, 1986, **29**, 1004 *(synth, pmr)*

4,5,6,7-Tetrahydroisoxazolo[5,4-*c*]pyridin-   T-60072
3-ol

*4,5,6,7-Tetrahydroisoxazolo[5,4-c]pyridin-3(2H)-one,
9CI. Gaboxadol. THIP*
[64603-91-4]

$C_6H_8N_2O_2$ M 140.141
GABA agonist. Mp 242-244° dec. Zwitterionic.

B, HBr: [65202-63-3]. Faint-reddish cryst.
   (MeOH/Et_2O). Mp 162-163°.

Krogsgaard-Larsen, P. *et al, Acta Chem. Scand., Ser. B*, 1977,
   31, 584 *(synth, uv, ir, pmr, use)*
Merck Index, 10th Ed., 9214.

1,2,3,4-Tetrahydro-1-methylenenaphtha-   T-60073
lene, 9CI

3,4-Dihydro-1(2H)-methylenenaphthalene
[25108-63-8]

$C_{11}H_{12}$ M 144.216
Oil. Bp_{16} 110°. n_D^{20} 1.575.

Christol, H. *et al, Bull. Soc. Chim. Fr.*, 1962, 1325 *(synth, uv)*
Buchanan, G.W. *et al, Can. J. Chem.*, 1979, **57**, 3028 *(synth,
cmr)*
Campbell, M.M. *et al, Tetrahedron*, 1985, **41**, 5637 *(synth, ir,
pmr)*

1,2,3,4-Tetrahydro-3-methyl-3-isoquinolin-   T-60074
ecarboxylic acid, 9CI

$C_{11}H_{13}NO_2$ M 191.229

(R)-form [105226-66-2]
   Mp 327° dec. $[\alpha]_D^{20} +31.6°$ (c, 1 in 6*M*HCl).

(±)-form
   B, HCl: [100486-34-8]. Solid. Mp >300°.

Skiles, J.W. *et al, J. Med. Chem.*, 1986, **29**, 784 *(synth)*
Schöllkopf, U. *et al, Angew. Chem., Int. Ed. Engl.*, 1987, **26**, 143
   (synth, pmr)

Tetrahydro-2-methylpyran, 9CI　　　T-60075

Updated Entry replacing T-00940

2-Methyltetrahydropyran. δ-Hexylene oxide

[10141-72-7]

(S)-form

$C_6H_{12}O$　　M 100.160

(S)-form [102208-92-4]

$[\alpha]_D$ +11.81° (c, 8.7 in $CHCl_3$).

(±)-form

Spar. sol. cold H_2O, less sol. hot. Bp_{749} 104.5-106°.

Hanschke, E., *Chem. Ber.*, 1955, **88**, 1053 (*synth*)
Keinan, E. *et al*, *J. Am. Chem. Soc.*, 1986, **108**, 3474 (*synth, pmr, ir, ms*)

3,6,7,8-Tetrahydro-2*H*-oxocin, 9CI　　　T-60076

$C_7H_{12}O$　　M 112.171

(Z)-form [38786-97-9]

Liq.

Paquette, L.A. *et al*, *J. Am. Chem. Soc.*, 1972, **94**, 6751 (*synth, pmr*)
Overman, L.E. *et al*, *J. Am. Chem. Soc.*, 1986, **108**, 3516 (*synth*)

1,2,3,4-Tetrahydro-2-oxo-4-quinolinecarboxylic acid　　　T-60077

$C_{10}H_9NO_3$　　M 191.186

(±)-form

Cryst. Mp 218°.

Me ester:
　$C_{11}H_{11}NO_3$　　M 205.213
　Cryst. (C_6H_6 or EtOH). Mp 164°.

Et ester:
　$C_{12}H_{13}NO_3$　　M 219.240
　Cryst. Mp 146°.

Aeschlimann, J.A., *J. Chem. Soc.*, 1926, 2902 (*synth, deriv*)
Julian, P.L. *et al*, *J. Am. Chem. Soc.*, 1953, 5305 (*synth, bibl*)

1,2,3,4-Tetrahydro-4-oxo-6-quinolinecarboxylic acid　　　T-60078

[51552-75-1]

$C_{10}H_9NO_3$　　M 191.186

Solid. No Mp given.

Me ester: [51552-76-2].
　$C_{11}H_{11}NO_3$　　M 205.213
　Yellow-brown cryst. (MeOH). Mp 157-158°.

Hirsch, J.A. *et al*, *J. Org. Chem.*, 1974, **39**, 2044 (*synth, ir, pmr*)

1,2,3,4-Tetrahydro-4-oxo-7-quinolinecarboxylic acid　　　T-60079

[19384-65-7]

$C_{10}H_9NO_3$　　M 191.186

Orange-yellow cryst. (H_2O). Mp ca. 370°.

Et ester: [22048-84-6].
　$C_{12}H_{13}NO_3$　　M 219.240
　Deep-yellow solid (AcOH aq.). Mp 110-111°.

Et ester, oxime: [22048-85-7].
　$C_{12}H_{14}N_2O_3$　　M 234.254
　Light-yellow cryst. (C_6H_6). Mp 170-172°.

Bekhli, A.F. *et al*, *J. Org. Chem. USSR* (*Engl. Transl.*), 1968, **4**, 2175 (*synth, uv*)

6,7,8,9-Tetrahydro-6-phenyl-5*H*-benzocyclohepten-5-one, 9CI　　　T-60080

2-Phenyl-1-benzosuberone

[51197-89-8]

$C_{16}H_{16}O$　　M 224.302

(±)-form

Oil.

McCague, R. *et al*, *J. Med. Chem.*, 1986, **29**, 2053 (*synth, pmr*)

1,2,3,6-Tetrahydro-4-phenylpyridine, 9CI　　　T-60081

Updated Entry replacing T-20070

4-Phenyl-Δ³-piperideine

[10338-69-9]

$C_{11}H_{13}N$　　M 159.230

$Bp_{1.5}$ 100-105°.

B,HCl: [43064-12-6]. Cryst. (Me_2CO/2-propanol). Mp 200-202°.

N-Me: [28289-54-5]. *1,2,3,6-Tetrahydro-1-methyl-4-phenylpyridine. 1-Methyl-4-phenyl-1,2,3,6-tetrahydropyridine. MPTP.*
　$C_{12}H_{15}N$　　M 173.257
　Produces permanent symptoms mimicking Parkinson's disease.

▷Highly toxic, cumulative poison causing irreversible Parkinsonism by inhalation or skin contact

N-Me; B,HCl: [23007-85-4]. Cryst. (Me_2CO). Mp 241-243°.

Ziering, A. *et al*, *J. Org. Chem.*, 1947, **12**, 894.
Schmidle, C.J. *et al*, *J. Am. Chem. Soc.*, 1956, **78**, 1702 (*synth*)
Markey, S.P. *et al*, *Chem. Eng. News*, Feb. 6, 1984, 2 (*tox*)
Flippen-Anderson, J.L. *et al*, *Acta Crystallogr.*, *Sect. C*, 1986, **42**, 1184 (*cryst struct*)
Fries, D.S. *et al*, *J. Med. Chem.*, 1986, **29**, 424 (*synth, props, bibl*)
Sayre, L.M. *et al*, *J. Am. Chem. Soc.*, 1986, **108**, 2464 (*bibl*)

Tetrahydro-2-(2-piperidinyl)furan   T-60082

2-Piperidinotetrahydrofuran

[99705-97-2]

C$_9$H$_{17}$NO   M 155.239

Bp$_6$ 65-68°.

Fuchigami, T. *et al, J. Org. Chem.*, 1986, **51**, 366 (*synth, pmr*)

4,5,9,10-Tetrahydropyrene, 9CI   T-60083

Updated Entry replacing T-20071

3,4,8,9-Tetrahydropyrene (*obsol.*)

[781-17-9]

C$_{16}$H$_{14}$ M 206.287

Flakes (EtOH). Mp 138°.

Coulson, E.A., *J. Chem. Soc.*, 1937, 1298 (*synth*)
Tintel, C. *et al, Recl. Trav. Chim. Pays-Bas*, 1983, **102**, 224 (*synth, pmr*)
Yamato, T. *et al, J. Chem. Soc., Perkin Trans. 1*, 1987, 1 (*synth*)

1,2,3,6-Tetrahydro-3-pyridinecarboxylic acid, 9CI   T-60084

[86447-23-6]

C$_6$H$_9$NO$_2$ M 127.143

(±)-*form*

Cryst. + ½H$_2$O. Mp 286-289° dec.

B,HBr: Cryst. (AcOH). Mp 210°.

N-*Me:* [86447-29-2].
   C$_7$H$_{11}$NO$_2$ M 141.169
   Cryst. (AcOH) (as hydrochloride). Mp 245-250° dec. (hydrochloride).

N-tert-*Butyloxycarbonyl:* Cryst. (EtOAc). Mp 142-144°.

Allan, R.D. *et al, Aust. J. Chem.*, 1987, **36**, 601 (*synth*)

7,8,9,10-Tetrahydropyrido[1,2-*a*]azepin-4(6*H*)-one   T-60085

[101773-63-1]

C$_{10}$H$_{13}$NO   M 163.219

Thomas, E.W., *J. Org. Chem.*, 1986, **51**, 2184 (*synth, pmr, cmr, ir, ms*)

5,6,9,10-Tetrahydro-4*H*,8*H*-pyrido[3,2,1-*ij*][1,6]naphthyridine, 9CI   T-60086

4,5,6,8,9,10-Hexahydropyrido[3,4,5-ij]quinolizine. Nor-dehydro-α-matrinidine

[6052-72-8]

C$_{11}$H$_{14}$N$_2$ M 174.245

Pale-yellow solid (pet. ether). Mp 62-66° (30-31°). Bp$_3$ 145°.

B,HBr: Needles (MeOH/EtOAc). Mp 275° (272°).

Tsuda, K. *et al, J. Org. Chem.*, 1956, **21**, 1481 (*synth, ir, uv*)
Sakamoto, T. *et al, Chem. Pharm. Bull.*, 1986, **34**, 2018 (*synth, pmr*)

5-(3,4,5,6-Tetrahydro-3-pyridylidene-methyl)-2-furanmethanol   T-60087

C$_{11}$H$_{13}$NO$_2$ M 191.229

(*E*)-*form*

Prod. of reacn. between glucose and lysine in sl. acid soln. Cryst. (toluene). Mp 104-105°.

Picrate: Mp 162-165.5°.

Miller, R. *et al, Acta Chem. Scand., Ser. B*, 1984, **38**, 689; 1987, **41**, 208 (*isol, ms, pmr, synth, cryst struct*)

Tetrahydro-2(1*H*)-pyrimidinone   T-60088

Updated Entry replacing T-50121

Trimethyleneurea

[1852-17-1]

C$_4$H$_8$N$_2$O   M 100.120

Cryst. Mp 260°. Bp$_{0.005}$ 130° subl.

1-β-D-Ribofuranosyl: [19149-48-5].
   C$_9$H$_{16}$N$_2$O$_5$ M 232.236
   Inhibitor of cytidine deaminase. Isomerises rapidly in acid and slowly in aqueous soln. to pyranose form.

1-β-D-Ribopyranosyl: [104051-86-7].
   C$_9$H$_{16}$N$_2$O$_5$ M 232.236
   Solid. Mp ca. 110°.

1-Me:
   C$_5$H$_{10}$N$_2$O   M 114.147
   Mp 91-92°.

1,3-Di-Me:
   C$_6$H$_{12}$N$_2$O   M 128.174
   Bp$_{0.05}$ 60-61°.

1,3-Di-Et: [30826-85-8].
   C$_8$H$_{16}$N$_2$O   M 156.227
   Bp$_{0.1}$ 58-61°.

▷UW7580000.

Behringer, H. *et al, Justus Liebigs Ann. Chem.*, 1957, **607**, 67 (*synth*)
Michels, J.G., *J. Org. Chem.*, 1960, **25**, 2246 (*synth, bibl*)
Sonoda, N. *et al, J. Am. Chem. Soc.*, 1971, **93**, 6344 (*synth*)
Li, C. *et al, J. Med. Chem.*, 1981, **24**, 1089 (*deriv*)

Bates, H.A. *et al*, *J. Org. Chem.*, 1986, **51**, 2228 (*synth*)
Kelley, J.A. *et al*, *J. Med. Chem.*, 1986, **29**, 2351 (*deriv, synth, cmr, ms, pmr, props*)

Tetrahydro-4(1*H*)-pyrimidinone, 9CI   T-60089

1,2,5,6-Tetrahydro-4-oxopyrimidine. 1,2,5,6-Tetrahydro-4-hydroxypyrimidine

[10167-09-6]

C$_4$H$_8$N$_2$O   M 100.120
Cryst. (EtOAc or by subl.). Mp 91-92°.
3-Me:
  C$_5$H$_{10}$N$_2$O   M 114.147
  Oil. Bp$_{0.03}$ 65°.

Škarić, V. *et al*, *Croat. Chim. Acta*, 1964, **36**, 87; 1966, **38**, 1 (*synth, ir, uv, pmr*)

Tetrahydro-2-(2-pyrrolidinyl)furan   T-60090

2-Pyrrolidinotetrahydrofuran

[99706-00-0]

C$_8$H$_{15}$NO   M 141.213
Bp$_{34}$ 90-93°.

Fuchigami, T. *et al*, *J. Org. Chem.*, 1986, **51**, 366 (*synth, pmr*)

Tetrahydro-1*H*-pyrrolizin-2(3*H*)-one, 9CI   T-60091

2-Pyrrolizidinone. 2-Oxopyrrolizidine. 2-Ketopyrrolizidine. 1-Azabicyclo[3.3.0]octan-3-one

[14174-86-8]

C$_7$H$_{11}$NO   M 125.170
(±)-*form*
  Liq. Bp$_1$ 78°. Rapidly forms cryst. dimer on standing at r.t.
  B,MeI: Cryst. (MeOH/Me$_2$CO). Mp 267-268°.
  Picrate: [14408-81-2]. Cryst. (EtOH). Mp 185°.
    Unstable.
  Picronolate: Cryst. Mp 212-213°.

Clemo, G. *et al*, *J. Chem. Soc.*, 1942, 424 (*synth*)
Aaron, H.S. *et al*, *J. Org. Chem.*, 1966, **31**, 3502 (*synth*)

6,7,8,9-Tetrahydro-4*H*-quinolizin-4-one   T-60092

[50720-19-9]

C$_9$H$_{11}$NO   M 149.192
Mp 45-47°.
Picrate: Mp 104.5-105.5°.

Wenkert, E. *et al*, *J. Am. Chem. Soc.*, 1973, **95**, 8427 (*synth*)

Thomas, E.W., *J. Org. Chem.*, 1986, **51**, 2184 (*synth, pmr, cmr, ir, uv, ms*)

Tetrahydro-2*H*-1,3-thiazine, 9CI   T-60093

Penthiazolidine. 1,3-Thiazane

[543-71-5]

C$_4$H$_9$NS   M 103.182
Liq.
▷XJ0776700.
  B,HCl: [79128-34-0]. Mp 225°.
  Picrate: Mp 147°.

Takata, Y., *J. Pharm. Soc. Jpn.*, 1952, **72**, 220; *CA*, **46**, 11182f (*synth*)
Cook, M.J. *et al*, *J. Chem. Soc., Perkin Trans. 2*, 1973, 325 (*conformn*)

Tetrahydro-2*H*-1,3-thiazine-2-thione, 9CI   T-60094

2-Thiotetrahydro-1,3-thiazine

[5554-48-3]

C$_4$H$_7$NS$_2$ M 133.226
Cryst. (MeOH). Mp 134-135.5°. Exists in the thione form in aq. soln.

Thione-form
  N-Me: [5554-52-9].
    C$_5$H$_9$NS$_2$ M 147.253
    Needles (EtOH). Mp 88°.
  N-Et: [64067-72-7].
    C$_6$H$_{11}$NS$_2$ M 161.280
    Cryst. (EtOH). Mp 68°.

SH-form
  S-Me: [58842-19-6]. *5,6-Dihydro-2-(methylthio)-4H-1,3-thiazine, 9CI.*
    C$_5$H$_9$NS$_2$ M 147.253
    Bp$_{50}$ 155-160°.
  S-Me; B,HI: Yellow cryst. Mp 156-158°.
  S-Me; B,MeI: Cryst. (EtOH). Mp 132°.
  S-Et: 5,6-Dihydro-2-(ethylthio)-4H-1,3-thiazine.
    C$_6$H$_{11}$NS$_2$ M 161.280
    Bp$_{40}$ 145-150°.

Gabriel, S. *et al*, *Ber.*, 1890, **23**, 91 (*synth*)
Hamer, F.M. *et al*, *J. Chem. Soc.*, 1943, 243 (*synth, deriv*)
McKay, A.F. *et al*, *J. Am. Chem. Soc.*, 1958, **80**, 3339 (*deriv, synth*)
Felder, E. *et al*, *Helv. Chim. Acta*, 1963, **46**, 752 (*synth*)
Garraway, J.L., *J. Chem. Soc. (B)*, 1966, 92 (*tautom*)
Owen, T.C. *et al*, *J. Chem. Soc. (C)*, 1967, 1373 (*synth*)
Pushtoshkin, G.I., *J. Org. Chem. USSR (Engl. Transl.)*, 1968, **4**, 761 (*synth*)
Fahey, J.L. *et al*, *J. Chem. Soc., Perkin Trans. 1*, 1977, 1117 (*deriv*)
Obata, N., *Bull. Chem. Soc. Jpn.*, 1977, **50**, 2187 (*synth*)

Tetrahydro-2*H*-1,2-thiazin-3-one, 9CI   T-60095

1,2-Thiazan-3-one

C_4H_7NOS M 117.165
Cryst. (H$_2$O). Mp 160°.
N-Ph: Tetrahydro-2-phenyl-2H-1,2-thiazin-3-one.
$C_{10}H_{11}NOS$ M 193.263
Needles (EtOH aq.). Mp 169-170°.

Kharasch, N. *et al*, *J. Org. Chem.*, 1963, **28**, 1901 (*synth*)
Luettringhaus, A. *et al*, *Chem. Ber.*, 1965, **98**, 1005 (*synth*)

Tetrahydro-2*H*-1,3-thiazin-2-one, 9CI   T-60096

2-Penthiazolidone
[14889-64-6]

C_4H_7NOS M 117.165
Cryst. Mp 85-88°.

Grisley, D.W. *et al*, *Synthesis*, 1972, 318 (*synth*)
Sonoda, N. *et al*, *Tetrahedron Lett.*, 1975, 1969 (*synth*)

2-(Tetrahydro-2-thienyl)pyridine, 9CI   T-60097

2-(2-Pyridyl)tetrahydrothiophene
[76732-76-8]

$C_9H_{11}NS$ M 165.253
(±)-***form***
Brown oil.

Aloup, J.-C. *et al*, *J. Med. Chem.*, 1987, **30**, 24 (*synth*)

2,3,4,5-Tetrahydrothiepino[2,3-*b*]pyridine, 9CI   T-60098

Thiepano[2,3-b]pyridine
[108005-21-6]

$C_9H_{11}NS$ M 165.253
Bp$_{0.4}$ 70°.

Seitz, G. *et al*, *Tetrahedron Lett.*, 1986, **27**, 2747 (*synth, pmr*)

Tetrahydro-1,2,4-triazine-3,6-dione, 9CI   T-60099

Hexahydro-1,2,4-triazine-3,6-dione
[19279-79-9]

$C_3H_5N_3O_2$ M 115.091
Cryst. (H$_2$O). Mp 227-228°. One claimed synthesis
(1964) was erroneous.

Lindemann, A. *et al*, *J. Am. Chem. Soc.*, 1952, **74**, 476 (*synth*)
Schwan, T.J. *et al*, *J. Heterocycl. Chem.*, 1983, **20**, 547 (*synth, pmr*)

2,3,5,6-Tetrahydroxybenzoic acid   T-60100

$C_7H_6O_6$ M 186.121
Tetra-Me ether: [6172-65-2]. *2,3,5,6-Tetramethoxybenzoic acid.*
$C_{11}H_{14}O_6$ M 242.228
Mp 144°.

Schäfer, W. *et al*, *Chem. Ber.*, 1971, **104**, 3211 (*synth, deriv*)

3,3′,4,4′-Tetrahydroxybibenzyl   T-60101

4,4′-(1,2-Ethanediyl)bis[1,2-benzenediol]. 1,2-Bis(3,4-dihydroxyphenyl)ethane

$C_{14}H_{14}O_4$ M 246.262
3,4;3′,4′-Bismethylene ether: [80784-19-6]. *5,5′-(1,2-Ethanediyl)bis-1,3-benzodioxole, 9CI. 3,4;3′,4′-Dimethylenedioxybibenzyl. 3,4;3′,4′-Bis(methylenedioxy)bibenzyl.*
$C_{16}H_{14}O_4$ M 270.284
Constit. of *Frullania* spp. Cryst. Mp 132-133°.

Asakawa, Y. *et al*, *Phytochemistry*, 1987, **26**, 1117.

3,3′,4,5-Tetrahydroxybibenzyl   T-60102

5-[2-[(3-Hydroxyphenyl)ethyl]]-1,2,3-benzenetriol
$C_{14}H_{14}O_4$ M 246.262
3,4-Methylene, 3′,5-Di-Me ether: [80357-96-6]. *3′,5-Dimethoxy-3,4-methylenedioxybibenzyl.*
$C_{17}H_{18}O_4$ M 286.327
Constit. of *Frullania* spp. Cryst. Mp 135-137°.
3,5-Di-Me ether: [108853-12-9]. *4-[2-(3-Hydroxyphenyl)ethyl]-2,6-dimethoxyphenol, 9CI. 3′,4-Dihydroxy-3,5-dimethoxybibenzyl. **Aloifol I**.*
$C_{16}H_{18}O_4$ M 274.316
Constit. of *Cymbidium aloifolium*. Viscous mass.

Asakawa, Y. *et al*, *Phytochemistry*, 1987, **26**, 1117 (*isol*)
Juneja, R.K. *et al*, *Phytochemistry*, 1987, **26**, 1123 (*isol*)

3',4',5,7-Tetrahydroxyflavan   T-60103

Updated Entry replacing T-30087

2-(3,4-Dihydroxyphenyl)-3,4-dihydro-2H-1-benzo-pyran-5,7-diol, 9CI

$C_{15}H_{14}O_5$ M 274.273

(*S*)-form

3',4'-Di-Me ether: [89289-92-9]. 5,7-Dihydroxy-3',4'-di-methoxyflavan. **Diffutidin**.
$C_{17}H_{18}O_5$ M 302.326
Constit. of *Canscora diffusa*. Cryst. Mp 195-200°.
[α]$_D$ −12.4° (MeOH).

3',4'-Di-Me ether, 5-O-β-D-glucopyranoside: [89289-91-8]. **Diffutin**.
$C_{23}H_{28}O_{10}$ M 464.468
Constit. of *C. diffusa*. Shows antistress and antianxiety activity. Cryst. Mp 144-145°. [α]$_D$ −46.3° (MeOH).

O^5-Xylopyranoside: [108403-45-8]. **Viscutin 3**.
$C_{20}H_{22}O_9$ M 406.388
Constit. of *Viscum tuberculatum*. Insect growth inhibitor.

O^5-(2-O-p-Hydroxybenzoylxylopyranoside): [108403-43-6]. **Viscutin 1**.
$C_{27}H_{26}O_{11}$ M 526.496
Constit. of *V. tuberculatum*. Insect growth inhibitor.

O^5-(2-O-Caffeoylxylopyranoside): [108403-44-7]. **Viscutin 2**.
$C_{29}H_{28}O_{12}$ M 568.533
Constit. of *V. tuberculatum*. Insect growth inhibitor.

Ghosal, S. *et al*, *J. Chem. Res. (S)*, 1983, 330 (*isol*)
Kubo, I. *et al*, *Tetrahedron Lett.*, 1987, **28**, 921 (*isol*)

3,4',5,7-Tetrahydroxyflavanone   T-60104

Updated Entry replacing A-00623

3,4-Dihydro-2-(4-hydroxyphenyl)-2H-1-benzopyran-3,5,7-triol, 9CI

(*2R,3R*)-form

$C_{15}H_{14}O_5$ M 274.273

(*2R,3R*)-form [24808-04-6]

(−)-*Epiafzelechin*

Constit. of afzelia hardwood, *Larix sibirica, Actinidia chinensis, Juniperis communis* and *Cassia javanica*. Needles (EtOH aq.). Mp 240-243° dec. (312°). [α]$_D^{19}$ −58.9° (c, 3 in EtOH), −4° (c, 1 in Me$_2$CO).

Tetra-Ac: Prisms (AcOH aq.). Mp 126-127°.

4',5,7-Tri-Me ether: Prisms (MeOH). Mp 110°. [α]$_D^{20}$ −67.4° (c, 2 in EtOH).

(*2R,3S*)-form [2545-00-8]

Afzelechin

Found in *Cochlospermum gillivraei, Desmoncus polycanthos, Eucalyptus calophylla*, etc. Mp 221-222° dec. [α]$_D^{20}$ +20.6° (5% Me$_2$CO aq.).

7-O-β-D-Apioside:
$C_{20}H_{22}O_9$ M 406.388

Constit. of rhizomes of *Polypodium glycyrrhiza*. Powder (CHCl$_3$). Mp 135-137°. [α]$_D$ −93.3° (c, 0.12 in MeOH).

(*2S,3S*)-form

(+)-*Epiafzelechin*

Constit. of *Livingstona chinensis*.

King, F.E. *et al*, *J. Chem. Soc.*, 1955, 2948 (*isol, struct*)
Hillis, W.E. *et al*, *Aust. J. Chem.*, 1960, **13**, 390 (*isol, struct*)
Korver, O. *et al*, *Tetrahedron*, 1971, **27**, 5459 (*cd*)
Monache, F.D. *et al*, *Phytochemistry*, 1972, **11**, 2333 (*isol, bibl*)
Friedrich, H. *et al*, *Planta Med.*, 1978, **33**, 251 (*isol*)
Watterman, P.G. *et al*, *Planta Med.*, 1979, **37**, 178 (*isol, cmr, pmr*)
Kim, J. *et al*, *Tetrahedron Lett.*, 1987, **28**, 3655 (*apioside*)

3,5,6,7-Tetrahydroxyflavanone   T-60105

$C_{15}H_{12}O_6$ M 288.256

6,7-Methylene ether: [110204-44-9]. 3,5-Dihydroxy-6,7-methylenedioxyflavanone.
$C_{16}H_{12}O_6$ M 300.267
Constit. of sugarbeet infected with *Rhizoctonia solani*. Yellow cryst. Mp 179-180°.

Takahashi, H. *et al*, *Bull. Chem. Soc. Jpn.*, 1987, **60**, 2261.

2',3',5,7-Tetrahydroxyflavone   T-60106

Updated Entry replacing T-40068

2-(2,3-Dihydroxyphenyl)-5,7-dihydroxy-4H-1-benzopyran-4-one, 9CI

[74805-70-2]

$C_{15}H_{10}O_6$ M 286.240
Constit. of *Scutellaria bicalensis*. Cryst. Mp >360°.

Tomimori, T. *et al*, *J. Pharm. Soc., Jpn.*, 1984, **104**, 529.
Tanaka, T. *et al*, *Yakugaku Zasshi*, 1987, **107**, 315 (*synth*)

2',5,6,7-Tetrahydroxyisoflavone   T-60107

$C_{15}H_{10}O_6$ M 286.240

2',6-Di-Me ether: [94285-21-9]. 5,7-Dihydroxy-2',6-dimethoxyisoflavone.
$C_{17}H_{14}O_6$ M 314.294
Constit. of *Iris missouriensis*. Pale-yellow prisms (MeOH/CHCl$_3$). Mp 190-193°.

6,7-Methylene ether: [97359-75-6]. 2',5-Dihydroxy-6,7-methylenedioxyisoflavone. **Irisone B**.
$C_{16}H_{10}O_6$ M 298.251
Constit. of *I. missouriensis*. Plates (EtOAc). Mp 230-233°.

6,7-Methylene ether, 2'-Me ether: [2652-16-6]. 5-Hydroxy-2'-methoxy-6,7-methylenedioxyisoflavone. **Irisone A**.
$C_{17}H_{12}O_6$ M 312.278
Constit. of *I. missouriensis*. Needles (EtOAc). Mp 186-189°.

Takahashi, H. *et al*, BSCJ, 1987, **60**, 2261.
Wong, S.-M. *et al*, *J. Nat. Prod.*, 1987, **50**, 178.

3′,4′,5,7-Tetrahydroxyisoflavone   T-60108

Updated Entry replacing T-01252
Orobol. *Norsantal*
[480-23-9]

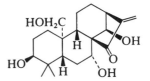

C₁₅H₁₀O₆   M 286.240

Isol. as glucoside (Oroboside) from *Orobus tuberosus* and *Baptisia lecontii* and constit. of *Bolusanthus speciosus*. Pale-yellow cryst. (AcOH). Mp 212°.

Tetra-Ac: [1061-93-4]. Cryst. (AcOH). Mp 212°.

3′-Me ether: [36190-95-1]. *4′,5,7-Trihydroxy-3′-methoxyisoflavone.*
C₁₆H₁₂O₆   M 300.267
Isol. from *Dalbergia inundata* and *Thermopsis* spp. Cryst. (MeOH). Mp 218-222°.

4′-Me ether: [2284-31-3]. *3′,5,7-Trihydroxy-4′-methoxyisoflavone.* **Pratensein**.
C₁₆H₁₂O₆   M 300.267
Constit. of *Bolusanthus speciosus* and *Trifolium pratense*. Cryst. (EtOH). Mp 272-273°.

7-Me ether: [529-60-2]. *3′,4′,5-Trihydroxy-7-methoxyisoflavone.* **Santal**.
C₁₆H₁₂O₆   M 300.267
Constit. of *Pterocarpus osun* and *P. soyauxi*. Cryst. (EtOH). Mp 222-223°.

7-Me ether, tri-Ac: Cryst. (C₆H₆). Mp 147°.

3′,4′-Di Me ether: *5,7-Dihydroxy-3′,4′-dimethoxyisoflavone. 3′-O-Methylpratensein.*
C₁₇H₁₄O₆   M 314.294
Constit. of *B. speciosus*.

3′,7-Di-Me ether: *4′,5-Dihydroxy-3′,7-dimethoxyisoflavone.*
C₁₇H₁₄O₆   M 314.294
Constit. of *P. soyauxii*.

3′,7-Di-Me ether, 4′,5-Di-Ac: Cryst. (EtOH). Mp 179°.

Charaux, H. *et al*, *Bull. Soc. Chim. Biol.*, 1939, **21**, 1330 (*isol*)
Akisanya, A. *et al*, *J. Chem. Soc.*, 1959, 2679 (*isol*)
Wong, E. *et al*, *J. Org. Chem.*, 1963, **28**, 2336 (*isol, struct, synth*)
Markham, K.R. *et al*, *Phytochemistry*, 1968, **7**, 791 (*isol*)
Dement, A.W. *et al*, *Phytochemistry*, 1972, **11**, 1089 (*isol*)
Adinarayana, D. *et al*, *Indian J. Chem.*, 1974, **12**, 911 (*synth*)
Leite de Almeida, M.E. *et al*, *Phytochemistry*, 1974, **13**, 751 (*isol*)
Asnes, K. *et al*, *Z. Naturforsch., C*, 1985, **40**, 617 (*isol*)
Bezuidenhoudt, B.C.B. *et al*, *Phytochemistry*, 1987, **26**, 531 (*isol*)

ent-1β,3α,7α,11α-Tetrahydroxy-16-kaurene-6,15-dione   T-60109

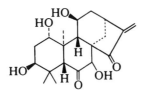

C₂₀H₂₈O₆   M 364.438

1,7,11-Tri-Ac: ent-*1β,7α,11α-Triacetoxy-3α-hydroxy-16-kaurene-6,15-dione.*
C₂₆H₃₄O₉   M 490.549
Constit. of *Rabdosia adenantha*. Cryst. Mp 255°. [α]₀¹³ −76° (c, 0.25 in CHCl₃).

Xu, Y. *et al*, *Tetrahedron Lett.*, 1987, **28**, 499.

ent-3α,7β,14α,20-Tetrahydroxy-16-kauren-15-one   T-60110

Coestinol

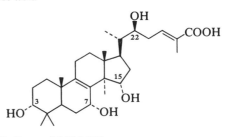

C₂₀H₃₀O₅   M 350.454
Constit. of *Plectranthus coesta*. Cryst. (MeOH). Mp 246-248°. [α]₀ −140.8° (c, 0.064 in MeOH).

Phadnis, A.P. *et al*, *Indian J. Chem., Sect. B*, 1987, **26**, 15.

3,7,15,22-Tetrahydroxy-8,24-lanostadien-26-oic acid   T-60111

C₃₀H₄₈O₆   M 504.706

(3α,5α,7α,15α,22S,24E)-form

3,15,22-Tri-Ac: *3α,15α,22S-Triacetoxy-7α-hydroxy-8,24E-lanostadien-26-oic acid. Ganoderic acid* O.
C₃₆H₅₄O₉   M 630.817
Constit. of cultured mycelium of *Ganoderma lucidum*. Needles. Mp 156-158°. Not the same as Granoderic acid *O* in 7,20-Dihydroxy-3,11,15,23-tetraoxo-8-lanosten-26-oic acid, D-60379 .

3,15,22-Tri-Ac, 7-Me ether: *3α,15α,22S-Triacetoxy-7α-methoxy-8,24E-lanostadien-26-oic acid. O-Methylganoderic acid* O.
C₃₇H₅₆O₉   M 644.844
Constit. of cultured mycelium of *G. lucidum*. Cryst. Mp 228-229.5°.

(3α,5α,7α,15α,22ξ,24E)-form

3,15,22-Tri-Ac: *3α,15α,22-Triacetoxy-7α-hydroxy-8,24E-lanostadien-26-oic acid. Ganoderic acid* Mb.
C₃₆H₅₄O₉   M 630.817
Metab. of *Ganoderma lucidum*. Syrup. [α]₀²⁷ −4.0° (c, 0.2 in MeOH).

3,7,22-Tri-Ac: *3α,7α,22-Triacetoxy-15α-hydroxy-8,24E-lanostadien-26-oic acid. Ganoderic acid* Mc.
C₃₆H₅₄O₉   M 630.817
Metab. of *G. lucidum*. Syrup. [α]₀²⁷ −23° (c, 0.2 in MeOH).

3,22-Di-Ac, 7-Me ether: *3α,22-Diacetoxy-15α-hydroxy-7α-methoxy-8,24E-lanostadien-26-oic acid. Ganoderic acid* Mg.
C₃₅H₅₄O₈   M 602.807
Metab. *G. lucidum*. Amorph. powder. Mp 126-129°. [α]₀²³ −23° (c, 0.2 in MeOH).

3,22-Di-Ac: *3α,22-Diacetoxy-7α,15α-dihydroxy-8,24E-lanostadien-26-oic acid. Ganoderic acid* Mh.
C₃₄H₅₂O₈   M 588.780
Syrup. [α]₀²³ +2° (c, 0.2 in MeOH).

Hirotani, H. et al, Phytochemistry, 1987, 26, 2797.
Nishitoba, T. et al, Agric. Biol. Chem., 1987, 51, 619, 1149.

1,3,4,5-Tetrahydroxy-2-methylanthraquin-one   T-60112

2-Hydroxyislandicin

[89701-80-4]

$C_{15}H_{10}O_6$ M 286.240

Constit. of root bark of *Ventilago calyculata*. Red cryst. (MeOH). Mp 233°.

4-Me ether: [93446-15-2]. *1,3,5-Trihydroxy-4-methoxy-2-methylanthraquinone.*
$C_{16}H_{12}O_6$ M 300.267
Constit. of root bark of *V. calyculata*. Orange-red needles. Mp 210°.

Rao, B.K. et al, Phytochemistry, 1983, 22, 2583; 1984, 23, 2104 (isol, uv, ir, pmr, ms, struct)

1,3,5,8-Tetrahydroxy-2-methylanthraquin-one   T-60113

1,3,5,8-Tetrahydroxy-2-methyl-9,10-anthracenedione, 9CI

[108027-02-7]

$C_{15}H_{10}O_6$ M 286.240

5-Me ether: [102786-92-5]. *1,3,8-Trihydroxy-5-methoxy-2-methylanthraquinone.*
$C_{16}H_{12}O_6$ M 300.267
Light-brown needles (Et$_2$O). Mp 130-132°.
5-Me ether, 8-O-α-L-rhamnopyranoside: [102786-89-0].
$C_{22}H_{22}O_{10}$ M 446.410
Isol. from *Acacia leucophloea*.

Saxena, M. et al, J. Nat. Prod., 1986, 49, 205.

1,3,5,8-Tetrahydroxy-2-(3-methyl-2-butenyl)xanthone   T-60114

$C_{18}H_{16}O_6$ M 328.321

3-Me ether: [110187-11-6]. *1,5,8-Trihydroxy-3-methyl-2-(3-methyl-2-butenyl)xanthone.*
$C_{19}H_{18}O_6$ M 342.348
Constit. of *Garcinia mangostana*. Cryst. Mp 193°.

Perveen, M. et al, Chem. Ind. (London), 1987, 418

1,2,5,6-Tetrahydroxyphenanthrene   T-60115

1,2,5,6-Phenanthrenetetrol

$C_{14}H_{10}O_4$ M 242.231

5,6-Di-Me ether, 1,2-Di-Ac: 1,2-Diacetoxy-5,6-dimethoxyphenanthrene.
$C_{20}H_{18}O_6$ M 354.359
Acetolysis prod. of Metaphanine, M-00437 . Pale-yellow cryst. (EtOH). Mp 150°.
Tetra-Me ether: 1,2,5,6-Tetramethoxyphenanthrene.
$C_{18}H_{18}O_4$ M 298.338
Needles (Et$_2$O). Mp 64-66°.

Kondo, H. et al, CA, 1953, 47, 5951a (synth, deriv)
Tomita, M. et al, Chem. Pharm. Bull., 1965, 13, 695 (deriv, pmr)

1,2,5,7-Tetrahydroxyphenanthrene   T-60116

Updated Entry replacing T-20096
1,2,5,7-Phenanthrenetetrol

$C_{14}H_{10}O_4$ M 242.231

1,5-Di-Me ether: [86630-47-9]. *1,5-Dimethoxy-2,7-phenanthrenediol. 2,7-Dihydroxy-1,5-dimethoxyphenanthrene.*
$C_{16}H_{14}O_4$ M 270.284
Constit. of *Oncidium cebolleta*.
Tetra-Me ether: [96754-01-7]. *1,2,5,7-Tetramethoxyphenanthrene.*
$C_{18}H_{18}O_4$ M 298.338
Cryst. (EtOAc/hexane). Mp 134°.
9,10-Dihydro, 2,7-Di-Me ether: [87402-72-0]. *9,10-Di-hydro-2,7-dimethoxy-1,5-phenanthrenediol, 9CI. 9,10-Dihydro-1,5-dihydroxy-2,7-dimethoxyphenanthrene.* **Eulophiol**.
$C_{16}H_{16}O_4$ M 272.300
Constit. of *Eulophia nuda* tubers. Cryst. (CHCl$_3$). Mp 202-203°.

Bhandari, S.R. et al, Phytochemistry, 1983, 22, 747 (isol)
Stermitz, F.R. et al, J. Nat. Prod., 1983, 46, 417 (isol)
Bhandari, S.R. et al, Indian J. Chem., Sect. B, 1985, 24, 204 (synth, deriv, uv, pmr)

1,2,6,7-Tetrahydroxyphenanthrene   T-60117

1,2,6,7-Phenanthrenetetrol

$C_{14}H_{10}O_4$ M 242.231

Tetra-Me ether: 1,2,6,7-Tetramethoxyphenanthrene.
$C_{18}H_{18}O_4$ M 298.338
Cryst. Mp 215-220°.

Kondo, H. et al, CA, 1953, 47, 5951a.

1,3,5,6-Tetrahydroxyphenanthrene   T-60118

1,3,5,6-Phenanthrenetetrol

$C_{14}H_{10}O_4$ M 242.231

Tetra-Me ether: 1,3,5,6-Tetramethoxyphenanthrene.
$C_{18}H_{18}O_4$ M 298.338
Cryst. Mp 107-108°.

Kondo, H. et al, J. Pharm. Soc. Jpn., 1952, 72, 834; CA, 1955, 49, 1075e.

1,3,6,7-Tetrahydroxyphenanthrene   T-60119

1,3,6,7-Phenanthrenetetrol

$C_{14}H_{10}O_4$ M 242.231

Tetra-Me ether: 1,3,6,7-Tetramethoxyphenanthrene.
$C_{18}H_{18}O_4$ M 298.338
Cryst. Mp 165-166°.

Kondo, H. et al, CA, 1955, 49, 1076b.

2,3,5,7-Tetrahydroxyphenanthrene   T-60120

Updated Entry replacing T-01301
2,3,5,7-Phenanthrenetetrol

$C_{14}H_{10}O_4$ M 242.231

5,7-Di-Me ether: [42050-16-8]. *5,7-Dimethoxy-2,3-phenanthrenediol, 9CI. 2,3-Dihydroxy-5,7-dimethoxyphenanthrene.*
$C_{16}H_{14}O_4$ M 270.284
Constit. of *Combretum hereroense*. Needles (MeOH/CHCl$_3$). Mp 213-214°.
3,5,7-Tri-Me ether: [39499-84-8]. *3,5,7-Trimethoxy-2-phenanthrenol, 9CI. 2-Hydroxy-3,5,7-trimethoxyphenanthrene.*
$C_{17}H_{16}O_4$ M 284.311

Isol. from *C. psidioides* and *Tamus communis*. Plates (CHCl₃/pet. ether). Mp 177-179°.
Tetra-Me ether: [22318-84-9]. *2,3,5,7-Tetramethoxyphenanthrene.*
$C_{18}H_{18}O_4$ M 298.338
Plates (MeOH). Mp 140-142°.
9,10-Dihydro, 3,5-di-Me ether: 9,10-Dihydro-2,7-dihydroxy-3,5-dimethoxyphenanthrene. 6-Methoxycoelonin.
$C_{16}H_{17}O_4$ M 273.308
Constit. of *Cymbidium aloifolium*. Cryst. (Me₂CO/hexane). Mp 118°.

Letcher, R.M. *et al*, *J. Chem. Soc.* (*C*), 1971, 3070 (*synth, deriv, uv, pmr*)
Letcher, R.M. *et al*, *J. Chem. Soc., Perkin Trans. 1*, 1972, 2941; 1973, 1179 (*isol, struct*))
Coxon, D.T. *et al*, *Phytochemistry*, 1982, **21**, 1389 (*deriv, pmr*)
Juneja, R.K. *et al*, *Phytochemistry*, 1987, **26**, 1123 (*isol*)

2,3,6,7-Tetrahydroxyphenanthrene   T-60121

2,3,6,7-Phenanthrenetetrol
[17425-64-8]
$C_{14}H_{10}O_4$ M 242.231
Cryst. Mp 262-266°.
Tetra-Me ether: [30269-34-2]. *2,3,6,7-Tetramethoxyphenanthrene.*
$C_{18}H_{18}O_4$ M 298.338
Cryst. (toluene/hexane). Mp 180-181°.

Horner, L. *et al*, *Chem. Ber.*, 1967, **100**, 2842 (*synth*)
Taylor, E.C. *et al*, *J. Am. Chem. Soc.*, 1980, **102**, 6513 (*synth, deriv*)
Nordlander, J.E. *et al*, *J. Org. Chem.*, 1987, **52**, 1627 (*synth, deriv, pmr, cmr*)

2,4,5,6-Tetrahydroxyphenanthrene   T-60122

2,4,5,6-Phenanthrenetetrol
[68570-34-3]
$C_{14}H_{10}O_4$ M 242.231
Isol. from rhizomes of *Dioscorea bulbifera*. Mp 220°.
Tetra-Ac: [68570-35-4].
$C_{22}H_{18}O_8$ M 410.379
Cryst. Mp 160-161°.
Tetra-Me ether: 2,4,5,6-Tetramethoxyphenanthrene.
$C_{18}H_{18}O_4$ M 298.338
Cryst. Mp 92-93°.

Kondo, H. *et al*, *CA*, 1955, **49**, 1076a (*synth, deriv*)
Wij, M. *et al*, *Indian J. Chem., Sect. B*, 1978, **16**, 643 (*isol, deriv, uv, pmr*)

3,4,5,6-Tetrahydroxyphenanthrene   T-60123

3,4,5,6-Phenanthrenetetrol
$C_{14}H_{10}O_4$ M 242.231
Tetra-Me ether: 3,4,5,6-Tetramethoxyphenanthrene.
$C_{18}H_{18}O_4$ M 298.338
Prisms (MeOH). Mp 114-115.5°.

Comin, J. *et al*, *J. Org. Chem.*, 1954, **19**, 1774.

2,3,12,16-Tetrahydroxy-4,7-pregnadien-20-one   T-60124

$C_{21}H_{30}O_5$ M 362.465

(2α,3β,12β,16α)-form
16-Ac: 16α-Acetoxy-2α,3β,12β-trihydroxy-4,7-pregnadien-20-one.
$C_{23}H_{32}O_6$ M 404.502
Constit. of *Stizophyllum riparium*. Amorph. Mp 90-92°. $[\alpha]_D^{25}$ +1.6° (c, 0.25 in CHCl₃).

Duh, C.-Y. *et al*, *J. Nat. Prod.*, 1987, **50**, 63.

3,4',5,7-Tetrahydroxy-8-prenylflavan   T-60125

3,4-Dihydro-3,5,7-trihydroxy-2-(4-hydroxyphenyl)-8-(3-methyl-2-butenyl)-2H-benzopyran

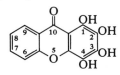

$C_{20}H_{22}O_5$ M 342.391
(2R,3S)-form
5-Me ether: 8-(3,3-Dimethylallyl)-5-methoxy-3,4',7-trihydroxyflavan.
$C_{21}H_{24}O_5$ M 356.418
Constit. of *Marshallia tenuifolia*. Gum. $[\alpha]_D^{20}$ −14° (c, 0.036 in CHCl₃).

Herz, W. *et al*, *Phytochemistry*, 1987, **26**, 1175.

1,2,3,4-Tetrahydroxyxanthone   T-60126

1,2,3,4-Tetrahydroxy-9H-xanthen-9-one, 9CI

$C_{13}H_8O_6$ M 260.203
1,2,4-Tri-Me ether: 3-Hydroxy-1,2,4-trimethoxyxanthone.
$C_{16}H_{14}O_6$ M 302.283
Constit. of *Psorospermum febrifugum*. Yellow needles (McOH). Mp 166-167°.
Tetra-Me ether: 1,2,3,4-Tetramethoxyxanthone.
$C_{17}H_{16}O_6$ M 316.310
Needles. Mp 82-84°.

Habib, A.M. *et al*, *J. Nat. Prod.*, 1987, **50**, 141.

1,3,6,8-Tetrahydroxyxanthone   T-60127

Updated Entry replacing T-01322
1,3,6,8-Tetrahydroxy-9H-xanthen-9-one, 9CI
[39731-51-6]
$C_{13}H_8O_6$ M 260.203
Cryst. (Py aq.). Mp 350° dec.
3,6-Di-Me ether: [20355-59-3]. *1,8-Dihydroxy-3,6-dimethoxyxanthone.*
$C_{15}H_{12}O_6$ M 288.256
Constit. of a *Diploschistes* sp. Pale-yellow cryst. (MeOH). Mp 193-194°.

Scott, A.I. *et al*, *Tetrahedron*, 1971, **27**, 3051 (*synth*)
Sundholm, E.G. *et al*, *Acta Chem. Scand., Ser. B*, 1978, **32**, 177 (*synth, cmr*)
Elix, J.A. *et al*, *Aust. J. Chem.*, 1987, **40**, 1031 (*isol, synth*)

2,3,4,7-Tetrahydroxyxanthone   T-60128

$C_{13}H_8O_6$ M 260.203

2,3,4-Tri-Me ether: 7-Hydroxy-2,3,4-trimethoxyxanth-one. 2-Hydroxy-5,6,7-trimethoxyxanthone.
$C_{16}H_{14}O_6$ M 302.283
Constit. of *Hypericum ericoides.* Yellow cryst. Mp 205-206°.

Cordona, M.L. *et al, J. Nat. Prod.,* 1982, **45,** 134 (*isol, struct*)
Gil, S. *et al, J. Nat. Prod.,* 1987, **50,** 301 (*synth*)

2,2,4,4-Tetrakis(trifluoromethyl)-1,3-dith-ietane, 9CI   T-60129

[791-50-4]

$C_6F_{12}S_2$ M 364.167
Dimer of 1,1,1,3,3,3-Hexafluoro-2-propanethione, H-40049 . Precursor to 1,1,1,3,3,3-Hexafluoro-2-propan-one, H-00631 and 1,1,1,3,3,3-Hexafluoro-2-propaneth-ione, H-40049 . Mp 24°. Bp 108-109°.
▷Toxic. JO4940000.
1-Oxide: [96025-84-2].
    $C_6F_{12}OS_2$ M 380.166
    Precursor to Bis(trifluoromethyl)sulfine, B-30194 .
1,1-Dioxide: [795-31-3].
    $C_6F_{12}O_2S_2$ M 396.166
    Mp 35°.
1,3-Dioxide: [87108-78-9].
    $C_6F_{12}O_2S_2$ M 396.166
    Precursor to Bis(trifluoromethyl)sulfine, B-30194 .
1,1,3-Trioxide: [96025-87-5].
    $C_6F_{12}O_3S_2$ M 412.165
    Mp 50°.
1,1,3,3-Tetraoxide: [96225-88-6].
    $C_6F_{12}O_4S_2$ M 428.164
    Mp 79°.

Martin, K.V., *J. Chem. Soc.,* 1964, 2944 (*synth*)
Dyatkin, B.L. *et al, Tetrahedron,* 1973, **29,** 2759 (*synth*)
Elsaesser, A. *et al, Chem. Ber.,* 1985, **118,** 116 (*oxides*)
Vanderluy, M. *et al, Org. Synth.,* 1985, **63,** 154 (*synth, bibl, F nmr, use*)

2,3,9,10-Tetrakis(trimethylsilyl)[5]-phenylene   T-60130

[111409-81-5]

$C_{42}H_{46}Si_4$ M 663.167
First example of a deriv. of this, the longest 'phenylene' system so far obtained. Deep-red cryst. Highly air-sensitive. Dec. on heating.

Blanco, L. *et al, Angew. Chem., Int. Ed. Engl.,* 1987, **26,** 1246 (*synth, uv, pmr*)

1,3,5,7-Tetramethyl-2,4-adamantanedione   T-60131

1,3,5,7-Tetramethyltricyclo[3.3.1.1^{3,7}]decane-2,4-dione
[52719-88-7]

$C_{14}H_{20}O_2$ M 220.311
Cryst. (pentane). Mp 138-139.5°.

Lenoir, D. *et al, J. Am. Chem. Soc.,* 1974, **96,** 2157 (*synth, ir, pmr*)

1,3,5,7-Tetramethyladamantanone   T-60132

1,3,5,7-Tetramethyltricyclo[3.3.1.1^{3,7}]decan-2-one
[52719-86-5]

$C_{14}H_{22}O$ M 206.327
Cryst. (pentane). Mp 109-110°.

Lenoir, D. *et al, J. Am. Chem. Soc.,* 1974, **96,** 2157 (*synth, ir, pmr*)
Raber, D.J. *et al, Tetrahedron,* 1986, **42,** 4347 (*pmr*)

2,2,6,6-Tetramethylcyclohexanone, 9CI, 8CI   T-60133

Updated Entry replacing T-01472
[1195-93-3]

$C_{10}H_{18}O$ M 154.252
Liq. Mp 11.2°. Bp 185°, Bp$_3$ 50-53°.
Oxime: [7007-40-1].
    $C_{10}H_{19}NO$ M 169.266
    Mp 151.5°.

Bory, S. *et al, Bull. Soc. Chim. Fr.,* 1965, 2541 (*synth*)
Ashby, E. *et al, J. Org. Chem.,* 1980, **45,** 1028 (*ir, pmr, ms*)
Lissel, M. *et al, Justus Liebigs Ann. Chem.,* 1987, 263 (*synth, ir, pmr, bibl*)

3,3,6,6-Tetramethyl-4-cyclohexene-1,2-dione   T-60134

[108586-95-4]

$C_{10}H_{14}O_2$ M 166.219

Mp 83.5-84.5°. Bp$_{0.02}$ 60° subl.

Schaltegger, A. *et al*, *Helv. Chim. Acta*, 1986, **69**, 1666 (*synth*, *ir*, *pmr*, *ms*)

5,5,6,6-Tetramethyl-2-cyclohexene-1,4-dione, 9CI   T-60135

[111192-78-0]

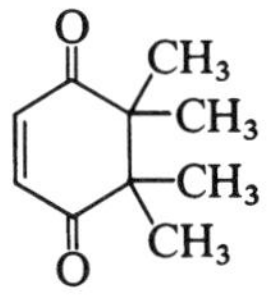

C$_{10}$H$_{14}$O$_2$ M 166.219
Mp 126-126.5°.

de Meijere, A. *et al*, *Tetrahedron*, 1986, **42**, 6487 (*synth*, *uv*, *ir*, *pmr*)

2,2,5,5-Tetramethyl-3-cyclohexen-1-one   T-60136

[13855-90-8]

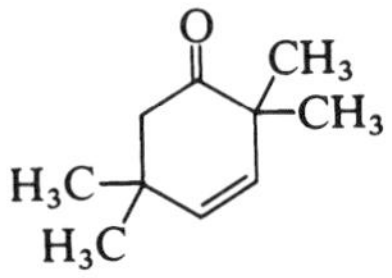

C$_{10}$H$_{16}$O   M 152.236
Bp$_{21}$ 73°. n_D^{21} 1.4489.

Gaoni, Y. *et al*, *J. Org. Chem.*, 1966, **31**, 3809 (*synth*, *uv*, *pmr*)

2,2,5,5-Tetramethylcyclopentaneselone, 9CI   T-60137

[87842-36-2]

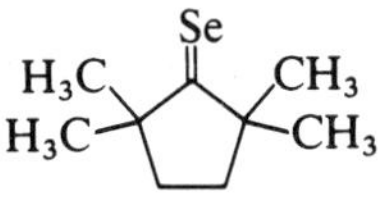

C$_9$H$_{16}$Se   M 203.185
Blue cryst. Mp 74-77°. Bp$_{0.5}$ 60°.

Guziec, F.S. *et al*, *J. Org. Chem.*, 1984, **49**, 189 (*synth*, *pmr*, *cmr*, *ir*, *ms*)
Guziec, F.S. *et al*, *J. Chem. Soc., Perkin Trans. 1*, 1985, 107 (*synth*, *ir*, *pmr*)

2,2,5,5-Tetramethylcyclopentanone, 9CI   T-60138

[4541-35-9]

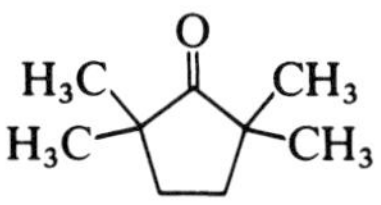

C$_9$H$_{16}$O   M 140.225
Bp 150-157°.

Lissel, M. *et al*, *Justus Liebigs Ann. Chem.*, 1987, 263 (*synth*)

2,2,5,5-Tetramethyl-3-cyclopentene-1-thione, 9CI   T-60139

[81396-39-6]

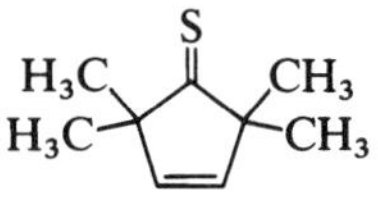

C$_9$H$_{14}$S   M 154.270
Mp 48-51°.

Krebs, A. *et al*, *Tetrahedron*, 1986, **42**, 1693 (*synth*, *ir*, *pmr*, *cmr*, *ms*)

2,2,5,5-Tetramethyl-3-cyclopenten-1-one, 9CI   T-60140

Updated Entry replacing T-20112
[81396-36-3]

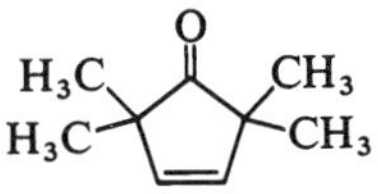

C$_9$H$_{14}$O   M 138.209
Solid by subl. Mp 44°.

Hydrazone:
C$_9$H$_{16}$N$_2$ M 152.239
Cryst. (hexane). Mp 72-74°.
2,4-Dinitrophenylhydrazone: [102521-12-0]. Large red needles (EtOH). Mp 131°.

Cullen, E.R. *et al*, *J. Org. Chem.*, 1982, **47**, 3563 (*synth*, *props*)
Adam, W. *et al*, *J. Am. Chem. Soc.*, 1986, **108**, 4556 (*synth*, *ir*, *pmr*, *cmr*,, *ms*)
Krebs, A. *et al*, *Tetrahedron*, 1986, **42**, 1693 (*synth*, *deriv*, *ir*, *pmr*, *cmr*, *ms*)

2,2,5,5-Tetramethyl-3-hexenedioic acid   T-60141

[108586-93-2]

$$HOOCC(CH_3)_2CH{=}CHC(CH_3)_2COOH$$

C$_{10}$H$_{16}$O$_4$ M 200.234
(E)-form

Cryst. (hexane). Mp 138-139°. A 1903 reference to this compd. was 3-methyl-2-butenoic acid.

Schaltegger, A. *et al*, *Helv. Chim. Acta*, 1986, **69**, 1666 (*synth*, *ir*, *pmr*, *cmr*, *ms*)

1,1,3,3-Tetramethyl-2-indaneselone   T-60142

1,3-Dihydro-1,1,3,3-2H-indene-2-selone, 9CI
[74768-64-2]

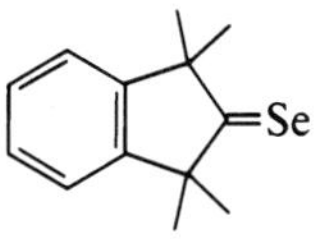

C$_{13}$H$_{16}$Se   M 251.229
Dark-blue cryst. Mp 40-43°.

Klages, C.P. *et al*, *Chem. Ber.*, 1980, **113**, 2255 (*synth*)
Guziec, F.S. *et al*, *J. Chem. Soc., Perkin Trans. 1*, 1985, 107 (*synth*, *ir*, *pmr*)

1,1,3,3-Tetramethyl-2-indanone   T-60143

*1,3-Dihydro-1,1,3,3-tetramethyl-2*H*-inden-2-one, 9CI*
[5689-12-3]

$C_{13}H_{16}O$ M 188.269
Cryst. by subl. Mp 75.0-76.5°.
Hydrazone: [74768-84-6].
  $C_{13}H_{18}N_2$ M 202.299
  Needles (hexane). Mp 106-106.5°.

Starr, J.E. *et al, J. Org. Chem.*, 1966, **31**, 1393 (*synth, uv, ir, pmr*)
Guziec, F.S. *et al, J. Chem. Soc., Perkin Trans. 1*, 1985, 107 (*deriv, synth, ir, pmr*)

1,1,3,3-Tetramethyl-2-methylenecyclohex-ane, 9CI   T-60144

[29779-78-0]

$C_{11}H_{20}$ M 152.279
Liq. Bp$_{92}$ 112°.

Fitjer, L. *et al, Chem. Ber.*, 1986, **119**, 1144 (*synth, ir, pmr, cmr*)

2′,3′,4′,5′-Tetramethyl-6′-nitroacetophen-one   T-60145

1-(2,3,4,5-Tetramethyl-6-nitrophenyl)ethanone. 1-Ace-tyl-2,3,4,5-tetramethyl-6-nitrobenzene
[103224-59-5]

$C_{12}H_{15}NO_3$ M 221.255
Mp 90-92°.

Keumi, T. *et al, J. Org. Chem.*, 1986, **51**, 3439 (*synth, pmr, ir*)

2′,3′,4′,6′-Tetramethyl-5′-nitroacetophen-one   T-60146

1-(2,3,4,6-Tetramethyl-5-nitrophenyl)ethanone. 1-Ace-tyl-2,3,4,6-tetramethyl-5-nitrobenzene
[78740-45-1]
$C_{12}H_{15}NO_3$ M 221.255
Mp 111-112°.

Keumi, T. *et al, J. Org. Chem.*, 1986, **51**, 3439 (*synth, pmr, ir*)

2′,3′,5′,6′-Tetramethyl-4′-nitroacetophen-one   T-60147

1-(2,3,5,6-Tetramethyl-4-nitrophenyl)ethanone. 1-Ace-tyl-2,3,5,6-tetramethyl-4-nitrobenzene
[64853-55-0]
$C_{12}H_{15}NO_3$ M 221.255
Mp 158-160°.

Keumi, T. *et al, J. Org. Chem.*, 1986, **51**, 3439 (*synth, pmr, ir*)

2,2,8,8-Tetramethyl-3,4,5,6,7-nonanepen-tone   T-60148

*Di-*tert-*butyl pentaketone*
[104779-79-5]

$$(H_3C)_3CCOCOCOCOCOC(CH_3)_3$$

$C_{13}H_{18}O_5$ M 254.282
First known aliphatic vicinal pentaketone. Deep-coloured.
λ_{max} 352sh (ϵ196), 436 (101), 559 (96) nm.
Covalent hydrate: [104779-77-3]. *2,2,8,8-Tetramethyl-5,5-dihydroxy-3,4,6,7-nonanetetrone.*
  $C_{13}H_{20}O_6$ M 272.297
  Yellow. Mp 64-65°.

Gleiter, R. *et al, Angew. Chem., Int. Ed. Engl.*, 1986, **25**, 999 (*synth, ir, cmr, uv*)

3,3,4,4-Tetramethyl-1,2-oxathietane, 9CI   T-60149

[102505-81-7]

$C_6H_{12}OS$ M 132.220
Pale-yellow oil.

Lown, J.W. *et al, J. Am. Chem. Soc.*, 1986, **108**, 3811 (*synth, pmr, ir, uv, ms*)

3,7,11,15-Tetramethyl-13-oxo-2,6,10,14-hexadecatetraenal   T-60150

(2E,6E,10E)-form

$C_{20}H_{30}O_2$ M 302.456
(2E,6E,10E)-form
  Eleganonal
  Constit. of *Cystoseira balearica*. Oil.
(2Z,6E,10E)-form
  Isoeleganonal
  From *C. balearica*. Oil.

Amico, V. *et al, Phytochemistry*, 1987, **26**, 2637.

1,3,6,8-Tetramethylphenanthrene, 9CI   T-60151

Updated Entry replacing T-01542
  [18499-99-5]
  $C_{18}H_{18}$ M 234.340
  Mp 164-165° (152.5-153.5°).
Monopicrate: [18500-00-0]. Mp 195-196°.

Canonne, P. *et al, Tetrahedron*, 1969, **25**, 2349.
Ho, T.I. *et al, Synthesis*, 1987, 795 (*synth, ir, pmr*)

2,4,5,7-Tetramethylphenanthrene, 9CI   T-60152

Updated Entry replacing T-01545
  [7396-38-5]

$C_{18}H_{18}$ M 234.340
Cryst. (hexane). Mp 110-113°.

1,3,5-Trinitrobenzene complex: Gold needles (EtOH).
  Mp 146.5-148°.

Wittig, G. *et al*, *Chem. Ber.*, 1953, **86**, 629.
Newman, M.S. *et al*, *J. Am. Chem. Soc.*, 1965, **87**, 5554.
Levi, E.J. *et al*, *J. Org. Chem.*, 1966, **31**, 4302 (*synth*)
Dougherty, R.C. *et al*, *Org. Mass Spectrom.*, 1977, **5**, 1321 (*ms*)
Ho, T.I. *et al*, *Synthesis*, 1987, 795 (*synth, ir, pmr*)

3,3,5,5-Tetramethyl-4-pyrazolidinone, 9CI   T-60153

[55790-79-9]

$C_7H_{14}N_2O$ M 142.200
Cryst. by subl. Mp 117-117.5°.

Crawford, R.J. *et al*, *Can. J. Chem.*, 1974, **52**, 4033 (*synth*)
Bushby, R.J. *et al*, *J. Chem. Soc., Perkin Trans. 1*, 1979, 2401
  (*synth, pmr, cmr*)

2,2,4,4-Tetramethyl-3-thietanone, 9CI   T-60154

[58721-01-0]

$C_7H_{12}OS$ M 144.231
Large needles (MeOH). Mp 102-107° (106-108°). Subl.
  >90°.

1,1-Dioxide: [58721-02-1].
  $C_7H_{12}O_3S$ M 176.230
  Cryst. (pet. ether). Mp 167.5-170°.

Claeson, G. *et al*, *Ark. Kemi*, 1964, **21**, 295 (*synth, ir, uv*)
Bushby, R.J., *J. Chem. Soc., Perkin Trans. 1*, 1975, 2513
  (*synth, ir*)
Furuhata, T. *et al*, *Tetrahedron*, 1986, **42**, 5301 (*synth*)

3,3,4,4-Tetramethyl-2-thietanone   T-60155

[42906-89-8]
$C_7H_{12}OS$ M 144.231
Cryst. Mp 66-68.5°.

Gotthardt, H., *Chem. Ber.*, 1974, **107**, 2544 (*synth, ir, pmr, ms*)

3,4,7,11-Tetramethyl-6,10-tridecadienal   T-60156

Updated Entry replacing T-20136
Faranal
[65395-77-9]

(*3S,4R*)-form

$C_{17}H_{30}O$ M 250.423
(*3S,4R,6E,10Z*)-*form*
  Trail pheromone of *Monomorium pharaonis*. Oil. $[\alpha]_D^{23}$
  +16.2° (c, 0.5 in hexane).

(*3R,4S,6E,10Z*)-*form*
  Oil. $[\alpha]_D^{23}$ −16.4° (c, 0.22 in hexane).

Ritter, F.J. *et al*, *Tetrahedron Lett.*, 1977, 2617 (*isol*)
Baker, R. *et al*, *J. Chem. Soc., Perkin Trans. 1*, 1983, 1387
  (*synth*)
Knight, D.W. *et al*, *J. Chem. Soc., Perkin Trans. 1*, 1983, 955
  (*synth, bibl*)
Poppe, L. *et al*, *Tetrahedron Lett.*, 1986, **27**, 5769 (*synth*)

2,4,6,8-Tetramethylundecanoic acid, 9CI   T-60157

Updated Entry replacing T-01600

$C_{15}H_{30}O_2$ M 242.401
(*2R,4R,6R,8R*)-*form* [10553-03-4]
  A main constit. of the preen gland wax of the common
    goose (*Anser anser*); also present in small amounts in
    the preen gland wax of other species. $[\alpha]_D^{23}$ −15.8°.
  Me ester: [2490-55-3]. Present in preen gland waxes of
    ruddy shelduck (*Tadorna ferruginea*) and common
    shelduck (*T. tadorna*). $[\alpha]_D^{21}$ −24° (c, 0.253 in CH$_3$).
    n_D^{22} 1.4317.

Murray, K.E., *Aust. J. Chem.*, 1962, **15**, 510 (*isol*)
Noble, R.E. *et al*, *Nature* (*London*), 1963, **199**, 600 (*biosynth*)
Odham, G., *Ark. Kemi*, 1964, **21**, 379; 1965, **23**, 431; 1966, **25**,
  543; 1967, **27**, 295 (*isol*)
Smith, C.R., *Top. Lipid Chem.*, 1970, **1**, 307 (*rev, bibl*)
Mori, K. *et al*, *Justus Liebigs Ann. Chem.*, 1987, 555 (*synth, abs
  config, ir, pmr, ms*)

Tetraneurin *D*   T-60158

Updated Entry replacing T-01612
[28587-47-5]

$C_{17}H_{24}O_6$ M 324.373
Constit. of *Parthenium lozanianum* and *P. fruticosum*.
  Cryst. (Me$_2$CO). Mp 203.5-205.5°. $[\alpha]_D^{25}$ −72.8° (c,
  0.54 in MeOH).
4-Ac: [28587-46-4]. **Tetraneurin *C*.**
  $C_{19}H_{26}O_7$ M 366.410
  Constit. of *P. alpinum* var. *tetraneuris* and *P.
  lozanianum*. Cryst. (CHCl$_3$/pet. ether). Mp 145°.
  $[\alpha]_D^{25}$ −109° (c, 0.05 in MeOH).
1-Deoxy, 14-Acetoxy: 14-Acetoxytetraneurin D.
  $C_{19}H_{26}O_7$ M 366.410
  Constit. of *P. lozanianum*. Oil. $[\alpha]_D^{24}$ −112° (c, 0.4 in
  CHCl$_3$). The published name for this compd. ignores
  the fact that there is an oxygen missing at C-1.
*15-Deacetyl, 15-(2-methylpropanoyl): Desacetyltetran-
  eurin* D *15-O-isobutyrate*.
  $C_{19}H_{28}O_6$ M 352.427
  Constit. of *P. lozanianum*. Cryst. Mp 213°. $[\alpha]_D^{24}$
  −42° (c, 1.5 in CHCl$_3$).
*15-Deacetyl, 4-(2-methylpropanoyl): Desacetyltetran-
  eurin* D *4-O-isobutyrate*.
  $C_{19}H_{28}O_6$ M 352.427
  Constit. of *P. lozanianum*. Oil. $[\alpha]_D^{24}$ −69° (c, 0.3 in
  CHCl$_3$).

15-Deacetyl, 4-Ac: [109291-64-7]. *15-Desacetyltetran-*
eurin C.
$C_{17}H_{24}O_6$ M 324.373
Constit. of *P. lozanianum*. Oil. $[\alpha]_D^{24}$ −31° (c, 1 in
CHCl₃).
15-Deacetyl, 4-Ac, 15-(2-methylpropanoyl): [109291-
66-9]. *15-Desacetyltetraneurin C isobutyrate.*
$C_{21}H_{30}O_7$ M 394.464
Constit. of *P. lozanianum*. Cryst. Mp 174°. $[\alpha]_D^{24}$
−46° (c, 0.7 in CHCl₃).

Rüesch, H. *et al, Tetrahedron*, 1969, **25**, 805 (*isol*)
Yoshioka, H. *et al, Tetrahedron*, 1970, **26**, 2167 (*struct*)
Jakupovic, J. *et al, Phytochemistry*, 1987, **26**, 761 (*cmr, isol,*
derivs)

2,3,4,6-Tetranitroaniline   T-60159

Updated Entry replacing T-01615
2,3,4,6-Tetranitrobenzenamine, 9CI
[3698-54-2]

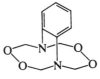

$C_6H_3N_5O_8$ M 273.118
Yellow cryst. (AcOH). Mp 220°. Slightly hygroscopic.
▷Explosive. BY9275000.
N-*Ac: 2,3,4,6-Tetranitroacetanilide.*
   $C_8H_5N_5O_9$ M 315.156
   Mp ca. 170° dec.
N-*Me:*
   $C_7H_5N_5O_8$ M 287.145
   Mp 127°.
N-*Di-Me:*
   $C_8H_7N_5O_8$ M 301.172
   Mp 153°.

v. Duin, C.F.V., *Recl. Trav. Chim. Pays-Bas*, 1917, **37**, 111
   (*synth*)
Forster, A. *et al, J. Chem. Soc.*, 1922, 1988 (*synth*)
U.S.P., 3 062 885, (*1962*); *CA*, **58**, 5572b (*synth*)
Kamlet, M.J., *J. Chem. Soc.* (B), 1968, 1147 (*uv*)
Atkins, R.L. *et al, J. Org. Chem.*, 1986, **51**, 2572 (*synth, ir,*
   pmr)

2,3,4,5-Tetranitro-1H-pyrrole   T-60160

$C_4HN_5O_8$ M 247.081
1-Me: [69726-67-6]. *1-Methyl-2,3,4,5-tetranitro-1H-*
pyrrole.
   $C_5H_3N_5O_8$ M 261.107
   Pale-orange cryst. (CH₂Cl₂). Mp 115° (101-102°).
▷Potentially toxic and explosive

Doddi, G. *et al, J. Org. Chem.*, 1979, **44**, 2321 (*synth, deriv*)
Cromer, D.T. *et al, Acta Crystallogr.*, *Sect. C*, 1986, **42**, 1428
   (*synth, cryst struct, deriv*)

1,7,10,16-Tetraoxa-4,13-diazacyclooctade-cane   T-60161

4,13-Diaza-18-crown-6
[23978-55-4]

$C_{12}H_{26}N_2O_4$ M 262.348
Cryst. (hexane). Mp 114-115°.

Dietrich, B. *et al, Tetrahedron*, 1973, **29**, 1629 (*synth, pmr,*
   conformn)
Buhleier, E. *et al, Justus Liebigs Ann. Chem.*, 1977, 1344
   (*synth*)
Desreux, J.F. *et al, J. Inorg. Nucl. Chem.*, 1977, **39**, 1587
   (*synth*)
Kulstad, S. *et al, Acta Chem. Scand., Ser. B*, 1979, **33**, 469
   (*synth*)
Biasius, E. *et al, Freseuius' Z. Anal. Chem.*, 1980, **304**, 10 (*ms*)
Gatto, V.J. *et al, J. Am. Chem. Soc.*, 1984, **106**, 8240 (*synth,*
   props, derivs)
Gatto, V.J. *et al, J. Org. Chem.*, 1986, **51**, 5373 (*synth, ir, pmr,*
   cmr)

10,11,14,15-Tetraoxa-1,8-diazatricyclo[6.4.4.0²,⁷]hexadeca-2(7),3,5-triene   T-60162

2H,5H-(Methanodioxymethano)-3,4,1,6-benzodioxa-
diazocine, 9CI. Benzenetetramethylene diperoxide
diamine
[105382-66-9]

$C_{10}H_{12}N_2O_4$ M 224.216
Prod. of reacn. of o-1,2-benzenediamine, HCHO and
H₂O₂. Cryst. (pentyl acetate).

Fourcas, J.T. *et al, Acta Crystallogr.*, *Sect. C*, 1986, **42**, 1395
   (*synth, cryst struct*)

1,3,7,9-Tetraoxaspiro[4,5]decane, 9CI   T-60163
[2812-66-0]

$C_6H_{10}O_4$ M 146.143
Bp₅ 45-50°.

Weiss, F. *et al, Bull. Soc. Chim. Fr.*, 1965, 1364 (*synth*)
Jones, R.A.Y. *et al, J. Chem. Soc.* (B), 1971, 1302 (*conformn*)

1,4,6,10-Tetraoxaspiro[4,5]decane, 9CI   T-60164

Ethylene 1,3-propylene orthocarbonate
[24472-05-7]

$C_6H_{10}O_4$ M 146.143
Bp₁ 70°.

Sakai, S. *et al*, *J. Org. Chem.*, 1970, **35**, 2347 (*synth*)
Endo, T. *et al*, *Synthesis*, 1984, 837 (*synth*)

3,3,6,6-Tetraphenyl-1,4-dioxane-2,5-dione   T-60165

Benzilide. Tetraphenylglycolide

[467-32-3]

C$_{28}$H$_{20}$O$_4$ M 420.464
Cryst. (EtOAc). Mp 196°.

Arnold, R.T. *et al*, *J. Am. Chem. Soc.*, 1949, **71**, 2439 (*synth*)
Wasserman, H.H. *et al*, *J. Am. Chem. Soc.*, 1950, **72**, 5787 (*ir*, *struct*)
Tanaka, M. *et al*, *J. Org. Chem.*, 1973, **38**, 1602 (*synth*)

4,4,5,5-Tetraphenyl-1,3-dithiolane, 9CI   T-60166

4,4,5,5-Tetraphenyltetramethylene disulfide

[88691-94-5]

C$_{27}$H$_{22}$S$_2$ M 410.591
Mp 207-209° (166-167°, 199-200°). Gives blue melt.

Bergmann, E. *et al*, *Ber.*, 1930, **63**, 2576 (*synth*)
Schönberg, A. *et al*, *Ber.*, 1931, **64**, 2577.
Kalwinsch, I. *et al*, *J. Am. Chem. Soc.*, 1981, **103**, 7032 (*synth*, *pmr*)

Tetraphenylethanone, 9CI   T-60167

2,2,2-Triphenylacetophenone, 8CI. β-Benzopinacolone.
Phenyl trityl ketone. Phenyl triphenylmethyl ketone.
Benzoyltriphenylmethane

[466-37-5]

Ph$_3$CCOPh

C$_{26}$H$_{20}$O   M 348.443
Cryst. (EtOH or C$_6$H$_6$). Mp 183-184° (181°).

Org. Synth., Coll. Vol., **2**, 73.
Kakis, F.J. *et al*, *J. Org. Chem.*, 1971, **36**, 4117 (*synth*)
Ando, W. *et al*, *Chem. Lett.*, 1980, 1255 (*synth*)
Lindner, E. *et al*, *Chem. Ber.*, 1981, **114**, 810 (*synth*)
Barton, D.H.R. *et al*, *J. Chem. Soc., Perkin Trans. 1*, 1985, 2667 (*synth*)

1,1,3,3-Tetraphenyl-2-propanone   T-60168

sym-*Tetraphenylacetone*

[7476-11-1]

Ph$_2$CHCOCHPh$_2$

C$_{27}$H$_{22}$O   M 362.470
Cryst. (EtOH). Mp 133-134°.

Dean, D.O. *et al*, *J. Am. Chem. Soc.*, 1950, **72**, 1740 (*synth*)
Kantor, S.W. *et al*, *J. Am. Chem. Soc.*, 1950, **72**, 3290 (*synth*)
Charumilind, P. *et al*, *J. Org. Chem.*, 1980, **45**, 4359 (*ir*, *pmr*, *cmr*, *ms*)
Barton, D.H.R. *et al*, *J. Chem. Soc., Perkin Trans. 1*, 1987, 241 (*synth*, *ir*, *pmr*, *ms*)

Tetra(2-triptycyl)ethylene   T-60169

2,2′,2″,2‴-(1,2-Ethenediylidene)tetrakis[9,10-dihydro-1′,2′-benzenoanthracene], 9CI. 1,1,2,2-Tetrakis(2-triptycyl)ethene

[106726-77-6]

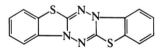

C$_{82}$H$_{52}$ M 1037.313
Pale-yellow cryst. Mp >400°. Strongly fluorescent.
Forms inclusion complexes with org. solvents.

Nakayama, J. *et al*, *J. Chem. Soc., Chem. Commun.*, 1986, 974 (*synth*, *uv*, *pmr*)

1,2,4,5-Tetrazine, 9CI   T-60170

Updated Entry replacing T-01760

[290-96-0]

C$_2$H$_2$N$_4$ M 82.065
Crimson cryst. Sol. H$_2$O, EtOH, Et$_2$O. Mp 99°. Subl.;
volatile at room temp.

Curtius, T. *et al*, *Ber.*, 1907, **40**, 84.
Wood, D. *et al*, *J. Am. Chem. Soc.*, 1933, **55**, 3649.
Bertinotte, F. *et al*, *Acta Crystallogr.*, 1955, **8**, 513 (*cryst struct*)
Marcelis, A.T.M. *et al*, *J. Heterocycl. Chem.*, 1987, **24**, 545 (*synth*)

[1,2,4,5]Tetrazino[3,4-*b*:6,1-*b′*]-   T-60171
bisbenzothiazole, 9CI

Bis(benzothiazolo)[3,2-b:3′,2″-e][1,2,4,5]tetrazine

[107550-91-4]

C$_{14}$H$_8$N$_4$S$_2$ M 296.364
Yellow-orange needles by subl. Mp 207-208.5°.

Eichenberger, T. *et al*, *Helv. Chim. Acta*, 1986, **69**, 1521 (*synth*, *uv*)

[1,2,4,5]Tetrazino[1,6-*a*:4,3-*a'*]- diisoquinoline, 9CI   T-60172

[226-65-3]

$C_{18}H_{12}N_4$ M 284.320
Brown cryst. Mp 263-265°.

Eichenberger, T. *et al*, *Helv. Chim. Acta*, 1986, **69**, 1521 (*synth*, *uv*, *ir*)

[1,2,4,5]Tetrazino[1,6-*a*:4,3-*a'*]diquinoline, 9CI   T-60173

[13090-26-1]

$C_{18}H_{12}N_4$ M 284.320
Fine brown needles. Mp 271-272°.

Eichenberger, T. *et al*, *Helv. Chim. Acta*, 1986, **69**, 1521 (*synth*, *ir*, *uv*)

Tetrazole-5-thione   T-60174

Updated Entry replacing T-01765
1,4-Dihydro-5H-tetrazole-5-thione, 9CI

1,4 Dihydro-*form*   1,3 Dihydro-*form*   5H-*form*

CH_2N_4S M 102.114

1,4-Dihydro-form
Needles. Mp 205° dec. Major tautomer.
1-Me:
  $C_2H_4N_4S$ M 116.140
  Cryst. (Me₂CO). Mp 125-126°.
1-Et:
  $C_3H_6N_4S$ M 130.167
  Cryst. (Me₂CO). Mp 50°.
1-Ph: [86-93-1].
  $C_7H_6N_4S$ M 178.211
  Faint-yellow cryst. (EtOH). Mp 150°.
  ▷XF7700000.
1,4-Di-Me: [54986-14-0].
  $C_3H_6N_4S$ M 130.167
  Cryst. (pet. ether). Mp 99-100°, 107°.
1-Me, 4-Ph: [1455-91-0].
  $C_8H_8N_4S$ M 192.238
  Cryst. Mp 47-49°.

1,3-Dihydro-form
Minor tautomer.
3-Et, 1-Ph: [62681-14-5].
  $C_9H_{10}N_4S$ M 206.265
  Yellowish-white prisms (EtOAc/hexane). Mp 114.5-115°.

(*SH*)-*form* [18686-81-2]
1H-Tetrazole-5-thiol. 5-Mercapto-1H-tetrazole
S-*Me:* [29914-17-8].
  $C_2H_4N_4S$ M 116.140
  Prisms (dioxan). Mp 144-146°, Mp 151° dec.
S-*Et:*
  $C_3H_6N_4S$ M 130.167
  Plates (toluene). Mp 86-88°.
S-*Ph:* [28986-48-3].
  $C_7H_6N_4S$ M 178.211
  Prisms (toluene). Mp 92-93°.
1-N-*Ph*, S-*Me:* [1455-92-1].
  $C_8H_8N_4S$ M 192.238
  Plates (EtOH). Mp 78.5-80°, Mp 84°.

Freund, M. *et al*, *Ber.*, 1901, **34**, 3110 (*synth*)
Lieber, E. *et al*, *Can. J. Chem.*, 1958, **36**, 801 (*ir*, *struct*)
Lieber, E. *et al*, *J. Org. Chem.*, 1961, **26**, 4472 (*deriv*)
Postovoskii, I.Ya. *et al*, *J. Gen. Chem. USSR (Engl. Transl.)*, 1964, **34**, 254 (*struct*, *deriv*)
Bartels-Keith, J.R. *et al*, *J. Org. Chem.*, 1977, **42**, 3725 (*deriv*, *cmr*, *struct*)

2-Tetrazolin-5-one   T-60175

Updated Entry replacing T-10156
1,4-Dihydro-5H-tetrazol-5-one
[16421-52-6]

CH_2N_4O M 86.053
Mp 243-245°.
1-Et: [69048-98-2].
  $C_3H_6N_4O$ M 114.107
  Cryst. (C₆H₆). Mp 79.7°. Bp₁ 133°.
1-Benzyl: [53798-95-1].
  $C_8H_8N_4O$ M 176.177
  Mp 145.2°.

Haines, D.R. *et al*, *J. Org. Chem.*, 1982, **47**, 474.
Janssens, F. *et al*, *J. Med. Chem.*, 1986, **29**, 2290 (*derivs*)

Tetrazolo[1,5-*a*]pyrimidine, 9CI   T-60176
2-Azidopyrimidine
[275-03-6]

$C_4H_3N_5$ M 121.101
Tautomeric system. Mp 121-123°.

Japan. Pat., 57 777, (*1957*); *CA*, **52**, 4699g (*synth*)
Temple, C. *et al*, *J. Org. Chem.*, 1965, **30**, 826 (*pmr*, *tautom*)
Pugmire, R.J. *et al*, *J. Heterocycl. Chem.*, 1987, **24**, 805 (*cmr*)

Tetrazolo[1,5-*c*]pyrimidine, 9CI   T-60177
4-Azidopyrimidine
[274-88-4]

$C_4H_3N_5$ M 121.101
Tautomeric, tetrazolo tautomer predominates. Mp 77-79°.

Temple, C. *et al*, *J. Org. Chem.*, 1965, **30**, 829 (*synth*, *uv*, *ir*, *tautom*)

Pugmire, R.J. *et al*, *J. Heterocycl. Chem.*, 1987, **24**, 805 (*cmr*)

Teucjaponin *A*   **T-60178**

Updated Entry replacing T-30140
  Montanin F
  [81422-44-8]

$C_{22}H_{28}O_7$ M 404.459
Constit. of *Teucrium japonicum* and *T. montanum*.
  Shows antifeedant activity for *Prodenia litura*. Cryst.
  (Me_2CO/C_6H_6). Mp 145-148°. $[\alpha]_D^{20}$ +38.8° (c, 3.84
  in $CHCl_3$).

6-Epimer: [72948-20-0]. ***Teucjaponin* B**.
  $C_{22}H_{28}O_7$ M 404.459
  Constit. of *T. japonicum*. Needles ($CHCl_3/MeOH$).
  Mp 255-258°. $[\alpha]_D^{19}$ +45.5° (c, 0.67 in $CHCl_3$).

Miyase, T. *et al*, *Chem. Pharm. Bull.*, 1981, **29**, 3561 (*isol, struct*)
Papanov, G.Y. *et al*, *Phytochemistry*, 1983, **22**, 2787 (*isol*)
Gács-Baitz, E. *et al*, *Phytochemistry*, 1987, **26**, 2110 (*struct*)

Teucretol   **T-60179**

$C_{24}H_{34}O_8$ M 450.528
Constit. of *Teucrium creticum*. Syrup. $[\alpha]_D^{27}$ -8.8° (c,
  0.341 in $CHCl_3$).

Savona, G. *et al*, *Phytochemistry*, 1987, **26**, 3285.

Teucrin *E*   **T-60180**

Updated Entry replacing T-50248
  [54927-87-6]

$C_{20}H_{24}O_6$ M 360.406
Constit. of *Teucrium chamaedrys*. Cryst. (Me_2CO/Et_2O). Mp 235-238°. $[\alpha]_D^{20}$ -25° (Me_2CO).

2,3-Didehydro: **2,3-Dehydroteucrin** E.
  $C_{16}H_{22}O_6$ M 310.346
  Constit. of *Teucrium scordium*. Cryst. Mp 249°.

Reinbol'd, A.M. *et al*, *Khim. Prir. Soedin.*, 1974, 589 (*isol*)
Jakupovic, J. *et al*, *Planta Med.*, 1985, 341 (*isol*)

Teumicropin   **T-60181**

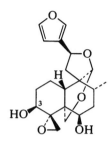

$C_{20}H_{26}O_6$ M 362.422
Constit. of *Teucrium micropodioides*. Cryst. (EtOAc/hexane). Mp 195-198°. $[\alpha]_D^{22}$ -22.9° (c, 0.088 in Py).

3-Ac: 3-Acetylteumicropin.
  $C_{22}H_{28}O_7$ M 404.459
  From *T. micropodioides*. Cryst. (EtOAc/hexane). Mp 256-259°. $[\alpha]_D^{20}$ +4.1° (c, 0.195 in $CHCl_3$).

De La Torre, M.C. *et al*, *Phytochemistry*, 1988, **27**, 213.

Teumicropodin   **T-60182**

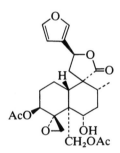

$C_{24}H_{30}O_9$ M 462.496
Constit. of *Teucrium micropodioides*. Cryst. (EtOAc/hexane). Mp 238-241°. $[\alpha]_D^{22}$ +42.2° (c, 0.303 in $CHCl_3$).

De La Torre, M.C. *et al*, *Phytochemistry*, 1988, **27**, 213.

Teupolin I   **T-60183**

Updated Entry replacing T-30148
  [72948-20-0]

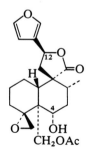

$C_{22}H_{28}O_7$ M 404.459
Constit. of *Teucrium polium*. Cryst. ($Et_2O/CHCl_3$). Mp 211-213°. $[\alpha]_D^{20}$ +60° (c, 0.2 in CH_2Cl_2).

Ac: [67987-83-1]. **Montanin** C.
  $C_{24}H_{30}O_8$ M 446.496

Constit. of *T. montanum*. Cryst. Mp 181-183°. $[\alpha]_D^{20}$ +8.4° (c, 0.262 in $CHCl_3$).

De-O-Ac, O⁴-Ac: [72948-21-1]. **Teupolin II**.
$C_{22}H_{28}O_7$ M 404.459
Isol. from *T. polium*. Cryst. (Et_2O/Me_2CO). Mp 187-189°. $[\alpha]_D^{20}$ +52° (c, 0.2 in CH_2Cl_2).

Malakov, P.Y. *et al*, *Z. Naturforsch., B*, 1978, **33**, 789; 1979, **34**, 1570 (*isol*)
Gács-Baitz, E. *et al*, *Heterocycles*, 1982, **19**, 539 (*cmr*)
Fayos, J. *et al*, *J. Org. Chem.*, 1984, **49**, 1789 (*struct*)
Gács-Baitz, E. *et al*, *Phytochemistry*, 1987, **26**, 2110 (*struct, bibl*)

Teupyrenone   T-60184
Updated Entry replacing T-20152
[85564-03-0]

$C_{22}H_{26}O_7$ M 402.443
Constit. of *Teucrium pyrenaicum*. Cryst. (EtOAc/hexane). Mp 213-215°. $[\alpha]_D^{18}$ −46.5° (c, 0.8 in $CHCl_3$).

Deacetyl: **Deacetylteupyrenone**.
$C_{20}H_{24}O_6$ M 360.406
Constit. of *T. micropodioides*. Cryst. (EtOAc/hexane). Mp 235-240°. $[\alpha]_D^{22}$ +1.6° (c, 0.183 in $CHCl_3$).

Garcia-Alvarez, M.C. *et al*, *Phytochemistry*, 1982, **21**, 2559 (*isol, struct*)
De La Torre, M.C. *et al*, *Phytochemistry*, 1988, **27**, 213 (*deriv*)

Theaspirone   T-60185
[36431-72-8]

$C_{13}H_{22}O$ M 194.316
Constit. of raspberry, yellow passion fruit and tea. Oil.

Ohloff, G., *Fortschr. Chem. Org. Naturst.*, 1978, **35**, 431 (*isol*)
Schulte-Elte, K.H. *et al*, *Helv. Chim. Acta*, 1978, **61**, 1125 (*synth*)
Okawara, H. *et al*, *Heterocycles*, 1979, **13**, 191 (*synth*)
Torii, S. *et al*, *Tetrahedron Lett.*, 1981, **22**, 2291 (*synth*)
Masuda, H. *et al*, *Agric. Biol. Chem.*, 1985, **49**, 861 (*synth*)

2-Thiabicyclo[2.2.1]heptan-3-one, 9CI   T-60186
[54396-36-0]

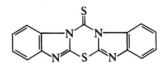

C_6H_8OS M 128.189
Solid by subl. Mp 88-89°.

Ueda, Y. *et al*, *Can. J. Chem.*, 1986, **64**, 2184 (*synth, ir, pmr*)

2-Thiabicyclo[2.2.2]octan-3-one, 9CI   T-60187
[105676-16-2]

$C_7H_{10}OS$ M 142.215
Cryst. (Et_2O/hexane). Mp 174-175°.

Ueda, Y. *et al*, *Can. J. Chem.*, 1986, **64**, 2184 (*synth, ir, pmr*)

2-Thiabicyclo[2.2.2]oct-5-en-3-one, 9CI   T-60188
3-Oxo-2-thiabicyclo[2.2.2]oct-5-ene
[40168-97-6]

C_7H_8OS M 140.200
Cryst. Mp 73-74° (61-63°).

Reich, H.J. *et al*, *J. Org. Chem.*, 1973, **38**, 2637 (*synth, pmr, ir*)
Harpp, D.N. *et al*, *Can. J. Chem.*, 1985, **63**, 951 (*synth, ir, pmr, ms*)
Ueda, Y. *et al*, *Can. J. Chem.*, 1986, **64**, 2184 (*synth, ir, pmr*)

1-Thia-2-cyclooctyne   T-60189
7,8-Didehydro-3,4,5,6-tetrahydro-2H-thiocin, 9CI
[103794-91-8]

$C_7H_{10}S$ M 126.216
Reacts readily with both nucleophiles and electrophiles due to polarised and strained triple band. Oily liq. with characteristic odour.

Meier, H. *et al*, *Angew. Chem., Int. Ed. Engl.*, 1986, **25**, 809 (*synth, pmr, cmr*)

1-Thia-3-cyclooctyne   T-60190
$C_7H_{10}S$ M 126.216

Meier, H. *et al*, *Chem. Ber.*, 1980, **113**, 2398.

**13H-[1,3,5]Thiadiazino[3,2-a:5,6-a']-   T-60191
bisbenzimidazole-13-thione, 9CI**
13H-Dibenzimidazo[2,1-b:1',2'-e][1,3,5]thiadiazine-13-thione
[34858-80-5]

$C_{15}H_8N_4S_2$ M 308.375
Yellow prismatic needles (DMF). Mp 189-190°.

Haugwitz, R.D. *et al*, *J. Org. Chem.*, 1972, **37**, 2776 (*synth*)
Hull, R., *Synth. Commun.*, 1979, **9**, 477 (*synth, pmr, cmr*)

Martin, D. *et al, J. Chem. Soc., Perkin Trans. 1*, 1985, 1007 (*synth*)

2-(1,3,4-Thiadiazol-2-yl)pyridine, 9CI   T-60192
2-(2-Pyridyl)-1,3,4-thiadiazole
[98273-52-0]

$C_7H_5N_3S$ M 163.197
Cryst. (pet. ether). Mp 83-84° (80-82°).

Hemmerich, P. *et al, Helv. Chim. Acta*, 1958, **41**, 2058.
Toyooka, K. *et al, Chem. Pharm. Bull.*, 1987, **35**, 1030.

4-(1,3,4-Thiadiazol-2-yl)pyridine, 9CI   T-60193
2-(4-Pyridyl)-1,3,4-thiadiazole
$C_7H_5N_3S$ M 163.197
Cryst. Mp 112-114°.

Toyooka, K. *et al, Chem. Pharm. Bull.*, 1987, **35**, 1030 (*synth, pmr*)

Thia[2.2]metacyclophane   T-60194
2-Thiatricyclo[9.3.1.1^{4,8}]hexadeca-1(15),4,6,8(16),11,13-hexaene, 9CI
[106484-46-2]

$C_{15}H_{14}S$ M 226.336
(+)-*form*
  $[\alpha]_D^{20}$ +152.73° (c, 1 in CHCl$_3$).
(±)-*form*
  Needles (heptane). Mp 108°.
  S-*Oxide:*
    $C_{15}H_{14}OS$ M 242.335
    Mp ~155°. Four diastereoisomers formed with one in excess. Sepd. by HPLC.
  S-*Dioxide:*
    $C_{15}H_{14}O_2S$ M 258.334
    Mp 200-222°.

Vögtle, F. *et al, J. Chem. Soc., Chem. Commun.*, 1986, 1248 (*synth, cryst struct, pmr, cmr, cd, abs config*)

2-Thiatricyclo[3.3.1.1^{3,7}]decane, 9CI   T-60195
Updated Entry replacing T-30163
2-Thiaadamantane, 8CI
[281-25-4]

$C_9H_{14}S$ M 154.270
Occurs in Middle East crude oil distillate. Sweet-smelling cryst. Mp 323°.

2,2-Dioxide: [23504-97-4].
  $C_9H_{14}O_2S$ M 186.268
  Mp 333.9-334.6° dec. (sealed tube).

Stetter, H. *et al, Chem. Ber.*, 1962, **95**, 1687 (*synth*)
Landa, S. *et al, Collect. Czech. Chem. Commun.*, 1969, **34**, 2014 (*synth*)
Duddeck, H. *et al, Org. Magn. Reson.*, 1976, **8**, 593 (*cmr*)
Hajek, M. *et al, Collect. Czech. Chem. Commun.*, 1976, **41**, 2533 (*pmr*)
Suginome, H. *et al, Synthesis*, 1986, 741 (*synth, ms, ir, pmr*)

1,4-Thiazepine   T-60196

C_5H_5NS M 111.161
Parent compd. currently unknown. 2,7-Di-*tert*-butyl-5-methoxy deriv. prepd.

Yamamoto, K. *et al, Angew. Chem., Int. Ed. Engl.*, 1986, **25**, 635.

2-Thiazoledinethione, 9CI   T-60197
[1437-88-3]

$C_3H_5NS_2$ M 119.199
Free compd. exists in thione form.

Thione-form
  Needles (H$_2$O). Mp 105-106°.
  N-*Me:*
    $C_4H_7NS_2$ M 133.226
    Mp 69.5°.
  N-*Ph:*
    $C_{10}H_9NS_2$ M 207.308
    Mp 129-130°.

Thiol-form
  S-*Me:* [19975-56-5]. *4,5-Dihydro-2-(methylthio)-thiazole. 2-Methyl-Δ^2-1,3-thiazoline.*
    $C_4H_7NS_2$ M 133.226
    Bp$_3$ 60°.
  S-*Me; B,HI:* [40836-94-0]. Mp 115-117°.

Knorr, L. *et al, Ber.*, 1903, **36**, 1278 (*synth*)
Dewey, C.S. *et al, J. Org. Chem.*, 1965, **30**, 491 (*synth, deriv, ir*)
Owen, T.C. *et al, J. Chem. Soc. (C)*, 1967, 1373 (*synth*)
Hirai, K. *et al, Tetrahedron Lett.*, 1971, 4359 (*synth*)
Bigg, D.C.H. *et al, J. Heterocycl. Chem.*, 1976, **13**, 977 (*synth*)
Dou, H.J.M. *et al, Helv. Chim. Acta*, 1978, **61**, 3143 (*synth, pmr*)
Fieser, M. *et al, Reagents for Organic Synthesis*, Wiley, 1967-84, **5**, 459.

4-Thiazolesulfonic acid, 9CI   T-60198
[82971-12-8]

$C_3H_3NO_3S_2$ M 165.182
Cryst. Mp 304-305°.

Erlenmeyer, H. *et al, Helv. Chim. Acta*, 1945, **28**, 985 (*synth*)

5-Thiazolesulfonic acid   T-60199

$C_3H_3NO_3S_2$ M 165.182
Needles (H_2O). Mp 304-305°.

Erlenmeyer, H. *et al*, *Helv. Chim. Acta*, 1945, **28**, 985 (*synth*)

2(3H)-Thiazolethione, 9CI   T-60200

4-Thiazoline-2-thione, 8CI. 2-Thiazolethiol. 2-Thiothia-zolone. 2-Mercaptothiazole
[5685-05-2]

$C_3H_3NS_2$ M 117.183
Exists as thione form in aq. soln. Cryst. (EtOH aq.). Mp 78-80°.

SH-form

S-*Me:* [5053-24-7]. *2-(Methylthio)thiazole*.
  $C_4H_5NS_2$ M 131.210
  Bp_2 68°.
S-*Et:* [14527-48-1]. *2-(Ethylthio)thiazole*.
  $C_5H_7NS_2$ M 145.237
  Bp_2 70°.
S-*Isopropyl:* [15545-30-9].
  $C_6H_9NS_2$ M 159.264
  Bp_5 66°.
Disulfide: [20362-54-3]. *2,2′-Dithiobisthiazole, 9CI*.
  $C_6H_4N_2S_4$ M 232.351
  Light-yellow needles (hexane). Mp 83°. Dec. slowly on storage.

Backer, H.J. *et al*, *Recl. Trav. Chim. Pays-Bas*, 1945, **64**, 102 (*synth*)
Mathes, R.A. *et al*, *J. Am. Chem. Soc.*, 1948, **70**, 1451 (*synth*)
Garraway, J.L., *J. Chem. Soc. (B)*, 1966, 92 (*ir, tautom*)
Bastionelli, P. *et al*, *Bull. Soc. Chim. Fr.*, 1967, 1948 (*synth, deriv, ir*)

Thiazolo[4,5-b]quinoline-2(3H)-thione, 9CI   T-60201

Quinrhodine
[495-35-2]

$C_{10}H_6N_2S_2$ M 218.291
Cryst. (AcOH). Mp 294-295° dec.
3-Ph:
  $C_{16}H_{10}N_2S_2$ M 294.388
  Cryst. (EtOH). Mp 110°.

Tanasescu, I. *et al*, *Chem. Ber.*, 1958, **91**, 160 (*synth*)

Thieno[2,3-c]cinnoline   T-60202

[96919-50-5]

$C_{10}H_6N_2S$ M 186.231
Pale-orange needles (cyclohexane). Mp 138-139°.

Barton, J.W. *et al*, *J. Chem. Soc., Perkin Trans. 1*, 1985, 131 (*synth, pmr*)

Thieno[3,2-c]cinnoline   T-60203

[60338-65-0]

$C_{10}H_6N_2S$ M 186.231
Orange cryst. (cyclohexane). Mp 133-134°.

Barton, J.W. *et al*, *J. Chem. Soc., Perkin Trans. 1*, 1985, 131 (*synth, pmr*)

Thieno[3,4-b]furan   T-60204

[18546-38-8]

C_6H_4OS M 124.157
Oil.

Moursounidis, J. *et al*, *Tetrahedron Lett.*, 1986, **27**, 3045 (*synth, uv, pmr*)

Thieno[3,2-d]pyrimidine   T-60205

Updated Entry replacing T-01956
[272-68-4]

$C_6H_4N_2S$ M 136.171
Cryst. (Et_2O). Mp 83°.

Robba, M. *et al*, *Tetrahedron*, 1971, **27**, 487.
Sakamoto, T. *et al*, *Chem. Pharm. Bull.*, 1986, **34**, 2719 (*synth, pmr*)

5H-Thieno[2,3-c]pyrrole   T-60206

[250-63-5]

C_6H_5NS M 123.172
Cryst. Mp 68.5-69.5°. Readily polym. in acid soln. Dark-purple melt.

Sha, C.-K. *et al*, *J. Chem. Soc., Chem. Commun.*, 1986, 310 (*synth, pmr, cmr, uv*)

1-(2-Thienyl)ethanol   T-60207

α-Methyl-2-thiophenemethanol, 9CI
[2309-47-9]

C_6H_8OS M 128.189
(R)-form
  Bp_7 87-89°. $[\alpha]_D^{20}$ +3.94° (neat).
(±)-form
  Bp_6 90-92°. n_D 1.5430.

Van Zyl, G. *et al*, *J. Am. Chem. Soc.*, 1956, **78**, 1955 (*synth*)
Červinka, O. *et al*, *Z. Chem.*, 1969, **9**, 448 (*synth, abs config*)

3-(2-Thienyl)-2-propenoic acid, 9CI   T-60208

3-(2-Thienyl)acrylic acid. 3-Thiopheneacrylic acid
[1124-65-8]

(*E*)-*form*

$C_7H_6O_2S$ M 154.183
(*E*)-**form** [15690-25-2]
Mp 146-147°.

Me ester:
$C_9H_8O_2S$ M 180.221
Cryst. (hexane). Mp 42°.

(*Z*)-**form** [51019-83-1]
Mp 136-137°.

King, W.J. *et al, J. Org. Chem.*, 1949, **14**, 405 (*synth*)
Heck, R.F., *J. Am. Chem. Soc.*, 1968, **90**, 5518 (*deriv, synth*)
Taniguchi, Y. *et al, Chem. Pharm. Bull.*, 1973, **21**, 2070 (*synth, ir, pmr, uv*)
Larock, R.C. *et al, J. Org. Chem.*, 1986, **51**, 5221 (*synth*)

2-Thiocyanato-1*H*-pyrrole   T-60209

Thiocyanic acid 1H-pyrrol-2-yl ester, 9CI. 2-Pyrrolyl thiocyanate
[66786-05-8]

$C_5H_4N_2S$ M 124.160
Originally thought to be the 3-isomer. Long needles (CH_2Cl_2/methylcyclohexane). Mp 40-44°.

▷Skin irritant, intense skin staining agent

Matteson, D.S. *et al, J. Org. Chem.*, 1957, **22**, 1500 (*synth, ir*)
Gronowitz, S. *et al, J. Org. Chem.*, 1961, **26**, 2615 (*pmr, struct*)
Olsen, R.K. *et al, J. Org. Chem.*, 1963, **28**, 3050 (*struct*)

1-Thiocyano-1,3-butadiene   T-60210

1,3-Butadienyl thiocyanate

$$H_2C=CHCH=CHSCN$$

C_5H_5NS M 111.161
(*E*)-**form**
Oil with onion odour. Bp_{30} 60°.
(*Z*)-**form**
Oil. Obt. only in admixture with (*E*)-form.

Huber, S. *et al, Helv. Chim. Acta*, 1986, **69**, 1898 (*synth, pmr, cmr, ir, ms*)

3-Thiopheneethanol, 9CI   T-60211

2-(3-Thienyl)ethanol. 3-(2-Hydroxyethyl)thiophene
[13781-67-4]

C_6H_8OS M 128.189
Liq. Bp_{14} 110-111°.

Gronowitz, S., *Ark. Kemi*, 1956, **8**, 441 (*synth*)

Thiophene-3-ol, 9CI   T-60212

3-Hydroxythiophene. 3-Thienol
[17236-59-8]
C_4H_4OS M 100.135
Free compd. not known.

Ac: 3-Acetoxythiophene.
$C_6H_6O_2S$ M 142.172
Liq. Bp_{10} 82-84°. n_D^{20} 1.5200.
Me ether: [17573-92-1]. 3-Methoxythiophene.
C_5H_6OS M 114.162
Liq. Bp_{65} 80-82°. n_D^{25} 1.5292.

Gronowitz, S., *Ark. Kemi*, 1958, **12**, 239.
Hörnfeldt, A.B. *et al, Acta Chem. Scand.*, 1962, **16**, 789.

2-Thiophenesulfonic acid, 9CI   T-60213

[79-84-5]

$C_4H_4O_3S_2$ M 164.194
Deliquescent cryst.
Ba salt: Cryst. + $3H_2O$ (H_2O). Sol. H_2O.
Et ester:
$C_6H_8O_3S_2$ M 192.247
Oil.
Chloride: [16629-19-9].
$C_4H_3ClO_2S_2$ M 182.640
Needles (pet. ether). Mp 31.5-33°. $Bp_{1.5}$ 94-98°.
Amide: [6339-87-3].
$C_4H_5NO_2S_2$ M 163.209
Cryst. (Et_2O). Mp 144-146°.
Anilide: [39810-46-3]. N-*Phenyl-2-thiophenesulfonamide, 9CI.*
$C_{10}H_9NO_2S_2$ M 239.307
Cryst. (EtOH). Mp 99.5-100°.

Weitz, L., *Chem. Ber.*, 1884, **17**, 792 (*synth*)
Steinkopf, W. *et al, Justus Liebigs Ann. Chem.*, 1933, **501**, 174 (*deriv*)
Terent'ev, A.P. *et al, Zh. Obshch. Khim.*, 1952, **22**, 153 (*synth*)
Cymerman-Craig, J. *et al, J. Chem. Soc.*, 1956, 4114 (*chloride*)

3-Thiophenesulfonic acid, 9CI   T-60214

$C_4H_4O_3S_2$ M 164.194
Deliquescent cryst. solid.
Na salt: [51175-72-5]. Hygroscopic solid.
Chloride: [51175-71-4].
$C_4H_3ClO_2S_2$ M 182.640
Cryst. (pet. ether). Mp 47°.
Amide: [64255-63-6].
$C_4H_5NO_2S_2$ M 163.209
Prisms (H_2O). Mp 148°.
Anilide: [51175-66-7]. N-*Phenyl-3-thiophenesulfonamide, 9CI.*
$C_{10}H_9NO_2S_2$ M 239.307
Cryst. (EtOH aq.). Mp 111-112°.

Langer, J., *Ber.*, 1884, **17**, 1566 (*synth, derivs*)
Arcoria, A., *J. Org. Chem.*, 1974, **12**, 1689 (*derivs*)

4-Thioxo-2-hexanone   T-60215

[81674-19-3]

$$H_3CCH_2CSCH_2COCH_3$$

$C_6H_{10}OS$ M 130.204

Chelated enol and ene-thiol forms prob. predominate.
Yellow oil. Bp$_9$ 67-68°.

Duus, F., *J. Am. Chem. Soc.*, 1986, **108**, 630 (*synth, uv, tautom*)

3-Thioxo-1,2-propadien-1-one   T-60216
Tricarbon oxide sulfide
[2219-62-7]

$$OC{=}C{=}CS$$

C$_3$OS   M 84.092
Gas. Mp ca. −70°.

Winnewisser, M. *et al*, *Chem. Phys. Lett.*, 1976, **37**, 270 (*synth, spectra*)
Bock, H. *et al*, *J. Am. Chem. Soc.*, 1986, **108**, 7844 (*synth, pe, ms, bibl*)

Threonine   T-60217
Updated Entry replacing A-01760
2-Amino-3-hydroxybutanoic acid, 9CI
[36676-50-3]

$$\begin{array}{c} COOH \\ | \\ H_2N{-}C{\blacktriangleleft}H \\ | \\ H{\blacktriangleright}C{\blacktriangleleft}OH \\ | \\ CH_3 \end{array}$$

(2S,3R)-form
Absolute
configuration

C$_4$H$_9$NO$_3$ M 119.120

(2S,3R)-form [72-19-5]
L-Threonine
From wide variety of protein hydrolysates. Cryst. (EtOH aq.). Mp 251-253°. [α]$_D^{26}$ −28.3° (c, 1.2 in H$_2$O). N-Protected derivatives useful in peptide synthesis are listed alphabetically elsewhere.

▷XO8590000.

Me ether: [4144-02-9].
C$_5$H$_{11}$NO$_3$ M 133.147
Cryst. (EtOH aq.). Mp 214-216°. [α]$_D$ −37.8°.
N-*Formyl:*
C$_5$H$_9$NO$_4$ M 147.130
Cryst. (H$_2$O). Mp 163-164°. [α]$_D^{26}$ +11.8°.
N-*Benzoyl:*
C$_{11}$H$_{13}$NO$_4$ M 223.228
Cryst. (EtOAc). Mp 147-148°. [α]$_D^{26}$ +25.1°.
N-*(2,4-Dinitrophenyl):* Cryst. (Et$_2$O/pet. ether). Mp 145°. [α]$_D$ +106° (4% NaHCO$_3$).
N-*Me:* [2812-28-4].
C$_5$H$_{11}$NO$_3$ M 133.147
Component of Stendomycin.

(2R,3S)-form [632-20-2]
D-Threonine
Cryst. (EtOH aq.). Mp 251-252°. [α]$_D^{26}$ +28.4° (H$_2$O).
▷XO8580000.
Me ether: Cryst. (EtOH aq.). Mp 214-216°. [α]$_D^{26}$ +38.2°.
N-*Formyl:* Mp 164.5°. [α]$_D$ −11.9°.
N-*Benzoyl:* Cryst. (EtOAc). Mp 147-148°. [α]$_D^{26}$ −25.5°.

(2RS,3SR)-form [80-68-2]
(±)-Threonine
Cryst. (EtOH aq.). Mp 234-235° dec.
Me ester: [53216-06-1].
C$_5$H$_{11}$NO$_3$ M 133.147

Mp 125° dec. (as hydrochloride).
Me ether: Cryst. (EtOH aq.). Mp 215-218°.
Formyl: Cryst. (H$_2$O). Mp 174-175°.
N-*Benzoyl:* Cryst. (H$_2$O). Mp 143-144°.
N-*2,4-Dinitrophenyl:* Mp 178°.

(2S,3S)-form [28954-12-3]
L-Allothreonine
▷BA4055000.
N-*Ac:*
C$_6$H$_{11}$NO$_4$ M 161.157
Mp 153-154°.
Ac, Me ether:
C$_7$H$_{13}$NO$_4$ M 175.184
Mp 153-155°.

(2RS,3RS)-form [144-98-9]
(±)-Allothreonine
Cryst. (EtOH aq.). Mp 242-243°.
Me ester; B,HCl: [76419-75-5]. Mp 150° dec.
Me ether: Cryst. (EtOH aq.). Mp 230-233°.
N-*Benzoyl:* Cryst. (H$_2$O). Mp 175-176°.

Org. Synth., Coll. Vol., 3, 813.
Winitz, M. *et al*, *J. Am. Chem. Soc.*, 1956, **78**, 2423.
Greenstein, J.P. *et al*, *Chemistry of the Amino Acids*, 1961, Wiley, N.Y., **3**, 2238.
Aruldhas, G., *Spectrochim. Acta, Part A*, 1967, **23**, 1345 (*pmr*)
Pachler, K.G.R. *et al*, *Spectrochim. Acta, Part A*, 1968, **24**, 1311 (*pmr*)
Mallikarjunan, M. *et al*, *Acta Crystallogr., Sect. B*, 1969, **25**, 220 (*abs config*)
Maldonado, P. *et al*, *Bull. Soc. Chim. Fr.*, 1971, 2933 (*synth*)
Swaminathan, P. *et al*, *Acta Crystallogr., Sect. B*, 1975, **31**, 217; *J. Cryst. Mol. Struct.*, 1975, **5**, 203 (*cryst struct*)
Soukup, M. *et al*, *Helv. Chim. Acta*, 1987, **70**, 232 (*synth*)
Seebach, D. *et al*, *Helv. Chim. Acta*, 1987, **70**, 237 (*synth*)

Thymifodioic acid   T-60218

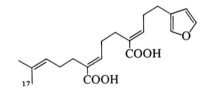

C$_{20}$H$_{26}$O$_5$ M 346.422
Constit. of *Baccharis thymifolia*. Oil.
17-Acetoxy: **17-Acetoxythymifodioic acid**.
C$_{22}$H$_{28}$O$_7$ M 404.459
From *B. thymifolia*.

Saad, J.R. *et al*, *Phytochemistry*, 1987, **26**, 3033.

Thymopoietin, 9CI   T-60219
Thymin (hormone)
[60529-76-2]

H-W-X-Phe-Leu-Glu-Asp-Pro-Ser-Val-Leu-Thr-Lys-Glu-Lys-Leu-Lys-Ser-Glu-Leu-Val-Ala-Asn-Asn-Val-Thr-Leu-Pro-Ala-Gly-Glu-Gln-Arg-Lys-Y-Val-Tyr-Val-Glu-Leu-Tyr-Leu-Gln-Z-Leu-Thr-Ala-Leu-Lys-Arg-OH. Thymopoietin I: W = Gly, X = Gln, Y = Asp, Z = His, II: W = Pro, X = Glu, Y = Asp, Z = Ser, III: W = Pro, X = Glu, Y = Glu, Z = His

Polypeptide hormone discovered by its effect on neuromuscular transmission. Induces T-lymphocyte differentiation and affects immunoregulatory balance.

Thymopoietin I [66943-28-0]

1-Glycine-2-glutamine-43-histidinethymopoietin II, 9CI

$C_{250}H_{410}N_{68}O_{75}$ M 5568.400

Isol. from bovine thymus. Fluffy powder. $[\alpha]_D^{21}$ −78.4° (c, 0.3 in AcOH aq.).

Thymopoietin II, 9CI [56996-26-0]

$C_{251}H_{412}N_{64}O_{78}$ M 5574.398

From bovine thymus.

Thymopoietin III [79103-34-7]

34-Glutamic acid-43-histidinethymopoietin II, 9CI

$C_{255}H_{416}N_{66}O_{77}$ M 5638.487

From bovine spleen. Fluffy powder. $[\alpha]_D^{21}$ −81.2° (c, 0.3 in AcOH aq.).

Audhya, T. et al, Biochemistry, 1981, **20**, 6195 (struct, bibl)
Abiko, T. et al, Chem. Pharm. Bull., 1985, **33**, 1583; 1986, **34**, 2133 (synth)

Thyrsiferol   T-60220

Updated Entry replacing T-50332

[66873-39-0]

$C_{30}H_{53}BrO_7$ M 605.649

Config. at C-14 and C-15 wrongly drawn in early references. Metab. of *Laurencia thyrsifera*.

23-Ac: [96304-95-9].

$C_{32}H_{55}BrO_8$ M 647.686

Constit. of *L. obtusa*. Cytotoxic agent. Cryst. (MeOH aq.). Mp 118-119°. $[\alpha]_D^{29}$ +1.99° (c, 4.4 in CHCl$_4$).

Blunt, J.W. et al, Tetrahedron Lett., 1978, 69 (isol, struct)
Suzuki, T. et al, Tetrahedron Lett., 1985, **26**, 1329 (isol)
Sakami, S. et al, Tetrahedron Lett., 1986, **27**, 4287 (isol, cryst struct)

7-Tirucallene-3,23,24,25-tetrol   T-60221

Updated Entry replacing T-10186

(3α,23R,24S)-form

$C_{30}H_{52}O_4$ M 476.738

(3α,23R,24S)-form [78739-37-4]

Hispidol A

Constit. of *Trichilia hispida*. Cryst. (MeOH/MeCN). Mp 118°. $[\alpha]_D^{25}$ −80° (Py).

(3β,23R,24S)-form [78739-39-6]

Hispidol B

From *T. hispida*. Cryst. (MeOH/CH$_2$Cl$_2$). Mp 252-253°. $[\alpha]_D^{25}$ −57° (Py).

Jolad, S.D. et al, J. Org. Chem., 1981, **46**, 4085 (isol)
Arisawa, M. et al, Phytochemistry, 1987, **26**, 3301 (cryst struct)

Tomenphantopin B   T-60222

[108864-23-9]

$C_{19}H_{22}O_7$ M 362.379

Constit. of *Elephantopus tomentosus*. Cytotoxin. Needles (CHCl$_3$). Mp 114-116°. $[\alpha]_D$ −11.1° (c, 0.4 in CHCl$_3$).

Me ether: [108864-22-8]. **Tomenphantopin A**.

$C_{20}H_{24}O_7$ M 376.405

From *E. tomentosus*. Cytotoxin. Powder. $[\alpha]_D$ −8.6° (c, 0.35 in CHCl$_3$).

Hayashi, T. et al, Phytochemistry, 1987, **26**, 1065 (isol, cryst struct)

Toosendantriol   T-60223

Updated Entry replacing A-00141

21,23R:24S,25-Diepoxyapotirucall-14-ene-3α,7α,21S-triol

$C_{30}H_{48}O_5$ M 488.706

21-Ac: [73114-34-8]. *21-Acetoxy-21,23:24,25-diepoxyapotirucall-14-ene-3,7-diol. 21-O-Acetyltoosendantriol*. **Chisocheton compound D**.

$C_{32}H_{50}O_6$ M 530.743

Constit. of fruit of *Melia toosendan* and of *Chisocheton paniculatus*. Powder. $[\alpha]_D^{25}$ −3.6° (c, 0.26 in CHCl$_3$).

3-Ketone, 21-Ac: [73114-31-5]. **Chisocheton compound A**.

$C_{32}H_{48}O_6$ M 528.728

Constit. of *C. paniculatus*. Cryst. (Et$_2$O/MeOH). Mp 209-211°.

Connolly, J.D. et al, J. Chem. Soc., Perkin Trans. 1, 1979, 2959 (isol)
Nakanishi, T. et al, Chem. Lett., 1986, 69 (isol, cryst struct)

Topazolin T-60224

5,7-Dihydroxy-2-(4-hydroxyphenyl)-3-methoxy-6-(3-methyl-2-butenyl)-4H-1-benzopyran-4-one, 9CI. 4',5,7-Trihydroxy-3-methoxy-6-prenylflavone

[109605-79-0]

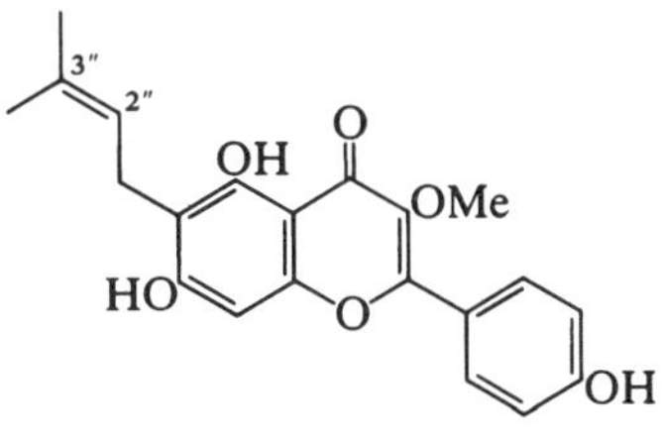

$C_{21}H_{20}O_6$ M 368.385

Constit. of *Lupinus luteus* cv. Topaz. Pale-yellow rods (Me$_2$CO/EtOAc). Mp 227-228.5°.

2'',3''-Dihydro, 3''-hydroxy: [109605-84-7]. **Topazolin hydrate**.
$C_{21}H_{22}O_7$ M 386.401
Constit. of *L. luteus* cv. Topaz. Pale-yellow rods (EtOAc/hexane). Mp 239-241°.

Tahara, S. *et al, Agric. Biol. Chem.*, 1987, **51**, 1039.

α-Torosol T-60225

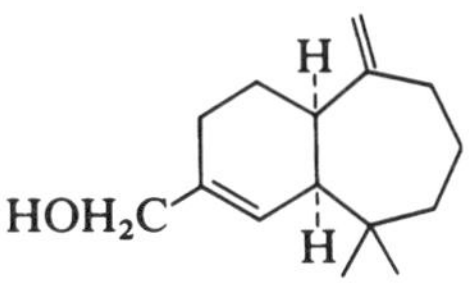

$C_{15}H_{24}O$ M 220.354
Constit. of *Cedrus libanotica*. Oil. $[\alpha]_D^{25}$ −180.2° (c, 2.7 in CHCl$_3$).

Avcibasi, H. *et al, Phytochemistry*, 1987, **26**, 2852.

β-Torosol T-60226

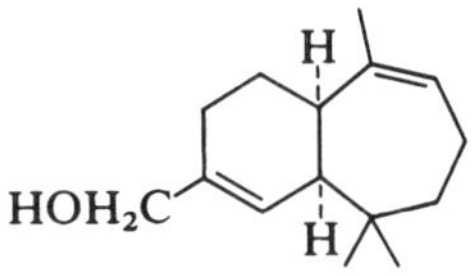

$C_{15}H_{24}O$ M 220.354
Constit. of *Cedrus libanotica*.

Avcibasi, H. *et al, Phytochemistry*, 1987, **26**, 2852.

Trechonolide A T-60227

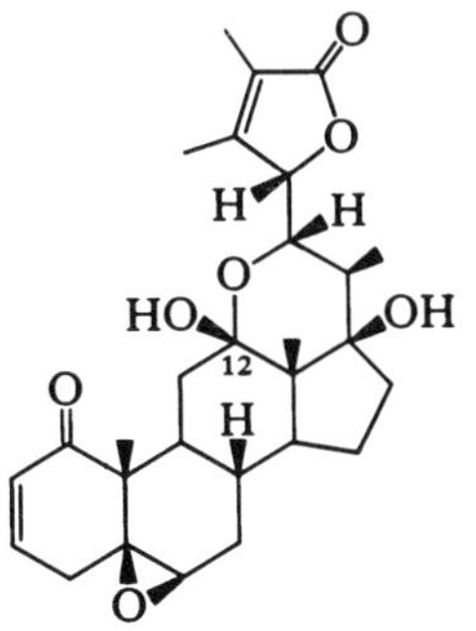

$C_{28}H_{36}O_7$ M 484.588
Constit. of *Trechnonaetes laciniata*. Cryst. (CH$_2$Cl$_2$/hexane). Mp 264-266°.

12-Me ether: Trechonolide B.
$C_{29}H_{38}O_7$ M 498.615

From *T. laciniata*. Prob. an artefact.

Lavie, D. *et al, Phytochemistry*, 1987, **26**, 1791 (*cryst struct*)

1,1,1-Triacetoxy-1,1-dihydro-1,2-benzio-doxol-3(1H)-one T-60228

Dess-Martin periodinane

[87413-09-0]

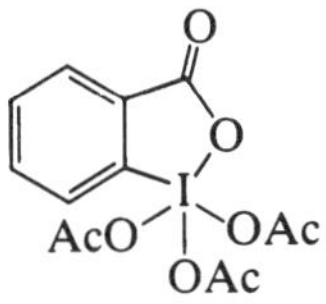

$C_{13}H_{13}IO_8$ M 424.145
Mild, selective reagent for the oxidation of primary and secondary alcohols to aldehydes and ketones. Mp 124-126° dec.

Dess, D.B. *et al, J. Org. Chem.*, 1983, **48**, 4155 (*synth, nmr, use*)
Becker, H.-D. *et al, J. Org. Chem.*, 1986, **51**, 2956 (*use*)

2,4,7-Triaminopteridine T-60229

2,4,7-Pteridinetriamine, 9CI

[14439-13-5]

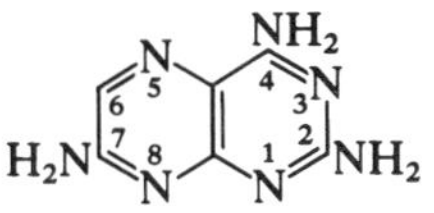

$C_6H_7N_7$ M 177.168
Yellow solid + 1H$_2$O. Mp >250°.

Albert, A. *et al, J. Chem. Soc.*, 1956, 4621.

4,6,7-Triaminopteridine T-60230

4,6,7-Pteridinetriamine, 9CI

[19167-62-5]
$C_6H_7N_7$ M 177.168
Pale-brown needles. Mp 179-181°.

Albert, A. *et al, J. Chem. Soc.*, 1956, 4621.

1,4,7-Triazacyclododecane, 9CI T-60231

[23635-83-8]

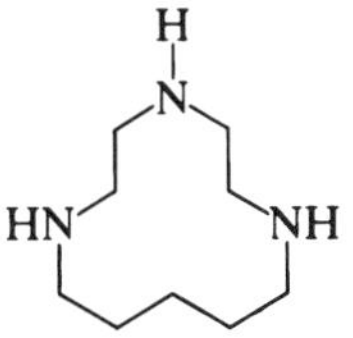

$C_9H_{21}N_3$ M 171.285
B,3HCl: [56187-03-2]. Cryst. + 1½H$_2$O. Mp 255° (230°).
Tris(4-methylbenzenesulfonyl): Mp 173-174°.

Richman, J.E. *et al, J. Am. Chem. Soc.*, 1974, **96**, 2268.
Briellmann, M. *et al, Helv. Chim. Acta*, 1987, **70**, 680 (*synth, pmr*)

1,5,9-Triazacyclododecane, 9CI   T-60232
[294-80-4]

$C_9H_{21}N_3$ M 171.285
Bp₃ 77-78°.

B,3HCl: Mp 286° (260°).
B,3HBr: [35980-62-2]. Mp 267.5-268.5°.
Tris(4-methylbenzenesulfonyl): Mp 172°.

Van Winkle, J.L. *et al, J. Org. Chem.*, 1966, **31**, 3300 (*synth, ir*)
Koyama, H. *et al, Bull. Chem. Soc. Jpn.*, 1972, **45**, 481 (*deriv*)
Zompa, L., *Inorg. Chem.*, 1978, **17**, 2531 (*synth*)
Briellmann, M. *et al, Helv. Chim. Acta*, 1987, **70**, 680 (*synth, pmr*)

1,5,9-Triazacyclotridecane, 9CI   T-60233
[54365-83-2]

$C_{10}H_{23}N_3$ M 185.312
B,3HCl: [70656-98-3]. Mp 282°.
Tris(4-methylbenzenesulfonyl): Monohydrate. Mp 232-235° (213-214°).

Weitl, F.L. *et al, J. Am. Chem. Soc.*, 1979, **101**, 2728 (*deriv*)
Kimura, E. *et al, Chem. Pharm. Bull.*, 1980, **28**, 994 (*deriv*)
Briellmann, M. *et al, Helv. Chim. Acta*, 1987, **70**, 680 (*synth, pmr*)

1,4,7-Triazacycloundecane   T-60234

$C_8H_{19}N_3$ M 157.258
B,3HCl: Mp 246°.
Tris(4-methylbenzenesulfonyl): Mp 172°.

Briellmann, M. *et al, Helv. Chim. Acta*, 1987, **70**, 680 (*synth, pmr*)

1,4,8-Triazacycloundecane, 9CI   T-60235
[35980-61-1]

$C_8H_{19}N_3$ M 157.258
B,3HCl: Mp 247° (243°).

Tris(4-methylbenzenesulfonyl): Mp 217° (213°).

Koyama, H. *et al, Bull. Chem. Soc. Jpn.*, 1972, **45**, 481 (*deriv*)
Zompa, L., *Inorg. Chem.*, 1978, **17**, 2531 (*synth*)
Briellmann, M. *et al, Helv. Chim. Acta*, 1987, **70**, 680 (*synth, pmr*)

19,20,21-Triazatetracyclo[13.3.1.1³,⁷.1⁹,¹³]-   T-60236
heneicosa-
1(19),3,5,7(21),9,11,13(20),15,17-non-
aene-2,8,14-trione, 9CI
1,3,5-Tri[2,6]pyridacyclohexaphane-2,4,6-trione
[68871-27-2]

$C_{18}H_9N_3O_3$ M 315.287
Yellow cryst. Mp 236.0-236.5°. Nonplanar.

Newkome, G.R. *et al, J. Am. Chem. Soc.*, 1986, **108**, 6074 (*synth, pmr, cmr, ir, uv, cryst struct*)

1,3,5-Triazine-2,4,6-tricarboxaldehyde   T-60237
Triformyl-s-triazine

$C_6H_3N_3O_3$ M 165.108
Free compd. unstable.

Trioxime:
   $C_6H_6N_6O_3$ M 210.152
   Dihydrate. Mp 284° dec.
Trisemicarbazone: Cryst. + 3H₂O (DMSO). Mp >320°.

Sumera, F. *et al, J. Heterocycl. Chem.*, 1987, **24**, 793.

[1,3,5]Triazino[1,2-a:3,4-a':5,6-a″]-   T-60238
trisbenzimidazole, 9CI
Tribenzimidazo[1,2-a:1',2'-c:1″,2″-e] [1,3,5]triazine
[32833-13-9]

$C_{21}H_{12}N_6$ M 348.366
Solid (DMF or toluene). Mp 380° (391-393°, 353-356°).

Gofen, G.I. *et al, Khim. Geterotsikl. Soedin.*, 1971, **7**, 282 (*synth*)
Lavagnino, E.R. *et al, J. Heterocycl. Chem.*, 1972, **9**, 149 (*synth, ir, pmr*)
Martin, D. *et al, J. Chem. Soc., Perkin Trans. 1*, 1985, 1007 (*synth*)

1*H*-1,2,3-Triazole-4-carboxylic acid, 9CI　　T-60239

v-*Triazole-4-carboxylic acid*, 8CI

[16681-70-2]

1*H*-*form*

C₃H₃N₃O₂　　M 113.076

2*H*-form probably predominates, in tautom. of acid and derivs.

(1*H*)-*form*

1-*Me:* [16681-71-3].
　C₄H₅N₃O₂　　M 127.102
　Cryst. (H₂O). Mp 218-219°, Mp 224° dec., Mp 240-242°.
1-*Me, Me ester:* [57362-82-0].
　C₅H₇N₃O₂　　M 141.129
　Cryst. (pet. ether or MeOH). Mp 159-161°.
1-*Me, Et ester:* [39039-48-0].
　C₆H₉N₃O₂　　M 155.156
　Cryst. (H₂O). Mp 93-94.5°.
1-*Me, amide:* [57362-84-2].
　C₄H₆N₄O　　M 126.118
　Cryst. (EtOH aq.). Mp 261-263°. Subl. at 150-220°.
1-*Ph:* see 1-*Phenyl-1H-1,2,3-triazole-4-carboxylic acid,*
　P-60138
1-*Benzyl, Me ester:* [76003-76-4].
　C₁₁H₁₁N₃O₂　　M 217.227
　Mp 104°.

(2*H*)-*form*

Plates (HCl aq.) or cryst. (H₂O). Mp 219° (211° dec.).
Me ester: [21977-71-9].
　C₄H₅N₃O₂　　M 127.102
　Cryst. (cyclohexane or MeOH). Mp 133-135°, Mp 145°.
Et ester: [40594-98-7].
　C₅H₇N₃O₂　　M 141.129
　Cryst. (CCl₄/CHCl₃). Mp 112-113° (117-118°).
Amide: [53897-99-7].
　C₃H₄N₄O　　M 112.091
　Cryst. (AcOH). Mp 261-263° (256-257°).
Nitrile: [18755-49-2]. 4-*Cyano-1,2,3-triazole*.
　C₃H₂N₄　　M 94.076
　Cryst. (C₆H₆). Mp 113-114°.
2-*Me:* [63062-88-4].
　C₄H₅N₃O₂　　M 127.102
　Cryst. (C₆H₆). Mp 141-142°.
2-*Me, Me ester:* [105020-39-1].
　C₅H₇N₃O₂　　M 141.129
　Cryst. (pet. ether). Mp 63-64°.
2-*Me, Et ester:*
　C₆H₉N₃O₂　　M 155.156
　Oil. Bp₆₀ 115°.
2-*Ph:* see 2-*Phenyl-2H-1,2,3-triazole-4-carboxylic acid,*
　P-60140

(3*H*)-*form*

1H-*1,2,3-Triazole-5-carboxylic acid*
3-*Me, Me ester:* [105020-38-0]: *Methyl 1-methyl-1H-*
　1,2,3-triazole-5-carboxylate.
　C₅H₇N₃O₂　　M 141.129
　Cryst. (pet. ether). Mp 55.5-56.5°.
3-*Ph:* see 1-*Phenyl-1H-1,2,3-triazole-5-carboxylic acid,*
　P-60139
3-*Benzyl, Me ester: Methyl 1-benzyl-1H-1,2,3-triazole-*
　5-carboxylate.
　C₁₁H₁₁N₃O₂　　M 217.227

Mp 44°.

Peratoner, A. *et al, Gazz. Chim. Ital.*, 1908, **38I**, 76 (*synth, derivs*)
Pedersen, C., *Acta Chem. Scand.*, 1959, **13**, 888 (*synth, derivs*)
Thurber, T.C. *et al, J. Org. Chem.*, 1976, **41**, 1041 (*synth, derivs*)
Huisgen, R. *et al, Chem. Ber.*, 1987, **130**, 159 (*derivs*)

[1,2,4]Triazolo[1,5-*b*]pyridazine, 9CI　　T-60240

[1,2,4]*Triazolo*[2,3-b]*pyridazine*

[40369-38-8]

C₅H₄N₄　　M 120.113
Cryst. (CHCl₃/hexane). Mp 138-142°.

Polanc, S. *et al, J. Org. Chem.*, 1974, **39**, 2143 (*synth*)
Pugmire, R.J. *et al, J. Heterocycl. Chem.*, 1987, **24**, 805 (*cmr*)

[1,2,3]Triazolo[1,5-*a*]pyridine, 9CI　　T-60241

2,3,4-*Pyridotriazole*

[274-59-9]

C₆H₅N₃　　M 119.126
Light-yellow oil or cryst. Bp₀.₀₉ 78°.

Boyer, J.H. *et al, J. Am. Chem. Soc.*, 1978, **79**, 678 (*synth*)
Pugmire, R.J. *et al, J. Heterocycl. Chem.*, 1987, **24**, 805 (*cmr*)

[1,2,4]Triazolo[1,5-*a*]pyrimidine, 9CI　　T-60242

Updated Entry replacing T-02387

[275-02-5]

C₅H₄N₄　　M 120.113
Complexing agent for transition metals. Mp 144-145°.

Allen, C.F.H., *J. Org. Chem.*, 1959, **24**, 796.
Tamura, Y. *et al, J. Heterocycl. Chem.*, 1975, **12**, 107.
Brown, D.J. *et al, Aust. J. Chem.*, 1977, **30**, 2515.
Biagini Cingi, M. *et al, Acta Crystallogr., Sect. C*, 1986, **42**, 427 (*cryst struct, complex*)
Pugmire, R.J. *et al, J. Heterocycl. Chem.*, 1987, **24**, 805 (*cmr*)

1,2,4-Triazolo[4,3-*c*]pyrimidine, 9CI　　T-60243

[274-81-7]

C₅H₄N₄　　M 120.113
Mp 165-167°.

Brown, D.J. *et al, Aust. J. Chem.*, 1978, **31**, 2505 (*synth, uv*)
Pugmire, R.J. *et al, J. Heterocycl. Chem.*, 1987, **24**, 805 (*cmr*)

Tribenzo[*b,n,pqr*]perylene   T-60244
[190-81-8]

$C_{30}H_{16}$ M 376.456
Plates or needles (xylene). Mp 388-389° (382-385°).

Clar, E. *et al, J. Chem. Soc.*, 1958, 1861 (*synth, uv*)
Burke, R. *et al, Chem. Ber.*, 1986, **119**, 3521 (*synth, pmr*)

2,4,5-Tribromo-1*H*-imidazole, 9CI   T-60245
Updated Entry replacing T-50374
Tribromoglyoxaline
[2034-22-2]

$C_3HBr_3N_2$ M 304.766
Silky needles (AcOH). Mp 221°.
▷Prob. neurotoxic. NI8660000.

Balaban, E. *et al, J. Chem. Soc.*, 1922, 947 (*synth*)
Stensiö, K.-E. *et al, Acta Chem. Scand.*, 1973, **27**, 2179 (*synth*)
Iddon, B. *et al, J. Chem. Soc., Chem. Commun.*, 1985, 1428
  (*deriv, synth*)
Iddon, B. *et al, Tetrahedron Lett.*, 1986, **27**, 1635 (*haz, use*)

3,4,5-Tribromo-1*H*-pyrazole   T-60246
[17635-44-8]

$C_3HBr_3N_2$ M 304.766
Cryst. (EtOH aq.). Mp 188°.
1-Nitro: [104599-40-8].
  $C_3Br_3N_3O_2$ M 349.764
  Yellowish solid. Mp 74°.

Juffermans, J.P.H. *et al, J. Org. Chem.*, 1986, **51**, 4656 (*synth,
  ir, ms*)

2,3,5-Tribromothiophene   T-60247
[3141-24-0]

C_4HBr_3S M 320.824
Mp 25-27°. Bp_{20} 136-138°, Bp_9 123-124°.

Org. Synth., Coll. Vol., **5**, 149 (*synth*)
Antionoletti, R. *et al, J. Chem. Soc., Perkin Trans. 1*, 1986,
  1755 (*pmr, ms*)

1,4,7-Tribromotriquinacene   T-60248
2a,4a,6a-*Tribromo*-2a,4a,6a,6b-
tetrahydrocyclopenta[cd]*pentalene*, 9CI
[91682-00-7]

$C_{10}H_7Br_3$ M 366.877
Cryst. (pentane). Mp 128°.

Butenschön, H. *et al, Chem. Ber.*, 1985, **118**, 2757 (*synth, ir,
  pmr, cmr, ms*)

Tri-*tert*-butylazete   T-60249
Tris(1,1-dimethylethyl)azete
[103794-87-2]

$(CH_3)_3C$ $C(CH_3)_3$
$(CH_3)_3C$ N

$C_{15}H_{27}N$ M 221.385
Red-brown needles (pentane). Mp 37°.

Vogelbacher, V.J. *et al, Angew. Chem., Int. Ed. Engl.*, 1986, **25**,
  842 (*synth, ir, uv, pmr, cmr, ms*)

Tri-*tert*-butyl-1,2,3-triazine   T-60250
Tris(1,1-dimethylethyl)-1,2,3-triazine, 9CI
[103794-88-3]

$(CH_3)_3C$ $C(CH_3)_3$
$(CH_3)_3C$ N
            N

$C_{15}H_{27}N_3$ M 249.398
Cryst. (pentane). Mp 130°.

Vogelbacher, V.J. *et al, Angew. Chem., Int. Ed. Engl.*, 1986, **25**,
  842.

Trichilinin   T-60251

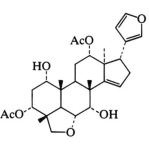

$C_{30}H_{40}O_8$ M 528.641
Constit. of *Trichilia roka*.

Nakatani, M. *et al, Heterocycles*, 1987, **26**, 43.

(2,2,2-Trichloroethyl)oxirane, 9CI   T-60252
1,1,1-Trichloro-3,4-epoxybutane, 8CI
[3083-25-8]

$C_4H_5Cl_3O$ M 175.442
▷EK3875000.

(±)-form
Intermed. in manuf. of agricultural and pharmaceutical chemicals and in prepn. of low flammability plastics. Much patented. Liq. Bp$_{100}$ 110°.

Dowbenko, R., *Tetrahedron*, 1965, **21**, 1647 (*synth*)
Bourgeois, M.J. *et al*, *Tetrahedron*, 1986, **42**, 5309 (*synth*)

5-(Trichloromethyl)isoxazole, 9CI   T-60253
[88283-10-7]

C$_4$H$_2$Cl$_3$NO   M 186.425
Mp 24-26°. Bp$_{40}$ 98°.

Spiegler, W. *et al*, *Synthesis*, 1986, 69 (*synth, ir, pmr*)

1,1,2-Trichloro-1-propene, 9CI   T-60254
[21400-25-9]

$$Cl_2C{=}CClCH_3$$

C$_3$H$_3$Cl$_3$ M 145.416
Bp 118°.

Kirrmann, A. *et al*, *Bull. Soc. Chim. Fr.*, 1948, 168 (*synth*)
Bauer, H. *et al*, *Chem. Ber.*, 1986, **119**, 1890 (*synth, ir, pmr, cmr*)

2,4,6-Trichloropyridine   T-60255
Updated Entry replacing T-02866
[16063-69-7]
C$_5$H$_2$Cl$_3$N   M 182.437
Mp 33°.
▷UU0875000.

Graf, R., *J. Prakt. Chem.*, 1932, **133**, 44.
Kaneko, C. *et al*, *Chem. Pharm. Bull.*, 1986, **34**, 3658 (*synth*)

1,2,3-Trichloro-4,5,6-trifluorobenzene, 9CI, 8CI   T-60256
[827-12-3]

C$_6$Cl$_3$F$_3$ M 235.420

Chambers, R.D. *et al*, *Tetrahedron*, 1963, **19**, 891 (*F nmr*)

1,3,5-Trichloro-2,4,6-trifluorobenzene, 9CI, 8CI   T-60257
[319-88-0]
C$_6$Cl$_3$F$_3$ M 235.420
Cryst. (EtOH). Mp 61-62°. Bp$_{12}$ 79.5°.
▷DC2450000.

Finger, G.C. *et al*, *J. Am. Chem. Soc.*, 1951, **73**, 149, 153 (*synth*)
Fuller, G., *J. Chem. Soc.*, 1965, 6264 (*synth*)
Gage, J.C., *Br. J. Ind. Med.*, 1970, **27**, 1 (*tox*)
Hitzke, J. *et al*, *Org. Mass Spectrom.*, 1972, **6**, 349 (*ms*)
Green, J.H.S. *et al*, *J. Mol. Spectrosc.*, 1976, **62**, 228 (*ir, raman*)
Chaplot, S.L. *et al*, *Acta Crystallogr.*, Sect. B, 1981, **37**, 1896 (*cryst struct*)
Yoshioka, Y. *et al*, *J. Mol. Struct.*, 1983, **111**, 195 (*F nmr, nqr*)
Yoshioka, Y. *et al*, *Z. Naturforsch.*, B, 1985, **40**, 137 (*F nmr*)

Trichotriol   T-60258

C$_{15}$H$_{24}$O$_4$ M 268.352
Isol. from *Fusarium sporotrichioides*. Mycotoxin. Oil.

Corley, D.G. *et al*, *J. Org. Chem.*, 1987, **52**, 4405.

9-Tricosene   T-60259
Updated Entry replacing T-20208
[52078-48-5]

$$H_3C(CH_2)_7CH{=}CH(CH_2)_{12}CH_3$$

C$_{23}$H$_{46}$ M 322.616
(E)-form [35857-62-6]
Bp$_{0.35}$ 164-165°. n$_D^{25}$ 1.4529.
(Z)-form [27519-02-4]
Muscalure
Sex attractant of the housefly. Oil. Bp$_{0.15}$ 170-172° (Bp$_{0.2}$ 160-161°). n$_D^{25}$ 1.4524.

Gribble, G.W. *et al*, *J. Chem. Soc., Chem. Commun.*, 1973, 735 (*synth*)
Küpper, F.-W. *et al*, *Chem.-Ztg.*, 1975, **99**, 464; *Z. Naturforsch., B*, 1976, **31**, 1256 (*synth*)
Rossi, R., *Chim. Ind. (Milan)*, 1975, **57**, 242 (*synth, pmr*)
Abe, K. *et al*, *Bull. Chem. Soc. Jpn.*, 1977, **50**, 2792 (*synth, ir, pmr*)
Cormier, R.A. *et al*, *J. Chem. Educ.*, 1979, **56**, 345 (*synth*)
Shani, A., *J. Chem. Ecol.*, 1979, **5**, 557 (*synth*)
Naoshima, Y. *et al*, *Agric. Biol. Chem.*, 1981, **45**, 1723 (*synth*)
Brown, H.C. *et al*, *J. Org. Chem.*, 1982, **47**, 3806 (*synth*)
Odinokov, V.N. *et al*, *Tetrahedron Lett.*, 1982, **23**, 1371 (*synth*)
Koumaglo, K. *et al*, *Tetrahedron Lett.*, 1984, **25**, 717 (*synth*)
Yadav, A.K. *et al*, *Helv. Chim. Acta*, 1984, **67**, 1698 (*synth*)
Wenkert, E. *et al*, *J. Org. Chem.*, 1985, **50**, 719 (*synth, pmr*)
Subramanian, G.B.V. *et al*, *Tetrahedron*, 1986, **42**, 3967 (*synth, ir, pmr*)

Tricyclo[5.2.1.0^{2,6}]dec-2(6)-ene   T-60260
2,3,4,5,6,7-Hexahydro-4,7-methano-1H-indene, 9CI
[87238-75-3]

C$_{10}$H$_{14}$ M 134.221
Oil.

Takaishi, N. *et al*, *J. Org. Chem.*, 1986, **51**, 4862 (*synth, ir, pmr, cmr, ms*)

Tricyclo[4.4.2.0^{1,6}]dodecane-2,7-dione   T-60261
Tetrahydro-4a,8a-ethanonaphthalene-1,5(2H,6H)-dione, 9CI
[42245-90-9]

C$_{12}$H$_{16}$O$_2$ M 192.257

Cryst. (hexane or by subl.). Mp 45-48°.

Peet, N.P. *et al*, *J. Org. Chem.*, 1973, **25**, 4281 (*synth, ir, uv, pmr*)

Jeffrey, D.A. *et al*, *J. Org. Chem.*, 1986, **51**, 3206 (*synth*)

Tricyclo[4.1.0.0^{1,3}]heptane   T-60262

[174-73-2]

C$_7$H$_{10}$ M 94.156
Bp 108°. n_D^{25} 1.4652.

Skattebøl, L., *J. Org. Chem.*, 1966, **31**, 2789 (*synth*)
Gleiter, R. *et al*, *J. Org. Chem.*, 1986, **51**, 2899 (*pe*)

Tricyclo[18.4.1.1^{8,13}]hexacosa-   T-60263
2,4,6,8,10,12,14,16,18,20,22,24-dode-
caene, 9CI

[105363-23-3]

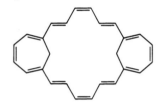

C$_{26}$H$_{24}$ M 336.476
Paratropic 24π electron system. Black-brown needles. Mp 200° dec.

Yamamoto, K. *et al*, *Tetrahedron Lett.*, 1986, **27**, 975 (*synth, pmr*)

Tricyclo[8.4.1.1^{3,8}]hexadeca-   T-60264
3,5,7,10,12,14-hexaene-2,9-dione, 9CI
Bishomoanthraquinone

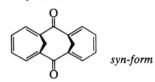

syn-form

C$_{16}$H$_{12}$O$_2$ M 236.270
syn-form [104713-91-9]
Yellow needles (CH$_2$Cl$_2$/Et$_2$O). Mp 198-199°.
anti-form [104713-92-0]
Colourless cryst. Mp 299-300°. Shows no pronounced oxidising activity.

Vogel, E. *et al*, *Angew. Chem., Int. Ed. Engl.*, 1966, **5**, 603; 1986, **25**, 1000 (*synth, pmr, cmr, uv*)

Tricyclo[3.1.1.1^{2,4}]octane, 9CI   T-60265
Diasterane
[51273-49-5]

C$_8$H$_{12}$ M 108.183
Cryst. Mp 96-97° (sealed tube). Readily subl.

Otterbach, A. *et al*, *Angew. Chem., Int. Ed. Engl.*, 1987, **26**, 554 (*synth, pmr, cmr*)

Tricyclo[5.1.0.0^{2,8}]octane, 9CI   T-60266

[36328-29-7]

C$_8$H$_{12}$ M 108.183
Liq.

Christl, M. *et al*, *Chem. Ber.*, 1986, **119**, 3045, 3067 (*synth, pmr, cmr*)

Tricyclo[3.2.1.0^{2,7}]oct-3-ene   T-60267
Tricyclo[2.2.2.0^{2,6}]oct-7-ene
[3725-23-3]

C$_8$H$_{10}$ M 106.167

Grob, C.A. *et al*, *Helv. Chim. Acta*, 1963, **46**, 1676 (*synth, uv, ir*)
Doering, W von E. *et al*, *Tetrahedron*, 1963, **19**, 715 (*synth*)
Sauers, R.R. *et al*, *J. Org. Chem.*, 1968, **33**, 799 (*synth*)

Tricyclo[5.1.0.0^{2,8}]oct-3-ene, 9CI   T-60268

[102575-26-8]

C$_8$H$_{10}$ M 106.167
Bp$_{30}$ 48°.

Christl, M. *et al*, *Chem. Ber.*, 1986, **119**, 3045, 3067 (*synth, pmr, cmr*)

Tricyclo[5.1.0.0^{2,8}]oct-4-ene, 9CI   T-60269

[102575-25-7]
C$_8$H$_{10}$ M 106.167
Liq.

Christl, M. *et al*, *Chem. Ber.*, 1986, **119**, 3045, 3067 (*synth, pmr, cmr*)

Tricyclopropylcyclopropenium(1+)   T-60270

C$_{12}$H$_{15}^{\oplus}$ M 159.251 (ion)
Chloride: [75094-00-7].
    C$_{12}$H$_{15}$Cl   M 194.704
    Cryst. (MeCN/Et$_2$O). Mp 89-90°.
Tetrafluoroborate: [75359-38-5].
    C$_{12}$H$_{15}$BF$_4$ M 246.054
    Cryst. (MeCN/Et$_2$O). Mp 137-140°.
Hexafluoroantimonate: [99310-21-1].
    C$_{12}$H$_{15}$F$_6$Sb   M 394.991
    Cryst. (Et$_2$O/MeCN). Mp 213-215°.

Komatsu, K. *et al*, *Tetrahedron Lett.*, 1980, **21**, 947 (*synth, ir, pmr, cmr, uv*)
Moss, R.A. *et al*, *J. Am. Chem. Soc.*, 1986, **108**, 134 (*synth, cryst struct, ir, uv, pmr*)

Tricycloquinazoline, 9CI   T-60271

Updated Entry replacing T-03010
Tetraazabenzo[a]*naphth*[*1,2,3*-de]*anthracene*
[195-84-6]

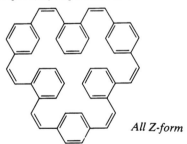

$C_{21}H_{12}N_4$ M 320.353
Fluffy yellow needles (toluene). Mp 322-323°.

▷Carcinogen. YD2800000.

Cooper, F.C. *et al, J. Chem. Soc.*, 1954, 3429 (*synth*)
Ball, J. *et al, Acta Crystallogr., Sect. B*, 1969, **25**, 882 (*cryst struct*)
Bailey, M.L. *et al, Acta Crystallogr., Sect. B*, 1970, **26**, 1622 (*cryst struct*)
Pakrashi, S.C. *et al, J. Org. Chem.*, 1971, **36**, 642 (*synth, ms*)
Yoneda, F. *et al, Chem. Pharm. Bull.*, 1973, **21**, 1610 (*synth*)
Yamaguchi, H. *et al, J. Chem. Soc., Faraday Trans. 2*, 1979, **75**, 1506 (*spectra, bibl*)
Sax, N.I., *Dangerous Properties of Industrial Materials*, 5th Ed., Van Nostrand-Reinhold, 1979, 1049.

6,7-Tridecadiene   T-60272

1,3-Dibutylallene

$C_{13}H_{24}$ M 180.333
(R)-form [93588-88-6]
$[\alpha]_{365}^{20} -67°$ (c, 0.14 in CH_2Cl_2). Enantiomeric purity 49 ± 1%.
(±)-form [104597-02-6]
No phys. props. given.

Mannschreck, A. *et al, Tetrahedron*, 1986, **42**, 399 (*synth, ir, pmr, ms*)

7,10:19,22:31,34-Triethenotribenzo[a,k,u]-   T-60273
cyclotriacontane, 9CI

[2₆](Orthopara)₃cyclophanehexaene

All Z-form

$C_{48}H_{36}$ M 612.812
(all-E)-form [104184-80-7]
Mp >370° dec.
(all-Z)-form [66726-67-8]
Mp 257-260°.

Thulin, B. *et al, Acta Chem. Scand., Ser. B*, 1978, **32**, 109 (*synth, pmr, ir, uv*)
Sundahl, M. *et al, Tetrahedron Lett.*, 1986, **27**, 1063 (*isom, uv, pmr, ir*)

Trifluoroacetyl nitrite, 8CI   T-60274

Trifluoroacetic acid, anhydride with nitrous acid, 9CI
[667-29-8]

$$F_3CCOONO$$

$C_2F_3NO_3$ M 143.022
Precursor to Trifluoronitrosomethane, T-20237 . Golden-yellow liq. d_4^{20} 1.607. Bp_{750} 101°, Bp_{66} 40.3°. n_D^{25} 1.3722.

▷Explosive when vapour heated above 100°C

Banks, R.E. *et al, J. Chem. Soc. (C)*, 1966, 1350 (*synth, haz, bibl*)
Banks, R.E. *et al, J. Chem. Soc., Perkin Trans. 1*, 1974, 2532 (*synth, ir*)
Umemoto, T. *et al, Bull. Chem. Soc. Jpn.*, 1983, **56**, 631 (*synth*)

1,1,1-Trifluoro-2-butyne, 9CI, 8CI   T-60275

[406-41-7]

$$F_3CC{\equiv}CCH_3$$

$C_4H_3F_3$ M 108.063
Gas. Bp 18-18.5° (extrap.).

Haszeldine, R.N. *et al, J. Chem. Soc.*, 1954, 1261 (*synth*)
Laurie, V.W., *J. Chem. Phys.*, 1959, **30**, 1101 (*struct*)
Tuazon, E.C. *et al, J. Chem. Phys.*, 1970, **53**, 3178 (*synth, ir, raman*)
Mielcarek, J.J. *et al, J. Fluorine Chem.*, 1978, **12**, 321 (*synth*)

1,4,4-Trifluorocyclobutene, 9CI   T-60276

[3932-66-9]

$C_4H_3F_3$ M 108.063
Volatile liq. d_4^{25} 1.190. Bp_{631} 26.8°. n_D^{25} 1.3170. Polymerises on standing.

Park, J.D. *et al, J. Org. Chem.*, 1960, **25**, 990 (*synth*)
Newark, R.A. *et al, J. Magn. Reson.*, 1969, **1**, 418 (*F nmr*)

Trifluoroethenesulfonic acid, 9CI   T-60277

Trifluorovinylsulfonic acid

$$F_2C{=}CFSO_3H$$

$C_2HF_3O_3S$ M 162.083
Fluoride: [684-10-6].
   $C_2F_4O_2S$ M 164.074
   Monomer. Liq. Bp 51.8-52.2°. $n_D^{25.4}$ 1.3237.

U.S.P., 3 041 317, (*1962*); *CA*, **58**, 451
Banks, R.E. *et al, J. Chem. Soc. (C)*, 1966, 1171 (*synth, F nmr*)

Trifluoroisocyanomethane, 9CI   T-60278

Trifluoromethyl isocyanide, 8CI
[19480-01-4]

$$F_3CNC$$

C_2F_3N M 95.024
Gas. Bp -80° (extrapol.). Polymerises at pressure >200 mm Hg or in condensed phase.

Makarov, S.P. *et al, Zh. Obshch. Khim.*, 1967, **37**, 2781 (*Engl. transl. p. 2667*) (*synth*)

Banks, R.E. *et al*, *J. Chem. Soc. (C)*, 1969, 2119 (*synth*)
Bock, H. *et al*, *Inorg. Chem.*, 1984, **23**, 1535 (*pe*)
Christen, D. *et al*, *J. Chem. Phys.*, 1984, **80**, 4020 (*ed*)
Lentz, D., *Ann. Chim. (Paris)*, 1984, **9**, 665 (*synth, bibl*)
Lentz, D. *et al*, *J. Fluorine Chem.*, 1984, **24**, 523 (*synth, F nmr, ir, ms*)

Trifluoromethanesulfenic acid, 9CI   T-60279

[306-79-6]

$$F_3CSOH$$

CHF_3OS M 118.074
Me ester: [1544-50-9]. *Methyl trifluoromethanesulfenate.*
$C_2H_3F_3OS$ M 132.100
Liq. Bp 25-26°.
Et ester: [691-04-3]. *Ethyl trifluoromethanesulfenate.*
$C_3H_5F_3OS$ M 146.127
Liq. Bp 47-49°.
2,2,2-Trifluoroethyl ester: Liq. Bp 44-50°.
Isopropyl ester:
$C_4H_7F_3OS$ M 160.154
Liq. Bp 68-70°.

Haszeldine, R.N. *et al*, *J. Chem. Soc.*, 1955, 2901.

Trifluoromethanesulfenyl bromide, 9CI   T-60280

[753-92-4]

$$F_3CSBr$$

$CBrF_3S$ M 180.970
Orange-red liq. d_4^{16} 1.770. Bp 36°. n_D^{16} 1.3855. Dec. by sunlight to Br_2 and CF_3SSCF_3.

Yarovenko, N.N. *et al*, *Zh. Obsch. Khim.*, 1959, **29**, 3786 (*Engl. transl. p. 3749*) (*synth*)
Ratcliffe, C.T. *et al*, *J. Am. Chem. Soc.*, 1968, **90**, 5403 (*synth, F nmr, ir*)
Altabef, A.B. *et al*, *Z. Anorg. Allg. Chem.*, 1983, **506**, 161 (*synth, ir, raman, props*)
Minkwitz, R. *et al*, *Z. Anorg. Allg. Chem.*, 1985, **527**, 161; **531**, 31 (*synth, ir, raman, F nmr, cmr, uv*)

Trifluoromethanesulfenyl fluoride, 9CI   T-60281

[17742-04-0]

$$F_3CSF$$

CF_4S M 120.065
Highly reactive substance. Forms dimer at low temp. Stored at ambient temp. <100 mm Hg.
▷Highly toxic
Dimer: [26391-89-9].
  Difluoro(trifluoromethanethiolato)(trifluoromethyl)-sulfur, 9CI, 8CI.
  $C_2F_8S_2$ M 240.129
  Dimer in equilibrium with CF_3SF. Has struct. $F_3CSF_2-S-CF_3$.

Stump, E.C., *Chem. Eng. News*, 1967, **45**, 16 (*tox*)
Seel, F. *et al*, *Angew. Chem., Int. Ed. Engl.*, 1969, **8**, 773 (*dimer, synth, fmr, props*)
Gombler, W. *et al*, *J. Fluorine Chem.*, 1974, **4**, 333 (*dimer, F nmr*)
Gombler, W. *et al*, *Z. Anorg. Allg. Chem.*, 1978, **439**, 193; 1980, **469**, 139 (*synth, ir, F nmr, ms, dimer*)
Oberhammer, H. *et al*, *J. Mol. Struct.*, 1981, **70**, 273 (*ed, ir, raman, bibl*)

Trifluoromethanesulfenyl iodide, 9CI   T-60282

[102127-62-8]

$$F_3CSI$$

CF_3IS M 227.971
Orange-yellow solid at −100°.

Minkwitz, R. *et al*, *Z. Anorg. Allg. Chem.*, 1985, **527**, 161 (*synth, F nmr, cmr, uv, raman*)

3-(Trifluoromethyl)benzenesulfonic acid   T-60283

[1643-69-2]

$C_7H_5F_3O_3S$ M 226.170
Mp 42-44°.
Et ester: [55400-68-5].
  $C_9H_9F_3O_3S$ M 254.224
  $Bp_{0.08}$ 66°.
Chloride: [777-44-6].
  $C_7H_4ClF_3O_2S$ M 244.616
  Bp_6 88-90°.
Amide:
  $C_7H_6F_3NO_2S$ M 225.185
  Mp 121-122°.
S-Benzythiuronium salt: Mp 138-139°.

Yagupol'skii, L.M. *et al*, *J. Gen. Chem. USSR (Engl. Transl.)*, 1959, **29**, 549 (*synth*)
Dannley, R.L. *et al*, *J. Org. Chem.*, 1975, **40**, 2278 (*deriv, synth*)
Crumrine, D.S. *et al*, *J. Org. Chem.*, 1986, **51**, 5013 (*synth, nmr*)

2-(Trifluoromethyl)benzenethiol, 9CI   T-60284

$α,α,α$-*Trifluoro-o-toluenethiol, 8CI. 2-(Trifluoromethyl)thiophenol. 1-Mercapto-2-(trifluoromethyl)benzene*
[13333-97-6]

$C_7H_5F_3S$ M 178.172
Bp_{10} 62-64°. n_D^{27} 1.4961.
S-Me: [322-58-7]. *1-(Methylthio)-2-(trifluoromethyl)-benzene, 9CI. 2-(Trifluoromethyl)phenyl methyl sulfide.* Liq. d^{25} 1.308. Mp 3.7-5.0°. Bp 208-209°, Bp_{10} 82-88°. n_D^{26} 1.5092.

McBee, E.T. *et al*, *J. Am. Chem. Soc.*, 1950, **72**, 4235 (*ethers*)
Sharghi, H. *et al*, *J. Chem. Eng. Data*, 1966, **11**, 612 (*synth, derivs*)

3-(Trifluoromethyl)benzenethiol, 9CI   T-60285

$α,α,α$-*Trifluoro-m-toluenethiol, 8CI. 3-(Trifluoromethyl)thiophenol. 1-Mercapto-3-(trifluoromethyl)benzene*
[937-00-8]
$C_7H_5F_3S$ M 178.172
Liq. Bp 161-163°. $η_D^{20}$ 1.4122.
S-Me: [328-98-3]. *1-(Methylthio)-3-(trifluoromethyl)-benzene, 9CI. 3-(Trifluoromethyl)phenyl methyl sulfide.*
$C_9H_7F_3S$ M 204.210

Liq. d^{25} 1.270. Bp 192-193°. η_D^{25} 1.497.

Soper, Q.T. *et al, J. Am. Chem. Soc.*, 1948, **70**, 2849 (*synth*)
McBee, E.T. *et al, J. Am. Chem. Soc.*, 1950, **72**, 4235 (*ethers*)
Stacey, G.W. *et al, J. Org. Chem.*, 1959, **24**, 1892 (*synth*)
Mindl, J. *et al, Collect. Czech. Chem. Commun.*, 1980, **45**, 3130 (*synth*)

4-(Trifluoromethyl)benzenethiol, 9CI   T-60286

α,α,α-Trifluoro-p-toluenethiol, 8CI. 4-(Trifluoromethyl)thiophenol. 1-Mercapto-4-(trifluoromethyl)benzene

[825-83-2]

$C_7H_5F_3S$ M 178.172
Liq. Bp$_{13}$ 60-61°.
S-Me: 1-(Methylthio)-4-(trifluoromethyl)benzene, 9CI. 4-(Trifluoromethyl)phenyl methyl sulfide.
$C_8H_7F_3S$ M 192.199
Mp 37°. Bp 198.5-199.0°.

Japan. Pat., 63 2 631, (*1963*); *CA*, **64**, 15858 (*synth*)

2-(Trifluoromethyl)-1H-benzimidazole, 9CI   T-60287

[312-73-2]

$C_8H_5F_3N_2$ M 186.136
Herbicide. Cryst. Mp 209-210°. pK_a 4.51 (5% EtOH aq., 30°).

Smith, W.T. *et al, J. Am. Chem. Soc.*, 1953, **75**, 1292 (*synth*)
Morgan, K.J., *J. Chem. Soc.*, 1961, 2343 (*ir*)
Rabiger, R.J. *et al, J. Org. Chem.*, 1964, **29**, 476 (*uv*)
Lopyrev, V.A. *et al, Org. Magn. Reson.*, 1981, **15**, 1219 (*cmr, pmr*)
Crank, G. *et al, Aust. J. Chem.*, 1982, **35**, 775 (*synth*)
Adamson, G.W. *et al, Pestic. Sci.*, 1984, **15**, 31 (*tox*)

(Trifluoromethyl)carbonimidic difluoride, 9CI   T-60288

(Trifluoromethyl)imidocarbonyl fluoride, 8CI. Perfluoroazapropene. Perfluoro(methylenemethylamine)

[371-71-1]

$$F_3CN{=}CF_2$$

C_2F_5N M 133.021
Gas. Bp −33.7° (extrap.). Readily hydrolysed.

Barr, D.A. *et al, J. Chem. Soc.*, 1955, 1881 (*synth*)
Young, J.A. *et al, J. Am. Chem. Soc.*, 1956, **78**, 5637 (*synth*)
Banks, R.E. *et al, J. Chem. Soc. (C)*, 1966, 901 (*synth*)

2-Trifluoromethyl-1H-imidazole   T-60289

[66675-22-7]

$C_4H_3F_3N_2$ M 136.076
Needles (CHCl$_3$). Mp 145-146°.

Kimoto, H. *et al, J. Org. Chem.*, 1978, **43**, 3403 (*synth, pmr*)

4(5)-Trifluoromethyl-1H-imidazole   T-60290

[33468-69-8]

$C_4H_3F_3N_2$ M 136.076
Cryst. (H$_2$O). Mp 148.5-149.5°.

Baldwin, J.E. *et al, J. Med. Chem.*, 1975, **18**, 895 (*synth*)

2-Trifluoromethyl-1H-indole, 9CI   T-60291

[51310-54-4]

$C_9H_6F_3N$ M 185.148
Plates (hexane). Mp 107-108° (102°).

Kobayashi, Y. *et al, J. Org. Chem.*, 1974, **39**, 1836 (*synth, ir, pmr*)
Girard, Y. *et al, J. Org. Chem.*, 1983, **48**, 3220 (*synth*)

3-Trifluoromethyl-1H-indole, 9CI   T-60292

[51310-55-5]
$C_9H_6F_3N$ M 185.148
Plates (hexane). Mp 110°.

Kobayashi, Y. *et al, J. Org. Chem.*, 1974, **39**, 1836 (*synth, ir, pmr*)
Girard, Y. *et al, J. Org. Chem.*, 1983, **48**, 3220 (*synth*)

6-Trifluoromethyl-1H-indole, 9CI   T-60293

[13544-43-9]
$C_9H_6F_3N$ M 185.148
Mp 109-110°.

Kalir, A. *et al, Isr. J. Chem.*, 1966, **4**, 155 (*synth*)
Raucher, S. *et al, J. Org. Chem.*, 1983, **48**, 2066 (*synth*)
Tischler, A.N. *et al, Tetrahedron Lett.*, 1986, **27**, 1653 (*synth*)

(Trifluoromethyl)oxirane, 9CI   T-60294

1,1,1-Trifluoro-2,3-epoxypropane, 8CI. 2,3-Epoxy-1,1,1-trifluoropropane

[359-41-1]

$C_3H_3F_3O$ M 112.051
(±)-form
Liq. d$_{20}^{20}$ 1.307. Bp$_{748.3}$ 39.1-39.3°. n_D^{20} 1.2997.

McBee, E.T. *et al, J. Am. Chem. Soc.*, 1952, **74**, 3902 (*synth*)
Politzer, P. *et al, Carcinogenesis*, 1984, **5**, 845; *CA*, **101**, 105531 (*tox*)

2-(Trifluoromethyl)propenoic acid, 9CI   T-60295

2-Trifluoromethylacrylic acid
[381-98-6]

$$H_2C{=}C(CF_3)COOH$$

$C_4H_3F_3O_2$ M 140.062
Mp 52.5-53° (50-51°). Bp 146-148°, Bp$_{28}$ 90°.
Me ester: [382-90-1].
$C_5H_5F_3O_2$ M 154.089
Liq. Bp 103.8-105°.
▷Lachrymator

Amide:
C₄H₄F₃NO   M 139.077
Cryst. (C₆H₆). Mp 104°.
Nitrile: [381-84-0]. *2-Trifluoromethylacrylonitrile. 2-Cyano-3,3-trifluoropropene.*
C₄H₂F₃N   M 121.062
Bp₇₅₉ 75.9-76.2°.
▷AT7050000.

Buxton, M.W., *J. Chem. Soc.*, 1954, 366.
Drakesmith, F.G. *et al*, *J. Org. Chem.*, 1968, **33**, 280 (*synth*)
Fuchikami, T. *et al*, *Synthesis*, 1984, 766 (*synth*)
Fuchikami, T. *et al*, *Tetrahedron Lett.*, 1986, **27**, 3173 (*deriv, use*)

2-(Trifluoromethyl)pyridine, 9CI   T-60296

[368-48-9]

C₆H₄F₃N   M 147.100
Liq. Bp 139-140°. n_D^{25} 1.4155, 1.4166.
1-Oxide: [22253-59-4].
C₆H₄F₃NO   M 163.099
Bp₂₀ 132-133°.

Janz, G.J. *et al*, *J. Am. Chem. Soc.*, 1956, **78**, 978 (*synth, ir*)
Raasch, M.S., *J. Org. Chem.*, 1962, **27**, 1406 (*synth*)
Kobayashi, Y. *et al*, *Chem. Pharm. Bull.*, 1967, **15**, 1896, 1901 (*synth, ir, ms*)
Kobayashi, Y. *et al*, *Chem. Pharm. Bull.*, 1969, **17**, 510 (*oxide*)
Matsui, K. *et al*, *Chem. Lett.*, 1981, 1719 (*synth*)

3-(Trifluoromethyl)pyridine, 9CI   T-60297

[3796-23-4]
C₆H₄F₃N   M 147.100
Bp 115-117° (113°). n_D^{25} 1.4150.
1-Oxide: [22253-72-1].
C₆H₄F₃NO   M 163.099
Mp 75-77°.

Janz, G.J. *et al*, *J. Am. Chem. Soc.*, 1956, **78**, 978 (*synth, ir*)
Kobayashi, Y. *et al*, *Chem. Pharm. Bull.*, 1967, **15**, 1896, 1901 (*synth, ms*)
Kobayashi, Y. *et al*, *Chem. Pharm. Bull.*, 1969, **17**, 510 (*oxide*)
Kobayashi, Y. *et al*, *J. Fluorine Chem.*, 1981, **18**, 533 (*synth*)

4-(Trifluoromethyl)pyridine, 9CI   T-60298

[3796-24-5]
C₆H₄F₃N   M 147.100
Bp 108-110°. n_D^{25} 1.4155.
1-Oxide: [22253-59-4].
C₆H₄F₃NO   M 163.099
Prisms (Me₂CO). Mp 175-177°.

Raasch, M.S., *J. Org. Chem.*, 1962, **27**, 1406 (*synth*)
Kobayashi, Y. *et al*, *Chem. Pharm. Bull.*, 1967, **15**, 1896, 1901 (*synth, ir, ms*)
Kobayashi, Y. *et al*, *Chem. Pharm. Bull.*, 1969, **17**, 510 (*oxide*)
Shustov, L.D. *et al*, *Zh. Obshch. Khim.*, 1983, **53**, 103 (*Engl. transl.* p. 85) (*synth*)

1,2,3-Trifluoro-4-nitrobenzene, 9CI   T-60299

[771-69-7]

C₆H₂F₃NO₂   M 177.083
Liq. Bp₂₀ 92°.

Finger, G.C. *et al*, *J. Am. Chem. Soc.*, 1959, **81**, 94 (*synth*)
Bolton, R. *et al*, *J. Chem. Soc., Perkin Trans. 2*, 1978, 141 (*F nmr*)

1,2,3-Trifluoro-5-nitrobenzene, 9CI   T-60300

[66684-58-0]
C₆H₂F₃NO₂   M 177.083

Bolton, R. *et al*, *J. Chem. Soc., Perkin Trans. 2*, 1978, 141 (*F nmr*)

1,2,4-Trifluoro-5-nitrobenzene, 9CI   T-60301

[2105-61-5]
C₆H₂F₃NO₂   M 177.083
d_4^{20} 1.54. Mp −11°. Bp 192°, Bp₂₀ 93.5°. n_D^{20} 1.4943.
▷Lachrymatory

Finger, G.C. *et al*, *J. Am. Chem. Soc.*, 1951, **73**, 145 (*synth*)
Weygand, F. *et al*, *Chem. Ber.*, 1951, **84**, 101 (*synth*)
Bolton, R. *et al*, *J. Chem. Soc., Perkin Trans. 2*, 1978, 141 (*F nmr*)
Aldrich Library of FT-IR Spectra, 1st Ed., **1**, 1379A.
Sigma-Aldrich Library of Chemical Safety Data, 1st Ed., 1769A.

1,2,5-Trifluoro-3-nitrobenzene, 9CI   T-60302

[66684-57-9]
C₆H₂F₃NO₂   M 177.083
Mp ca. −20°. Bp 187°. n_D^{20} 1.4873.

Finger, G.C. *et al*, *J. Am. Chem. Soc.*, 1951, **73**, 152 (*synth*)
Bolton, R. *et al*, *J. Chem. Soc., Perkin Trans. 2*, 1978, 141 (*F nmr*)

1,2,5-Trifluoro-4-nitrobenzene, 9CI   T-60303

C₆H₂F₃NO₂   M 177.083
n_D^{20} 1.4985.

Finger, G.C. *et al*, *J. Am. Chem. Soc.*, 1956, **78**, 6034 (*synth*)
Bolton, R. *et al*, *J. Chem. Soc., Perkin Trans. 2*, 1978, 141 (*F nmr*)

1,3,5-Trifluoro-2-nitrobenzene, 9CI   T-60304

[315-14-0]
C₆H₂F₃NO₂   M 177.083
Shows nematocidal properties. Pale-yellow liquid. Mp ca. 3.5°. Bp 172°, Bp₂₀ 81°. n_D^{20} 1.4783.

Finger, G.C. *et al*, *J. Am. Chem. Soc.*, 1951, **73**, 153 (*synth*)
U.S.P., 3 294 629, (*1966*); CA, **66**, 54566 (*use*)
Biemond, J. *et al*, *Chem. Phys.*, 1973, **1**, 335 (*pmr, F nmr, nmr*)
Correll, T. *et al*, *J. Mol. Struct.*, 1980, **65**, 43 (*microwave*)

Trifluorooxirane, 9CI   T-60305

Epoxytrifluoroethane. Trifluoroethylene oxide

[2925-24-8]

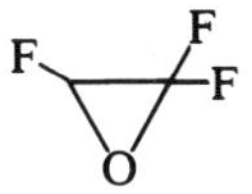

C_2HF_3O M 98.025

(±)-form

Gas. Bp −42°.

U.S.P., 3 775 440, (1973); CA, **80**, 59846 (*synth*)
Agopovich, J.W. et al, J. Am. Chem. Soc., 1983, **105**, 5047
  (*synth, pmr, F nmr, ir*)

Trifluoropyrazine, 9CI   T-60306

[55215-60-6]

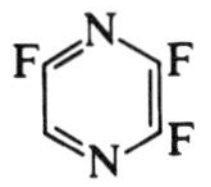

$C_4HF_3N_2$ M 134.061

Liq. Mp 0-1°. Bp 89°.

Chambers, R.D. et al, J. Chem. Soc., Perkin Trans. 1, 1974,
  2580 (*synth, F nmr*)

2,4,6-Trifluoropyrimidine, 9CI   T-60307

[696-82-2]

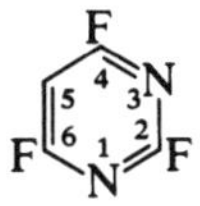

$C_4HF_3N_2$ M 134.061

Liq. Bp 98°.

Bailey, R.T. et al, Spectrochim. Acta, Part A, 1967, **23**, 2989
  (*ir, raman*)
Dewar, M.J. et al, J. Chem. Phys., 1968, **49**, 499 (*pmr*)
Shkurko, O.P. et al, Khim. Geterotsikl. Soedin, 1972, 1281
  (*synth*)
Hitzke, J. et al, Org. Mass Spectrom., 1976, **11**, 20 (*F nmr, ms*)

4,5,6-Trifluoropyrimidine, 8CI   T-60308

[17573-78-3]

$C_4HF_3N_2$ M 134.061

Banks, R.E. et al, J. Chem. Soc. (C), 1967, 1822 (*synth, F nmr*)

4,4,4-Trifluoro-3-(trifluoromethyl)-2-bu-   T-60309
tenal, 9CI

[104291-39-6]

$$(F_3C)_2C=CHCHO$$

$C_5H_2F_6O$ M 192.061

Yellow liq. Bp 71°.

Abele, H. et al, Chem. Ber., 1986, **119**, 3502 (*synth, pmr, ir, F nmr*)

1,1,1-Trifluoro-2-(trifluoromethyl)-3-bu-   T-60310
tyn-2-ol, 9CI

[646-72-0]

$$(F_3C)_2C(OH)C≡CH$$

$C_5H_2F_6O$ M 192.061

Liq. Bp 77°.

Abele, H. et al, Chem. Ber., 1986, **119**, 3502 (*synth, ir, pmr, F nmr, cmr*)

4-[2,2,2-Trifluoro-1-(trifluoromethyl)-   T-60311
ethylidene]-1,3,2-dioxathietane 2,2-diox-
ide, 9CI

Hexafluoroisobutylidene sulfate

[36638-46-7]

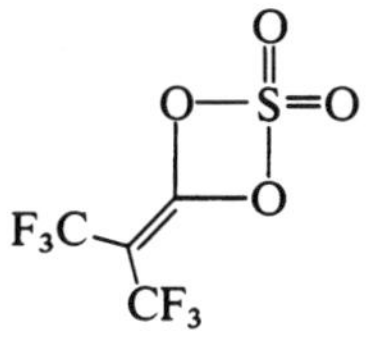

$C_4F_6O_4S$ M 258.092

Obt. by SO_3 addition to $(CF_3)_2C=C=O$.

  Sulfotrioxidising agent. Liq., which fumes in moist air.
  d_4^{20} 1.791. Bp 49°. n_D^{20} 1.3382.

Pavlov, V.M. et al, Khim. Geterotsikl. Soedin., 1971, 1645;
  1973, 13 (Engl. transl. pp. 1529, 10) (*synth, F nmr*)
Sokol'skii, G.A. et al, Khim. Geterotsikl. Soedin, 1973, 178
  (Engl. transl. p. 164) (*use*)

11,12,16-Trihydroxy-5,8,11,13-abietate-   T-60312
traen-7-one

Anhydronellionol

[108646-47-5]

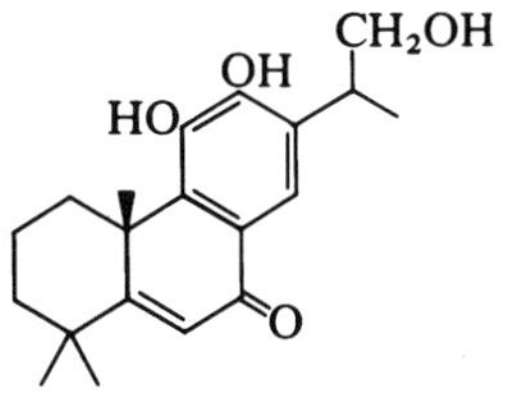

$C_{20}H_{26}O_4$ M 330.423

Constit. of *Premna integrifolia*. Cryst. ($CHCl_3$/MeOH).
  Mp 251-252°.

Rao, P.V.S. et al, Indian J. Chem., Sect. B, 1987, **26**, 191.

11,12,14-Trihydroxy-8,11,13-abietatriene-   T-60313
3,7-dione

Candelabrone

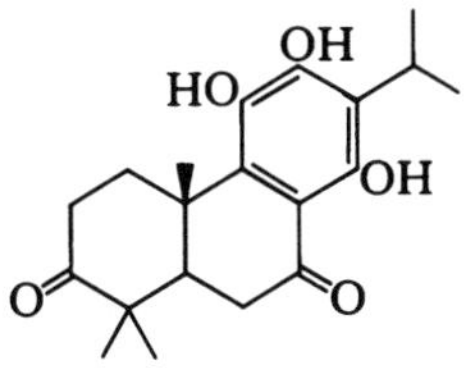

$C_{20}H_{26}O_5$ M 346.422

Constit. of *Salvia candelabrum*. Cryst. (MeOH). Mp
  224-226°.

Canigueral, S. et al, Phytochemistry, 1988, **27**, 221.

3,11,12-Trihydroxy-8,11,13-abietatrien-7-one   T-60314

$C_{20}H_{28}O_4$ M 332.439

3β-form

3β-Hydroxydemethylcryptojaponol
Constit. of *Salvia pubescens*. Cryst. Mp 233-235°.

Galicia, M.A. *et al*, *Phytochemistry*, 1988, **27**, 217.

3′,4′,5′-Trihydroxyacetophenone, 8CI   T-60315

Updated Entry replacing T-03242
1-(3,4,5-Trihydroxyphenyl)ethanone, 9CI. 5-Acetopyrogallol

[33709-29-4]

$C_8H_8O_4$ M 168.149
Needles (H_2O). Mp 187-188°.

▷AN0529000.

Tri-Ac:
  $C_{14}H_{14}O_7$ M 294.260
  Needles (ligroin). Mp 111-112°.

3′,5′-Di-Me ether: [2478-38-8]. *1-(4-Hydroxy-3,5-dimethoxyphenyl)ethanone, 9CI. 4′-Hydroxy-3′,5′-dimethoxyacetophenone. Acetosyringone.*
  $C_{10}H_{12}O_4$ M 196.202
  Found in kraft mill effluent. Plant stress hormone, insect attractant. Prisms (Et_2O). Mp 120-121°.

▷AM8440000.

Tri-Me ether: [1136-86-3]. *3′,4′,5′-Trimethoxyacetophenone.*
  $C_{11}H_{14}O_4$ M 210.229
  Needles (ligroin). Mp 78°.

Tri-Me ether, oxime:
  $C_{11}H_{15}NO_4$ M 225.244
  Needles (H_2O), prisms (EtOH). Mp 102-103°.

Tri-Me ether, semicarbazone: Cryst. (H_2O). Mp 178-179°.

Bogert, M.T. *et al*, *J. Am. Chem. Soc.*, 1914, **36**, 523.
Mauthner, F., *J. Prakt. Chem.*, 1927, **115**, 137.
Kurosawa, K. *et al*, *Bull. Chem. Soc. Jpn.*, 1978, **51**, 3612 (*synth, Acetosyringone*)

3,3′,4-Trihydroxybibenzyl   T-60316

4-[2-(3-Hydroxyphenyl)ethyl]-1,2-benzenediol. 1-(3,4-Dihydroxyphenyl)-2-(3-hydroxyphenyl)ethane

$C_{14}H_{14}O_3$ M 230.263

3′,4-Di-Me ether: [63367-96-4]. *3-Hydroxy-3′,4-dimethoxybibenzyl.*
  $C_{16}H_{18}O_3$ M 258.316
  Constit. of *Frullania falciloba*. Cryst. (hexane). Mp 81-82°.

3,4-Methylene ether, 3′-Me ether: [80357-95-5]. *5-[2-(3-Methoxyphenyl)ethyl]-1,3-benzodioxole, 9CI. 3′-Methoxy-3,4-methylenedioxybibenzyl.*
  $C_{16}H_{16}O_3$ M 256.301

Cryst. of *F. falciloba*. Cryst. Mp 49-50°.

Asakawa, Y. *et al*, *Phytochemistry*, 1987, **26**, 1023.

3,11,15-Trihydroxycholanic acid   T-60317

$C_{24}H_{40}O_5$ M 408.577

(3α,5β,11β,15β)-form
Transformation prod. of lithocholic acid by by fungus *Cunninghamella blakesleeana* ST 22. Cryst. (2-Propanol/hexane). Mp 236-237°. $[\alpha]_D^{25}$ +20.8° (c, 0.31 in MeOH).

(3α,5β,11α,15β)-form
Transformation prod. of lithocholic acid by fungus *C. blakesleeana* ST 22. Cryst. (EtOAc). Mp 153-155°.

Jodoi, Y. *et al*, *Tetrahedron*, 1987, **43**, 487.

3,15,18-Trihydroxycholanic acid   T-60318

$C_{24}H_{40}O_5$ M 408.577 ,

(3α,5β,15β)-form
Transformation prod. of lithocholic acid by fungus *Cunninghamella blakesleeana*.

Me ester: Cryst. (EtOAc/hexane). Mp 191-192°. $[\alpha]_D^{15}$ +10.2° (c, 0.46 in MeOH).

Jodoi, Y. *et al*, *Tetrahedron*, 1987, **43**, 487.

12,20,25-Trihydroxy-23-dammaren-3-one   T-60319

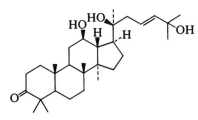

$C_{30}H_{50}O_4$ M 474.723

(12β,20S,23E)-form [105798-67-2]
Constit. of *Betula pendula* leaves. $[\alpha]_D^{20}$ +32.2° (c, 0.5 in $CHCl_3$).

Pokhilo, N.D. *et al*, *Khim. Prir. Soedin.*, 1986, **22**, 166.

5,14,15-Trihydroxy-6,8,10,12-eicosatetraenoic acid, 9CI   T-60320

[92950-25-9]

(5S,6E,8Z,10E,12E,14R,15S)-form

$C_{20}H_{32}O_5$ M 352.470

(5S,6E,8Z,10E,12E,14R,15S)-form [106062-83-3]
Lipoxin B. LX-B
Formed in human leukocytes from arachadonic acid. Major biol. isomer.

(5S,6E,8E,10E,12E,14R,15S)-form
8-trans-Lipoxin B. 8-trans-LXB
Minor biologically derived isomer.

(5S,6E,8E,10E,12E,14S,15S)-form
(14S)-8-trans-Lipoxin B. (14S)-8-trans-LXB
Minor biologically derived isomer.

Serhan, C.N. *et al*, *Proc. Natl. Acad. Sci. USA*, 1984, **81**, 5335; 1986, **83**, 1983 (*biosynth, isol, ms, uv, stereochem*)

Nicolaou, K.C. *et al*, *Synthesis*, 1986, 453 (*synth, ir, uv, pmr*)

5,7,8-Trihydroxyflavanone   T-60321

Updated Entry replacing T-20253

$C_{15}H_{12}O_5$ M 272.257

7-Me ether: 5,8-Dihydroxy-7-methoxyflavanone.
$C_{16}H_{14}O_5$ M 286.284
Constit. of *Notholaena neglecta*. Light-yellow cryst.
Mp 220-225°.

7,8-Di-Me ether: 5-Hydroxy-7,8-dimethoxyflavone.
$C_{17}H_{16}O_5$ M 300.310
Constit. of *Andrographis paniculata*. Cryst.
(Me$_2$CO/pet. ether). Mp 98-99°.

Tri-Me ether: 5,7,8-Trimethoxyflavone.
$C_8H_{16}O_5$ M 192.211
Cryst. (MeOH). Mp 156-158°.

Gupta, K.K. *et al*, *Phytochemistry*, 1983, **22**, 314.
Scheele, C. *et al*, *J. Nat. Prod.*, 1987, **50**, 181 (*isol*)

5,6,7-Trihydroxyflavone   T-60322

Updated Entry replacing T-20255

*5,6,7-Trihydroxy-2-phenyl-4H-1-benzopyran-4-one,
9CI. Baicalein*

[491-67-8]

$C_{15}H_{10}O_5$ M 270.241
Mp 223-226°, 263-264°.

6,7-Di-Me ether: [740-33-0]. *5-Hydroxy-6,7-
dimethoxyflavone.*
$C_{17}H_{14}O_5$ M 298.295
Isol. from stems of *Popowia cauliflora*. Yellow plates
(CHCl$_3$/pet. ether). Mp 150°.

Tri-Me ether: [973-67-1]. *5,6,7-Trimethoxyflavone.*
$C_{18}H_{16}O_5$ M 312.321
Isol. from *Colebrookia oppositifolia, Callicarpa
japonica* and *Zeyhera tuberulosa*. Piscicide. Mp 165-
167°.

*6,7-Methylene ether: 5-Hydroxy-6,7-
methylenedioxyflavone.*
$C_{16}H_{10}O_5$ M 282.252
Constit. of sugarbeet infected with *Rhizoctonia solani*.
Yellow cryst. Mp 205° dec.

Kutney, J.P. *et al*, *Phytochemistry*, 1971, **10**, 3298 (*isol*)
Ahmed, S.A. *et al*, *Indian J. Chem.*, 1974, **12**, 1327 (*isol*)
Panchipol, K. *et al*, *Phytochemistry*, 1978, **17**, 1363 (*isol*)
Takahashi, H. *et al*, *Bull. Chem. Soc. Jpn.*, 1987, **60**, 2261 (*isol*)

8,9,14-Trihydroxy-1(10),4,11(13)-germa-   T-60323
cratrien-12,6-olide

Updated Entry replacing T-50469

$C_{15}H_{20}O_4$ M 264.321

($6\alpha,8\beta,9\beta$)-form illus.

($6\alpha,8\beta,9\beta$)-form

*8-(2-Methylpropanoyl), 14-Ac: 14-Acetoxy-9β-
hydroxy-8β-(2-methylpropanoyloxy)-1(10),4,11(13)-
germacratrien-12,6α-olide.*
$C_{21}H_{28}O_7$ M 392.448
Constit. of *Blainvillea latifolia*. Cryst. (Me$_2$CO/pet.
ether). Mp 159-160°. [α]$_D$ −47.8° (CHCl$_3$).

($6\beta,8\alpha,9\beta$)-form

8-Angeloyl: **Gigantanolide A.**
$C_{20}H_{26}O_6$ M 362.422
Constit. of *Montanoa gigas*. Cryst. Mp 118-122°. [α]$_D$
−152° (CHCl$_3$).

9-Angeloyl: **Gigantanolide B.**
$C_{20}H_{26}O_6$ M 362.422
From *M. gigas*.

9-(2-Methylbutanoyl): **Gigantanolide C.**
$C_{20}H_{28}O_6$ M 364.438
From *M. gigas*.

Rojatkar, S.R. *et al*, *J. Chem. Res. (S)*, 1986, 272 (*isol, cryst
struct*)
Quijano, L. *et al*, *Phytochemistry*, 1987, **26**, 2589 (*derivs*)

4′,5,7-Trihydroxyisoflavone   T-60324

Updated Entry replacing T-10289

*5,7-Dihydroxy-3-(4-hydroxyphenyl)-4H-1-benzopyran-
4-one, 9CI. Differenol A. Genistein. Prunetol. Sophoricol*

[446-72-0]

$C_{15}H_{10}O_5$ M 270.241
Widely distributed in plants such as clover and alfalfa
and prod. by *Aspergillus* spp., *Micromonospora* spp.,
Pseudomonas spp. and *Streptomyces* spp. Weak
oestrogen. Shows insect antifeedant and weak
antibacterial activity against *E. coli* and *Xanthomonas
oryzae*. Induces cell differentiation and inhibits DOPA
carboxylase. Prisms (EtOH aq.). Mp 301-302° dec.
pK_{a1} 8.6, pK_{a2} 10.7.

▷Exp. carcinogen. NR2392000.

4-β-D-Glucoside: [152-95-4]. **Sophicoroside.**
$C_{21}H_{20}O_{10}$ M 432.383
Constit. of *Sophora japonica*. Prisms (EtOH). Mp
297°. [α]$_D^{20}$ −46.7°.

7-β-D-Glucoside: [529-59-9]. **Genistin.**
$C_{21}H_{20}O_{10}$ M 432.383
Isol. from *Genista tinctoria* and soybean. Leaflets
(EtOH). Mp 254-256° dec. [α]$_D^{21}$ −27.7° (MeOH aq.).

7-β-D-Rutinosyl: [14988-20-6]. **Sphaerobioside.**
$C_{27}H_{30}O_{14}$ M 578.526
Isol. from *Baptisia sphaerocarpa* and *B. lecontii*.
Needles (MeOH). Mp 203-204°. [α]$_D^{20}$ −73.3° (c, 9.9
in Py).

4′,7-Bis(4-O-β-D-glucopyranosyl-β-D-apiofuranoside):
[78693-95-5]. **Sarothamnoside.**
$C_{37}H_{46}O_{23}$ M 858.757
Obt. from seeds of *Sarothamnus* spp. Mp 136-138°.
[α]$_D^{20}$ −128° (c, 0.1 in H$_2$O).

7-O-(Apiosyl-(1→6)-β-D-glucopyranoside): [108044-
05-9]. **Ambocin.**
$C_{26}H_{28}O_{14}$ M 564.499
Constit. of *Neorautanenia amboensis*. Glass. [α]$_D$
−36.5° (c, 0.011 in H$_2$O).

4'-Apioside, 7-β-D-glucopyranoside: [108069-00-7].
Neobacin.
$C_{26}H_{28}O_{14}$ M 564.499
From *N. amboensis*. Glass. $[\alpha]_D$ −38.2° (c, 0.011 in H_2O).

Bognár, R. *et al, Chem. Ind.* (*London*), 1954, 518
Rosler, H. *et al, Chem. Ber.*, 1965, **98**, 2193 (*deriv*)
Wagner, H. *et al, Chem. Ber.*, 1967, **100**, 101 (*synth*)
Markham, K.R. *et al, Phytochemistry*, 1968, **7**, 791 (*isol*)
Ganguly, A.K. *et al, Chem. Ind.* (*London*), 1970, 201 (*isol*)
Umezawa, H. *et al, J. Antibiot.*, 1975, **28**, 947 (*isol*)
Wagner, H. *et al, Tetrahedron Lett.*, 1976, 1799 (*cmr*)
Hazeto, T. *et al, J. Antibiot.*, 1979, **32**, 217 (*isol*)
Asahi, K. *et al, J. Antibiot.*, 1981, **34**, 919 (*uv, ir*)
Brum-Bousquet, M. *et al, Tetrahedron Lett.*, 1981, 1223 (*Sarothamnoside*)
Breytenbach, J.C., *J. Nat. Prod.*, 1986, **49**, 1003 (*isol, deriv*)
Jain, A.C. *et al, J. Chem. Soc., Perkin Trans. 1*, 1986, 215 (*synth*)
Ogawara, H. *et al, J. Antibiot.*, 1986, **39**, 606 (*Ambocin, Neobacin*)

4',6,7-Trihydroxyisoflavone   T-60325

Updated Entry replacing T-10290
[17817-31-1]
$C_{15}H_{10}O_5$ M 270.241
Isol. from fermented soybeans. Cryst. (MeOH). Mp 322°.
Tri-Ac: Cryst. (EtOH). Mp 217°.
4',7-Di-Me ether: [550-79-8]. *6-Hydroxy-4',7-dimethoxyisoflavone.* **Afromosin**. Afromosin.
$C_{17}H_{14}O_5$ M 298.295
Constit. of *Afrormosia elata, Castanospermum australe* and *Myrocarpus* spp. Needles (MeOH or EtOH). Mp 236-237° (228-229°).
4',7-Di-Me ether, 6-β-D-glucopyranosyl: [19046-26-5].
Wistin.
$C_{23}H_{24}O_9$ M 444.437
Constit. of *Wistaria floribunda*. Needles (MeOH aq. or Me_2CO aq.). Mp 209-210°. $[\alpha]_D^{12}$ −67.15° (c, 1.43 in AcOH).
Tri-Me ether: [798-61-8]. *4',6,7-Trimethoxyisoflavone.*
$C_{18}H_{16}O_5$ M 312.321
Cryst. Mp 179-180°.

McMurry, T.B.H. *et al, J. Chem. Soc.*, 1960, 1491 (*isol, struct*)
Fujita, M. *et al, Chem. Pharm. Bull.*, 1963, **11**, 382 (*deriv*)
Harbourne, J. *et al, J. Org. Chem.*, 1963, **28**, 881 (*isol*)
Bevan, C.W.L. *et al, J. Chem. Soc.* (*C*), 1966, 509 (*isol, struct*)
Ikehata, H. *et al, Agric. Biol. Chem.*, 1968, **32**, 740 (*isol*)
Dewick, P.M. *et al, Phytochemistry*, 1978, **17**, 249 (*biosynth*)
Caballero, P. *et al, J. Nat. Prod.*, 1986, **49**, 1126 (*cryst struct*)

ent-3α,7β,14α-Trihydroxy-16-kauren-15-one   T-60326

[84294-86-0]

$C_{20}H_{30}O_4$ M 334.455
Constit. of *Isodon amethystoides* and *Plectranthus coesta*. Oil. $[\alpha]_D$ −80° (c, 0.03 in MeOH).
Tri-Ac: Cryst. (EtOH). Mp 249-250°. $[\alpha]_D$ +20.74° (c, 0.0675 in MeOH).

Wang, X. *et al, Zhongcaoyao*, 1982, **13**, 12; *CA*, **98**, 50318d (*isol*)
Phadnis, A.P. *et al, Indian J. Chem., Sect. B*, 1987, **26**, 15 (*isol*)

3,7,15-Trihydroxy-8,24-lanostadien-26-oic acid   T-60327

$C_{30}H_{48}O_5$ M 488.706
(3α,5α,7α,15α,24E)-form
3,7-Di-Ac: 3α,7α-Diacetoxy-15α-hydroxy-8,24E-lanostadien-26-oic acid. **Ganoderic acid Ma**.
$C_{34}H_{52}O_7$ M 572.781
Metab. of *Ganoderma lucidum*. Syrup. $[\alpha]_D^{24}$ −16° (c, 0.3 in MeOH).
3-Ac, 7-Me ether: 3α-Acetoxy-15α-hydroxy-22-methoxy-8,24-lanostadien-26-oic acid. **Ganoderic acid Mi**.
$C_{33}H_{52}O_6$ M 544.770
Metab. of *G. lucidum*. Syrup. $[\alpha]_D^{23}$ −11° (c, 0.2 in MeOH).

Nishitoba, T. *et al, Agric. Biol. Chem.*, 1987, **51**, 619, 1149.

3,7,22-Trihydroxy-8,24-lanostadien-26-oic acid   T-60328

$C_{30}H_{48}O_5$ M 488.706
(3α,5α,7α,22ξ,24E)-form
3,22-Di-Ac, 7-Me ether: 3α,22-Diacetoxy-7α-methoxy-8,24E-lanostadien-26-oic acid. **Ganoderic acid Md**.
$C_{35}H_{54}O_7$ M 586.807
Metab. of *Ganoderma lucidum*. Cryst. Mp 181-182°. $[\alpha]_D^{24}$ −20° (c, 0.33 in MeOH).
22-Ac, 7-Me ether: 22-Acetoxy-3α-hydroxy-7α-methoxy-8,24E-lanostadien-26-oic acid. **Ganoderic acid Mj**.
$C_{33}H_{52}O_6$ M 544.770
Metab. of *G. lucidum*. Syrup. $[\alpha]_D^{23}$ −8° (c, 0.05 in MeOH).

Nishitoba, T. *et al, Agric. Biol. Chem.*, 1987, **51**, 619, 1149.

3,15,22-Trihydroxy-7,9(11),24-lanostatrien-26-oic acid   T-60329

$C_{30}H_{46}O_5$ M 486.690
(3α,15α,22S,24E)-form
15,22-Di-Ac: 15α,22S-Diacetoxy-3α-hydroxy-7,9(11),24E-lanostatrien-26-oic acid. Ganoderic acid P.
$C_{34}H_{50}O_7$ M 570.765
Constit. of the cultured mycelium of *Ganoderma lucidum*. Cryst. Mp 211-212.5°.
3,22-Di-Ac: 3α,22S-Diacetoxy-15α-hydroxy-7,9(11),24E-lanostatrien-26-oic acid. Ganoderic acid Q.
$C_{34}H_{50}O_7$ M 570.765
Constit. of the cultured mycelium of *G. lucidum*. Cryst. Mp 131-132°.

Tri-Ac: [103992-91-2]. *3α,15α,22S-Triacetoxy-7,9(11),24-lanostatrien-26-oic acid. Ganoderic acid* T.
$C_{36}H_{52}O_8$ M 612.802
Constit. of cultured mycelium of *G. lucidum.* Cryst.
Mp 200-202°.

(3α,5α,22ξ,24E)-form

3,22-Di-Ac: Ganoderic acid MK.
$C_{34}H_{50}O_7$ M 570.765
Constit. of *G. lucidum.* Syrup. $[\alpha]_D^{23}$ +23° (c, 0.2 in MeOH).

Hirotani, M. *et al, Phytochemistry,* 1987, **26**, 2797.
Nishitoba, T. *et al, Agric. Biol. Chem.,* 1987, **51**, 1149.

15,26,27-Trihydroxy-7,9(11),24-lanosta-trien-3-one   T-60330

$C_{30}H_{46}O_4$ M 470.691
15α-form [106518-62-1]
Ganoderiol B
Constit. of *Ganoderma lucidum.* Amorph. powder.

Sato, H. *et al, Agric. Biol. Chem.,* 1986, **50**, 2887.

1,11,20-Trihydroxy-3-lupanone   T-60331

$C_{30}H_{50}O_4$ M 474.723
(1β,11α)-form
Constit. of *Salvia deserta.* Cryst. (MeOH). Mp 260-264°. $[\alpha]_D^{18}$ +69° (c, 0.121 in CHCl$_3$).

Savona, G. *et al, Phytochemistry,* 1987, **26**, 3305.

3,6,8-Trihydroxy-1-methylanthraquinone-2-carboxylic acid   T-60332

Updated Entry replacing T-03400
9,10-Dihydro-3,6,8-trihydroxy-1-methyl-9,10-dioxo-2-anthracenecarboxylic acid, 9CI. Laccaic acid D. *Xantho-kermesic acid. Flavokermesic acid*
[18499-84-8]
$C_{16}H_{10}O_7$ M 314.251
Constit. of Rangini stick lac (from host trees *Butea monosperma* and *Zizyphus mauritiana*), rhubarb and senna and from the insects *Kermes ilicis.* Yellow needles (H$_2$O). Dec. >300°.
Me ester: [53254-85-6].
$C_{17}H_{12}O_7$ M 328.278
Isol. from subterranean stems of *Aloe saponaria.* Orange needles (MeOH). Mp 270-275°.
Me ester, tri-Me ether: Mp 226°.

Gadgil, D.D. *et al, Tetrahedron Lett.,* 1968, 2223 (*synth*)
Mehandale, A.R. *et al, Tetrahedron Lett.,* 1968, 2231 (*isol*)
Yagi, A. *et al, Chem. Pharm. Bull.,* 1974, **22**, 1159 (*ms, ir, pmr*)
Cameron, D.W. *et al, J. Chem. Soc., Chem. Commun.,* 1978, 688 (*synth*)
Yagi, A. *et al, Phytochemistry,* 1978, **17**, 895 (*biosynth*)
Wouters, J. *et al, Tetrahedron Lett.,* 1987, **28**, 1199 (*isol*)

Handle all chemicals with care

2,5,8-Trihydroxy-3-methyl-6,7-methylene-dioxy-1,4-naphthoquinone   T-60333

4,6,9-Trihydroxy-7-methylnaphtho[2,3-d]-1,3-dioxole-5,8-dione, 9CI

$C_{12}H_8O_7$ M 264.191
5- or 8-Me ether: [75628-34-1]. **Nepenthone A**. *2,5-Di-hydroxy-8-methoxy-3-methyl-6,7-methylenedioxy-1,4-naphthoquinone.*
$C_{13}H_{10}O_7$ M 278.218
Isol. from roots of *Nepenthes rafflesiana.* Feathery orange needles (Me$_2$CO). Mp 255-260° dec.
Tri-Me ether: [75628-93-2].
$C_{15}H_{14}O_7$ M 306.271
Yellow needles (Me$_2$CO/pet. ether). Mp 165-167°.

Cannon, J.R. *et al, Aust. J. Chem.,* 1980, **33**, 1073 (*isol, struct*)
Rizzacasa, M.A. *et al, J. Chem. Soc., Perkin Trans. 1,* 1987, 2017 (*struct, synth, deriv*)

9,10,18-Trihydroxyoctadecanoic acid   T-60334

Updated Entry replacing T-03434
[496-86-6]

$$(CH_2)_7COOH$$
$$H\!-\!\overset{|}{C}\!-\!OH$$
$$HO\!-\!\overset{|}{C}\!-\!H$$
$$(CH_2)_7CH_2OH$$

(9S,10S)-form

$C_{18}H_{36}O_5$ M 332.479
(9S,10S)-form [17705-68-9]
Phloinolic acid. Phloionolic acid. Floionolic acid
Constit. of cork, cutins, and *Chamaepeuce afra* seeds.
Cryst. (MeOH). Mp 104-105°. $[\alpha]_{589}$+22.75°. The (9R,10R)-(−) form has also been prepared.
Me ester: [36693-05-7].
$C_{19}H_{38}O_5$ M 346.506
Mp 79-80.5°. $[\alpha]_D$ +21.52°.
(9RS,10RS)-form [583-86-8]
(±)-threo-*form*
Cryst. (EtOAc). Mp 104°.
(9RS,10SR)-form [17673-81-3]
erythro-*form. Isophloionolic acid*
Cryst. (EtOAc). Mp 132-133°. Opt. inactive (*meso-*).

Ames, D.E. *et al, J. Chem. Soc. (C),* 1967, 1556 (*synth*)
Eglinton, G. *et al, Phytochemistry,* 1968, **7**, 313 (*ms*)
McGhie, J.F. *et al, Chem. Ind. (London),* 1972, 463 (*stereochem*)

2,3,24-Trihydroxy-12-oleanen-28-oic acid   T-60335

Updated Entry replacing T-50497
$C_{30}H_{48}O_5$ M 488.706
(2α,3α)-form
Constit. of *Prunella vulgaris.*
Me ester: Cryst. (MeOH). Mp 280-282°. $[\alpha]_D^{27}$ +59.6° (c, 1 in CHCl$_3$).
(2β,3β)-form
Arboreic acid
Constit. of the bark of *Myrianthus arboreus.*

Me ester: Cryst. (MeOH). Mp 239-240°.

Kojima, H. *et al, Phytochemistry*, 1986, **25**, 729 (*isol, struct*)
Ngounou, F.N. *et al, Phytochemistry*, 1987, **26**, 3080 (*isol, struct*)

2,3,24-Trihydroxy-11,13(18)-oleonadien-28-oic acid   T-60336

$C_{30}H_{46}O_5$ M 486.690

(2α,3α)-form

Constit. of *Prunella vulgaris.*

Me ester: Needles (MeOH). Mp 267-269°. $[\alpha]_D^{24}$ −156° (c, 0.3 in CHCl$_3$).

Kojima, H. *et al, Phytochemistry*, 1987, **26**, 1107.

7,11,12-Trihydroxy-6-oxo-8,11,13-abieta-trien-20-oic acid   T-60337

7-Hydroxy-6-oxocarnosic acid

$C_{20}H_{26}O_6$ M 362.422

7α-form [110201-67-7]

Constit. of *Salvia canariensis.*

6-Deoxo, 7-ketone: 11,12-Dihydroxy-7-oxo-8,11,13-abietatrien-20-oic acid. 7-Oxocarnosic acid.
$C_{20}H_{26}O_5$ M 346.422
Constit. of *S. canariensis.*

7β-form [110201-66-6]

Constit. of *S. canariensis.*

González, A.G. *et al, Phytochemistry*, 1987, **26**, 1471.

(3,4,5-Trihydroxy-6-oxo-1-cyclohexen-1-yl)methyl 2-butenoate, 9CI   T-60338

Updated Entry replacing T-03458

2-Crotonyloxymethyl-4,5,6-trihydroxy-2-cyclohexenone

[57449-30-6]

$C_{11}H_{14}O_6$ M 242.228

Isol. from *Streptomyces griseosporeus* and *S. filipinensis.* Glyoxalase I inhibitor, antitumor antibiotic. Needles (CHCl$_3$/MeOH). Mp 181°. $[\alpha]_D^{24}$ −109° (c, 1.5 in EtOH).

Chimura, H. *et al, J. Antibiot.*, 1975, **28**, 743 (*struct*)
Takeuchi, T. *et al, J. Antibiot.*, 1975, **28**, 737 (*isol*)
Japan. Pat., 77 71 448, (*1977*); *CA*, **87**, 184080b (*synth*)
Mirza, S. *et al, Helv. Chim. Acta*, 1985, **68**, 988 (*synth, uv, ir, pmr*)
Takayama, H. *et al, Tetrahedron Lett.*, 1986, **27**, 5509 (*synth*)

2,3,7-Trihydroxyphenanthrene   T-60339

2,3,7-Phenanthrenetriol

$C_{14}H_{10}O_3$ M 226.231

3,7-Di-Me ether: [110202-83-0]. *3,7-Dimethoxy-2-phenanthrenol, 9CI. 2-Hydroxy-3,7-dimethoxyphenanthrene.*
$C_{16}H_{14}O_3$ M 254.285
Constit. of *Marchantia polymorpha.* Powder. Mp 159-160°.

Tri-Me ether: [64701-00-4]. *2,3,7-Trimethoxyphenanthrene.*
$C_{17}H_{16}O_3$ M 268.312
Cryst. Mp 114-116°.

Asakawa, Y. *et al, Phytochemistry*, 1987, **26**, 1811.

2,3,4-Trihydroxyphenylacetic acid   T-60340

2,3,4-Trihydroxybenzeneacetic acid, 9CI

$C_8H_8O_5$ M 184.148

Free acid unknown.

Tri-Me ether: [22480-91-7]. *2,3,4-Trimethoxybenzeneacetic acid, 9CI. 2,3,4-Trimethoxyphenylacetic acid.*
$C_{11}H_{14}O_5$ M 226.229
Cryst. (EtOH aq. or Et$_2$O/pet. ether). Mp 101-102°.

Tri-Me ether, Me ester: [22480-88-2].
$C_{12}H_{16}O_5$ M 240.255
Bp$_{22}$ 170-174°.

Tri-Me ether, nitrile: [68913-85-9].
$C_{11}H_{13}NO_3$ M 207.229
Bp$_{14}$ 170-175°.

Quelet, R. *et al, Bull. Soc. Chim. Fr.*, 1953, C46.
Kametani, T. *et al, Yakugaku Zasshi*, 1969, **89**, 279; *CA*, **70**, 106356.
Ahluwalia, V.K. *et al, Indian J. Chem., Sect. B*, 1978, **16**, 372.
Lenz, G.R. *et al, J. Heterocycl. Chem.*, 1981, **18**, 691.
Patra, A. *et al, J. Indian Chem. Soc.*, 1983, **60**, 265 (*cmr*)

2,3,5-Trihydroxyphenylacetic acid   T-60341

2,3,5-Trihydroxybenzeneacetic acid, 9CI

[57154-98-0]

$C_8H_8O_5$ M 184.148

Metab. of Zearalenone by *Fusarium roseum.* No phys. props. reported.

Steele, J.A. *et al, J. Agric. Food Chem.*, 1976, **24**, 89 (*isol*)

2,4,5-Trihydroxyphenylacetic acid   T-60342

2,4,5-Trihydroxybenzeneacetic acid, 9CI

[51109-27-4]

$C_8H_8O_5$ M 184.148

Needles + ½H$_2$O. Mp 164-165°.

Tri-Me ether: [4463-16-5]. *2,4,5-Trimethoxybenzeneacetic acid, 9CI. 2,4,5-Trimethoxyphenylacetic acid. Homoasaronic acid.*
$C_{11}H_{14}O_5$ M 226.229
Cryst. Mp 84-87°.

Tri-Me ether, Me ester: [2638-15-5].
$C_{12}H_{16}O_5$ M 240.255
Mp 44-46°.

Tri-Me ether, nitrile: [38444-50-7].
$C_{11}H_{13}NO_3$ M 207.229
Cryst. (MeOH). Mp 84°. Bp$_1$ 160-165°.

Gottlieb, O.R. *et al, J. Org. Chem.*, 1961, **26**, 2449 (*deriv*)
Short, J.H. *et al, Tetrahedron*, 1973, **29**, 1931 (*deriv*)
Wada, G.H. *et al, Biochemistry*, 1973, **12**, 5212 (*synth*)

Maruyama, K. *et al, Bull. Chem. Soc. Jpn.*, 1978, **51**, 3586 (*deriv*)
Patra, A. *et al, J. Indian Chem. Soc.*, 1983, **60**, 265 (*cmr, deriv*)

3,4,5-Trihydroxyphenylacetic acid   T-60343

3,4,5-Trihydroxybenzeneacetic acid, 9CI
[29511-09-9]
$C_8H_8O_5$ M 184.148
Cryst. (EtOAc/pet. ether). Mp 161° (158°).

Me ester:
$C_9H_{10}O_5$ M 198.175
Bp_1 139-140°.

tert-*Butyl ester:*
$C_{12}H_{16}O_5$ M 240.255
Needles (pet. ether). Mp 66-67°.

Tri-Ac:
$C_{14}H_{14}O_8$ M 310.260
Cryst. (EtOH). Mp 196°.

Tri-Ac, Me ester:
$C_{15}H_{16}O_8$ M 324.287
Plates. Mp 138.8-139.7°.

3-Me ether: [2989-10-8]. *3,4-Dihydroxy-5-methoxyben-zeneacetic acid, 9CI. 3,4-Dihydroxy-5-methoxyphenyl-lacetic acid.*
$C_9H_{10}O_5$ M 198.175
Needles (Et$_2$O/CHCl$_3$). Mp 160°.

4-Me ether: [34021-73-3]. *3,5-Dihydroxy-4-methoxy-benzeneacetic acid, 9CI. 3,5-Dihydroxy-4-methoxy-phenylacetic acid.*
$C_9H_{10}O_5$ M 198.175
Needles (Et$_2$O). Mp 130°.

Tri-Me ether: [951-82-6]. *3,4,5-Trimethoxybenzeneace-tic acid, 9CI. 3,4,5-Trimethoxyphenylacetic acid.*
$C_{11}H_{14}O_5$ M 226.229
Mp 121°.

Tri-Me ether, nitrile: [13338-63-1]. *3,4,5-Trimethoxy-benzeneacetonitrile, 9CI. 3,4,5-Trimethoxyphenylacetonitrile.*
$C_{11}H_{13}NO_3$ M 207.229
Cryst. (Et$_2$O). Mp 77-78°. $Bp_{0.1}$ 140-155°.

▷AM2475000.

Kamal, A. *et al, Tetrahedron*, 1965, **21**, 1411 (*deriv*)
Smith, G.E. *et al, Biochem. J.*, 1972, **130**, 141 (*synth*)
Short, J.H. *et al, Tetrahedron*, 1973, **29**, 1931 (*deriv*)
Becker, D. *et al, J. Chem. Soc., Perkin Trans. 1*, 1977, 1674 (*deriv*)
Patra, A. *et al, Indian J. Chem., Sect. B*, 1982, **21**, 173 (*deriv*)

1-(2,3,4-Trihydroxyphenyl)-2-(3,4,5-trihydroxyphenyl)ethylene   T-60344

2,3,3′,4,4′,5′-Stilbenehexol. 2,3,3′,4,4′,5′-Hexahydroxystilbene

$C_{14}H_{12}O_6$ M 276.245
(**Z**)-**form**

3′,4,4′,5′-Tetra-Me ether: 3-Methoxy-6-[2-(3,4-5-trimethoxyphenyl)ethenyl]-1,2-benzenediol, 9CI. 1-(2,3-Dihydroxy-4-methoxyphenyl)-2-(3,4,5-trimethoxyphenyl)ethylene. **Combretastatin A1**.
$C_{18}H_{20}O_6$ M 332.352
Constit. of *Combretum caffrum*. Antileukaemic agent. Cryst. (CHCl$_3$/hexane). Mp 113-115°.

3′,4,4′,5′-Tetra-Me ether, dihydro: [109971-64-4]. *3-Methoxy-6-[2-(3,4,5-trimethoxyphenyl)ethyl]1,2-benzenediol, 9CI. 1-(2,3-Dihydroxy-4-methoxyphenyl)-2-(3,4,5-trimethoxyphenyl)ethane. 2,3-Dihydroxy-3,4,4′,5-tetramethoxybibenzyl.* **Combretastatin B1**.
$C_{18}H_{22}O_6$ M 334.368
From *C. caffrum*. Antileukaemic agent. Gum.

Pettit, G.R. *et al, J. Nat. Prod.*, 1987, **50**, 119.

2,3,12-Trihydroxy-4,7-pregnadien-20-one   T-60345

$C_{21}H_{30}O_4$ M 346.466
(**2α,3β,12β**)-**form** [109237-01-6]
Constit. of *Stizophyllum riparium*. Amorph. Mp 115-118°. $[\alpha]_D^{25}$ +16.5° (c, 0.33 in CHCl$_3$).

Duh, C.-Y. *et al, J. Nat. Prod.*, 1987, **50**, 63.

2,3,12-Trihydroxy-4,7,16-pregnatrien-20-one   T-60346

$C_{21}H_{28}O_4$ M 344.450
(**2α,3β,12β**)-**form** [109237-00-5]
Constit. of *Stizophyllum riparium*. Cytotoxic agent. Amorph. Mp 194-197°. $[\alpha]_D^{25}$ +60.5° (c, 0.54 in CHCl$_3$).

Duh, C.-Y. *et al, J. Nat. Prod.*, 1987, **50**, 63.

3,23,25-Trihydroxy-7-tiracallen-24-one   T-60347

$C_{30}H_{50}O_4$ M 474.723
(**3β,23R**)-**form**
Constit. of *Simaba multiflora*. Cryst. (Me$_2$CO). Mp 208-213°.

Arisawa, M. *et al, Phytochemistry*, 1987, **26**, 3301.

3,19,24-Trihydroxy-12-ursen-28-oic acid   T-60348

Updated Entry replacing B-00083

$C_{30}H_{48}O_5$ M 488.706
(**3α,19α**)-**form** [64199-78-6]
Barbinervic acid
Constit. of *Clethra barbinervis*. Cryst. (MeOH). Mp 298°. $[\alpha]_D^{31}$ +18° (c, 1 in EtOH).

(*3β,19α*)-*form*

24-(3-Hydroxy-4-methoxyphenyl)-2E-propanoyl ester:
3β,19α-Dihydroxy-24-trans-ferulyloxy-12-ursen-28-
oic acid.
$C_{40}H_{56}O_8$ M 664.878
Constit. of *Stizophyllum riparium*. Amorph. Mp 195-
197°. $[\alpha]_D^{25}$ +1.58° (c, 0.38 in CHCl$_3$).

Takani, M. *et al, Chem. Pharm. Bull.*, 1977, **25**, 981 (*isol*)
Duh, C.-Y. *et al, J. Nat. Prod.*, 1987, **50**, 63 (*isol*)

1,5,6-Trihydroxyxanthone   T-60349

Updated Entry replacing T-03546
1,5,6-Trihydroxy-9H-xanthen-9-one, 9CI. **Mesuxanth-**
one B
[5042-03-5]
$C_{13}H_8O_5$ M 244.203
Constit. of the woods of *Symphonia globulifera*, *Mesua*
ferrea, *Calophyllum scriblitifolium*, *C. calaba* and
Garcinia buchananii. Yellow needles or plates
(EtOAc/CHCl$_3$ or EtOH). Mp 286-288°.

1-Me ether: [93930-21-3]. *5,6-Dihydroxy-1-*
methoxyxanthone.
$C_{14}H_{10}O_5$ M 258.230
Isol. from *Tovomita excelsa*. Cryst. (MeOH). Mp 258-
260°.

5-Me ether: [20081-65-6]. *1,6-Dihydroxy-5-methoxyx-*
anthone. **Buchanaxanthone**.
$C_{15}H_{12}O_5$ M 272.257
From *G. buchananii* and *C.* spp. Pale-yellow cryst.
(C$_6$H$_6$). Mp 243-246°.

6-Me ether: [20081-69-0]. *1,5-Dihydroxy-6-*
methoxyxanthone.
$C_{14}H_{10}O_5$ M 258.230
Constit. of *Tovomita excelsa*. Cryst. (MeOH aq.). Mp
247-248°.

5,6-Di-Me ether: [5042-07-9]. *1-Hydroxy-5,6-*
dimethoxyxanthone.
$C_{15}H_{12}O_5$ M 272.257
Cream-coloured or pale-yellow needles (pet. ether or
MeOH). Mp 184°.

Govindachari, T.R. *et al, Tetrahedron*, 1967, **23**, 243 (*isol, uv,*
ir)
Jackson, B. *et al, J. Chem. Soc.* (*C*), 1968, 2579 (*isol, synth,*
deriv)
Carpenter, I. *et al, J. Chem. Soc.* (*C*), 1969, 2421 (*isol*)
Locksley, H.D. *et al, J. Chem. Soc.* (*C*), 1969, 1567 (*isol, deriv*)
Somanathan, R. *et al, J. Chem. Soc., Perkin Trans. 1*, 1972,
1935 (*isol, deriv*)
Quillinan, A.J. *et al, J. Chem. Soc.* (*C*), 1973, 1329 (*synth, uv,*
pmr)
Gunaskera, S.P. *et al, J. Chem. Soc., Perkin Trans. 1*, 1975,
2215 (*isol*)
De Olivera, W.G. *et al, Phytochemistry*, 1984, **23**, 2390 (*isol*)
Patnaik, M. *et al, Acta Chem. Scand., Ser. B*, 1987, **41**, 210
(*synth*)

3,3′,5-Triiodothyronine   T-60350

Updated Entry replacing L-00355
O-(4-Hydroxy-3-iodophenyl)-3,5-diiodotyrosine, 9CI. 3-
[*4-(4-Hydroxy-3-iodophenoxy)-3,5-diiodophenyl*]-
alanine. 3′-Deiodothyroxine. Triothyrone
[327-86-0]

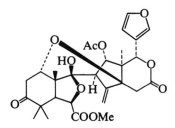

$C_{15}H_{12}I_3NO_4$ M 650.978

(*R*)-*form* [5714-08-9]
D-form. Detrothyronine, INN
Hypocholesterolaemic agent.
(*S*)-*form* [6893-02-3]
L-form. Liothyronine. T$_3$
Thyroid hormone. Mp 236-237° dec. $[\alpha]_D^{30}$ +21.5° (c,
4.75 in EtOH/1*M* HCl).
▷AY6750000.
B,HCl: [6138-47-2]. *Trionine. Thybon.* Long needles.
Mp 202-203° dec.
Na salt: [55-06-1]. *Cynomel. Cytobin. Cytomel. Cyto-*
mine. Tertroxin.

Gross, J. *et al, Lancet*, 1952, 439; *Biochem. J.*, 1953, **53**, 645.
Roche, J. *et al, Bull. Soc. Chim. Fr.*, 1957, 462.
Goodman, H.M. *et al, Med. Physiol. 13th Ed.*, 1974, **2**, 1632
(*rev*)
Block, P., *J. Med. Chem.*, 1976, **19**, 1067 (*synth*)
Mazzocchi, P.H. *et al, Org. Magn. Reson.*, 1978, **11**, 143 (*cmr*)
Chopra, I.J. *et al, Monogr. Endocrinol.*, 1981, **18**, 1-125 (*revs*)
Okabe, N. *et al, Biochim. Biophys. Acta*, 1982, **717**, 179 (*cryst*
struct)

Trijugin *A*   T-60351

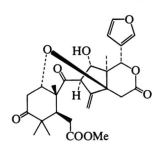

$C_{29}H_{34}O_{11}$ M 558.581
Constit. of *Heynea trijuga*. Cryst. (Me$_2$CO/hexane). Mp
265-266°. $[\alpha]_D^{30}$ −32.5° (c, 2 in CHCl$_3$).

Purushothaman, K.K. *et al, Can. J. Chem.*, 1987, **65**, 35.

Trijugin *B*   T-60352

$C_{27}H_{32}O_9$ M 500.544
Constit. of *Heynea trijuga*. Cryst. (Me$_2$CO/hexane). Mp
230-231°. $[\alpha]_D^{30}$ −15° (c, 2 in CHCl$_3$).

Purushothaman, K.K. *et al, Can. J. Chem.*, 1987, **65**, 35.

1,2,4-Trimethoxy-3*H*-phenoxazin-3-one   T-60353
AV toxin E

C$_{15}$H$_{13}$NO$_5$ M 287.271
Prod. by leaf spot fungus *Acrospermum viticola*.
Phytotoxin. Red needles (MeOH). Mp 160-162°.
Related to Questiomycin *A* and Michigazone.

*4-Demethoxy: 1,2-Dimethoxy-3*H-*phenoxazin-3-one.*
AV toxin D.
C$_{14}$H$_{11}$NO$_4$ M 257.245
From *A. viticola*. Phytotoxin. Yellow-brown plates
(MeOH). Mp 148-150°.

Kinjo, J. *et al, Tetrahedron Lett.*, 1987, **28**, 3697 (*isol, struct*)

Trimethylamine oxide, 8CI   T-60354
Updated Entry replacing T-03647
N,N-*Dimethylmethanamine* N-*oxide, 9CI.*
Trimethyloxamine
[1184-78-7]

$$Me_3N^{\oplus}{-}O^{\ominus}$$

C$_3$H$_9$NO   M 75.110
Widely distrib. in animal tissues, esp. fish. Oxidising
agent used in synthesis. Cleaves Si—C bonds. Large
cryst. (DMF). Sol. H$_2$O, MeOH. Mp 224-226° dec.
Hygroscopic.
▷Potentially explosive
Dihydrate: Cryst. (H$_2$O). Mp 96°.
B,HCl: [7651-88-9]. Needles (EtOH). Sol. H$_2$O, hot
MeOH. Mp 218°.
B,HI: Prisms (EtOH). Sol. H$_2$O, EtOH. Mp 130° dec.

Dunstan, W.R. *et al, J. Chem. Soc.*, 1899, **75**, 792 (*synth*)
Hickinbottom, W.J., *Reactions of Organic Compounds*, 1936,
  Longmans, New York, 277 (*synth*)
Craig, J.C. *et al, J. Org. Chem.*, 1970, **35**, 1721 (*synth, props*)
Sakurai, H. *et al, Tetrahedron Lett.*, 1986, **27**, 75 (*use*)
Soderquist, J.A. *et al, Tetrahedron Lett.*, 1986, **27**, 3961 (*synth*)
Fieser, M. *et al, Reagents for Organic Synthesis*, Wiley, 1967-
  84, **8**, 507.
Bretherick, L., *Handbook of Reactive Chemical Hazards*, 2nd
  Ed., Butterworths, London and Boston, 1979, 452.
Hazards in the Chemical Laboratory, (Bretherick, L., Ed.), 3rd
  Ed., Royal Society of Chemistry, London, 1981, 522.

3,4,5-Trimethylbenzoic acid   T-60355
Updated Entry replacing T-03695
α-*Isodurylic acid*
[1076-88-6]
C$_{10}$H$_{12}$O$_2$ M 164.204
Needles (H$_2$O). Mp 215-216°.
Me ester: [13544-66-6].
  C$_{11}$H$_{14}$O$_2$ M 178.230
  Cryst. (MeOH aq.). Mp 40.5-41.5°.

Jacobsen, O., *Ber.*, 1882, **15**, 1855 (*synth*)
Jannasch, P., *Ber.*, 1894, **27**, 3441 (*synth*)
Porowska, N., *Rocz. Chem.*, 1957, **31**, 677 (*deriv*)
Cano, F., *Acta Crystallogr., Sect. B*, 1970, **26**, 972 (*struct*)
Wadsten, T., *J. Chem. Soc., Chem. Commun.*, 1973, 4 (*struct*)

Hartshorn, M.P. *et al, Aust. J. Chem.*, 1985, **38**, 587 (*ir, pmr*)
Baciocchi, E. *et al, J. Org. Chem.*, 1986, **51**, 4544 (*deriv, synth,
  pmr, ms*)

2,5,8-Trimethylbenzotriimidazole   T-60356

C$_{12}$H$_{12}$N$_6$ M 240.267
Yellowish cryst. Mp >350°. Triethyl and tripentyl homo-
logues also prepd. Cryst. struct. determination on the
tripentyl compd.

Kohne, B. *et al, Angew. Chem., Int. Ed. Engl.*, 1986, **25**, 650
  (*synth, pmr, ms, cryst struct*)

1,2,3-Trimethyl-9*H*-carbazole   T-60357
[74404-32-3]

C$_{15}$H$_{15}$N   M 209.290
Mp 127.5-128.5°.

Carlin, R.B. *et al, J. Am. Chem. Soc.*, 1962, **84**, 4107 (*synth, uv,
  ir*)
Miller, B. *et al, J. Am. Chem. Soc.*, 1980, **102**, 4772 (*synth,
  pmr*)

1,2,4-Trimethyl-9*H*-carbazole   T-60358
[97218-74-1]
C$_{15}$H$_{15}$N   M 209.290
Pale-yellow cryst. (pet. ether). Mp 101.5-102°.

Takeuchi, H. *et al, J. Chem. Soc., Perkin Trans. 1*, 1986, 611
  (*synth, ir, pmr, cmr*)

2,3,4-Trimethyl-2-cyclohexen-1-one   T-60359
[22070-24-2]

C$_9$H$_{14}$O   M 138.209
(±)-*form*
Oil. Bp$_8$ 93-98°.
Semicarbazone: Mp 208-210°.

Ananthanarayan, K.A. *et al, Can. J. Chem.*, 1972, **50**, 3550
  (*synth, pmr*)
Safaryn, J.E. *et al, Tetrahedron*, 1986, **42**, 2635 (*synth, ir, pmr*)

2,2,5-Trimethylcyclopentanecarboxylic acid, 9CI
T-60360

$C_9H_{16}O_2$ M 156.224

(1RS,5SR)-form [85545-06-8]
(±)-trans-*form*
Oil. Bp_{14} 130° (bulb).
Amide: [85545-07-9].
  $C_9H_{17}NO$ M 155.239
  Needles (hexane/Me_2CO/Et_2O). Mp 153-155°.

Dreiding, A.S. *et al, Helv. Chim. Acta*, 1982, **65**, 2413; 1986, **69**, 1163 (*synth, ir, pmr, ms*)

1,2,5-Trimethyl-1-cyclopentanecarboxylic acid
T-60361

[88907-88-4]

$C_9H_{16}O_2$ M 156.224

(1r,2RS,5SR)-form
Prisms (pentane at −30°). Mp 71-73°.

Šolaja, B. *et al, Helv. Chim. Acta*, 1986, **69**, 1163 (*synth, ir, pmr, cmr, ms*)

2,6,10-Trimethyl-6,11-dodecadiene-2,3,10-triol
T-60362

$C_{15}H_{28}O_3$ M 256.384

(3S,10S)-form
Oil. $[\alpha]_D^{25}$ −13.4° (c, 1.08 in MeOH).
3-O-β-D-Glucopyranoside: [108906-50-9]. **Icariside C_1**.
  $C_{21}H_{38}O_8$ M 418.526
  Constit. of *Epimedium grandiflorum* var. *thunbergianum*. Amorph. powder. $[\alpha]_D^{25}$ −22.5° (c, 1.00 in MeOH).
2-O-β-D-glucopyranoside: **Icariside C_2**.
  $C_{21}H_{38}O_8$ M 418.526
  Constit. of *E. grandiflorum* var. *thunbergianum*. Amorph. powder. $[\alpha]_D^{25}$ −19.3° (c, 0.96 in MeOH).
11-O-β-D-Glucopyranoside: **Icariside C_3**.
  $C_{21}H_{38}O_8$ M 418.526
  Constit. of *E. grandiflorum* var. *thunbergianum*. Amorph. powder. $[\alpha]_D^{25}$ −34.7° (c, 0.88 in MeOH).
(3R,10S)-form
Oil. $[\alpha]_D^{25}$ +44.5° (c, 0.64 in MeOH).
3-O-β-D-Glucopyranoside: **Icariside C_4**.
  $C_{21}H_{38}O_8$ M 418.526
  Constit. of *Epimedium grandiflorum* var. *thunbergianum*. Amorph. powder. $[\alpha]_D^{25}$ +3.4° (c, 0.87 in MeOH).

Miyase, T. *et al, Chem. Pharm. Bull.*, 1987, **35**, 1109.

2,6,10-Trimethyl-2,6,10-dodecatrien-1,5,8,12-tetrol
T-60363

$C_{15}H_{26}O_4$ M 270.368

(2E,6E,10E)-form
5,8,12-Trihydroxyfarnesol
Constit. of *Cousinia adenostica*. Oil.
1-Ac: **12-Acetoxy-5,8-dihydroxyfarnesol**.
  $C_{17}H_{28}O_5$ M 312.405
  From *C. adenostica*. Oil.

Rustaiyan, A. *et al, Phytochemistry*, 1987, **26**, 2635.

3,7,11-Trimethyl-2,6,10-dodecatrienoic acid
T-60364

Updated Entry replacing T-10328
Farnesenic acid. Farnesic acid. Farnesylic acid
[7548-13-2]

$C_{15}H_{24}O_2$ M 236.353
All isomers (*E,E; E,Z; Z,E;* and *Z,Z*) are known. Oil. Bp_{16} 202-206°.
Me ester: [10485-70-8].
  $C_{16}H_{26}O_2$ M 250.380
  Juvenile hormone. Oil. Bp_{10} 177-185°.
2,3-Dihydroxypropyl ester:
  $C_{18}H_{30}O_4$ M 310.433
  Constit. of *Archidoris odhneri*. Oil.
2-Acetoxy-3-hydroxypropyl ester:
  $C_{20}H_{32}O_5$ M 352.470
  Constit. of *A. odhneri*. Oil.
3-Acetoxy-2-hydroxypropyl ester: Constit. of *A. odhneri*. Oil.
Me ester, 10,11-epoxide: [22963-93-5]. *Juvenile hormone III.*
  $C_{16}H_{26}O_3$ M 266.380
  Oil. $Bp_{0.08}$ 125-126°.

Ohki, M. *et al, Agric. Biol. Chem.*, 1974, **38**, 175 (*synth*)
Crombie, L. *et al, J. Chem. Soc., Perkin Trans. 1*, 1975, 913 (*cmr*)
Katzenellenbogen, J.A. *et al, J. Am. Chem. Soc.*, 1976, **98**, 4925 (*synth*)
Pitzele, B.S. *et al, Tetrahedron*, 1976, **32**, 1347 (*synth*)
Anderson, R.J. *et al, Tetrahedron Lett.*, 1980, 797 (*isol, deriv*)
Kuhnz, W. *et al, Org. Magn. Reson.*, 1981, **16**, 138 (*cmr*)
Mori, K. *et al, Tetrahedron*, 1987, **43**, 4097 (*synth*)

2,2,6-Trimethyl-5-hepten-3-one, 9CI
T-60365
tert-*Butyl isobutenyl ketone*
[84352-58-9]

$$(H_3C)_3CCOCH_2CH{=}C(CH_3)_2$$

$C_{10}H_{18}O$ M 154.252
Bp_{18} 100°.

Bretsch, W. *et al, Justus Liebigs Ann. Chem.*, 1987, 175 (*synth, pmr*)

2,2,5-Trimethyl-4-hexen-3-one, 9CI
T-60366
tert-*Butyl isopropyl ketone*
[14705-30-7]

$$(H_3C)_3CCOCH{=}C(CH_3)_2$$

$C_9H_{16}O$ M 140.225
Liq. Bp_{40} 110°.

Bretsch, W. et al, Justus Liebigs Ann. Chem., 1987, 175 (synth, pmr, ir)

3,4,4-Trimethyl-5(4H)-isoxazolone, 9CI   T-60367
[15731-98-3]

$C_6H_9NO_2$ M 127.143
Bp_{15} 103°, Bp_{11} 96-97°.

Billon, P., Ann. Chim. (Paris), 1927, 7, 357 (synth)
Jacquier, R. et al, Bull. Soc. Chim. Fr., 1970, 2690 (synth, pmr, uv, ir)
Cannone, P. et al, Tetrahedron, 1986, 42, 4203 (synth, ir, pmr)

7,11,15-Trimethyl-3-methylene-1,2-hexa-decanediol, 9CI   T-60368
[100605-94-5]

$C_{20}H_{40}O_2$ M 312.535
Constit. of Senecio gallicus. Oil. $[\alpha]_D$ −1.2° (c, 0.5 in $CHCl_3$).

Di-Ac:
   $C_{24}H_{44}O_4$ M 396.609
   Constit. of S. gallicus. Oil.

Urones, J.G. et al, Phytochemistry, 1987, 26, 1113.

2,3,4-Trimethyl-5-nitrobenzoic acid   T-60369
[13667-30-6]

$C_{10}H_{11}NO_4$ M 209.201
Cryst. (EtOH aq.). Mp 177°.

Kubota, T. et al, Tetrahedron Lett., 1967, 745.
Okukado, N. et al, CA, 1969, 70, 3406t (synth)

2,4,6-Trimethyl-3-nitrobenzoic acid   T-60370
3-Nitromesitoic acid
[106567-41-3]
$C_{10}H_{11}NO_4$ M 209.201
Yellowish needles (EtOH aq.). Mp 183.5-184°. pK_a 4.34
(55% EtOH aq., 22°).

Beringer, F.M. et al, J. Am. Chem. Soc., 1953, 75, 3319 (synth)
Cuyegkeng, M.A. et al, Chem. Ber., 1987, 120, 803 (synth, pmr)

3,4,5-Trimethylphenol, 9CI, 8CI   T-60371
Updated Entry replacing T-04033
sym-Hemimellitenol. 5-Hydroxyhemimellitene
[527-54-8]

$C_9H_{12}O$ M 136.193
Needles with bluish fluor. (hexane). Mp 108°. Bp 248-249°.

Ac: [6719-74-0].
   $C_{11}H_{14}O_2$ M 178.230
   Cryst. Mp 59-60°.
Phenylurethane: Mp 148-149°.
Me ether: [21573-41-1]. 5-Methoxy-1,2,3-trimethylbenzene.
   $C_{10}H_{14}O$ M 150.220
   Bp_{22} 120°.

Banwell, T. et al, J. Am. Chem. Soc., 1977, 99, 3042 (synth)
Netzel, D.A., Org. Magn. Reson., 1978, 11, 58 (nmr)
Baciocchi, E. et al, J. Org. Chem., 1986, 51, 4544 (deriv, synth, pmr, ms)

4,6,8-Trimethyl-2-undecanol, 9CI   T-60372
Updated Entry replacing T-20298
[83474-31-1]

1R,3R,5R,7R-form

$C_{14}H_{30}O$ M 214.390
Formyl: [83540-84-5]. **Lardolure**.
   $C_{15}H_{30}O_2$ M 242.401
   Aggregation pheromone of the Acarid mite, Lardoglyphus konoi.

Kiswahara, Y. et al, Agric. Biol. Chem., 1982, 46, 2283 (isol, struct)
Mori, K. et al, Tetrahedron, 1986, 42, 5539, 5545 (synth, abs config)

2,3,4-Trinitro-1H-pyrrole   T-60373

$C_4H_2N_4O_6$ M 202.083
1-Me: [69726-53-0]. 1-Methyl-2,3,4-trinitro-1H-pyrrole.
   $C_5H_4N_4O_6$ M 216.110
   Used in explosives. Mp 111-112°.
▷Potentially toxic and explosive

Doddi, G. et al, J. Org. Chem., 1979, 44, 2321 (synth, pmr)

2,3,5-Trinitro-1H-pyrrole   T-60374
$C_4H_2N_4O_6$ M 202.083
1-Me: [69726-47-2]. 1-Methyl-2,3,5-trinitro-1H-pyrrole.
   $C_5H_4N_4O_6$ M 216.110
   Mp 49-50°.
▷Potentially toxic and explosive

Doddi, G. et al, J. Org. Chem., 1979, 44, 2321 (synth, pmr)

4,7,13-Trioxa-1,10-diazabicyclo[8.5.5]-icosane   T-60375

[72640-82-5]

C₁₅H₃₀N₂O₃   M 286.414
Cryptand. Oil.

Lincoln, S.F. *et al, J. Am. Chem. Soc.*, 1986, **108**, 8134 (*use*)
Lincoln, S.F. *et al, J. Chem. Soc., Dalton Trans.*, 1986, 1075 (*synth, pmr, ir, ms*)

1,4,10-Trioxa-7,13-diazacyclopentadecane, 9CI   T-60376

4,10-Diaza-15-crown-5

[31249-95-3]

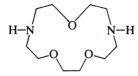

C₁₀H₂₂N₂O₃   M 218.295
Mp 89-90°. Bp₀.₁ 120-122°.

4,10-Dibenzyl: [94195-16-1].
   C₂₄H₃₄N₂O₃   M 398.544
   Oil.

Dietrich, B. *et al, Tetrahedron*, 1973, **29**, 1629, 1647 (*synth, pmr, conformn, props*)
Gatto, V.J. *et al, J. Org. Chem.*, 1986, **51**, 5373 (*synth, pmr, ir, cmr*)
Gatto, V.J. *et al, Tetrahedron Lett.*, 1986, **27**, 327 (*synth, use*)

2,4,10-Trioxatricyclo[3.3.1.1³,⁷]decane   T-60377

2,4,10-Trioxaadamantane

[281-32-3]

C₇H₁₀O₃   M 142.154
Cryst. (pet. ether). Mp 219-220°. Hygroscopic.

Stetter, H. *et al, Chem. Ber.*, 1953, **86**, 790 (*synth*)

Triphenodioxazine, 9CI   T-60378

Updated Entry replacing T-04265
5,12-Dioxa-7,14-diazapentacene. [1,4]Benzoxazino[2,3-b]phenoxazine

[258-72-0]

C₁₈H₁₀N₂O₂   M 286.289
One of the prods. of condensation of 2-aminophenol with 1,4-benzoquinone. Red, blue or violet needles (PhNO₂, xylene or by subl.). Mp >360°.

Kehrmann, F. *et al, Helv. Chim. Acta*, 1924, **7**, 973; 1925, **8**, 223 (*synth*)

Musso, H. *et al, Chem. Ber.*, 1965, **98**, 3937 (*synth, ir*)
Bolognese, A. *et al, J. Heterocycl. Chem.*, 1986, **23**, 1003 (*synth, uv, pmr*)

3,4,4-Triphenyl-2-cyclohexen-1-one   T-60379

[103367-47-1]

C₂₄H₂₀O   M 324.421
Solid (EtOH). Mp 146-147°.

Tosylhydrazone: [103367-48-2]. Solid (EtOH). Mp 205-206°.

Zimmerman, H.E. *et al, J. Am. Chem. Soc.*, 1968, **90**, 954; 1986, **108**, 6276 (*synth, pmr, ir*)

3,5,5-Triphenyl-2-cyclohexen-1-one   T-60380

[103367-54-0]
C₂₄H₂₀O   M 324.421
Solid (EtOH). Mp 117.0-117.5°.

Zimmerman, H.E. *et al, J. Am. Chem. Soc.*, 1986, **108**, 6276 (*synth, ir, pmr, uv*)

4,5,5-Triphenyl-2-cyclohexen-1-one, 9CI   T-60381

[105555-58-6]

C₂₄H₂₀O   M 324.421
(±)-*form* [103367-50-6]
   Solid (EtOAc). Mp 215.0-216.0°.

Zimmerman, H.E. *et al, J. Am. Chem. Soc.*, 1986, **108**, 6276 (*synth, ir, pmr, uv*)

2-Triphenylenecarboxaldehyde   T-60382

2-Formyltriphenylene

[96404-79-4]

C₁₉H₁₂O   M 256.303
Cryst. Mp 155-156°.

Tanga, M.J. *et al, J. Heterocycl. Chem.*, 1987, **24**, 39 (*synth, uv, ir, pmr, cmr*)

1-Triphenylenecarboxylic acid

T-60383

$C_{19}H_{12}O_2$ M 272.303
Cryst. (EtOH). Mp 245°.

Me ester:
  $C_{20}H_{14}O_2$ M 286.329
  Yellow cryst. (EtOH). Mp 150°.

Plieninger, H. *et al, Chem. Ber.*, 1963, **96**, 1610 (*synth, deriv*)

2-Triphenylenecarboxylic acid

T-60384

[109534-58-9]
$C_{19}H_{12}O_2$ M 272.303
Cryst. (1-pentanol). Mp 336-338°.

Monohydrate: Mp 157-158°.
Me ester:
  $C_{20}H_{14}O_2$ M 286.329
  Cryst. (EtOH). Mp 171-172°.
Chloride:
  $C_{19}H_{11}ClO$ M 290.748
  Yellow cryst. (pet. ether). Mp 130-131°.
Nitrile: [34177-31-6]. *2-Cyanotriphenylene.*
  $C_{19}H_{11}N$ M 253.303
  Needles (EtOH). Mp 220-221°.

Barker, C.C. *et al, J. Chem. Soc.*, 1955, 4482 (*synth, deriv*)
Sato, T. *et al, Bull. Chem. Soc. Jpn.*, 1971, **44**, 2484 (*synth, deriv*)
Tanga, M.J. *et al, J. Heterocycl. Chem.*, 1987, **24**, 39 (*synth, uv, ir, pmr, cmr*)

Triphenylene-1,2-oxide

T-60385

1,2-Epoxy-1,2-dihydrotriphenylene

$C_{18}H_{12}O$ M 244.292
Racemic form obtained from both chiral and achiral
precursors. Needles (Me$_2$CO at low temp.). Mp 167-
168°.

Boyd, D.R. *et al, J. Chem. Soc., Perkin Trans. 1*, 1987, 369
(*synth, pmr*)

1,3,6-Triphenylhexane

T-60386

1,1',1"-(1,3,6-Hexanetriyl)trisbenzene, 9CI
[70547-85-2]

$$PhCH_2CH_2CHPhCH_2CH_2CH_2Ph$$

$C_{24}H_{26}$ M 314.469
(±)-form
Oil. Bp$_{0.1}$ 110-115°.

Bergman, J. *et al, Tetrahedron*, 1986, **42**, 763 (*synth, ir, cmr*)

1,2,3-Triphenyl-1H-indene, 9CI

T-60387

[38274-35-0]

$C_{27}H_{20}$ M 344.455
(±)-form
Needles. Mp 136°.

Maroni, R. *et al, J. Chem. Soc., Perkin Trans. 1*, 1974, 353
  (*synth*)
Yamamura, K. *et al, Bull. Chem. Soc. Jpn.*, 1986, **59**, 3699
  (*synth, pmr*)

(Triphenylmethyl)hydrazine

T-60388

Tritylhydrazine. Hydrazinotriphenylmethane
[104933-75-7]

$$Ph_3CNHNH_2$$

$C_{19}H_{18}N_2$ M 274.365
B,HCl: [104370-29-8]. Solid. Mp 109-112°.

Senior, J.K., *J. Am. Chem. Soc.*, 1916, **38**, 2718 (*synth*)
Baldwin, J.E. *et al, Tetrahedron*, 1986, **42**, 4235 (*synth, use*)

1,2,2-Triphenyl-1-propanone

T-60389

*2,2-Diphenylpropiophenone. 1-Benzoyl-1,1-
diphenylethane*
[36504-01-5]

$$H_3CCPh_2COPh$$

$C_{21}H_{18}O$ M 286.373
Cryst. (EtOH). Mp 91-92°.

Kharasch, M.S. *et al, J. Org. Chem.*, 1951, **16**, 1458 (*synth*)
Borowitz, I.J. *et al, J. Org. Chem.*, 1972, **37**, 3873 (*synth, ir, uv,
  pmr, ms*)

1,2,3-Triphenyl-1-propanone

T-60390

Benzyldesoxybenzoin. 2,3-Diphenylpropiophenone
[4842-45-9]

$$PhCH_2CHPhCOPh$$

$C_{21}H_{18}O$ M 286.373
(±)-form
Needles. Mp 120°.

Meyer, V. *et al, Ber.*, 1888, **21**, 1300 (*synth*)
Cragoe, E.J. *et al, J. Org. Chem.*, 1958, **23**, 971 (*synth*)
Miyano, S. *et al, Chem. Pharm. Bull.*, 1965, **13**, 1372 (*synth*)
Makosza, M. *et al, Rocz. Chem.*, 1973, **47**, 77 (*synth*)

1,3,3-Triphenyl-1-propanone, 9CI

T-60391

3,3-Diphenylpropiophenone, 8CI
[606-86-0]

$$Ph_2CHCH_2COPh$$

$C_{21}H_{18}O$ M 286.373
Mp 96°.

Oxime: [1669-73-4].
  $C_{21}H_{19}NO$ M 301.387
  Needles (EtOH aq.). Mp 131°.
Phenylhydrazone: Pale-yellow cryst. (EtOH). Mp 137°.

Kohler, E.P., *Am. Chem. J.*, 1904, **31**, 642 (*synth*)

Chemisat, B., *Bull. Soc. Chim. Fr.*, 1972, 3415 (*synth, pmr*)
Jayamani, M. *et al*, *Tetrahedron*, 1986, **42**, 4325 (*synth*)

Triptofordin *A*   T-60392

R^1 = H, R^2 = OCCH=CHPh

C$_{31}$H$_{36}$O$_6$ M 504.622
Constit. of *Tripterygium wilfordii* var. *regelii*. Cryst.
(MeOH). Mp 220-221°. $[\alpha]_D^{28}$ +191.7° (c, 0.23 in
MeOH).

Takaishi, Y. *et al*, *Phytochemistry*, 1987, **26**, 2325.

Triptofordin *B*   T-60393

As Triptofordin *A*, T-60392 with

R^1 = OH, R^2 = COPh

C$_{29}$H$_{34}$O$_7$ M 494.583
Constit. of *Tripterygium wilfordii* var. *regelii*. Cryst.
(MeOH). Mp 193-195°. $[\alpha]_D^{27}$ +75.6° (c, 0.23 in
MeOH).

Takaishi, Y. *et al*, *Phytochemistry*, 1987, **26**, 2325.

Triptofordin *C*-2   T-60394

C$_{33}$H$_{38}$O$_{11}$ M 610.657
Constit. of *Tripterygium wilfordii* var. *regelii*. Amorph.
powder. Mp 128-131°. $[\alpha]_D^{22}$ −44.7° (c, 0.27 in
MeOH). MF incorrectly given as C$_{33}$H$_{34}$O$_{11}$.

2-Ketone: Triptofordin C-1.
C$_{33}$H$_{36}$O$_{11}$ M 608.641
From *T. wilfordii* var. *regelii*. Cryst. (MeOH). Mp
249-251°. $[\alpha]_D^{28}$ −22.6° (c, 0.22 in MeOH).

Takaishi, Y. *et al*, *Phytochemistry*, 1987, **26**, 2325.

Triptofordin *D*-1   T-60395

R^1 = H, R^2 = −COCH=CHPh, R^3 = O

C$_{35}$H$_{38}$O$_{11}$ M 634.679
Constit. of *Tripterygium wilfordii*. Needles (MeOH).
Mp 224-226°. $[\alpha]_D^{27}$ +56.1° (c, 0.25 in MeOH).
Takaishi, Y. *et al*, *Phytochemistry*, 1987, **26**, 2581.

Triptofordin *D*-2   T-60396

As Triptofordin *D*-1, T-60395 with

R^1 = H, R^2 = COCH=CHPh, R^3 = αOAc, H

C$_{37}$H$_{42}$O$_{12}$ M 678.732
Constit. of *Tripterygium wilfordii*. Needles (MeOH).
Mp 103-109°. $[\alpha]_D^{27}$ −3.5° (c, 0.28 in MeOH).

Takaishi, Y. *et al*, *Phytochemistry*, 1987, **26**, 2581.

Triptofordin *E*   T-60397

As Triptofordin *D*-1, T-60395 with

R^1 = OAc, R^2 = −COPh, R^3 = O

C$_{35}$H$_{38}$O$_{13}$ M 666.677
Constit. of *Tripteryguim willfordii*. Granules (MeOH).
Mp 116-118°. $[\alpha]_D^{27}$ +47.4° (c, 0.25 in MeOH).

Takaishi, Y. *et al*, *Phytochemistry*, 1987, **26**, 2581.

1,8,11-Triptycenetricarboxylic acid   T-60398

*9,10-Dihydro-9,10[1',2']benzenoanthracene-1,8,16-tri-
carboxylic acid, 9CI*

[103259-15-0]

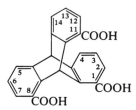

C$_{23}$H$_{14}$O$_6$ M 386.360
Solid. Mp 426-428°. The numbering used in the paper is
based on the 9CI name which is not the usual
triptycene numbering (the two systems give the same
result for mono- and disubst. triptycenes).

Tri-Me ester: [103259-19-4].
C$_{26}$H$_{20}$O$_6$ M 428.440
Cryst. (CH$_2$Cl$_2$/EtOAc). Mp 287-289°.

Rogers, M.E. *et al*, *J. Org. Chem.*, 1986, **51**, 3308 (*synth, pmr,
cmr, ms*)

1,8,14-Triptycenetricarboxylic acid   T-60399

*9,10-Dihydro-9,10[1',2']benzenoanthracene-1,8,13-tri-
carboxylic acid, 9CI*

[103259-13-8]
C$_{23}$H$_{14}$O$_6$ M 386.360
See note under 1,8,11-Triptycenetricarboxylic acid, T-
60398 . Solid. Mp 432-435°.

Tri-Me ester: [103259-20-7].
C$_{26}$H$_{20}$O$_6$ M 428.440
Cryst. (CH$_2$Cl$_2$/EtOAc). Mp 256-257°.

Rogers, M.E. *et al*, *J. Org. Chem.*, 1986, **51**, 3308 (*synth, pmr,
cmr, ms*)

Trisabbreviatin BBB   T-60400
[84633-05-6]

$C_{34}H_{40}O_{12}$ M 640.683
Isol. from *Dryopteris abbreviata*. Pale-yellow powder
(cyclohexane/EtOH).

Coskun, M. *et al, Chem. Pharm. Bull.*, 1982, **30**, 4102 (*isol*)

1,3,5-Tris(bromomethyl)benzene, 9CI   T-60401
[18226-42-1]

$C_9H_9Br_3$ M 356.882
Cryst. (CHCl_3/pet. ether). Mp 94°.

Ried, W. *et al, Chem. Ber.*, 1959, **92**, 2532 (*synth*)
Cochrane, W.P. *et al, J. Chem. Soc. (C)*, 1968, 630 (*synth, pmr*)
Vogtle, F. *et al, Chem. Ber.*, 1973, **106**, 717 (*synth, pmr*)

1,3,5-Tris(chloromethyl)benzene, 9CI   T-60402
[17299-97-7]

$C_9H_9Cl_3$ M 223.529
Cryst. (pet. ether). Mp 64°.

Cochrane, W.P. *et al, J. Chem. Soc. (C)*, 1968, 630 (*synth, pmr*)

2,3,5-Tris(methylene)bicyclo[2.2.1]heptane, T-60403
9CI
2,3,5-Trimethylidenebicyclo[2.2.1]heptane. 2,3,5-
Trimethylenebicyclo[2.2.1]heptane
[86963-85-1]

$C_{10}H_{12}$ M 132.205
Oil. Bp_{0.2} 35°.

Burnier, G. *et al, Helv. Chim. Acta*, 1986, **69**, 1310 (*synth, uv, pmr, cmr*)

20,29,30-Trisnor-3,19-lupanediol   T-60404
20,29,30-Trinorlupane-3,19-diol, 9CI

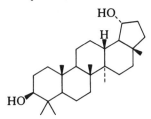

$C_{27}H_{46}O_2$ M 402.659
(3β,19α)-form [109795-03-1]
Metab. of *Pseudocyphellaria rubella*. Cryst. Mp 234-
236°.

Corbett, R.E. *et al, Aust. J. Chem.*, 1987, **40**, 461.

Tris(pentachlorophenyl)methane   T-60405
1,1',1''-Methylidynetris[2,3,4,5,6-pentachlorobenzene],
9CI. Perchlorotriphenylmethane
[33240-61-8]

$$(C_6Cl_5)_3CH$$

$C_{19}HCl_{15}$ M 761.012
Mp 320° dec.
Radical: [4070-01-3]. *Tris(pentachlorophenyl)methyl,*
9CI.
$C_{19}Cl_{15}$ M 760.004
Inert free radical. Deep-red solid. Mp 305° dec.
Carbanion, tetraethylammonium salt: Tetraethylam-
monium tris(pentachlorophenyl)methylide.
$C_{27}H_{20}Cl_{15}N$ M 890.257
Garnet cryst. powder.
Carbonium ion hexachloroantimonate: [30482-37-2].
Tris(pentachlorophenyl)methyl
hexachloroantimonate.
$C_{19}Cl_{21}Sb$ M 1094.472
Dark-green cryst.

Ballester, M. *et al, Tetrahedron Lett.*, 1970, 3615, 4509
(*carbonium ion, carbanion*)
Ballester, M. *et al, J. Am. Chem. Soc.*, 1971, **93**, 2215 (*synth, pmr, esr, uv, ir, use*)
Ballester, M. *et al, Synthesis*, 1986, 64 (*synth*)

Tris(trifluoromethyl)amine   T-60406
1,1,1-Trifluoro-N,N-bis(trifluoromethyl)methanamine,
9CI. Nonafluorotrimethylamine.
Perfluorotrimethylamine
[432-03-1]

$$(F_3C)_3N$$

C_3F_9N M 221.025
Gas. Mp −116°. Bp_{760} −10.1°.

Haszeldine, R.N., *J. Chem. Soc.*, 1951, 102 (*synth*)
Burger, H. *et al, J. Mol. Struct.*, 1979, **54**, 159 (*ir, raman, ed*)
Burger, H. *et al, J. Fluorine Chem.*, 1981, **17**, 65 (*synth, spectra*)

b,b',b''-Tritriptycene T-60407

5,7,9,14,16,18,28,33-Octahydro-28,33[1,2']benzeno-7,16[2',3']anthraceno-5,18[1,2']:9,14[1'',2'']-dibenzeno-heptacene, 9CI

[103960-12-9]

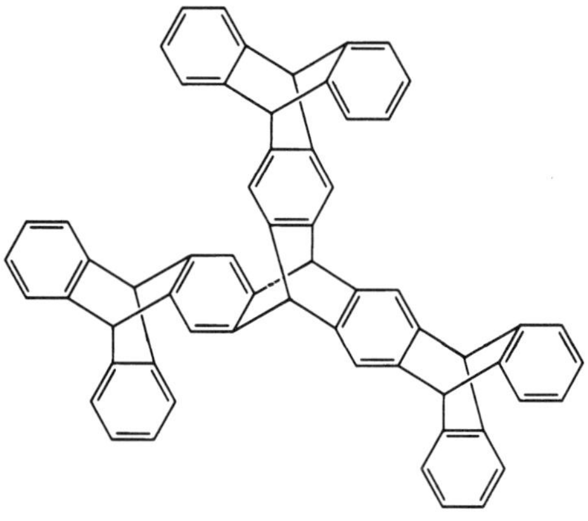

$C_{62}H_{38}$ M 782.982
Mp >550°.

Bashir-Hashemi, A. *et al, J. Am. Chem. Soc.*, 1986, **108**, 6675 (*synth, pmr, cmr, ir, cryst struct*)

Trunculin *A* T-60408

[105969-64-0]

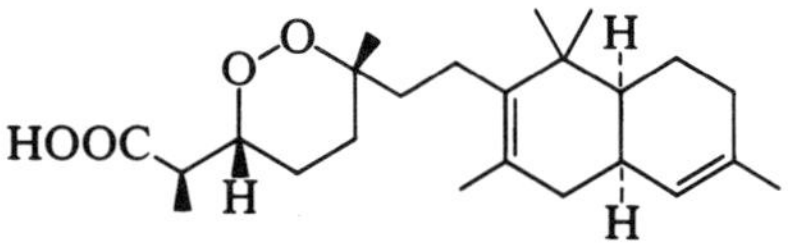

$C_{24}H_{38}O_4$ M 390.562
Constit. of *Latrunculia brevis.*

Me ester: Cryst. (MeOH aq.). Mp 66.5-67.5°. $[\alpha]_D$ +158.3° (c, 1.02 in $CHCl_3$).

Capon, R.J. *et al, J. Org. Chem.*, 1987, **52**, 339 (*cryst struct*)

Trunculin *B* T-60409

[105969-65-1]

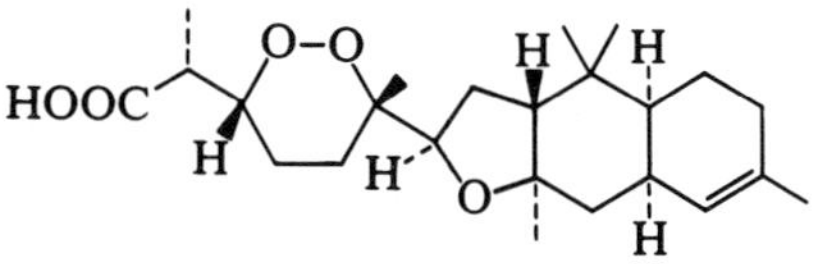

$C_{24}H_{38}O_5$ M 406.561
Constit. of *Latrunculia brevis.*

Me ester: Cryst. (MeOH aq.). Mp 91-93°. $[\alpha]_D$ +13.1° (c, 1.54 in $CHCl_3$).

Capon, R.J. *et al, J. Org. Chem.*, 1987, **52**, 339 (*cryst struct*)

Tucumanoic acid T-60410

Updated Entry replacing T-50568
ent-2β,3β,4α-Trihydroxy-13-cleroden-15-oic acid

[97165-43-0]

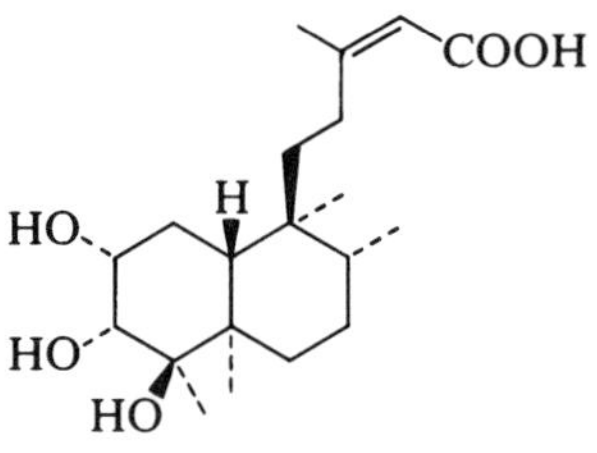

$C_{20}H_{34}O_5$ M 354.486
Constit. of *Baccharis tucumanensis.*

Me ester: Cryst. Mp 190-191°. $[\alpha]_D^{20}$ −5.6° (c, 0.48 in MeOH).

13,14-Dihydro: ent-2β,3β,4α-Trihydroxy-15-clerodan-oic acid. **Dihydrotucumanoic acid.**
$C_{20}H_{36}O_5$ M 356.501
From *B. pedicellata* and *B. marginalis.* Cryst. (Me_2CO). Mp 190-191°. $[\alpha]_D^{25}$ −16.0° (c, 0.1 in MeOH).

Rossomando, P.C. *et al, Phytochemistry*, 1985, **24**, 787 (*isol, struct*)
Fiani, F. *et al, Phytochemistry*, 1987, **26**, 3281 (*deriv*)

N-Tyrosylalanine T-60411

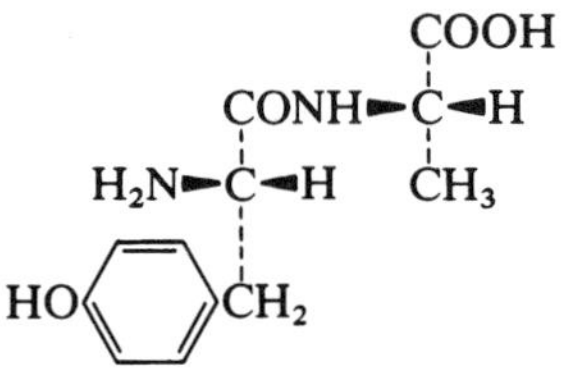

L-L-form

$C_{12}H_{16}N_2O_4$ M 252.269

L-D-form
Boc-Tyr-D-Ala-OH: Cryst. (EtOAc). Mp 232-234°.
Boc-Tyr-D-Ala-OMe: Cryst. (EtOAc). Mp 78-80°.

L-L-form [730-08-5]
Cryst. (EtOH aq.).
Boc-Tyr-Ala-OBzl: [96523-73-8]. Mp 120°. $[\alpha]_D^{20}$ −15.8° (c, 1 in MeOH).
Z-Tyr-Ala-OMe: Mp 135-136°. $[\alpha]_D$ −18.5° (c, 1 in MeOH).
Z-Tyr(OZ)-Ala-OBzl: Cryst. (EtOH). Mp 145°.

Dukler, S. *et al, Tetrahedron*, 1971, **27**, 607 (*synth*)
Tomatis, R. *et al, Farmaco, Ed. Sci.*, 1979, **34**, 496 (*deriv*)
Gacel, G. *et al, J. Med. Chem.*, 1981, **24**, 1119 (*deriv*)
Loennechen, T. *et al, Acta Chem. Scand., Ser. B*, 1984, **38**, 647 (*synth, derivs*)
Omodei-Sale, A. *et al, J. Chem. Res. (S)*, 1984, 50 (*synth*)

N-Tyrosylphenylalanine, 9CI T-60412

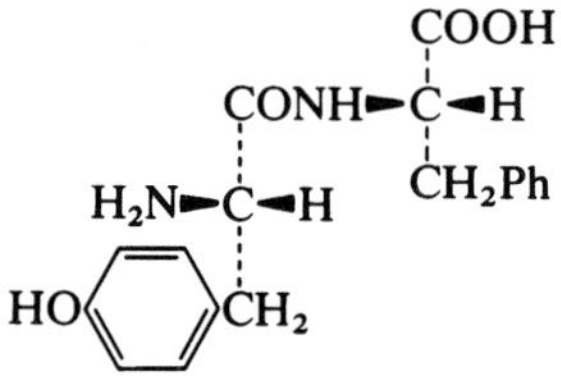

(S,S)-form

$C_{18}H_{20}N_2O_4$ M 328.367
(S,S)-form [17355-11-2]
L-L-form
Needles + $1H_2O$ (H_2O). Mp 310-312° dec. $[\alpha]_D^{21}$ +17.7° (c, 0.5 in 2*M* HCl).

N-*Benzyloxycarbonyl:* Cryst. (EtOH aq.). Mp 175-176°. $[\alpha]_D^{22}$ +2.1° (c, 1 in 1*M* NaOH).

N,O-*Bis(benzyloxycarbonyl):* Cryst. (EtOH). Mp 182-184°. $[\alpha]_D^{21}$ +6.1° (c, 1 in AcOH).

N-*Formyl:* Needles (EtOH aq.). Mp 247-248° dec.

Stewart, F.H.C., *J. Org. Chem.*, 1960, **25**, 1828 (*synth*)

Dale, B.J. *et al*, *J. Chem. Soc., Perkin Trans. 2*, 1976, 91 (*pmr*)

Murali, R. *et al*, *Int. J. Pept. Protein Res.*, 1987, **29**, 187 (*cryst struct*)

U

Umbellifolide   U-60001

Updated Entry replacing U-30003

Tetrahydro-6-methyl-3-methylene-6-(4-oxopentyl)-2,5-
(3H,4H)-benzofurandione, 9CI. 4,5-Dioxo-4,5-seco-
11(13)-eudesmen-12,8β-olide

[89026-40-4]

$C_{15}H_{20}O_4$ M 264.321

Constit. of *Artemisia umbelliformis*. Cryst.
(EtOAc/Et$_2$O). Mp 113°. $[\alpha]_D^{25}$ +60.8° (c, 0.5 in
CHCl$_3$).

Appendino, G. *et al, J. Chem. Soc., Perkin Trans. 1*, 1983, 2705
 (*isol, cryst struct*)
Marco, J.A. *et al, Tetrahedron*, 1987, **43**, 2523 (*synth*)

Undecafluoropiperidine   U-60002

Perfluoropiperidine

[836-77-1]

$C_5F_{11}N$ M 283.044

Mild fluorinating agent. Liq. Bp 49°.

Banks, R.E. *et al, J. Chem. Soc.*, 1962, 3407 (*synth*)
Lee, J. *et al, Trans. Faraday Soc.*, 1967, **63**, 16 (*F nmr*)

Undecamethylcyclohexanol   U-60003

[103457-84-7]

$C_{17}H_{34}O$ M 254.455

Mp >300°. Readily dehydrates.

Wehle, D. *et al, Chem. Ber.*, 1986, **119**, 3127 (*synth, pmr, cmr*)

6-Undecyn-1-ol   U-60004

[69222-10-2]

$$H_3C(CH_2)_3C{\equiv}C(CH_2)_4CH_2OH$$

$C_{11}H_{20}O$ M 168.278

Liq. Bp$_2$ 108-110°, Bp$_{0.03}$ 90-92°.

Svirskaya, P.I. *et al, J. Chem. Eng. Data*, 1979, **24**, 152 (*synth*)
Brown, H.C. *et al, J. Org. Chem.*, 1986, **51**, 4518 (*synth, ir, pmr,
 cmr*)

12-Ursene-1,2,3,11,20-pentol   U-60005

$C_{30}H_{50}O_5$ M 490.722

(1β,2α,3β,11α,20β)-form

3-Ac: 3β-Acetoxy-12-ursene-1β,2α,11α,20β-tetrol.
  $C_{32}H_{52}O_6$ M 532.759
  Constit. of *Salvia argentea*. Cryst. (EtOAc/hexane).
  Mp 112-114°. $[\alpha]_D^{20}$ +33.1° (c, 0.136 in CHCl$_3$).

Bruno, M. *et al, Phytochemistry*, 1987, **26**, 497.

12-Ursene-1,2,3,11-tetrol   U-60006

$C_{30}H_{50}O_4$ M 474.723

(1β,2α,3β,11α)-form

3-Ac: 3β-Acetoxy-12-ursene-1β,2α,11α-triol.
  $C_{32}H_{52}O_5$ M 516.760
  Constit. of *Salvia agentea*. Amorph. powder. $[\alpha]_D^{20}$
  +33.9° (c, 0.106 in CHCl$_3$).

Bruno, M. *et al, Phytochemistry*, 1987, **26**, 497.

12-Ursene-2,3,11,20-tetrol   U-60007

$C_{30}H_{50}O_4$ M 474.723

(2α,3β,11α,20β)-form

3-Ac: 3β-Acetoxy-12-ursene-2α,11α,20β-triol.
  $C_{32}H_{52}O_5$ M 516.760
  Constit. of *Salvia argentea*. Cryst. (EtOAc/hexane).
  Mp 116-120°. $[\alpha]_D^{20}$ +49.4° (c, 0.155 in CHCl$_3$).

Bruno, M. *et al, Phytochemistry*, 1987, **26**, 497.

12-Ursene-2,3,11-triol   U-60008

$C_{30}H_{50}O_3$ M 458.723

(2α,3β,11α)-form

3-Ac: 3β-Acetoxy-12-ursene-2α,11α-diol.
  $C_{32}H_{52}O_4$ M 500.760
  Constit. of *Salvia argentea*. Cryst. (EtOAc/hexane).
  Mp 205-208°. $[\alpha]_D^{18}$ +20.3° (c, 0.118 in CHCl$_3$).

Bruno, M. *et al, Phytochemistry*, 1987, **26**, 497.

V

N-Valylleucine, 9CI   **V-60001**

$(H_3C)_2CHCH(NH_2)CONHCH(COOH)CH_2CH(CH_3)_2$

$C_{11}H_{22}N_2O_3$ M 230.306

D-L-form

Boc-Val-Leu-OMe: [78981-67-6]. Prisms (hexane). Mp 87-88°. $[\alpha]_D$ −7.7° (c, 1 in MeOH).

L-D-form [22906-55-4]

Amorph. solid + 1H$_2$O. Mp 280-285° dec. $[\alpha]_D^{20}$ +72.1° (c, 1.9 in H$_2$O).

Z-Val-Leu-OBzl: Cryst. (CHCl$_3$/Et$_2$O). Mp 132°.

L-L-form [3989-97-7]

B,HCl: Cryst. (MeOH/Et$_2$O). Mp 217-219°. $[\alpha]_D$ −6.0° (c, 1 in 6M HCl).

H-Val-Leu-OMe; B,HCl: [23365-04-0]. Mp 174-176°. $[\alpha]_D$ −12.6° (c, 1 in H$_2$O).

Z-Val-Leu-OH: [17708-79-1]. Cryst. (EtOAc/pet. ether). Mp 135-137°. $[\alpha]_D^{20}$ −24.0° (c, 0.49 in CH$_2$Cl$_2$).

Z-Val-Leu-OBzl: [95303-77-8]. Mp 115-116°. $[\alpha]_D$ −10.7° (c, 1 in DMF).

Z-Val-Leu-OMe: [4817-93-0]. Mp 134-135°. $[\alpha]_D^{25}$ −25.5° (c, 2 in DMF).

Z-Val-Leu-NHNH$_2$: [17137-03-0]. Mp 124-138°. $[\alpha]_D$ −42.6° (c, 1 in MeOH).

DL-DL-form [3929-62-2]

Mp 247-248°.

Luebke, K. *et al, Justus Liebigs Ann. Chem.*, 1963, **665**, 205 (*synth*)

Barrett, G.C., *J. Chem. Soc. (C)*, 1969, 1123 (*synth*)

Bodanszky, M. *et al, J. Am. Chem. Soc.*, 1975, **97**, 2857 (*synth*)

Oya, M. *et al, Bull. Chem. Soc. Jpn.*, 1981, **54**, 2705 (*synth*)

Suzuki, K. *et al, Chem. Pharm. Bull.*, 1981, **29**, 233 (*deriv*)

Ariyashi, Y., *Bull. Chem. Soc. Jpn.*, 1984, **57**, 3197 (*deriv*)

Okada, Y. *et al, Chem. Pharm. Bull.*, 1984, **32**, 4608 (*deriv*)

Schmidt, U. *et al, Synthesis*, 1987, 236 (*deriv*)

Vaticaffinol   **V-60002**

Updated Entry replacing V-40002

[81344-96-9]

$C_{56}H_{42}O_{12}$ M 906.941

Constit. of *Vatica affinis*. Cryst. (Me$_2$CO/C$_6$H$_6$). Mp 285° dec. $[\alpha]_D$ −22.5° (MeOH).

Sotheeswaran, S. *et al, J. Chem. Soc., Perkin Trans. 1*, 1985, 159 (*isol*)

Sotheeswaran, S. *et al, Phytochemistry*, 1987, **26**, 1505 (*struct*)

Vebraside   **V-60003**

$C_{16}H_{24}O_{10}$ M 376.360

Constit. of *Verbena brasiliensis*. Mp 131-133°. $[\alpha]_D^{23}$ +80.7° (c, 0.195 in CHCl$_3$).

Franke, A. *et al, Phytochemistry*, 1987, **26**, 3015.

Ventilone C   **V-60004**

1,3,4,9-Tetrahydro-8-hydroxy-5-methoxy-6,7-methylenedioxy-1-methylnaphtho[2,3-c]furan-4,9-dione

[96385-81-8]

$C_{15}H_{12}O_7$ M 304.256

Constit. of root bark of *Ventilago maderaspatana*. Red needles (C$_6$H$_6$/pet. ether). Mp 196°. $[\alpha]_D^{22}$ +144° (c, 0.16 in CHCl$_3$).

8-Me ether: [96385-83-0]. **Ventilone** E.

   $C_{16}H_{14}O_7$ M 318.282

   Constit. of *V. maderaspatana*. Yellow microcryst. (MeOH). Mp 180-181°. $[\alpha]_D^{22}$ +174° (c, 0.13 in CHCl$_3$).

8-Me ether, 5-O-De-Me: [96385-82-9]. **Ventilone** D.

   $C_{15}H_{12}O_7$ M 304.256

   Constit. of *V. maderaspatana*. Dark-red needles (C$_6$H$_6$/pet. ether). Mp 177-178°. $[\alpha]_D$ +240° (c, 0.10 in CHCl$_3$).

Hanumaiah, T. *et al, Tetrahedron*, 1985, **41**, 635 (*cryst struct*)

Verecynarmin A   **V-60005**

$C_{22}H_{28}O_4$ M 356.461

Constit. of *Armina maculata* and *Veretillum cynomorium*. Foam. Mp 127-135°. $[\alpha]_D^{20}$ −255.7° (c, 1.13 in EtOH).

Guerriero, A. *et al*, *Helv. Chim. Acta*, 1987, **70**, 984.

Verrucosidin   V-60006

Updated Entry replacing V-20010
[88389-71-3]

$C_{24}H_{32}O_6$ M 416.513

Tremorgen from *Penicillium verrucosum* var. *cyclopium*. Potent neurotoxin. Plates (Et$_2$O). Mp 90-91°. $[\alpha]_D^{26}$ +92.4° (c, 0.25 in MeOH).

Ganguli, M. *et al*, *J. Org. Chem.*, 1984, **49**, 3762 (*isol, struct*)
Nishiyama, S. *et al*, *Tetrahedron Lett.*, 1986, **27**, 723 (*synth, abs config*)
Cha, J.K. *et al*, *Tetrahedron Lett.*, 1987, **28**, 5473 (*synth, abs config*)

Vetidiol   V-60007

$C_{15}H_{24}O_2$ M 236.353

Constit. of vetiver oil. Cryst. Mp 170°. $[\alpha]_D^{30}$ −140°.

Kalsi, P.S. *et al*, *Tetrahedron*, 1987, **43**, 2985.

4-Vinyl-2-azetidinone, 8CI   V-60008

[7486-94-4]

C_5H_7NO M 97.116

(±)-*form*
Bp$_{0.3}$ 67-68°.

Moriconi, E.J. *et al*, *Tetrahedron Lett.*, 1968, 3823 (*synth*)
Durst, T. *et al*, *J. Org. Chem.*, 1970, **35**, 2043 (*synth*)

3-Vinylcyclohexene   V-60009

3-Ethenylcyclohexene, 9CI
[766-03-0]

(*R*)-*form*

C_8H_{12} M 108.183

(*R*)-*form* [95421-88-8]
   $[\alpha]_D^{25}$ −139° (c, 1 in toluene).
(*S*)-*form* [76152-63-1]
   $[\alpha]_D^{25}$ +250° (c, 1 in toluene).
(±)-*form*
   Bp 129°, Bp$_{17}$ 30°.

Rice, F.O. *et al*, *J. Am. Chem. Soc.*, 1944, **66**, 765.

Breil, H. *et al*, *Makromol. Chem.*, 1963, **69**, 18 (*synth*)
Cywinski, N.F., *J. Org. Chem.*, 1965, **30**, 361 (*synth, pmr*)
Paquette, L.A. *et al*, *J. Am. Chem. Soc.*, 1972, **94**, 7771 (*synth*)
Beger, J. *et al*, *J. Prakt. Chem.*, 1974, **316**, 449 (*synth*)
Peiffer, G. *et al*, *Bull. Soc. Chim. Fr.*, 1979, 415 (*synth*)
Buono, G. *et al*, *J. Org. Chem.*, 1985, **50**, 1781 (*synth*)
Goering, H.L. *et al*, *J. Org. Chem.*, 1986, **51**, 2884 (*synth*)

5-Vinyl-1-cyclopentenecarboxylic acid   V-60010

5-Ethenyl-1-cyclopentenecarboxylic acid

$C_8H_{10}O_2$ M 138.166

(±)-*form*

Me ester: [102979-48-6]. *2-Carbomethoxy-3-vinylcyclopentene.*
$C_9H_{12}O_2$ M 152.193
Characterised spectroscopically.

Nugent, W.A. *et al*, *J. Org. Chem.*, 1986, **51**, 3376 (*synth, pmr*)

2-Vinyl-1,1-cyclopropanedicarboxylic acid   V-60011

Updated Entry replacing D-10264
2-Ethenyl-1,1-cyclopropanedicarboxylic acid, 9CI
[7686-78-4]

(*R*)-*form*

$C_7H_8O_4$ M 156.138

(*R*)-*form*
Di-Me ester:
   $C_9H_{12}O_4$ M 184.191
   Chiral synthon for steroid total synth. $[\alpha]_D^{25}$ +54.6° (c, 0.984 in CCl$_4$).
(±)-*form*
Me ester: [63364-52-3]. Bp$_{0.05}$ 55°.

Braun, H. *et al*, *Tetrahedron Lett.*, 1976, 2121 (*synth, pmr*)
Cho, I. *et al*, *J. Polym. Sci., Polym. Chem. Ed.*, 1979, **17**, 3169 (*synth, pmr*)
Quinkert, G. *et al*, *Justus Liebigs Ann. Chem.*, 1981, 2335; 1982, 1999 (*synth, pmr, use*)

3-Vinylcyclopropene   V-60012

3-Ethenylcyclopropene, 9CI
[61082-23-3]

C_5H_6 M 66.102

Stable only <−70°, readily dimerises.

Billups, W.E. *et al*, *Tetrahedron*, 1986, **42**, 1575 (*synth, pmr*)

2-Vinyl-1*H*-pyrrole V-60013

*2-Ethenyl-1*H-*pyrrole, 9CI*

[2433-66-1]

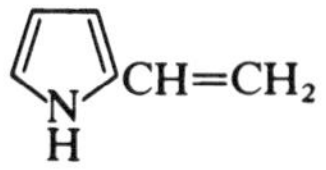

C_6H_7N M 93.128

Pale-yellow oil. Bp$_{18}$ 64-67°.

N-*Me:* [2540-06-9].

C_7H_9N M 107.155

Bp$_{16-18}$ 62-64°.

Herz, W. *et al, J. Am. Chem. Soc.,* 1954, **76**, 576 (*synth, ir*)
Jones, R.A. *et al, Aust. J. Chem.,* 1965, **18**, 875 (*synth, deriv, ir*)

4-Vinyl-1,2,3-thiadiazole V-60014

4-Ethenyl-1,2,3-thiadiazole

[101541-87-1]

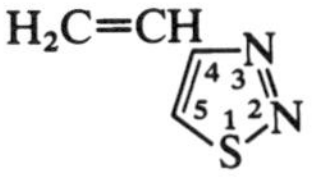

$C_4H_4N_2S$ M 112.149

Oil. Other 4-vinyl-1,2,3-thiadiazoles also prepd.

Hanold, N. *et al, Justus Liebigs Ann. Chem.,* 1986, 1344 (*synth, ir, ms, pmr, cmr*)

5-Vinyl-1,2,3-thiadiazole V-60015

5-Ethenyl-1,2,3-thiadiazole

[101055-88-3]

$C_4H_4N_2S$ M 112.149

Bp$_{0.5}$ 43-45°. Other 5-alkenyl-1,2,3-thiadiazoles also prepd.

Hanold, N. *et al, Justus Liebigs Ann. Chem.,* 1986, 1334 (*synth, ir, ms, pmr, cmr*)

Volkensiachromone V-60016

[110024-18-5]

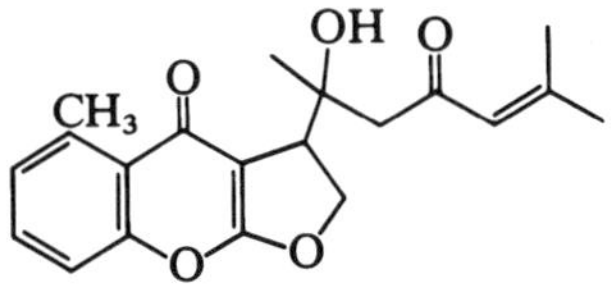

$C_{20}H_{22}O_5$ M 342.391

Constit. of *Bothriocline ripensis.* Oil.

Jakupovic, J. *et al, Phytochemistry,* 1987, **26**, 1069.

Vulgarin V-60017

Updated Entry replacing V-20038

Tauremisin A. *Barrelin. Judaicin*

[3162-56-9]

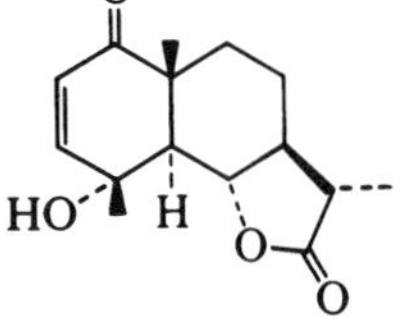

Absolute
configuration

$C_{15}H_{20}O_4$ M 264.321

Constit. of *Artemisia vulgaris.* Oral hypoglycemic agent. Cryst. (EtOH). Mp 174-175°. [α]$_{546}^{27}$+48.7° (c, 3.86 in CHCl$_3$).

▷LE3170000.

4-Epimer: [66289-87-0]. *4-Epivulgarin.*

$C_{15}H_{20}O_4$ M 264.321

From *A. judaica* and *A. canariensis.* Cryst. (C$_6$H$_6$/hexane)(also descr. as gum). Mp 192-194°. [α]$_D$ +77.5° (c, 5.7 in CHCl$_3$).

4-Deoxy,4α-hydroperoxy: [72505-77-2]. *4α-Hydroperoxydesoxyvulgarin.*

$C_{15}H_{20}O_5$ M 280.320

From *A. judaica.* Gum.

Geissman, T.A. *et al, J. Org. Chem.,* 1962, **27**, 1855 (*isol, struct*)
Ando, M. *et al, Bull. Chem. Soc. Jpn.,* 1978, **51**, 283 (*synth*)
González, G. *et al, J. Chem. Soc., Perkin Trans. 1,* 1978, 1243 (*synth*)
Ando, M. *et al, Bull. Chem. Soc. Jpn.,* 1979, **52**, 2737 (*synth*)
Gonzalez, A.G. *et al, Phytochemistry,* 1983, **22**, 1509 (*isol*)
Metwally, M.A. *et al, Phytochemistry,* 1985, **24**, 1103 (*derivs*)
Abagaz, B. *et al, Tetrahedron,* 1986, **42**, 6003 (*cryst struct*)
Arias, J.M. *et al, J. Chem. Soc., Perkin Trans. 1,* 1987, 471 (*synth*)

W

Warburganal W-60001

Updated Entry replacing W-10002

1,4,4a,5,6,7,8,8a-Octahydro-1-hydroxy-5,5,8a-tri-methyl-1,2-naphthalenedicarboxaldehyde, 9CI

[62994-47-2]

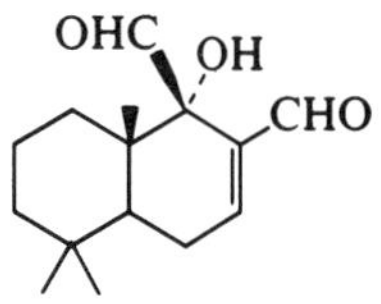

$C_{15}H_{22}O_3$ M 250.337

Isol. from *Warburgia* spp. Antifeeding compd. Cryst. (hexane). Mp 106-107°. $[\alpha]_D^{22}$ +260° (c, 0.45 in $CHCl_3$).

Kubo, I. *et al, J. Chem. Soc., Chem. Commun.*, 1976, 1013 (*isol, struct*)
Nakanishi, K. *et al, Isr. J. Chem.*, 1977, **16**, 28 (*isol, struct*)
Hollinshead, D.M. *et al, J. Chem. Soc., Perkin Trans. 1*, 1983, 1579 (*synth, bibl*)
Razmilic, I. *et al, Chem. Lett.*, 1985, 1113 (*synth*)
Manna, S. *et al, J. Chem. Soc., Chem. Commun.*, 1987, 1324 (*synth*)

Wasabidienone A W-60002

Updated Entry replacing W-30003

[90052-97-4]

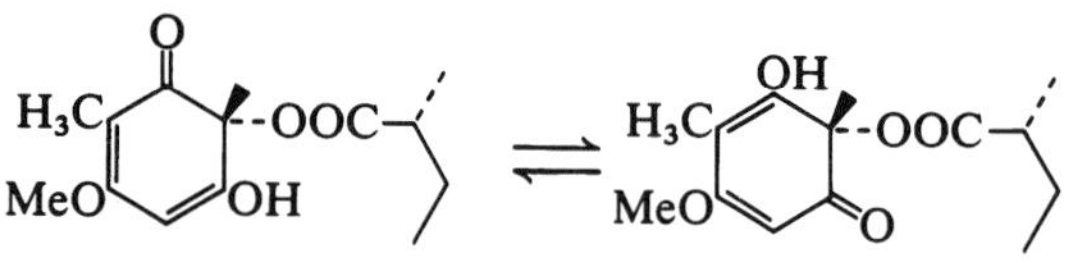

$C_{14}H_{20}O_5$ M 268.309

Tautomeric mixture. Pigment from potato culture solution of *Phoma wasabiae*. Oil.

p-*Nitrobenzoyl:* Pale-yellow prisms. Mp 114-114.5°. $[\alpha]_D$ +159° (c, 0.55 in $CHCl_3$).

Soga, O. *et al, Chem. Lett.*, 1984, 339; *Agric. Biol. Chem.*, 1987, **51**, 283 (*isol, struct*)

Wasabidienone E W-60003

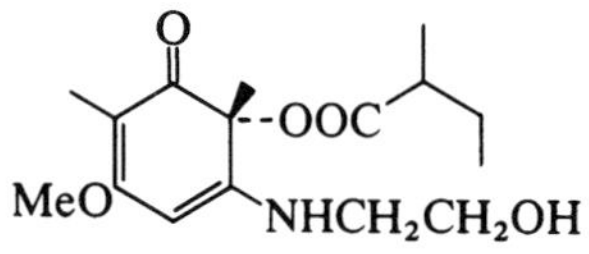

$C_{16}H_{25}NO_5$ M 311.377

Constit. of potato culture solution of *Phoma wasabiae*. Yellow-orange prisms (Et_2O/hexane). Mp 132-134°. $[\alpha]_D$ +50.5° (c, 1.1 in $CHCl_3$).

Soga, O. *et al, Chem. Lett.*, 1987, 815.

Wedeliasecokaurenolide W-60004

Updated Entry replacing W-00010

[85552-23-4]

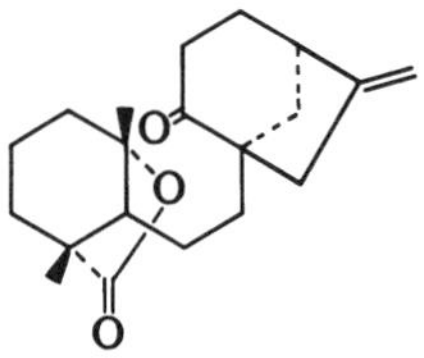

$C_{20}H_{28}O_3$ M 316.439

Constit. of *Wedelia trilobata*. Gum. $[\alpha]_D^{24}$ −44.9° (c, 3.2 in $CHCl_3$).

3β-Acetoxy: 3β-Acetoxywedeliasecokaurenolide.
$C_{22}H_{30}O_5$ M 374.476
Constit. of *Alepidea amatynsia*. Cryst. (pet. ether). Mp 120°. $[\alpha]_D^{24}$ −25° (c, 0.1 in $CHCl_3$).

Bohlmann, F. *et al, Phytochemistry*, 1981, **20**, 751 (*isol, struct*)
Rustaiyan, A. *et al, Phytochemistry*, 1987, **26**, 2106 (*deriv*)

Withametelin W-60005

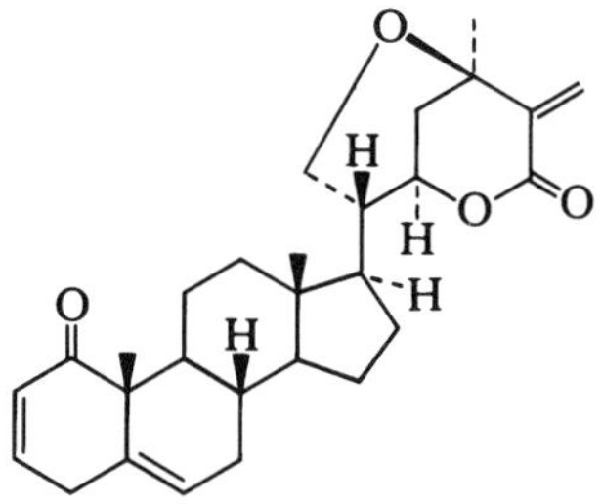

$C_{28}H_{36}O_4$ M 436.590

Constit. of *Datura metel*. Needles (EtOAc). Mp 210°. $[\alpha]_D$ −64.4° (c, 0.45 in $CHCl_3$).

Oshima, Y. *et al, Tetrahedron Lett.*, 1987, **28**, 2025.

Withaminimin W-60006

15α-Acetoxy-5α,6β,14α-trihydroxy-1-oxo-20S,22R-witha-2,16,24-trienolide

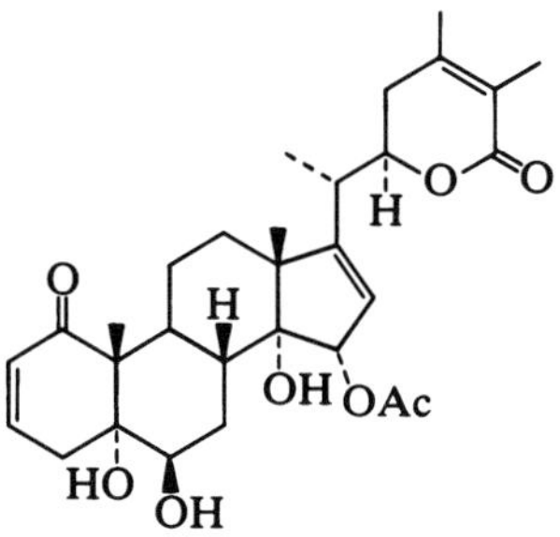

$C_{30}H_{40}O_8$ M 528.641

Constit. of *Physalis minima*. Amorph. powder. Mp 208°.

Gottlieb, H.E. *et al, Phytochemistry* 1987, **26**, 1801.

Withanolide *Y* W-60007

5α,6α-Epoxy-7α,17α,20R-trihydroxy-1-oxo-22R-6α,7α-witha-2,24-dienolide

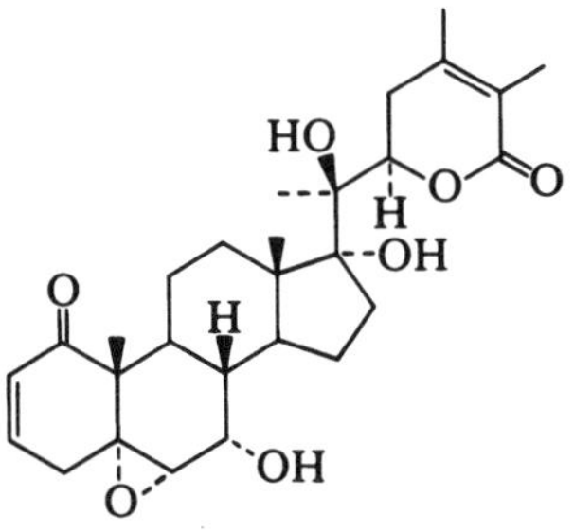

$C_{28}H_{38}O_7$ M 486.604

Constit. of *Withania somnifera*. Cryst. (EtOAc/CH$_2$Cl$_2$). Mp 270-273°.

Bessalle, R. *et al, Phytochemistry*, 1987, **26**, 1797.

Withaphysacarpin W-60008

5β,6β-Epoxy-4β,16β,20R-trihydroxy-1-oxo-22R,24R,25R-with-2-enolide

[41929-21-9]

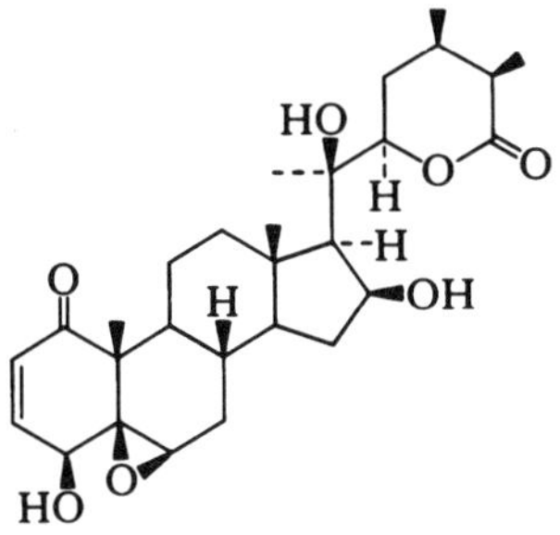

$C_{28}H_{40}O_7$ M 488.620

Constit. of *Physalis ixocarpa*. Cryst. Mp 275-278°. [α]$_D$ +20° (CHCl$_3$).

Subramanian, S.S. *et al, Indian J. Pharm.*, 1973, **35**, 36.

Withaphysalin *E* W-60009

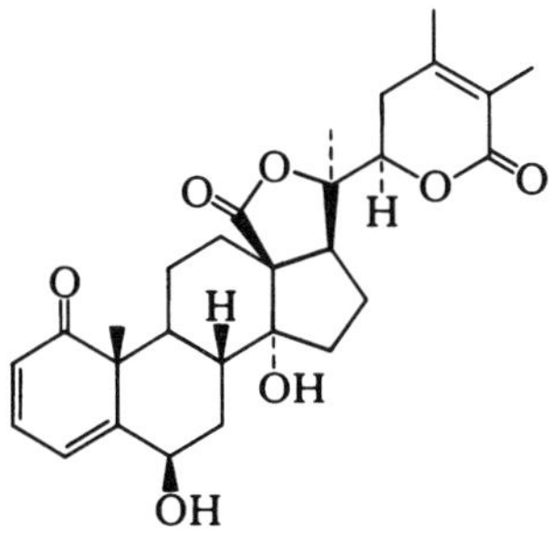

$C_{28}H_{34}O_7$ M 482.572

Constit. of *Physalis minima* var. *indica*. Amorph. powder. Mp 311-312°. [α]$_D$ +61.4° (c, 0.51 in Py).

Sinha, S.C. *et al, Phytochemistry*, 1987, **26**, 2115.

Withaphysanolide W-60010

4β,14α,17β,20R-Tetrahydroxy-1-oxo-20R-witha-2,5,24-trienolide

[74799-65-8]

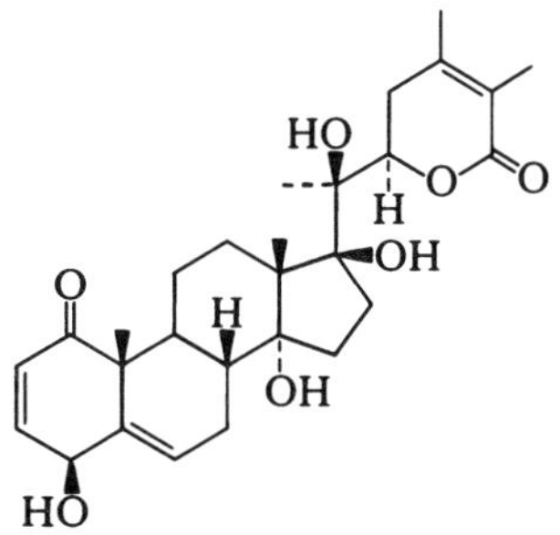

$C_{28}H_{38}O_7$ M 486.604

Constit. of *Physalis viscosa*. Cryst. (MeOH). Mp 215°. [α]$_D$ +95° (c, 2.63 in dioxane).

Maslennikova, V.A. *et al, Khim. Prir. Soedin.*, 1980, **16**, 167 (*isol*)

Abdullaev, N.D. *et al, Khim. Prir. Soedin.*, 1984, **20**, 182 (*struct*)

X

Xanthotoxin   X-60001

*9-Methoxy-7H-furo[3,2-g] [1]benzopyran-7-one, 9CI,
8CI. 9-Methoxyfuro[3,2-g]chromen-7-one. 8-Methoxy-
4′,5′:6,7-furocoumarin. 9-Methoxypsoralen. Methoxsa-
len, BAN. Zanthotoxin. Ammoidin. Meloxine. Methoxa-
Dome*

[298-81-7]

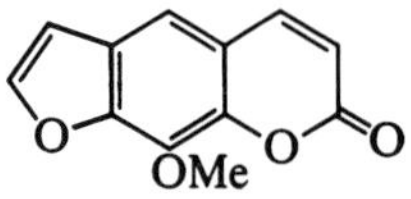

$C_{12}H_8O_4$ M 216.193

Isol. from *Cnidium dubium, Cryptodiscus didymus,
Evodia hupehensis, Selinum tenuifolium* and *Aegle
marmelos*. Aid to dermal pigmentation. Cryst. (EtOH
aq.). Mp 148°. Bitter tasting with tingling sensation.

▷LV1400000.

Späth, E. *et al, Ber.*, 1936, **69**, 769 (*isol*)
Schonberg, J.R. *et al, Nature* (*London*), 1947, **168**, 468 (*isol*)
Sommers, A.H. *et al, J. Am. Chem. Soc.*, 1957, **79**, 3491 (*synth*)
Scheel, W. *et al, Biochemistry*, 1963, **12**, 1127 (*isol*)
Chatterjee, A. *et al, Tetrahedron*, 1972, **28**, 5175 (*synth*)
Austin, D.J. *et al, Phytochemistry*, 1973, **12**, 1657 (*biosynth*)
Mikayado, M. *et al, Phytochemistry*, 1978, **17**, 143 (*synth*)
Lin, Y.-Y. *et al, J. Heterocycl. Chem.*, 1979, **16**, 799 (*synth*)
Loutfy, M.A. *et al, Anal. Profiles Drug. Subst.*, 1980, **9**, 427
(*rev*)
Nore, P. *et al, J. Heterocycl. Chem.*, 1980, **17**, 985 (*synth, bibl*)
Dall'Acqua, F. *et al, Cryst. Struct. Commun.*, 1981, **10**, 505
(*cryst struct*)

Xanthotoxol   X-60002

9-Hydroxy-7H-furo[3,2-g] [1]benzopyran-7-one, 9CI

[2009-24-7]

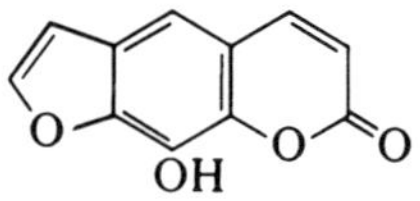

$C_{11}H_{16}O_4$ M 212.245

Isol. from *Cnidium dubium, Cryptodiscus didymus,
Evodia hupehensis, Selinum tenuifolium* and *Aegele
marmelos*. Cryst. (EtOH aq.). Mp 251-252°.

▷Photocarcinogen

O-(3-Methyl-2-butenyl): [55747-78-9]. **Ferulidene**.
$C_{16}H_{14}O_4$ M 270.284
Constit. of *Prangos ferulacea*. Cryst. Mp 119-120°.

O-(2-Hydroxy-3-methyl-3-butenyl): [53319-52-1]. **Iso-
gosferol**. *Isogospherol*.
$C_{16}H_{14}O_5$ M 286.284
Constit. of *P. lophoptera* and *Heracleum granatense*.
Cryst. (Et₂O). Mp 72-73.5°.

Me ether: see Xanthotoxin, X-60001

Kommissarenko, N.F. *et al, Khim. Prir. Soedin.*, 1966, **2**, 375
(*isol*)
Nikonov, G.K. *et al, Khim. Prir. Soedin.*, 1971, **7**, 115 (*isol*)
Perel'son, M.E. *et al, Khim. Prir. Soedin.*, 1971, **7**, 576 (*pmr*)
Gellert, M. *et al, Phytochemistry*, 1972, **11**, 2894 (*isol*)
Stemple, N.R. *et al, Acta Crystallogr., Sect. B*, 1972, **28**, 2485
(*cryst struct*)
Abyshev, A.Z., *Khim. Prir. Soedin.*, 1974, **10**, 83, 568
(*Ferulidene*)
González, A.G. *et al, An. Quim.*, 1974, **70**, 369 (*Isogosferol*)
Sharma, B.R. *et al, Indian J. Chem., Sect. B*, 1980, **19**, 162
(*isol*)
Ceska, O. *et al, Experientia*, 1986, **42**, 1302 (*occur, tox*)

Y

Yadanzioside *K* Y-60001

[101559-98-2]

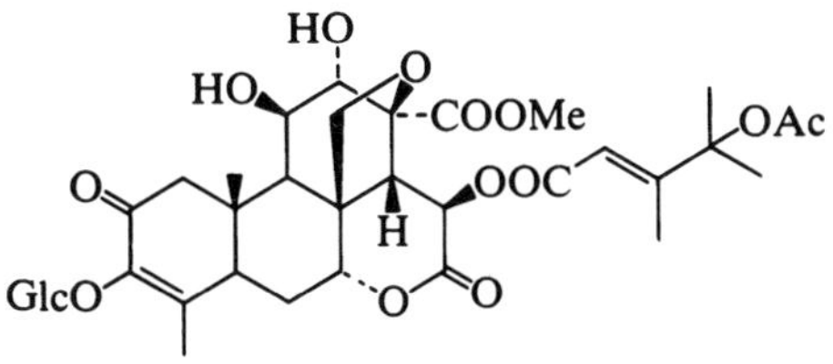

HO
HO
--COOMe
OAc
O
OOC
GlcO
H
O
O

$C_{36}H_{48}O_{18}$ M 768.764

Constit. of *Brucea javanica*. Cryst. (MeOH). Mp 214.5-
216.5°. $[\alpha]_D^{23}$ +15° (c, 1 in EtOH).

Sakaki, T. *et al*, *Bull. Chem. Soc. Jpn.*, 1986, **59**, 3541.

Yessotoxin Y-60002

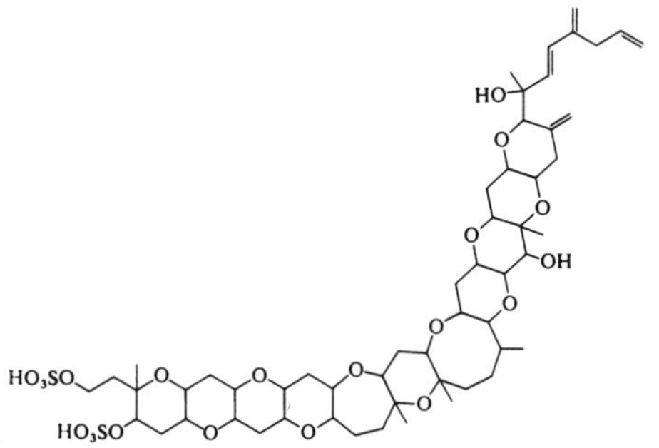

HO
O
O
OH
O
O
HO$_3$SO
O
O
O
O
O
HO$_3$SO

$C_{55}H_{82}O_{21}S_2$ M 1143.360

Toxic constit. of scallops *Patinopecten yessoensis*.
Amorph. solid (as Di-Na salt). $[\alpha]_D^{20}$ +3.01° (c, 0.45 in
MeOH).

Murata, M. *et al*, *Tetrahedron Lett.*, 1987, **28**, 5869.

Yezoquinolide Y-60003

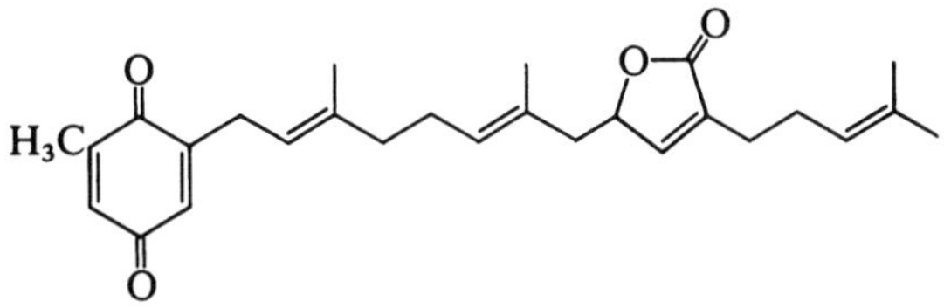

O
H$_3$C
O
O
O

$C_{27}H_{34}O_4$ M 422.563

Plastoquinone constit. of brown alga *Sargassum saga-
mianium* var. *yezoense*. Pale-yellow oil. $[\alpha]_D^{24}$ −23.5°
(c, 0.52 in CHCl$_3$).

Segawa, M. *et al*, *Chem. Lett.*, 1987, 1365.

Z

Zaluzanin C Z-60001

Updated Entry replacing Z-40001

[16838-87-2]

$C_{15}H_{18}O_3$ M 246.305

Constit. of *Zaluzania* spp. Cryst. (Me$_2$CO/isopropyl ether). Mp 103-104°. [α]$_D$ ±0° (CHCl$_3$).

Ac: [16838-85-0]. **Zaluzanin D**.
$C_{17}H_{20}O_4$ M 288.343
Constit. of *Z.* spp. Cryst. (EtOAc/isopropyl ether). Mp 103-104°. [α]$_D$ ±0° (CHCl$_3$).

(3-Methyl-2-butenoyl) ester: [57576-43-9]. **Vernoflexin**.
$C_{20}H_{24}O_4$ M 328.407
Constit. of the roots of *Veronia flexuosa*. Bitter taste. Cryst. (EtOH). Mp 73-75°.

β-D-Glucoside: [57576-33-7]. **Vernoflexuoside**. *Glucozaluzanin* C.
$C_{21}H_{28}O_8$ M 408.447
Constit. of the roots of *V. flexuosa*, *Ainsliaea acerifolia*, *Macroclinidium trilobum* and *Brachylaena* spp. Cryst. (MeOH aq. or H$_2$O) with bitter taste. Mp 104-105°. [α]$_D^{21}$ −18.4° (c, 0.93 in MeOH).

Angeloyl:
$C_{20}H_{24}O_4$ M 328.407
Constit. of *Zinnia multiflora*. Oil.

11β,13-Dihydro: **11β,13-Dihydrozaluzanin** C.
$C_{15}H_{20}O_3$ M 248.321
Constit. of *Brachylaena transvaalensis*. Gum.

11β,13-Dihydro, Ac: Gum. [α]$_D^{24}$ +11° (c, 0.23 in CHCl$_3$).

3-Epimer: [67667-64-5]. **3-Epizaluzanin** C. *Isozaluzanin* C.
$C_{15}H_{18}O_3$ M 246.305
Constit. of *V. anisochaetoides* and *Saussurea lappa*. Cryst. (also descr. as oil). Mp 143°. [α]$_D^{24}$ −55.6° (c, 0.24 in CHCl$_3$).

11α,13-Dihydro: **11α,13-Dihydrozaluzanin** C.
$C_{15}H_{20}O_3$ M 248.321
Constit. of *Ainsliaea fragrans*. Gum. [α]$_D^{24}$ −12° (c, 0.25 in CHCl$_3$).

8α-Hydroxy, 11α,13-dihydro: **8α-Hydroxy-11α,13-dihydrozaluzanin** C.
$C_{15}H_{20}O_4$ M 264.321
Constit. of *A. fragrans*. Gum. [α]$_D^{24}$ −24° (c, 0.25 in CHCl$_3$).

8β-Hydroxy, 3-ketone: **8β-Hydroxydehydrozaluzanin** C.
$C_{15}H_{16}O_4$ M 260.289
Constit. of *Andryala pinnatifida*. Gum. [α]$_D^{24}$ +10° (c, 0.1 in CHCl$_3$).

7α-Hydroxy,3-deoxy: **7α-Hydroxy-3-deoxyzaluzanin** C.
$C_{15}H_{18}O_3$ M 246.305
Constit. of *Podachaenium reminens*. Cryst. Mp 132-133°.

3-O-(2-O-Caffeoylglucoside): [93236-48-7]. **Ainsliaside A**.
$C_{30}H_{34}O_{11}$ M 570.592
Constit. of *A. acerifolia*. Amorph. powder. [α]$_D^{19}$ +59.7° (c, 0.50 in MeOH).

β-D-Glucosido(3→1)glucoside: [94474-62-1]. **Macrocliniside B**.
$C_{27}H_{38}O_{13}$ M 570.589
Constit. of *Macroclinidium trilobum*. Amorph. powder. [α]$_D^{25}$ −18.3° (c, 1.15 in H$_2$O).

3-O-(β-D-Glucopyranosyl-(1→4)-β-D-glycopyranosyl-(1→3)-β-D-glucopyranoside: [100187-63-1]. **Macrocliniside I**.
$C_{33}H_{48}O_{18}$ M 732.731
Constit. of *M. trilobum*. Amorph. powder. [α]$_D^{25}$ −13.0° (c, 1.35 in H$_2$O).

3-O-β-D-(6-O-p-Hydroxyphenylacetyl)-glucopyranoside: [100202-28-6]. **Crepiside A**.
$C_{29}H_{34}O_{10}$ M 542.582
Constit. of *Crepis japonica*. Amorph. powder. [α]$_D^{20}$ +12.3° (c, 0.65 in Py).

3-O-β-D-(4-O-p-hydroxyphenylacetylglucopyranoside): [100202-26-4]. **Crepiside B**.
$C_{29}H_{34}O_{10}$ M 542.582
Constit. of *C. japonica*. Amorph. powder. [α]$_D^{18}$ −6.7° (c, 0.52 in MeOH).

9α-Hydroxy: **9α-Hydroxyzaluzanin B**. Constit. of *Lactuca laciniata*. Cryst. (MeOH aq.). Mp 76-78°. [α]$_D^{25}$ +13.0° (c, 0.58 in MeOH).

9α-Hydroxy, 11β,13-Dihydro: **11β,13-Dihydro-9α-hydroxyzaluzanin** C. *9α-Hydroxy-11β,13-dihydrozaluzanin* C.
$C_{15}H_{20}O_4$ M 264.321
From *L. laciniata*. Amorph. [α]$_D^{25}$ +50.7° (c, 0.75 in MeOH).

11β,13-Dihydro, β-D-Glucoside:
$C_{21}H_{30}O_8$ M 410.463
Constit. of *B. nereifolia*. Cryst. Mp 96°.

8α-Hydroxy, 11β,13-dihydro, 3-β-D-glucoside: **11β,13-Dihydro-8α-hydroxyglucozaluzanin** C.
$C_{21}H_{30}O_9$ M 426.463
Constit. of *B. nereifolia*.

8α-Acetoxy, 11β,13-dihydro, 3-β-D-glucoside, tetra-Ac: Cryst. Mp 130°.

Romo de Vivar, A. *et al*, *Tetrahedron*, 1967, **23**, 3903 (isol, struct)
Bohlmann, F. *et al*, *Phytochemistry*, 1978, **17**, 475; 1979, **18**, 1343 (isol)
Ando, M. *et al*, *Chem. Lett.*, 1982, 501 (synth)
Bohlmann, F. *et al*, *Phytochemistry*, 1982, **21**, 647, 1799, 2120 (isol, derivs)
Kalsi, P.S. *et al*, *Phytochemistry*, 1983, **22**, 1993 (isol)
Fronczek, F.R. *et al*, *J. Nat. Prod.*, 1984, **47**, 1036 (deriv)
Miyase, T. *et al*, *Chem. Pharm. Bull.*, 1984, **32**, 3043, 3912 (derivs)
Miyase, T. *et al*, *Chem. Pharm. Bull.*, 1985, **33**, 4445, 4451 (isol)
Nishimura, K. *et al*, *Phytochemistry*, 1986, **25**, 2375 (derivs)
Zdero, C. *et al*, *Phytochemistry*, 1987, **26**, 2597 (deriv)

Zearalenone   Z-60002

Updated Entry replacing Z-00007
*3,4,5,6,9,10-Hexahydro-14,16-dihydroxy-3-methyl-1H-
2-benzoxacyclotetradecin-1,7(8H)-dione, 9CI. Mycotoxin F2*

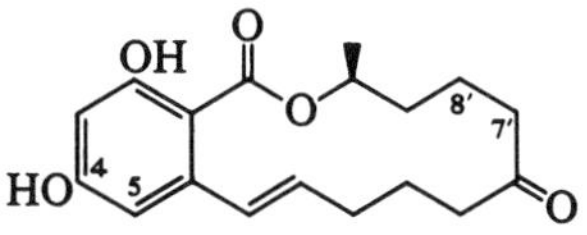

$C_{18}H_{22}O_5$ M 318.369

▷DM2550000.

(S)-form [17924-92-4]

Fungal metab. of *Fusarium* spp. Has weak oestrogenic
activity and causes physiol. changes when ingested by
animals as foodstuffs contaminant. Mp 164-165° (157-
159°). $[\alpha]_D$ −190° (CHCl₃).

▷DM2550000.

4-O-Me: Mp 120-122°. $[\alpha]_D$ −177° (CHCl₃).
Di-Ac: Mp 123-125°.
7′,8′-Didehydro (E-): **7′-Dehydrozearalenone.** Minor
metab. of *G. zeae.* Needles (Me₂CO/Et₂O). Mp 197-
200°. $[\alpha]_D^{24}$ −133.8° (CHCl₃). 7′,8′-bond has (*E*)-
config.
8′-Hydroxy: [40785-64-6]. **8′-Hydroxyzearalenone.** *F-5-
3.*
   $C_{18}H_{22}O_6$ M 334.368
   Cryst. (MeOH/CHCl₃). Mp 210-212° (198-199°).
   $[\alpha]_D^{24}$ −149.1° (c, 1 in Me₂CO). 8′-Config. unknown.
Epi-8′-hydroxy: [40785-65-7]. **8′-epi-Hydroxyzearalen-
one.** *F-5-4.* Cryst. (MeOH/CHCl₃). Mp 172-174°
   (168-169°). $[\alpha]_D^{24}$ −53.1° (c, 1 in Me₂CO).
5-Formyl: **5-Formylzearalenone.**
   $C_{19}H_{22}O_6$ M 346.379
   Metab. of *G. zeae.* Platelets. Mp 188-190°. $[\alpha]_D^{24}$
   −42.0° (CHCl₃).

(±)-form

Prismatic needles (MeNO₂). Mp 188-190°.

4-O-Me: Mp 108-111°.

Stob, M. *et al, Nature* (*London*), 1962, **196**, 1318 (*isol*)
Urry, W.H. *et al, Tetrahedron Lett.*, 1966, 3109 (*uv, pmr, ms,
   struct*)
Taub, D. *et al, Tetrahedron,* 1968, **24**, 2443 (*synth, resoln, abs
   config*)
Girotra, N.N. *et al, J. Org. Chem.,* 1969, **34**, 3192 (*synth*)
Bolliger, G. *et al, Helv. Chim. Acta,* 1972, **55**, 3030 (*deriv*)
Jackson, R.A. *et al, J. Agric. Food Chem.,* 1974, **22**, 1015 (*ms*)
Ciegler, A., *Lloydia,* 1975, **38**, 21 (*rev*)
Shipchandler, M.T., *Heterocycles,* 1975, **3**, 471 (*rev, bibl*)
Pohland, A.E. *et al, Pure Appl. Chem.,* 1982, **54**, 2219 (*rev, uv,
   ir, pmr, ms*)
Rao, A.V.R. *et al, Tetrahedron,* 1987, **43**, 779 (*synth*)
Cole, R.J. *et al, Handbook of Toxic Fungal Metabolites,*
   Academic Press, N.Y. 1981, 902.

Zuonin A   Z-60003

[79120-58-4]

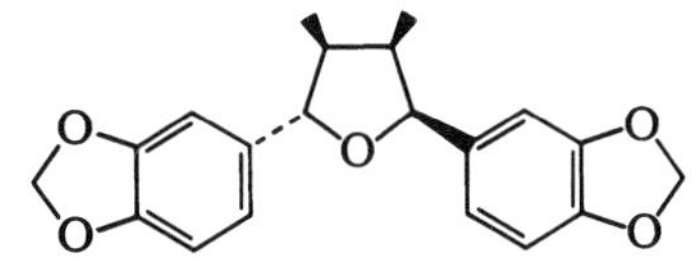

$C_{20}H_{20}O_5$ M 340.375

Constit. of *Aristolochia chilensis.* Cryst. (EtOH). Mp
   119-121°. $[\alpha]_D$ −139.4° (c, 1 in CHCl₃).

Urzúa, A. *et al, Phytochemistry,* 1987, **26**, 1509.

Name Index

This index becomes invalid after publication of the Seventh Supplement.

The Name Index lists in alphabetical order all Names and Synonyms contained in this Supplement. Names contained in Supplements 1 to 5 are listed in the cumulative Index Volume published with the Fifth Supplement.

Each index term refers the user to a DOC Number consisting of a single letter of the alphabet followed by five digits. The letter is the first letter of the relevant DOC Name.

The first digit of the DOC Number (printed in bold type) indicates the number of the Supplement in which the entry is printed. In this Sixth Supplement the first digit is invariably 6.

A DOC Number which follows immediately upon an index term means that the term is itself used as the Entry Name.

A DOC Number which is preceded by the word '*see*' means that the term is a synonym to an Entry Name.

A DOC Number which is preceded by the word '*in*' means that the term is embedded within an Entry, usually as a synonym to a particular stereoisomeric form or to a derivative.

The symbol ▷ preceding an index term indicates that the DOC Entry contains information on toxic or hazardous properties of the compound.

Name Index

AAP, *see* A-60283

Abbreviatin PB, A-60001

(5*S*)-9(10→20)-Abeo-1(10),8,11,13-abietatetraene-11,12-diol, *see* B-60007

Abeoanticopalic acid, A-60002

19(4→3)-Abeo-*O*-demethylcryptojaponol, *see* A-60003

19(4→3)-Abeo-11,12-dihydroxy-4(18),8,11,13-abietatetraen-7-one, A-60003

19(4β→3β)-Abeo-6,11-epoxy-6,12-dihydroxy-6,7-seco-4(18),-8,11,13-abietatetraen-7-al, A-60004

Abhexone, *see* E-60051

Abiesonic acid, A-60005

6,8,11,13-Abietatetraene-11,12,14-triol, A-60006

8,11,13-Abietatrien-19-ol, A-60007

8,11,13-Abietatrien-12,16-oxide, A-60008

8(14)-Abieten-18-oic acid 9,13-endoperoxide, A-60009

Absinthifolide, *in* H-60138

Acalycixeniolide *A*, A-60010

Acalycixeniolide *B*, A-60011

Acanthoglabrolide, *see* A-60012

Acanthospermal *A*, A-60012

3,4-Acenaphthenedicarboxylic acid, A-60013

4-Acenaphthenol, *see* H-60099

Acenaphtho[5,4-*b*]furan, A-60014

Acenaphtho[5,4-*b*]thiophene, A-60015

5,6-Acepleiadylenedione, *see* C-60194

5,8-Acepleiadylenedione, *see* C-60195

2-Acetamido-3*H*-phenoxazin-3-one, *in* A-60237

▷Acetamidoxime, A-60016

Acetoguanamine, *see* D-60040

Acetonediacetic acid, *see* O-60071

Acetonedicarboxylic acid, *see* O-60084

3-Acetonyl-5,8-dihydroxy-2-(hydroxymethyl)-6-methoxy-1,4-naphthoquinone, *see* F-60108

Acetonylmercaptan, *see* M-60028

2-Acetonylpyridine, *see* P-60229

3-Acetonylpyridine, *see* P-60230

4-Acetonylpyridine, *see* P-60231

▷Acetophenone, A-60017

▷5-Acetopyrogallol, *see* T-60315

Acetosyringone, *in* T-60315

4-Acetoxy-2-amino-3,5,14-trihydroxy-6-eicosenoic acid, *see* F-60082

2-Acetoxy-1,3-benzenedicarboxylic acid, *in* H-60101

ent-18-Acetoxy-15-beyeren-19-oic acid, *in* H-60106

4-Acetoxy-2-bromo-5,6-epoxy-2-cyclohexen-1-one, *in* B-60239

17α-Acetoxy-6α-butanoyloxy-15,17-oxido-16-spongianone, *in* D-60365

3-Acetoxy-3-(2-chloro-2-isocyanovinyl)-2-oxoindole, *see* I-60028

9β-Acetoxy-4,5-dehydro-4(15)-dihydrocostic acid, *in* H-60135

4-Acetoxy-2,6-dibromo-1,5-dihydroxy-2-cyclohexen-1-one, *in* D-60097

4-Acetoxy-2,6-dibromo-5-hydroxy-2-cyclohexen-1-one, *in* D-60097

21-Acetoxy-21,23:24,25-diepoxyapotirucall-14-ene-3,7-diol, *in* T-60223

2α-Acetoxy-11α,13-dihydroconfertin, *in* D-60211

7α-Acetoxy-12,20-dihydroxy-8,12-abietadiene-11,14-dione, *in* R-60015

2′-Acetoxy-3,5-dihydroxy-7,8-dimethoxyflavone, *in* P-60045

12-Acetoxy-5,8-dihydroxyfarnesol, *in* T-60363

3β-Acetoxydrimenin, *in* D-60519

6β-Acetoxy-1α,4α-epoxyeudesmane, *in* E-60019

19-Acetoxy-15,16-epoxy-13,17-spatadien-5α-ol, *in* E-60029

3β-Acetoxy-21,23-epoxy-7,24-tirucalladiene, *in* E-60032

9β-Acetoxy-4,11(13)-eudesmadien-12-oic acid, *in* H-60135

3β-Acetoxy-4,11(13)-eudesmadien-12,8β-olide, *in* H-60136

1β-Acetoxy-3,7(11),8-eudesmatrien-12,8-olide, *in* H-60140

1β-Acetoxy-4(15),7(11),8-eudesmatrien-12,8-olide, *in* H-60141

6β-Acetoxy-4(15)-eudesmene-1β,5α-diol, *in* E-60068

3β-Acetoxyhaageanolide acetate, *in* H-60001

6β-Acetoxyhebemacrophyllide, *in* H-60010

1-Acetoxy-7-hydroperoxy-3,7-dimethyl-2*E*,5*E*-octadien-4-one, *in* D-60318

4-Acetoxy-6-(4-hydroxy-2-cyclohexenyl)-2-(4-methyl-3-pentenyl)-2-heptenoic acid, A-60018

1-Acetoxy-7-hydroxy-3,7-dimethyl-2*E*,5*E*-octadien-4-one, *in* D-60318

1β-Acetoxy-8β-hydroxy-3,7(11)-eudesmadien-12,8-olide, *in* D-60322

3β-Acetoxy-2α-hydroxy-1(10),4,11(13)-germacratrien-12,6α-olide, *in* H-60006

3α-Acetoxy-15α-hydroxy-7,9(11),24*E*-lanostatrien-26-oic acid, *in* D-60344

3α-Acetoxy-15α-hydroxy-22-methoxy-8,24*E*-lanostadien-26-oic acid, *in* T-60327

22-Acetoxy-3α-hydroxy-7α-methoxy-8,24*E*-lanostadien-26-oic acid, *in* T-60328

14-Acetoxy-9β-hydroxy-8β-(2-methylpropanoyloxy)-1(10),-4,11(13)-germacratrien-12,6α-olide, *in* T-60323

6α-Acetoxy-17β-hydroxy-15,17-oxido-16-spongianone, *in* D-60365

3α-Acetoxy-15β-hydroxy-7,16-seco-7,11-trinervitadiene, *see* S-60019

2-Acetoxyisopropenyl-6-acetyl-5-hydroxybenzofuran, *see* A-60065

ent-3β-Acetoxy-16β,17-kauranediol, *in* K-60005

ent-3β-Acetoxy-15-kauren-17-oic acid, *in* H-60167

ent-3β-Acetoxy-15-kauren-17-ol, *in* K-60007

ent-6β-Acetoxy-7,12*E*,14-labdatrien-17-oic acid, *in* L-60004

ent-6β-Acetoxy-7,12*E*,14-labdatrien-17-ol, *in* L-60004

15-Acetoxy-7-labden-17-oic acid, *in* H-60168

3β-Acetoxy-14(26),17*E*,21-malabaricatriene, *in* M-60008

1-Acetoxy-6-(2-methoxy-4,5-methylenedioxyphenyl)-2-(3,4-methylenedioxyphenoxy)-3,7-dioxabicyclo[3.3.0]octane, *see* P-60142

3β-Acetoxy-12-oleanene-2α,11α-diol, *in* O-60037

3β-Acetoxy-12-oleanene-1β,2α,11α-triol, *in* O-60034

9β-Acetoxy-3-oxo-1,4(15),11(13)-eudesmatrien-12,6-olide, *in* H-60202

1-Acetoxypinoresinol, *in* H-60221

1-Acetoxypinoresinol 4′-*O*-β-D-glucopyranoside, *in* H-60221

▷3-Acetoxy-1,2-propanediol, *see* G-60029

4-Acetoxy-10-puteninone, *see* L-60009

7α-Acetoxyroyleanone, *in* D-60304

7β-Acetoxyroyleanone, *in* D-60304

14-Acetoxytetraneurin *D*, *in* T-60158

3-Acetoxythiophene, *in* T-60212

17-Acetoxythymifodioic acid, *in* T-60218

9β-Acetoxytournefortiolide, *in* H-60137

15α-Acetoxy-5α,6β,14α-trihydroxy-1-oxo-20*S*,22*R*-witha-2,16,24-trienolide, *see* W-60006

16α-Acetoxy-2α,3β,12β-trihydroxy-4,7-pregnadien-20-one, *in* T-60124

3β-Acetoxy-28,20β-ursanolide, *in* H-60233

3β-Acetoxy-12-ursene-2α,11α-diol, *in* U-60008

3β-Acetoxy-12-ursene-1β,2α,11α,20β-tetrol, *in* U-60005

3β-Acetoxy-12-ursene-1β,2α,11α-triol, *in* U-60006

3β-Acetoxy-12-ursene-2α,11α,20β-triol, *in* U-60007

3β-Acetoxywedeliasecokaurenolide, *in* W-60004

▷Acetylacetone, *see* P-60059

501

Name Index

Tetrahydro-2*H*-1,3-thiazine — 2,4,6,8-Tetramethylundecano . . .

Molecular Formula Index

This index becomes invalid after publication of the Seventh Supplement.

The Molecular Formula Index lists the molecular formulae of compounds in this Supplement which occur as Entry Names or as important derivatives. Molecular formulae of compounds contained in Supplements 1 to 5 are listed in the cumulative Index Volume published with the Fifth Supplement.

The first digit of the DOC Number (printed in bold type) refers to the number of the Supplement in which the Entry appears. In this Sixth Supplement the first digit is invariably 6.

Where a molecular formula applies to a derivative the DOC Number is prefixed by the word '*in*'.

The symbol ▷ preceding an Index Entry indicates that the DOC Entry contains information on toxic or hazardous properties of the compound.

Molecular Formula Index

CBrF₃S
Trifluoromethanesulfenyl bromide, T-60280

CClNO₃S
▷Sulfuryl chloride isocyanate, S-60062

CF₃IS
Trifluoromethanesulfenyl iodide, T-60282

CF₄S
▷Trifluoromethanesulfenyl fluoride, T-60281

CF₈S
Pentafluoro(trifluoromethyl)sulfur, P-60041

CHClOS
Carbonochloridothioic acid, C-60013

CHF₃OS
Trifluoromethanesulfenic acid, T-60279

CH₂BrClO₂S
Bromomethanesulfonic acid; Chloride, *in* B-60291

CH₂BrIO₂S
Iodomethanesulfonic acid; Bromide, *in* I-60048

CH₂Br₂O₂S
Bromomethanesulfonic acid; Bromide, *in* B-60291

CH₂N₄O
2-Tetrazolin-5-one, T-60175

CH₂N₄S
Tetrazole-5-thione, T-60174

CH₃BrO₃S
Bromomethanesulfonic acid, B-60291

CH₃IO₃S
Iodomethanesulfonic acid, I-60048

CH₃N₃O₂S
▷Methanesulfonyl azide, M-60033

CH₄N₂O₃
O,O'-Carbonylbis(hydroxylamine), C-60015

CH₄N₂O₃S
Aminoiminomethanesulfonic acid, A-60210

CH₄N₂S₂
Dithiocarbazic acid, D-60509

CH₅NO₃S
Aminomethanesulfonic acid, A-60215

C₂ClF₃O
Chlorotrifluorooxirane, C-60140

C₂Cl₂F₂O
2,2-Dichloro-3,3-difluorooxirane, D-60128

C₂F₂I₂
1,1-Difluoro-2,2-diiodoethylene, D-60183

C₂F₃N
Trifluoroisocyanomethane, T-60278

C₂F₃NO₃
▷Trifluoroacetyl nitrite, T-60274

C₂F₄O
▷Tetrafluorooxirane, T-60049

C₂F₄OS₂
2,2,4,4-Tetrafluoro-1,3-dithietane; 1-Oxide, *in* T-60048

C₂F₄O₂S
Trifluoroethenesulfonic acid; Fluoride, *in* T-60277

C₂F₄O₂S₂
2,2,4,4-Tetrafluoro-1,3-dithietane; 1,1-Dioxide, *in* T-60048

C₂F₄O₃S₂
2,2,4,4-Tetrafluoro-1,3-dithietane; 1,1,3-Trioxide, *in* T-60048

C₂F₄O₄S₂
2,2,4,4-Tetrafluoro-1,3-dithietane; 1,1,3,3-Tetroxide, *in* T-60048

C₂F₄S₂
2,2,4,4-Tetrafluoro-1,3-dithietane, T-60048

C₂F₅N
(Trifluoromethyl)carbonimidic difluoride, T-60288

C₂F₆N₂O₂
1,1,1-Trifluoro-*N*-(nitrosooxy)-*N*-(trifluoromethyl)-
methanamine, *in* B-60180

C₂F₈S₂
Difluoro(trifluoromethanethiolato)(trifluoromethyl)sulfur, *in*
T-60281

C₂HF₃O
Trifluorooxirane, T-60305

C₂HF₃O₃S
Trifluoroethenesulfonic acid, T-60277

C₂HF₆NO
N,N-Bis(trifluoromethyl)hydroxylamine, B-60180

C₂HN₅S
3-Azido-1,2,4-thiadiazole, A-60351

C₂H₂F₂O
2,3-Difluorooxirane, D-60187

C₂H₂F₂O₃S
4,4-Difluoro-1,2-oxathietane 2,2-dioxide, D-60186

C₂H₂N₂O₂
1,3-Diazetidine-2,4-dione, D-60057

C₂H₂N₂O₄
2*H*-1,5,2,4-Dioxadiazine-3,6(4*H*)dione, D-60466

C₂H₂N₄
1,2,4,5-Tetrazine, T-60170

C₂H₂S₄
Ethanebis(dithioic)acid, E-60040

C₂H₂Se₂
1,2-Diselenete, D-60498

C₂H₃ClOS
Carbonochloridothioic acid; *O*-Me, *in* C-60013

C₂H₃FN₂O
▷3-Fluoro-3-methoxy-3*H*-diazirine, F-60050

C₂H₃F₃OS
Methyl trifluoromethanesulfenate, *in* T-60279

C₂H₃N
Ethynamine, E-60056

C₂H₃NS₂
Methanesulfenyl thiocyanate, M-60032

C₂H₃NSe
Methyl selenocyanate, M-60127

C₂H₃N₃S
▷1,2-Dihydro-3*H*-1,2,4-triazole-3-thione, D-60298

C₂H₄BrNO₂
1-Bromo-2-nitroethane, B-60303

C₂H₄ClI
1-Chloro-1-iodoethane, C-60106

C₂H₄IN₃
1-Azido-2-iodoethane, A-60346

C₂H₄N₂O₄
▷Oxalohydroxamic acid, O-60048

C₂H₄N₂S₂
▷Ethanedithioamide, E-60041

529

C₂H₄N₄S
2,5-Diamino-1,3,4-thiadiazole, D-60049
Tetrazole-5-thione; 1-Me, *in* T-60174
Tetrazole-5-thione; *S*-Me, *in* T-60174

C₂H₄O₂
▷1,2-Dioxetane, D-60469

C₂H₅N₃O₂
Biuret, B-60190
▷Semioxamazide, S-60020

C₂H₆N₂O
▷Acetamidoxime, A-60016

C₂H₆N₂O₃S
Aminoiminomethanesulfonic acid; *N*-Me, *in* A-60210

C₂H₆N₂S₂
Dithiocarbazic acid; Me ester, *in* D-60509

C₂H₆O₂S₂
2-Mercaptoethanesulfinic acid, M-60022

C₂H₇NO₂
O-(2-Hydroxyethyl)hydroxylamine, H-60134

C₂H₇NO₃S
(Methylamino)methanesulfonic acid, *in* A-60215

C₂H₁₀N₄
1,2-Dihydrazinoethane, D-60192

C₂N₂Se₃
Dicyanotriselenide, D-60165

C₃Br₃N₃O₂
3,4,5-Tribromo-1*H*-pyrazole; 1-Nitro, *in* T-60246

C₃ClF₂N₃
2-Chloro-4,6-difluoro-1,3,5-triazine, C-60052
6-Chloro-3,5-difluoro-1,2,4-triazine, C-60053

C₃F₆N₂
3,3-Bis(trifluoromethyl)-3*H*-diazirine, B-60178

C₃F₇NO
3,3,4,4-Tetrafluoro-2-(trifluoromethyl)-1,2-oxazetidine, T-60051

C₃F₉N
Tris(trifluoromethyl)amine, T-60406

C₃HBrINS
2-Bromo-4-iodothiazole, B-60287
2-Bromo-5-iodothiazole, B-60288
4-Bromo-2-iodothiazole, B-60289
5-Bromo-2-iodothiazole, B-60290

C₃HBr₂NS
2,4-Dibromothiazole, D-60109
2,5-Dibromothiazole, D-60110
4,5-Dibromothiazole, D-60111

C₃HBr₃N₂
▷2,4,5-Tribromo-1*H*-imidazole, T-60245
3,4,5-Tribromo-1*H*-pyrazole, T-60246

C₃HCl₂NS
2,4-Dichlorothiazole, D-60154
2,5-Dichlorothiazole, D-60155
4,5-Dichlorothiazole, D-60156

C₃HI₂NS
2,4-Diiodothiazole, D-60388
2,5-Diiodothiazole, D-60389

C₃H₂BrN₃O₂
4-Bromo-3(5)-nitro-1*H*-pyrazole, B-60304
3(5)-Bromo-5(3)-nitro-1*H*-pyrazole, B-60305

C₃H₂Br₂
1,2-Dibromocyclopropene, D-60094

C₃H₂Br₂N₂
4,5-Dibromo-1*H*-imidazole, D-60100

C₃H₂ClF₃
2-Chloro-3,3,3-trifluoro-1-propene, C-60141

C₃H₂Cl₄
▷1,1,2,3-Tetrachloro-1-propene, T-60029

C₃H₂INS
2-Iodothiazole, I-60064
4-Iodothiazole, I-60065
5-Iodothiazole, I-60066

C₃H₂N₄
4-Cyano-1,2,3-triazole, *in* T-60239
4-Diazo-4*H*-imidazole, D-60063

C₃H₃ClN₂O
4-Chloro-5-methylfurazan, C-60114

C₃H₃ClN₂O₂
2-Chloro-2,3-dihydro-1*H*-imidazole-4,5-dione, C-60054
4-Chloro-3-methylfuroxan, *in* C-60114

C₃H₃Cl₂F₃O
2,2-Dichloro-3,3,3-trifluoro-1-propanol, D-60161

C₃H₃Cl₂N₂O₂
3-Chloro-4-methylfuroxan, *in* C-60114

C₃H₃Cl₃
1,1,2-Trichloro-1-propene, T-60254

C₃H₃F₃O
(Trifluoromethyl)oxirane, T-60294

C₃H₃F₆NO
1,1,1,1′,1′,1′-Hexafluoro-*N*-methoxydimethylamine, *in* B-60180

C₃H₃NO₃S₂
4-Thiazolesulfonic acid, T-60198
5-Thiazolesulfonic acid, T-60199

C₃H₃NS₂
2(3*H*)-Thiazolethione, T-60200

C₃H₃N₃O₂
1*H*-1,2,3-Triazole-4-carboxylic acid, T-60239

C₃H₃N₃O₃
▷Cyanuric acid, C-60179

C₃H₄I₂N₆
1,3-Diiodo-2,2-diazidopropane, D-60387

C₃H₄N₂
Diazocyclopropane, D-60060

C₃H₄N₂O₂
2-Propyn-1-amine; *N*-Nitro, *in* P-60184

C₃H₄N₂O₃
2,3-Dihydro-2-hydroxy-1*H*-imidazole-4,5-dione, D-60243
5-Hydroxy-2,4-imidazolidinedione, H-60156

C₃H₄N₄O
1*H*-1,2,3-Triazole-4-carboxylic acid; Amide, *in* T-60239

C₃H₄O₂
1,3-Dioxole, D-60473

C₃H₄O₃
3-Oxopropanoic acid, O-60090

C₃H₅ClOS
▷Carbonochloridothioic acid; *O*-Et, *in* C-60013
Carbonochloridothioic acid; *S*-Et, *in* C-60013

C₃H₅F₂NO₂
2-Amino-3,3-difluoropropanoic acid, A-60135

C₃H₅F₃OS
Ethyl trifluoromethanesulfenate, *in* T-60279

C₃H₅N
▷2-Propyn-1-amine, P-60184

C₃H₅NO₃
Isonitrosopropanoic acid, *in* O-60090
▷3-Nitropropanal, N-60037

C₃H₅NS₂
2-Thiazolidinethione, T-60197

C₃H₅N₃O₂
Dihydro-1,3,5-triazine-2,4(1*H*,3*H*)-dione, D-60297

Tetrahydro-1,2,4-triazine-3,6-dione, T-60099

$C_3H_6Br_2$
▷1,2-Dibromopropane, D-60105

$C_3H_6ClNO_2$
▷2-Chloro-2-nitropropane, C-60120

$C_3H_6FNO_2$
3-Amino-2-fluoropropanoic acid, A-60162

$C_3H_6IN_3$
1-Azido-3-iodopropane, A-60347

$C_3H_6N_2$
4,5-Dihydro-1*H*-imidazole, D-60245

$C_3H_6N_2O_2$
Acetamidoxime; *O*-Formyl, *in* A-60016

$C_3H_6N_4O$
2-Tetrazolin-5-one; 1-Et, *in* T-60175

$C_3H_6N_4O_2$
2-Azido-2-nitropropane, A-60349

$C_3H_6N_4S$
Tetrazole-5-thione; 1-Et, *in* T-60174
Tetrazole-5-thione; 1,4-Di-Me, *in* T-60174
Tetrazole-5-thione; *S*-Et, *in* T-60174

$C_3H_6N_6$
2,2-Diazidopropane, D-60058

$C_3H_6N_6O_3$
▷Hexahydro-1,3,5-triazine; 1,3,5-Trinitroso, *in* H-60060

C_3H_6OS
1-Mercapto-2-propanone, M-60028

$C_3H_6O_2S$
▷2-Mercaptopropanoic acid, M-60027

C_3H_7NOS
2-Mercaptopropanoic acid; Amide, *in* M-60027

$C_3H_7NO_3$
▷3-Nitro-1-propanol, N-60039

$C_3H_7NO_4$
▷2-Nitro-1,3-propanediol, N-60038

$C_3H_7N_3O_2$
Allophanic methylamide, *in* B-60190

$C_3H_8N_2O_2$
N,N-Dimethylcarbamohydroxamic acid, D-60414

$C_3H_8N_2O_3S$
Aminoiminomethanesulfonic acid; *N,N*-Di-Me, *in* A-60210

$C_3H_8S_2$
2,2-Propanedithiol, P-60175

C_3H_9NO
▷Trimethylamine oxide, T-60354

$C_3H_9NO_3S$
(Dimethylamino)methanesulfonic acid, *in* A-60215

$C_3H_9N_3$
Hexahydro-1,3,5-triazine, H-60060

C_3OS
3-Thioxo-1,2-propadien-1-one, T-60216

C_4Br_4O
Tetrabromofuran, T-60023

$C_4ClF_3N_2$
3-Chloro-4,5,6-trifluoropyridazine, C-60142
4-Chloro-3,5,6-trifluoropyridazine, C-60143
5-Chloro-2,4,6-trifluoropyrimidine, C-60144

C_4F_6
▷(Difluoromethylene)tetrafluorocyclopropane, D-60185

$C_4F_6O_2$
1,1,1,4,4,4-Hexafluoro-2,3-butanedione, H-60043

$C_4F_6O_4S$
4-[2,2,2-Trifluoro-1-(trifluoromethyl)ethylidene]-1,3,2-dioxathietane 2,2-dioxide, T-60311

C_4F_6S
Bis(trifluoromethyl)thioketene, B-60187

$C_4F_6S_2$
3,4-Bis(trifluoromethyl)-1,2-dithiete, B-60179

$C_4F_8O_3S$
4,4-Difluoro-3,3-bis(trifluoromethyl)-1,2-oxathietane 2,2-dioxide, D-60181

C_4F_9NO
Nonafluoromorpholine, N-60056

C_4HBr_3S
2,3,5-Tribromothiophene, T-60247

$C_4HF_3N_2$
Trifluoropyrazine, T-60306
2,4,6-Trifluoropyrimidine, T-60307
4,5,6-Trifluoropyrimidine, T-60308

C_4HF_5
3,3,4,4,4-Pentafluoro-1-butyne, P-60032
1,3,3,4,4-Pentafluorocyclobutene, P-60033

C_4HF_8NO
2,2,3,3,5,5,6,6-Octafluoromorpholine, O-60009

$C_4HN_5O_8$
2,3,4,5-Tetranitro-1*H*-pyrrole, T-60160

$C_4H_2Br_2N_2O$
4,5-Dibromo-1*H*-imidazole-2-carboxaldehyde, D-60101

$C_4H_2Br_2N_2O_2$
4,5-Dibromo-1*H*-imidazole-2-carboxylic acid, D-60102

$C_4H_2ClNO_2$
5-Isoxazolecarboxylic acid; Chloride, *in* I-60142

$C_4H_2Cl_2N_2O$
4,5-Dichloro-1*H*-imidazole-2-carboxaldehyde, D-60134

$C_4H_2Cl_2N_2O_2$
4,5-Dichloro-1*H*-imidazole-2-carboxylic acid, D-60135

$C_4H_2Cl_3NO$
5-(Trichloromethyl)isoxazole, T-60253

$C_4H_2F_2N_2$
2,3-Difluoropyrazine, D-60188
2,6-Difluoropyrazine, D-60189

$C_4H_2F_3N$
▷2-Trifluoromethylacrylonitrile, *in* T-60295

$C_4H_2F_4$
3,3,4,4-Tetrafluorocyclobutene, T-60043

$C_4H_2N_2O$
5-Cyanoisoxazole, *in* I-60142

$C_4H_2N_2S_2$
(Dimercaptomethylene)malononitrile, *in* D-60395

$C_4H_2N_4O_6$
2,3,4-Trinitro-1*H*-pyrrole, T-60373
2,3,5-Trinitro-1*H*-pyrrole, T-60374

$C_4H_2O_2$
2-Butynedial, B-60351

$C_4H_3BrN_2O_2$
▷5-Bromo-2,4-(1*H*,3*H*)-pyrimidinedione, B-60318
6-Bromo-2,4-(1*H*,3*H*)-pyrimidinedione, B-60319

$C_4H_3ClN_2O_2$
▷5-Chloro-2,4-(1*H*,3*H*)-pyrimidinedione, C-60130

$C_4H_3ClO_2S_2$
2-Thiophenesulfonic acid; Chloride, *in* T-60213
3-Thiophenesulfonic acid; Chloride, *in* T-60214

C_4H_3ClS
3-Chlorothiophene, C-60136

$C_4H_3F_3$
1,1,1-Trifluoro-2-butyne, T-60275
1,4,4-Trifluorocyclobutene, T-60276

$C_4H_3F_3N_2$
2-Trifluoromethyl-1*H*-imidazole, T-60289

4(5)-Trifluoromethyl-1*H*-imidazole, T-60290

C$_4$H$_3$F$_3$O$_2$
2-(Trifluoromethyl)propenoic acid, T-60295

C$_4$H$_3$IN$_2$O$_2$
▷5-Iodo-2,4(1*H*,3*H*)-pyrimidinedione, I-60062

C$_4$H$_3$IS
3-Iodothiophene, I-60067

C$_4$H$_3$NO$_3$
5-Isoxazolecarboxylic acid, I-60142

C$_4$H$_3$N$_3$
4(5)-Cyanoimidazole, *in* I-60004

C$_4$H$_3$N$_3$O$_2$
2-Nitropyrimidine, N-60043

C$_4$H$_3$N$_3$O$_3$
6-Formyl-1,2,4-triazine-3,5(2*H*,4*H*)-dione, F-60075

C$_4$H$_3$N$_3$O$_4$
2,3-Dinitro-1*H*-pyrrole, D-60461
2,4-Dinitro-1*H*-pyrrole, D-60462
2,5-Dinitro-1*H*-pyrrole, D-60463
3,4-Dinitro-1*H*-pyrrole, D-60464

C$_4$H$_3$N$_5$
Tetrazolo[1,5-*a*]pyrimidine, T-60176
Tetrazolo[1,5-*c*]pyrimidine, T-60177

C$_4$H$_4$BrN$_3$O$_2$
4-Bromo-3(5)-nitro-1*H*-pyrazole; 1-Me, *in* B-60304

C$_4$H$_4$Br$_2$O
2,2-Dibromocyclopropanecarboxaldehyde, D-60092

C$_4$H$_4$Br$_2$O$_2$
2,2-Dibromocyclopropanecarboxylic acid, D-60093

C$_4$H$_4$Cl$_2$O$_2$
2,2-Dichlorocyclopropanecarboxylic acid, D-60127

C$_4$H$_4$F$_3$NO
2-(Trifluoromethyl)propenoic acid; Amide, *in* T-60295

C$_4$H$_4$N$_2$O$_2$
1*H*-Imidazole-4-carboxylic acid, I-60004
5-Isoxazolecarboxylic acid; Amide, *in* I-60142
3-Nitropyrrole, N-60044

C$_4$H$_4$N$_2$S
4-Vinyl-1,2,3-thiadiazole, V-60014
5-Vinyl-1,2,3-thiadiazole, V-60015

C$_4$H$_4$N$_6$
4,4′-Bi-4*H*-1,2,4-triazole, B-60189

C$_4$H$_4$OS
Thiophene-3-ol, T-60212

C$_4$H$_4$O$_2$
1,2-Cyclobutanedione, C-60183

C$_4$H$_4$O$_3$S$_2$
2-Thiophenesulfonic acid, T-60213
3-Thiophenesulfonic acid, T-60214

C$_4$H$_4$O$_4$
Diformylacetic acid, D-60190

C$_4$H$_4$O$_4$S$_2$
(Dimercaptomethylene)propanedioic acid, D-60395

C$_4$H$_5$Cl$_2$NO$_2$
2-Amino-4,4-dichloro-3-butenoic acid, A-60125

C$_4$H$_5$Cl$_3$O
▷(2,2,2-Trichloroethyl)oxirane, T-60252

C$_4$H$_5$F$_6$NO
N-Ethoxy-1,1,1,1′,1′,1′-hexafluorodimethylamine, *in* B-60180

C$_4$H$_5$IO$_2$
Dihydro-3-iodo-2(3*H*)-furanone, D-60252

C$_4$H$_5$NO$_2$
1-Amino-2-cyclopropene-1-carboxylic acid, A-60122
1-Nitro-1,3-butadiene, N-60023

C$_4$H$_5$NO$_2$S$_2$
2-Thiophenesulfonic acid; Amide, *in* T-60213
3-Thiophenesulfonic acid; Amide, *in* T-60214

C$_4$H$_5$NO$_4$
2-Oxo-4-oxazolidinecarboxylic acid, O-60082

C$_4$H$_5$NS
4-Mercapto-2-butenoic acid; Nitrile, *in* M-60018

C$_4$H$_5$NS$_2$
2-(Methylthio)thiazole, *in* T-60200

C$_4$H$_5$N$_3$O
1*H*-Imidazole-4-carboxylic acid; Amide, *in* I-60004

C$_4$H$_5$N$_3$O$_2$
6-Methyl-1,3,5-triazine-2,4(1*H*,3*H*)-dione, M-60132
1*H*-1,2,3-Triazole-4-carboxylic acid; 1-Me, *in* T-60239
1*H*-1,2,3-Triazole-4-carboxylic acid; Me ester, *in* T-60239
1*H*-1,2,3-Triazole-4-carboxylic acid; 2-Me, *in* T-60239

C$_4$H$_5$N$_3$S$_2$
5-Amino-2-thiazolecarbothioamide, A-60260
5-Amino-4-thiazolecarbothioamide, A-60261

C$_4$H$_6$ClF$_2$NO$_2$
2-Amino-4-chloro-4,4-difluorobutanoic acid, A-60107

C$_4$H$_6$I$_2$
▷3-Iodo-2-(iodomethyl)-1-propene, I-60047

C$_4$H$_6$N$_2$O
4,5-Dihydro-3(2*H*)-pyridazinone, D-60272
3,4-Dihydro-2(1*H*)-pyrimidinone, D-60273
5,6-Dihydro-4(1*H*)-pyrimidinone, D-60274

C$_4$H$_6$N$_2$O$_2$
1,3-Diazetidine-2,4-dione; 1,3-Di-Me, *in* D-60057

C$_4$H$_6$N$_2$O$_3$
2-Oxo-4-imidazolidinecarboxylic acid, O-60074

C$_4$H$_6$N$_4$O
1*H*-1,2,3-Triazole-4-carboxylic acid; 1-Me, amide, *in* T-60239

C$_4$H$_6$O$_2$
2-Methylene-1,3-dioxolane, M-60075
4-Methylene-1,3-dioxolane, M-60076

C$_4$H$_6$O$_2$S
4-Mercapto-2-butenoic acid, M-60018

C$_4$H$_6$O$_3$
Dihydro-4-hydroxy-2(3*H*)-furanone, D-60242
1-Hydroxycyclopropanecarboxylic acid, H-60117
Methylene-1,3,5-trioxane, M-60081
3-Oxopropanoic acid; Me ester, *in* O-60090

C$_4$H$_6$S
2,3-Dihydrothiophene, D-60290

C$_4$H$_6$S$_4$
Ethanebis(dithioic)acid; Di-Me ester, *in* E-60040

C$_4$H$_7$ClFNO$_2$
2-Amino-4-chloro-4-fluorobutanoic acid, A-60108

C$_4$H$_7$ClOS
Carbonochloridothioic acid; *O*-Propyl, *in* C-60013
Carbonochloridothioic acid; *O*-Isopropyl, *in* C-60013
▷Carbonochloridothioic acid; *S*-Propyl, *in* C-60013

C$_4$H$_7$F$_2$NO$_2$
2-Amino-3,3-difluorobutanoic acid, A-60133
3-Amino-4,4-difluorobutanoic acid, A-60134

C$_4$H$_7$F$_3$OS
Trifluoromethanesulfenic acid; Isopropyl ester, *in* T-60279

C$_4$H$_7$IO
2-Iodobutanal, I-60040
3-Iodo-2-butanone, I-60041
2-Iodo-2-methylpropanal, I-60051

C$_4$H$_7$N
2-Propyn-1-amine; *N*-Me; B,HI, *in* P-60184

C$_4$H$_7$NOS
Tetrahydro-2*H*-1,2-thiazin-3-one, T-60095

Tetrahydro-2*H*-1,3-thiazin-2-one, T-60096

C₄H₇NO₂
3-Aminodihydro-2(3*H*)-furanone, A-60138
5-Methyl-2-oxazolidinone, M-60107
3-Morpholinone, M-60147
Nitrocyclobutane, N-60024

C₄H₇NS₂
4,5-Dihydro-2-(methylthio)thiazole, *in* T-60197
Tetrahydro-2*H*-1,3-thiazine-2-thione, T-60094
2-Thiazoledinethione; *N*-Me, *in* T-60197

C₄H₇N₃O₂
Biuret; *N*-Ac, *in* B-60190

C₄H₇N₅
2,4-Diamino-6-methyl-1,3,5-triazine, D-60040

C₄H₇N₅O
2,4-Diamino-6-methyl-1,3,5-triazine; *N*³-Oxide, *in* D-60040
2,4-Diamino-6-methyl-1,3,5-triazine; *N*⁵-Oxide, *in* D-60040

C₄H₈BrNO₂
2-Amino-3-bromobutanoic acid, A-60097
2-Amino-4-bromobutanoic acid, A-60098
4-Amino-2-bromobutanoic acid, A-60099

C₄H₈ClNO₂
2-Amino-3-chlorobutanoic acid, A-60103
2-Amino-4-chlorobutanoic acid, A-60104
4-Amino-2-chlorobutanoic acid, A-60105
4-Amino-3-chlorobutanoic acid, A-60106

C₄H₈FNO₂
2-Amino-3-fluorobutanoic acid, A-60154
2-Amino-4-fluorobutanoic acid, A-60155
3-Amino-4-fluorobutanoic acid, A-60156
4-Amino-2-fluorobutanoic acid, A-60157
4-Amino-3-fluorobutanoic acid, A-60158

C₄H₈I₂O
1,1′-Oxybis[2-iodoethane], O-60096

C₄H₈NO₆P
Antibiotic SF 2312, A-60288

C₄H₈N₂O
Tetrahydro-2(1*H*)-pyrimidinone, T-60088
Tetrahydro-4(1*H*)-pyrimidinone, T-60089

C₄H₈N₂O₂
Acetamidoxime; *O*-Ac, *in* A-60016
4-Amino-5-methyl-3-isoxazolidinone, A-60221

C₄H₈N₄O₂
▷Hexahydropyrimidine; *N*,*N*′-Dinitroso, *in* H-60056

C₄H₈N₄O₄
▷Hexahydropyrimidine; *N*,*N*′-Dinitro, *in* H-60056

C₄H₈OS
3-Mercaptocyclobutanol, M-60019

C₄H₈O₂
▷2-Methyl-1,3-dioxolane, M-60063

C₄H₈O₂S
2-(Methylthio)propanoic acid, *in* M-60027

C₄H₈O₄
2,3-Dihydroxybutanoic acid, D-60312
3,4-Dihydroxybutanoic acid, D-60313

C₄H₈S₂
1,3-Cyclobutanedithiol, C-60184

C₄H₉NO₃
3-Amino-2-hydroxybutanoic acid, A-60183
Threonine, T-60217

C₄H₉NS
▷Tetrahydro-2*H*-1,3-thiazine, T-60093

C₄H₉N₃O₃
3-Amino-2-ureidopropanoic acid, A-60268

C₄H₁₀N₂
Hexahydropyrimidine, H-60056

C₄H₁₀O₂
2-Methyl-1,3-propanediol, M-60120

C₄H₁₀O₃
▷1-Ethoxy-1-hydroperoxyethane, E-60044

C₄H₁₀S₂
1,3-Butanedithiol, B-60339
2,2-Butanedithiol, B-60340
2-Methyl-1,3-propanedithiol, M-60121

C₄H₁₂N₄
Octahydro-1,2,5,6-tetrazocine, O-60020

C₄H₁₃N₃O
2-[(2-Hydrazinoethyl)amino]ethanol, H-60091

C₅BrF₄N
4-Bromo-2,3,5,6-tetrafluoropyridine, B-60328

C₅BrF₅N
2-Bromo-3,4,5,6-tetrafluoropyridine, B-60326
3-Bromo-2,4,5,6-tetrafluoropyridine, B-60327

C₅F₆
1,2,3,4,5,5-Hexafluoro-1,3-cyclopentadiene, H-60044

C₅F₁₀N₂O₂
2,2,3,3,4,4,5,5,6,6-Decafluoropiperidine; *N*-Nitro, *in* D-60006

C₅F₁₁N
Undecafluoropiperidine, U-60002

C₅HF₁₀N
2,2,3,3,4,4,5,5,6,6-Decafluoropiperidine, D-60006

C₅H₂Cl₂N₂O₂
2,3-Dichloro-5-nitropyridine, D-60140
2,4-Dichloro-3-nitropyridine, D-60141
2,4-Dichloro-5-nitropyridine, D-60142
2,5-Dichloro-3-nitropyridine, D-60143
2,6-Dichloro-3-nitropyridine, D-60144
2,6-Dichloro-4-nitropyridine, D-60145
3,4-Dichloro-5-nitropyridine, D-60146
3,5-Dichloro-4-nitropyridine, D-60147

C₅H₂Cl₂N₂O₃
2,6-Dichloro-4-nitropyridine; 1-Oxide, *in* D-60145
3,5-Dichloro-4-nitropyridine; 1-Oxide, *in* D-60147

C₅H₂Cl₃N
▷2,4,6-Trichloropyridine, T-60255

C₅H₂F₄N₂
4-Amino-2,3,5,6-tetrafluoropyridine, A-60257

C₅H₂F₆O
4,4,4-Trifluoro-3-(trifluoromethyl)-2-butenal, T-60309
1,1,1-Trifluoro-2-(trifluoromethyl)-3-butyn-2-ol, T-60310

C₅H₂INS
2-Cyano-3-iodothiophene, *in* I-60075
2-Cyano-4-iodothiophene, *in* I-60076
2-Cyano-5-iodothiophene, *in* I-60078
3-Cyano-2-iodothiophene, *in* I-60074
3-Cyano-4-iodothiophene, *in* I-60077
4-Cyano-2-iodothiophene, *in* I-60079

C₅H₃BrClN
3-Bromo-5-chloropyridine, B-60223
4-Bromo-3-chloropyridine, B-60224

C₅H₃BrFN
2-Bromo-5-fluoropyridine, B-60252
3-Bromo-2-fluoropyridine, B-60253
3-Bromo-5-fluoropyridine, B-60254
4-Bromo-3-fluoropyridine, B-60255
5-Bromo-2-fluoropyridine, B-60256

C₅H₃BrIN
2-Bromo-4-iodopyridine, B-60280
2-Bromo-5-iodopyridine, B-60281
3-Bromo-4-iodopyridine, B-60282
3-Bromo-5-iodopyridine, B-60283

4-Bromo-2-iodopyridine, B-60284
4-Bromo-3-iodopyridine, B-60285
5-Bromo-2-iodopyridine, B-60286

C$_5$H$_3$BrO$_2$
3-Bromo-2-furancarboxaldehyde, B-60257
4-Bromo-2-furancarboxaldehyde, B-60258
5-Bromo-2-furancarboxaldehyde, B-60259

C$_5$H$_3$ClFN
2-Chloro-3-fluoropyridine, C-60081
2-Chloro-4-fluoropyridine, C-60082
2-Chloro-5-fluoropyridine, C-60083
2-Chloro-6-fluoropyridine, C-60084
3-Chloro-2-fluoropyridine, C-60085
4-Chloro-2-fluoropyridine, C-60086
4-Chloro-3-fluoropyridine, C-60087
5-Chloro-2-fluoropyridine, C-60088

C$_5$H$_3$ClFNO
2-Chloro-3-fluoropyridine; N-Oxide, in C-60081

C$_5$H$_3$ClIN
2-Chloro-3-iodopyridine, C-60107
2-Chloro-4-iodopyridine, C-60108
2-Chloro-5-iodopyridine, C-60109
3-Chloro-2-iodopyridine, C-60110
3-Chloro-5-iodopyridine, C-60111
4-Chloro-2-iodopyridine, C-60112
4-Chloro-3-iodopyridine, C-60113

C$_5$H$_3$FIN
2-Fluoro-4-iodopyridine, F-60045
3-Fluoro-4-iodopyridine, F-60046

C$_5$H$_3$IOS
2-Iodo-3-thiophenecarboxaldehyde, I-60068
3-Iodo-2-thiophenecarboxaldehyde, I-60069
4-Iodo-2-thiophenecarboxaldehyde, I-60070
4-Iodo-3-thiophenecarboxaldehyde, I-60071
5-Iodo-2-thiophenecarboxaldehyde, I-60072
5-Iodo-3-thiophenecarboxaldehyde, I-60073

C$_5$H$_3$IO$_2$S
2-Iodo-3-thiophenecarboxylic acid, I-60074
3-Iodo-2-thiophenecarboxylic acid, I-60075
4-Iodo-2-thiophenecarboxylic acid, I-60076
4-Iodo-3-thiophenecarboxylic acid, I-60077
5-Iodo-2-thiophenecarboxylic acid, I-60078
5-Iodo-3-thiophenecarboxylic acid, I-60079

C$_5$H$_3$NS
2-Ethynylthiazole, E-60061
4-Ethynylthiazole, E-60062

C$_5$H$_3$N$_5$O$_8$
▷1-Methyl-2,3,4,5-tetranitro-1H-pyrrole, in T-60160

C$_5$H$_4$BrNO
▷2-Bromo-3-hydroxypyridine, B-60275
2-Bromo-5-hydroxypyridine, B-60276
3-Bromo-5-hydroxypyridine, B-60277
4-Bromo-3-hydroxypyridine, B-60278

C$_5$H$_4$BrNO$_2$
2-Bromo-3-hydroxypyridine; N-Oxide, in B-60275

C$_5$H$_4$BrNS
3-Bromo-2(1H)-pyridinethione, B-60315
3-Bromo-4(1H)-pyridinethione, B-60316
5-Bromo-2(1H)-pyridinethione, B-60317

C$_5$H$_4$ClNO
▷2-Chloro-3-hydroxypyridine, C-60100
2-Chloro-5-hydroxypyridine, C-60101
3-Chloro-5-hydroxypyridine, C-60102

C$_5$H$_4$ClNS
3-Chloro-2(1H)-pyridinethione, C-60129

C$_5$H$_4$ClN$_5$
▷2-Amino-6-chloro-1H-purine, A-60116

C$_5$H$_4$Cl$_4$
1-Chloro-1-(trichlorovinyl)cyclopropane, C-60138

C$_5$H$_4$FN$_5$
8-Amino-6-fluoro-9H-purine, A-60163

C$_5$H$_4$INO
3-Hydroxy-2-iodopyridine, H-60160

C$_5$H$_4$INOS
2-Iodo-3-thiophenecarboxaldehyde; Oxime, in I-60068
3-Iodo-2-thiophenecarboxaldehyde; Oxime, in I-60069
4-Iodo-2-thiophenecarboxaldehyde; Oxime, in I-60070
4-Iodo-3-thiophenecarboxaldehyde; Oxime, in I-60071
5-Iodo-2-thiophenecarboxaldehyde; Oxime, in I-60072
5-Iodo-3-thiophenecarboxaldehyde; Oxime, in I-60073
2-Iodo-3-thiophenecarboxylic acid; Amide, in I-60074
3-Iodo-2-thiophenecarboxylic acid; Amide, in I-60075
4-Iodo-2-thiophenecarboxylic acid; Amide, in I-60076
4-Iodo-3-thiophenecarboxylic acid; Amide, in I-60077
5-Iodo-2-thiophenecarboxylic acid; Amide, in I-60078
5-Iodo-3-thiophenecarboxylic acid; Amide, in I-60079

C$_5$H$_4$N$_2$O$_3$
1,2,3,4-Tetrahydro-2,4-dioxo-5-pyrimidinecarboxaldehyde, T-60064

C$_5$H$_4$N$_2$S
▷2-Thiocyanato-1H-pyrrole, T-60209

C$_5$H$_4$N$_4$
Pyrazolo[1,5-b][1,2,4]triazine, P-60207
[1,2,4]Triazolo[1,5-b]pyridazine, T-60240
[1,2,4]Triazolo[1,5-a]pyrimidine, T-60242
1,2,4-Triazolo[4,3-c]pyrimidine, T-60243

C$_5$H$_4$N$_4$O
1,5-Dihydro-6H-imidazo[4,5-c]pyridazin-6-one, D-60248
2H-Imidazo[4,5-b]pyrazin-2-one, I-60006

C$_5$H$_4$N$_4$O$_2$
1H-Pyrazolo[3,4-d]pyrimidine-4,6(5H,7H)-dione, P-60202

C$_5$H$_4$N$_4$O$_6$
▷1-Methyl-2,3,4-trinitro-1H-pyrrole, in T-60373
▷1-Methyl-2,3,5-trinitro-1H-pyrrole, in T-60374

C$_5$H$_4$N$_4$S
1,3-Dihydro-2H-imidazo[4,5-b]pyrazine-2-thione, D-60247
▷1,7-Dihydro-6H-purine-6-thione, D-60270

C$_5$H$_5$BF$_5$N
1-Fluoropyridinium; Tetrafluoroborate, in F-60066

C$_5$H$_5$BrN$_2$O
2-Amino-5-bromo-3-hydroxypyridine, A-60101

C$_5$H$_5$Cl
1-Chloro-1-ethynylcyclopropane, C-60068

C$_5$H$_5$ClFNO$_4$
1-Fluoropyridinium; Perchlorate, in F-60066

C$_5$H$_5$ClN$_2$O
2-Amino-5-chloro-3-hydroxypyridine, A-60111

C$_5$H$_5$ClO
2-Chloro-2-cyclopenten-1-one, C-60051

C$_5$H$_5$Cl$_2$F$_3$O$_2$
2,2-Dichloro-3,3,3-trifluoro-1-propanol; Ac, in D-60161

C$_5$H$_5$FN$^\oplus$
1-Fluoropyridinium, F-60066

C$_5$H$_5$F$_3$O$_2$
Dihydro-4-(trifluoromethyl)-2(3H)-furanone, D-60299
Dihydro-5-(trifluoromethyl)-2(3H)-furanone, D-60300
▷2-(Trifluoromethyl)propenoic acid; Me ester, in T-60295

C$_5$H$_5$F$_7$NSb
1-Fluoropyridinium; Hexafluoroantimonate, in F-60066

C$_5$H$_5$NOS
3-Hydroxy-2(1H)-pyridinethione, H-60222
▷2(1H)-Pyridinethione; 1-Hydroxy, in P-60224
4(1H)-Pyridinethione; 1-Hydroxy, in P-60225

C$_5$H$_5$NO$_2$
3-Hydroxy-2(1H)-pyridinone, H-60223
3-Hydroxy-4(1H)-pyridinone, H-60227
4-Hydroxy-2(1H)-pyridinone, H-60224

5-Hydroxy-2(1*H*)-pyridinone, H-60225
6-Hydroxy-2(1*H*)-pyridinone, H-60226

$C_5H_5NO_3$
5-Isoxazolecarboxylic acid; Me ester, *in* I-60142

C_5H_5NS
2-Pyridinethiol, P-60223
▷2(1*H*)-Pyridinethione, P-60224
▷4(1*H*)-Pyridinethione, P-60225
1,4-Thiazepine, T-60196
1-Thiocyano-1,3-butadiene, T-60210

$C_5H_5NS_2$
2,3-Pyridinedithiol, P-60218
2,4-Pyridinedithiol, P-60219
2,5-Pyridinedithiol, P-60220
3,4-Pyridinedithiol, P-60221
3,5-Pyridinedithiol, P-60222

$C_5H_5N_3O_2$
4-Methyl-5-nitropyrimidine, M-60104

$C_5H_5N_3O_4$
1-Methyl-2,3-dinitro-1*H*-pyrrole, *in* D-60461
1-Methyl-2,4-dinitro-1*H*-pyrrole, *in* D-60462
1-Methyl-2,5-dinitro-1*H*-pyrrole, *in* D-60463
1-Methyl-3,4-dinitro-1*H*-pyrrole, *in* D-60464

$C_5H_5N_5$
▷2-Aminopurine, A-60243
8-Aminopurine, A-60244
9-Aminopurine, A-60245

$C_5H_5N_5O$
6-Amino-1,3-dihydro-2*H*-purin-2-one, A-60144
5-Aminopyrazolo[4,3-*d*]pyrimidin-7(1*H*,6*H*)-one, A-60246

C_5H_6
3-Vinylcyclopropene, V-60012

$C_5H_6Br_2O_2$
2,2-Dibromocyclopropanecarboxylic acid; Me ester, *in* D-60093

$C_5H_6Cl_2O_2$
2,2-Dichlorocyclopropanecarboxylic acid; Me ester, *in* D-60127

$C_5H_6N_2O$
2-Amino-3-hydroxypyridine, A-60200
2-Amino-5-hydroxypyridine, A-60201
4-Amino-3-hydroxypyridine, A-60202
5-Amino-3-hydroxypyridine, A-60203
1-Amino-2(1*H*)-pyridinone, A-60248
2-Amino-4(1*H*)-pyridinone, A-60253
3-Amino-2(1*H*)-pyridinone, A-60249
3-Amino-4(1*H*)-pyridinone, A-60254
4-Amino-2(1*H*)-pyridinone, A-60250
5-Amino-2(1*H*)-pyridinone, A-60251
6-Amino-2(1*H*)-pyridinone, A-60252

$C_5H_6N_2O_2$
4-Amino-2(1*H*)-pyridinone; *N*-Oxide, *in* A-60250
1*H*-Imidazole-4-carboxylic acid; Me ester, *in* I-60004

$C_5H_6N_2S$
▷3-Amino-2(1*H*)-pyridinethione, A-60247

$C_5H_6N_6O$
2,8-Diamino-1,7-dihydro-6*H*-purin-6-one, D-60035

C_5H_6O
2-Cyclopenten-1-one, C-60226

C_5H_6OS
3-Methoxythiophene, *in* T-60212

$C_5H_6O_2$
4-Hydroxy-2-cyclopenten-1-one, H-60116
▷5-Methyl-2(5*H*)-furanone, M-60083
4-Pentene-2,3-dione, P-60061

$C_5H_6O_3$
2-Cyclopropyl-2-oxoacetic acid, C-60231
5-Hydroxymethyl-2(5*H*)-furanone, H-60182

$C_5H_6O_4$
Methyl diformylacetate, *in* D-60190

$C_5H_6O_5$
3-Oxopentanedioic acid, O-60084

C_5H_7Br
1-Bromo-2,3-dimethylcyclopropene, B-60235
1-Bromo-1,3-pentadiene, B-60311

C_5H_7Cl
5-Chloro-1,3-pentadiene, C-60121

$C_5H_7ClO_2$
1-Chloro-2,3-pentanedione, C-60123

C_5H_7FO
2-Fluorocyclopentanone, F-60022

C_5H_7IO
2-Iodocyclopentanone, I-60043

$C_5H_7IO_2$
Tetrahydro-3-iodo-2*H*-pyran-2-one, T-60069

C_5H_7NO
2-Cyclopenten-1-one; Oxime, *in* C-60226
4-Vinyl-2-azetidinone, V-60008

$C_5H_7NO_2S$
3-(Ethylthio)-2-propenoic acid; Nitrile, *S*-dioxide, *in* E-60055

$C_5H_7NO_4$
2-Amino-4-hydroxypentanedioic acid; Lactone, *in* A-60195
5-Methyl-2-oxo-4-oxazolidinecarboxylic acid, M-60108
3-Oxopentanedioic acid; Amide, *in* O-60084

$C_5H_7NO_5$
3-Oxopentanedioic acid; Oxime, *in* O-60084

C_5H_7NS
3-(Ethylthio)-2-propenoic acid; Nitrile, *in* E-60055

$C_5H_7NS_2$
2-(Ethylthio)thiazole, *in* T-60200

$C_5H_7N_3O$
2-Amino-6-methyl-4(1*H*)-pyrimidinone, A-60230
4-Hydrazino-2(1*H*)-pyridinone, H-60093
1*H*-Imidazole-4-carboxylic acid; Methylamide, *in* I-60004

$C_5H_7N_3O_2$
1,6-Dimethyl-5-azauracil, *in* M-60132
3,6-Dimethyl-5-azauracil, *in* M-60132
Methyl 1-methyl-1*H*-1,2,3-triazole-5-carboxylate, *in* T-60239
1*H*-1,2,3-Triazole-4-carboxylic acid; 1-Me, Me ester, *in* T-60239
1*H*-1,2,3-Triazole-4-carboxylic acid; Et ester, *in* T-60239
1*H*-1,2,3-Triazole-4-carboxylic acid; 2-Me, Me ester, *in* T-60239

$C_5H_7N_3O_3$
Cyanuric acid; 1,3-Di-Me, *in* C-60179

$C_5H_8Br_2$
1,1-Dibromo-3-methyl-1-butene, D-60104

$C_5H_8ClNO_2$
1-Chloro-2,3-pentanedione; 2-Oxime, *in* C-60123

$C_5H_8Cl_2O$
▷3,3-Bis(chloromethyl)oxetane, B-60149

$C_5H_8FNO_2$
4-Amino-5-fluoro-2-pentenoic acid, A-60161

$C_5H_8N_2O$
3,4-Dihydro-2(1*H*)-pyrimidinone; 1-Me, *in* D-60273
5,6-Dihydro-4(1*H*)-pyrimidinone; 3-Me, *in* D-60274

$C_5H_8N_2O_2$
▷*N*-(2-Cyanoethyl)glycine, *in* C-60016
Squamolone, S-60049

$C_5H_8N_2O_7$
3,3-Bis(nitratomethyl)oxetane, *in* O-60056

C_5H_8O
▷1-Penten-3-one, P-60062

$C_5H_8O_2$
2,6-Dioxaspiro[3.3]heptane, D-60468
2-Hydroxycyclopentanone, H-60114
3-Hydroxycyclopentanone, H-60115
4-Oxopentanal, O-60083
▷2,4-Pentanedione, P-60059

$C_5H_8O_2S$
3-(Ethylthio)-2-propenoic acid, E-60055
4-Mercapto-2-butenoic acid; Me ester, *in* M-60018

$C_5H_8O_3S$
2-Mercaptopropanoic acid; *S*-Ac, *in* M-60027

C_5H_9ClOS
Carbonochloridothioic acid; *O*-Butyl, *in* C-60013
Carbonochloridothioic acid; *O-tert*-Butyl, *in* C-60013
Carbonochloridothioic acid; *S*-Butyl, *in* C-60013

$C_5H_9ClO_3$
(1-Chloroethyl) ethyl carbonate, C-60067

$C_5H_9FN_2$
2-Amino-3-fluoro-3-methylbutanoic acid; Nitrile, *in* A-60159

C_5H_9IO
2-Iodo-3-methylbutanal, I-60049
2-Iodopentanal, I-60054
5-Iodo-2-pentanone, I-60057
5-Iodo-4-penten-1-ol, I-60058

C_5H_9N
2-Propyn-1-amine; *N*-Di-Me, *in* P-60184

C_5H_9NO
4-Imino-2-pentanone, *in* P-60059

$C_5H_9NO_2$
2-Amino-3-methyl-3-butenoic acid, A-60218
5-(Aminomethyl)dihydro-2(3*H*)-furanone, *in* A-60196
3,5-Dimethyl-2-oxazolidinone, *in* M-60107
2-Hydroxycyclopentanone; Oxime, *in* H-60114

$C_5H_9NO_3$
2-Amino-2-(hydroxymethyl)-3-butenoic acid, A-60187
3,4-Dihydro-3,4-dihydroxy-2-(hydroxymethyl)-2*H*-pyrrole, D-60222

$C_5H_9NO_4$
3-(Carboxymethylamino)propanoic acid, C-60016
3-Nitro-1-propanol; Ac, *in* N-60039
Threonine; *N*-Formyl, *in* T-60217

$C_5H_9NO_5$
2-Amino-4-hydroxypentanedioic acid, A-60195

$C_5H_9NS_2$
5,6-Dihydro-2-(methylthio)-4*H*-1,3-thiazine, *in* T-60094
Tetrahydro-2*H*-1,3-thiazine-2-thione; *N*-Me, *in* T-60094

$C_5H_9N_3O$
3,5-Dihydro-3,5,5-trimethyl-4*H*-triazol-4-one, D-60303

$C_5H_9N_3O_2$
3-Azido-3-methylbutanoic acid, A-60348

$C_5H_{10}ClNO_2$
2-Amino-3-chlorobutanoic acid; Me ester, *in* A-60103
2-Amino-4-chlorobutanoic acid; Me ester, *in* A-60104

$C_5H_{10}FNO_2$
2-Amino-3-fluoro-3-methylbutanoic acid, A-60159
2-Amino-3-fluoropentanoic acid, A-60160

$C_5H_{10}N_2$
2,5-Diazabicyclo[4.1.0]heptane, D-60053

$C_5H_{10}N_2O$
Tetrahydro-2(1*H*)-pyrimidinone; 1-Me, *in* T-60088
Tetrahydro-4(1*H*)-pyrimidinone; 3-Me, *in* T-60089

$C_5H_{10}N_2O_2$
4-Oxopentanal; Dioxime, *in* O-60083
2,4-Pentanedione; Dioxime, *in* P-60059

$C_5H_{10}O_2S$
2-Mercapto-3-methylbutanoic acid, M-60023
2-Mercaptopropanoic acid; Et ester, *in* M-60027

2-Mercaptopropanoic acid; Me ester, *S*-Me ether, *in* M-60027

$C_5H_{10}O_3$
3,3-Oxetanedimethanol, O-60056

$C_5H_{10}O_4$
2,3-Dihydroxybutanoic acid; Me ester, *in* D-60312
▷Glycerol 1-acetate, G-60029

$C_5H_{10}S_2$
1,1-Cyclopentanedithiol, C-60222
5-Methyl-1,3-dithiane, M-60067

$C_5H_{11}BrO$
5-Bromo-2-pentanol, B-60313

$C_5H_{11}IO$
1-Iodo-3-pentanol, I-60055
5-Iodo-2-pentanol, I-60056

$C_5H_{11}NO_3$
2-Amino-4-hydroxy-2-methylbutanoic acid, A-60186
5-Amino-4-hydroxypentanoic acid, A-60196
Threonine; Me ether, *in* T-60217
Threonine; *N*-Me, *in* T-60217
Threonine; Me ester, *in* T-60217

$C_5H_{11}NO_4$
1,1-Dimethoxy-3-nitropropane, *in* N-60037

$C_5H_{11}N_3$
1-Azido-2,2-dimethylpropane, A-60344

$C_5H_{11}N_3O_2$
Biuret; 1,3,5-Tri-Me, *in* B-60190

$C_5H_{12}N_2O$
3,3-Bis(aminomethyl)oxetane, B-60140

$C_5H_{12}N_2O_2S$
S-(2-Aminoethyl)cysteine, A-60152

$C_5H_{12}N_2O_3S$
S-(2-Aminoethyl)cysteine; *S*-Oxide, *in* A-60152
Aminoiminomethanesulfonic acid; *N-tert*-Butyl, *in* A-60210

$C_5H_{12}N_2O_4S$
S-(2-Aminoethyl)cysteine; *S,S*-Dioxide, *in* A-60152

$C_5H_{12}S_4$
2,2-Bis(mercaptomethyl)-1,3-propanedithiol, B-60164

$C_5H_{13}NO$
3-Amino-3-methyl-2-butanol, A-60217

$C_5H_{13}NO_2$
1-Amino-3-methyl-2,3-butanediol, A-60216

$C_6BrF_4NO_2$
1-Bromo-2,3,4,5-tetrafluoro-6-nitrobenzene, B-60323
1-Bromo-2,3,4,6-tetrafluoro-5-nitrobenzene, B-60324
1-Bromo-2,3,5,6-tetrafluoro-4-nitrobenzene, B-60325

$C_6Br_2F_4$
1,2-Dibromo-3,4,5,6-tetrafluorobenzene, D-60106
1,3-Dibromo-2,4,5,6-tetrafluorobenzene, D-60107
1,4-Dibromo-2,3,5,6-tetrafluorobenzene, D-60108

$C_6Cl_2F_4$
1,2-Dichloro-3,4,5,6-tetrafluorobenzene, D-60151
1,3-Dichloro-2,4,5,6-tetrafluorobenzene, D-60152
1,4-Dichloro-2,3,5,6-tetrafluorobenzene, D-60153

$C_6Cl_3F_3$
1,2,3-Trichloro-4,5,6-trifluorobenzene, T-60256
▷1,3,5-Trichloro-2,4,6-trifluorobenzene, T-60257

$C_6F_4I_2$
1,2,3,4-Tetrafluoro-5,6-diiodobenzene, T-60044
1,2,3,5-Tetrafluoro-4,6-diiodobenzene, T-60045
▷1,2,4,5-Tetrafluoro-3,6-diiodobenzene, T-60046

C_6F_5IO
Pentafluoroiodosobenzene, P-60034

C_6F_5NO
Pentafluoronitrosobenzene, P-60039

3,5-Hexadiyn-2-one, H-60041

C₆H₄OS
Thieno[3,4-*b*]furan, T-60204

C₆H₄O₄
3,6-Dihydroxy-1,2-benzoquinone, D-60309

C₆H₄O₆
3,5-Dihydroxy-4-oxo-4*H*-pyran-2-carboxylic acid, D-60368

C₆H₄O₈
Ethylenetetracarboxylic acid, E-60049

C₆H₅BrIN
5-Bromo-2-iodoaniline, B-60279

C₆H₅BrO₃
2-Bromo-1,3,5-benzenetriol, B-60199
4-Bromo-1,2,3-benzenetriol, B-60200
5-Bromo-1,2,3-benzenetriol, B-60201
2-Bromo-5,6-epoxy-4-hydroxy-2-cyclohexen-1-one, B-60239

C₆H₅Cl₂N
2,3-Dichloro-5-methylpyridine, D-60138
2,5-Dichloro-3-methylpyridine, D-60139

C₆H₅FO₂
4-Fluoro-1,2-benzenediol, F-60020

C₆H₅F₄NO₃S
1-Fluoropyridinium; Trifluoromethanesulfonate, *in* F-60066

C₆H₅NO₂
Amino-1,4-benzoquinone, A-60089

C₆H₅NS
5*H*-Thieno[2,3-*c*]pyrrole, T-60206

C₆H₅N₃
Imidazo[1,2-*a*]pyrazine, I-60005
Imidazo[1,2-*b*]pyridazine, I-60007
▷Imidazo[4,5-*c*]pyridine, I-60008
5*H*-Pyrrolo[2,3-*b*]pyrazine, P-60248
[1,2,3]Triazolo[1,5-*a*]pyridine, T-60241

C₆H₅N₃O
4-Amino-5-ethynyl-2(1*H*)-pyrimidinone, A-60153
1,3-Dihydro-2*H*-imidazo[4,5-*b*]pyridin-2-one, D-60249
1,3-Dihydro-2*H*-imidazo[4,5-*c*]pyridin-2-one, D-60250
4-Hydroxyimidazo[4,5-*b*]pyridine, H-60157

C₆H₅N₅O
2-Amino-4(1*H*)-pteridinone, A-60242

C₆H₅N₅O₂
2-Amino-4(1*H*)-pteridinone; 8-Oxide, *in* A-60242
▷Isoxanthopterin, I-60141

C₆H₆BrNO
2-Bromo-3-hydroxy-6-methylpyridine, B-60267
2-Bromo-3-methoxypyridine, *in* B-60275
3-Bromo-5-methoxypyridine, *in* B-60277

C₆H₆BrNO₂
2-Bromo-3-hydroxy-6-methylpyridine; *N*-Oxide, *in* B-60267
2-Bromo-3-hydroxypyridine; Me ether, *N*-Oxide, *in* B-60275

C₆H₆BrNS
3-Bromo-2-(methylthio)pyridine, *in* B-60315

C₆H₆Br₂N₂O₂
4,5-Dibromo-1*H*-imidazole-2-carboxylic acid; Et ester, *in* D-60102

C₆H₆Br₂O₃
2,6-Dibromo-4,5-dihydroxy-2-cyclohexen-1-one, D-60097

C₆H₆ClNO
2-Chloro-5-hydroxy-6-methylpyridine, C-60099
2-Chloro-3-methoxypyridine, *in* C-60100
3-Chloro-5-methoxypyridine, *in* C-60102

C₆H₆ClNO₂S
3-Chloro-2-(methylsulfonyl)pyridine, *in* C-60129

C₆H₆ClNS
3-Chloro-2-(methylthio)pyridine, *in* C-60129

C₆H₆Cl₂O
2,4-Dichloro-3,4-dimethyl-2-cyclobuten-1-one, D-60130

4,4-Dichloro-2,3-dimethyl-2-cyclobuten-1-one, D-60131

C₆H₆INO
3-Hydroxy-2-iodo-6-methylpyridine, H-60159

C₆H₆N₂
2-Methyl-4-cyanopyrrole, *in* M-60125

C₆H₆N₂O
2-Acetylpyrimidine, A-60044
4-Acetylpyrimidine, A-60045
5-Acetylpyrimidine, A-60046
1,4-Dihydrofuro[3,4-*d*]pyridazine, D-60241

C₆H₆N₂O₂
5-Acetyl-2(1*H*)-pyrimidinone, A-60048
N-Hydroxy-*N*-nitrosoaniline, H-60196

C₆H₆N₂O₃
5-Acetyl-2,4(1*H*,3*H*)-pyrimidinedione, A-60047

C₆H₆N₂S
1,4-Dihydrothieno[3,4-*d*]pyridazine, D-60289

C₆H₆N₂S₂
[Bis(methylthio)methylene]propanedinitrile, *in* D-60395
1,4-Dithiocyano-2-butene, D-60510

C₆H₆N₄
2,2′-Bi-1*H*-imidazole, B-60107
1,3(5)-Bi-1*H*-pyrazole, B-60119

C₆H₆N₄O
2-Aminopyrrolo[2,3-*d*]pyrimidin-4-one, A-60255
1,3-Dihydro-1-methyl-2*H*-imidazo[4,5-*b*]pyrazin-2-one, *in* I-60006

C₆H₆N₄O₂
3,9-Dihydro-9-methyl-1*H*-purine-2,6-dione, D-60257

C₆H₆N₄S
1,3-Dihydro-1-methyl-2*H*-imidazo[4,5-*b*]pyrazine-2-thione, *in* D-60247
1,7-Dihydro-6*H*-purine-6-thione; 1-Me, *in* D-60270
1,7-Dihydro-6*H*-purine-6-thione; 7-Me, *in* D-60270
1,7-Dihydro-6*H*-purine-6-thione; 9-Me, *in* D-60270
1,7-Dihydro-6*H*-purine-6-thione; 3-Me, *in* D-60270
▷1,7-Dihydro-6*H*-purine-6-thione; *S*-Me, *in* D-60270
2-(Methylthio)-1*H*-imidazo[4,5-*b*]pyrazine, *in* D-60247

C₆H₆N₆
2,4-Diaminopteridine, D-60042
4,6-Diaminopteridine, D-60043
4,7-Diaminopteridine, D-60044
6,7-Diaminopteridine, D-60045

C₆H₆N₆O₂
2,4-Diaminopteridine; 5,8-Dioxide, *in* D-60042
2,6-Diamino-4,7(3*H*,8*H*)-pteridinedione, D-60047
2,7-Diamino-4,6(3*H*,5*H*)-pteridinedione, D-60046

C₆H₆N₆O₃
1,3,5-Triazine-2,4,6-tricarboxaldehyde; Trioxime, *in* T-60237

C₆H₆O
2,3-Dihydro-2,3-bis(methylene)furan, D-60208
4,5-Dihydrocyclobuta[*b*]furan, D-60212

C₆H₆O₂
3-Furanacetaldehyde, F-60083
1,5-Hexadiene-3,4-dione, H-60039
3-Oxo-1-cyclopentenecarboxaldehyde, O-60062

C₆H₆O₂S
3-Acetoxythiophene, *in* T-60212

C₆H₆O₄
3-Acetyl-4-hydroxy-2(5*H*)-furanone, A-60030
1,2,3,4-Benzenetetrol, B-60014

C₆H₇ClO
2-Chloro-2-cyclohexen-1-one, C-60050

C₆H₇N
2-Vinyl-1*H*-pyrrole, V-60013

C₆H₇NO
4-(2-Propynyl)-2-azetidinone, P-60185

C_6H_7NOS

2-Pyridinethiol; *S*-Me, *N*-Oxide, *in* P-60223

$C_6H_7NO_2$

1-Azabicyclo[3.2.0]heptane-2,7-dione, A-60330
3-Hydroxy-2(1*H*)-pyridinone; *N*-Me, *in* H-60223
3-Hydroxy-4(1*H*)-pyridinone; *N*-Me, *in* H-60227
4-Hydroxy-2(1*H*)-pyridinone; *N*-Me, *in* H-60224
5-Hydroxy-2(1*H*)-pyridinone; *N*-Me, *in* H-60225
6-Hydroxy-2(1*H*)-pyridinone; *N*-Me, *in* H-60226
3-Methoxy-4(1*H*)-pyridinone, *in* H-60227
6-Methoxy-2(1*H*)-pyridinone, *in* H-60226
5-Methyl-1*H*-pyrrole-2-carboxylic acid, M-60124
5-Methyl-1*H*-pyrrole-3-carboxylic acid, M-60125

$C_6H_7NO_3$

3-Amino-5-hydroxy-7-oxabicyclo[4.1.0]hept-3-en-2-one, A-60194
5-Isoxazolecarboxylic acid; Et ester, *in* I-60142

C_6H_7NS

▷2-Pyridinethiol; *S*-Me, *in* P-60223
▷2(1*H*)-Pyridinethione; *N*-Me, *in* P-60224
4(1*H*)-Pyridinethione; *N*-Me, *in* P-60225

$C_6H_7NS_2$

4-(Methylthio)-2(1*H*)-pyridinethione, *in* P-60219

$C_6H_7N_3$

2-Amino-5-vinylpyrimidine, A-60269

$C_6H_7N_3OS_2$

5-Amino-2-thiazolecarbothioamide; N^5-Ac, *in* A-60260
5-Amino-4-thiazolecarbothioamide; N^5-Ac, *in* A-60261

$C_6H_7N_5$

6-Amino-7-methylpurine, A-60226
2-(Methylamino)purine, *in* A-60243
8-(Methylamino)purine, *in* A-60244

$C_6H_7N_5O$

6-Amino-1,3-dihydro-1-methyl-2*H*-purine-2-one, *in* A-60144
2,6-Diamino-1,5-dihydro-4*H*-imidazo[4,5-*c*]pyridin-4-one, D-60034

$C_6H_7N_7$

2,4,7-Triaminopteridine, T-60229
4,6,7-Triaminopteridine, T-60230

$C_6H_8F_4$

1,1,2,2-Tetrafluoro-3,4-dimethylcyclobutane, T-60047

$C_6H_8N_2$

▷2-(Aminomethyl)pyridine, A-60227
3-(Aminomethyl)pyridine, A-60228
4-(Aminomethyl)pyridine, A-60229
3,4,5,6-Tetrahydro-4,5-bis(methylene)pyridazine, T-60062

$C_6H_8N_2O$

2-Amino-3-hydroxy-6-methylpyridine, A-60192
5-Amino-3-hydroxy-2-methylpyridine, A-60193
2-Amino-4-methoxypyridine, *in* A-60253
2-Amino-6-methoxypyridine, *in* A-60252
3-Amino-2-methoxypyridine, *in* A-60249
3-Amino-4-methoxypyridine, *in* A-60254
4-Amino-2-methoxypyridine, *in* A-60250
5-Amino-2-methoxypyridine, *in* A-60251
3-Amino-1-methyl-2(1*H*)-pyridinone, *in* A-60249
3-Amino-1-methyl-4(1*H*)-pyridinone, *in* A-60254
5-Amino-1-methyl-2(1*H*)-pyridinone, *in* A-60251
6-Amino-1-methyl-2(1*H*)-pyridinone, *in* A-60252
7-Hydroxy-6,7-dihydro-5*H*-pyrrolo[1,2-*a*]imidazole, H-60118

$C_6H_8N_2O_2$

5-Ethyl-2,4(1*H*,3*H*)-pyrimidinedione, E-60054
1*H*-Imidazole-4-carboxylic acid; Et ester, *in* I-60004
4,5,6,7-Tetrahydroisoxazolo[4,5-*c*]pyridin-3-ol, T-60071
4,5,6,7-Tetrahydroisoxazolo[5,4-*c*]pyridin-3-ol, T-60072

$C_6H_8N_2S$

4,6-Dimethyl-2(1*H*)-pyrimidinethione, D-60450

$C_6H_8N_2S_2$

1,4-Diamino-2,3-benzenedithiol, D-60030

$C_6H_8N_2S_4$

1,4-Diamino-2,3,5,6-benzenetetrathiol, D-60031

$C_6H_8N_4O_4$

5-Nitrohistidine, N-60028

C_6H_8O

Bicyclo[3.1.0]hexan-2-one, B-60088
3-Cyclohexen-1-one, C-60211
2,3-Dimethyl-2-cyclobuten-1-one, D-60415
7-Oxabicyclo[4.1.0]hept-3-ene, O-60047
2-Oxatricyclo[4.1.0^{1,6}.0^{3,5}]heptane, O-60052

C_6H_8OS

2-Thiabicyclo[2.2.1]heptan-3-one, T-60186
1-(2-Thienyl)ethanol, T-60207
3-Thiopheneethanol, T-60211

$C_6H_8O_2$

Bicyclo[1.1.1]pentane-1-carboxylic acid, B-60101
3,3-Dimethyl-1-cyclopropene-1-carboxylic acid, D-60417
3,4-Dimethyl-2(5*H*)-furanone, D-60423
▷2-Hydroxy-3-methyl-2-cyclopenten-1-one, H-60174
4-Hydroxy-2-methyl-2-cyclopenten-1-one, H-60175

$C_6H_8O_3$

2,5-Bis(hydroxymethyl)furan, B-60160
3,4-Bis(hydroxymethyl)furan, B-60161
2-Cyclopropyl-2-oxoacetic acid; Me ester, *in* C-60231
5,5-Dimethyl-4-methylene-1,2-dioxolan-3-one, D-60428
4-Hydroxy-2,5-dimethyl-3(2*H*)-furanone, H-60120
5-Hydroxymethyl-4-methyl-2(5*H*)-furanone, H-60187

$C_6H_8O_3S_2$

2-Thiophenesulfonic acid; Et ester, *in* T-60213

$C_6H_8O_4$

Ethyl diformylacetate, *in* D-60190

C_6H_9FO

2-Fluorocyclohexanone, F-60021

C_6H_9IO

2-Iodocyclohexanone, I-60042

C_6H_9NO

3-Cyclohexen-1-one; Oxime, *in* C-60211
1,2-Dihydro-2,2-dimethyl-3*H*-pyrrol-3-one, D-60230
4-(2-Propenyl)-2-azetidinone, *in* A-60172

$C_6H_9NO_2$

3-(1-Aminocyclopropyl)-2-propenoic acid, A-60123
4-Amino-2,5-hexadienoic acid, A-60166
1,2,3,6-Tetrahydro-3-pyridinecarboxylic acid, T-60084
3,4,4-Trimethyl-5(4*H*)-isoxazolone, T-60367

$C_6H_9NO_3$

3-Aminodihydro-2(3*H*)-furanone; *N*-Ac, *in* A-60138
2-Amino-2-(hydroxymethyl)-4-pentynoic acid, A-60191
4-Oxo-2-piperidinecarboxylic acid, O-60089
1,2,3,6-Tetrahydro-3-hydroxy-2-pyridinecarboxylic acid, T-60068

$C_6H_9NO_4$

2-Amino-3-methylenepentanedioic acid, A-60219

$C_6H_9NS_2$

2(3*H*)-Thiazolethione; *S*-Isopropyl, *in* T-60200

$C_6H_9N_3O$

2-Amino-6-methyl-4(1*H*)-pyrimidinone; 2-*N*-Me, *in* A-60230

$C_6H_9N_3O_2$

α-Amino-1*H*-imidazole-1-propanoic acid, A-60204
α-Amino-1*H*-imidazole-2-propanoic acid, A-60205
2,4-Dimethoxy-6-methyl-1,3,5-triazine, *in* M-60132
4-Methoxy-1,6-dimethyl-1,3,5-triazin-2(1*H*)-one, *in* M-60132
1*H*-1,2,3-Triazole-4-carboxylic acid; 1-Me, Et ester, *in* T-60239
1*H*-1,2,3-Triazole-4-carboxylic acid; 2-Me, Et ester, *in* T-60239
1,3,6-Trimethyl-5-azauracil, *in* M-60132

$C_6H_9N_3O_3$

Trimethyl isocyanurate, *in* C-60179

C₆H₁₀Br₄
2,3-Bis(bromomethyl)-1,4-dibromobutane, B-60144

C₆H₁₀ClF₂NO₂
2-Amino-4-chloro-4,4-difluorobutanoic acid; Et ester, *in* A-60107

C₆H₁₀ClNO₃
2-Amino-5-chloro-6-hydroxy-4-hexenoic acid, A-60110

C₆H₁₀Cl₄
2,3-Bis(chloromethyl)-1,4-dichlorobutane, B-60148

C₆H₁₀N₂
2-Amino-5-hexenoic acid; Nitrile, *in* A-60169
3-Amino-2-hexenoic acid; Nitrile, *in* A-60170

C₆H₁₀N₂O₃
2-Amino-3-methylenepentanedioic acid; Amide, *in* A-60219

C₆H₁₀N₂O₄
2,5-Piperazinedicarboxylic acid, P-60157

C₆H₁₀N₄
1,2,3,5-Tetraaminobenzene, T-60012
1,2,4,5-Tetraaminobenzene, T-60013

C₆H₁₀N₄S
1-(4,5-Dihydro-1*H*-imidazol-2-yl)-2-imidazolidinethione, D-60246

C₆H₁₀O
▷2-Hexenal, H-60076

C₆H₁₀OS
4-Thioxo-2-hexanone, T-60215

C₆H₁₀O₂
2-Methoxycyclopentanone, *in* H-60114
4-Methoxy-3-penten-2-one, *in* P-60059
4-Oxohexanal, O-60073

C₆H₁₀O₂S
3-(Acetylthio)cyclobutanol, *in* M-60019
2,5-Dihydro-2,5-dimethylthiophene; 1,1-Dioxide, *in* D-60233
3-Mercaptocyclopentanecarboxylic acid, M-60021

C₆H₁₀O₃
Epiverrucarinolactone, *in* D-60353
6-(Hydroxymethyl)tetrahydro-2*H*-pyran-2-one, *in* D-60329
3-Oxopropanoic acid; Isopropyl ester, *in* O-60090
Tetrahydro-3-hydroxy-4-methyl-2*H*-pyran-2-one, *in* D-60353
3,4,5,6-Tetrahydro-4-hydroxy-6-methyl-2*H*-pyran-2-one, T-60067

C₆H₁₀O₄
4-Hydroxy-3,3-dimethyl-2-oxobutanoic acid, H-60124
1,3,7,9-Tetraoxaspiro[4,5]decane, T-60163
1,4,6,10-Tetraoxaspiro[4,5]decane, T-60164

C₆H₁₀S
2,3-Dihydro-2,2-dimethylthiophene, D-60232
2,5-Dihydro-2,5-dimethylthiophene, D-60233

C₆H₁₀S₄
2,2′-Bi-1,3-dithiolane, B-60103
Hexahydro-1,4-dithiino[2,3-*b*]-1,4-dithiin, H-60051

C₆H₁₁Br
5-Bromo-1-hexene, B-60263
6-Bromo-1-hexene, B-60264
5-Bromo-3-methyl-1-pentene, B-60299

C₆H₁₁BrO
5-Bromo-2-hexanane, B-60262
1-Bromo-4-methyl-2-pentanone, B-60298

C₆H₁₁Br₂NO₂
6-Amino-2,2-dibromohexanoic acid, A-60124

C₆H₁₁Cl
5-Chloro-3-methyl-1-pentene, C-60117

C₆H₁₁ClFNO₂
2-Amino-4-chloro-4-fluorobutanoic acid; Et ester, *in* A-60108

C₆H₁₁Cl₂NO₂
6-Amino-2,2-dichlorohexanoic acid, A-60126

C₆H₁₁F₂NO₂
2-Amino-3,3-difluorobutanoic acid; Et ester, *in* A-60133

C₆H₁₁IO
1-Iodo-3,3-dimethyl-2-butanone, I-60044

C₆H₁₁N
3,4-Dihydro-2,2-dimethyl-2*H*-pyrrole, D-60229

C₆H₁₁NO
3,4-Dihydro-2,2-dimethyl-2*H*-pyrrole; *N*-Oxide, *in* D-60229

C₆H₁₁NOS
3-Mercaptocyclopentanecarboxylic acid; Amide, *in* M-60021

C₆H₁₁NO₂
2-Amino-2-ethyl-3-butenoic acid, A-60150
1-Amino-2-ethylcyclopropanecarboxylic acid, A-60151
2-Amino-2-hexenoic acid, A-60167
2-Amino-3-hexenoic acid, A-60168
2-Amino-5-hexenoic acid, A-60169
3-Amino-2-hexenoic acid, A-60170
3-Amino-4-hexenoic acid, A-60171
3-Amino-5-hexenoic acid, A-60172
6-Amino-2-hexenoic acid, A-60173
2-Amino-3-methyl-3-butenoic acid; Me ester, *in* A-60218
2-Methyl-2-pyrrolidinecarboxylic acid, M-60126

C₆H₁₁NO₃
2-Amino-1-hydroxy-1-cyclobutaneacetic acid, A-60184
2-Amino-2-(hydroxymethyl)-4-pentenoic acid, A-60190

C₆H₁₁NO₄
2-Amino-3,3-dimethylbutanedioic acid, A-60147
2-Amino-4-methylpentanedioic acid, A-60223
Threonine; *N*-Ac, *in* T-60217

C₆H₁₁NS₂
5,6-Dihydro-2-(ethylthio)-4*H*-1,3-thiazine, *in* T-60094
Tetrahydro-2*H*-1,3-thiazine-2-thione; *N*-Et, *in* T-60094

C₆H₁₁N₃O₂
Dihydro-1,3,5-triazine-2,4(1*H*,3*H*)-dione; 1,3,5-Tri-Me, *in* D-60297

C₆H₁₂BrNO₂
6-Amino-2-bromohexanoic acid, A-60100

C₆H₁₂ClNO₂
6-Amino-2-chlorohexanoic acid, A-60109

C₆H₁₂I₂O₂
1,2-Bis(2-iodoethoxy)ethane, B-60163

C₆H₁₂NO₂⊕
4-Azoniaspiro[3.3]heptane-2,6-diol, A-60354

C₆H₁₂N₂
2-Dimethylamino-3,3-dimethylazirine, D-60402
5-Hexyne-1,4-diamine, H-60080

C₆H₁₂N₂O
3-Amino-2-hexenoic acid; Amide, *in* A-60170
Tetrahydro-2(1*H*)-pyrimidinone; 1,3-Di-Me, *in* T-60088

C₆H₁₂N₂O₃S
N-Alanylcysteine, A-60069

C₆H₁₂N₄
1,2,3,4,5,6,7,8-Octahydropyridazino[4,5-*d*]pyridazine, O-60018

C₆H₁₂N₄O₂
2,5-Piperazinedicarboxylic acid; Diamide, *in* P-60157

C₆H₁₂N₆
▷Benzenehexamine, B-60012

C₆H₁₂O
4-Hexen-3-ol, H-60077
▷4-Methyl-2-pentanone, M-60109
Tetrahydro-2-methylpyran, T-60075

C₆H₁₂OS
3,3,4,4-Tetramethyl-1,2-oxathietane, T-60149

C₆H₁₂O₂
β,β-Dimethyloxiraneethanol, D-60431

C₇H₅FO₂

2-Fluoro-3-hydroxybenzaldehyde, F-60035
2-Fluoro-5-hydroxybenzaldehyde, F-60036
4-Fluoro-3-hydroxybenzaldehyde, F-60037

C₇H₅FO₃

2-Fluoro-5-hydroxybenzoic acid, F-60038
2-Fluoro-6-hydroxybenzoic acid, F-60039
3-Fluoro-2-hydroxybenzoic acid, F-60040
3-Fluoro-4-hydroxybenzoic acid, F-60041
4-Fluoro-2-hydroxybenzoic acid, F-60042
4-Fluoro-3-hydroxybenzoic acid, F-60043
5-Fluoro-2-hydroxybenzoic acid, F-60044

C₇H₅FO₄

3-Fluoro-2,6-dihydroxybenzoic acid, F-60023
4-Fluoro-3,5-dihydroxybenzoic acid, F-60024
5-Fluoro-2,3-dihydroxybenzoic acid, F-60025

C₇H₅F₃O₃S

3-(Trifluoromethyl)benzenesulfonic acid, T-60283

C₇H₅F₃S

2-(Trifluoromethyl)benzenethiol, T-60284
3-(Trifluoromethyl)benzenethiol, T-60285
4-(Trifluoromethyl)benzenethiol, T-60286

C₇H₅IO₃

▷1-Hydroxy-1,2-benziodoxol-3(1H)-one, H-60102

C₇H₅IO₄

2,3-Dihydroxy-6-iodobenzoic acid, D-60330
2,4-Dihydroxy-3-iodobenzoic acid, D-60331
2,4-Dihydroxy-5-iodobenzoic acid, D-60332
2,5-Dihydroxy-4-iodobenzoic acid, D-60333
2,6-Dihydroxy-3-iodobenzoic acid, D-60334
3,4-Dihydroxy-5-iodobenzoic acid, D-60335
3,5-Dihydroxy-2-iodobenzoic acid, D-60336
3,5-Dihydroxy-4-iodobenzoic acid, D-60337
4,5-Dihydroxy-2-iodobenzoic acid, D-60338

C₇H₅NO

Benzonitrile N-oxide, B-60035
Pentafluoroisocyanatobenzene, P-60035

C₇H₅NO₂

4-Nitrosobenzaldehyde, N-60046

C₇H₅NO₃

4-Hydroxy-3-nitrosobenzaldehyde, H-60197

C₇H₅NO₆

4,5-Dihydroxy-2-nitrobenzoic acid, D-60360

C₇H₅NS₂

Benzenesulfenyl thiocyanate, B-60013

C₇H₅N₃OS

2,3-Dihydro-2-thioxopyrido[2,3-d]pyrimidin-4(1H)-one, D-60291
2,3-Dihydro-2-thioxopyrido[3,2-d]pyrimidin-4(1H)-one, D-60292
2,3-Dihydro-2-thioxopyrido[3,4-d]pyrimidin-4(1H)-one, D-60293
2,3-Dihydro-2-thioxo-4H-pyrido[1,2-a]-1,3,5-triazin-4-one, D-60294

C₇H₅N₃O₂

Pyrido[2,3-d]pyrimidine-2,4(1H,3H)-dione, P-60236
Pyrido[3,4-d]pyrimidine-2,4(1H,3H)-dione, P-60237

C₇H₅N₃O₂S

3-Amino-5-nitro-2,1-benzisothiazole, A-60232

C₇H₅N₃O₅

3,5-Dinitrobenzamide, in D-60458

C₇H₅N₃S

1,2,3-Benzotriazine-4(3H)-thione, B-60050
2-(1,3,4-Thiadiazol-2-yl)pyridine, T-60192
4-(1,3,4-Thiadiazol-2-yl)pyridine, T-60193

C₇H₅N₅O₈

▷2-Methyl-3,4,5,6-tetranitroaniline, M-60129
▷3-Methyl-2,4,5,6-tetranitroaniline, M-60130
▷4-Methyl-2,3,5,6-tetranitroaniline, M-60131

C₇H₆BrFO

3-Bromo-4-fluorobenzyl alcohol, B-60244
5-Bromo-2-fluorobenzyl alcohol, B-60245
2-Bromo-1-fluoro-3-methoxybenzene, in B-60247
2-Bromo-4-fluoro-1-methoxybenzene, in B-60248
4-Bromo-2-fluoro-1-methoxybenzene, in B-60251

C₇H₆Br₂

1,3-Dibromo-2-methylbenzene, D-60103

C₇H₆ClFO

1-Chloro-2-fluoro-4-methoxybenzene, in C-60080
1-Chloro-4-fluoro-2-methoxybenzene, in C-60074
2-Chloro-4-fluoro-1-methoxybenzene, in C-60073
4-Chloro-2-fluoro-1-methoxybenzene, in C-60079

C₇H₆ClNO₂

3-Chloro-4-hydroxybenzoic acid; Amide, in C-60091

C₇H₆ClNO₃

4-Chloro-2-methoxy-1-nitrobenzene, in C-60119
2-Chloro-3-methyl-4-nitrophenol, C-60115
3-Chloro-4-methyl-5-nitrophenol, C-60116

C₇H₆Cl₂S

1,2-Dichloro-4-(methylthio)benzene, in D-60123
1,4-Dichloro-2-(methylthio)benzene, in D-60121

C₇H₆FNO₂

4-Fluoro-2-hydroxybenzoic acid; Amide, in F-60042

C₇H₆FNO₃

4-Fluoro-2-methoxy-1-nitrobenzene, in F-60061

C₇H₆F₂S

1,3-Difluoro-2-(methylthio)benzene, in D-60174
1,3-Difluoro-5-(methylthio)benzene, in D-60176
1,4-Difluoro-2-(methylthio)benzene, in D-60173

C₇H₆F₃NO₂S

3-(Trifluoromethyl)benzenesulfonic acid; Amide, in T-60283

C₇H₆N₂

▷1H-Pyrrolo[3,2-b]pyridine, P-60249

C₇H₆N₂O

▷2-Aminobenzoxazole, A-60093

C₇H₆N₂O₄

▷2-Methyl-1,3-dinitrobenzene, M-60061

C₇H₆N₂O₅

2,6-Dinitrobenzyl alcohol, D-60459

C₇H₆N₄

4-Amino-1,2,3-benzotriazine, A-60092

C₇H₆N₄O

4-Amino-1,2,3-benzotriazine; 2-Oxide, in A-60092
4-Amino-1,2,3-benzotriazine; 3-Oxide, in A-60092
4-Hydroxylamino-1,2,3-benzotriazine, in A-60092

C₇H₆N₄O₂

2,4(1H,3H)-Pteridinedione; 1-Me, in P-60191
2,4(1H,3H)-Pteridinedione; 3-Me, in P-60191

C₇H₆N₄S

▷Tetrazole-5-thione; 1-Ph, in T-60174
Tetrazole-5-thione; S-Ph, in T-60174

C₇H₆O

2,4,6-Cycloheptatrien-1-one, C-60204
Dimethylenebicyclo[1.1.1]pentanone, D-60422
4-Methylene-2,5-cyclohexadien-1-one, M-60071
6-Methylene-2,4-cyclohexadien-1-one, M-60072

C₇H₆O₂S

3-(2-Thienyl)-2-propenoic acid, T-60208

C₇H₆O₃

2,7-Dihydroxy-2,4,6-cycloheptatrien-1-one, D-60314

C₇H₆O₄S

2-Carboxy-3-thiopheneacetic acid, C-60017
4-Carboxy-3-thiopheneacetic acid, C-60018

C₇H₆O₆

Pentahydroxybenzaldehyde, P-60042

2,3,5,6-Tetrahydroxybenzoic acid, T-60100

$C_7H_6O_7$
Pentahydroxybenzoic acid, P-60043

C_7H_6S
2,4,6-Cycloheptatriene-1-thione, C-60203

$C_7H_7BrO_3$
4-Bromo-2,6-dimethoxyphenol, *in* B-60201

C_7H_7BrSe
Bromomethyl phenyl selenide, B-60300

$C_7H_7Br_2N$
▷ 2,6-Bis(bromomethyl)pyridine, B-60146
3,5-Bis(bromomethyl)pyridine, B-60147

C_7H_7FO
2-Fluoro-3-methylphenol, F-60052
2-Fluoro-4-methylphenol, F-60053
2-Fluoro-5-methylphenol, F-60054
2-Fluoro-6-methylphenol, F-60055
3-Fluoro-2-methylphenol, F-60056
3-Fluoro-4-methylphenol, F-60057
4-Fluoro-2-methylphenol, F-60058
4-Fluoro-3-methylphenol, F-60059
5-Fluoro-2-methylphenol, F-60060

C_7H_7IO
4-Iodo-3-methylphenol, I-60050

C_7H_7NO
2,3-Dihydro-1*H*-pyrrolizin-1-one, D-60277

$C_7H_7NO_2S$
(Phenylthio)nitromethane, P-60137
2(1*H*)-Pyridinethione; 1-Acetoxy, *in* P-60224

$C_7H_7NO_3$
3-Hydroxy-2(1*H*)-pyridinone; 3-Ac, *in* H-60223
5-Hydroxy-2(1*H*)-pyridinone; 5-Ac, *in* H-60225

$C_7H_7NO_3S$
4-Carboxy-3-thiopheneacetic acid; Acetamide, *in* C-60018

$C_7H_7NO_4S$
(Nitromethyl)sulfonylbenzene, *in* P-60137

$C_7H_7N_3$
5*H*-Pyrrolo[2,3-*b*]pyrazine; 5-Me, *in* P-60248

$C_7H_7N_3O$
1,3-Dihydro-2*H*-imidazo[4,5-*c*]pyridin-2-one; 1-Me, *in* D-60250

$C_7H_7N_5O$
2-Amino-1-methyl-4(1*H*)-pteridinone, *in* A-60242
2-(Methylamino)-4(1*H*)-pteridinone, *in* A-60242

C_7H_8
4-Methylenebicyclo[3.1.0]hex-2-ene, M-60069

$C_7H_8BrF_3NO_2$
4-Bromo-1-nitro-2-(trifluoromethyl)benzene, B-60309

C_7H_8BrNO
2-Bromo-3-ethoxypyridine, *in* B-60275
2-Bromo-3-methoxy-6-methylpyridine, *in* B-60267

$C_7H_8N_2$
N-Phenylformamidine, P-60101

$C_7H_8N_2O$
4-Oxoheptanedinitrile, *in* O-60071

$C_7H_8N_2O_2$
1-Amino-2(1*H*)-pyridinone; *N*-Ac, *in* A-60248
2,4-Diaminobenzoic acid, D-60032
1,2-Diamino-4,5-methylenedioxybenzene, D-60038
1,3-Diisocyanatocyclopentane, D-60390

$C_7H_8N_2O_3$
3-Amino-4-methyl-5-nitrophenol, A-60222
3-Amino-5-nitrobenzyl alcohol, A-60233

$C_7H_8N_2O_3S$
Aminoiminomethanesulfonic acid; *N*-Ph, *in* A-60210

$C_7H_8N_2O_5$
3-Amino-2(1*H*)-pyridinethione; N^3-Ac, *in* A-60247

$C_7H_8N_4$
1′-Methyl-1,5′-bi-1*H*-pyrazole, *in* B-60119

$C_7H_8N_4O$
1,3-Dihydro-1,3-dimethyl-2*H*-imidazo[4,5-*b*]pyrazin-2-one, *in* I-60006
1,4-Dihydro-1,4-dimethyl-2*H*-imidazo[4,5-*b*]pyrazin-2-one, *in* I-60006
2-Methoxy-4-methyl-4*H*-imidazo[4,5-*b*]pyrazine, *in* I-60006

$C_7H_8N_4O_2$
1*H*-Pyrazolo[3,4-*d*]pyrimidine-4,6(5*H*,7*H*)-dione; 1,5-Di-Me, *in* P-60202
1*H*-Pyrazolo[3,4-*d*]pyrimidine-4,6(5*H*,7*H*)-dione; 5,7-Di-Me, *in* P-60202

$C_7H_8N_4S$
1,3-Dihydro-1,3-dimethyl-2*H*-imidazo[4,5-*b*]pyrazine-2-thione, *in* D-60247
1,7-Dihydro-6*H*-purine-6-thione; 1,9-Di-Me, *in* D-60270
1,7-Dihydro-6*H*-purine-6-thione; 3,7-Di-Me, *in* D-60270
1,7-Dihydro-6*H*-purine-6-thione; 3,9-Di-Me, *in* D-60270
1,7-Dihydro-6*H*-purine-6-thione; *S*,3*N*-Di-Me, *in* D-60270
1,7-Dihydro-6*H*-purine-6-thione; *S*,7*N*-Di-Me, *in* D-60270
1,7-Dihydro-6*H*-purine-6-thione; *S*,9*N*-Di-Me, *in* D-60270
1-Methyl-2-methylthio-1*H*-imidazo[4,5-*b*]pyrazine, *in* D-60247
4-Methyl-2-methylthio-4*H*-imidazo[4,5-*b*]pyrazine, *in* D-60247

C_7H_8OS
2,5-Dimethyl-3-thiophenecarboxaldehyde, D-60452
2-Thiabicyclo[2.2.2]oct-5-en-3-one, T-60188

$C_7H_8O_2$
3-Oxo-1-cyclohexene-1-carboxaldehyde, O-60060
α-Oxo-3-cyclopentene-1-acetaldehyde, O-60061

$C_7H_8O_4$
5-Hydroxymethyl-2(5*H*)-furanone; Ac, *in* H-60182
2-Vinyl-1,1-cyclopropanedicarboxylic acid, V-60011

$C_7H_8O_5$
4-Oxo-2-heptenedioic acid, O-60072
1,4,8-Trioxaspiro[4.5]decan-8-one, *in* O-60084

$C_7H_9BrO_2$
7-Bromo-5-heptynoic acid, B-60261

C_7H_9ClO
2-Chloro-2-cyclohepten-1-one, C-60049

C_7H_9N
2-Propyn-1-amine; *N*-Di-Et, *in* P-60184
2-Vinyl-1*H*-pyrrole; *N*-Me, *in* V-60013

C_7H_9NO
7-Azabicyclo[4.2.0]oct-3-en-8-one, A-60331

$C_7H_9NO_2$
3-Amino-5-hydroxybenzyl alcohol, A-60182
2,3-Dimethoxypyridine, *in* H-60223
2,4-Dimethoxypyridine, *in* H-60224
2,6-Dimethoxypyridine, *in* H-60226
3,4-Dimethoxypyridine, *in* H-60227
3-Ethoxy-4(1*H*)-pyridinone, *in* H-60227
6-Hydroxy-2(1*H*)-pyridinone; Me ether, *N*-Me, *in* H-60226
2-Methoxy-1-methyl-4(1*H*)-pyridinone, *in* H-60224
3-Methoxy-1-methyl-4(1*H*)-pyridinone, *in* H-60227
4-Methoxy-1-methyl-2(1*H*)-pyridinone, *in* H-60224
5-Methyl-1*H*-pyrrole-2-carboxylic acid; Me ester, *in* M-60124
5-Methyl-1*H*-pyrrole-3-carboxylic acid; Me ester, *in* M-60125

$C_7H_9NO_3$
3-Hydroxy-4(1*H*)-pyridinone; Di-Me ether, 1-oxide, *in* H-60227
4-Hydroxy-2(1*H*)-pyridinone; Di-Me ether, 1-oxide, *in* H-60224

$C_7H_9NO_4S_2$
2,3-Bis(methylsulfonyl)pyridine, *in* P-60218
2,5-Bis(methylsulfonyl)pyridine, *in* P-60220
3,5-Bis(methylsulfonyl)pyridine, *in* P-60222

C₇H₉NS
4-Amino-5,6-dihydro-4*H*-cyclopenta[*b*]thiophene, A-60137
2-(2-Pyridyl)ethanethiol, P-60239
2-(4-Pyridyl)ethanethiol, P-60240

C₇H₉NS₂
2,3-Bis(methylthio)pyridine, *in* P-60218
2,4-Bis(methylthio)pyridine, *in* P-60219
2,5-Bis(methylthio)pyridine, *in* P-60220
3,5-Bis(methylthio)pyridine, *in* P-60222
1-Methyl-4-(methylthio)-2(1*H*)-pyridinethione, *in* P-60219

C₇H₉N₅
3,7-Dimethyladenine, D-60400
2-(Dimethylamino)purine, *in* A-60243
8-(Dimethylamino)purine, *in* A-60244

C₇H₉O₃
4-Hydroxy-2-cyclopenten-1-one; Ac, *in* H-60116

C₇H₁₀
Bicyclo[3.1.1]hept-2-ene, B-60087
1,2-Bis(methylene)cyclopentane, B-60169
1,3-Bis(methylene)cyclopentane, B-60170
Tricyclo[4.1.0.0¹,³]heptane, T-60262

C₇H₁₀N₂O₂
4,5,6,7-Tetrahydro-3-methoxyisoxazolo[4,5-*c*]pyridine, *in* T-60071

C₇H₁₀N₂O₄
5-Methyl-3,6-dioxo-2-piperazineacetic acid, M-60064

C₇H₁₀N₂S
4,6-Dimethyl-2(1*H*)-pyrimidinethione; 1-Me, *in* D-60450

C₇H₁₀O
1-Acetylcyclopentene, A-60028
2,4-Heptadienal, H-60020
7-Oxatricyclo[4.1.1.0²,⁵]octane, O-60053
8-Oxatricyclo[3.3.0.0²,⁷]octane, O-60054

C₇H₁₀OS
2-Thiabicyclo[2.2.2]octan-3-one, T-60187

C₇H₁₀O₂
3,3-Dimethyl-1-cyclopropene-1-carboxylic acid; Me ester, *in* D-60417
2,4-Heptadienoic acid, H-60021
2-Methoxy-3-methyl-2-cyclopenten-1-one, *in* H-60174

C₇H₁₀O₃
5-Ethyl-3-hydroxy-4-methyl-2(5*H*)-furanone, E-60051
2-Hydroxycyclopentanone; Ac, *in* H-60114
7-Hydroxy-5-heptynoic acid, H-60150
4-Methoxy-2,5-dimethyl-3(2*H*)-furanone, *in* H-60120
2,4-Pentanedione; Ac, *in* P-60059
2,4,10-Trioxatricyclo[3.3.1.1³,⁷]decane, T-60377

C₇H₁₀O₄
Ethyl 3-acetoxyacrylate, *in* O-60090

C₇H₁₀O₅
4-Oxoheptanedioic acid, O-60071
3-Oxopentanedioic acid; Di-Me ester, *in* O-60084

C₇H₁₀O₆
1,3-Dioxolane-2,2-diacetic acid, *in* O-60084

C₇H₁₀S
1-Thia-2-cyclooctyne, T-60189
1-Thia-3-cyclooctyne, T-60190

C₇H₁₁ClO
3,3-Dimethyl-4-pentenoic acid; Chloride, *in* D-60437

C₇H₁₁NO
4,4-Dimethylglutaraldehydonitrile, *in* D-60434
Hexahydro-1*H*-pyrrolizin-1-one, H-60057
Hexahydro-3*H*-pyrrolizin-3-one, H-60058
Tetrahydro-1*H*-pyrrolizin-2(3*H*)-one, T-60091

C₇H₁₁NO₂
1,2,3,6-Tetrahydro-3-pyridinecarboxylic acid; *N*-Me, *in* T-60084

C₇H₁₁NO₄
2-Amino-3-heptenedioic acid, A-60164

2-Amino-5-heptenedioic acid, A-60165

C₇H₁₁NO₅
4-Oxoheptanedioic acid; Oxime, *in* O-60071

C₇H₁₁NS
4-Mercaptocyclohexanecarbonitrile, *in* M-60020

C₇H₁₂
4,4-Dimethyl-2-pentyne, D-60438

C₇H₁₂Br₂
1,1-Dibromocycloheptane, D-60091

C₇H₁₂Br₂O
2,4-Dibromo-2,4-dimethyl-3-pentanone, D-60099

C₇H₁₂N₂O
3,5-Dihydro-3,3,5,5-tetramethyl-4*H*-pyrazol-4-one, D-60286
Hexahydro-1*H*-pyrrolizin-1-one; Oxime, *in* H-60057

C₇H₁₂N₂OS
3,5-Dihydro-3,3,5,5-tetramethyl-4*H*-pyrazole-4-thione; *S*-Oxide, *in* D-60285

C₇H₁₂N₂O₄
2,6-Diamino-3-heptene-1,2-dioic acid, D-60036

C₇H₁₂N₂S
3,5-Dihydro-3,3,5,5-tetramethyl-4*H*-pyrazole-4-thione, D-60285

C₇H₁₂N₄O
Caffeidine, C-60003

C₇H₁₂O
1-Methoxycyclohexene, M-60037
3,6,7,8-Tetrahydro-2*H*-oxocin, T-60076

C₇H₁₂OS
2-Hexyl-5-methyl-3(2*H*)furanone; *S*-Oxide (*exo*-), *in* H-60079
2,2,4,4-Tetramethyl-3-thietanone, T-60154
3,3,4,4-Tetramethyl-2-thietanone, T-60155

C₇H₁₂O₂
3,3-Dimethyl-4-pentenoic acid, D-60437
4-Oxoheptanal, O-60070

C₇H₁₂O₂S
3-(Ethylthio)-2-propenoic acid; Et ester, *in* E-60055
4-Mercaptocyclohexanecarboxylic acid, M-60020

C₇H₁₂O₃
▷Botryodiplodin, B-60194
3,3-Dimethyl-4-oxopentanoic acid, D-60433
4,4-Dimethyl-5-oxopentanoic acid, D-60434
3-Oxopropanoic acid; *tert*-Butyl ester, *in* O-60090

C₇H₁₂O₃S
2,2,4,4-Tetramethyl-3-thietanone; 1,1-Dioxide, *in* T-60154

C₇H₁₂O₅
1,3-Diacetylglycerol, D-60023
2-Hydroxy-2-isopropylbutanedioic acid, H-60165
2-Hydroxy-3-isopropylbutanedioic acid, H-60166

C₇H₁₂S
Hexahydro-2*H*-cyclopenta[*b*]thiophene, H-60049

C₇H₁₃Br
7-Bromo-1-heptene, B-60260

C₇H₁₃FO
2-Fluoroheptanal, F-60034

C₇H₁₃IO₂
5-Iodo-2-pentanol; Ac, *in* I-60056
2-(3-Iodopropyl)-2-methyl-1,3-dioxolane, *in* I-60057

C₇H₁₃NO
4,4-Dimethyl-5-oxopentanoic acid; Oxime, *in* D-60434

C₇H₁₃NOS
4-Mercaptocyclohexanecarboxylic acid; Amide, *in* M-60020

C₇H₁₃NOS₂
1-Isothiocyanato-5-(methylsulfinyl)pentene, *in* I-60140

C$_8$H$_5$F$_3$N$_2$
2-(Trifluoromethyl)-1H-benzimidazole, T-60287

C$_8$H$_5$F$_5$O
1-(Pentafluorophenyl)ethanol, P-60040

C$_8$H$_5$NO
3H-Indol-3-one, I-60035

C$_8$H$_5$NO$_2$
Furo[2,3-b]pyridine-2-carboxaldehyde, F-60093
Furo[2,3-c]pyridine-2-carboxaldehyde, F-60096
Furo[2,3-b]pyridine-3-carboxaldehyde, F-60094
Furo[3,2-b]pyridine-2-carboxaldehyde, F-60095
Furo[3,2-c]pyridine-2-carboxaldehyde, F-60097
Isatogen, in I-60035

C$_8$H$_5$NO$_3$
Furo[2,3-b]pyridine-2-carboxylic acid, F-60098
Furo[2,3-c]pyridine-2-carboxylic acid, F-60102
Furo[2,3-b]pyridine-3-carboxylic acid, F-60099
Furo[2,3-c]pyridine-3-carboxylic acid, F-60103
Furo[3,2-b]pyridine-2-carboxylic acid, F-60100
Furo[3,2-c]pyridine-2-carboxylic acid, F-60104
Furo[3,2-b]pyridine-3-carboxylic acid, F-60101
Furo[3,2-c]pyridine-3-carboxylic acid, F-60105

C$_8$H$_5$NS$_3$
3-Phenyl-1,4,2-dithiazole-5-thione, P-60088

C$_8$H$_5$N$_2$O
3-Hydroxyiminoindole, in I-60035

C$_8$H$_5$N$_3$O$_9$
3-Hydroxy-5-methyl-2,4,6-trinitrobenzoic acid, H-60193

C$_8$H$_5$N$_5$O$_9$
2,3,4,6-Tetranitroacetanilide, in T-60159

C$_8$H$_6$BrFO$_2$
2-Bromo-4-fluorobenzoic acid; Me ester, in B-60240

C$_8$H$_6$BrNO$_4$
4-(Bromomethyl)-2-nitrobenzoic acid, B-60297

C$_8$H$_6$Cl$_2$O$_3$
2,3-Dichloro-4-methoxybenzoic acid, in D-60133

C$_8$H$_6$FN$_3$O$_5$
6-Fluoro-3,4-dinitroaniline; N-Ac, in F-60032

C$_8$H$_6$F$_3$NO$_3$
1-Methoxy-2-nitro-4-(trifluoromethyl)benzene, in N-60048
1-Methoxy-3-nitro-5-(trifluoromethyl)benzene, in N-60052
1-Methoxy-4-nitro-2-(trifluoromethyl)benzene, in N-60053
4-Methoxy-1-nitro-2-(trifluoromethyl)benzene, in N-60054

C$_8$H$_6$N$_2$OS
2,3-Dihydro-2-thioxo-4(1H)-quinazolinone, D-60296
2,3-Dihydro-4-thioxo-2(1H)-quinazolinone, D-60295

C$_8$H$_6$N$_2$OSe
3,4-Dihydro-4-selenoxo-2(1H)-quinazolinone, D-60280
2,3-Dihydro-2-selenoxo-4(1H)-quinazolinone, 9Ci, D-60281

C$_8$H$_6$N$_2$O$_2$
1,4-Dihydro-2,3-quinoxalinedione, D-60279
Furo[2,3-b]pyridine-2-carboxaldehyde; Oxime, in F-60093
Furo[2,3-c]pyridine-2-carboxaldehyde; Oxime, in F-60096
Furo[3,2-b]pyridine-2-carboxaldehyde; Oxime, in F-60095
Furo[3,2-c]pyridine-2-carboxaldehyde; Oxime, in F-60097
Furo[2,3-b]pyridine-3-carboxylic acid; Amide, in F-60099
Furo[2,3-c]pyridine-3-carboxylic acid; Amide, in F-60103
Furo[3,2-b]pyridine-3-carboxylic acid; Amide, in F-60101
Furo[3,2-c]pyridine-3-carboxylic acid; Amide, in F-60105
3H-Indol-3-one; 1-Oxide, oxime, in I-60035
5,8-Quinazolinediol, Q-60002
5,8-Quinoxalinediol, Q-60008
6,7-Quinoxalinediol, Q-60009

C$_8$H$_6$N$_2$O$_4$
DDED, in E-60049

C$_8$H$_6$N$_2$O$_6$
2,6-Dinitrobenzoic acid; Me ester, in D-60457
3,5-Dinitrobenzoic acid; Me ester, in D-60458

C$_8$H$_6$OS
Benzo[b]thiophen-3(2H)-one, B-60048

C$_8$H$_6$O$_3$
4H-1,3-Benzodioxin-2-one, B-60028

C$_8$H$_6$O$_3$S
Benzo[b]thiophen-3(2H)-one; 1,1-Dioxide, in B-60048

C$_8$H$_6$O$_4$
5,7-Dihydroxy-1(3H)-isobenzofuranone, D-60339

C$_8$H$_6$O$_5$
2-Formyl-3,4-dihydroxybenzoic acid, F-60070
2-Formyl-3,5-dihydroxybenzoic acid, F-60071
4-Formyl-2,5-dihydroxybenzoic acid, F-60072
6-Formyl-2,3-dihydroxybenzoic acid, F-60073
2-Hydroxy-1,3-benzenedicarboxylic acid, H-60101
2-Hydroxy-3,4-methylenedioxybenzoic acid, H-60177
3-Hydroxy-4,5-methylenedioxybenzoic acid, H-60178
4-Hydroxy-2,3-methylenedioxybenzoic acid, H-60179
6-Hydroxy-2,3-methylenedioxybenzoic acid, H-60180
6-Hydroxy-3,4-methylenedioxybenzoic acid, H-60181

C$_8$H$_6$O$_6$
4,5-Dihydroxy-1,3-benzenedicarboxylic acid, D-60307

C$_8$H$_6$S
Cyclopenta[b]thiapyran, C-60224

C$_8$H$_6$S$_2$
2-(2,4-Cyclopentadienylidene)-1,3-dithiole, C-60221

C$_8$H$_7$BrO
2-Bromo-3-methylbenzaldehyde, B-60292

C$_8$H$_7$BrO$_2$
5′-Bromo-2′-hydroxyacetophenone, B-60265
2-Bromo-3-methylbenzoic acid, B-60293
3-Bromo-4-methylbenzoic acid, B-60294

C$_8$H$_7$BrO$_4$
4-Acetoxy-2-bromo-5,6-epoxy-2-cyclohexen-1-one, in B-60239

C$_8$H$_7$Br$_2$NO$_2$
1,3-Bis(bromomethyl)-5-nitrobenzene, B-60145

C$_8$H$_7$Br$_2$NO$_3$
3,5-Dibromo-1,6-dihydroxy-4-oxo-2-cyclohexene-1-acetonitrile, D-60098

C$_8$H$_7$ClO$_2$
▷Benzyloxycarbonyl chloride, B-60071
3-Chloro-4-methoxybenzaldehyde, in C-60089
4-Chloro-3-methoxybenzaldehyde, in C-60090

C$_8$H$_7$ClO$_3$
3-Chloro-2,6-dihydroxy-4-methylbenzaldehyde, C-60059
3-Chloro-4,6-dihydroxy-2-methylbenzaldehyde, C-60060
3-Chloro-4-hydroxybenzoic acid; Me ester, in C-60091
3-Chloro-6-hydroxy-2-methoxybenzaldehyde, in C-60058
2-Chloro-3-hydroxy-5-methylbenzoic acid, C-60095
3-Chloro-4-hydroxy-5-methylbenzoic acid, C-60096
5-Chloro-4-hydroxy-2-methylbenzoic acid, C-60097
3-Chloro-4-methoxybenzoic acid, in C-60091
3-Chloro-5-methoxybenzoic acid, in C-60092
4-Chloro-3-methoxybenzoic acid, in C-60093

C$_8$H$_7$FO
2-Fluoro-2-phenylacetaldehyde, F-60062

C$_8$H$_7$FO$_3$
3-Fluoro-2-hydroxybenzoic acid; Me ester, in F-60040
3-Fluoro-4-hydroxybenzoic acid; Me ester, in F-60041
4-Fluoro-2-hydroxybenzoic acid; Me ester, in F-60042
5-Fluoro-2-hydroxybenzoic acid; Me ester, in F-60044
2-Fluoro-5-methoxybenzoic acid, in F-60038
3-Fluoro-4-methoxybenzoic acid, in F-60041
5-Fluoro-2-methoxybenzoic acid, in F-60044

C$_8$H$_7$F$_3$S
1-(Methylthio)-4-(trifluoromethyl)benzene, in T-60286

C$_8$H$_7$IO
2-Iodoacetophenone, I-60039

C$_8$H$_7$IO$_3$
2-Iodosophenylacetic acid, I-60063

1-Methoxy-1,2-benziodoxol-3(1*H*)one, *in* H-60102

C_8H_7N
Indolizine, I-60033

C_8H_7NO
6-Hydroxyindole, H-60158
2-Methylfuro[2,3-*b*]pyridine, M-60084
3-Methylfuro[2,3-*b*]pyridine, M-60085

$C_8H_7NO_3$
Amino-1,4-benzoquinone; *N*-Ac, *in* A-60089

$C_8H_7NO_4$
3-Amino-7-oxabicyclo[4.1.0]hept-3-ene-2,5-dione, *in* A-60194

$C_8H_7NO_5$
2′,3′-Dihydroxy-6′-nitroacetophenone, D-60357
3′,6′-Dihydroxy-2′-nitroacetophenone, D-60358
4′,5′-Dihydroxy-2′-nitroacetophenone, D-60359

$C_8H_7NO_6$
4-Hydroxy-5-methoxy-2-nitrobenzoic acid, *in* D-60360
5-Hydroxy-4-methoxy-2-nitrobenzoic acid, *in* D-60360

$C_8H_7N_3O$
2′-Azidoacetophenone, A-60334
3′-Azidoacetophenone, A-60335
4′-Azidoacetophenone, A-60336
Furo[2,3-*b*]pyridine-3-carboxaldehyde; Hydrazone, *in* F-60094

$C_8H_7N_3S$
1,2,3-Benzotriazine-4(3*H*)-thione; 3-*N*-Me, *in* B-60050
1,2,3-Benzotriazine-4(3*H*)-thione; *S*-Me, *in* B-60050
4-Mercapto-2-methyl-1,2,3-benzotriazinium hydroxide inner salt, *in* B-60050

$C_8H_7N_5O_2$
2-Amino-4(1*H*)-pteridinone; N^2-Ac, *in* A-60242

$C_8H_7N_5O_8$
2,3,4,6-Tetranitroaniline; *N*-Di-Me, *in* T-60159

C_8H_8
4-Octene-1,7-diyne, O-60028

$C_8H_8Br_2O_4$
4-Acetoxy-2,6-dibromo-5-hydroxy-2-cyclohexen-1-one, *in* D-60097
2,6-Dibromo-4,5-dihydroxy-2-cyclohexen-1-one; 4-Ac, *in* D-60097

C_8H_8ClN
4-Chloro-2,3-dihydro-1*H*-indole, C-60055

$C_8H_8ClNO_3$
4-Chloro-2-ethoxy-1-nitrobenzene, *in* C-60119
1-Chloro-5-methoxy-2-methyl-3-nitrobenzene, *in* C-60116
2-Chloro-1-methoxy-3-methyl-4-nitrobenzene, *in* C-60115

C_8H_8INO
2-Iodoacetophenone; Oxime (*Z*-), *in* I-60039

$C_8H_8INO_2$
3-Hydroxy-2-iodo-6-methylpyridine; *O*-Ac, *in* H-60159

$C_8H_8N_2O$
▷1-(Diazomethyl)-4-methoxybenzene, D-60064
3,4-Dihydro-2(1*H*)-quinazolinone, D-60278

$C_8H_8N_2O_2S$
Benzo[*b*]thiophen-3(2*H*)-one; Hydrazone, 1,1-dioxide, *in* B-60048

$C_8H_8N_4$
2,4-Diaminoquinazoline, D-60048
▷1-Hydrazinophthalazine, H-60092

$C_8H_8N_4O$
2-Tetrazolin-5-one; 1-Benzyl, *in* T-60175

$C_8H_8N_4O_2$
2,4(1*H*,3*H*)-Pteridinedione; 1,3-Di-Me, *in* P-60191

$C_8H_8N_4S$
Tetrazole-5-thione; 1-Me, 4-Ph, *in* T-60174
Tetrazole-5-thione; 1-*N*-Ph, *S*-Me, *in* T-60174

C_8H_8O
▷Acetophenone, A-60017

Bicyclo[2.2.1]hepta-2,5-diene-2-carboxaldehyde, B-60083
Bicyclo[3.3.0]oct-1(2)-en-3-one; Dimer, *in* B-60098
4,7-Dihydroisobenzofuran, D-60253
▷(4-Hydroxyphenyl)ethylene, H-60211
1-Phenylethenol, P-60092

C_8H_8OS
4′-Mercaptoacetophenone, M-60017

$C_8H_8O_2$
Bicyclo[2.2.1]hepta-2,5-diene-2-carboxylic acid, B-60084
Bicyclo[4.2.0]oct-7-ene-2,5-dione, B-60097
4-(2-Furanyl)-3-buten-2-one, F-60086
2-(Hydroxymethyl)benzaldehyde, H-60171
2-Hydroxyphenylacetaldehyde, H-60207
3-Hydroxyphenylacetaldehyde, H-60208
4-Hydroxyphenylacetaldehyde, H-60209
2-Hydroxy-2-phenylacetaldehyde, H-60210

$C_8H_8O_2S$
2-Mercapto-2-phenylacetic acid, M-60024

$C_8H_8O_3$
2,4-Dihydroxyphenylacetaldehyde, D-60369
2,5-Dihydroxyphenylacetaldehyde, D-60370
3,4-Dihydroxyphenylacetaldehyde, D-60371

$C_8H_8O_4$
3,6-Dimethoxy-1,2-benzoquinone, *in* D-60309
▷3′,4′,5′-Trihydroxyacetophenone, T-60315

$C_8H_8O_5$
2,3,4-Trihydroxyphenylacetic acid, T-60340
2,3,5-Trihydroxyphenylacetic acid, T-60341
2,4,5-Trihydroxyphenylacetic acid, T-60342
3,4,5-Trihydroxyphenylacetic acid, T-60343

$C_8H_8S_2$
2-(2,4-Cyclopentadien-1-ylidene)-1,3-dithiolane, *in* C-60221

$C_8H_9BrO_3$
3-Bromo-2,6-dimethoxyphenol, *in* B-60200

C_8H_9ClO
2-Chloro-2-phenylethanol, C-60128

C_8H_9F
2-Fluoro-1,3-dimethylbenzene, F-60026

C_8H_9FO
1-Fluoro-2-methoxy-4-methylbenzene, *in* F-60054
1-Fluoro-3-methoxy-6-methylbenzene, *in* F-60057
1-Fluoro-4-methoxy-3-methylbenzene, *in* F-60059
2-Fluoro-1-methoxy-4-methylbenzene, *in* F-60053

$C_8H_9FO_2$
4-Fluoro-1,2-dimethoxybenzene, *in* F-60020

C_8H_9N
N-Benzylidenemethylamine, B-60070
6,7-Dihydro-5*H*-1-pyrindine, D-60275
6,7-Dihydro-5*H*-2-pyrindine, D-60276

C_8H_9NO
Acetophenone; (*E*)-Oxime, *in* A-60017
2,3-Dihydro-5(1*H*)-indolizinone, D-60251
6,7-Dihydro-5*H*-1-pyrindine; *N*-Oxide, *in* D-60275
N-Methylbenzenecarboximidic acid, M-60045
1-(2-Pyridinyl)-2-propanone, P-60229
1-(3-Pyridinyl)-2-propanone, P-60230
1-(4-Pyridinyl)-2-propanone, P-60231

$C_8H_9NO_2$
4-Hydroxyphenylacetaldoxime, *in* H-60209

$C_8H_9NO_3S$
4-Carboxy-3-thiopheneacetic acid; *N*-Methylacetamide, *in* C-60018

$C_8H_9NO_4$
Antibiotic MT 35214, *in* A-60194
N-(5-Hydroxy-2-oxo-7-oxobicyclo[4.1.0]hept-3-en-3-yl)-acetamide, *in* A-60194

$C_8H_9N_5O_2$
N-(2,9-Dihydro-1-methyl-2-oxo-1*H*-purin-6-yl)acetamide, *in* A-60144

C$_8$H$_{10}$

Bicyclo[4.1.1]octa-2,4-diene, B-60090
Tricyclo[3.2.1.0^{2,7}]oct-3-ene, T-60267
Tricyclo[5.1.0.0^{2,8}]oct-3-ene, T-60268
Tricyclo[5.1.0.0^{2,8}]oct-4-ene, T-60269

C$_8$H$_{10}$Br$_2$O$_4$

4-Acetoxy-2,6-dibromo-1,5-dihydroxy-2-cyclohexen-1-one, *in* D-60097

C$_8$H$_{10}$N$_2$O

Acetamidoxime; *N*-Ph, *in* A-60016
1-(3-Pyridinyl)-2-propanone; Oxime, *in* P-60230

C$_8$H$_{10}$N$_2$O$_2$

3-Amino-4(1*H*)-pyridinone; 1-Me, *N*^{3}-Ac, *in* A-60254

C$_8$H$_{10}$N$_2$O$_3$

5-Ethyl-2,4(1*H*,3*H*)-pyrimidinedione; 1-Ac, *in* E-60054
5-Methoxy-2-methyl-3-nitroaniline, *in* A-60222

C$_8$H$_{10}$N$_2$O$_3$S

Aminoiminomethanesulfonic acid; *N*-Benzyl, *in* A-60210

C$_8$H$_{10}$N$_4$O$_2$

1*H*-Pyrazolo[3,4-*d*]pyrimidine-4,6(5*H*,7*H*)-dione; 1,5,7-Tri-Me, *in* P-60202
1*H*-Pyrazolo[3,4-*d*]pyrimidine-4,6(5*H*,7*H*)-dione; 2,5,7-Tri-Me, *in* P-60202

C$_8$H$_{10}$N$_4$O$_5$

5-Nitrohistidine; *N*$^\alpha$-Ac, *in* N-60028

C$_8$H$_{10}$O

Bicyclo[3.3.0]oct-6-en-2-one, B-60099
Bicyclo[3.3.0]oct-1(2)-en-3-one, B-60098

C$_8$H$_{10}$O$_2$

Bicyclo[4.2.0]octane-2,5-dione, B-60093
4,4-Dimethyl-3-oxo-1-cyclopentene-1-carboxaldehyde, D-60432
3-Ethynyl-3-methyl-4-pentenoic acid, E-60059
5-Vinyl-1-cyclopentenecarboxylic acid, V-60010

C$_8$H$_{10}$O$_3$

Halleridone, H-60004

C$_8$H$_{10}$O$_4$

3,6-Dimethoxy-1,2-benzenediol, *in* B-60014
4-Hydroxy-2,5-dimethyl-3(2*H*)-furanone; Ac, *in* H-60120

C$_8$H$_{10}$O$_5$

5-Acetyl-2,2-dimethyl-1,3-dioxane-4,6-dione, A-60029

C$_8$H$_{10}$S$_2$

2,5-Dimethyl-1,4-benzenedithiol, D-60406

C$_8$H$_{11}$BrO$_2$

2-Bromobicyclo[2.2.1]heptane-1-carboxylic acid, B-60210
7-Bromo-5-heptynoic acid; Me ester, *in* B-60261

C$_8$H$_{11}$BrO$_3$

(3-Bromomethyl)-2,4,10-trioxatricyclo[3.3.1.1^{3,7}]decane, B-60301

C$_8$H$_{11}$ClO$_5$

2,3-Dihydroxybutanoic acid; Di-Ac, chloride, *in* D-60312

C$_8$H$_{11}$NO$_2$

5-Methyl-1*H*-pyrrole-3-carboxylic acid; Et ester, *in* M-60125

C$_8$H$_{11}$NO$_3$

4-Oxoheptanedioic acid; Me ester, mononitrile, *in* O-60071

C$_8$H$_{11}$N$_2$

Acetophenone; Hydrazone, *in* A-60017

C$_8$H$_{11}$N$_3$

2-Amino-5-vinylpyrimidine; *N*^{2},*N*^{2}-Di-Me, *in* A-60269

C$_8$H$_{11}$N$_3$O$_3$

α-Amino-1*H*-imidazole-2-propanoic acid; *N*$^\alpha$-Ac, *in* A-60205

C$_8$H$_{12}$

Bicyclo[4.1.1]oct-2-ene, B-60095
Bicyclo[4.1.1]oct-3-ene, B-60096
1,2-Bis(methylene)cyclohexane, B-60166
1,3-Bis(methylene)cyclohexane, B-60167

1,4-Bis(methylene)cyclohexane, B-60168
2,5-Dimethyl-1,3,5-hexatriene, D-60424
Tricyclo[3.1.1.1^{2,4}]octane, T-60265
Tricyclo[5.1.0.0^{2,8}]octane, T-60266
3-Vinylcyclohexene, V-60009

C$_8$H$_{12}$Cl$_2$N$_2$O$_2$

4,5-Dichloro-1*H*-imidazole-2-carboxaldehyde; Di-Et acetal, *in* D-60134

C$_8$H$_{12}$N$_2$O

2-*tert*-Butyl-4(3*H*)-pyrimidinone, B-60349

C$_8$H$_{12}$N$_2$O$_2$

▷1,3-Diazaspiro[4.5]decane-2,4-dione, D-60056
4,4-Dimethyl-6-nitro-5-hexenenitrile, *in* D-60429
3-Ethoxy-4,5,6,7-tetrahydroisoxazolo[4,5-*c*]pyridine, *in* T-60071

C$_8$H$_{12}$N$_2$S$_2$

2,3-Bis(methylthio)-1,4-benzenediamine, *in* D-60030

C$_8$H$_{12}$O

1-Acetyl-2-methylcyclopentene, A-60043
Bicyclo[3.2.1]octan-8-one, B-60094

C$_8$H$_{12}$O$_2$

Bicyclo[2.2.1]heptane-1-carboxylic acid, B-60085
Bicyclo[3.1.1]heptane-1-carboxylic acid, B-60086
5,6-Dimethoxy-1,3-cyclohexadiene, D-60397
1,4-Dioxaspiro[4,5]dec-7-ene, *in* C-60211

C$_8$H$_{12}$O$_3$

4,4-Diethoxy-2-butynal, *in* B-60351
7-Hydroxy-5-heptynoic acid; Me ester, *in* H-60150

C$_8$H$_{12}$O$_4$S$_2$

(Dimercaptomethylene)propanedioic acid; Di-*S*-Me, di-Me ester, *in* D-60395

C$_8$H$_{12}$O$_6$

2,3-Dihydroxybutanoic acid; Di-Ac, *in* D-60312

C$_8$H$_{13}$BF$_4$O$_2$

3,4,4*a*,5,6,7-Hexahydro-2*H*-pyrano[2,3-*b*]pyrilium; Tetrafluoroborate, *in* H-60055

C$_8$H$_{13}$Br$_2$NO$_3$

6-Amino-2,2-dibromohexanoic acid; *N*-Ac, *in* A-60124

C$_8$H$_{13}$Cl$_2$NO$_3$

6-Amino-2,2-dichlorohexanoic acid; *N*-Ac, *in* A-60126

C$_8$H$_{13}$NO

Bicyclo[2.2.1]heptane-1-carboxylic acid; Amide, *in* B-60085

C$_8$H$_{13}$NO$_2$

7-Aminobicyclo[4.1.0]heptane-7-carboxylic acid, A-60095

C$_8$H$_{13}$NO$_3$

4-Oxo-2-piperidinecarboxylic acid; *N*-Me, Me ester, *in* O-60089

C$_8$H$_{13}$NO$_4$

1-Amino-1,4-cyclohexanedicarboxylic acid, A-60117
2-Amino-1,4-cyclohexanedicarboxylic acid, A-60118
3-Amino-1,2-cyclohexanedicarboxylic acid, A-60119
4-Amino-1,1-cyclohexanedicarboxylic acid, A-60120
4-Amino-1,3-cyclohexanedicarboxylic acid, A-60121
4,4-Dimethyl-6-nitro-5-hexenoic acid, D-60429

C$_8$H$_{13}$O$_2$$^\oplus$

3,4,4*a*,5,6,7-Hexahydro-2*H*-pyrano[2,3-*b*]pyrilium, H-60055

C$_8$H$_{14}$

Bicyclo[4.1.1]octane, B-60091
1-Ethylcyclohexene, E-60046
3-Ethylcyclohexene, E-60047
4-Ethylcyclohexene, E-60048

C$_8$H$_{14}$BrNO$_3$

6-Amino-2-bromohexanoic acid; *N*-Ac, *in* A-60100

C$_8$H$_{14}$ClNO$_3$

6-Amino-2-chlorohexanoic acid; *N*-Ac, *in* A-60109

C$_8$H$_{14}$N$_2$

4,5-Dihydro-3,3,5,5-tetramethyl-4-methylene-3*H*-pyrazole, D-60283

$C_8H_{14}N_2O_2$
1,3-Diazetidine-2,4-dione; 1,3-Diisopropyl, *in* D-60057

$C_8H_{14}N_2O_4$
2,6-Diamino-4-methyleneheptanedioic acid, D-60039
2,5-Piperazinedicarboxylic acid; Di-Me ester, *in* P-60157
N-Prolylserine, P-60173
N-Serylproline, S-60025

$C_8H_{14}N_4S_2$
1,6-Diallyl-2,5-dithiobiurea, D-60026

$C_8H_{14}O$
2,2-Dimethyl-5-hexen-3-one, D-60425
4-Methyl-3-hepten-2-one, M-60087
4-Octen-2-one, O-60030

$C_8H_{14}OS$
Dihydro-3,3,4,4-tetramethyl-2(3*H*)-furanthione, D-60282
4,5-Dihydro-3,3,4,4-tetramethyl-2(3*H*)-thiophenone, D-60288

$C_8H_{14}O_2$
4,4-Dimethoxycyclohexene, *in* C-60211
4-Hexen-3-ol; Ac, *in* H-60077
3-Methyl-2-heptenoic acid, M-60086
4-Oxooctanal, O-60081

$C_8H_{14}O_2S$
4-Mercapto-2-butenoic acid; *tert*-Butyl ester, *in* M-60018
Octahydrobenzo[*b*]thiophene; 1,1-Dioxide, *in* O-60012

$C_8H_{14}O_2S_2$
Bis(3-hydroxycyclobutyl)disulfide, *in* M-60019

$C_8H_{14}O_3$
2-Cyclohexyl-2-hydroxyacetic acid, C-60212
4,4-Dimethyl-5-oxopentanoic acid; Me ester, *in* D-60434
4-Hydroxy-4-(2-hydroxyethyl)cyclohexanone, H-60153

$C_8H_{14}S$
Octahydrobenzo[*b*]thiophene, O-60012

$C_8H_{14}S_2$
4,5-Dihydro-3,3,4,4-tetramethyl-2(3*H*)-thiophenethione, D-60287

$C_8H_{15}BrO_2$
5-Bromo-2-hexanane; Ethylene acetal, *in* B-60262

$C_8H_{15}ClO$
2,2-Diethylbutanoic acid; Chloride, *in* D-60171

$C_8H_{15}IO$
3-Iodo-2-octanone, I-60053

$C_8H_{15}NO$
▷Physoperuvine, P-60151
Tetrahydro-2-(2-pyrrolidinyl)furan, T-60090

$C_8H_{15}NO_2$
3-Amino-2-hexenoic acid; Et ester, *in* A-60170
2-Methyl-2-piperidinecarboxylic acid; Me ester, *in* M-60119
4-(Tetrahydro-2-furanyl)morpholine, T-60065

$C_8H_{15}NO_4$
2-Amino-3-butylbutanedioic acid, A-60102

$C_8H_{16}BrNO_2$
2-Amino-4-bromobutanoic acid; *tert*-Butyl ester, *in* A-60098

$C_8H_{16}N_2O$
▷Tetrahydro-2(1*H*)-pyrimidinone; 1,3-Di-Et, *in* T-60088

$C_8H_{16}N_2O_3S$
N-Methionylalanine, M-60035

$C_8H_{16}N_2O_4S$
N-Serylmethionine, S-60024

$C_8H_{16}O$
2-Octen-1-ol, O-60029

$C_8H_{16}OS$
2-Methyl-4-propyl-1,3-oxathiane, M-60122

$C_8H_{16}O_2$
2,2-Diethylbutanoic acid, D-60171
3,4-Dimethylpentanoic acid; Me ester, *in* D-60435

5-Hydroxy-4-methyl-3-heptanone, H-60183
Pityol, P-60160

$C_8H_{16}O_3$
1-(2-Hydroxyethyl)-1,4-cyclohexanediol, H-60133
Isorengyol, *in* H-60133

$C_8H_{16}O_4$
Methyl 3,3-diethoxypropionate, *in* O-60090

$C_8H_{16}O_5$
5,7,8-Trimethoxyflavone, *in* T-60321

$C_8H_{16}S_2$
1,2-Dithiecane, D-60507

$C_8H_{17}NO$
2,2-Diethylbutanoic acid; Amide, *in* D-60171

$C_8H_{17}N_3O_3$
N-Ornithyl-β-alanine, O-60041

$C_8H_{18}N_2$
Hexahydropyrimidine; 1,3-Di-Et, *in* H-60056

$C_8H_{18}N_2O_2$
1,7-Dioxa-4,10-diazacyclododecane, D-60465

$C_8H_{19}N$
2-Octylamine, O-60031

$C_8H_{19}N_3$
1,4,7-Triazacycloundecane, T-60234
1,4,8-Triazacycloundecane, T-60235

$C_8H_{22}N_4$
1,2-Dihydrazinoethane; $N^\alpha,N^{\alpha'},N^\beta,N^\beta,N^{\beta'},N^{\beta'}$-Hexa-Me, *in* D-60192

$C_8O_2S_6$
2,6-Dithioxobenzo[1,2-*d*:4,5-*d'*]bis[1,3]dithiole-4,8-dione, D-60512

$C_8O_4S_2$
2,7-Dithiatricyclo[6.2.0.0^{3,6}]deca-1(8),3(6)-diene-4,5,9,10-tetrone, D-60506

$C_9H_3Cl_2NO_2$
6,7-Dichloro-5,8-isoquinolinedione, D-60137
6,7-Dichloro-5,8-quinolinedione, D-60149

$C_9H_3Cl_2NO_3$
6,7-Dichloro-5,8-quinolinedione; *N*-Oxide, *in* D-60149

$C_9H_4N_4O_3$
Imidazo[4,5-*g*]quinazoline-4,8,9(3*H*,7*H*)-trione, I-60010

$C_9H_5Cl_2F_3N_2$
5,6-Dichloro-2-(trifluoromethyl)-1*H*-benzimidazole; *N*-Me, *in* D-60160

$C_9H_5Cl_2N$
6,7-Dichloroisoquinoline, D-60136

$C_9H_5NO_2$
▷5,8-Isoquinolinequinone, I-60131

$C_9H_5N_3O_2$
[1]-Benzopyrano[2,3-*d*]-1,2,3-triazol-9(1*H*)-one, B-60043

C_9H_6ClNO
▷5-Chloro-8-hydroxyquinoline, C-60103
8-Chloro-2(1*H*)-quinolinone, C-60131
8-Chloro-4(1*H*)-quinolinone, C-60132

$C_9H_6ClNO_2$
5-Chloro-8-hydroxyquinoline; 1-Oxide, *in* C-60103

$C_9H_6Cl_2O_2$
1,3,5-Cycloheptatriene-1,6-dicarboxylic acid; Dichloride, *in* C-60202

$C_9H_6F_3N$
2-Trifluoromethyl-1*H*-indole, T-60291
3-Trifluoromethyl-1*H*-indole, T-60292
6-Trifluoromethyl-1*H*-indole, T-60293

$C_9H_6N_2O_2$
5,8-Isoquinolinequinone; 8-Oxime, *in* I-60131
1-Nitroisoquinoline, N-60029
4-Nitroisoquinoline, N-60030

5-Nitroisoquinoline, N-60031
6-Nitroisoquinoline, N-60032
7-Nitroisoquinoline, N-60033
8-Nitroisoquinoline, N-60034
5-Phenyl-1,2,4-oxadiazole-3-carboxaldehyde, P-60122

C$_9$H$_6$N$_2$O$_3$

5-Nitroisoquinoline; 2-Oxide, *in* N-60031
8-Nitroisoquinoline; 2-Oxide, *in* N-60034

C$_9$H$_6$N$_2$S

3-Amino-2-cyanobenzo[*b*]thiophene, *in* A-60091

C$_9$H$_6$N$_4$

4-Cyano-1-phenyl-1,2,3-triazole, *in* P-60138
4-Cyano-2-phenyl-1,2,3-triazole, *in* P-60140

C$_9$H$_6$N$_4$O

3,8-Dihydro-9*H*-pyrazolo[4,3-*f*]quinazolin-9-one, D-60271
Imidazo[4,5-*h*]quinazolin-6-one, I-60013
Imidazo[4,5-*g*]quinazolin-8-one, I-60012
Imidazo[4,5-*f*]quinazolin-9(8*H*)-one, I-60011
Pyrazolo[3,4-*f*]quinazolin-9(8*H*)-one, P-60204
Pyrazolo[4,3-*g*]quinazolin-5(6*H*)-one, P-60205

C$_9$H$_6$N$_4$O$_2$

Imidazo[4,5-*g*]quinazoline-6,8(5*H*,7*H*)-dione, I-60009
1*H*-Pyrazolo[4,3-*g*]quinazoline-5,7(6*H*,8*H*)-dione, P-60203

C$_9$H$_6$N$_4$S

Pyrazino[2,3-*b*]pyrido[3',2'-*e*][1,4]thiazine, P-60199

C$_9$H$_6$O$_2$

1*H*-2-Benzopyran-1-one, B-60041

C$_9$H$_6$O$_3$

3-Benzofurancarboxylic acid, B-60032

C$_9$H$_6$O$_4$

4-Hydroxy-5-benzofurancarboxylic acid, H-60103

C$_9$H$_7$ClO$_3$

1,4-Benzodioxan-2-carboxylic acid; Chloride, *in* B-60024
1,4-Benzodioxan-6-carboxylic acid; Chloride, *in* B-60025

C$_9$H$_7$ClO$_4$

3-Hydroxy-4,5-methylenedioxybenzoic acid; Me ether, chloride, *in* H-60178
6-Hydroxy-3,4-methylenedioxybenzoic acid; Me ether, chloride, *in* H-60181

C$_9$H$_7$F$_3$S

1-(Methylthio)-3-(trifluoromethyl)benzene, *in* T-60285

C$_9$H$_7$IO$_4$

1-Hydroxy-1,2-benziodoxol-3(1*H*)-one; Ac, *in* H-60102

C$_9$H$_7$N

Cyclopent[*b*]azepine, C-60225

C$_9$H$_7$NO

Cyclohepta[*b*]pyrrol-2(1*H*)-one, C-60201
Cyclopent[*b*]azepine; *N*-Oxide, *in* C-60225
5,6-Epoxyquinoline, E-60027
7,8-Epoxyquinoline, E-60028
1*H*-Indole-5-carboxaldehyde, I-60029
1*H*-Indole-6-carboxaldehyde, I-60030
1*H*-Indole-7-carboxaldehyde, I-60031
5-Phenyloxazole, P-60125

C$_9$H$_7$NO$_2$

1,4-Benzodioxan-2-carbonitrile, *in* B-60024
1,3-Benzodioxole-4-acetonitrile, *in* B-60029

C$_9$H$_7$NO$_2$S

▷3-Amino-2-benzo[*b*]thiophenecarboxylic acid, A-60091

C$_9$H$_7$NO$_3$

1*H*-Isoindolin-1-one-3-carboxylic acid, I-60102
3-Methoxy-4,5-methylnedioxybenzonitrile, *in* H-60178

C$_9$H$_7$NS

1(2*H*)-Isoquinolinethione, I-60132
3(2*H*)-Isoquinolinethione, I-60133
3-Quinolinethiol, Q-60003
5-Quinolinethiol, Q-60004
6-Quinolinethiol, Q-60005
4(1*H*)-Quinolinethione, Q-60006

C$_9$H$_7$N$_3$OS

1,2,3-Benzotriazine-4(3*H*)-thione; *N*-Ac, *in* B-60050

C$_9$H$_7$N$_3$O$_2$

6-Amino-8-nitroquinoline, A-60234
8-Amino-6-nitroquinoline, A-60235
5-Phenyl-1,2,4-oxadiazole-3-carboxaldehyde; Oxime, *in* P-60122
1-Phenyl-1*H*-1,2,3-triazole-4-carboxylic acid, P-60138
1-Phenyl-1*H*-1,2,3-triazole-5-carboxylic acid, P-60139
2-Phenyl-2*H*-1,2,3-triazole-4-carboxylic acid, P-60140

C$_9$H$_7$N$_5$

8-Aminoimidazo[4,5-*g*]quinazoline, A-60207
9-Aminoimidazo[4,5-*f*]quinazoline, A-60206

C$_9$H$_7$N$_5$O

7-Aminoimidazo[4,5-*f*]quinazolin-9(8*H*)-one, A-60208
6-Aminoimidazo[4,5-*g*]quinolin-8(7*H*)-one, A-60209

C$_9$H$_8$ClNO$_5$

4,5-Dihydroxy-2-nitrobenzoic acid; Di-Me ether, chloride, *in* D-60360

C$_9$H$_8$Cl$_3$NO$_2$

2-Amino-3-(2,3,4-trichlorophenyl)propanoic acid, A-60262
2-Amino-3-(2,3,6-trichlorophenyl)propanoic acid, A-60263
2-Amino-3-(2,4,5-trichlorophenyl)propanoic acid, A-60264

C$_9$H$_8$N$_2$O

5-Amino-3-phenylisoxazole, A-60240
3-Amino-2(1*H*)-quinolinone, A-60256
5-Methyl-3-phenyl-1,2,4-oxadiazole, M-60110
1*H*-Pyrrolo[3,2-*b*]pyridine; *N*-Ac, *in* P-60249

C$_9$H$_8$N$_2$OS

3-Amino-2-benzo[*b*]thiophenecarboxylic acid; Amide, *in* A-60091

C$_9$H$_8$N$_2$O$_2$

1,4-Dihydro-2,3-quinoxalinedione; 1-Me, *in* D-60279
6-Hydroxy-7-methoxyquinoxaline, *in* Q-60009

C$_9$H$_8$N$_2$O$_3$

5-Phenyl-1,2,4-oxadiazole-3-carboxaldehyde; Covalent hydrate, *in* P-60122

C$_9$H$_8$N$_2$O$_4$

1-Cyano-4,5-dimethoxy-2-nitrobenzene, *in* D-60360

C$_9$H$_8$N$_2$O$_6$

3,5-Dinitrobenzoic acid; Et ester, *in* D-60458

C$_9$H$_8$N$_2$S

4-Methyl-5-phenyl-1,2,3-thiadiazole, M-60116
5-Methyl-4-phenyl-1,2,3-thiadiazole, M-60117

C$_9$H$_8$N$_4$O

1-Phenyl-1*H*-1,2,3-triazole-4-carboxylic acid; Amide, *in* P-60138
1-Phenyl-1*H*-1,2,3-triazole-5-carboxylic acid; Amide, *in* P-60139
2-Phenyl-2*H*-1,2,3-triazole-4-carboxylic acid; Amide, *in* P-60140

C$_9$H$_8$N$_6$O$_3$

Lepidopterin, L-60020

C$_9$H$_8$O

3-Phenyl-2-propyn-1-ol, P-60130

C$_9$H$_8$OS

1,4-Dihydro-3*H*-2-benzothiopyran-3-one, D-60206

C$_9$H$_8$OSe

1,4-Dihydro-3*H*-2-benzoselenin-3-one, D-60204

C$_9$H$_8$OTe

1,4-Dihydro-3*H*-2-benzotellurin-3-one, D-60205

C$_9$H$_8$O$_2$

Cubanecarboxylic acid, C-60174
3-(4-Hydroxyphenyl)-2-propenal, H-60213
2-Methyl-1,3-benzenedicarboxaldehyde, M-60046
2-Methyl-1,4-benzenedicarboxaldehyde, M-60047
4-Methyl-1,2-benzenedicarboxaldehyde, M-60048
4-Methyl-1,3-benzenedicarboxaldehyde, M-60049

5-Methyl-1,3-benzenedicarboxaldehyde, M-60050

$C_9H_8O_2S$

3,4-Dihydro-2H-1,5-benzoxathiepin-3-one, D-60207
3-(2-Thienyl)-2-propenoic acid; Me ester, *in* T-60208

$C_9H_8O_3$

1,4-Benzodioxan-2-carboxaldehyde, B-60022
1,4-Benzodioxan-6-carboxaldehyde, B-60023
4H-1,3-Benzodioxin-6-carboxaldehyde, B-60026
2,3-Dihydro-2-benzofurancarboxylic acid, D-60197

$C_9H_8O_4$

1,4-Benzodioxan-2-carboxylic acid, B-60024
1,4-Benzodioxan-6-carboxylic acid, B-60025
4H-1,3-Benzodioxin-6-carboxylic acid, B-60027
1,3-Benzodioxole-4-acetic acid, B-60029
1,3,5-Cycloheptatriene-1,6-dicarboxylic acid, C-60202
5,7-Dihydroxy-6-methyl-1(3H)-isobenzofuranone, D-60349
5-Hydroxy-7-methoxyphthalide, *in* D-60339
7-Hydroxy-5-methoxyphthalide, *in* D-60339

$C_9H_8O_5$

2-Formyl-3,5-dihydroxybenzoic acid; Me ester, *in* F-60071
2-Hydroxy-1,3-benzenedicarboxylic acid; Mono-Me ester, *in* H-60101
2-Methoxy-1,3-benzenedicarboxylic acid, *in* H-60101
5-Methoxy-1,3-benzodioxole-4-carboxylic acid, *in* H-60180
6-Methoxy-1,3-benzodioxole-5-carboxylic acid, *in* H-60181
7-Methoxy-1,3-benzodioxole-4-carboxylic acid, *in* H-60179
2-Methoxy-3,4-methylenedioxybenzoic acid, *in* H-60177
3-Methoxy-4,5-methylenedioxybenzoic acid, *in* H-60178

C_9H_9BrO

3-(Bromomethyl)-2,3-dihydrobenzofuran, B-60295

$C_9H_9BrO_2$

1-(5-Bromo-2-methoxyphenyl)ethanone, *in* B-60265

$C_9H_9Br_3$

1,3,5-Tris(bromomethyl)benzene, T-60401

C_9H_9ClO

2-Chloro-4,6-dimethylbenzaldehyde, C-60061
4-Chloro-2,6-dimethylbenzaldehyde, C-60062
4-Chloro-3,5-dimethylbenzaldehyde, C-60063
5-Chloro-2,4-dimethylbenzaldehyde, C-60064

$C_9H_9ClO_3$

3-Chloro-4-hydroxybenzoic acid; Me ester, Me ether, *in* C-60091
2-Chloro-3-methoxy-5-methylbenzoic acid, *in* C-60095
3-Chloro-4-methoxy-5-methylbenzoic acid, *in* C-60096
5-Chloro-4-methoxy-2-methylbenzoic acid, *in* C-60097

$C_9H_9Cl_2NO_2$

2-Amino-3-(2,3-dichlorophenyl)propanoic acid, A-60127
2-Amino-3-(2,4-dichlorophenyl)propanoic acid, A-60128
2-Amino-3-(2,5-dichlorophenyl)propanoic acid, A-60129
2-Amino-3-(2,6-dichlorophenyl)propanoic acid, A-60130
2-Amino-3-(3,4-dichlorophenyl)propanoic acid, A-60131
2-Amino-3-(3,5-dichlorophenyl)propanoic acid, A-60132

$C_9H_9Cl_3$

1,3,5-Tris(chloromethyl)benzene, T-60402

C_9H_9FO

2-Fluoro-2-phenylpropanal, F-60063
2-Fluoro-3-phenylpropanal, F-60064
2-Fluoro-1-phenyl-1-propanone, F-60065

$C_9H_9FO_2$

3-Fluoro-2-methylphenol; Ac, *in* F-60056

$C_9H_9FO_4$

3-Fluoro-2,6-dimethoxybenzoic acid, *in* F-60023
5-Fluoro-2,3-dimethoxybenzoic acid, *in* F-60025

$C_9H_9F_3O_3S$

3-(Trifluoromethyl)benzenesulfonic acid; Et ester, *in* T-60283

$C_9H_9IO_4$

2-Iodo-4,5-dimethoxybenzoic acid, *in* D-60338
3-Iodo-2,4-dimethoxybenzoic acid, *in* D-60331
3-Iodo-2,6-dimethoxybenzoic acid, *in* D-60334

3-Iodo-4,5-dimethoxybenzoic acid, *in* D-60335
4-Iodo-2,5-dimethoxybenzoic acid, *in* D-60333
5-Iodo-2,4-dimethoxybenzoic acid, *in* D-60332
6-Iodo-2,3-dimethoxybenzoic acid, *in* D-60330

C_9H_9N

6-Methyl-1H-indole, M-60088
3-Phenyl-2-propyn-1-amine, P-60129

C_9H_9NO

3-Amino-1-indanone, A-60211
7,8-Dihydro-5(6H)-isoquinolinone, D-60254
1H-Indole-6-methanol, I-60032
6-Methoxyindole, *in* H-60158

$C_9H_9NO_2$

4-Phenyl-2-oxazolidinone, P-60126

$C_9H_9NO_3$

1,4-Benzodioxan-6-carboxaldehyde; Oxime, *in* B-60023
1,4-Benzodioxan-2-carboxylic acid; Amide, *in* B-60024

$C_9H_9NO_4$

3-Hydroxy-4,5-methylenedioxybenzoic acid; Me ether, amide, *in* H-60178

$C_9H_9NO_6$

4,5-Dihydroxy-2-nitrobenzoic acid; 4-Me ether, Me ester, *in* D-60360
4,5-Dimethoxy-2-nitrobenzoic acid, *in* D-60360

$C_9H_9N_3O_3$

Biuret; N-Benzoyl, *in* B-60190

$C_9H_9N_3S$

1,2,3-Benzotriazine-4(3H)-thione; 2-N-Et, *in* B-60050

$C_9H_9N_5$

▷2,4-Diamino-6-phenyl-1,3,5-triazine, D-60041

C_9H_{10}

7-Methylene-1,3,5-cyclooctatriene, M-60074

$C_9H_{10}ClN$

4-Chloro-2,3-dihydro-1H-indole; 1-Me, *in* C-60055

$C_9H_{10}ClNO$

2′-Amino-2-chloro-3′-methylacetophenone, A-60112
2′-Amino-2-chloro-4′-methylacetophenone, A-60113
2′-Amino-2-chloro-5′-methylacetophenone, A-60114
2′-Amino-2-chloro-6′-methylacetophenone, A-60115
2-Chloro-4,6-dimethylbenzaldehyde; Oxime, *in* C-60061

$C_9H_{10}N_2O$

1,3,4,5-Tetrahydro-2H-1,3-benzodiazepin-2-one, T-60058

$C_9H_{10}N_2O_3$

1,4-Benzodioxan-6-carboxylic acid; Hydrazide, *in* B-60025
2,4-Diaminobenzoic acid; 2-N-Ac, *in* D-60032
2,4-Diaminobenzoic acid; 4-N-Ac, *in* D-60032

$C_9H_{10}N_2S$

2-Amino-4,5,6,7-tetrahydrobenzo[b]thiophene-3-carboxylic acid; Nitrile, *in* A-60258

$C_9H_{10}N_4$

4-Dimethylamino-1,2,3-benzotriazine, *in* A-60092

$C_9H_{10}N_4O$

4-Amino-1,2,3-benzotriazine; N,N(4)-Di-Me, 2-Oxide, *in* A-60092

$C_9H_{10}N_4S$

Tetrazole-5-thione; 3-Et, 1-Ph, *in* T-60174

$C_9H_{10}O$

1-Methoxy-1-phenylethylene, *in* P-60092

$C_9H_{10}OS$

1-[4-(Methylthio)phenyl]ethanone, *in* M-60017

$C_9H_{10}O_2$

Bicyclo[2.2.1]hepta-2,5-diene-2-carboxylic acid; Me ester, *in* B-60084
3,4-Dihydro-2H-1-benzopyran-2-ol, D-60198
3,4-Dihydro-2H-1-benzopyran-3-ol, D-60199
3,4-Dihydro-2H-1-benzopyran-4-ol, D-60200
3,4-Dihydro-2H-1-benzopyran-5-ol, D-60201
3,4-Dihydro-2H-1-benzopyran-7-ol, D-60202

$C_9H_{15}NO_5$

4-Oxoheptanedioic acid; Di-Me ester, oxime, *in* O-60071

$C_9H_{15}N_3O_2$

Histidine trimethylbetaine, H-60083

$C_9H_{15}N_3O_3$

Triethyl isocyanurate, *in* C-60179

$C_9H_{15}N_3O_8$

Hexahydro-1,3,5-triazine; 1,3,5-Tri-Ac, *in* H-60060

C_9H_{16}

1-Isopropylcyclohexene, I-60118
3-Isopropylcyclohexene, I-60119
4-Isopropylcyclohexene, I-60120

$C_9H_{16}N_2$

2-Diazo-1,1,3,3-tetramethylcyclopentane, D-60066
2,2,5,5-Tetramethyl-3-cyclopenten-1-one; Hydrazone, *in* T-60140

$C_9H_{16}N_2O_4S$

S-(2-Aminoethyl)cysteine; N^α,N^ϵ-Di-Ac, *in* A-60152

$C_9H_{16}N_2O_5$

Tetrahydro-2(1H)-pyrimidinone; 1-β-D-Ribofuranosyl, *in* T-60088
Tetrahydro-2(1H)-pyrimidinone; 1-β-D-Ribopyranosyl, *in* T-60088

$C_9H_{16}O$

2-Methylcyclooctanone, M-60059
2-Methyl-2-octenal, M-60106
2,2,5,5-Tetramethylcyclopentanone, T-60138
2,2,5-Trimethyl-4-hexen-3-one, T-60366

$C_9H_{16}O_2$

3,3-Dimethyl-4-pentenoic acid; Et ester, *in* D-60437
2-Hydroxycyclononanone, H-60112
3-Hydroxycyclononanone, H-60113
2-Methyl-1,6-dioxaspiro[4.5]decane, M-60062
2,2,5-Trimethylcyclopentanecarboxylic acid, T-60360
1,2,5-Trimethyl-1-cyclopentanecarboxylic acid, T-60361

$C_9H_{16}O_3$

2-Cyclohexyl-2-hydroxyacetic acid; Me ester, *in* C-60212
2-Cyclohexyl-2-methoxyacetic acid, *in* C-60212
3,3-Dimethyl-4-oxopentanoic acid; Et ester, *in* D-60433

$C_9H_{16}Se$

2,2,5,5-Tetramethylcyclopentaneselone, T-60137

$C_9H_{17}Br$

1-Bromo-3-nonene, B-60310

$C_9H_{17}NO$

Tetrahydro-2-(2-piperidinyl)furan, T-60082
2,2,5-Trimethylcyclopentanecarboxylic acid; Amide, *in* T-60360

$C_9H_{17}NO_2$

2-Methyl-2-piperidinecarboxylic acid; Et ester, *in* M-60119
Stachydrine ethyl ester, *in* S-60050

$C_9H_{17}NO_3$

2-Amino-3-hydroxy-4-methyl-6-octenoic acid, A-60189

$C_9H_{17}NO_4$

3-(Carboxymethylamino)propanoic acid; Di-Et ester, *in* C-60016

$C_9H_{18}N_2$

1,5-Diazabicyclo[5.2.2]undecane, D-60055

$C_9H_{18}N_2O_2$

▷3-[(3-Methylbutylnitrosoamino]-2-butanone, M-60055

$C_9H_{18}N_2O_4S$

N-Serylmethionine; Me ester, *in* S-60024

$C_9H_{18}O$

1-Nonen-3-ol, N-60058

$C_9H_{18}O_3$

6-Hydroxy-2,4-dimethylheptanoic acid, H-60121

$C_9H_{18}O_4$

Ethyl 3,3-diethoxypropionate, *in* O-60090

$C_9H_{20}ClNO_2$

▷Muscarine; Chloride, *in* M-60153

$C_9H_{20}INO_2$

Muscarine; Iodide, *in* M-60153

$C_9H_{20}NO_2^\oplus$

Muscarine, M-60153

$C_9H_{20}N_2O_3$

N,N'-[Carbonylbis(oxy)]bis[2-methyl-2-propanamine], *in* C-60015

$C_9H_{20}O_4S_4$

Tetrakis(methylsulfinylmethyl)methane, *in* B-60164

$C_9H_{20}S_4$

Tetrakis(methylthiomethyl)methane, *in* B-60164

$C_9H_{21}N_3$

▷Hexahydro-1,3,5-triazine; 1,3,5-Tri-Et, *in* H-60060
1,4,7-Triazacyclododecane, T-60231
1,5,9-Triazacyclododecane, T-60232

$C_{10}H_4^{\ominus\ominus}$

Dihydrocyclopenta[c,d]pentalene(2−), D-60216

$C_{10}H_4Cl_2FNO_2$

3,4-Dichloro-1-(4-fluorophenyl)-1H-pyrrole-2,5-dione, D-60132

$C_{10}H_4K_2$

Dihydrocyclopenta[c,d]pentalene(2−); Di-K salt, *in* D-60216

$C_{10}H_4O_2S_2$

Benzo[1,2-b:4,5-b']dithiophene-4,8-dione, B-60031

$C_{10}H_5BrO_3$

2-Bromo-3-hydroxy-1,4-naphthoquinone, B-60268
2-Bromo-5-hydroxy-1,4-naphthoquinone, B-60269
2-Bromo-6-hydroxy-1,4-naphthoquinone, B-60270
2-Bromo-7-hydroxy-1,4-naphthoquinone, B-60271
2-Bromo-8-hydroxy-1,4-naphthoquinone, B-60272
7-Bromo-2-hydroxy-1,4-naphthoquinone, B-60273
8-Bromo-2-hydroxy-1,4-naphthoquinone, B-60274

$C_{10}H_5NO_4$

6-Nitro-1,4-naphthoquinone, N-60035
7-Nitro-1,2-naphthoquinone, N-60036

$C_{10}H_5NO_5$

2-Hydroxy-3-nitro-1,4-naphthoquinone, H-60195

$C_{10}H_5N_3O_2$

1H-Naphtho[2,3-d]triazole-4,9-dione, N-60011

$C_{10}H_6F_3NO_2$

7-Amino-4-(trifluoromethyl)-2H-1-benzopyran-2-one, A-60265

$C_{10}H_6N_2O$

2-Cyano-3-hydroxyquinoline, *in* H-60228

$C_{10}H_6N_2OS$

[1,4]Oxathiino[3,2-b:5,6-c']dipyridine, O-60050

$C_{10}H_6N_2O_2S$

[1,4]Oxathiino[3,2-b:5,6-c']dipyridine; 8-Oxide, *in* O-60050

$C_{10}H_6N_2S$

Thieno[2,3-c]cinnoline, T-60202
Thieno[3,2-c]cinnoline, T-60203

$C_{10}H_6N_2S_2$

Thiazolo[4,5-b]quinoline-2(3H)-thione, T-60201

$C_{10}H_6N_4O_2$

Pyrazino[2,3-g]quinazoline-2,4-(1H,3H)-dione, P-60200
Pyrazino[2,3-f]quinazoline-8,10-(7H,9H)-dione, P-60201

$C_{10}H_6O_3$

4-Hydroxy-1,2-naphthoquinone, H-60194

$C_{10}H_6O_4$

2,5-Dihydroxy-1,4-naphthoquinone, D-60354
2,8-Dihydroxy-1,4-naphthoquinone, D-60355
▷5,8-Dihydroxy-1,4-naphthoquinone, D-60356

C₁₀H₆O₈
Hexahydroxy-1,4-naphthoquinone, H-60065

C₁₀H₆S₂
▷Benzo[1,2-*b*:4,5-*b*′]dithiophene, B-60030

C₁₀H₇Br₃
1,4,7-Tribromotriquinacene, T-60248

C₁₀H₇ClO₃
4-Hydroxy-5-benzofurancarboxylic acid; Me ether, chloride, *in* H-60103

C₁₀H₇ClO₄
6-Chloro-1,2,3,4-naphthalenetetrol, C-60118

C₁₀H₇N
2-Ethynylindole, E-60057
3-Ethynylindole, E-60058

C₁₀H₇NO₂
4-Benzoylisoxazole, B-60062
5-Benzoylisoxazole, B-60063

C₁₀H₇NO₃
3-Hydroxy-2-quinolinecarboxylic acid, H-60228

C₁₀H₇NS
2-(1,2-Propadienyl)benzothiazole, P-60174

C₁₀H₇N₃
Pyrazolo[3,4-*c*]quinoline, P-60206

C₁₀H₇N₃O
2,5-Dihydro-1*H*-dipyrido[4,3-*b*:3′,4′-*d*]pyrrol-1-one, D-60237

C₁₀H₈
Fulvalene, F-60080
Naphthvalene, N-60012

C₁₀H₈N₂
▷2,3′-Bipyridine, B-60120
2,4′-Bipyridine, B-60121
3,3′-Bipyridine, B-60122
3,4′-Bipyridine, B-60123

C₁₀H₈N₂O
2-Acetylquinoxaline, A-60056
5-Acetylquinoxaline, A-60057
6-Acetylquinoxaline, A-60058
4(5)-Benzoylimidazole, B-60057
4-Benzoylpyrazole, B-60065
2,3′-Bipyridine; 1′-Oxide, *in* B-60120
2,4′-Bipyridine; 1-Oxide, *in* B-60121
3,3′-Bipyridine; Mono-*N*-oxide, *in* B-60122
2,2′-Oxybispyridine, O-60098
2,3′-Oxybispyridine, O-60099
3,3′-Oxybispyridine, O-60100
4,4′-Oxybispyridine, O-60101
2-Phenyl-1*H*-imidazole-4(5)-carboxaldehyde, P-60111

C₁₀H₈N₂OS
2,3-Dihydro-6-phenyl-2-thioxo-4(1*H*)-pyrimidinone, D-60268

C₁₀H₈N₂O₂
5-Amino-3-phenylisoxazole; *N*-Formyl, *in* A-60240
4-Benzoylisoxazole; Oxime, *in* B-60062
2,3′-Bipyridine; 1,1′-Dioxide, *in* B-60120
2,4′-Bipyridine; 1,1′-Dioxide, *in* B-60121
3-Hydroxy-2-quinolinecarboxylic acid; Amide, *in* H-60228
3,3′-Oxybispyridine; *N*-Oxide, *in* O-60100

C₁₀H₈N₂O₃
2-Acetylquinoxaline; 1,4-Dioxide, *in* A-60056
3,3′-Oxybispyridine; *N*,*N*′-Dioxide, *in* O-60100

C₁₀H₈N₂O₃S
Di-2-pyridyl sulfite, D-60497

C₁₀H₈N₄
Dipyrido[1,2-*b*:1′,2′-*e*][1,2,4,5]tetrazine, D-60494

C₁₀H₈OS
1-Benzothiepin-5(4*H*)-one, B-60047

C₁₀H₈O₂
1,2-Di(2-furanyl)ethylene, D-60191
2-(3-Oxo-1-propenyl)benzaldehyde, O-60091

3-Phenyl-2(5*H*)furanone, P-60102
4-Phenyl-2(5*H*)-furanone, P-60103

C₁₀H₈O₃
3-Benzofurancarboxylic acid; Me ester, *in* B-60032
5-Hydroxy-4-methyl-2*H*-1-benzopyran-2-one, H-60173
1,4,5-Naphthalenetriol, N-60005

C₁₀H₈O₃S
2-Oxo-4-phenylthio-3-butenoic acid, O-60088

C₁₀H₈O₄
Albidin, A-60070
4,7-Dihydroxy-5-methyl-2*H*-1-benzopyran-2-one, D-60348
4-Hydroxy-5-benzofurancarboxylic acid; Me ester, *in* H-60103
4-Methoxy-5-benzofurancarboxylic acid, *in* H-60103

C₁₀H₈O₆
2-Acetoxy-1,3-benzenedicarboxylic acid, *in* H-60101

C₁₀H₈S₈
2-(5,6-Dihydro-1,3-dithiolo[4,5-*b*][1,4]dithiin-2-ylidene)-5,6-dihydro-1,3-dithiolo[4,5-*b*][1,4]dithiin, D-60238

C₁₀H₉Br
3-Bromo-1-phenyl-1-butyne, B-60314

C₁₀H₉BrO
7-Bromo-3,4-dihydro-1(2*H*)-naphthalenone, B-60234

C₁₀H₉IO₂
1,3-Diacetyl-5-iodobenzene, D-60024

C₁₀H₉NO
2-Acetylindole, A-60032
4-Acetylindole, A-60033
▷5-Acetylindole, A-60034
6-Acetylindole, A-60035
7-Acetylindole, A-60036
1-(2-Furanyl)-2-(2-pyrrolyl)ethylene, F-60088
1-(4-Hydroxyphenyl)pyrrole, H-60216
2-(2-Hydroxyphenyl)pyrrole, H-60217

C₁₀H₉NO₂S
α-Aminobenzo[*b*]thiophene-3-acetic acid, A-60090
3-Amino-2-benzo[*b*]thiophenecarboxylic acid; Me ester, *in* A-60091

C₁₀H₉NO₂S₂
N-Phenyl-2-thiophenesulfonamide, *in* T-60213
N-Phenyl-3-thiophenesulfonamide, *in* T-60214

C₁₀H₉NO₃
2,3-Dihydro-2-oxo-1*H*-indole-3-acetic acid, D-60259
2-Nitro-3,4-dihydro-1(2*H*)naphthalenone, N-60026
1,2,3,4-Tetrahydro-2-oxo-4-quinolinecarboxylic acid, T-60077
1,2,3,4-Tetrahydro-4-oxo-6-quinolinecarboxylic acid, T-60078
1,2,3,4-Tetrahydro-4-oxo-7-quinolinecarboxylic acid, T-60079

C₁₀H₉NS
3(2*H*)-Isoquinolinethione; *S*-Me; B,HCl, *in* I-60133
1-Methylthioisoquinoline, *in* I-60132
4-Methylthioquinoline, *in* Q-60006
5-Methylthioquinoline, *in* Q-60004
6-Methylthioquinoline, *in* Q-60005
1-(2-Pyrrolyl)-2-(2-thienyl)ethylene, P-60250
3-Quinolinethiol; *S*-Me, *in* Q-60003
4(1*H*)-Quinolinethione; *N*-Me, *in* Q-60006

C₁₀H₉NS₂
2-Thiazoledinethione; *N*-Ph, *in* T-60197

C₁₀H₉N₃
6-Methyl-3-phenyl-1,2,4-triazine, M-60118
3-(Phenylazo)-2-butenenitrile, P-60079

C₁₀H₉N₃O
2-Acetylquinoxaline; Oxime, *in* A-60056

C₁₀H₉N₃O₂
1-Phenyl-1*H*-1,2,3-triazole-4-carboxylic acid; Me ester, *in* P-60138
1-Phenyl-1*H*-1,2,3-triazole-5-carboxylic acid; Me ester, *in* P-60139

2-Phenyl-2*H*-1,2,3-triazole-4-carboxylic acid; Me ester, *in* P-60140

$C_{10}H_9N_3O_9$

3-Hydroxy-5-methyl-2,4,6-trinitrobenzoic acid; Me ether, Me ester, *in* H-60193

$C_{10}H_{10}$

Bi-2,4-cyclopentadien-1-yl, B-60100
9-Methylene-1,3,5,7-cyclononatetraene, M-60073
1-Methyleneindane, M-60078

$C_{10}H_{10}N_2O$

5-Acetylindole; Oxime, *in* A-60034
5-Amino-3-phenylisoxazole; *N*-Me, *in* A-60240

$C_{10}H_{10}N_2O_2$

1,4-Dihydro-2,3-quinoxalinedione; 1,4-Di-Me, *in* D-60279
2,3-Dimethoxyquinoxaline, *in* D-60279
5,8-Dimethoxyquinoxaline, *in* Q-60002
6,7-Dimethoxyquinoxaline, *in* Q-60009
3-Ethoxy-2-quinoxalinol, *in* D-60279
3-Ethoxy-2(1*H*)quinoxalinone, *in* D-60279
5,8-Quinoxalinediol; Di-Me ether, *in* Q-60008

$C_{10}H_{10}O$

2,3-Dihydro-6(1*H*)-azulenone, D-60193
4,7-Dihydro-4,7-ethanoisobenzofuran, D-60240
Pentacyclo[5.3.0.0^{2,5}.0^{3,9}.0^{4,8}]decan-6-one, P-60027
1-Phenyl-3-buten-1-one, P-60085
2-(2-Propenyl)benzaldehyde, P-60176
4-(2-Propenyl)benzaldehyde, P-60177

$C_{10}H_{10}O_2$

Dispiro[2.0.2.4]dec-8-ene-7,10-dione, D-60501
(4-Hydroxyphenyl)ethylene; Ac, *in* H-60211
3-(4-Methoxyphenyl)-2-propenal, *in* H-60213
2-Phenylcyclopropanecarboxylic acid, P-60087

$C_{10}H_{10}O_2S$

4'-Mercaptoacetophenone; Ac, *in* M-60017

$C_{10}H_{10}O_3$

Hexahydrocyclopenta[*cd*]pentalene-1,3,5(2*H*)-trione, H-60048
2-Hydroxy-2-phenylacetaldehyde; Ac, *in* H-60210
3-Oxopropanoic acid; Benzyl ester, *in* O-60090

$C_{10}H_{10}O_3S$

2-Mercaptopropanoic acid; *S*-Benzoyl, *in* M-60027

$C_{10}H_{10}O_4$

1,4-Benzodioxan-2-carboxylic acid; Me ester, *in* B-60024
1,4-Benzodioxan-6-carboxylic acid; Me ester, *in* B-60025
4*H*-1,3-Benzodioxin-6-carboxylic acid; Me ester, *in* B-60027
1,3,5-Cycloheptatriene-1,6-dicarboxylic acid; Mono-Me ester, *in* C-60202
3,4-Dihydroxyphenylacetaldehyde; 3-Me ether, Ac, *in* D-60371
5,7-Dimethoxyphthalide, *in* D-60339
5-Hydroxy-7-methoxy-6-methylphthalide, *in* D-60349

$C_{10}H_{10}O_5$

2-Formyl-3,4-dimethoxybenzoic acid, *in* F-60070
4-Formyl-2,5-dimethoxybenzoic acid, *in* F-60072
6-Formyl-2,3-dimethoxybenzoic acid, *in* F-60073
2-Hydroxy-1,3-benzenedicarboxylic acid; Di-Me ester, *in* H-60101
2-Hydroxy-3,4-methylenedioxybenzoic acid; Me ether, Me ester, *in* H-60177
3-Hydroxy-4,5-methylenedioxybenzoic acid; Me ether, Me ester, *in* H-60178
4-Hydroxy-2,3-methylenedioxybenzoic acid; Me ether, Me ester, *in* H-60179

$C_{10}H_{10}O_6$

4,5-Dihydroxy-1,3-benzenedicarboxylic acid; Di-Me ester, *in* D-60307
4,5-Dimethoxy-1,3-benzenedicarboxylic acid, *in* D-60307

$C_{10}H_{11}BrO_2$

3-Bromo-2,4,6-trimethylbenzoic acid, B-60331

$C_{10}H_{11}ClO_2$

2-Chloro-2-phenylethanol; Ac, *in* C-60128

$C_{10}H_{11}ClO_3$

5-Chloro-4-hydroxy-2-methylbenzoic acid; Me ether, Me ester, *in* C-60097

$C_{10}H_{11}FO_4$

Methyl 4-fluoro-3,5-dimethoxybenzoate, *in* F-60024

$C_{10}H_{11}IO$

2-Iodo-1-phenyl-1-butanone, I-60060

$C_{10}H_{11}IO_4$

2,3-Dihydroxy-6-iodobenzoic acid; Di-Me ether, Me ester, *in* D-60330
2,4-Dihydroxy-5-iodobenzoic acid; Di-Me ether, Me ester, *in* D-60332
2,6-Dihydroxy-3-iodobenzoic acid; Di-Me ether, Me ester, *in* D-60334
3,4-Dihydroxy-5-iodobenzoic acid; Di-Me ether, Me ester, *in* D-60335
Methyl 2-iodo-3,5-dimethoxybenzoate, *in* D-60336
Methyl 4-iodo-3,5-dimethoxybenzoate, *in* D-60337

$C_{10}H_{11}N$

1-Amino-1,4-dihydronaphthalene, A-60139
1-Amino-5,8-dihydronaphthalene, A-60140
1,6-Dimethyl-1*H*-indole, *in* M-60088
2,2-Dimethyl-3-phenyl-2*H*-azirine, D-60440
2-(4-Methylphenyl)propanoic acid; Nitrile, *in* M-60114

$C_{10}H_{11}NOS$

4-Methyl-5-phenyl-2-oxazolidinethione, M-60111
Tetrahydro-2-phenyl-2*H*-1,2-thiazin-3-one, *in* T-60095

$C_{10}H_{11}NO_2$

2-Amino-3-phenyl-3-butenoic acid, A-60238
2-Amino-4-phenyl-3-butenoic acid, A-60239
5-Methyl-3-phenyl-2-oxazolidinone, *in* M-60107
4-Phenyl-3-morpholinone, *in* M-60147
1,2,3,4-Tetrahydro-1-isoquinolinecarboxylic acid, T-60070

$C_{10}H_{11}NO_3$

1-Amino-2-(4-hydroxyphenyl)cyclopropanecarboxylic acid, A-60197
3-Amino-4-oxo-4-phenylbutanoic acid, A-60236
4-Hydroxyphenylacetaldehyde; Ac, oxime, *in* H-60209

$C_{10}H_{11}NO_4$

2,3,4-Trimethyl-5-nitrobenzoic acid, T-60369
2,4,6-Trimethyl-3-nitrobenzoic acid, T-60370

$C_{10}H_{11}NO_5$

2',3'-Dimethoxy-6'-nitroacetophenone, *in* D-60357
3',6'-Dimethoxy-2'-nitroacetophenone, *in* D-60358
4',5'-Dimethoxy-2'-nitroacetophenone, *in* D-60359

$C_{10}H_{11}N_3O$

3,4-Diamino-1-methoxyisoquinoline, D-60037

$C_{10}H_{11}N_5$

2,4-Diamino-6-methyl-1,3,5-triazine; *N*-Ph, *in* D-60040

$C_{10}H_{11}N_5O_3$

6-Amino-1,3-dihydro-2*H*-purin-2-one; 6,9-Di-Ac, 1-Me, *in* A-60144

$C_{10}H_{12}$

5-Decene-2,8-diyne, D-60015
[4]Metacyclophane, M-60029
[4]Paracyclophane, P-60006
2,3,5-Tris(methylene)bicyclo[2.2.1]heptane, T-60403

$C_{10}H_{12}FN_5O_4$

8-Amino-6-fluoro-9*H*-purine; 9-β-D-Ribofuranosyl, *in* A-60163

$C_{10}H_{12}N_2O_3$

3-Amino-4(1*H*)-pyridinone; 1-*N*-Me, N^3,N^3-di-Ac, *in* A-60254

$C_{10}H_{12}N_2O_4$

10,11,14,15-Tetraoxa-1,8-diazatricyclo[6.4.4.0^{2,7}]-hexadeca-2(7),3,5-triene, T-60162

$C_{10}H_{12}N_4O_3$

Leucettidine, L-60023

$C_{10}H_{12}N_4O_4S$
▷1,7-Dihydro-6*H*-purine-6-thione; 9-β-D-Ribofuranosyl, *in* D-60270

$C_{10}H_{12}O$
1-Ethoxy-1-phenylethylene, *in* P-60092

$C_{10}H_{12}O_2$
3,4-Dihydro-2*H*-1-benzopyran-7-ol; Me ether, *in* D-60202
3,4-Dihydro-8-methoxy-2*H*-1-benzopyran, *in* D-60203
Dispiro[cyclopropane-1,5′-[3,8]dioxatricyclo[5.1.0.$0^{2,4}$]-octane-6′,1″-cyclopropane], D-60500
2-(2-Methylphenyl)propanoic acid, M-60113
2-(4-Methylphenyl)propanoic acid, M-60114
3,4,5-Trimethylbenzoic acid, T-60355

$C_{10}H_{12}O_2S$
2-Mercapto-3-phenylpropanoic acid; Me ester, *in* M-60026

$C_{10}H_{12}O_3$
2,5-Dimethoxybenzeneacetaldehyde, *in* D-60370
3,4-Dimethoxybenzeneacetaldehyde, *in* D-60371

$C_{10}H_{12}O_4$
Hallerone, H-60005
▷1-(4-Hydroxy-3,5-dimethoxyphenyl)ethanone, *in* T-60315

$C_{10}H_{12}O_5$
2,5-Bis(hydroxymethyl)furan; Di-Ac, *in* B-60160
3,4-Bis(hydroxymethyl)furan; Di-Ac, *in* B-60161
Gelsemide, G-60010

$C_{10}H_{13}I$
1-*tert*-Butyl-2-iodobenzene, B-60346
1-*tert*-Butyl-3-iodobenzene, B-60347
1-*tert*-Butyl-4-iodobenzene, B-60348

$C_{10}H_{13}NO$
2-(4-Methylphenyl)propanoic acid; Amide, *in* M-60114
7,8,9,10-Tetrahydropyrido[1,2-*a*]azepin-4(6*H*)-one, T-60085

$C_{10}H_{13}NO_2$
2-Amino-2-methyl-3-phenylpropanoic acid, A-60225
3-Amino-2,4,6-trimethylbenzoic acid, A-60266
5-Amino-2,3,4-trimethylbenzoic acid, A-60267

$C_{10}H_{13}NO_3$
2-Amino-2-methyl-3-(4-hydroxyphenyl)propanoic acid, A-60220

$C_{10}H_{13}NO_4$
2-Amino-3-hydroxy-4-(4-hydroxyphenyl)butanoic acid, A-60185

$C_{10}H_{13}N_5O_3$
2-Amino-6-(1,2-dihydroxypropyl)-3-methylpterin-4-one, A-60146

$C_{10}H_{13}N_5O_5$
6-Amino-1,3-dihydro-2*H*-purin-2-one; 9-(β-D-Arabinofuranosyl), *in* A-60144

$C_{10}H_{14}$
1,2,3,5,8,8*a*-Hexahydronaphthalene, H-60053
1,2,3,7,8,8*a*-Hexahydronaphthalene, H-60054
2-Methyl-6-methylene-1,3,7-octatriene, M-60089
Tricyclo[5.2.1.$0^{2,6}$]dec-2(6)-ene, T-60260

$C_{10}H_{14}N_2O_2$
2,4-Dihydroxyphenylacetaldehyde; Dimethylhydrazone, *in* D-60369
Octahydro-5*H*,10*H*-dipyrrolo[1,2-*a*:1′,2′-*d*]pyrazine-5,10-dione, O-60016

$C_{10}H_{14}N_2O_3$
Octahydro-2-hydroxy-5*H*,10*H*-dipyrrolo[1,2-*a*:1′,2′-*d*]-pyrazine-5,10-dione, O-60017

$C_{10}H_{14}N_2O_4$
Isoporphobilinogen, I-60114
Octahydro-2,7-dihydroxy-5*H*,10*H*-dipyrrolo[1,2-*a*:1′,2′-*d*]-pyrazine-5,10-dione, O-60013
Octahydro-5*a*,10*a*-dihydroxy-5*H*,10*H*-dipyrrolo[1,2-*a*:1′,2′-*d*]-pyrazine-5,10-dione, O-60014

$C_{10}H_{14}N_6O_5$
8-Aminoguanosine, *in* D-60035

$C_{10}H_{14}O$
3-Caren-5-one, C-60019
2,7-Cyclodecadien-1-one, C-60185
3,7-Cyclodecadien-1-one, C-60186
2-Cyclopentylidenecyclopentanone, C-60227
3,3-Dimethylbicyclo[2.2.2]oct-5-en-2-ene, D-60412
5-Methoxy-1,2,3-trimethylbenzene, *in* T-60371

$C_{10}H_{14}O_2$
4,4-Dimethylbicyclo[3.2.1]octane-2,3-dione, D-60410
3-Ethynyl-3-methyl-4-pentenoic acid; Et ester, *in* E-60059
▷1-(3-Furanyl)-4-methyl-1-pentanone, F-60087
2-Isopropyl-5-methyl-1,3-benzenediol, I-60122
2-Isopropyl-6-methyl-1,4-benzenediol, I-60123
3-Isopropyl-6-methyl-1,2-benzenediol, I-60124
5-Isopropyl-2-methyl-1,3-benzenediol, I-60125
5-Isopropyl-3-methyl-1,2-benzenediol, I-60126
5-Isopropyl-4-methyl-1,3-benzenediol, I-60127
Nepetalactone, N-60020
4-Oxatricyclo[4.3.1.$1^{3,8}$]undecan-5-one, O-60055
3,3,6,6-Tetramethyl-4-cyclohexene-1,2-dione, T-60134
5,5,6,6-Tetramethyl-2-cyclohexene-1,4-dione, T-60135

$C_{10}H_{14}O_3$
2-(Dimethoxymethyl)benzenemethanol, *in* H-60171
α-(Dimethoxymethyl)benzenemethanol, *in* H-60210

$C_{10}H_{14}O_4$
1,2,3,4-Tetramethoxybenzene, *in* B-60014

$C_{10}H_{14}S_2$
1,4-Dimethyl-2,5-bis(methylthio)benzene, *in* D-60406

$C_{10}H_{15}Cl$
1-Chloroadamantane, C-60039
2-Chloroadamantane, C-60040

$C_{10}H_{15}NO$
2-Cyclopentylidenecyclopentanone; Oxime, *in* C-60227
3-Ethynyl-3-methyl-4-pentenoic acid; *N*,*N*-Dimethylamide, *in* E-60059

$C_{10}H_{15}NO_4$
α-Allokainic acid, *in* K-60003
▷Kainic acid, K-60003

$C_{10}H_{15}N_3$
2-Azidoadamantane, A-60337

$C_{10}H_{16}N_2S_4$
2,3,5,6-Tetrakis(methylthio)-1,4-benzenediamine, *in* D-60031

$C_{10}H_{16}O$
1-Acetylcyclooctene, A-60027
3,3-Dimethylbicyclo[2.2.2]octan-2-one, D-60411
2-Isopropyl-5-methyl-4-cyclohexen-1-one, I-60128
2,2,5,5-Tetramethyl-3-cyclohexen-1-one, T-60136

$C_{10}H_{16}O_2$
2-Hydroperoxy-2-methyl-6-methylene-3,7-octadiene, H-60097
3-Hydroperoxy-2-methyl-6-methylene-1,7-octadiene, H-60098
Iridodial, I-60084
Lineatin, L-60027
Spiro[bicyclo[3.2.1]octane-8,2′-[1,3]dioxolane], *in* B-60094

$C_{10}H_{16}O_3$
1,7-Dihydroxy-3,7-dimethyl-2,5-octadien-4-one, D-60318

$C_{10}H_{16}O_4$
Gelsemiol, G-60011
7-Hydroxy-1-hydroxy-3,7-dimethyl-2*E*,5*E*-octadien-4-one, *in* D-60318
2,2,5,5-Tetramethyl-3-hexenedioic acid, T-60141

$C_{10}H_{16}S_2$
1,3-Bis(allylthio)cyclobutane, *in* C-60184

$C_{10}H_{17}N$
4-Azatricyclo[4.3.1.$1^{3,8}$]undecane, A-60333

$C_{10}H_{18}$
1-*tert*-Butylcyclohexene, B-60341
3-*tert*-Butylcyclohexene, B-60342
4-*tert*-Butylcyclohexene, B-60343

$C_{10}H_{18}N_2O_4$
2*H*-1,5,2,4-Dioxadiazine-3,6(4*H*)dione; 2,4-Di-*tert*-butyl, *in* D-60466

$C_{10}H_{18}N_2O_6S_2$
γ-Glutamylmarasmine, G-60026

$C_{10}H_{18}N_4O_4$
1,2-Dihydrazinoethane; $N^{\alpha},N^{\alpha'},N^{\beta},N^{\beta'}$-Tetra-Ac, in D-60192

$C_{10}H_{18}O$
3,7-Dimethyl-6-octenal, D-60430
Hexamethylcyclobutanone, H-60069
2,2,6,6-Tetramethylcyclohexanone, T-60133
2,2,6-Trimethyl-5-hepten-3-one, T-60365

$C_{10}H_{18}OS$
Di-tert-butylthioketene; S-Oxide, in D-60114

$C_{10}H_{18}O_2$
▷5-Hexyldihydro-2(3H)-furanone, H-60078
2-Octen-1-ol; Ac, in O-60029

$C_{10}H_{18}O_3$
10-Oxodecanoic acid, O-60063

$C_{10}H_{18}S$
Di-tert-butylthioketene, D-60114

$C_{10}H_{19}Br$
5-Bromo-5-decene, B-60225

$C_{10}H_{19}NO$
2,2,6,6-Tetramethylcyclohexanone; Oxime, in T-60133

$C_{10}H_{19}NO_2$
3-Amino-2-hexenoic acid; tert-Butyl ester, in A-60170

$C_{10}H_{19}NO_3$
2-Amino-1-hydroxy-1-cyclobutaneacetic acid; tert-Butyl ester, in A-60184
2-Amino-3-hydroxy-4-methyl-6-octenoic acid; N-Me, in A-60189
10-Oxodecanoic acid; Oxime, in O-60063

$C_{10}H_{20}$
Methylcyclononane, M-60058

$C_{10}H_{20}N_2$
1,5-Diazabicyclo[5.2.2]undecane; N-Me, in D-60055

$C_{10}H_{20}O$
5-Decanone, D-60014
2-Decen-4-ol, D-60016
2,2,3,3,4,4-Hexamethylcyclobutanol, H-60068

$C_{10}H_{20}O_4$
3,4-Dihydroxybutanoic acid; 4-O-tert-Butyl, Et ester, in D-60313

$C_{10}H_{22}N_2$
Hexahydropyrimidine; 1,3-Diisopropyl, in H-60056

$C_{10}H_{22}N_2O_3$
1,4,10-Trioxa-7,13-diazacyclopentadecane, T-60376

$C_{10}H_{22}N_4O_2S_2$
Alethine, A-60073

$C_{10}H_{22}O_2$
5,6-Decanediol, D-60013

$C_{10}H_{23}N_3$
1,5,9-Triazacyclotridecane, T-60233

$C_{10}H_{25}N_5$
1,4,7,10,13-Pentaazacyclopentadecane, P-60025

$C_{11}H_6N_4$
9-Diazo-9H-cyclopenta[1,2-b:4,3-b']dipyridine, D-60059

$C_{11}H_6N_6$
Pyrimido[4,5-i]imidazo[4,5-g]cinnoline, P-60244

$C_{11}H_7BrO$
1-Bromo-2-naphthalenecarboxaldehyde, B-60302

$C_{11}H_7BrO_3$
3-Bromo-5-hydroxy-2-methyl-1,4-naphthoquinone, B-60266
2-Bromo-6-methoxy-1,4-naphthaquinone, in B-60270
2-Bromo-3-methoxy-1,4-naphthoquinone, in B-60268
2-Bromo-5-methoxy-1,4-naphthoquinone, in B-60269

2-Bromo-7-methoxy-1,4-naphthoquinone, in B-60271
2-Bromo-8-methoxy-1,4-naphthoquinone, in B-60272

$C_{11}H_7ClO_3$
3-Chloro-5-hydroxy-2-methyl-1,4-naphthoquinone, C-60098
2-Oxo-2H-benzopyran-4-acetic acid; Chloride, in O-60057

$C_{11}H_7NO$
9H-Pyrrolo[1,2-a]indol-9-one, P-60247

$C_{11}H_7NOSe$
[1,4]Benzoxaselenino[3,2-b]pyridine, B-60054

$C_{11}H_7NO_2$
4-(Cyanomethyl)coumarin, in O-60057
4-Oxo-4H-1-benzopyran-3-acetic acid; Nitrile, in O-60059

$C_{11}H_7NS$
Azuleno[1,2-d]thiazole, A-60356
Azuleno[2,1-d]thiazole, A-60357
2-(Phenylethynyl)thiazole, P-60099
4-(Phenylethynyl)thiazole, P-60100

$C_{11}H_7N_3O_2$
Pyrimido[4,5-b]quinoline-2,4(3H,10H)-dione, P-60245

$C_{11}H_8Br_2$
▷1-Bromo-2-(bromomethyl)naphthalene, B-60212
1-Bromo-4-(bromomethyl)naphthalene, B-60213
1-Bromo-5-(bromomethyl)naphthalene, B-60214
1-Bromo-7-(bromomethyl)naphthalene, B-60215
1-Bromo-8-(bromomethyl)naphthalene, B-60216
2-Bromo-3-(bromomethyl)naphthalene, B-60217
2-Bromo-6-(bromomethyl)naphthalene, B-60218
3-Bromo-1-(bromomethyl)naphthalene, B-60219
6-Bromo-1-(bromomethyl)naphthalene, B-60220
7-Bromo-1-(bromomethyl)naphthalene, B-60221
1,2-Dibromo-1,4-dihydro-1,4-methanonaphthalene, D-60095
1,3-Dibromo-1,4-dihydro-1,4-methanonaphthalene, D-60096

$C_{11}H_8N_2O_2$
5-Benzoyl-2(1H)-pyrimidinone, B-60066

$C_{11}H_8N_2O_2S$
1,1'-Carbonothioylbis-2(1H)pyridinone, C-60014

$C_{11}H_8O_2$
Homoazulene-1,5-quinone, H-60085
Homoazulene-1,7-quinone, H-60086
Homoazulene-4,7-quinone, H-60087

$C_{11}H_8O_3$
5-Hydroxy-7-methyl-1,2-naphthoquinone, H-60189
8-Hydroxy-3-methyl-1,2-naphthoquinone, H-60190
4-Methoxy-1,2-naphthoquinone, in H-60194
Pentacyclo[6.3.0.0^{2,6}.0^{5,9}]undecane-4,7,11-trione, P-60031

$C_{11}H_8O_4$
3,5-Dihydroxy-2-methyl-1,4-naphthoquinone, D-60350
5,6-Dihydroxy-2-methyl-1,4-naphthoquinone, D-60351
5,8-Dihydroxy-2-methyl-1,4-naphthoquinone, D-60352
2-Hydroxy-8-methoxy-1,4-naphthoquinone, in D-60355
2-Oxo-2H-benzopyran-4-acetic acid, O-60057
1-Oxo-1H-2-benzopyran-3-acetic acid, O-60058
4-Oxo-4H-1-benzopyran-3-acetic acid, O-60059

$C_{11}H_8O_5$
4-Hydroxy-5-benzofurancarboxylic acid; Ac, in H-60103
2-Hydroxy-5-methoxy-1,4-naphthoquinone, in D-60354

$C_{11}H_8O_8$
2,3,5,6,8-Pentahydroxy-7-methoxy-1,4-naphthoquinone, in H-60065

$C_{11}H_9BF_4N_4$
3-Phenyltetrazolo[1,5-a]pyridinium; Tetrafluoroborate, in P-60134

$C_{11}H_9Br$
1-Bromo-1,4-dihydro-1,4-methanonaphthalene, B-60231
2-Bromo-1,4-dihydro-1,4-methanonaphthalene, B-60232
9-Bromo-1,4-dihydro-1,4-methanonaphthalene, B-60233

$C_{11}H_9BrN_4$
3-Phenyltetrazolo[1,5-a]pyridinium; Bromide, in P-60134

C₁₁H₉NO

1-Acetylisoquinoline, A-60039
3-Acetylisoquinoline, A-60040
4-Acetylisoquinoline, A-60041
5-Acetylisoquinoline, A-60042
2-Acetylquinoline, A-60049
3-Acetylquinoline, A-60050
4-Acetylquinoline, A-60051
5-Acetylquinoline, A-60052
6-Acetylquinoline, A-60053
7-Acetylquinoline, A-60054
8-Acetylquinoline, A-60055
2-Phenyl-1,3-oxazepine, P-60123
5-Phenyl-1,4-oxazepine, P-60124

C₁₁H₉NO₂

4-Acetylquinoline; 1-Oxide, *in* A-60051
Cyclohepta[*b*]pyrrol-2(1*H*)-one; *N*-Ac, *in* C-60201

C₁₁H₉NO₃

2-Oxo-2*H*-benzopyran-4-acetic acid; Amide, *in* O-60057
4-Oxo-4*H*-1-benzopyran-3-acetic acid; Amide, *in* O-60059

C₁₁H₉N₃

1-Methyl-8*H*-benzo[*cd*]triazirino[*a*]indazole, M-60052
9*H*-Pyrrolo[1,2-*a*]indol-9-one; Hydrazone, *in* P-60247

C₁₁H₉N₃O₃

8-Amino-6-nitroquinoline; 8-*N*-Ac, *in* A-60235

C₁₁H₉N₄⊕

3-Phenyltetrazolo[1,5-*a*]pyridinium, P-60134

C₁₁H₁₀

1,4-Dihydro-1,4-methanonaphthalene, D-60256

C₁₁H₁₀IN₃

6*H*-Dipyrido[1,2-*a*:2′,1′-*d*][1,3,5]triazin-5-ium; Iodide, *in* D-60495

C₁₁H₁₀N₂

3,4-Dihydro-β-carboline, D-60209
Di-2-pyridylmethane, D-60496

C₁₁H₁₀N₂O

1-Acetylisoquinoline; Oxime, *in* A-60039
4-Acetylisoquinoline; Oxime, *in* A-60041
2-Acetylquinoline; Oxime, *in* A-60049
5-Acetylquinoline; Oxime, *in* A-60052
6-Acetylquinoline; Oxime, *in* A-60053
7-Acetylquinoline; Oxime, *in* A-60054
8-Acetylquinoline; Oxime, *in* A-60055
1-Methyl-2-phenyl-1*H*-imidazole-4-carboxaldehyde, *in* P-60111
1-Methyl-2-phenyl-1*H*-imidazole-5-carboxaldehyde, *in* P-60111

C₁₁H₁₀N₂OS₂

Spirobrassinin, S-60041

C₁₁H₁₀N₂O₂

5-Amino-3-phenylisoxazole; *N*-Ac, *in* A-60240
3-Amino-2(1*H*)-quinolinone; 3-*N*-Ac, *in* A-60256

C₁₁H₁₀N₃⊕

6*H*-Dipyrido[1,2-*a*:2′,1′-*d*][1,3,5]triazin-5-ium, D-60495

C₁₁H₁₀N₄O₂

lin-Benzotheophylline, *in* I-60009

C₁₁H₁₀O

Hexacyclo[5.4.0.0²,⁶.0³,¹⁰.0⁵,⁹.0⁸,¹¹]undecan-4-one, H-60035
2-Naphthalenemethanol, N-60004

C₁₁H₁₀O₂

2-Methyl-1,3-naphthalenediol, M-60091
2-Methyl-1,5-naphthalenediol, M-60092
3-Methyl-1,2-naphthalenediol, M-60093
4-Methyl-1,2-naphthalenediol, M-60094
6-Methyl-1,2-naphthalenediol, M-60095
Pentacyclo[5,4.0.0²,⁶.0³,¹⁰.0⁵,⁹]undecane-1,11-dione, P-60030

C₁₁H₁₀O₃

7-Hydroxy-3,4-dimethyl-2*H*-1-benzopyran-2-one, H-60119
5-Methoxy-4-methylcoumarin, *in* H-60173

C₁₁H₁₀O₄

8,13-Dioxapentacyclo[6.5.0.0²,⁶.0⁵,¹⁰.0³,¹¹]tridecane-9,12-dione, D-60467
Methyl 3-benzoyloxyacrylate, *in* O-60090

C₁₁H₁₀O₅

2,7-Dihydroxy-2,4,6-cycloheptatrien-1-one; Di-Ac, *in* D-60314

C₁₁H₁₁BrN₂O₂

2-Bromotryptophan, B-60332
5-Bromotryptophan, B-60333
6-Bromotryptophan, B-60334
7-Bromotryptophan, B-60335

C₁₁H₁₁ClN₂O₂

2-Chlorotryptophan, C-60145
6-Chlorotryptophan, C-60146

C₁₁H₁₁N

▷2-Methyl-5-phenyl-1*H*-pyrrole, M-60115
3-(2-Propenyl)indole, P-60183

C₁₁H₁₁NO

1-(2-Furanyl)-2-(2-pyrrolyl)ethylene; *N*-Me, *in* F-60088
1-(4-Methoxyphenyl)pyrrole, *in* H-60216
2-(2-Methoxyphenyl)pyrrole, *in* H-60217

C₁₁H₁₁NO₂

3,3-Dimethyl-2,4(1*H*,3*H*)-quinolinedione, D-60451

C₁₁H₁₁NO₂S

α-Aminobenzo[*b*]thiophene-3-acetic acid; Me ester, *in* A-60090
3-Amino-2-benzo[*b*]thiophenecarboxylic acid; Et ester, *in* A-60091

C₁₁H₁₁NO₃

3-Aminodihydro-2(3*H*)-furanone; *N*-Benzoyl, *in* A-60138
2,3-Dihydro-2-oxo-1*H*-indole-3-acetic acid; Me ester, *in* D-60259
3-Morpholinone; 4-Benzoyl, *in* M-60147
1,2,3,4-Tetrahydro-2-oxo-4-quinolinecarboxylic acid; Me ester, *in* T-60077
1,2,3,4-Tetrahydro-4-oxo-6-quinolinecarboxylic acid; Me ester, *in* T-60078

C₁₁H₁₁NS

1-(2-Pyrrolyl)-2-(2-thienyl)ethylene; *N*-Me, *in* P-60250

C₁₁H₁₁N₃O₂

Methyl 1-benzyl-1*H*-1,2,3-triazole-5-carboxylate, *in* T-60239
1-Phenyl-1*H*-1,2,3-triazole-4-carboxylic acid; Et ester, *in* P-60138
1-Phenyl-1*H*-1,2,3-triazole-5-carboxylic acid; Et ester, *in* P-60139
2-Phenyl-2*H*-1,2,3-triazole-4-carboxylic acid; Et ester, *in* P-60140
1*H*-1,2,3-Triazole-4-carboxylic acid; 1-Benzyl, Me ester, *in* T-60239

C₁₁H₁₂

1,4-Methano-1,2,3,4-tetrahydronaphthalene, M-60034
1,2,3,4-Tetrahydro-1-methylenenaphthalene, T-60073

C₁₁H₁₂N₂O

5-Amino-3-phenylisoxazole; *N*-Et, *in* A-60240

C₁₁H₁₂N₂O₄

2,4-Diaminobenzoic acid; 2,4-*N*-Di-Ac, *in* D-60032

C₁₁H₁₂N₂O₅

2′-Deoxy-5-ethynyluridine, *in* E-60060

C₁₁H₁₂N₂O₆

5-Ethynyluridine, *in* E-60060

C₁₁H₁₂O

2,4,6-Cycloheptatrien-1-ylcyclopropylmethanone, C-60206
2,3-Dihydro-2,2-dimethyl-1*H*-inden-1-one, D-60227
1-Phenyl-3-penten-1-one, P-60127
1,2,3,4-Tetrahydro-7*H*-benzocyclohepten-7-one, T-60057

C₁₁H₁₂O₂

2-Phenylcyclopropanecarboxylic acid; Me ester, *in* P-60087

C$_{11}$H$_{12}$O$_2$S
4-Mercapto-2-butenoic acid; Benzyl ester, *in* M-60018

C$_{11}$H$_{12}$O$_3$
3-Benzoyl-2-methylpropanoic acid, B-60064
3,4-Dihydro-2*H*-1-benzopyran-2-ol; Ac, *in* D-60198
3,4-Dihydro-2*H*-1-benzopyran-4-ol; Ac, *in* D-60200

C$_{11}$H$_{12}$O$_4$
1,3,5-Cycloheptatriene-1,6-dicarboxylic acid; Di-Me ester, *in* C-60202
5,7-Dimethoxy-6-methylphthalide, *in* D-60349
Pyrenocin *A*, P-60213

C$_{11}$H$_{12}$O$_5$
2-Formyl-3,5-dihydroxybenzoic acid; Di-Me ether, Me ester, *in* F-60071

C$_{11}$H$_{13}$BrO$_2$
3-Bromo-2,4,6-trimethylbenzoic acid; Me ester, *in* B-60331

C$_{11}$H$_{13}$IO$_4$
2,5-Dihydroxy-4-iodobenzoic acid; Di-Me ether, Et ester, *in* D-60333

C$_{11}$H$_{13}$N
1,2,3,6-Tetrahydro-4-phenylpyridine, T-60081

C$_{11}$H$_{13}$NO
3,4-Dihydro-3,3-dimethyl-2(1*H*)quinolinone, D-60231
4-(Phenylimino)-2-pentanone, *in* P-60059

C$_{11}$H$_{13}$NO$_2$
2-Amino-3-phenyl-3-butenoic acid; Me ester, *in* A-60238
1,2,3,4-Tetrahydro-4-hydroxyisoquinoline; *N*-Ac, *in* T-60066
1,2,3,4-Tetrahydro-3-methyl-3-isoquinolinecarboxylic acid, T-60074
5-(3,4,5,6-Tetrahydro-3-pyridylidenemethyl)-2-furanmethanol, T-60087

C$_{11}$H$_{13}$NO$_3$
1-Amino-2-(4-hydroxyphenyl)cyclopropanecarboxylic acid; Me ester, *in* A-60197
1-Amino-2-(4-hydroxyphenyl)cyclopropanecarboxylic acid; Me ether, *in* A-60197
2,3,4-Trihydroxyphenylacetic acid; Tri-Me ether, nitrile, *in* T-60340
2,4,5-Trihydroxyphenylacetic acid; Tri-Me ether, nitrile, *in* T-60342
▷3,4,5-Trimethoxybenzeneacetonitrile, *in* T-60343

C$_{11}$H$_{13}$NO$_4$
2-Amino-2-benzylbutanedioic acid, A-60094
2-Amino-3-phenylpentanedioic acid, A-60241
3,5-Pyridinedipropanoic acid, P-60217
Threonine; *N*-Benzoyl, *in* T-60217

C$_{11}$H$_{13}$N$_3$O$_4$
2'-Deoxy-5-ethynylcytidine, *in* A-60153
Imidazo[4,5-*c*]pyridine; 1-β-D-Ribofuranosyl, *in* I-60008

C$_{11}$H$_{13}$N$_3$O$_5$
5-Ethynylcytidine, *in* A-60153

C$_{11}$H$_{14}$NO$_3$
3-Amino-2-hydroxybutanoic acid; *N*-Benzoyl, *in* A-60183

C$_{11}$H$_{14}$N$_2$
5,6,9,10-Tetrahydro-4*H*,8*H*-pyrido[3,2,1-*ij*][1,6]-naphthyridine, T-60086

C$_{11}$H$_{14}$N$_4$
4-(Butylimino)-3,4-dihydro-1,2,3-benzotriazine (incorr.), *in* A-60092

C$_{11}$H$_{14}$N$_4$O$_4$S
▷1,7-Dihydro-6*H*-purine-6-thione; 5-Me, 9-β-D-ribofuranosyl, *in* D-60270

C$_{11}$H$_{14}$O
2,2-Dimethyl-1-phenyl-1-propanone, D-60443
Spiro[5.5]undeca-1,3-dien-7-one, S-60046

C$_{11}$H$_{14}$O$_2$
Andirolactone, A-60271

3,4,5-Trimethylbenzoic acid; Me ester, *in* T-60355
3,4,5-Trimethylphenol; Ac, *in* T-60371

C$_{11}$H$_{14}$O$_4$
2-(3,4-Dihydroxyphenyl)propanoic acid; Di-Me ether, *in* D-60372
3',4',5'-Trimethoxyacetophenone, *in* T-60315

C$_{11}$H$_{14}$O$_5$
Pyrenocin B, *in* P-60213
2,3,4-Trimethoxybenzeneacetic acid, *in* T-60340
2,4,5-Trimethoxybenzeneacetic acid, *in* T-60342
3,4,5-Trimethoxybenzeneacetic acid, *in* T-60343

C$_{11}$H$_{14}$O$_6$
2,3,5,6-Tetramethoxybenzoic acid, *in* T-60100
(3,4,5-Trihydroxy-6-oxo-1-cyclohexen-1-yl)methyl 2-butenoate, T-60338

C$_{11}$H$_{15}$Br
(Bromomethylidene)adamantane, B-60296
Bromopentamethylbenzene, B-60312

C$_{11}$H$_{15}$Cl
Chloropentamethylbenzene, C-60122

C$_{11}$H$_{15}$N
2-Phenylcyclopentylamine, P-60086

C$_{11}$H$_{15}$NO$_2$
2-Amino-2-methyl-3-phenylpropanoic acid; Me ester, *in* A-60225
1,2,3,4,5-Pentamethyl-6-nitrobenzene, P-60057

C$_{11}$H$_{15}$NO$_2$S
2-Amino-4,5,6,7-tetrahydrobenzo[*b*]thiophene-3-carboxylic acid; Et ester, *in* A-60258

C$_{11}$H$_{15}$NO$_3$
2-Amino-3-(4-hydroxyphenyl)-3-methylbutanoic acid, A-60198
4-Amino-3-hydroxy-5-phenylpentanoic acid, A-60199

C$_{11}$H$_{15}$NO$_4$
3',4',5'-Trihydroxyacetophenone; Tri-Me ether, oxime, *in* T-60315

C$_{11}$H$_{15}$N$_5$O
1'-Methylzeatin, M-60133

C$_{11}$H$_{15}$N$_5$O$_4$
Euglenapterin, E-60070

C$_{11}$H$_{15}$N$_5$O$_5$
Ara-doridosine, *in* A-60144
Doridosine, D-60518

C$_{11}$H$_{16}$
(2,2-Dimethylpropyl)benzene, D-60445
Methyleneadamantane, M-60068
Tetracyclo[4.4.1.0^{3,11}.0^{9,11}]undecane, T-60037

C$_{11}$H$_{16}$N$_2$O$_5$
▷2'-Deoxy-5-ethyluridine, *in* E-60054

C$_{11}$H$_{16}$N$_2$O$_6$
5-Ethyluridine, *in* E-60054

C$_{11}$H$_{16}$O
5,6,7,8,9,10-Hexahydro-4*H*-cyclonona[*c*]furan, H-60047
1-Methyladamantanone, M-60040
5-Methyladamantanone, M-60041
Pentamethylphenol, P-60058

C$_{11}$H$_{16}$O$_4$
▷Xanthotoxol, X-60002

C$_{11}$H$_{16}$O$_8$
Ranunculin, *in* H-60182

C$_{11}$H$_{17}$NO$_7$
2-(Dimethylamino)benzaldehyde; Di-Me acetal, *in* D-60401

C$_{11}$H$_{18}$
2,2-Dimethyl-3-methylenebicyclo[2.2.2]octane, D-60427

C$_{11}$H$_{18}$N$_2$O$_2$
Hexahydro-3-(1-methylpropyl)pyrrolo[1,2-*a*]pyrazine-1,4-dione, H-60052

C₁₁H₁₈N₂O₃

Cyclo(hydroxyprolylleucyl), C-60213

C₁₁H₁₈O₂

Bicyclo[2.2.2]octane-1-carboxylic acid; Et ester, *in* B-60092
2-Hexyl-5-methyl-3(2*H*)furanone, H-60079

C₁₁H₁₈O₄

Citreoviral, C-60150
Phaseolinic acid, P-60067

C₁₁H₁₈O₅

4-Oxoheptanedioic acid; Di-Et ester, *in* O-60071

C₁₁H₂₀

1,1,3,3-Tetramethyl-2-methylenecyclohexane, T-60144

C₁₁H₂₀O

6-Undecyn-1-ol, U-60004

C₁₁H₂₀O₂

2,4-Heptadienal; Di-Et acetal, *in* H-60020

C₁₁H₂₀O₃

10-Oxodecanoic acid; Me ester, *in* O-60063

C₁₁H₂₁Br

11-Bromo-1-undecene, B-60336

C₁₁H₂₂

1,1-Di-*tert*-butylcyclopropane, D-60112

C₁₁H₂₂N₂O₃

N-Valylleucine, V-60001

C₁₁H₂₄

2,2,3,3,4,4-Hexamethylpentane, H-60071

C₁₁H₂₇N₅

1,4,7,10,13-Pentaazacyclohexadecane, P-60021

C₁₂H₄Cl₆

▷2,2′,4,4′,5,5′-Hexachlorobiphenyl, H-60034

C₁₂H₆Cl₂N₂O₂

[2,2′-Bipyridine]-4,4′-dicarboxylic acid; Dichloride, *in* B-60126
[2,2′-Bipyridine]-6,6′-dicarboxylic acid; Dichloride, *in* B-60128

C₁₂H₆Cl₁₂

Hexakis(dichloromethyl)benzene, H-60067

C₁₂H₆N₂

2,3-Dicyanonaphthalene, *in* N-60003

C₁₂H₆N₂O₂

Benzo[*g*]quinazoline-6,9-dione, B-60044
Benzo[*g*]quinoxaline-6,9-dione, B-60046

C₁₂H₆N₄

5,5′-Dicyano-2,2′-bipyridine, *in* B-60127
Pyrazino[2′,3′:3,4]cyclobuta[1,2-*g*]quinoxaline, P-60198
1,2,4,5-Tetracyano-3,6-dimethylbenzene, *in* D-60407

C₁₂H₇BrO₄

2-Bromo-5-hydroxy-1,4-naphthoquinone; Ac, *in* B-60269
2-Bromo-6-hydroxy-1,4-naphthoquinone; Ac, *in* B-60270
2-Bromo-8-hydroxy-1,4-naphthoquinone; Ac, *in* B-60272

C₁₂H₇NOS

1*H*-Phenothiazin-1-one, P-60078

C₁₂H₇NO₂

10*H*-[1]Benzopyrano[3,2-*c*]pyridin-10-one, B-60042
1*H*-Carbazole-1,4(9*H*)-dione, C-60012
3-Cyano-2-naphthoic acid, *in* N-60003
2,3-Naphthalenedicarboxylic acid; Imide, *in* N-60003

C₁₂H₇NO₃

2-Hydroxy-3*H*-phenoxazin-3-one, H-60206

C₁₂H₇N₅

7*H*-2,3,4,6,7-Pentaazabenz[*de*]anthracene, P-60017

C₁₂H₈ClNOS

3-Chloro-10*H*-phenothiazine; 5-Oxide, *in* C-60126
4-Chloro-10*H*-phenothiazine; 5-Oxide, *in* C-60127

C₁₂H₈ClNO₂S

3-Chloro-10*H*-phenothiazine; 5,5-Dioxide, *in* C-60126
4-Chloro-10*H*-phenothiazine; 5,5-Dioxide, *in* C-60127

C₁₂H₈ClNS

1-Chloro-10*H*-phenothiazine, C-60124
▷2-Chloro-10*H*-phenothiazine, C-60125
3-Chloro-10*H*-phenothiazine, C-60126

C₁₂H₈Cl₂

5,6-Dichloroacenaphthene, D-60115

C₁₂H₈F₃NO₃

7-Amino-4-(trifluoromethyl)-2*H*-1-benzopyran-2-one; *N*-Ac, *in* A-60265

C₁₂H₈N₂

Benzo[*c*]cinnoline, B-60020

C₁₂H₈N₂O

Benzo[*c*]cinnoline; *N*-Oxide, *in* B-60020
2-(2-Furanyl)quinoxaline, F-60089

C₁₂H₈N₂O₂

▷2-Amino-3*H*-phenoxazin-3-one, A-60237
Benzo[*c*]cinnoline; 5,6-Di-*N*-oxide, *in* B-60020

C₁₂H₈N₂O₄

[2,2′-Bipyridine]-3,3′-dicarboxylic acid, B-60124
[2,2′-Bipyridine]-3,5′-dicarboxylic acid, B-60125
[2,2′-Bipyridine]-4,4′-dicarboxylic acid, B-60126
[2,2′-Bipyridine]-5,5′-dicarboxylic acid, B-60127
[2,2′-Bipyridine]-6,6′-dicarboxylic acid, B-60128
[2,3′-Bipyridine]-2,3′-dicarboxylic acid, B-60129
[2,4′-Bipyridine]-2′,6′-dicarboxylic acid, B-60130
[2,4′-Bipyridine]-3,3′-dicarboxylic acid, B-60131
[2,4′-Bipyridine]-3′,5-dicarboxylic acid, B-60132
[3,3′-Bipyridine]-2,2′-dicarboxylic acid, B-60133
[3,3′-Bipyridine]-4,4′-dicarboxylic acid, B-60134
[3,4′-Bipyridine]-2′,6′-dicarboxylic acid, B-60135
[4,4′-Bipyridine]-2,2′-dicarboxylic acid, B-60136
[4,4′-Bipyridine]-3,3′-dicarboxylic acid, B-60137
2,2′-Dinitrobiphenyl, D-60460

C₁₂H₈N₄

Pyrido[2″,1″:2′,3′]imidazo[4′,5′:4,5]imidazo[1,2-*a*]pyridine, P-60234

C₁₂H₈N₁₀O₁₂

3,3′,5,5′-Tetraamino-2,2′,4,4′,6,6′-hexanitrobiphenyl, T-60014

C₁₂H₈O

Naphtho[2,3-*c*]furan, N-60008
1-Phenyl-1,4-hexadiyn-3-one, P-60108
▷1-Phenyl-2,4-hexadiyn-1-one, P-60109
6-Phenyl-3,5-hexadiyn-2-one, P-60110

C₁₂H₈OS

4-Phenylthieno[3,4-*b*]furan, P-60135

C₁₂H₈O₂

1,2-Dihydrocyclobuta[*a*]naphthalene-3,4-dione, D-60213
1,2-Dihydrocyclobuta[*b*]naphthalene-3,8-dione, D-60214
1,2-Naphthalenedicarboxaldehyde, N-60001
1,3-Naphthalenedicarboxaldehyde, N-60002

C₁₂H₈O₃

α-Oxo-1-naphthaleneacetic acid, O-60078
α-Oxo-2-naphthaleneacetic acid, O-60079

C₁₂H₈O₄

2,3-Naphthalenedicarboxylic acid, N-60003
▷Xanthotoxin, X-60001

C₁₂H₈O₇

2,5,8-Trihydroxy-3-methyl-6,7-methylenedioxy-1,4-naphthoquinone, T-60333

C₁₂H₈S₂

2-Dibenzothiophenethiol, D-60078
4-Dibenzothiophenethiol, D-60079

C₁₂H₉BrO

1-Acetyl-3-bromonaphthalene, A-60021
1-Acetyl-4-bromonaphthalene, A-60022

$C_{12}H_{14}O_3$

Senkyunolide F, *in* L-60026
(Z)-6,7-Epoxyligustilide, *in* L-60026

$C_{12}H_{14}O_3S$

2-Mercapto-3-methylbutanoic acid; *S*-Benzoyl, *in* M-60023

$C_{12}H_{14}O_4$

4-Hydroxy-3-(2-hydroxy-3-methyl-3-butenyl)benzoic acid, H-60154

$C_{12}H_{14}O_5$

2-Hydroxy-1,3-benzenedicarboxylic acid; Di-Et ester, *in* H-60101

$C_{12}H_{14}O_6$

4,5-Dihydroxy-1,3-benzenedicarboxylic acid; Di-Me ether, di-Me ester, *in* D-60307

$C_{12}H_{15}^{\oplus}$

Tricyclopropylcyclopropenium(1+), T-60270

$C_{12}H_{15}BF_4$

Tricyclopropylcyclopropenium(1+); Tetrafluoroborate, *in* T-60270

$C_{12}H_{15}Cl$

Tricyclopropylcyclopropenium(1+); Chloride, *in* T-60270

$C_{12}H_{15}F_6Sb$

Tricyclopropylcyclopropenium(1+); Hexafluoroantimonate, *in* T-60270

$C_{12}H_{15}N$

▷1,2,3,6-Tetrahydro-1-methyl-4-phenylpyridine, *in* T-60081

$C_{12}H_{15}NO_2$

2-Amino-3-methyl-3-butenoic acid; Benzyl ester, *in* A-60218
1,2,3,4-Tetrahydro-1-isoquinolinecarboxylic acid; Et ester, *in* T-60070

$C_{12}H_{15}NO_3$

2-Amino-2-methyl-3-phenylpropanoic acid; *N*-Ac, *in* A-60225
2′,3′,4′,5′-Tetramethyl-6′-nitroacetophenone, T-60145
2′,3′,4′,6′-Tetramethyl-5′-nitroacetophenone, T-60146
2′,3′,5′,6′-Tetramethyl-4′-nitroacetophenone, T-60147

$C_{12}H_{16}$

Tetracyclo[6.2.1.1³,⁶.0²,⁷]dodec-2(7)-ene, T-60033

$C_{12}H_{16}BrClO$

Furocaespitane, F-60090

$C_{12}H_{16}BrClO_2$

5-(3-Bromo-4-chloro-4-methylcyclohexyl)-5-methyl-2(5*H*)-furanone, B-60222

$C_{12}H_{16}N_2O_4$

N-Tyrosylalanine, T-60411

$C_{12}H_{16}O$

4-(4-Methylphenyl)-2-pentanone, M-60112

$C_{12}H_{16}O_2$

2-(4-Methylphenyl)propanoic acid; Et ester, *in* M-60114
Tricyclo[4.4.2.0¹,⁶]dodecane-2,7-dione, T-60261

$C_{12}H_{16}O_4$

Senkyunolide H, *in* L-60026

$C_{12}H_{16}O_5$

2,3,4-Trihydroxyphenylacetic acid; Tri-Me ether, Me ester, *in* T-60340
2,4,5-Trihydroxyphenylacetic acid; Tri-Me ether, Me ester, *in* T-60342
3,4,5-Trihydroxyphenylacetic acid; *tert*-Butyl ester, *in* T-60343

$C_{12}H_{16}O_6$

Pentamethoxybenzaldehyde, *in* P-60042

$C_{12}H_{16}O_7$

Pentamethoxybenzoic acid, *in* P-60043

$C_{12}H_{17}NO_2$

2-Amino-2-methyl-3-phenylpropanoic acid; Et ester, *in* A-60225

$C_{12}H_{17}NO_3$

2-Amino-3-(4-methoxyphenyl)-3-methylbutanoic acid, *in* A-60198

$C_{12}H_{17}NO_7$

Epivolkenin, *in* T-60007
Taraktophyllin, T-60007

$C_{12}H_{18}$

2-Isopropyl-1,3,5-trimethylbenzene, I-60130

$C_{12}H_{18}O$

Methoxypentamethylbenzene, *in* P-60058

$C_{12}H_{18}O_2$

[1,1′-Bicyclohexyl]-2,2′-dione, B-60089
Clavularin *A*, C-60153
Clavularin B, *in* C-60153
6-(1-Heptenyl)-5,6-dihydro-2*H*-pyran-2-one, H-60025
5-Isopropyl-2-methyl-1,3-benzenediol; Di-Me ether, *in* I-60125

$C_{12}H_{18}O_3$

7-Isojasmonic acid, *in* J-60002
Jasmonic acid, J-60002

$C_{12}H_{18}O_4$

1-Acetoxy-7-hydroxy-3,7-dimethyl-2*E*,5*E*-octadien-4-one, *in* D-60318
3,4-Dihydroxyphenylacetaldehyde; Di-Me ether, di-Me acetal, *in* D-60371

$C_{12}H_{18}O_5$

1-Acetoxy-7-hydroperoxy-3,7-dimethyl-2*E*,5*E*-octadien-4-one, *in* D-60318

$C_{12}H_{20}$

1-*tert*-Butyl-1,2-cyclooctadiene, B-60344
Cyclododecyne, C-60192
1,1′,2,2′,3,4,5,5′,6,6′-Decahydrobiphenyl, D-60007
1,1′,2,3,4,4′,5,5′,6,6′-Decahydrobiphenyl, D-60008
1,2,3,3′,4,4′,5,5′,6,6′-Decahydrobiphenyl, D-60009

$C_{12}H_{20}N_2O_7$

2′-Deoxymugeneic acid, *in* M-60148

$C_{12}H_{20}N_2O_8$

Isomugeneic acid, *in* M-60148
Mugineic acid, M-60148

$C_{12}H_{20}N_2O_9$

3-Hydroxymugeneic acid, *in* M-60148

$C_{12}H_{20}N_4O_4$

Octahydro-1,2,5,6-tetrazocine; Tetra-Ac, *in* O-60020

$C_{12}H_{20}O_2$

10-Dodecen-12-olide, *in* H-60126

$C_{12}H_{20}O_4$

9,12-Dioxododecanoic acid, D-60471

$C_{12}H_{20}O_8$

Parasorboside, *in* T-60067

$C_{12}H_{21}NO_4$

2-Amino-1,4-cyclohexanedicarboxylic acid; Di-Et ester, *in* A-60118
3-Amino-1,2-cyclohexanedicarboxylic acid; Di-Et ester, *in* A-60119
4-Amino-1,1-cyclohexanedicarboxylic acid; Di-Et ester, *in* A-60120
4-Amino-1,3-cyclohexanedicarboxylic acid; Di-Et ester, *in* A-60121

$C_{12}H_{22}N_2O_2$

3,6-Bis(2-methylpropyl)-2,5-piperazinedione, B-60171

$C_{12}H_{22}O_2$

Invictolide, I-60038

$C_{12}H_{22}O_3$

10-Hydroxy-11-dodecenoic acid, H-60125
12-Hydroxy-10-dodecenoic acid, H-60126
12-Oxododecanoic acid, O-60066

$C_{12}H_{22}O_4$

Talaromycin *A*, T-60003

$C_{13}H_{12}$

Benz[f]indane, B-60015
3-Methylbiphenyl, M-60053
Tetracyclo[5.5.1.0^{4,13}.0^{10,13}]trideca-2,5,8,11-tetraene, T-60036

$C_{13}H_{12}N_2$

N,N'-Diphenylformamidine, D-60479

$C_{13}H_{12}N_2O$

2-(Aminomethyl)pyridine; N-Benzoyl, in A-60227
4-(Aminomethyl)pyridine; N-Benzoyl, in A-60229

$C_{13}H_{12}N_2O_3S$

Aminoiminomethanesulfonic acid; N,N-Di-Ph, in A-60210

$C_{13}H_{12}O$

4-Methoxyacenaphthene, in H-60099

$C_{13}H_{12}O_2$

2-(Benzyloxy)phenol, B-60072
3-(Benzyloxy)phenol, B-60073
▷4-(Benzyloxy)phenol, B-60074
▷Goniothalamin, G-60031
4-Methyl-1-azulenecarboxylic acid; Me ester, in M-60042

$C_{13}H_{12}O_3$

6-Acetyl-5-hydroxy-2-isopropenylbenzo[b]furan, in A-60065
Goniothalamin oxide, in G-60031

$C_{13}H_{12}O_4$

6-Acetyl-5-hydroxy-2-(1-hydroxymethylvinyl)benzo[b]furan, in A-60065
Altholactone, A-60083
5,8-Dihydroxy-2,3,6-trimethyl-1,4-naphthoquinone, D-60380
5,6-Dimethoxy-2-methyl-1,4-naphthoquinone, in D-60351
1,4,5-Naphthalenetriol; 5-Me ether, 1-Ac, in N-60005
2-Oxo-2H-benzopyran-4-acetic acid; Et ester, in O-60057
Platypterophthalide, P-60161

$C_{13}H_{12}O_5$

Acuminatolide, A-60063

$C_{13}H_{13}BrN_2O_3$

5-Bromotryptophan; N^α-Ac, in B-60333
6-Bromotryptophan; N^α-Ac, in B-60334
7-Bromotryptophan; N^α-Ac, in B-60335

$C_{13}H_{13}BrO_2$

1-Bromo-(2-dimethoxymethyl)naphthalene, in B-60302

$C_{13}H_{13}IO_8$

1,1,1-Triacetoxy-1,1-dihydro-1,2-benziodoxol-3(1H)-one, T-60228

$C_{13}H_{13}N$

2-(2-Phenylethenyl)-1H-pyrrole; N-Me, in P-60094
2-(2-Phenylethenyl)-1H-pyrrole; N-Me, in P-60094

$C_{13}H_{13}NO$

1,3,4,10-Tetrahydro-9(2H)-acridinone, T-60052

$C_{13}H_{13}NO_2$

5-Methyl-1H-pyrrole-2-carboxylic acid; Benzyl ester, in M-60124

$C_{13}H_{13}N_3O_5$

Antibiotic PDE I, A-60286

$C_{13}H_{14}$

2,3,5,6,7-Pentamethylenebicyclo[2.2.2]octane, P-60056

$C_{13}H_{14}N_2O$

1-Benzyl-1,4-dihydronicotinamide, B-60067

$C_{13}H_{14}N_2O_4S_3$

Gliotoxin E, G-60024

$C_{13}H_{14}O$

1-(1-Naphthyl)-2-propanol, N-60015

$C_{13}H_{14}O_2$

7-Phenyl-5-heptynoic acid, P-60106

$C_{13}H_{14}O_3$

2,2-Dimethyl-2H-1-benzopyran-6-carboxylic acid; Me ester, in D-60408
7-Ethoxy-3,4-dimethylcoumarin, in H-60119
1,4,5-Trimethoxynaphthalene, in N-60005

$C_{13}H_{14}O_4$

6-Acetyl-5-hydroxy-2-hydroxymethyl-2-methylchromene, A-60031
Anaphatol, in D-60339
4-Hydroxy-2-isopropyl-5-benzofurancarboxylic acid; Me ester, in H-60164

$C_{13}H_{14}O_5$

Diaporthin, D-60052

$C_{13}H_{15}Cl_2NO_3$

6-Amino-2,2-dichlorohexanoic acid; N-Benzoyl, in A-60126

$C_{13}H_{16}Cl_2O_4$

1-(3,5-Dichloro-2,6-dihydroxy-4-methoxyphenyl)-1-hexanone, D-60129

$C_{13}H_{16}NO_3$

3-Benzoyl-2-methylpropanoic acid; Et ester, in B-60064

$C_{13}H_{16}N_2$

2-Diazo-1,1,3,3-tetramethylindane, D-60067

$C_{13}H_{16}N_2O_4$

Bursatellin, B-60338

$C_{13}H_{16}O$

7-Phenyl-3-heptyn-2-ol, P-60107
1,1,3,3-Tetramethyl-2-indanone, T-60143

$C_{13}H_{16}O_2$

3,4-Dihydro-3,3,8a-trimethyl-1,6(2H,8aH)-naphthalenedione, D-60301

$C_{13}H_{16}O_3$

1-(Dimethoxymethyl)-2,3,5,6-tetrakis(methylene)-7-oxabicyclo[2.2.1]heptane, D-60399
Flossonol, F-60014

$C_{13}H_{16}O_4$

4-Hydroxy-3-(2-hydroxy-3-methyl-3-butenyl)benzoic acid; Me ester, in H-60154

$C_{13}H_{16}Se$

1,1,3,3-Tetramethyl-2-indaneselone, T-60142

$C_{13}H_{17}NO_3S$

2-Amino-4,5,6,7-tetrahydrobenzo[b]thiophene-3-carboxylic acid; Et ester, N-Ac, in A-60258

$C_{13}H_{17}NO_4$

2-Amino-3-(4-hydroxyphenyl)-3-methylbutanoic acid; N-Ac, in A-60198

$C_{13}H_{18}BrClO_3$

Methyl 3-(3-bromo-4-chloro-4-methylcyclohexyl)-4-oxo-2-pentenoate, M-60054

$C_{13}H_{18}N_2$

1,1,3,3-Tetramethyl-2-indanone; Hydrazone, in T-60143

$C_{13}H_{18}N_4O_3$

Citrulline; α-N-Benzoyl, amide, in C-60151

$C_{13}H_{18}O_2$

Pentamethylphenol; Ac, in P-60058

$C_{13}H_{18}O_5$

2,2,8,8-Tetramethyl-3,4,5,6,7-nonanepentone, T-60148

$C_{13}H_{18}O_7$

Pentahydroxybenzoic acid; Penta-Me ether, Me ester, in P-60043

$C_{13}H_{20}$

2-Isopropylideneadamantane, I-60121
Tetracyclo[5.5.1.0^{4,13}.0^{10,13}]tridecane, T-60035

$C_{13}H_{20}O$

γ-Ionone, I-60080

$C_{13}H_{20}O_3$

Methyl jasmonate, in J-60002

$C_{13}H_{20}O_6$

2,2,8,8-Tetramethyl-5,5-dihydroxy-3,4,6,7-nonanetetrone, in T-60148

$C_{13}H_{22}O$

1-Cyclododecenecarboxaldehyde, C-60191

Theaspirone, T-60185

$C_{13}H_{22}O_4$

9,12-Dioxododecanoic acid; Me ester, *in* D-60471

$C_{13}H_{24}$

Octamethylcyclopentene, O-60026
6,7-Tridecadiene, T-60272

$C_{13}H_{24}O$

Octamethylcyclopentanone, O-60025

$C_{13}H_{24}O_3$

10-Hydroxy-11-dodecenoic acid; Me ester, *in* H-60125
12-Hydroxy-10-dodecenoic acid; Me ester, *in* H-60126
7-Megastigmene-5,6,9-triol, M-60012
12-Oxododecanoic acid; Me ester, *in* O-60066

$C_{13}H_{25}N_3$

11-Methylene-1,5,9-triazabicyclo[7.3.3]pentadecane, M-60080

$C_{13}H_{26}O$

2,2,3,3,4,4,5,5-Octamethylcyclopentanol, O-60024

$C_{13}H_{26}O_4$

10-Oxodecanoic acid; Me ester, Di-Me acetal, *in* O-60063

$C_{13}H_{31}N_5$

1,4,7,11,15-Pentaazacyclooctadecane, P-60023
1,4,8,11,15-Pentaazacyclooctadecane, P-60024

$C_{14}H_4Cl_2O_4$

2,3-Dichloro-1,4,9,10-anthracenetetrone, D-60117

$C_{14}H_5ClO_4$

2-Chloro-1,4,9,10-anthracenetetrone, C-60041

$C_{14}H_6N_4O_{11}$

3,5-Dinitrobenzoic acid; Anhydride, *in* D-60458

$C_{14}H_6O_8$

▷Ellagic acid, E-60003

$C_{14}H_8Br_2$

9,10-Dibromoanthracene, D-60085

$C_{14}H_8Cl_2$

9,10-Dichloroanthracene, D-60116

$C_{14}H_8F_2O_4$

4,4′-Difluoro-[1,1′-biphenyl]-3,3′-dicarboxylic acid, D-60178
4,5-Difluoro-[1,1′-biphenyl]-2,3-dicarboxylic acid, D-60179
6,6′-Difluoro-[1,1′-biphenyl]-2,2′-dicarboxylic acid, D-60180

$C_{14}H_8N_2$

Pyrido[2′,3′:3,4]cyclobuta[1,2-*g*]quinoline, P-60232
Pyrido[3′,2′:3,4]cyclobuta[1,2-*g*]quinoline, P-60233

$C_{14}H_8N_2O_2$

[1,4]Benzoxazino[3,2-*b*][1,4]benzoxazine, B-60055

$C_{14}H_8N_4S_2$

[1,2,4,5]Tetrazino[3,4-*b*:6,1-*b*′]bisbenzothiazole, T-60171

$C_{14}H_8O$

Acenaphtho[5,4-*b*]furan, A-60014

$C_{14}H_8OS$

Phenanthro[4,5-*bcd*]thiophene; 4-Oxide, *in* P-60077

$C_{14}H_8O_2$

Benz[*a*]azulene-1,4-dione, B-60011
1,8-Biphenylenedicarboxaldehyde, B-60118

$C_{14}H_8O_2S$

Phenanthro[4,5-*bcd*]thiophene; 4,4-Dioxide, *in* P-60077

$C_{14}H_8O_3$

3,4-Acenaphthenedicarboxylic acid; Anhydride, *in* A-60013

$C_{14}H_8S$

Acenaphtho[5,4-*b*]thiophene, A-60015
Phenanthro[4,5-*bcd*]thiophene, P-60077

$C_{14}H_8S_2$

[1]Benzothiopyrano[6,5,4-*def*][1]benzothiopyran, B-60049

$C_{14}H_9Cl_2NO_2$

2,2′-Iminodibenzoic acid; Dichloride, *in* I-60014

$C_{14}H_9NO$

2-Phenyl-3*H*-indol-3-one, P-60114

$C_{14}H_9NO_2$

3-(Phenylimino)-1(3*H*)-isobenzofuranone, P-60112
2-Phenylisatogen, *in* P-60114
2-(2-Pyridyl)-1,3-indanedione, P-60241
2-(3-Pyridyl)-1,3-indanedione, P-60242
2-(4-Pyridyl)-1,3-indanedione, P-60243

$C_{14}H_9NO_3$

1-Amino-2-hydroxyanthraquinone, A-60174
▷1-Amino-4-hydroxyanthraquinone, A-60175
1-Amino-5-hydroxyanthraquinone, A-60176
1-Amino-8-hydroxyanthraquinone, A-60177
2-Amino-1-hydroxyanthraquinone, A-60178
2-Amino-3-hydroxyanthraquinone, A-60179
3-Amino-1-hydroxyanthraquinone, A-60180
9-Amino-10-hydroxy-1,4-anthraquinone, A-60181

$C_{14}H_9N_3$

4,4′-Iminobisbenzonitrile, *in* I-60018
5*H*-Indolo[2,3-*b*]quinoxaline, I-60036
Pyrido[2′,1′:2,3]imidazo[4,5-*c*]isoquinoline, P-60235

$C_{14}H_9N_3OS$

1,2,3-Benzotriazine-4(3*H*)-thione; *N*-Benzoyl, *in* B-60050

$C_{14}H_9N_3O_2$

5*H*-Indolo[2,3-*b*]quinoxaline; 5,11-Dioxide, *in* I-60036

$C_{14}H_{10}$

Cyclohepta[*de*]naphthalene, C-60198

$C_{14}H_{10}N_2O$

3,5-Diphenyl-1,2,4-oxadiazole, D-60484
3-Nitroso-2-phenylindole, *in* P-60114
1*H*-Pyrrolo[3,2-*b*]pyridine; *N*-Benzoyl, *in* P-60249

$C_{14}H_{10}N_2O_2$

1,3-Diazetidine-2,4-dione; 1,3-Di-Ph, *in* D-60057
3,5-Diphenyl-1,2,4-oxadiazole; 4-Oxide, *in* D-60484
2-Phenylisatogen oxime, *in* P-60114

$C_{14}H_{10}N_2O_3$

N-(3-Oxo-3*H*-phenoxazin-2-yl)acetamide, *in* A-60237

$C_{14}H_{10}O_2$

3,4-Heptafulvalenedione, H-60022

$C_{14}H_{10}O_3$

2,4,5-Phenanthrenetriol, P-60071
2,3,7-Trihydroxyphenanthrene, T-60339

$C_{14}H_{10}O_4$

3,4-Acenaphthenedicarboxylic acid, A-60013
Moracin *M*, M-60146
1,2,5,6-Tetrahydroxyphenanthrene, T-60115
1,2,5,7-Tetrahydroxyphenanthrene, T-60116
1,2,6,7-Tetrahydroxyphenanthrene, T-60117
1,3,5,6-Tetrahydroxyphenanthrene, T-60118
1,3,6,7-Tetrahydroxyphenanthrene, T-60119
2,3,5,7-Tetrahydroxyphenanthrene, T-60120
2,3,6,7-Tetrahydroxyphenanthrene, T-60121
2,4,5,6-Tetrahydroxyphenanthrene, T-60122
3,4,5,6-Tetrahydroxyphenanthrene, T-60123

$C_{14}H_{10}O_5$

1,5-Dihydroxy-6-methoxyxanthone, *in* T-60349
5,6-Dihydroxy-1-methoxyxanthone, *in* T-60349
1,2,5,6,7-Pentahydroxyphenanthrene, P-60050

$C_{14}H_{10}O_6$

2,5-Dihydroxy-1,4-naphthoquinone; Di-Ac, *in* D-60354
2,8-Dihydroxy-1,4-naphthoquinone; Di-Ac, *in* D-60355
5,8-Dihydroxy-1,4-naphthoquinone; Di-Ac, *in* D-60356
1,2,3,5,6,7-Hexahydroxyphenanthrene, H-60066

$C_{14}H_{11}ClN_2O_3$

N-Methylindisocin, *in* I-60028

$C_{14}H_{11}FO$

2-Fluoro-2,2-diphenylacetaldehyde, F-60033
3-Fluoro-4-methylphenol; Benzoyl, *in* F-60057

C$_{14}$H$_{11}$FO$_2$

2-Fluoro-3-hydroxybenzaldehyde; Benzyl ether, *in* F-60035
2-Fluoro-5-hydroxybenzaldehyde; Benzyl ether, *in* F-60036
4-Fluoro-3-hydroxybenzaldehyde; Benzyl ether, *in* F-60037

C$_{14}$H$_{11}$IO$_3$S

2-Iodo-1-phenyl-2-(phenylsulfonyl)ethanone, I-60061

C$_{14}$H$_{11}$N

6-Phenyl-1*H*-indole, P-60113

C$_{14}$H$_{11}$NO$_2$

2-Hydroxycarbazole; Ac, *in* H-60109

C$_{14}$H$_{11}$NO$_4$

1,2-Dimethoxy-3*H*-phenoxazin-3-one, *in* T-60353
▷2,2'-Iminodibenzoic acid, I-60014
2,3'-Iminodibenzoic acid, I-60015
2,4'-Iminodibenzoic acid, I-60016
3,3'-Iminodibenzoic acid, I-60017
4,4'-Iminodibenzoic acid, I-60018

C$_{14}$H$_{12}$

9,10-Dihydrophenanthrene, D-60260
9*b*-Methyl-9*bH*-benz[*cd*]azulene, M-60044

C$_{14}$H$_{12}$N$_2$O$_2$

5*a*,6,11*a*,12-Tetrahydro[1,4]benzoxazino[3,2-*b*][1,4]-
benzoxazine, T-60059

C$_{14}$H$_{12}$N$_2$O$_4$

[2,2'-Bipyridine]-3,3'-dicarboxylic acid; Di-Me ester, *in* B-60124
[2,2'-Bipyridine]-3,5'-dicarboxylic acid; Di-Me ester, *in* B-60125
[2,2'-Bipyridine]-4,4'-dicarboxylic acid; Di-Me ester, *in* B-60126
[2,2'-Bipyridine]-5,5'-dicarboxylic acid; Di-Me ester, *in* B-60127
[3,3'-Bipyridine]-2,2'-dicarboxylic acid; Di-Me ester, *in* B-60133
[3,3'-Bipyridine]-4,4'-dicarboxylic acid; Di-Me ester, *in* B-60134

C$_{14}$H$_{12}$N$_2$S

2,5-Dihydro-2,2-diphenyl-1,3,4-thiadiazole, D-60236

C$_{14}$H$_{12}$N$_4$

4-Amino-1,2,3-benzotriazine; *N*(4)-Benzyl, *in* A-60092
4-Amino-1,2,3-benzotriazine; 2-Me, *N*(4)-Ph, *in* A-60092
4-Amino-1,2,3-benzotriazine; 3-Me, *N*(4)-Ph, *in* A-60092
4-Amino-1,2,3-benzotriazine; 3-Benzyl, *in* A-60092

C$_{14}$H$_{12}$N$_4$S

Hector's base, H-60012

C$_{14}$H$_{12}$O$_2$

4-Hydroxyacenaphthene; Ac, *in* H-60099
2,3-Naphthalenedicarboxylic acid; Di-Me ester, *in* N-60003
1-Phenoxy-1-phenylethylene, *in* P-60092
1,2,3,4-Tetrahydroanthraquinone, T-60054

C$_{14}$H$_{12}$O$_3$

α-Oxo-1-naphthaleneacetic acid; Et ester, *in* O-60078
α-Oxo-2-naphthaleneacetic acid; Et ester, *in* O-60079

C$_{14}$H$_{12}$O$_5$

▷Khellin, K-60010

C$_{14}$H$_{12}$O$_6$

1-(2,3,4-Trihydroxyphenyl)-2-(3,4,5-trihydroxyphenyl)-
ethylene, T-60344

C$_{14}$H$_{12}$O$_8$

Fulvic acid, F-60081
Polivione, P-60167

C$_{14}$H$_{13}$N

9-Amino-9,10-dihydroanthracene, A-60136
2-Amino-9,10-dihydrophenanthrene, A-60141
4-Amino-9,10-dihydrophenanthrene, A-60142
9-Amino-9,10-dihydrophenanthrene, A-60143

C$_{14}$H$_{13}$NO

2-Ethoxycarbazole, *in* H-60109

C$_{14}$H$_{13}$NO$_4$

6-Amino-2-hexenoic acid; *N*-Phthalimido, *in* A-60173

C$_{14}$H$_{13}$N$_3$O$_2$

2,2'-Iminobisbenzamide, *in* I-60014

C$_{14}$H$_{14}$

1,2,3,4-Tetrahydroanthracene, T-60053

C$_{14}$H$_{14}$N$_2$

2,5-Dimethyl-3-(2-phenylethenyl)pyrazine, D-60442
N,*N*'-Diphenylformamidine; *N*-Me, *in* D-60479

C$_{14}$H$_{14}$N$_2$O

2,2'-Diaminobiphenyl; 2-*N*-Ac, *in* D-60033

C$_{14}$H$_{14}$O$_2$

1-Methoxy-2-(phenylmethoxy)benzene, *in* B-60072
1-Methoxy-3-(phenylmethoxy)benzene, *in* B-60073
1-Methoxy-4-(phenylmethoxy)benzene, *in* B-60074
2-(2-Phenylethyl)-1,4-benzenediol, P-60095
4-(2-Phenylethyl)-1,2-benzenediol, P-60096
4-(2-Phenylethyl)-1,3-benzenediol, P-60097
5-(2-Phenylethyl)-1,3-benzenediol, P-60098

C$_{14}$H$_{14}$O$_3$

2-(Dimethoxymethyl)-1-naphthalenecarboxaldehyde, *in*
N-60001
3,3',4-Trihydroxybibenzyl, T-60316

C$_{14}$H$_{14}$O$_4$

5,8-Dihydroxy-2,3,6,7-tetramethyl-1,4-naphthoquinone,
D-60377
1,4,5-Naphthalenetriol; 1,5-Di-Me ether, Ac, *in* N-60005
1,4,5-Naphthalenetriol; 4,5-Di-Me ether, Ac, *in* N-60005
Phthalidochromene, P-60143
3,3',4,4'-Tetrahydroxybibenzyl, T-60101
3,3',4,5-Tetrahydroxybibenzyl, T-60102

C$_{14}$H$_{14}$O$_5$

3,3',4,4',5-Pentahydroxybibenzyl, P-60044

C$_{14}$H$_{14}$O$_6$

3,3',4,4',5,5'-Hexahydroxybibenzyl, H-60062

C$_{14}$H$_{14}$O$_7$

3',4',5'-Trihydroxyacetophenone; Tri-Ac, *in* T-60315

C$_{14}$H$_{14}$O$_8$

1,2,3,4-Benzenetetrol; Tetra-Ac, *in* B-60014
Hexahydroxy-1,4-naphthoquinone; Tetra-Me ether, *in* H-60065
3,4,5-Trihydroxyphenylacetic acid; Tri-Ac, *in* T-60343

C$_{14}$H$_{14}$Se$_2$

Dibenzyl diselenide, D-60082

C$_{14}$H$_{15}$ClO$_5$

Mikrolin, M-60135

C$_{14}$H$_{15}$NO

1,3,4,10-Tetrahydro-9(2*H*)-acridinone; *N*-Me, *in* T-60052

C$_{14}$H$_{15}$N$_3$O$_6$

Antibiotic FR 900482, A-60285

C$_{14}$H$_{15}$N$_5$O$_4$

8-Aminoimidazo[4,5-*g*]quinazoline; 1-(β-D-Ribofuranosyl), *in*
A-60207
lin-Benzoadenosine, *in* A-60207

C$_{14}$H$_{16}$

1,3-Diethynyladamantane, D-60172

C$_{14}$H$_{16}$N$_2$

2,2'-Bis(methylamino)biphenyl, *in* D-60033
1-(1-Naphthyl)piperazine, N-60013
1-(2-Naphthyl)piperazine, N-60014

C$_{14}$H$_{16}$N$_4$

5,6,8,9-Tetraaza[3.3]paracyclophane, T-60018

C$_{14}$H$_{16}$O$_2$

7-Phenyl-5-heptynoic acid; Me ester, *in* P-60106

C$_{14}$H$_{16}$O$_4$

Pyriculol, P-60216

C$_{14}$H$_{16}$O$_5$

Dechloromikrolin, *in* M-60135

$C_{15}H_{11}ClN_2O_2$
2-Chloro-2,3-dihydro-1H-imidazole-4,5-dione; 1,3-Di-Ph, *in* C-60054

$C_{15}H_{11}ClO_4$
Dibenzo[a,d]cycloheptenylium; Perchlorate, *in* D-60071

$C_{15}H_{11}NO$
6-Benzoylindole, B-60058
8-Hydroxy-7-phenylquinoline, H-60218
2-Phenyl-4(1H)-quinolinone, P-60132
4-Phenyl-2(1H)-quinolinone, P-60131

$C_{15}H_{11}NOS$
▷2-Phenyl-1,5-benzothiazepin-4(5H)-one, P-60081

$C_{15}H_{11}NO_2$
Dibenz[b,g]azocine-5,7(6H,12H)-dione, D-60068
2,4-Diphenyl-5(4H)-oxazolone, D-60485

$C_{15}H_{11}NO_3$
▷1-Amino-2-methoxyanthraquinone, *in* A-60174
▷1-Amino-4-methoxyanthraquinone, *in* A-60175
1-Amino-5-methoxyanthraquinone, *in* A-60176
9-Amino-10-methoxy-1,4-anthraquinone, *in* A-60181
1-Hydroxy-4-(methylamino)anthraquinone, *in* A-60175

$C_{15}H_{11}NS$
2-(2-Phenylethenyl)benzothiazole, P-60093

$C_{15}H_{12}I_3NO_4$
3,3′,5-Triiodothyronine, T-60350

$C_{15}H_{12}N_2O$
5-Amino-3-phenylisoxazole; N-Ph, *in* A-60240

$C_{15}H_{12}N_2O_3$
2,3-Dihydro-2-hydroxy-1H-imidazole-4,5-dione; 1,3-Di-Ph, *in* D-60243

$C_{15}H_{12}O$
10,11-Dihydro-5H-dibenzo[a,d]cyclohepten-5-one, D-60218

$C_{15}H_{12}OS$
3-Hydroxymethyldibenzo[b,f]thiepin, H-60176

$C_{15}H_{12}O_2$
2,7-Dimethylxanthone, D-60453
(4-Hydroxyphenyl)ethylene; Benzoyl, *in* H-60211

$C_{15}H_{12}O_3$
2,3-Dihydro-2-phenyl-2-benzofurancarboxylic acid, D-60262
4-Methoxy-2,5-phenanthrenediol, *in* P-60071

$C_{15}H_{12}O_3S$
3-Hydroxymethyldibenzo[b,f]thiepin; 5,5-Dioxide, *in* H-60176
2-Mercapto-2-phenylacetic acid; S-Benzyl, *in* M-60024

$C_{15}H_{12}O_4$
3,9-Dihydroxypterocarpan, D-60373

$C_{15}H_{12}O_5$
1,6-Dihydroxy-5-methoxyxanthone, *in* T-60349
1-Hydroxy-5,6-dimethoxyxanthone, *in* T-60349
3-(4-Hydroxyphenyl)-1-(2,4,6-trihydroxyphenyl)-2-propen-1-one, H-60220
Protosappanin A, P-60186
5,7,8-Trihydroxyflavanone, T-60321

$C_{15}H_{12}O_6$
1,8-Dihydroxy-3,6-dimethoxyxanthone, *in* T-60127
3,5-Dihydroxy-2-methyl-1,4-naphthoquinone; Di-Ac, *in* D-60350
3,5,6,7-Tetrahydroxyflavanone, T-60105

$C_{15}H_{12}O_7$
Nectriafurone, N-60016
2′,3,5,7,8-Pentahydroxyflavanone, P-60045
3′,4′,5,6,7-Pentahydroxyflavanone, P-60046
1,2,5,6,8-Pentahydroxyxanthone, P-60052
1,5,8-Trihydroxy-2,6-dimethoxyxanthone, *in* P-60052
Ventilone C, V-60004
Ventilone D, *in* V-60004

$C_{15}H_{13}Br$
1-Bromo-1,2-diphenylpropene, B-60237

$C_{15}H_{13}NO$
N-(Diphenylmethylene)acetamide, *in* D-60482
6-Hydroxyindole; Benzyl ether, *in* H-60158
2-Octylamine; N-Benzoyl, *in* O-60031

$C_{15}H_{13}NOS$
2-Phenyl-1,5-benzothiazepin-4(5H)-one; 2,3-Dihydro, *in* P-60081

$C_{15}H_{13}NO_2$
3,4-Diphenyl-2-oxazolidinone, *in* P-60126

$C_{15}H_{13}NO_5$
1,2,4-Trimethoxy-3H-phenoxazin-3-one, T-60353

$C_{15}H_{14}$
10,11-Dihydro-5H-dibenzo[a,d]cycloheptene, D-60217
1,1-Diphenylcyclopropane, D-60478
1,1-Diphenylpropene, D-60491

$C_{15}H_{14}ClN$
4-Chloro-2,3-dihydro-1H-indole; 1-Benzyl, *in* C-60055

$C_{15}H_{14}N_2$
10,11-Dihydro-5H-dibenzo[a,d]cyclohepten-5-one; Hydrazone, *in* D-60218

$C_{15}H_{14}N_4$
4-Amino-1,2,3-benzotriazine; 2-Et, N(4)-Ph, *in* A-60092
4-Amino-1,2,3-benzotriazine; 3-Et, N(4)-Ph, *in* A-60092

$C_{15}H_{14}O$
Di(2,4,6-cycloheptatrien-1-yl)ethanone, D-60166

$C_{15}H_{14}OS$
Thia[2.2]metacyclophane; S-Oxide, *in* T-60194

$C_{15}H_{14}O_2S$
Thia[2.2]metacyclophane; S-Dioxide, *in* T-60194

$C_{15}H_{14}O_3$
4-(Benzyloxy)phenol; Ac, *in* B-60074
Cyclolongipesin, C-60214
Dehydroosthol, D-60020
cis-Dehydroosthol, *in* D-60020

$C_{15}H_{14}O_4$
Allopteroxylin, A-60075
Helicquinone, H-60015
2-Methyl-1,3-naphthalenediol; Di-Ac, *in* M-60091
4-Methyl-1,3-naphthalenediol; Di-Ac, *in* M-60094

$C_{15}H_{14}O_5$
Ageratone, A-60065
Altholactone; Ac, *in* A-60083
3-(4-Hydroxyphenyl)-1-(2,4,6-trihydroxyphenyl)-1-propanone, H-60219
3′,4′,5,7-Tetrahydroxyflavan, T-60103
3,4′,5,7-Tetrahydroxyflavanone, T-60104

$C_{15}H_{14}O_7$
Fusarubin, F-60108
2,5,8-Trihydroxy-3-methyl-6,7-methylenedioxy-1,4-naphthoquinone; Tri-Me ether, *in* T-60333

$C_{15}H_{14}S$
Thia[2.2]metacyclophane, T-60194

$C_{15}H_{15}N$
1,2,3-Trimethyl-9H-carbazole, T-60357
1,2,4-Trimethyl-9H-carbazole, T-60358

$C_{15}H_{15}NO$
2-Amino-1,3-diphenyl-1-propanone, A-60149

$C_{15}H_{15}NO_4$
6-Amino-2-hexenoic acid; N-Phthalimido, Me ester, *in* A-60173

$C_{15}H_{15}NO_6$
Ascorbigen, A-60309

$C_{15}H_{16}N_{10}O_2$
Drosopterin, D-60521
Isodrosopterin, *in* D-60521
Neodrosopterin, *in* D-60521

$C_{15}H_{16}O$
1,3,5,7(11),9-Guaiapentaen-14-al, G-60039

8-Isopropyl-5-methyl-2-naphthalenecarboxaldehyde, I-60129

$C_{15}H_{16}O_3$
Gnididione, G-60030
Heritol, H-60028

$C_{15}H_{16}O_4$
Cedrelopsin, C-60026
8β-Hydroxydehydrozaluzanin C, in Z-60001
9-Hydroxy-3-oxo-1,4(15),11(13)-eudesmatrien-12,6-olide, H-60202
Longipesin, L-60034
Murraol, M-60150

$C_{15}H_{16}O_7$
Dihydrofusarubin A, in F-60108
Dihydrofusarubin B, in F-60108

$C_{15}H_{16}O_8$
3,4,5-Trihydroxyphenylacetic acid; Tri-Ac, Me ester, in T-60343

$C_{15}H_{17}BrN_2O_3$
6-Bromotryptophan; N^α-Ac, Et ester, in B-60334

$C_{15}H_{18}$
1-(2,2-Dimethylpropyl)naphthalene, D-60446
2-(2,2-Dimethylpropyl)naphthalene, D-60447
2,3,4,5,6,7,8,9-Octahydro-1H-triindene, O-60021

$C_{15}H_{18}N_2$
Aurantioclavine, A-60316

$C_{15}H_{18}O$
3,4-Dihydro-8-isopropyl-5-methyl-2-naphthalenecarboxaldehyde, in I-60129

$C_{15}H_{18}O_2$
Tannunolide A, T-60005
Tannunolide B, in T-60005

$C_{15}H_{18}O_3$
Bullerone, B-60337
3-Epizaluzanin C, in Z-60001
7α-Hydroxy-3-deoxyzaluzanin C, in Z-60001
1-Hydroxy-3,7(11),8-eudesmatrien-12,8-olide, H-60140
1-Hydroxy-4(15),7(11),8-eudesmatrien-12,8-olide, H-60141
3-Oxo-4,11(13)-eudesmadien-12,8β-olide, in H-60136
9-Oxo-4,11(13)-eudesmadien-12,16β-olide, in H-60137
Zaluzanin C, Z-60001

$C_{15}H_{18}O_4$
2,8-Dihydroxy-3,10(14),11(13)-guaiatrien-12,6-olide, D-60327
$1\beta,10\beta$-Epoxy-8α-hydroxyachillin, in H-60100
8-Hydroxyachillin, H-60100
Pentalenolactone E, P-60055

$C_{15}H_{18}O_5$
Artelin, A-60304
2,9-Dihydroxy-8-oxo-1(10),4,11(13)-germacratrien-12,6-olide, D-60366
2,3-Epoxy-1,4-dihydroxy-7(11),8-eudesmadien-12,8-olide, E-60012

$C_{15}H_{18}O_6$
Araneophthalide, A-60292
Chilenone B, C-60034

$C_{15}H_{18}O_9$
5,7-Dihydroxy-1(3H)-isobenzofuranone; 5-Me ether, glucoside, in D-60339

$C_{15}H_{19}N_5O_5$
3,4-Dihydro-4,6,7-trimethyl-3-β-D-ribofuranosyl-9H-imidazo[1,2-a]purin-9-one, D-60302

$C_{15}H_{20}BrClO_2$
Epoxyisodihydrorhodophytin, E-60024

$C_{15}H_{20}Br_2O_2$
Dehydrochloroprepacifenol, D-60018

$C_{15}H_{20}O$
Herbacin, H-60027

$C_{15}H_{20}O_2$
Aristolactone, A-60296
Cantabradienic acid, C-60009
6-Deoxyilludin M, in I-60001
8,12-Epoxy-3,7,11-eudesmatrien-1-ol, E-60020
Glechomanolide, G-60022
Isomyomontanone, in M-60155
Mokkolactone, M-60142
Myomontanone, M-60155
Pinguisanin, P-60155

$C_{15}H_{20}O_3$
Artesovin, A-60306
Bisabolangelone, B-60138
4,8-Bis-epi-inuviscolide, in I-60037
Cantabrenonic acid, in C-60010
Cyclodehydromyopyrone A, C-60188
Cyclodehydromyopyrone B, C-60189
Dehydromyoporone, D-60019
6-Deoxyilludin S, in I-60002
11β,13-Dihydrozaluzanin C, in Z-60001
11α,13-Dihydrozaluzanin C, in Z-60001
1-epi-Inuviscolide, in I-60037
8-epi-Inuviscolide, in I-60037
1,10-Epoxy-4,11(13)-germacradien-12,8-olide, E-60022
Eumorphistonol, E-60073
Haageanolide, H-60001
Hanphyllin, H-60006
3-Hydroxy-4,11(13)-eudesmadien-12,8-olide, H-60136
9-Hydroxy-4,11(13)-eudesmadien-12,6-olide, H-60137
15-Hydroxy-4,11(13)-eudesmadien-12,8-olide, H-60138
▷Illudin M, I-60001
Inuviscolide, I-60037
9-Oxo-4,11(13)-eudesmadien-12-oic acid, in H-60135
1-Oxo-7(11)-eudesmen-12,8-olide, O-60067
1-Oxo-4,10(14)-germacradien-12,6α-olide, in O-60068
α-Pipitzol, P-60158
Quadrangolide, Q-60001
Santamarine, S-60007

$C_{15}H_{20}O_4$
Artecalin, A-60302
Confertin, in D-60211
2,3-Dehydro-11α,13-dihydroconfertin, in D-60211
11,13-Dehydroeriolin, in E-60036
11β,13-Dihydro-9α-hydroxyzaluzanin C, in Z-60001
1,8-Dihydroxy-3,7(11)-eremophiladien-12,8-olide, D-60320
1,5-Dihydroxyeriocephaloide, D-60321
1,8-Dihydroxy-3,7(11)-eudesmadien-12,8-olide, D-60322
8,9-Dihydroxy-1(10),4,11(13)-germacratrien-12,6-olide, D-60323
9,15-Dihydroxy-1(10),4,11(13)-germacratrien-12,6-olide, D-60324
4,10-Dihydroxy-2,11(13)-guaiadien-12,6-olide, D-60326
3,5-Dihydroxy-4(15),10(14)-guaiadien-12,8-olide, D-60325
9,10-Dihydroxy-7-marasmen-5,13-olide, D-60346
4,5-Dioxo-1(10)-xanthen-12,8-olide, D-60474
4-Epivulgarin, in V-60017
Epoxycantabronic acid, in C-60010
$1\beta,10\alpha$-Epoxyhaageanolide, in H-60001
$4\alpha,5\alpha$-Epoxy-3α-hydroxy-11(13)-eudesmen-12,8β-olide, in H-60136
$4\alpha,5\alpha$-Epoxy-3β-hydroxy-11(13)-eudesmen-12,8β-olide, in H-60136
3α-Hydroperoxy-4,11(13)-eudesmadien-12,8β-olide, in H-60136
8α-Hydroxybalchanin, in S-60007
8α-Hydroxy-11α,13-dihydrozaluzanin C, in Z-60001
9α-Hydroxysantamarine, in S-60007
▷Illudin S, I-60002
Schkuhridin B, S-60012
8,9,14-Trihydroxy-1(10),4,11(13)-germacratrien-12,6-olide, T-60323
Umbellifolide, U-60001
▷Vulgarin, V-60017

$C_{15}H_{20}O_5$
3,5-Dihydroxy-4(15),10(14)-guaiadien-12,8-olide; 10α,14-Epoxide, in D-60325

569

C$_{15}$H$_{20}$O$_5$

4,5-Epoxy-2,8-dihydroxy-1(10),11(13)-germacradien-12,6-
olide, E-60013
4α-Hydroperoxydesoxyvulgarin, *in* V-60017

C$_{15}$H$_{20}$O$_6$

Ajafinin, A-60068
2,3-Epoxy-1,4,8-trihydroxy-7(11)-eudesmen-12,8-olide, E-60033

C$_{15}$H$_{20}$O$_8$

Laurencia Polyketal, P-60168

C$_{15}$H$_{21}$NO$_4$

3,5-Pyridinedipropanoic acid; Di-Et ester, *in* P-60217

C$_{15}$H$_{22}$O

Elvirol, E-60004
Isobicyclogermacrenal, I-60088
Penlanpallescensin, P-60015
Perforenone, P-60063
3-Silphinenone, S-60033

C$_{15}$H$_{22}$O$_2$

Capsenone, C-60011
Drimenin, D-60519
4,5-Epoxy-1(10),7(11)-germacradien-8-one, E-60023
Helminthosporal, H-60016
6-Hydroxy-2,7,10-bisabolatrien-9-one, H-60107
6-Hydroxy-1,4-eudesmadien-3-one, H-60139

C$_{15}$H$_{22}$O$_3$

Bedfordiolide, B-60008
Cantabrenolic acid, C-60010
Dihydroconfertin, D-60211
11β,13-Dihydroconfertin, *in* D-60211
11β,13-Dihydro-8-epi-confertin, *in* D-60211
11β,13-Dihydro-1-epi-inuviscolide, *in* I-60037
11β,13-Dihydroinuviscolide, *in* I-60037
3,4-Epoxy-11(13)-eudesmen-12-oic acid, E-60021
9-Hydroxy-4,11(13)-eudesmadien-12-oic acid, H-60135
4-Hydroxy-11(13)-eudesmen-12,8-olide, H-60142
Myoporone, M-60156
1-Oxo-4-germacren-12,6-olide, O-60068
7-Oxo-11-nordrim-8-en-12-oic acid; Me ester, *in* O-60080
Pentalenic acid, P-60054
Santamarine; 11β,13-Dihydro, *in* S-60007
Sterpuric acid, S-60053
Warburganal, W-60001

C$_{15}$H$_{22}$O$_4$

Artapshin, A-60301
Avenaciolide, A-60325
8,10-Dihydroxy-3-guaien-12,6-olide, D-60328
Eriolin, E-60036
11β-Hydroxy-11,13-dihydro-8-epi-confertin, *in* D-60211
9α-Hydroxy-11β,13-dihyrosantamarine, *in* S-60007
13-Hydroxysterpuric acid, *in* S-60053
Sydonic acid, S-60065

C$_{15}$H$_{22}$O$_5$

2α-Hydroxy-11α,13-dihydroconfertin, *in* D-60211
4-Oxo-3,4-secoambrosan-12,6-olid-3-oic acid, O-60093

C$_{15}$H$_{22}$O$_9$

Eranthemoside, E-60034
Galiridoside, G-60002

C$_{15}$H$_{23}$ClO$_2$

2-Chloro-3,7-epoxy-9-chamigranone, C-60066
2-Chloro-3-hydroxy-7-chamigren-9-one, C-60094

C$_{15}$H$_{24}$

ε-Cadinene, C-60001
1,1,3,3,5,5-Hexamethyl-2,4,6-tris(methylene)cyclohexane,
H-60072
1,2,3,4,4a,5,6,7-Octahydro-4a,5-dimethyl-2-(1-
methylethenyl)naphthalene, O-60015
β-Patchoulene, P-60010
Precapnelladiene, P-60171
Sinularene, S-60034

C$_{15}$H$_{24}$O

α-Biotol, B-60112

C$_{15}$H$_{24}$O

β-Biotol, B-60113
2,3-Epoxy-7,10-bisaboladiene, E-60009
3,5-Eudesmadien-1-ol, E-60064
5,7-Eudesmadien-11-ol, E-60065
5,7(11)-Eudesmadien-15-ol, E-60066
Fervanol, F-60007
α-Torosol, T-60225
β-Torosol, T-60226

C$_{15}$H$_{24}$O$_2$

10(14)-Aromadendrene-4,8-diol, A-60299
Artemone, A-60305
cis-9,10-Dihydrocapsenone, *in* C-60011
7,12-Dihydroxysterpurene, D-60374
4,7(11)-Eudesmadiene-12,13-diol, E-60063
11-Hydroxy-4-guaien-3-one, H-60149
11-Hydroxy-1(10)-valencen-2-one, H-60234
Kurubasch aldehyde, K-60014
6-Oxocyclonerolidol, *in* H-60111
Tanavulgarol, T-60004
3,7,11-Trimethyl-2,6,10-dodecatrienoic acid, T-60364
Vetidiol, V-60007

C$_{15}$H$_{24}$O$_3$

Dihydromyoporone, *in* M-60156
FS-2, F-60078
Lapidol, L-60014
Punctaporonin *A*, P-60197
Punctaporonin D, *in* P-60197

C$_{15}$H$_{24}$O$_4$

Trichotriol, T-60258

C$_{15}$H$_{24}$O$_{11}$

Avicennioside, A-60326

C$_{15}$H$_{25}$BrO

Brasudol, B-60197
Isobrasudol, *in* B-60197

C$_{15}$H$_{25}$NS

6-Isothiocyano-4(15)-eudesmene, *in* F-60069

C$_{15}$H$_{26}$

Cyclopentadecyne, C-60220

C$_{15}$H$_{26}$O

9(11)-Drimen-8-ol, D-60520
11-Eudesmen-5-ol, E-60069

C$_{15}$H$_{26}$O$_2$

Alloaromadendrane-4,10-diol, A-60074
1,4-Epoxy-6-eudesmanol, E-60019
11-Eudesmene-1,5-diol, E-60067
1(10),4-Germacradiene-6,8-diol, G-60014
6-Hydroxycyclonerolidol, H-60111

C$_{15}$H$_{26}$O$_3$

8-Carotene-4,6,10-triol, C-60021
8-Daucene-3,6,14-triol, D-60004
8(14)-Daucene-4,6,9-triol, D-60005
4(15)-Eudesmene-1,5,6-triol, E-60068
Fexerol, F-60009

C$_{15}$H$_{26}$O$_4$

Fercoperol, F-60001
2,6,10-Trimethyl-2,6,10-dodecatrien-1,5,8,12-tetrol, T-60363

C$_{15}$H$_{26}$O$_5$

9,12-Dioxododecanoic acid; Me ester, ethylene acetal, *in*
D-60471
Lapidolinol, L-60015

C$_{15}$H$_{26}$O$_7$

ent-5α,11-Epoxy-1β,4α,6α,8α,9β,14-eudesmanehexol, E-60018

C$_{15}$H$_{27}$N

7-Amino-3,10-bisaboladiene, A-60096
Tri-*tert*-butylazete, T-60249

C$_{15}$H$_{27}$N$_3$

Tri-*tert*-butyl-1,2,3-triazine, T-60250

C$_{15}$H$_{28}$O

▷Cyclopentadecanone, C-60219

$C_{15}H_{28}O_3$
12-Hydroxy-13-tetradecenoic acid; Me ester, *in* H-60229
2,6,10-Trimethyl-6,11-dodecadiene-2,3,10-triol, T-60362

$C_{15}H_{29}NO$
Cyclopentadecanone; Oxime, *in* C-60219

$C_{15}H_{30}$
3-*tert*-Butyl-2,2,4,5,5-tetramethyl-3-hexene, B-60350

$C_{15}H_{30}N_2O_3$
4,7,13-Trioxa-1,10-diazabicyclo[8.5.5]icosane, T-60375

$C_{15}H_{30}O_2$
Lardolure, *in* T-60372
2,4,6,8-Tetramethylundecanoic acid, T-60157

$C_{15}H_{30}O_4$
12-Oxododecanoic acid; Me ester, Di-Me acetal, *in* O-60066

$C_{15}H_{33}N_3$
Hexahydro-1,3,5-triazine; 1,3,5-Tri-*tert*-butyl
, *in* H-60060

$C_{15}H_{35}N_5$
1,5,9,13,17-Pentaazacycloeicosane, P-60018

$C_{16}H_4N_4S_2$
2,2'-(4,8-Dihydrobenzo[1,2-*b*:5,4-*b'*]dithiophene-4,8-
diylidene)bispropanedinitrile, D-60196

$C_{16}H_8Cl_2O_2$
1,8-Anthracenedicarboxylic acid; Dichloride, *in* A-60281

$C_{16}H_8Cl_4$
1,4,5,8-Tetrachloro-9,10-anthraquinodimethane, T-60026

$C_{16}H_8F_8$
4,5,7,8,12,13,15,16-Octafluoro[2.2]paracyclophane, O-60010

$C_{16}H_8N_2$
1,8-Dicyanoanthracene, *in* A-60281

$C_{16}H_8N_2O_4S_2$
Dithiobisphthalimide, D-60508

$C_{16}H_8O_2$
Cyclohept[*fg*]acenaphthylene-5,6-dione, C-60194
Cyclohept[*fg*]acenaphthylene-5,8-dione, C-60195

$C_{16}H_8O_3$
Benzo[*b*]naphtho[2,1-*d*]furan-5,6-dione, B-60033
Benzo[*b*]naphtho[2,3-*d*]furan-6,11-dione, B-60034
3,4-Phenanthrenedicarboxylic acid; Anhydride, *in* P-60068

$C_{16}H_8S_2$
Fluorantheno[3,4-*cd*]-1,2-dithiole, F-60017

$C_{16}H_8Se_2$
Fluorantheno[3,4-*cd*]-1,2-diselenole, F-60015

$C_{16}H_8Te_2$
Fluorantheno[3,4-*cd*]-1,2-ditellurole, F-60016

$C_{16}H_9ClO_2S$
1-Pyrenesulfonic acid; Chloride, *in* P-60210

$C_{16}H_9NO_2$
▷1-Nitropyrene, N-60040
2-Nitropyrene, N-60041
4-Nitropyrene, N-60042

$C_{16}H_{10}$
Benzo[*a*]biphenylene, B-60016
Benzo[*b*]biphenylene, B-60017
Dicyclopenta[*ef,kl*]heptalene, D-60167

$C_{16}H_{10}F_3NO_2$
7-Amino-4-(trifluoromethyl)-2*H*-1-benzopyran-2-one; *N*-Ph, *in*
A-60265

$C_{16}H_{10}F_6I_2O_5$
µ-Oxodiphenylbis(trifluoroacetato-*O*)diiodine, O-60065

$C_{16}H_{10}N_2O_2$
Quino[7,8-*h*]quinoline-4,9(1*H*,12*H*)dione, Q-60007

$C_{16}H_{10}N_2S_2$
Thiazolo[4,5-*b*]quinoline-2(3*H*)-thione; 3-Ph, *in* T-60201

$C_{16}H_{10}N_6$
2,2'-Azodiquinoxaline, A-60353

$C_{16}H_{10}O$
Phenanthro[9,10-*b*]furan, P-60073
Phenanthro[9,10-*c*]furan, P-60074

$C_{16}H_{10}O_2$
3-Phenyl-1,2-naphthoquinone, P-60120
4-Phenyl-1,2-naphthoquinone, P-60121

$C_{16}H_{10}O_3$
5-Formyl-4-phenanthrenecarboxylic acid, F-60074

$C_{16}H_{10}O_3S$
4-Oxo-2-phenyl-4*H*-1-benzothiopyran-3-carboxylic acid,
O-60087
1-Pyrenesulfonic acid, P-60210
2-Pyrenesulfonic acid, P-60211
4-Pyrenesulfonic acid, P-60212

$C_{16}H_{10}O_4$
1,8-Anthracenedicarboxylic acid, A-60281
4-Oxo-2-phenyl-4*H*-1-benzopyran-3-carboxylic acid, O-60085
4-Oxo-3-phenyl-4*H*-1-benzopyran-2-carboxylic acid, O-60086
3,4-Phenanthrenedicarboxylic acid, P-60068

$C_{16}H_{10}O_5$
ψ-Baptigenin, B-60006
5-Hydroxy-6,7-methylenedioxyflavone, *in* T-60322

$C_{16}H_{10}O_6$
2',5-Dihydroxy-6,7-methylenedioxyisoflavone, *in* T-60107

$C_{16}H_{10}O_7$
3,6,8-Trihydroxy-1-methylanthraquinone-2-carboxylic acid,
T-60332

$C_{16}H_{10}O_8$
Nasutin C, *in* E-60003

$C_{16}H_{11}N$
Indolo[1,7-*ab*][1]benzazepine, I-60034

$C_{16}H_{11}NO$
1-Benzoylisoquinoline, B-60059
3-Benzoylisoquinoline, B-60060
4-Benzoylisoquinoline, B-60061

$C_{16}H_{11}NO_2$
2-Phenyl-4-(phenylmethylene)-5(4*H*)-oxazolone, P-60128

$C_{16}H_{11}NO_3$
1-Amino-8-hydroxyanthraquinone; *N*-Ac, *in* A-60177

$C_{16}H_{11}NO_4$
1-Amino-2-hydroxyanthraquinone; *N*-Ac, *in* A-60174
1-Amino-5-hydroxyanthraquinone; *N*-Ac, *in* A-60176

$C_{16}H_{12}$
Cyclohepta[*ef*]heptalene, C-60197
1,10-Dihydrodicyclopenta[*a,h*]naphthalene, D-60220
3,8-Dihydrodicyclopenta[*a,h*]naphthalene, D-60221

$C_{16}H_{12}F_2O_4$
4,4'-Difluoro-[1,1'-biphenyl]-3,3'-dicarboxylic acid; Di-Me ester,
in D-60178
4,5-Difluoro-[1,1'-biphenyl]-2,3-dicarboxylic acid; Di-Me ester,
in D-60179
6,6'-Difluoro-[1,1'-biphenyl]-2,2'-dicarboxylic acid; Di-Me ester,
in D-60180

$C_{16}H_{12}F_4$
4,5,7,8-Tetrafluoro[2.2]paracyclophane, T-60050

$C_{16}H_{12}N_2$
2,2'-Bis(cyanomethyl)biphenyl, *in* B-60115
4,4'-Bis(cyanomethyl)biphenyl, *in* B-60116

$C_{16}H_{12}N_2O$
1-Benzoylisoquinoline; Oxime, *in* B-60059

$C_{16}H_{12}O$
9-Anthraceneacetaldehyde, A-60280
1-Phenyl-2-naphthol, P-60115
3-Phenyl-2-naphthol, P-60116
4-Phenyl-2-naphthol, P-60117
5-Phenyl-2-naphthol, P-60118
8-Phenyl-2-naphthol, P-60119

5-[2-(3-Methoxyphenyl)ethyl]-1,3-benzodioxole, *in* T-60316
9-Methylcyclolongipesin, *in* C-60214

$C_{16}H_{16}O_4$
9,10-Dihydro-2,7-dimethoxy-1,5-phenanthrenediol, *in* T-60116
Gleinadiene, *in* G-60023
Homocyclolongipesin, H-60088
Perforatin A, *in* A-60075

$C_{16}H_{16}O_5$
Deoxyaustrocortilutein, *in* A-60321
1-(2,6-Dihydroxy-4-methoxyphenyl)-3-(4-hydroxyphenyl)-1-propanone, *in* H-60219
3-(4-Hydroxyphenyl)-1-(2,4-dihydroxy-6-methoxyphenyl)-1-propanone, *in* H-60220

$C_{16}H_{16}O_6$
Altersolanol B, *in* A-60080
Austrocortilutein, A-60321
Deoxyaustrocortirubin, *in* A-60322

$C_{16}H_{16}O_7$
Altersolanol C, *in* A-60080
Austrocortirubin, A-60322
Fusarubin; O^9-Me, *in* F-60108
Fusarubin methyl acetal, *in* F-60108

$C_{16}H_{16}O_8$
Altersolanol *A*, A-60080

$C_{16}H_{16}S_2$
2,11-Dithia[3.3]paracyclophane, D-60505

$C_{16}H_{17}O_4$
9,10-Dihydro-2,7-dihydroxy-3,5-dimethoxyphenanthrene, *in* T-60120

$C_{16}H_{18}N_2$
[3.3][2.6]Pyridinophane, P-60226
[3](2.2)[3](5.5)Pyridinophane, P-60227
[3](2.5)[3](5.2)Pyridinophane, P-60228

$C_{16}H_{18}O_2$
Methyl 8-isopropyl-5-methyl-2-naphthalenecarboxylate, *in* I-60129

$C_{16}H_{18}O_3$
3-Hydroxy-3',4-dimethoxybibenzyl, *in* T-60316

$C_{16}H_{18}O_4$
Gleinene, G-60023
4-[2-(3-Hydroxyphenyl)ethyl]-2,6-dimethoxyphenol, *in* T-60102
O-Methylcedrelopsin, *in* C-60026

$C_{16}H_{18}O_5$
Skimminin, S-60035

$C_{16}H_{18}O_7$
3-O-Methyldihydrofusarubin A, *in* F-60108

$C_{16}H_{18}O_8$
3,6-Dimethyl-1,2,4,5-benzenetetracarboxylic acid; Tetra-Me ester, *in* D-60407

$C_{16}H_{20}ClN_3$
Bindschedler's green; Chloride, *in* B-60110

$C_{16}H_{20}N_2$
2,2'-Bis(dimethylamino)biphenyl, *in* D-60033

$C_{16}H_{20}N_3^{\oplus}$
Bindschedler's green, B-60110

$C_{16}H_{20}O_2$
Methyl 3,4-dihydro-8-isopropyl-5-methyl-2-naphalenecarboxylate, *in* I-60129

$C_{16}H_{20}O_6$
Orthopappolide, O-60045

$C_{16}H_{22}$
[3^{4,10}][7]Metacyclophane, M-60030

$C_{16}H_{22}O_6$
2,3-Dehydroteucrin E, *in* T-60180
2α,3α-Epoxy-1β,4α-dihydroxy-8β-methoxy-7(11)-eudesmen-12,8-olide, *in* E-60033

$C_{16}H_{22}O_8$
Synrotolide, S-60066

$C_{16}H_{22}O_{10}$
Gelsemide 7-glucoside, *in* G-60010

$C_{16}H_{23}N$
Axisonitrile-4, *in* A-60328

$C_{16}H_{23}NS$
Axisothiocyanate-4, *in* A-60329

$C_{16}H_{23}N_5O_5$
1'-Methylzeatin; 9-β-D-Ribofuranosyl, *in* M-60133

$C_{16}H_{24}O$
5-Methyl-2-(1-methyl-1-phenylethyl)cyclohexanol, M-60090

$C_{16}H_{24}O_9$
Semperoside, S-60021

$C_{16}H_{24}O_{10}$
9-Hydroxysemperoside, *in* S-60021
Vebraside, V-60003

$C_{16}H_{24}O_{11}$
Shanzhiside, S-60028

$C_{16}H_{25}N$
Axisonitrile-1, *in* A-60328
Axisonitrile-2, *in* A-60328
Axisonitrile-3, *in* A-60328
10α-Isocyanoalloaromadendrane, *in* A-60328
3-Isocyano-7,9-bisaboladiene, I-60091
7-Isocyano-3,10-bisaboladiene, I-60092
11-Isocyano-5-eudesmene, *in* F-60068
6-Isocyano-4(15)-eudesmene, *in* F-60069

$C_{16}H_{25}NO$
7-Isocyanato-2,10-bisaboladiene, I-60089

$C_{16}H_{25}NO_5$
Wasabidienone *E*, W-60003

$C_{16}H_{25}NS$
Axisothiocyanate-1, *in* A-60329
Axisothiocyanate-2, *in* A-60329
Axisothiocyanate-3, *in* A-60329
10α-Isothiocyanatoalloaromadendrane, *in* A-60329
11-Isothiocyano-5-eudesmene, *in* F-60068

$C_{16}H_{26}O_2$
3,7,11-Trimethyl-2,6,10-dodecatrienoic acid; Me ester, *in* T-60364

$C_{16}H_{26}O_3$
ent-12-Hydroxy-13,14,15,16-tetranor-1(10)-halimen-18-oic acid, H-60231
Juvenile hormone III, *in* T-60364

$C_{16}H_{26}O_9$
Gelsemiol 1-glucoside, *in* G-60011
Gelsemiol 3-glucoside, *in* G-60011
Gibboside, G-60020

$C_{16}H_{27}NO$
Axamide-2, A-60327
10α-Formamidoalloaromadendrane, *in* A-60327
11-Formamido-5-eudesmene, F-60068
6-Formamido-4(15)-eudesmene, F-60069

$C_{16}H_{28}N_4O_8$
1,4,7,10-Tetraazacyclododecane-1,4,7,10-tetraacetic acid, T-60015

$C_{16}H_{28}O$
Dodecahydro-3a,6,6,9a-tetramethylnaphtho[2,1-*b*]furan, D-60515

$C_{16}H_{28}O_2$
Isoambrettolide, I-60086

$C_{16}H_{30}$
5-Hexadecyne, H-60037

$C_{16}H_{30}O$
Decamethylcyclohexanone, D-60011
2-Hexadecenal, H-60036

C$_{16}$H$_{32}$N$_2$O$_5$
4,7,13,16,21-Pentaoxa-1,10-diazabicyclo[8.8.5]tricosane,
P-60060

C$_{17}$H$_8$N$_2$O
10-(Dicyanomethylene)anthrone, D-60164

C$_{17}$H$_{10}$F$_3$NO$_3$
7-Amino-4-(trifluoromethyl)-2H-1-benzopyran-2-one; N-
Benzoyl, in A-60265

C$_{17}$H$_{10}$N$_2$
9-(Dicyanomethyl)anthracene, D-60163

C$_{17}$H$_{10}$O$_5$
1,5-Diphenylpentanepentone, D-60488

C$_{17}$H$_{12}$
3H-Cyclonona[def]biphenylene, C-60215
1-Methylazupyrene, M-60043

C$_{17}$H$_{12}$OS
2,6-Diphenyl-4H-thiopyran-4-one, D-60492

C$_{17}$H$_{12}$O$_3$S
2,6-Diphenyl-4H-thiopyran-4-one; 1,1-Dioxide, in D-60492

C$_{17}$H$_{12}$O$_6$
3,3-Dihydroxy-5,5-diphenyl-1,2,4,5-pentanetetrone, in D-60488
5-Hydroxy-2′-methoxy-6,7-methylenedioxyisoflavone, in
T-60107

C$_{17}$H$_{12}$O$_7$
3,6,8-Trihydroxy-1-methylanthraquinone-2-carboxylic acid; Me
ester, in T-60332

C$_{17}$H$_{12}$O$_8$
Nasutin B, in E-60003

C$_{17}$H$_{13}$NO$_2$S
2-Phenyl-1,5-benzothiazepin-4(5H)-one; N-Ac, in P-60081

C$_{17}$H$_{14}$
Bicyclo[5.3.1]undeca-1,3,5,7,9-pentaene, B-60102

C$_{17}$H$_{14}$N$_2$O$_4$
2,3-Dihydro-2-hydroxy-1H-imidazole-4,5-dione; 1,3-Di-Ph, O-
Ac, in D-60243

C$_{17}$H$_{14}$O
2-Methoxy-1-phenylnaphthalene, in P-60115
2-Methoxy-5-phenylnaphthalene, in P-60118
7-Methoxy-1-phenylnaphthalene, in P-60119

C$_{17}$H$_{14}$O$_2$
1,2-Dihydro-1-phenyl-1-naphthalenecarboxylic acid, D-60264
2-(4-Hydroxyphenyl)-5-(1-propenyl)benzofuran, H-60214

C$_{17}$H$_{14}$O$_3$
1,5-Diphenyl-1,3,5-pentanetrione, D-60489
4-[5-(1-Propenyl)-2-benzofuranyl]-2,3-benzenediol, in H-60214

C$_{17}$H$_{14}$O$_4$
Bonducellin, B-60192

C$_{17}$H$_{14}$O$_5$
5-Hydroxy-6,7-dimethoxyflavone, in T-60322
6-Hydroxy-4′,7-dimethoxyisoflavone, in T-60325

C$_{17}$H$_{14}$O$_6$
4′,5-Dihydroxy-3′,7-dimethoxyisoflavone, in T-60108
5,7-Dihydroxy-2′,6-dimethoxyisoflavone, in T-60107
5,7-Dihydroxy-3′,4′-dimethoxyisoflavone, in T-60108
Sophorocarpan B, S-60038

C$_{17}$H$_{14}$O$_7$
3,4′,7-Trihydroxy-5,6-dimethoxyflavone, in P-60047
3′,4′,5-Trihydroxy-6,7-dimethoxyisoflavone, in P-60049

C$_{17}$H$_{15}$N
2-Methyl-1,5-diphenyl-1H-pyrrole, in M-60115

C$_{17}$H$_{15}$NO$_2$S
2-Phenyl-1,5-benzothiazepin-4(5H)-one; 2,3-Dihydro, N-Ac, in
P-60081

C$_{17}$H$_{16}$N$_2$O
1-[1-Methoxy-1-(2-pyridinyl)ethyl]isoquinoline, in I-60134

C$_{17}$H$_{16}$N$_2$O$_3$
3-Oxopentanedioic acid; Dianilide, in O-60084

C$_{17}$H$_{16}$O$_2$
1,5-Bis(4-hydroxyphenyl)-1,4-pentadiene, B-60162

C$_{17}$H$_{16}$O$_3$
2,3,7-Trimethoxyphenanthrene, in T-60339

C$_{17}$H$_{16}$O$_4$
3,5,7-Trimethoxy-2-phenanthrenol, in T-60120

C$_{17}$H$_{16}$O$_5$
2′,4-Dihydroxy-4′,6′-dimethoxychalcone, in H-60220
5-Hydroxy-7,8-dimethoxyflavone, in T-60321
Methylsainfuran, in S-60002
Sophorocarpan A, S-60037
1,5,6-Trimethoxy-2,7-phenanthrenediol, in P-60050

C$_{17}$H$_{16}$O$_6$
1,5-Bis(3,4-dihydroxyphenyl)-4-pentyne-1,2-diol, B-60155
5,8-Dihydroxy-2,3,6-trimethyl-1,4-naphthoquinone; Di-Ac, in
D-60380
1,2,3,4-Tetramethoxyxanthone, in T-60126

C$_{17}$H$_{16}$O$_7$
2′,3,5-Trihydroxy-7,8-dimethoxyflavanone, in P-60045
1,3,8-Trihydroxy-4,7-dimethoxyxanthone, in P-60052

C$_{17}$H$_{18}$O$_4$
3′,5-Dimethoxy-3,4-methylenedioxybibenzyl, in T-60102

C$_{17}$H$_{18}$O$_5$
9$β$-Acetoxy-3-oxo-1,4(15),11(13)-eudesmatrien-12,6-olide, in
H-60202
5,7-Dihydroxy-3′,4′-dimethoxyflavan, in T-60103
4′-Hydroxy-3′,5-dimethoxy-3,4-methylenedioxybibenzyl, in
P-60044
Longipesin; 9-Ac, in L-60034

C$_{17}$H$_{18}$O$_6$
Pandoxide, P-60002

C$_{17}$H$_{18}$O$_7$
Fusarubin; O^3-Et, in F-60108

C$_{17}$H$_{20}$O$_2$
1,5-Diphenyl-1,3-pentanediol, D-60487

C$_{17}$H$_{20}$O$_3$
1$β$-Acetoxy-4(15),7(11),8-eudesmatrien-12,8-olide, in H-60141

C$_{17}$H$_{20}$O$_4$
1$β$-Acetoxy-3,7(11),8-eudesmatrien-12,8-olide, in H-60140
Zaluzanin D, in Z-60001

C$_{17}$H$_{20}$O$_5$
Cavoxinine, C-60024
Cavoxinone, C-60025
4,4′-Dihydroxy-3,3′,5-trimethoxybibenzyl, in P-60044
8-Hydroxyachillin; Ac, in H-60100
Murraxocin, M-60151

C$_{17}$H$_{20}$O$_6$
Achillolide A, in A-60061
Tarchonanthus lactone, T-60009

C$_{17}$H$_{20}$O$_7$
3-O-Ethyldihydrofusarubin A, in F-60108

C$_{17}$H$_{21}$ClO$_6$
Arctodecurrolide, in D-60325

C$_{17}$H$_{22}$O$_3$
8,12-Epoxy-3,7,11-eudesmatrien-1-ol; Ac, in E-60020
6-Methoxy-2-(3,7-dimethyl-2,6-octadienyl)-1,4-
benzoquinone, M-60038
Nimbidiol, N-60021

C$_{17}$H$_{22}$O$_4$
3$β$-Acetoxy-4,11(13)-eudesmadien-12,8$β$-olide, in H-60136
9$β$-Acetoxytournefortiolide, in H-60137
Subexpinnatin C, S-60058

C$_{17}$H$_{22}$O$_5$
1$β$-Acetoxy-8$β$-hydroxy-3,7(11)-eudesmadien-12,8-olide, in
D-60322

3β-Acetoxy-2α-hydroxy-1(10),4,11(13)-germacratrien-12,6α-olide, *in* H-60006
Ergolide, E-60035

$C_{17}H_{22}O_6$
Neovasinone, N-60019

$C_{17}H_{24}O_4$
3β-Acetoxydrimenin, *in* D-60519
9β-Acetoxy-4,11(13)-eudesmadien-12-oic acid, *in* H-60135
2-(7-Hydroxy-3,7-dimethyl-2-octenyl)-6-methoxy-1,4-benzoquinone, H-60123
Sterpuric acid; 13-Hydroxy, 3,13-ethylidene acetal, *in* S-60053

$C_{17}H_{24}O_5$
Nitrosin, N-60045

$C_{17}H_{24}O_6$
2α-Acetoxy-11α,13-dihydroconfertin, *in* D-60211
15-Desacetyltetraneurin C, *in* T-60158
Tetraneurin D, T-60158

$C_{17}H_{26}O_2$
Coralloidin C, *in* E-60066
Coralloidin E, *in* E-60065

$C_{17}H_{26}O_{11}$
Shanzhiside; Me ester, *in* S-60028

$C_{17}H_{28}O_2$
5,6-Decanediol; Benzyl ether, *in* D-60013

$C_{17}H_{28}O_3$
6β-Acetoxy-1α,4α-epoxyeudesmane, *in* E-60019

$C_{17}H_{28}O_4$
6β-Acetoxy-4(15)-eudesmene-1β,5α-diol, *in* E-60068

$C_{17}H_{28}O_5$
12-Acetoxy-5,8-dihydroxyfarnesol, *in* T-60363

$C_{17}H_{30}O$
3,4,7,11-Tetramethyl-6,10-tridecadienal, T-60156

$C_{17}H_{32}$
1,1,2,2,3,3,4,4,5,5-Decamethyl-6-methylenecyclohexane, D-60012

$C_{17}H_{32}O_2$
Cyclopentadecanone; Ethylene acetal, *in* C-60219

$C_{17}H_{32}O_3$
2-Tetradecyloxiranecarboxylic acid, T-60041

$C_{17}H_{34}O$
Undecamethylcyclohexanol, U-60003

$C_{18}H_6N_6O_{12}$
Benzo[1,2-b:3,4-b':5,6-b'']tripyrazine-2,3,6,7,10,11-tetracarboxylic acid, B-60051

$C_{18}H_9N_3O_3$
19,20,21-Triazatetracyclo[13.3.1.1³,⁷.1⁹,¹³]heneicosa-1(19),3,5,7(21),9,11,13(20),15,17-nonaene-2,8,14-trione, T-60236

$C_{18}H_{10}$
▷Cyclopenta[cd]pyrene, C-60223

$C_{18}H_{10}N_2O_2$
Triphenodioxazine, T-60378

$C_{18}H_{10}O$
Pyreno[4,5-b]furan, P-60214

$C_{18}H_{10}O_2$
Naphth[2,3-a]azulene-5,12-dione, N-60006

$C_{18}H_{10}O_6$
Bicoumol, B-60082

$C_{18}H_{10}S_2$
Benzo[1,2-b:4,5-b']bis[1]benzothiophene, B-60018
Benzo[1,2-b:5,4-b']bis[1]benzothiophene, B-60019
Phenanthro[1,10-bc:8,9-b',c']bisthiopyran, P-60072

$C_{18}H_{12}$
Δ¹,¹'-Biindene, B-60109
Cyclohepta[a]phenalene, C-60199

$C_{18}H_{12}N_4$
[1,2,4,5]Tetrazino[1,6-a:4,3-a']diisoquinoline, T-60172
[1,2,4,5]Tetrazino[1,6-a:4,3-a']diquinoline, T-60173

$C_{18}H_{12}N_{12}O_6$
Benzo[1,2-b:3,4-b':5,6-b'']tripyrazine-2,3,6,7,10,11-tetracarboxylic acid; Hexaamide, *in* B-60051

$C_{18}H_{12}O$
Triphenylene-1,2-oxide, T-60385

$C_{18}H_{12}O_2$
1-Pyreneacetic acid, P-60208
4-Pyreneacetic acid, P-60209

$C_{18}H_{12}O_2S_2$
5a,5b,11a,11b-Tetrahydrocyclobuta[1,2-b:4,3-b']-benzothiopyran-11,12-dione, T-60063

$C_{18}H_{12}O_4$
Glabone, G-60021

$C_{18}H_{12}O_6$
Grevilline A, G-60038

$C_{18}H_{12}O_7$
Grevilline B, *in* G-60038

$C_{18}H_{12}O_8$
Grevilline C, *in* G-60038
Grevilline D, *in* G-60038

$C_{18}H_{12}O_9$
Substictic acid, S-60060

$C_{18}H_{14}$
1,1'-Bi-1H-indene, B-60108
[2.2.2](1,2,3)Cyclophane-1,9-diene, C-60229
7,12-Dihydropleiadene, D-60269

$C_{18}H_{14}N_6$
1,4-Bis(2-pyridylamino)phthalazine, B-60172

$C_{18}H_{14}O$
2,7-Diphenyloxepin, D-60486

$C_{18}H_{14}O_4$
3,4-Acenaphthenedicarboxylic acid; Di-Et ester, *in* A-60013
1,8-Anthracenedicarboxylic acid; Di-Me ester, *in* A-60281
4-Oxo-2-phenyl-4H-1-benzopyran-3-carboxylic acid; Et ester, *in* O-60085
3,4-Phenanthrenedicarboxylic acid; Di-Me ester, *in* P-60068

$C_{18}H_{15}N_3$
Amino-1,4-benzoquinone; Dianil, *in* A-60089

$C_{18}H_{16}$
1,2-Dibenzylidenecyclobutane, D-60083
1,2-Diphenyl-1,4-cyclohexadiene, D-60477

$C_{18}H_{16}O_2$
Eupomatenoid 6, E-60075

$C_{18}H_{16}O_3$
2-(2-Hydroxy-4-methoxyphenyl)-5-(1-propenyl)benzofuran, *in* H-60214
2-(4-Hydroxyphenyl)-7-methoxy-5-(1-propenyl)benzofuran, *in* H-60214

$C_{18}H_{16}O_4$
Danshexinkun A, D-60002
2-(2,4-Dihydroxyphenyl)-7-methoxy-5-(1-propenyl)-benzofuran, *in* H-60214

$C_{18}H_{16}O_5$
8-Methoxybonducellin, *in* B-60192
5,6,7-Trimethoxyflavone, *in* T-60322
4',6,7-Trimethoxyisoflavone, *in* T-60325

$C_{18}H_{16}O_6$
Isoamericanin A, I-60087
Scleroderolide, S-60013
1,3,5,8-Tetrahydroxy-2-(3-methyl-2-butenyl)xanthone, T-60114
2',5',6-Trihydroxy-3,5,7-trimethoxyflavone, *in* H-60064

$C_{18}H_{16}O_7$
3,4'-Dihydroxy-5,6,7-trimethoxyflavone, *in* P-60047
3,5-Dihydroxy-4',6,7-trimethoxyflavone, *in* P-60047

4′,5-Dihydroxy-3,6,7-trimethoxyflavone, *in* P-60047
4′,5-Dihydroxy-3,6,8-trimethoxyflavone, *in* P-60048
5,6-Dihydroxy-3,4′,8-trimethoxyflavone, *in* P-60048

C$_{18}$H$_{16}$O$_{10}$
3′,4′,5,5′,7-Pentahydroxy-3,6,8-trimethoxyflavone, *in* O-60023

C$_{18}$H$_{17}$NO$_5$
2-Amino-3-phenylpentanedioic acid; *N*-Benzoyl, *in* A-60241

C$_{18}$H$_{18}$
[2.2.2](1,2,3)Cyclophane, C-60228
1,3,6,8-Tetramethylphenanthrene, T-60151
2,4,5,7-Tetramethylphenanthrene, T-60152

C$_{18}$H$_{18}$O$_2$
Conocarpan, C-60161

C$_{18}$H$_{18}$O$_3$
Olmecol, O-60040

C$_{18}$H$_{18}$O$_4$
1,2,5,6-Tetramethoxyphenanthrene, *in* T-60115
1,2,5,7-Tetramethoxyphenanthrene, *in* T-60116
1,2,6,7-Tetramethoxyphenanthrene, *in* T-60117
1,3,5,6-Tetramethoxyphenanthrene, *in* T-60118
1,3,6,7-Tetramethoxyphenanthrene, *in* T-60119
2,3,5,7-Tetramethoxyphenanthrene, *in* T-60120
2,3,6,7-Tetramethoxyphenanthrene, *in* T-60121
2,4,5,6-Tetramethoxyphenanthrene, *in* T-60122
3,4,5,6-Tetramethoxyphenanthrene, *in* T-60123

C$_{18}$H$_{18}$O$_5$
5,6-Dimethoxy-3-(4-methoxybenzyl)phthalide, D-60398
Echinofuran, E-60002
Homocyclolongipesin; Ac, *in* H-60088
2′-Hydroxy-4,4′,6′-trimethoxychalcone, *in* H-60220
4-Hydroxy-2′,4′,6′-trimethoxychalcone, *in* H-60220

C$_{18}$H$_{18}$O$_6$
5,8-Dihydroxy-2,3,6,7-tetramethyl-1,4-naphthoquinone; Di-Ac, *in* D-60377
Monocillin I, *in* N-60060
3,4,7,8-Tetramethoxy-2,6-phenanthrenediol, *in* H-60066

C$_{18}$H$_{18}$O$_7$
1,2,3,4,8-Pentamethoxyxanthone, *in* P-60051
Senepoxide, S-60023
β-Senepoxide, *in* S-60023
3′,4′,5,6-Tetrahydroxy-7-methoxyflavanone, *in* P-60046

C$_{18}$H$_{19}$NO
2-Phenylcyclopentylamine; *N*-Benzoyl, *in* P-60086
2-Phenylcyclopentylamine; Benzoyl, *in* P-60086

C$_{18}$H$_{19}$NO$_3$S
2-Amino-4,5,6,7-tetrahydrobenzo[*b*]thiophene-3-carboxylic acid; Et ester, *N*-benzoyl, *in* A-60258

C$_{18}$H$_{20}$
3,3-Dimethyl-4,4-diphenyl-1-butene, D-60418
5,13-Dimethyl[2.2]metacyclophane, D-60426

C$_{18}$H$_{20}$N$_2$O$_4$
N-Tyrosylphenylalanine, T-60412

C$_{18}$H$_{20}$O$_2$
Pentamethylphenol; Benzoyl, *in* P-60058

C$_{18}$H$_{20}$O$_3$
2,3-Dihydro-2-(4-hydroxyphenyl)-5-(3-hydroxypropyl)-3-methylbenzofuran, D-60244
1-(4-Hydroxyphenyl)-2-(4-propenylphenoxy)-1-propanol, H-60215
Tetrahydrobis(4-hydroxyphenyl)-3,4-dimethylfuran, T-60061

C$_{18}$H$_{20}$O$_5$
Monocillin II, *in* N-60060

C$_{18}$H$_{20}$O$_6$
3-Methoxy-6-[2-(3,4-5-trimethoxyphenyl)ethenyl]-1,2-benzenediol, *in* T-60344
Monocillin III, *in* N-60060
Syringopicrogenin *A*, S-60067

C$_{18}$H$_{20}$O$_7$
Syringopicrogenin B, *in* S-60067

C$_{18}$H$_{21}$N$_3$O$_6$
10-Deazariboflavin, *in* P-60245

C$_{18}$H$_{22}$N$_2$O$_3$S$_2$
Dithiosilvatin, D-60511

C$_{18}$H$_{22}$O$_5$
Monocillin IV, *in* N-60060
▷Zearalenone, Z-60002

C$_{18}$H$_{22}$O$_6$
8′-Hydroxyzearalenone, *in* Z-60002
3-Methoxy-6-[2-(3,4,5-trimethoxyphenyl)ethyl]1,2-benzenediol, *in* T-60344
Monocillin V, *in* N-60060

C$_{18}$H$_{24}$
1,2,3,4,5,6,7,8,9,10,11,12-Dodecahydrotriphenylene, D-60516

C$_{18}$H$_{24}$O$_2$
3-(2-Hydroxy-4,8-dimethyl-3,7-nonadienyl)benzaldehyde, H-60122

C$_{18}$H$_{24}$O$_5$
Nordinone, N-60060

C$_{18}$H$_{24}$O$_7$
Nordinonediol, *in* N-60060

C$_{18}$H$_{26}$O$_8$
Boronolide, B-60193

C$_{18}$H$_{28}$
9,9′-Bi(bicyclo[3.3.1]nonylidene), B-60079

C$_{18}$H$_{28}$O$_3$
ent-6β,17-Dihydroxy-14,15-bisnor-7,11*E*-labdadien-13-one, D-60311

C$_{18}$H$_{28}$O$_4$
ent-12-Hydroxy-13,14,15,16-tetranor-1(10)-halimen-18-oic acid; Ac, *in* H-60231

C$_{18}$H$_{30}$
▷Hexaethylbenzene, H-60042

C$_{18}$H$_{30}$O$_3$
Sterebin *D*, S-60052

C$_{18}$H$_{30}$O$_4$
Sterebin A, *in* S-60052
3,7,11-Trimethyl-2,6,10-dodecatrienoic acid; 2,3-Dihydroxypropyl ester, *in* T-60364

C$_{18}$H$_{30}$O$_5$
Gloeosporone, G-60025

C$_{18}$H$_{30}$S$_3$
7,14,21-Trithiatrispiro[5.1.5.1.5.1]heneicosane, *in* C-60209

C$_{18}$H$_{32}$N$_4$O$_8$
1,4,8,11-Tetraazacyclotetradecane-1,4,8,11-tetraacetic acid, T-60017

C$_{18}$H$_{32}$O$_5$
Aspicillin, A-60312

C$_{18}$H$_{34}$O
2,4-Octadecadien-1-ol, O-60004

C$_{18}$H$_{34}$O$_2$
1-Tetradecylcyclopropanecarboxylic acid, T-60040

C$_{18}$H$_{34}$O$_3$
2-Tetradecyloxiranecarboxylic acid; Me ester, *in* T-60041

C$_{18}$H$_{34}$O$_5$
Aspicillin; Dihydro, *in* A-60312

C$_{18}$H$_{36}$N$_2$O$_6$
▷4,7,13,16,21,24-Hexaoxa-1,10-diazabicyclo[8.8.8]-hexacosane, H-60073

C$_{18}$H$_{36}$O
2-Octadecen-1-ol, O-60007

C$_{18}$H$_{36}$O$_2$
3-Octadecene-1,2-diol, O-60005
9-Octadecene-1,12-diol, O-60006

C$_{18}$H$_{36}$O$_5$
9,10,18-Trihydroxyoctadecanoic acid, T-60334

C$_{18}$H$_{38}$N$_2$
1,1,1′,1′-Tetra-*tert*-butylazomethane, T-60024

C$_{18}$N$_6$O$_9$
Benzo[1,2-*b*:3,4-*b*′:5,6-*b*″]tripyrazine-2,3,6,7,10,11-tetracarboxylic acid; Trianhydride, *in* B-60051

C$_{18}$N$_{12}$
Benzo[1,2-*b*:3,4-*b*′:5,6-*b*″]tripyrazine-2,3,6,7,10,11-tetracarboxylic acid; Hexanitrile, *in* B-60051

C$_{19}$Cl$_{15}$
Tris(pentachlorophenyl)methyl, *in* T-60405

C$_{19}$Cl$_{21}$Sb
Tris(pentachlorophenyl)methyl hexachloroantimonate, *in* T-60405

C$_{19}$HCl$_{15}$
Tris(pentachlorophenyl)methane, T-60405

C$_{19}$H$_{11}$ClO
2-Triphenylenecarboxylic acid; Chloride, *in* T-60384

C$_{19}$H$_{11}$N
2-Cyanotriphenylene, *in* T-60384

C$_{19}$H$_{12}$O
2-Triphenylenecarboxaldehyde, T-60382

C$_{19}$H$_{12}$O$_2$
1-Triphenylenecarboxylic acid, T-60383
2-Triphenylenecarboxylic acid, T-60384

C$_{19}$H$_{14}$N$_4$
4-Amino-1,2,3-benzotriazine; 3,*N*(4)-Di-Ph, *in* A-60092

C$_{19}$H$_{14}$O$_2$
1-Pyreneacetic acid; Me ester, *in* P-60208
4-Pyreneacetic acid; Me ester, *in* P-60209

C$_{19}$H$_{14}$O$_4$
4′,7-Dimethoxyisoflavone, *in* D-60340
Ochromycinone, O-60003

C$_{19}$H$_{14}$O$_6$
4′,7-Dihydroxyisoflavone; Di-Ac, *in* D-60340

C$_{19}$H$_{14}$O$_7$
6-Aldehydoisoophiopogone *A*, A-60071

C$_{19}$H$_{14}$O$_{10}$
Constictic acid, C-60163

C$_{19}$H$_{14}$S$_2$
2-Dibenzothiophenethiol; *S*-Benzyl, *in* D-60078
4-Dibenzothiophenethiol; *S*-Benzyl, *in* D-60079

C$_{19}$H$_{15}$N$_3$
Azidotriphenylmethane, A-60352

C$_{19}$H$_{16}$
4-(2,4,6-Cycloheptatrien-1-ylidene)bicyclo[5.4.1]dodeca-2,5,7,9,11-pentaene, C-60207

C$_{19}$H$_{16}$O$_4$
2-(4-Hydroxyphenyl)-7-methoxy-5-(1-propenyl)-3-benzofurancarboxaldehyde, H-60212
Moracin *D*, M-60143
Moracin *E*, M-60144
Moracin *G*, M-60145

C$_{19}$H$_{16}$O$_5$
3′,4′-Deoxypsorospermin, D-60021

C$_{19}$H$_{16}$O$_6$
6-Aldehydoisoophiopogone *B*, A-60072
Psorospermin, P-60189

C$_{19}$H$_{17}$ClO$_6$
3′,4′-Deoxy-4′-chloropsorospermin-3′-ol, *in* D-60021

C$_{19}$H$_{17}$NO$_2$
2-Hydroxycyclononanone; Oxime, *in* H-60112

C$_{19}$H$_{18}$N$_2$
(Triphenylmethyl)hydrazine, T-60388

C$_{19}$H$_{18}$O$_3$
Eupomatenoid 5, *in* E-60075
Kachirachirol A, *in* K-60001

C$_{19}$H$_{18}$O$_4$
Demethylfruticulin A, *in* F-60076
2-(2-Hydroxy-4-methoxyphenyl)-7-methoxy-5-(1-propenyl)-benzofuran, *in* H-60214
Isotanshinone IIB, I-60138
6-Methoxy-2-[2-(4-methoxyphenyl)ethyl]-4*H*-1-benzopyran-4-one, M-60039
Moracin C, *in* M-60146

C$_{19}$H$_{18}$O$_6$
Fruticulin *B*, F-60077
1,5,8-Trihydroxy-3-methyl-2-(3-methyl-2-butenyl)xanthone, *in* T-60114

C$_{19}$H$_{18}$O$_7$
3′,4′-Deoxypsorospermin-3′,4′-diol, *in* D-60021
5-Hydroxy-3,4′,6,7-tetramethoxyflavone, *in* P-60047
6-Hydroxy-3,4′,5,7-tetramethoxyflavone, *in* P-60047

C$_{19}$H$_{18}$O$_8$
2′-Acetoxy-3,5-dihydroxy-7,8-dimethoxyflavone, *in* P-60045
3′,5-Dihydroxy-4′,6,7,8-tetramethoxyflavone, D-60375
5,7-Dihydroxy-3′,4′,6,8-tetramethoxyflavone, D-60376
5′,6-Dihydroxy-2′,3,5,7-trimethoxyflavone, *in* H-60064

C$_{19}$H$_{18}$O$_9$
3,4′,5-Trihydroxy-3′,6,7,8-tetramethoxyflavone, *in* H-60024

C$_{19}$H$_{18}$O$_{10}$
2′,3,4′,5,5′,6,7,8-Octahydroxyflavone; 2′,3,6,8-Tetra-Me ether, *in* O-60022
2′,4′,5,7-Tetrahydroxy-3,5′,6,8-tetramethoxyflavone, *in* O-60022
2′,5,5′,7-Tetrahydroxy-3,4′,6,8-tetramethoxyflavone, *in* O-60022

C$_{19}$H$_{20}$
1-Methyl-4,4-diphenylcyclohexene, M-60065
3-Methyl-4,4-diphenylcyclohexene, M-60066

C$_{19}$H$_{20}$O$_2$
3,3-Dimethyl-5,5-diphenyl-4-pentenoic acid, D-60420
Hermosillol, H-60029

C$_{19}$H$_{20}$O$_3$
Cryptotanshinone, C-60173
2,3-Dihydro-2-(4-hydroxy-3-methoxyphenyl)-3-methyl-5-(1-propenyl)benzofuran, *in* C-60161
2,3-Dihydro-2-(4-hydroxyphenyl)-7-methoxy-3-methyl-5-(1-propenyl)benzofuran, *in* C-60161

C$_{19}$H$_{20}$O$_4$
Galipein, G-60001
Kachirachirol B, K-60001

C$_{19}$H$_{20}$O$_5$
Agasyllin, A-60064
Homocyclolongipesin; Propanoyl, *in* H-60088
3-(4-Hydroxyphenyl)-1-(2,4,6-trihydroxyphenyl)-2-propen-1-one; Tetra-Me ether, *in* H-60220

C$_{19}$H$_{20}$O$_6$
Asadanin, A-60308
1(10)E,8E-Millerdienolide, *in* M-60136

C$_{19}$H$_{20}$O$_7$
Annulin *A*, A-60276
4β,15-Epoxy-1(10)E,8E-millerdienolide, *in* M-60136
Isoelephantopin, I-60099

C$_{19}$H$_{22}$O$_3$
Gravelliferone, G-60037

C$_{19}$H$_{22}$O$_4$
Chalepin, C-60031
2,3-Dihydro-2-(4-hydroxy-3-methoxyphenyl)-5-(3-hydroxypropyl)-3-methylbenzofuran, *in* D-60244
1,5-Dihydroxy-1,7-diphenyl-3-heptanone, D-60480
1-(4-Hydroxyphenyl)-2-(2-methoxy-4-propenylphenoxy)-1-propanol, *in* H-60215
Neocryptotanshinone, N-60017

Tetrahydrobis(4-hydroxyphenyl)-3,4-dimethylfuran; 3'-Methoxy, *in* T-60061
Tetrahydro-2-(4-hydroxyphenyl)-5-(4-hydroxy-3-methoxyphenyl)-3,4-dimethylfuran, *in* T-60061
Yashabushiketodiol B, *in* D-60480

C$_{19}$H$_{22}$O$_6$
Antheridic acid, A-60278
Asadanol, *in* A-60308
5-Formylzearalenone, *in* Z-60002
Isotaxiresinol, I-60139
9-Millerenolide, M-60137
1(10)-Millerenolide, M-60136
Strigol, S-60057

C$_{19}$H$_{22}$O$_7$
4β,15-Epoxy-9E-millerenolide, *in* M-60137
4β,15-Epoxy-9Z-millerenolide, *in* M-60137
Tomenphantopin B, T-60222

C$_{19}$H$_{22}$O$_8$
Syringopicrogenin C, *in* S-60067

C$_{19}$H$_{22}$O$_{10}$
Aranochromanophthalide, A-60293

C$_{19}$H$_{23}$ClO$_7$
Chlororepdiolide, C-60133
Repensolide, R-60005

C$_{19}$H$_{24}$O$_3$
1,7-Diphenyl-1,3,5-heptanetriol, D-60480

C$_{19}$H$_{24}$O$_4$
Gibberellin A$_9$, G-60018

C$_{19}$H$_{24}$O$_5$
Gibberellin A$_{40}$, *in* G-60018
Gibberellin A$_{45}$, *in* G-60018
Gibberellin A$_{51}$, *in* G-60018

C$_{19}$H$_{24}$O$_6$
3β-Acetoxyhaageanolide acetate, *in* H-60001
Caleine *E*, C-60005
Desacetylacanthospermal A, *in* A-60012
9β,15-Diacetoxy-1(10),4,11(13)-germacratrien-12,6α-olide, *in* D-60324
Gibberellin A$_{55}$, *in* G-60018
Gibberellin A$_{63}$, *in* G-60018

C$_{19}$H$_{24}$O$_7$
Achillolide *B*, A-60061

C$_{19}$H$_{26}$O$_2$
Acalycixeniolide *B*, A-60011

C$_{19}$H$_{26}$O$_6$
Gibberellin A$_2$, G-60017

C$_{19}$H$_{26}$O$_7$
14-Acetoxytetraneurin D, *in* T-60158
Resinoside, R-60006
Tetraneurin C, *in* T-60158

C$_{19}$H$_{26}$O$_{10}$
Ptelatoside A, *in* H-60211

C$_{19}$H$_{28}$O$_2$
Acalycixeniolide *A*, A-60010

C$_{19}$H$_{28}$O$_4$
Coralloidin D, *in* E-60063

C$_{19}$H$_{28}$O$_5$
1β,10α; 4α,5β-Diepoxy-8α-isobutoxyglechomanolide, *in* G-60022
1β,10α; 4α,5β-Diepoxy-8β-isobutoxyglechomanolide, *in* G-60022

C$_{19}$H$_{28}$O$_6$
Desacetyltetraneurin *D* 4-*O*-isobutyrate, *in* T-60158
Desacetyltetraneurin *D* 15-*O*-isobutyrate, *in* T-60158

C$_{19}$H$_{28}$O$_7$
Tagitinin *A*, T-60002

C$_{19}$H$_{28}$O$_{12}$
6-*O*-Acetylshanghiside methyl ester, *in* S-60028

Barlerin, *in* S-60028

C$_{19}$H$_{30}$N$_6$O$_8$
Antiarrhythmic peptide (ox atrium), A-60283

C$_{19}$H$_{30}$O$_2$
Gracilin *F*, G-60035

C$_{19}$H$_{30}$O$_3$
Norjuslimdiolone, N-60061
17-Nor-7-oxo-8-labden-15-oic acid, *in* H-60198

C$_{19}$H$_{30}$O$_4$
ent-6α,15,16-Trihydroxy-19-nor-4-rosen-3-one, *in* N-60061

C$_{19}$H$_{32}$O$_2$
15-Hydroxy-17-nor-8-labden-7-one, H-60198

C$_{19}$H$_{34}$O$_5$
3,6-Epidioxy-6-methoxy-4-octadecenoic acid, E-60008

C$_{19}$H$_{37}$NO$_2$
14-Azaprostanoic acid, A-60332

C$_{19}$H$_{38}$O$_5$
9,10,18-Trihydroxyoctadecanoic acid; Me ester, *in* T-60334

C$_{20}$H$_{12}$
Benz[*d*]aceanthrylene, B-60009
Benz[*k*]aceanthrylene, B-60010

C$_{20}$H$_{12}$Br$_2$
2,2'-Dibromo-1,1'-binaphthyl, D-60089
4,4'-Dibromo-1,1'-binaphthyl, D-60090

C$_{20}$H$_{12}$I$_2$
2,2'-Diiodo-1,1'-binaphthyl, D-60385
4,4'-Diiodo-1,1'-binaphthyl, D-60386

C$_{20}$H$_{12}$O$_5$
Halenaquinone, *in* H-60002

C$_{20}$H$_{12}$O$_6$
Altertoxin III, A-60082

C$_{20}$H$_{12}$S
Benzo[3,4]phenanthro[1,2-*b*]thiophene, B-60039
Benzo[3,4]phenanthro[2,1-*b*]thiophene, B-60040

C$_{20}$H$_{14}$
9,10-Dihydrodicyclopenta[*c,g*]phenanthrene, D-60219

C$_{20}$H$_{14}$O
1-Acetyltriphenylene, A-60059
2-Acetyltriphenylene, A-60060

C$_{20}$H$_{14}$O$_2$
1-Triphenylenecarboxylic acid; Me ester, *in* T-60383
2-Triphenylenecarboxylic acid; Me ester, *in* T-60384

C$_{20}$H$_{14}$O$_5$
Halenaquinol, H-60002

C$_{20}$H$_{14}$O$_6$
Alterperylenol, A-60078
▷Altertoxin II, A-60081
7,7'-Dimethoxy-[6,6'-bi-2*H*-benzopyran]-2,2'-dione, *in* B-60082

C$_{20}$H$_{14}$O$_8$S
Halenaquinol; O^{16}-Sulfate, *in* H-60002

C$_{20}$H$_{14}$Se
9-(Phenylseleno)phenanthrene, P-60133

C$_{20}$H$_{15}$NO
2-Acetyltriphenylene; Oxime, *in* A-60060

C$_{20}$H$_{16}$
▷7,12-Dimethylbenz[*a*]anthracene, D-60405

C$_{20}$H$_{16}$N$_2$
2,2'-Dimethyl-4,4'-biquinoline, D-60413

C$_{20}$H$_{16}$N$_2$O$_3$
1-[4-[6-(Diethylamino)-2-benzofuranyl]phenyl]-1*H*-pyrrole-2,5-dione, D-60170

C$_{20}$H$_{16}$O$_3$
3-(Benzyloxy)phenol; Benzoyl, *in* B-60073

C$_{20}$H$_{16}$O$_4$
Ochromycinone methyl ether, *in* O-60003

$C_{20}H_{16}O_5$

2,5-Bis(hydroxymethyl)furan; Dibenzoyl, *in* B-60160
Citrusinol, C-60152

$C_{20}H_{16}O_6$

Crotarin, C-60172
Dihydroalterperylenol, *in* A-60078
7,6-(2,2-Dimethylpyrano)-3,4′,5-trihydroxyflavone, D-60449

$C_{20}H_{18}$

5,6-Didehydro-1,4,7,10-tetramethyldibenzo[*ae*]cyclooctene, D-60169

$C_{20}H_{18}N_2O$

3-Amino-1,2,3,4-tetrahydrocarbazole; N^3-Benzoyl, *in* A-60259

$C_{20}H_{18}O_4$

3,4-Dibenzoyl-2,5-hexanedione, D-60081
4,5-Phenanthrenedimethanol; Di-Ac, *in* P-60070

$C_{20}H_{18}O_5$

Crotalarin, C-60171
Moracin H, *in* M-60145

$C_{20}H_{18}O_6$

1,2-Diacetoxy-5,6-dimethoxyphenanthrene, *in* T-60115
Garveatin *A* quinone, G-60006

$C_{20}H_{18}O_8$

Arborone, A-60294

$C_{20}H_{18}O_{10}$

Conphysodalic acid, C-60162

$C_{20}H_{19}N_3$

▷Rosaniline, R-60010

$C_{20}H_{20}$

▷7,12-Dimethylbenz[*a*]anthracene; 1,2,3,4-Tetrahydro, *in* D-60405
Dodecahedrane, D-60514

$C_{20}H_{20}O_3$

Eupomatenoid 4, *in* E-60075

$C_{20}H_{20}O_4$

Eupamatenoid 7, *in* E-60075
Fruticulin *A*, F-60076
Moracin I, *in* M-60146
Psorolactone, P-60188

$C_{20}H_{20}O_5$

7,8-(2,2-Dimethylpyrano)-3,4′,5-trihydroxyflavan, D-60448
3-(4-Hydroxyphenyl)-1-(2,4,6-trihydroxyphenyl)-2-propen-1-one; 4′-(3-Methyl-2-butenyl), *in* H-60220
10-Isopentenylemodinanthran-10-ol, I-60111
Machilin *B*, M-60002
Moracin J, *in* M-60146
Zuonin *A*, Z-60003

$C_{20}H_{20}O_6$

2-Hydroxygarveatin A, *in* H-60148
3-(4-Hydroxyphenyl)-1-(2,4,6-trihydroxyphenyl)-2-propen-1-one; 2′,4′,6′-Tri-Me, 4-Ac, *in* H-60220
3-(4-Hydroxyphenyl)-1-(2,4,6-trihydroxyphenyl)-2-propen-1-one; 4,4′,6′-Tri-Me, 2-Ac, *in* H-60220
O^5-Methyl-3′,4′-deoxypsorospermin-3′-ol, *in* D-60021
Rhynchosperin *A*, R-60007

$C_{20}H_{20}O_7$

3,4′,5,6,7-Pentamethoxyflavone, *in* P-60047
Rhynchosperin B, *in* R-60007
Sigmoidin *D*, S-60031

$C_{20}H_{20}O_8$

6-Hydroxy-2′,3,5,5′,7-pentamethoxyflavone, *in* H-60064

$C_{20}H_{20}O_{10}$

4′,5,5′,7-Tetrahydroxy-2′,3,6,8-tetramethoxyflavone, *in* O-60022
3′,5,7-Trihydroxy-3,4′,5′,6,8-pentamethoxyflavone, *in* O-60023
4′,5,5′-Trihydroxy-2′,3,6,7,8-pentamethoxyflavone, *in* O-60022
4′,5,7-Trihydroxy-2′,3,5′,6,8-pentamethoxyflavone, *in* O-60022

4′,5,7-Trihydroxy-3,3′,5′,6,8-pentamethoxyflavone, *in* O-60023

$C_{20}H_{22}$

2,5-Dimethyl-3,4-diphenyl-2,4-hexadiene, D-60419

$C_{20}H_{22}ClNO_2$

Fontonamide, F-60067

$C_{20}H_{22}O_2$

3,3-Dimethyl-5,5-diphenyl-4-pentenoic acid; Me ester, *in* D-60420

$C_{20}H_{22}O_4$

6-(3-Methyl-2-butenyl)allopteroxylin, *in* A-60075
Polemannone, P-60166

$C_{20}H_{22}O_5$

Ethuliacoumarin, E-60045
2-Hydroxygarveatin *B*, H-60148
3,4′,5,7-Tetrahydroxy-8-prenylflavan, T-60125
Volkensiachromone, V-60016

$C_{20}H_{22}O_6$

Acuminatin, A-60062
Columbin, C-60160
2-Deoxychamaedroxide, *in* C-60032

$C_{20}H_{22}O_7$

Chamaedroxide, C-60032
Desacylisoelephantopin senecioate, *in* I-60099
Desacylisoelephantopin tiglate, *in* I-60099
8β-Hydroxycolumbin, *in* C-60160
1-Hydroxypinoresinol, H-60221
Isojateorin, *in* C-60160
Jateorin, *in* C-60160

$C_{20}H_{22}O_9$

3,4′,5,7-Tetrahydroxyflavanone; 7-*O*-β-D-Apioside, *in* T-60104
Viscutin 3, *in* T-60103

$C_{20}H_{23}NO_5$

Fuligorubin *A*, F-60079

$C_{20}H_{24}O_3$

Gravelliferone; Me ether, *in* G-60037
Rubifolide, R-60016

$C_{20}H_{24}O_4$

Austrobailignan-6, A-60320
Brayleanin, *in* C-60026
Coralloidolide *A*, C-60164
3,6-Dioxo-4,7,11,15-cembratetraen-10,20-olide, D-60470
12,16-Epoxy-11,14-dihydroxy-5,8,11,13-abietatetraen-7-one, E-60011
Gersemolide, G-60015
Gersolide, G-60016
Gravelliferone; 8-Methoxy, *in* G-60037
Hebeclinolide, H-60009
Vernoflexin, *in* Z-60001
Zaluzanin C; Angeloyl, *in* Z-60001

$C_{20}H_{24}O_5$

Epoxylaphodione, *in* D-60470
3β-Hydroxyhebeclinolide, *in* H-60009
4-Hydroxyisobacchasmacranone, H-60161
Machilin *C*, M-60003
Machilin D, *in* M-60003

$C_{20}H_{24}O_6$

Deacetylsessein, *in* S-60027
Deacetylteupyrenone, *in* T-60184
Isolariciresinol, *in* I-60139
ent-Isolariciresinol, *in* I-60139
Isotaxiresinol; 7-Me ether, *in* I-60139
Ladibranolide, L-60008
Sanguinone *A*, S-60006
Teucrin *E*, T-60180

$C_{20}H_{24}O_7$

Melampodin *D*, M-60013
Orthopappolide methacrylate, *in* O-60045
Tomenphantopin A, *in* T-60222

$C_{20}H_{24}O_8$

4β,15-Epoxy-1β-methoxy-9Z-millerenolide, *in* M-60137

$C_{20}H_{24}S_6$

2,3,11,12-Dibenzo-1,4,7,10,13,16-hexathia-2,11-cyclooctadecadiene, D-60073

$C_{20}H_{26}O_3$

19(4→3)-Abeo-11,12-dihydroxy-4(18),8,11,13-abietatetraen-7-one, A-60003

6,7-Dehydroroyleanone, *in* R-60015

3-(2-Hydroxy-4,8-dimethyl-3,7-nonadienyl)benzaldehyde; Ac, *in* H-60122

Kahweol, K-60002

ent-6-Oxo-7,12E,14-labdatrien-17,11α-olide, O-60076

$C_{20}H_{26}O_4$

19(4β→3β)-Abeo-6,11-epoxy-6,12-dihydroxy-6,7-seco-4(18),-8,11,13-abietatetraen-7-al, A-60004

Anisomelic acid, A-60275

Desoxyarticulin, *in* A-60307

11,12-Dihydroxy-6,8,11,13-abietatetraen-20-oic acid, D-60305

15,16-Epoxy-12-oxo-8(17),13(16),14-labdatrien-19-oic acid, *in* L-60011

Hardwickiic acid; 19-Oxo, *in* H-60007

17-Oxohebemacrophyllide, *in* H-60010

11,12,16-Trihydroxy-5,8,11,13-abietatetraen-7-one, T-60312

$C_{20}H_{26}O_5$

Articulin, A-60307

8,9-Dihydroxy-1(10),4,11(13)-germacratrien-12,6-olide; 9-(2R,3R-Epoxy-2-methylbutanoyl), *in* D-60323

11,12-Dihydroxy-7-oxo-8,11,13-abietatrien-20-oic acid, *in* T-60337

Feruginin, F-60004

Gibberellin A_{24}, G-60019

Senecioodontol, S-60022

Thymifodioic acid, T-60218

11,12,14-Trihydroxy-8,11,13-abietatriene-3,7-dione, T-60313

$C_{20}H_{26}O_6$

Caleine F, *in* C-60005

Coleon *C*, C-60157

Coralloidolide *B*, C-60165

4,5-Epoxy-2,8-dihydroxy-1(10),11(13)-germacradien-12,6-olide; 8-(3-Methyl-2-butenoyl), *in* E-60013

Gibberellin A_{36}, *in* G-60019

Gigantanolide A, *in* T-60323

Gigantanolide B, *in* T-60323

Jurinelloide, J-60003

Sesebrinic acid, S-60026

Teumicropin, T-60181

7,11,12-Trihydroxy-6-oxo-8,11,13-abietatrien-20-oic acid, T-60337

$C_{20}H_{26}O_7$

20-Hydroxyelemajurinelloide, H-60132

20-Hydroxyjurinelloide, *in* J-60003

$C_{20}H_{26}O_8$

Specionin, S-60039

$C_{20}H_{26}O_{10}$

Anamarine, A-60270

$C_{20}H_{28}N_2O_3$

Oxyphencyclimine, O-60102

$C_{20}H_{28}O$

8,11,13-Abietatrien-12,16-oxide, A-60008

8,11,13-Cleistanthatrien-19-al, *in* C-60154

$C_{20}H_{28}O_2$

Barbatusol, B-60007

Cedronellone, C-60027

3,7,11,15(17)-Cembratetraen-16,2-olide, C-60028

8,11,13-Cleistanthatrien-19-oic acid, *in* C-60154

Taonianone, T-60006

$C_{20}H_{28}O_3$

6,8,11,13-Abietatetraene-11,12,14-triol, A-60006

3,12-Dihydroxy-8,11,13-abietatrien-1-one, D-60306

Hardwickiic acid, H-60007

Hebemacrophyllide, H-60010

6-Hydroxy-5-oxo-7,15-fusicoccadien-15-al, H-60203

Isolobophytolide, I-60106

Lambertianic acid, L-60011

ent-15-Oxo-16-kauren-18-oic acid, O-60075

Royleanone, R-60015

Wedeliasecokaurenolide, W-60004

$C_{20}H_{28}O_4$

Brevifloralactone, B-60198

Curculathyrane *A*, C-60176

Curculathyrane *B*, C-60177

Demethylpinusolide, *in* P-60156

2β,12-Dihydroxy-8,12-abietadiene-11,14-dione, *in* R-60015

7,12-Dihydroxy-8,12-abietadiene-11,14-dione, D-60304

3,10-Dihydroxy-5,11-dielmenthadiene-4,9-dione, D-60316

Ephemeric acid, E-60006

6,11-Epoxy-6,12-dihydroxy-6,7-seco-8,11,13-abietatrien-7-al, E-60014

Hautriwaic acid, *in* H-60007

17-Hydroxyhebemacrophyllide, *in* H-60010

Isospongiadiol, I-60137

3,11,12-Trihydroxy-8,11,13-abietatrien-7-one, T-60314

$C_{20}H_{28}O_5$

8,9-Dihydroxy-1(10),4,11(13)-germacratrien-12,6-olide; 8-(2-Methylbutanoyl), *in* D-60323

2-Hydroxyhautriwaic acid, *in* H-60007

$C_{20}H_{28}O_6$

Amarolide, A-60085

Gigantanolide C, *in* T-60323

Schkuhridin A, *in* S-60012

ent-1β,3α,7α,11α-Tetrahydroxy-16-kaurene-6,15-dione, T-60109

$C_{20}H_{28}O_{10}$

Ptelatoside B, *in* H-60211

$C_{20}H_{30}$

Pentaisopropylidenecyclopentane, P-60053

$C_{20}H_{30}O$

8,11,13-Abietatrien-19-ol, A-60007

8,11,13-Cleistanthatrien-19-ol, C-60154

$C_{20}H_{30}O_2$

Abeoanticopalic acid, A-60002

Auricularic acid, A-60317

Cycloanticopalic acid, C-60181

3,4-Epoxy-13(15),16,18-sphenolobatrien-5-ol, E-60031

18-Hydroxy-8,15-isopimaradien-7-one, H-60162

ent-15-Kauren-17-oic acid, K-60008

Perrottetianal *A*, P-60064

Reiswigin B, *in* R-60004

Sanadaol, S-60005

3,7,11,15-Tetramethyl-13-oxo-2,6,10,14-hexadecatetraenal, T-60150

$C_{20}H_{30}O_3$

Agrostistachin, A-60067

5,6-Epoxy-7,9,11,14-eicosatetraenoic acid, E-60015

14,15-Epoxy-8(17),12-labdadien-16-oic acid, E-60025

15,16-Epoxy-13,17-spatadiene-5,19-diol, E-60029

2-[(3,4,4a,5,6,8a-Hexahydro-2H-1-benzopyran-2-yl)-ethylidene]-6-methyl-5-heptenoic acid, H-60046

2-[(2,3,3a,4,5,7a-Hexahydro-3,6-dimethyl-2-benzofuranyl)-ethylidene]-6-methyl-5-heptenoic acid, H-60050

ent-7α-Hydroxy-15-beyeren-19-oic acid, H-60104

ent-12β-Hydroxy-15-beyeren-19-oic acid, H-60105

ent-18-Hydroxy-15-beyeren-19-oic acid, H-60106

ent-3β-Hydroxy-15-kauren-17-oic acid, H-60167

ent-5-Oxo-1(10),13-halimadien-18-oic acid, O-60069

2-Oxokolavenic acid, *in* K-60013

16-Oxo-17-spongianal, O-60094

$C_{20}H_{30}O_4$

8(14)-Abieten-18-oic acid 9,13-endoperoxide, A-60009

β-Cyclohallerin, *in* C-60193

α-Cyclohallerin, C-60193

Cymbodiacetal, C-60233

580

ent-3β,19-Dihydroxy-15-kauren-17-oic acid, D-60341
ent-1(10)13E-Halimadiene-15,18-dioic acid, *in* O-60069
ent-1(10),13Z-Halimadiene-15,18-dioic acid, *in* O-60069
17-Hydroxy-15,17-oxido-16-spongianone, H-60201
Isohallerin, I-60100
Kurubashic acid angelate, *in* K-60014
Lapidin, *in* L-60014
ent-3α,7β,14α-Trihydroxy-16-kauren-15-one, T-60326

$C_{20}H_{30}O_5$
6,17-Dihydroxy-15,17-oxido-16-spongianone, D-60365
1β,10α-Epoxykurubashic acid angelate, *in* K-60014
6-Oxo-7-labdene-15,17-dioic acid, O-60077
ent-3α,7β,14α,20-Tetrahydroxy-16-kauren-15-one, T-60110

$C_{20}H_{30}O_7S$
Hymatoxin *A*, H-60235

$C_{20}H_{30}O_8$
▷Ptaquiloside, P-60190

$C_{20}H_{32}$
Laurenene, L-60018

$C_{20}H_{32}Br_2O_2$
12-Hydroxybromosphaerol, H-60108

$C_{20}H_{32}O$
Dictymal, D-60162
ent-11-Kaurene-16β,18-diol, K-60006
18-Oxo-3-virgene, O-60095

$C_{20}H_{32}O_2$
Dilophic acid, D-60394
12-Isoagathen-15-oic acid, I-60085
8(14),15-Isopimaradiene-2,18-diol, I-60113
ent-15-Kaurene-3β,17-diol, K-60007
Kolavenic acid, K-60013
ent-7,12E,14-Labdatriene-6β,17-diol, L-60004
Reiswigin *A*, R-60004

$C_{20}H_{32}O_3$
8,9-Epoxy-5,11,14-eicosatrienoic acid, E-60016
11,12-Epoxy-5,8,14-eicosatrienoic acid, E-60017
3,4-Epoxy-13(15),16-sphenolobadiene-5,18-diol, E-60030
5-Hydroxy-6,8,11,14-eicosatetraenoic acid, H-60127
8-Hydroxy-5,9,11,14-eicosatetraenoic acid, H-60128
12-Hydroxy-5,8,10,14-eicosatetraenoic acid, H-60131
9-Hydroxy-5,7,11,14-icosatetraenoic acid, H-60129
11-Hydroxy-5,8,12,14-icosatetraenoic acid, H-60130
2-(6-Hydroxy-4-methyl-4-hexenylidene)-6,10-dimethyl-7-oxo-9-undecenal, H-60186
ent-7,11E,14-Labdatriene-6β,13ξ,17-triol, L-60005
Pilosanone *A*, P-60153
ent-5-Rosene-3α,15,16,18-tetrol, *in* R-60011

$C_{20}H_{32}O_4$
Chatferin, *in* F-60009
12,19-Dihydroxy-5,8,10,14-eicosatetraenoic acid, D-60319
10-Hydroxy-2-(6-hydroxy-4-methyl-4-hexylidene)-6,10-dimethyl-7-oxo-8-undecenal, *in* H-60186
7-Labdene-15,17-dioic acid, L-60006
7-Labdene-15,18-dioic acid, L-60007
Oliveric acid, O-60039
Pilosanone *B*, *in* P-60153
Salvicin, S-60003
Stolonidiol, S-60055
ent-2α,3α,16-Trihydroxy-8(14)-pimaren-15-one, *in* P-60154

$C_{20}H_{32}O_5$
11,15-Dihydroxy-9-oxo-5,13-prostadienoic acid, D-60367
Methyl 3,6-epidioxy-6-methoxy-4,14,16-octadecatrienoate, *in* E-60008
Sterebin *B*, *in* S-60052
Sterebin *C*, *in* S-60052
Tetradecahydro-4,6-dihydroxy-9,9,12a-trimethyl-6*H*-phenanthro[1,10a-c]furan-3-carboxylic acid, T-60038
5,14,15-Trihydroxy-6,8,10,12-eicosatetraenoic acid, T-60320
3,7,11-Trimethyl-2,6,10-dodecatrienoic acid; 2-Acetoxy-3-hydroxypropyl ester, *in* T-60364

$C_{20}H_{32}O_6$
Lascrol, *in* L-60016

$C_{20}H_{34}O$
8-Hydroxyisopimar-15-ene, H-60163
Obscuronatin, O-60002

$C_{20}H_{34}O_2$
Chromophycadiol, C-60148
ent-7,13E-Labdadiene-3β,15-diol, L-60001

$C_{20}H_{34}O_3$
7,8-Epoxy-2,11-cembradiene-4,6-diol, E-60010
15-Hydroxy-7-labden-17-oic acid, H-60168
ent-2α,16β,17-Kauranetriol, K-60004
ent-3β,16β,17-Kauranetriol, K-60005
8,13-Labdadiene-6,7,15-triol, L-60003
ent-5-Rosene-15,16,18-triol, *in* R-60012
ent-5-Rosene-15,16,19-triol, R-60012

$C_{20}H_{34}O_4$
6,15-Dihydroxy-7-labden-17-oic acid, D-60342
8,13-Labdadiene-2,6,7,15-tetrol, L-60002
ent-8(14)-Pimarene-2α,3α,15R,16-tetrol, P-60154
ent-5-Rosene-3α,15,16,19-tetrol, R-60011

$C_{20}H_{34}O_5$
Methyl 3,6-epidioxy-6-methoxy-4,16-octadecadienoate, *in* E-60008
Tucumanoic acid, T-60410

$C_{20}H_{36}N_4O_8$
1,5,9,13-Tetraazacyclohexadecane-1,5,9,13-tetraacetic acid, T-60016

$C_{20}H_{36}O_5$
Methyl 3,6-epidioxy-6-methoxy-4-octadecenoate, *in* E-60008
ent-2β,3β,4α-Trihydroxy-15-clerodanoic acid, *in* T-60410

$C_{20}H_{37}NO_3$
N-(Tetrahydro-2-oxo-3-furanyl)hexadecanamide, *in* A-60138

$C_{20}H_{38}O$
3,7,11,15-Tetramethyl-2-hexadecenal, *in* P-60152

$C_{20}H_{38}O_3$
Gyplure, *in* O-60006

$C_{20}H_{40}O$
Phytol, P-60152

$C_{20}H_{40}O_2$
7,11,15-Trimethyl-3-methylene-1,2-hexadecanediol, T-60368

$C_{21}H_{11}F_5O_3$
9-Fluorenylmethyl pentafluorophenyl carbonate, F-60019

$C_{21}H_{12}N_4$
▷Tricycloquinazoline, T-60271

$C_{21}H_{12}N_6$
[1,3,5]Triazino[1,2-a:3,4-a':5,6-a'']trisbenzimidazole, T-60238

$C_{21}H_{12}O_2$
7*H*-Dibenzo[c,h]xanthen-7-one, D-60080

$C_{21}H_{13}N$
Phenanthro[9,10-g]isoquinoline, P-60075

$C_{21}H_{15}NO_2$
Dibenz[b,g]azocine-5,7(6*H*,12*H*)-dione; *N*-Ph, *in* D-60068

$C_{21}H_{16}O_5$
Calopogonium isoflavone *B*, C-60006

$C_{21}H_{16}O_6$
Gerberinol 1, G-60013
Justicidin *B*, *in* D-60493

$C_{21}H_{16}O_7$
▷Diphyllin, D-60493

$C_{21}H_{16}O_{11}$
α-Acetylconstictic acid, *in* C-60163

$C_{21}H_{18}O$
Maximaisoflavone *B*, *in* B-60006
1,2,2-Triphenyl-1-propanone, T-60389
1,2,3-Triphenyl-1-propanone, T-60390
1,3,3-Triphenyl-1-propanone, T-60391

C$_{21}$H$_{18}$O$_3$
6-Methylene-2,4-cyclohexadien-1-one; Trimer, *in* M-60072

C$_{21}$H$_{18}$O$_6$
Pipoxide, P-60159

C$_{21}$H$_{18}$O$_7$
Hildecarpidin, H-60081

C$_{21}$H$_{19}$NO
1,3,3-Triphenyl-1-propanone; Oxime, *in* T-60391

C$_{21}$H$_{20}$O$_6$
Angeloylprangeline, *in* P-60170
Garvin *B*, G-60009
Kwakhurin, K-60016
Topazolin, T-60224

C$_{21}$H$_{20}$O$_7$
2-Hydroxygarvin B, *in* G-60009
3-(4-Hydroxyphenyl)-1-(2,4,6-trihydroxyphenyl)-2-propen-1-one; 4′,6′-Di-Me, 2′,4-di-Ac, *in* H-60220
Podoverine *A*, P-60164

C$_{21}$H$_{20}$O$_9$
Daidzin, *in* D-60340

C$_{21}$H$_{20}$O$_{10}$
Genistin, *in* T-60324
Sophicoroside, *in* T-60324

C$_{21}$H$_{21}$ClN$_2$O
Anhydrohapaloxindole, A-60273

C$_{21}$H$_{22}$O$_4$
Eupomatenoid 12, *in* E-60075

C$_{21}$H$_{22}$O$_5$
Aristotetralone, A-60298
7,8-(2,2-Dimethylpyrano)-3,4′-dihydroxy-5-methoxyflavan, *in* D-60448
Garveatin *D*, G-60007

C$_{21}$H$_{22}$O$_7$
Annulin *B*, A-60277
Praeruptorin *A*, P-60169
Topazolin hydrate, *in* T-60224

C$_{21}$H$_{22}$O$_9$
3-Hydroxy-3′,4′,5,6,7,8-hexamethoxyflavone, *in* H-60024
5-Hydroxy-3,3′,4′,6,7,8-hexamethoxyflavone, *in* H-60024

C$_{21}$H$_{22}$O$_{10}$
5,7-Dihydroxy-2′,3,4′,5′,6,8-hexamethoxyflavone, *in* O-60022
5,7-Dihydroxy-3,3′,4′,5′,6,8-hexamethoxyflavone, *in* O-60023

C$_{21}$H$_{22}$O$_{12}$
Lanceoside, *in* P-60052

C$_{21}$H$_{24}$O$_4$
6-(3-Methyl-2-butenyl)allopteroxylin methyl ether, *in* A-60075

C$_{21}$H$_{24}$O$_5$
8-(3,3-Dimethylallyl)-5-methoxy-3,4′,7-trihydroxyflavan, *in* T-60125
5-Methylethuliacoumarin, *in* E-60045
Rutamarin, *in* C-60031

C$_{21}$H$_{24}$O$_6$
Phillygenin, P-60141
▷Phloridzin, *in* H-60219

C$_{21}$H$_{24}$O$_7$
1-Hydroxypinoresinol; 4″-Me ether, *in* H-60221

C$_{21}$H$_{26}$O$_5$
Aristolignin, A-60297
3-*O*-Methyl-2,5-dehydrosenecioodentol, *in* S-60022

C$_{21}$H$_{26}$O$_6$
Cordatin, C-60166
Isolariciresinol 4′-methyl ether, *in* I-60139
Subexpinnatin B, *in* S-60058

C$_{21}$H$_{26}$O$_7$
1-Ethoxy-9Z-millerenolide, *in* M-60137
Orthopappolide senecioate, *in* O-60045
Orthopappolide tiglate, *in* O-60045

C$_{21}$H$_{26}$O$_8$
4β,15-Epoxy-1β-ethoxy-4Z-millerenolide, *in* M-60137

C$_{21}$H$_{28}$O$_4$
Membranolide, M-60016
2,3,12-Trihydroxy-4,7,16-pregnatrien-20-one, T-60346

C$_{21}$H$_{28}$O$_5$
3-*O*-Methylsenecioodentol, *in* S-60022
7-O-Formylhorminone, *in* D-60304

C$_{21}$H$_{28}$O$_7$
14-Acetoxy-9β-hydroxy-8β-(2-methylpropanoyloxy)-1(10),-4,11(13)-germacratrien-12,6α-olide, *in* T-60323
9-Desacetoxymelcanthin F, *in* M-60014

C$_{21}$H$_{28}$O$_8$
Vernoflexuoside, *in* Z-60001

C$_{21}$H$_{29}$NO$_3$
Smenospongine, S-60036

C$_{21}$H$_{30}$O$_3$
12-Methoxy-8,11,13-abietatrien-20-oic acid, M-60036

C$_{21}$H$_{30}$O$_4$
Pinusolide, P-60156
2,3,12-Trihydroxy-4,7-pregnadien-20-one, T-60345

C$_{21}$H$_{30}$O$_5$
Microglossic acid, M-60134
Spongionellin, S-60048
2,3,12,16-Tetrahydroxy-4,7-pregnadien-20-one, T-60124

C$_{21}$H$_{30}$O$_6$
11-Dihydro-12-norneoquassin, D-60258

C$_{21}$H$_{30}$O$_7$
15-Desacetyltetraneurin *C* isobutyrate, *in* T-60158

C$_{21}$H$_{30}$O$_8$
Brachynereolide, B-60196
Zaluzanin *C*; 11β,13-Dihydro, β-D-Glucoside, *in* Z-60001

C$_{21}$H$_{30}$O$_9$
11β,13-Dihydro-8α-hydroxyglucozaluanin C, *in* Z-60001

C$_{21}$H$_{30}$O$_{13}$
Acetylbarlerin, *in* S-60028

C$_{21}$H$_{31}$N
8-Isocyano-10,14-amphilectadiene, I-60090
7-Isocyano-1-cycloamphilectene, I-60093
7-Isocyano-11-cycloamphilectene, I-60094
8-Isocyano-1(12)-cycloamphilectrene, I-60095

C$_{21}$H$_{32}$O$_2$
1-(3,4-Methylenedioxyphenyl)-1-tetradecene, M-60077

C$_{21}$H$_{32}$O$_3$
Gracilin E, *in* G-60035
Methyl 14ξ,15-epoxy-8(17),12E-labdadien-16-oate, *in* E-60025

C$_{21}$H$_{32}$O$_4$
Lycopersiconolide, L-60042

C$_{21}$H$_{32}$O$_5$
10,11-Dihydromicroglossic acid, *in* M-60134

C$_{21}$H$_{32}$O$_8$
11β,13-Dihydrobrachynereolide, *in* B-60196

C$_{21}$H$_{32}$O$_{10}$
Ebuloside, E-60001

C$_{21}$H$_{34}$O$_2$
Kolavenic acid; Me ester, *in* K-60013

C$_{21}$H$_{34}$O$_3$
5-Hydroxy-6,8,11,14-eicosatetraenoic acid; Me ester, *in* H-60127
8-Hydroxy-5,9,11,14-eicosatetraenoic acid; Me ester, *in* H-60128
12-Hydroxy-5,8,10,14-eicosatetraenoic acid; Me ester, *in* H-60131
9-Hydroxy-5,7,11,14-icosatetraenoic acid; Me ester, *in* H-60129
11-Hydroxy-5,8,12,14-icosatetraenoic acid; Me ester, *in* H-60130

$C_{21}H_{34}O_5$
3,6-Epidioxy-6-methoxy-4,16,18-eicosatrienoic acid, E-60007

$C_{21}H_{34}O_{10}$
7,7-O-Dihydroebuloside, *in* E-60001

$C_{21}H_{38}O_8$
Icariside C_1, *in* T-60362
Icariside C_2, *in* T-60362
Icariside C_3, *in* T-60362
Icariside C_4, *in* T-60362

$C_{22}H_{12}N_2$
Dibenz[*b,h*]indeno[1,2,3-*de*][1,6]naphthyridine, D-60069
▷Dibenzo[*c,f*]indeno[1,2,3-*ij*][2,7]naphthyridine, D-60074
2,2'-Diisocyano-1,1'-binaphthyl, D-60391

$C_{22}H_{12}N_2O$
Benzo[*a*]benzofuro[2,3-*c*]phenazine, *in* B-60033
▷Dibenzo[*c,f*][1]benzopyrano[2,3,4-*ij*][2,7]naphthyridine, D-60070

$C_{22}H_{12}N_4$
Dibenzo[*cd:c'd'*][1,2,4,5]tetrazino[1,6-*a*:4,3-*a'*]diindole, D-60076

$C_{22}H_{12}O$
▷Indeno[1,2,3-*cd*]pyren-1-ol, I-60022
▷Indeno[1,2,3-*cd*]pyren-2-ol, I-60023
▷Indeno[1,2,3-*cd*]pyren-6-ol, I-60024
▷Indeno[1,2,3-*cd*]pyren-7-ol, I-60025
▷Indeno[1,2,3-*cd*]pyren-8-ol, I-60026

$C_{22}H_{12}S_2$
Naphtho[2,1-*b*:6,5-*b'*]bis[1]benzothiophene, N-60007

$C_{22}H_{14}O_6$
3,3'-Bi[5-hydroxy-2-methyl-1,4-naphthoquinone], B-60106

$C_{22}H_{14}O_9$
Aurintricarboxylic acid, A-60318

$C_{22}H_{16}$
1,8-Diphenylnaphthalene, D-60483

$C_{22}H_{16}O_2$
1,3-Diphenyl-1*H*-indene-2-carboxylic acid, D-60481

$C_{22}H_{17}N$
3,5-Dihydro-4*H*-dinaphth[2,1-*c*:1',2'-*e*]azepine, D-60234

$C_{22}H_{17}NO_2$
Dibenz[*b,g*]azocine-5,7(6*H*,12*H*)-dione; *N*-Benzyl, *in* D-60068

$C_{22}H_{18}O_7$
Justicidin A, *in* D-60493

$C_{22}H_{18}O_8$
Desertorin *A*, D-60022
2,4,5,6-Tetrahydroxyphenanthrene; Tetra-Ac, *in* T-60122

$C_{22}H_{18}O_{14}$
Hexahydroxy-1,4-naphthoquinone; Hexa-Ac, *in* H-60065

$C_{22}H_{20}$
3-(1,3,6-Cycloheptatrien-1-yl-2,4,6-cycloheptatrien-1-ylidenemethyl)-1,3,5-cycloheptatriene, C-60205
[2.2](4,7)(7,4)Indenophane, I-60021

$C_{22}H_{20}N_2S_2$
1,4-Bis(ethylthio)-3,6-diphenylpyrrolo[3,4-*c*]pyrrole, B-60158

$C_{22}H_{20}O_{10}$
ψ-Baptisin, *in* B-60006

$C_{22}H_{20}O_{13}$
Ellagic acid; 3,3'-Di-Me ether, 4-glucoside, *in* E-60003

$C_{22}H_{22}O_4$
Paralycolin *A*, P-60009

$C_{22}H_{22}O_8$
4,5;4',5'-Bismethylenedioxypolemannone, *in* P-60166
Dalbergia Rotenolone, R-60014

$C_{22}H_{22}O_{10}$
1,3,5,8-Tetrahydroxy-2-methylanthraquinone; 5-Me ether, 8-*O*-α-L-rhamnopyranoside, *in* T-60113

$C_{22}H_{24}O_6$
Ramosissin, R-60001

$C_{22}H_{24}O_7$
Machilin E, *in* M-60003

$C_{22}H_{24}O_8$
1-Acetoxypinoresinol, *in* H-60221

$C_{22}H_{24}O_{10}$
Helichrysin, *in* H-60220

$C_{22}H_{25}N_3O_3S$
Antibiotic FR 900452, A-60284

$C_{22}H_{25}N_3O_{10}$
▷Aluminon, *in* A-60318

$C_{22}H_{26}O_4$
21-Ethyl-2,6-epoxy-17-hydroxy-1-oxacyclohenicosa-2,5,14,18,20-pentaen-11-yn-4-one, E-60050

$C_{22}H_{26}O_5$
Isodidymic acid, I-60098
Licarin *C*, L-60024

$C_{22}H_{26}O_6$
Asebotin, *in* H-60219
Henricine, H-60017

$C_{22}H_{26}O_7$
Laferin, L-60009
Sessein, S-60027
Teupyrenone, T-60184

$C_{22}H_{26}O_8$
Abbreviatin PB, A-60001
3,6-Bis(3,4-dimethoxyphenyl)tetrahydro-1*H*,3*H*-furo[3,4-*c*]-furan-1,4-diol, B-60156
Isopicropolin, I-60112
7-Oxodihydrogmelinol, O-60064

$C_{22}H_{26}O_9$
1-Hydroxysyringaresinol, *in* H-60221

$C_{22}H_{28}O_2$
Fervanol benzoate, *in* F-60007

$C_{22}H_{28}O_3$
Fervanol *p*-hydroxybenzoate, *in* F-60007
Kurubasch aldehyde benzoate, *in* K-60014

$C_{22}H_{28}O_4$
Kurubashic acid benzoate, *in* K-60014
Verecynarmin *A*, V-60005

$C_{22}H_{28}O_5$
1α,10β-Epoxykurubaschic acid benzoate, *in* K-60014
1β,10α-Epoxykurubaschic acid benzoate, *in* K-60014
Hardwickiic acid; 19-Acetoxy, 1,2-didehydro, *in* H-60007

$C_{22}H_{28}O_6$
Articulin acetate, *in* A-60307
8,9-Dihydroxy-1(10),4,11(13)-germacratrien-12,6-olide; 9-Ac, 8-(2*R*,3*R*-epoxy-2-methylbutanoyl), *in* D-60323
Homoheveadride, H-60089

$C_{22}H_{28}O_7$
17-Acetoxythymifodioic acid, *in* T-60218
3-Acetylteumicropin, *in* T-60181
Teucjaponin *A*, T-60178
Teucjaponin B, *in* T-60178
Teupolin I, T-60183
Teupolin II, *in* T-60183

$C_{22}H_{28}O_8$
Gracilin *B*, G-60034
Gracilin C, *in* G-60034

$C_{22}H_{30}$
Biadamantylideneethane, B-60075

$C_{22}H_{30}O_3$
Siccanin, S-60029
Siccanochromene *E*, S-60030

$C_{22}H_{30}O_4$
Ferolin, *in* G-60014

C$_{22}$H$_{30}$O$_5$

6β-Acetoxyhebemacrophyllide, *in* H-60010
7α-Acetoxyroyleanone, *in* D-60304
7β-Acetoxyroyleanone, *in* D-60304
3β-Acetoxywedeliasecokaurenolide, *in* W-60004
Brevifloralactone acetate, *in* B-60198
3,5-Di-*O*-methylsenecioodontol, *in* S-60022
5-epi-6β-acetoxyhebemacrophyllide, *in* H-60010
Feraginidin, *in* D-60004
Ferugin, *in* D-60005
Hautriwaic acid acetate, *in* H-60007
Lasianthin, L-60017

C$_{22}$H$_{30}$O$_6$

7α-Acetoxy-12,20-dihydroxy-8,12-abietadiene-11,14-dione, *in* R-60015

C$_{22}$H$_{32}$O$_2$

Dehydroabietinol acetate, *in* A-60007

C$_{22}$H$_{32}$O$_3$

Acetylsanadaol, *in* S-60005
3,4-Epoxy-13(15),16,18-sphenolobatrien-5-ol; Ac, *in* E-60031
2-Hydroxy-6-(8,11-pentadecadienyl)benzoic acid, H-60205

C$_{22}$H$_{32}$O$_4$

ent-18-Acetoxy-15-beyeren-19-oic acid, *in* H-60106
19-Acetoxy-15,16-epoxy-13,17-spatadien-5α-ol, *in* E-60029
ent-3β-Acetoxy-15-kauren-17-oic acid, *in* H-60167
ent-6β-Acetoxy-7,12*E*,14-labdatrien-17-oic acid, *in* L-60004
Iloprost, I-60003
Isoiloprost, *in* I-60003
Lagerstronolide, L-60010

C$_{22}$H$_{32}$O$_5$

ent-6β,17-Diacetoxy-14,15-bisnor-7,11*E*-labdadien-13-one, *in* D-60311
7-Hydroxy-8(17),13-corymbidienolide, H-60110

C$_{22}$H$_{32}$O$_6$

6α-Acetoxy-17β-hydroxy-15,17-oxido-16-spongianone, *in* D-60365

C$_{22}$H$_{34}$O$_3$

ent-3β-Acetoxy-15-kauren-17-ol, *in* K-60007
ent-6β-Acetoxy-7,12*E*,14-labdatrien-17-ol, *in* L-60004
2-Hydroxy-6-(8-pentadecenyl)benzoic acid, *in* H-60205
2-Hydroxy-6-(10-pentadecenyl)benzoic acid, *in* H-60205

C$_{22}$H$_{34}$O$_4$

Maesanin, M-60004

C$_{22}$H$_{34}$O$_5$

4-Acetoxy-6-(4-hydroxy-2-cyclohexenyl)-2-(4-methyl-3-pentenyl)-2-heptenoic acid, A-60018
Cornudentanone, C-60167
3,4-Epoxy-13(15),16-sphenolobadiene-5,18-diol; 5-Ac, *in* E-60030
Stolonidiol acetate, *in* S-60055

C$_{22}$H$_{34}$O$_7$

Coleonol *B*, C-60158
Coleonol C, *in* C-60158

C$_{22}$H$_{36}$O$_3$

Chromophycadiol monoacetate, *in* C-60148
2-Hydroxy-6-pentadecylbenzoic acid, *in* H-60205
Secotrinervitane, S-60019

C$_{22}$H$_{36}$O$_4$

ent-3β-Acetoxy-16β,17-kauranediol, *in* K-60005
15-Acetoxy-7-labden-17-oic acid, *in* H-60168

C$_{22}$H$_{36}$O$_5$

Methyl 3,6-epidioxy-6-methoxy-4,16,18-eicosatrienoate, *in* E-60007

C$_{22}$H$_{36}$O$_{16}$

Shanzhisin methyl ester gentiobioside, *in* S-60028

C$_{22}$H$_{40}$O$_4$

9-Octadecene-1,12-diol; Di-Ac, *in* O-60006

C$_{22}$H$_{41}$NO$_7$

Fumifungin, F-60082

C$_{23}$H$_{14}$O

7-Methoxyindeno[1,2,3-*cd*]pyrene, *in* I-60025
8-Methoxyindeno[1,2,3-*cd*]pyrene, *in* I-60026

C$_{23}$H$_{14}$O$_6$

1,8,11-Triptycenetricarboxylic acid, T-60398
1,8,14-Triptycenetricarboxylic acid, T-60399

C$_{23}$H$_{18}$O$_2$

1,3-Diphenyl-1*H*-indene-2-carboxylic acid; Me ester, *in* D-60481

C$_{23}$H$_{20}$O$_8$

Desertorin B, *in* D-60022

C$_{23}$H$_{22}$N$_3$O$_3$

2,5-Dihydro-2,2,5,5-tetramethyl-3-[[[(2-phenyl-3*H*-indol-3-ylidene)amino]oxy]carbonyl]-1*H*-pyrrol-1-yloxy, D-60284

C$_{23}$H$_{22}$O$_{10}$

Phrymarolin II, P-60142

C$_{23}$H$_{24}$O$_9$

Wistin, *in* T-60325

C$_{23}$H$_{26}$O$_6$

Garvin *A*, G-60008

C$_{23}$H$_{26}$O$_8$

4,5-Dimethoxy-4′,5′-methylenedioxypolemannone, *in* P-60166
1-Hydroxypinoresinol; 4″-Me ether, 1-Ac, *in* H-60221
Sikkimotoxin, S-60032

C$_{23}$H$_{26}$O$_{11}$

Nyasicaside, *in* B-60155

C$_{23}$H$_{28}$N$_4$O$_7$

Biphenomycin B, *in* B-60114

C$_{23}$H$_{28}$N$_4$O$_8$

Biphenomycin *A*, B-60114

C$_{23}$H$_{28}$O$_3$

Citreomontanin, C-60149

C$_{23}$H$_{28}$O$_7$

Isosphaeric acid, I-60136
Palliferinin, *in* L-60014

C$_{23}$H$_{28}$O$_{10}$

Diffutin, *in* T-60103

C$_{23}$H$_{29}$NO$_6$

Fusarin *A*, F-60107

C$_{23}$H$_{29}$NO$_7$

Fusarin D, *in* F-60107

C$_{23}$H$_{30}$O$_4$

Fervanol vanillate, *in* F-60007
Guayulin D, *in* A-60299

C$_{23}$H$_{30}$O$_5$

Kurubasch aldehyde vanillate, *in* K-60014

C$_{23}$H$_{30}$O$_8$

Acanthospermal *A*, A-60012
Gracilin D, *in* G-60034

C$_{23}$H$_{32}$O$_4$

8,11,13-Abietatrien-19-ol; 19-(Carboxyacetyl), *in* A-60007
Methyl 11,12-dimethoxy-6,8,11,13-abietatrien-20-oate, *in* D-60305

C$_{23}$H$_{32}$O$_5$

Chimganidin, *in* G-60014

C$_{23}$H$_{32}$O$_6$

16α-Acetoxy-2α,3β,12β-trihydroxy-4,7-pregnadien-20-one, *in* T-60124

C$_{23}$H$_{32}$O$_9$

Absinthifolide, *in* H-60138

C$_{23}$H$_{34}$O$_4$

7,13-Corymbidienolide, C-60169

C$_{23}$H$_{34}$O$_5$

▷Coroglaucigenin, C-60168
Gracilin *A*, G-60033

C$_{25}$H$_{24}$O$_2$S
3-[(Triphenylmethyl)thio]cyclopentanecarboxylic acid, *in* M-60021

C$_{25}$H$_{24}$O$_7$
Laserpitinol, *in* L-60016

C$_{25}$H$_{26}$O$_5$
Cajaflavanone, C-60004

C$_{25}$H$_{26}$O$_6$
Orotinin, O-60044

C$_{25}$H$_{26}$O$_7$
Garvalone *B*, G-60005

C$_{25}$H$_{26}$O$_8$
Rhynchosperin C, *in* R-60007

C$_{25}$H$_{28}$O$_4$
1-[2,4-Dihydroxy-3,5-bis(3-methyl-2-butenyl)phenyl]-3-(4-hydroxyphenyl)-2-propen-1-one, D-60310

C$_{25}$H$_{28}$O$_5$
Lespedazaflavone *B*, L-60021
Lonchocarpol *A*, L-60030

C$_{25}$H$_{28}$O$_6$
Lonchocarpol *C*, L-60031
Lonchocarpol *D*, L-60032

C$_{25}$H$_{28}$O$_7$
Lonchocarpol *E*, L-60033

C$_{25}$H$_{30}$N$_8$O$_9$
7-Hydro-8-methylpteroylglutamylglutamic acid, H-60096

C$_{25}$H$_{30}$O$_5$
Ircinin 1, I-60083
Ircinin 2, *in* I-60083

C$_{25}$H$_{30}$O$_7$
Lonchocarpol B, *in* L-60030

C$_{25}$H$_{31}$N$_3$O$_5$
1-[*N*-[3-(Benzoylamino)-2-hydroxy-4-phenylbutyl]alanyl]-proline, B-60056

C$_{25}$H$_{32}$O$_4$
Ircinianin, I-60081
Ircinic acid, I-60082

C$_{25}$H$_{32}$O$_6$
3-*O*-Angeloylsenecioodontol, *in* S-60022

C$_{25}$H$_{32}$O$_{11}$
Melcanthin *F*, M-60014

C$_{25}$H$_{32}$O$_{12}$
Isoligustroside, I-60104

C$_{25}$H$_{32}$O$_{13}$
Syringopicroside C, *in* S-60067

C$_{25}$H$_{33}$NO
Aurachin *D*, A-60315

C$_{25}$H$_{33}$NO$_2$
Aurachin *B*, A-60314
Aurachin *C*, *in* A-60315

C$_{25}$H$_{33}$NO$_3$
Aurachin *A*, A-60313

C$_{25}$H$_{34}$O$_3$
Apo-12′-violaxanthal, *in* P-60065

C$_{25}$H$_{36}$O$_3$
Persicaxanthin, P-60065

C$_{25}$H$_{36}$O$_4$
24-Methyl-25-nor-12,24-dioxo-16-scalaren-22-oic acid, M-60105

C$_{25}$H$_{36}$O$_6$
Pseudopterosin *A*, P-60187

C$_{25}$H$_{38}$O$_3$
Palauolide, P-60001

C$_{25}$H$_{38}$O$_4$
Spongiolactone, S-60047

C$_{25}$H$_{38}$O$_5$
Pallinin, *in* C-60021

C$_{25}$H$_{38}$O$_7$
Laserpitine, L-60016
Methyl 15α,17β-diacetoxy-15,16-dideoxy-15,17-oxido-16-spongianoate, *in* T-60038

C$_{25}$H$_{40}$O$_3$
3-Octadecene-1,2-diol; 1-Benzoyl, *in* O-60005

C$_{25}$H$_{40}$O$_6$
Salvisyriacolide, S-60004

C$_{25}$H$_{42}$N$_4$O$_{14}$
Allosamidin, A-60076

C$_{25}$H$_{42}$O$_4$
6β-Isovaleroyloxy-8,13*E*-labdadiene-7α,15-diol, *in* L-60003

C$_{25}$H$_{45}$FeN$_6$O$_8$
Ferrioxamine *B*, F-60003

C$_{26}$H$_{14}$
Bisbenzo[3,4]cyclobuta[1,2-*c*;1′,2′-*g*]phenanthrene, B-60143
Naphtho[1,2,3,4-*ghi*]perylene, N-60009

C$_{26}$H$_{14}$O$_2$
6,15-Hexacenedione, H-60033

C$_{26}$H$_{16}$
9,9′-Bifluorenylidene, B-60104

C$_{26}$H$_{18}$
1,3-Di-(1-naphthyl)benzene, D-60454
1,3-Di(2-naphthyl)benzene, D-60455

C$_{26}$H$_{18}$N$_6$
7,11:20,24-Dinitrilodibenzo[*b,m*][1,4,12,15]-tetraazacyclodocosine, D-60456

C$_{26}$H$_{20}$O
Tetraphenylethanone, T-60167

C$_{26}$H$_{20}$O$_6$
1,8,11-Triptycenetricarboxylic acid; Tri-Me ester, *in* T-60398
1,8,14-Triptycenetricarboxylic acid; Tri-Me ester, *in* T-60399

C$_{26}$H$_{24}$
Tricyclo[18.4.1.1^{8,13}]hexacosa-2,4,6,8,10,12,14,16,18,20,22,24-dodecaene, T-60263

C$_{26}$H$_{24}$O$_2$
[2.2][2.2]Paracyclophane-5,8-quinone, P-60007
[2.2][2.2]Paracyclophane-12,15-quinone, P-60008

C$_{26}$H$_{26}$
[2.2]Paracyclo(4,8)[2.2]metaparacyclophane, P-60005

C$_{26}$H$_{26}$O$_2$S
4-[(Triphenylmethyl)thio]cyclohexanecarboxylic acid, *in* M-60020

C$_{26}$H$_{26}$O$_9$
Dukunolide *A*, D-60522

C$_{26}$H$_{26}$O$_{10}$
Dukunolide B, *in* D-60522

C$_{26}$H$_{28}$O$_6$
Orotinichalcone, O-60043
Orotinin; 5-Me ether, *in* O-60044

C$_{26}$H$_{28}$O$_8$
Dukunolide *D*, D-60523

C$_{26}$H$_{28}$O$_9$
Dukunolide E, *in* D-60523
Dukunolide F, *in* D-60523
Ephemeroside, *in* E-60006

C$_{26}$H$_{28}$O$_{13}$
Ambonin, *in* D-60340
Neobanin, *in* D-60340

C$_{26}$H$_{28}$O$_{14}$
Ambocin, *in* T-60324
Neobacin, *in* T-60324

$C_{26}H_{30}O_6$
Lespedezaflavanone A, L-60022

$C_{26}H_{30}O_8$
Dysoxylin, D-60525

$C_{26}H_{30}O_{10}$
Graucin A, G-60036

$C_{26}H_{30}O_{13}$
3-(4-Hydroxyphenyl)-1-(2,4,6-trihydroxyphenyl)-2-propen-1-one; 2'-(O-Rhamnosyl(1→4)xyloside), *in* H-60220

$C_{26}H_{32}O_5$
Licoricidin, L-60025

$C_{26}H_{32}O_{12}$
1-Hydroxypinoresinol; 1-O-β-D-Glucopyranoside, *in* H-60221
1-Hydroxypinoresinol; 4-O-β-D-Glucopyranoside, *in* H-60221

$C_{26}H_{32}O_{13}$
Decumbeside C, D-60017
Decumbeside D, *in* D-60017

$C_{26}H_{33}ClO_9$
12-Ptilosarcenol; 12-Ac, *in* P-60193

$C_{26}H_{34}O_6$
Nimolicinoic acid, N-60022

$C_{26}H_{34}O_8$
Agrimophol, A-60066

$C_{26}H_{34}O_9$
Deoxyhavannahine, *in* H-60008
ent-1β,7α,11α-Triacetoxy-3α-hydroxy-16-kaurene-6,15-dione, *in* T-60109

$C_{26}H_{34}O_{10}$
9α,14-Diacetoxy-1α-benzoyloxy-4β,6β,8β-trihydroxydihydro-β-agarofuran, *in* E-60018
Eumaitenin, E-60071
Havannahine, H-60008
Rhynchospermoside A, R-60008
Rhynchospermoside B, *in* R-60008

$C_{26}H_{36}O_9$
Caesalpin F, C-60002

$C_{26}H_{36}O_{10}$
Melianolone, M-60015

$C_{26}H_{38}O_4$
12-Hydroxy-24-methyl-24-oxo-16-scalarene-22,25-dial, H-60191

$C_{26}H_{38}O_5$
Chinensin II, C-60036
12α-Hydroxy-24-methyl-24,25-dioxo-14-scalaren-22-oic acid, *in* H-60191

$C_{26}H_{38}O_7$
17α-Acetoxy-6α-butanoyloxy-15,17-oxido-16-spongianone, *in* D-60365

$C_{26}H_{40}O_{14}$
6'-O-Apiosylebuloside, *in* E-60001

$C_{26}H_{44}N_4O_{14}$
Methylallosamidin, *in* A-60076

$C_{26}H_{48}O_3$
Ficulinic acid A, F-60011

$C_{26}H_{50}O_2$
Aparjitin, A-60291

$C_{27}H_{20}$
1,2,3-Triphenyl-1H-indene, T-60387

$C_{27}H_{20}Cl_{15}N$
Tetraethylammonium tris(pentachlorophenyl)methylide, *in* T-60405

$C_{27}H_{20}O_7$
Mulberrofuran R, M-60149

$C_{27}H_{22}O$
1,1,3,3-Tetraphenyl-2-propanone, T-60168

$C_{27}H_{22}S_2$
4,4,5,5-Tetraphenyl-1,3-dithiolane, T-60166

$C_{27}H_{26}O_{11}$
Cleistanthoside B, *in* D-60493
Viscutin 1, *in* T-60103

$C_{27}H_{30}O_{14}$
4',7-Dihydroxyisoflavone; 4',7-Di-O-β-D-glucopyranosyl, *in* D-60340
Sphaerobioside, *in* T-60324

$C_{27}H_{32}O_7$
Garvalone A, G-60004

$C_{27}H_{32}O_9$
Pedonin, P-60014
Trijugin B, T-60352

$C_{27}H_{32}O_{10}$
Baccharinoid B25, B-60004
Eumaitenol, E-60072

$C_{27}H_{33}N_3O_5$
Andrimide, A-60272

$C_{27}H_{34}O_4$
Yezoquinolide, Y-60003

$C_{27}H_{34}O_5$
5-O-Methyllicoricidin, *in* L-60025
Scopadulcic acid B, S-60016

$C_{27}H_{34}O_6$
Scopadulcic acid A, S-60015

$C_{27}H_{34}O_9$
Isocyclocalamin, I-60096

$C_{27}H_{34}O_{11}$
Phillyrin, *in* P-60141

$C_{27}H_{35}ClO_9$
12-Ptilosarcenol; 12-Propanoyl, *in* P-60193

$C_{27}H_{36}O_7$
12α-Hydroxy-4,4,14α-trimethyl-3,7,11,15-tetraoxo-5α-chol-8-en-24-oic acid, *in* L-60037
Hyperlatolic acid, H-60236
Isohyperlatolic acid, I-60101

$C_{27}H_{38}O_2$
Sargaquinone, S-60010

$C_{27}H_{38}O_4$
8',9'-Dihydroxysargaquinone, *in* S-60010
Sargahydroquinoic acid, S-60009

$C_{27}H_{38}O_5$
Amentadione, A-60086
Amentaepoxide, A-60087
Amentol, A-60088
Bifurcarenone, B-60105

$C_{27}H_{38}O_7$
3β,12β-Dihydroxy-4,4,14α-trimethyl-7,11,15-trioxo-5α-chol-8-en-24-oic acid, *in* L-60037
3β-Hydroxy-4α-hydroxymethyl-4β,14α-dimethyl-7,11,15-trioxo-5α-chol-8-en-24-oic acid, *in* L-60036
Pseudopterosin B, *in* P-60187
Pseudopterosin C, *in* P-60187
Pseudopterosin D, *in* P-60187

$C_{27}H_{38}O_8$
3β,12β-Dihydroxy-4α-hydroxymethyl-4β,14α-dimethyl-7,11,15-trioxo-5α-chol-8-en-24-oic acid, *in* L-60036

$C_{27}H_{38}O_{13}$
Macrocliniside B, *in* Z-60001

$C_{27}H_{40}O_5$
Chinensin I, C-60035

$C_{27}H_{40}O_7$
Lucidenic acid H, L-60036

$C_{27}H_{40}O_9$
Methyl $6\alpha,15\alpha,17\beta$-triacetoxy-15,16-dideoxy-15,17-oxido-16-spongianoate, *in* T-60038

$C_{27}H_{42}O$
Papakusterol, P-60004

$C_{27}H_{42}O_3$
5-Spirosten-3-ol, S-60045

$C_{27}H_{42}O_4$
Deacetoxybrachycarpone, *in* B-60195
Nuatigenin, N-60062
5-Spirostene-3,25-diol, S-60044

$C_{27}H_{42}O_6$
Lucidenic acid *M*, L-60037

$C_{27}H_{42}O_{15}$
Penstebioside, P-60016

$C_{27}H_{44}O$
$(24S,25S)$-24,26-Cyclo-5α-cholest-22E-en-3β-ol, *in* P-60004

$C_{27}H_{44}O_2$
3-Hydroxy-25,26,27-trisnor-24-cycloartanal, H-60232

$C_{27}H_{44}O_3$
Lansilactone, L-60013

$C_{27}H_{44}O_5$
1,2,3-Spirostanetriol, S-60043

$C_{27}H_{44}O_7$
Antibiotic FR 900406, *in* H-60063
▷2,3,14,20,22,25-Hexahydroxy-7-cholesten-6-one, H-60063
Pterosterone, P-60192

$C_{27}H_{44}O_8$
2,3,5,14,20,22,25-Heptahydroxycholest-7-en-6-one, H-60023

$C_{27}H_{46}O_2$
5-Cholestene-3,26-diol, C-60147
20,29,30-Trisnor-3,19-lupanediol, T-60404

$C_{27}H_{47}FeN_6O_9$
Ferrioxamine D_1, *in* F-60003

$C_{27}H_{54}$
13-Heptacosene, H-60019

$C_{28}H_{14}$
Benzo[*a*]coronene, B-60021

$C_{28}H_{20}O_4$
3,3,6,6-Tetraphenyl-1,4-dioxane-2,5-dione, T-60165

$C_{28}H_{22}O_{10}$
Cephalochromin, C-60029

$C_{28}H_{24}$
[2.0.2.0]Metacyclophane, M-60031

$C_{28}H_{24}O_4$
Isomarchantin *C*, I-60107
Isoriccardin *C*, I-60135

$C_{28}H_{28}O_{11}$
Dukunolide C, *in* D-60522

$C_{28}H_{30}O_{10}$
Physalin *G*, P-60146

$C_{28}H_{32}O_6$
Garcinone *E*, G-60003

$C_{28}H_{32}O_{10}$
Physalin *L*, P-60147

$C_{28}H_{32}O_{11}$
Physalin *D*, P-60145

$C_{28}H_{34}O_7$
Withaphysalin *E*, W-60009

$C_{28}H_{34}O_{11}$
$6\beta,9\alpha,14$-Triacetoxy-1α-benzoyloxy-4β-hydroxy-8-oxodihydro-β-agarofuran, *in* E-60018

$C_{28}H_{34}O_{13}$
1-Acetoxypinoresinol 4'-*O*-β-D-glucopyranoside, *in* H-60221

$C_{28}H_{36}O_4$
Daturilin, D-60003
Withametelin, W-60005

$C_{28}H_{36}O_6$
Jaborol, J-60001

$C_{28}H_{36}O_7$
Trechonolide *A*, T-60227

$C_{28}H_{36}O_{11}$
$6\beta,9\alpha,14$-Triacetoxy-1α-benzoyloxy-4$\beta,8\beta$-dihydroxydihydro-β-agarofuran, *in* E-60018

$C_{28}H_{36}O_{16}$
Shanzhiside; 6-(4-Hydroxy-3,5-dimethoxybenzoyl), 8-Ac, Me ester, *in* S-60028

$C_{28}H_{37}ClO_{10}$
Ptilosarcone, P-60196

$C_{28}H_{37}NO_4$
Cytochalasin *O*, C-60235

$C_{28}H_{38}N_2O_{10}$
1,2-Bis(4'-benzo-15-crown-5)diazene, B-60141

$C_{28}H_{38}N_4O_6$
Chlamydocin, C-60038

$C_{28}H_{38}O_4$
Cystoketal, C-60234
Isocystoketal, *in* C-60234

$C_{28}H_{38}O_7$
Physanolide, P-60149
Withanolide *Y*, W-60007
Withaphysanolide, W-60010

$C_{28}H_{38}O_{10}$
Lapidolinin, *in* L-60015

$C_{28}H_{39}ClO_8$
Physalolactone, P-60148

$C_{28}H_{39}ClO_{10}$
Ptilosarcol, P-60195

$C_{28}H_{39}NO$
Emindole *DA*, E-60005
Emindole SA, *in* E-60005

$C_{28}H_{40}$
1,3,5,7-Tetra-*tert*-butyl-*s*-indacene, T-60025

$C_{28}H_{40}BrN_3O_6$
Geodiamolide B, *in* G-60012

$C_{28}H_{40}IN_3O_6$
Geodiamolide *A*, G-60012

$C_{28}H_{40}O_3$
9'-Methoxysargaquinone, *in* S-60010
11'-Methoxysargaquinone, *in* S-60010

$C_{28}H_{40}O_4$
8',9'-Dihydroxy-5-methylsargaquinone, *in* S-60010

$C_{28}H_{40}O_5$
Amentol 1'-methyl ether, *in* A-60088
Balearone, B-60005
Bifurcarenone; 2'-Me ether, *in* B-60105
12-Hydroxy-24-methyl-24-oxo-16-scalarene-22,25-dial; O^{12}-Ac, *in* H-60191
Isobalearone, *in* B-60005
Strictaepoxide, S-60056

$C_{28}H_{40}O_6$
Ixocarpanolide, I-60143

$C_{28}H_{40}O_7$
$6\alpha,7\alpha$-Epoxy-5$\alpha,14\alpha,20R$-trihydroxy-1-oxo-22$R,24S,25R$-with-2-enolide, *in* I-60143
Withaphysacarpin, W-60008

$C_{28}H_{44}O$
24-Methyl-5,22,25-cholestatrien-3-ol, M-60057

$C_{28}H_{46}O_4$
24-Methylene-5-cholestene-3,4,7,20-tetrol, M-60070

$C_{28}H_{46}O_7$
Polypodoaurein, *in* H-60063

$C_{28}H_{49}NO_8$
Kayamycin, K-60009

$C_{28}H_{52}O_3$
Ficulinic acid B, F-60012

$C_{29}H_{18}O$
Di-9-anthracenylmethanone, D-60050

$C_{29}H_{20}$
Tetrabenzotetracyclo[5.5.1.0^{4,13}.0^{10,13}]tridecane, T-60019

$C_{29}H_{24}O_{10}$
Chaetochromin C, *in* C-60029

$C_{29}H_{28}O_{12}$
Viscutin 2, *in* T-60103

$C_{29}H_{34}O_7$
Triptofordin *B*, T-60393

$C_{29}H_{34}O_{10}$
Crepiside A, *in* Z-60001
Crepiside B, *in* Z-60001

$C_{29}H_{34}O_{11}$
Physalin I, *in* P-60145
Trijugin *A*, T-60351

$C_{29}H_{34}O_{12}$
Acetyleumaitenol, *in* E-60072

$C_{29}H_{36}O_{10}$
Baccharinoid B27, *in* B-60002

$C_{29}H_{36}O_{13}$
1-Hydroxypinoresinol; 4″-Me ether, 1-Ac, 4′-*O*-β-D-glucopyranoside, *in* H-60221

$C_{29}H_{38}O_7$
Isonimolide, I-60110
Trechonolide *B*, *in* T-60227

$C_{29}H_{38}O_9$
Scutellone *A*, S-60017
Scuterivulactone C₂, *in* S-60017

$C_{29}H_{38}O_{10}$
Baccharinoid B9, B-60001
Baccharinoid B10, *in* B-60001
Baccharinoid B13, B-60002
Baccharinoid B14, *in* B-60002
Baccharinoid B16, B-60003

$C_{29}H_{40}O_7$
Superlatolic acid, S-60063

$C_{29}H_{40}O_8$
Physalactone, P-60144

$C_{29}H_{40}O_{10}$
Baccharinoid B20, *in* B-60001
Baccharinoid B23, *in* B-60003
Baccharinoid B24, *in* B-60003

$C_{29}H_{44}O_3$
3-Hydroxy-30-nor-12,20(29)-oleanadien-28-oic acid, H-60199
18β-Hydroxy-28-nor-3,16-oleanenedione, *in* D-60363

$C_{29}H_{44}O_6$
Brachycarpone, B-60195
Heteronemin, H-60030

$C_{29}H_{44}O_9$
Coroglaucigenin; 3-*O*-Rhamnoside, *in* C-60168
▷Frugoside, *in* C-60168
Methyl 15α,17β-diacetoxy-6α-butanoyloxy-15,16-dideoxy-15,17-oxido-16-spongianate, *in* T-60038

$C_{29}H_{45}N_3O_3$
Blastmycetin *D*, B-60191

$C_{29}H_{46}O_3$
3,18-Dihydroxy-28-nor-12-oleanen-16-one, D-60363

$C_{29}H_{46}O_7$
2α,7α-Diacetoxy-6β-isovaleroyloxy-8,13*E*-labdadien-15-ol, *in* L-60002

$C_{29}H_{46}O_8$
Viticosterone E, *in* H-60063

$C_{29}H_{47}N_5O_5$
Stacopin P1, S-60051

$C_{29}H_{47}N_5O_6$
Stacopin P2, *in* S-60051

$C_{29}H_{48}O$
Ficisterol, F-60010
Hebesterol, H-60011

$C_{29}H_{50}O$
Petrostanol, P-60066

$C_{29}H_{50}O_2$
4-Stigmastene-3,6-diol, S-60054

$C_{29}H_{50}O_3$
24,24-Dimethoxy-25,26,27-trisnor-3-cycloartanol, *in* H-60232

$C_{29}H_{52}N_{10}O_6$
Argiopine, A-60295

$C_{30}H_{14}$
[5]Phenylene, P-60090

$C_{30}H_{16}$
Tribenzo[*b,n,pqr*]perylene, T-60244

$C_{30}H_{16}O_2$
7,16-Heptacenedione, H-60018

$C_{30}H_{18}$
Di-9-anthrylacetylene, D-60051

$C_{30}H_{18}O_{10}$
Hinokiflavone, H-60082

$C_{30}H_{20}O_4$
Tetrabenzoylethylene, T-60021

$C_{30}H_{22}O_4$
1,1,2,2-Tetrabenzoylethane, T-60020

$C_{30}H_{22}O_8$
Lophirone *A*, L-60035

$C_{30}H_{22}O_9$
Maackiasin, M-60001

$C_{30}H_{22}O_{13}$
Chiratanin, C-60037

$C_{30}H_{24}O_4$
Tecomaquinone I, T-60010

$C_{30}H_{24}O_{10}$
Chaetochromin D, *in* C-60029

$C_{30}H_{24}O_{13}$
[2′,2′]-Catechin-taxifolin, C-60023

$C_{30}H_{26}O_{10}$
Chaetochromin, *in* C-60029
Chaetochromin B, *in* C-60029

$C_{30}H_{30}O_8$
▷Gossypol, G-60032

$C_{30}H_{34}O_{11}$
Ainsliaside A, *in* Z-60001

$C_{30}H_{36}O_8$
Artelein, A-60303

$C_{30}H_{36}O_9$
Isonimolicinolide, I-60109

$C_{30}H_{37}N_5O_5$
Avellanin *B*, A-60324

$C_{30}H_{38}O_4$
Cochloxanthin, C-60156

$C_{30}H_{38}O_9$
Isolimbolide, I-60105

$C_{30}H_{38}O_{12}$
6β,8β,9α,14-Tetraacetoxy-1α-benzoyloxy-4β-hydroxydihydro-β-agarofuran, *in* E-60018

$C_{31}H_{24}$
4-Methylene-1,2,3,5-tetraphenylbicyclo[3.1.0]hex-2-ene,
M-60079

$C_{31}H_{32}O_8$
Gossypol; 6-Me ether, *in* G-60032

$C_{31}H_{36}O_6$
Triptofordin *A*, T-60392

$C_{31}H_{39}N_5O_5$
Avellanin *A*, A-60323

$C_{31}H_{41}NO_6$
Pyrichalasin *H*, P-60215

$C_{31}H_{42}O_8$
Fevicordin *A*, F-60008

$C_{31}H_{42}O_{11}$
Euphorianin, E-60074

$C_{31}H_{46}O_3$
Disidein, D-60499

$C_{31}H_{46}O_7$
Heteronemin; 12-Ac, *in* H-60030
Heteronemin; 12-Epimer, 12-Ac, *in* H-60030

$C_{31}H_{46}O_{13}$
Chamaepitin, C-60033

$C_{31}H_{52}O$
Cyclopterospermol, C-60232

$C_{32}H_{14}$
Ovalene, O-60046

$C_{32}H_{22}O_{10}$
Chamaecyparin, *in* H-60082
Cryptomerin B, *in* H-60082

$C_{32}H_{24}$
3,6-Bis(diphenylmethylene)-1,4-cyclohexadiene, B-60157

$C_{32}H_{26}O_{13}$
Alterporriol *B*, A-60079
Alterporriol A, *in* A-60079

$C_{32}H_{28}O_2$
Dypnopinacol, D-60524

$C_{32}H_{30}N_4$
13,15-Ethano-17-ethyl-2,3,12,18-tetramethylmonobenzo[g]-
porphyrin, E-60043

$C_{32}H_{30}O_8$
2′,7-Dihydroxy-4′-methoxy-4-(2′,7-dihydroxy-4′-
methoxyisoflavan-5′-yl)isoflavan, D-60347

$C_{32}H_{30}O_9$
2′,7-Dihydroxy-4′-methoxy-4-(2′,7-dihydroxy-4′-
methoxyisoflavan-5′-yl)isoflavan; 3′-Hydroxy, *in* D-60347

$C_{32}H_{34}O_8$
Gossypol; 6,6′-Di-Me ether, *in* G-60032

$C_{32}H_{36}O_8$
Artanomaloide, A-60300

$C_{32}H_{38}O_{12}$
Luminamicin, L-60038

$C_{32}H_{38}O_{15}$
Neoleuropein, N-60018

$C_{32}H_{44}O_4$
3-Octadecene-1,2-diol; Dibenzoyl, *in* O-60005

$C_{32}H_{48}O_5$
3α-Acetoxy-15α-hydroxy-7,9(11),24E-lanostatrien-26-oic
acid, *in* D-60344

$C_{32}H_{48}O_6$
Chisocheton compound A, *in* T-60223

$C_{32}H_{50}O_3$
3β-Acetoxy-21,23-epoxy-7,24-tirucalladiene, *in* E-60032

$C_{32}H_{50}O_4$
3β-Acetoxy-28,20β-ursanolide, *in* H-60233

$C_{32}H_{50}O_6$
21-Acetoxy-21,23:24,25-diepoxyapotirucall-14-ene-3,7-diol, *in*
T-60223

$C_{32}H_{52}O_2$
3β-Acetoxy-14(26),17E,21-malabaricatriene, *in* M-60008

$C_{32}H_{52}O_4$
3β-Acetoxy-12-oleanene-2α,11α-diol, *in* O-60037
3β-Acetoxy-12-ursene-2α,11α-diol, *in* U-60008

$C_{32}H_{52}O_5$
3β-Acetoxy-12-oleanene-1β,2α,11α-triol, *in* O-60034
3β-Acetoxy-12-ursene-1β,2α,11α-triol, *in* U-60006
3β-Acetoxy-12-ursene-2α,11α,20β-triol, *in* U-60007

$C_{32}H_{52}O_6$
3β-Acetoxy-12-ursene-1β,2α,11α,20β-tetrol, *in* U-60005

$C_{32}H_{52}O_9$
Neotokoronin, *in* S-60043
1,2,3-Spirostanetriol; 1-O-α-L-Arabinopyranoside, *in* S-60043
Tokoronin, *in* S-60043

$C_{32}H_{54}O_{13}$
ent-2α,16β,17-Kauranetriol; O^2,O^{17}-Bis-β-D-glucopyranoside, *in*
K-60004

$C_{32}H_{55}BrO_8$
Thyrsiferol; 23-Ac, *in* T-60220

$C_{33}H_{28}O_{10}$
Podocarpusflavanone, P-60163

$C_{33}H_{32}N_4$
13,15-Ethano-3,17-diethyl-2,12,18-trimethylmonobenzo[g]-
porphyrin, E-60042

$C_{33}H_{32}N_4O_2$
$13^2,17^3$-Cyclopheophorbide enol, C-60230

$C_{33}H_{32}O_9$
2′,7-Dihydroxy-4′-methoxy-4-(2′,7-dihydroxy-4′-
methoxyisoflavan-5′-yl)isoflavan; 4′-Methoxy, *in* D-60347
2′,7-Dihydroxy-4′-methoxy-4-(2′,7-dihydroxy-4′-
methoxyisoflavan-5′-yl)isoflavan; 3′-Hydroxy, 2′-Me ether, *in*
D-60347

$C_{33}H_{36}O_{11}$
Triptofordin C-1, *in* T-60394

$C_{33}H_{38}O_2$
▷α-Guttiferin, G-60040

$C_{33}H_{38}O_{11}$
9α,14-Diacetoxy-1α,8β-dibenzoyloxy-4β,8β-dihydroxydihydro-
β-agarofuran, *in* E-60018
Pringleine, P-60172
Triptofordin *C*-2, T-60394

$C_{33}H_{40}BNO_2$
Muscarine; Tetraphenylborate, *in* M-60153

$C_{33}H_{42}O_8$
Sarothralen *B*, S-60011

$C_{33}H_{48}O_{18}$
Macrocliniside I, *in* Z-60001

$C_{33}H_{52}O_6$
3α-Acetoxy-15α-hydroxy-22-methoxy-8,24E-lanostadien-26-
oic acid, *in* T-60327
22-Acetoxy-3α-hydroxy-7α-methoxy-8,24E-lanostadien-26-oic
acid, *in* T-60328

$C_{33}H_{52}O_8$
5-Spirosten-3-ol; 3-O-β-D-Glucopyranoside, *in* S-60045
Trillin, *in* S-60045

$C_{33}H_{54}O_{12}$
Sileneoside, *in* H-60063

$C_{34}H_{16}N_4$
29,29,30,30-Tetracyanobianthraquinodimethane, T-60030

C$_{34}$H$_{18}$
Anthra[9,1,2-*cde*]benzo[*rst*]pentaphene, A-60279
Benzo[*rst*]phenaleno[1,2,3-*de*]pentaphene, B-60036
Benzo[*rst*]phenanthro[1,10,9-*cde*]pentaphene, B-60037
Benzo[*rst*]phenanthro[10,1,2-*cde*]pentaphene, B-60038
Dibenzo[*a,rst*]naphtho[8,1,2-*cde*]pentaphene, D-60075

C$_{34}$H$_{20}$
5,10-Dihydroanthra[9,1,2-*cde*]benzo[*rst*]pentaphene, *in* A-60279
9,18-Dihydrobenzo[*rst*]phenanthro[10,1,2-*cde*]pentaphene, *in* B-60038

C$_{34}$H$_{36}$N$_4$
13^1-Methyl-13,15-ethano-13^2,17-prop-13^2(15^2)-enoporphyrin, M-60082

C$_{34}$H$_{38}$O$_{16}$
Cleistanthoside A, *in* D-60493

C$_{34}$H$_{38}$O$_{17}$
Aloenin B, A-60077

C$_{34}$H$_{40}$O$_{12}$
Trisabbreviatin BBB, T-60400

C$_{34}$H$_{44}$O$_{13}$
Taccalonolide B, T-60001

C$_{34}$H$_{48}$O$_8$
Ecdysterone 22-*O*-benzoate, *in* H-60063

C$_{34}$H$_{50}$O$_6$
Caloverticillic acid A, C-60007
Caloverticillic acid C, C-60008
Caloverticillic acid B, *in* C-60007
3α,15α-Diacetoxy-7,9(11),24*E*-lanostatrien-26-oic acid, *in* D-60344

C$_{34}$H$_{50}$O$_7$
3α,22*S*-Diacetoxy-15α-hydroxy-7,9(11),24*E*-lanostatrien-26-oic acid, *in* T-60329
15α,22*S*-Diacetoxy-3α-hydroxy-7,9(11),24*E*-lanostatrien-26-oic acid, *in* T-60329
Ganoderic acid *MK*, *in* T-60329

C$_{34}$H$_{52}$O$_7$
3α,7α-Diacetoxy-15α-hydroxy-8,24*E*-lanostadien-26-oic acid, *in* T-60327

C$_{34}$H$_{52}$O$_8$
3α,22-Diacetoxy-7α,15α-dihydroxy-8,24*E*-lanostadien-26-oic acid, *in* T-60111

C$_{34}$H$_{55}$BrO$_8$
15-Anhydrothyrsiferyl diacetate, *in* A-60274
15(28)-Anhydrothyrsiferyl diacetate, *in* A-60274

C$_{35}$H$_{36}$N$_4$O$_4$
N-Methylprotoporphyrin IX, M-60123

C$_{35}$H$_{38}$O$_{11}$
Triptofordin *D*-1, T-60395

C$_{35}$H$_{38}$O$_{13}$
Triptofordin E, T-60397

C$_{35}$H$_{40}$O$_{12}$
Acetylpringleine, *in* P-60172

C$_{35}$H$_{42}$O$_{11}$
1-Cinnamoylmelianone, *in* M-60015

C$_{35}$H$_{50}$N$_8$O$_6$S$_2$
Patellamide A, P-60011

C$_{35}$H$_{54}$O$_7$
3α,22-Diacetoxy-7α-methoxy-8,24*E*-lanostadien-26-oic acid, *in* T-60328

C$_{35}$H$_{54}$O$_8$
3α,22-Diacetoxy-15α-hydroxy-7α-methoxy-8,24*E*-lanostadien-26-oic acid, *in* T-60111

C$_{35}$H$_{62}$O$_{12}$
Indicoside A, I-60027

C$_{36}$H$_{24}$N$_2$O$_2$
2,2′-(1,4-Phenylene)bis[5,1-[1,1′-biphenyl]-4-yl]oxazole, P-60091

C$_{36}$H$_{30}$O$_{14}$
Podoverine C, P-60165

C$_{36}$H$_{30}$O$_{15}$
Podoverine B, *in* P-60165

C$_{36}$H$_{44}$O$_4$
Panduratin B, P-60003

C$_{36}$H$_{46}$O$_{14}$
Taccalonolide A, *in* T-60001

C$_{36}$H$_{48}$O$_{18}$
Yadanzioside K, Y-60001

C$_{36}$H$_{50}$O$_8$
β-Ecdysone 2-cinnamate, *in* H-60063
2,3,14,20,22,25-Hexahydroxy-7-cholesten-6-one; 2-Cinnamoyl, *in* H-60063

C$_{36}$H$_{50}$O$_9$
β-Ecdysone 3-*p*-coumarate, *in* H-60063
2,3,5,14,20,22,25-Heptahydroxycholest-7-en-6-one; 2-Cinnamoyl, *in* H-60023

C$_{36}$H$_{52}$O$_8$
3α,15α,22*S*-Triacetoxy-7,9(11),24-lanostatrien-26-oic acid, *in* T-60329

C$_{36}$H$_{54}$O$_9$
3α,7α,22-Triacetoxy-15α-hydroxy-8,24*E*-lanostadien-26-oic acid, *in* T-60111
3α,15α,22-Triacetoxy-7α-hydroxy-8,24*E*-lanostadien-26-oic acid, *in* T-60111
3α,15α,22*S*-Triacetoxy-7α-hydroxy-8,24*E*-lanostadien-26-oic acid, *in* T-60111

C$_{36}$H$_{56}$O$_{11}$
Ilexsaponin A1, *in* D-60382

C$_{36}$H$_{58}$O$_9$
Corchorusin B, *in* S-60001

C$_{36}$H$_{60}$O$_8$
Corchorusin A, *in* O-60038

C$_{36}$H$_{60}$O$_9$
Corchorusin C, *in* O-60035

C$_{37}$H$_{42}$O$_{12}$
Triptofordin *D*-2, T-60396

C$_{37}$H$_{46}$N$_8$O$_6$S$_2$
Patellamide C, P-60013

C$_{37}$H$_{46}$O$_{23}$
Sarothamnoside, *in* T-60324

C$_{37}$H$_{51}$BrO$_6$
6′-Bromodisidein, *in* D-60499

C$_{37}$H$_{51}$ClO$_6$
6′-Chlorodisidein, *in* D-60499

C$_{37}$H$_{52}$O$_{13}$
Fevicordin *A* glucoside, *in* F-60008

C$_{37}$H$_{53}$N$_3$O$_9$
30-Demethoxycurromycin A, *in* C-60178

C$_{37}$H$_{56}$O$_9$
3α,15α,22*S*-Triacetoxy-7α-methoxy-8,24*E*-lanostadien-26-oic acid, *in* T-60111

C$_{38}$H$_{18}$
Dibenzo[*jk,uv*]dinaphtho[2,1,8,7-*defg*:2′,1′,8′,7′-*opqr*]pentacene, D-60072

C$_{38}$H$_{26}$
Cycloheptatrienylidene(tetraphenylcyclopentadenylidene)ethylene, C-60208

C$_{38}$H$_{28}$
[1,1′-Biphenyl]-4,4′-diylbis[diphenylmethyl], B-60117

C$_{38}$H$_{35}$NO$_{10}$
Scleroderris green, S-60014

C$_{38}$H$_{42}$O$_{12}$
Esulone *A*, E-60038
Esulone *C*, E-60039

Carnosifloside IV, *in* C-60175

$C_{48}H_{80}O_{18}$
Carnosifloside V, *in* C-60175
Carnosifloside VI, *in* C-60175

$C_{50}H_{63}BrO_{16}$
Bromothricin, *in* C-60137

$C_{50}H_{63}ClO_{16}$
Chlorothricin, C-60137

$C_{50}H_{63}ClO_{17}$
2‴-Hydroxychlorothricin, *in* C-60137

$C_{50}H_{80}O_{21}$
Balanitin 2, *in* S-60045

$C_{50}H_{80}O_{22}$
▷Funkioside F, *in* S-60045

$C_{51}H_{72}Cl_2O_{18}$
Lipiarmycin A_4, *in* L-60028

$C_{51}H_{82}O_{22}$
Aculeatiside A, *in* N-60062
Balanitin 1, *in* S-60045
Funkioside E, *in* S-60045

$C_{51}H_{82}O_{23}$
Aculeatiside B, *in* N-60062
Avenacoside A, *in* N-60062

$C_{52}H_{74}Cl_2O_{18}$
Lipiarmycin, L-60028
Lipiarmycin A_3, *in* L-60028

$C_{52}H_{82}O_{20}$
Guaianin D, *in* H-60199

$C_{53}H_{84}O_{24}$
Camellidin II, *in* D-60363

$C_{55}H_{82}O_{21}S_2$
Yessotoxin, Y-60002

$C_{55}H_{86}O_{25}$
Camellidin I, *in* D-60363

$C_{56}H_{40}O_{12}$
Miyabenol *B*, M-60140

$C_{56}H_{42}O_{12}$
Miyabenol *A*, M-60139
Vaticaffinol, V-60002

$C_{57}H_{92}O_{28}$
Avenacoside B, *in* N-60062

$C_{58}H_{78}N_4O_{22}S_4$
Esperamicin A_{1b}, *in* E-60037

$C_{58}H_{92}O_{52}$
Guaianin E, *in* H-60199

$C_{59}H_{80}N_4O_{22}S_4$
Esperamicin A_1, *in* E-60037
Esperamicin A_2, *in* E-60037

$C_{59}H_{90}O_{26}$
▷Funkioside G, *in* S-60045

$C_{62}H_{38}$
b,b′,b″-Tritriptycene, T-60407

$C_{63}H_{98}N_{18}O_{13}S$
Substance *P*, S-60059

$C_{73}H_{126}O_{20}$
Bistheonellide B, *in* M-60138

$C_{74}H_{128}O_{20}$
Misakinolide *A*, M-60138

$C_{82}H_{52}$
Tetra(2-triptycyl)ethylene, T-60169

$C_{100}H_{164}O$
Dolichol, D-60517

$C_{122}H_{210}N_2O_{16}$
[3][14-Acetyl-14-azacyclohexacosanone][25,26,53,54,55,56-hexaacetoxytricyclo[49.3.1.1^{24,28}]hexapentaconta-1(55),-24,26,28(56),51,53-hexaene][14-acetyl-14-azacyclohexacosanone]catenane, A-60020

$C_{250}H_{410}N_{68}O_{75}$
Thymopoietin I, *in* T-60219

$C_{251}H_{412}N_{64}O_{78}$
Thymopoietin II, 9CI, *in* T-60219

$C_{255}H_{416}N_{66}O_{77}$
Thymopoietin III, *in* T-60219

Chemical Abstracts Service Registry Number Index

This index becomes invalid after publication of the Seventh Supplement.

The CAS Registry Number Index lists in ascending numerical order all CAS Registry Numbers recorded in this Supplement. CAS Registry numbers of compounds contained in Supplements 1 to 5 are listed in the cumulative Index Volume published with the Fifth Supplement.

Each CAS Registry Number listed refers the user to a chemical name (with stereochemical or derivative descriptors where relevant) and a DOC number.

The first digit of the DOC Number (printed in bold type) refers to the number of the Supplement in which the Entry appears. In this Sixth Supplement the first digit is invariably 6.

A DOC Number which follows immediately upon a chemical name means that the name is the DOC Name and it is to this name that the CAS Registry Number refers.

A DOC Number which is preceded by the word '*in*' means that CAS Registry Number refers to the specified stereoisomer or derivative which is to be found embedded within the particular Entry.

The symbol ▷ preceding an index term indicates that the DOC Entry contains information on toxic or hazardous properties of the compound.

CAS Registry Number Index

312-73-2   2-(Trifluoromethyl)-1*H*-benzimidazole, T-60287
315-14-0   1,3,5-Trifluoro-2-nitrobenzene, T-60304
319-88-0 ▷1,3,5-Trichloro-2,4,6-trifluorobenzene, T-60257
322-58-7   1-(Methylthio)-2-(trifluoromethyl)benzene, *in* T-60284
327-86-6   3,3′,5-Triiodothyronine, T-60350
328-98-3   1-(Methylthio)-3-(trifluoromethyl)benzene, *in* T-60285
337-65-5   Spiro[bicyclo[3.2.1]octane-8,2′-[1,3]dioxolane], *in* B-60094
341-27-5   3-Fluoro-2-hydroxybenzoic acid, F-60040
342-69-8 ▷1,7-Dihydro-6*H*-purine-6-thione; 6-*Thiol-form*, 5-Me, 9-β-D-ribofuranosyl, *in* D-60270
344-03-6   1,4-Dibromo-2,3,5,6-tetrafluorobenzene, D-60108
344-38-7   4-Bromo-1-nitro-2-(trifluoromethyl)benzene, B-60309
344-39-8   4-Methoxy-1-nitro-2-(trifluoromethyl)benzene, *in* N-60054
345-16-4   5-Fluoro-2-hydroxybenzoic acid, F-60044
345-29-9   4-Fluoro-2-hydroxybenzoic acid, F-60042
348-62-9   4-Chloro-2-fluorophenol, C-60079
350-29-8   3-Fluoro-4-hydroxybenzoic acid, F-60041
359-41-1   (Trifluoromethyl)oxirane, T-60294
359-63-7   *N,N*-Bis(trifluoromethyl)hydroxylamine, B-60180
359-75-1   1,1,1-Trifluoro-*N*-(nitrosooxy)-*N*-(trifluoromethyl)-methanamine, *in* B-60180
360-11-2   Difluorodiphenylmethane, D-60184
360-91-8   3,4-Bis(trifluoromethyl)-1,2-dithiete, B-60179
363-24-6 ▷11,15-Dihydroxy-9-oxo-5,13-prostadienoic acid; (5*Z*,8*R*,11*R*,12*R*,13*E*,15*S*)-*form*, *in* D-60367
367-32-8   4-Fluoro-1,2-benzenediol, F-60020
367-67-9   1-Bromo-4-nitro-2-(trifluoromethyl)benzene, B-60307
367-78-2   2-Fluoro-4,6-dinitroaniline, F-60028
367-81-7   5-Fluoro-2,4-dinitroaniline, F-60031
367-83-9   2-Fluoro-5-methoxybenzoic acid, *in* F-60038
368-48-9   2-(Trifluoromethyl)pyridine, T-60296
371-71-1   (Trifluoromethyl)carbonimidic difluoride, T-60288
372-75-8   Citrulline; (*S*)-*form*, *in* C-60151
373-80-8   Pentafluoro(trifluoromethyl)sulfur, P-60041
374-31-2   1,3,3,4,4-Pentafluorocyclobutene, P-60033
378-94-9   Nonafluoromorpholine, N-60056
381-84-0 ▷2-Trifluoromethylacrylonitrile, *in* T-60295
381-98-6   2-(Trifluoromethyl)propenoic acid, T-60295
382-90-1 ▷2-(Trifluoromethyl)propenoic acid; Me ester, *in* T-60295
386-72-1   2-Nitro-3-(trifluoromethyl)phenol, N-60047
391-92-4   5-Fluoro-2-hydroxybenzoic acid; Me ester, *in* F-60044
392-57-4 ▷1,2,4,5-Tetrafluoro-3,6-diiodobenzene, T-60046
394-04-7   5-Fluoro-2-methoxybenzoic acid, *in* F-60044
394-25-2   1-Methoxy-2-nitro-4-(trifluoromethyl)benzene, *in* N-60048
394-28-5   2-Bromo-5-fluorobenzoic acid, B-60241
398-62-9   4-Fluoro-1,2-dimethoxybenzene, *in* F-60020
399-55-3   2-Fluoro-1-methoxy-4-methylbenzene, *in* F-60053
400-99-7 ▷2-Nitro-4-(trifluoromethyl)phenol, N-60048
401-53-6   2-Amino-4-fluorobutanoic acid, A-60155
402-31-3   1,3-Bis(trifluoromethyl)benzene, B-60176
403-01-0   3-Fluoro-4-hydroxybenzoic acid; Me ester, *in* F-60041
403-20-3   3-Fluoro-4-methoxybenzoic acid, *in* F-60041
404-71-7 ▷1-Fluoro-3-isocyanatobenzene, F-60048
406-41-7   1,1,1-Trifluoro-2-butyne, T-60275
427-77-0   Gibberellin *A*₉, G-60018
432-03-1   Tris(trifluoromethyl)amine, T-60406
433-19-2   1,4-Bis(trifluoromethyl)benzene, B-60177
433-69-2   4-Amino-5-methyl-3-isoxazolidinone, A-60221
433-95-4   1,2-Bis(trifluoromethyl)benzene, B-60175
443-88-9   2-Fluoro-1,3-dimethylbenzene, F-60026
443-90-3   2-Fluoro-6-methylphenol, F-60055
446-36-6 ▷5-Fluoro-2-nitrophenol, F-60061
446-72-0 ▷4′,5,7-Trihydroxyisoflavone, T-60324
450-89-5   1-Chloro-4-fluoro-2-methoxybenzene, *in* C-60074
451-25-2   4-(4-Methylphenyl)-2-pentanone, M-60112
452-06-2 ▷2-Aminopurine, A-60243

452-08-4   2-Bromo-4-fluoro-1-methoxybenzene, *in* B-60248
452-09-5   4-Chloro-2-fluoro-1-methoxybenzene, *in* C-60079
452-70-0   4-Fluoro-3-methylphenol, F-60059
452-72-2   4-Fluoro-2-methylphenol, F-60058
452-78-8   3-Fluoro-4-methylphenol, F-60057
452-81-3   2-Fluoro-4-methylphenol, F-60053
452-85-7   5-Fluoro-2-methylphenol, F-60060
454-99-9   2,4-Bis(trifluoromethyl)pyridine, B-60182
455-00-5   2,6-Bis(trifluoromethyl)pyridine, B-60184
465-94-1   12-Oleanene-3,16,28-triol; (3β,16β)-*form*, *in* O-60038
465-95-2   12-Oleanene-3,16,28-triol; (3β,16α)-*form*, *in* O-60038
466-37-5   Tetraphenylethanone, T-60167
467-32-3   3,3,6,6-Tetraphenyl-1,4-dioxane-2,5-dione, T-60165
468-19-9 ▷Coroglaucigenin, C-60168
470-30-4   4-Hydroxy-3,3-dimethyl-2-oxobutanoic acid, H-60124
471-32-9   Dithiocarbazic acid, D-60509
471-87-4   Stachydrine, S-60050
473-84-7   2-Hydroxycyclopentanone, H-60114
475-38-7 ▷5,8-Dihydroxy-1,4-naphthoquinone, D-60356
476-37-9   Hexahydroxy-1,4-naphthoquinone, H-60065
476-66-4 ▷Ellagic acid, E-60003
478-40-0   3,5-Dihydroxy-2-methyl-1,4-naphthoquinone, D-60350
479-11-8   Benzo[*b*]naphtho[2,3-*d*]furan-6,11-dione, B-60034
479-66-3   Fulvic acid, F-60081
480-23-9   3′,4′,5,7-Tetrahydroxyisoflavone, T-60108
481-40-3   1,4,5-Naphthalenetriol, N-60005
483-85-2   2-Formyl-3,4-dimethoxybenzoic acid, *in* F-60070
484-32-2   2-Methoxy-3,4-methylenedioxybenzoic acid, *in* H-60177
484-51-5   Khellinone, K-60011
484-92-4   1-(4,5-Dihydro-1*H*-imidazol-2-yl)-2-imidazolidinethione, D-60246
485-38-1   4,5-Dimethoxy-1,3-benzenedicarboxylic acid, *in* D-60307
486-66-8   4′,7-Dihydroxyisoflavone, D-60340
487-21-8   2,4(1*H*,3*H*)-Pteridinedione, P-60191
487-39-8   Phillygenin; (+)-*form*, *in* P-60141
487-41-2   Phillyrin, *in* P-60141
487-42-3   3-(Phenylimino)-1(3*H*)-isobenzofuranone, P-60112
487-56-9   4-Hydroxy-5-benzofurancarboxylic acid, H-60103
487-79-6 ▷Kainic acid, K-60003
490-06-2   3-Isopropyl-6-methyl-1,2-benzenediol, I-60124
491-31-6   1*H*-2-Benzopyran-1-one, B-60041
491-40-7   Octahydro-4*H*-quinolizin-4-one, O-60019
491-67-8   5,6,7-Trihydroxyflavone, T-60322
492-98-8   2,2′-Bi-1*H*-imidazole, B-60107
494-21-3   2-(2-Furanyl)quinoxaline, F-60089
494-67-7   2,2′-(1,4-Phenylene)bis[5,1-[1,1′-biphenyl]-4-yl]-oxazole, P-60091
495-35-2   Thiazolo[4,5-*b*]quinoline-2(3*H*)-thione, T-60201
495-52-3   *s*-Hydrindacene, H-60095
495-74-9 ▷1-Phenyl-2,4-hexadiyn-1-one, P-60109
496-69-5   2-Bromo-4-fluorophenol, B-60248
496-83-3   2-Hydroxycyclononanone, H-60112
496-86-6   9,10,18-Trihydroxyoctadecanoic acid, T-60334
497-26-7 ▷2-Methyl-1,3-dioxolane, M-60063
501-32-6   α-Amino-1*H*-imidazole-1-propanoic acid, A-60204
501-53-1 ▷Benzyloxycarbonyl chloride, B-60071
502-50-1   4-Oxoheptanedioic acid, O-60071
502-51-2   4-Oxo-2-heptenedioic acid, O-60072
502-65-8   Lycopene, L-60041
502-72-7 ▷Cyclopentadecanone, C-60219
502-87-4   4-Methylene-2,5-cyclohexadien-1-one, M-60071
503-53-7   4,4-Dimethyl-5-oxopentanoic acid, D-60434
504-75-6   4,5-Dihydro-1*H*-imidazole, D-60245
505-19-1   Hexahydropyrimidine, H-60056
505-57-7 ▷2-Hexenal, H-60076
505-72-6   3-(Carboxymethylamino)propanoic acid, C-60016
507-29-9   Histidine trimethylbetaine, H-60083
512-04-9 ▷5-Spirosten-3-ol; (3β,25*R*)-*form*, *in* S-60045
514-51-2   β-Patchoulene, P-60010
515-85-5   3,3,4,4-Tetrafluoro-2-(trifluoromethyl)-1,2-oxazetidine, T-60051
515-96-8 ▷Semioxamazide, S-60020

13055-36-2	3,4-Acenaphthenedicarboxylic acid, A-60013
13058-73-6	7-Nitroisoquinoline, N-60033
13090-26-1	[1,2,4,5]Tetrazino[1,6-*a*:4,3-*a'*]diquinoline, T-60173
13136-51-1	2-Mercapto-2-phenylacetic acid; (*S*)-form, *S*-Benzyl, *in* M-60024
13136-52-2	2-Mercapto-2-phenylacetic acid; (*R*)-form, *S*-Benzyl, *in* M-60024
13161-85-8	1,3,4,10-Tetrahydro-9(2*H*)-acridinone, T-60052
13164-04-0	Chalepin, C-60031
13171-59-0	1,2,3,4,5-Pentamethyl-6-nitrobenzene, P-60057
13177-73-6	7-Hydroxyfuro[2,3-*d*]pyridazine, H-60147
13186-45-3	C.I. Mordant Violet 39, *in* A-60318
13200-02-7	7,16-Heptacenedione, H-60018
13214-71-6	6,15-Hexacenedione, H-60033
13261-50-2	2-Hydroxy-8-methoxy-1,4-naphthoquinone, *in* D-60355
13274-42-5	1,5,9,13,17-Pentaazaheptadecane, P-60026
13306-99-5	2-Phenyl-2*H*-1,2,3-triazole-4-carboxylic acid, P-60140
13325-14-9	3-Amino-3-methyl-2-butanol, A-60217
13333-97-6	2-(Trifluoromethyl)benzenethiol, T-60284
13338-63-1	▷3,4,5-Trimethoxybenzeneacetonitrile, *in* T-60343
13344-76-8	2-Fluoro-2-phenylacetaldehyde, F-60062
13401-18-8	2,4(1*H*,3*H*)-Pteridinedione; 1,3-Di-Me, *in* P-60191
13464-19-2	Carbonochloridothioic acid; *S*-Ph, *in* C-60013
13466-30-3	Acetophenone; Hydrazone, *in* A-60017
13472-57-6	2,6-Diethoxypyridine, *in* H-60226
13484-76-9	*N*-Phenylformamidine, P-60101
13511-13-2	1-(2-Propenyl)cyclohexene, P-60179
13538-66-4	9-Methylene-1,3,5,7-cyclononatetraene, M-60073
13544-43-9	6-Trifluoromethyl-1*H*-indole, T-60293
13544-66-6	3,4,5-Trimethylbenzoic acid; Me ester, *in* T-60355
13577-71-4	6-Chloro-2-(trifluoromethyl)-1*H*-imidazo[4,5-*b*]pyridine, C-60139
13580-54-6	2,2,3,3,5,5,6,6-Octafluoromorpholine, O-60009
13602-68-1	4-Amino-2(1*H*)-pyridinone; *NH-form*, *N*-Oxide; B,HCl, *in* A-60250
13602-69-2	4-Amino-2(1*H*)-pyridinone; *NH-form*, *N*-Oxide, *in* A-60250
13606-71-8	3,4-Diphenyl-2-oxazolidinone, *in* P-60126
13659-21-7	3-Bromo-2,4-dichlorophenol, B-60228
13659-22-8	2-Bromo-3,5-dichlorophenol, B-60226
13667-30-6	2,3,4-Trimethyl-5-nitrobenzoic acid, T-60369
13726-16-4	4-Chloro-3-methoxybenzaldehyde, *in* C-60090
13728-34-2	2,3-Naphthalenedicarboxylic acid; Di-Me ester, *in* N-60003
13754-69-3	3,7,11,15-Tetramethyl-2-hexadecenal, *in* P-60152
13781-67-4	3-Thiopheneethanol, T-60211
13849-32-6	3,4-Dihydro-2*H*-1-benzopyran-5-ol, D-60201
13855-90-8	2,2,5,5-Tetramethyl-3-cyclohexen-1-one, T-60136
13881-91-9	Aminomethanesulfonic acid, A-60215
13885-13-7	2-Cyclopropyl-2-oxoacetic acid, C-60231
13889-92-4	▷Carbonochloridothioic acid; *S*-Propyl, *in* C-60013
13889-94-6	Carbonochloridothioic acid; *S*-Butyl, *in* C-60013
13906-09-7	2,3-Dihydro-2-thioxo-4(1*H*)-quinazolinone, D-60296
13980-04-6	▷Hexahydro-1,3,5-triazine; 1,3,5-Trinitroso, *in* H-60060
14033-65-9	4-(1-Propenyl)cyclohexene, P-60181
14072-82-3	4-Isopropylcyclohexene, I-60120
14072-87-8	3-*tert*-Butylcyclohexene, B-60342
14090-99-4	2,5,7,8-Tetrahydroxy-3,6-dimethoxy-1,4-naphthoquinone, *in* H-60065
14144-06-0	Trillin, *in* S-60045
14164-34-2	4,6-Dimethyl-2-phenylpyrimidine, D-60444
14174-83-5	Hexahydro-1*H*-pyrrolizin-1-one, H-60057
14174-86-8	Tetrahydro-1*H*-pyrrolizin-2(3*H*)-one, T-60091
14189-85-6	2-Methoxy-3-methyl-2-cyclopenten-1-one, *in* H-60174
14252-67-6	2-Chloro-2-phenylethanol; (*S*)-form, *in* C-60128
14255-88-0	▷Phenyl 5,6-dichloro-2-(trifluoromethyl)-1*H*-benzimidazole-1-carboxylate, *in* D-60160
14289-45-3	α-Oxo-2-naphthaleneacetic acid, O-60079
14307-88-1	Alethine; B,2HCl, *in* A-60073
14309-25-2	Azidotriphenylmethane, A-60352
14346-19-1	3-Amino-5-nitro-2,1-benzisothiazole, A-60232
14352-55-7	2-Hydroxycyclopentanone; (±)-form, Oxime, *in* H-60114
14353-90-3	Pentafluoroiodosobenzene, P-60034
14356-59-3	Saikogenin *F*, S-60001
14377-11-8	1-Acetylcycloheptene, A-60026
14408-81-2	Tetrahydro-1*H*-pyrrolizin-2(3*H*)-one; (±)-form, Picrate, *in* T-60091
14439-13-5	2,4,7-Triaminopteridine, T-60229
14491-02-2	2,2-Dimethyl-3-phenyl-2*H*-azirine, D-60440
14496-24-3	3,4-Bis(hydroxymethyl)furan, B-60161
14527-48-1	2-(Ethylthio)thiazole, *in* T-60200
14531-52-3	5-(2-Phenylethyl)-1,3-benzenediol, P-60098
14554-09-7	5,8-Dihydroxy-2-methyl-1,4-naphthoquinone, D-60352
14561-37-6	2-Amino-3-chlorobutanoic acid; (2*RS*,3*RS*)-form, *in* A-60103
14561-56-9	2-Amino-3-chlorobutanoic acid, A-60103
14578-68-8	3,4-Dihydro-4-phenyl-1(2*H*)-naphthalenone, D-60267
14671-09-1	Hexahydro-1*H*-pyrrolizin-1-one; (±)-form, Picrate, *in* H-60057
14696-36-7	*ent*-15-Kauren-17-oic acid, K-60008
14699-32-2	8-Hydroxyisopimar-15-ene; 8β-form, *in* H-60163
14705-30-7	2,2,5-Trimethyl-4-hexen-3-one, T-60366
14730-25-7	3-Phenyl-1,4,2-dithiazole-5-thione, P-60088
14757-77-8	4-Hydroxyfuro[2,3-*d*]pyridazine, H-60146
14757-78-9	3-Bromo-2-furancarboxaldehyde, B-60257
14787-34-9	5-Hydroxy-3,4′,6,7-tetramethoxyflavone, *in* P-60047
14802-18-7	2-Phenyl-4(1*H*)-quinolinone, P-60132
14804-37-6	Methoxypentamethylbenzene, *in* P-60058
14811-73-5	10-Oxodecanoic acid; Me ester, *in* O-60063
14836-73-8	Ferrioxamine *B*, F-60003
14882-94-1	Rutamarin, *in* C-60031
14889-64-6	Tetrahydro-2*H*-1,3-thiazin-2-one, T-60096
14944-26-4	3,4-Dihydro-3-phenyl-1(2*H*)-naphthalenone, D-60266
14958-06-6	2-Methyl-1,3-naphthalenediol; Di-Ac, *in* M-60091
14988-20-6	Sphaerobioside, *in* T-60324
15055-49-9	5-Amino-3-phenylisoxazole; *N*-Ph, *in* A-60240
15055-81-9	5-Isoxazolecarboxylic acid; Me ester, *in* I-60142
15089-43-7	2,2-Butanedithiol, B-60340
15123-45-2	13,19:14,18-Dimethenoanthra[1,2-*a*]benzo[*o*]pentaphene, D-60396
15149-00-5	4-Bromo-2-fluorophenol; Methanesulfonate, *in* B-60251
15158-36-8	4-Methyl-1,2-benzenedicarboxaldehyde, M-60048
15159-65-6	2-Amino-4-bromobutanoic acid; (*S*)-form, B,HBr, *in* A-60098
15176-29-1	▷2′-Deoxy-5-ethyluridine, *in* E-60054
15232-95-8	3-(2-Propenyl)cyclohexene, P-60180
15250-38-1	5,8-Quinoxalinedione, Q-60010
15300-62-6	Benzo[*a*]biphenylene; 2,4,7-Trinitrofluorenone complex, *in* B-60016
15301-54-9	2-Phenylcyclopentylamine, P-60086
15308-22-2	2,5,6,8-Tetrahydroxy-3,7-dimethoxy-1,4-naphthoquinone, *in* H-60065
15308-24-4	2,3,5,6,8-Pentahydroxy-7-methoxy-1,4-naphthoquinone, *in* H-60065
15383-91-2	2-Amino-2-butylbutanedioic acid; (±)-form, *in* A-60102
15414-36-5	4-(2-Propenyl)cyclohexene, P-60182
15430-52-1	2-Propyn-1-amine; B,HCl, *in* P-60184
15462-43-8	2-Cyano-3-hydroxyquinoline, *in* H-60228
15462-44-9	3-Hydroxy-2-quinolinecarboxylic acid; Amide, *in* H-60228
15462-45-0	3-Hydroxy-2-quinolinecarboxylic acid, H-60228
15537-53-8	Benzylidenecycloheptane, B-60068
15540-18-8	2-Cyclohexyl-2-methoxyacetic acid, *in* C-60212
15545-30-9	2(3*H*)-Thiazolethione; *SH-form*, *S*-Isopropyl, *in* T-60200
15567-77-8	2-(Hydroxymethyl)-1,4-benzodioxan; Carbamate, *in* H-60172

22395-22-8 7-Methoxyflavone, *in* H-60144
22436-26-6 9-Bromo-1,4-dihydro-1,4-methanonaphthalene; *syn-form, in* B-60233
22480-88-2 2,3,4-Trihydroxyphenylacetic acid; Tri-Me ether, Me ester, *in* T-60340
22480-91-7 2,3,4-Trimethoxybenzeneacetic acid, *in* T-60340
22503-68-0 4-Oxo-2-heptenedioic acid; Di-Me ester, 2,4-Dinitrophenylhydrazone, *in* O-60072
22503-69-1 4-Oxo-2-heptenedioic acid; Di-Me ester, *in* O-60072
22557-12-6 4,5,7,8-Tetrafluoro[2.2]paracyclophane, T-60050
22621-26-7 3-Acetyl-4-hydroxy-2(5*H*)-furanone, A-60030
22627-00-5 Diphenylmethaneimine; *N*-Me, *in* D-60482
22634-92-0 4-Oxoheptanedioic acid; Di-Me ester, *in* O-60071
22644-96-8 2-Hexadecenal; (*E*)-*form, in* H-60036
22680-62-2 4-Methoxy-2-phenylquinoline, *in* P-60132
22721-28-4 2-Amino-4,5,6,7-tetrahydrobenzo[*b*]thiophene-3-carboxylic acid; Hydrazide, *in* A-60258
22733-60-4 Siccanin, S-60029
22756-44-1 Asadanin, A-60308
22776-74-5 4-Azatricyclo[4.3.1.1³,⁸]undecane, A-60333
22800-71-1 *N*-(Diphenylmethylene)acetamide, *in* D-60482
22804-63-3 1,2,3,4,8-Pentamethoxyxanthone, *in* P-60051
22810-28-2 2-Phenyl-4*H*-1-benzothiopyran-4-one; *S*,*S*-Dioxide, *in* P-60082
22839-47-0 ▷Aspartame; L-L-*form, in* A-60311
22839-65-2 Aspartame; L-D-*form, in* A-60311
22856-30-0 2,3-Dicyanonaphthalene, *in* N-60003
22856-34-4 4-Oxo-2-heptenedioic acid; 2,4-Dinitrophenyl-hydrazone, *in* O-60072
22906-55-4 *N*-Valylleucine; L-D-*form, in* V-60001
22918-03-2 4-Chloro-2-iodopyridine, C-60112
22934-58-3 3-Hydroxy-4,5-methylenedioxybenzoic acid; Me ether, Me ester, *in* H-60178
22940-91-6 2,2-Dichloro-3,3-difluorooxirane, D-60128
22963-93-5 Juvenile hormone III, *in* T-60364
22993-77-7 6-Amino-2,2-dichlorohexanoic acid, A-60126
23003-22-7 3-Hydroxy-2(1*H*)-pyridinethione, H-60222
23003-30-7 3-Hydroxy-2-iodo-6-methylpyridine, H-60159
23003-35-2 2-Bromo-3-hydroxy-6-methylpyridine, B-60267
23007-85-4 1,2,3,6-Tetrahydro-4-phenylpyridine; *N*-Me; B,HCl, *in* T-60081
23020-15-7 2-Phenylcyclopropanecarboxylic acid; (1*S*,2*S*)-*form, in* P-60087
23020-18-0 2-Phenylcyclopropanecarboxylic acid; (1*S*,2*R*)-*form, in* P-60087
23038-58-6 4-Methyl-1,3-benzenedicarboxaldehyde, M-60049
23043-62-1 ▷2,9-Diaminoacridine, D-60027
23109-05-9 ▷Amanitins; α-Amanitin, *in* A-60084
23130-58-7 3-Oxopropanoic acid; *Oxo-form*, 2,4-Dinitrophenyl-hydrazone, *in* O-60090
23141-17-5 Isotaxiresinol; 7-Me ether, *in* I-60139
23145-06-4 5,7-Dichlorobenzofuran, D-60126
23145-08-6 5,7-Dibromobenzofuran, D-60087
23156-68-5 *O*-(2-Hydroxyethyl)hydroxylamine; B,HCl, *in* H-60134
23239-35-2 2-Amino-2-methyl-3-phenylpropanoic acid; (*S*)-*form, in* A-60225
23245-77-4 [4,4′-Bipyridine]-3,3′-dicarboxylic acid, B-60137
23245-99-0 Acetophenone; (*E*)-2,4-Dinitrophenylhydrazone, *in* A-60017
23304-25-8 ▷1-(Diazomethyl)-4-methoxybenzene, D-60064
23311-86-6 2-Amino-3-methyl-3-butenoic acid, A-60218
23334-72-7 2,3-Dihydroxybutanoic acid; (2*S*,3*S*)-*form, in* D-60312
23351-09-9 1-(4-Hydroxyphenyl)pyrrole, H-60216
23365-04-0 *N*-Valylleucine; L-L-*form*, H-Val-Leu-OMe; B,HCl, *in* V-60001
23369-74-6 Jateorin, *in* C-60160
23402-19-9 Agasyllin, A-60064
23461-67-8 3,3-Dimethyl-4-oxopentanoic acid, D-60433
23479-73-4 Elvirol, E-60004
23500-57-4 3,3-Bis(aminomethyl)oxetane, B-60140
23504-97-4 2-Thiatricyclo[3.3.1.1³,⁷]decane; 2,2-Dioxide, *in* T-60195

23527-24-4 Kolavenic acid; Me ester, *in* K-60013
23535-17-3 4-Hydroxy-2-methyl-2-cyclopenten-1-one, H-60175
23537-79-3 2-Bromo-1,4-dihydro-1,4-methanonaphthalene, B-60232
23537-80-6 1-Bromo-1,4-dihydro-1,4-methanonaphthalene, B-60231
23592-45-2 (Methylamino)methanesulfonic acid, *in* A-60215
23635-83-8 1,4,7-Triazacyclododecane, T-60231
23658-61-9 2-(Dimethylamino)purine, *in* A-60243
23658-67-5 8-(Methylamino)purine, *in* A-60244
23664-28-0 4,7-Dihydroxy-5-methyl-2*H*-1-benzopyran-2-one, D-60348
23687-23-2 8-(Dimethylamino)purine, *in* A-60244
23720-86-7 Phrymarolin II, P-60142
23724-57-4 7-Methoxy-1,3-benzodioxole-4-carboxylic acid, *in* H-60179
23731-78-4 2-Hydroxy-3,4-methylenedioxybenzoic acid; Me ether, Me ester, *in* H-60177
23732-22-1 4-Methyl-3-hepten-2-one; (*E*)-*form, in* M-60087
23732-23-2 4-Methyl-3-hepten-2-one; (*Z*)-*form, in* M-60087
23745-82-6 4-Hydroxyphenylacetaldoxime, *in* H-60209
23785-21-9 1*H*-Imidazole-4-carboxylic acid; Et ester, *in* I-60004
23812-55-7 4-Hydroxy-2,3-methylenedioxybenzoic acid; Me ether, Me ester, *in* H-60179
23827-93-2 *N*-Serylproline, S-60025
23833-96-7 8-Chloro-4(1*H*)-quinolinone, C-60132
23978-09-8 ▷4,7,13,16,21,24-Hexaoxa-1,10-diazabicyclo[8.8.8]-hexacosane, H-60073
23978-55-5 1,7,10,16-Tetraoxa-4,13-diazacyclooctadecane, T-60161
23981-25-1 8-Chloro-2(1*H*)-quinolinone, C-60131
24007-54-3 2-Hydroxyphenylacetaldehyde; 2,4-Dinitrophenyl-hydrazone, *in* H-60207
24034-22-8 2-Bromo-1-nitro-3-(trifluoromethyl)benzene, B-60308
24035-43-6 8,11,13-Abietatrien-19-ol, A-60007
24100-18-3 2-Bromo-3-methoxypyridine, *in* B-60275
24190-32-7 γ-Ionone, I-60080
24207-04-3 2-Bromo-3-hydroxy-6-methylpyridine; *N*-Oxide, *in* B-60267
24207-22-5 2-Bromo-3-methoxy-6-methylpyridine, *in* B-60267
24211-27-6 Siccanochromene *E*, S-60030
24259-39-0 3-(Phenylimino)-1(3*H*)-isobenzofuranone; B,HClO₄, *in* P-60112
24271-83-8 5,8-Quinazolinediol, Q-60002
24308-04-1 1,4,5-Naphthalenetriol; Tri-Ac, *in* N-60005
24330-52-7 1,3-Butanedithiol, B-60339
24330-53-8 2-Methyl-1,3-propanediol; Bis(4-methyl-benzenesulfonyl), *in* M-60120
24462-15-5 Dehydroabietinol acetate, *in* A-60007
24470-47-1 Hardwickiic acid; (+)-*form, in* H-60007
24472-05-7 1,4,6,10-Tetraoxaspiro[4.5]decane, T-60164
24509-73-7 1*H*-Pyrrolo[3,2-*b*]pyridine; *N*-Ac, *in* P-60249
24513-65-3 2,3-Dihydro-3-(phenylmethylene)-4*H*-1-benzopyran-4-one; (*Z*)-*form, in* D-60263
24513-66-4 2,3-Dihydro-3-(phenylmethylene)-4*H*-1-benzopyran-4-one; (*E*)-*form, in* D-60263
24514-98-5 3-(Phenylazo)-2-butenenitrile; (*E*,*E*)-*form, in* P-60079
24515-82-0 3-(Phenylazo)-2-butenenitrile; (*Z*,*E*)-*form, in* P-60079
24526-70-3 2-Oxo-2*H*-benzopyran-4-acetic acid; Chloride, *in* O-60057
24526-71-4 2-Oxo-2*H*-benzopyran-4-acetic acid; Et ester, *in* O-60057
24526-73-6 2-Oxo-2*H*-benzopyran-4-acetic acid, O-60057
24526-74-7 4-(Cyanomethyl)coumarin, *in* O-60057
24570-49-8 Muscarine; (2*S*,4*R*,5*S*)-*form*, Iodide, *in* M-60153
24653-75-6 1-Mercapto-2-propanone, M-60028
24656-54-0 [2.0.2.0]Metacyclophane, M-60031
24769-97-9 6-Amino-2,2-dichlorohexanoic acid; *N*-Benzoyl, *in* A-60126
24778-20-9 Artecalin, A-60302
24808-04-6 3,4′,5,7-Tetrahydroxyflavanone; (2*R*,3*R*)-*form, in* T-60104

614

53342-16-8	Chlamydocin, C-60038
53342-27-1	2-Acetylpyrimidine, A-60044
53377-54-1	4,7-Dimethoxy-5-methyl-2*H*-1-benzopyran-2-one, *in* D-60348
53449-12-0	8-Aminoimidazo[4,5-*g*]quinazoline, A-60207
53449-13-1	8-Aminoimidazo[4,5-*g*]quinazoline; B,HCl, *in* A-60207
53449-18-6	Imidazo[4,5-*g*]quinazolin-8-one, I-60012
53449-43-7	9-Aminoimidazo[4,5-*f*]quinazoline, A-60206
53449-49-3	Imidazo[4,5-*h*]quinazolin-6-one, I-60013
53449-52-8	Imidazo[4,5-*f*]quinazolin-9(8*H*)-one, I-60011
53515-94-9	4-(2-Phenylethyl)-1,2-benzenediol, P-60096
53518-15-3	7-Amino-4-(trifluoromethyl)-2*H*-1-benzopyran-2-one, A-60265
53537-91-0	4,4-Dimethylbicyclo[3.2.1]octane-2,3-dione; (±)-form, *in* D-60410
53549-12-5	4,4-Dimethylbicyclo[3.2.1]octane-2,3-dione; (1*S*)-form, *in* D-60410
53578-15-7	Cubanecarboxylic acid, C-60174
53585-93-6	2-Cyclohexyl-2-hydroxyacetic acid; (*R*)-form, *in* C-60212
53601-99-3	4,5,6,7-Tetrahydroisoxazolo[4,5-*c*]pyridin-3-ol; B,HBr, *in* T-60071
53602-00-9	4,5,6,7-Tetrahydroisoxazolo[4,5-*c*]pyridin-3-ol, T-60071
53636-08-1	Dithiocarbazic acid; Hydrazine salt, *in* D-60509
53658-78-9	3,3-Dimethylbicyclo[2.2.2]oct-5-en-2-ene, D-60412
53681-67-7	4′,7-Dihydroxyisoflavone; 4′,7-Di-*O*-β-D-glucopyranosyl, *in* D-60340
53755-85-4	2-Phenyl-1,3-benzoxathiol-1-ium; Perchlorate, *in* P-60084
53798-51-9	2,5-Dihydroxy-3-methylpentanoic acid; (2*S*,3*R*)-form, *in* D-60353
53798-95-1	2-Tetrazolin-5-one; 1-Benzyl, *in* T-60175
53822-96-1	Axisonitriles; Axisonitrile-1, *in* A-60328
53822-97-2	Axisothiocyanates; Axisothiocyanate-1, *in* A-60329
53870-24-9	2,3-Dichloroquinoxaline; 1-Oxide, *in* D-60150
53873-12-4	1-(3-Pyridinyl)-2-propanone; Picrate, *in* P-60230
53897-99-7	1*H*-1,2,3-Triazole-4-carboxylic acid; (2*H*)-form, Amide, *in* T-60239
53978-11-3	2-Amino-1,4-cyclohexanedicarboxylic acid; (1*RS*,2*RS*,4*RS*)-form, Di-Et ester, *in* A-60118
53978-12-4	2-Amino-1,4-cyclohexanedicarboxylic acid; (1*RS*,2*RS*,4*SR*)-form, Di-Et ester, *in* A-60118
53978-13-5	2-Amino-1,4-cyclohexanedicarboxylic acid; (1*RS*,2*SR*,4*SR*)-form, Di-Et ester, *in* A-60118
53978-17-9	2-Amino-1,4-cyclohexanedicarboxylic acid; (1*RS*,2*SR*,4*RS*)-form, Di-Et ester, *N*-benzoyl, *in* A-60118
53978-31-7	2-Amino-1,4-cyclohexanedicarboxylic acid; (1*RS*,2*RS*,4*RS*)-form, Di-Et ester, B,HCl, *in* A-60118
53978-32-8	2-Amino-1,4-cyclohexanedicarboxylic acid; (1*RS*,2*RS*,4*SR*)-form, Di-Et ester, B,HCl, *in* A-60118
53978-33-9	2-Amino-1,4-cyclohexanedicarboxylic acid; (1*RS*,2*SR*,4*SR*)-form, Di-Et ester, B,HCl, *in* A-60118
53984-36-4	3-Chloro-5-hydroxybenzoic acid, C-60092
53990-67-3	Octahydro-2,7-dihydroxy-5*H*,10*H*-dipyrrolo[1,2-*a*:1′,2′-*d*]pyrazine-5,10-dione; (2*R*,5a*R*,7*R*,10a*S*)-form, *in* O-60013
53990-68-4	Octahydro-2,7-dihydroxy-5*H*,10*H*-dipyrrolo[1,2-*a*:1′,2′-*d*]pyrazine-5,10-dione; (2*R*,5a*R*,7*R*,10a*R*)-form, *in* O-60013
53990-71-9	Octahydro-5*H*,10*H*-dipyrrolo[1,2-*a*:1′,2′-*d*]-pyrazine-5,10-dione; (5a*R*,10a*R*)-form, *in* O-60016
53990-72-0	Octahydro-5*H*,10*H*-dipyrrolo[1,2-*a*:1′,2′-*d*]-pyrazine-5,10-dione; (5a*R*,10a*S*)-form, *in* O-60016
54002-75-4	Octahydro-2,7-dihydroxy-5*H*,10*H*-dipyrrolo[1,2-*a*:1′,2′-*d*]pyrazine-5,10-dione; (2*R*,5a*S*,7*S*,10a*R*)-form, *in* O-60013
54022-19-4	Diazocyclopropane, D-60060
54053-93-9	2,3,14,20,22,25-Hexahydroxy-7-cholesten-6-one; (2β,3α,5β,2o*R*,22*R*)-form, *in* H-60063
54060-73-0	2-(9-Anthracenyl)ethanol, A-60282
54070-65-4	Biuret; 1,1,3,5-Tetra-Me, *in* B-60190
54147-74-9	2,2′-Diaminobiphenyl; 2-*N*-Ac, *in* D-60033
54166-20-0	3,3-Diphenylcyclobutanone, D-60476
54255-13-9	1-(4,5-Dihydro-1*H*-imidazol-2-yl)-2-imidazolidinethione; B,2HCl, *in* D-60246
54290-13-0	2,5-Dimethyl-3-(2-phenylethenyl)pyrazine; (*E*)-form, *in* D-60442
54302-42-0	Gossypol; (±)-form, 6-Me ether, *in* G-60032
54338-82-8	Bicyclo[4.2.0]octane-2,5-dione; (1*RS*,2*SR*)-form, *in* B-60093
54338-83-9	Bicyclo[4.2.0]oct-7-ene-2,5-dione; (1*RS*,5*SR*)-form, *in* B-60097
54354-35-7	1-(1-Propenyl)cyclohexene; (*E*)-form, *in* P-60178
54365-83-2	1,5,9-Triazacyclotridecane, T-60233
54378-77-7	1,3-Difluoro-5-(methylthio)benzene, *in* D-60176
54378-78-8	1,4-Difluoro-2-(methylthio)benzene, *in* D-60173
54396-36-0	2-Thiabicyclo[2.2.1]heptan-3-one, T-60186
54397-83-0	12-Hydroxy-5,8,10,14-eicosatetraenoic acid; (5*Z*,8*Z*,10*E*,12*S*,14*Z*)-form, *in* H-60131
54415-44-0	5-Acetylisoquinoline, A-60042
54532-45-5	Amanitins; Amanullinic acid, *in* A-60084
54629-30-0	10*H*-[1]Benzopyrano[3,2-*c*]pyridin-10-one, B-60042
54707-49-2	Grevilline D, *in* G-60038
54730-18-6	Bromomethanesulfonic acid; Bromide, *in* B-60291
54808-16-1	8-Bromo-2-hydroxy-1,4-naphthoquinone, B-60274
54808-30-9	2-Hydroxy-3-nitro-1,4-naphthoquinone, H-60195
54819-17-9	1*H*-Phenothiazin-1-one, P-60078
54856-83-6	2-Dimethylamino-3,3-dimethylazirine, D-60402
54863-75-1	13-Heptacosene; (*Z*)-form, *in* H-60019
54927-87-6	Teucrin E, T-60180
54931-11-2	1-Amino-2(1*H*)-pyridinone, A-60248
54980-22-2	Physalin D, P-60145
54986-14-0	Tetrazole-5-thione; 1,4-*Dihydro-form*, 1,4-Di-Me, *in* T-60174
55035-85-3	Gibberellin A₅₅, *in* G-60018
55215-60-6	Trifluoropyrazine, T-60306
55304-68-2	Senepoxide; (±)-form, *in* S-60023
55400-68-5	3-(Trifluoromethyl)benzenesulfonic acid; Et ester, *in* T-60283
55479-94-2	2-(Hydroxymethyl)benzaldehyde, H-60171
55483-01-7	5,7-Dihydroxy-6-methyl-1(3*H*)-isobenzofuranone, D-60349
55504-08-0	*N*-Methylbenzenecarboximidic acid; (*E*)-form, Me ester, *in* M-60045
55532-07-5	2,2-Dimethyl-5-hexen-3-one, D-60425
55679-31-7	Bicyclo[3.2.1]octan-8-one, B-60094
55704-78-4	2,5-Dimethyl-1,4-dithiane-2,5-diol, D-60421
55717-45-8	2-Bromo-5-hydroxypyridine, B-60276
55717-46-9	2-Amino-5-hydroxypyridine, A-60201
55747-78-9	Ferulidene, *in* X-60002
55790-78-8	4,5-Dihydro-3,3,5,5-tetramethyl-4-methylene-3*H*-pyrazole, D-60283
55790-79-9	3,3,5,5-Tetramethyl-4-pyrazolidinone, T-60153
55812-47-0	Gibberellin A₄₅, *in* G-60018
55890-24-9	Austrobailignan-6; (2*R*,3*R*)-form, *in* A-60320
55901-80-9	2,3-Dichloro-4-methoxybenzoic acid, *in* D-60133
55903-97-4	Octahydro-2-hydroxy-5*H*,10*H*-dipyrrolo[1,2-*a*:1′,2′-*d*]pyrazine-5,10-dione; (2*S*,5a*R*,10a*S*)-form, *in* O-60017
55907-33-0	Axisonitriles; Axisonitrile-2, *in* A-60328
55993-21-0	9,9′-Bi(bicyclo[3.3.1]nonylidene), B-60079
56003-01-1	5,7-Dihydroxy-3′,4′,6,8-tetramethoxyflavone, D-60376
56012-79-4	Disidein, D-60499
56012-89-6	Axamide-2, A-60327
56012-90-9	Axisothiocyanates; Axisothiocyanate-2, *in* A-60329
56187-03-2	1,4,7-Triazacyclododecane; B,3HCl, *in* T-60231
56221-26-2	2,3-Dihydroxy-6-iodobenzoic acid; Di-Me ether, Me ester, *in* D-60330
56221-41-1	6-Iodo-2,3-dimethoxybenzoic acid, *in* D-60330
56234-20-9	8-Acetylquinoline, A-60055
56257-59-1	▷Altertoxin II, A-60081
56283-44-4	1(10),4-Germacradiene-6,8-diol; (1*E*,4*E*,6*R*,8*R*)-form, *in* G-60014

59989-18-3	5-Ethynyl-2,4(1*H*,3*H*)-pyrimidinedione, E-60060
59995-47-0	4-Hydroxy-2-cyclopenten-1-one; (*R*)-*form*, *in* H-60116
59995-48-1	4-Hydroxy-2-cyclopenten-1-one; (*R*)-*form*, Ac, *in* H-60116
59995-49-2	4-Hydroxy-2-cyclopenten-1-one; (*S*)-*form*, *in* H-60116
60049-36-7	2-Amino-3-methyl-3-butenoic acid; (±)-*form*, A-60218
60064-29-1	6-Aminoimidazo[4,5-*g*]quinolin-8(7*H*)-one, A-60209
60077-57-8	4-Hydroxy-5-benzofurancarboxylic acid; Me ester, *in* H-60103
60103-01-7	2-Amino-3-methyl-3-butenoic acid; (*R*)-*form*, *in* A-60218
60189-62-0	*lin*-Benzoadenosine, *in* A-60207
60189-64-2	Imidazo[4,5-*g*]quinazoline-6,8(5*H*,7*H*)-dione, I-60009
60189-88-0	8-Aminoimidazo[4,5-*g*]quinazoline; 1-(β-D-Ribofuranosyl), *in* A-60207
60239-18-1	1,4,7,10-Tetraazacyclododecane-1,4,7,10-tetraacetic acid, T-60015
60239-22-7	1,4,8,11-Tetraazacyclotetradecane-1,4,8,11-tetraacetic acid, T-60017
60268-40-8	Hanphyllin, H-60006
60297-83-8	Licarin *C*, L-60024
60302-27-4	2,2,3,3,4,4-Hexamethylpentane, H-60071
60338-65-0	Thieno[3,2-*c*]cinnoline, T-60203
60454-77-5	▷Funkioside C, *in* S-60045
60454-78-6	▷Funkioside D, *in* S-60045
60454-79-7	Funkioside E, *in* S-60045
60454-80-0	▷Funkioside F, *in* S-60045
60454-81-1	▷Funkioside G, *in* S-60045
60466-50-4	Bromomethyl phenyl selenide, B-60300
60498-89-7	Gnididione, G-60030
60529-76-2	Thymopoietin, T-60219
60549-48-6	[2.2](2,6)Azulenophane; *anti-form*, *in* A-60355
60583-16-6	5,8-Dihydroxy-2,3,6,7-tetramethyl-1,4-naphthoquinone, D-60377
60595-16-6	(Phenylthio)nitromethane, P-60137
60705-54-6	1-Bromo-3-nonene; (*Z*)-*form*, *in* B-60310
60723-27-5	▷Physoperuvine, P-60151
60727-70-0	1-Acetylcyclooctene; (*E*)-*form*, *in* A-60027
60811-22-5	4-Chloro-3-fluorobenzenethiol, C-60071
60811-23-6	3-Chloro-4-fluorobenzenethiol, C-60069
60811-24-7	3,4-Difluorobenzenethiol, D-60175
60814-30-4	4-Acetylquinoline, A-60051
60828-09-3	13(18)-Oleanene-2,3-diol; (2β,3β)-*form*, *in* O-60033
60958-71-6	Mikrolin, M-60135
60958-72-7	Dechloromikrolin, *in* M-60135
61047-38-9	10,11-Dihydro-5*H*-dibenzo[*a,d*]cyclohepten-5-one; Hydrazone, *in* D-60218
61062-50-8	4-*tert*-Butylcyclohexene; (*R*)-*form*, *in* B-60343
61082-23-3	3-Vinylcyclopropene, V-60012
61117-54-2	Hexahydro-3-(1-methylpropyl)pyrrolo[1,2-*a*]pyrazine-1,4-dione; (3*R*,8a*S*,1′*S*)-*form*, *in* H-60052
61117-55-3	Hexahydro-3-(1-methylpropyl)pyrrolo[1,2-*a*]pyrazine-1,4-dione; (3*R*,8a*S*,1′*R*)-*form*, *in* H-60052
61135-33-9	2′-Deoxy-5-ethynyluridine, *in* E-60060
61203-48-3	2-Iodo-4,5-dimethoxybenzoic acid, *in* D-60338
61276-36-6	2,8-Dihydroxy-1,4-naphthoquinone; Di-Ac, *in* D-60355
61305-27-9	4-Hydroxy-2-cyclopenten-1-one, H-60116
61348-75-2	2-Amino-3-methyl-3-butenoic acid; (±)-*form*, B,HCl, *in* A-60218
61348-76-3	2-Amino-3-methyl-3-butenoic acid; (*R*)-*form*, B,HCl, *in* A-60218
61348-78-5	2-Amino-3-methyl-3-butenoic acid; (*R*)-*form*, Me ester, *in* A-60218
61366-76-5	2-Mercaptopropanoic acid; (±)-*form*, Me ester, *S*-Me ether, *in* M-60027
61376-23-6	2-Amino-3-methyl-3-butenoic acid; (*S*)-*form*, *in* A-60218
61376-24-7	2-Amino-3-methyl-3-butenoic acid; (*S*)-*form*, B,HCl, *in* A-60218
61456-87-9	4-Cyano-1-phenyl-1,2,3-triazole, *in* P-60138
61468-81-3	4,5-Dihydro-3(2*H*)-pyridazinone, D-60272
61481-40-1	γ-Glutamylmarasmine, G-60026
61485-47-0	Ethanebis(dithioic)acid; Di-Me ester, *in* E-60040
61494-52-8	1-Pyrenesulfonic acid; Chloride, *in* P-60210
61521-39-9	2-Amino-3,3-dimethylbutanedioic acid; (±)-*form*, *in* A-60147
61566-58-3	2,4-Diaminobenzoic acid; B,HCl, *in* D-60032
61650-52-0	2-(3-Oxo-1-propenyl)benzaldehyde; (*E*)-*form*, *in* O-60091
61719-58-2	1,3-Dihydro-2*H*-imidazo[4,5-*c*]pyridin-2-one; 1-Benzyl, *in* D-60250
61740-29-2	4-Hydroxy-2-cyclopenten-1-one; (±)-*form*, *in* H-60116
61752-45-2	1-Phenyl-3-penten-1-one; (*Z*)-*form*, *in* P-60127
61760-11-0	2-(2,2-Dimethylpropyl)naphthalene, D-60447
61782-63-6	2,2-Dibromocyclopropanecarboxaldehyde, D-60092
61826-89-9	Helichrysin, *in* H-60220
61836-40-6	4,5-Dimethoxy-1-naphthol, *in* N-60005
61836-58-6	3,5-Dihydroxy-2-methyl-1,4-naphthoquinone; Di-Ac, *in* D-60350
61871-80-5	2′-Amino-2-chloro-5′-methylacetophenone, A-60114
61883-14-5	3-Amino-1,2-cyclohexanedicarboxylic acid; (1*RS*,2*RS*,3*RS*)-*form*, Di-Et ester, *in* A-60119
61883-62-3	3-Morpholinone; 4-Benzoyl, *in* M-60147
61885-53-8	Bicyclo[4.1.1]octa-2,4-diene, B-60090
61885-54-9	Bicyclo[4.1.1]oct-3-ene, B-60096
61906-95-4	13-Heptacosene, H-60019
61915-20-6	3-Amino-1,2-cyclohexanedicarboxylic acid; (1*RS*,2*SR*,3*RS*)-*form*, Di-Et ester, *in* A-60119
61915-22-8	3-Amino-1,2-cyclohexanedicarboxylic acid; (1*RS*,2*SR*,3*SR*)-*form*, Di-Et ester, *in* A-60119
61978-16-3	3-Methyl-1,2-naphthalenediol, M-60093
61978-33-4	6-Methyl-1,2-naphthalenediol, M-60095
61985-32-8	4(5)-Benzoylimidazole, B-60057
62008-04-2	Heteronemin, H-60030
62012-57-1	4,4′-Diiodo-1,1′-binaphthyl, D-60386
62013-43-8	5-Nitrohistidine; (*S*)-*form*, Me ester; B,HCl, *in* N-60028
62023-30-7	3-Amino-2-hydroxy-5-methylhexanoic acid, A-60188
62076-67-9	2-Amino-3-chlorobutanoic acid; (2*RS*,3*SR*)-*form*, Me ester, *in* A-60103
62076-72-6	2-Amino-3-chlorobutanoic acid; (2*RS*,3*SR*)-*form*, *N*-Benzoyl, *in* A-60103
62076-81-7	2-Amino-3-chlorobutanoic acid; (2*RS*,3*RS*)-*form*, Me ester, *in* A-60103
62078-10-8	Axisonitriles; Axisonitrile-4, *in* A-60328
62078-11-9	Axisothiocyanates; Axisothiocyanate-4, *in* A-60329
62114-58-3	Dibenzo[*a,d*]cycloheptenylium; Perchlorate, *in* D-60071
62171-59-9	1-*tert*-Butyl-2-iodobenzene, B-60346
62289-79-6	2-Phenyl-2*H*-1,2,3-triazole-4-carboxylic acid; Me ester, *in* P-60140
62316-29-4	3-Chloro-4-methoxy-5-methylbenzoic acid, *in* C-60096
62322-48-9	5-Methyl-2(5*H*)-furanone; (*R*)-*form*, *in* M-60083
62348-13-4	5-Isoxazolecarboxylic acid; Chloride, *in* I-60142
62365-78-0	3-Ethynylindole, E-60058
62393-64-0	4-Isopropenylcyclohexene; (*R*)-*form*, *in* I-60117
62438-05-5	1-Amino-2(1*H*)-pyridinone; B,HCl, *in* A-60248
62452-74-8	2-Bromo-3-ethoxy-1,4-naphthoquinone, *in* B-60268
62497-62-5	Antibiotic PDE I, A-60286
62502-14-1	Calopogonium isoflavone *B*, C-60006
62681-14-5	Tetrazole-5-thione; 1,3-*Dihydro-form*, 3-Et, 1-Ph, *in* T-60174
62690-65-7	1,2,3,5,8,8a-Hexahydronaphthalene, H-60053
62690-66-8	1,2,3,7,8,8a-Hexahydronaphthalene, H-60054
62708-42-3	2-(2-Propenyl)benzaldehyde, P-60176
62785-91-5	Furo[2,3-*d*]pyrimidin-2(1*H*)-one, F-60106
62835-95-4	2-(2-Methylphenyl)propanoic acid, M-60113
62928-74-9	Pentacyclo[5.3.0.0²,⁵.0³,⁹.0⁴,⁸]decan-6-one; (+)-*form*, *in* P-60027
62952-40-3	3-Oxo-1-cyclohexene-1-carboxaldehyde, O-60060

620

84633-06-7	Abbreviatin PB, A-60001
84673-59-6	2-Amino-5-hexenoic acid; Nitrile, *in* A-60169
84743-77-1	2-Bromo-1,3,5-benzenetriol, B-60199
84800-12-4	2-Mercapto-3-phenylpropanoic acid; (*R*)-*form*, *in* M-60026
84871-06-7	Papakusterol, P-60004
84996-45-2	1,3-Dihydro-1-methyl-2*H*-imidazo[4,5-*b*]pyrazin-2-one, *in* I-60006
84996-46-3	1,3-Dihydro-1-methyl-2*H*-imidazo[4,5-*b*]pyrazine-2-thione, *in* D-60247
84996-47-4	1-Methyl-2-methylthio-1*H*-imidazo[4,5-*b*]pyrazine, *in* D-60247
84996-48-5	2-(Methylthio)-1*H*-imidazo[4,5-*b*]pyrazine, *in* D-60247
84996-49-6	4-Methyl-2-methylthio-4*H*-imidazo[4,5-*b*]pyrazine, *in* D-60247
84996-50-9	1,3-Dihydro-1,3-dimethyl-2*H*-imidazo[4,5-*b*]-pyrazine-2-thione, *in* D-60247
84996-53-2	1,3-Dihydro-1,3-dimethyl-2*H*-imidazo[4,5-*b*]-pyrazin-2-one, *in* I-60006
84996-54-3	2-Methoxy-4-methyl-4*H*-imidazo[4,5-*b*]pyrazine, *in* I-60006
84996-55-4	1,4-Dihydro-1,4-dimethyl-2*H*-imidazo[4,5-*b*]-pyrazin-2-one, *in* I-60006
85179-07-3	Lapidolinin, *in* L-60015
85179-08-4	Lapidolin, *in* L-60015
85202-11-5	Ochromycinone methyl ether, *in* O-60003
85202-12-6	Lapidolinol, L-60015
85216-79-1	2-Tetradecyloxiranecarboxylic acid; (±)-*form*, Na salt, *in* T-60041
85282-84-4	Bis(quinuclidine)bromine(1+); Bromide, *in* B-60173
85282-86-6	Bis(quinuclidine)bromine(1+); Tetrafluoroborate, *in* B-60173
85288-97-7	10-Hydroxy-11-dodecenoic acid, H-60125
85288-98-8	12-Hydroxy-13-tetradecenoic acid, H-60229
85288-99-9	12-Hydroxy-10-dodecenoic acid, H-60126
85289-00-5	14-Hydroxy-12-tetradecenoic acid, H-60230
85330-62-7	2,5-Bis(methylthio)pyridine, *in* P-60220
85330-63-8	2,5-Bis(methylsulfonyl)pyridine, *in* P-60220
85330-79-6	2,3-Bis(methylsulfonyl)pyridine, *in* P-60218
85385-68-8	1,6:7,12-Bismethano[14]annulene; *syn-form*, *in* B-60165
85386-94-3	2-Chloro-3-fluoropyridine; *N*-Oxide, *in* C-60081
85440-62-6	1,6:7,12-Bismethano[14]annulene; *anti-form*, *in* B-60165
85452-72-8	3,4-Dihydroxyphenylacetaldehyde; Di-Me ether, di-Me acetal, *in* D-60371
85515-06-6	6-Bromotryptophan; (±)-*form*, *N*α-Ac, *in* B-60334
85526-81-4	10(14)-Aromadendrene-4,8-diol; (4β,8α)-*form*, *in* A-60299
85531-49-3	[4,4'-Bipyridine]-2,2'-dicarboxylic acid, B-60136
85545-06-8	2,2,5-Trimethylcyclopentanecarboxylic acid; (1*RS*,5*SR*)-*form*, *in* T-60360
85545-07-9	2,2,5-Trimethylcyclopentanecarboxylic acid; (1*RS*,5*SR*)-*form*, Amide, *in* T-60360
85552-23-4	Wedeliasecokaurenolide, W-60004
85564-03-0	Teupyrenone, T-60184
85570-50-9	Dimethylenebicyclo[1.1.1]pentanone, D-60422
85620-94-6	2-Azido-2-nitropropane, A-60349
85620-95-7	2,2-Diazidopropane, D-60058
85814-86-4	3-Hydroxycyclononanone, H-60113
85846-83-9	5-Hydroxymethyl-2(5*H*)-furanone; (*S*)-*form*, Ac, *in* H-60182
85960-33-4	Cyclohept[*fg*]acenaphthylene-5,6-dione, C-60194
85960-34-5	Cyclohept[*fg*]acenaphthylene-5,8-dione, C-60195
85963-75-3	2,2,4-Tetrafluoro-1,3-dithietane; 1-Oxide, *in* T-60048
85963-76-4	2,2,4,4-Tetrafluoro-1,3-dithietane; 1,1,3-Trioxide, *in* T-60048
86229-97-2	RA-VII, *in* R-60002
86379-88-6	Octahydro-7*H*-1-benzopyran-7-one; (4a*RS*,8a*RS*)-*form*, *in* O-60011
86402-39-3	Cleistanthoside A, *in* D-60493
86447-23-6	1,2,3,6-Tetrahydro-3-pyridinecarboxylic acid, T-60084
86447-29-2	1,2,3,6-Tetrahydro-3-pyridinecarboxylic acid; (±)-*form*, *N*-Me, *in* T-60084
86450-76-2	Shanzhiside; 6-(4-Hydroxy-3,5-dimethoxybenzoyl), 8-Ac, Me ester, *in* S-60028
86456-69-1	6-Bromo-1-(bromomethyl)naphthalene, B-60220
86474-77-3	5-Methylnaphtho[2,1-*b*]thiophene, M-60099
86474-78-4	5-Methylnaphtho[2,1-*b*]thiophene; Picrate, *in* M-60099
86474-80-8	4-Methylnaphtho[2,1-*b*]thiophene; Picrate, *in* M-60098
86474-81-9	2-Methylnaphtho[2,1-*b*]thiophene; Picrate, *in* M-60097
86475-22-1	Pentamethoxybenzaldehyde, *in* P-60042
86544-37-8	2-Hydroxy-2-phenylacetaldehyde; (*S*)-*form*, Me ether, *in* H-60210
86546-83-0	9(11)-Drimen-8-ol; (5*R*,8*S*,10*R*)-*form*, *in* D-60520
86546-84-1	9(11)-Drimen-8-ol; (5*R*,8*R*,10*R*)-*form*, *in* D-60520
86582-92-5	Clavularin *A*, C-60153
86630-47-9	1,5-Dimethoxy-2,7-phenanthrenediol, *in* T-60116
86632-29-3	2,2'-Diiodo-1,1'-binaphthyl; (±)-*form*, *in* D-60385
86635-85-0	2,4-Dihydroxy-3-iodobenzoic acid, D-60331
86685-13-4	5-Amino-3-phenylisoxazole; *N*-Formyl, *in* A-60240
86685-14-5	5-Amino-3-phenylisoxazole; *N*-Me, *in* A-60240
86685-93-0	5-Amino-3-phenylisoxazole; *N*-Et, *in* A-60240
86688-06-4	2,2'-Diiodo-1,1'-binaphthyl; (*R*)-*form*, *in* D-60385
86688-08-6	2,2'-Dibromo-1,1'-binaphthyl; (*R*)-*form*, *in* D-60089
86690-14-4	Halenaquinone, *in* H-60002
86702-02-5	Magnoshinin, M-60007
86746-90-9	Clavularin B, *in* C-60153
86747-02-6	β-Senepoxide, *in* S-60023
86810-04-0	*N*-Alanylcysteine; L-L-*form*, Boc-Ala-Cys-OMe, *in* A-60069
86849-13-0	RA-IV, *in* R-60002
86963-85-1	2,3,5-Tris(methylene)bicyclo[2.2.1]heptane, T-60403
86988-90-1	4-Chloro-3-methylfuroxan, *in* C-60114
86989-08-4	Isomyomontanone, *in* M-60155
86989-09-5	Myomontanone, M-60155
87108-78-9	2,2,4,4-Tetrakis(trifluoromethyl)-1,3-dithietane; 1,3-Dioxide, *in* T-60129
87121-83-3	Benz[*a*]azulene-1,4-dione, B-60011
87206-01-7	*N*-(Tetrahydro-2-oxo-3-furanyl)hexadecanamide, *in* A-60138
87216-46-4	1,4-Dihydro-3*H*-2-benzothiopyran-3-one, D-60206
87216-47-5	1,4-Dihydro-3*H*-2-benzoselenin-3-one, D-60204
87216-48-6	1,4-Dihydro-3*H*-2-benzotellurin-3-one, D-60205
87219-30-5	3-Aminodihydro-2(3*H*)-furanone; (*S*)-*form*, *N*-Benzoyl, *in* A-60138
87238-75-3	Tricyclo[5.2.1.0²,⁶]dec-2(6)-ene, T-60260
87255-31-0	3-Amino-5-hexenoic acid, A-60172
87402-72-0	9,10-Dihydro-2,7-dimethoxy-1,5-phenanthrenediol, *in* T-60116
87413-09-0	1,1,1-Triacetoxy-1,1-dihydro-1,2-benziodoxol-3(1*H*)-one, T-60228
87441-73-4	11,13-Dehydroeriolin, *in* E-60036
87462-15-5	5-Methyl-1*H*-pyrrole-2-carboxylic acid; Benzyl ester, *in* M-60124
87480-58-8	1-Amino-2-ethylcyclopropanecarboxylic acid, A-60151
87512-32-1	3-Amino-2-hexenoic acid; *tert*-Butyl ester, *in* A-60170
87533-53-7	1,3-Diacetyl-5-iodobenzene, D-60024
87573-88-4	5-Acetyl-2(1*H*)-pyrimidinone, A-60048
87596-55-2	Nectriafurone, N-60016
87619-34-9	1-Bromo-2,3-dimethylcyclopropene, B-60235
87625-62-5	▷Ptaquiloside, P-60190
87656-32-4	2-(Dimethoxymethyl)benzenemethanol, *in* H-60171
87667-46-7	Fusarubin methyl acetal, *in* F-60108
87701-33-5	Ptaquiloside; Tetra-Ac, *in* P-60190
87842-36-2	2,2,5,5-Tetramethylcyclopentaneselone, T-60137
87999-44-8	1-Azidodiamantane, A-60341
87999-45-9	4-Azidodiamantane, A-60343

103258-84-0	2H-1,5,2,4-Dioxadiazine-3,6(4H)dione; 2,4-Di-*tert*-butyl, *in* D-60466
103259-13-8	1,8,14-Triptycenetricarboxylic acid, T-60399
103259-15-0	1,8,11-Triptycenetricarboxylic acid, T-60398
103259-19-4	1,8,11-Triptycenetricarboxylic acid; Tri-Me ester, *in* T-60398
103259-20-7	1,8,14-Triptycenetricarboxylic acid; Tri-Me ester, *in* T-60399
103322-24-3	2,6-Dihydro-2,6-dimethyleneazulene, D-60226
103348-91-0	7-Aminobicyclo[4.1.0]heptane-7-carboxylic acid; *trans-form*, *in* A-60095
103367-32-4	5,6-Dihydroxyhexanoic acid; (S)-*form*, Me ester, *in* D-60329
103367-47-1	3,4,4-Triphenyl-2-cyclohexen-1-one, T-60379
103367-48-2	3,4,4-Triphenyl-2-cyclohexen-1-one; Tosylhydrazone, *in* T-60379
103367-50-6	4,5,5-Triphenyl-2-cyclohexen-1-one; (±)-*form*, *in* T-60381
103367-54-0	3,5,5-Triphenyl-2-cyclohexen-1-one, T-60380
103367-60-8	3-Methyl-4,4-diphenylcyclohexene; (±)-*form*, *in* M-60066
103438-45-5	2,6-Diamino-1,5-dihydro-4H-imidazo[4,5-c]pyridin-4-one, D-60034
103438-47-7	2,6-Diamino-1,5-dihydro-4H-imidazo[4,5-c]pyridin-4-one; Methanesulfonyl, *in* D-60034
103438-84-2	2-Fluoro-5-hydroxybenzaldehyde, F-60036
103438-85-3	4-Fluoro-3-hydroxybenzaldehyde, F-60037
103438-86-4	2-Fluoro-3-hydroxybenzaldehyde, F-60035
103438-90-0	2-Fluoro-3-hydroxybenzaldehyde; Benzyl ether, *in* F-60035
103438-91-1	4-Fluoro-3-hydroxybenzaldehyde; Benzyl ether, *in* F-60037
103438-92-2	2-Fluoro-5-hydroxybenzaldehyde; Benzyl ether, *in* F-60036
103457-83-6	Nonamethylcyclopentanol, N-60057
103457-84-7	Undecamethylcyclohexanol, U-60003
103457-85-8	1,1,2,2,3,3,4,4-Octamethyl-5-methylenecyclopentane, O-60027
103457-86-9	1,1,2,2,3,3,4,4,5,5-Decamethyl-6-methylenecyclohexane, D-60012
103495-77-8	1,1,3,3,5,5-Hexamethyl-2,4,6-tris(methylene)cyclohexane, H-60072
103495-83-6	1,1,2,2,3,3-Hexamethylcyclohexane, H-60070
103499-08-7	5-Phenyl-1,2,4-oxadiazole-3-carboxaldehyde; Oxime, *in* P-60122
103499-59-8	2-Mercapto-3-phenylpropanoic acid; (R)-*form*, Me ester, *in* M-60026
103499-61-2	2-Mercapto-3-methylbutanoic acid; (R)-*form*, S-Benzoyl, *in* M-60023
103500-26-1	3-(1-Aminocyclopropyl)-2-propenoic acid; (E)-*form*, N-Benzyloxycarbonyl, *in* A-60123
103500-27-2	3-(1-Aminocyclopropyl)-2-propenoic acid; (E)-*form*, B,HCl, *in* A-60123
103528-06-9	Antibiotic SF 2339, A-60289
103538-56-3	13², 17³-Cyclopheophorbide enol, C-60230
103547-83-7	2,2,3,3,4,4-Hexamethylcyclobutanol, H-60068
103547-87-1	2,2,3,3,4,4,5,5-Octamethylcyclopentanol, O-60024
103553-93-1	Ebuloside, E-60001
103562-47-6	3-Mercaptocyclobutanol; *trans-form*, *in* M-60019
103562-48-7	3-Mercaptocyclobutanol; *cis-form*, *in* M-60019
103562-49-8	3-Mercaptocyclobutanol; *trans-form*, O-(4-Methylbenzenesulfonyl), *in* M-60019
103562-50-1	3-Mercaptocyclobutanol; *cis-form*, O-(4-Methylbenzenesulfonyl), *in* M-60019
103562-67-0	3-(Acetylthio)cyclobutanol, *in* M-60019
103562-70-5	Bis(3-hydroxycyclobutyl)disulfide, *in* M-60019
103562-73-2	1,3-Bis(allylthio)cyclobutane, *in* C-60184
103562-76-1	1,3-Cyclobutanedithiol; *cis-form*, *in* C-60184
103616-07-5	2-Mercaptopropanoic acid; (R)-*form*, Et ester, *in* M-60027
103616-08-6	2-Mercapto-2-phenylacetic acid; (S)-*form*, *in* M-60024
103620-96-8	4,5-Dihydro-3,3,4,4-tetramethyl-2(3H)-thiophenone, D-60288
103620-99-1	Dihydro-3,3,4,4-tetramethyl-2(3H)-furanthione, D-60282
103621-02-9	4,5-Dihydro-3,3,4,4-tetramethyl-2(3H)-thiophenethione, D-60287
103654-23-5	1,5-Dihydroxy-1,7-diphenyl-3-heptanone, *in* D-60480
103654-24-6	Yashabushiketodiol B, *in* D-60480
103654-25-7	1,7-Diphenyl-1,3,5-heptanetriol; (1R,3R,5S)-*form*, *in* D-60480
103654-30-4	Artelein, A-60303
103654-31-5	8-Carotene-4,6,10-triol, C-60021
103665-30-1	[2.2](4,7)(7,4)Indenophane, I-60021
103668-59-3	1-Bromo-(2-dimethoxymethyl)naphthalene, *in* B-60302
103668-60-6	2-(Dimethoxymethyl)-1-naphthalenecarboxaldehyde, *in* N-60001
103692-62-2	9,18-Diphenylphenanthro[9,10-b]triphenylene, D-60490
103722-81-2	5-Decene-2,8-diyne; (E)-*form*, *in* D-60015
103733-33-1	1,2,3,4-Tetrahydro-1-isoquinolinecarboxylic acid; (±)-*form*, Et ester; B, HCl, *in* T-60070
103744-81-6	Rehmaglutin C, R-60003
103782-08-7	Allosamidin, A-60076
103794-87-2	Tri-*tert*-butylazete, T-60249
103794-88-3	Tri-*tert*-butyl-1,2,3-triazine, T-60250
103794-91-8	1-Thia-2-cyclooctyne, T-60189
103884-21-5	7-Aminoimidazo[4,5-f]quinazolin-9(8H)-one, A-60208
103884-30-6	7-Aminoimidazo[4,5-f]quinazolin-9(8H)-one; B,2HCl, *in* A-60208
103960-12-9	b,b',b''-Tritriptycene, T-60407
103980-84-3	1-Bromo-1,3-pentadiene; (E,E)-*form*, *in* B-60311
103992-91-2	3α,15α,22S-Triacetoxy-7,9(11),24-lanostatrien-26-oic acid, *in* T-60329
104007-33-2	1,6-Dihydro-1,6-dimethyleneazulene, D-60225
104014-34-8	Dibenz[b,g]azocine-5,7(6H,12H)-dione; N-Benzyl, *in* D-60068
104014-40-6	Dibenz[b,g]azocine-5,7(6H,12H)-dione; N-Me, *in* D-60068
104014-78-0	3,5-Dihydro-3,5,5-trimethyl-4H-triazol-4-one, D-60303
104014-86-0	2,5-Dimethyl-1,4-benzenedithiol, D-60406
104019-20-7	7-Acetylindole, A-60036
104051-86-7	Tetrahydro-2(1H)-pyrimidinone; 1-β-D-Ribopyranosyl, *in* T-60088
104070-31-7	5-Chloro-3-methyl-1-pentene; (±)-*form*, *in* C-60117
104090-87-1	1-(2-Naphthyl)piperazine; B,HCl, *in* N-60014
104091-52-3	2,2-Dimethyl-5-cycloheptene-1,3-dione, D-60416
104113-37-3	2-Bromobicyclo[2.2.1]heptane-1-carboxylic acid; (1R,2R)-*form*, *in* B-60210
104113-38-4	2-Bromobicyclo[2.2.1]heptane-1-carboxylic acid; (1S,2S)-*form*, *in* B-60210
104113-47-5	5a,5b,11a,11b-Tetrahydrocyclobuta[1,2-b:4,3-b']benzothiopyran-11,12-dione; (5aα,5bα,11aα,11bβ)-*form*, *in* T-60063
104113-71-5	1-(1-Naphthyl)piperazine; B,HCl, *in* N-60013
104144-79-8	2,2'-Diisocyano-1,1'-binaphthyl; (R)-*form*, *in* D-60391
104144-80-1	2,2'-Diisocyano-1,1'-binaphthyl; (S)-*form*, *in* D-60391
104157-40-6	Methanesulfenyl thiocyanate, M-60032
104170-63-0	9-Benzylidene-1,3,5,7-cyclononatetraene, B-60069
104184-80-7	7,10:19,22:31,34-Triethenotribenzo[a,k,u]cyclotriacontane; (all-E)-*form*, *in* T-60273
104241-29-4	2-Amino-3-hydroxy-4-(4-hydroxyphenyl)butanoic acid; (2S,3R)-*form*, *in* A-60185
104286-26-2	Pentacyclo[6.4.0.0²,⁷.0³,¹².0⁶,⁹]dodeca-4,10-diene, P-60028
104291-39-6	4,4,4-Trifluoro-3-(trifluoromethyl)-2-butenal, T-60309

631

109163-59-9	*ent*-15-Oxo-1(10),13-halimadien-18-oic acid; (13*E*)-*form*, Me ester, *in* O-60069
109163-61-3	ent-1(10),13Z-Halimadiene-15,18-dioic acid, *in* O-60069
109163-62-4	ent-1(10)13E-Halimadiene-15,18-dioic acid, *in* O-60069
109163-64-6	*ent*-12-Hydroxy-13,14,15,16-tetranor-1(10)-halimen-18-oic acid; Ac, *in* H-60231
109163-65-7	*ent*-15-Oxo-1(10),13-halimadien-18-oic acid; (13*Z*)-*form*, *in* O-60069
109178-63-4	7*H*-2,3,4,6,7-Pentaazabenz[*de*]anthracene, P-60017
109179-31-9	2-Bromo-3-methylbenzaldehyde, B-60292
109181-77-3	*ent*-15-Oxo-1(10),13-halimadien-18-oic acid; (13*E*)-*form*, *in* O-60069
109181-97-7	6α-Butanoyloxy-17β-hydroxy-15,17-oxido-16-spongianone, *in* D-60365
109181-98-8	6α-Acetoxy-17β-hydroxy-15,17-oxido-16-spongianone, *in* D-60365
109181-99-9	Methyl 6α,15α,17β-triacetoxy-15,16-dideoxy-15,17-oxido-16-spongianoate, *in* T-60038
109194-67-4	2,3-Dihydro-2-(4-hydroxyphenyl)-5-(3-hydroxypropyl)-3-methylbenzofuran; (2*R*,3*R*)-*form*, *in* D-60244
109194-68-5	2-(4-Hydroxyphenyl)-7-methoxy-5-(1-propenyl)-3-benzofurancarboxaldehyde, H-60212
109194-69-6	2-(4-Hydroxyphenyl)-5-(1-propenyl)benzofuran, H-60214
109194-70-9	2-(4-Hydroxyphenyl)-7-methoxy-5-(1-propenyl)-benzofuran, *in* H-60214
109194-71-0	4-[5-(1-Propenyl)-2-benzofuranyl]-2,3-benzenediol, *in* H-60214
109194-72-1	2-(2,4-Dihydroxyphenyl)-7-methoxy-5-(1-propenyl)benzofuran, *in* H-60214
109194-73-2	1-(4-Hydroxyphenyl)-2-(4-propenylphenoxy)-1-propanol, H-60215
109194-74-3	Tetrahydrobis(4-hydroxyphenyl)-3,4-dimethylfuran; (2*R*,3*R*,4*S*,5*R*)-*form*, *in* T-60061
109194-75-4	Tetrahydro-2-(4-hydroxyphenyl)-5-(4-hydroxy-3-methoxyphenyl)-3,4-dimethylfuran, *in* T-60061
109202-01-9	4-Mercapto-2-butenoic acid; (*E*)-*form*, Benzyl ester, *in* M-60018
109202-04-2	4-Mercapto-2-butenoic acid; (*E*)-*form*, Nitrile, *in* M-60018
109217-15-4	Fontonamide, F-60067
109217-16-5	Anhydrohapaloxindole, A-60273
109225-42-5	2,3-Dihydro-2-(4-hydroxy-3-methoxyphenyl)-5-(3-hydroxypropyl)-3-methylbenzofuran, *in* D-60244
109225-43-6	1-(4-Hydroxyphenyl)-2-(2-methoxy-4-propenylphenoxy)-1-propanol, *in* H-60215
109237-00-5	2,3,12-Trihydroxy-4,7,16-pregnatrien-20-one; (2α,3β,12β)-*form*, *in* T-60346
109237-01-6	2,3,12-Trihydroxy-4,7-pregnadien-20-one; (2α,3β,12β)-*form*, *in* T-60345
109237-38-9	Chiratanin, C-60037
109274-90-0	3-Methylfuro[2,3-*b*]pyridine, M-60085
109274-92-2	Furo[2,3-*b*]pyridine-2-carboxaldehyde, F-60093
109274-95-5	2-Cyanofuro[2,3-*b*]pyridine, *in* F-60098
109274-96-6	3-Cyanofuro[2,3-*b*]pyridine, *in* F-60099
109274-98-8	Furo[2,3-*b*]pyridine-3-carboxylic acid; Amide, *in* F-60099
109274-99-9	Furo[2,3-*b*]pyridine-3-carboxaldehyde, F-60094
109275-00-5	Furo[2,3-*b*]pyridine-3-carboxaldehyde; Hydrazone, *in* F-60094
109280-45-7	Tetrahydrobis(4-hydroxyphenyl)-3,4-dimethylfuran; (2*R*,3*R*,4*S*,5*S*)-*form*, 3'-Methoxy, *in* T-60061
109291-64-7	15-Desacetyltetraneurin *C*, *in* T-60158
109291-66-9	15-Desacetyltetraneurin *C* isobutyrate, *in* T-60158
109296-83-5	3,4-Heptafulvalenedione, H-60022
109305-25-1	5,6,8,9-Tetraaza[3.3]paracyclophane, T-60018
109360-94-3	Alloaromadendrane-4,10-diol; (4α,10β)-*form*, *in* A-60074
109517-68-2	Crotalarin, C-60171
109517-69-3	Crotarin, C-60172
109517-70-6	Cavoxinone, C-60025
109517-71-7	Cavoxinine, C-60024
109517-72-8	Ferugin, *in* D-60005
109532-22-1	2'-Amino-2-chloro-3'-methylacetophenone, A-60112
109532-23-2	2'-Amino-2-chloro-4'-methylacetophenone, A-60113
109532-24-3	2'-Amino-2-chloro-6'-methylacetophenone, A-60115
109534-58-9	2-Triphenylenecarboxylic acid, T-60384
109575-71-5	Indicoside *A*, I-60027
109605-79-0	Topazolin, T-60224
109605-84-7	Topazolin hydrate, *in* T-60224
109621-33-2	Hymatoxin *A*, H-60235
109664-01-9	Isotanshinone IIB, I-60138
109664-02-0	Neocryptotanshinone, N-60017
109667-12-1	Phaseolinic acid, P-60067
109681-73-4	[1,4]Benzoxaselenino[3,2-*b*]pyridine, B-60054
109703-33-5	4,4-Dimethyl-3-oxo-1-cyclopentene-1-carboxaldehyde, D-60432
109719-06-4	Dihydro-4-(trifluoromethyl)-2(3*H*)-furanone; (*S*)-*form*, *in* D-60299
109719-07-5	Dihydro-4-(trifluoromethyl)-2(3*H*)-furanone; (*R*)-*form*, *in* D-60299
109795-03-1	20,29,30-Trisnor-3,19-lupanediol; (3β,19α)-*form*, *in* T-60404
109803-47-6	5-Chloro-4-methoxy-2-methylbenzoic acid, *in* C-60097
109803-48-7	5-Chloro-4-hydroxy-2-methylbenzoic acid; Me ether, Me ester, *in* C-60097
109803-52-3	1-Bromo-2,3-dichloro-4-methoxybenzene, *in* B-60229
109872-63-1	Nordinone, N-60060
109872-64-2	Nordinonediol, *in* N-60060
109894-10-2	2-Hydroxygarveatin A, *in* H-60148
109894-11-3	2-Hydroxygarveatin *B*, H-60148
109905-56-8	Epivolkenin, *in* T-60007
109905-96-6	2-Amino-3-phenylpentanedioic acid; (2*S*,3*S*)-*form*, *in* A-60241
109918-65-2	2-Amino-4-hydroxy-2-methylbutanoic acid; (*R*)-*form*, *in* A-60186
109918-71-0	2-Amino-2-ethyl-3-butenoic acid; (*R*)-*form*, *in* A-60150
109948-61-0	2-(1,2-Propadienyl)benzothiazole, P-60174
109954-46-3	2-(7-Hydroxy-3,7-dimethyl-2-octenyl)-6-methoxy-1,4-benzoquinone; (*Z*)-*form*, *in* H-60123
109954-47-4	2-(7-Hydroxy-3,7-dimethyl-2-octenyl)-6-methoxy-1,4-benzoquinone; (*E*)-*form*, *in* H-60123
109954-48-5	6-Methoxy-2-(3,7-dimethyl-2,6-octadienyl)-1,4-benzoquinone; (*Z*)-*form*, *in* M-60038
109971-64-4	3-Methoxy-6-[2-(3,4,5-trimethoxyphenyl)ethyl]-1,2-benzenediol, *in* T-60344
109986-01-8	*cis*-9,10-Dihydrocapsenone, *in* C-60011
110024-18-5	Volkensiachromone, V-60016
110024-19-6	Cyclolongipesin, C-60214
110024-20-9	Homocyclolongipesin, H-60088
110024-35-6	Longipesin; 9-Ac, *in* L-60034
110024-46-9	9-Methylcyclolongipesin, *in* C-60214
110024-47-0	Homocyclolongipesin; Ac, *in* H-60088
110024-48-1	Homocyclolongipesin; Propanoyl, *in* H-60088
110042-12-1	Longipesin; 9-Ac, 4-Me ether, *in* L-60034
110044-90-1	6-Isocyano-4(15)-eudesmene, *in* F-60069
110044-92-3	6-Formamido-4(15)-eudesmene; (6α,10α)-*form*, *in* F-60069
110064-50-1	7-Hydroxy-3-(4-hydroxybenzylidene)-4-chromanone, H-60152
110064-64-7	Blastmycetin *D*, B-60191
110064-65-8	Isolecanoric acid, I-60103
110066-02-9	12,19-Dihydroxy-5,8,10,14-eicosatetraenoic acid; (5*Z*,8*Z*,10*E*,12*S*,14*Z*,19*R*)-*form*, *in* D-60319
110066-03-0	12,19-Dihydroxy-5,8,10,14-eicosatetraenoic acid; (5*Z*,8*Z*,10*E*,12*S*,14*Z*,19*S*)-*form*, *in* D-60319